COASTAL, ESTUARIAL AND HARBOUR ENGINEERS' REFERENCE BOOK

COASTAL, ESTUARIAL AND HARBOUR ENGINEERS' REFERENCE BOOK

Edited by

M.B. Abbott

Professor, International Institute for Infrastructural, Hydraulic and Environmental Engineering, Delft, The Netherlands

and

W.A. Price

Coastal and Port Engineering Consultant, formerly Deputy Director, Hydraulics Research Ltd, Wallingford, UK

CRC Press
Taylor & Francis Group
Boca Raton London New York

CRC Press is an imprint of the
Taylor & Francis Group, an **informa** business
A TAYLOR & FRANCIS BOOK

CRC Press
Taylor & Francis Group
6000 Broken Sound Parkway NW, Suite 300
Boca Raton, FL 33487-2742

First issued in paperback 2019

ISBN-13: 978-0-419-15430-3 (hbk)
ISBN-13: 978-0-367-86464-4 (pbk)

Typeset in 9/10 Palatino by Best-set Typesetter Ltd., Hong Kong

A catalogue record for this book is available from the British Library

Library of Congress Cataloging-in-Publication data
Coastal, estuarial, and harbour engineer's reference book / edited by
M.B. Abbott and W.A. Price. – 1st ed.
p. cm.
Includes index.
ISBN 0–419–15430–2 (alk. paper)
1. Coastal engineering. 2. Harbors – Design and construction.
3. Estuaries. I. Abbott, Michael B. II. Price, W. A. (W. Alan)
TC205.C56 1993
627 – dc20 93–6888
CIP

Visit the Taylor & Francis Web site at
http://www.taylorandfrancis.com

and the CRC Press Web site at
http://www.crcpress.com

CONTENTS

CONTRIBUTORS

PROFESSOR M.B. ABBOTT
International Institute for Infrastructural,
Hydraulic and Environmental Engineering
Delft
The Netherlands

DR P.C. BARBER
Shoreline Management Partnership
Chester
Clwyd
UK

PROFESSOR F.B.J. BARENDS
Delft Geotechnics
Delft
The Netherlands

A.D. BATES
Anthony D Bates Partnership
Dredging and Coastal Consulting Engineers
Axbridge
Somerset
UK

K.W. BEDFORD
Professor of Civil Engineering
Ohio State University
Columbus
Ohio
USA

Rene du Bois D'Oingt
6932 Mirwart
Belgium

A.G. BRINSON
Rendel Palmer & Tritton
London
UK

PROFESSOR H.F. BURCHARTH
Department of Civil Engineering
University of Aalborg
Aalborg
Denmark

J.A. CUNGE
Director and Senior Engineer
Laboratoire d'Hydraulic de France
Echirolles
France

R.A. DALRYMPLE
Professor and Director
Centre of Applied Coastal Research
University of Delaware
Delaware
USA

A.M. DAVIES
Proudman Oceanographic Laboratory
Bidston Observatory
Birkenhead
Merseyside
UK

R.A. FALCONER
Professor of Water Engineering
Department of Civil and Structural Engineering
University of Bradford
Bradford
UK

C.A. FLEMING
Sir William Halcrow & Partners Ltd
Wiltshire
Swindon
UK

PROFESSOR J. FREDSOE
Institute for Hydrodynamics and
Hydraulic Engineering
Technical University of Denmark
Lyngby
Denmark

PROFESSOR A. FÜHRBÖTER
Leichtweiss-Institut für Wasserbau
Technical University of Braunschweig
Braunschweig
Germany

MR L. HAMM
SOGREAH, Consulting Engineers
Grenoble
France

DR.-ING. G. HEERTEN
Naue-Fasertechnik
Gmbh & Co KG
Lübbecke
Germany

F.F.M. HOLLY JR
Professor, Department of Civil
and Environmental Engineering
Institute of Hydraulic Research
University of Iowa
Iowa
USA

J.C. HUIS IN T'VELD
Rijkswaterstaat
Afdeling Sluizen en Stuwen
Rijskantoorgebouw 'Westhaven'
Utrecht
The Netherlands

O.J. JENSEN
Danish Hydraulic Institute
Hørsholm
Denmark

M.P. KENDRICK
Hurdlemakers
Crazies Hill
Wargrave
Berkshire
UK

J.L.M. KONTER
Rijkswaterstaat
Afdeling Sluizen en Stuwen
Rijskantoorgebouw 'Westhaven'
Utrecht
The Netherlands

P. LACEY
Ove Arup & Partners
London
UK

J.-P. LATHAM
Geomaterials: School of Engineering
Queen Mary and Westfield College
University of London
London
UK

PER MADSEN
Danish Hydraulic Institute
and Environmental Engineering
Hørsholm
Denmark

A.J. MEHTA
Coastal and Oceanographic Engineering Department
University of Florida
Gainesville
Florida
USA

C. MIGNIOT
SOGREAH, Consulting Engineers
Grenoble
France

D. MILES
Glovers (Solicitors)
London
UK

P. NEVILLE-JONES
Binnie & Partners
Consulting Engineers
Redhill
Surrey
UK

A. HASLE NIELSEN
Director, Danish Maritime Institute
Lyngby
Denmark

B.A. O'CONNOR
Professor of Maritime Civil Engineering
Department of Civil Engineering
University of Liverpool
Liverpool
UK

J.F. O'HARA
Principal Engineer
Costain Engineering and Construction
International Division
Maidenhead
Berkshire
UK

W.R. PARKER
Blackdown Consultants Ltd
Wellington
Somerset
UK

J. PARKS
President Dyneqs Ltd
Tampa
Florida
USA

A.B. POOLE
Geomaterials: School of Engineering
Queen Mary and Westfield College
University of London
London
UK

W.A. PRICE
Coastal and Port Engineering Consultant
Didcot
Oxfordshire
UK

PROFESSOR J.L. RAHUEL
Laboratoire d'Hydraulique de France
Echirolles
France

E.B. RASSMUSSEN
Senior Engineer
Danish Hydraulic Institute
Hørsholm
Denmark

DR. R.F. RAYNER
Wimpey Environmental Ltd
Swindon
Wiltshire
UK

N.P. REILLY
Mott MacDonald
Altrincham
Manchester
UK

A.G. ROBERTS
Canterbury City Council
Canterbury
Kent
UK

G.S. RODENHUIS
Director, Research and Marketing
Delft Hydraulics
Delft
The Netherlands

W. RODI
Professor, Institute for Hydromechanics
University of Karlsruhe
Karlsruhe
Germany

S.E. SANDS
Danish Maritime Institute
Lyngby
Denmark

J.H. SARGENT
Chairman, HR Wallingford Group Ltd
Oxfordshire
UK
Vice-President, PIANC

J.D. SIMM
Robert West & Partners
Chartered Consulting Engineers
Orpington
Kent
UK

P.G. SPRAGG
Rendal Palmer and Tritton
London
UK

J. SÜNDERMANN
Professor and Director
Institute of Oceanography
University of Hamburg
Hamburg
Germany

G. THOMPSON
Binnie & Partners
Consulting Engineers
Redhill
Surrey
UK

M.F.C. THORN
Hydraulics Research
Wallingford
Oxon
UK

R.G. TICKELL
Department of Civil Engineering
University of Liverpool
Liverpool
UK

I.H. TOWNEND
Sir William Halcrow and Partners Ltd
Swindon
Wiltshire
UK

J.V. TOWNER
Environmental Assessment Group Ltd
31 Dover Street
London W1X 3RA

J.R. VIEIRA
Senior Engineer
Danish Hydraulic Institute
Hørsholm
Denmark

H. DE VLIEGER
Director and Head of Research
N.V. Baggerwerken de Clodt en Zoon
Belgium

H.J. DE VRIEND
Head, Coastal Engineering Department
Delft Hydraulics
Emmeloord
The Netherlands

H. WANG
Geomaterials: School of Engineering
Queen Mary and Westfield College
University of London
London
UK

PREFACE

The aim of this book is to give the civil engineer, and especially the marine engineer, in whatever capacity they are acting, a concise presentation of the fundamentals of the theory and practice of all branches of coastal, estuarial and harbour engineering. We have concentrated on aspects relating to the marine environment itself and the coastal structures which interact directly with it. Our contributors have tried to convey the fundamentals of their subjects concisely yet authoritatively.

We believe this book will therefore be of value to many categories of reader. It will provide the first point of contact for those seeking a general introduction to the main topics covered, by means of the reviews contained in each chapter, coupled with the extensive references provided, which guide the reader to the most important sources of further information. For students of civil and marine engineering, environmental science, geology and geophysics, it will help them to appreciate the complex interactions between the individual aspects of coastal engineering. Consultants and contractors often work in very diverse fields and they will be able to supplement their own basic engineering knowledge and skills with specialized aspects relevant to a particular project. An appreciation will also be gained of the power and limitations of the computational techniques available to other specialists and this is especially relevant in view of the rapid development of modelling and computational methods at the present time.

It is worth pointing out the international nature of the subject. Our contributors are drawn from many leading research, consulting and contracting organizations and they have experience of engineering projects throughout the developing and developed world. This emphasizes the need to take account not only of technical means of solving coastal engineering problems, but also of the economic constraints on the design and construction approaches, which arise from the local conditions and resources applicable to any site.

We have endeavoured to achieve a balanced spread of subject coverage, though with a subject of such breadth and complexity, some overlap between chapters contributed by different authors has been inevitable. With the rapid changes in our subject, this reflects the diversity of development underway throughout the engineering and research community. By use of the references provided, both to other chapters within the book and external sources, the reader will be able to find the information he needs.

It only remains for us to thank the contributors for the tremendous efforts that they have made in the preparation of this comprehensive approach to the solution of coastal, estuarial and harbour engineering problems.

It is very sad that during the preparation of this book we heard of the death of Professor Dr Ing Alfred Fuhrböter (1931–1992). Apart from enjoying his close friendship he was the role model of a research based practical engineer. In music a reviewer would say that this rendering of a particular work is definitive. What he means is that he doubts whether it could be performed any better. The same could be said about all of Alfred's research projects.

But it is as a man we will miss him – brilliant but humble, and as Professor Bijker has said, having authority without being authoritative.

Responsibility for editing the book has been divided between us, and we have each introduced our own sections at the relevant point in the book; The Dynamic Environment (Parts One to Four, M.B. Abbott) and The Physical Environment (Parts Five to Nine, W.A. Price).

DISCLAIMER

THE DYNAMIC ENVIRONMENT

THE DYNAMIC ENVIRONMENT: AN INTRODUCTION

M.B. Abbott

THE STRUCTURES OF INDUSTRY IN THIS FIELD

There must be very few current branches of engineering that have been subjected to such rapid and drastic changes as coastal and marine engineering. Whereas, only twenty or so years ago, the engineer working in this field could find in one reference book almost all the knowledge required to do his job, he is nowadays dependent upon a veritable industry, composed of consulting companies, technological institutes, laboratories, survey companies, instrument suppliers and many other kinds of specialized organizations, which offer him an immense range and variety of services and products. Correspondingly, a reference book that only twenty or so years ago could provide a 'do-it-yourself kit' of equations, tables and graphs, has now to become, in the first place, an introduction to the services and products that are currently available.

The changes that have occurred in the nature and availability of knowledge have clearly indicated the necessity for preparing a new kind of engineers' reference book. However, the shear immensity of the field and the speed of its transformation have made such preparation a particularly daunting task. What could be most usefully included? How should it all be arranged? These and many other questions have had to be answered in the preparation of this work. This introduction provides an opportunity to outline the reasoning behind the choices made and introduces the new kind of knowledge industry that now serves the engineer working in these fields.

As in most branches of technology, the changes that have occurred have been driven, in the first place, by the needs of society. Modern society continues to place new and ever more exigent demands upon the engineer, whether for protection from flooding, for providing beaches suitable for recreation, for providing acceptable cargo-handling conditions in harbours, or for answering any number of environmental concerns. In order to satisfy these demands, the engineer working in this area needs knowledge of an ever greater range, depth and level of detail. This is to say, however, that he must now make recourse to such sources of knowledge as are ordered and organized in our current knowledge industries. He has then to make recourse to research within his areas of interest, but he has to do this by using certain types of organization that are themselves organized into the industry that is to be described here.

The first impression by an outsider would probably be that of great complexity: there are so many elements and so many interactions between them. Whenever we have to react to complex situations, our most natural tendency is to consult with others, and preferably those with an experience of similar situations. Those with whom the coastal, estuarial and harbour engineers consult are of course, in the first line, those **consulting engineers** who are specialized in the relevant areas. The division of activities between the client (who will usually be represented by his own professional engineer, who usually stands in an immediate and direct relation to one particular problem) and the consulting engineer (whose expertise extends over a whole range of problems of this class) corresponds to a natural division of labour within knowledge industries generally. The consulting engineer draws upon a wide range of experience in a particular field and applies this experience to a particular problem within this field, in support of the client, as usually represented by his engineer. Of course, by the very practice of his profession, the consulting engineer further strengthens his abilities. In short, the client wishes only to solve a problem, but this may necessitate the construction of works or the introduction of new operating procedures, or whatever, and in order to obtain advice about how to proceed in these matters, he turns to the consultant. The consultant thus serves as a sort of 'midwife' at the delivery of a new design or procedure or whatever else is required to solve the problem.

This position of the consultant, as one who helps his client to manage complex situations or solve complex problems, can be exercised in a number of different ways. The first, and most direct way is in the form of an independent private consulting company that has accumulated a particular kind of expertise, and a corresponding reputation. A certain differentiation then naturally occurs between these companies, as a function of their different areas of specialization. However, there is usually such an overlap of expertise between these companies that an element of competition exists between them. There is accordingly a market for specialized knowledge at this level and in this form.

It has been a feature of a world where almost everything becomes more complex and more interconnected that individual consulting companies have tended to extend themselves over ever greater areas of expertise and have attempted to integrate these areas together in various ways for the purpose of performing projects. As a consequence, many consulting companies have become very large: many now employ some hundreds of engineers and applied scientists. On the one hand, this has the consequence that these companies are able to concentrate knowledge resources that far exceed those normally available to the individual client and his engineering staff. On the other hand, of course, there can be a commercial temptation to use such resources more prodigally than is really justified by the job in hand, a tendency that the consulting company is naturally obliged to resist on ethical grounds. In any event there still remain good grounds for employing the services of smaller consulting companies, and even single individuals in many cases.

There are, of course, other kinds of consultants than private consulting companies that are active in this field. The nature of these organizations and their relative importance, however, commonly varies quite widely from one country to another. A country like The Netherlands, the very existence of which depends upon the mobilization of expertise in these fields, has its State Public Works (Rijkswaterstaat), which combines the functions of consultant and contractor and, in some instances, research organization as well. Similarly, in the USA, the US Army Corps of Engineers fulfils many functions that are in other countries situated in the private domain. Other countries have professional associations which provide a range of consulting services.

To the extent that these activities are seen as components of one part of a more general knowledge industry, then we must identify other kinds of organizations again that support these activities with even more specialized knowledge. The most significant among these are technological-service organizations that were originally called 'laboratories'. In most cases these were established to carry out physical-model testing of proposed works and operating methodologies. In some cases these originated as extensions of consulting activities, but in most instances they were established as state or semi-state organizations. Over recent years, however, they have passed, for the greater part, into the private domain. Moreover, as their main field of activity has passed from physical modelling to numerical modelling, so they have for the most part discarded the name of 'laboratory' altogether.

In addition to the technological-service institutes, we may identify other organizations that supply such specialized services as surveying, instrumentation and telemetry.

The other main class of organization to which the client and his engineer must usually turn in their search for solutions is the civil engineering contractor. Whereas the consultant serves at the birth of a design, it is the contractor who has to bring this newborn design to maturity, as a construction, an on-line system, or whatever other physical entity. From this point of view, and to mix our metaphors a little, the consulting engineering is only the half-way-house between the desires and aims of the client and the satisfaction of these desires and the fulfilment of these aims in the material world. In order to carry out his function, the contractor also draws upon knowledge, albeit often a different kind of knowledge from that of the consultant. Thus, to take closely-related examples, the consultant may be particularly interested in those current and wave conditions that his new structure will have to resist during the course of its lifetime, while the contractor may be more interested in the wave and current conditions that can be expected at a particular period during his construction programme. More generally, the consultant in this field will usually be interested more in managing natural processes, while for the contractor the process aspect is of a lesser importance as compared with, for example, the construction-organizational aspects. As has been argued elsewhere (Abbott, 1991), whereas the consultant tends to concentrate on the description of processes, the contractor is more interested in material, plant and work flow descriptions, which come down to descriptions of data.

All of this has to do with the 'supply side' of the industry. Meanwhile, however, we may identify a considerable number of organizations that we can better place on the 'demand side', in that they promote and canalize political action, which leads to legislation and associated institutional arrangements (e.g. regulating and other agencies). These include any number of environmental pressure groups and associations.

THE CENTRAL ROLE OF MODELLING IN THE TRANSFORMATION OF THE INDUSTRY

By a **model** of any entity or concept we mean a representation of that entity or concept, and so **its presentation in another shape or form**. Thus, for example, a physical model of a harbour is a reproduction of a real-world harbour at a much smaller scale, and so with an appropriate scale distortion. Similarly, a conceptual model of a morphological process may be a particular or more definite representation in thought of what is otherwise a rather undifferentiated and less structured impression of the process of interaction between fluid and sediment.

As our society has become increasingly information-dependent, so modelling has attained ever greater importance. This modelling is then carried out for much the greater part using computers, so that it constitutes mathematical modelling or, more precisely, numerical modelling. In numerical modelling, the real-world entity, the prototype, is described by a set of numbers, while the behaviour of that entity in space and time is described by sets of operations on these numbers. In the fields of interest here, numerical modelling has attained a central role in the design of works, in the elaboration of operating rules and policies, in the optimization of management practices, and indeed in almost all other activities. At the same time, however, this increased emphasis on numerical modelling, and indeed on the use of computers generally, has had a considerable influence upon the structure of the industry in this area, as indeed it has done in other areas as well.

From the point of view of its applications, the model has one basic purpose, namely that of indicating to us the consequences of any proposed course of action before we become committed irreversibly to that course. The purpose of the model is to answer the question of 'What would happen if . . . ?' Thus, what would happen if we continue the dumping of a certain chemical at a particular off-shore site? What would happen if we replenished the sand on a particular beach? What would happen if we introduced tertiary treatment on storm-sewer effluents in a particular estuary? What would happen if we designed a new container terminal in this way rather than that? And so on indefinitely. Models help us to see ahead, so that they give us a **prevision** of the consequences of our proposed actions. To the extent that our futures depend upon our own actions, so the model guides our decisions, shaping our actions towards our more-desired futures.

Now even as models have become the key elements in decision making in almost all parts of society, so the demands placed upon these models have also increased. Whereas only some twenty years ago a few relatively simple models sufficed to suggest the most likely consequences of a restricted number of basic interventions in the courses of nature, today large numbers of models are commonly coupled together to predict the consequences of a complex combination of interventions. A similar increase in the requirements of accuracy and reliability has occurred. This has had the consequence that, for the most part, modelling has become a complicated and quite highly-specialized activity, and, certainly until very recently, preferably pursued in a few major centres, where the required level and variety of expertise could be mobilized. These centres, however, developed in different parts of the industry, while their location and their development varied considerably from one national industrial environment to another. Thus, although end-users of models are only now beginning to do their own modelling, modelling centres have developed in some consulting engineering companies, in some industry associations, in almost all research and development institutes, in many government agencies and in some other places besides. Moreover, and especially in the case of numerical modelling, there appears to be little correlation between the industrial environments within which modelling has developed and the quality of the modelling work itself. Although the research and development organizations would appear to be the best suited to this kind of activity, in point of fact some European consulting engineers, in particular, have developed very advanced and capable modelling groups, while some research and development organizations have done very little in this area. In the USA, a number of government agencies and private, non-profit foundations have provided extensive numerical modelling facilities, while consulting engineers have tended for the most part only to adapt and apply these modelling facilities to particular problems.

Thus although there has been a great increase in the amount of modelling activity, so that in the present areas alone it now employs some thousands of engineers and applied scientists, its organization within the industry structure has become quite disconnected, diffused and altogether unclear. It is nowadays often by no means obvious which kind of organization is best suited to a particular task. Most private consulting companies and institutes will claim to have models available for almost every purpose, while several university departments, often under the pressure of reduced central-government funding, will offer to provide similar modelling services, but at a much lower price, while, again, government agencies and other organizations will offer to provide the models themselves for next to nothing. And then, beyond this again, there are some contractors who will often carry out the whole modelling exercise without any charge at all. The client for modelling services is thus confronted with a plethora of models offered by all manner of kinds of modelling-service organizations. How is he to choose in this situation? The answer to this must always be based upon the total cost that is associated with the modelling work.

After all, if the cost of a certain inaccuracy is a very expensive failure of works or operating procedures, savings made at the expense of this accuracy at the modelling end of the work will not be justified. Similarly, if a modelling study carried out by a contractor leads to the use of a methodology that is a speciality of that contractor, the loss in flexibility of approach and competitive pricing will quickly cancel out any savings made at the level of the modelling. The 'cost of the modelling work' must always be seen within this kind of frame. Thus, to continue along these lines, a model may be made available very cheaply, or at no charge at all, but it may be an exceedingly expensive model when seen in the wider perspective. Not only may it give rise to inaccuracies and thus faulty designs, but it may also consume a very great deal of the engineer's time – commonly several years – and even then lack the potential to be developed further. The model may, similarly, itself be made available at a nominal cost, but then necessitate the employment of a great deal of expensive consulting time in order to make it operational, and to adapt it for further use, and so on indefinitely. Indeed, there are some individuals and organizations which operate deliberately in just this manner, by making their models 'freely available', and then earning on the subsequent model-consulting side. Simply because so much depends upon the quality of the modelling work, however, this is not the place to trade-off quality for what must always be relatively small savings in total cost. Of course, if a failure of works, operating policies, or whatever else, occurs as a result of inadequate modelling, the costs will usually exceed any possible cost of a sound modelling investment by a large margin.

There is a further dimension in this total-cost picture that is particularly relevant to the coastal, estuarial and harbour engineer. This has to do with the expenses that are imposed upon the end-user of models, 'the clients', by community, national and other legislative and licensing bodies, usually on the basis of model studies. In order to evaluate the true relevance of the measures that are imposed, as a means of evaluating and possibly avoiding often punitive expenditures, this client and his engineer need in their turn to call upon professional modelling counsel. It rarely pays, or is otherwise a mere gamble, to sacrifice model quality for small savings in the modelling studies.

The real requirement of the engineer and his client can perhaps be best expressed as one of 'value for money'. This comes down, for most practical purposes, to the requirement for an efficient industry structure. As the industry is presently organized, the consulting engineer usually stands in the first and most immediate relation to the client and the client's own engineering staff. In some cases the consultant may be able to perform the required modelling studies in house, but in many if not most cases he will need to employ the services of an institute specializing in modelling. Such institutes, of course, usually carry out both physical and numerical modelling, while they commonly have survey teams, instrumentation departments and other such related services in house. Over the last few years, these specialist institutes have begun to sell modelling tools, which the client's engineering staff and the consultant can use in their investigations. Some industrial and professional associations have recently begun to evaluate the quality of these tools on behalf of their members. The use of the same or similar tools for legislative, licensing and monitoring bodies similarly leads to the introduction of some practical, albeit still in most cases *de facto* standards. Some consultants and most technological-service institutes in turn sponsor and otherwise promote research in universities, thus ensuring that this research is more relevant to the needs of practice while alleviating the financial pressure otherwise placed on university departments to compete with private consultants and technological-service institutes. Of course the flow of models from university departments, government agencies and other organizations can be expected to continue. Some of these may be taken up and adapted by the clients own engineering staff, by consultants and others, but the overall tendency appears to be in the direction of providing specialized modelling services and tool products on a professional basis.

THE OUTLOOK FOR THE 1990s AND BEYOND

The engineer is now becoming a major user of tools. Whereas it may still be possible even today for an engineer to work only with pen and paper, in a few years this engineer will be dependent on a whole range of computer-based design and management aids. At the moment the tools concerned are essentially stand-alone devices, even though the output from one may be formatted so as to serve as input to another. Only a few such tools are mentioned in this volume. These correspond to what are usually called 'fourth-generation' modelling systems. In this nomenclature, first generation modelling had to do with simple computerized equation solving and other operations that could be done just as well manually as by using a computer, except that the computer was faster and usually more reliable. In second-generation modelling, attention shifted to methods that were more specifically suited to digital computation, which was usually synonymous with their being unsuited for manual computation. These methods (finite differences, finite elements, boundary elements, etc. etc.) were 'computer friendly', even if 'human unfriendly'. Second-generation modelling was concerned with building models one at a time with each model dedicated to a particular geographical area. In third-generation modelling, on the other hand, 'modelling systems' were evolved that were generic to a whole class of models and which could be used to generate and run models of any particular geographical region simply by inputting the description of that region in a standardized form. Third-generation modelling offered great advantages in reducing lead times and costs and generally increasing reliability in modelling studies.

However, third-generation modelling became an increasingly esoteric and hermetic activity, practised by only a few consulting engineering companies, technological-service institutes, government agencies and other such organizations. Through second- and third-generation modelling, a hydraulics suited to digital computer applications, called **computational hydraulics**, evolved, but the result was that second- and third-generation modelling could only be pursued by computational-hydraulic experts. However, during the 1980s a demand arose from the end-users of modelling studies, the clients, for the modelling systems themselves. At about the same time, the proliferation of personal computers and work stations brought the cost of computing power down to such a level that real-world modelling could be carried out much more economically. The result was an attempt to design modelling systems that could be used by end-users, and this constituted the fourth-generation of modelling. Very briefly, fourth-generation modelling can be characterized as follows (Abbott *et al.*, 1991):

1. The modelling system becomes a design and management tool suited to use by persons who are not computational-hydraulics specialists.
2. In fourth-generation modelling, a differentiation correspondingly occurs between tool makers, who are computational-hydraulics specialists, and tool users, who usually are not.
3. The fourth-generation tool is a product, so that computational hydraulics specialists have to become, in part, producers of products, rather than performers of projects.
4. The principal line of development of fourth-generation modelling is in the direction of making computational hydraulics much more useful to a much wider range of end-users.
5. Research in this area thus moves from purely scientific research into research on the application of science. It is thus technological rather than scientific research in the usually understood sense.
6. When seen from the point of view of computer-science, research in fourth-generation modelling is concentrated on the control aspects of the code, with the operational aspects developing rather little and the data-base aspects only beginning to become significant.

Fourth-generation modelling tools are currently operated through standard computer terminals, using a mouse, a keyboard and a reasonably high-resolution screen. As professional design and management tools, they require a very great deal of research and development, over and above that expended upon the third-generation systems that are their natural progenitors.

Although the fourth-generation tools are only now coming into widespread use in engineering practice, fifth-generation modelling is already being actively researched. In brief, whereas the fourth-generation modelling system does not incorporate intelligent expert-advice-serving facilities and equivalent intelligent interrogation facilities, the fifth generation does. The difference between these generations is in the extent to which they make use of artificial intelligence (AI). Fourth-generation systems make very little use of this; fifth-generation systems come to depend upon it.

From the side of practical applications, fourth-generation modelling has to do primarily with equipping the hydraulic engineer with specific 'closed' tools and tool sets: in effect the engineer-user only 'parameterizes' these tools, so as to provide modelling and other facilities. Fifth-generation modelling, on the other hand, has more to do with providing the engineer, chemist, biologist, sedimentologist or whatever other user with a more general, 'open' **working environment**, within which this user is able to assemble his or her own tools and tool sets. Corresponding to this, fourth-generation systems are written in only one, or at most two languages, whereas fifth-generation environments link together many diverse language-like elements (e.g. various shell-built expert systems, structured query-language (SQL) data bases, object-orientated data bases, lists and list processors, windowing systems and graphics facilities, i.e. elements which may be regarded as constituting statements written in a considerable variety of different languages). In these systems, the numerical models serve only as means for 'encapsulating' what is known about the physical world. Whereas fourth-generation modelling is restricted primarily to the simulation of natural processes for the purposes of making designs and elaborating operating rules, fifth-generation modelling is also directed to diagnostic systems, planning and scheduling systems, on-line control systems, alarm and maintenance systems, advice-serving, and many other kinds of application besides.

More generally still, both fourth-generation and fifth-generation modelling together constitute just one part of that area of technology for which the name **Hydroinformatics** has been proposed and is already quite widely accepted (Abbott, 1991). Hydroinformatics has to do with the whole gamut of information technology applied in hydraulics and water resources, and in this total picture simulation and encapsulation themselves constitute only relatively limited parts.

INTRODUCTION TO SECTION 1: THE DYNAMIC ENVIRONMENT

This reference book is directed primarily to a younger generation of coastal, estuarial and harbour engineers, who will have received a different education from that of their seniors. Although possibly less broadly based, it is considerably more mathematically orientated. Whereas the engineer of twenty or thirty years ago finished his education with only some linear algebra and classical analysis in Cartesian co-ordinate frames, today's young engineer has usually followed courses in vector and more general tensor analysis, numerical analysis and some aspects of computer science. This is the background that is assumed here.

THE SCIENTIFIC BACKGROUND

The volume opens with a contribution by Sündermann which is quite rigorous in this respect. It sets out the origins of the tides, storm surges and short-period waves that act upon the areas that are of most direct interest in this book in their most general mathematical forms. The discussion is, however, restricted to the continuum view of nature, with its mathematics of differential and integral equations. The various modes of motion so introduced are then treated one at a time in the following sections. The first mode is that which is most immediately associated with tidal and storm-surge effects, which is flow that is predominantly along the direction of the earth's surface, so that it is **nearly-horizontal**. Falconer outlines the theory in this case, introducing the main concepts and their mathematical description, but using a simpler mathematical apparatus. Once again, the description is given entirely in continuum terms, which is much the most convenient way from our own point of view, even though the corresponding equations cannot be solved, as they stand, using a digital machine. The continuum equations thus serve as the conceptual foundations in the elaboration of models and modelling systems.

Following the descriptions of those nearly-horizontal flows that are most suited to tidal, storm-surge and other such processes, Dalrymple contributes Chapter 3 on short-period waves. This bridges the gap between classical wave theory, as originated during the last century and as still active today, and those modern computational theories which are more suited to digital computer applications, being incorporated in numerical models.

All motion on our planet tends to die away due to resistance of one or the other form. Tidal waves are damped, short-period waves may break and their energy may be otherwise dissipated; in almost all cases we observe how these directed motions pass into those less directed and confused motions that we call, generally, 'turbulence'. It is this turbulence that is the prime agent for carrying sediments, and so forming the sea bed in the shallower areas of interest, for diffusing and dispersing our waste products, and for many other purposes. In Chapter 4, Rodi introduces the main ideas that underlie current research, development and tool making in the area of turbulence modelling. Both the current further developments of traditional approaches and the elaboration of new and more computational-hydraulic approaches are considered.

Turbulence is the name that we give to confused motion, but science still tries to find some order even in this confusion. It accordingly posits a certain 'absolute state of confusion', in which there is no semblance of order whatsoever in that there is no local causal relation at all between a present state and a prior one. The corresponding process of 'completely disordered' motion is called **diffusion**, and science writes a very neat and simple differential equation to describe it under the continuum assumption. On the other hand, in the real world, such an absolute state of confusion rarely obtains. In practice, fluid-borne materials will be **dispersed** by turbulence, and there will commonly be some trends or tendencies which lead to a certain semblance of order in the process. Thus dispersion is by no means identical with diffusion, even though it is often described and treated in the same way. It has been one of the achievements of computational hydraulics, and indeed of computational fluid dynamics generally, that it has been able to demarcate some of the limits between these two concepts and to relate their relative influences to the modelling procedure used. Bedford describes this relation and its modelling consequences in a particularly searching and complete chapter (Chapter 5).

Nearly-horizontal flows and waves generate turbulence, which picks up sediment for net transport by currents and waves. In many practical situations, however, it is the sediment transport and subsequent morphological changes that are of the greatest interest to the coastal, estuarial and harbour engineer. A very important part in this process is then played by non-cohesive sediments, such as sands and silts. Fredsøe deals with these non-cohesive sediments in Chapter 6. On the one hand he outlines how classical, pre-computational approaches have been adapted to serve in numerical modelling applications, while on the other hand he introduces the notion that more specifically computational hydraulic methods can be employed here. After all, the hydrodynamic 'platforms' of models – those which describe the nearly-horizontal and short-period-wave flows – provide distributions in space and time of such eminently sediment-transport-relevant quantities as the turbulent-energy level and the specific rate of turbulent-energy dissipation, as introduced earlier by Rodi, and it is only natural to include these as the main driving forces in sediment-transport computations.

The computation of non-cohesive sediment transport and corresponding morphological changes has already become established in the repertoire of the modelling community, but the corresponding computations of cohesive sediments are at an earlier stage in their development. They are so important for practice, however, and especially in relation to environmental studies, that a considerable effort is now being made to elaborate working methodologies in this direction. In Chapter 7, Hamm and Migniot draw upon their considerable experience to outline the progress that is being made in this area and to indicate the possibilities that are now appearing.

HYDRODYNAMIC MODELS

The contributions up to this point are intended to introduce the foundations of the subject and to give it a certain mathematical form. This mathematical form is primarily that of algebraic and differential equations, which are much the simplest forms from the point of view of our own understanding. However, as they stand, they are mostly quite useless, for the purposes of digital computation. In order that they may be accommodated on a digital machine they must be cast into other forms, such as those that we call 'difference equations', 'finite element equations' and so on. In effect, we have here to do with two quite different **languages**, the first being called the **language of the continuum** and the second the **language of arithmetic** (Abbott and Basco, 1989). The language of arithmetic that we use on our digital machines is what is called in mathematical logic, and thus in computer science generally, a **first-order language**, which we represent by L_1Ar, while the language of the continuum, which is typically that of algebraic equations, trigonometric expressions, differential equations, and so forth, is definitely **not** a first-order language. It can however be represented for most practical purposes by a **second-order language of real numbers**, which we represent by L_2Real. In computational hydraulics and most other computational subjects we **translate**, backward and forward, between statements written in L_1Ar and L_2Real. Thus, for example, we usually write our descriptions of phenomena in partial differential equations, which can be regarded as statements in L_2Real, but we translate these into other forms, such as difference equations, which are statements in L_1Ar, for the purposes of realizing codes that will run on our digital machines. Similarly, our machines produce millions of numbers which we cannot possibly comprehend in their entirety, but we can make continuum solutions of our difference schemes so as to describe numerically-induced errors in terms that we can ourselves best understand. In this case we translate from statements written in L_1Ar into statements written, in effect, in L_2Real. There is accordingly a considerable jump involved in moving from the second part of this work, which has to do almost entirely with descriptive statements written in L_2Real, to the third part, where we have to consider computational processes, invariably expressed in L_1Ar.

We commonly divide our models up on the basis of their **dimensionality**. The most complete descriptions should then be provided by three-dimensional models. These are descriptions in terms of three independent space variables, to which a variation in time is usually conjoined. Of course the completeness may well be illusory if the model is not well designed and constructed, which is to say, among other things, that the translation from L_2Real to L_1Ar may not be well done. However, to the extent that this translation is well done, the three-dimensional model is the most sensitive of all models, at least hydrodynamically. The three-dimensional hydrodynamic model can most easily produce incorrect solutions on the basis of apparently quite small faults in schematization, it can most easily pass over into numerical instability and is generally the most demanding of all the productions of computational hydraulics. Correspondingly there have been many failures in three-dimensional modelling studies. Chapter 8 on this subject, has been written by Rasmussen, who has had much practical experience in three-dimensional modelling.

The difficulties that have been experienced when using full three-dimensional models have been so great that approaches intermediate between the two- and three-dimensional cases have been quite widely explored. In models of this type a rather complete description of a nearly-horizontal flow field is given in the horizontal, while the nearly-horizontal-velocity distribution in the vertical is represented by analytic functions characterized by only a few variables. Davies, who has contributed the chapter on this class of model here, is one of the leading exponents of this intermediate class of model. There can be little doubt that this is a much easier way to arrive at working models than the full three-dimensional approach, but the resulting models will usually be more restricted in their range of applications. Taking account of the resources that have to be mobilized in order to produce even a third-generation modelling capability in full three-dimensional modelling (a broadly based and competent design time, a lead time of at least five years and a cost of several millions of dollars), there may be no alternative to this intermediate approach for many organizations.

In most practical situations, the vertical variation of the nearly-horizontal flow is of little or no consequence, so that two-dimensional nearly-horizontal flow models can be used. These are, so to say, the 'workhorses' of modelling in coastal, estuarial and harbour engineering: the number and variety of applications are legion, and there have been few major constructions or other interventions in the course of nature over the last decade or so that have been planned and designed without them. This is an area where very heavy investments have been made over the last twenty or so years, and where fourth-generation tools are currently entering service. The two-dimensional models are themselves commonly divided into nearly-horizontal flow models and short-period-wave models. In his contribution to this book, Rodenhuis concentrates on nearly-horizontal flow, this being the more widely applied at present.

As the dimensionality is reduced, so the effects of flow in the neglected unresolved directions have to be introduced parametrically in the resolved directions. The problem of modelling thus shifts from one of describing a rather complete flow system to one of accounting for the effects of the unresolved parts on the resolved parts of the flow. As the modelling process becomes easier in one respect, it becomes more difficult in another. Thus when we arrive at one-dimensional nearly-horizontal flow models we arrive at models which are the simplest of all to construct in principle and the most difficult of all to use effectively in practice. This has had the consequence that although fourth-generation modelling has made its greatest advances in this area, the quality of the resulting modelling tools also varies greatly. The main point about these tools is that they should allow the professional engineer who is not a computational-hydraulics expert to build and run models of a quality commensurate with the requirements placed upon those models. The computational-hydraulics expert will usually be able to build a good model even with a very low quality modelling system, albeit still with some difficulties, while the less experienced user can easily be led into incorrect assumptions, settings, parameterizations and other model building errors by such a system. In his contribution on this subject, Cunge outlines some of the problems of one-dimensional modelling and introduces some typical fourth-generation tools that are available in this area. This part of the book is completed by a brief sketch of short-period wave modelling by Madsen and Abbott.

TRANSPORT DISPERSION AND WATER QUALITY MODELS

So far in our description of models we have only considered the hydrodynamic components, or hydrodynamic 'stages', and indeed we have even then kept mainly to the nearly-horizontal flow models. These are, however, rarely sufficient in themselves for engineering studies. In most cases they serve more as 'platforms' for a considerable number of 'applications superstructures'. Thus they may be used to provide the extents of flooded and dried areas and velocity fields for advection-diffusion models, and these last may be coupled together with water-quality and morphological models. We here again begin at the top end of the scale of dimensionality with the three-dimensional advection-diffusion models. Rasmussen continues his earlier introduction to the three-dimensional hydrodynamic platform and outlines, in Chapter 13, how this is used to drive transport-diffusion superstructures.

We continue to the two-dimensional advection-diffusion models, which are again the most widely applied and correspondingly most highly developed models of their class. Here Vieira provides an overview of some of the modelling problems, which connects with Chapter 5 by Bedford.

The reduction in dimensionality is completed by a description of one-dimensional transport-diffusion models, but in the more general context of transport and dispersion and water quality applications. Holly provides a critical evaluation of model requirements in this area, which again extends the earlier contribution by Bedford.

SEDIMENT TRANSPORT MODELS

In many cases the advection-diffusion stage is itself coupled to a sediment transport stage, and this may even provide a complete morphological stage. We begin with the full three-dimensional case, even though this is still at an early stage for practical applications. In Chapter 16 O'Connor explains the approaches currently followed in this case and introduces some of the second-generation models and third-generation modelling systems that are currently being used.

With the two-dimensional and quasi-three-dimensional sediment transport models we return to 'workhorses' again, in that these are the main application tools used in this field. There are, however, a number of mechanisms involved, and a very careful analysis has often to be made in order to separate out the different influences. This is one of several areas where physical laboratory tests, field measurements and numerical computations are used interactively in order to come up with realistic predictions. This topic is very completely and critically covered by de Vriend in Chapter 17.

Returning to one-dimensional models, but now for sediment transport, Cunge and Rahuel outline the approaches followed when the sediment transport is due to currents alone. This restriction is often acceptable in most of the reaches of estuaries and in many harbour areas, where the effects of short-period wave motions are small. They also introduce some fourth-generation modelling tools.

PHYSICAL MODELS

The more senior engineer may well have asked what has happened to the physical models of his own early years. It is now time to say that these are still alive and well, although they have on the whole come to occupy a number of niches, where they live in a sort of symbiosis with the numerical models. As a major example of such a situation, Juul Jensen explains how short-period wave modelling has progressed between physical and numerical models.

A somewhat similar symbiosis between physical and numerical models has developed in the case of studies of fluid–structures interaction. This aspect is covered here by Huis in't Veld and Kunter.

REFERENCES

Abbott, M.B. (1991) *Hydroinformatics: Information Technology and the Aquatic Environment*, Gower, Aldershot, UK.

Abbott, M.B. and Basco, D.R. (1989) *Computational Fluid Dynamics, an Introduction for Engineers*, Longman, London, and Wiley, New York.

Abbott, M.B., Harnø, K. and Lindberg, S. (1991) The fourth generation of modelling in hydraulics. *Journal of Hydraulic Research*, **29** (5), 581–600.

PART ONE

THE COASTAL, ESTUARIAL AND HARBOUR ENVIRONMENT

THE DEEP-WATER ORIGINS OF THE PHYSICAL ENVIRONMENTS OF COASTAL, ESTUARIAL AND HARBOUR REGIONS 1

J. Sündermann

We begin with a thoroughgoing mathematical description of that physical environment in deep water which is the primary source of the currents, waves and other driving forces that act upon coastal regions, estuaries and harbours. We adopt the conventional division of flows into currents and waves, only preceeding descriptions of these with a description of the theoretical foundation that is common to both. The treatment is restricted to the hydrodynamic aspects, with sedimentation, chemical, biological and general-ecological aspects being put to one side. We make the usual assumption that the oceans and the deep seas act as almost homogeneous and effectively-infinite sources and sinks so far as the shallower regions are concerned.

CONTENTS

1.1 THEORETICAL FOUNDATIONS

1.1.1 BASIC PRINCIPLES

The physical state of the ocean is described by seven macroscopic quantities, all of which are functions of space $\mathbf{x}^1$ and time t: the components of the velocity vector $\mathbf{v}$, pressure p, salinity S, temperature T and density ρ. Their inter-relation follows from the basic conservation laws of mechanics and thermodynamics (usually expressed in differential terms).

Conservation of momentum (Reynolds equation of motion):

$$\frac{\partial \mathbf{v}}{\partial t} + (\mathbf{v} \cdot \nabla)\mathbf{v} + 2\mathbf{\Omega} \times \mathbf{v} - \nabla \cdot (A \nabla \mathbf{v}) = \frac{1}{\rho}(\mathbf{a} - \nabla p) \quad (1.1)$$

where $\mathbf{\Omega}$ is the angular velocity of the earth's rotation, $\mathbf{a}$ is the resultant of the external accelerations and A is the coefficient of the turbulent inner friction. Depending upon whether Cartesian space co-ordinates (x, y, z) or geographical co-ordinates (λ, θ, z) are used, so (x, y) denote the horizontal directions in a tangential plane at the globe, and (λ, θ) the geographical longitude and latitude. The z-axis is directed upwards. The origin of co-ordinates is situated at the undisturbed sea surface. The surface $z = \eta(x, y, t)$ describes the actual sea level, the surface $z = -h(x, y)$ the sea bottom. The velocity vector $\mathbf{v}$ has the components (u, v, w).

Conservation of total mass (continuity equation):

$$\frac{\partial \rho}{\partial t} + \nabla \cdot (\rho \mathbf{v}) = 0 \quad \text{or equivalent}$$

$$\frac{\partial \rho}{\partial t} + \rho \nabla \cdot \mathbf{v} = 0 \quad (1.2a)$$

In most cases the water is assumed to be incompressible; then from Equation (1.2a) it follows that:

$$\nabla \cdot \mathbf{v} = 0 \quad (1.2b)$$

The conservation of salt mass (continuity equation of salt) is:

$$\frac{\partial S}{\partial t} + \mathbf{v} \cdot \nabla S - \nabla \cdot (D_S \nabla S) = Q_S \quad (1.3)$$

where D_S is the coefficient of turbulent salt diffusion and Q_S are the external sources and sinks of

Coastal, Estuarial and Harbour Engineer's Reference Book. Edited by M.B. Abbott and W.A. Price. Published in 1993 by Chapman & Hall, 2–6 Boundary Row, London SE1 8HN. ISBN 0 419 15430 2.

salinity (e.g. by evaporation or precipitation at the sea surface).

The conservation of energy (heat transport equation) is:

$$\frac{\partial T}{\partial t} + \mathbf{v} \cdot \nabla T - \nabla \cdot (D_T \nabla T) = Q_T \qquad (1.4)$$

where D_T is the coefficient of turbulent heat diffusion and Q_T are the external sources and sinks of heat (e.g. by heating or cooling at the sea surface).

The equation of state is given by

$$\rho = \rho(S, T, p) \qquad (1.5)$$

This is an empirical relation which is generally taken today as the **UNESCO formulation** (1981). Equations (1.1) and (2.2) are often referred to as **primitive equations.** Detailed presentations of them have been given by Krauss (1973) and Pedlosky (1982).

1.1.2 BOUNDARY CONDITIONS

Equations (1.1)–(1.5) form a system of second-order non-linear partial differential equations. Its integration requires the prescription of boundary conditions. These are all deduced from the **Kotchine theorem** (Krauss, 1973). Especially at the surface (index s) and at the bottom (index b) these follow two classes of condition.

Kinematic boundary conditions (the index h denotes the horizontal component):

$$\omega^s = \frac{\partial \eta}{\partial t} + \mathbf{v}_h \cdot \nabla_h \eta \qquad (1.6a)$$

$$\omega^b = -\mathbf{v}_h \cdot \nabla_h h \qquad (1.6b)$$

Generally $v_n = 0$ holds for a solid boundary, where n represents the normal direction (**no-flow condition**). For $\partial\eta/\partial t = 0$, Equation (1.6a) is called the **rigid lid condition.**

The dynamic boundary conditions are:

$$A_z \frac{\partial \mathbf{v}_h}{\partial z}\bigg|^s = \tau_h^s \qquad (1.7a)$$

$$A_z \frac{\partial \mathbf{v}_h}{\partial z}\bigg|^b = \tau_h^b \qquad (1.7b)$$

where A_z is the coefficient of vertical inner friction and τ_h is the horizontal external stress at the boundary surfaces. Commonly, the expressions

$$\tau_h^s = \lambda|\mathbf{V}_h^s|V_h^s \qquad (1.8a)$$

$$\tau_h^b = r|\mathbf{v}_h^b|v_h^b \qquad (1.8b)$$

are used, where λ denotes the wind drag coefficient (typical value: 1.75×10^{-6}), r is the bottom friction coefficient (typical value: 2.5×10^{-3}) and $\mathbf{V}_h^s$ is the horizontal wind velocity at the sea surface. Further boundary conditions have also been given by Krauss (1973) and Pedlosky (1982).

1.1.3 SCALE ANALYSIS

The dynamics of the real ocean are governed in most of its regions by simple balances allowing for simplified equations. These are derived by normalizing the state quantities with typical space and time scales. Equation (1.1) then takes the dimensionless form:

$$\frac{\omega_0}{f_0}\frac{\partial \hat{\mathbf{v}}_h}{\partial \hat{t}} + Ro\left(\hat{\mathbf{v}}_h \cdot \hat{\nabla}_h \hat{v}_h + \hat{\omega}\frac{\partial \hat{\mathbf{v}}_h}{\partial z}\right) + \hat{\mathbf{z}} \times \hat{\mathbf{v}}_h$$
$$-E_h \hat{\nabla}_h^2 \hat{\mathbf{v}}_h - E_z \frac{\partial^2 \hat{v}_h}{\partial \hat{z}^2} = -\hat{\nabla}_h \hat{p} \qquad (1.9a)$$

$$\left(\frac{H}{L}\right)^2 \left\{\frac{\omega_0}{f_0}\frac{\partial \hat{\omega}}{\partial \hat{t}} + Ro\left(\hat{\mathbf{v}}_h \cdot \nabla_h \hat{\omega} + \hat{\omega}\frac{\partial \hat{\omega}}{\partial \hat{z}}\right)\right\}$$
$$-\left(\frac{H}{L}\right)^2 \left\{E_h \hat{\nabla}_h^2 \hat{\omega} + E_z \frac{\partial^2 \hat{\omega}}{\partial z^2}\right\}$$
$$= -\frac{H}{f_0 L V_0} g - \frac{\partial \hat{p}}{\partial \hat{z}} \qquad (1.9b)$$

where typical scales are used for normalization: L for the horizontal, H for the vertical dimension, θ_0 for geographical latitude, ω_0 for frequency and V_0 for the speed of the horizontal current. By means of the relations $\omega_0 = (H/L)V_0$, $f_0 = 2|\mathbf{\Omega}| \sin \theta_0$ and $p_0 = \rho f_0 V_0 L$, characteristical values for the vertical velocity ω_0, the Coriolis parameter f_0 and the pressure p_0 are derived.

The resulting dimensionless variables are marked by the circumflex symbol '^': $\hat{x} = \frac{x}{L}$, $\hat{y} = \frac{y}{L}$, $\hat{z} = \frac{z}{H}$, $\hat{t} = \omega_0 t$, $\hat{\mathbf{v}}_h = \frac{\mathbf{v}_h}{V_0}$, $\hat{\omega} = \frac{\omega}{\omega_0}$, $\hat{p} = \frac{p}{p_0}$.

Ro denotes the dimensionless **Rossby number**:

$$Ro = \frac{V_0}{f_0 L}$$

which represents the quotient of the centrifugal and the Coriolis force in an ocean current. It characterizes the relative importance of the non-linear inertia terms.

E_h and E_z are the dimensionless **Ekman numbers** in the horizontal and vertical directions, respectively

$$E_h = \frac{A_h}{f_0 L^2}, \quad E_z = \frac{A_z}{f_0 H^2}$$

They represent the quotient of frictional and Coriolis force and they are a measure of the relative importance of the turbulent inner friction.

The acceleration of gravity, g, which is assumed to

be the only external acceleration: $\mathbf{a} = (0, 0, -g)$. By estimating the coefficients of Equation (1.9) for real ocean conditions, fundamental balances are achieved characterizing different regimes of the sea. For the general ocean circulation the following values can be assumed: $L = 10^6\,\mathrm{m}$, $H = 10^3\,\mathrm{m}$, $f_0 = 10^{-4}\,\mathrm{s}^{-1}$, $\omega_0 = 10^{-6}\,\mathrm{s}^{-1}$, $V_0 = 10^{-1}\,\mathrm{m\,s}^{-1}$, $A_h = 10^4\,\mathrm{m}^2\,\mathrm{s}^{-1}$, $A_z = 10^{-1}\,\mathrm{m}^2\,\mathrm{s}^{-1}$. From this the following numbers are derived: $H/L = 10^{-3}$, $\omega_0/f_0 = 10^{-2}$, $Ro = 10^{-3}$, $E_h = 10^{-4}$, $E_z = 10^{-3}$.

Neglecting small terms in Equation (1.9), the most important basic balances in the ocean are obtained:

$$\hat{\mathbf{z}} \times \mathbf{v}_h = -\hat{\boldsymbol{\nabla}}_h \hat{p} \tag{1.10a}$$

$$0 = -\frac{H}{f_0 L V_0} g - \frac{\partial \hat{p}}{\partial \hat{z}} \tag{1.10b}$$

Rewriting Equation (1.10) in terms of acceleration we obtain the **geostrophic equilibrium***,

$$-f \underset{\neg h}{v} + \frac{1}{\rho} \boldsymbol{\nabla}_h p = 0 \tag{1.11}$$

and the **hydrostatic equilibrium**;

$$\frac{1}{\rho}\frac{\partial p}{\partial z} = -g \tag{1.12}$$

The **hydrostatic pressure** distribution follows from Equation (12) by vertical integration:

$$p(z) - p(\eta) = \rho_0 g(\eta - z) + \int_z^{\eta} g(\rho - \rho_0)\mathrm{d}z \tag{1.13}$$

Here $p(\eta)$ denotes the air pressure at the surface (mostly neglected), the first term on the right-hand side the barotropic part of the pressure $\bar{p}$ and the second term the **baroclinic** part p'. ρ_0 is a constant mean density. The barotropic pressure characterizes the basic state of the homogeneous ocean, the baroclinic the deviation from this state due to a density anomaly, $(\rho - \rho_0)$.

The hydrostatic approximation which is generally used in oceanography implies a significant separation of horizontal and vertical motions and its calculations. Another important simplification is the **Boussinesq approximation** whereby variations of the density can be neglected with the exception of the gravity acceleration term, i.e. the buoyancy.

*The symbol '$\neg$' denotes a horizontal operator, rotating a vector by 90° in the mathematically negative sense: $(u, v) \rightarrow (v, -u)$. The following rules are valid: $\underset{\neg h}{\underset{\neg h}{v}} = -\mathbf{v}_h$, $\mathbf{v}_h \cdot \mathbf{v}_h = \underset{\neg h}{v} \cdot \underset{\neg h}{v}$, $\mathbf{v}_h \cdot \underset{\neg h}{v} = \underset{\neg h}{v} \cdot \mathbf{v}_h = 0$.

1.1.4 BAROTROPIC AND BAROCLINIC CURRENTS

For small Rossby numbers, five equations are obtained from Equations (1.1), (1.6a), (1.3) and (1.4), for the five **prognostic**, (i.e. locally fast varying) quantities $\mathbf{v}_h$ (with two components), η, S and T:

$$\frac{\partial \mathbf{v}_h}{\partial t} - f \underset{\neg h}{v} + \frac{1}{\rho_0} \boldsymbol{\nabla}_h p = \mathbf{q}_h \tag{1.14a}$$

$$\frac{\partial \eta}{\partial t} - \omega^s = 0 \tag{1.14b}$$

$$\frac{\partial}{\partial t}\begin{Bmatrix} S \\ T \end{Bmatrix} + \left(\mathbf{v}_h \cdot \boldsymbol{\nabla}_h + \omega \frac{\partial}{\partial z}\right)\begin{Bmatrix} S \\ T \end{Bmatrix} = \begin{Bmatrix} Q_S \\ Q_T \end{Bmatrix} \tag{1.14c}$$

whereby, in $\mathbf{q}_h$, the external and the frictional accelerations are included. The three **diagnostic** variables, ω, p and ρ, are determined from the prognostic ones by applying Equations (1.5) and (1.13) and by vertical integration of Equation (1.2b):

$$\omega(z) = -\boldsymbol{\nabla}_h \cdot \int_{-h}^{z} \mathbf{v}_h \mathrm{d}z \quad \text{with } \omega(-h) = 0 \tag{1.15}$$

Introducing, by vertical integration, the (barotropic) velocities (strictly speaking, the upper limit is $z = \eta$, but this difference is neglected here.)

$$\bar{\mathbf{v}}_h = \frac{1}{h}\int_{-h}^{0} \mathbf{v}_h \mathrm{d}z \tag{1.16a}$$

$$\bar{\omega} = -\boldsymbol{\nabla}_h \cdot \int_{-h}^{z} \bar{\mathbf{v}}_h \mathrm{d}z \tag{1.16b}$$

the flow can be separated into a barotropic and a baroclinic subsystem:

$$\mathbf{v}_h = \bar{\mathbf{v}}_h + \mathbf{v}_h'$$

$$\omega = \bar{\omega} + \omega'$$

$$p = \bar{p} + p'$$

The prognostic variables of the barotropic subsystem are $\bar{\mathbf{v}}_h$ and η and they are determined by

$$\frac{\partial \bar{\mathbf{v}}_h}{\partial t} - f \underset{\neg h}{\bar{v}} + \frac{1}{\rho_0} \boldsymbol{\nabla}_h \bar{p} + \left[\frac{1}{h\rho_0}\int_{-h}^{0} \boldsymbol{\nabla}_h p' \mathrm{d}z\right] = \bar{\mathbf{q}}_h \tag{1.17a}$$

$$\frac{\partial \eta}{\partial t} - \bar{\omega}_s = 0 \tag{1.17b}$$

The baroclinic subsystem is described by

$$\frac{\partial \mathbf{v}'_h}{\partial t} - f \underset{\neg h}{v'} + \frac{1}{\rho_0} \boldsymbol{\nabla}_h p' - \frac{1}{h\rho_0}\int_{-h}^{0} \boldsymbol{\nabla}_h p' \mathrm{d}z = \mathbf{q}_h'$$

$$\frac{\partial}{\partial t}\begin{Bmatrix} S \\ T \end{Bmatrix} + \left[(\bar{\mathbf{v}}_h + \mathbf{v}_h') \cdot \nabla_h + (\bar{\omega} + \omega') \frac{\partial}{\partial z} \right] \begin{Bmatrix} S \\ T \end{Bmatrix} = \begin{Bmatrix} Q_S \\ Q_T \end{Bmatrix}$$

The interactions of both systems are given by the terms in square brackets: the barotropic currents can also be forced by baroclinic pressure gradients, and the density field is advected by barotropic velocities. This coupling between both systems can be neglected in the case of free linear disturbances of a horizontally homogeneous, stabily stratified ocean.

The barotropic system has much shorter time scales and accordingly much higher phase velocities c then the baroclinic system:

$$c_1 = \sqrt{gh} \quad \text{barotropic} \tag{1.18a}$$

$$c_2 = c_1 \sqrt{\frac{h}{\rho_0} \frac{\partial \rho}{\partial z}} \quad \text{baroclinic} \tag{1.18b}$$

The typical length scale is the **Rossby radius of deformation** $R_i = c/f$, which characterizes the horizontal orbital motion caused by the equilibrium of centrifugal and Coriolis forces. It ranges from 100 to 1000 km for barotropic and from 5 to 50 km for baroclinic motions.

For quasi-stationary conditions, it follows from Equations (1.16b) and (1.17b) that:

$$-\nabla_h \cdot (h\bar{\mathbf{v}}_h) = \bar{\omega}^s = 0$$

(for the rigid lid condition, see section 1.1.2). This means that the divergence of the flow vanishes and there exists a **stream function** ψ with:

$$\bar{\mathbf{v}}_h = \frac{1}{h} \underset{\neg h}{\nabla} \psi \tag{1.19}$$

Instead of the primitive equations (1.17a), in oceanography the scalar **vorticity equation** is often used after the vorticity operator $\underset{\neg h}{\nabla}$. This equation, in the quasi-stationary case, is:

$$\underset{\neg h}{\nabla} \left(\frac{f}{h}\right) \cdot \nabla_h \psi = -\underset{\neg h}{\nabla} \cdot \left(\frac{1}{h\rho_0} \int_{-h}^{0} \nabla_h p' \mathrm{d}z \right) + \underset{\neg h}{\nabla} \cdot \bar{\mathbf{q}}_h = Q \tag{1.20}$$

where Q denotes the vertical component of the acting external torque. For $Q = 0$ follows:

$$\bar{\mathbf{v}}_h \cdot \nabla_h \left(\frac{f}{h}\right) = 0 \tag{1.21}$$

Equation (1.21) means that the vertically-averaged currents (corresponding to the mass transport) tend to follow the isolines f/h = const, and in smaller regions ($f \sim$ const) they simply follow the isobathic lines.

In the non-stationary case, the horizontal flow is generally divergent and no stream function can be defined. In the concept of **quasi-geostrophy** which is broadly used in oceanography, the variations are assumed to be sufficiently slow in the sense that a stream function exists, while the cinematic boundary condition (Equation (1.17b)) holds. Then the vorticity equation for instationary quasi-geostrophic currents is derived from Equation (1.17a) (Hasselmann, 1982):

$$\frac{\partial}{\partial t}\left\{ \nabla_h \cdot \left(\frac{1}{h}\nabla_h \psi\right) - \frac{f}{h}\eta \right\} + \underset{\neg h}{\nabla} \left(\frac{f}{h}\right) \cdot \nabla_h \psi = Q \tag{1.22}$$

This equation yields, for vanishing Q, the law of conservation of potential vorticity:

$$\frac{\mathrm{d}}{\mathrm{d}t}\left(\frac{\zeta + f}{h}\right) = 0 \tag{1.23}$$

with $\zeta = -\underset{\neg h}{\nabla} \cdot \bar{\mathbf{v}}_h$ representing the **relative vorticity** of the flow. Equation (1.23) expresses the fact that any horizontal flow in the ocean undergoes a deflection, if there are changes in bathymetry or latitude. Northward currents (the **planetary vorticity** f **increases**) gain a negative relative vorticity, i.e. they are deflected clockwise. For southward currents the opposite is true.

1.2 CURRENTS IN THE OCEAN

1.2.1 CLASSIFICATION

Looking at the external forcing it is possible to distinguish between wind-, thermohaline- and tidal-driven currents (see Muller and Willebrand, 1989, for this entire chapter). The first requirement is for the turbulent inner friction for the vertical transmission of the entrained momentum; it is limited to a surface boundary layer (see section 1.2.3). Nevertheless, the associated deformation of the sea surface causes barotropic pressure gradients acting in the whole water column.

The thermohaline circulation is generated by the inhomogeneous structure of water masses in the world ocean which is a result of heat and salt fluxes at the sea surface. The resulting baroclinic pressure gradients drive currents practically everywhere in the ocean, but especially in the deep waters. With the exception of a narrow equatorial belt (±5°) and small friction layers near the boundaries, these flows are geostrophically balanced (see section 1.2.2). The tidal motions are periodic and are treated in section 1.3.2. For the global circulation, a fair approximation is that these are stationary, i.e. $\omega_0/f_0 \ll 1$. This is not the case for wavelike motions (see Chapter 3).

1.2.2 GEOSTROPHIC FLOW

The basic equation is given by Equations (1.11) and (1.12). The continuity of the flow can then be deduced, i.e. non-divergent, horizontal flow, and therefore the existence of a stream function ψ with

$$\mathbf{v}_h = \frac{1}{h}\underline{\nabla}_h \psi$$

With (1.11), the equation

$$\nabla_h\left(\psi + \frac{h}{\rho f}p\right) = 0 \tag{1.24}$$

can be derived. It expresses the fact that the streamlines and isobars are identical, i.e. the currents are perpendicular to the pressure gradients.

For a given region of high pressure at the northern hemisphere the geostrophic current flows around clockwise, and anticlockwise on the southern hemisphere. This orientation is termed **anticyclonic**, and its opposite is called **cyclonic** (see Figure 1.1).

From Equations (1.11) and (1.12), by elimination of the pressure terms, with a good approximation, the **thermal wind equation**:

$$\frac{\partial \underline{v}_h}{\partial z} = -\frac{g}{\rho f}\nabla_h p$$

is determined.

Vertical integration assuming the Boussinesq approximation yields:

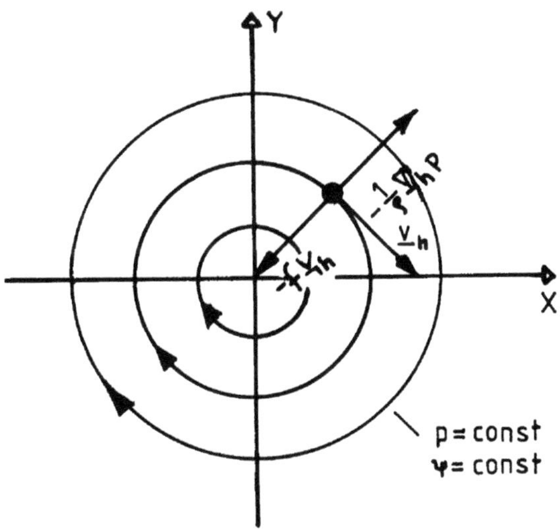

Figure 1.1 Geostrophical equilibrium on northern hemisphere with high pressure in origin.

$$\underline{v}_h(z) - \underline{v}_h(z_0) = -\frac{g}{\rho_0 f}\int_{z_0}^{z}\nabla_h \rho dz$$

This equation forms the basis of the important **dynamical method** for the calculation of thermohaline currents from given density gradients. If a reference velocity $\underline{v}_h(z_0)$ is not known from direct measurements, the integration is started from a **layer of no motion** $z = z_0$ (Defant, 1961) with

$$\underline{v}_h(z_0) = \mathbf{0}$$

This reference layer has a depth of the order of 1000 m.

1.2.3 FLOW IN THE EKMAN LAYER

The vertical profile of the wind-driven horizontal currents in the upper boundary layer is deduced by the Ekman theory. With the assumptions of stationarity ($\omega_0/f_0 \ll 1$), small Rossby and horizontal Ekman numbers (Ro, $E_h \ll 1$) and an infinitely extended ocean (no horizontal pressure gradients), it follows Equation (1.9a) in terms of acceleration that:

$$-f\underline{v}_h - A_z\frac{\partial^2 \mathbf{v}_h}{\partial z^2} = \mathbf{0}$$

With the boundary conditions (1.6b) and (1.7a) and prescribing a southerly wind with stress τ_y^s, the following current field is derived by integration

$$u(z) = \pm|\mathbf{v}_h|\cos\pi\left(\frac{z}{D} + \frac{1}{4}\right)$$
$$v(z) = |\mathbf{v}_h|\sin\pi\left(\frac{z}{D} + \frac{1}{4}\right) \tag{1.26}$$

(the positive sign is for the northern hemisphere, the negative for the southern). The speed of the current is given by

$$|\mathbf{v}_h| = \frac{\tau_y^s}{\sqrt{A_z|f|}}e^{\pi z/D}$$

where D denotes the **Ekman's friction depth**:

$$D = \pi\sqrt{\frac{2A_z}{|f|}} \tag{1.27}$$

Equation (1.26) describes the **Ekman Spiral**. The wind driven current at the surface is deflected by 45° against the wind direction, to the right in the northern and to the left in the southern hemisphere. The rotation is continued into deeper water (see Figure 1.2).

The current speed is highest at the surface:

$$|\mathbf{v}_h^s| = \frac{\tau_y^s}{\sqrt{A_z|f|}} \tag{1.28}$$

and decreases exponentially with depth. At $z = -D$

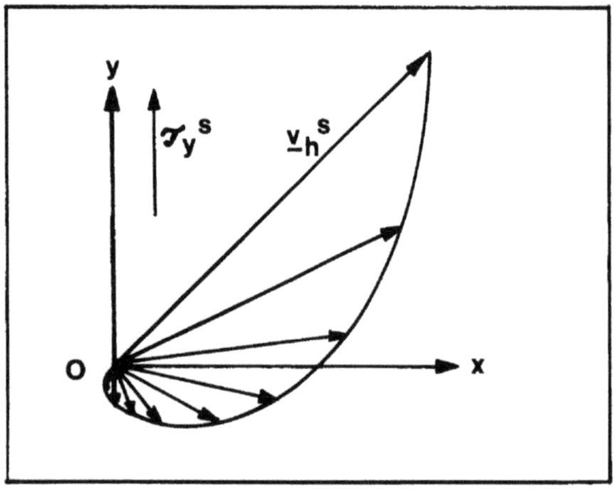

Figure 1.2 Ekman spiral. $\mathbf{v}_h^s$ is surface vector, further vectors represent deeper layers.

the flow direction is opposite to the surface one and the speed is

$$|\mathbf{v}_h| = |\mathbf{v}_h^s|e^{-\pi} \approx 0.043|\mathbf{v}_h^s|$$

At this depth the wind driven current has practically disappeared. D is of the order of 100 m, and decreases according to Equation (1.27) with higher latitudes. From Equation (1.28) it appears that, for typical values of A_z and f, the current speed at the surface amounts to a few percent of the wind speed.

The mass transport $\mathbf{M}_h = (M_x, M_y)$ in the Ekman layer is obtained by vertical integration of Equation (1.26):

$$M_x = \int_{-\infty}^{0} u(z)\mathrm{d}z = \pm\frac{\tau_y^s}{|f|}$$

$$M_y = \int_{-\infty}^{0} v(z)\mathrm{d}z = 0 \qquad (1.29)$$

This means that the total transport is directed perpendicularly to the wind stress, to the right on the northern and the left on the southern hemisphere. In the case of finite water depth, the Ekman spiral is modified (Krauss, 1973).

The Ekman theory can be extended to the frictional layer near the bottom. The result is, that a (e.g. geostrophic) deep-water flow approaching the bottom is deflected to the left in the northern and to the right in the southern hemisphere (Krauss, 1973).

In this case the Ekman transport is given by

$$\mathbf{M}_h = \frac{D}{2\pi}\underline{v}_g$$

where $\mathbf{v}_g$ denotes the basic geostrophic current. Generally the transport has a non-zero divergence:

$$\nabla_h \cdot \mathbf{M}_h = -\frac{D}{2\pi}\nabla_h \cdot \underline{v}_g = -\frac{D}{2\pi}\zeta$$

By means of the continuity equation, (Equation (1.16b)), and the condition $\omega^b = 0$, integration over the bottom layer yields

$$\omega^D = -\nabla_h \cdot \mathbf{M}_h = \frac{D}{2\pi}\zeta$$

This means, that the vorticity of the basic horizontal flow generates a vertical mass flux at the interface. Anticyclonic (negative) vorticity causes an inflow into the boundary layer (**Ekman suction**), cyclonic vorticity an outflow (**Ekman pumping**).

The Ekman transport essentially explains the **upwelling** phenomena in the ocean, i.e. the flow of water from a few hundred metres depth towards the surface. The trade winds which are blowing parallel to the coast cause a seaward transport of surface water at the eastern sides of the oceans. This divergence is compensated by an upwelling of deeper water. Along the equator a divergent flow is caused by different signs of the Coriolis parameter, and again upwelling appears.

1.2.4 THE GLOBAL WIND-DRIVEN CIRCULATION

For a homogeneous ocean without tides the external accelerations are:

$$\mathbf{q}_h = A_z\frac{\partial^2\mathbf{v}_h}{\partial z^2} + A_h\nabla_h^2 v_h$$

Vertical integration from the surface to the bottom with boundary conditions (1.7) yields

$$\bar{q}_h = \frac{1}{h}(\tau_h^s - \tau_h^b) + A_h\nabla_h^2\bar{\mathbf{v}}_h) \qquad (1.30)$$

First, it is assumed that there is no bottom friction or turbulent viscosity in the water. Then from Equation (1.20) it follows that:

$$\underline{\nabla}_h f \cdot \nabla_h\psi = \underline{\nabla}_h \cdot \tau_h^s \qquad (1.31)$$

In a Cartesian system the dependency of the Coriolis parameter on the latitude can be expressed approximately by

$$f(y) \approx f_{(0)} + \frac{\mathrm{d}f}{\mathrm{d}y}y = f_0 + \beta y$$

with $\beta = \mathrm{d}f/\mathrm{d}y > 0$. This relation is called the **β – plane approximation**. From Equation (1.31) it then follows that:

$$\beta\frac{\partial\psi}{\partial x} = \underline{\nabla}_h \cdot \tau_h^s \quad \text{or} \quad \beta M_y = -\underline{\nabla}_h \cdot \tau_h^s \qquad (1.32)$$

This so-called **Sverdrup balance** means that the vorticity of the meridional transport is balanced by the vorticity of the wind stress. Assuming

$$\tau_x^s \sim \sin y, \quad \tau_y^s = 0, \quad \text{so that} \quad -\underset{h}{\nabla} \cdot \tau_h^s \sim -\cos y$$

as an approximation of the major zonal wind belts on the northern hemisphere (westerlies and trades), Equation (1.32) describes the eastern parts of the large anticyclonic gyres. It is also valid for the equatorial dynamics (while geostrophic or Ekman conditions fail because $f = 0$).

The western boundary currents, such as the Gulf stream and Kuroshio, cannot be explained by the Sverdrup balance. They can be justified, however, by introducing a horizontal friction into Equation (1.31). With a linear approach

$$A_h \nabla_h^2 \underset{h}{\mathbf{v}} = -R \underset{h}{\mathbf{v}} = -\frac{R}{h} \underset{h}{\nabla} \psi$$

($R \approx 10^{-6}\,s^{-1}$ is the linear friction factor), which is called the **Stommel balance**,

$$R\nabla_h^2\psi + \beta\frac{\partial\psi}{\partial x} = \underset{h}{\nabla} \cdot \tau_h^s \qquad (1.33)$$

As a solution for a schematic North Atlantic Ocean, with $\psi = 0$ at the boundary, concentric stream lines are obtained with a high density in the extreme western part (**westward intensification**): see Figure 1.3.

The transports are, in Sverdrup's case, $Sv = 10^6\,m^3\,s^{-1}$. This picture reflects qualitatively and – up to a factor of 2 – also quantitatively, the global ocean circulation in the northern hemisphere. In the southern hemisphere, the rotation is anti-clockwise. The observed asymmetry in the ocean general circulation (narrow and strong westerly boundary currents and broad, weak recirculation in the interior and eastern regions) is finally caused by the β effect, i.e. the curvature of the globe. More sophisticated solutions are obtained by introducing higher-order friction terms, additional non-linear inertia terms and a realistic bottom topography.

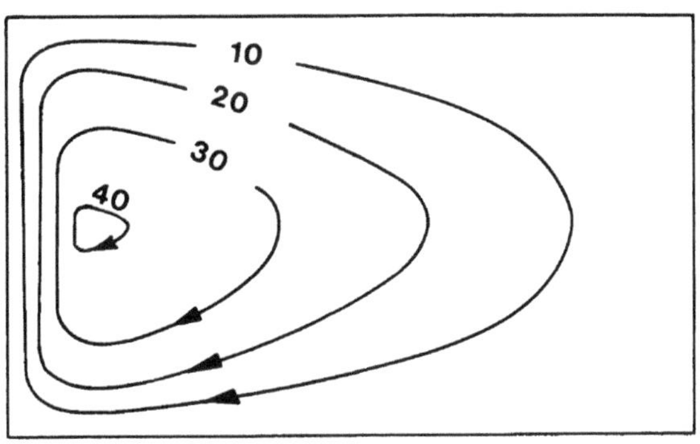

Figure 1.3 Streamlines (in *Sv*) in a schematic North Atlantic Ocean.

1.3 WAVES IN THE OCEAN

1.3.1 CLASSIFICATION

Ocean waves can be classified according to the restoring accelerations (Magaard and Mysak, 1986).

1. Gravity. It is expressed in the homogeneous ocean by the (external) earth's acceleration g. It controls all barotropic **gravitational waves**, from the long tidal waves to the short period wind waves. In the stratified ocean, additionally, the reduced gravity or **buoyancy** is acting according to Equation (1.13):

$$g' = \frac{\rho - \rho_0}{\rho_0} g$$

controlling the **internal waves**.
2. β-effect. According to Equation (1.23), this forces a periodic exchange of relative and planetary vorticity. The corresponding long waves are called **planetary** or **rossby waves**.
3. Surface tension. This causes the short **capillary waves**.
4. Compressibility of the water. This is normally neglected in oceanography but it naturally provides sound waves in the ocean, the most important means of undersea communication.

This classification can be completed by further differentiations:

1. Free or forced oscillations. Eigen oscillations (**seiches**) and ocean swell are belonging to the first group; tides and wind waves to the latter one.
2. Waves without and with the influence of topography. The first are deep-water waves, the second are shallow-water waves. Strong depth gradients cause topographically trapped waves, e.g. **Double Kelvin** waves and **topographical Rossby waves**.

First, the gravity waves are considered. They are described for barotropic conditions by Equations (1.14a,b) and (1.2b). Assuming a non-rotational motion,

$$\nabla \times \mathbf{v} = \mathbf{0}$$

a **velocity potential** ϕ exists with

$$\mathbf{v} = -\nabla\phi$$

It then follows, using Equation (1.13) and neglecting the Coriolis and the external accelerations, that:

$$\frac{\partial^2\phi}{\partial t^2} + g\frac{\partial\phi}{\partial z} = 0 \qquad (1.34a)$$

$$\nabla^2\phi = 0 \qquad (1.34b)$$

With the boundary condition, (1.6b), taking the form:

$$\left.\frac{\partial\phi}{\partial z}\right|_{-h} = 0 \qquad (1.34c)$$

an equation system is obtained for ϕ. The harmonic approach:

$$\phi(\mathbf{x}, t) = \hat{\phi}(z)e^{i(\boldsymbol{\kappa}\cdot\mathbf{x}-\omega t)}$$

with $\boldsymbol{\kappa}$ a wave number vector, then yields a non-trivial solution of Equation (1.34) only if a frequency–wave number relation holds:

$$\tanh \kappa h = \frac{\omega^2}{g\kappa}$$

with $\kappa = |\boldsymbol{\kappa}|$. The **phase velocity** $c = \frac{\omega}{\kappa}$ of gravity waves is then

$$c = \sqrt{\frac{g}{\kappa}\tanh \kappa h} \qquad (1.35)$$

This expression is termed the **dispersion relation.** For deep-water waves ($\kappa h \gg 1$), it follows from Equation (1.35) that:

$$c = \sqrt{\frac{g}{\kappa}} \qquad (1.36)$$

Long waves are propagated faster than short ones; i.e. normal dispersion occurs as in the case of ocean swell and wind waves.

For shallow-water waves ($\kappa h \ll 1$), from Equation (1.35):

$$c = \sqrt{gh} \qquad (1.37)$$

is deduced. These waves are not dispersive and their phase velocity depends only on depth. Seiches, tides and earthquake waves (**tsunamis**) belong to this class.

1.3.2 LONG WAVES

These are represented in the first place by the barotropic shallow-water waves. From Equations (1.13), (1.14a,b) and (1.15) the equation system

$$\frac{\partial v_h}{\partial t} - f\underline{v}_{\urcorner h} + g\nabla_h\eta = \mathbf{q}_h$$

$$\frac{\partial\eta}{\partial t} + \nabla_h\cdot(h\mathbf{v}_h) = 0$$

is achieved, and with a harmonical approach ($\mathbf{v}_h = \hat{\mathbf{v}}_h e^{i\omega t}$, $\eta = \hat{\eta}e^{i\omega t}$) and the assumption $\mathbf{q}_h = 0$, the **wave equation**:

$$\nabla_h^2\eta + \frac{\omega^2 - f^2}{gh}\eta = 0 \qquad (1.38)$$

is found. The solution of Equation (1.38) depends on the coefficient $(\omega^2 - f^2)/gh$, in particular on its sign. For $\omega \leqslant f$ only one solution exists, the **Kelvin wave.** It propagates with a phase velocity (Equation (1.37)) with a perpendicular exponential slope, rising to the right in the northern hemisphere. For $\omega > f$, an infinite number of solutions exist, the **Poincaré waves,** which are undamped in both horizontal directions.

For a closed basin of length l and for $f = 0$, eigen oscillations occur with a period defined by the **Merian formula**:

$$T_m = \frac{2l}{mc} \quad (m = 1, 2, 3, \ldots)$$

Introducing the tidal generating accelerations:

$$\mathbf{q}_h = \nabla_h\Phi$$

where Φ denotes the tidal generating potential, we obtain:

$$\Phi = CgR\cos^2\phi\cos(\sigma t + s\lambda)$$

Thus a class of forced oscillations, the ocean tides, are obtained. Here C, $= 0.263 \times 10^{-7}$, is a tidal constant, $R \approx 6370$ km, the earth's radius, σ is the angular velocity and s the index of the partial tide (for the most important tide, the semi-diurnal principal lunar tide, M_2, $\sigma = 28.984°/\text{h}$, $s = 2$).

Internal waves which are conducted by internal interfaces instead of by the sea surface (maximum density gradient) have a similar horizontal shape as

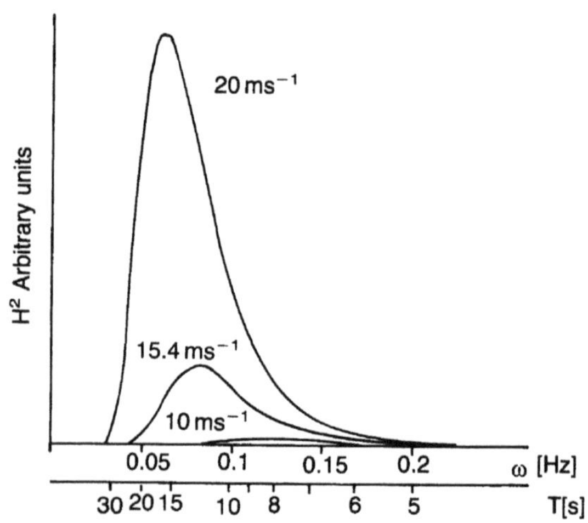

Figure 1.4 The Pierson–Moskowitz spectrum.

long surface waves. Their vertical elevation η obeys the wave equation

$$\frac{d^2\eta}{dt^2} + N^2(z)\eta = 0$$

Here

$$N = \sqrt{-\frac{g}{\rho}\frac{d\rho}{dz}}$$

denotes the **Brunt–Väisälä frequency**. Since only the reduced gravity acts as a restoring force, the amplitudes of internal waves are much higher than those of the surface waves (order of magnitude: 100 m).

The Rossby waves, also long waves, are obtained as solutions of the instationary vorticity Equation (1.22) with $Q = 0$:

$$\frac{\partial}{\partial t}\nabla^2\psi + \beta\frac{\partial\psi}{\partial x} = 0$$

Their phase velocity is

$$c = -\frac{\beta}{\kappa^2}$$

showing that they are always westward directed and normally dispersive.

1.3.3 SHORT WAVES

The short surface waves obey the dispersion relation (Equation (1.36)). The water particles are performing orbital motions which are exponentially damped with depth. Thus, the wind waves are penetrating typically only a few tens of metres.

A fully developed wind-forced sea state is characterized by the **Pierson–Moskowitz spectrum**, (Pierson *et al.*, 1955), see (Figure 1.4). It shows the square of the wave height H (as a measure of energy) depending on wave frequency ω (resp. period $T = 2\pi/\omega$) for different wind speeds. It can be seen that the wave energy increases exponentially with increasing wind velocity. Simultaneously, the energy maximum is shifted to lower frequencies, i.e. to longer periods. For the initial regime of wind-forced surface waves, a modified picture has been deduced, namely the JONSWAP spectrum (Hasselmann *et al.*, 1973).

REFERENCES

Defant, A. (1961) *Physical Oceanography* (2 Volumes), Pergamon, Oxford.

Hasselmann, K. (1982) An ocean model for climate variability studies, *Progress in Oceanography* **11**, 69–92.

Hasselmann, K., Barnett, T.P., Bouws, E. *et al.* (1973) Measurements of wind-wave growth and swell during the Joint North Sea Wave Project (JONSWP). *Deutsche Hydr Zertschrift A (8°)* **12**.

Krauss, W. (1973) *Methods and Results of Theoretical Oceanography. Dynamics of the Homogeneous and the Quasihomogeneous Ocean*, Gbr. Bornträger, Berlin-Stuttgart.

Muller, P. and Willebrand, J. (1989) Ocean circulation: equations for oceanic motions, in *Landolt–Börnshein: Numerical data and functional relationships in science and technology*, Vol V 3b Oceanography (ed. J. Sündermann), Springer-Verlag, Berlin, Chap. 4.1, pp. 1–14.

Magaard, L. and Mysak, L.A. (1986) Ocean waves. Classification and basic features, in *Landolt–Börnstein: Numerical data and functional relationships in science and technology*, Vol V 3c Oceanography (ed. J. Sündermann), Springer-Verlag, Berlin, Chap. 6.1, pp. 1–16.

Pedlosky, J. (1982) *Geophysical Fluid Dynamics*, Springer Verlag, Berlin.

Pierson, W.J., Neumann, G. and James, R.W. (1955) Practical methods for observing and forecasting ocean waves. *US Naval Oceanographic Office Publ. 603.*

UNESCO, ICES, SCOR, IAPSO (1981), Background papers and supporting data on the international equation of state of sea water, 1980. Unesco Technical Paper on Marine Science (Paris) *38*.

PART TWO

THE SCIENTIFIC BACKGROUND

AN INTRODUCTION TO NEARLY-HORIZONTAL FLOWS

2

R.A. Falconer

The flows that occur most commonly in coastal areas, estuaries and harbours are nearly horizontal, that is, they closely follow the earth's surface and so are subject to negligible accelerations in the vertical direction. Pressure distributions in their verticals are thus nearly hydrostatic. These are the flows of tides and storm surges, of salinity intrusions and short-wave-induced currents. Indeed, they are so ubiquitous in their applications that they have come to constitute the most important single class of flows studied in this area. Correspondingly, considerable effort has been expended on their modelling. This section is devoted to a simple mathematical exposition of the theory of nearly horizontal flows, with the aim of deriving the basic equations that are employed in their modelling.

CONTENTS

2.1 INTRODUCTION

This chapter is devoted to the establishment of the governing equations of fluid flow in coastal and estuarine waters, with particular reference to the flow associated with tidal, or long period, wave propagation. The corresponding equations are based on the conservation laws of mass (continuity equation) and momentum (Newton's second law of motion), with the required unknown variables being the water elevation above datum ζ and the velocity components u, v, w as shown in Figure 2.1. These equations can be defined in words for an imaginary control volume as, respectively:

1. The mass inflow per unit time equals the mass outflow per unit time plus the change in mass within the control volume per unit time.
2. The rate of change of momentum is proportional to the resultant applied force and is along the line of action of the force.

2.2 CONTINUITY EQUATION

In deriving the continuity equation for three-dimensional unsteady flow, an infinitesimal control volume of sides Δx, Δy, Δz is considered, with the velocity components being u, v, w at the centre of the control volume and with all derivatives being assumed known at this point. By applying Taylor's series, defined for some fluid characteristic ϕ (e.g. velocity, pressure etc.) in the x-direction as:

$$\phi\left[x \pm \frac{\Delta x}{2}\right] = \phi(x) \pm \frac{1}{2}\left[\frac{\partial\phi(x)}{\partial x}\frac{\Delta x}{1!}\right] + \frac{1}{4}\left[\frac{\partial^2\phi(x)}{\partial x^2}\frac{\Delta x^2}{2!}\right] \pm \frac{1}{8}\left[\frac{\partial^3\phi(x)}{\partial x^3}\frac{\Delta x^3}{3!}\right] + \ldots$$

and assuming that terms higher than third order, i.e. $0(\Delta x^3)$, are small, then the mass fluxes per unit area in the x-direction for the control volume are as shown in Figure 2.2. Likewise, similar mass flux components can be expressed for the y and z directions, respectively.

For flow in the x-direction, the mass influx across side ABCD in time interval Δt is given as

$$M_{\text{in}} = \left[\rho u - \frac{\partial \rho u}{\partial x}\frac{\Delta x}{2}\right]\Delta y\,\Delta z\,\Delta t$$

where ρ is the fluid density. Similarly, the mass efflux across the side EFGH in time interval Δt is given as

Coastal, Estuarial and Harbour Engineer's Reference Book. Edited by M.B. Abbott and W.A. Price. Published in 1993 by Chapman & Hall, 2–6 Boundary Row, London SE1 8HN. ISBN 0 419 15430 2.

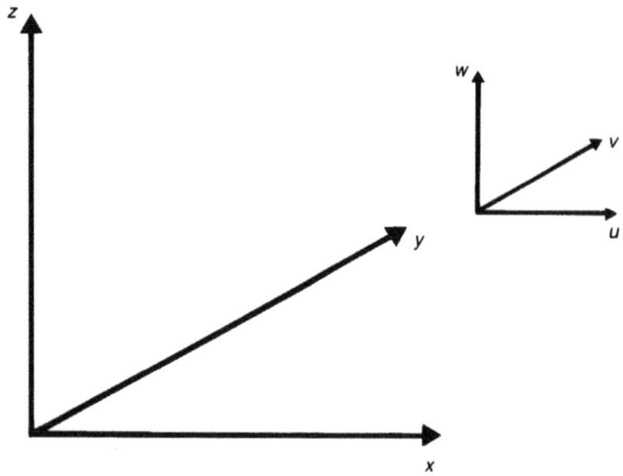

Figure 2.1 Co-ordinate system.

$$M_{out} = \left[\rho u + \frac{\partial \rho u}{\partial x}\frac{\Delta x}{2}\right]\Delta y\,\Delta z\,\Delta t$$

with the resulting net x-direction mass flux being given as

$$\begin{aligned} M_{net} &= M_{in} - M_{out} \\ &= -\frac{\partial \rho u}{\partial x}\Delta x\,\Delta y\,\Delta z\,\Delta t \end{aligned}$$

Similar expressions can be obtained for the y and z directions giving, for the net mass into the control volume:

$$\text{Net mass influx} = \left[\frac{\partial \rho u}{\partial x} + \frac{\partial \rho v}{\partial y} + \frac{\partial \rho w}{\partial z}\right]\Delta x\,\Delta y\,\Delta z\,\Delta t \qquad (2.1)$$

For the corresponding change of mass within the control volume, if the fluid elemental mass at time t is given as $\rho\,\Delta x\,\Delta y\,\Delta z$, the masses at times $t + \Delta t/2$ and $t - \Delta t/2$, as given by Taylor's series, are respectively

$$M_t + \frac{\Delta t}{2} = \rho\,\Delta x\,\Delta y\,\Delta z + \frac{\partial}{\partial t}(\rho\,\Delta x\,\Delta y\,\Delta z)\frac{\Delta t}{2}$$

$$M_t - \frac{\Delta t}{2} = \rho\,\Delta x\,\Delta y\,\Delta z - \frac{\partial}{\partial t}(\rho\,\Delta x\,\Delta y\,\Delta z)\frac{\Delta t}{2}$$

Hence, the net mass increase in time Δt can be expressed as:

$$\text{Net mass increase} = M_{\Delta t} = \frac{\partial}{\partial t}(\rho\,\Delta x\,\Delta y\,\Delta z)\Delta t \qquad (2.2)$$

Equating the net mass influx (Equation (2.1)) with the net mass increase within the control volume in time Δt gives the three-dimensional continuity equation as:

$$\frac{\partial \rho}{\partial t} + \frac{\partial \rho u}{\partial x} + \frac{\partial \rho v}{\partial y} + \frac{\partial \rho w}{\partial z} = 0 \qquad (2.3)$$

This equation applies for all types of flow, including unsteady, turbulent compressible flow.

For steady flows both compressible and incompressible, Equation (2.3) reduces to:

$$\frac{\partial \rho u}{\partial x} + \frac{\partial \rho v}{\partial y} + \frac{\partial \rho w}{\partial z} = 0 \qquad (2.4)$$

where $u, v, w = \phi(x, y, z)$ but $\neq \phi(t)$, and for incompressible flows (both steady and unsteady), Equation (2.3) reduces to:

$$\frac{\partial u}{\partial x} + \frac{\partial v}{\partial y} + \frac{\partial w}{\partial z} = 0 \qquad (2.5)$$

where $u, v, w = \phi(x, y, z, t)$.

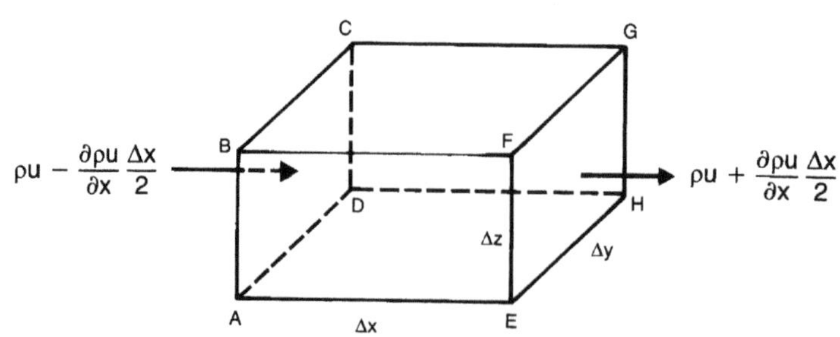

Figure 2.2 Flow into element.

2.3 MOMENTUM EQUATIONS

2.3.1 GENERAL FORMULATION

In considering the momentum equations for three-dimensional unsteady flow, Newton's second law of motion states that the sum of the external forces acting on a unit mass must equal the rate of change of linear momentum, that is:

$$\Sigma \mathbf{F} = \frac{d(m\mathbf{V})}{dt} = m\frac{d\mathbf{V}}{dt} + \mathbf{V}\frac{dm}{dt}$$

where F is the resultant force, m is the mass and $\mathbf{V}$ is the velocity. From the statement of mass conservation, $dm/dt = 0$, giving

$$\Sigma \mathbf{F} = m\frac{d\mathbf{V}}{dt} \tag{2.6}$$

This equation can now be expanded by considering the force components illustrated for the infinitesimal control volume of sides Δx, Δy, Δz shown in Figure 2.3. Using the notation that the first subscript defines the plane normal to the subscript and the second subscript defines the stress direction on the plane, then the x-direction forces on the fluid element include the shear stress components τ_{yz} and τ_{zx}, the normal stress component σ_{xx} and the body force per unit mass X. If the elemental volume is assumed to remain constant, then application of Equation (2.6) to the control volume gives:

$$\rho\,\Delta x\,\Delta y\,\Delta z\frac{du}{dt} = \rho\,X\,\Delta x\,\Delta y\,\Delta z + \frac{\partial\sigma_{xx}}{\partial x}\Delta x\,\Delta y\,\Delta z + \frac{\partial\tau_{yx}}{\partial y}\Delta x\,\Delta y\,\Delta z + \frac{\partial\tau_{zx}}{\partial z}\Delta x\,\Delta y\,\Delta z \tag{2.7}$$

The tensile fluid stress (σ_{xx}) is composed of two components: firstly, the hydrostatic pressure P, which is a negative stress, and secondly the component σ'_{xx}, which is proportional to the time rate of change of strain. This normal stress can be defined as (Schlichting, 1979)

$$\left.\begin{aligned}\sigma_{xx} &= \bar{\sigma} + \sigma'_{xx}\\ \sigma_{yy} &= \bar{\sigma} + \sigma'_{yy}\\ \sigma_{zz} &= \bar{\sigma} + \sigma'_{zz}\end{aligned}\right\} \tag{2.8}$$

where $\bar{\sigma} = 1/3\,(\sigma_{xx} + \sigma_{yy} + \sigma_{zz}) = -P$.

Substituting Equation (2.8) in (2.7) and dividing by $\rho\Delta x\,\Delta y\,\Delta z$ gives the momentum equation in the x-direction for laminar or turbulent flow:

$$\frac{du}{dt} = X - \frac{1}{\rho}\frac{\partial P}{\partial x}$$

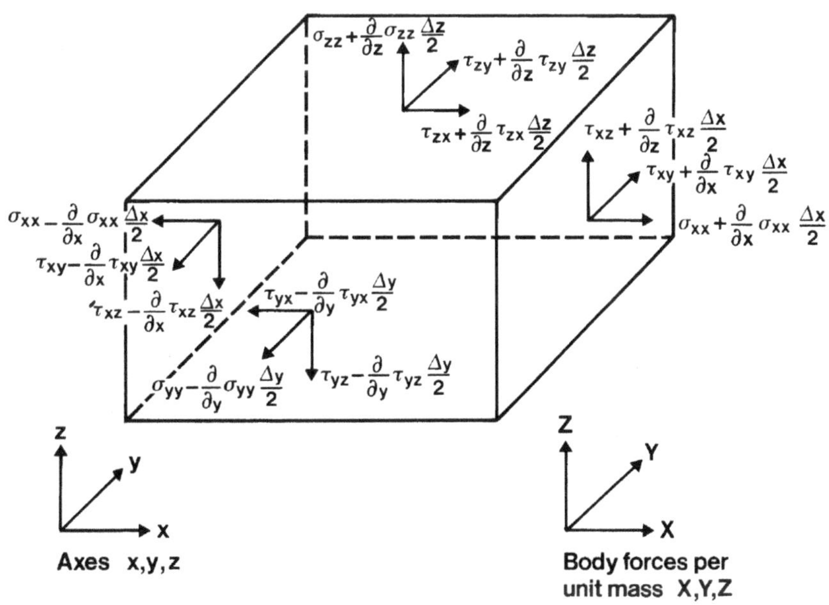

Figure 2.3 Forces acting on fluid element.

$$+\frac{1}{\rho}\left\{\frac{\partial\sigma'_{xx}}{\partial x}+\frac{\partial\tau_{yx}}{\partial y}+\frac{\partial\tau_{zx}}{\partial z}\right\} \qquad (2.9)$$

However, for three-dimensional unsteady flow, $u = \phi(x, y, z, t)$, and the total acceleration can be decomposed into its local and advective components, giving:

$$\frac{du}{dt}=\frac{\partial u}{\partial t}+\frac{\partial u}{\partial x}\frac{dx}{dt}+\frac{\partial u}{\partial y}\frac{dy}{dt}+\frac{\partial u}{\partial z}\frac{dz}{dt}$$

or

$$\frac{du}{dt}=\underbrace{\frac{\partial u}{\partial t}}_{\text{Local acceleration}}+\underbrace{u\frac{\partial u}{\partial x}+v\frac{\partial u}{\partial y}+w\frac{\partial u}{\partial z}}_{\text{Advective acceleration}} \qquad (2.10)$$

2.3.2 FLOW CLASSIFICATION

Fluid flow can be classified into two broad categories, laminar and turbulent flow. Turbulent flow is characterized by mixing action throughout the flow field due to eddies of various scales within the flow. An example is the gusting of wind which causes the wind speed to fluctuate around the mean velocity in an apparently random and chaotic manner. However, laminar flow does not exhibit this intense mixing process and the flow has a smooth appearance, with fluid 'particles' flowing in discrete layers or laminae. The principle differences between the two types of flow are illustrated for open channel flow in Figure 2.4.

Three points are significant from the illustrated profiles in Figure 2.4. Firstly, the eddies create a mixing across the depth of flow resulting in a more uniform velocity profile for the case of turbulent flow; secondly, the turbulence introduces a continuously fluctuating velocity component to the flow; and thirdly, turbulent flow could be described as unsteady flow since at any point the velocity varies continuously with time. However, if the discharge is maintained at a constant rate, and the temporal or time average velocity $\bar{u}$ is considered, then the velocity is virtually constant and the flow is defined as steady turbulent flow.

An index relating the flow conditions to the flow category is the Reynolds number (Re) defined as:

$$\text{Re}=\rho\frac{LU}{\mu}\quad\text{or}\quad\frac{LU}{\nu}$$

where U is a characteristic velocity, L is a characteristic length scale (taken to be the hydraulic radius, or in wide channels the depth of flow), ρ is the fluid density and μ or ν is the absolute or kinematic viscosity, respectively. For free-surface flows, the critical Reynolds number dividing the flow categories is generally assumed to be 500, with turbulent free-surface flow occurring at values of Re greater than 500 and laminar flow at Reynolds numbers less than 500. Another distinction regarding the differences between laminar and turbulent flow occurs immediately adjacent to any solid-wall boundaries. For the 'no-slip' condition to apply to any boundary this requires the velocity immediately adjacent to the boundary to be zero, with the velocity having to increase with distance from the boundary until the free-stream velocity is achieved. The distance between the boundary and the elevation where the free-stream temporal-mean velocity occurs (or nearly occurs) defines the boundary layer thickness. From the definition of the Reynolds number, it is evident that part of the boundary layer must be laminar, termed the laminar sub-layer for turbulent flow. This observation has important ramifications in sediment transport processes and resistance characteristics for all types of flows. A more complete description of this influence of the laminar sub-layer has been given by Schlichting (1979).

2.3.3 LAMINAR FLOW

For the case of laminar flow the normal stress components can be defined as (Schlichting, 1979):

$$\left.\begin{aligned}\sigma'_{xx}&=\mu\left[2\frac{\partial u}{\partial x}-\frac{2}{3}\left[\frac{\partial u}{\partial x}+\frac{\partial v}{\partial y}+\frac{\partial w}{\partial z}\right]\right]\\ \sigma'_{yy}&=\mu\left[2\frac{\partial v}{\partial y}-\frac{2}{3}\left[\frac{\partial u}{\partial x}+\frac{\partial v}{\partial y}+\frac{\partial w}{\partial z}\right]\right]\\ \sigma'_{zz}&=\mu\left[2\frac{\partial w}{\partial z}-\frac{2}{3}\left[\frac{\partial u}{\partial x}+\frac{\partial v}{\partial y}+\frac{\partial w}{\partial z}\right]\right]\end{aligned}\right\} \qquad (2.11)$$

where μ is the coefficient of viscosity.

Similarly, the shear stresses in Equation (2.8) can be expressed in terms of rates of change of strain as:

$$\left.\begin{aligned}\tau_{yx}&=\mu\left[\frac{\partial v}{\partial x}+\frac{\partial u}{\partial y}\right]\\ \tau_{zx}&=\mu\left[\frac{\partial w}{\partial x}+\frac{\partial u}{\partial z}\right]\end{aligned}\right\} \qquad (2.12)$$

with similar shear stress components in the y and z directions. For a full derivation of the shear stress relationships given in Equation (2.12), see Schlichting (1979).

Substitution of Equations (2.12), (2.11) and (2.10) in (2.9) gives the Navier–Stokes equation for compressible flow in the x-direction:

$$\frac{\partial u}{\partial t}+u\frac{\partial u}{\partial x}+v\frac{\partial u}{\partial y}+w\frac{\partial u}{\partial z}=X$$

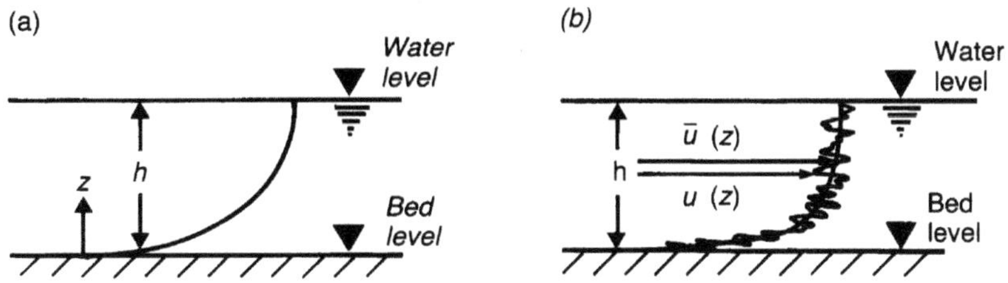

Figure 2.4 Comparison of (a) laminar and (b) turbulent velocity profiles.

$$-\frac{1}{\rho}\frac{\partial P}{\partial x} + \nu\left[\frac{\partial^2 u}{\partial x^2} + \frac{\partial^2 u}{\partial y^2} + \frac{\partial^2 u}{\partial z^2}\right] + \frac{\nu}{3}\left[\frac{\partial^2 u}{\partial x^2} + \frac{\partial^2 v}{\partial x \partial y} + \frac{\partial^2 w}{\partial x \partial z}\right] \qquad (2.13)$$

For incompressible flow, substitution of the corresponding continuity equation (Equation (2.5)) in Equation (2.13) gives:

$$\frac{\partial u}{\partial t} + u\frac{\partial u}{\partial x} + v\frac{\partial u}{\partial y} + w\frac{\partial u}{\partial z} = X - \frac{1}{\rho}\frac{\partial P}{\partial x} + \nu\left[\frac{\partial^2 u}{\partial x^2} + \frac{\partial^2 u}{\partial y^2} + \frac{\partial^2 w}{\partial z^2}\right] \qquad (2.14)$$

with similar equations being obtained for the y and z directions.

2.3.4 TURBULENT FLOW

In practical model studies of coastal and estuarine flows, the flow is almost always highly turbulent and Equation (2.14) has to be modified accordingly. In deriving the x-direction momentum equation for incompressible turbulent flow, the corresponding continuity equation, (2.5) is first multiplied by u, giving:

$$u\frac{\partial u}{\partial x} + u\frac{\partial v}{\partial y} + u\frac{\partial w}{\partial z} = 0$$

and then added to Equation (2.14) to give

$$\frac{\partial u}{\partial t} + \frac{\partial u^2}{\partial x} + \frac{\partial uv}{\partial y} + \frac{\partial uw}{\partial z} = X - \frac{1}{\rho}\frac{\partial P}{\partial x} + \nu\left[\frac{\partial^2 u}{\partial x^2} + \frac{\partial^2 u}{\partial y^2} + \frac{\partial^2 u}{\partial z^2}\right] \qquad (2.15)$$

As indicated in Figure 2.5 the turbulent instantaneous velocity component u can be expressed as

$$u = \bar{u} + u'$$

where

$\bar{u}$ = temporal average velocity component

$$= \frac{1}{\Delta t}\int_t^{t+\Delta t} u\,dt$$

and

u' = fluctuating velocity component where $\bar{u}' = 0$, or

$$\frac{1}{\Delta t}\int_t^{t+\Delta t} u'dt = 0$$

Substituting for the time average fluctuating components in Equation (2.15) gives

$$\frac{\partial}{\partial t}(\bar{u} + u') + \frac{\partial}{\partial x}(\bar{u} + u')(\bar{u} + u') + \frac{\partial}{\partial y}(\bar{u} + u')(\bar{v} + v') + \frac{\partial}{\partial z}(\bar{u} + u')(\bar{w} + w')$$
$$= \bar{X} + X' - \frac{1}{\rho}\frac{\partial}{\partial x}(\bar{P} + P') + \nu\left[\frac{\partial^2}{\partial x^2}(\bar{u} + u') + \frac{\partial^2}{\partial y^2}(\bar{u} + u') + \frac{\partial^2}{\partial z^2}(\bar{u} + u')\right]$$

Expanding and time-averaging the whole equation gives

$$\frac{\partial}{\partial t}\overline{(\bar{u} + u')} + \frac{\partial}{\partial x}\overline{(\bar{u}\bar{u} + 2u'\bar{u} + u'^2)} + \frac{\partial}{\partial y}\overline{(\bar{u}\bar{v} + u'\bar{v} + v'\bar{u} + u'v')} + \frac{\partial}{\partial z}\overline{(\bar{u}\bar{w} + u'\bar{w} + w'\bar{u} + u'w')}$$
$$= \overline{\bar{x} + x'} - \frac{1}{\rho}\frac{\partial}{\partial x}(\overline{\bar{P} + P'}) + \nu\left[\frac{\partial^2}{\partial x^2}\overline{(\bar{u} + u')} + \frac{\partial^2}{\partial y^2}\overline{(\bar{u} + u')} + \frac{\partial^2}{\partial z^2}\overline{(\bar{u} + u')}\right] \qquad (2.16)$$

However,

$$\overline{\bar{u} + u'} = \frac{1}{\Delta t}\int_t^{t+\Delta t}(\bar{u} + u')dt = \bar{u} + \frac{1}{\Delta t}\int_t^{t+\Delta t} u'\,dt = \bar{u},$$

since

$$\frac{1}{\Delta t}\int_t^{t+\Delta t} u'\,dt = 0$$

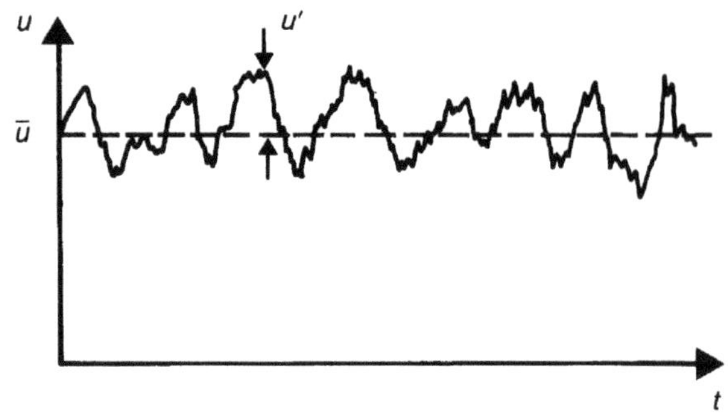

Figure 2.5 Typical turbulent velocity record.

Hence $\overline{\bar{u} + u'} = \bar{u}$, and similarly for all other terms of this form. Likewise, other derivates can be expanded, giving

$$\overline{\bar{u}\bar{u} + 2u'\bar{u} + u'^2} = \frac{1}{\Delta t}\int_t^{t+\Delta t} (\bar{u}\bar{u} + 2u'\bar{u} + u'^2)dt = \bar{u}\bar{u} + \overline{u'u'}$$

since $\bar{u}\bar{u}$ = constant and:

$$\frac{1}{\Delta t}\int_t^{t+\Delta t} 2u'\bar{u} = \frac{2\bar{u}}{\Delta t}\int_t^{t+\Delta t} u'\,dt = 0$$

with similar expressions being obtained for the remaining components of the advective accelerations.

Substituting these time integration formulae in Equation (2.16) gives

$$\frac{\partial \bar{u}}{\partial t} + \frac{\partial \bar{u}\bar{u}}{\partial x} + \frac{\partial \bar{u}\bar{v}}{\partial y} + \frac{\partial \bar{u}\bar{w}}{\partial z} = \bar{X} - \frac{1}{\rho}\frac{\partial \bar{P}}{\partial x} + \nu\left[\frac{\partial^2 \bar{u}}{\partial x^2} + \frac{\partial^2 \bar{u}}{\partial y^2} + \frac{\partial^2 \bar{u}}{\partial z^2}\right] - \left[\frac{\partial \overline{u'u'}}{\partial x} + \frac{\partial \overline{u'v'}}{\partial y} + \frac{\partial \overline{u'w'}}{\partial z}\right] \quad (2.17)$$

As before, multiplying the incompressible continuity equation (2.5) by $\bar{u}$ and subtracting from Equation (2.17), and recalling that $\nu = \mu/\rho$, gives the x-direction Navier–Stokes equation for incompressible turbulent flow:

$$\frac{\partial \bar{u}}{\partial t} + \bar{u}\frac{\partial \bar{u}}{\partial x} + \bar{v}\frac{\partial \bar{u}}{\partial y} + \bar{w}\frac{\partial \bar{u}}{\partial z} = \bar{X} - \frac{1}{\rho}\frac{\partial \bar{P}}{\partial x} + \frac{1}{\rho}\frac{\partial}{\partial x}\left[\mu\frac{\partial \bar{u}}{\partial x} - \rho\overline{u'u'}\right] + \frac{1}{\rho}\frac{\partial}{\partial y}\left[\mu\frac{\partial \bar{u}}{\partial y} - \rho\overline{u'v'}\right] + \frac{1}{\rho}\frac{\partial}{\partial z}\left[\mu\frac{\partial \bar{u}}{\partial z} - \rho\overline{u'w'}\right] \quad (2.18)$$

with similar equations for the y and z directions.

Comparison of Equation (2.18) with the original Navier–Stokes equation (2.14) in the x-direction shows that the instantaneous values of velocity and pressure have been replaced by the corresponding temporal average values, together with the addition of three further terms on the right-hand side of Equation (2.18), namely:

$$-\frac{1}{\rho}\frac{\partial}{\partial x}\rho\overline{u'u'}, \quad -\frac{1}{\rho}\frac{\partial}{\partial x}\rho\overline{u'v'}, \quad -\frac{1}{\rho}\frac{\partial}{\partial x}\rho\overline{u'w'}$$

The corresponding terms $-\rho\overline{u'u'}$, $-\rho\overline{u'v'}$ and $-\rho\overline{u'w'}$ are known as the Reynolds (or apparent) stresses, with their form being similar to the viscous shear stresses. These additional terms arise from the time averaging of the turbulence and are zero for laminar flow.

In solving for these Reynolds stresses, Boussinesq (see Goldstein, 1938) proposed that they could be represented in a diffusive manner whereby:

$$\left.\begin{aligned} -\rho\overline{u'u'} &= \eta\left[\frac{\partial \bar{u}}{\partial x} + \frac{\partial \bar{u}}{\partial x}\right] \\ -\rho\overline{u'v'} &= \eta\left[\frac{\partial \bar{u}}{\partial y} + \frac{\partial \bar{v}}{\partial x}\right] \\ -\rho\overline{u'w'} &= \eta\left[\frac{\partial \bar{u}}{\partial z} + \frac{\partial \bar{w}}{\partial x}\right] \end{aligned}\right\} \quad (2.19)$$

where η is the absolute eddy viscosity, or, as for laminar viscosity, ε is the kinematic eddy viscosity which equals η/ρ. In general, for turbulent flow, $\eta \gg \mu$, or $\varepsilon \gg \nu$.

To determine the Reynolds stresses given in Equation (2.19), many different and varyingly complicated expressions exist for ε (or η), which can be summarized as follows.

1. The simplest method of closing the equation by evaluating ε is to assume a constant value based on field data. Typically $\varepsilon = 10\,\text{m}^2\,\text{s}^{-1}$ (see Bengtsson, 1978).
2. The representation most widely used in coastal and estuarine flows is to apply a zero-equation turbulence model similar to that prescribed by Prandtl's mixing length hypothesis (Goldstein, 1938) where, for the x-direction:

$$\varepsilon = \ell^2 J$$

where ℓ is some characteristic mixing length and

$$J = \left[2\left[\frac{\partial \bar{u}}{\partial x}\right]^2 + 2\left[\frac{\partial \bar{v}}{\partial y}\right]^2 + 2\left[\frac{\partial \bar{w}}{\partial z}\right]^2 + \left[\frac{\partial \bar{w}}{\partial y} + \frac{\partial \bar{v}}{\partial z}\right]^2 + \left[\frac{\partial \bar{u}}{\partial z} + \frac{\partial \bar{w}}{\partial x}\right]^2 + \left[\frac{\partial \bar{v}}{\partial x} + \frac{\partial \bar{u}}{\partial y}\right]^2\right]^{1/2}$$

although this expression can usually be simplified due to the dominance of one or more velocity gradients. In determing the mixing length ℓ, Prandtl suggested $\ell = \kappa y$ near a wall, where κ is the von Kármán constant (= 0.4) and y is the co-ordinate perpendicular to the wall, with $y = 0$ at the wall. Other expressions are available for ℓ including, for example, the von Kármán representation given as

$$\ell = \kappa \left| \frac{\partial \bar{u}}{\partial y} \Big/ \frac{\partial^2 \bar{u}}{\partial y^2} \right|$$

3. For more refined turbulence models, transport equations exist for the turbulent kinetic energy and the energy dissipation, i.e. the $\kappa - \varepsilon$ model, or for the Reynolds stresses directly, i.e. algebraic stress models. Further details of these refined models are given elsewhere in this volume and by Rodi (1984).

In modelling large-scale coastal and estuarine flows it is often convenient to assume that the flow is inviscid (i.e. non-viscous) and that the turbulent stresses are insignificant in comparison with other forces, such as the pressure gradient and bed resistance. For such flow approximations the Navier–Stokes equations reduce to the Euler equations, giving for the x-direction:

$$\frac{\partial u}{\partial t} + u\frac{\partial u}{\partial x} + v\frac{\partial u}{\partial y} + w\frac{\partial u}{\partial z} = X - \frac{1}{\rho}\frac{\partial P}{\partial x} \qquad (2.20)$$

2.3.5 BODY FORCES AND CORIOLIS ACCELERATION

If the effects of the earth's rotation are neglected and assuming that z is the vertical co-ordinate then the body forces are given as

$$\left.\begin{aligned} X &= 0 \\ Y &= 0 \\ Z &= -g \end{aligned}\right\} \qquad (2.21)$$

where g is the acceleration due to gravity. However, in modelling coastal and estuarine flows the effect of the earth's rotation cannot be ignored and an additional component of acceleration arises due to relative motion. This additional acceleration component is known as the Coriolis acceleration and can be included in the body force components as:

$$\left.\begin{aligned} X &= +2\bar{v}\omega \sin\phi \\ Y &= -2\bar{u}\omega \sin\phi \\ Z &= -g \end{aligned}\right\} \qquad (2.22)$$

where ω is the earth's rotation $\simeq 7.3 \times 10^{-5}$ rad s^{-1}, i.e. $2\pi/(24 \times 36\,000)$, and ϕ is the earth's latitude at the site of interest. The effects of the Coriolis acceleration are related to various hydrodynamic phenomena such as;

1. The transverse surface slopes occurring in open channels and coastal seas;
2. The bending of streamlines and the deflection of jets, including tidal jets;
3. Amphidromal tidal systems, where high water rotates about points of no tidal amplitude.

Such flow modelling is also affected by the influence of the Coriolis acceleration on air flows, such as tornadoes. Further details of the Coriolis acceleration and a full derivation of the acceleration components have been given by Dronkers (1964).

2.4 DEPTH INTEGRATED EQUATIONS

For many coastal- and estuarine-flow problems the vertical velocity component $\bar{w}$ is relatively small in comparison with the horizontal velocity components $\bar{u}$, $\bar{v}$. Hence the continuity equation and the Navier–Stokes equations can be integrated over the depth and solved numerically to give the depth-averaged velocity fields. Such flow fields can generally be assumed to occur in unstratified, or weakly stratified, estuaries, harbours, bays and coastal seas where the wind shear is also relatively small.

Using the notation given in Figure 2.6 and dropping the overbar for convenience, depth integration of the continuity equation (2.5) gives

$$\int_{-h}^{\zeta}\left[\frac{\partial u}{\partial x} + \frac{\partial v}{\partial y} + \frac{\partial w}{\partial z}\right]\mathrm{d}z = \int_{-h}^{\zeta}\left[\frac{\partial u}{\partial x} + \frac{\partial v}{\partial y}\right]\mathrm{d}z + w_\zeta - w_{-h} \qquad (2.23)$$

Using the Leibnitz rule (Sokolnikoff and Redheffer, 1966), given as:

$$\int_a^b \frac{\partial}{\partial y} f(x,y)\mathrm{d}z = \frac{\partial}{\partial y}\int_a^b f(x,y)\mathrm{d}z - f(b,y)\frac{\partial b}{\partial y} + f(a,y)\frac{\partial a}{\partial y}$$

Equation (2.23) can be expanded to give

$$\frac{\partial}{\partial x}\int_{-h}^{\zeta} u\mathrm{d}z + \frac{\partial}{\partial y}\int_{-h}^{\zeta} v\mathrm{d}z - u_\zeta \left.\frac{\partial \zeta}{\partial x}\right|_\zeta + u_{-h}\left.\frac{\partial(-h)}{\partial x}\right|_{-h} - v_\zeta\left.\frac{\partial \zeta}{\partial y}\right|_\zeta + v_{-h}\left.\frac{\partial(-h)}{\partial y}\right|_{-h} + w_\zeta - w_h = 0 \qquad (2.24)$$

At the bed, the no-slip boundary condition gives $u_{-h} = v_{-h} = w_{-h} = 0$, and at the surface the kinematic free

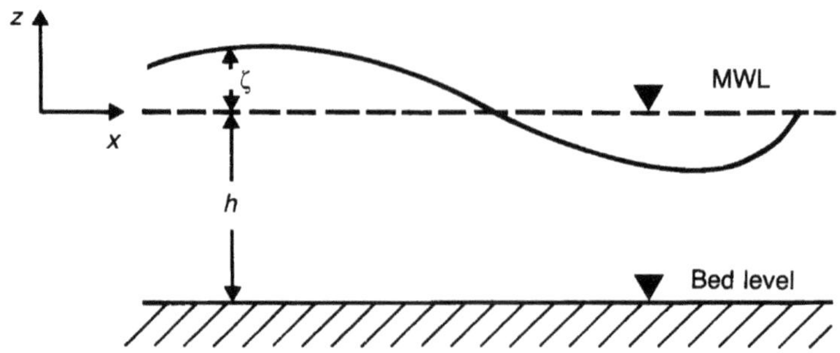

Figure 2.6 Co-ordinate system for depth integrated equations.

surface condition, which assumes that a fluid particle remains on the surface (Vallentine, 1969), gives

$$\left.\frac{d\zeta}{dt}\right|_\zeta = \left[\frac{\partial\zeta}{\partial t} + \frac{\partial\zeta}{\partial x}\frac{dx}{dt} + \frac{\partial\zeta}{\partial y}\frac{dy}{dt}\right]_\zeta$$

$$= \left[\frac{\partial\zeta}{\partial t} + u\frac{\partial\zeta}{\partial x} + v\frac{\partial\zeta}{\partial y}\right]_\zeta = w_\zeta$$

Substituting in Equation (2.24) gives, for the depth integrated continuity equation:

$$\frac{\partial\zeta}{\partial t} + \frac{\partial UH}{\partial x} + \frac{\partial VH}{\partial y} = 0 \tag{2.25}$$

where U, V = depth average velocity components, defined as

$$U = \frac{1}{H}\int_{-h}^{\zeta} u\mathrm{d}z, \quad V = \frac{1}{H}\int_{-h}^{\zeta} v\mathrm{d}z$$

and H = total depth of water column = $h + \zeta$.

Similarly, for incompressible turbulent flow, substitution of Equations (2.19) and (2.22) in (2.17) and integration over the depth gives:

$$\int_{-h}^{\zeta}\left[\frac{\partial u}{\partial t} + \frac{\partial u^2}{\partial x} + \frac{\partial uv}{\partial y} + \frac{\partial uw}{\partial z}\right]\mathrm{d}z = \int_{-h}^{\zeta} fv\mathrm{d}z - \frac{1}{\rho}\int_{-h}^{\zeta}\frac{\partial P}{\partial x}\mathrm{d}z$$
$$+ \int_{-h}^{\zeta}\left[\frac{\partial}{\partial x}(\varepsilon+\nu)\left[2\frac{\partial u}{\partial x}\right] + \frac{\partial}{\partial y}(\varepsilon+\nu)\left[\frac{\partial u}{\partial y} + \frac{\partial v}{\partial x}\right]\right.$$
$$\left. + \frac{\partial}{\partial z}(\varepsilon+\nu)\left[\frac{\partial u}{\partial z} + \frac{\partial w}{\partial x}\right]\right]\mathrm{d}z \tag{2.26}$$

Applying the kinematic free surface condition, assuming $\varepsilon \gg \nu$ and making the approximation that:

$$\frac{\partial}{\partial x}\int_{-h}^{\zeta}\varepsilon\frac{\partial u}{\partial x} = \frac{\partial}{\partial x}\left[\bar{\varepsilon}H\frac{\partial U}{\partial x}\right]$$

where $\bar{\varepsilon}$ is the depth mean eddy viscosity (Falconer, 1976), this gives, for the x-direction momentum equation (2.26):

$$\frac{\partial UH}{\partial t} + \frac{\partial}{\partial x}(\beta U^2H) + \frac{\partial}{\partial y}(\beta UVH) = fVH$$
$$- \frac{1}{\rho}\int_{-h}^{\zeta}\frac{\partial P}{\partial x}\mathrm{d}z + \frac{\tau_{xw}}{\rho} - \frac{\tau_{xb}}{\rho} + 2\frac{\partial}{\partial x}\left[\bar{\varepsilon}H\frac{\partial U}{\partial x}\right]$$
$$+ \frac{\partial}{\partial y}\left[\bar{\varepsilon}H\left[\frac{\partial U}{\partial y} + \frac{\partial V}{\partial x}\right]\right] \tag{2.27}$$

where β is the momentum correction factor, τ_{xw} is the surface wind shear stress component, τ_{xb} is the bed shear stress component and f is the Coriolis parameter (= $2\omega \sin\theta$).

In practical model studies, and in the absence of extensive field data, the momentum correction factor β is either set to unity or a specific vertical velocity profile is assumed. For an assumed logarithmic vertical velocity profile of the form

$$u = \frac{U_*}{k}\log_e(h+z) + U_*C_1$$

where U_* is the shear velocity (equal to $\sqrt{\tau_o/\rho}$, where τ_o is the boundary shear stress) it can readily be shown that

$$\beta = \left[1 + \frac{g}{C^2\kappa^2}\right] \tag{2.28}$$

where C is the de Chezy bed roughness coefficient.

Alternatively, for a wind generated flow field, a parabolic velocity distribution can be assumed of the form (Koutitas and Cousidou–Koutita, 1986):

$$u = \left[\frac{3a_x}{4} - \frac{3U}{2}\right]\left[\frac{z^2}{h} - 1\right] + a_x\left[\frac{z}{h} + 1\right] \tag{2.29}$$

where $a_x = H\tau_{xw}/\rho\varepsilon_\zeta$ and ε_ζ is the eddy viscosity at the free surface, giving, for β

$$\beta = 1.2 \tag{2.29}$$

For the pressure term and for nearly horizontal flows, the vertical acceleration due to hydrodynamic processes is generally small in comparison with gravity effects. Consequently, in the z-direction momentum equation corresponding to Equation (2.14), the vertical accelerations and shear stresses are neglected, such that a hydrostatic pressure variation results (Falconer, 1976), giving:

$$\frac{\partial P}{\partial z} + \rho g = 0 \qquad (2.30)$$

Integrating Equation (2.30) gives a linear pressure distribution with depth, that is:

$$P(z) = \int -\rho g dz = -\rho g z + \text{constant} \qquad (2.31)$$

Using the boundary condition that the pressure at the free surface, i.e. $z = \zeta$, is at atmospheric pressure, Pa, then Equation (2.31) becomes:

$$P(z) = \rho g(\zeta - z) + Pa$$

The derivatives of the pressure in any horizontal direction can be expressed in terms of the water elevation giving for the x-direction

$$\frac{\partial P}{\partial x} = \rho g \frac{\partial \zeta}{\partial x} + \frac{\partial Pa}{\partial x} \qquad (2.32)$$

Differences in atmospheric pressure at the water surface can give rise to water level variations which may be of significance if the area being modelled is relatively large. However, in most coastal and estuarine engineering studies the barometric pressure gradient is insignificant and is not included in the governing equations.

For the surface shear stress due to wind action, resolving forces horizontally for a steady uniform flow gives for the x-direction:

$$\tau_{xw} = \gamma \rho_a W_x W_s \qquad (2.33)$$

where γ is the air-water resistance coefficient, ρ_a is the air density, W_x is the wind velocity component in the x-direction and W_s is the wind speed. Various constant values and formulae have been proposed for the air–water resistance coefficient, with the most widely used representation being that attributed to Munk (Dronkers, 1964) whereby $\gamma = 0.0026$.

In most coastal and estuarine flow studies the bed shear stress is represented in a manner similar to that for steady, uniform open-channel flow (Henderson, 1966). Thus, for the x-direction bed shear stress

$$\tau_{xb} = \frac{\rho g U\sqrt{U^2 + V^2}}{C^2} \qquad (2.34)$$

To determine the Chezy value either a constant value can be included directly for C (typically $30\,\text{m}^{1/2}\,\text{s}^{-1} < C < 100\,\text{m}^{1/2}\,\text{s}^{-1}$ for such flows) or C can be evaluated from the Manning equation, that is:

$$C = H^{1/6} n^{-1} \qquad (2.35)$$

where n is the Manning roughness coefficient (typically $0.015 < n < 0.04$). Alternatively, the Colebrook–White equation can be used giving

$$C = \left[\frac{2g}{f}\right]^{1/2} = -\sqrt{32g}\,\log_{10}\left[\frac{k_s}{14.84H} + \frac{1.255C}{(4UH/\nu)\sqrt{2g}}\right] \qquad (2.36)$$

where f is the Darcy–Weisbach friction coefficient and k_s is the Nikuradse equivalent sand roughness size. The advantage of using the Colebrook–White equation to calculate C is that the value of the roughness coefficient k_s can be more closely related to bed features such as ripples or dunes and the representation of the Chezy value can include transitional turbulent flow (or turbulent rough flow that is not fully developed). This refinement can be particularly important when modelling the flooding and drying of tidal flood plains, where the Reynolds number effects may be significant (Falconer and Owens, 1987).

Finally, the turbulent diffusion terms in Equation (2.27) can be simplified by neglecting, firstly the gradients of $\bar{\varepsilon}H$, and secondly, the divergence of the mean flow (Kuipers and Vreugdenhil, 1973), giving:

$$2\frac{\partial}{\partial x}\left[\bar{\varepsilon}H\frac{\partial U}{\partial x}\right] + \frac{\partial}{\partial y}\left[\bar{\varepsilon}H\left[\frac{\partial U}{\partial y} + \frac{\partial V}{\partial x}\right]\right] = 2\bar{\varepsilon}H\frac{\partial^2 U}{\partial x^2} + \bar{\varepsilon}H\frac{\partial}{\partial y}\left[\frac{\partial U}{\partial y} + \frac{\partial V}{\partial x}\right] + 2\frac{\partial u}{\partial x}\frac{\partial \bar{\varepsilon}H}{\partial x} + \left[\frac{\partial U}{\partial y} + \frac{\partial V}{\partial x}\right]\frac{\partial \bar{\varepsilon}H}{\partial y} \simeq \bar{\varepsilon}H\left[\frac{\partial^2 U}{\partial x^2} + \frac{\partial^2 U}{\partial y^2}\right] \qquad (2.37)$$

Values of β and $\bar{\varepsilon}$ can either be estimated from field data or, assuming that bed generated turbulence dominates over free shear layer turbulence, a logarithmic velocity profile can be given, such as (Elder, 1959):

$$\bar{\varepsilon} = \kappa/6U_*H \qquad (2.38)$$

However, field data by Fischer (1973) for turbulent diffusion in rivers has shown that the value for $\bar{\varepsilon}$ is generally greater than that given by Equation (2.38), and is more accurately given by

$$\bar{\varepsilon} = 0.15U_*H \qquad (2.39)$$

Alternatively, and particularly where free-shear-layer turbulence and turbulent advection are important (e.g. as in tidal jets), then the two-equation depth-integrated k–ε turbulence model can be used to evaluate $\bar{\varepsilon}$.

Hence, in modelling coastal and estuarine flows where the vertical accelerations are small and the flow is well mixed vertically, the governing momentum equations can be obtained by substituting Equations (2.28), (2.32)–(2.34) and (2.37) in Equation (2.27), giving for the x-direction:

$$\frac{\partial UH}{\partial t} + \frac{\partial}{\partial x}(\beta U^2 H) + \frac{\partial}{\partial y}(\beta UVH) = fVH - gH\frac{\partial \zeta}{\partial x} - \frac{H}{\rho}\frac{\partial P_a}{\partial x} + \frac{\gamma \rho_a W_x W_s}{\rho} - \frac{gU\sqrt{U^2 + V^2}}{C^2} + \bar{\varepsilon} H\left[\frac{\partial^2 U}{\partial x^2} + \frac{\partial^2 U}{\partial y^2}\right] \quad (2.40)$$

and similarly for the y-direction:

$$\frac{\partial VH}{\partial t} + \frac{\partial}{\partial x}(\beta UVH) + \frac{\partial}{\partial y}(\beta V^2 H) = fUH - gH\frac{\partial \zeta}{\partial y} - \frac{H}{\rho}\frac{\partial P_a}{\partial x} + \frac{\gamma \rho_a W_y W_s}{\rho} - \frac{gV\sqrt{U^2 + V^2}}{C^2} + \bar{\varepsilon} H\left[\frac{\partial^2 V}{\partial x^2} + \frac{\partial^2 V}{\partial y^2}\right] \quad (2.41)$$

When Equations (2.40) and (2.41) are combined with Equation (2.25), the three equations can be solved numerically, for prescribed open boundary conditions, to give the time and space varying values for the three unknown variables, namely ζ, U and V. Details of the solution of these equations are given elsewhere in this volume.

REFERENCES

Bengtsson, L. (1978) Wind-induced circulation in lakes. *Nordic Hydrology*, **9**, 75–94.

Dronkers, J.J. (1964) *Tidal Computations in Rivers and Coastal Waters*, North Holland, Amsterdam.

Elder, J.W. (1959) The dispersion of marked fluid in turbulent shear flow. *Journal of Fluid Mechanics*, **5** (4), 544–560.

Falconer, R.A. (1976) Mathematical modelling of jet-forced circulation in reservoirs and harbours. *PhD thesis* University of London, UK.

Falconer, R.A. and Owens, P.H. (1987) Numerical simulation of flooding and drying in a depth-averaged tidal flow model. *Proceedings of the Institution of Civil Engineers*, **83** (2), 161–180.

Fischer, H.B. (1973) Longitudinal dispersion and turbulent mixing in open channel flow. *Annual Review of Fluid Mechanics*, **5**, 59–78.

Goldstein, S. (1938) *Modern Developments in Fluid Dynamics*, Vol 1, Oxford University Press, Oxford.

Henderson, F.M. (1966) *Open Channel Flow*, Collier MacMillan, London.

Koutitas, C. and Cousidou-Koutita, M. (1986) A comparative study of three mathematical models for wind generated circulation in coastal areas. *Coastal Engineering*, **10**, 127–138.

Kuipers and Vrengdenhil (1973) Calculations of two-dimensional horizontal flow. *Delft Hydraulics Laboratory Tech. Rep. S163*, Part 1 pp. 1–44.

Rodi, W. (1984) *Turbulence Models and their Application in Hydraulics*, 2nd edn, International Association for Hydraulic Research, Delft, The Netherlands.

Schlichting, H. (1979) *Boundary Layer Theory*, 6th edn, McGraw-Hill, New York.

Sokolnikoff, I.S. and Redheffer, R.M. (1966) *Mathematics of Physics and Modern Engineering*, 2nd edn, McGraw-Hill, New York.

Vallentine, H.R. (1969) *Applied Hydrodynamics*, 2nd edn, Butterworth, London.

THE THEORY OF SHORT-PERIOD WAVES 3

R.A. Dalrymple

The next most important class of flow for the engineer working in the coastal, estuarial and harbour zones is that of short-period waves. Waves of this type have been studied for more than 150 years by mathematicians and physicists, but it is only recently, with the advent of the digital computer, that they have been studied extensively from the point of view of engineering applications. This chapter provides an introduction to the classical mathematical theory, which still has its following, while also introducing the way in which this theory is being modified and developed in order to meet the requirements of practice.

CONTENTS

3.1 INTRODUCTION

Water waves are complicated phenomena, involving water motions created by a single wave train, as in the laboratory, or the superimposition of many different wave trains having a variety of shapes, speeds, and directions. Some of the analytical and numerical theories for water waves will be discussed briefly here.

This chapter will primarily discuss the Airy wave theory which is the simplest wave theory available. While it has numerous drawbacks, due to the assumptions made in its derivation, it has a surprisingly wide range of applications, even being used for the design of large offshore structures.

Non-linear theories, including numerical models, will then be presented. It is important to be aware that there are models which are valid in two dimensions and those that are capable of treating three dimensions. Two-dimensional models have been used in the past for offshore design; three-dimensional models are necessary to predict wave behaviour over large areas where the bathymetry is variable.

3.2 AIRY WAVE THEORY

Waves are primarily created by the wind blowing over the ocean surface and they serve as a mechanism to transport wind energy (impressed on the water) away from the generation area. Because of the large extent of the oceans and the fact that waves can propagate over great distances without significant damping, waves are almost always present at a given site. During a storm, the sea surface can be very confused, referred to as 'sea', since the wind is generating waves in many directions and sizes. Far from the storm area, the waves are more regular (swell) and often can be considered as coming from a single direction.

A given wave can be described geometrically by a wave height, H, which is the distance between a wave crest and the previous (or trailing) wave trough, as shown in Figure 3.1. For small waves, the crest and trough elevations above the mean sea level are equal, while for large waves, the crest elevation is greater than that of the trough, due to non-linear effects. The wave length, L, is twice the distance from the trough to the trailing (or leading) crest. The wave steepness is defined as the ratio of these two wave characteristics: H/L. Waves are steep when the wave height is large with respect to the wave length; the maximum wave steepness for a deep water wave is approximately 1/7. Waves steeper than 1/7 will break.

Another wave property which can easily be measured is the wave period, T, which is the time taken for two crests to pass a given point. The speed at which the wave propagates is the wave celerity, C, which is defined as the ratio of the distance the wave travels in one wave period divided by the period ($C = L/T$). This is sometimes referred to as the wave phase speed.

Coastal, Estuarial and Harbour Engineer's Reference Book. Edited by M.B. Abbott and W.A. Price. Published in 1993 by Chapman & Hall, 2–6 Boundary Row, London SE1 8HN. ISBN 0 419 15430 2.

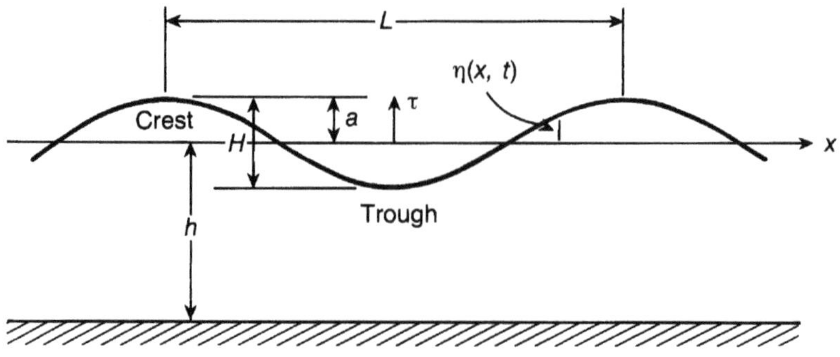

Figure 3.1 Geometrical parameters for waves.

The simplest description of waves assumes that we can describe the wave form with a sinusoidal curve, which presumes that the crest and trough elevations are similar in shape. At a fixed instant in time, the wave form can be described as a function of distance x (x is assumed to lie in the direction of the waves) as:

$$\eta = \frac{H}{2}\cos kx \tag{3.1}$$

Here η determines the elevation of the wave form above the still water level, and k is the wave number, defined as $k = 2\pi/L$. The k ensures that as we move a distance equal to a wavelength in the wave direction (say from $x = 0$ to $x = L$), the wave repeats (since $\cos 0 = \cos 2\pi$).

To permit the wave form to move, we allow time to advance, while requiring that the wave form repeats in time every T seconds.

$$\eta = \frac{H}{2}\cos k(x - Ct) = \frac{H}{2}\cos(kx - \sigma t) \tag{3.2}$$

where σ is the angular frequency of the wave ($\sigma = 2\pi/T$).

A mathematical derivation of the Airy theory, with all of the assumptions needed to determine the relevant governing mathematical equation and appropriate conditions that apply at the water surface and the bottom, is provided in numerous texts (e.g. Dean and Dalrymple, 1984).

3.2.1 WAVE KINEMATICS AND ENERGY

The wave motion is determined from the velocity potential, $\phi(x, y, z, t)$, a mathematical function of space (x, y, z), and time t, which permits the determination of velocities, accelerations and pressures under the wave form. The existence of this potential function is assured by the assumptions that the fluid is incompressible and the wave motion is irrotational. This last assumption means that fluid particles do not rotate about an axis as they move, which might seem to be a very dubious assumption in the presence of viscosity and bottom friction; however, in practice, only in the vicinity of the bottom and the surface is the rotation of the fluid important. These two assumptions lead to the governing Laplace equation for the fluid motion.

The velocity potential, satisfying the Laplace equation, is given by

$$\phi(x, z, t) = \frac{Hg}{2\sigma}\frac{\cosh k(h + z)}{\cosh kh}\sin(kx - \sigma t) \tag{3.3}$$

where g is the acceleration of gravity, h is the water depth and z is the elevation above the still water line. The following requirement also results from the wave theory

$$\sigma^2 = gk \tanh kh \tag{3.4}$$

which is the **dispersion relationship**, which fixes the dependency of the wave number k on the angular frequency σ. Here dispersion means that a group of waves of different frequencies will disperse in space as the longer period waves travel faster than the shorter period waves. By substituting the definitions of k and σ, this expression can be rewritten as

$$L = \frac{gT^2}{2\pi}\tanh kh \tag{3.5}$$

The disadvantage of this equation is that it must be solved by trial and error or graphically.

There are some simplifications of the dispersion relationship available for shallow and deep water. Defining shallow water as requiring a relative water depth, $h/L < 1/20$, we have

$$L = \sqrt{gh}T \tag{3.6}$$

In deep water, with an associated relative water depth of $h/L > 1/2$, we have

$$L = \frac{gT^2}{2\pi} \tag{3.7}$$

The water particle velocities and accelerations as well as the pressure under the waves can be obtained from derivatives of the velocity potential. The horizontal and vertical velocities are

$$u(x, z, t) = \frac{gHk}{2\sigma} \frac{\cosh k(h + z)}{\cosh kh} \cos(kx - \sigma t) \quad (3.8)$$

$$w(x, z, t) = \frac{gHk}{2\sigma} \frac{\sinh k(h + z)}{\cosh kh} \sin(kx - \sigma t) \quad (3.9)$$

The trajectories of these water particles can be shown to be elliptical paths about the mean particle position. At the bottom, in shallow water, these paths collapse to a horizontal line parallel to the bottom, while in deep water the ellipses tend to circular orbits, decaying exponentially with depth.

The accelerations according to the Airy theory are

$$\dot{u}(x, z, t) = \frac{gHk}{2} \frac{\cosh k(h + z)}{\cosh kh} \sin(kx - \sigma t) \quad (3.10)$$

$$\dot{w}(x, z, t) = -\frac{gHk}{2} \frac{\sinh k(h + z)}{\cosh kh} \cos(kx - \sigma t) \quad (3.11)$$

The pressure is a combination of the hydrostatic pressure of the water and the dynamic pressure due to the presence of the waves

$$p = \rho g z + \rho g \frac{H}{2} \frac{\cosh k(h + z)}{\cosh kh} \cos(kx - \sigma t) \quad (3.12)$$

This wave pressure will lead to wave forces on submerged obstacles.

For waves travelling in directions other than the x direction, the following potential is used

$$\phi(x, y, z, t) = \frac{Hg}{2\sigma} \frac{\cosh k(h + z)}{\cosh kh} \times \sin(k \cos\theta x + k \sin\theta y - \sigma t) \quad (3.13)$$

where θ is the angle made by the wave direction with the x axis. The argument of the sine function, $S(x, y, t) \equiv (k \cos\theta x + k \sin\theta y - \sigma t)$ is often written as $S = (\mathbf{k} \cdot \mathbf{x} - \sigma t)$, where S is the **phase function**. The wave number vector, $\mathbf{k} \equiv \nabla S$, where ∇ is defined as the horizontal gradient

$$\nabla \equiv \left(\frac{\partial}{\partial x}\mathbf{i}, \frac{\partial}{\partial y}\mathbf{j} \right)$$

where the unit vectors, $\mathbf{i}, \mathbf{j}$ point in the x and y directions, respectively.

The average energy per unit surface area is $\frac{1}{8}(\rho g H^2)$, which clearly will increase rapidly with wave height. The energy is transported by the waves with a speed equal to or slower than that of the wave form; this energy-related velocity is called the **group velocity**

$$C_g = \frac{\sigma}{2k} \left(1 + \frac{2kh}{\sinh 2kh} \right) \quad (3.14)$$

3.2.2 WAVE TRANSFORMATIONS

As waves propagate from deep to shallow water and then break, they undergo a number of transformations due to the decreasing water depth and the interaction with structures and/or mean currents. These transformations are shoaling, refraction, wave–current interaction, breaking and diffraction.

Shoaling is the change in wave height as the waves encounter shallower water. Since the wave length becomes shorter in shallow water and the wave celerity decreases, in order that the wave may transport energy at the same rate, the wave height must increase, as the higher the wave the more energy it contains. Waves propagating from deep to shallow water first decrease in height (to a maximum of 9% less), then begin to increase in height. The expression for the wave height at any location on a simple coast characterized by straight and parallel contours due to shoaling is

$$H = H_o K_s \quad (3.15)$$

where H_o is the deep water wave height and K_s is the shoaling coefficient. From Airy theory:

$$K_s = \sqrt{\frac{C_{g0}}{C_g}} \quad (3.16)$$

where the C_{g0} is the deep water value of the group velocity and C_g is the local value.

Wave refraction, associated with shoaling is the change in wave direction as waves encounter shallow water. As a rule, wave crests tend to become parallel with the bottom contours, due to the fact that waves in shallow water travel slower than waves in deeper water. Several equations can show the wave direction change, such as:

$$\frac{\partial\theta}{\partial s} = -\frac{1}{C} \frac{\partial C}{\partial n} \quad (3.17)$$

where the s, n directions lie in the directions tangent to and perpendicular to the local wave direction. Alternatively, the irrotationality of the wave number, defined identically by

$$\nabla \times \mathbf{k} = 0, \quad (3.18)$$

where $\mathbf{k} = (k \cos\theta\, \mathbf{i} + k \sin\theta\, \mathbf{j})$, also leads to an equation for θ.

The wave direction change also causes wave-height changes, so we add another term to our wave height equation.

$$H = H_o K_s K_r \quad (3.19)$$

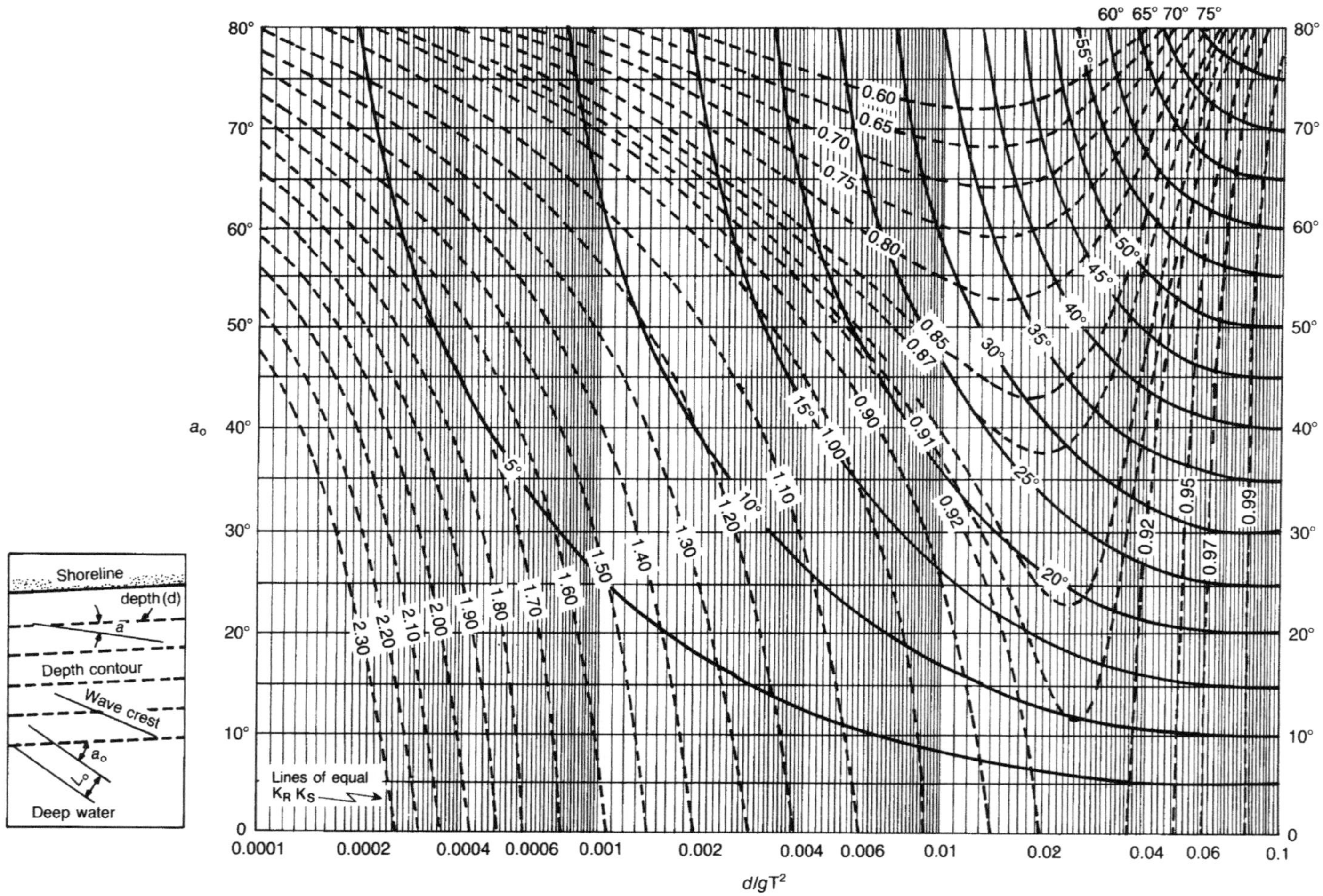

Figure 3.2 Change in wave direction and height for straight and parallel contours (*Shore Protection Manual*).

and K_r is the refraction coefficient, defined as

$$K_r = \sqrt{\frac{\cos\theta_0}{\cos\theta}} \tag{3.20}$$

Both coefficients and their variation with relative water depth are shown in Figure 3.2, for the case of straight and parallel contours.

Numerous computer programs exist to calculate the behaviour of waves from deep to shallow water over more realistic bathymetry than straight and parallel contours. The major classifications are ray tracing programs, which in effect, plot the trajectories of portions of the wave crests, showing where wave energy will concentrate (e.g. Noda, 1974) and wave energy (or action) programs (e.g. Dalrymple, 1988), which solve for the wave height and directions over a bathymetric grid. Later chapters will discuss these types of models in detail.

Wave–current interaction can provide analogous changes in wave direction and wave height as changes in water depth, because the presence of a current changes the wave length. Currents flowing in the same direction as the wave cause the wave to lengthen and decrease in height, while currents opposing the waves cause them to become shorter and higher. Waves also change directions on a current. Peregrine (1976) has provided a good review of wave–current interaction.

When the wave becomes too steep, either by encountering shallow water or by having too much energy input to it from the wind, the wave will break. **Wave breaking** in deep water occurs when the wave steepness becomes too large, while in shallow water, the maximum wave height is directly related to the water depth:

$$H = \kappa h \tag{3.21}$$

where κ is nearly 1, depending on the bottom slope, reefs, etc. This relationship provides an upper limit for the height of waves at a shallow water site.

Wave diffraction occurs when there is a variation of wave height along the wave crest, this might occur when waves pass an offshore island or pass through a gap in a breakwater at a harbour. Waves tend to turn away from regions of high wave height, so for the case of waves through a breakwater gap, the wave crests tend to become circular, as if they originated at a point. The importance of wave diffraction results from this direction changes, as it introduces wave energy into regions, say, behind the breakwater, where one might otherwise assume there were no waves.

For simple calculations for breakwaters, the US Army *Shore Protection Manual* (1973) provides plots of wave height as a function of distance from the breakwater tip. For more elaborate calculations, such as in the vicinity of offshore islands, this is a more difficult calculation, requiring sophisticated computer models.

3.2.3 MILD-SLOPE EQUATION

The mild-slope equation (Berkhoff, 1972) is obtained by vertically integrating the Laplace equation. This two-dimensional equation is capable of predicting refraction and diffraction and has proved to be very popular. Numerous applications in finite element and finite difference formulation of this model have been carried out (Bettess and Zienkiewicz, 1977; Houston, 1981; Ito and Tanimoto, 1972). Parabolic representations provide a rapid solution of the mild-slope equation and eliminate the need for a downwave boundary condition, as shown by Radder (1979). The development of parabolic models lies in assuming that the waves travel nearly in the x direction and

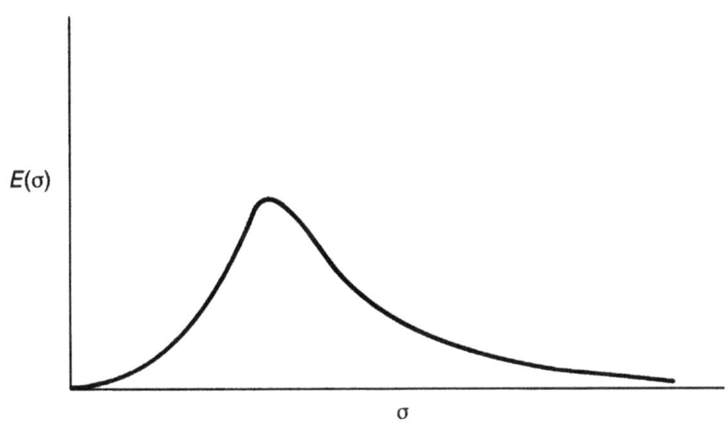

Figure 3.3 Wave spectrum.

factoring out the rapidly varying x component. Kirby and Dalrymple (1983) provide the weakly non-linear form of the mild-slope equation in parabolic form, incorporating a more correct wave speed in the model, permitting the accurate calculation of diffraction over a large area.

Kirby (1986) has provided a modified mild-slope equation to include the effect of rapidly varying bottom features and wave–current interaction has been included by Booij (1981) and Kirby (1984).

One disadvantage of parabolic modelling of the mild-slope equation is that there is an angular restriction on the waves; they must primarily travel near the x direction. Recently, angular spectrum modelling has been discussed, which requires less restriction on the wave direction (Dalrymple and Kirby, 1988).

3.3 WAVE STATISTICS AND SPECTRA

The surface of the ocean is clearly very irregular and not described by a single sinusoidal function, as described in the earlier sections. The water surface is made up of a number of wave trains with different sizes and frequencies travelling in different directions. This superposition of waves is the manifestation of the wave spectrum. A frequency spectrum, as shown in Figure 3.3, is the amount of energy per frequency plotted versus frequency. It does not contain any information as to the wave direction. As can be seen from the figure, the low- and high-frequency ends of the spectrum have far less energy than the portion of the energy spectrum associated with the wind waves.

The shape of the wave spectrum, or, more correctly, the energy density spectrum, depends on the factors that govern wave growth: wind speed and duration, and the water depth. The overall shape of the spectrum, however, is common enough to have a generic description. There are numerous common spectral shapes; here are some of the more widely used:

1. The Pierson–Moscowitz spectrum;
2. The Bretschneider spectrum;
3. The JONSWOP spectrum.

In order to introduce wave directionality, the wave spectrum is multiplied by a direction factor, resulting in a directional wave spectrum. The most widely used factor is the Mitsuyasu spreading factor, which spreads the wave energy about a mean wave direction, θ_o. The spreading factor is defined as

$$D(\theta) = A \cos 2s\left(\frac{\theta - \theta_o}{2}\right) \tag{3.22}$$

where A is a normalization parameter to force the $\int D(\theta)d\theta$ to be unity. Larger values of the spreading parameter s imply that the waves travel more nearly in the θ_o direction. This s value is obtained from wave data. Further information on spectral method has been given by Goda (1985).

The shoaling and refraction of spectra is a problem of current interest. Often spectral data is known offshore and must be transformed to a nearshore site. An example of this type of model is provided by Booij *et al.* (1985).

3.4 NON-LINEAR AND HIGHER-ORDER WAVE THEORIES

The Airy wave theory is a linear wave theory in the sense that mathematical assumptions were made which involved neglecting the non-linear (and hence mathematically more difficult) terms in the boundary conditions. This is valid when the velocities are small, and, consequently, the Airy theory is limited to situations when the waves are very small. Since it is a simple theory to use, it remains a good first estimate of wave properties; however, for design purposes, a more complete (mathematically) non-linear theory must be used. One physical justification of the non-linear theories is that high steep waves have high narrow crests and shallow wide troughs, which cannot be described by a simple sine or cosine function.

Numerous analytical non-linear wave theories have been developed over the past 150 years, but, in general, they can be classified as shallow water or deep water theories. Examples of shallow water theories are the cnoidal (Korteweg and DeVries, 1895) and solitary (Boussinesq, 1877) wave theories, while in deeper water the Stokes (1847) theory is the most widely used. These theories are often only valid for their region of applicability: cnoidal theory cannot be used in deep water and the Stokes theory cannot be used in shallow water. Numerical, or computer enhanced, theories, such as the stream function wave theory (Dean, 1965, with least-squares fit to the boundary conditions; Fenton and Rienecker, 1982, with a collocation fit), can be valid for all water depths. Dean (1974) has tabulated 40 example waves, which permits the convenient calculation of wave properties. Cokelet (1977), using an accelerated perturbation method, has produced the most accurate method of computing waves of constant form.

More recently, time-marching solutions of waves have been carried out using boundary element methods, which permit the calculation of unsteady waves and their characteristics, Longuet–Higgins and Cokelet (1976) and Dold and Peregrine (1984).

The propagation of long waves in three dimensions can be computed using the long wave equations, as in tidal propagation models, or through the use of newer Fourier models or the K–P equation, both of which are discussed in Liu *et al.* (1985).

The brevity of this chapter has resulted in the omission of vast amounts of information concerning waves. The interested reader is referred to texts such as Mei (1983), Whitham (1974) and Dean and Dalrymple (1984) for further details.

ACKNOWLEDGEMENT

This work was supported in part by the Chevron Oil Field Research Company, La Habra, California.

REFERENCES

Berkhoff, J.C.W. (1972) Computation of combined refraction-diffraction, *Proceedings of the 13th International Coastal Engineering Conference, Vanconver, Canada.* ASCE, New York pp. 471–90.

Bettess, P. and Zienkiewicz, O.C. (1977) Diffraction and refraction of surface waves using finite and infinite elements. *International Journal on Numerical Methods in Engineering*, **11**, 1271–90.

Booij, N. (1981) Gravity waves on water with non-uniform depth and current. Department of Civil Engineering, Techical University of Delft, *Report 81–1*.

Booij, N., Holthuijsen, L.H. and Herbers, T.H.C. (1985) A numerical model for wave boundary conditions in port design. *International Conference* on Numerical and Hydraulic Modelling of Ports and Harbours, Birmingham, UK. BHRA, Cranfield, UK. pp. 263–8.

Boussinesq, J. (1877) Essai sur la thèorie des eaux courantes. *Mémoires Présentés Par Divers Savants à L'Académie des Sciences de L'Institute National de France*, (Ser. 2), **23**, 1–680.

Cokelet, E.D. (1977) Steep gravity waves in water of arbitrary uniform depth. *Philosophical Transactions of the Royal Society of London, A*, **286**, 183–230.

Dalrymple, R.A. (1988) A model for the refraction of water waves. *Journal of Waterway, Port, Coastal and Ocean Engineering, ASCE*, **114**, (4), 423–35.

Dalrymple, R.A. and Kirby, J.T. (1988) Models for very wide-angle water waves and wave diffraction. *Journal of Fluid Mechanics*, **192**, 33–50.

Dean, R.G. (1965) Stream function representation of ocean waves. *Journal of Geophysical Research*, **70**, (18), 4561–72.

Dean, R.G. (1974) Evaluation and development of water wave theories for engineering application, US Army Coastal Engineering Research Center, *Special Report 1* (2 Vols).

Dean, R.G. and Dalrymple, R.A. (1984) *Water Wave Mechanics for Engineers and Scientists*, Prentice-Hall, Englewood Cliffs.

Dold, J.W. and Peregrine, D.H. (1984) Steep unsteady water waves: an efficient computational scheme, *Proceedings 19th International Coastal Engineering Conference*, Houston, Texas, USA, (ed B.L. Edge), ASCE, New York, pp. 955–67.

Fenton, J.D. and Rienecker, M.M. (1982) A numerical method for solving nonlinear water-wave problems: application to solitary-wave interactions. *Journal of Fluid Mechanics*, **118**, 411–43.

Goda, Y. (1985) *Random Seas and Design of Maritime Structures*, University of Tokyo Press, Tokyo.

Houston, J.R. (1981) Combined refraction and diffraction of short waves using the finite element method. *Applied Ocean Research*, **3**, 163–70.

Ito, Y. and Tanimoto, K. (1972) A method of numerical analysis of wave propagation: application to wave diffraction and refraction, *Proceedings of the 13th International Coastal Engineering Conference, Vanconver*. ASCE, New York, pp. 503–22.

Kirby, J.T. (1984) A note on linear surface wave-current interaction over slowly varying topography. *Journal Geophysical Research*, **89** (C1), 745–7.

Kirby, J.T. (1986) A general wave equation for waves over rippled beds. *Journal of Fluid Mechanics*, **162**, 171–86.

Kirby, J.T. and Dalrymple, R.A. (1983) A parabolic equation for the combined refraction-diffraction of Stokes waves by mildly varying topography. *Journal of Fluid Mechanics*, **136**, 453–66.

Korteweg, D.J. and DeVries, G. (1895) On the change of form of long waves advancing in a rectangular canal and on a new type of long stationary wave. *Philosophical Magazine*, **39**, 422–43.

Liu, P.L.-F., Yoon, S.B. and Kirby, J.T. (1985) Nonlinear refraction-diffraction of waves in shallow water. *Journal of Fluid Mechanics*, **153**, 185–201.

Longuet-Higgins, M.S. and Cokelet, E.D. (1976) The deformation of steep surface waves on water. I. A numerical method of computation. *Proceedings of the Royal Society of London, A*, **350**, 1–26.

Mei, C.C. (1983) *The Applied Dynamics of Ocean Surface Waves*, John Wiley New York.

Noda, E.K. (1974) Wave-induced nearshore circulation. *Journal of Geophysical Research*, **79**, (27), 4097–4106.

Peregrine, D.H. (1976) Interaction of waves and currents. *Advances in Applied Mathematics*, **16**, 9–117.

Radder, A.C. (1979) On the parabolic equation method for water-wave propagation. *Journal of Fluid Mechanics*, **95**, 159–76.

Stokes, G.G. (1847) On the theory of oscillatory waves. *Transactions of the Cambridge Philosophical Society*, **8**, 441–55.

US Army, Coastal Engineering Research Center (1973) *Shore Protection Manual*, US Government Printing Office.

Whitham, G.B. (1974) *Linear and Nonlinear Waves*, Wiley-Interscience, New York.

ELEMENTS OF THE THEORY OF TURBULENCE 4

W. Rodi

After the nearly-horizontal flows and the short-period wave motions, the next important aspect that has to be considered is turbulence. It is turbulence that carries sediments, forms the beds of coastal and estuarial regions, diffuses and disperses all manner of waste products of industry, urban settlements and agriculture, leads to variations in loadings on pipelines and drives any number of other phenomena. As corresponds to the importance of this subject, considerable effort has been invested in the development of turbulence modelling tools for practical applications. The intention here is to introduce the foundations of these modelling tools.

CONTENTS

4.1 INTRODUCTION

4.1.1 PHYSICAL MECHANISMS

Like most other flows of practical interest, the water flows in coastal areas, estuaries and harbours are usually turbulent, i.e. they involve an irregularly fluctuating, eddying motion. This turbulent motion contributes significantly to the transport of momentum, heat and mass and therefore has a determining influence on the velocity, temperature and concentration distributions over the water body. Hence, the calculation of these fields usually requires a realistic simulation of the turbulent transport processes.

The turbulent fluctuating motion is basically always three-dimensional and consists of a wide spectrum of eddy sizes, associated with a wide range of fluctuation frequencies. The size of the largest eddies with low frequency, at which the energy input from the mean motion occurs, is of the same order of magnitude as the flow domain. The smallest eddies with high-frequency fluctuations, at which kinetic energy is dissipated, are determined by viscous forces. A special feature of flow in shallow water bodies is that the depth is much smaller than the horizontal dimensions of the flow, resulting in a two-range spectrum of the turbulent motions. Large-scale, virtually-horizontal eddying motions exist which are basically two-dimensional and which are generated by and hence receive their energy input from horizontal shear layers near banks, irregularities and abrupt changes in flow geometry causing separation (as in bays), and obstructions such as islands. When periodic vortices are shed, as for example past obstructions, the question arises whether these should be considered as turbulence. On a much smaller scale, three-dimensional turbulence is generated by bottom and wind shear, providing energy input at a completely different range of the spectrum. At even smaller scales, dissipation takes place which is mainly determined by the presence of the bottom and wind-shear generated turbulence, but there must of course be an interaction between the large-scale two-dimensional turbulent motion and the small-scale three-dimensional turbulence. This interaction is mainly due to the bed shear damping the large-scale two-dimensional motion. The bed shear itself is effected by the small-scale three-dimensional turbulent motion.

Another important mechanism is the influence of density stratification, which is present in many coastal waters and estuaries. Stratification is caused by fresh water flowing over salt water in estuaries, by solar heating of surface layers in coastal waters, leading to the formation of mixed layers and thermoclines, or by man-made discharges of lighter or heavier water. Stable stratification damps the turbulent fluctuating motion (in particular the vertical fluctuations) and hence the vertical mixing and can lead to the complete collapse of turbulence. On the other hand, unstable

Coastal, Estuarial and Harbour Engineer's Reference Book. Edited by M.B. Abbott and W.A. Price. Published in 1993 by Chapman & Hall, 2–6 Boundary Row, London SE1 8HN. ISBN 0 419 15430 2.

stratification, which can for example occur due to heat loss from the water surface, amplifies turbulence and mixing. Under stratified situations, a very complex vertical structure of the flow can emerge, with several layers and complex exchange mechanisms at their interfaces. Further information on turbulence and the effects of stratification in water bodies, has been given by Nezu and Nakagawa (1990) and a special issue of the *Journal of Geophysical Research* (1987).

4.1.2 POSSIBLE WAYS OF SIMULATING TURBULENCE

The turbulent motions of all scales are governed by the exact time-dependent Navier–Stokes and continuity equations, and in stratified situations also by the corresponding equations determining the local density. Numerical methods are now available for solving these equations in principle. However, owing to the discretization of the flow field in these methods, only a limited resolution is possible, and in spite of the great advances in computer technology the resolution of all scales in the complete water body is far beyond the capacity of currently available computers. In recognition of this, large eddy simulation (LES) techniques have been developed, which simulate directly only the large-scale motion that can be resolved by a given numerical grid, while the small-scale, unresolvable motion is simulated by a sub-grid-scale model. Basically, large eddy simulations are three-dimensional, and if the turbulence responsible for the vertical mixing is to be captured in the simulation, then the fairly fine vertical discretization necessary also has to be applied in the horizontal directions. This is, at present, not feasible for the complete water body so the LES approach is only suitable for studying local phenomena. Such LES calculations are certainly of considerable fundamental interest, providing important information on the vertical structure of the flow. In shallow water bodies, something like a two-dimensional large-eddy simulation appears possible in which only the large-scale horizontal turbulent motion is resolved. The interaction with the smaller-scale three-dimensional turbulent motion then needs to be modelled. The inclusion of bottom-friction terms in the two-dimensional equations may go a long way towards this, and numerical dissipation present in virtually all the methods employed may also help. Realistic calculations of this kind seem possible as has been shown by Abbott and Larsen (1985) and Abbott *et al.* (1985). However, it is not entirely clear at present whether an additional sub-grid-scale model is necessary in such two-dimensional calculations, especially when turbulent heat and mass transfer is to be calculated. Further clarifying studies are necessary on this issue.

When vertical exchange processes are to be simulated in an engineering calculation or when the horizontal turbulent motion cannot be resolved numerically, statistical calculation procedures are used in which averaged mean-flow equations are solved. The averaging is now carried out over all scales, thereby introducing turbulence correlations into the equations governing the distribution of the averaged quantities. These correlations represent the turbulent transport of momentum, heat and mass, and a turbulence model has to be introduced for determining these correlations – which is now a model for the whole spectrum of the turbulent motion. The rest of this chapter considers turbulence models of this type.

When the vertical structure is important and needs to be resolved, especially in situations with stratification, a model for the vertical turbulent transport has to be employed. A brief review of such models is given in section 4.2. These vertical-structure models can also be used in connection with a three-dimensional time-dependent calculation in which the large-scale horizontal turbulent motion is resolved (Oey *et al.*, 1985). In many cases, the vertical turbulent mixing is so strong that the flow quantities are distributed nearly uniformly over the depth. Under such circumstances, the vertical structure has no influence or is not of interest and hence does not have to be calculated, so that a two-dimensional depth-average calculation, determining only the horizontal distribution, is sufficient. The vertical turbulent transport enters into such calculations only through bottom and surface-shear-stress and heat-transfer terms. In large-water-body calculations, the horizontal turbulent momentum flux can usually be neglected and the heat and mass transfer can be simulated with the aid of a constant eddy diffusivity. However, this rather crude treatment is not sufficient for near-field regions such as near banks, structures and discharges, where refined modelling of both turbulent momentum and heat/mass transfer is necessary so as to allow the calculation of the variation of the eddy viscosity/diffusivity over the flow field. An example of a special turbulence model for depth-averaged calculations is presented in section 4.3.

Only a brief account of turbulence modelling can be given in this chapter. A more complete treatment of the subject has been given elsewhere (Rodi, 1984, 1987).

4.2 VERTICAL-STRUCTURE MODELS

4.2.1 MEAN-FLOW EQUATIONS

When turbulence is averaged out completely, the temporal and spatial distribution of the mean quantities of the water body is governed by the following equations, given here in tensor notation:

Continuity equation:

$$\frac{\partial U_i}{\partial x_i} = 0 \tag{4.1}$$

Momentum equations:

$$\frac{\partial U_i}{\partial t} + U_j \frac{\partial U_i}{\partial x_j} = -\frac{1}{\rho_r}\frac{\partial P}{\partial x_i} + \frac{\partial}{\partial x_j}\left(\nu \frac{\partial U_i}{\partial x_j} - \overline{u_i u_j}\right) + g_i \frac{\rho - \rho_r}{\rho_r} \tag{4.2}$$

Scalar transport equation:

$$\frac{\partial \phi}{\partial t} + U_i \frac{\partial \phi}{\partial x_i} = \frac{\partial}{\partial x_i}\left(\lambda \frac{\partial \phi}{\partial x_i} - \overline{u_i \varphi}\right) + S_\phi \tag{4.3}$$

Equation of state:

$$\rho = \rho(\phi) \tag{4.4}$$

The symbols are defined in Figure 4.1 and in the Nomenclature. The mean quantities are statistical averages over a time which is long compared with the time scale of the turbulent fluctuations but short compared with that of the mean-flow variations (e.g. tidal period).

Equations (4.2) and (4.3) contain correlations between fluctuating quantities, indicated by an overbar. These quantities represent the transport due to turbulence in the mean-flow equations and need to be simulated with the aid of a turbulence model. In this section, we are concerned only with the vertical turbulent transport of momentum and scalar quantities due to the vertical fluctuations. If we change to the x, y, z notation also given in Figure 4.1, the correlations of interest are the velocity correlations $\overline{uw}$ and $\overline{vw}$ appearing as shear stresses in the momentum equations (4.2) and the velocity–scalar correlation $\overline{w\varphi}$ appearing as heat or mass flux in equation (4.3). Models will now be reviewed for simulating these turbulent stresses and fluxes.

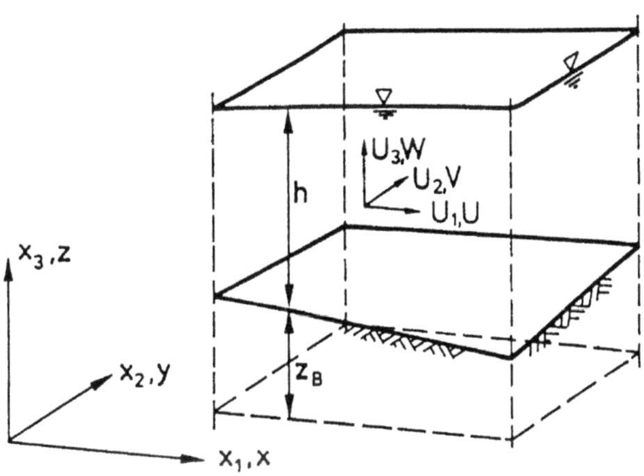

Figure 4.1 Co-ordinate system.

4.2.2 TURBULENCE MODELS

Types of turbulence model

Existing models for the turbulent transport can be grouped into two categories: those which employ the turbulent (or eddy) viscosity/diffusivity concept and those which do not. The turbulent viscosity/diffusivity concept assumes that, in analogy to the viscous stresses and fluxes in laminar flow, the turbulent stresses and fluxes are proportional to the gradients of the mean velocity and scalar quantities, respectively. For the turbulent transport terms of interest here, this concept provides:

$$\overline{uw} = -\nu_t \frac{\partial U}{\partial z}, \quad \overline{vw} = -\nu_t \frac{\partial V}{\partial z}, \quad \overline{w\varphi} = -D_t \frac{\partial \phi}{\partial z} \tag{4.5}$$

where ν_t and D_t are the turbulent (or eddy) viscosity and diffusivity, respectively. In contrast to their molecular counterparts, these quantities are not fluid properties but depend strongly on the state of the turbulence. The turbulent diffusivity D_t is usually assumed to be proportional to the turbulent viscosity ν_t:

$$D_t = \frac{\nu_t}{\sigma_t} \tag{4.6}$$

where the proportionality factor σ_t is the turbulent Prandtl or Schmidt number. The assumption of a constant turbulent viscosity/diffusivity is not adequate in vertical-structure models, as it is well known that these quantities have a nearly parabolic distribution with depth in open channel flows (Nezu and Nakagawa, 1993). The distribution of ν_t and D_t must therefore also be provided by the turbulence model, and various models for this are briefly described below.

The second group of turbulence models does not use the eddy viscosity/diffusivity concept but solves transport equations for the shear stresses $\overline{uw}$ and $\overline{vw}$ and for the scalar flux $\overline{w\varphi}$. These models are usually referred to as second-order closure schemes, and they will also be discussed briefly below.

Mixing-length models

Prandtl (1925) proposed to relate the eddy viscosity to the mean velocity gradients as follows:

$$\nu_t = l_m^2\left[\left(\frac{\partial U}{\partial z}\right)^2 + \left(\frac{\partial V}{\partial z}\right)^2\right]^{1/2} \tag{4.7}$$

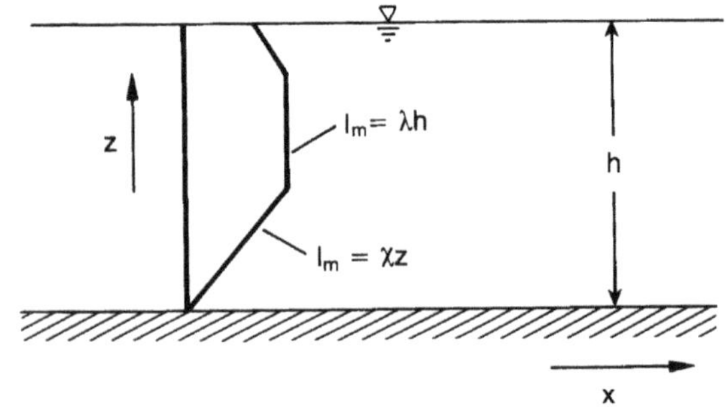

Figure 4.2 Mixing length distribution in open channel flow.

where l_m is the mixing length whose distribution over the flow field has to be determined empirically. In channel flow, a mixing length distribution as shown in Figure 4.2 has been found suitable, with the von Kármán constant $\kappa \approx 0.4$ and $\lambda = 0.1$. As indicated in Figure 4.2, the mixing length is often reduced near the water surface in order to account for its restricting effect on the size of the turbulent eddies.

Buoyancy forces may significantly alter the turbulent exchange processes and hence also the eddy viscosity and the mixing length distribution. This effect can be accounted for by empirical formulae in which the influence of buoyancy is characterized by a Richardson number. These formulae have been given elsewhere (Rodi, 1984).

The mixing length model has the advantage of being simple and economical, and in many cases reasonable results can be obtained provided the proper empirical constants are chosen. However, extensive testing has also brought to light the limitations of the model, in particular the lack of universality of the empirical input, especially in buoyant situations. This can be traced to the implied assumption in the model that turbulence is in a state of local equilibrium everywhere, which means the turbulence produced at a certain point (mainly by interaction of shear stresses and velocity gradients) is dissipated at this point at the same rate. Thus the mixing length model does not account for the transport of turbulence (from regions of high to those of low production), nor for history effects, and is therefore not suitable for situations where transport and history effects are important. In tidal channel flow, for example, the model grossly underpredicts the turbulent diffusivity for times near slack water, when the velocity gradients are low, but turbulence which was produced earlier still persists.

One-equation models

The transport and history effects of turbulence can be accounted for in an eddy viscosity/diffusivity model by solving a transport equation for a suitable parameter characterizing the turbulence. The turbulent kinetic energy, k, is such a parameter, as it characterizes the intensity of the fluctuating motion. The following semi-empirical transport equation for k is usually employed in models of this type (Rodi, 1984):

$$\frac{\partial k}{\partial t} + U\frac{\partial k}{\partial x} + V\frac{\partial k}{\partial y} + W\frac{\partial k}{\partial z} = \frac{\partial}{\partial z}\left(\frac{\nu_t}{\sigma_k}\frac{\partial k}{\partial z}\right) + \nu_t\left[\left(\frac{\partial U}{\partial z}\right)^2 + \left(\frac{\partial V}{\partial z}\right)^2\right] + \beta g\frac{\nu_t}{\sigma_t}\frac{\partial \phi}{\partial z} - \varepsilon \quad (4.8)$$

Rate of change	Convective transport by the mean motion	Diffusive transport by the turbulent motion	P = Shear production	G = buoyancy production/ destruction	viscous dissipation

History effects are accounted for in this equation by the rate of change term, and the transport effects by the convective and diffusive transport terms. The distribution of k is further influenced by the shear production term extracting energy from the mean motion, by viscous dissipation into thermal energy and, in general, also by an interchange between turbulent kinetic energy and potential energy via the action of buoyancy. The mixing length model implies that the last three terms on the right-hand side balance each other. Equation (4.8) is a modelled form of the exact k-equation that can be derived from the Navier–Stokes equations. In order to obtain a complete turbulence model, further assumptions are necessary. The eddy viscosity ν_t and the dissipation rate ε need to be related to k and to a second parameter characterizing the turbulent motion, here chosen as the length scale L. From dimensional analysis it follows that:

$$\nu_t = c'_\mu \sqrt{k}L, \quad \varepsilon = c_D\frac{k^{3/2}}{L} \quad (4.9)$$

where c'_μ and c_D are empirical constants. Only the product of these constants which has a value of $c'_\mu c_D \approx 0.08$ is important. Relations (4.9) together with the k-equation (4.8) form a so-called 'one-equation' model.

In one-equation turbulence models the distribution of the length scale L has to be determined empirically. Similar distributions are used as for the mixing length l_m. In open-channel flow, the following distribution with depth appears to work reasonably well:

$$\frac{L}{h} = c_L \operatorname{erf}(4y/h) \quad (4.10)$$

where h is the local water depth and $c_L \approx 0.105$. In buoyant situations the modifications introduced by

Rodi (1984) for the mixing length can be used also to account for the effect of buoyancy on the length scale L.

Energy-equation models are superior to the mixing length model whenever transport and history effects are important. An example is oscillating channel flow during times of slack water. Smith and Takhar (1977) concluded that one-equation models represent the optimum level of complexity for channel flows, since the length–scale distribution can be specified empirically to sufficient accuracy in this case. This conclusion was reached for non-buoyant channel flow situations, and it should be emphasized that the effect of buoyancy on the length scale L has to be accounted for empirically. Furthermore, the coefficients c'_μ and σ_t are also dependent on buoyancy effects, and possibilities to express this dependence are discussed in the next section.

Two-equation models

The characteristic length scale of the turbulent motion is also subject to transport and history effects. In order to account for these effects, models have been developed which use a transport equation also for the length scale L, and are therefore more general than the simpler models. The difficulties in finding widely valid formulae for prescribing or calculating L have stimulated the use of such models. The length-scale determining equation does not need to have the length scale L itself as the dependent variable; any combination with k will suffice, since k is know from the solution of the k-equation. The combination $k^{3/2}/L$ has become most popular so that, according to Equation (4.9), an equation is solved for the dissipation rate ε of the turbulent kinetic energy. This ε-equation reads for the flows of interest here:

$$\frac{\partial \varepsilon}{\partial t} + U\frac{\partial \varepsilon}{\partial x} + v\frac{\partial \varepsilon}{\partial y} + w\frac{\partial \varepsilon}{\partial z} = \frac{\partial}{\partial z}\left(\frac{\nu_t}{\sigma_\varepsilon}\frac{\partial \varepsilon}{\partial z}\right) + c_{1\varepsilon}\frac{\varepsilon}{k}[P + (1 - c_{3\varepsilon})G] - c_{2\varepsilon}\frac{\varepsilon^2}{k} \qquad (4.11)$$

where P and G are the stress and buoyancy production, respectively, of the turbulent kinetic energy defined in the k-equation (4.8). The ε-equation is considerably more empirical than the k-equation: all terms on the right-hand side of Equation (4.11) are based on model assumptions and therefore contain empirical constants. Together with the k-equation (4.8) and the eddy-viscosity expression resulting from combining the two relations in Equation (4.9)

$$\nu_t = c_\mu \frac{k^2}{\varepsilon} \qquad (4.12)$$

the ε-equation forms the so-called k–ε model. For non-buoyant situations, the empirical constants c_μ, σ_ε, $c_{1\varepsilon}$, $c_{2\varepsilon}$ have the values 0.09, 1.3, 1.44 and 1.92, respectively. Unfortunately, the value of the empirical constant $c_{3\varepsilon}$ associated with the buoyancy term in the ε-equation has been found to depend on the flow situation considered. In unstably stratified flows, where G is a source term, $c_{3\varepsilon}$ should take a value of 1, while in stably stratified flow, where G is a sink term, $c_{3\varepsilon}$ should be chosen near zero (0–0.2).

In stratified flows under local equilibrium conditions, the empirical constant c_μ and the turbulent Prandtl/Schmidt number σ_t can be shown to depend on the buoyancy influence. Further, the k–ε model introduced so far does not account for the influence of a free surface on the turbulence, which is mainly to damp the vertical fluctuations and thereby reduce the coefficient c_μ. The model has therefore been extended to account for these effects by making c_μ and σ_t functions of buoyancy and surface-damping parameters. These functions were not obtained directly from experiments but were derived by simplifying a more complex stress/flux equation model introduced briefly in the next section. The actual c_μ and σ_t functions can be found in Rodi (1985).

Two-equation models like the k–ε model are the simplest models that promise success for flows for which the length scale cannot easily be prescribed empirically. The k–ε model is one of the most widely tested turbulence models in engineering fluid mechanics and is also gaining popularity for geophysical flow calculations. A few examples of k–ε-model calculations will be presented in section 4.2.3. Further examples can be found in Rodi (1984, 1987).

Stress/flux equation models and algebraic derivatives

The turbulence models discussed so far are all based on the eddy viscosity/diffusivity concept involving a close tie-up of the stresses and fluxes to the velocity and temperature or concentration gradients. These models clearly have their limitations, especially in certain geophysical flows with counter-gradient transport and directional influence on turbulence due to gravitational effects. Models have therefore been developed which do not use the eddy viscosity/diffusivity concept but solve transport equations for the individual stress and flux components. Such models take direct account of the transport and history effects of these components and also of the anisotropy of the turbulent transport in complex flows and thus simulate the physical processes most realistically. The equations can be derived in exact form, but they contain higher-order correlations for which model approximations have to be introduced in order to obtain a closed system. A particular advantage of starting the model derivation from exact equations is that terms accounting for the effect of buoyancy are introduced automatically. The stress/flux equation models are the most complex models and so far have found little use in practical water flow calculations. The actual model equations are therefore not presented here and the reader is referred to more comprehensive reviews (Rodi, 1984, 1987).

In order to derive simpler models for practical calculations, so-called algebraic stress/flux models have been developed by simplifying the differential transport equations so that they are reduced to algebraic expressions but retain most of their basic features. To this end, the rate of change, convective and diffusive transport terms in the equations need to be replaced by model approximations. The simplest approximation is, of course, to neglect these terms, but a more generally applicable one is to relate the terms to the corresponding ones in the k-equation (Rodi, 1984). In this way the buoyancy terms in the original equations are retained in the resulting algebraic expressions. The same is true for surface proximity terms when these were introduced into the original stress transport model. For the stresses $\overline{uw}$, $\overline{vw}$, and the flux $\overline{w\varphi}$, the resulting algebraic expressions can be written in form of the eddy viscosity/diffusivity relations appearing in the k–ε model, but with c_μ and σ_t now being functions of various parameters, such as the ratio of production to dissipation, buoyancy and also the surface proximity. With these functions, the k–ε model is therefore extended to account for buoyancy and surface-proximity effects. As has already been mentioned, the actual functional dependences have been given by Rodi (1985). It was found that in shear-layer-type flows, i.e. in situations with a predominant flow direction, this algebraic-stress extension of the k–ε model yields results which are not markedly inferior to results obtained with the full stress-equation model.

Boundary conditions

In the turbulence-model equations given above, viscous terms have been neglected, so that the equations are not applicable to the viscous sub-layer near walls, if such a layer exists. The viscous sub-layer is usually bridged by the so-called wall function approach as described by Launder and Spalding (1974), which relates the resultant velocity parallel to the bottom, U_r at a point z_c just outside the viscous sub-layer to the resultant wall shear stress via the logarithmic law of the wall:

$$\frac{U_{rc}}{U_*} = \frac{1}{\kappa}\ln\left(\frac{z_c U_*}{\nu}E\right) \tag{4.13}$$

in which U_* is the bottom friction velocity, κ is the von Kármán constant ($\approx$0.4) and E is a roughness coefficient (=9 for hydraulically smooth walls). It is further assumed that at this point local equilibrium prevails, which leads to the following boundary conditions for k and ε to be applied at the nearest bottom point in models which solve the k- and ε-equations:

$$\frac{k_c}{U_*} = \frac{1}{\sqrt{c_\mu}}, \quad \varepsilon_c = \frac{U_*^3}{\kappa z_c} \tag{4.14}$$

The wall-function relations (4.13) and (4.14) are also useful as boundary conditions for rough bottoms, where the first grid point is placed above the actual roughness elements. The roughness coefficient E accounts for the roughness effect, and lowering E increases the ratio between the velocity at the first grid point and the friction velocity.

In the absence of wind-induced shear stresses and heat exchange with the atmosphere, a free surface can be considered to a first approximation as a symmetry plane. When a shear layer is created by wind forces, the boundary conditions (4.13) and (4.14) for wall boundaries seem appropriate; that is, the surface can be considered as a moving wall. More refined conditions for intermediate situations are discussed in Rodi (1984).

4.2.3 CALCULATION EXAMPLES

This sub-section presents some applications of the turbulence models introduced above to the simulation of the vertical flow structure in water bodies. All examples are at least nominally two-dimensional so that the velocity component V is zero and the flow development in only one vertical plane is considered.

The first example concerns developed flow in an open channel, and Figure 4.3 compares calculated vertical velocity and eddy viscosity distributions with measurements. The calculations were obtained with the k–ε model extended by the surface-proximity dependence of the coefficient c_μ as described in Rodi (1985). The flow is non-buoyant so that the buoyancy extension also described there is not effective. The agreement of the mean-velocity and eddy-viscosity distributions with experimental data can be seen to be good. The reduction of the eddy viscosity near the free surface is reproduced well by the present model. This reduction is due to the surface-proximity correction of c_μ. The difference between the commonly employed parabolic ν_t-distribution and the experimentally observed one in the mid-depth region is also noteworthy. The parabolic ν_t-distribution results from assuming a logarithmic velocity distribution up to the free surface. In close agreement with the experiments, the calculated eddy viscosity also falls below the parabolic distribution. Celik and Rodi (1988) have shown that, with the same turbulence model, suspended sediment concentration in channel flow can also be calculated realistically for various conditions.

Smith and Takhar (1977) have studied the performance of various turbulence models when applied to oscillatory channel flow as this occurs in estuaries. The most complex model employed transport equations for the shear stress $\overline{uw}$, the kinetic energy k and the dissipation rate ε. For various amplitudes and periods of oscillations, the depth-mean phase difference between the shear stress $\overline{uw}$, and the velocity gradient $\partial U/\partial z$ was determined with the aid of this model. The difference was found to be small, especially for larger periods, so that the eddy viscosity assumption $\overline{uw} \propto \partial U/\partial z$ is sufficiently accurate for tidal channel flow. Smith and Takhar made calculations with two eddy-viscosity models. One was the standard k–ε model and the other the one-equation model described above, using an empirical length-scale distribution according to Equation (4.10). As this distribution was close to that obtained from the k–ε model, they concluded that the one-equation model is the optimum model for unsteady channel flow. Figure 4.4 shows the vertical variation of the amplitude of the velocity oscillations predicted with this model in the Humber estuary in the north of England and also compares it with measurements and predictions obtained with uniform and parabolic eddy-viscosity distributions. It is obvious from Figure 4.4 that the one-equation model, which accounts for the history effects of turbulence, predicts the oscillations well, while the parabolic eddy-viscosity distribution suitable for steady channel flow does not do so well. This is understandable when one looks at the eddy-viscosity profiles in Figure 4.5b which were predicted for the Humber estuary at various time intervals during the tidal cycle by Celik and Rodi (1985) with the standard k–ε model. It is clear that ν_t can neither be prescribed as time invariant, nor can it be approximated by a simple function of the vertical distance. Figure 4.5a compares Celik and Rodi's predicted variations of the relative amplitude of the semi-diurnal tidal current with observations by Johns (1966) for the relatively low tidal frequency ($\sigma = 1.45 \times 10^{-4}\,s^{-1}$). The agreement can be seen to be excellent. The predicted amplitude of the surface current was, however, about 30% less than the observed one, which is not so surprising in view of the considerable uncertainty about the bottom roughness parameter entering the calculations.

The last calculation example concerns the development of stratified flow, and in particular the influence of both stable and unstable stratification on the turbulent mixing in a mixing layer. This flow situation is of great importance for the dynamics of thermoclines. Viollet (1985) simulated a particular flow situation that was also studied experimentally (Viollet, 1980) and this situation is sketched in the top part of Figure 4.6. The calculations were obtained both with the standard k–ε model and with an algebraic stress/flux version (modifying the coefficients c_μ and σ_t) as de-

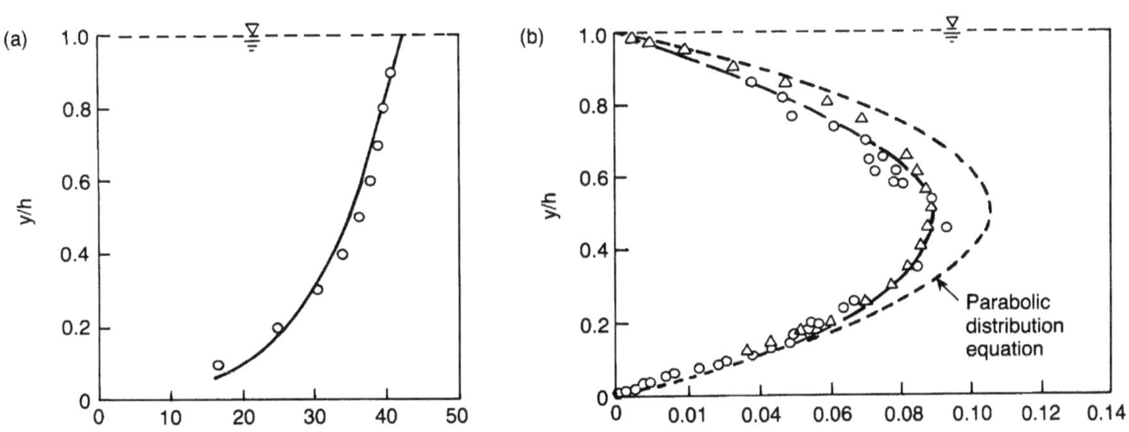

Figure 4.3 (a) Velocity distribution in open channel flow; (○) experiments, Jobson and Sayre (1970); (——), predictions. (b), eddy–viscosity distribution in open channel flow, experiments: (○), Ueda *et al.*, 1976; (△), Nezu and Rodi (1986); (——), predictions. Taken from Celik and Rodi (1988).

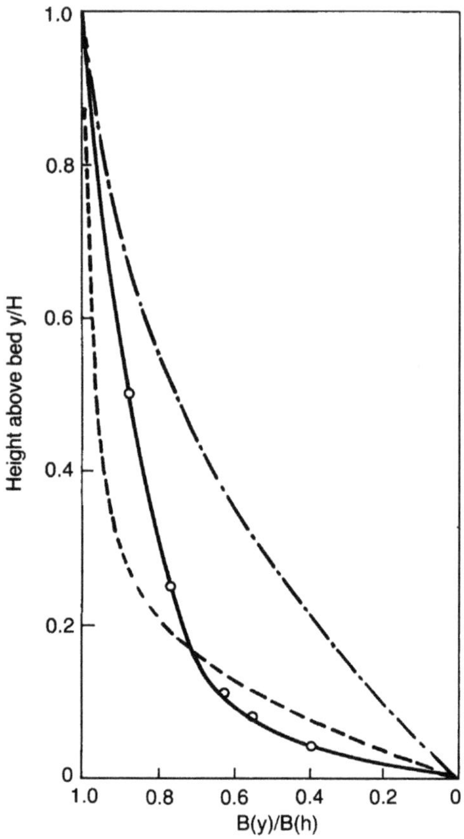

Figure 4.4 Vertical variation of velocity oscillation of tidal flow in Humber estuary, after Smith and Takhar (1977). (——), one-equation model; (– – –) Johns, parabolic eddy viscosity; (–·–), Johns, constant eddy viscosity; (○), Johns, observed values.

scribed above. The lower parts of Figure 4.6 compare the downstream development of the temperature profile under stable and unstable stratification for two densimetric Froude numbers (F) each. The results show clearly that the mixing is significantly reduced by the stable stratification and is virtually absent for the low Froude-number case. In this situation, there is little difference between the results obtained with the standard k–ε model and the algebraic stress/flux extension, and both agree quite well with the measurements. In the unstable cases, it is only the algebraic stress/flux version that is able to predict the very strong mixing when the Froude number is low. It appears that with this model the influence of stratification on the mixing in shear layers can be described fairly well. Further examples of model applications to stratified flow can be found in Rodi (1987).

4.3 DEPTH-AVERAGE HORIZONTAL TRANSPORT MODELS

4.3.1 MEAN-FLOW EQUATIONS

Depth-averaged forms of the mean-flow equations may be obtained by integration of the three-dimensional equations (4.1) to (4.3) over the depth, assuming hydrostatic pressure distribution as introduced by Falconer in Chapter 2 of this volume. The resulting equations read:

$$\frac{\partial h}{\partial t} + \frac{\partial}{\partial x}(h\bar{U}) + \frac{\partial}{\partial y}(h\bar{V}) = 0 \tag{4.15}$$

$$\begin{aligned}
&\frac{\partial \bar{U}}{\partial t} + \bar{U}\frac{\partial \bar{U}}{\partial x} + \bar{V}\frac{\partial \bar{U}}{\partial y} \\
&= -g\frac{\partial}{\partial x}(h + z_B) + \frac{1}{\rho h}\frac{\partial(h\bar{\tau}_{xx})}{\partial x} \\
&\quad + \frac{1}{\rho h}\frac{\partial(h\bar{\tau}_{xy})}{\partial y} + \frac{\tau_{sx} - \tau_{bx}}{\rho h} \\
&\quad \left.\begin{aligned}
&+ \frac{1}{\rho h}\frac{\partial}{\partial x}\int_{z_B}^{z_B+h} \rho(U - \bar{U})^2 \mathrm{d}z \\
&+ \frac{1}{\rho h}\frac{\partial}{\partial y}\int_{z_B}^{z_B+h} \rho(U - \bar{U})(V - \bar{V}) \mathrm{d}z
\end{aligned}\right\}\text{Dispersion}
\end{aligned} \tag{4.16}$$

$$\begin{aligned}
&\frac{\partial \bar{V}}{\partial t} + \bar{U}\frac{\partial \bar{V}}{\partial x} + \bar{V}\frac{\partial \bar{V}}{\partial y} \\
&= -g\frac{\partial}{\partial y}(h + z_B) + \frac{1}{\rho h}\frac{\partial(h\bar{\tau}_{xy})}{\partial x} \\
&\quad + \frac{1}{\rho h}\frac{\partial(h\bar{\tau}_{yy})}{\partial y} + \frac{\tau_{sy} - \tau_{by}}{\rho h} \\
&\quad \left.\begin{aligned}
&+ \frac{1}{\rho h}\frac{\partial}{\partial x}\int_{z_B}^{z_B+h} \rho(U - \bar{U})(V - \bar{V}) \mathrm{d}z \\
&+ \frac{1}{\rho h}\frac{\partial}{\partial y}\int_{z_B}^{z_B+h} \rho(V - \bar{V})^2 \mathrm{d}z
\end{aligned}\right\}\text{Dispersion}
\end{aligned} \tag{4.17}$$

$$\begin{aligned}
&\frac{\partial \bar{\phi}}{\partial t} + \bar{U}\frac{\partial \bar{\phi}}{\partial x} + \bar{V}\frac{\partial \bar{\phi}}{\partial y} \\
&= \frac{1}{\rho h}\frac{\partial(h\bar{J}x)}{\partial x} + \frac{1}{\rho h}\frac{\partial(h\bar{J}y)}{\partial y} + \frac{q_s}{\rho h} \\
&\quad \left.\begin{aligned}
&+ \frac{1}{\rho h}\frac{\partial}{\partial x}\int_{z_B}^{z_B+h} \rho(U - \bar{U})(\phi - \bar{\phi}) \mathrm{d}z \\
&+ \frac{1}{\rho h}\frac{\partial}{\partial y}\int_{z_B}^{z_B+h} \rho(V - \bar{V})(\phi - \bar{\phi}) \mathrm{d}z
\end{aligned}\right\}\text{Dispersion}
\end{aligned} \tag{4.18}$$

These equations govern the distribution of the water depth h and of the depth-averaged horizontal velocities $\bar{U}$ and $\bar{V}$ as well as of the scalar quantity $\bar{\phi}$. τ_b and τ_s are, the bottom and surface shear stresses, respectively, and q_s is the heat flux through the surface. Simple empirical laws involving friction and heat-

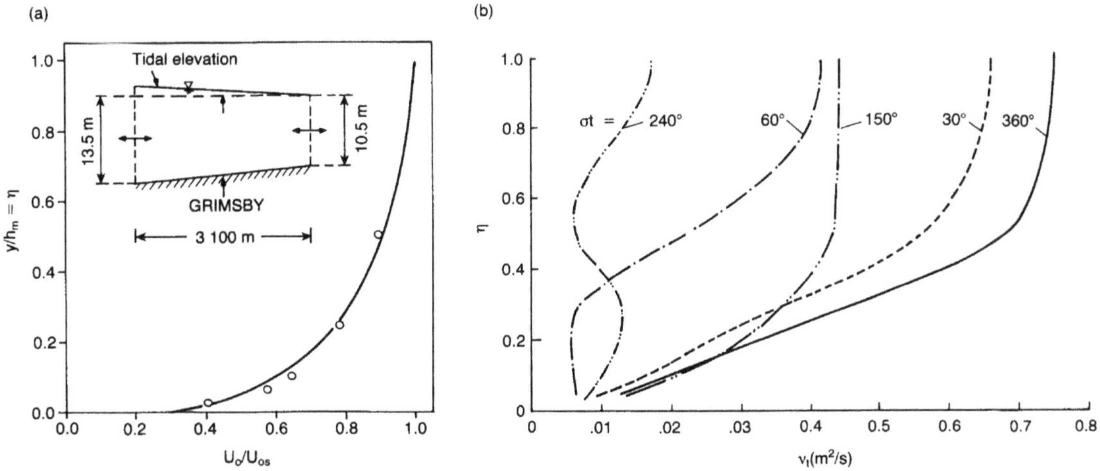

Figure 4.5 Distribution of relative amplitude of: (a) tidal current; (b) eddy viscosity in Humber estuary. Lines are predictions, symbols are measurements. From Celik and Rodi (1985).

transfer coefficients are used for calculating these terms, e.g. the quadratic friction law for the bottom shear stress. The task of a turbulence model is now to determine the depth-averaged stresses $\tilde{\tau}_{ij}$ and scalar fluxes $\bar{J}_i$. In many large-scale situations, the stresses $\tilde{\tau}_{ij}$ are negligible compared with the other terms in Equations (4.16) and (4.17), and the only influence of turbulence is through the bottom shear stress τ_b. In contrast, the turbulent heat and mass fluxes $\bar{J}_i$ are always important in Equation (4.18) and require a model assumption. Equations (4.16) to (4.18) also contain so-called dispersion terms which are due to vertical non-uniformities of the mean-flow quantities and are especially important under unsteady conditions. These terms have nothing to do with turbulence *per se* because they are solely a product of the depth-averaging process. In general, models also have to be introduced for these terms, but such models are not turbulence models. In the calculations shown in section 4.3.3, these terms have been neglected.

4.3.2 DEPTH-AVERAGE TURBULENCE MODELS

All depth-average calculations to date involving a turbulence representation employed the eddy viscosity/diffusivity concept reading (in tensor notation):

$$\frac{\tilde{\tau}_{ij}}{\rho} = \tilde{\nu}_t\left(\frac{\partial \bar{U}_i}{\partial x_j} + \frac{\partial \bar{U}_j}{\partial x_i}\right) - \frac{2}{3}\tilde{k}\delta_{ij}, \quad \frac{\bar{J}_i}{\rho} = \tilde{D}_t \frac{\partial \bar{\phi}}{\partial x_i} \quad (4.19)$$

In many practical depth-average calculations, a constant eddy diffusivity has been assumed while the turbulent stresses (and hence the eddy viscosity) have been neglected. A suitable value for the mean-depth diffusivity is usually found by calibration calculations, but it should be realized that the value chosen also accounts for the effects of dispersion and numerical diffusion.

Depth-average κ–ε model

When the near-field in the vicinity of banks, structures, discharges, etc. needs to be resolved, a model is required that calculates a realistic distribution of the eddy viscosity and diffusivity. Such a model has been developed by Rastogi and Rodi (1978) who adopted the k–ε model described earlier for use in depth-average calculations. They assumed that the local depth-averaged state of turbulence can be characterized by the energy and dissipation parameters $\tilde{k}$ and $\tilde{\varepsilon}$ and that the mean-depth eddy viscosity and diffusivity can be related to these parameters in the same way as given by Equations (4.12) and (4.6). The distribution of $\tilde{k}$ and $\tilde{\varepsilon}$ is determined by the following transport equations:

$$\bar{U}\frac{\partial \tilde{k}}{\partial x} + \bar{V}\frac{\partial \tilde{k}}{\partial y} = \frac{\partial}{\partial x}\left(\frac{\tilde{\nu}_t}{\sigma_k}\frac{\partial \tilde{k}}{\partial x}\right) + \frac{\partial}{\partial y}\left(\frac{\tilde{\nu}_t}{\sigma_k}\frac{\partial \tilde{k}}{\partial y}\right) + P_h + P_{kV} - \tilde{\varepsilon} \quad (4.20)$$

$$\bar{U}\frac{\partial \tilde{\varepsilon}}{\partial x} + \bar{V}\frac{\partial \tilde{\varepsilon}}{\partial y} = \frac{\partial}{\partial x}\left(\frac{\tilde{\nu}_t}{\sigma_\varepsilon}\frac{\partial \tilde{\varepsilon}}{\partial x}\right) + \frac{\partial}{\partial y}\left(\frac{\tilde{\nu}_t}{\sigma_\varepsilon}\frac{\partial \tilde{\varepsilon}}{\partial y}\right) + c_{1\varepsilon}\frac{\tilde{\varepsilon}}{\tilde{k}}P_h + P_{\varepsilon V} - c_{2\varepsilon}\frac{\tilde{\varepsilon}^2}{\tilde{k}} \quad (4.21)$$

Here,

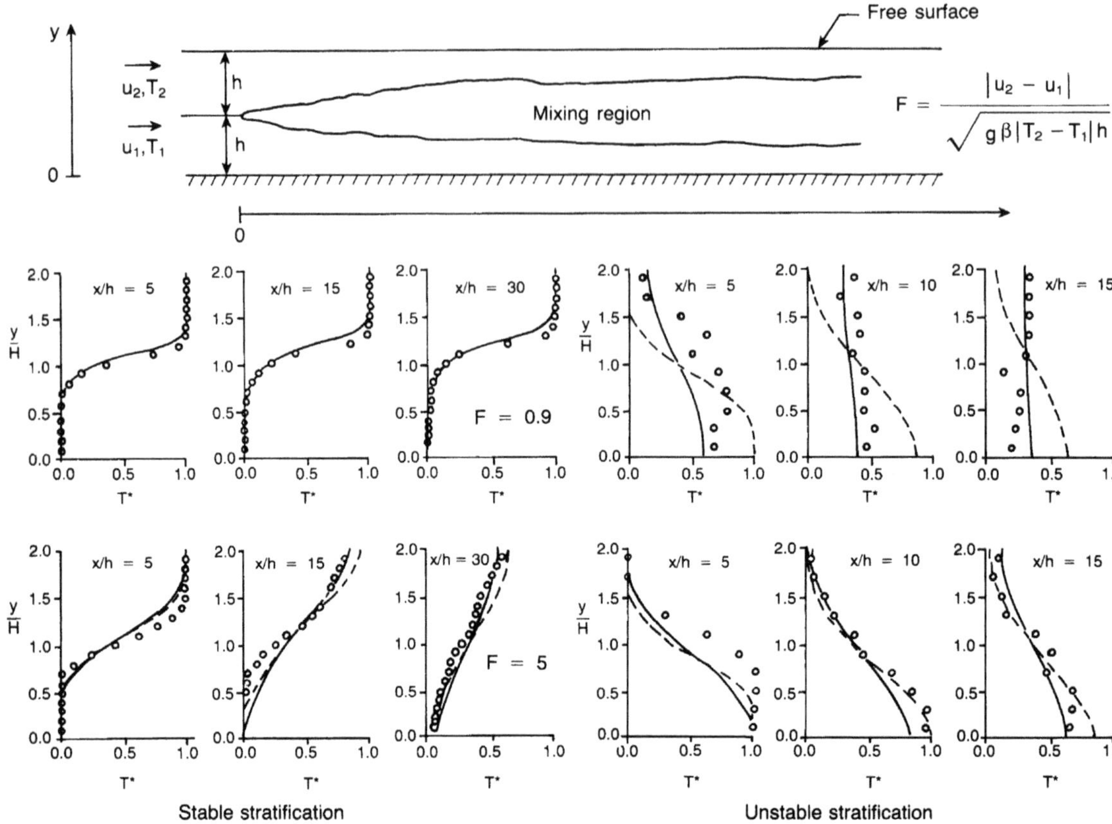

Figure 4.6 Development of temperature profiles in stratified mixing layers. Calculations by Viollet (1985). (---), standard k-ε model; (——) with algebraic stress/flux extension; (O) data of Viollet (1980).

$$P_h = \tilde{\nu}_t\left[2\left(\frac{\partial \bar{U}}{\partial x}\right)^2 + 2\left(\frac{\partial \bar{V}}{\partial y}\right)^2 + \left(\frac{\partial \bar{U}}{\partial y} + \frac{\partial \bar{V}}{\partial x}\right)^2\right] \quad (4.22)$$

is the production of k due to the interactions of turbulent stresses with horizontal mean velocity gradients. The terms P_{kv} and $P_{\varepsilon v}$ account for the terms originating from the non-uniformity of the vertical profiles. The main contribution to these terms stems from the significant vertical velocity gradients near the bottom which, by interaction with the relatively large turbulent shear stresses in this region, produce turbulence. This production is in addition to the production of P_h due to horizontal velocity gradients and depends strongly on the bottom roughness. The source terms $P_{\kappa v}$ and $P_{\varepsilon v}$ are therefore related to the friction velocity U_* by writing:

$$P_{kV} = c_k \frac{U_*^3}{h}, \quad P_{\varepsilon V} = c_\varepsilon \frac{U_*^4}{h^2} \quad (4.23)$$

where U_* is determined from the quadratic friction law $U_* = \sqrt{c_f(\bar{U}^2 + \bar{V}^2)}$, with c_f being the friction coefficient. The empirical constants c_k and c_ε were determined from undisturbed normal channel flow, where the dissipation rate ε and the friction velocity U_* can be directly related to the energy slope and where fairly reliable measurements of the depth mean diffusivity $\tilde{D}_t$ are available:

$$c_k = \frac{1}{\sqrt{c_f}}, \quad C_\varepsilon = \frac{c_{2\varepsilon}\, c_\mu^{1/2}}{(e^*\sigma_t)^{1/2} c_f^{3/4}} \quad (4.24)$$

The other constants in the depth-average k–ε model have the same value as given in section 4.2.2 for the vertical-structure k–ε model. The coefficient $e^* = \tilde{D}_t/(U_* h)$ is the dimensionless diffusivity for which Fischer *et al.* (1979) recommend a value of 0.15 for wide laboratory flumes, while $\varepsilon^* = 0.6$ is more suitable for natural rivers.

The adaptation of the k–ε model for depth-average calculations is certainly of a rather empirical nature, but calculations performed so far, show encouraging results. Examples will be presented in the next section. The depth-average k–ε model is best suited for near-field applications. In far-field calculations, refined turbulence models seem to not be warranted because of the relative unimportance of the turbulent stresses, the slow variation of all time-averaged quantities in the horizontal directions and last, but not least, the uncertainties due to dispersion and numerical diffusion effects and in specifying the boundary conditions.

4.3.3 CALCULATION EXAMPLES

Two examples of applications of the depth-average k–ε model will be presented here. The first example concerns a side discharge of heated water into steady, rectangular, open channel flow which demonstrates the ability of the model to work for rather complex flow situations. McGuirk and Rodi (1978) have calculated the near-field of such a discharge with the depth-average k–ε model. The flow situation is illustrated in Figure 4.7a together with the predicted streamline pattern. Owing to the restriction of jet entrainment by the near-bank, a recirculation region with decreased surface elevation is formed behind the jet. This region is predicted correctly and so are the jet trajectories for various ratios R of discharge-to-channel velocity (McGuirk and Rodi, 1978). Isotherms are of particular interest to environmentalists since they show the thermally polluted zones and also the extent to which they are polluted. Figure 4.7b compares predicted and depth-averaged isotherms for two velocity ratios R with measurements. There is little agreement on such details as the contorted nature of the isotherms, but the general behaviour of the isotherms is certainly predicted correctly and so is the extent of the thermally polluted region in the streamwise and lateral directions.

The second example concerns the mass exchange between a channel flow and an adjacent dead-water zone such as may occur in a harbour. The flow situation is sketched in Figure 4.8a, which also gives the calculated streamlines. In the shallow-water situation considered here, the turbulence causing the mass exchange is of particularly complex nature because it stems partly from turbulence produced in the free mixing layer springing off the edge of the expansion and partly from the shear production near the bottom. The depth-average k–ε model accounts for both of these production processes. A steady flow field was calculated, but the development of the mass concentration field with time was simulated starting with an initially uniform concentration in the harbour. The process simulated was that of a washing out of pollutants in the harbour, as effected by the mass exchange. Except at the beginning, the calculated pollutant concentration decays exponentially with time, in agreement with the experiment. The calculations were carried out for harbour situations studied experimentally by Nestmann (1977) who kept the channel and harbour geometries constant but varied the water depth and the inlet velocity. Figure 4.8b compares the calculated and measured half-life time $t_{1/2}$ of the average concentration in the harbour, and it can be seen that the mass exchange is reasonably well predicted by the model for all depths studied, encompassing both cases of predominant mixing-layer generation of turbulence and predominant bottom generation. It should be noted that none of the empirical constants in the model have been adjusted to obtain this result.

4.4 CONCLUSIONS

This chapter has discussed the current possibilities of and practices in simulating the effects of turbulence in calculations of the flow, heat and mass transfer in coastal and estuarial waters as well as in harbours. With the presently available computing power, three-dimensional large-eddy simulations can be used only for calculating local phenomena but not the flow in the whole water body; they are more suitable for fundamental studies which aid both our understanding of the flow phenomena and the development of simpler statistical models. The latter do not resolve any details of the turbulent motion but account only for its global effects via a turbulence model. When the flow is not well mixed vertically and the vertical structure needs to be resolved, as for example under stratified conditions, refined turbulence modelling is necessary. For this, eddy-viscosity/diffusivity models appear adquate for most problems, except when counter-gradient transport occurs in larger areas of the flow. The use of a constant eddy viscosity/diffusivity over the flow field is, however, not sufficient, but the distribution of these turbulence quantities must also be determined by the model. The simple mixing-length models work quite well for near-equilibrium shear layers but are not suitable for strongly transient situations as they neglect transport and history effects on turbulence. Models employing transport equations for turbulence quantities account for these effects and therefore have greater generality. One-equation models employing a transport equation for the kinetic energy of turbulence appear to be sufficiently refined for those situations where the flow has such a shear-layer character that the length scale can be prescribed with confidence. Two-equation models, which also solve an equation for the length scale of turbulence, are more generally applicable and

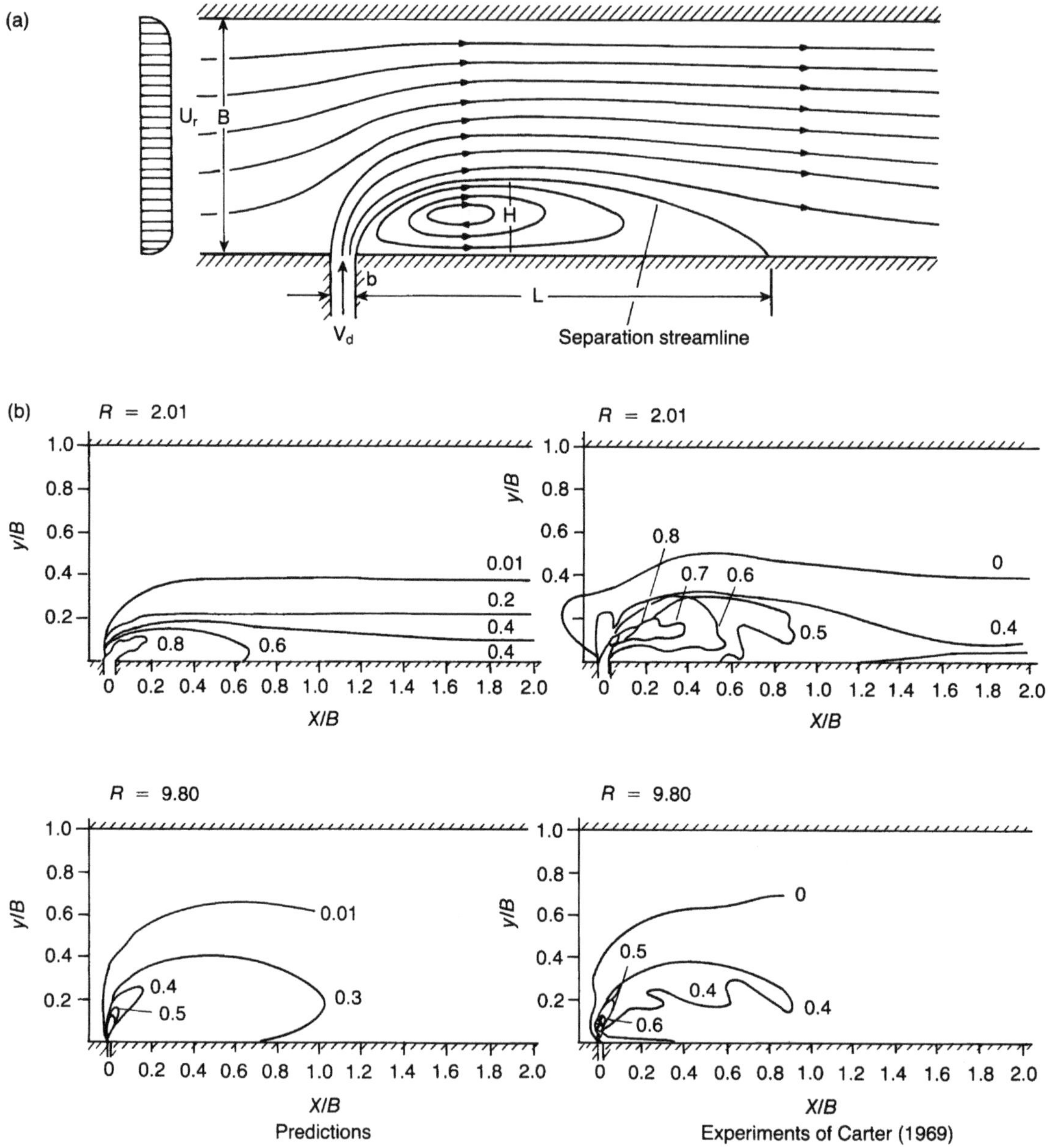

Figure 4.7 Side discharge into rectangular open channel flow. (a) predicted streamlines; (b) isotherms; depth-average calculations, $R = V_d/U_r$ (from McGuirk and Rodi (1978)).

are superior particularly in buoyant situations. The k–ε model is the most widely tested of the two-equation models, and an algebraic stress/flux extension is available that takes better account of the influence of stratification and also allows the effects of a free surface on the turbulence to be simulated. This model seems to be sufficient for most situations. The more complex stress/flux-equation models certainly do more justice to the physics and simulate the details better, but for obtaining global information of

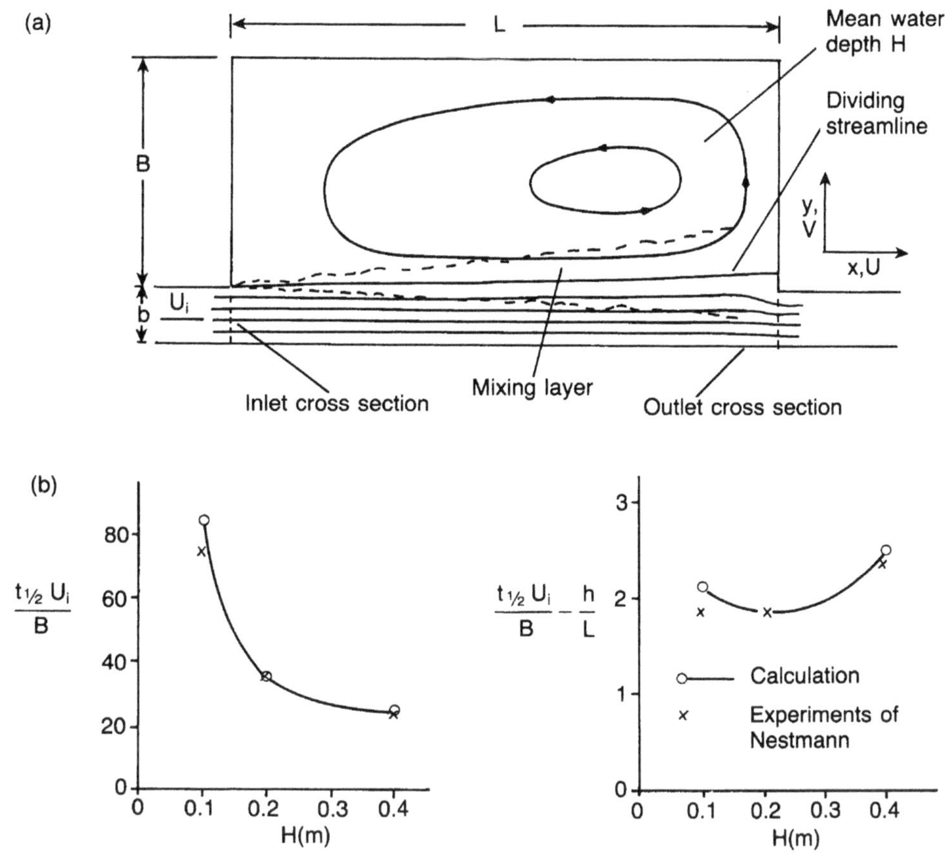

Figure 4.8 Depth-average calculation of unsteady mass exchange between channel flow and a dead-water zone (from McGuirk and Rodi, 1979). (a) Flow configuration and calculated streamline pattern for $h = 0.4$ m, $b = 0.4$ m; (b) variation of dimensionless half-life parameters with depth.

practical interest, simpler models are often adequate.

When the flow is vertically fairly well mixed, only the horizontal transport needs to be simulated in a two-dimensional depth-average calculation. Something like a two-dimensional large-eddy simulation seems possible in this case, resolving the large-scale horizontal turbulent motion. The interaction with the smaller-scale bed-generated three-dimensional turbulence and its modelling is not very clear at present, and further studies are necessary to examine whether the inclusion of the bed-shear terms is sufficient or whether a sub-grid-scale model is, in general, necessary. When the horizontal turbulent motion is not resolved in the numerical calculation, a statistical turbulence model has to be used. In far-field calculations, a refined model is now not necessary. In fact, the turbulent shear-stress terms appearing in the depth-averaged momentum equations can usually be neglected compared with the bed-shear terms and a constant eddy diffusivity for heat and mass transfer, obtained from calibration calculations, is usually employed. On the other hand, near-field situations require more refined modelling in which a realistic distribution of the eddy viscosity and diffusivity over the flow field has to be provided by the turbulence model. A depth-average version of the k–ε model has been presented which allows such refined modelling. This model has been found to describe the flow field and also the distribution of scalar quantities reasonably well for a number of relatively complex situations, including recirculating flows.

NOMENCLATURE

B	Channel width
	Oscillation amplitude

c's	Constants in turbulence models
c_f	Friction coefficient
$\tilde{D}_t$	Eddy diffusivity
E	Roughness coefficient in log law (Equation (4.13))
e^*	$= \tilde{D}_t(U_* h)$ = dimensionless, depth-average diffusivity
F	Densimetric Froude number (defined in Figure 4.6)
G	Buoyancy production/destruction of k, defined in Equation (4.8)
g	Gravitational acceleration
H, h	Water depth
J_i	Turbulent flux of ϕ in direction x_i
k	Turbulent kinetic energy
L	Length scale of turbulence
l_m	Prandtl mixing length
P	Mean pressure, Shear production of k, defined in Equation (4.8)
q_s	Heat flux at surface
R	Velocity ratio (defined in Figure 4.7)
T	Temperature
t	Time
U, V, W	Mean velocity components in x, y, z-direction
u, v, w	Fluctuating velocity components in x, y, z-direction
U_i	Mean velocity component in x_i-direction
u_i	Fluctuating velocity component in x_i-direction
U_*	Friction velocity
x, y, z	Co-ordinates (see Figure 4.1)
x_i	Co-ordinates in tensor notation
z_B	Height above datum

GREEK SYMBOLS

β	Volumetric expansion coefficient
δ_{ij}	Kronecker delta, = 1 for $i = j$ and = 0 for $i \neq j$
ε	Dissipation rate of k
η	$= y/h$ relative height above bottom
κ	von Kármán constant in log law (Equation 4.13))
λ	Molecular diffusivity of scalar quantity ϕ, Constant in specifying mixing length distribution (see Figure 4.2).
ν	Kinematic molecular viscosity
ν_t	Eddy viscosity
ρ	Fluid density
σ	Oscillation frequency of tidal current
σ_t	Turbulent Prandtl/Schmidt number
$\sigma_{k,\varepsilon}$	Diffusion constants in k, ε equations
τ_{ij}	Stress acting in direction x_i on a face perpendicular to x_j
ϕ	Mean temperature or concentration
φ	Fluctuating temperature or concentration

SUBSCRIPTS

B, b	Bottom
h	Horizontal
r	Reference Resultant
s	Surface
t	Turbulent
v	Vertical

REFERENCES

Abbott, M.B., Larsen, J. and Jianhua Tao (1985) Modelling circulation in depth-integrated flows, part I: the accumulation of evidence. *Journal of Hydraulic Research*, **23** (4), 309–26.

Abbott, M.B. and Larsen, J. (1985) Modelling circulation in depth-integrated flows, part II: a reconciliation. *Journal of Hydraulic Research*, **23** (5), 397–420.

Carter, H.H. (1969) A preliminary report on the characteristics of a heated jet discharged horizontally into a transverse current, part I, constant depth. Chesapeake Bay Institute, Johns Hopkins University, *Tech. Rep. 61*.

Celik, I. and Rodi, W. (1985) Calculation of wave-induced turbulent flows in estuaries. *Ocean Engineering*, **12** (6), 531–42.

Celik, I. and Rodi, W. (1988) Modeling suspended sediment transport in non-equilibrium situations. *Journal of Hydraulic Engineering*, **114** (10), 1157–91.

Fischer, H.D., List, E.J., Koh, R.C.Y., *et al.* (1979) *Mixing in Inland and Coastal Waters*, Academic Press, London.

Jobson, H.E. and Sayre, W.W. (1970) Vertical transfer in open channel flow. *Journal of the Hydraulics/Division, ASCE*, **HY3**, 703–24.

Johns, B. (1966) On the vertical structure of tidal flow in river estuaries. *Geophysical Journal of the Royal Astronomical Society*, **12**, 103–10.

Journal of Geophysical Research (1987) (Special issue containing papers from the IUTAM symposium on mixing in stratified fluids) **92** (C5), 5229–495.

Launder, B.E. and Spalding, D.B. (1974) The numerical computation of turbulent flow. *Computer Methods in Applied Mechanics and Engineering*, **3**, 269–89.

McGuirk, J.J. and Rodi, W. (1978) A depth-averaged mathematical model for the near field of side discharge into open channel flow. *Journal of Fluid Mechanics*, **86**, 761–81.

McGuirk, J.J. and Rodi, W. (1979) Calculation of unsteady mass exchange between a main stream and a dead water zone, Proceedings of the 18th IAHR Congress, Cagliari, Italy.

Nestmann, F. (1977) Strömungsmechanische Kriterien zur Klassifizierung und Beurteilung der Durchströmungsverhältnisse in Wasserbehältern. Vertieferarbeit am Institut für Hydromechanik, University of Karlsruhe.

Nezu, I. and Nakagawa, H. (1993) *Turbulence in Open Channel Flows*, IAHR monograph, A.A. Balkema, Rotterdam.

Nezu, I. and Rodi, W. (1986) Open channel flow measurements with a laser-doppler anemometer. *Journal of Hydraulic Engineering*, **112** (5), 335–55.

Oey, L.Y., Mellor, G. and Hires, R. (1985) A three-dimensional simulation of the Hudson-Raritan estuary, part I: description of the model and model simulations, part II: comparison with observation, part III: salt flux analyses. *Journal of Physical Oceanography*, **15**, 1676–1720.

Prandtl, L. (1925) Über die ausgebildete Turbulenz. *Zeitschrift für Angewandte Mathematikund Mechanik*, **5**, 136.

Rastogi, A.K. and Rodi, W. (1978) Predictions of heat and mass transfer in open channels. *Journal of the Hydraulics Division, ASCE.* **HY3**, 397–420.

Rodi, W. (1984) *Turbulence Models and their Application in Hydraulics*, International Association for Hydraulic Research, Delft, The Netherlands.

Rodi, W. (1985) Calculation of stably stratified shear-layer flows with a buoyancy-extended k–ε turbulence model, in *Turbulence and Diffusion in a Stable Environment* (ed. J.C.R. Hunt), Oxford University Press, Oxford, pp. 111–40.

Rodi, W. (1987) Examples of calculation methods for flow and mixing in stratified fluids. *Journal of Geophysical Research*, **92** (C5), 5305–28.

Smith, T.J. and Takhar, H.S. (1977) The calculation of oscillatory flow in open channels using mean turbulence energy models. Simon Engineering Laboratories, University of Manchester, Manchester, UK, *Rep. No. M 139 PL.*

Ueda, H., Möller, R., Komori, S. *et al.* (1976) Eddy diffusivity near the free surface of open channel flow. *International Journal of Heat and Mass Transfer*, **20** (11), 1127–36.

Viollet, P.L. (1980) Turbulent mixing in two-layer stratified shear flow. *Proceeding of the 2nd Symposium on Stratified Flows*, Tapir, Trondheim, Norway, pp. 315–25.

Viollet, P.L. (1985) The modelling of turbulent recirculating flows for the purpose of reactor thermal-hydraulic analysis. Electricité de France, Chatou, France, *Rep. E44/R05.*

DIFFUSION, DISPERSION AND SUB-GRID PARAMETERIZATION 5

K.W. Bedford

In the practice of modelling, the results of the continuum theory of turbulence are introduced for the most part through the inclusion of diffusion and dispersion terms in the model equations. However, the way in which this is done in any particular model must depend upon the degree to which the model itself resolves diffusion and discretization processes at the scale of its numerical grid. Correspondingly, the results of turbulence theory must be translated primarily into methodologies of sub-grid-scale parameterization. This translation process is by no means simple, however, and the aim of this, very ambitious chapter is to expose the principal difficulties and to indicate how these are currently being overcome in the main areas of modelling practice.

CONTENTS

5.1 INTRODUCTION

It is particularly difficult to review and summarize the literature in a field so rich in history as this one. In fact, the subject of turbulent diffusion is as old as Reynolds' (1882) original averaging definitions for turbulence. The earliest parameterizations for turbulent diffusion originated with work by Taylor (1921) based upon homogeneous isotopic turbulence in the absence of boundaries. He said that, essentially, after an initial portion of time, the turbulent flux of concentration was approximated by a gradient diffusion term. The coefficient of proportionality has been termed the eddy diffusivity (Fischer *et al.*, 1979). In the presence of boundaries which introduce spatial gradients in the mean velocity and concentration gradients, Taylor (1953, 1954) clearly demonstrated (via pipe-flow analyses) that the effect of these gradients (and hence boundaries) on the turbulent flux could also be modelled by a gradient-diffusion approximation. The proportionality coefficient was labelled the dispersion coefficient (Fischer *et al.*, 1979). Elder (1959) extended Taylor's analysis to a shallow shear flow down an infinitely wide plane and assumed that the vertical velocity variations dominated the dispersion coefficient. He demonstrated that the dispersion coefficient was roughly 100 times larger than the eddy diffusivity which, in turn, was roughly five orders of magnitude greater than the typical moleculer diffusivity.

Fischer (1967) reasoned that for rivers and streams with steady shear flow, the vertical variations in velocity were not nearly so important as the transverse variations. Field data summarized in Table 5.3 of Fischer *et al.* (1979) show that the dispersion coefficient in streams is quite strong, approximately 20 times (or more) greater than the dispersion coefficient estimated by Elder. Fischer's incorporation of transverse variations went quite some way towards explaining these field data.

The dispersion coefficient has therefore clearly been demonstrated to be an important transport agent and its parameterization, as well as its relationship to turbulent diffusion, have been studied extensively. While the importance of a proper parameterization of diffusion and dispersion has been

Coastal, Estuarial and Harbour Engineer's Reference Book. Edited by M.B. Abbott and W.A. Price. Published in 1993 by Chapman & Hall, 2–6 Boundary Row, London SE1 8HN. ISBN 0 419 15430 2.

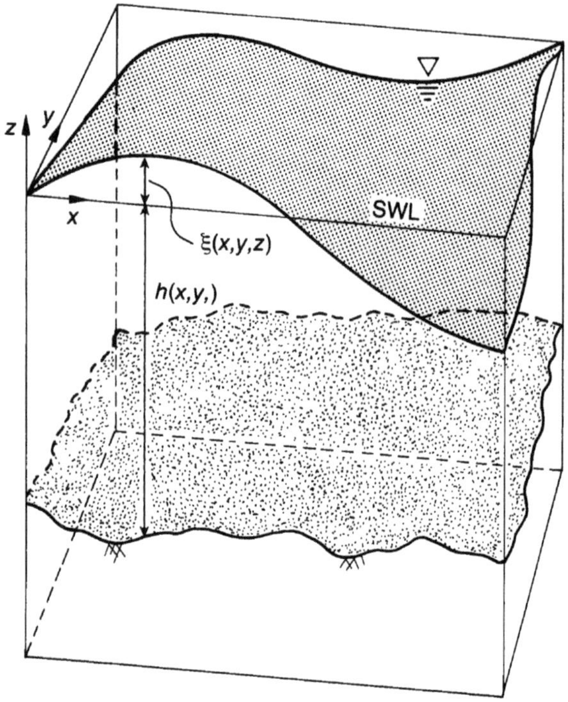

Figure 5.1 Co-ordinate notation schematic.

identified in the literature, a universal form of such parameterization has not. The objective of this chapter is to provide the scientific setting for the parameterization of dispersion and diffusion, especially as regards the two- and three-dimensional modelling of shallow estuaries, coastal seas harbours, etc. Recent efforts at reviewing this highly complex field include Csanady (1972), Fischer *et al.* (1979) and review articles by Fischer (1976), Holley *et al.* (1970, 1972), Holly (1985), Chatwin and Allan (1985) and the unfortunately brief, but interesting, review by Smith (1989). It is tempting to simply recast the information in these review articles to meet this objective. However, the complexity of the water bodies considered in this volume, coupled with the highly limiting assumptions used to derive most of the reviewed parameterizations, often yield improperly selected model formulations, inappropriate to the task. It is therefore the present purpose to expose a number of considerations which take the form of consistency requirements between the physics, governing equations and the grid resolution which can provide a procedure for identifying adequate parameterizations of processes which, although not resolved by the grid and model equations, nevertheless affect the model results. Chief amongst these sub-grid-scale parameterizations are, of course, diffusion and dispersion.

This chapter is based on the following simple ideas. Firstly, the flow and transport physics are assumed to vary continuously in time and space over as many as eight decades of variability. Continuous dynamic complexity has been clearly documented (Monin *et al.*, 1974; Monin and Ozmidov, 1985; Nihoul and Djenidi, 1987; and Woods, 1977) and must be explicitly dealt with in the resulting model structure. Secondly, the use of a finite grid representation of the continuum suggests that all the physics variability at the Nyquist limit of twice the grid size or smaller will not be resolved in the model but, because it affects the physics, must be parameterized by a bulk representation. Finally, the interface between the continuum (analogue) physics and the digital representation is a set of governing differential equations which have been properly averaged or prepared such that the information removed by the grid has been commensurately filtered or averaged out from the governing equations. It is implicit in these assumptions that averaging of the equations in space or time gives rise to terms requiring closure which, for a number of simple flows, have been adequately parameterized by elementary diffusion and dispersion descriptions. However, due to their basis in averaging, the historically available diffusion and dispersion relations are only valid if the conditions under which they were derived (i.e. averaging conditions) obtain in the field data and model. If not, then errors occur and the results are reduced to a library of site- and simulation-specific results. Diffusion and dispersion parameterizations can no longer be viewed as separate from the physics, which is intuitive and well known, but further, it can no longer be seen as separate from the digital model structure, i.e., grid size, arrangement and resulting resolution. The purpose of this chapter is to point out these consistency and sub-grid-scale considerations as a means of generalizing the evaluation of the diffusion and dispersion parameterizations.

5.2 THE AVERAGE EQUATIONS AND DIFFUSION AND DISPERSION TERMS

For the remainder of this chapter attention is directed to the concentration or transport equation as it is the source of the diffusity and dispersion terms. The basic starting point is the three-dimensional continuum equation. With reference to the notation in Figure 5.1 the three-dimensional transport equation is

$$\frac{\partial c}{\partial t} + (u_i c)_{,i} = D c_{,jj} \tag{5.1a}$$

$$\frac{\partial c}{\partial t} + \frac{\partial (uc)}{\partial x} + \frac{\partial (vc)}{\partial y} + \frac{\partial (wc)}{\partial z} = D\left(\frac{\partial^2 c}{\partial x^2} + \frac{\partial^2 c}{\partial y^2} + \frac{\partial^2 c}{\partial z^2}\right) \quad (5.1b)$$

$$\frac{\partial c}{\partial t} + (\bar{v}\nabla)c = D\nabla^2 c \quad (5.1c)$$

In Equation (5.1a,b,c), D is the molecular diffusion coefficient and $\bar{v} = u_i + v_j + w_k$ are the velocity components in the $\mathbf{x} = x_i + y_j + z_k$ directions, respectively.

5.2.1 AVERAGING

The traditional definition for equation averaging and the subsequent analysis procedures have recently been recast in the larger context of generalized convolution averaging (Leonard, 1974; Bedford *et al.*, 1987; Yeo and Bedford, 1988) with a rather different set of averaging rules. The convolution definition of averaging is:

$$\bar{\alpha}(\mathbf{x}, t) = \int_{-\infty}^{\infty} G(\mathbf{x} - \mathbf{x}', t - t')\,\alpha(\mathbf{x}', t')\,d\mathbf{x}'dt' \quad (5.2)$$

where the filter function G is defined such that

$$\int_{-\infty}^{\infty} G(\mathbf{x}, t)\,d\mathbf{x}\,dt = 1 \quad (5.3)$$

The original averaging definition for turbulence, taken from Reynold's original work, used a uniform, weight function, fixed interval average for G over a time period Δ_t. This averaging is defined (Hinze, 1975) as

$$\bar{\alpha}(\mathbf{x}, t) = \frac{1}{\Delta_t}\int_t^{t+\Delta t} \alpha(\mathbf{x}, t)\,dt \quad (5.4)$$

and it assumes that the turbulent quantity $\alpha'(\mathbf{x}, t)$ is defined relative to $\bar{\alpha}$ as

$$\alpha' = \alpha(\mathbf{x}, t) - \alpha(\mathbf{x}, t) \quad (5.5)$$

From the moving average perspective, various forms of $G(\mathbf{x}, t)$ have been used, one popular form being the Gaussian spatial filter

$$G(x, t) = \left(\frac{\gamma}{\pi}\right)^{1/2} \sum_{i=1}^{3} \frac{1}{\Delta_i} e^{-\sigma x_i/\Delta_i^2} \quad (5.6)$$

where γ is a filter coefficient (usually taken as 6.0) and Δ_i is the averaging length scale in each co-ordinate direction.

Whatever the form of $G(\mathbf{x}, t)$, the central question in these averages is what is the effective averaging scale (Δ_t, Δ_i). As the filtering or averaging removes all processes and information at scales smaller than these cut-off scales, it is essential to specify and select these scales consistently.

5.2.2 THREE-DIMENSIONAL TRANSPORT

Using the overbar to denote the average value of the variable, the governing equation is averaged using rules appropriate to either the fixed (Reynolds) or moving average (Bedford *et al.*, 1987) and the following three-dimensional equation results:

$$\frac{\partial \bar{c}}{\partial t} + (\bar{u}_{,i}\bar{c})_{,i} = \bar{Q}_{i,i} + D\nabla^2\bar{c} + (\overline{u', c'})_{,i} \quad (5.7)$$

In Equation (5.7) the quantity $Q_{i,i}$ results from the use of the higher-order moving average, which for the Gaussian space filter has the form

$$Q_{i,i} = \frac{\Delta_i^2}{2\gamma}(\bar{u}_{i,k}\bar{c}_{,k})_{,i} \quad (5.8)$$

Space-time filters have been derived (Bedford and Dakhoul, 1982) and tested (Dakhoul and Bedford, 1986; Aldama, 1990). Equations (5.7) and (5.8) have formed the bases for a number of detailed turbulence models, with their use in surface water models being first due to Babajimopoulos and Bedford (1980) and Bedford and Babajimopoulos (1980).

The ubiquitous Reynolds-averaged forms of the three-dimensional equations result from the averaging defined in Equation (5.4) and, when combined with the proper averaging rules, are

$$\frac{\partial \bar{c}}{\partial t} + (\bar{u}_i\bar{c})_{,i} = D\bar{c}_{,jj} - (\overline{u_i' c'})_{,i} \quad (5.9)$$

The last terms in Equation (5.9) are called the Reynolds fluxes and require closure with mean values of the dependent variables.

5.2.3 DEPTH-AVERAGE TWO-DIMENSIONAL TRANSPORT

Due to the shallowness of most flows and the consequent rapid vertical mixing relative to the horizontal components, the three-dimensional equations are often further averaged by vertical or depth averaging to form a two-dimensional model. The depth averaging operation is a fixed interval average based upon the local total depth i.e. $H = h + \xi$. Therefore, at each instant of time, the averaging scale (depth) is continuously varying. The average is defined as

$$\beta = \bar{\beta}^* + \beta'^* \quad (5.10)$$

where (*) denotes a definition base on

$$\bar{\beta}^* = \frac{1}{(h + \xi)}\int_{-h}^{\xi} \beta dz \quad (5.11)$$

If the definition in Equation (5.11) is applied to the three-dimensional equations (Equations (5.7) or (5.9)) then:

$$\begin{aligned}\frac{\partial \bar{\bar{c}}^*}{\partial t} &+ \frac{\partial}{\partial x}(H\bar{\bar{u}}^*\,\bar{\bar{c}}^*) + \frac{\partial}{\partial y}(H\bar{\bar{u}}^*\,\bar{\bar{c}}^*)\\ &= D\nabla^2(H\bar{\bar{c}}^*) + \bar{A}^* \ldots.(3)\\ &- \left[\frac{\partial}{\partial x}\int_{-h}^{\xi}\overline{u'c'}\,\mathrm{d}z + \frac{\partial}{\partial y}\int_{-h}^{\xi}\overline{v'c'}\,\mathrm{d}z\right]\ldots.(1)\\ &- \left[\frac{\partial}{\partial x}\int_{-h}^{\xi}\left\{(\bar{u}')^*(\bar{c}')^*\mathrm{d}z + \frac{\partial}{\partial y}\int_{-h}^{\xi}(v')^*(c')^*\right\}\mathrm{d}z\right]\ldots.(2)\end{aligned} \quad (5.12)$$

In Equation (5.12), term (3) is the vertically averaged term resulting from the use of higher-order averaging and is defined from Equation (5.8) as:

$$-\frac{\Delta_i^2}{2\gamma}\int_{-h}^{\xi}\{(\bar{u}_{i,k}, \bar{c}_{,k})_{,i}\}\mathrm{d}z \quad (5.13)$$

Term (1) represents the vertical averaging of the Reynolds flux terms while term (2) results from the vertical integration of the temporally averaged non-linear advection term. Note that $(\bar{\alpha})'^*$ is the departure of the temporal mean of α from the vertical average value of α, i.e. $(\bar{\alpha})'^* = \bar{\alpha} - \bar{\bar{\alpha}}^*$.Term (2) is called the dispersion term and closure must be specified in terms of the space–time average values of $\bar{\bar{c}}^*$, and possibly $\bar{\bar{v}}^*$. It should be noted that the depth averaging need not start from the turbulent form, but since the models discussed in this volume are turbulent, it is appropriate to start with this as a basis for the derivation.

Finally, from these equations, we can, for comparative analysis, identify a number of key length scales. Firstly, the horizontal extent of the basin, L, and the water depths (H, h and ξ) are intuitive. Secondly, the averaging scales in either time, Δ_t, or space, Δ_i, are critical because, from a Nyquist sampling theorem viewpoint, the minimum wave numbers, k_m, or frequency, ω_m, of a physical process resolved after the averaging process are π/Δ_i and π/Δ_t, respectively. The next question to be explored concerns how these length and time scales compare to the physics at hand.

5.3 THE PHYSICAL SETTING

A number of chapters in this volume fully explain the detailed physics of the settings, yet it is necessary to try and enumerate and order all the physical processes in terms of size and possibly amplitude so that our desired grid and filtering scales can be compared to the physical scales of the various processes.

5.3.1 ENUMERATION OF PROCESSES

In the crudest sense, a water body consists of two basic motions; wave-like coherent motions occurring in distinct narrow-frequency or wavenumber bands, and turbulent eddies. Turbulent eddies are broad-band and can occur over the entire spectrum whose width is governed by the Reynolds number of the flow. As is described in most works on turbulence, such as Tennekes and Lumley (1972), Hinze (1975) or Monin and Ozmidov (1985), a non-linear interaction of vortex stretching, shear, stratification and vorticity results in the existance of a continuous range of eddy sizes bounded from above by the geometry of the problem, i.e. basin size (L, H), and from below by the molecular viscosity. The effective extent of the horizontal dimension is roughly (Woods, 1977) 1 cm to 100 km for estuaries giving a factor of 10^7 – an enormously broad spectrum. In the absence of any energy inputs, the wave number or frequency distribution of the eddy kinetic energy is roughly a red spectrum, which decreases monotonically from the largest to the smallest scales of eddies. The energy cascade (Tennekes and Lumley, 1972) resulting from vortex stretching summarizes this behaviour from a continuous point of view, while a more quantized discrete approach is offered in the oceanographic literature (see Woods, 1977) for an early overview. In the energy-cascade viewpoint, full three-dimensional turbulence energy flows continuously from the largest to the smallest scales. A more specialized two-dimensional turbulence (Rhines, 1977) may exist at larger scales, and in this energy may flow to both larger and smaller scales.

Against the broad-band backdrop of turbulent eddies is superimposed a variety of time varying wave motions whose energy is concentrated in narrow wavelength and frequency bands but whose effects broadly spread to both larger and smaller wave-number bands about the central band. These wave-like motions are often the means by which kinetic energy is injected into the turbulence spectrum, as, for example, through surface gravity waves, internal waves via Kelvin–Helmholtz instabilities or the baroclinic instability mechanism of converting potential to kinetic energy. There is a rather large number of coherent motions in the limited-area models considered in this volume. Table 5.1 contains a brief summary of many of these for estuaries while Figure 5.2 portrays a space–time display for many of these motions.

5.3.2 PROCESS ORDERING

A variance spectrum is one of the more fundamental procedures for mapping the scale, size and magni-

Table 5.1 Summary of length and time scales associated with various activities and influences within estuaries (Bedford, 1985)

No. Activity/process	*Time/length scale*	*Reference*
1 Wind stress	$T = 3$ d $D = 6.7\,W/\sqrt{\sin\phi}$	Elliot and Wang (1978) Eckman depth of influence
2 Far-field forcing (coastal current, storm surge, etc.)	$T = 10–20$ d	Elliot *et al.* (1978)
3 Tidal activity	Diurnal/semi-diurnal ranges variable (>1 to >40 ft)	
4 Local turbulence (Brunt–Väisälä)	$T = 1$ min–1 h $T = 1$ min $L = 1$ m (vertical scale)	Cannon (1971) Nihoul (1978)
5 Residual circulation	$T = 20$ d (overall) $T = 1–2.5$ d (type)	Elliot and Wang (1978)
6 River inflow	$T = 5$ days–10 d Coherent with seal level at periods of 2.5–20 d	Elliot and Wang (1978) Elliot and Wang (1978)
7 Surface slope (set up)	4.2 days (barotropic)	Kjerfve *et al.* (1978)
8 Sea-level	(a) Coherent with wind at 9.2 days (surges) creation time ~6 h. (b) Coherent with surface pressure at 6 d (pressure jumps) creation time ~6 h.	Kjerfve *et al.* (1978) N. Smith (1978)
9 Velocities (a) Tidal (b) Residual (c) Turbulent	$T = 2–10$ d $T = 18$ d $u' = u\,(10^4)$	Elliot (1978) Weisberg (1976)
10 Salinity instruction	$10^2 < L/H < 10^4$ (ratio of salt intrusion distance to estuary depth)	Rattray and Mitsuda (1974)
11 Internal waves	$T = 2–11$ h $L = 10–1000$ m $H = 1$ m	Mortimer (1968) Garrett and Munk (1979)
12 Kelvin, Topographic waves	$T = 14–30$ h $L = 10–1000$ m	Mortimer (1968) Pedlosky (1979)
13 (a) coastal jets (b) Upselling	$T = 1–2$ d $L = 100–1000$ km	Csanady (1982)
14 Thermal stratification	$D = 10–100$ m $T = 100–200$ d	Turner (1973)
15 Vertical shear	$D = 10–100$ m $T = 10$ s	Fischer *et al.* (1979)
16 Inertial motions	$D = 6.7\,W\sqrt{\sin\phi}$ $T = 14$ h	Pedlosky (1979)

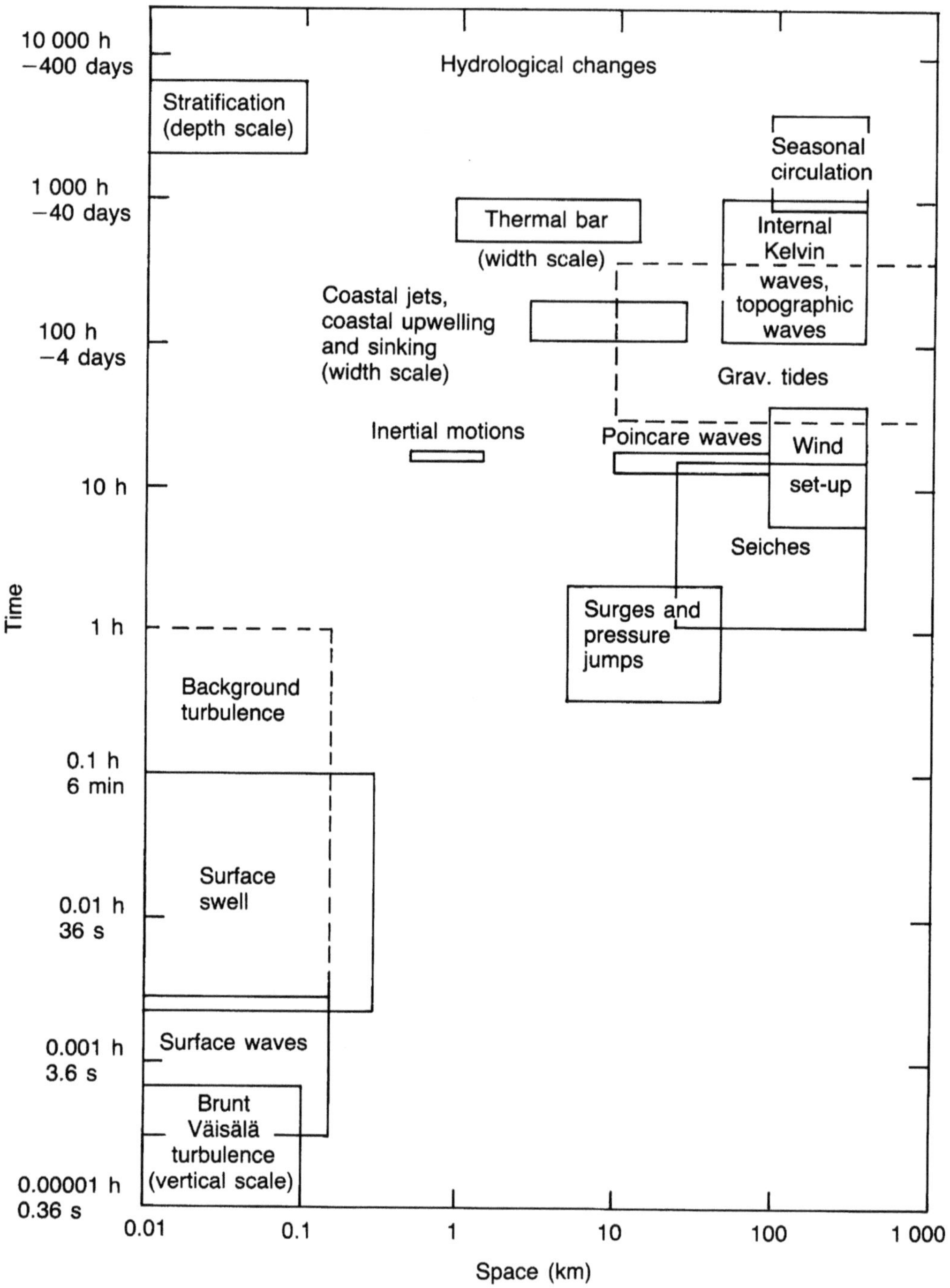

Figure 5.2 Typical space-time process diagram.

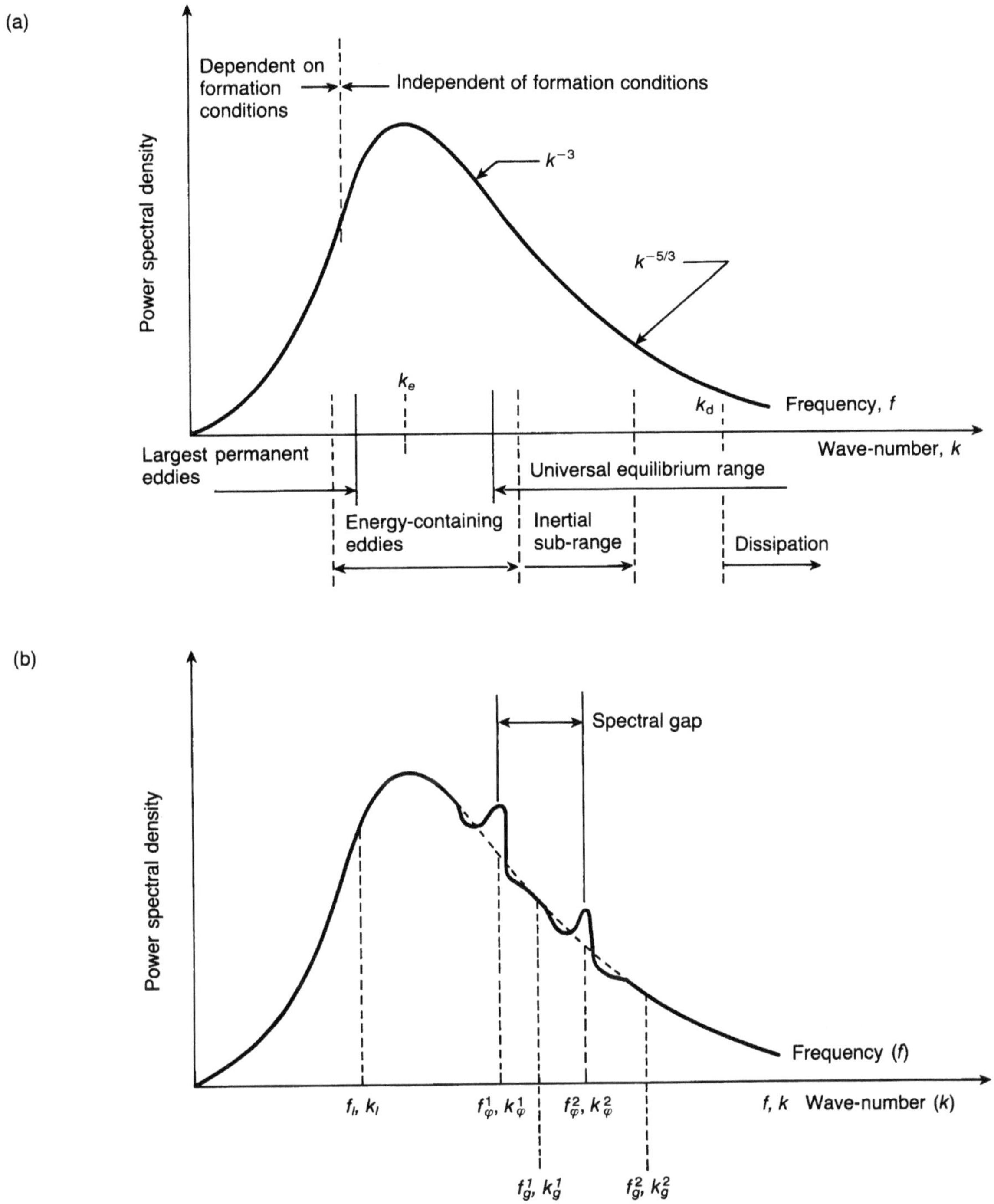

Figure 5.3 (a) Typical kinetic energy spectrum for pure turbulence; (b) kinetic energy spectrum including wave motions.

tude of a broad-band system. A log-log plot of the variance (per wavelength) against wavenumber or frequency is typical. Figure 5.3a contains a general schematic of the typical kinetic-energy (velocity variance) spectral density as assembled for pure turbulent motions. With the imposition of wave motion, spectral 'spikes' occur at the typical frequency or wavenumber of that motion. Near the spectral 'spike', energy is locally flowing both up and down the spectra and, consequently, the energy transfer is not in equilibrium i.e. production does not equal dissipation.

Within the kinetic energy variance spectrum there are four different ranges where the turbulence eddies are distinctly different, and it is important to specify the ranges. Table 5.3 has been compiled from Nihoul and Djenidi (1987) and Woods (1977), summarizing the various ranges as based on the non-dimensional numbers in Table 5.2.

The space–time assemblage of data is currently undergoing considerable refinement, especially as new instrumentation techniques become available. The literature on comprehensive space-time diagrams is sparse and contributing papers are scattered through a number of journals. The recent compilation of papers from the IUTAM Mixing in Stratified Fluids Conference (Imberger, 1987 and *Journal of Geophysical Research*, 1987, 5229–495) provides an extremely intense dose of field-based reality concerning the non-linear complexity of surface water physics. The papers by Thorpe (1987), Hopfinger (1987) and Gregg (1987) are particularly noteworthy.

Table 5.2 Definitions of dimensionless parameters

Ro, Rossby no.	$Uf^{-1}L^{-1}$
Ri, Richardson no.	$H^2N^2U^{-1}$
Re, Reynolds no.	LU/ν

where:
U = Horizontal velocity scale
f = Coriolis frequency
L = Horizontal length scale
H = Depth scale
ν = Kinematic viscosity
N = Brunt–Väisälä frequency
$= -\frac{g}{\rho}\frac{\partial \rho_0}{\partial z}^{1/2}$
ρ = Local density
ρ_0 = Average density
g = Acceleration due to gravity

5.4 RECONCILING THE GRID, PHYSICS AND AVERAGE EQUATIONS

5.4.1 NYQUIST OR MICRO-SCALE CONSIDERATIONS

The relationship of the grid size to the averaged differential equations has been explored in a number of ways. From the meteorological modelling work of Smagorinsky *et al.* (1963), Haltiner and Williams (1982), Mesinger and Arakawa (1976) and Deardorff (1970) to the more recent large-eddy simulation turbulence modelling work by Leonard (1974) or Rogallo and Moin (1984), the rule of thumb is fairly clear: the averaging or cut-off scale (Δ_i or Δ_t) in the differential equations must be, for a Nyquist sampling theorem, at least twice as large as the time (δt), or space (δx) step in the grid, i.e. two grid points per minimum resolved wavelength or frequency:

$$\Delta_t \geqslant 2\,\delta t$$
$$\Delta_i \geqslant 2\,\delta x_i \qquad (5.14)$$

In practice (Haltiner and Williams, 1982; Oran and Boris, 1987; Fletcher, 1988), due to Gibb's phenomena and run-off error it is necessary to have more than two grid points for two minimum wavelengths to be resolved in the model. Therefore, consistency between the minimally resolved wavenumber and frequency in the gridded model and the averaging scale in the equations must be ensured. This consistency leaves the attendant caution that the model calculations between $4\,\delta x_i$ (or δt) required for including the quantization errors give information that has no physical basis as the equations have been filtered essentially to resolve only down to Δ_t and Δ_i. In this wavenumber region, aliasing difficulties must therefore be expected, and if not explicitly accounted for will therefore show up as low-frequency errors that are then in practice always handled by 'tuning' the diffusion or dispersion coefficients. Resulting diffusion and dispersion models rarely agree with theoretically expected values, and inadvertant aliasing reduction is one principal reason why this is so.

The relationship of the Nyquist boundary and averaging scales to the physics is not nearly so well explored in the literature. In surface water flows, Woods (1977) provides a good starting point, which is essentially based upon the experience of the meteorological community (Stewart, 1972). Bedford (1981) discusses the effect of time and space-grid sizes and the question of whether two- or three-dimensional flows have been properly resolved. Abbott *et al.* (1985) and Abbott and Larsen (1985) discuss additional aspects, especially the inadvertant physical distortion of the information between k_{ci} $(=2\pi/\Delta x_i)$ and k_{gi} $(=2\pi/\delta x_i)$. Essentially, three points apply. Firstly, if the Nyquist (i.e. grid) boundary is selected to occur at k_{gi} (or ω_g) on the full spectrum then the physics not resolved by k_{gi} (or ω_g) must be par-

Table 5.3 Summary of turbulent eddy types

Eddy class	*Governing dimensionless number*	*Range*	*Eddy lifetime (Te)*	*Eddy size (Le)*	*Ratio vertical to horizontal eddy size*	*Comments*
Rotation range (no overturn)	Rossby no.	Ro < 1	$Te > f^{-1}$	$Le = L$	$\frac{He}{Le} =$ e.g. 1/500	(a) Quasi-qeostrophic (b) Two-dimensional isotropic (c) Axes in plane of density surface, therefore, two-dimensional turbulence does not apply
Buoyancy range (no overturn)	Rossby no. Richardson no	Ro > 1 Ri > 1	$f^{-1} < Te < N^{-1}$	$Le = (\varepsilon_*/f^3)^{1/2}$	$\frac{He}{Le} = \left(\frac{f}{N}\right) \mathrm{Ro} \cdot \mathrm{Ri}^{1/2}$	Same comments as above
Billow turbulence (overturn)	Richardson no	Ri < 1	$Te < N^{-1}$	$Le = (\varepsilon_*/N^3)^{1/2}$	$\frac{He}{Le} = 1/4$ (limit)	(a) Overturning of density surfaces (b) Statistically resembles three-dimensional isotropy on average
Inertial subrange (overturn)	Richardson no Reynolds no	Ri < 1 Re > 1	$T_e \sim (\varepsilon_*/\nu)^{-1/2}$	$Le = (\varepsilon_* T_e^3)^{1/2}$	$\frac{He}{Le} = 1$ only in average sense	Same as billow turbulence

ε_* is the dissipation rate of kinetic energy

ameterized, as it is this information that must be supplied as the Reynolds fluxes or the dispersion terms.

Secondly, as discussed by Townsend (1976) and subsequently by many others, the parameterization of the unresolved motions must be achieved by variables that remain resolved by the grid, which means that the sub-grid-scale parameterization of the remaining spectra can only be done with a length scale controlled by the largest unresolved motion in the portion of the spectra to be parameterized. When the spectra contains no spikes or peaks then Townsend's arguments can be used for generating parameterizations, and many important ones have been found and reported upon and will be reviewed here.

Thirdly, when spectral spikes or peaks exist, then the equilibrium condition required by Townsend does not obtain near the spike; hence neither k_{gi} (ω_g) nor Δ_i (Δ_t) should be selected near a spectral peak. Figure 5.3b depicts proper and improper grid wavenumbers. Finally, placing the grid wavenumber near or at a boundary between the four types of turbulent eddy should also be avoided, for similar reasons.

5.4.2 MACROSCALE CONSIDERATIONS

With sub-grid parameterizations now seen in the context of an unresolved spectrum at the high frequency (or wavenumber) microscale portion of the spectrum, it should come as no surprise that there are limits to resolution in the low-wavenumber (frequency) region. Due primarily to round-off error and computer size, it is the case that most digital models only span 2 to 2.5 decades of resolution in their solution. For example, in the author's Great Lakes forecasting system (Bedford *et al.*, 1990), Lake Erie is being modelled with a three-dimensional system, with the operational horizontal grid size being approximately 1 km and the horizontal extent of the lake being approximately 200 km. Resolution bandwidths greater than this introduce truncation errors that are difficult to separate from actual calculations. Woods (1977) has graphically summarized this bandwidth limit and its effect on defining sub-grid-scale parameterizations. His plot has been redrawn in Figure 5.4.

Clearly, the positioning of the model resolution bandwidth relative to the typical variance spectra for the site has a profound implication for the way in which the sub-grid-scale Reynolds fluxes and dispersion terms are modelled and in deciding which processes the parameterizations are to describe. In the literature, the most ubiquitous form of representation is the gradient–diffusion model of diffusion and dispersion. The background of these forms will therefore be reviewed in the next section.

5.5 TRADITIONAL DIFFUSION AND DIFFUSIVITY ANALYSES

5.5.1 BASIC CONCEPTS

The location of the Nyquist/grid boundary away from spectral spikes and regime boundaries allows, in a simple sense, the physics of the resulting flow field to reach an 'equilibrium' wherein variance is governed by relatively simple and balanced processes. Mathematical representations are therefore 'easy' to find. Historically, diffusion and dispersion are among the most frequently studied transport processes and their parameterizations are generally derived from such 'equilibrium' approaches. The previously cited works by Taylor form the bases for these studies, with reviews by Fischer *et al.* (1979), Holly (1985), Csanady (1972), Chatwin and Allen (1985) and Smith (1989) forming the basis for the summary information presented here.

Identification of the stages of evolution in the diffusion process forms a basis for categorizing various parameterizations. If the contaminant is assumed to be a cloud of 'particles' placed at a point far away from any boundaries in homogeneous, isotropic (stationary) turbulence (no stratification and velocity shear implied), then the Lagrangian correlation coefficient for the particle displacements in the x direction is defined as (Holly, 1985)

$$R_x = \frac{\overline{\overline{u'(t)u'(t+\tau)}}}{\overline{\overline{u'^2}}} \qquad (5.15)$$

In Equation (5.15) the velocity u'_i stands for the instantaneous velocity of the ith particle and $\overline{\overline{u'^2}}$ stands for the mean square velocity fluctuation of the entire ensemble of particles (the double overbar). The variance growth with time of the distribution in the x direction is then:

$$\sigma_x(t) = 2\overline{\overline{u'^2}}\int_0^t (t-\tau)\,R_x\,d\tau \qquad (5.16)$$

Taylor demonstrated the existence of two distinct regions in the integration of Equation (5.16) and Csanady (1972) has generalized these further.

Table 5.4 summarizes the results where the variable T_L is called the Lagrangian integral time scale, and as long as $R_x(t)$ decreases monotonically it is a measure of how long it takes the particles to 'forget' the original velocity distribution. In three-dimensional diffusion, the integral time scale is the average of the integral scales in each co-ordinate direction. The corresponding Lagrangian length scale is

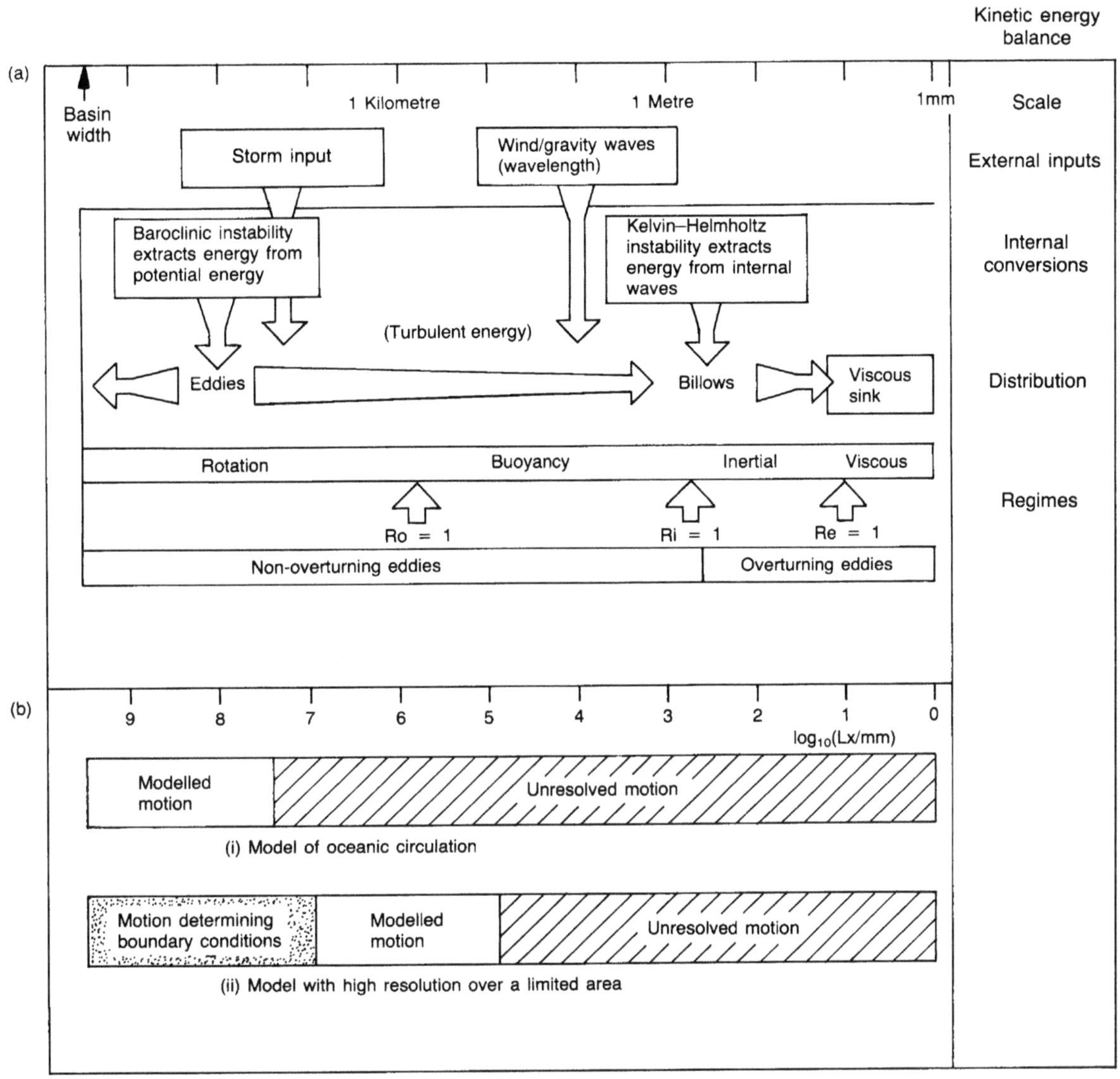

Figure 5.4 Model bandwith/energy process resolution diagram (redrawn from Woods 1977). (a) Characteristics of oceanic turbulence; (b) spectral windows of primitive equation models.

$$l_L^2 = \overline{u'^2}\, T_L^2 \qquad (17)$$

and the equivalent length scale regimes in Table 5.4 come from setting the cloud size, L^2, proportional to σ_x^2, i.e. $L^2 \simeq \overline{2u'^2}\, T_L t$.

Regime 3 (Table 5.4) is a most important region, as Taylor was able to show that linear cloud growth (variance) with time is a necessary condition for the diffusion equation and gradient–diffusion approximation to apply.

For time periods less than T_x (regime 1), the analyses of Richardson (1926) and Batchelor (1952) apply. Here the time variation of the ensemble average particle separation, $\langle s^2 \rangle$, is presumed to be affected by the initial separation, s_o, and the Kolmogorov scale $(\nu^3/\varepsilon_*)^{1/4}$; here ε_* is the dissipation. After analysis, it was found that

$$\frac{d}{dt}\langle s^2 \rangle \sim \varepsilon_*^{1/3}[\langle s^2 \rangle]^{2/3} \qquad (5.17)$$

Table 5.4 Diffusion variance solution regimes

Region	*Criteria (time)*	*Criteria (length)*	*Variance growth*
1	$t \rightarrow 0, t < T_x$	$L^2 < 2l_L{}^2$	$\sigma_x^2(t) = \bar{\bar{u}}'^2 t^2$
2	$t \rightarrow T_x$	$L^2 \sim 2l_L{}^2$	$\sigma_x^2(t) = \overline{\overline{2u'^2}} t\, T_x - \overline{\overline{\partial u'^2}} \int_0^t \tau\, R_x(\tau)\, d\tau$
3	$t \rightarrow \infty, t > T_x$	$L^2 > 2l_L{}^2$	$\sigma_x^2(t) = \overline{\overline{2u'^2}} t\, T_x$

1. $T_x = \int_0^\infty R_x(\tau)\, d\tau$

2. $T_L = (T_x + T_y + T_z)/3$

which corresponds to the most elementary form of the famous four-thirds law of relative particle diffusion.

5.5.2 THE GRADIENT-DIFFUSION APPROXIMATION AND EDDY DIFFUSIVITIES

In complete analogy to Fickian diffusion theory (Csanady, 1972; Fischer *et al.*, 1979), if a cloud has dispersed for sufficient time (regime 3) in homogeneous isotopic turbulence, the distribution of variance (σ_x^2) becomes a Gaussian normal distribution and a turbulent eddy diffusivity can be defined in each direction as

$$\varepsilon_x = \frac{1}{2}\frac{d}{dt}(\sigma_x^2) = \overline{\overline{u'^2}}\, T_x = (\overline{\overline{u'^2}})^{1/2} l_{Lx} \quad (5.18a)$$

$$\varepsilon_y = \frac{1}{2}\frac{d}{dt}(\sigma_y^2) = \overline{\overline{v'^2}}\, T_y = (\overline{\overline{v'^2}})^{1/2} l_{Ly} \quad (5.18b)$$

$$\varepsilon_z = \frac{1}{2}\frac{d}{dt}(\sigma_z^2) = \overline{\overline{w'^2}}\, T_z = (\overline{\overline{w'^2}})^{1/2} l_{Lz} \quad (5.18c)$$

The governing transport equation then becomes a diffusion equation with constant (though unequal) eddy diffusivities. Therefore,

$$\overline{u'c'} = -\varepsilon_x \frac{\partial \bar{c}}{\partial x}; \; \overline{v'c'} = -\varepsilon_x \frac{\partial \bar{c}}{\partial y} \text{ and } \overline{w'c'} = -\varepsilon_z \frac{\partial \bar{c}}{\partial z} \quad (5.19)$$

It is essential to understand that this formulation obtains for a flow oriented along the principal flow axis, i.e. the streamline. This is a consequence of arguments presented by Kolmogorov (1931, 1933), wherein cross-axis terms for closure are required for axes not aligned with the streamline. These terms are always neglected, with the restriction that the forms in Equation (5.19) are only valid for $t > T_L$ which (Fisher *et al.*, 1979) allows time for smoothing to occur.

Insofar as short time (regime 1) diffusion is concerned, if time progressed so far that the initial distribution is forgotten, then Batchelor's theory allows the distribution of concentration to be predicted by a diffusion equation with spatial gradients measured relative to the displacement from the centre of mass of the cloud. The temporally-and spatially-varying eddy diffusivity is now proportional to the cloud size to the 'four-third power'; i.e.

$$\varepsilon = \frac{1}{2}\frac{d\langle \chi^2 \rangle}{dt} = \alpha \langle \chi^2 \rangle^{2/3} \quad (5.20)$$

For a point source, the ensemble average cloud variance $\langle \chi^2 \rangle$ grows as:

$$\langle \chi^2 \rangle = \left(\frac{2}{3}\alpha\, t\right)^3 \quad (5.21)$$

Okubo (1974) has shown that this behaviour is operable in the oceans up to 10^3 km and is not necessarily confined to small-scale 'turbulence.' The unified diffusion diagram given by Okubo (1974) or Fischer *et al.* (1979) indicates this range of validity, and also indicates that α ranges between 0.002 (cm$^{2/3}$) s^{-1} to 0.01 (cm$^{2/3}$) s^{-1}. While this is more a conjecture than an experimental observation, the occurrence of 'diffusion-like' turbulence at large scales is then not surprising if one accepts that equilibrium conditions at the four scale ranges in Table 5.3 exist and that a diffusion solution is an excellent representation for equilibrium.

Again, for the Fickian diffusion concept to apply, the co-ordinate system must be based on the centre of mass of the cloud and by implication its subsequent advection by the currents. This is a small but crucial consideration, almost completely overlooked in most model adaptations of the eddy-diffusivity formulations for regime 1. Values for ε_x, ε_y, ε_z, etc. are parameterized by relation to some presumably more readily measured group of variables. However, while the prescriptions offered for full three-dimensional homogeneous isotropic turbulence are straightforward, the presence of mitigating circumstances quickly overwhelms the functional simplicity. Essentially, the presence of walls or density interfaces which introduce shear, irregular basin geometries, irregular bottom roughnesses, and dynamic time varying recirculation, all combine to quickly confuse the parameterization formulations.

Several simple diffusivity parameterizations have been discussed by Fischer *et al.* (1979) and Holly (1985), starting out from Elder's (1959) analysis, made for an infinitely wide channel of constant depth. Here, the depth h is supposed to be the only significant scale in parameterizing how a point source fills the depth and subsequently diffuses. Using a simple boundary layer theory, the friction velocity u_* is defined from the weak bottom slope (S) as

$$u_* = \sqrt{ghS} \tag{5.22}$$

and the vertical eddy diffusivity is defined from the vertical eddy viscosity as

$$\varepsilon_v = khu^*\left(\frac{z}{h}\right)\left(1-\frac{z}{h}\right) \tag{5.23}$$

which gives a vertically averaged value of

$$\bar{\varepsilon}_v = 0.067\,hu_* \tag{5.24}$$

Transverse eddy diffusivities in the absence of walls, have been inferred from laboratory experiments (Lau and Krishnappen, 1977; Table 5.1 in Fischer *et al.*, 1979) and have a range of values approximated (with a 50% error bound, Fischer *et al.*, 1979) by

$$\varepsilon_t \cong 0.15\,hu_* \tag{5.25}$$

The effect of including river cross-channel geometry is that both lateral and vertical scales have to be contended with. Holley *et al.* (1972) have investigated the effect of cross channel depth variations on the transverse mixing coefficient and Table 5.2 in Fischer *et al.* (1979) summarizes these and other data. The explicit resolution of deterministic circulation motions perpendicular to the streamline is intuitively seen as being in direct conflict with the gradient–diffusion approximation.

5.5.3 HIGHER-ORDER EDDY DIFFUSIVITY AND FLUX MODELS

While the confining nature of rivers gave a historical boost to the possibility of exact parameterizations for gradient–diffusion-based eddy diffusivities, the model resolution of two- and three-dimensional activity containing multiple physical processes and corresponding length scales creates transport fields that are further from the required equilibrium circumstances than can be handled by simple gradient-diffusion models. Consequently, a number of higher-order procedures have been created to provide for these circumstances.

While these are the provinces of other chapters in this volume, higher-order gradient-diffusion models exist for predicting aspects of non-local, multiple length scale eddy diffusivities. These models have been reviewed by Rodi in this volume, and elsewhere (Rodi, 1987; ASCE, 1988; Mellor and Yamada, 1982). These resulting one- and two-equation turbulence models have been used to determine eddy diffusivities but have been shown to suffer when anisotropies occur in the plane perpendicular to the streamline. Higher-order theories based upon second-moment closure theories (Chou, 1945) eliminate the gradient-diffusion concept altogether by solving transport equations and simple algebraic expressions for Reynolds fluxes directly. However, such precision comes at an intense computational cost. These issues will be dealt with in subsequent chapters.

5.5.4 SUB-GRID, VOLUME-AVERAGE DIFFUSION AND DIFFUSIVITY-EQUILIBRIUM APPROACHES

Meteorologists have long recognized the explicit dependence of the grid-wave-number resolution, the resulting process-spectra cut-offs, and the necessary sub-grid-scale parameterizations. In the three-dimensional inertial sub-range region (Figure 5.3a), Leith (1972) states that turbulent production equals dissipation and this fact is exploited to devise eddy viscosities which, although they fully reproduce the required energy cascade, do so only in an average sense over the grid volume.

In the inertial sub-range, the turbulent kinetic energy distribution is given by

$$E(k) = C_1\,\varepsilon_*^{2/3}\,k^{-5/3} \tag{5.26}$$

Over the grid volume characterized by the grid vector size, δx, the eddy viscosity or diffusivity is proportional to

$$\varepsilon = \mathrm{Pr}C_2\,\varepsilon_*^{1/3}\,k_g^{-4/3} = \mathrm{Pr}\nu_{t3} \tag{5.27}$$

where Pr is the Prandtl number, ν_t is the eddy viscosity and k_g is the grid wave number. Using the local grid-based velocity deformation based on volume averaged velocities, Smagorinsky *et al.* (1963) stated that

$$\varepsilon = \mathrm{Pr}\sqrt{2}C_2 k_g^{-2}|D_{ij}|^3 \tag{5.28}$$

where $D_{ij} = u_{i,j} + u_{j,i}$.

Further down the spectrum, as $k \rightarrow$ o, at scales greater than the depth, elements of two-dimensional turbulence exist, and, if the equilibrium range in this portion of the spectrum is found, similar eddy diffusivities can in turn be found for this region. Leith (1972) and Kraichnan (1967), clarified the existence of two equilibrium ranges and noted that, for the two-dimensional region, not only energy but also entropy was conserved.

If k_g lies in the two-dimensional range, then the energy cascade is given by

$$E(k) = C_3\eta^{2/3}k^{-3} \tag{5.29}$$

where η is the entropy cascade rate, which is constant. The accompanying eddy diffusivity is found from dimensional arguments for the given grid cell as

$$\varepsilon = \mathrm{Pr}C_4\eta^{1/3}k_g^{-2} = \mathrm{Pr}\,\nu_{t2} \tag{5.30}$$

which, if η is defined as

$$\eta = \nu_{t2}|\nabla\omega|^2$$

gives

$$\varepsilon = \mathrm{Pr}C_5|\nabla\omega|^3(\delta x)^3 \tag{5.31}$$

where ω is the vorticity. Grid dependent, spectral diffusivity parameterizations are ubiquitous in the meteorological literature and one cannot help but wonder why they have gained so little favour in the surface water community. Clearly, the problems are physically similar.

5.5.5 CAVEATS

The assumptions of homogeneous, isotropic and stationary turbulence along with the implication of relatively equilibrium conditions are difficult to meet in most shallow-water three-dimensional models and the gradient diffusion-based model with its characteristic broad-based smoothing tends to mask a number of important features of turbulent flow. Inadvertent grid placement also results in the resolution of spectral bands which are not governed by equilibrium parameterizations. As computers become more powerful and the grid sizes decline correspondingly, a number of these heretofore neglected processes must be more fully incorporated to achieve adequate parameterizations. These processes include incorporation of not only diffusive-like turbulence but also transport due to surface and internal gravity waves, intermittency and specific mixing resulting from the interaction of individual eddies of various sizes, fontal upwelling including stratification and inclined isopycnals, and billow turbulence.

5.6 DISPERSION

5.6.7 INITIAL WORK-CHANNELS

Taylor was able to demonstrate, in (1953) for laminar flow and in (1954) for turbulent flow, that, for a frame of reference shifted to the mean flow and after a sufficiently long start-up or adjustment period, the mass flux of concentration could be described (relative to the velocity/streamline) by a gradient of the area-average concentration times a dispersion coefficient, K. The dispersion coefficient varied only down the axis of the pipe and the adjustment time (of the order of $0.4\,r/\varepsilon_x$) allowed for the velocity deviations about the cross-section mean to be normally distributed. The bulk-dispersion process was shown to be of far greater magnitude than the turbulent-diffusivity-based processes in comprising the average.

The results by Elder (1959) for infinitely wide flow down an inclined plane were similar. As mentioned earlier, vertical variations were deemed essential and the dispersion coefficient, K, was found by vertical integration to be close to

$$K = \frac{0.404}{k}hu_* \tag{5.32}$$

where k is the von Kármán coefficient and u_* is defined in Equation (5.22). Again, after a brief adjustment period to attain normality, the parametric result is that the vertically-averaged distribution of c down the plane could be predicted with a one-dimensional advection–diffusion equation with K as the dispersion coefficient.

Fischer (1969) showed that, for straight river channels with proximate side banks, the transverse velocity variations from the cross-section average were more important than the vertical variations, and he found the dispersion coefficient, after the normality adjustment time, was equal to

$$K = \frac{IW^2\overline{u'^2}^*}{\varepsilon_t} \tag{5.33}$$

where W is the river width, $\overline{u'^2}^*$ is the cross-section average of the square of the velocity derivation from the cross-section average, ε_t is the transverse diffusivity and I is the integral found in Equation (4.48) in Fischer *et al.* (1979). Fischer found that the initial adjustment to normality took place over a time period given by $T = 0.4\,l^2/E_t$; where l is the distance from the maximum velocity streamline to the farthest bank. Often this time period is quite large in comparison to the period of interest. Holly (1985) gives the example that the Missouri River would have $T = 13\,\mathrm{h}$ or a distance of 50 miles before the dispersion formulation applied. Smith (1989) suggests that this time period corresponds to a longitudinal distance of upwards of 100 channel breadths downstream, which gives a 10 km distance for a 100 m wide river. Clearly, transverse mixing takes much longer to complete than vertical mixing, and that given the frequent location of various outfalls, the required transverse diffusivity equilibrium adjustment required for Taylor's analysis might never obtain in a typical river.

Table 5.3 in Fischer *et al.* (1979) presents various experimental results for K in channels.

5.6.2 EXTENSIONS TO COMPLEX GEOMETRIES: CHANNELS

The introduction of more complex, large-scale geometrical features, such as channel bends, and improvements in small-scale roughness resolution, resulting from the use of more powerful computers and so the employment of more refined grids, creates difficult problems for channel dispersion parameterization. Introducing bends and irregular side shapes is difficult because of the introduction of secondary circulations introduced by centrifugal accelerations. The necessary transverse eddy viscosities are difficult to determine because of the formation of an advective component cross-wise of the channel. Two-dimensional formulations which partially alleviate this problem are reviewed in the next section. Holly (1985) has articulated the arguments and uncertainties pertaining to one-dimensional dispersion models in his review article.

Irregular side-wall geometries have been considered to cause dead zones in channels. These, in turn, result in such complex release characteristics that the longitudinal variance does not vary linearly with time. Consequently, the Taylor analysis does not apply (Chatwin and Allan, 1985). Valentine and Wood (1977) were among the first to parameterize these effects. Methodologies based on dead-zone or exact-solution approaches have also been developed (Wallis *et al.* 1989; Beer and Young 1983).

5.6.3 TRANSITION REGIME: CHANNELS

In an excellent, but too brief review, Smith (1989) has summarized the research involved in describing methods for adapting Taylor's dispersion analysis to time periods before normality adjustment has been achieved. As indicated earlier, this transition region is crucial, as for many rivers with frequent pollution outfalls, the entire diffusion/dispersion scenario might have to be described by transition dynamics, as Taylor equilibrium is never achieved.

For a point far from the boundaries, governed by turbulent diffusivities ε_x, ε_y and containing a downstream advection u, Smith (1989) noted several important results for the developing dispersion region. He referred back, in the first place, to Sullivan and Yip (1985) who used the Gaussian diffusion model as the basis for a power series expansion procedure for including non-uniformities in the longitudinal velocity and diffusivities. This work in turn extended work by Townsend (1951), who showed that, even for the smallest times, the concentration distribution in a shear flow was Gaussian (the centroid and variance, however, were not so well behaved). He showed that for linear shear flow

$$u_0 + \alpha y \tag{5.34}$$

that:

$$\sigma_x^2 = 2\varepsilon_x t + \frac{1}{6}\varepsilon_y \alpha^2 t^3, \quad \sigma_y^2 = 2\varepsilon_y t \tag{5.35}$$

Smith (1989) then provided a more general relation for the cloud variance, as

$$\frac{\mathrm{d}}{\mathrm{d}t}\sigma_x^2 = 2\left(\varepsilon_x + \frac{\alpha^2\sigma_y^4}{16\varepsilon_y}\right) \tag{5.36}$$

which shows the growth of σ_x accelerates as lateral cloud growth occurs. He also demonstrated that when σ_y has grown to half of the channel width, a, the channel dispersion coefficient, K, is of the order of

$$K \simeq \frac{\bar{u}^2 a}{64\varepsilon_y} \tag{5.37}$$

Extension of these concepts to arbitrary shear distributions is more complex but Smith (1981) has presented a ray-tracing model to analyse this case for a variety of nearly exact outfall velocity fields.

Townsend also adapted the distorted Gaussian procedure for the initial period to time varying shear flow, and work by Young *et al.* (1982), Smith (1982b), and Smith (1985) clearly demonstrate that, during the transition stage, effective dispersion can be quite large compared to ε_x and that the timing of the discharge relative to a periodic velocity variation can introduce a factor of three into the dispersion value.

Using the method due to Aris (1956), progress has been made in adapting the Gaussian channel dispersion model to times between the Townsend initial period and the Taylor equilibrium period. The procedure by Smith (1982a) is for steady flows while Yasuda (1984) has offered a procedure for oscillatory flows.

5.6.4 TWO-DIMENSIONAL DISPERSION

One of the major requirements for extending the Fickian gradient-diffusion concept to turbulent diffusion and dispersion was a viewpoint attached to the principal velocity axes or the streamline. This fact is central in adapting a proper 'closure' for the dispersion terms in the two-dimensional vertically averaged equation (Equations 5.12 and 5.13). Originally (see Leendertse (1970) for one of many examples), closure was achieved by a simple gradient-diffusion expression of the form

$$\int_{-h}^{\xi}\left(\overline{u'c'} + \bar{u}'^{*}\bar{ci}^{*}\right)\mathrm{d}z = (\bar{\varepsilon}_x{}^{*} + K_x)\frac{\partial \bar{\bar{c}}^{*}}{\partial x} \tag{5.38a}$$

$$\int_{-h}^{\xi} (\overline{v'c'} + \overline{v'}^* \overline{c'}^*)\,dz = (\bar{\varepsilon}_y{}^* + K_y)\frac{\partial \bar{\bar{c}}^*}{\partial y} \qquad (5.38b)$$

where K is the dispersion coefficient in the x or y direction. No consideration of whether the x or y axis corresponded to the streamline co-ordinate was given until the paper by Fischer (1978) on the bulk dispersion tensor, and the metric coefficient approach by Yotsukura and Sayre (1976). Both seek to address this problem.

Yotsukura and Sayre (see Holly, 1985) defined the metric coefficients, $m_1(x, y)$ and $m_2(x, y)$ which define distances along an arbitrary curve between points A and B, ΔL_x, and the corresponding change in the x co-ordinate; i.e.

$$m_1(x, y) = \frac{\Delta L_x}{\Delta x}, \quad m_2(x, y) = \frac{\Delta L_y}{\Delta x} \qquad (39)$$

Originally adapted to river channels, y = constant is defined as a plane aligned parallel to the depth-average velocity streamline. With the metric coefficients usually being placed between 0.8 and 1.2 (1.0 corresponding to a straight co-ordinate system), the resulting depth and Reynolds-average equations become

$$\frac{\partial \bar{\bar{c}}^*}{\partial t} + \frac{\bar{\bar{u}}^*}{m_1}\frac{\partial \bar{\bar{c}}^*}{\partial x} + \frac{\bar{\bar{\omega}}^*}{m_2}\frac{\partial \bar{\bar{c}}^*}{\partial y} = \frac{1}{Hm_1m_2}\frac{\partial}{\partial x}\left(\frac{m3}{m_1}H\bar{\varepsilon}_x^*\frac{\partial \bar{\bar{c}}^*}{\partial x}\right)$$
$$+ \frac{1}{Hm_1m_2}\frac{\partial}{\partial_y}\left(\frac{m_3}{m_1}H\bar{\varepsilon}_y^*\right) + \frac{1}{Hm_1m_2}\frac{\partial}{\partial x}\left(\frac{m_3}{m_1}HK_x\frac{\partial \bar{\bar{c}}^*}{\partial x}\right)$$
$$+ \frac{1}{Hm_1m_2}\frac{\partial}{\partial y}\left(\frac{m_3}{m_1}HK_y\frac{\partial \bar{\bar{c}}^*}{\partial y}\right) \qquad (5.40)$$

In Equation (5.40), the vertically averaged diffusivities $(\bar{\varepsilon}_x^*, \bar{\varepsilon}_y^*)$ are usually added to the dispersion coefficients (K_x, K_y) to form an overall bulk dispersion coefficient, $D_x = \bar{\varepsilon}_x^* + K_x$ and $D_y = \bar{\varepsilon}_y^* + D_y$.

Fischer (1978), and later Holly and Usseglio-Polatera (1984), and Falconer (1984, 1986), have presented forms for the bulk-dispersion tensor formulation which are more appropriate to recirculation zones with time varying streamline trajectories. If s refers to the local co-ordinate tangent to the streamline (two-dimensional in plan) and n represents the local normal to that streamline, then the following bulk dispersion equation obtains:

$$\frac{\partial \bar{\bar{c}}^*}{\partial t} + \bar{\bar{u}}^*\frac{\partial \bar{\bar{c}}^*}{\partial x} + \bar{\bar{w}}^*\frac{\partial \bar{\bar{c}}^*}{\partial z}$$
$$= \frac{1}{H}\frac{\partial}{\partial x}\left(HD_{xx}\frac{\partial \bar{\bar{c}}^*}{\partial x}\right) + \frac{1}{H}\frac{\partial}{\partial x}\left(HD_{xy}\frac{\partial \bar{\bar{c}}^*}{\partial y}\right)$$
$$+ \frac{1}{H}\frac{\partial}{\partial y}\left(HD_{yx}\frac{\partial \bar{\bar{c}}^*}{\partial x}\right) + \frac{1}{H}\frac{\partial}{\partial y}\left(HD_{yy}\frac{\partial \bar{\bar{c}}^*}{\partial y}\right) \qquad (5.41)$$

where

$$D_{xx} = D_s\cos^2\theta + D_n\sin^2\theta$$
$$D_{yx} = D_{xy} = (D_s - D_n)\sin\theta\cos\theta$$
$$D_{yy} = D_s\sin^2\theta + D_n\cos^2\theta \qquad (5.42)$$

In turn, D_s and D_n are the bulk 'longitudinal' (streamline direction) and 'transverse' bulk dispersion coefficient, defined as the sum of the vertically-averaged turbulent diffusivity and dispersion coefficients, $D_n = \bar{\varepsilon}_n^* + K_n$ and $D_x = \bar{\varepsilon}_s^* + K_s$, respectively. In Equation (5.42), θ is the angle between the streamline and the x-axis.

Falconer's references report a boundary layer formulation for D_{xx}, D_{yy} and D_{yx} which takes the form

$$D_{xx} = \frac{\{[\bar{\varepsilon}_n^*(\bar{\bar{u}}^*)^2 + \bar{\varepsilon}_s^*(\bar{\bar{v}}^*)^3]H\sqrt{g}\}}{\{[(\bar{\bar{u}}^*)^2 + (\bar{\bar{v}}^*)^2]^{1/2}C_h\}} \qquad (5.43a)$$

$$D_{yy} = \frac{\{[\bar{\varepsilon}_n^*(\bar{\bar{v}}^*)^2 + \bar{\varepsilon}_s^*(\bar{\bar{u}}^*)^2]H\sqrt{g}\}}{\{[(\bar{\bar{u}}^*)^2 + (\bar{\bar{v}}^*)^2]^{1/2}C_h\}} \qquad (5.43b)$$

$$D_{xy} = D_{yx} = \left\{\frac{(\bar{\varepsilon}_n^* - \bar{\varepsilon}_s^*)\,\bar{\bar{u}}^*\,\bar{\bar{v}}^*\,H\sqrt{g}}{[(\bar{\bar{u}}^*)^2 + (\bar{\bar{v}}^*)^2]^{1/2}C_h}\right\} \qquad (5.43c)$$

In Equation (5.43a,b,c), C_h is the Chezy coefficient.

The influence of the bulk dispersion formulation centres on improving the spatial structure of the concentration distribution. As an example, the author and a co-worker have been examining the response of Sandusky Bay (Lake Erie, North America) to storm effects, principally increased tributary inflow from the Sandusky River (Figure 5.5) and wind driven circulation. Chloride concentrations were computed (Bedford and Mark, 1988) for a six-week time period (1 June–15 July 1981) when a detailed series of data were collected at 18 stations within the Bay (Figure 5.5). Skill test scores were computed and for the initial calculations a simple Leendertse-type (Equation 5.38) closure with constant dispersion coefficients of 40 and $4\,m^2\,d^{-1}$ gave the best scores. Figure 5.6a contains a calculated chloride concentration map resulting from a storm-induced Sandusky River freshwater inflow. These calculations were redone (Lee and Bedford, 1990, work in progress) using the bulk-dispersion tensor formulation (Fischer, 1978) and the boundary layer friction formulation for the bulk-dispersion coefficients. Figure 5.6b clearly portrays the stretching, re-orienting and sharpening of the plume due to the fluid shear. This more refined spatial structure agrees much more directly with remote sensing data (Lyons *et al.*, 1988), especially in the Upper Bay where the tributary inflow and Bay recirculation are governed by processes which can be described by two-dimensional vertically averaged representations.

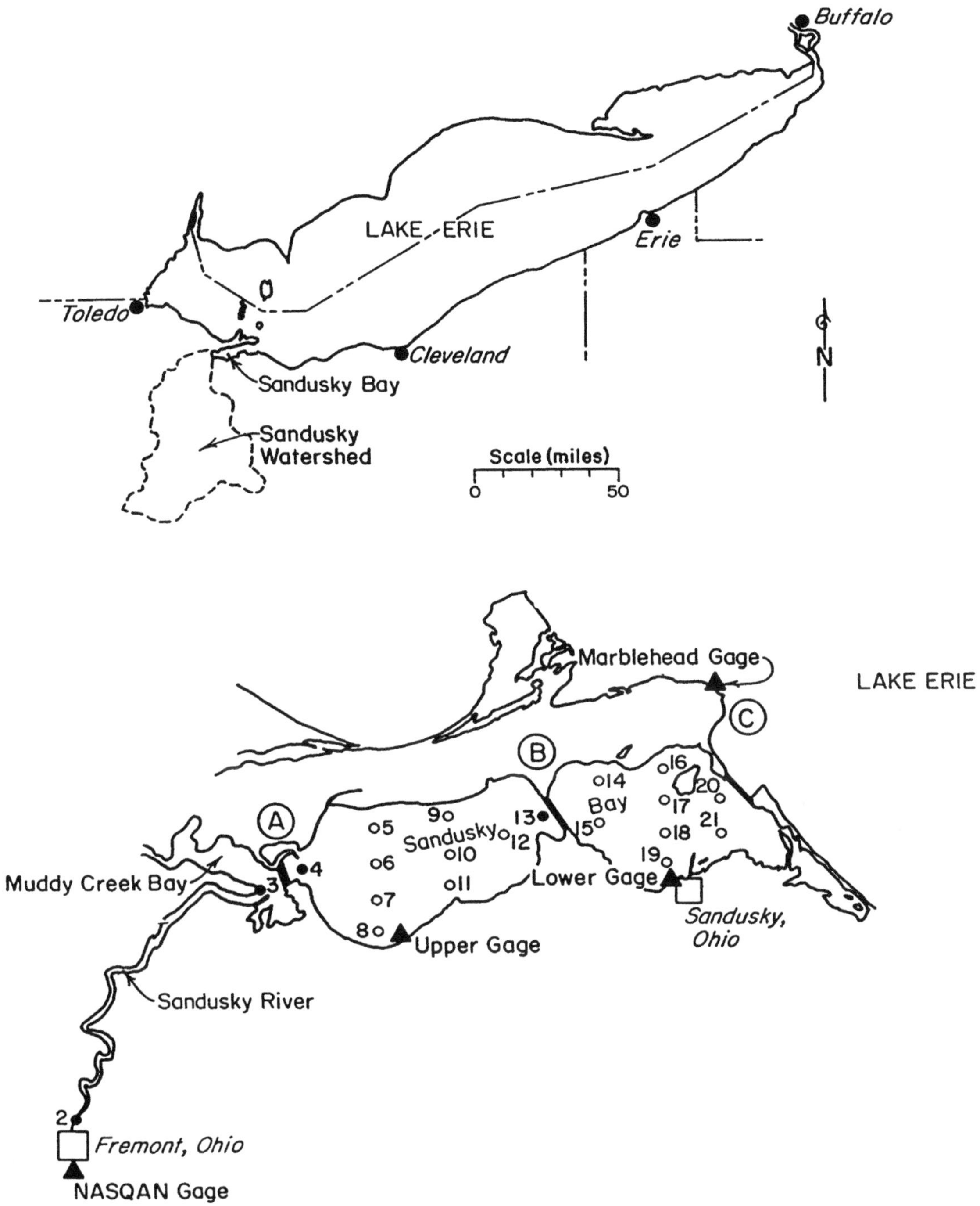

Figure 5.5 Sandusky River/Bay system, Lake Erie, North America.

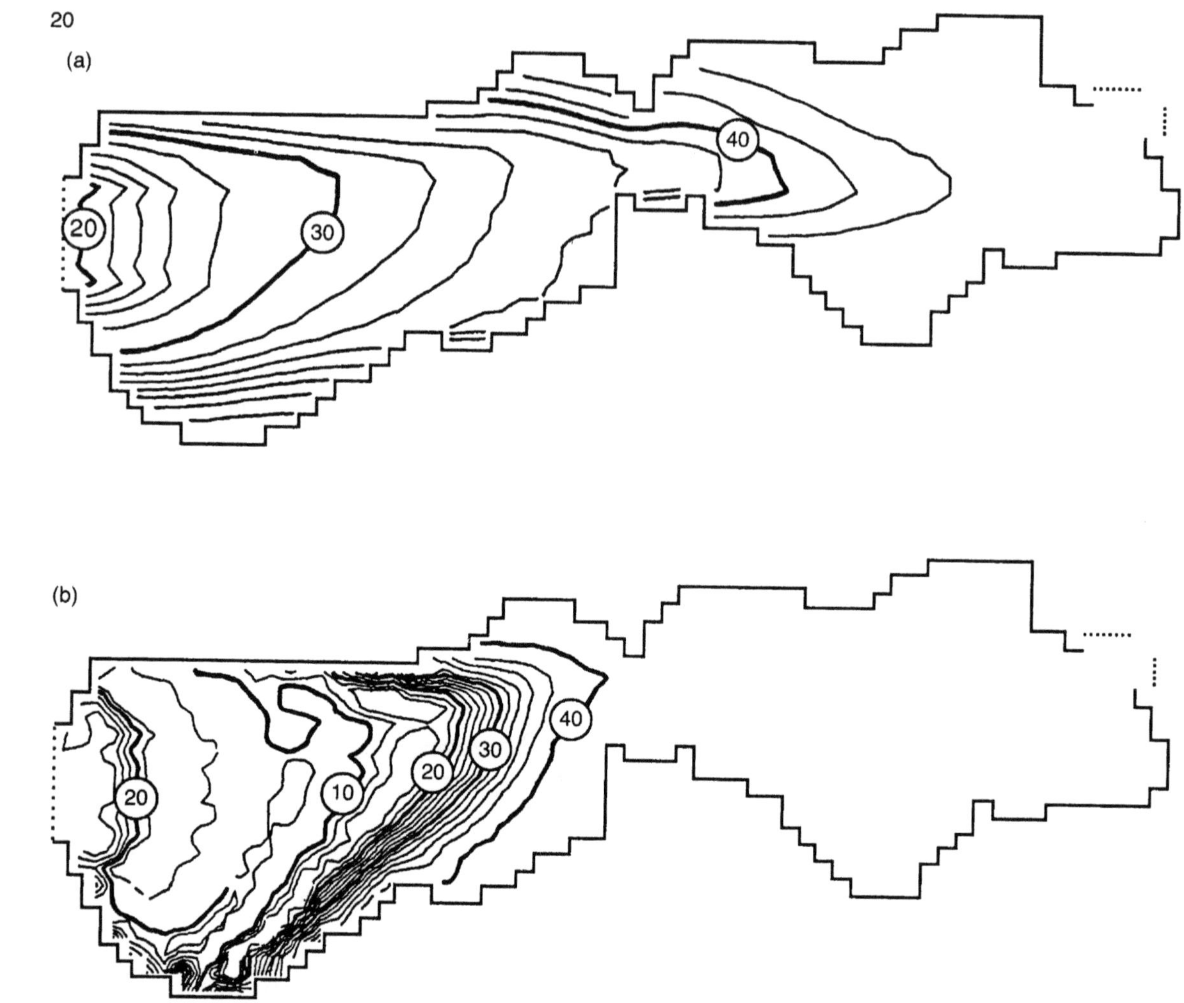

Figure 5.6 Chloride distributions: (——) 2 $mg\,l^{-1}$ contour interval; (–○–), chloride concentration in $mg\,l^{-1}$. (a) 16 June 1981, constant bulk dispersion coefficients; $D_x = 40\,m^2\,d^{-1}$; $D_y = 4\,m^2\,d^{-1}$. (b) 16 June 1981, Fischer (1978) dispersion tensor, Falconer (1984) formulation.

While the predicted chloride structure for the Upper Bay was improved considerably, the limits to reproducing spatial structures governed by three-dimensional processes with a two-dimensional model quickly appear in the Lower Bay. Here (as in most simulations of embayments or harbours) the mixing zone between the Lake and Lower Bay waters introduces a density interface (Bedford, 1989) which can be clearly seen in a Landsat image taken on 25 June 1981, (Figure 5.7) (also Lyons *et al.*, 1988). This interface is not only governed by many of the three-dimensional mechanisms which give rise to the marine turbidity maxima (Dyer, 1986) but also coincides with the null oscillation line of the second mode of the Bay and the furthest bayward excursion distance of the second free mode of Lake Erie (Bedford, 1989). Clearly, such non-linear interactions will provide serious challenges to those wishing to model these mixing zones with anything less than a three-dimensional model. Dispersion approaches will not work in such cases.

5.7 DEVELOPMENTS IN PROGRESS

A number of research areas have not been addressed here and may warrant attention in the future. Firstly, a

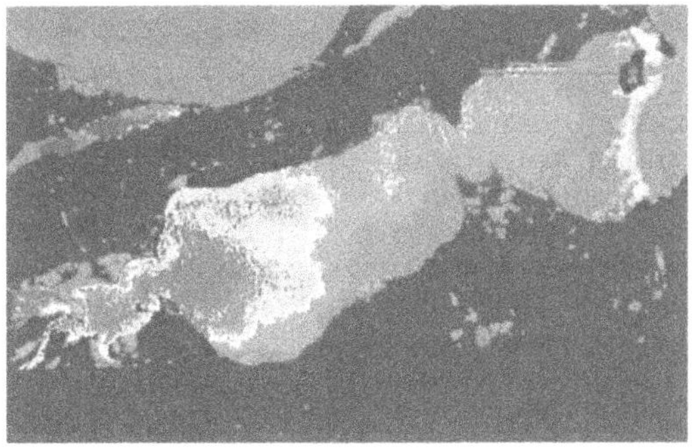

Figure 5.7 Sandusky Bay satellite image, 25 June 1981, enhanced to show turbidity distributions (see also Lyons *et al.* 1988).

small body of research (Gill and Sankarasubramanian, 1970; Smith, 1981) have attempted, in a complete analogy to second-moment closure theory for Reynold's fluxes, to write one- and two-equation models for the velocity and concentration departure from the local area or depth mean. These are then carried along as extra transport calculations and seem quite promising in describing the time varying dispersion coefficients without resorting to local, equilibrium descriptions (e.g. Equation (5.43)).

Secondly, it can be seen from this review that local mesh refinement, done indiscriminately without regard to the physics thereby resolved, can introduce transport processes not supported in other portions of the grid. Such highly refined mesh regions are often introduced to refine solutions locally near tributary confluences, outfalls, sharp platforms, rapid changes in bottom topography, and other such discontinuous features. Many of these regions have well known, exact solutions and Lick (1989) has presented a quite interesting procedure for embedding these exact solutions directly into the solution algorithm. Such procedures might be useful in retaining global consistency in the grid resolution and they could thereby eliminate the need for excessively high dispersion coefficients in the *ad hoc* filtering schemes required to deal with the errors introduced by overly-refined meshes.

Finally, there is a small but growing literature on methods for better averaging of the equations governing these two- and three-dimensional models. Low-pass, higher-order filters have been used in surface water flow models since work by Babajimopoulos and Bedford (1980) and Bedford (1981). Recently, space-time filters developed by Dakhoul and Bedford (1986) as implemented by Aldama (1990) have been used in a two-dimensional dispersion model by Rossman (1987). Since all the terms requiring closure are remnants of the averaging procedure, it is this author's opinion that a small amount of research should continue in the area of 'how to average.'

5.8 CONCLUSIONS

Given the limited scope of this chapter, comprehensive generalized conclusions are neither warranted nor justified. The literature on both the physics and its appropriate parameterization provides a welter of information and it is hoped that the modelling considerations suggested here provide a reference to the appropriate sources for more focused information. Several points need to be made.

It is this author's belief that the literature clearly documents the conditions under which the Fickian diffusion analogy will work in describing turbulent diffusion. The conditions for use of the gradient diffusion analogy for parameterizing the Reynolds fluxes and dispersion fluxes are also well known and documented. These parameterizations will work quite well as long as the conditions under which the parameterizations were derived are appropriate for the physics being described in the model and consistent with the discretizations and grid structure employed. Careful adherence to such features as short-time and long-time equilibrium constraints, placement of grid Nyquist boundaries within the equilibrium ranges, and streamline following co-ordinate representations all help ensure a successful model adaptation of the available theoretical parameterizations.

All too often, when modellers analyse results the question arises – especially when the results are not 'good' – as to whether the closure approximation was adequate to the task. Quite often the reasons offered for the inadequacy of these parameterizations is that the field data were too sparse, or that the physics was more complicated than the parameterizations could handle. While these are always in some sense correct assertions, what is suggested here is that a number of seemingly innocuous consistency requirements, if not satisfactorily addressed, could be a major source of the perceived discrepancies in model performance.

The increasing availability of ultra-fast supercomputers will exacerbate these problems in two ways. The enormous storage capacity allows model discretization to become extremely small. While, in a fractal sense, it seems that the closer we look, the more complex things become, the pragmatic concern is that the powerful adaptive grid generators developed to provide the grids for these models indiscriminately resolve spatial mesh sizes without regard to the impact on the physics. A second supercomputer-derived caution is that we can now calculate results that are so spatially dense that we can no longer provide anywhere near the amount of data necessary to 'validate' the model. Therefore, while the grid densities have increased by several orders of magnitude over the last decade alone, the instrumental technology has not kept pace. In fact, the bulk of our useful data is still collected by time series at selected points. Therefore, the information content disparity between model and field data becomes, a most serious obstacle, and this must also be overcome before introducing any further complexity in the numerical structure or sub-grid-scale parameterizations.

ACKNOWLEDGEMENTS

This contribution was originally prepared for the Great Lakes Forecasting System project, which is financed by a consortium of sponsors. This support, and particularly that given by CRAY Research Inc., is gratefully acknowledged.

REFERENCES

Abbott, M. and Larsen, J. (1985) *Journal of Hydraulic Research*, **23** (5), 397–420.

Abbott, M., Larsen, J. and Tao, J. (1985) *Journal of Hydraulic Research*, **23** (4), 309–26.

Aldarma, A. (1990) *Filtering Techniques for Turbulent Flow Simulation*, Springer Verlag, Berlin.

Aris, R. (1982) *Proceedings of the Royal Society of London*, A **381**, 179–94.

ASCE Task Committee (1988) *Journal of Hydraulic Engineering*, **114** (9), 970–1073.

Babajimopoulos, C. and Bedford, K. (1980) *Journal of the Hydraulics Division, ASCE*, **106** (1), 1–20.

Batchelor, G. (1952) *Proceedings of the Cambridge Philosophical Society*, **48**, 345–62.

Bedford, K. (1981) in *Transport Models for Inland and Coastal Waters*, (ed H. Fischer), Academic Press, pp. 172–221.

Bedford, K. (1985) *Selection of Turbulence and Mixing Parameterization for Estuary Water Quality Models*, U.S. Army Corps of Engrs., Environmental Impact Research Prog, paper EL-85-2, Vicksburg, MS., p. 153.

Bedford, K. (1989) in *Sediment Transport Modeling*, (ed S. Wang), American Society of Civil Engineers, New York, 106–111.

Bedford, K. and Babajimopoulos, C. (1980) *Journal of the Hydraulics Division, ASCE*, **106** (1), 21–39.

Bedford, K. and Dakhoul, Y. (1982) in *Applying Research to Hydraulic Practice*, (ed P. Smith) American Society of Civil Engineers, New York, 32–43.

Bedford, K., Dingman, J. and Yeo, W. (1987) in *Three-Dimensional Models of Marine and Estuary Physics*, (eds J.C.J. Nihoul and B. Jamart), Elsevier, Amsterdam, 113–26.

Bedford, K. and Mark, D. (1988) *Journal of Environmental Engineering*, **114** (2), 352–67.

Bedford, K., Yen, C.C., Kempf, J. *et al.* (1990) in: *Estuarine and Coastal Modeling*, (ed M. Spaldung), American Society of Civil Engineers, 248–57.

Beer, T. and Young, P. (1983) *Journal Environmental Engineering*, **109**, 1049–67.

Brown, W.S. and Trask, R.P. (1980) *Journal of the Physics of the Ocean*, **10**, 1742–54.

Cannon, G. (1971) *Journal of Geophysical Research*, **76** (C24), 5852–8.

Chatwin, P. and Allen, C. (1985) *Annual Review of Fluid Mechanics.*, **17**, 117–49.

Chou, P. (1945) *Quarterly Journal of Applied Mathematics*, **3** (38), 38–54.

Csanady, G. (1973) *Turbulent Diffusion in the Environment*, Reidel Dordrecht, Holland.

Dakhoul, Y. and Bedford, K. (1986) *International Journal of Numerical Methods in Fluids*, **6** (2), 49–64, 65–82.

Deardorff, J. (1970) *Journal of Fluid Mechanics*, **41** (2), 453–83.

Dyer, K. (1986) *Coastal and Estuarine Sediment Dynamics*, John Wiley, Chichester, UK.

Elder, J. (1959) *Journal of Fluid Mechanics*, **5**, 544–60.

Elliot, A. (1978) *Estuary and Coastal Marine Science*, **6**, 285–99.

Elliot, A. and Wang, D.-P. (1978) *Hydrodynamics of Estuaries and Fjords*, (ed J.C.J. Nihoul) Elsevier, Amsterdam.

Falconer, R. (1984) *Journal of Environmental Engineering*, **110** (6), 1099–116.

Falconer, R. (1986) *Journal of Waterways, Port, Coastal and Ocean Engineering*, **112** (1), 15–34.

Fischer, H. (1969) *Water Resources Research*, **5**, 496–506.

Fischer, H. (1967) *Journal of the Hydraulics Division, ASCE*, **93**, 187–216.

Fischer, H. (1976) *Annual Review of Fluid Mechanics*, **8**, 107–33.

Fischer, H. (1978) *Journal of Geophysical Research*, **83** (C5), 2373–5.

Fischer, H., List, E.J., Koh, R. *et al.* (1979) *Mixing in Inland and Coastal Waters*, Academic Press, New York.

Fletcher, C. (1988) *Computational Techniques for Fluid Dynamics*, Springer Verlag, Berlin.

Garrett, C. and Munk, W. (1979) *Annual Review of Fluid Mechanics*, **11**, 265–85.

Gill, W. and Sankarasubramanian, R. (1970) *Proceedings of the Royal Society of London*, A **316**, 341–50.

Gregg, M. (1987) *Journal of Geophysical Research*, **92** (C5), 5249–86.

Haltiner, G. and Williams, R. (1980) *Numerical Prediction and Dynamic Meteorology*, John Wiley, New York.

Hamming, R. (1983) *Digital Filters*, Prentice Hall, Englewood Cliffs, NJ.

Hinze, J. (1975) *Turbulence*, McGraw Hill, New York.

Holly, F. (1985) *Reviews in Hydraulics*, BARA Publications, London, pp. 1–37.

Holly, F. Jr. and Usseglio-Polatera, J. (1984) *Journal of Hydraulic Engineering*, **110** (7), 905–26.

Holley, E.R., Harleman, D.R.F. and Fischer, H.B. (1970) *Journal of the Hydraulics Division, ASCE*, 96, 1691–709.

Holley, E., Siemons, J. and Abraham, G. (1972) *Journal of Hydraulic Research*, **10**, 27–57.

Hopfinger, E. (1987) *Journal of Geophysical Research*, 92 (C5), 5287–303.

Imberger, J. (1987) *Journal of Geophysical Research*, 92 (C5), 5229.

Kent, R. and D. Pritchard. (1959) *J. Marine Res.*, **18**, 62–72.

Kjerfve, B., Greer, J.E. and Crout, R.L. (1978) in *Estuarine Interactions*, (ed M. Wiley), Academic Press, New York, pp. 497–514.

Kolmogorov, A. (1931) *Math. Ann.*, **104**, 415–58.

Kolmogorov, A. (1933) *Math. Ann.*, **108**, 149–60.

Kraichnan, R. (1967) *Physics of Fluids*, **10**, 513–24.

Lau, Y. and Krishnappan, B. (1977) *Journal of the Hydraulics Division, ASCE*, **103**, 1173–1189.

Lee, K. and Bedford, K. (1990) Thesis work in progess.

Leendertse, J. (1970) *A Water Quality Simulation Model for Well-Mixed Estuaries and Coastal Seas.*, Rept. Rand Corp., RM-6230-RC.

Leith, C. (1972) in *Parameterization of Sub-Grid Scale Processes*, (ed Joint Organizing Committee, R. Stewart Chairman) *GARP, Rep. 8*, World Meteorological Organization, Geneva, Switzerland.

Leonard, A. (1974) *Advances in Geophysics*, **18** A, 237.

Lyons, J. Bedford, K., Yen, C.C. *et al.* (1988) *Remote Sensing of the Environment*, **25**, 107–15.

Mellor, G.L. and Yamada, T. (1982) *Reviews of Geophysics and Space Physics*, **20** (4), 851–75.

Messinger, F. and Arakawa, A. (1976) Numerical methods used in atmospheric models, *Global Atmospheric Research Programme No 17 Part 1*, World Meteorological Organization, Geneva, Switzerland.

Monin, A., Kamenkavich, V. and Kort, V. (1974) *Variability of the Oceans*, Wiley Interscience, New York.

Monin, A. and Ozmidov, R. (1985) *Turbulence in the Ocean*, Reidel Dordrecht, Holland.

Mortimer, C.H. (1968) *Internal Waves and Associated Currents Observed in Lake Michigan During The Summer of 1963*, No. 1, Center for Great Lakes Studies, University of Wisconsin, Milwaukee.

Nihoul, J.C.J., (ed) (1978) *Hydrodynamics of Estuaries and Fjords*, Elsevier Scientific; Amsterdam.

Nihoul, J.C.J. and Djenidi, S. (1987) in *Three-Dimensional Models of Marine and Estuarine Dynamics*, (eds J.C.J Nihoul and B. Jamart) Elsevier, Amsterdam, pp. 1–36.

Okubo, A. (1974) *Rapp. P.-V. Reun. Cons. unt Explor Mer.*, **167**, pp. 77–85. See also Okubo, A. (1971), *Deep Sea Research*, **18**, 789–802.

Okubo, A. (1976) *Deep Sea Research*, **23**, 1213–14.

Oran, E. and Boris, J. (1987) *Numerical Simulation of Reactive Flow*, Elsevier, New York.

Pedlosky, J. (1979) *Geophysical Fluid Dynamics*, Springer-Verlag, Berlin.

Rattray, M., Jr. and Mitsuda, E. (1974) *Estuary and Coastal Marine Science*, **2**, 375–94.

Reynolds, Osborne (1892) *Philosophical Transactions of the Royal Society*, **186**, 123–64.

Rhines, P. (1977) in *The Sea, Vol 6*, (eds E. Goldberg, I. McCave, J. O'Brien *et al.*), Wiley Interscience, New York, 189–318.

Richardson, L. (1926) *Proceedings of the Royal Society of London*, A **110**, 709–37.

Rodi, W. (1980) *Turbulence Models and Their Application in Hydraulics – A State of the Art Review, International Association for Hydraulic Research*, Delft, The Netherlands.

Rodi, W. (1987) *Journal of Geophysical Research*, **92** (C5), 5305–5328.

Rogallo, R. and Moin, P. (1984) *Annual Review of Fluid Mechanics*, **16**, 99–137.

Rossman, P. (1990) ScD MIT (unpublished)

Smagorinsky, J., Manobe, S. and Holloway, J. (1963) *Monthly Weather Review*, **93** (12), 99–164.

Smith, N.P. (1978) in Nihoul, pp. 147–60.

Smith, R. (1981) *Journal of Fluid Mechanics*, pp. 469–486.

Smith, R. (1982a) *Quarterly Journal of Mechanics and Applied Mathematics*, **35**, 345–66.

Smith, R. (1982b) *Journal of Fluid Mechanics*, **114**, 379–98.

Smith, R. (1985) *Journal of Fluid Mechanics*, **153**, 479–99.

Smith, R. (1989) in: *Hydraulic and Environmental Modeling of Coastal, Estuarine and River Waters*, ed Falconer, R., Goodwin, P. and Matthew, R., Gower Aldershot UK, pp. 277–90.

Stewart, R., Joint Organizing Committee (1972) *Parameterization of Subgrid Scale Processes, Global Atmospheric Research Program, Publ. 8*, World Meterological Organization, Geneva, Switzerland.

Sullivan, P. and Yip, H. (1985) ZAMP, **36**, 596–608.

Taylor, G. (1921) *Proceedings of the London Mathematical Society*, A **20**, 196–211.

Taylor, G. (1953) *Proceedings of the Royal Society of London*, A **219**, 186–203.

Taylor, G. (1954) *Proceedings of the Royal Society of London*, A **223**, 446–68.

Tennekes, H. and Lumley, J. (1972) *A First Course in Turbulence*, MIT Press, Cambridge, MA.

Thorpe, S. (1987) *Journal of Geophysical Research*, **92** (C5), 5231–5248.

Townsend, A. (1951) *Proceedings of the Royal Society of London*, A **209**, 418–30.

Townsend, A. (1976) *The Structure of Turbulent Shear Flow*, Cambridge University Press, Cambridge, UK.

Turner, J.S. (1973) *Buoyancy Effects in Fluids*, Cambridge University Press, Cambridge, UK.

Valentine, E. and Wood, I. (1977) *Journal of the Hydraulics Division*, ASCE, **103**, 975–90.

Wallis, S., Guymen, I. and Biligi, A. (1989) in *Hydraulic and Environmental Modeling of Coastal Estuarine and River Water*, (eds R. Falconer, P. Goodwin and R. Matthew), Gower Technical, Aldershot, UK.

Weisberg, R.H. (1976) *Journal of Marine Science*, **34**, 387–94.

Woods, J. (1977) in *Modeling and Prediction of the Upper Layers of the Ocean*, (ed E. Kraus), Pergammon Press, Oxford, pp. 118–42.

Yasuda, H. (1984) *Journal of Fluid Mechanics*, **148**, 383–403.

Yeo, W. and Bedford, K. (1988) in *Computational Methods in Flow Analysis*, (eds H. Niki and M. Kawahara), Okayama U. Science Pub. Co, Okayama, Japan, pp. 844–51.

Yotsukura, N. and Sayre, W. (1977) *Water Resources Research*, **12** (4), 695–704.

Young, W., Rhines, P. and Garrett, C. (1982) *Journal of Physical Oceanography*, **12**, 515–27.

ELEMENTS OF NON-COHESIVE SEDIMENT TRANSPORT 6

J. Fredsøe

The turbulence models enter into the nearly-horizontal flow and short-wave models through diffusion, dispersion and other terms, together with their parameterizations. At the same time, however, it is these nearly-horizontal flow and short-wave models that provide the energy sources and sinks to the turbulence models. So far as the use of such models is concerned, all of this is still only the 'hydrodynamic platform' upon which the models of direct interest are constructed, as 'superstructures'. One of the most important classes of superstructure models is that of sediment transport. However, so different are the cases of non-cohesive sediment transport and cohesive sediment transport, that they are treated here in two separate contributions.

This chapter outlines methods based upon classical (pre-computational) correlations of sediment motion with mean or local velocities and methods that are more computational hydraulic, in that they make use of the turbulent energy level and rate of turbulent energy dissipation, which arise quite naturally from the hydrodynamic models.

CONTENTS

6.1 INTRODUCTION

Some basic elements of sediment transport are described, starting from the classical steady-current case, which has been treated extensively for most of this century. Next, sediment transport in oscillatory flow is described and the influences of current and breaking waves are discussed.

A considerable number of formulae exist which attempt to quantify the sediment transport in waves and currents. The basic hydrodynamics and sediment behaviour as described in this section should properly be included in these formulae in order to be a reliable tool for engineering predictions.

6.2 SEDIMENT TRANSPORT IN STEADY CURRENT

6.2.1 BASIC CONCEPTS

The transport of sediment is usually split into three parts: bed load, suspended load and wash load.

Bed load is that part of the total sediment load which occurs very close to the bed: see Figure 6.1a. Usually, the bed load is taken as the sediment which is rolling, sliding or jumping immediately above the bed. At low fluid shear stresses, this process occurs in a thin layer of the order of 1–2 grain diameters in thickness (Einstein, 1950). At high fluid-shear stresses, the bed-load transport can occur in a layer of thickness equal to several grain diameters: see Figure 5.1b, also Engelund (1981) and Wilson (1988). This is because, at high concentrations, the bed-load particles are kept moving by intergranular collisions.

In general, the mobility of a bed-load particle is determined as the ratio between the driving forces and the stabilizing forces (frictional forces). The driving forces are proportional to shear stress τ times surface area d^2, d being the grain diameter. The stabilizing forces are proportional to the submerged weight $(s-1)gd^3$, where s is the relative density of sediment and g is the acceleration due to gravity. The ratio between driving and stabilizing forces is known as the Shields parameter θ:

$$\theta = \frac{\tau}{(s-1)gd} \tag{6.1}$$

Coastal, Estuarial and Harbour Engineer's Reference Book. Edited by M.B. Abbott and W.A. Price. Published in 1993 by Chapman & Hall, 2–6 Boundary Row, London SE1 8HN. ISBN 0 419 15430 2.

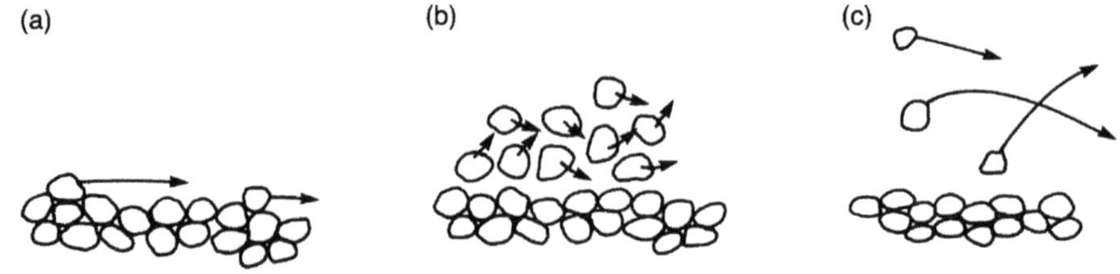

Figure 6.1 Different modes of sediment transport; (a) bed load; (b) sheet flow; (c) suspended load.

If θ is below a certain critical value, θ_c, the bed-load particles will not move, θ_c varies between 0.03 and 0.05, being slightly dependent on the grain Reynolds number. For $\theta > \theta_c$, the grains move, and a sediment transport q_b occurs, q_b is frequently taken to be the volumetric transport of solid material (i.e. without pores!) per unit time and width. The Einstein number:

$$\Phi_b = q_b/\sqrt{(s-1)gd^3} \tag{6.2}$$

is then the dimensionless form of the bed-load discharge. It is possible to show, by more or less refined mathematical models (Einstein, 1950; Engelund and Fredsøe, 1976) that the dimensionless bed-load discharge is mainly a function of the Shields parameter. The best known relationship is the Meyer–Peter formula (Meyer–Peter and Müller, 1948):

$$\Phi_b = 8(\theta - \theta_c)^{3/2} \tag{6.3}$$

This formula is based on an extensive series of experiments. Engelund and Fredsøe (1976) obtained a similar expression derived by expressing equilibrium between the driving and frictional forces on a moving sand particle.

At high shear stresses, Equation (6.3) should be slightly modified because the bed load occurs in several layers, especially if the sediment is so coarse that it is difficult to get into suspension. This type of flow is called sheet flow. Engelund (1981) suggested that Equation (6.3) be modified to

$$\Phi_b = 4.5\, s\, \theta^{3/2} \tag{6.4}$$

for high values of θ.

Suspended load is that part of the load which is moving without continuous contact with the bed, due to the turbulence in the flow. In the case of a plane bed, the vertical distribution of sediment is usually calculated by adapting the diffusion concept, in a similar way to that used to describe the exchange of momentum by turbulent eddies. Taking the diffusion coefficient, ε_s, for suspended sediment to be proportional to the eddy viscosity for the flow

$$\varepsilon_s = \beta\varepsilon \tag{6.5}$$

the vertical distribution of suspended sediment can be found by expressing an equilibrium between settling and diffusion. This gives

$$\frac{c}{c_a} = \left(\frac{D-y}{y}\frac{a}{D-a}\right)^z,$$

$$z = \frac{w}{\beta\kappa U_f} \tag{6.6}$$

where c is the concentration at a distance y above the bed, c_a is the concentration of suspended material at a distance a above the bed, D is the water depth, w is the fall velocity and κ is the von Kármán constant. Equation (6.6) is the so-called Rouse–Vanoni distribution. Several suggestions have been proposed for the bed concentration c_a. For fine sediment, where the bed load only occurs in a layer of one or two grain diameters, Engelund and Fredsøe (1976) gave analytical expressions for c_a. Their expression is based on Bagnold's concept for the dispersive stresses (Bagnold, 1954), the main idea being that the fluid shear stress close to the bottom is transferred to the bed by grain-to-grain collisions. In the case of sheet flow, other expressions have been suggested for the bed concentration (Wilson and Pugh, 1988). This problem, however, seems not to be fully solved, even though it is generally agreed today that c_a is mainly a function of the Shields parameter θ and the ratio w/U_f.

Finally, **wash load** is that part of the sediment load where the particle size is so small that it never will be in contact with the bed. This load cannot be predicted without supplementary field measurements, and is usually not accounted for in predictive sediment-transport models.

6.2.2 BEDFORMS

The bed is usually not plane but breaks up, forming different kinds of sandwaves on the bottom. A complete discussion of these bed-forms can be found in Guy and Richardson (1961). At low flow velocities, where the shear stress is just above the critical level, small current ripples will be formed. At higher shear

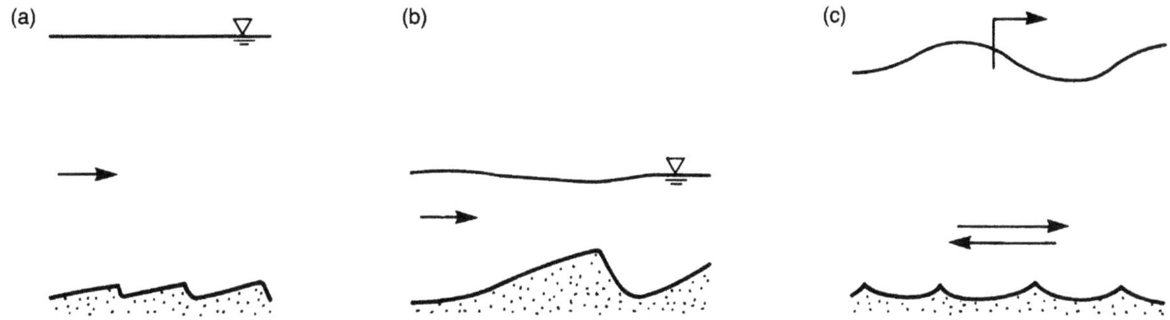

Figure 6.2 Typical bed forms; (a) current ripples; (b) current dunes; (c) wave ripples.

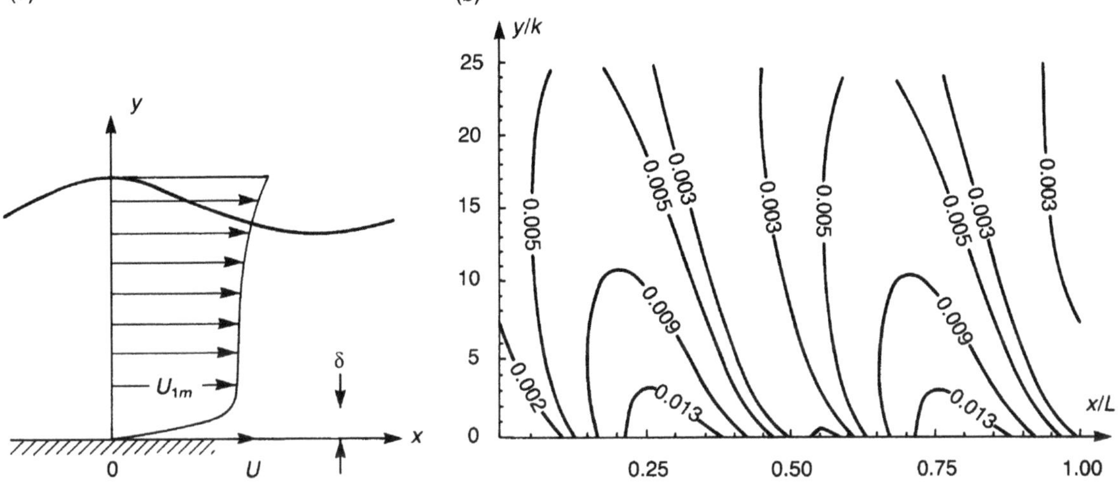

Figure 6.3 (a) sketch of waveboundary layer δ; and (b) contour plot of turbulent kinetic energy in the wave boundary layer for $a/k = 1000$.

stresses, these ripples will be substituted by dunes, which are much larger in size than the current ripples (see Figure 6.2). At even higher shear stresses, the dunes will be washed out due to the action of suspended sediment, and the bottom will become plane.

The presence of bed-forms complicates the description of sediment transport. The bed shear stress will now be composed of two contributions, one originating from skin friction along the bed wave surface, τ', and the other, τ'', originating from the form drag acting on the dunes due to downstream flow separation.

In the case of the presence of bed-forms, the bed load is usually related to the skin friction, τ'. It is still uncertain whether the vertical distribution of suspended sediment should be related to the total U_f or only to U_f'.

6.3 OSCILLATORY FLOW

6.3.1 BED SHEAR STRESS AND TURBULENCE IN NON-BREAKING WAVES

The movement of sediment in an oscillatory flow is, just as in the case of steady flow, determined by the bed shear stress and the near-bed turbulence. In purely oscillating flow without any resulting current, the turbulence is restricted to the thin, wave boundary layer just above the bed, in which the flow velocity drops from the outer wave-induced flow velocity U_w to zero at the bed (see Figure 6.3).

Because of the large velocity gradient in the wave boundary layer, the bed shear stress in waves becomes much larger than that originating from a current with the same mean flow velocity V as U_{1m}, U_{1m}

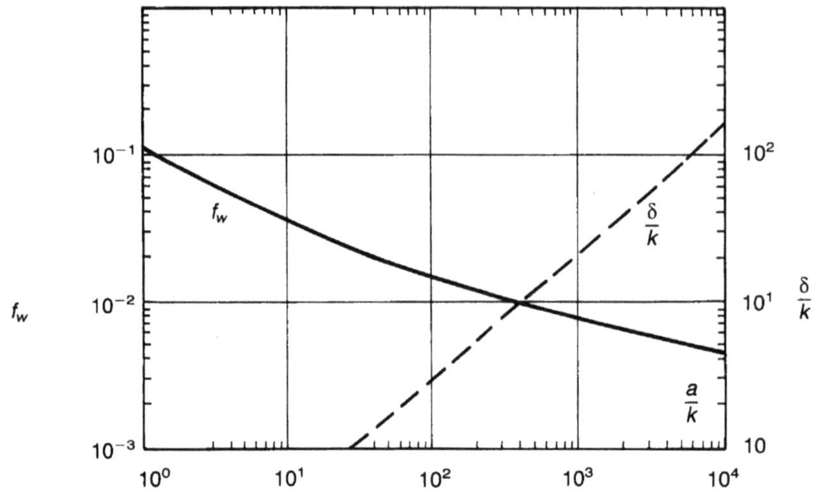

Figure 6.4 Wavefriction factor and wave boundary layer thickness in case of a rough bed.

being the maximum wave-induced flow velocity just outside the wave boundary layer (see Figure 6.3). Figure 6.4 shows the thickness δ of the turbulent region just above the bed, made dimensionless with the bottom roughness k. δ/k is mainly a function of the dimensionless quantity a/k, where a is the free-stream particle amplitude.

As can be seen from Figure 6.4, the thickness of the wave boundary layer is only of the order of 2–10 cm (with a roughness equal 0.1–1 mm). Figure 6.4 further shows the variation in the wave friction factor f_w defined by

$$\tau_{w,\max} = f_w 1/2\, \rho\, U_{1m}{}^2 \tag{6.7}$$

As can be seen from Figure 6.4, f_w is much larger than in ordinary channel flow, where the friction factor is usually in the range 0.002–0.005, and is dependent on the water depth and the bed roughness.

The turbulence level, and thus the eddy viscosity, changes with time as shown in Figure 6.5, which shows the result of a theoretical solution obtained by a turbulence model (Justesen and Fredsøe, 1985).

6.3.2 BED-LOAD TRANSPORT IN OSCILLATORY FLOW

The transport of the bed-load can in most cases be found by a quasi-steady application of bed-load formulae like Equation 6.3 as suggested by Madsen and Grant (1976). They integrated a relation quite similar to Equation 6.3 over half a wave period and found that the average rate of sediment transport back or forth is given by

$$\bar{\Phi} = 12.5\, \theta_m{}^3 \tag{6.8}$$

where θ_m is the maximum dimensionless shear stress, determined by $\tau_{w,\max}$.

In order to use such a quasi-steady model, it is assumed that the inertia of the bed-load particles is negligible. It can be shown (Fredsøe and Rasmussen, 1980) that the sediment size must be larger than several millimeters before the inertia forces have any influence.

6.3.3 DISTRIBUTION OF SUSPENDED SEDIMENT IN WAVES

In general, the transport of suspended sediment constitutes the major part of the total sediment load due to waves. In this unsteady case it becomes much more complicated to estimate the vertical distribution of the suspended sediment, because the level of turbulence is rising and falling all the time (see Figure 6.5). As long as the bed is plane (see below), it is still relevant to consider the distribution of suspended sediment as a result of a diffusion process due to the turbulence. The diffusion equation now reads

$$\frac{dc}{dt} = w\frac{\partial c}{\partial y} + \frac{\partial}{\partial y}\left(\varepsilon_s \frac{\partial c}{\partial y}\right) \tag{6.9}$$

which is a continuity equation between settling and diffusion. Figure 6.7 presents the solution of Equation (6.9), as solved numerically. The eddy viscosity is calculated by means of more-or-less advanced turbulent models, as indicated in the figure.

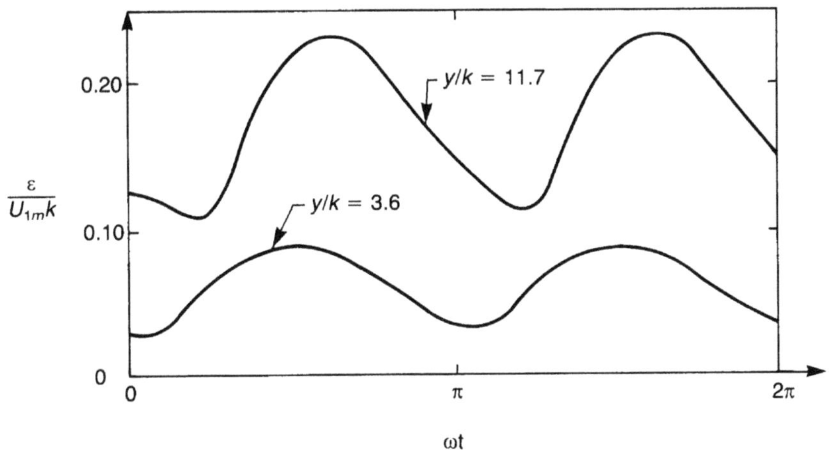

Figure 6.5 Time variation in eddy viscosity in turbulent wave boundary layer. Predicted by one-equation model. $a/k = 10^3$.

Effect of a current superimposed on waves on the amount of suspended sediment

When waves are superimposed on a current, the turbulence is no longer restricted to the thin wave boundary layer, but will extend to cover the whole water depth, whereby sediment in suspension can now be lifted further away from the sea bed.

Figure 6.8, shows an example of the vertical distribution of suspended sediment in combined wave–current motion (Fredsøe *et al.*, 1985). Close to the bottom, the distribution is mainly determined by the waves, due to the large friction factor in wave motion. Further away from the bottom, the current-generated current takes over, and is responsible for the presence of suspended sediment far from the bed.

In the numerical example shown in Figure 6.8, the direction of the current is the same as the direction of wave propagation. More sediment can be kept in suspension if the current is perpendicular to the direction of propagation of waves than if it is in the same direction.

The resulting sediment transport of suspended sediment in wave–current motion is the product of the concentration of sediment and the flow velocity at the same level, integrated over the depth. In this connection it is important to take into account that, in combined wave–current motion, the presence of waves is felt by the current as an increased bed friction. This is because of the large shear stresses formed by the wave boundary layer. Figure 6.9 shows an example of how the current friction increases, V being the mean current velocity. This larger friction implies that if the current water discharge is kept constant, the velocity is redistributed over the vertical when waves are present (Bakker and Doorn, 1978). Close to the bottom, the presence of the waves will reduce the current velocity (and thus the amount of suspended sediment transport).

Effect of wave breaking

Like the current, wave breaking implies that more sediment can be kept in suspension because of the increase in the level of turbulence.

When waves break, a large amount of turbulent energy is produced. While production of turbulent energy due to a current is mainly concentrated close to the bed, the production of turbulent energy in breaking and broken waves of the spilling breaker type is concentrated near the water surface. From here, the turbulent energy is transported downwards in the fluid by turbulent diffusion. The vertical distribution of turbulence can be found by the use of turbulence models as described earlier in this volume (see also Deigaard *et al.*, 1986a,b), from which the vertical distribution of sediment can be found from the diffusion equation (Equation (6.9)). Figure 6.10 shows an example of the results of such a calculation. It can be seen that close to the bottom, the wave breaking does not influence the predicted amount of suspended sediment very much. This is because at this level it is mainly the turbulence from the wave boundary layer which keeps the sediment in suspension. Further away, however, the sediment is kept in suspension mainly by the turbulence originating from the wave breaking.

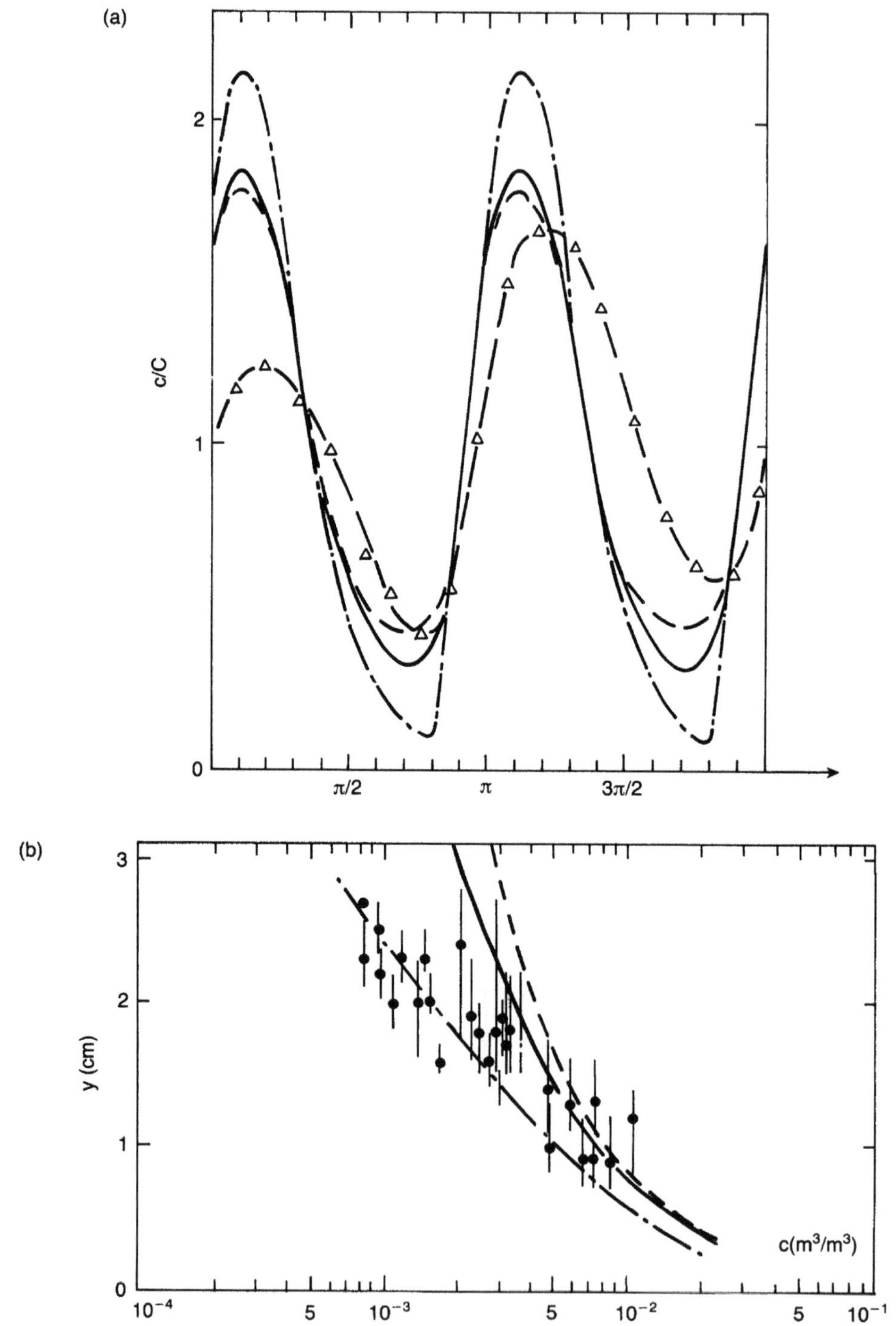

Figure 6.6 Variation with time in instantaneous value of concentration *c* of suspended sediment (a) and predicted vertical mean concentration distribution C (b). (——) one equation model; (– – –), mixing length theory; (– · –), momentum method; (–△–), measurements by Staub *et al.*; $y/k = 37.9$.

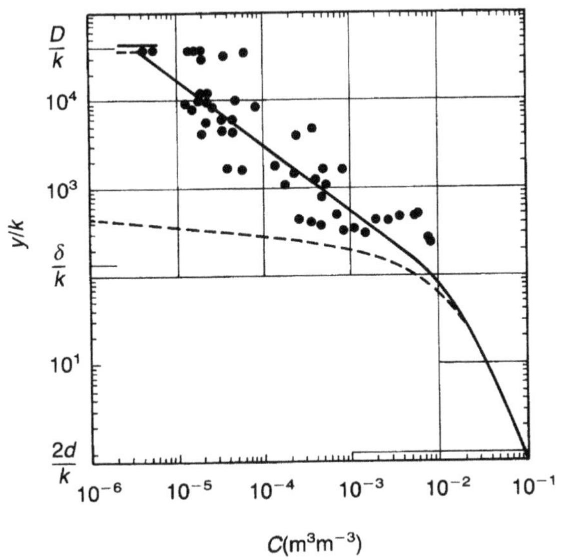

Figure 6.7 Distribution of suspended sediment in combined wave–current motion. $D = 10\,\text{m}$, $H = 1.7\,\text{m}$, $T = 12\,\text{s}$, $U = 0.20\,\text{m}\,\text{s}^{-1}$ (– – –), mean concentration in pure wave form; (——) mean concentration in combined wave–current motion; (●), field measurements. From Fredsøe *et al.* (1985).

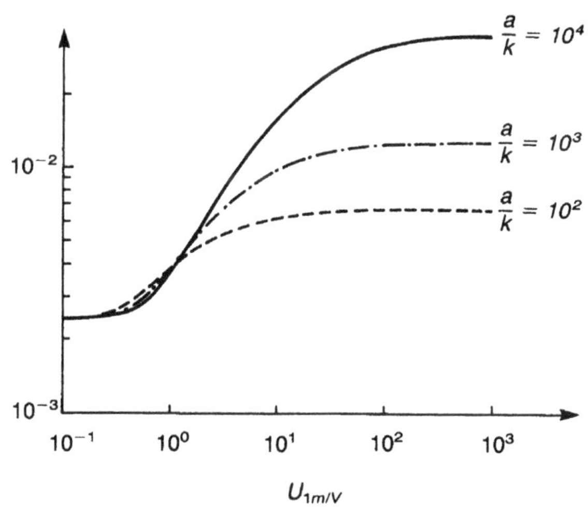

Figure 6.8 Increase in current-resistance due to presence of waves.

6.3.4 BEDFORMS IN WAVES

As in steady flow, the bed will not remain plane in the oscillatory case for small values of the Shields parameter θ. For values of θ smaller than around 1, the bed is covered by wave ripples, as sketched in

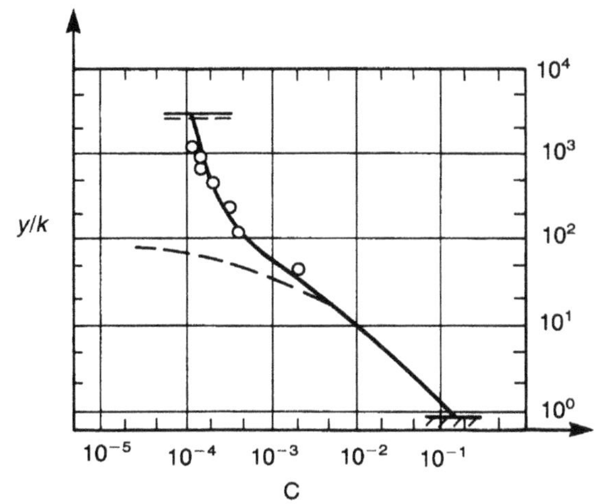

Figure 6.9 Distribution of suspended sediment in spilling breaker. From Deigaard *et al.* (1986a) (——), theory including wavebreaking; (– – –), theory without wavebreaking; (○), field measurements.

Figure 6.1c. At small values of θ, the length λ of these ripples is correlated to a, the free stream particle amplitude. Usually a good fit is given by

$$\lambda = 1.2a \tag{6.10}$$

The steepness of the ripples, η/λ, in which η is the ripple height, is nearly constant for small values of θ, but decreases with increasing θ-values until the bed becomes flat for $\theta' \sim 0.8 - 1.0$ (see Figure 6.10), which is reproduced from Nielsen (1979).

In the case of ripples, an organized pattern of eddies is formed close to the bottom, due to flow separation behind the ripple crest. These eddies are able to bring considerable amounts of sediment into suspension away from the bed (see Figure 6.11). This process is not a diffusion process because it is the result of an organized motion. Thus, both coarse and fine materials are lifted nearly the same distance away from the bed, so that the vertical distribution of sediment will not be affected by the actual value of the fall velocity (or grain size) to the same degree as in the plane-bed case.

So far, the vertical distribution of sediments over ripples has not been solved theoretically, but many field and laboratory measurements are available.

6.3.5 FORMULATION OF SEDIMENT TRANSPORT FORMULAE

In principle, the resulting average transport of suspended sediment during a certain time period, T, can be found from the expression:

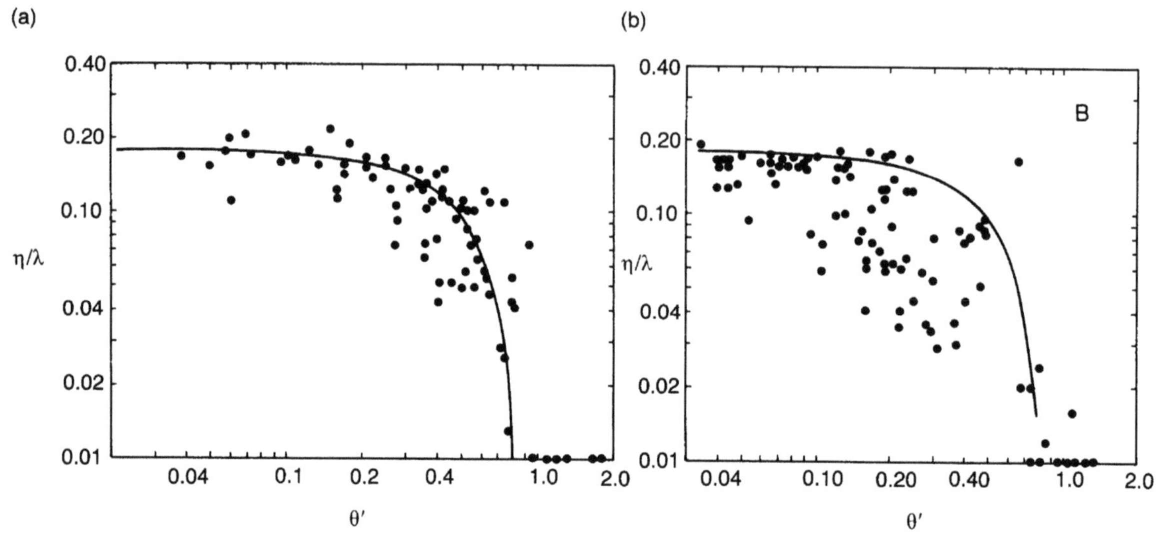

Figure 6.10 Ripple steepness versus dimensionless shear stress for laboratory data (A) and field data (B). After Nielsen (1979).

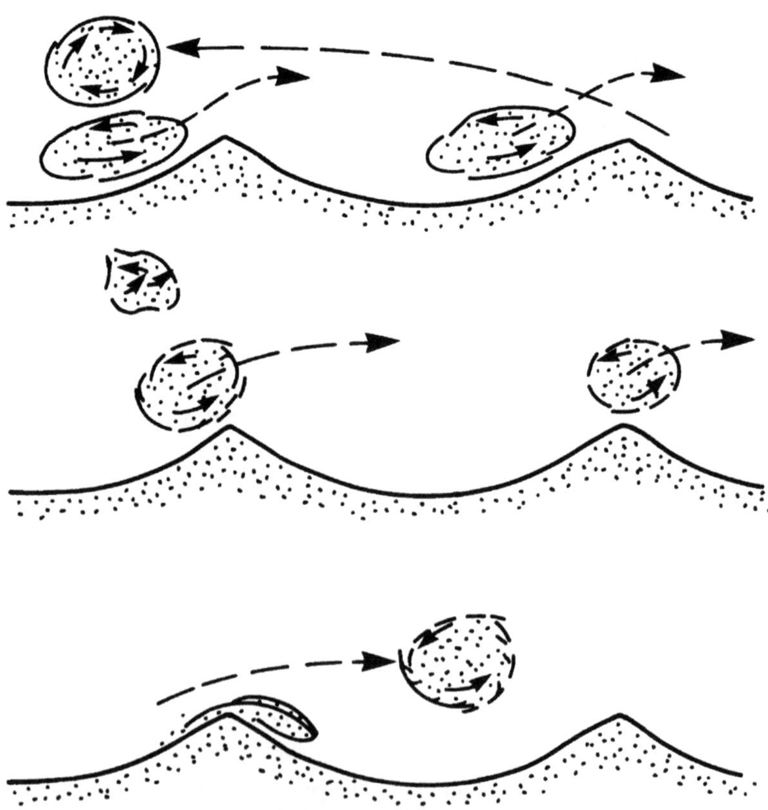

Figure 6.11 Three different stages in the organized motion over a wave-generated ripple. From Nielsen (1979).

$$q_s = \frac{1}{T}\int_0^T \int_0^D uc\,\mathrm{d}y\,\mathrm{d}t \qquad (6.11)$$

in which u and c are the instantaneous values of velocity and concentration, respectively. In steady currents, Equation (6.11) is reduced to

$$q_s = \int_0^D UC\,\mathrm{d}y \qquad (6.12)$$

where U is the average velocity and C is the average concentration.

In the wave case, it is still common to apply Equation (6.12), taking U as the average current velocity over one wave period, and C as the average concentration. This corresponds to the usual notion that waves are responsible for bringing the sediment into suspension, while the current is responsible for the net transportation of the sediment.

However, in unsteady flow the instantaneous amount of sediment in suspension is not determined by the instantaneous value of the bed shear stress, because the sediment takes some time to settle after being picked up from the bed. In the case of wave ripples, this effect can be so pronounced that the resulting sediment transport direction can be the opposite of that of a weak current superposed upon a wave motion (Inman and Bowen, 1963).

An example of a sediment transport formula based on Equation (6.2) for the steady-current case is that developed by Engelund and Fredsøe (1976). Deigaard *et al.* (1986b) have constructed a formula for the longshore sediment transport in the surf zone, based on the complete Equation (6.11). However, the presence of wave ripples was not included, which restricted the model's validity to regions with high shear stresses.

6.4 ENERGETIC APPROACHES

In the above description, the processes responsible for sediment transport are based on a traction approach, in which the hydrodynamics of the flow must be known in detail in order to calculate the resulting sediment transport.

Another school of thought within sediment transport believes in a more integrated approach, where the transport of sediment is related to the rate of energy dissipation ω of the stream. This approach was originally introduced by Bagnold (1966) for sediment transport in streams. The argument behind his model was that a certain amount of energy is necessary, partly to keep the bed-load moving, and partly to keep the centre of mass of the suspended sediment at a certain height above the bed. Bagnold developed the following equation for the total load i of sediment transport (in total immersed weight)

$$i = \left(\frac{\varepsilon_b}{\tan\phi} + \frac{\varepsilon_s}{w/V}\right)\omega \qquad (6.13)$$

which is valid for a stream bed with small slopes. In Equation (6.13), ϕ is the angle of repose of the sediment and V is the mean flow velocity. The two terms on the right-hand side represent bed load and suspended load, respectively.

Bagnold assumed that the bed load and suspended load efficiencies were constant. His approach has been widely used in rivers by, among others, Yang (1972) and McDowell (1989), and in wave–current motion by, for instance, Komar and Inman (1970) and Bailard (1981).

In general, it is inevitable that a kind of proportionality exists between ω and the sediment transport rate.

In the steady-current case, where the dissipation of energy occurs close to the bed, i.e. at the same location where the sediment transport occurs, the efficiency factor becomes quite large. But in cases like breaking and broken waves, the energy dissipation mainly takes place close to the water surface, far from the sediment in transportation. In such a case, the efficiency factor must be quite low. Hence, in general, the efficiency factor cannot be constant, but must depend on the flow situation, so in all cases a refined hydrodynamic description will be necessary.

However, for a preliminary investigation of morphological modelling, the energetic models are often favourable because of their simplicity.

REFERENCES

Bagnold, R.A. (1954) Experiments on a gravity-free dispersion of large solid spheres in a Newtonian fluid under shear. *Proceedings of the Royal Society of London*, A **225**, 49–63.

Bagnold, R.A. (1966) An approach to the sediment transport problem from general physics. US Geological Survey *Professional paper* 422-I.

Bailard, J.A. (1981) An energetics total load test. trans. model for a plane sloping beach. *Journal of Geophysical Research*, **86**, (C11), 10938–54,

Bakker, W.T. and Doorn, Th. (1978) Near-bottom velocities in waves with a current, *16th Coastal Engineering Conference*, Hamburg 1978, **2**, 1394–1413.

Deigaard, R., Fredsøe, J. and Hedegaard, I.B. (1986a) Suspended sediment in the surf zone. *Journal of Waterway, Port, Coastal and Ocean Engineering, ASCE*, **112**, (1), 115–28.

Deigaard, R., Fredsøe, J. and Hedegaard, I.B. (1986b) Mathematical model for littoral drift. *Journal of Waterway, Port, Coastal and Ocean Engineering, ASCE*, **112**, (3), 351–69.

Einstein, H.A. (1950) The bed-load function for sediment transport in open channel flows. US Department of Agriculture, *Tech. Bull. 1026*.

Engelund, F. (1981) *Transport of bed load at high shear stress.* Institute of Hydrodynamics and Hydraulic Engineering, ISVA, Technical University of Denmark, Progress Rep. 53, pp. 31–5.

Engelund, F. and Fredsøe, J. (1976) A sediment transport model for straight alluvial channels. *Nordic Hydrology* **7**, 293–306.

Fredsøe, J., Andersen, O.H. and Silberg, S. (1985) Distribution of suspended sediment in large waves. *Journal of Waterway, Port, Coastal and Ocean Engineering, ASCE*, **111**, (6), 1041–59.

Fredsøe, J. and Rasmussen, P. (1980) *Inertia of bed load in oscillatory flow*. Institute of Hydrodynamics and Hydraulic Engineering, ISVA, Technical University of Denmark, Progress Rep. 51, pp. 29–35.

Guy, H.P., and Richardson, E.V. (1961) Forms of bed roughness in alluvial channels. *Journal of Hydraulic Division, ASCE*, **87**, (HY3), 87–105.

Inman, D.L. and Bowen, A.J. (1962) Flume experiments on sand transport by waves and current. *Proceedings of the 8th International Coastal Engineering Conference*, Mexico City 1962, ASCE, 1963, pp. 137–50.

Justesen, P. and Fredsøe, J. (1985) *Distribution of turbulence and suspended sediment in the wave boundary layer*. Institute of Hydrodynamics and Hydraulic Engineering, ISVA, Technical University of Denmark, Progress Rep. 62 pp. 61–7.

Komar, P.D. and Inman, D.L. (1970) Longshore sand transport on beaches. *Journal of Geophysical Research*, **75**, 5914–27.

Madsen, O.S. and Grant, W.D. (1976) Sediment transport in the coastal environment. *MIT Rep*. 209.

McDowell, D.M. (1989) A general formula for estimation of the rate of transport of non-cohesive bed-load. *Journal of Hydraulic Research*, **27**, (3), 355–62.

Meyer-Peter, E. and Müller, R. (1948) Formulas for bed load transport. *Proceedings of 2nd Meeting, IAHSR*, Stockholm 1948, pp. 39–64.

Nielsen, P. (1979) *Some basic concepts of wave sediment transport*. Institute of Hydrodynamics and Hydraulic Engineering, ISVA, Technical University of Denmark. Ser. Paper 20.

Wilson, K.C. (1988) *Frictional behaviour of sheet flow*. Institute of Hydrodynamics and Hydraulic Engineering, ISVA, Technical University of Denmark, Progress Rep. 67, pp. 11–22.

Wilson, K.C. and Pugh, F.J. (1988) Dispersive force basis for concentration profiles. *Journal of Hydraulic Division, ASCE*, **114**, (7), 806–10.

Yang, C.T. (1972) Unit stream power and sediment transport. *Journal of Hydraulic Division, ASCE*, **98**, (HY10), 1805–26.

ELEMENTS OF COHESIVE SEDIMENT DEPOSITION, CONSOLIDATION AND EROSION 7

L. Hamm and C. Migniot

The contribution describes methodologies currently employed in studying cohesive sediment deposition, consolidation and erosion. In this case also, experimental investigations must be properly integrated with the numerical modelling studies. Similarly, the formation and subsequent behaviour of sediments is influenced by biochemical factors, which enter here, for the first time in this book. Although studies of cohesive sediments are currently not so common as studies of non-cohesive sediments, their frequency is increasing under the twin impacts of an accelerating research effort and an increased awareness of the needs for such studies.

CONTENTS

7.1 INTRODUCTION: SPECIFIC FEATURES OF COHESIVE SEDIMENTS

The dynamics of cohesive sediments are distinguished from those of sands, gravels and pebbles by the overriding importance of electrochemical forces acting between the particles and generating the cohesion that hydrodynamic forces have difficulty in overcoming. The intensity of these electrochemical forces varies considerably and their range of variations exceeds that of hydrodynamic forces so that all manner of situations may be encountered in nature, from widespread re-suspension of sediment to consolidated deposits that are impossible to scour. Cohesion is a function of a large number of parameters, involving in particular the composition of the sediment and the physico-chemical composition of the water. The following parameters may be mentioned in particular:

1. Sediment composition
   (a) **Sediment particle size** A cohesive sediment may consist of varying proportions of very fine sands, silts and clays. It is mainly the clays that ensure overall cohesion.
   (b) **Mineralogy of clay** The three main types of clay (montmorillonite, illite and kaolinite) have very different cohesive properties, which are generally represented by a characteristic parameter that may be the measured specific surface or cation exchange capacity (CEC).
   (c) **Organic matter content** (biological detritus, micro-organisms).
   (d) **Water content** related to the dry sediment density.
2. Composition of free and interstitial water
   (a) **Salinity** This is an important factor, especially in estuaries, as it contributes to the rate of flocculation.
   (b) **Temperature, pH** Concentration of the main cations and anions, including Na^+, Ca^+, Mg^{++}, Cl^- and $SO4^{--}$. The three cations usually met with may be represented by a characteristic parameter referred to as the sodium adsorption ratio (SAR). Other parameters derived from soil mechanics and normally applied to

Coastal, Estuarial and Harbour Engineer's Reference Book. Edited by M.B. Abbott and W.A. Price. Published in 1993 by Chapman & Hall, 2–6 Boundary Row, London SE1 8HN. ISBN 0 419 15430 2.

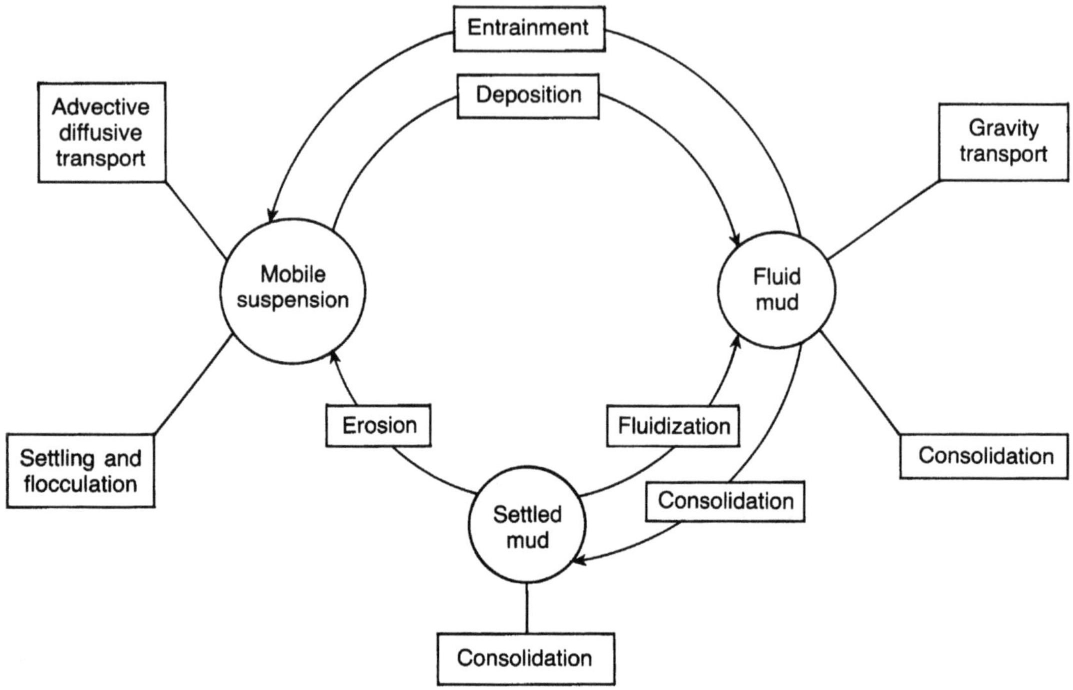

Figure 7.1 Cohesive sediment behavioural model (derived from Kirby, 1988).

non-cohesive sediments may also be used. It is therefore essential at the start of the study of a particular site to take *in situ* samples and carry out laboratory tests to determine the relative importance of the above parameters.

7.2 MAIN CONCEPTS CONCERNING THE DYNAMICS OF COHESIVE SEDIMENTS

7.2.1 GENERAL OUTLINE

In the case of non-cohesive sediments, a distinction is usually made between bed load and suspended load. The situation is more complex with cohesive sediments, whose behaviour varies considerably depending on the dry sediment concentration.

Several theoretical models have been proposed to attempt to describe the dynamics of cohesive sediments (Mehta *et al.*, 1982; Parker, 1984; Kirby, 1988; Mehta, 1989). Figure 7.1 shows one such model derived from Kirby (1988) in which three physical states of sediment occurrence may be distinguished. These are;

1. Mobile suspension;
2. Fluid mud;
3. Bed deposit or settled mud.

The transition from one state to another occurs as a result of different mechanisms, namely deposition, consolidation, erosion, re-suspension and fluidization.

In addition, the various states described are not static but possess time-dependent properties that also form part of the dynamic cycle, such as settlement, flocculation, transport by advection–diffusion and gravity transport (density currents). These have also been introduced into the figure.

7.2.2 FLUID MUD

One major difficulty in understanding the mechanisms involved is due to the frequent occurrence, especially in estuaries, of fluid mud. This corresponds to an area in which there is a gradual increase in the dry sediment between mobile suspension (concentration below $10\,g\,l^{-1}$) and bed deposit (concentration above $250\,g\,l^{-1}$). Hydrographs detect this area, which may be several metres thick, when they obtain a double echo on the sounding tape.

Fluid mud has special properties that are linked to its rheological behaviour, and is often described as a complex plastic fluid. In practice, two lines of research are of particular interest to harbour engineers in this

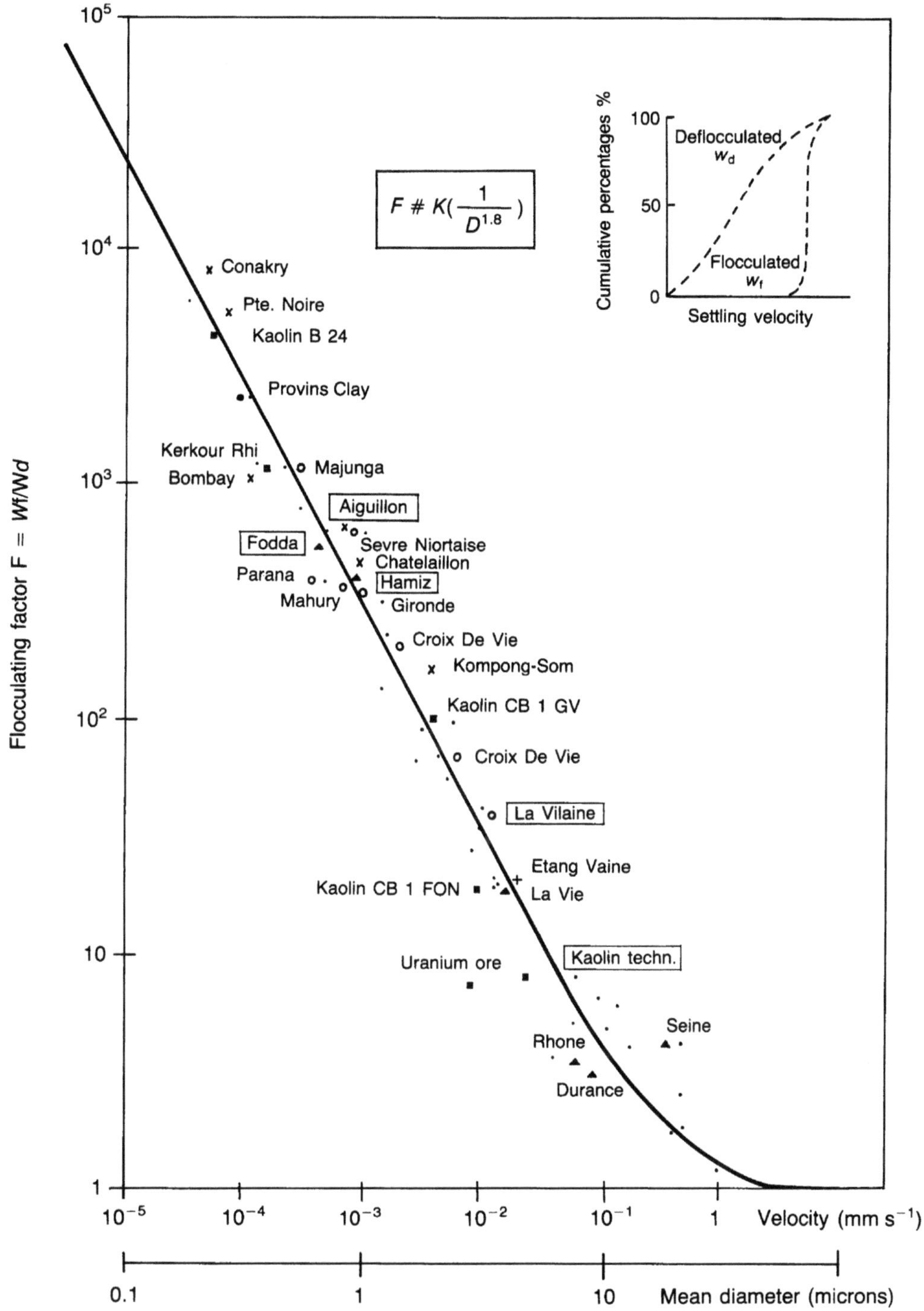

Figure 7.2 Effect of particle size on flocculation of natural muds (after Migniot, 1968). w = 50%; (×), marine silt; (○), estuary silt; (▲), river silt; (*), lake silt; (■), muds and powders; (...), other sediments obtained by separation.

respect: sedimentation in basins and channels as a result of density currents set up by the high concentrations of sediment near the bed (Lin *et al.*, 1986), and the possibility of navigating in such an environment, which means that drafts may be increased and maintenance dredging reduced (Brossard *et al.*, 1990).

7.2.3 DEVELOPMENT OF AN EMPIRICAL APPROACH

At present, it is impossible to describe all the processes involved, taking into account all the parameters relating to the behaviour of a particular type of mud, and because of this researchers have had to adopt an empirical approach in modelling the principal phenomena.

With this in view, suitable measuring methods have been developed *in situ* and in the laboratory, enabling the properties of cohesive sediments to be characterized by a small number of parameters. The work of Migniot (1968, 1989) is fully representative of this approach, which is the only one that is operational at the present time in harbour engineering.

7.3 DEPOSITION OF SUSPENDED SEDIMENTS

7.3.1 PHYSICAL DESCRIPTION OF COHESIVE SEDIMENT SETTLEMENT

Importance of flocculation

The settlement of cohesive sediments is a complex phenomenon, owing to the cohesive properties of the elementary particles, which tend to stick together when suspended: this is the phenomenon known as flocculation. Migniot (1968) showed that this tended to accelerate settlement to a considerable extent (Figure 7.2).

Factors favouring or restricting flocculation

Flocculation is the result of a combination of two types of phenomena (Leussen, 1988);

1. Mechanical mechanisms favouring collision between particles;
2. Mechanisms encouraging particles that have met to adhere to one another.

The two main mechanisms responsible for collision are:

1. The current velocity gradient (turbulence);
2. The differential settling of the suspended particles.

In addition, turbulence may become a limiting factor for flocculation, especially near the bed, where the flocs may be broken and resuspended. The three main mechanisms favouring adhesion of the particles are:

1. The presence of flocculating salts (electrochemical phenomenon);
2. Microbiological activity;
3. Bio-deposition.

The first mechanism is the most well known and was for a long time considered to be the main factor responsible for sedimentation in estuaries. The second mechanism is being increasingly studied. Microzoobenthos activity tends to cause mineral particles and organic matter to stick together and form flocs, even in fresh water. Lastly, biodeposition resulting from mollusc digestion processes (in oyster beds) also tends to cause particles to stick to one another (Diara *et al.*, 1984).

7.3.2 SETTLEMENT IN CALM WATER: ESTIMATION OF FALL VELOCITY

The rate of settlement is defined as being equal to the flow of sediment passing through the area separating the bed from the liquid medium. In calm water this flow is:

$$D = W_s C \tag{7.1}$$

in which: D is the rate of deposition in $\mathrm{kg\,m^{-2}\,s^{-2}}$; W_s is the fall velocity in $\mathrm{m\,s^{-1}}$; and C is the sediment concentration in $\mathrm{kg\,m^{-3}}$. The values of W_s and C are generally averaged along a vertical for either the entire depth of water or for only a part.

Fall velocity of elementary particles

In the absence of flocculation, the fall velocity of fine particles (i.e. those with a diameter of less than 0.1 mm) may be connected to an equivalent diameter by applying the Stokes formula:

$$W_s = \frac{\Delta g d^2}{18\gamma} \tag{7.2}$$

in which:

$$\Delta = \frac{\rho s - \rho w}{\rho w}$$

ρs is the density of sediment in $\mathrm{kg\,m^{-3}}$; ρw is the density of water in $\mathrm{kg\,m^{-3}}$; d is the equivalent diameter of particle in m; and γ is the kinematic viscosity of water in $\mathrm{m^2\,s^{-1}}$ ($10^{-6}\,\mathrm{m^2\,s^{-1}}$ in pure water at 20°C, $1.6 \times 10^{-6}\,\mathrm{m^2\,s^{-1}}$ at 4°C).

In principle, a direct measurement of fall velocity due to settlement in a laboratory tube defines an equivalent diameter. These velocities are extremely low and are practically never encountered in real

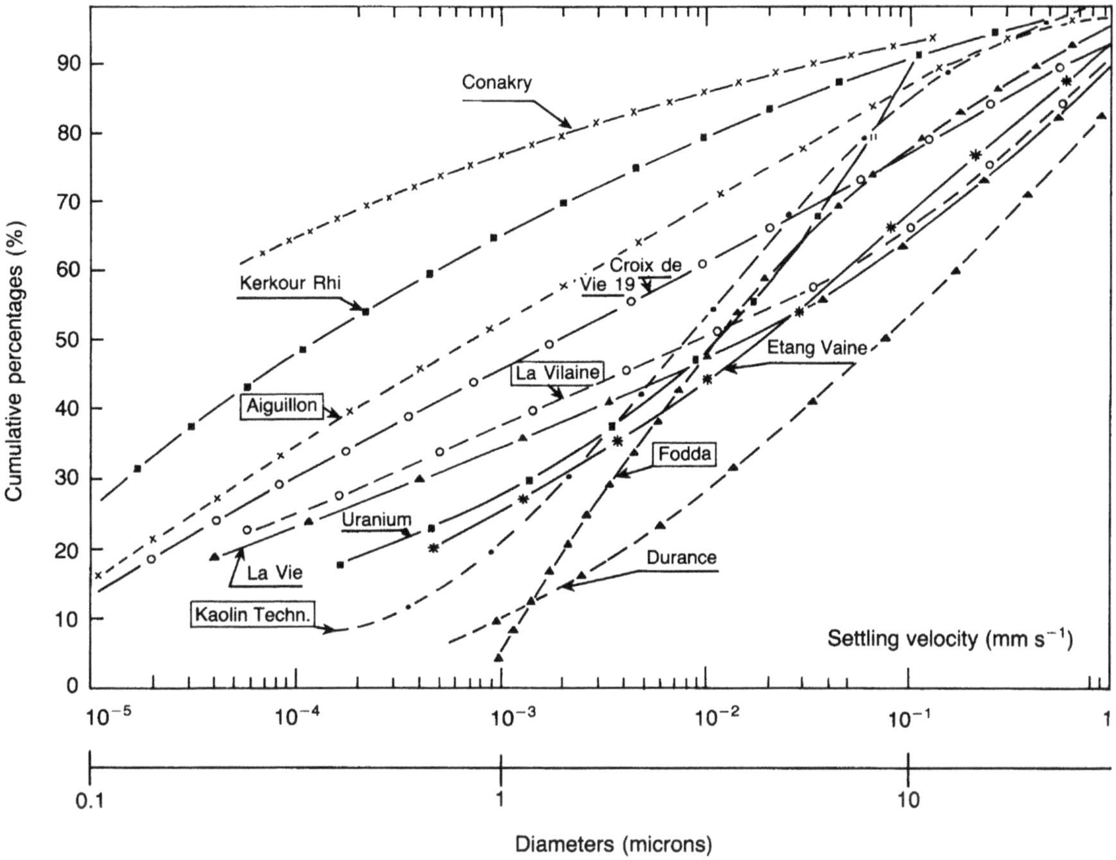

Figure 7.3 Fall velocity of elementary particles of natural muds (after Migniot, 1968).

conditions (Figure 7.3). However, these measurements are used to compare muds (Migniot, 1989).

Fall velocity induced by flocculation: methods of measurement

The method of measurement is similar to the previous one. It involves measuring the settlement of sediment in a tube and deducing the fall velocity distribution. Two methodologies are distinguished in the literature: laboratory measurements and *in situ* measurements.

These two methods are distinguished by the way in which the flocs are formed. In the laboratory, samples of mud come from natural deposits which have been disturbed. Flocs form under the effect of water salinity and the frequency of collision, which depends on the initial concentration. The main drawback with this method is that the turbulence that exists in nature is not reproduced, and this tends to cause larger flocs and accelerate settlement. For this reason, various instruments have been designed for taking sediment-bearing samples of water in nature without destroying the flocs. The best known of these is the Owen tube. However, this method does not yet give complete satisfaction, as the fall velocities appear to accelerate in the tube because of exaggerated differential settlement (Puls *et al.*, 1988).

Main parameters affecting flocculation

Particle size

Flocculation increases as the size of the particles decreases. This has been demonstrated experimentally

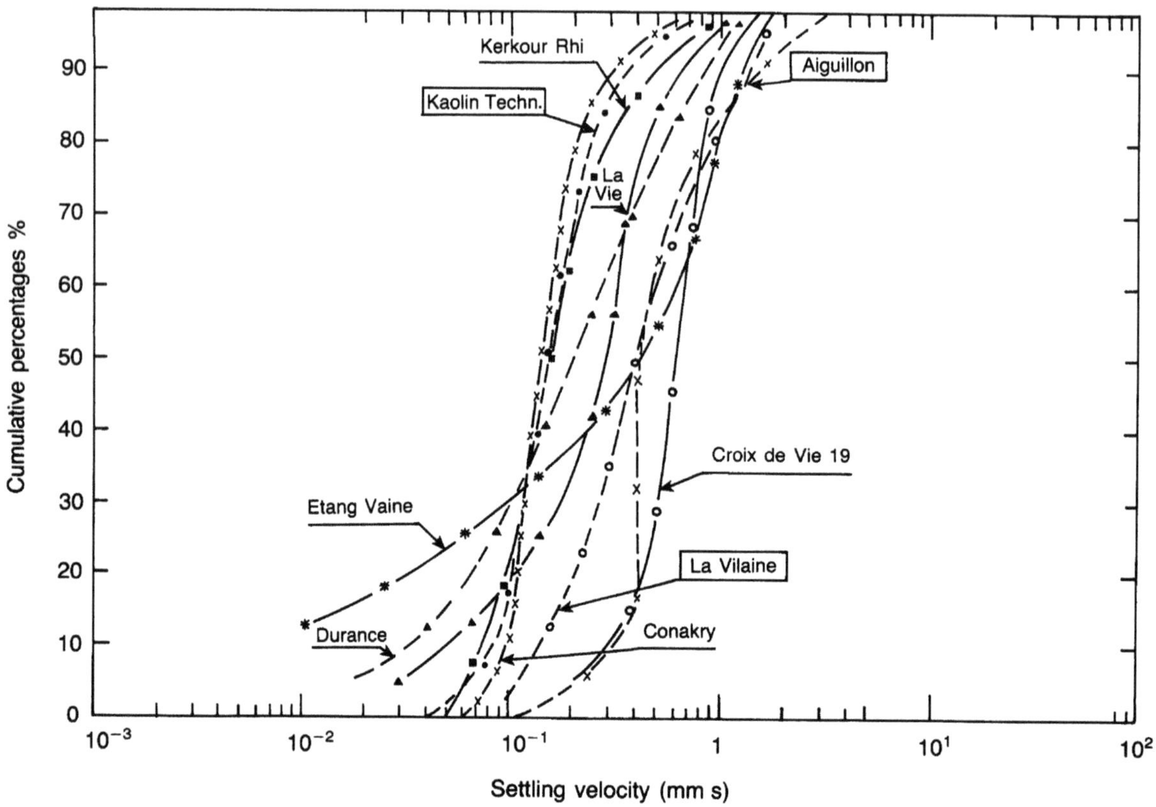

Figure 7.4 Fall velocity of flocculated particles of natural muds (after Migniot, 1968).

by Migniot (Figure 7.2). As the size of particles decreases, there is an increase in their specific surface area.

Salinity

Laboratory tests (Krone, 1962; Migniot, 1968) have shown the importance of salinity in the phenomenon of flocculation, once it rises above 0.3% (Figures 7.3 and 7.4).

In contrast to the laboratory results, however, *in situ* measurements do not reveal any marked effect of salinity. Measurements carried out with an Owen tube in the River Thames do not reveal any effect from salinity. This difference between the two types of measurement (laboratory and *in situ*) is explained by the fact that in nature the sediment remains in suspension for a much longer period than in the laboratory, thus enabling a large amount of flocculation to take place, even with low levels of salinity (Burt, 1984).

In addition, *in situ* measurements taken in the Elbe estuary showed that fall velocities were higher in fresh water than in saline water, for the same concentration. This may be explained by the fact that flocculation occurs in the river part of the Elbe as a result of the cohesive properties of organic matter. This flocculation is then destroyed by the action of the saline water, which decreases the fall velocities (Puls and Kuehl, 1986). A similar phenomenon has also been observed at Rotterdam (Leussen, 1988).

Concentration

The effect of concentration on fall velocity in calm water has been demonstrated by many authors, both in the laboratory and in nature. Three concentration ranges must be considered;

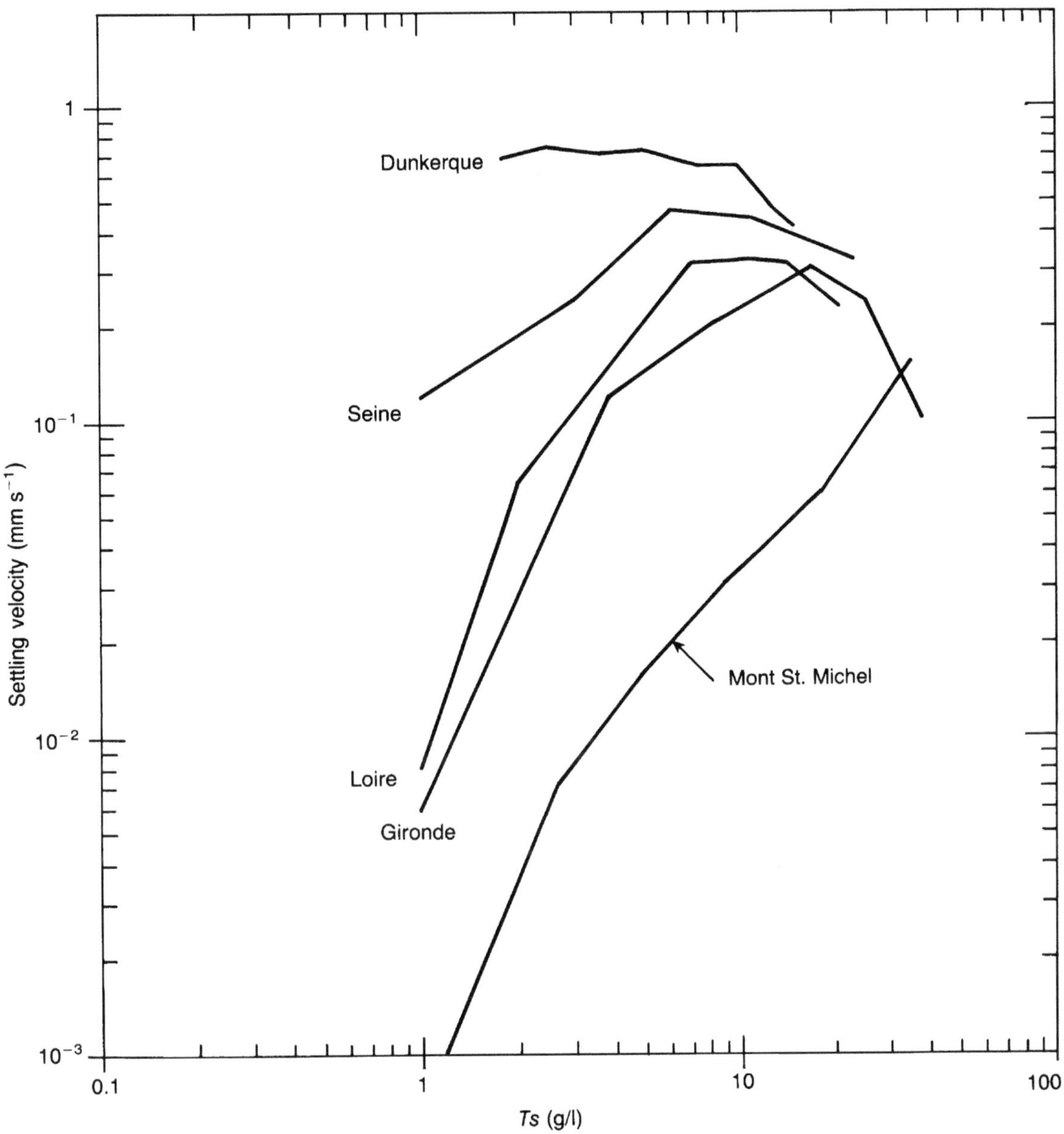

Figure 7.5 Medium settlement velocities of mud flocs as function of concentration, laboratory experiments.

1. Below 20–50 $mg\,l^{-1}$, measurement is not possible;
2. From 20–50 $mg\,l^{-1}$ to a few grams per litre, a relation of the form $W = KC^n$; is generally found, with n varying between 1 and 2;
3. Above this, the fall velocity reaches a maximum and then decreases; this is the phenomenon known as hindered settling (Figure 7.5).

Turbulence

The effect of this parameter is very difficult to estimate quantitatively. An indirect method involves taking *in situ* measurements at spring and neap tides. In 1970, Owen demonstrated higher fall velocities during neap tides than during spring tides (for equal concentrations), which he explained by the destruction of flocs as a result of turbulence. Measurements carried out in 1983 contradict this conclusion and do not reveal any effect from the tide coefficient on the fall velocity (Burt, 1984). It should be recalled that only the effect of turbulence on fall velocity in calm water is being considered here. In the presence of a current, the turbulence plays a fundamental role, which will now be discussed.

7.3.3 SETTLEMENT IN THE PRESENCE OF A CURRENT

In the presence of a current, turbulence will cause a shear in the fluid and a vertical diffusion of momentum acting in an opposite direction to the force of gravity. It is usual to distinguish between two regions in studying this phenomenon, namely the bed boundary layer and the rest of the column of water.

The bed boundary layer involves 10–20% of the total depth and is characterized by a rapid variation in current velocity, with zero velocity on the bed itself. There is considerable turbulence and only those flocs that are sufficiently cohesive can cross the area and be deposited on the bed. Turbulence in the rest of the column of water is less intense but is nevertheless sufficient to prevent the settlement of fine sediment, and also to encourage flocculation. Deposition is therefore controlled by friction on the bed, which may be represented by the shear stress τ_f or the shear velocity u_*.

The settlement rate is generally represented by the expression:

$$D = W_s C\left(1 - \frac{\tau_f}{\tau_d}\right) \qquad (7.3)$$

in which: D is the rate of deposition in $\mathrm{kg\,m^{-2}\,s^{-2}}$; W_s is the fall velocity in calm water in $\mathrm{m\,s^{-1}}$; C is the concentration of suspended sediment in $\mathrm{kg\,m^{-3}}$; τ_f is the shear stress on the bed in $\mathrm{N\,m^{-2}}$; and τ_d is the critical shear stress beyond which there is no further deposition.

The term W_sC, which represents the rate of settlement in calm water, must be estimated near the bed. However, it is often replaced by a value averaged over the entire depth.

The concept of critical shear stress, beyond which there is no further deposition, was introduced by Krone (1962) and has been used without any notable modification since then. It should be noted that, in the case of a wide range of particle sizes, a critical shear stress must be defined for each class of fall velocity. In this way, it is possible to take into account the sorting of particle sizes that has sometimes been observed (Mehta and Lott, 1987). Values for this critical stress vary from 0.06 to $0.1\,\mathrm{N\,m^{-2}}$ according to Krone (1962) and from 0.04 to $1\,\mathrm{N\,m^{-2}}$ according to Mehta and Lott (1987).

7.4 CONSOLIDATION OF A MUD DEPOSIT

7.4.1 GENERAL DESCRIPTION OF CONSOLIDATION

A recent mud deposit occurs in the form of a mixture of water and sediment, in which the sediment concentration is of the order of a few tens of grams per litre. This mixture will develop in three stages, as follows;

1. Settlement of flocs for the first few hours, thus forming fluid mud;
2. Elimination of interstitial water (one to two days);
3. Elimination of water connected with micelles (very slow consolidation).

The dry sediment density will then progress, resulting in the settlement of the deposit and consolidation, seen in the form of greater cohesion and thus increased resistance to erosion. For this reason, it is very important to study the consolidation of a deposit in order to evaluate possible future erosion.

The two fundamental parameters to be considered in defining the state of a deposit are the dry sediment density (T_s) and the initial rigidity or yield value (τ_y). These two parameters change with time and along the vertical through the deposit. Many parameters are involved in the change (Migniot, 1968) and only an empirical approach can enable them to be quantified for a particular site.

7.4.2 CHANGE IN SEDIMENT CONTENT

Migniot (1989) has summarized many results obtained from laboratory studies using a standardized methodology. These demonstrate that each type of mud has its own consolidation curve, which depends on its particle size and mineralogical characteristics (Figure 7.6). The same is true of the concentration gradient in the deposit (Figure 7.7). In addition, sedimentation conditions play an important role, such as, for example, the initial concentration of suspended sediment and its initial thickness.

The change in sediment content has also been studied empirically by many authors (Hayter (1983)

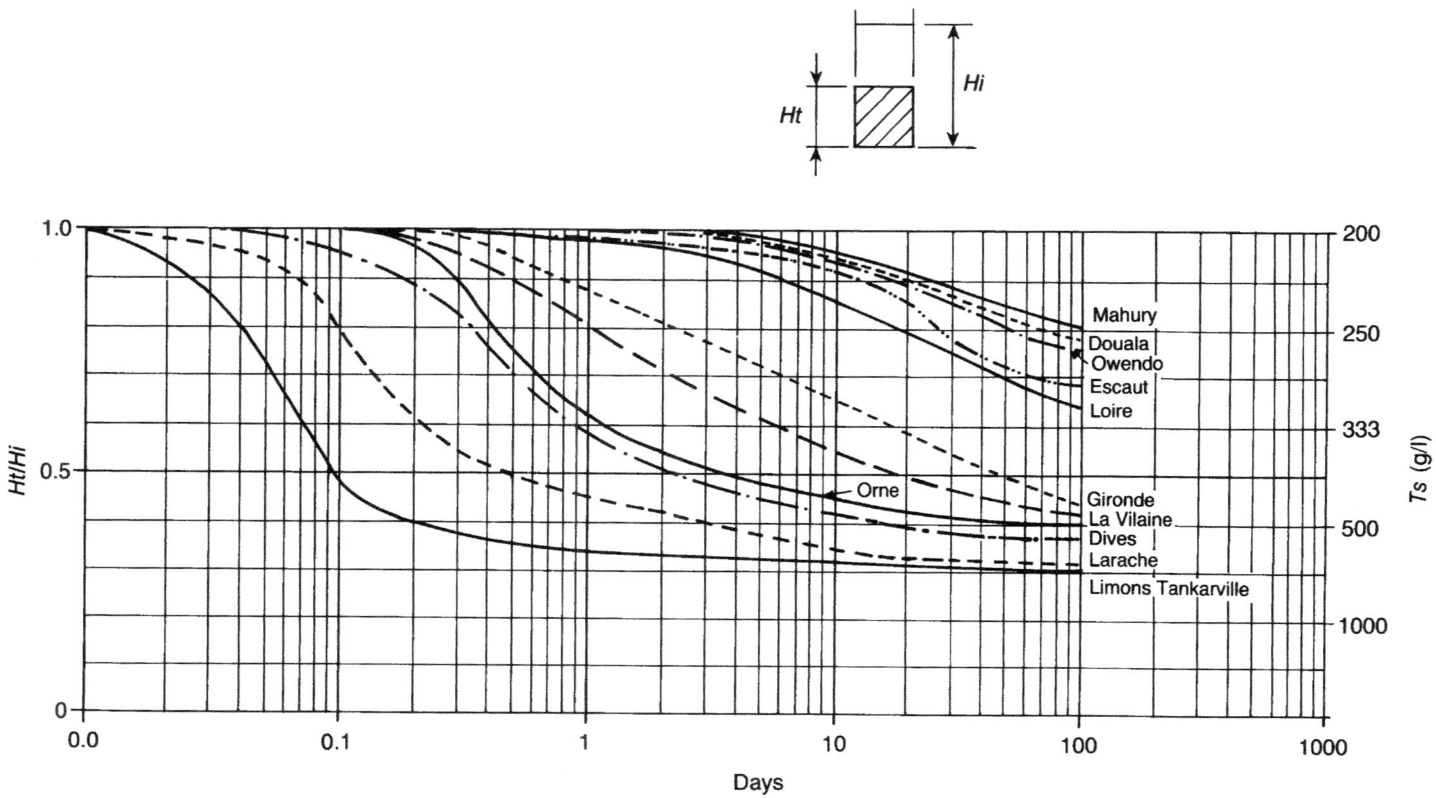

Figure 7.6 Consolidation rate of estuarine muds (after Migniot, 1989) – laboratory experiments (initial thickness: 1 m).

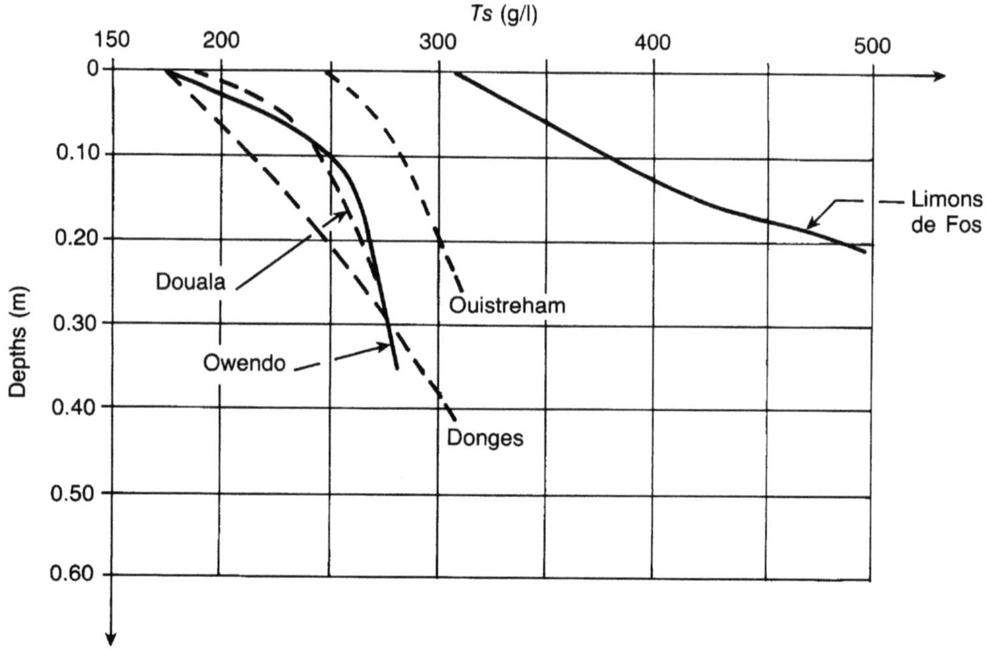

Figure 7.7 Dry sediment content profile in mud deposits laboratory experiments (after Migniot, 1989).

gives a summary) but there is no clear consensus of opinion. However, it should be noted that concentration profiles all show a rapid variation at the surface, which is then attenuated. This is an important fact in understanding the method by which a deposit is eroded.

7.4.3 CHANGE IN INITIAL RIGIDITY

The rheological properties of a mud vary in parallel with the change in sediment content. In practice, this means that it will be more and more difficult for a consolidated deposit to be eroded.

Migniot (1989) has proposed the initial rigidity as characterizing the resistance of a deposit to erosion. This is defined as the torque needed to cause a rotor immersed in the mud to begin turning. A Brookfield LVT viscosimeter is used to take measurements, following a very precise and repetitive procedure. Figure 7.8 gives the results obtained for various muds. The initial rigidity, τ_y, is shown as a function of the dry sediment density.

These results demonstrate a relation of the form:

$$\tau_y = bT_s^a \tag{7.4}$$

between the initial rigidity in $\mathrm{N\,m^{-2}}$ and the dry sediment density in $\mathrm{kg\,m^{-3}}$.

It should also be observed that there is a critical value of the dry sediment density T_{sc} and hence of the critical initial rigidity τ_{yc}, on either side of which the values of coefficients a and b are different. This boundary marks the separation between fluid or hardly-concentrated and plastic or concentrated muds. The transition from fluid to plastic states is usually between 1 and $3\,\mathrm{N\,m^{-2}}$. The coefficient a is around 3 for a fluid mud and 6 for a plastic one.

Thorn and Parsons (1980) and Mehta *et al.* (1982) prefer to characterize resistance to erosion directly in terms of the critical shear value of the fluid generating the erosion (shear strength). Tests must therefore be carried out systematically in a straight or circular flume.

7.4.4 THE ROLE OF EMERSION

When muds are deposited on the foreshore and exposed to the air, consolidation will accelerate at the surface because of the following phenomena (Diara *et al.*, 1984);

1. Reduction in water content: runoff along the convex slopes of mud flats, inter-particle flow in the surface film of the mud, evaporation during the summer, in particular;

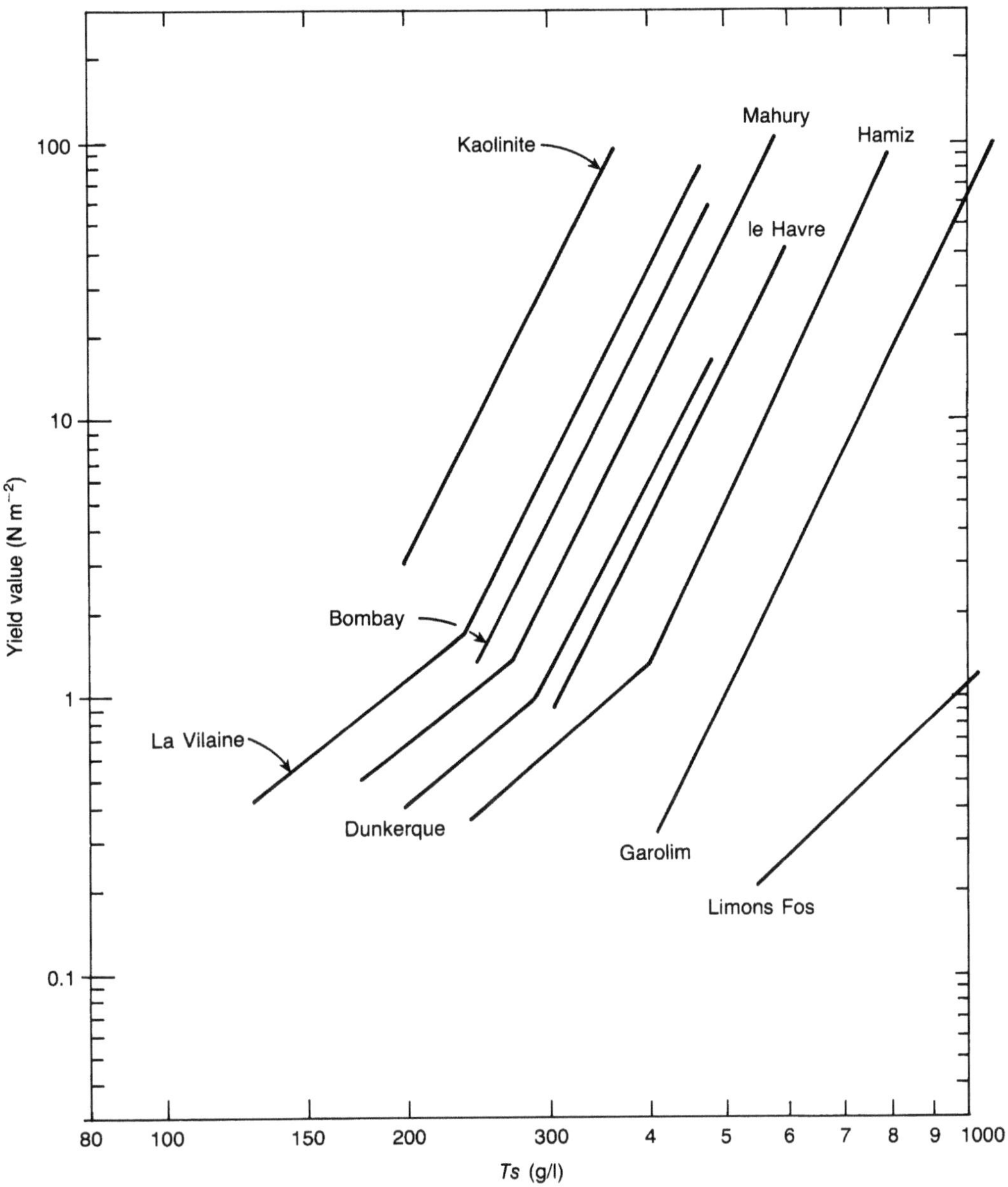

Figure 7.8 Yield values of mud deposits as function of dry sediment content, laboratory experiments (after Migniot, 1989).

2. Improvement of rheological and mechanical properties, increase in initial rigidity, cohesion and shear rate, resulting from loss of water;
3. Capillary evaporation-ascension phenomena: during emersion, water that has evaporated is progressively replaced through capillary action by interstitial water, which contains dissolved salts (silica, ammonia); these modify the development of micro-organisms living at the surface of the mud flats.

The role of emersion will be even greater in shellfish breeding areas, where biodeposits consisting of pellets with a high organic matter content will tend to consolidate.

7.5 EROSION OF MUD DEPOSITS BY CURRENTS

7.5.1 PHYSICAL DESCRIPTION OF PHENOMENA

Mud is eroded by currents in different ways, depending on the state of consolidation of the deposit. Three fields may be distinguished: unconsolidated deposits (fluid mud), consolidating deposits (plastic mud) and old consolidated deposits (solid mud).

Migniot (1989) fixes the limit between fluid and plastic muds at an initial rigidity of $3\,\mathrm{N\,m^{-2}}$ measured with a Brookfield viscosimeter and that between plastic and solid muds at $75\,\mathrm{N\,m^{-2}}$. Migniot (1968) and Périgaud (1983) have described the different methods of erosion for fluid and plastic muds. In the case of a fluid mud, there is an undulation of the interface, which is gradually accentuated, increasingly deforming the layer of sediments. In the end, these are carried away in the form of mud streaks, and diluted in the current water. In the case of a plastic mud, the surface creases and the surface film is torn. The mud is eroded in the form of 'flakes', which may or may not be diluted in the current.

From the hydrodynamic point of view, bursts of turbulence near the bed appear to play a dominant role in the erosion process (Périgaud, 1983). Here again, there is no established theory for calculating the rate of erosion of a mud deposit and it is current practice to use empirical relations.

7.5.2 GENERAL PRINCIPLES OF EROSION EXPERIMENTS

Tests are carried out in an open or recirculating flume, which may be straight or circular. The rate of erosion is calculated on the basis of measurements of suspended sediment concentration over time. The following relation may be written:

$$E = h\frac{\mathrm{d}\bar{C}}{\mathrm{d}t} = Ts(zf)\frac{\mathrm{d}zf}{\mathrm{d}t} \qquad (7.5)$$

in which: E is the rate of erosion ($\mathrm{kg\,m^{-2}\,s^{-1}}$); h is the water depth (m); $\bar{C}$ is the mean concentration of suspended sediment ($\mathrm{kg\,m^{-2}}$); $T_s(zf)$ is the dry sediment density ($\mathrm{kg\,m^{-3}}$); and zf is the elevation of the eroded bed (m). The rate of erosion is then related to the flow characteristics (shear velocity or stress) and characteristics of the deposit (dry sediment density, initial rigidity). It is usual to observe that erosion can only occur beyond a certain velocity threshold. In this case, it is common to refer to a critical shear stress or velocity corresponding to a given deposit.

7.5.3 CRITICAL EROSION VELOCITY

This is the velocity beyond which erosion begins for a well-defined mud. Migniot (1968) relates the shear velocity u_* to the mean dry sediment density T_s of the deposit and to the initial rigidity measured in a test tube using a Brookfield viscosimeter. From this observation, he deduces the following two empirical relations (Migniot, 1989):

Fluid mud $\quad u_{*c} = 0.013\,\tau_y^{1/4} \quad \tau_y < 3\,\mathrm{N\,m^{-2}} \qquad (7.6)$

Plastic mud $\quad u_{*c} = 0.009\,\tau_y^{1/2} \quad \tau_y > 3\,\mathrm{N\,m^{-2}} \qquad (7.7)$

Here again, there is the qualitative threshold between fluid and plastic muds, already observed quantitatively. In practice, *in situ* measurements of dry sediment density in the deposit, combined with laboratory measurements of initial rigidity, enable u_*c to be estimated in each particular case.

7.5.4 EROSION RELATIONS

Non-stratified deposit

In the case of a deposit with no vertical stratification of the initial rigidity concentration (in general, a consolidated deposit), the following two equivalent expressions are the most commonly used:

$$E = M_1\left(\frac{\tau f}{\tau c} - 1\right) \quad \text{(Kandiah, 1974)} \qquad (7.8)$$

or

$$E = M_2(\tau f - \tau c) \quad \text{(Odd, 1988)} \qquad (7.9)$$

In the second expression, M_2 is estimated at $2 - 4 \times 10^{-3}\,\mathrm{kg\,m^{-2}\,s^{-1}}$.

The critical shear stress τ_c is deduced from the expressions given in section 7.5.3, using the relation $\tau = \rho_s u_*^2$, in which ρ_s is the density of the water. The shear stress caused by the fluid at the bottom (τ_f) is calculated by assuming that flow is smooth (i.e. with a Strickler coefficient of the order of 80–100). These expressions indicate that once erosion has begun, it continues as long as the hydraulic conditions remain the same.

Partially consolidated deposits

In the case of a partially consolidated deposit, in which there is considerable vertical stratification of rigidity values, the phenomena are more complex, as erosion will often start at low speeds, and stop when it reaches the layer of sediment where the critical entrainment velocity is equal to the shear stress of the fluid.

A precise experimental procedure has been used by Thorn and Parsons (1980), Mehta *et al.* (1982), Parchure and Mehta (1985) and Kuijper *et al.* (1989) to study this phenomenon. The method used to exploit the results differs from one author to another, however, and no common formulation has been detected. In practice, this means that systematic flume tests have to be carried out when designing a project for a particular site, where such a phenomenon has to be taken into account.

Resuspension of fluid mud

The resuspension of fluid mud occurs when the shear stress on the bed exceeds the critical stress at the surface of the deposit (Hayter, 1983). The layer of fluid mud is resuspended almost instantaneously down to a depth at which the critical stress is equal to the shear stress of the fluid.

7.6 EROSION RESULTING FROM WAVE AGITATION

The action of wave agitation on a mud deposit is an important phenomenon, the role of which was stressed as far back as 1966 by Krone, in his study of San Francisco Bay. The interaction between wave disturbance and a muddy bed causes fluidization of the mud deposit and damps the wave agitation (Mehta, 1989). This damping effect has been observed particularly on the coasts of Surinam, Guiana and in the Gulf of Mexico, and theoretical studies have been carried out taking into account rheological models of the mud deposit of varying complexity (Maa and Mehta, 1987).

Fluidization of the deposit has been studied far less. Initial flume test results (Mehta, 1989) show that this is also an important phenomenon that can cause rapid erosion when currents are superimposed on wave disturbance, particularly in shallow areas. Mention should also be made of erosion phenomena caused by the breaking of waves on muddy banks, in tidal seas.

7.7 MOBILITY OF MUDDY SEDIMENTS

7.7.1 SUSPENDED TRANSPORT

Huge quantities of fine sediments move around in estuaries or along coasts, usually greatly exceeding the quantities of sand. In particular, it is estimated that at the mouth of a river, solid transport consists of 90% mud and 10% sand. These fine sediments are the result of soil erosion and form wash load in rivers. Continuous removal down to the sea has formed a permanent stock of suspended sediment along coasts, which is deposited offshore during calms periods or in sheltered areas, especially those created by harbour development schemes.

These sediments are transported by advection–diffusion by general tidal currents or wind-induced currents. In estuaries, the alternation of spring- and neap-tides gives rise to the well known phenomenon of turbidity maximum, where the concentrations of sediment increase considerably in specific areas (Migniot, 1983).

Along coasts, the phenomena are less spectacular but water turbidity measurements should be made over a complete year when drawing up a harbour development project, even if it is located far from the mouth of a river. Throughout the world, there are indeed many examples of harbours situated along sandy and rocky coasts that have suffered from serious siltation.

7.7.2 MOBILITY OF FLUID MUD

The movement of fluid mud in the form of density currents is also an important phenomenon in the silting of harbour areas. Fluid mud is likely to travel when the equilibrium slope of the deposit is exceeded or when there is a steep spatial gradient of the dry sediment density.

Migniot (1968) studied the equilibrium slope of a mud deposit in calm water in a flume and proposed the following relation:

$$tg\alpha = 0.01\,\tau_y \quad (\tau_y > 3\,\mathrm{N\,m^{-2}})$$

for an immersed deposit.

In the case of a deposit on a sloping bed or subject to a spatial pressure gradient, Odd (1988) calculated the threshold thickness for the deposit to remain stable by making the induced tractive force (gravitation, pressure gradient) equal to the initial rigidity of the deposit.

Lin *et al.* (1986) has presented a method for calculating a density current set up by a major variation in concentration between the water with a high sediment load circulating in the channel of an estuary and that with a relatively low load in a harbour basin. A density current of this kind could cause major sedimentation in harbour basins.

REFERENCES

Brossard, C., Delouis, A., Galichon, P. *et al.* (1990) Navigability of mud-silted channels. Scale model experiments. *Proceedings of the 22nd International Conference on Coastal Engineering*, July 2–6, 1990, Delft, The Netherlands. ASCE, New York, pp. 3088–101.

Burt, T.N. (1984) Field settling velocities of estuary muds, in *Estuarine Cohesive Sediment Dynamics*, Springer-Verlag, Berlin, pp. 126–50.

Diara, M., Gouleau, D., Nikodic, J. *et al.* (1984) Les mécanismes d'envasement de certaines zones littorales. *Proc. 18e journées de l'hydraulique*, SHF, II.2.1–II.2.6 (in French).

Hayter, E.J. (1983) Prediction of cohesive sediment movement in estuarial waters. *PhD dissertation*. University of Florida, Gainesville, Florida.

Kandiah, A. (1974) Fundamental aspects of surface erosion of cohesive soils. *PhD dissertation*. University of California, Davis, California.

Kirby, R. (1988) High concentration suspension (fluid mud) layers in estuaries, in *Physical Processes in Estuaries*, (eds J. Dronkers and W. van Leussen), Springer-Verlag, Berlin, pp. 463–87.

Krone, R.R. (1962) Flume studies of the transport of sediment in estuarial shoaling processes. *Internal Report*. Hydraulic Engineering Laboratory University of Berkeley, California.

Kuijper, C., Cornellise, J.M. and Winterwerp, J.C. (1989) Research on erosive properties of cohesive sediments. *Journal of Geophysical Research*, **94** (C10), 14341–50.

Leussen, W. van (1988) Aggregation of particles, settling velocity of mud flocs – a review, in *Physical Processes in Estuaries*, Springer-Verlag, Berlin, pp. 347–403.

Lin, C.P., Lott, J.W. and Mehta, A.J. (1986) Turbidity sedimentation in closed-end channels. Proceedings of the 20th International Conference on Coastal Engineering, November 9–14, 1986, Taipei, Taiwan. ASCE, New York, pp. 1336–50.

Maa, Y.P. and Mehta, A.J. (1987) Mud erosion by waves: a laboratory study. *Continental Shelf Research*, **7** (11–12), 1269–84.

Mehta, A.J., Parchure, T.M., Dixit, J.G. and Ariathurai, R. (1982) Resuspension potential of deposited cohesive sediment beds, in *Estuarine Comparison*, (ed. V.S. Kennedy), Academic Press, New York, pp. 591–609.

Mehta, A.J. and Lott, J.W. (1987) Sorting of fine sediment during deposition. *Proceedings of the International Conference on Coastal Sediments 87*, May 12–14, 1987, Louisiana, USA. ASCE, New York, pp. 348–62.

Mehta, A.J. (1989) On estuarine cohesive sediment suspension behaviour. *Journal of Geophysical Research*, **34** (010), 14303–14.

Migniot, C. (1968) Etude des propriétés physiques de différents sédiments très fins et de leur comportement sous les actions hydrodynamiques. *La Houille Blanche*, **7**, 591–620 (in French).

Migniot, C. (1983) Les matières en suspension dans les estuaires. *Bulletin de l'Association Internationale de Géologie de l'ingénieur*. **28**, 61–75, (in French).

Migniot, C. (1989) Tassement et rhéologie des vases – Parties 1 et 2. *La Houille Blanche*, 1 and 2, (in French).

Odd, N.V.M. (1988) Mathematical modelling of mud transport in estuaries, in *Physical Processes in Estuaries*, Springer-Verlag, Berlin, pp. 503–31.

Parchure, T.M. and Mehta, A.J. (1985) Erosion of soft cohesive sediment deposits. *Journal of Hydraulic Engineering, ASCE*, **111** (10), 1308–26.

Parker, W.R. (1984) On the observation of cohesive sediment behaviour for engineering purposes, in *Estuarine Cohesive Sediment Dynamics* (ed. A.J. Mehta), Springer-Verlag, Berlin, pp. 270–89.

Puls, W. and Kuehl, H. (1986) Field measurements of the settling velocities of estuarine flocs, in *Proceedings of the Third International Symposium On River Sedimentation*, March 31–April 14, 1986, Jackson, Mississipi. ASCE, New York, pp. 525–36.

Puls, W., Kuehl, H. and Heymann, K. (1988) Settling veloclty of mud flocs – results of field measurement in the Elbe and the Weser estuary, in *Physical Processes in Estuaries*, Springer Verlag, Berlin, pp. 404–24.

Perigaud, C. (1983) Mécanique de l'érosion des vases. *La Houille Blanche*, **7/8**, 501–12 (in French).

Thorn, M.F.C. and Parsons, J.G. (1980) Erosion of cohesive sediments in estuaries: an engineering guide, in *Proceedings of the Third International Symposium on Dredging Technology*, Bordeaux, BHRA, pp. 349–58.

PART THREE

NUMERICAL TOOLS AND THEIR APPLICATIONS

3.1 HYDRODYNAMIC MODELS

THREE-DIMENSIONAL HYDRODYNAMIC MODELS 8

E.B. Rasmussen

We begin our description of engineering modelling tools with an outline of the most general of these, namely the three-dimensional flow-simulation models. Despite several earlier, pioneering engineering investigations, it is really only over the last few years that models of this type have been applied routinely to hydraulic engineering and aquatic-system management problems. Such applications have been made possible by dramatic reductions in the cost of computer processing capacity and memory, on the one hand, and less visible but equally significant developments in computational hydraulics and computational fluid dynamics, generally, on the other hand. A basic problem is that in full three-dimensional models there is a minimal filtering-out of unresolvable features, so that there is a minimum of diffusion, dispersion and other such terms that would normally contribute to the stability of the models. Models of this class, if they are to be accurate at all, must in practice run at the outermost limits of their stability, so that they remain sensitive and fragile instruments, and are correspondingly difficult to use. As much effort is currently being invested in modelling capabilities in this area, it is to be expected that full three-dimensional simulation will become a consistent, reliable and economically viable aid to practice during the 1990s.

CONTENTS

8.1 INTRODUCTION

Three-dimensional models have been used successfully in the meteorological community for weather prediction. Compared with meteorology the history of three-dimensional models in oceanography and, especially, traditional hydraulics is quite young. The first three-dimensional models were developed in the 1960s as an attempt to create coupled atmosphere–ocean models to study the world's climate (*cf.* Bryan and Cox, 1968). These ocean models were, however, based on the so-called rigid-lid condition (i.e. the top vertical velocity is identically zero) in order to use larger time steps than the physically correct free surface condition would allow. The rigid-lid condition, however, is insufficient for most engineering

Coastal, Estuarial and Harbour Engineer's Reference Book. Edited by M.B. Abbott and W.A. Price. Published in 1993 by Chapman & Hall, 2–6 Boundary Row, London SE1 8HN. ISBN 0 419 15430 2.

problems concerning free surface flows and, consequently, this approach was left behind in the early 1970s with the development of three-dimensional models for free surface flows in estuaries and coastal regions (e.g. Leendertse *et al.*, 1973).

Practical applications with three-dimensional modelling is a fairly new branch of hydraulic engineering and may today still be considered as a niche activity, due to the computational time and storage involved. In recent years, however, the rapidly growing computer industry has provided much more powerful and less expensive computers so that three-dimensional models are beginning to play a larger role in the analysis and design of applications in hydraulics. So far the number of practical applications has been very limited and restricted to areas where the two-dimensional, depth-integrated models are insufficient i.e. mainly to the modelling of stratified regions and to environmental modelling where details on the vertical velocity field are required.

The class of three-dimensional models is quite broad in the sense that the mathematical bases of these models might differ considerably from one model to another, ranging from the 'complete' Navier–Stokes equations to narrower sub-systems like the Euler equations. The terms 'complete Navier–Stokes equations' or 'exact Navier–Stokes equations' are often seen in the literature for an averaged form of the Navier–Stokes equations, whereas this is really something quite different.

8.2 CONSERVATION LAWS

The equation for conservation of mass and the Navier–Stokes equations usually form the basis of a general continuum model for fluid flow:

$$\frac{\partial \rho}{\partial t} + \frac{\partial}{\partial x_j}(\rho u_j) = 0 \tag{8.1}$$

$$\frac{\partial}{\partial t}(\rho u_i) + \frac{\partial}{\partial x_j}(\rho u_i u_j) = -\frac{\partial P}{\partial x_i} + \rho g_i + \frac{\partial}{\partial x_j}(\sigma_{ij}) \tag{8.2}$$

in which ρ is the instantaneous local density of the fluid, u_j is the instantaneous velocity in the x_j-direction and t denotes the time. P is the fluid pressure, g_i is the x_i-component of the gravitational acceleration and σ_{ij} is the instantaneous deformation stress, which for Newtonian fluids is given by

$$\sigma_{ij} = \mu\left(\frac{\partial u_i}{\partial x_j} + \frac{\partial u_j}{\partial x_i}\right) + \left(\mu^b - \frac{2}{3}\mu\right)\delta_{ij}\frac{\partial u_m}{\partial x_m} \tag{8.3}$$

μ is the dynamic viscosity, and μ^b is a bulk viscosity of the same order as μ. The value of μ^b is very difficult to determine experimentally and is usually unknown, δ_{ij} is Kronecker's delta.

For inhomogeneous fluids, two additional equations, for the conservation of the scalar quantities salt and temperature (heat), are required, together with an equation of state to close the system expressed through Equations (8.1)–(8.3). The equation for conservation of salinity reads

$$\frac{\partial}{\partial t}(\rho S) + \frac{\partial}{\partial x_j}(\rho S u_j) = \frac{\partial}{\partial x_j}\left(\rho \kappa_D \frac{\partial S}{\partial x_j}\right) \tag{8.4}$$

in which S denotes the salinity and κ_D the diffusivity of salt in water. A similar equation may be derived for the temperature through expressing the law of heat conservation:

$$\frac{\partial}{\partial t}(\rho T) + \frac{\partial}{\partial x_j}(\rho T u_j) = \frac{\partial}{\partial x_j}\left(\rho \kappa \frac{\partial T}{\partial x_j}\right) + \frac{\alpha T}{c_p}\left(\frac{\partial P}{\partial t} + u_j \frac{\partial P}{\partial x_j}\right) \tag{8.5}$$

where T is the temperature (in degrees Kelvin), κ the heat diffusivity coefficient, α the thermal expansion coefficient and c_p the specific heat capacity. The local density is related to salinity, temperature and fluid pressure through an equation of state of the type

$$\rho = \rho(P, S, T) \tag{8.6}$$

Such a system constitutes the mathematical representation of fluid flow under the assumption that the fluid is a continuum, that is Newtonian, and that the flow can be described by the Navier–Stokes equations.

For turbulent flow, these equations represent instantaneous values of the flow properties. Since it is impossible to solve these equations for the small scales which are involved in the turbulent motions, the quantities are decomposed into a mean value and a random fluctuation with zero mean about that mean value; this is followed by an averaging process usually performed in the time domain. This decomposition of the instantaneous variables into time-averaged quantities and fluctuations causes a new set of equations with a number of unknown correlations (the Reynolds stresses) that are all related to the turbulent fluctuations. This is known as the closure problem, and is described by Rodi in Chapter 4 of this volume.

From an engineering point of view, the best bases for the modelling of turbulence are the so-called turbulent-energy transport models, most of which are based on the Boussinesq eddy-viscosity concept,

$$-\overline{u_i' u_j'} = \nu^T\left(\frac{\partial u_i}{\partial x_j} + \frac{\partial u_j}{\partial x_i}\right) - \frac{2}{3}\delta_{ij}k \tag{8.7}$$

in which k is the turbulent kinetic energy defined as

$$k = \frac{1}{2}\overline{u_i' u_i'} \tag{8.8}$$

and ν^T is the turbulent eddy viscosity. In Equation (8.7) and below, the bars over the time-averaged quantities have been omitted for convenience.

The argument for expressing the Reynolds stresses in terms of time-averaged quantities through the eddy-viscosity concept is that two of the most important effects of turbulence are included in the viscosity in laminar flow. The first effect is that turbulence absorbs kinetic energy from the mean flow and converts it into thermal energy while the second effect is that turbulence increases the rate of transport of such quantities as mass, momentum and energy across the averaged 'streamlines' of the flow.

This approach leads to the Reynolds-averaged Navier–Stokes equations

$$\frac{\partial u_i}{\partial t} + \frac{\partial}{\partial x_j}(u_i u_j) = -\frac{1}{\rho}\frac{\partial P}{\partial x_i} - g_i + \frac{1}{\rho}\frac{\partial}{\partial x_j}\left\{\rho\nu^T\left(\frac{\partial u_i}{\partial x_j} + \frac{\partial u_j}{\partial x_i}\right) - \frac{2}{3}\delta_{ij}k\rho\right\} \tag{8.9}$$

which in conjunction with the divergence-free continuity equation

$$\frac{\partial u_j}{\partial x_j} = 0 \tag{8.10}$$

forms the mathematical basis for most three-dimensional models. For compressible flows, Equation (8.10) reads

$$\frac{\partial \rho}{\partial t} + \frac{\partial}{\partial x_j}(\rho u_j) = D_{\text{mass}}\frac{\partial \rho}{\partial x_j} \tag{8.11}$$

where a conventional gradient-diffusion model has been applied to describe the turbulent density fluctuations. D_{mass} is a diffusivity coefficient similar to the eddy viscosity, but for weak density fluctuations this term can be omitted.

To close the system it is necessary either to express the eddy viscosity in terms of the time-averaged quantities (zero-equation models) or to provide equations that govern the evolution of the eddy viscosity (one-equation and two-equation models), as described by Rodi in Chapter 4 of this volume. One-equation models focus on the turbulent kinetic energy for which an equation can be derived from the Navier–Stokes equations. Dimensional analysis of the eddy viscosity necessitates a length scale which in a one-equation model is prescribed. Two-equation models seek to overcome this problem by use of an additional equation. The most successful two-equation turbulence model is probably the k–ε model where ε denotes the isotropic energy dissipation rate. As explained by Rodi earlier, this additional equation for the dissipation rate can be derived from the Navier–Stokes equations in a similar manner to the k-equation. Further reviews of the various turbulence models have been given by Markatos (1986) and Ferziger (1987) among others.

Alternative ways of modelling the turbulence, such as, large eddy simulation and direct numerical simulation, have been devised but have only a restricted range of application in three-dimensional modelling applications as they require so much computational time and storage.

8.3 ASPECTS OF THE COMPUTATIONAL METHODS

Equations (8.9)–(8.11) are, along with an appropriate turbulence model, the mathematical formulation of homogeneous or inhomogeneous three-dimensional flow expressed in a conservation form in a Cartesian co-ordinate system. These partial differential equations (PDEs) cannot, however, be solved by a computer as they express a continuous change in both time and space. Thus, the PDEs need to be re-formulated into a discrete set of equations which can be solved numerically. Methods like finite difference (FD), finite element (FE), finite volume (FV), boundary element (BE), marker and cell (MAC), integral (I), etc. have been devised for this conversion, but, because of the complexity of the problems involved and the need for high accuracy, finite difference methods are applied almost exclusively in hydraulics. It is thus natural that this chapter should concentrate on aspects related to the FD method, with only a few sporadic references to other methods. An outline of alternative methods has been given by Abbott and Basco (1989).

Most FD methods are expressed within an Eulerian formulation imposed on a rectangular Cartesian grid, or, at most, on a classical curvilinear or conformal grid that is transformed into a rectangular Cartesian grid prior to computation (see Figure 8.1).

In general, the differences in methods have only limited effects on the numerical results. The differences between FD and FV, for instance, only play a role for a non-Cartesian grid in the sense of providing different ways of handling the grid geometry (e.g. Vinokur, 1989).

Mathematically, the equations are usually characterized through their properties as elliptic, parabolic or hyperbolic (e.g. Abbott and Basco, 1989). The classification of the equations into these three classes is essential to the computational algorithm being adopted. Unfortunately many practical problems governed by Equations (8.1)–(8.3) are both elliptic and hyperbolic, that is, they are elliptic in some

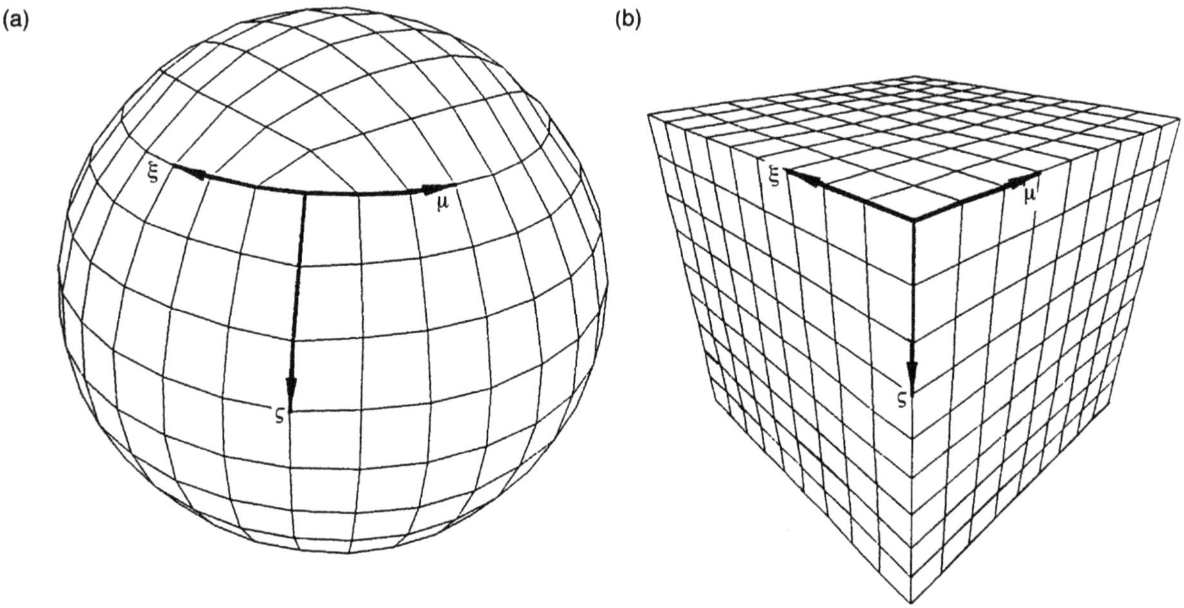

Figure 8.1 Classical curvilinear grid and corresponding transformed grid on which computations are performed.

part(s) of the computational domain and hyperbolic elsewhere. As no generally accessible mathematical theory has been proposed for such mixed-type equations, simplifications have often been introduced in the governing equations, such as neglecting the effects of viscosity, time derivatives etc., so that more suitable equations, in a mathematical and numerical sense, can be derived. An example of such a simplification arises in steady-state models. The steady-state solution, if such such exists in nature, is, however, reached, if at all, asymptotically in fully dynamic models.

From an engineering point of view, though, it can often be more feasible to adopt a method based on a general theory (e.g. the Navier–Stokes equations), rather than to adopt a more detailed method based on a narrower theory with a limited application, even though it requires more computer power and coding.

The primitive equations as given by Equations (8.9)–(8.11) will form a mathematically ill-posed problem whenever the fluid pressure and the velocities constitute the prognostic (i.e. dependent) variables due to a weak coupling between the pressure and the velocities. **This is the key element in three-dimensional modelling**. The system is said to be 'stiff', as both slow and fast processes are present which inherently cause difficulties in the numerical algorithm. In free-surface flows, however, only the slow processes are of interest, and the fast processes (such as shock waves) usually have no substantial influence on the slow processes (such as the free-surface waves), suggesting that they may be removed without loss of information. The fast processes are easily eliminated by replacing the time derivative of the density in Equation (8.11) with the pressure term in the equation of state (Equation (8.6)), whereby a compressibility of the fluid is introduced. Similarly, a pseudocompressibility term can be added in Equation (8.10). The fast processes are then subsequently eliminated through an artificial compressibility whereby the system has become hyperbolic-dominated. This approach known as 'the artificial compressibility approach' was first proposed by Chorin (1967).

Alternatively, so-called pressure correcting methods can be applied in which the divergence-free continuity equation is enforced through the solution of a Poisson equation for the pressure (*cf.* Ferziger, 1987; Patankar, 1980).

8.4 NUMERICAL STABILITY AND ACCURACY

The problems of recreating nature in a model arise from both those approximations that alter the physical system and those approximations involved in converting the continuous equations (PDEs), that describe that physical system, into a discrete set of equations which can be solved numerically.

The FD conversion technique is based upon a Taylor-series expansion of each of the terms in the PDEs. Such an expansion will lead to functions each of which has an infinite number of terms, and, of course, they cannot be solved either, as they are fully equivalent to the PDEs. The equations are made solvable by discarding an infinite number of terms above a certain level, thus providing the FD equations. The level of the discarded terms determines the order of accuracy of the FD scheme.

An exact solution to a problem represented by the PDEs has thus become an approximate solution and the difference between the true and approximate solution is usually called the truncation error.

Given the FD equations, it is anticipated that the approximate solution will converge towards the true solution for decreasing time step and grid spacing. Under such circumstances the FD scheme is regarded as convergent. Under certain general conditions it can be shown that if the FD scheme is stable it will also be convergent, implying that the stability of an FD scheme is a necessary and sufficient condition for convergence. Further details on this subject have been given by Richtmyer and Morton, (1967) and Abbott (1979).

As a low order of accuracy is equivalent to the introduction of a numerical dissipation, it is essential for the reliability of the results that the FD scheme has a high order of accuracy. A high accuracy, however, brings the FD scheme closer to the region of numerical instability. Hence, on the one hand, a high accuracy makes possible a good representation of the physical system, but, on the other hand, it tends to move the FD scheme closer to the limits of numerical stability. A low accuracy can always be used to ensure stability, but will also tend to obscure the details of the physical system. Thus, the proper combination of high accuracy and numerical stability is an essential aspect in numerical modelling.

A stability problem can be alleviated in many ways by filtering out the numerical instabilities, but this again is equivalent to introducing numerical dissipation and is thus equivalent to introducing a lower accuracy. Thus, filtering processes should be avoided unless these have a definite physical and parameterization significance, as explained by Bedford in Chapter 5 of this volume.

8.5 SOLUTION TECHNIQUES

Similar to the computational methods, several techniques for discretizing the PDEs into FD equations and solution algorithms have been devised. Once given the adopted grid, a spatial definition of the variables is required. Essentially, the spatial finite-difference approximations can be divided into the three classes: backward, central and forward differences, of which the central difference is often preferable as this class has a second-order truncation error on the first derivatives whereas the two others have a first-order accuracy only. Similarly, the time derivatives imply the definition of certain time levels leading to either explicit or implicit schemes. In general, an explicit method will require more computational time compared with implicit methods as the distinction between the particle speed and the celerity becomes more pronounced. It is in this vicinity that the implicit methods are superior to the explicit methods (Abbott and Basco (1989)).

Usually, all prognostic variables in implicit schemes are defined on the same time level and an iterative technique is applied to inverse the set of equations to advance the solution one time step. This inversion may be performed on the entire matrix in one step which, due to the size of the matrix, is an expensive approach. Alternatively, a dimensional decomposition may be applied whereby the inversion is split into three operations according to the three co-ordinate directions. In each operation, only the prognostic variables directly associated with the directions are considered as actively prognostic, whereas the other direction variables are 'locked'. This technique is known as the alternating-direction-implicit (ADI) technique. The successive-over-relaxation (SOR) technique is also widely used. An outline of these techniques has been given by Abbott (1979) and Abbott and Basco (1989).

8.6 THE FREE SURFACE

The numerical treatment of discontinuities such as free surfaces and fronts is essential to the computational algorithms in three-dimensional models, in which the fluid pressure and velocity are the prognostic variables. The position of the free surface does not alter the governing equations as given by Equations (8.9)–(8.11). The free surface is described in a physically correct manner through the kinematic boundary condition, but this gives restrictions on the time step that can be used because of the presence of external gravity waves. This has led to a mode-splitting technique and the introduction of the rigid-lid condition whereby much larger time steps are allowed (*cf.* Simons, 1974; Madala and Piacsek, 1977).

Most three-dimensional models for free surface flows assume hydrostatic pressure, whereby the pressure term in the two remaining momentum equations is replaced by a variation of the free surface. In three-dimensional models in which the full momentum equation in the vertical is solved, the free surface variation is correctly given by the kinematic

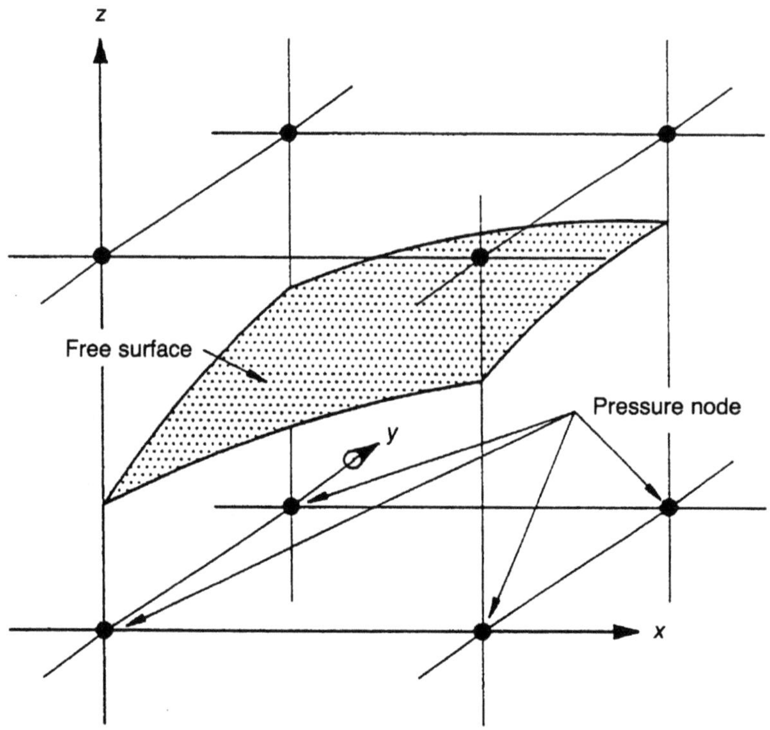

Figure 8.2 Representation of free surface on a rectangular grid. In three-dimensional models where the full momentum equation in the vertical direction is solved a relation between the surface position and the fluid pressure at the uppermost computational point is required.

boundary condition. However, this necessitates a relation between the position of the free surface and the fluid pressure at the uppermost computational point when applied on a rectangular grid (see Figure 8.2). Typically, the hydrostatic pressure assumption is applied in this uppermost region, which will then provide the required relation.

This assumption, of course, limits the applications to problems where the vertical acceleration close to the surface is negligible. Adaptive grids have attractive properties in this respect compared with rectangular grids, but they may reduce the computational speed considerably and, moreover, may introduce difficulties regarding numerical stability and accuracy. It is emphasized that an adaptive grid does not eliminate the need for the relation between the surface variation and the fluid pressure, but it may substantially reduce the size of the region for which this assumption is made.

8.7 ASPECTS OF THREE-DIMENSIONAL MODELLING

Although every flow in nature is three-dimensional, it can often be regarded as nearly-horizontal and two-dimensional models can be applied with excellent results (see Chapter 2 in this volume). There are still, however, a large number of flow problems requiring a numerical three-dimensional model in order to simulate the physical laws concerned.

An important aspect of three-dimensional modelling is the need for accurate boundary data that must be known prior to computation. In practice, however, these are not known or can only be poorly estimated. Inconsistent boundary data will inherently lead to numerical instability (blow-up) in the models. Hence, data gathering and assimilating techniques are aspects that are closely linked to the practical applications of numerical models. Attempts to overcome this problem have been made through the use of boundary generated data from two-dimensional models (*cf.* Rasmussen, 1989) but this technique is not yet generally applicable.

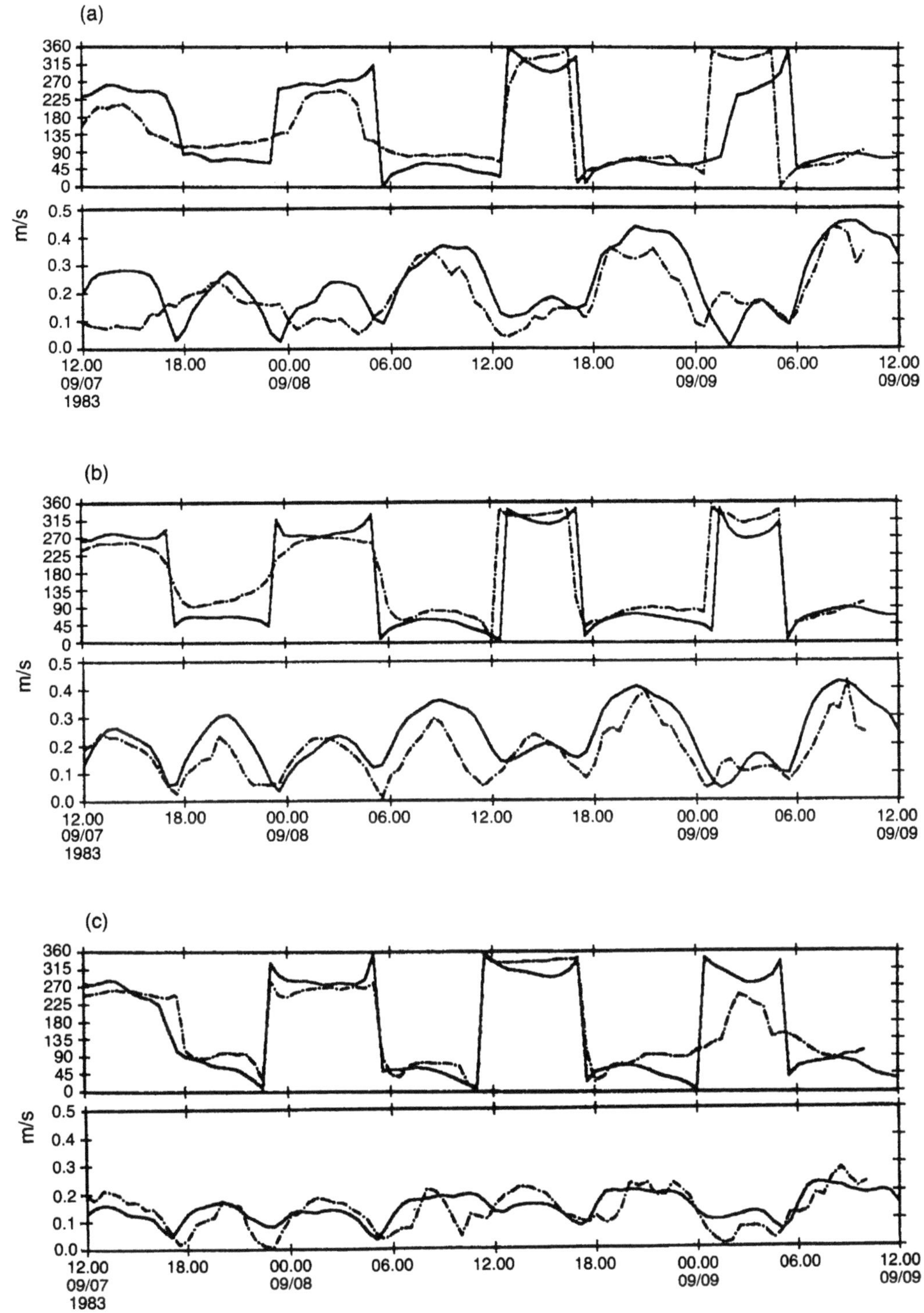

Figure 8.3 Time series of current speeds and directions at location in the North Sea for three levels during a storm in September 1983. (---), measured speeds; (——), calculated speeds. Total water depth ≏ 40 m.

However, three-dimensional models have received much attention recently, and the successful use of these models in related branches such as hydrodynamics, meteorology and aerodynamics will inevitably affect hydraulics. Problems involving the three-dimensional velocity field are becoming more technically and, particularly, economically tractable. An early example of such an application has been described by Daubert and Cahouet (1984).

Boundary-fitted co-ordinates may greatly facilitate such applications, in that the flow can then be accurately resolved with a smaller number of computational points. Estuarial and coastal engineering, however, only rarely involves such complex geometries that conventional rectangular co-ordinate systems are excluded. For the moment, a modelling system using a rectangular grid, and so with a high computational speed and high numerical accuracy, is considered as the best tool for an engineering approach.

REFERENCES

Abbott, M.B. (1979) *Computational Hydraulics: Elements of the Theory of Free Surface Flows*, Pitman, London.

Abbott, M.B. and Basco, D.R. (1989) *Computational Fluid Dynamics: An Introduction For Engineers*, Longman, Harlow, UK.

Bryan, K. and Cox, M.D. (1968) A non-linear model of an ocean driven by wind and differential heating: parts I and II. *Journal of Atmosphere Science*, **25**, 945–78.

Chorin, A.J. (1967) A numerical method for solving incompressible viscous flow problems. *Journal of Computational Physics*, **2**, 12.

Daubert, O. and Cahouet, J. (1984) Approche Numerique de la Houle par les Equations de Navier Stokes. 1[re] Partie, *Annales des Ponts et Chaussees* – 4[e] trimestre, 3–25.

Ferziger, J.H. (1987) Simulation of incompressible turbulent flows. *Journal of Computational Physics*, **69**, 1–48.

Leendertse, J.J., Alexander, R.C. and Liu, S.K. (1973) A three-dimensional model for estuaries and coastal seas: volume I, principles of computation. The Rand Corporation, *Rep. R-1417-OWRR*.

Madala, R.V. and Piacsek, S.A. (1977) A semi-implicit numerical model for baroclinic oceans. *Journal of Computational Physics*, **23**, 167–78.

Markatos, N.C. (1986) The mathematical modelling of turbulent flows. *Applied Mathematical Modelling*, **10**, 190–220.

Patankar, S.V. (1980) *Numerical Heat Transfer And Fluid Flow*, McGraw-Hill, New York.

Patankar, S.V. and Spalding, D.B. (1972) A calculation procedure for heat, mass and momentum transfer in three dimensional parabolic flows. *International Journal of Heat and Mass Transfer*, **15**, 1787.

Rasmussen, E.B. (1989) Ultilization of 2-D results in 3-D modelling – a new approach, in *Proceeding of the International Conference on Hydraulic and Environmental Modelling of Coastal, Estuarine and River Waters*, (eds R.A. Falconer, P. Goodwin and R.G.S. Mathew) Gower, Aldershot, UK.

Richtmyer, R.D. and Morton, K.W. (1967) *Difference Methods for Initial Value Problems*, 2nd edn, Interscience, New York.

Simons, T.J. (1974) Verification of numerical models of Lake Ontario, part I, circulation in spring and early summer. *Journal of Physical Oceanography*, **4**, 507–23.

Vinokur, M. (1989) An analysis of finite-difference and finite-volume formulations of conservation laws. *Journal of Computational Physics*, **81**, 1–52.

QUASI-THREE-DIMENSIONAL MODELLING USING MIXED FINITE-DIFFERENCE AND SPECTRAL MODELS 9

A.M. Davies

As explained in Chapter 8, full three-dimensional modelling has long been plagued with difficulties and, even today, it still requires much care and attention if accurate results are to be obtained while maintaining numerical stability. As a consequence of this situation considerable effort has been expended on a range of models intermediate between the two-dimensional, nearly-horizontal kind and the fully three-dimensional. This chapter is concerned with such models, in which the capabilities of a nearly-horizontal flow model are enhanced by providing a profile of velocity over depth at each grid point in the horizontal.

CONTENTS

9.1 INTRODUCTION

The most common method of solving the three-dimensional hydrodynamic equations describing flow in a sea region is to use finite difference methods in all three spatial dimensions, in the manner described in Chapter 8. As explained there, this method is usually applied over a regular finite difference grid fixed in the horizontal and vertical domain. Using such a grid system it is necessary to approximate lateral coastlines by a series of line elements with corners (Figure 9.1). Similarly the sea-bed is approximated by a staircase of grid boxes, although in some cases an irregular grid is used close to the bed in order to resolve variations in bottom topography, which are known to be important in determining the speed of propagation of waves, more accurately. The application of a fixed grid in the vertical causes some problems at the sea-surface, when account has to be taken of the movement of the free surface through the grid. This problem does not occur when a 'rigid lid' is applied in which the free surface is removed from the problem. However, such a surface boundary condition is only appropriate in deep ocean situations, and is generally not applicable in shallow water where gravity waves are important.

The limitations of using regular grids in the horizontal for resolving near-coastal features, have been recognized for some time, and have lead to the use of irregular grids (Thacker, 1977) or alternative approaches using finite element methods (Werner and Lynch, 1987, 1989; Walters, 1989; Werner, 1987; Lynch and Werner, 1987; Lynch and Officer, 1985). Other methods of refining the grid in near-coastal regions, using transformation techniques (namely boundary fitted co-ordinates) have recently been applied (see Chapter 8 and Spaulding (1984)).

The application of transformation techniques in the vertical, namely 'sigma co-ordinates' which follow the sea-surface and sea-bed has been known for some time, and has been applied by many authors (Davies, 1986; James, 1989, 1990; Noye and Stevens, 1987; Owen, 1980; Spaulding and Isaji, 1987; Lardner and Cekirge, 1988). The principal advantage of this method, as we

Coastal, Estuarial and Harbour Engineer's Reference Book. Edited by M.B. Abbott and W.A. Price. Published in 1993 by Chapman & Hall, 2–6 Boundary Row, London SE1 8HN. ISBN 0 419 15430 2.

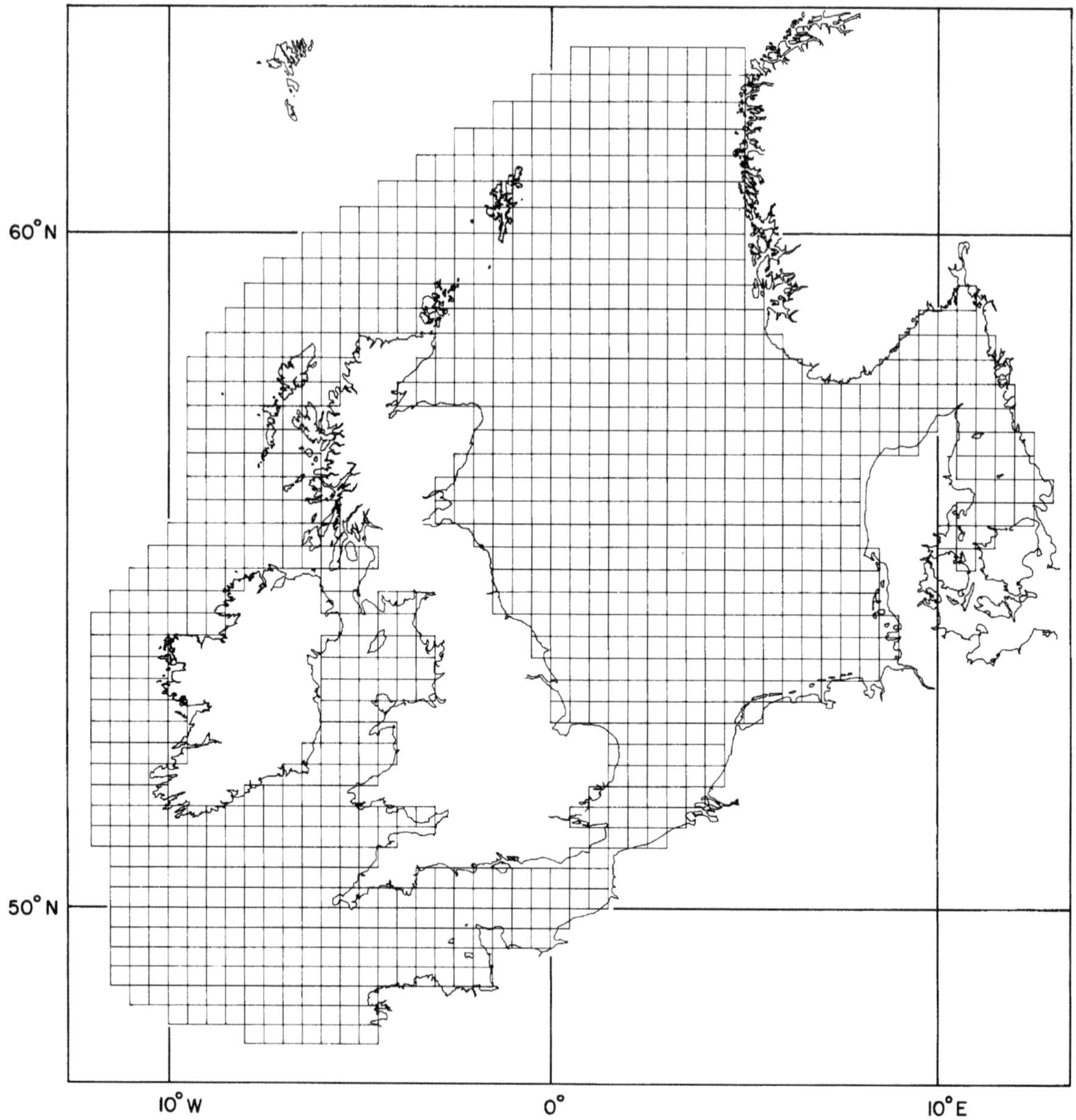

Figure 9.1 Typical finite difference grid used in horizontal.

will demonstrate, is that it can accurately follow the sea-surface and sea-bed, and yield high resolution in these boundary regions.

In this chapter we consider the solution of the hydrodynamic equations using a sigma co-ordinate transformation in the vertical. However, instead of applying a finite difference method through the vertical, the solution using the Galerkin method with various basis functions is considered. Discretization in the horizontal is briefly described using a uniform finite difference grid, although the method developed here, namely a mixed finite difference – Galerkin approach, can be readily applied using horizontally transformed grids, or finite elements in the horizontal domain.

In the last decade, the solution of the three-dimensional equations using a functional expansion through the vertical has become more popular, and a large number of tidal and wind driven circulation problems in homogeneous seas have been solved using this method (Choi, 1984; Davies, 1986, 1987a; Davies and Furnes, 1980; Furnes, 1983; Gordon and Spaulding, 1987; Heaps and Jones, 1981; Owen, 1980; Proctor, 1987; Wolf, 1983).

Solutions using linear elements through the vertical (finite elements) have also been developed. The method has also been extended to stratified regions (Furnes and Mork, 1987; Davies, 1983a,c). Davies (1987b, 1988) has demonstrated that the Galerkin approach can be used to develop 'super two-dimensional models' in which the bed stress in a vertically integrated two-dimensional hydrodynamic model is no longer determined from the depth-mean current (a major deficiency of conventional two-dimensional models, particularly for near-coastal wind driven flow problems). Calculations (Davies, 1987b, 1988) have shown that the bed stress in these 'super two-dimensional models' is in close agreement with that found in full three-dimensional models. Also by using convolution methods, (Jelesnianski, 1970; Forristall, 1974, 1980) current profiles can be extracted from these two-dimensional models which are in good agreement with those obtained using full three-dimensional models (Davies, 1987b, 1988).

In this chapter we briefly review the development of a mixed finite difference – Galerkin type of model. The use of a range of basis functions through the vertical is considered, with particular reference to the computational advantages of a basis set of eigenfunctions of the eddy viscosity profile. The solution of these equations on multiprocessor vector computers such as the CRAY X-MP is also discussed together with their computational advantages (minimization of computer time and memory) for small personal computers. The development of 'super two-dimensional models' and the means of extracting current profiles are also considered.

Recent improvements in the accuracy of the eigenfunction (modal) method for wind-driven problems are subsequently described. Suggestions for the likely future development of new methods of this kind are also discussed.

9.2 SOLUTION OF THE HYDRODYNAMIC EQUATIONS USING A GALERKIN APPROACH IN THE VERTICAL

The Galerkin method for solving partial differential equations, and its associated numerical analysis, is now well established and will not be discussed here. The approach is, in essence, a special case of the method of weighted residuals in which the weighting functions are also the functions used in the expansion. The finite element method in which discontinuous functions (elements) are used to approximate the solution is a common application of the method. For a detailed analysis of these various methods the reader is referred to Finlayson (1972), Strang and Fix (1973) and Gottlieb and Orszag (1977). In this chapter we will, of course, only be concerned with the application of the method rather than a detailed analysis of its numerical properties.

9.2.1 HYDRODYNAMIC EQUATIONS

In order to illustrate the major steps in solving the hydrodynamic equations using the Galerkin approach, we consider, for clarity, the solution using Cartesian co-ordinates. However, in many problems (Davies, 1986), polar co-ordinates have been used to account for the curvature of the earth.

Thus, using Cartesian co-ordinates, including the non-linear advective terms, the working equations in sigma co-ordinates (σ), with

$$\sigma = (a - b)\left(\frac{z + \zeta}{h + \zeta}\right) + b \tag{9.1}$$

are given by

$$\frac{\partial \zeta}{\partial t} + \frac{\partial}{\partial x}\left\{\frac{(h+\zeta)}{(a-b)}\int_b^a u \, d\sigma\right\} + \frac{\partial}{\partial y}\left\{\frac{(h+\zeta)}{(a-b)}\int_b^a v \, d\sigma\right\} = 0 \tag{9.2}$$

$$\frac{\partial u}{\partial t} + u\frac{\partial u}{\partial x} + v\frac{\partial u}{\partial y} + w^*\frac{\partial u}{\partial \sigma} - \gamma v = -g\frac{\partial \zeta}{\partial x} + \left(\frac{a-b}{h+\zeta}\right)^2 \frac{\partial}{\partial \sigma}\left(\mu \frac{\partial u}{\partial \sigma}\right) \tag{9.3}$$

$$\frac{\partial v}{\partial t} + u\frac{\partial v}{\partial x} + v\frac{\partial v}{\partial y} + w^*\frac{\partial v}{\partial \sigma} + \gamma u = -g\frac{\partial \zeta}{\partial y} + \left(\frac{a-b}{h+\zeta}\right)^2 \frac{\partial}{\partial \sigma}\left(\mu \frac{\partial v}{\partial \sigma}\right) \tag{9.4}$$

with

$$w^* = \frac{1}{h+\zeta}\left\{\frac{\partial \zeta}{\partial t}(a - \sigma)\right\} + \frac{1}{h+\zeta}\frac{\partial}{\partial x}\left\{(h+\zeta)\int_\sigma^a u \, d\sigma\right\} + \frac{1}{h+\zeta}\frac{\partial}{\partial y}\left\{(h+\zeta)\int_\sigma^a v \, d\sigma\right\} \tag{9.5}$$

The sigma transformation (Equation (9.1)) maps the interval from the sea-surface ζ to the sea-bed h, which changes with horizontal position and time, onto the constant interval a to b. The constants a and b can then be chosen to correspond to the interval over which the basis set of functions are defined, such as Chebyshev polynomials over the interval +1 to −1, eigenfunctions over 0 to +1.

In Equations (9.1)–(9.5), t denotes time, x, y, z Cartesian co-ordinates, with corresponding velocity components u, v, w^*, with w^* the vertical velocity in sigma co-ordinates. Acceleration due to gravity is denoted by g, with γ the geostrophic coefficient, and μ the vertical eddy viscosity.

Surface boundary conditions in sigma co-ordinates are given by

$$-\rho\left(\mu\frac{\partial u}{\partial\sigma}\right)_b = \frac{(h+\zeta)}{(a-b)}\tau_x^{\,s},$$
$$-\rho\left(\mu\frac{\partial v}{\partial\sigma}\right)_b = \frac{(h+\zeta)}{(a-b)}\tau_y^{\,s} \qquad (9.6)$$

with bed conditions given by

$$-\rho\left(\mu\frac{\partial u}{\partial\sigma}\right)_a = \frac{(h+\zeta)}{(a-b)}\tau_x^{\,h},$$
$$-\rho\left(\mu\frac{\partial v}{\partial\sigma}\right)_b = \frac{(h+\zeta)}{(a-b)}\tau_y^{\,h} \qquad (9.7)$$

In these equations, ρ is the density, $\tau_x^{\,s}$, $\tau_y^{\,s}$ are components of the surface wind stress (zero in tidal problems), and $\tau_x^{\,h}$, $\tau_y^{\,h}$ are components of bed stress.

Relating the bed stress to the bottom current, using a quadratic law, gives

$$\tau_x^{\,h} = K\rho u_h(u_h^2 + v_h^2)^{1/2},$$
$$\tau_y^{\,h} = K\rho v_h(u_h^2 + v_h^2)^{1/2} \qquad (9.8)$$

with K a quadratic bottom friction coefficient, and u_h, v_h current velocities at the sea bed.

An alternative is to apply a no-slip condition at the bed, namely

$$u_h = v_h = 0 \qquad (9.9)$$

9.2.2 SOLUTION USING THE GALERKIN APPROACH

We now consider the major steps in solving Equations (9.2)–(9.5) using the Galerkin approach in the vertical. Thus, expanding the two velocity components in terms of m functions f_r (σ) (the basis functions) through the vertical and coefficients A_r (x, y, t) and B_r (x, y, t), gives

$$u = \sum_{r=1}^{m} A_r f_r(\sigma), \quad v = \sum_{r=1}^{m} B_r f_r(\sigma) \qquad (9.10)$$

Substituting expansions (9.10) into Equations (9.2) and (9.5) gives

$$\frac{\partial\zeta}{\partial t} + \frac{1}{(a-b)}\left(\sum_{r=1}^{m}\frac{\partial}{\partial x}\left\{A_r(h+\zeta)\int_b^a f_r\,\mathrm{d}\sigma\right\} + \frac{\partial}{\partial y}\left\{B_r(h+\zeta)\int_b^a f_r\,\mathrm{d}\sigma\right\}\right) = 0 \qquad (9.11)$$

$$w^* = \frac{1}{h+\zeta}\left\{\frac{\partial\zeta}{\partial t}(a-\sigma) + \frac{1}{h+\zeta}\sum_{r=1}^{m}\left[\frac{\partial}{\partial x}\left\{A_r(h+\zeta)q_r\right\} + \frac{\partial}{\partial y}\left\{B_r(h+\zeta)q_r\right\}\right]\right. \qquad (9.12)$$

where

$$q_r = \int_\sigma^a f_r\,\mathrm{d}\sigma$$

Applying the Galerkin method to Equation (9.3), this equation is multiplied by f_k and integrated from a to b, as above. By integrating the term involving μ, by parts, the boundary conditions (9.6) and (9.7) can be included (see Davies, 1983b, 1987a for details).

Thus, from Equation (9.3), we obtain using expansions (9.10), the Galerkin form of the u equation:

$$\sum_{r=1}^{m}\frac{\partial A_r}{\partial t}\int_b^a f_r f_k\,\mathrm{d}\sigma = \gamma\sum_{r=1}^{m} B_r\int_b^a f_r f_k\,\mathrm{d}\sigma - g\frac{\partial\zeta}{\partial x}\int_b^a f_r\,\mathrm{d}\sigma - \frac{(a-b)f_k(a)\tau_x^{\,h}}{\rho(h+\zeta)} + \frac{(a-b)f_k(b)\tau_x^{\,s}}{\rho(h+\zeta)} - \left(\frac{a-b}{h+\zeta}\right)^2\sum_{r=1}^{m}A_r\int_b^a \mu\frac{\mathrm{d}f_r}{\mathrm{d}\sigma}\frac{\mathrm{d}f_k}{\mathrm{d}\sigma}\mathrm{d}\sigma - S_{uk} \qquad (9.13)$$

where

$$S_{uk} = \int_b^a u\frac{\partial u}{\partial x}f_k\,\mathrm{d}\sigma + \int_b^a v\frac{\partial u}{\partial y}f_k\,\mathrm{d}\sigma + \int_b^a w^*\frac{\partial u}{\partial\sigma}f_k\,\mathrm{d}\sigma \qquad (9.14)$$

is the non-linear advective term. An equation similar to Equation (9.13) can be derived for the v component of velocity.

The solution of these equations in the horizontal can be accomplished on a uniform or transformed finite difference grid, or on a finite element lattice. We will not discuss this in detail here, but rather concentrate on the solution in the vertical and the choice of basis functions in expansions (9.10).

9.2.3 CHOICE OF BASIS FUNCTIONS

In theory the choice of basis function is arbitrary, provided that they satisfy essential boundary condi-

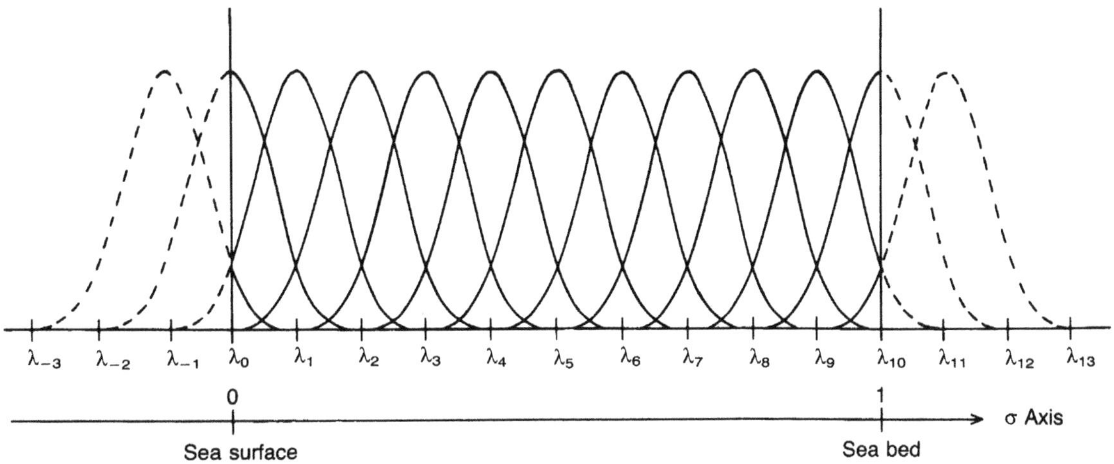

Figure 9.2 Vertical distribution of fourth order B-splines and associated knots.

tions, such as $u_h = v_h = 0$. As the number of functions in expansions (9.10) increases, the expansion rapidly converges to the true solution. Also spectral methods are known to converge more rapidly than finite difference methods.

In practice computational and physical considerations often influence the choice of basis functions, which can take two distinct forms, namely continuous functions (spectral methods) or discontinuous functions (finite element methods). We will briefly consider both methods here.

Finite element methods in the vertical

Davies (1977) used fourth-order B-splines as basis functions in solving the problem of wind induced flow in homogeneous (1977) and stratified (1983a) sea regions. These functions are piecewise polynomials which are only non-zero over a finite interval of space (i.e. a finite element). The positions at which the spline changes from zero to non-zero are termed knots, and are denoted by λ_r (see Figure 9.2), and a fourth-order B-spline is non-zero only over the interval $\lambda_{r-4} \leqslant \sigma \leqslant \lambda_r$. The positioning of the knots is arbitrary and in practice a higher knot spacing would be applied in regions where the current profile changes most rapidly. For example, for wind induced flow in a sea region, with a no-slip condition at the sea-bed, high shear surface and bed layers would arise, requiring a finer knot spacing in these regions if this shear was not to be artifically reduced by poor resolution.

Expressing the linear part (i.e. set $S_{uk} = 0$) of (9.13) in matrix form, gives

$$[\mathbf{M}]\frac{\mathrm{d}\mathbf{A}}{\mathrm{d}t} = \gamma[\mathbf{M}](\mathbf{B}) - \left(\frac{a-b}{(h+\zeta)}\right)^2[\mathbf{K}](\mathbf{A}) + (\mathbf{f}) \quad (9.15)$$

where **[M]** is an $m \times m$ matrix of integrals $\int_b^a f_r f_k \mathrm{d}\sigma$, with **[K]** an $m \times m$ matrix of integrals $\int_b^a \mu(\mathrm{d}f_r/\mathrm{d}\sigma)(\mathrm{d}f_r/\mathrm{d}\sigma)\mathrm{d}\sigma$, and **(A)**, **(B)**, are vectors of length m, of coefficients A_r and B_r. The vector (f) of length m containing the forcing and retarding terms is shown in Equation (9.13).

It is evident from Figure 9.2, that for fourth-order B-splines, their piecewise nature means that the matrices **(M)** and **(K)** in Equation (9.15) are diagonal, with a bandwidth of nine. Consequently the number of arithmetic operations per time step associated with the matrix vector multiplications involving **[M]** and **[K]** required to integrate Equation (9.15) is of order ($3 \times 9m$).

The algorithm used by Davies (1977) to compute these B-splines could compute splines of arbitrary order. Considering 2nd-order splines (linear elements, see Figure 9.3), it is evident that the matrices in Equation (9.15) are now tridiagonal and hence the computational overhead is reduced, of order ($3 \times 3m$). Obviously as the order of the spline is increased, the bandwidth of the matrix increases and the computational effort rises. It is, however, important to note that, in general, the higher the order of the spline, the greater the accuracy of the solution, and hence for a given accuracy the number of terms m in the expansion will be less for higher-order splines. The computational effort associated with various methods will be discussed in more detail later in this chapter. It is interesting to note the connection between the

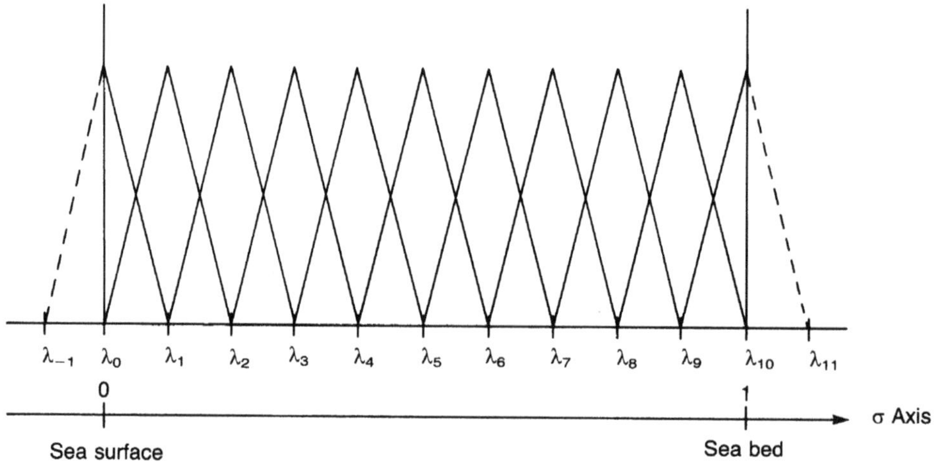

Figure 9.3 Vertical distribution of linear elements and associated knots.

Galerkin approach developed here, with low-order elements, and the conventional finite difference method, which also yields a tridiagonal matrix **K** when a central finite difference method is used to approximate the term involving eddy viscosity. In fact it can be readily shown that by using the method of weighted residuals with various expansion and weighting functions of various orders of polynomial, it is possible to generate a range of finite difference methods, with the order of the finite differencing increasing with the order of the polynomial. A detailed discussion of this is beyond the scope of this chapter.

Spectral methods in the vertical

In this section we briefly consider the computational effort involved in using continuous functions as a basis set within the Galerkin approach (a spectral method). Davies and Owen (1979), showed that for wind driven problems, with high shear surface and bottom boundary layers, an expansion of Chebyshev polynomials, because of their high rates of convergence in the boundary layers, has an optimal level of accuracy. However, the nature of these functions was found to be such that the matrices [**M**] and [**K**] in Equation (9.15) were full; consequently the computational effort was of order ($3m^2$). By using orthogonal functions (e.g. Legendre polynomials, see Davies and Owen, 1979), the matrix **M** is diagonal, although matrix **K** is in general full.

Modal models in the vertical

An alternative to using arbitrary continuous functions through the vertical is to choose the basis set to be eigenfunctions of the eddy viscosity profile (a modal approach). In physically realistic simulation models, the eddy viscosity is normally related to the flow field, using empirical equations of a form (Bowden 1978) suggested by shallow water measurements: $\mu = K_1 hu$ or $\mu = K_2 u^2/\sigma_1$ with σ_1 a characteristic frequency, or $\mu = K_3 \Delta u$, with $\Delta = CU_*/\gamma$, with U_* frictional velocity and C of order 0.2 to 0.4, a fixed constant.

In order to allow for such time and space variations of eddy viscosity, it is advantageous to write

$$\mu = \alpha(x, y, t)\psi(\sigma) \tag{9.16}$$

with $\psi(\sigma)$ a fixed function, representing the vertical variation of μ.

The basis functions f_r are then chosen as eigenfunctions of an eigenvalue problem (Davies, 1983b, 1985b), namely

$$\frac{d}{d\sigma}\left[\psi \frac{df}{d\sigma}\right] = -\varepsilon f \tag{9.17}$$

with ε_r the eigenvalues. The boundary conditions on Equation (9.17) can vary, but for tidal problems, where there is a zero surface stress, taking $df/d\sigma = 0$ at the surface is appropriate, with $df/d\sigma = 0$ in the case of quadratic bottom friction or $f_r = 0$ at the sea-bed for a no-slip bottom boundary condition.

The orthogonality property of the eigenfunctions, and the fact that they are eigenfunctions of the eddy viscosity profile, means that the matrices [**M**] and [**K**] in Equation (9.15) are diagonal, leading to a significant reduction in computational effort (Davies and Stephens, 1983).

Also by choosing surface and bed boundary conditions, such that

$$\left.\frac{df_r}{d\sigma}\right|_0 = \left.\frac{df_r}{d\sigma}\right|_1 = 0 \tag{9.18}$$

(i.e. a and b in the transformation are taken as 0 and 1), it is possible to make f_1 = constant with $\varepsilon_1 = 0$, and

$$\int_0^1 f_r \, d\sigma = 0, \quad r = 2,3, \ldots m \tag{9.19}$$

Consequently the gradient term only forces the first mode and does not influence the other modes. This together with the orthogonality property of the eigenfunctions enables the equations (omitting the advective terms for clarity) to be written as

$$\frac{\partial \zeta}{\partial t} + \frac{\partial}{\partial x}\left\{a_1(h+\zeta)\Phi_1 \int_0^1 f_1 d\sigma\right\} + \frac{\partial}{\partial y}\left\{b_1(h+\zeta)\Phi_1 \int_0^1 f_1 d\sigma\right\} = 0 \tag{9.20}$$

$$\frac{\partial a_1}{\partial t} = \gamma b_1 - g\frac{\partial \zeta}{\partial x} - \frac{\tau_x^h}{\rho(h+\zeta)} + \frac{\tau_x^s}{\rho(h+\zeta)} \tag{9.21}$$

$$\frac{\partial b_1}{\partial t} = -\gamma a_1 - g\frac{\partial \zeta}{\partial y} - \frac{\tau_y^h}{\rho(h+\zeta)} + \frac{\tau_y^s}{\rho(h+\zeta)} \tag{9.22}$$

$$\frac{\partial a_k}{\partial t} = \gamma b_k - \frac{\tau_x^h}{\rho(h+\zeta)} f_k(1) + \frac{\tau_x^s}{\rho(h+\zeta)} - \frac{\alpha \varepsilon_k}{(h+\zeta)^2} a_k \tag{9.23}$$

where

$$k = 2, \ldots, m$$

$$\frac{\partial b_k}{\partial t} = -\gamma a_k - \frac{\tau_y^h}{\rho(h+\zeta)} f_k(1) + \frac{\tau_y^s}{\rho(h+\zeta)} - \frac{\alpha \varepsilon_k}{(h+\zeta)^2} b_k \tag{9.24}$$

where

$$k = 2, \ldots, m$$

In deriving Equations (9.20)–(9.24), we have normalized the modes, such that $f_r(0) = 1$ for all r, and for reasons of computational economy we have expressed expansions (9.10) in the form

$$u = \sum_{r=1}^{m} A_r f_r(\sigma) = \sum_{r=1}^{m} a_r \Phi_r f_r(\sigma),$$
$$v = \sum_{r=1}^{m} B_r f_r(\sigma) = \sum_{r=1}^{m} b_r \Phi_r f_r(\sigma) \tag{9.25}$$

with

$$\Phi_r = 1 \Big/ \int_0^1 f_r f_r \, d\sigma \tag{9.26}$$

In the absence of the advective terms, Equations (9.20)–(9.24) are a set of semi-coupled equations, with coupling coming through the bottom stress term, Equation (9.8), which depends upon the sum of all modes.

In essence, Equations (9.20)–(9.22) correspond to the conventional vertically integrated hydrodynamic equations which can be integrated forward in time using either an implicit, or semi-implicit scheme such as ADI (as was described in Chapter 8), or an explicit method, restricted in its size of time step by the CFL condition.

If the viscosity term in Equations (9.23) and (9.24) is evaluated at the lower time step, then the time step τ is restricted to

$$\tau \leqslant \frac{(h+\zeta)^2}{2\alpha\varepsilon_k} \tag{9.27}$$

which can be of the order of a few seconds.

However, by centring the viscosity term such that

$$\left(\frac{a_k^{t+\tau} - a_k^t}{\tau}\right) = -\frac{\alpha\varepsilon_k}{(h+\zeta)^2}\left(\frac{a_k^{t+\tau} + a_k^t}{2}\right) \tag{9.28}$$

an unconditionally stable scheme can be obtained (Davies 1987a). By using Equation (9.28), a forward-time-stepping method having significant computational advantages on multiprocessor, parallel-vector supercomputers can be developed.

9.3 THE FORMULATION OF 'SUPER TWO-DIMENSIONAL MODELS' AND CURRENT PROFILE EXTRACTION FROM VERTICALLY INTEGRATED MODELS

9.3.1 GENERALIZED TIME VARYING FORCING

We now briefly consider how the modal approach developed previously can be used to enhance the bed stress formulation in two-dimensional vertically integrated models, and to extract current profiles from these models.

Using complex notation, the linear hydrodynamic equations at a point (the linear form of Equations (9.3) and (9.4)), can be written as

$$\frac{\partial q}{\partial t} + i\gamma q = -S + \frac{1}{h^2}\frac{\partial}{\partial \sigma}\left(\mu \frac{\partial q}{\partial \sigma}\right) \tag{9.29}$$

where $q = u + iv$, $S = g((\partial\zeta/\partial x) + i(\partial\zeta/\partial y))$. A normalized linear sigma co-ordinate $\sigma = z/h$ has been introduced with $i = \sqrt{-1}$.

Expressing $q = \bar{q} + q'$, with $\bar{q}$ the depth mean current and q' the deviation, i.e. the current profile (and substituting μ from Equation (9.16)), Equation (9.29) can be written as two separate equations:

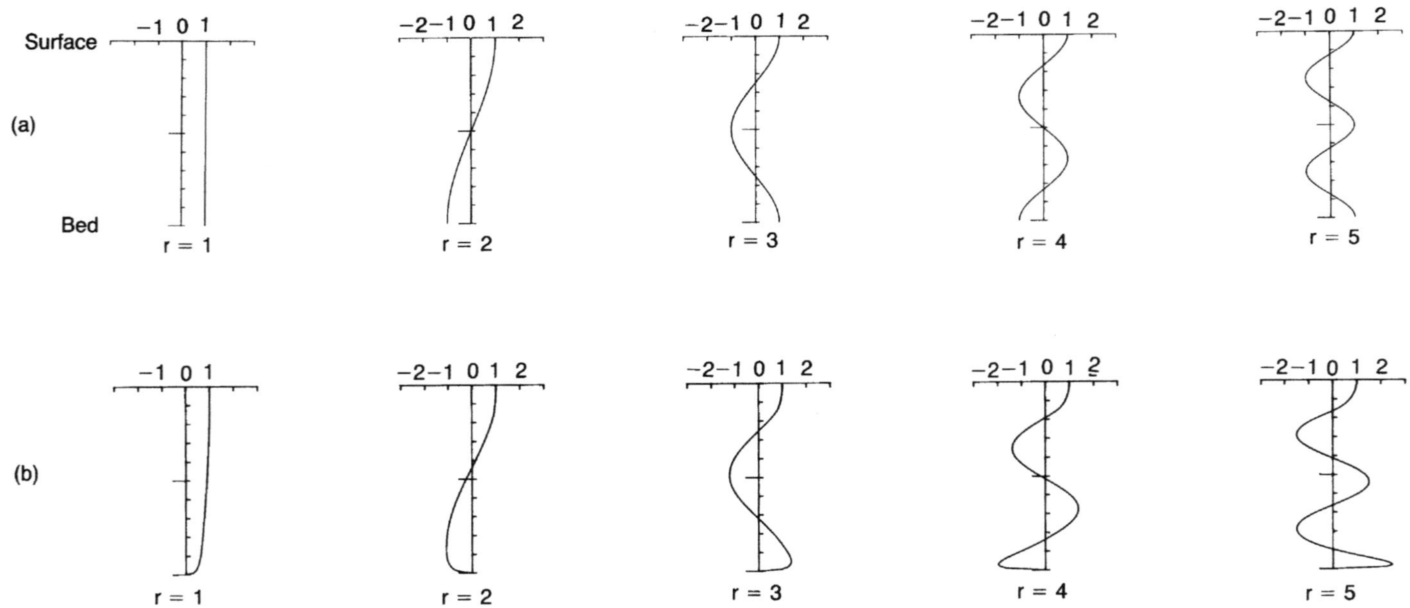

Figure 9.4 Profile of first five modes (a) using a slip bottom boundary condition; (b) a no-slip condition at the sea-bed.

$$\frac{\partial \bar{q}}{\partial t} + i\gamma\bar{q} = -S + \frac{F_s}{\rho h} - \frac{F_B}{\rho h} \quad (9.30)$$

$$\frac{\partial q'}{\partial t} + i\gamma q' = \frac{\alpha}{h^2}\frac{\partial}{\partial \sigma}\left(\psi \frac{\partial q'}{\partial \sigma}\right) - \frac{F_s}{\rho h} + \frac{F_B}{\rho h} \quad (9.31)$$

In Equations (9.30) and (9.31), the surface and bed stresses F_s and F_B are given by

$$F_s = \tau_x{}^s + i\tau_y{}^s, \quad F_B = \tau_x{}^h + i\tau_y{}^h \quad (9.32)$$

It is important to note that Equation (9.30) only gives the depth mean current, with Equation (9.31) giving no net transport, but only determining the current's variation through the vertical.

Considering the solution of Equation (9.31) using the Galerkin method through the vertical, in order to compute the current's profile and the associated bed stress, we express

$$q' = \sum_{r=2}^{m} A_r f_r(\sigma) \quad (9.33)$$

with the f_r taken as the modes of the eddy viscosity profile, omitting the first mode, which does not contribute to the current profile but only determines the depth-mean flow (see Figure 9.4, profiles (a)). Following the Galerkin method presented earlier in this chapter, multiplying Equation (9.31) by $f_k(\sigma)$, and integrating, then substituting in expansion (9.33), gives

$$\frac{dA_r}{dt} = -\left(i\gamma + \frac{\alpha\varepsilon_r}{h^2}\right)A_r + \frac{F_s}{\rho h}(f_r(0)\Phi_r - \bar{f}_r\Phi_r)$$
$$- \frac{F_B}{\rho h}(f_r(1)\Phi_r - \bar{f}_r\Phi_r) \quad (9.34)$$

where

$$\bar{f}_r = \int_0^1 f_r \, d\sigma$$

Considering now the solution of the general form of Equation (9.34), namely

$$\frac{dA}{dt} = -(i\gamma + b)A + cF \quad (9.35)$$

with γ, b and c arbitrary constants and F a time varying forcing. The solution of Equation (9.35) can be expressed as a convolution integral (Jelesnianski, 1970):

$$A = c\int_0^t F(t - \tau)e^{-(i\gamma+b)t}d\tau \quad (9.36)$$

Equation (9.36) can be readily solved using a simple recursion formula to progress from the (M) to the $(M + 1)$ time level:

$$A^{(M+1)} = \alpha^{(1)}A^{(M)} + \alpha^{(2)}F^{(M)} + \alpha^{(3)}F^{(M+1)} \quad (9.37)$$

with coefficients $\alpha^{(1)}$, $\alpha^{(2)}$ and $\alpha^{(3)}$ time-invariant functions of the parameters c, γ and b (see Davies 1987b, 1988 for details).

By solving Equation (9.31) using this convolution method, the current q_h at the sea-bed can be readily determined by adding the depth-mean current to expansion (9.33) computed at the bed:

$$q_h = \bar{q} + \sum_{\tau=2}^{m} A_r f_r(1) \quad (9.38)$$

Having computed q_h from (9.38), the bed stress components $\tau_x{}^h$ and $\tau_y{}^h$ can be obtained using Equation (9.8), or a linear friction law if preferred.

Using this approach the bed stress used in the two-dimensional model (Equation 9.30) is not determined from the depth mean current, a major deficiency of vertically integrated two-dimensional models, but is computed from the bottom current. Davies (1987b, 1988) found that by using four terms in Equation (9.33) the bed stress computed in a 'super two-dimensional model' was equivalent to that determined with a full three-dimensional model. The computational overhead introduced through solving the recurrence relationship (Equation (9.37)) is small, and the computer memory requirement is significantly less than that required for a full three-dimensional model. Also the two-dimensional, vertically integrated model need not be restricted to the linear form given in Equation (9.30), but can include the vertically integrated, non-linear advective terms. The inclusion of the recursion equation (9.37) into an established fully non-linear, two-dimensional model is particularly easy and has recently been performed for an existing fine resolution, non-linear model of the Celtic Sea (Davies and Jones, 1993), with a significant improvement in bed-stress representation.

Davies (1987b) has shown that by saving a time history of elevation gradient, bed stress and surface stress at points in the horizontal where current profiles were required, these profiles could be extracted from the two-dimensional model, and were in excellent agreement with those computed using a full three-dimensional model. A simplified version of this approach was used by Davies and Flather (1987) to derive current profiles for 50-year extreme currents, for a range of eddy viscosity distributions.

9.3.2 OSCILLATORY FORCING AT A SINGLE FREQUENCY (TIDAL MOTION)

By considering oscillatory motion (Davies 1985a) at a single frequency ω, q and S can be written as

$$q = R_+e^{i\omega t} + R_-e^{-i\omega t} \quad (9.39)$$

$$S = S_+e^{i\omega t} + S_-e^{-i\omega t} \quad (9.40)$$

with R_+ the amplitude of the anti-clockwise rotating component, and R_- that for the clockwise component.

Substituting Equations (9.39) and (9.40) into (9.29) and writing it in terms of anti-clockwise and clockwise components, gives

$$i(\gamma + \omega)R_+ = \frac{\alpha}{h^2}\frac{\partial}{\partial\sigma}\left(\psi\frac{\partial R_+}{\partial\sigma}\right) - S_+ \qquad (9.41)$$

$$i(\gamma - \omega)R_- = \frac{\alpha}{h^2}\frac{\partial}{\partial\sigma}\left(\psi\frac{\partial R_-}{\partial\sigma}\right) - S_- \qquad (9.42)$$

It is important to note that by considering a single frequency, time dependence has been removed, giving two time-independent Equations (9.41) and (9.42).

The solution of these equations can now be accomplished using the Galerkin approach, thus expanding R_+ and R_- in terms of the modes f_r, and coefficients a_r, b_r, gives

$$R_+ = \sum_{r=1}^{m} a_r\Phi_r f_r, \quad R_- = \sum_{r=1}^{m} b_r\Phi_r f_r \qquad (9.43)$$

Substituting Equation (9.43) into (9.41) and (9.42), gives

$$\left[i(\gamma + \omega) + \frac{\partial}{h^2}\varepsilon_r\right]a_r = -S_+\bar{f}_r \qquad (9.44)$$

$$\left[i(\gamma - \omega) + \frac{\partial}{h^2}\varepsilon_r\right]b_r = -S_-\bar{f}_r \qquad (9.45)$$

with $\Phi_r = 1/\int_0^1 f_r^2 d\sigma$ and $\bar{f}_r = \int_0^1 f_r d\sigma$.

In Equations (9.44) and (9.45), S_+ and S_- are the rotary components of the gradients of tidal elevation, which can be readily obtained from a co-tidal chart computed with a standard two-dimensional vertically integrated model. Having computed these input forcing terms, Equations (9.44) and (9.45) can be readily solved to yield coefficients a_r, b_r and hence tidal-current profiles.

Davies (1990a, 1991a,b) successfully applied this method with a basis set of functions satisfying a no-slip bottom boundary condition (see Figure 9.4, profiles (b)). He found that using the Galerkin approach it was possible to generate a basis set of modes that were highly sheared in the bottom boundary layer (see profiles (b) in Figure 9.4). The highly sheared nature of these modes meant that a small number of them (of order five to ten) could accurately resolve the current profile, compared to the use of the order of 50 grid boxes on a logarithmically transformed grid (Davies, 1991a).

9.3.3 COMPUTATIONAL ASPECTS

The method of enhancing the bed stress in a conventional vertically integrated two-dimensional model yielding a 'super two-dimensional model', with the associated ability to extract current profiles at selected points (Davies, 1987b, 1988) is ideally suited to applications where computer time and main memory are limited. Also the method of deriving tidal current profiles (Davies, 1990a,b) from co-tidal charts determined using conventional two-dimensional models is computationally very fast (of the order of two to three minutes for a thousand profiles on a computer having a processing speed of 3 mips (million instructions per second) a typical mini-computer speed). However, as was pointed out by Davies (1990b), it is important to accurately compute and store the elevation gradient determined by these models.

9.4 METHODS OF IMPROVING THE RATE OF CONVERGENCE OF THE MODAL METHOD FOR WIND-DRIVEN FLOWS

In the previous sections we have clearly shown the computational advantages, that appear even more emphatically on multi-processors, of using the Galerkin method with an eigenfunction (modal) basis set, and how this method can be used to derive 'super' two-dimensional models. Although the 'classical' method of determining the modes (Heaps, 1972), subject to a zero derivative boundary condition at the sea-surface (Equation (9.18)), is ideal and highly accurate for tidal and wind wave problems (Davies, 1991a), it has been shown (Davies, 1983b), that this leads to a slow rate of convergence of the modes in the near-surface layer for wind driven problems. This slow rate of convergence gives an appreciable error (of the order of 20%) in the surface wind driven current when a small number (of order ten) modes is used to solve the hydrodynamic equations with an eddy viscosity of the order of $0.0130\,m^2\,s^{-1}$. The reason for this is that for wind driven problems the surface boundary condition is one of non-zero surface stress, with an associated highly sheared (for the case of low surface viscosity) surface layer. Davies and Owen (1979) showed that an accurate solution could be obtained using Chebyshev or Legendre polynomials, although the equations were coupled and consequently would not be so readily amenable to parallel solution on multiprocessing computers. The solution of the eigenvalue problem subject to a derivative surface boundary condition (Davies, 1983c), a 'stressed' modal approach, did improve the rate of convergence of the modes. However, this method also coupled the equations and hence is not so attractive for solution on a multiprocessor.

Recently (Davies, 1991c) a mixed basis set approach has been developed using a fixed function ψ (an 'enhancement' function) and an expansion of 'classical' modes computed subject to boundary conditions (9.18) or satisfying exactly a linear slip condition (Davies, 1991c). This 'enhanced' eigenfunction approach gives an uncoupled set of equations, ideal for parallel solution, and leads to an accurate wind driven solution, even at the sea surface, with the order of six modes in the vertical.

9.5 CONCLUSIONS

It is clear from the mathematical development presented in this chapter that the Galerkin method with a basis set of modes is a computationally efficient alternative to a finite difference representation in the vertical. Also, depending upon the basis functions and test functions used, the method can be shown to be equivalent to a finite element or a finite difference scheme through the vertical. The approach can be readily applied with a regular or irregular finite difference or finite element method in the horizontal.

The computational advantages of the Galerkin-modal method on multi-processing vector computers is clearly evident, with an optimal design appearing to be one in which vectorization is accomplished in the horizontal domain, with each mode being integrated in parallel. The high rate of convergence of the modes in shallow water also means that the method is computationally efficient in terms of both computer time and memory, making it ideally suitable for the smaller range of computers (e.g. personal computers).

The development of enhanced bed stress two-dimensional models using the convolution approach, based upon a modal expansion, with the ability to recover current profiles, clearly has major computational advantages in engineering problems where cost-effective means of achieving a solution are required. Also the means to determine tidal current profiles using computer power available on personal computers from co-tidal charts computed using 'conventional' two-dimensional models is particularly valuable.

The ability of the modes to include the high-shear bottom boundary layer within their vertical structure makes them ideally suitable for engineering problems, such as sediment transport where a detailed description of the physics of the flow in the bottom boundary layer is required. At present a fine horizontal grid, enhanced bed stress, Celtic Sea model, including a wave friction effect is being used to examine the physics of the near-bed layer. A turbulence energy model is also being used to determine an appropriate viscosity profile and time variation to use in a modal model. Initial results (Davies, 1991b) suggest that a modal model designed in this way contains the majority of the physics included within the turbulence model, with a very significant reduction in computational time (of the order of 100), compared to full three-dimensional turbulence energy models (Davies and Jones, 1990).

From developments over the last decade, it appears clear that modal models will continue to be used as alternatives to finite difference methods. From an engineering viewpoint, the enhanced-bed-stress, vertically integrated models with profile recovery may prove to be a useful alternative to a full three-dimensional model. The accuracy of the 'enhanced' three-dimensional modal models in determining surface currents makes them particularly attractive for engineering problems involving surface currents (e.g. oil spill problems) where computational costs must be minimized.

ACKNOWLEDGEMENTS

The author is indebted to Mr R.A. Smith for preparing the diagrams and Mrs J. Hardcastle for typing the text.

REFERENCES

Bowden, K.F. (1978) Physical problems of the benthic boundary layer. *Geophysical Survey*, **3**, 255–96.

Choi, B.H. (1984) A three-dimensional model of the East China sea, in *Ocean Hydrodynamics of the Japan and East China Seas*, Elsevier, Amsterdam, pp. 209–24.

Davies, A.M. (1977) The numerical solution of the three-dimensional hydrodynamic equations using a B-spline representation of the vertical current profile, in *Bottom Turbulence*. Proceedings of the 8th Liege Colloquium on Ocean Hydrodynamics, (ed. J.C.J. Nihoul) Elsevier, New York, pp. 1–25.

Davies, A.M. (1983a) Numerical modelling of stratified flow: a spectral approach. *Continental Shelf Research*, **2**, 275–300.

Davies, A.M. (1983b) Formulation of a linear three-dimensional hydrodynamic sea model using a Galerkin-eigenfunction method. *International Journal for Numerical Methods in Fluids*, **3**, 33–60.

Davies, A.M. (1983c) Application of a Galerkin-eigenfunction method to computing currents in homogeneous and stratified seas, in *Numerical Methods for Fluid Dynamics*, (eds K.W. Morton and M.J. Baines) Academic Press, New York, pp. 287–301.

Davies, A.M. (1985a) On determining current profiles in oscillatory flows. *Applied Mathematical Modelling*, **9**, 419–28.

Davies, A.M. (1985b) A three-dimensional modal model of wind induced flow in a sea region. *Progress in Oceanography*, **15**, 71–128.

Davies, A.M. (1986) A three-dimensional model of the north west European Continental Shelf, with application to the M_4 tide. *Journal of Physical Oceanography*, **16**, 797–813.

Davies, A.M. (1987a) Spectral models in Continental Shelf sea oceanography, in *Three-Dimensional Coastal Ocean Models*, (ed. N.S. Heaps) American Geophysical Union, Washington DC, pp. 71–106 .

Davies, A.M. (1987b) On extracting current profiles from vertically integrated numerical models. *Coastal Engineering*, **11**, 445–77.

Davies, A.M. (1988) On formulating two-dimensional vertically integrated hydrodynamic numerical models with an enhanced representation of bed stress. *Journal of Geophysical Research*, **93**, 1241–63.

Davies, A.M. (1991a) On the accuracy of finite difference and spectral methods for computing tidal and wind wave current profiles. *International Journal for Numerical Methods in Fluids*, **12**, 101–24.

Davies, A.M. (1990) On extracting tidal current profiles from vertically integrated two dimensional hydrodynamic tidal models. *Journal of Geophysical Research*, **95**, 18,317–18,342.

Davies, A.M. (1991) On using turbulence energy models to develop spectral viscosity models. *Continental Shelf Research*, **11**, 1313–53.

Davies, A.M. (1990) On the importance of time varying eddy viscosity in generating higher tidal harmonics. *Journal of Geophysical Research*, **95**, 20,287–20,312.

Davies, A.M. (1991) Solution of the three dimensional hydrodynamic equations using an enhanced eigenfunction approach. *International Journal for Numerical Methods in Fluids*, **13**, 235–50.

Davies, A.M. and Jones, J.E. (1993) On improving the bed stress formulation in storm surge models. *Journal of Geophysical Research*, in press.

Davies, A.M. and Owen, A. (1979) Three-dimensional numerical sea model using the Galerkin method with a polynomial basis set. *Applied Mathematical Modelling*, **3**, 421–8.

Davies, A.M. and Furnes, G.K. (1980) Observed and computed M_2 tidal currents in the North Sea. *Journal of Physical Oceanography*, **10**, 237–57.

Davies, A.M. and Stephens, C.V. (1983) Comparison of the finite difference and Galerkin methods as applied to the solution of the hydrodynamic equations. *Applied Mathematical Modelling*, **7**, 226–40.

Davies, A.M. and Flather, R.A. (1987) On computing extreme meteorologically induced currents, with application to the north-west European Continental Shelf. *Contintental Shelf Research*, **7**, 643–83.

Davies, A.M. and Jones, J.E. (1990) Application of a three dimensional turbulence energy model to the determination of tidal currents on the Northwest European Continental shelf. *Journal of Geophysical Research*, **95**, 18,143–18,162.

Finlayson, B.A. (1972) *The Method of Weighted Residuals and Variational Principles*, Academic Press, New York.

Forristall, G.Z. (1974) Three-dimensional structure of storm-generated current, *Journal of Geophysical Research*, **79**, 2721–29.

Forristall, G.Z. (1980) A two-layer model for hurricane-driven currents on an irregular grid, *Journal of Physical Oceanography*, **9**, 1417–38.

Furnes, G.K. (1983) A three dimensional numerical sea model with eddy viscosity varying piecewise linearly in the vertical. *Continental Shelf Research*, **2**, 231–42.

Furnes, G.K. and Mork, M. (1987) Formulation of a continuously stratified sea model with three-dimensional representation of the upper layer. *Coastal Engineering*, **11**, 415–45.

Gordon, R.B. and Spaulding, M.L. (1987) Numerical simulations of the tidal- and wind-driven circulation in Narragansett Bay. *Estuarine, Coastal and Shelf Science*, **24**, 611–36.

Gottlieb, D. and Orszag, S. (1977) A numerical analysis of spectral methods. Society for Industrial and Applied Mathematics, Philadelphia, *NSF-CBMS Monograph*, 26.

Heaps, N.S. (1972) On the numerical solution of the three dimensional hydrodynamical equations for tides and storm surges. *Memoires de la Société Royal des Sciences de Liege Ser 6*, **2**, 143–80.

Heaps, N.S. and Jones, J.E. (1981) Three-dimensional model for tides and surges with vertical eddy viscosity prescribed in two layers – II. Irish Sea with bed friction layer. *Geophysical Journal of the Royal Astronomical Society*, **64**, 303–20.

James, I.D. (1989) A three-dimensional model of circulation in a frontal region of the North Sea. *Deutsche Hydrographische Zeitschrift*, **42**, 231–47.

James, I.D. (1990) Numerical modelling of density-driven circulation in shelf seas, in *Modelling Marine Systems, Vol. 2*, (ed. A.M. Davies), CRC Press, Boca Raton, FL, USA, pp. 345–372.

Jelesnianski, C.P. (1970) Bottom stress time history in linearized equations of motion for storm surges, *Monthly Weather Review*, **98**, 462–78.

Lardner, R.W. and Cekirge, H.M. (1988) A new algorithm for three-dimensional tidal and storm surge computations. *Applied Mathematical Modelling*, **12**, 471–81.

Lynch, D.R. and Officer, C.B. (1985) Analytic test cases for three-dimensional hydrodynamic models. *International Journal for Numerical Methods in Fluids*, **5**, 529–43.

Lynch, D.R. and Werner, F.E. (1987) Three-dimensional hydrodynamics on finite elements: part 1: linearised harmonic mode. *International Journal for Numerical Methods in Fluids*, **7**, 871–909.

Noye, J. and Stevens, M. (1987) A three-dimensional model of tidal propagation using transformations and variable grids, in *Three-Dimensional Coastal Ocean Models*, (ed. N.S. Heaps) American Geophysical Union, Washington DC, pp. 41–70.

Owen, A. (1980) A three-dimensional model of the Bristol Channel. *Journal of Physical Oceanography*, **10**, 1290–302.

Proctor, R. (1987) A three-dimensional numerical model of the eastern Irish Sea, in *Numerical Modelling: Applications to Marine Systems*, (ed. J. Noye) North Holland, Amsterdam, pp. 25–45.

Spaulding, M.L. (1984) A vertically averaged circulation model using boundary fitted coordinates. *Journal of Physical Oceanography*, **14**, 973–82.

Spaulding, M.L. and Isaji, T. (1987) Three dimensional continental shelf hydrodynamic model including wave current interaction, in *Three-dimensional models of marine and estuarine dynamics*, (eds J.C.J. Nihoul and B.M. Jamart) Elsevier, Amsterdam, pp. 405–26.

Strang, G. and Fix, G.J. (1973) *An Analysis of the Finite Element Method*, Prentice Hall, Englewood Cliffs NJ.

Thacker, W.C. (1977) Irregular grid finite-difference techniques: Simulations of oscillations in shallow circular basins. *Journal of Physical Oceanography*, **7**, 284–92.

Walters, R.A. (1989) A model for tides and currents in the English Channel and Southern North Sea, *Advances in Water Resources*, **10**, 138–48.

Werner, F.E. (1987) A numerical study of secondary flows over continental shelf edges. *Continental Shelf Research*, **7**, 379–409.

Werner, F.E. and Lynch, D.R. (1987) Field verification of wave equation tidal dynamics in the English Channel and southern North Sea. *Advances in Water Resources*, **10**, 115–30.

Werner, F.E. and Lynch, D.R. (1989) Harmonic structure of English Channel/Southern Bight tides from a Wave Equation simulation. *Advances in Water Resources*, **12**, 121–42.

Wolf, J. (1983) A comparison of a semi-implicit with an explicit scheme in a three-dimensional hydrodynamic model. *Continental Shelf Research*, **2**, 215–29.

TWO-DIMENSIONAL NEARLY-HORIZONTAL FLOW MODELS 10

G.S. Rodenhuis

As explained in the introduction to this book, these are the hydrodynamic 'workhorses' of modelling throughout the areas of coastal, estuarial and harbour engineering. Correspondingly, more effort has gone into developing these models than other kinds of model. A considerable variety of types of model now exists, and it can be difficult to choose between them. In this section, the main kinds of two-dimensional, nearly-horizontal flow models are described and their operating characteristics are compared.

CONTENTS

10.1 PURPOSE

In this chapter we seek to offer some help to the practical engineer faced with the task of judging the appropriateness of a given numerical model to a given application. The situation we have in mind is one where the engineer, having to find a solution to a specific problem, invites one or more specialists to help him. Each of the specialists proposes to apply a numerical model, each offering his own particular type. This is, for example, often the situation in international contracting, where specialist consultants, including hydraulic institutes, submit proposals for modelling in response to certain terms of reference. It applies equally to the situation where local institutes or organizations or even an in-house computer department, offer to run a particular model. In any case our engineer, not being a specialist himself, has to evaluate what is offered to him. There are several ways open to him to do this; faith in the specialist is one of them, just going for the lowest offer is another. Hopefully, however, he is looking for some cost-effectiveness criteria.

This chapter should make it possible for him to establish such criteria. We shall discuss briefly, the historical development of numerical models, give the present state-of-the-art and describe trends that may be indicative for future developments. We then continue with an outline of the various types of model, indicating their particular features in relation to the types of problem to which they can be applied. The problems and pitfalls that may be encountered in model applications are also reviewed.

The numerical models that we shall discuss, it should be noted, are the so-called two-dimensional models for nearly-horizontal flows, the theory of which has been developed in a previous chapter. Basically, they provide flow velocities, integrated over depth and water levels, at the nodes of a computational grid. These models find application in problems that are governed by the large-scale behaviour of the flow. To set the stage for our later discussion, a few examples may be useful.

When certain engineering works are to be constructed, be it a bridge across an estuary, a closure dam or the reclamation of tidal flats, the problem is often to determine existing flow patterns and the subsequent changes that may result from the construction. This problem frequently leads to another problem, that of determining changes in the morphology under the new hydrodynamic conditions. This type of model application is also often encountered in offshore engineering projects, where two-dimensional hydrodynamic models find increasing application in connection with the construction of marine pipelines, offshore platforms and subsea completions in shallow coastal seas.

Coastal, Estuarial and Harbour Engineer's Reference Book. Edited by M.B. Abbott and W.A. Price. Published in 1993 by Chapman & Hall, 2–6 Boundary Row, London SE1 8HN. ISBN 0 419 15430 2.

For coastal defence works, the obvious problem is that of determining safe levels against flooding under storm surges. This is often done using statistics, but when data is too scarce to provide a sound basis for these statistics, or when freak incidents such as hurricanes are the determining factor, deterministic simulation of certain scenarios with a two-dimensional model may provide a better alternative. A similar demand for modelling will perhaps develop under a climatic change as is currently predicted. Changes in storm patterns as a result of a change in the dynamics of the climate may make statistics based on data from the past unreliable, and in that case one could with advantage revert to model simulation of selected storms.

Often the hydrodynamic model is only a first step in a sequence of model applications; as has already been noted, it may provide basic input data for a computation of morphological changes. At present two-dimensional hydrodynamic models are much used in studies of the transport and dispersion of pollutants, the hydrodynamics being the basic carrier. An important example is the spreading of hot water discharged from a power-plant. The engineering problem here is to balance the cost of construction work to reduce recirculation, for example by increasing the distance between intake and outlet, against the benefit to the plant operation of low intake temperatures, while at the same time complying with environmental standards. Another example in this category is that of determining the distance from the coast of an offshore spoil dumping site. The costs of longer travel times for the dumping barges have to be balanced against the benefits of lower concentration levels of the dumped material at the shore.

10.2 PAST, PRESENT AND FUTURE

At the beginning of the 1970s, a request for a hydraulic model study, generally constituted a request for a physical model. It was the period during which laboratories directed their main investments towards large model halls. In this situation, to offer a numerical model as an alternative (and this author has been in that situation) came close to an act of masochism. It usually involved an uphill battle against established practice and preconceived concepts. Gradually, towards the end of the 1970s, and in increasing numbers, numerical models took over from physical models, and today we have, in fact, reached the reverse situation where, justified or not, the request generally is for a numerical model.

The reason for the success of the numerical models was not essentially that they provided more accurate results. The whole issue of which technique provides the more accurate result is still, in general, unresolved and left to 'engineering judgement'. The true reasons were mostly a combination of the following:

1. A numerical model is often cheaper, in particular if little problem-specific model development is required and use can be made of an already operational modelling system. (A modelling system is here understood as a generic software code that is formulated in general terms, which through specific problem data can generate a model for a specific situation. Modelling with such codes is often called 'third-generation modelling'. Thus such a system can generate, for instance, without the need for changes in the source code, a two-dimensional model of the North Sea as well as a model of the Persian Gulf.)
2. It is usually faster in several ways. Especially when a modelling system is employed, the time from study request to study result can be very much shorter. A numerical model can be set up in weeks rather than the months required for a physical model and alternative problem formulations can be generated and computed within hours or days.
3. In general a numerical model provides more information. A numerical model basically provides velocities and water-levels at each grid point and time step. This information can easily be organized in flow patterns and contour maps for areas of special interest at selected times, or can be processed into time series at selected locations. The physical model on the other hand is usually limited in the volume of information that it can provide through limitations in instrumentation. In a two-dimensional physical model, data is usually obtained by placing velocity meters or water-level followers at a limited number of locations in the model.
4. A numerical model can simulate situations which are difficult or impossible to reproduce in a physical model. An example is the case where water movement is determined predominantly by wind or by barometric pressure variations. Another example is the case where the area to be modelled is simply too large to be reproduced in a physical model, e.g. in a model of the North Sea or the Bay of Bengal.
5. Sometimes a numerical model is requested merely to follow the current trend. There are, in fact, situations where a physical model is still the more appropriate instrument. This is often the case in problems where the general behaviour of the flow is determined by turbulence processes, as for example in the case of flowlines and eddies in a

stilling basin in front of a power plant, or in the case of the formation and exchange of a pollutant plume discharged into a bathymetry of a complex form or into an irregular ambient current.

The early models mostly dealt with large-scale, two-dimensional water motion, i.e. tidal motion and storm surges in coastal seas or estuaries. They were mostly built on a rectangular grid, and their accuracy was limited. Water levels were usually predicted fairly accurately. These models were mostly calibrated on this data. Velocities, however, could be grossly incorrect. But in those early days we, the modellers, were already quite happy if our model remained stable, water-levels corresponded to some selected tidal curves and the flow patterns more or less resembled what we expected. These models were adequate for relative comparisons of various alternative solutions to an engineering problem and, when calibrated carefully for particular situations, sufficiently accurate to form the basis for decisions.

With their increasing success, however, certain shortcomings of the fixed grid models became apparent. Attention in hydraulic engineering and as a consequence in modelling, moved from the simulation of overall flow behaviour to detailed simulation of the flow. We now want to compute in detail the changes in sedimentation and erosion resulting from the construction of a closure dam. With increasing emphasis on the environment, we want accurate knowledge of the expected pollution levels at specific locations along the shoreline. Since an increase in dike levels costs millions of dollars, we want to know with great accuracy the high-water levels to be expected along a shoreline or a river bank.

The early fixed rectangular grid models now showed an important shortcoming in the form of an inherent conflict in resolution. Their mesh size is fixed. Therefore a mesh that adequately resolves the features in large open sea areas – and for obvious reasons of computational expense there is always a tendency to let the resolution requirements in these areas dominate the choice of mesh size – is in most cases too coarse to resolve the more detailed features near the coast. In addition to this lack of an adaptive mesh size, there is another important shortcoming; the rectangular grid has severe limitations when describing flow lines, channels and shorelines. These features have, by the essence of their nature, a curved, streamline form. In a rectangular grid they can only be described by a step-wise 'staircase' form.

In response, rectangular grid models were developed with a grid nesting capability (Figure 10.1). Grids with a finer mesh can be nested within a grid with a coarser net. This introduces the problem of how the various grids can be adequately coupled and, as we shall see later this may be a source of errors. So-called 'telescoping' grid and 'stretched' grid models were also developed (Figure 10.2). These models still employ grids formed by the perpendicular intersection of straight lines, but the line spacing in each direction can be varied. The control over resolution is, however, still inadequate. They have the further problem of a dependency of the computational accuracy on the direction along which one intersects the grid. Both the nested grid models and the telescoping grid models have the same limitation as fixed grid rectangular models when describing curved forms.

The above models are essentially finite difference models. During the 1970s, however, the finite element method became very popular in engineering. Because of its success in other areas and its ability to allow the computational grid to be adapted to the irregular form of a natural bathymetry, finite element methods were introduced into computational hydraulics. Finite element models have now matured and are in use at several professional hydraulic laboratories.

But the early finite element models did not perform well. Since the nature of the method has, in essence, the characteristics of a steady-state (strictly 'elliptic') approximation, results in terms of water levels were usually acceptable, in particular when the method was applied to slowly varying or steady-state problems. However, the method could be horrendously incorrect for velocities, in particular in problems of a fairly dynamic character. Often large internal dampening by unnaturally high viscosity or bottom friction had to be employed to bring results more or less in line with calibration data. Usually, the method was also very costly due to the form of the matrix that had to be inverted and solved for each time step. By nature of the irregular grid form, this matrix is unstructured and is therefore less amenable to fast inversion algorithms. Many of these earlier shortcomings have, however, been overcome in modern finite element models.

The tendency towards more requirements for computing the detailed flow behaviour had, by the end of the 1980s, led to refined numerical models. This refinement is in two areas, in the numerical formulation and in the formulation of the physical processes. As an improvement in numerical formulation we see models being developed that are based on curvilinear co-ordinates or so-called boundary fitted co-ordinates. As this last name indicates, a grid built on such co-ordinates can be fitted to the particular geometric shapes of a certain problem area, be it a shoreline or a deep water channel. Also the gridlines can be concentrated in areas where either the flow or

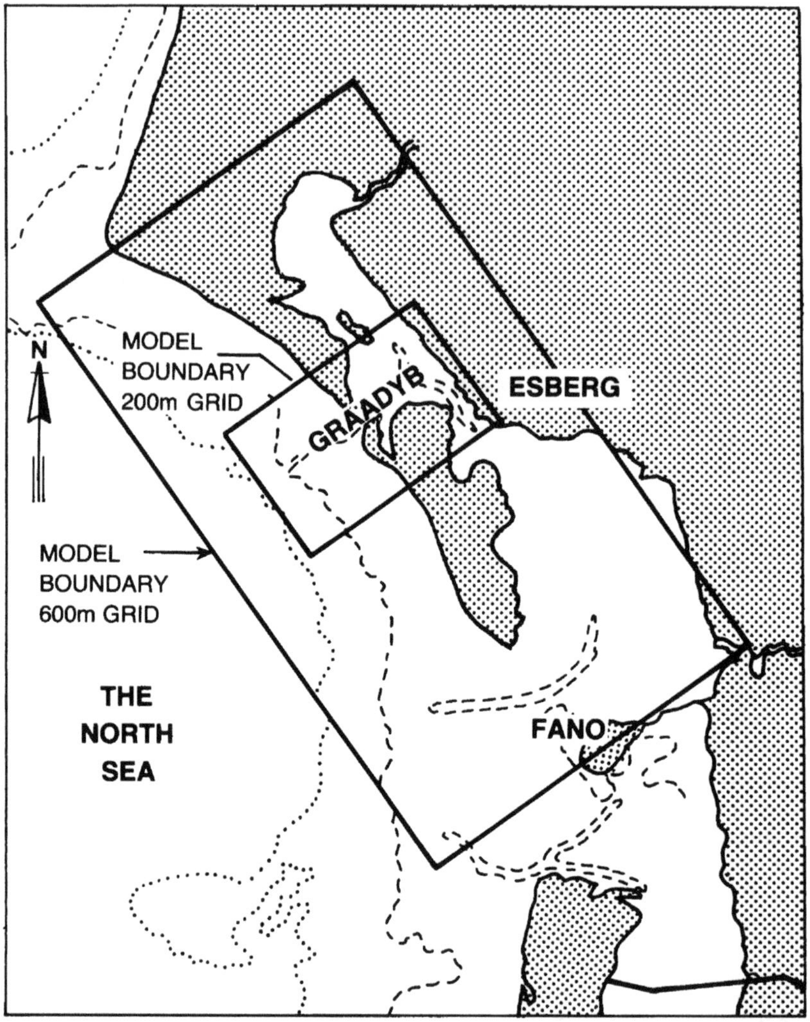

Figure 10.1 Example of model application using nested grids. Grådby, the entrance to Esbjerg, Danish west coast. Two-way interactive coupling is required. Flow plot shows velocities during flood (courtesy of Danish Hydraulic Institute).

the geometry or both vary rapidly and can be spread out in areas with little variation, for example over shallow tidal flats. There are various types of such grid systems. They may differ in the mathematical formation of the grid, in the orientation of the co-ordinates – tangential to the streamlines or normal to gridcell borderlines – and, most importantly, in whether the co-ordinates are orthogonal or non-orthogonal to each other. All of these models have, in addition to the normal inaccuracies related to solving continuous differential equations on a discrete grid, inaccuracies due to the geometry of the grid. When the differential equations are transformed to equations in curvilinear co-ordinates, certain 'geometrical terms' enter into the equations. These terms describe the variation of the curvilinear co-ordinates with respect to the Cartesian co-ordinates. They must themselves be approximated on the discrete grid. If the formation of the curvilinear grid is not well controlled, these terms may vary irregularly over the grid and introduce 'numerical noise' into the solution. The further refinement of numerical models in terms of the formulation of the physical processes is generally through the introduction of more complete

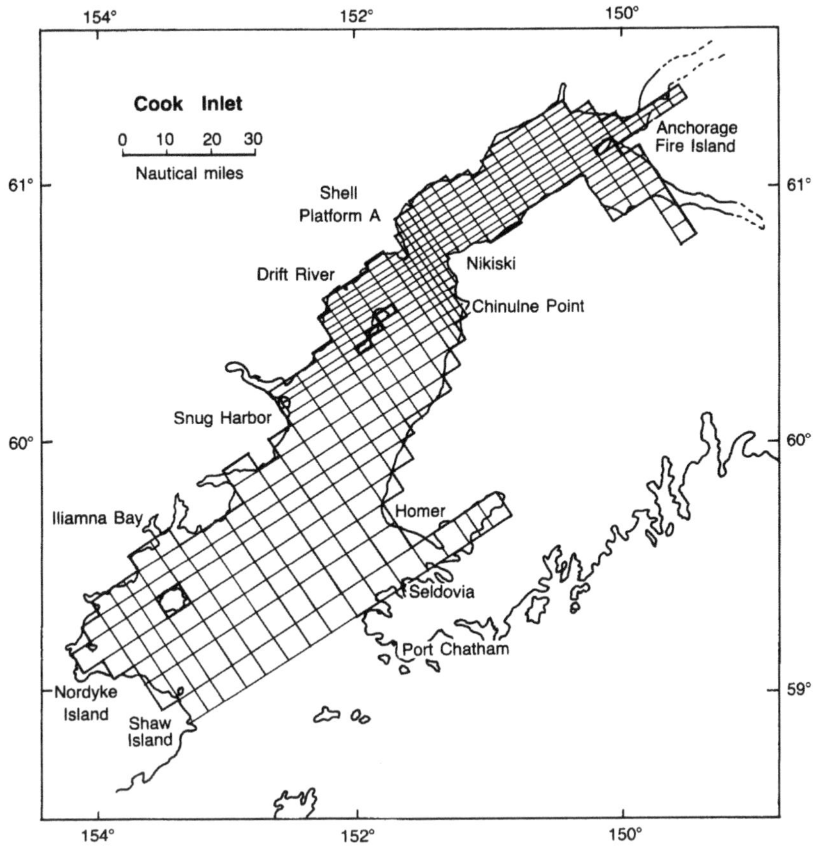

Figure 10.2 Example of a telescoping grid: Schematization of Cook Inlet, Alaska, Canada (after Mungall and Matthews).

descriptions of the internal structure of the flow, i.e. through improved formulations of turbulence and an extension from two-dimensional to three-dimensional processes. We meet here some fundamental modelling problems, fundamental because the need for a further refinement of the description conflicts with the essence of numerical modelling, i.e. a description on a discrete grid. So present research is directed to, among other things, the representation of sub-grid-scale processes on a discrete grid with a certain mesh size.

What then can be expected in the future? Since, as the saying goes, 'all prediction is difficult, especially prediction of the future', I shall be brief. A few trends can, however, be discerned that may at least give an indication of the developments in the near future. First a few remarks about the relation between numerical modelling and physical modelling. The numerical models have, at least for certain types of problems, reached maturity. Does that mean that physical modelling is to be dismissed? Not entirely, in my opinion. Although the large-scale water motion can be, and should be modelled numerically, detailed flow processes (especially when the influence of turbulence is important and when the geometries involved are complicated) are best modelled in a physical model. In some cases the two techniques can be combined with advantage: wave formation over a large sea area can be modelled in a wind–wave model, and the detailed wave action on a coastal or offshore structure can then be modelled in a physical model, where wave–structure interactions can also be included; general flow patterns in an estuary or coastal sea can be modelled in a two- or three-dimensional numerical flow model, and the detailed motions near an outfall or a lock can then be modelled in a physical model. We must expect, however, that the borderline between what can best be done in a numerical model and what should be done in a physical model, will increasingly shift towards numerical modelling when confidence in such models for particular types of problems increases.

The physical modelling that will take place in future will increasingly be the modelling of processes rather than the modelling of an actual problem situation. It will be the modelling of 'generic' processes, and, in direct relation to this, the mathematical and numerical formulation of these processes and the subsequent testing of this formulation against the physical results, that must help us to build more realistic and accurate numerical models. This work requires specialized facilities to study, for example, fresh water–salt water interaction, water–silt interaction, water–sand interaction and, possibly, the deposition and uptake of pollutants near the bottom of a water course.

Studies of processes will be part of what this author expects to be the main issue in numerical modelling in future, which is that of model validation. As in other areas of industry, quality assurance and control is presently being introduced into software development for hydraulic engineering. The first contracts in which a software quality assurance plan is included as a requirement have been signed in our branch of industry. This development of software quality assurance proceeds along two distinct lines: one is directed towards internal model quality, the other to external model quality. To start with the last, the activities under this assurance task deal with issues such as ensuring the integrity of the computer code being used. Rather than having the situation where somebody uses his personal version of the XYZ model, the client (and for that matter the management of the executing organization) needs to be assured that a well-defined release of the XYZ code is being used for his problem. Only in that case can one be sure that the quality of the model, established in an earlier validation of this particular XYZ release, is transferred to the application for which the code is used.

The task of ensuring model integrity relates to the organizational framework in which model development and model application takes place. This framework in turn has two sides: a 'soft' side, i.e. organization of people, operational procedures, quality audits etc., and a 'hard' side, i.e. the computer system on which the code, the data and the results are resident. An important point in relation to quality control on the hardware side has to do with the manner of access of users to information. This can be direct, like access to 'personal' memory on a PC, or indirect, as access to shared memory on a distributed system with certain computers configured as file servers. In particular, in the case of operating with personal memory, measures have to be taken to ensure code and information integrity, as this may quickly deteriorate through the insertion of uncontrolled, 'personal' versions of parameters and modules.

The internal side of software quality is directed to such issues as: are the equations embedded in the code a good approximation to the phenomenon to be studied, and does the numerical technique used to solve these equations introduce errors, (errors that sometimes lead to effects in the model results that are difficult to distinguish from effects caused by physical phenomena)? A common example in the latter category is numerical diffusion in transport–dispersion models.

Information to answer such questions should be compiled in a 'program validation document'. This document may include test results on certain 'standard' cases, comparisons with analytical computations, with physical model results or with field data, as well as information on model accuracy as a function of model parameters, organized in so-called 'model performance diagrams'. It must be said that the presence of a validation document as part of the program documentation is at this moment more the exception than the rule. However, certain organizations are beginning to introduce them and there are also cases where a number of users of model results, for example oil companies with interests offshore, join a model validation study. It is expected that there will be an increase in validation activities in the future.

10.3 MODEL TYPES: CHARACTERISTICS AND APPLICATIONS

In the discussion of the historical development of the various types of numerical model we have already briefly indicated their respective advantages and disadvantages. Often shortcomings provided the incentive to further model development. We shall now discuss the various types of model and their characteristics in more detail, still from the point of view of the user who has to evaluate the model for a particular application. In this discussion we shall deal with numerical as well as physical aspects for each model type. Under physical aspects we shall consider those aspects that are associated with the question of whether the model accurately represents certain physical features of the problem. Of course, being a numerical model it is only through the numerical results that we can find out to what extent it represents the physical features. Under numerical aspects we shall deal with questions such as model accuracy, stability, complexity of the algorithms and numerical dissipation, but always in terms of their implications for the application. If the reader wishes to go into the mathematical details of these subjects, he is referred to, for example, Abbott (1979) Abbott and Basco (1989) and Rodenhuis (1985).

10.4 RECTILINEAR FINITE DIFFERENCE MODELS

Probably still the most commonly used type of model, rectilinear models come in a variety of types, with uniform square or rectangular grid, an explicit or implicit time integration, with or without grid telescoping or nesting capabilities, etc.

10.4.1 UNIFORM GRID MODELS: NUMERICAL ASPECTS

By uniform grid models we shall understand models that have one fixed mesh size in the x-direction and one fixed mesh size in the y-direction; the mesh sizes in the two dimensions do not need to be the same, but they usually are.

The most important distinction when considering these types of model is the distinction between explicit and implicit models. Explicit models use a time integration technique where values at the upper time level, $t + \Delta t$, are entirely computed from values at the lower time level, t. However, the timestep, Δt, with which the computation can advance in time is limited by the time it takes for a physical disturbance, propagating along a characteristic in the time–space plane with celerity c, to reach the particular gridpoint at time $t + \Delta t$. This limitation is expressed in terms of the so-called Courant number,

$$C_r = -\frac{c\Delta t}{\Delta x}$$

For explicit models C_r must be kept less than or equal to 1. When this limit is exceeded, the solution generally becomes unstable. In that case the model is expected to calculate new values at a forward point in the space–time plane, while the information has not had sufficient time to physically propagate from points with known values to the forward point. (It should be realized in this that stability is a matter of the ratio between time step and grid size; refining the discretization while still violating the stability relation, will not stabilize a model.)

In an implicit time integration technique there is no such stability constraint. Courant numbers of up to 10 are quite possible, the upper limit being determined by accuracy considerations rather then by stability considerations. On the other hand, the implicit technique is more complex, as a matrix has to be built up, introducing a row for each unknown value at a grid point at time $t + \Delta t$, until the boundary of the computational domain is reached where the known boundary value is used to complete the matrix. In principle, only after completion of the whole matrix can it be solved.

What does this mean for the user and for model applications? This is often a question of costs and accuracy. Although modern computers are so powerful that available computer capacity should not limit the number of grid points and the number of timesteps for a particular application, in practice, there are limitations. The present development to implement even rather large models on personal computers and workstations, in fact, sets new limits, albeit perhaps only for the next few years. The limitations are sometimes in terms of limitations in computer memory, but most often in terms of cost and the duration of a computer run. When in an explicit model the mesh size is reduced by a certain factor, then the duration of a run increases almost directly with the third power of this factor, since a reduction in mesh size almost always requires a corresponding reduction in time step. In an implicit model the increase in duration will usually be smaller, with the factor to the second power as the lower limit, as there is more freedom in the choice of the time step. Although, because of the simplicity of its algorithmic structure, an explicit model is usually faster in terms of the number of computational points (points in the space–time grid) computed per CPU- second, in overall run time it is usually slower, since it may have to compute to up to something like ten times more computational points in an application.

Because of the direct relation between the number of gridpoints and the run time, there is always the temptation for the modeller to choose a rather large mesh size, i.e. to choose a rather coarse grid, where for reasons of accuracy he should perhaps choose a finer grid. Clearly the penalty to be paid for grid refinement in terms of run time is higher for explicit models than for implicit models.

This brings us to the point of resolution and accuracy: with what mesh size are features in the topography of the area to be modelled and, in the expected solution, to be reasonably resolved? And, correspondingly, for what time step are time variations in water levels and currents reasonably resolved? The answer depends on the quality of the model. The numerical model, being a discretization of a set of continuum equations, has certain truncation errors. By an elaborate construction of the computational molecule of the model, balancing positive discretization errors against negative discretization errors, truncation errors of second- or third-order accuracy can be obtained. Usually the developer of a model seeks to build so-called centred schemes, i.e. schemes in which the discretization is symmetrical in space and time relative to the centre point of the computational molecule. However, such a delicately balanced high-accuracy scheme may in practice turn out to be just a little too delicate. It would work correctly in a perfect world, but due to various reasons, non-linearities in equation coefficients being one of

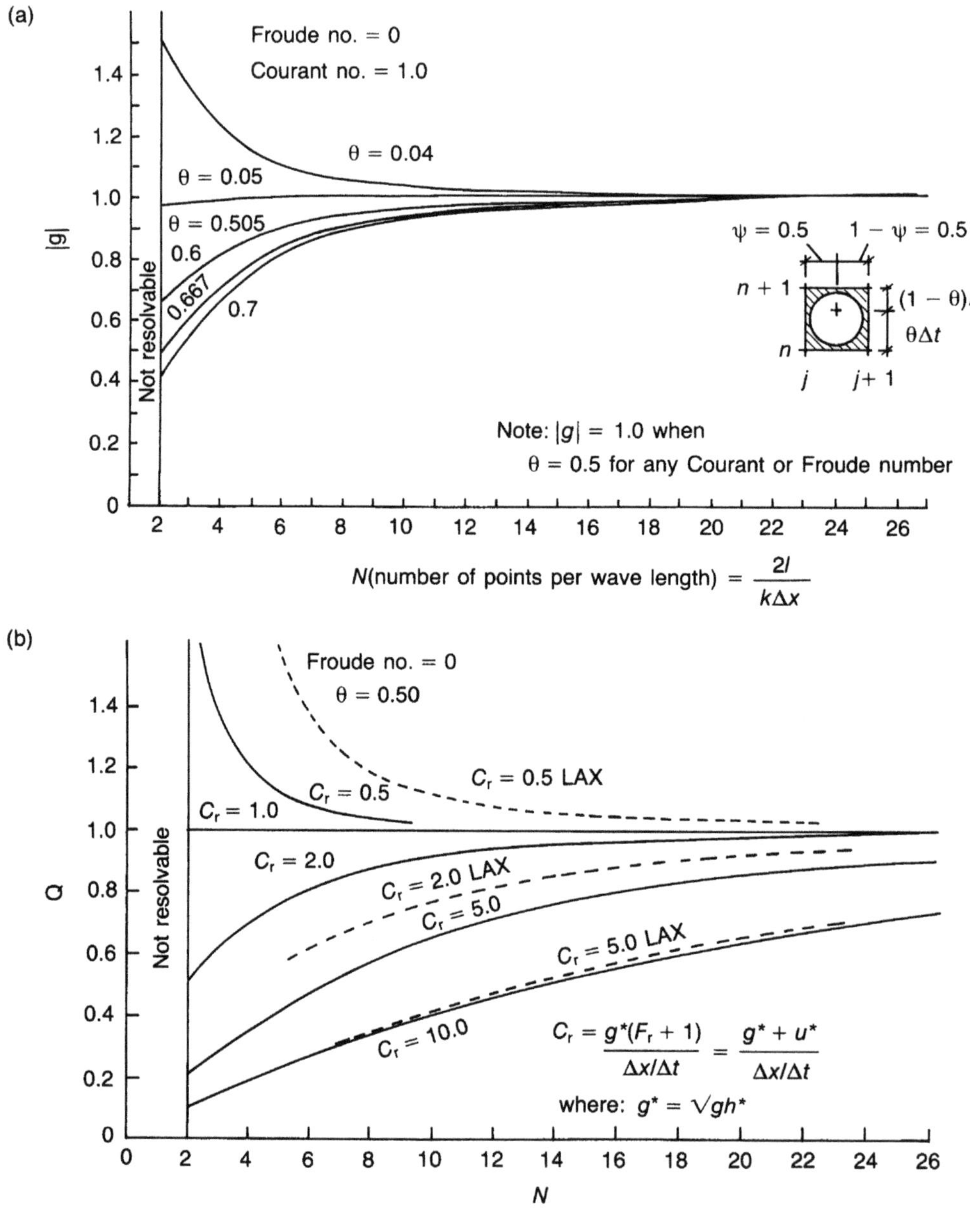

Figure 10.3 Typical amplitude and phase portraits (from Abbott and Basco, 1989, p. 257).

them, it may not work so well. Irregularities in data for example, may amplify out of bounds and destroy the solution. Therefore, instead of the perfectly centred schemes, sometimes off-centred schemes are used. The off-centring introduces some dissipation into the model, usually in the form of a dissipative term in the truncation error, which may serve to dissipate oscillations above certain frequencies.

The quality of the model – how well it represents the solution as a function of the number of grid points per 'wavelength' – can be expressed in diagrams, for example in a so-called amplitude portrait or phase portrait. Wavelength is here to be understood as the length of a sine wave approximating some characteristic wave-like feature in topography or in the expected solution. Figure 10.3 reproduces some typical amplitude and phase portraits. It can be seen that when the grid size is chosen so that a characteristic wavelength is represented by 10 or more gridpoints, features of this wavelength or larger are reproduced without error, while features with a wavelength shorter than this characteristic wavelength will be dampened.

It should be noted that diagrams such as those in Figure 10.3, are the result of theoretical calculations, usually a linearized stability analysis for a one-dimensional situation. Since the actual equations are non-linear and the situation to be modelled is two-dimensional, the model performance may in practice differ from the theoretical performance predicted by the diagrams.

Two-dimensional nearly-horizontal flow problems are so-called hyperbolic problems: a disturbance propagates in space and time along characteristics. For this reason there is a correspondence between resolution in space and resolution in time and a resolution of, say, 10 points per wavelength in space, often translates into a similar figure for the resolution in time, i.e. the number of time steps per wave period.

There may be other sources of errors than those introduced by the choice of mesh size and timestep. The whole process of discretization and numerical integration is an elaborate effort to, on the one hand, achieve high accuracy while, on the other hand maintaining reasonable computational speed. In developing a model, the developer may emphasise one or the other aspect. We shall not go into the details of the various possibilities here, but only mention those aspects that may show up in the quality of the model application.

An essential problem in all schemes is how to approximate the non-linear coefficients of the equations of nearly-horizontal flow and how to account for the fact that the equations are two-dimensional. One method, with emphasis on high accuracy, is to use a fully implicit scheme in which the full two-dimensional matrix is developed and solved by a matrix manipulation technique. The accuracy of the coefficients is achieved by extensive iteration at each time level until a certain prescribed accuracy is achieved. At the other end of the spectrum of methods is one in which the equations are, effectively, 'one-dimensionalized': solved at each time level or sometimes even at successive time levels alternatively first for the x-direction and then for the y-direction, and in which the non-linear coefficients are approximated in a multi-level difference scheme. The approximation in this case can be achieved without iteration; errors in the coefficients at one time level are compensated by an opposite error in the coefficients used at the next time level.

In the extreme case, the first method should give highly accurate and smooth solutions at each time level. The second method would be accurate only in an overall manner, being slightly incorrect at each time level. When used with a large time step the solution may show abrupt changes in flow patterns, especially near corners, and fairly large jumps in flow patterns between time levels. Inadequate approximation of coefficients (in particular the coefficients of the convective momentum terms) show up through zig-zag patterns in current vectors. Not surprisingly, the first method results in a model that may require fairly long run times, and the second method should give a rather fast model.

It should be noted at this point that model inaccuracies usually show up much more pronouncedly in flow patterns than in water-levels. This implies that where the application requires accurate current predictions, a model constructed along the ideas explained above for the first method might well be used with advantage. When the interest is in good predictions of water-levels, a model constructed using the second method would usually suffice.

There are many more numerical aspects to be considered. Of great importance is the treatment of boundary conditions and, for models of tidal estuaries, the treatment of flooding and drying of tidal flats. We shall only briefly indicate here the most important modelling problems. A full discussion is difficult as the treatment of these problems is very dependent on specific model characteristics such as the numerical scheme and the time integration techniques employed.

Conditions at the model boundaries, and more specifically the open model boundaries, are frequently the source of modelling problems. In principle, the following conditions must be specified: at an inflow boundary two conditions, for example the water-level and the direction of flow; at outflow boundaries, only one condition, usually taken to be the water-level. However, often only water-levels are available and the modeller must in some way or the other 'create' the other boundary condition at inflow boundaries. This is often done by either choosing the grid orientation so as to have flow vectors perpendicular to the inflow boundaries, or by creating an artificial model extension upstream of the actual model boundary where the flow can be assumed to enter perpendicular to the artificial boundary.

The relative importance of providing two boundary conditions at the inflow is determined by the relative magnitude of the velocity u in comparison with the characteristic celerity $\sqrt{gh}$, where g is the acceleration due to gravity and h is the fluid depth. If $u \ll \sqrt{gh}$, then specifying only water-levels does not create problems. This is, for example, often the case in most tidal applications.

Another boundary problem is often related to errors in the levelling datum used across an open boundary such as a sea strait. Due to Coriolis effects, there is often a slight tilt in the water-level across a strait. If this is not represented in the levelling data and the modeller applies only amplitudes and phases referenced to a fixed horizontal datum, errors in the flow pattern will result.

In general, errors in boundary data show up in the form of local circulation patterns at the boundary, sometimes to the extent that water flows in through one segment of the open boundary while at the same time flowing out through another segment of the boundary.

As already mentioned, problems may arise also from the model formulation of the flooding and drying of tidal flats. When, in a tidal estuary, the tidal flats dry out, part of the computational grid has to become inoperative. Later, during the flood, these inoperative regions have to be included again in the computation. In the numerical treatment of this feature there is the risk that numerically mass is erroneously added or subtracted. This error may not be so important with large tidal amplitudes entering through a boundary near to the area of tidal flats. It may, however, spoil results at the tidal-flat region itself, and in particular if the tidal amplitude is modest and the boundary some distance away. The error reveals itself by a book-keeping exercise or a test with an oscillation in a closed basin: one will observe how the mean water level goes down gradually.

Applications

Uniform grid model can best be applied in areas where the features to be modelled, both in terms of the topography and the expected solution, are fairly uniformly distributed. So, in open seas or fairly regular bays, good results can be expected. That is, water-levels should be predicted to within 5% of the amplitude, with currents usually less accurately predicted. When the topographical features become more complex, as is the case in a tidal inlet, an estuary with deep channels or near an irregular shore, then the accuracy of the uniform grid model deteriorates. Of course this is all related to whether the grid size is chosen so as to resolve the detailed features or to give a resolution 'on average' (and due to the cost involved the latter is often chosen). Again, the model inaccuracies reveal themselves first in velocity magnitudes and directions. Water-levels, even over a complex topography which is inadequately resolved on the grid, may still come out fairly accurately. So, also here, the requirements of the application – whether the interest is in water-levels or currents – should be taken into account when selecting a numerical model. When there is a requirement for a better resolution of local complex features and when the interest is in a good representation of the currents (for example in morphological studies or studies concerning the transportation of pollutants), then the use of non-uniform grid models should be considered. In the following we briefly review some of the features of non-uniform rectilinear grid models, before we continue with curvilinear models.

10.4.2 TELESCOPING GRID MODELS: NUMERICAL ASPECTS AND APPLICATION

By a 'telescoping' grid we shall understand a grid in which the grid spacing can be gradually increased or reduced. The grid spacing can be increased or reduced in one or in two directions. In two dimensions the grid is an intersection of two one-dimensional meshes with varying spacings. An area of high resolution is created by the intersection of bands of high resolution in each direction. These bands extend outside the area of interest, thus introducing an unnecessary large number of grid points into the equation. An example has been shown in Figure 10.2.

The drawback of this technique is that it becomes difficult to construct on such a grid a properly balanced centred difference scheme with a high order of accuracy. As a result phase and amplitude dissipation or amplification errors of varying magnitude may exist, depending on the direction of propagation relative to the grid. For this reason, this author would be doubtful of the results if the model was used to create the hydrodynamic basis for a study of the spreading of pollutants. For water-level predictions, good results have been reported.

10.4.3 NESTED GRIDS

Numerical aspects

To overcome the resolution problem of the uniform grid methods, the technique of nesting of a grid with smaller mesh size within a grid of larger mesh size has been developed. A number of different versions exist: coupled nested grids and uncoupled nested grids, and under each of these explicit or implicit time-integration methods.

When an uncoupled grid-nesting technique is employed, basically a computation on the coarse mesh is executed over the whole domain of the problem and the entire period of simulation. Results are stored and used to create boundary conditions for a sub-domain of the problem, where the grid with fine mesh size is used. In a separate run a solution is computed for the sub-domain with the new boundary data.

The problem in this technique is that important features of the sub-domain and their effect on the solution may not be well represented on the coarse grid (and the fine grid was introduced specifically to improve this). If this effect propagates outside the region of the sub-domain, the boundary conditions created on the coarse grid will be incorrect. A way around this problem is to schematize the features of the sub-domain on the coarse grid so that their overall effect on the solution is (hopefully) correctly represented. We shall give an illustration of this problem further below.

This problem of incorrect representation of the features of the sub-domain is avoided when the grids of the domain and sub-domain are coupled in one and the same computation. The computation proceeds on the coarse grid, starting from the boundary of the domain. When the boundary of the sub-domain is reached, the mesh size is changed, interpolations for intermediate points on or adjacent to the sub-domain boundary are made and the computation proceeds on the sub-domain grid. This is, in principle, carried out at each time step of the computation (interpolations in time at the sub-domain boundary and intermediate sub-domain computations could be introduced, but this would lead to a rather complicated model).

The coupled and uncoupled technique of nesting can be carried out with an explicit as well as an implicit time integration method. In the opinion of this author, implicit time integration is to be preferred. This is quickly understood if one recalls the limitations set to the time step by the choice of the mesh size in explicit methods. The time step in the sub-domain computation will be governed by the mesh size of the sub-domain; in the coarse-grid computation, the time step can be larger, corresponding to the larger mesh size. In the uncoupled application one can, in principle, run the coarse-grid computation with the coarse-grid time step, and the fine-grid computation with the fine-grid time step. However, the problem of the representation of the sub-domain features and their effect on the solution, discussed above for the spatial resolution, also has a translation in time. At the coarse-grid time step, variations at the boundary of the sub-domain at the time scale of the sub-domain are averaged out. When working with coupled grids, the explicit time-integration method forces one to use the smaller time step of the fine grid (unless, as mentioned earlier, one introduces elaborate intermediate computations on the sub-domain).

The implicit time-integration method allows more freedom of choice, for the technique with coupled as well as with uncoupled grids. Since one is not rigidly restricted by the magnitude of the mesh size, an average time step that can be used for the domain as well as the sub-domain calculations can be chosen. The implicit method with uncoupled grids is fairly straightforward. A model with implicit time integration and coupled grids becomes highly complicated. On the other hand, the structure of the implicit time integration in which model results at various time levels can be introduced, allows a good coupling between domain and sub-domain to be constructed. There are not many of these coupled, implicit models. The author is, in fact, aware of only one, S21.MK3, the model that he developed himself in 1973 (Rodenhuis, 1985).

When considering the use of a nested model, there is of course the question of where to place the borders of the sub-domain. To answer this question one has

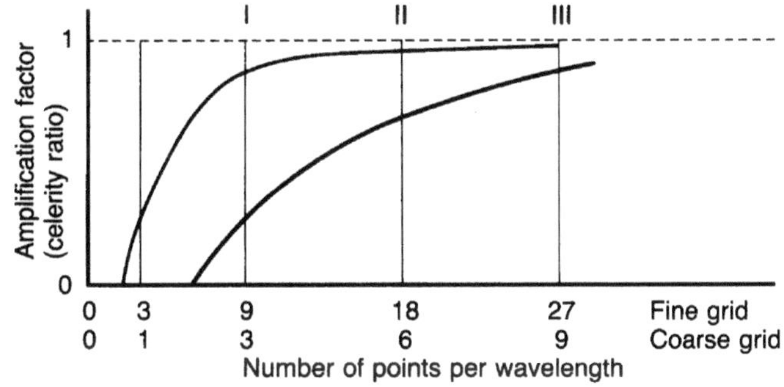

Figure 10.4 Typical amplification factor (celerity ratio) for two different grid resolutions.

to consider again the error-propagation qualities of the model as expressed, for example, in amplitude and phase portraits. This is explained with the aid of the schematic diagram (Figure 10.4) that could be a typical phase or amplitude portrait.

The two curves in Figure 10.4 are identical; the curve to the right has only been redrawn in order to correspond correctly to the number of grid points per wavelenth of the coarse grid. The factor of grid reduction is 3. It is clear from this diagram that placing the border line of the sub-domain in location I, that is, in a location where there are still features on the scale of the sub-domain that are resolved only by 3 points per wavelength on the coarse grid, will be wrong. The sub-domain features will be incorrectly represented in the solution. The correct choice of border location would be in location III.

This can be expressed another way: a grid transition should be made in the area where the topography and the flow are still of the scale of the coarse grid. Often this condition is violated, the demand of computational economy forcing the modeller to restrict the mesh refinement to as small an area as possible. Also here the observation made earlier that water-levels are still reasonably represented even on a grid that does not adequately resolve the topography, implies that the condition, introduced for the location of the grid transition, can be somewhat relaxed for a model application where the interest is only in water levels.

Incorrect positioning of the grid transition may show up in several ways: waves outside the sub-domain may reflect from the interior transition boundary, spurious waves may develop that are trapped inside the sub-domain or waves passing from domain to sub-domain may become stretched. The form in which these errors occur in an actual application will depend, among other things, on the error propagation characteristics of the numerical model employed.

Application

Clearly nested grid models find application where there is a need for a local grid refinement in order to resolve local topographical features better or to allow a better representation of detailed features in the solution, such as local eddies or rapid flow transitions. We shall here consider only the question of when to use coupled grids or uncoupled grids. The matter is best explained with the use of the two situations sketched in Figure 10.5. The situation on the left represents a problem where the possible recirculation of the cooling water from a power-plant must be studied in detail near the coast, including the effect of local topographical features such as coral reefs, discharge points etc. This problem can be solved using the two grids uncoupled from each other. It may be expected that the grid transition can be placed sufficiently far from the dominant sub-domain features so that these features do not influence the overall behaviour of the flow at the transition borders.

The situation on the right is the problem of a tidal inlet, with strong local features at the throat section of the inlet, deep channels, banks etc. This problem is best solved using the two grids coupled. It will be difficult to resolve the features of the throat section on the coarse grid adequately. There is the risk that the sea and the tidal basin will be badly connected, and that the boundary data created on the coarse grid will be inadequate.

The nesting of grids may be repeated by introducing further sub-domains of a second degree of refinement into the sub-domain at the first degree of refinement. The refinement itself may be by different factors. In coupled grid methods factors of two or three can be used; in uncoupled grid methods the reduction factor need not even be a whole number. The implicit coupled grid model, S21.MK3, allows grid nesting to several degrees of refinement deep. Several sub-domains of the same degree of refinement can be introduced at the same time. The refinement factor is 3.

10.5 CURVILINEAR FINITE DIFFERENCE MODELS

10.5.1 BASIC PRINCIPLES

To overcome the various resolution problems in rectilinear models, curvilinear models have been developed over the last decade. Their basic principle is that the resolution problem and the computational problem are separated. The physical space, with all its resolution problems, is represented on a grid of varying mesh size, so that a dense grid spacing can be obtained where needed, while in areas that are uninteresting and not important for the detailed flow behaviour (for example tidal flats) the grid can be coarse. This grid is then mapped into the computational space on a rectilinear grid. In this computational space one can then use all the tricks developed in rectilinear numerical model construction to build high accuracy difference schemes.

10.5.2 NUMERICAL ASPECTS

As nothing in life comes free, there are a number of snags that may cause inaccuracies that may ultimately turn up in the solution. When the equations of nearly-horizontal flow are expressed on a curvilinear grid, new terms need to be introduced into the equations. These so-called geometrical terms are usually combinations of derivatives of the co-ordinates in phy-

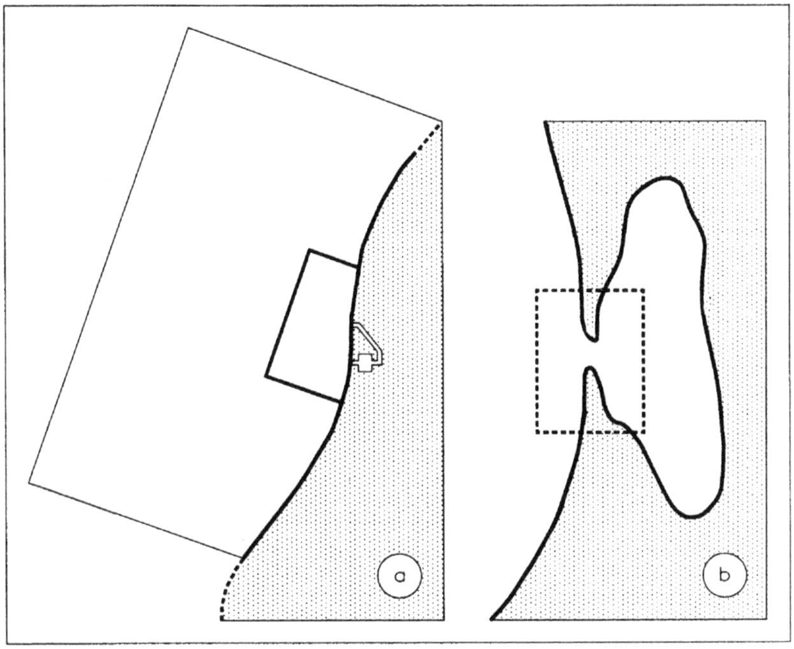

Figure 10.5 Typical modelling situations. (a) can be modelled with uncoupled grids; (b) should be modelled with coupled grids.

sical space with respect to the co-ordinates in the computational space and their inverse relations. The curvilinear grid can be orthogonal or non-orthogonal, and in particular in the latter case these terms become fairly complex. Errors are introduced into these terms when the curvilinear grid on which they are based is not smooth. The geometrical terms appear in many terms of the flow equations as coefficients. When they contain 'noise' produced as a result of an inadequate grid-generation process, this will spread throughout the solution procedure and, regardless of the accuracy of the numerical integration procedure, will disturb the results.

There are two methods of grid generation, which differ in the amount of 'noise' that is produced in the geometrical terms. One method, called the elliptic method, generates the grid as the solution to an elliptical equation. This method generally produces the smoothest grid. The other method, the algebraic method, generates the grid through a process of interpolation. First a 'skeleton' grid is sketched over the problem domain, and by successive interpolation and approximation, often using splines, the final grid is developed. This procedure does not necessarily result in a smooth grid, although it may look smooth on paper. There may be errors, in the form of discontinuities or large gradients in the derivatives of the physical co-ordinates with respect to the computational co-ordinates. A good check is to plot the respective derivatives as function of the computational co-ordinates in three dimensions. This may show a mountainous landscape with high ridges and peaks, which is not very good, or smooth, gentle terrain which is as it should be.

As has been already mentioned, curvilinear grids come in orthogonal and non-orthogonal versions. The elliptic method of grid generation produces a grid that is, by definition, orthogonal. The algebraic method produces a grid that is usually non-orthogonal. The flow equations on a non-orthogonal grid become very complicated. On the other hand, the algebraic method gives considerable flexibility and control to the modeller. One may thus be tempted to use an algebraic generation method and orthogonal equation form, but a significant degree of non-orthogonality may then spoil the results.

These are the main points that a non-specialist user of curvilinear models should be aware of. There are more esoteric items, such as whether expansion or contraction in the mapping of regions of the physical space to regions of the computational space effects stability or the error propagation qualities of the nu-

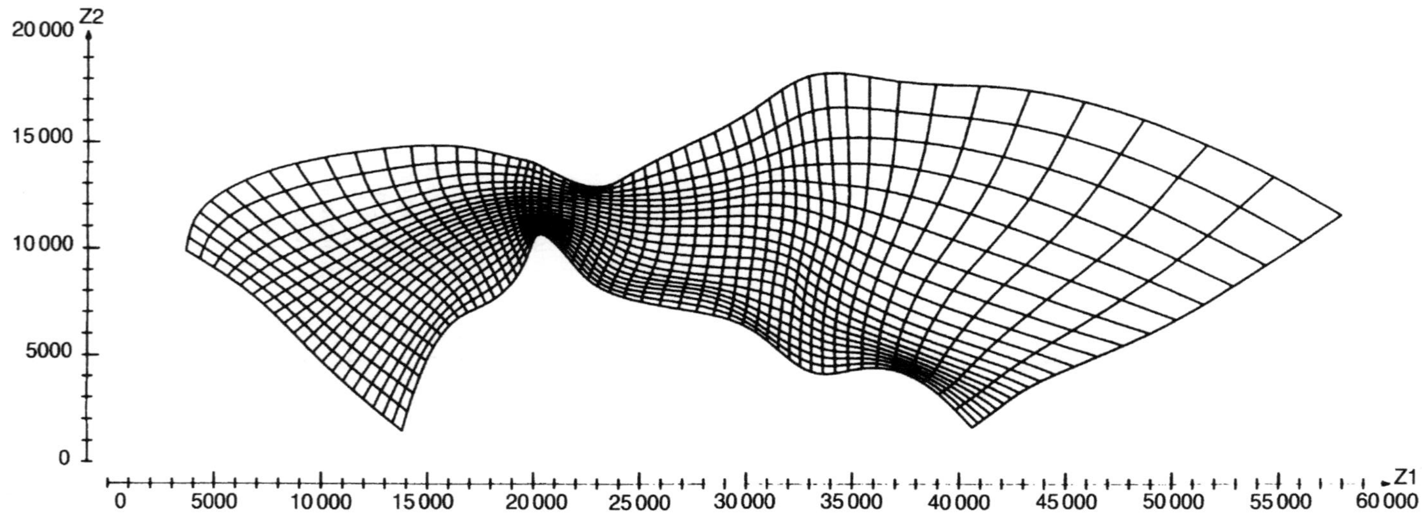

Figure 10.6 Final smoothed grid.

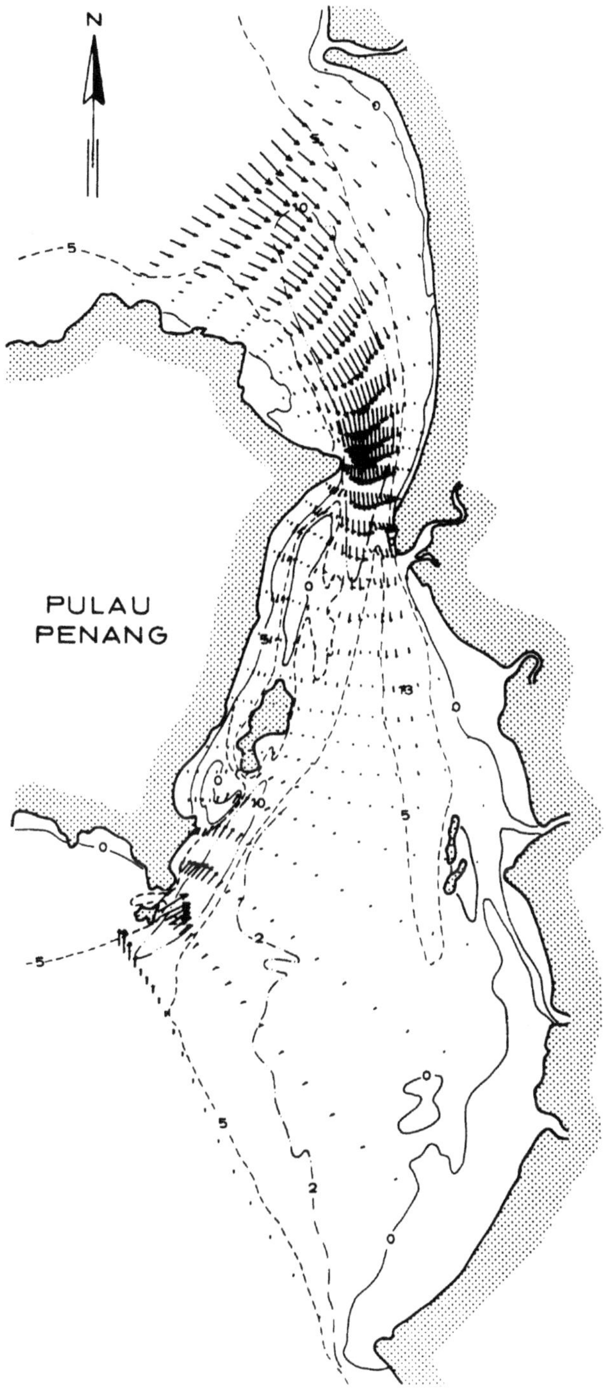

Figure 10.7 Penang Strait, velocities during flood.

merical model, or what is the effect of large model distortions on the results, but not even the specialists have answered these questions yet.

10.5.3 APPLICATIONS

There is not so much to say here. The areas where a curvilinear grid can be used with advantage follow, almost by default, from the discussions above about the resolution problems of the more traditional techniques. It should, however, be realized that when one choses a curvilinear model for an application, there should be a reason. Water-levels in an open sea can perfectly well be predicted by a rectilinear uniform grid model.

Economy may also be a reason to consider when a curvilinear grid application is contemplated. The model is more complex than the rectilinear models, requiring more effort per grid point. But this may be outweighed by the fact that the curvilinear model needs less grid points to describe the same domain better. Moreover, a bathymetry that is well described by a curvilinear grid at low tide for example, may not be so well described at high tide, so that in principle a further development to flow – an adaptive grid – is required. So far, these have been very little applied in practice.

More on curvilinear modelling may be found in Rodenhuis (1985), from which the two diagrams, Figures 10.6 and 10.7, are reproduced.

REFERENCES

Abbott, N.B. (1979) *Computational Hydraulics, Elements at the Theory of Free-surface Flows*, Longman, London.

Abbott, M.B. and Basco, D.R. (1989) *Computational Fluid Dynamics, an Introduction for Engineers*, Longman, London and Wiley, New York.

Rodenhuis, G.S. (1985) Toward a general curvilinear model for nearly-horizontal flows in two horizontal dimensions. Delft Hydraulics, Delft.

Mungall, J.C.H. and Matthews, J.B. (1973) Numerical tidal model with unequal grid spacing. Institute of Marine Science, University of Alaska, Canada. *Rep.* R73-2.

ONE-DIMENSIONAL NEARLY-HORIZONTAL FLOW MODELS 11

J.A. Cunge and J.L. Rahuel

From a superficial point of view, the one-dimensional models appear as the simplest kinds of model, since the number of governing equations is minimal and the set of numbers required to describe a one-dimensional flow is usually considerably less than that required to describe a two-dimensional or three-dimensional flow. However, on closer aquaintance, this class of model is seen to necessitate a much more complicated and delicate process of schematization of flow features, as it introduces quite different problems of connectivity (of dendritic and looped networks), while one-dimensional models are commonly used in estuarial and delta regions with a multiple of disparate, and even contradictory user requirements. Thus, on the one hand, the construction of simulation tools has progressed rapidly in this area, but, on the other hand, very heavy demands are placed upon both the tools themselves and their users. In this chapter, the difficulties are placed alongside the potentialities, and some of the pitfalls of model schematization are exposed.

CONTENTS

11.1 INTRODUCTION

Numerical simulation of one-dimensional unsteady free-surface flow in fixed-bed open channels is a well established technique and one of the best explored among the various tools described in other chapters of this volume. Thus an engineer can simulate complex topographies of tidal channels and rivers, including inundated areas and multi-connected deltaic networks of streams. Existing tools permit the simulation of occurrence, growth and propagation of tidal bores, as well as the evolution of salt concentration along channels and with time. Transitions, either natural (narrowings and widenings, local head-losses, rapid-slope variations, bifurcations, confluences, etc.) or man-made (dams, weirs, gates, dredging, etc.) are simulated with engineering accuracy by a number of available software tools. These tools also enable simulation and testing of management devices, such as feed-back controlled structures, and policies, such as the best strategies to protect estuaries from storm surges by activating gates and using polders.

The difficulty lies in the choice, as not all available tools can do all of this. Those which can often do it in different ways, some of which may be preferred by a designer or a manager as compared to others. At the end of the chapter some references concerning the tools available at a number of major institutes are given, but, in general, they are far from 'off-the-shelf' products. A few of the available off-the-shelf systems still call for considerable competence, and their application field is usually limited.

The equations describing the physical process of one-dimensional unsteady open channel flow were published by De Saint-Venant (1871) and the first numerical solution method was presented 18 years later (Massau, 1889). Since the first successful attempt using electronic computers (Isaacson *et al.*, 1954), hundreds of models have been built and run, either for inland or coastal water applications. This rich past does not imply the existence of a standard software

Coastal, Estuarial and Harbour Engineer's Reference Book. Edited by M.B. Abbott and W.A. Price. Published in 1993 by Chapman & Hall, 2–6 Boundary Row, London SE1 8HN. ISBN 0 419 15430 2.

tool used by everybody; on the contrary, whoever looks for one-dimensional flow simulation software will find too many on offer, rather than too few. Thus the purpose of the present chapter is to consider essential features which make a tool reliable, and to bring to the user's attention the difficulties of applying existing tools. Indeed, it is well known that a skilled and qualified modeller can obtain useful indications and correct conclusions using rather primitive software tools. A newcomer to the profession, on the other hand, may well obtain nothing, or even worse, wrong results, while applying a sophisticated simulation system. Given the stakes and liability problems on the one hand and the impossibility of a user knowing the software thoroughly on the other, a number of meaningful points should be considered by engineers applying simulation tools to design, management, control and other similar problems.

11.2 THEORETICAL BACKGROUND

One-dimensional gradually varied open-channel unsteady flow of density-constant water can be described by two equations expressing conservation laws. There are four hypotheses (De Saint-Venant, 1871) which permit us to describe such flow.

1. The velocity is uniform over the cross-section and the water level across the section is horizontal.
2. The pressure distribution is hydrostatic (streamline curvature is sufficiently weak to not induce pressure distribution variations).
3. All dissipative effects (boundary friction, turbulence) can be accounted for through global resistance laws such as those used in steady flows (e.g. Manning–Strickler, Chézy, etc.).
4. The bed slope is small, the flow is nearly horizontal.

These hypotheses enable the definition of flow in a cross-section located at a distance x at the time t by two dependent variables, e.g. discharge $Q(x, t)$ and elevation $y(x, t)$, consistent with the statement that two equations in terms of these variables can describe the flow. If the variability of water density is taken into account (an important feature for coastal areas), salt concentration $s(x, t)$ is the third unknown dependent variable; hence a third equation (salt mass conservation) must be added to the system of conservation laws.

Two possible sets of laws may in practice be chosen: either the triplet (water mass, salt mass and momentum conservation) or the triplet (water mass, salt mass and energy conservation). If a finite reach Δx of a river is considered, the resulting equations (integral form of the conservation laws applied to a control volume) are not the same and are not equivalent for the two triplets. Their solutions are also different. For steady flow, the momentum conservation equation must be applied rather than a Bernoulli-like energy conservation equation, to deal with the hydraulic jump. In unsteady flow, discontinuities such as moving hydraulic jumps, bores, etc., also occur and the mechanical–energy conservation equation is not suitable for such situations as its solution is not physically sound. The two laws of conservation (momentum and energy) are only equivalent when all variables (discharge, free-surface elevation, velocity, cross-sectional area, width, conveyance) are continuous, differentiable functions, and so only for infinitely small control volume (dx, dt).

Simulation tools based on the energy conservation equation, which cannot deal with surface discontinuity, should not be used to model a tidal river along which a bore can appear. Confusion between the two sets of equations is common in many publications.

There are two forms of basic equations obtained from the De Saint-Venant hypotheses (Cunge *et al.*, 1980): integral and differential. Nearly all existing tools refer to the differential form, which is given below for the constant density case:

Water mass continuity equation

$$\frac{\partial A}{\partial t} + \frac{\partial Q}{\partial x} = 0 \tag{11.1}$$

Dynamic momentum conservation equation

$$\frac{\partial Q}{\partial t} + \frac{\partial}{\partial x}\left(\frac{Q^2}{A}\right) + gA\frac{\partial y}{\partial x} + gAS_f = 0 \tag{11.2}$$

where $y(x, t)$ is the free surface elevation above datum, $A(y) = A(x, t)$ is the wetted cross-sectional area, $Q(x, t) = Au$, is the discharge through the cross-section, $u(x, t)$ is the uniform water velocity in the cross-section, S_f is the friction slope which represents all dissipative effects and is defined by the steady state flow formula:

$$Q = K\sqrt{S_f} \tag{11.3}$$

where $K(y) = K(x, t)$ is the steady-state defined conveyance based on the Manning–Strickler or a similar formula. The dynamic energy conservation equation:

$$\frac{\partial u}{\partial t} + u\frac{\partial u}{\partial x} + g\frac{\partial y}{\partial x} + gS_f = 0 \tag{11.4}$$

is equivalent to Equation (11.2) only if all variables are continuous. It must be remembered that the system of hyperbolic partial differential non-linear Equations (11.1) and (11.2) may have discontinuous solutions, the physical manifestation of which are bores (i.e. mobile hydraulic jumps). The system of Equations (11.1) and (11.4) may also have discontinuous solutions, but they do not correspond to physical reality, i.e. the celerity and height of a jump found from Equations (11.1) and (11.4) will not correspond to the real ones occurring in nature.

The hypothesis of uniform velocity within a cross-section is only seldom kept in mind by users, probably because it so obviously looks wrong. Firstly, it is wrong so far as the vertical distribution of horizontal velocities is concerned. Secondly, it is obviously wrong when the depth of water varies across the section. A hundred years of observations, tests and comparisons have proved that Equations (11.1) and (11.2) are sufficiently representative of physical reality for engineering practice in spite of the rather drastic hypotheses. Nevertheless, corrections are often introduced into the equations, such as the Boussinesq corrective coefficient for vertical momentum distribution or the 'live cross-sectional area' concept which, for each cross-section, makes a difference between the area involved in storage and that involved in the dynamics of the flow. While these modifications increase modelling flexibility, they introduce an arbitrary element into the equations. Note that in Equations (11.1) and (11.2) the only empirical coefficient involved is the friction factor encompassed by the conveyance K. Since it is empirical, it is arbitrary only within the range defined by experience and is definitely linked to the type of river bed and banks.

Where one-dimensional coastal problems are concerned, the system of Equations (11.1) and (11.2) must be used as such and must not be simplified. Indeed, there are many software tools offered for unsteady flow simulation which are based on the simplified dynamic equation. The simplification usually consists of eliminating the first left-hand-side term (temporal inertia term) in Equation (11.4). Equations (11.1) and (11.4) thus lead to a diffusive wave system, which is solved in lieu of the full system of Equations (11.1) and (11.2). The temporal inertia term is essential for tidal flow and, if eliminated, the resulting equations loose physical sense for coastal applications, although they can still be used for certain flood propagation problems (Cunge *et al.*, 1980).

11.3 NUMERICAL METHODS

The system of Equations (11.1) and (11.2) cannot be solved analytically in a closed form and approximate numerical methods are used for that purpose. Although there are three principal classes of methods known (characteristics, finite differences and finite elements), finite differences have found the widest use in industrial codes and only that approach is considered here. Indeed, the method of characteristics is used essentially for didactic and interpretation purposes. The most advantageous practical feature of finite elements is their ability to follow complicated (in plan-view) topographies, an advantage which is lost in one-dimensional applications.

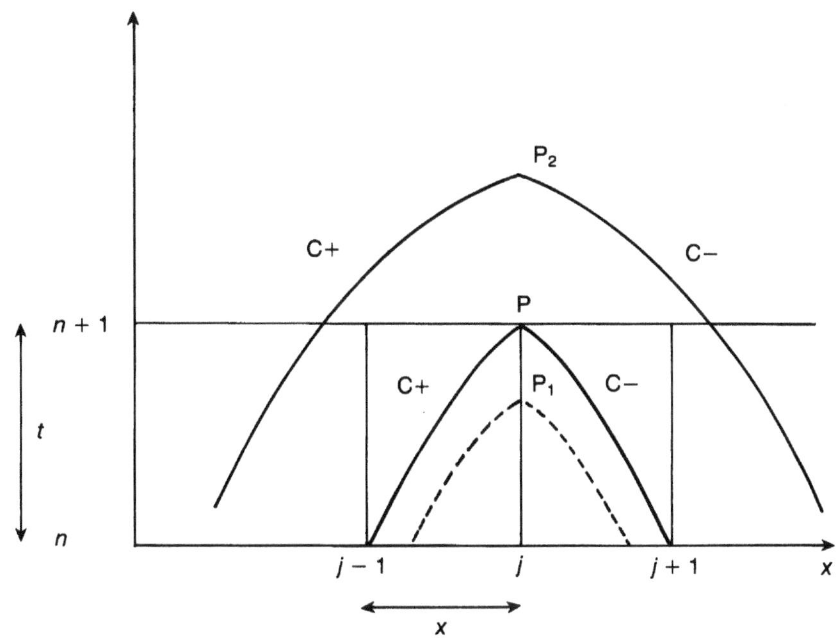

Figure 11.1 Illustration of modified lax scheme in characteristics plan.

A finite difference approach consists in schematizing the water course by a number of representative cross-sections (or computational points) separated from one another by a finite distance Δx. Then 'divided difference' formulae are applied to replace the derivatives in the original differential equations.

Considering a very simple way of replacing derivatives of a function $f(x, t)$ by finite differences, the so-called modified Lax scheme:

$$\frac{\partial f}{\partial t} = \left\{f_j^{n+1} - \frac{1}{2}[\alpha(f_{j+1}^n + f_{j-1}^n) + (1 - \alpha)f_j^n]\right\}\frac{1}{\Delta t} \quad (11.5a)$$

$$\frac{\partial f}{\partial x} = \{f_{j+1}^n - f_{j-1}^n\}\frac{1}{2\Delta x} \quad (11.5b)$$

Applied to Equation (11.1) this produces:

$$A_j^{n+1} = \frac{1}{2}\left[\alpha\left(A_{j+1}^n + A_{j-1}^n\right) + (1 - \alpha)A_j^n\right] - \frac{\Delta t}{2\Delta x}(Q_{j+1}^n - Q_{j-1}^n) \quad (11.6)$$

Assuming that the initial state $(y(x), A(x), Q(x))$ at time t_n is known along the channel, it is possible to compute values A_j, $(j = 2, \ldots, N - 1)$ for N computational points along the x-axis at time $t_{n+1} = t_n + \Delta t$. This method is termed explicit because each new value at the time t_{n+1} is computed separately, using only a limited number of neighbouring points. New values of the discharges Q_j at time level t_{n+1} can be computed in the same way by replacing the derivatives in Equation (11.2) by the formulae of Equation (11.6) and solving for Q^{n+1}. In Figure 11.1 two trajectories C_- and C_+ are shown going backwards from point P. Their equations in the (x, t) plane are, respectively

$$\frac{dx}{dt} = u \pm c; \quad c = \sqrt{g\frac{A}{b}} \quad (11.7)$$

They are termed characteristics and play a very important role. Let:

$$\mathrm{Cr} = \frac{\Delta t}{\Delta x}\max\,[|u + c|, |u - c|] \quad (11.8)$$

be called the Courant number. It can be proved that for all explicit schemes a limitation is imposed on the time step (the so-called Courant–Friedrichs–Lewy condition):

$$\mathrm{Cr} \leqslant 1 \Rightarrow \Delta t \leqslant \frac{\Delta x}{\max(|u \pm c|)} \quad (11.9)$$

If this limitation is violated, the solution is impossible to obtain (the computations 'blow up'). This situation is shown in Figure 11.1: the solution can still be obtained at the point P_1 with the aid of the segment $(j - 1, j + 1)$. Solving for point P_2 needs information lying beyond this segment. This limitation is very serious. Suppose there is a weir and a bridge, 50 m apart, along a tidal channel of average depth A/b = 4 m. When using an explicit difference scheme the time step must be smaller than 12.5 s while everywhere else a space step of 500 m (which would have led to a time step of 125 s) could be adopted. That is why most industrialized codes use implicit finite difference schemes which are not subject to such limitations.

It should be noted, that numerical solution is always approximate, never exact. The degree of accuracy obtained during simulation depends essentially upon the value of the Courant number, Cr. The value of Cr actually used in computations is variable (because the celerity c depends on local depth and also because the time step Δt and space step Δx are seldom constant). Thus the problem is not to obtain the exact solution but rather to know, as accurately as possible, how far the solution offered by the simulation is from the exact one. This vital information (for the user) is never explicitly given by the computer output and qualified advice is necessary to obtain it.

11.4 INITIAL AND BOUNDARY CONDITIONS

Models need both initial and boundary conditions, which must be supplied by the user. Two initial conditions are required at every computational point, most often discharge Q and elevation y at initial time. For one-dimensional flow, the most often used and required boundary conditions are natural ones;

1. Free surface water elevation as a function of time, $y(t)$.
2. Discharge hydrograph, $Q(t)$.
3. Discharge/elevation relationship, $Q = f(y)$, such as a weir formula.

One of these conditions should be imposed at each end of the model. Some rules should, however, be respected in this matter. If the flow is supercritical at a boundary, **two** conditions are to be imposed at the **inflow** section and **none** at the **outflow**. The discharge/elevation condition must not and cannot be imposed at an inflow section. Finally, if discharge hydrographs are imposed at all limits of the model, the water volume which was stored initially within the model will have no chance to change. In other words, an elevation or rating curve condition permits the system to adapt with time to this condition and the initial state loses its importance with time. Because of this it is possible to start the computations with inaccurate initial conditions and still obtain correct results after a few tidal cycles ('cold start'). If all boundary conditions are expressed as a $Q(t)$ relationship, the initial state is of utmost importance and must be given accurately.

Tools available on the market should of course accept natural initial conditions and deal correctly, from the numerical point of view, with all physically sound boundary conditions. Unfortunately, some of them do not live up to expectations. In some numerical methods it is very difficult to implement all desirable conditions accurately. In most explicit finite difference schemes, the only correct way to do so is through a locally-applied method of characteristics (Cunge *et al.*, 1980). Since this is complicated and costly, shortcuts are used which are often questionable. In 'staggered' schemes, which are schemes in which the discharges Q and elevations y are not computed at the same computational points, interpolations have to be used, and these may introduce serious errors unless they are done carefully.

The inland boundary condition may be a practical problem. The most common condition used is the discharge time-varying hydrograph. This condition, however, must not be used at the boundary which is affected by the tide. Indeed, if the discharge is imposed at such a boundary lying within the region of tidal influence, it acts as a rigid wall (or a dam) and provokes an artificial reflection of the tidal wave. The remedy is to extend the model as far as the section lying beyond tidal influence. If this is impossible or impracticable, a **radiation (or weakly-reflective) boundary condition** should be used, which lets all waves from inside the model cross the boundary with negligible reflection. Such conditions can often only be attained approximately.

A seaward boundary condition is usually imposed in the form of a water stage hydrograph, i.e. either a recorded tidal level or an artificial tide (e.g. only the main, M_2, sinusoidal component). Care must be taken to ensure that this condition is really independent of conditions inside the modelled area. A typical example of an erroneous approach is to calibrate a model using a tidal record near the river mouth and then to change considerably the topography of the river (diking, dredging, etc.) and to run the modified model with the same boundary condition. Indeed, a tidal level recorded before modification of the estuary may well be (actually always will be, the question is how much) affected by these modifications. The remedy here is to extend the model seaward (possibly to build a two-dimensional model which will compute the boundary condition for the nested inner one-dimensional model) or, again, to accept the approximation of a radiating weakly-reflective boundary condition.

11.5 VALIDATION OF SIMULATION TOOLS

There is a distinction to be made between validation and calibration. The latter can be considered as a part of the former. The validation of a tool has a general meaning: a tool is designed to be used in numerous situations for various rivers and purposes. Its validation implies that the limits of application (with definition of accuracy) have been explored and that within these limits the user can be confident of the results obtained, provided that the rules defined in the user's manual and other documentation have been followed. Calibration means repetitive simulation of past, observed events for a given water course, while the empirical and otherwise inaccurately defined coefficients (e.g. friction coefficients) are modified until an acceptable coincidence between the computed and observed results is reached. Calibration problems are dealt with later. As far as validation is concerned, only very few of the available tools are accompanied by a validation file. Validation is usually considered effective and proven on the basis of a few scientific publications, some of them dealing with the numerical characteristics of the finite difference scheme or algorithms, some of them showing that the software has been used to simulate observed real-life situations accurately.

While such data are important, it is extremely dangerous to consider positive comparisons of computed and observed results as software validation. Firstly, field-measured observations, except for extremely detailed laboratory data, are never sufficiently numerous, repetitive and accurate to be representative of all phenomena whose simulation is sought. Secondly, such a process implies more calibration than validation. It is nearly always possible, for a skilful modeller who knows his software, to reproduce observed data. Thus there is not much reason to be confident in the results of one's specific simulation simply because a correct coincidence between simulation and observation has been shown in some other case, run with the same software.

Proper validation should comprise a very detailed analysis of all aspects of the numerical method used. It should also prove that the results, in all possible admitted cases, satisfy the basic physical laws such as conservation of mass and momentum, on which they are founded, and that all boundary conditions are really well imposed (e.g. that the output discharges at a limit with an imposed $Q(t)$ hydrograph correspond at a given time exactly to the imposed values), and that the steady state solution obtained after keeping boundary conditions frozen coincides with the known analytical solution, etc.

We strongly advise using models of simple topographical and geometrical situations for validation purposes and multiplyir simulations of the cases for which solutions are known or are physically obvious. The most interesting features and predictive capability of a given software can be seen from such models. A comparison of observed and computed results is interesting *per se*, but it is not a proof of software quality, and at any rate not a proof of its predictive capacity.

A proper validation file, possibly completed with specific tests taking into account the user's own aims related to the applications in which he is interested, obviously increases the software cost, but should be mandatory. It should be profitable for the user to pay more for a tool in order to know the limits of its application precisely.

11.6 REPRESENTATION OF TOPOGRAPHIC, HYDRAULIC AND TOPOLOGICAL FEATURES: NECESSARY DATA

Anyone using a numerical modelling tool, be he a consultant, designer, manager of industrial installations or of a coastal area, planner, etc. needs it to simulate either existing situations or modified, future situations, or unobserved, exceptional situations. One-dimensional simulations are concerned with rivers which may or may not have adjacent inundated plains, with man-made canals, harbour basins whose elongated shapes make it possible to use a one-dimensional approach, or locks. In order to simulate flow in such water courses or man-made installations, it is necessary to describe their topography (geometry) as well as their hydraulic characteristics.

The definition of necessary data stems from the hypotheses upon which the flow equations are based. Any water course or canal must be first conceptually divided into reaches along which the De Saint-Venant hypotheses apply. Typically, such reaches should neither encompass rapid changes of cross-section nor singular head-losses or control structures. This implies that numerical tools should accept the concept of reaches along which gradually-varied unsteady flow equations are numerically solved and which are connected together through laws such as local head-losses, sudden cross-section variations, weirs (free-flowing or submerged), gates, dams etc. Quite obviously any tool which is not flexible enough to accept such sub-division of a simulated water course cannot be recommended. This flexibility concerns not only existing situations which are to be simulated, but also possible future situations which might subsequently be studied. A user must be able to insert at any particular cross-section of his choice a weir, a dam, a diversion, a widening, a defluent or a tributary, etc. These general requirements are satisfied by most reliable tools found on the market and at the same time they specify the topographical–geometrical data which the model builder needs.

For the reaches along which the De Saint-Venant hypotheses apply, it is necessary to have the longitudinal bed profile, a number of cross-sectional profiles sufficient to describe the reach, and information on inundated areas or inundation conditions. The latter include maps of inundated areas and longitudinal profiles of dikes. All these data must be tied in to the same geodetic altitude network and must be drawn (or otherwise presented) to a sufficiently accurate scale. There is no rule concerning the scale: engineering judgment and data availability decide this, but the quality of simulation is closely related to it.

For the special points (singular head-losses, weirs, etc.) which are also called internal boundary conditions and which link the above-mentioned reaches, as detailed a description as possible of their geometry and purpose is needed. Thus for a sudden widening of a canal, two cross-sections will be required, one on each side of the widening, and an estimate of the head-loss coefficient is required, especially for expanding flow. For a weir, a definition of the weir type will be required (practical shape, broad-crested, etc.) in order to chose an appropriate flow formula, and the weir width, elevation and discharge coefficients will also be required.

One must ask at this point whether or not a selected tool can actually simulate the internal boundary conditions required by a user. If the latter is interested in simulating changes in the flow regime of a tidal stream introduced by a weir, the question is whether a proposed tool can simulate weirs at all and, if so, what kind of weirs? Available tools usually offer a number of typical structures and the possibility of introducing new structures if the ones desired by a user are not included in the system. This introduction may be straightforward or complicated and costly. If the structures designed or to be tested by the user do not belong to the set included in the software, then the user can find himself facing an alternative: to pay (usually a high price) for modification of the software or to accept that the structure modelled will not be the one he would like to use.

Representation of anything more complicated than a simple one-reach canal brings in the problem of inundated areas and the topology of the network. Any useful software tool should be able to simulate accumulation basins separated from the main stream by weirs with or without gates. This does not mean that such a tool can readily simulate polders (used for flood alleviation purposes) or the plains inundated by floods or by tides. Flow over such plains is not really two-dimensional; it is more like the flow filling up a network of cells (reservoirs) interconnected by dis-

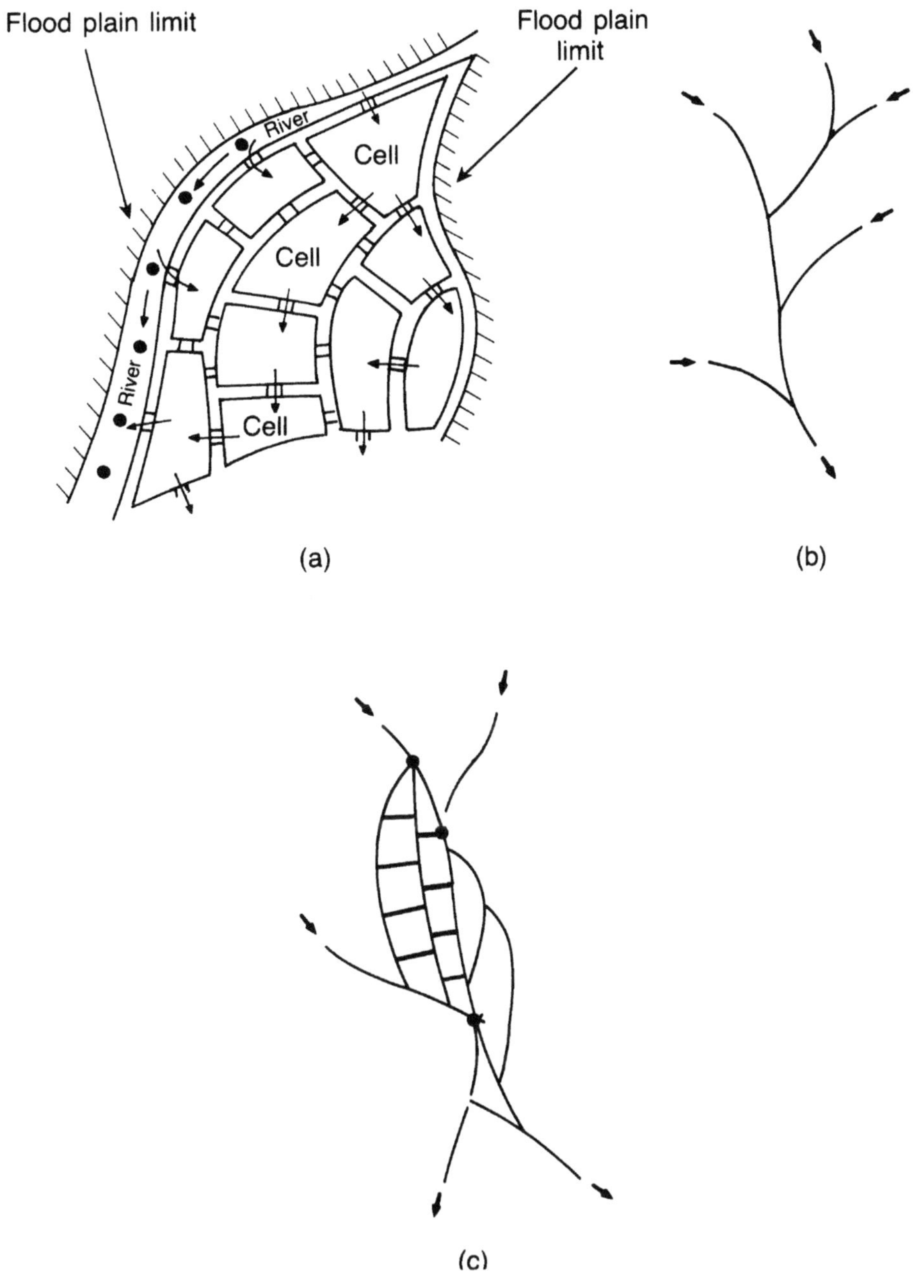

Figure 11.2 (a) schematic river-plain interactions (model has both riverine and flood plain flow); (b) dendritic and multi-connected canal networks; (c) real-life model of looped network of rivers and plain cells, under tidal influence, built to simulate Mekong Delta.

charge links (dikes, roads, gates, culverts). Such a network is obviously multi-connected and the flow in it cannot be described by the De Saint-Venant equations. In a correct software tool, the inundated plains and rivers (or canals) must be algorithmically interconnected: the flows from the plain into the river and vice versa, depending on varying unsteady conditions, have to be simulated. It is up to the user to define the topology of such a network (limits of cells, type of links, location of links with a river etc.) but it is not acceptable for the flow direction to be predefined (e.g. from the river towards the plain only) or disconnected (e.g. with the river levels being computed for the new time step first and then flow from the river to the polders or the plain in a second separate step).

Topologically speaking, the river (or canal) network conveying unsteady flows described by the De Saint-Venant equations can be dendritic (branched, tree-like) or multi-connected (looped). For many applications, the dendritic network may be sufficient, but to have a tool which cannot simulate looped networks is a limitation which must be kept in mind and which may be serious, as most deltaic regions are covered by networks of tidal rivers or canals which are looped.

A complete tool should be able to describe looped networks connected with an inundated plain cell network, with the possibility of placing internal boundary conditions (structures) anywhere. The user must be able to supply all necessary data describing topography, geometry and topology of the overall network. The various features discussed above are shown in Figure 11.2.

From the hydraulics point of view, the user must supply the first estimate of roughness coefficients and head losses. These values are then calibrated (see Section 11.7) but it is important for the user to know exactly how the tool takes these data into account. Some tools consider only double trapezoidal sections (main bed and overflow bed) and represent energy dissipation by composite formulae with single-valued but depth-variable Manning coefficients. Such a representation would not be convenient for larger rivers or for extrapolation beyond observed situations for which the conveyance factors are known. Nevertheless it is quite representative for trapezoidal man-made canals. A tool can be useful, but its application limited.

Obviously the user must supply all boundary conditions, both for calibration periods and for simulating future situations. It must be stressed that the definition of some boundary conditions may require hydraulic expertise and very flexible tools (*viz.* weakly-reflecting boundary conditions). The user must also supply all the data necessary to calibrate a model, namely the free-surface elevations recorded at a number of sections during the same tide for which boundary conditions are known.

11.7 CALIBRATION AND EVALUATION OF RESULTS

A mathematical model is a simplified, discrete representation of a complex physical flow situation. Three-dimensional terrain features are represented by one-dimensional elements and the flow is assumed to obey differential equations in which empirical coefficients appear. Strictly speaking, model calibration should be the process of adjusting the values of empirical coefficients so that flow events simulated in the model will reproduce as faithfully as possible the comparable natural events. According to this ideal situation, once a model is built (i.e. all geometric features along a modelled river are introduced into the database of a modelling system) it is enough to have recorded a number of water-stage hydrographs along the river, including those at the limits of the model. The simulation can then be run with boundary limits imposed and the changes to be made in the conveyance factors at the computational points can be deduced by comparing recorded and computed intermediate hydrographs. Indeed, the only empirical coefficient in Equations (11.1) and (11.2) is the roughness coefficient included in the conveyance factors. Such an approach is acceptable if the topography is represented with 100% accuracy, if the measurements of boundary conditions and intermediate hydrographs are accurate and if the hypotheses on which the equations are based are never violated anywhere.

Such an ideal situation does not exist and model calibration often requires adjustment of the dimensions of geometric elements and modification of boundary conditions. A model's potential for reproducing and predicting real flow events, and the potential quality of its calibration, depends on the amount and quality of topographical, topological and hydraulic data available for the watercourse. The questions of model calibration and data needs cannot be separated. Figure 11.3 gives a comparison of computed and observed hydrographs during the calibration process for a one-dimensional model. Is such a coincidence acceptable? There is no straightforward answer to this question. Indeed, only the user who knows the model's objectives and the hydraulics of the watercourse can answer such a question. But if the answer is negative and further calibration is required, then a number of alternative possibilities arise.

The first alternative is to modify the roughness coefficients. To modify the Strickler–Manning or Chézy

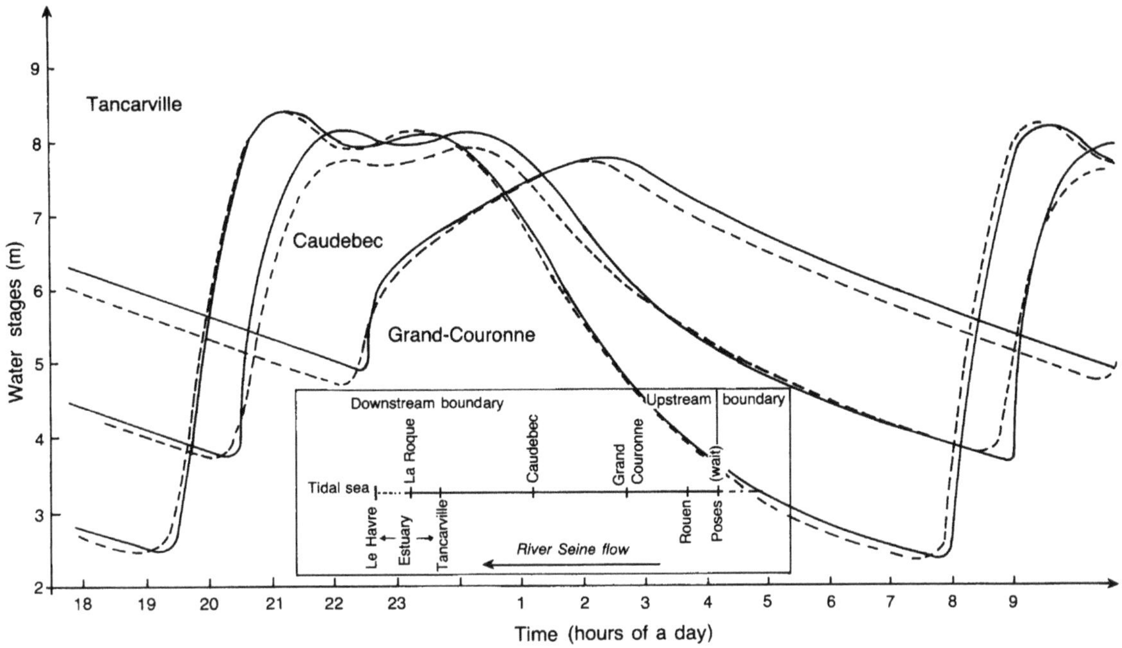

Figure 11.3 Comparison of observed and computed hydrographs for a tidal section of River Seine; (——) observed hydrographs; (– – –), computed hydrographs.

coefficients may or may not significantly affect the results. In tidal areas it generally only has a moderate effect unless the coefficient values used are physically absurd. An example of the differences is shown in Figure 11.5. Physically, the reason for such behaviour is clear: inertia terms have more weight than resistance terms in the equations.

The lack of coincidence may also be due to discretization, either in time or in space. Thus the time step can be adjusted, presumably by decreasing it, e.g. dividing it by two and repeating the computation in order to see if the result is better. It should be borne in mind, however, that systematically decreasing Δt does not enable the convergence process to be verified, unless the space step Δx is also decreased. During the refining process the Courant number Cr must be kept constant. It must be chosen with a view to correct reproduction of the most important phenomena for the study. If the important feature is the M_2 main component of the tide, then a Cr very different from 1.0 can be chosen. This is possible, since the designer is not interested in detailed modelling of steep fronts. If, however, the appearance of a bore and its propagation is important, then Cr must be as near to unity as is possible. An example of the differences is shown in Figure 11.7.

Another reason is inadequate representation of the geometry and topology. It is often claimed that resistance to flow in tidal reaches is not symmetric: different coefficients should be used for the flow and ebb phases. While this statement is disputed in the published literature, it is possible to show that the introduction of an accumulation area (or a polder) into the model can have effects comparable to an increase in roughness during ebb flow.

Yet another explanation for the lack of coincidence may be the inadequacy of the basic hypotheses. It is possible that the cross-sectional form is too complex to be approximated by the one-dimensional hypotheses. Sometimes the results can be improved by introducing 'live' cross-sections or accumulation areas, which enable the simulation of storage, or by introducing the Boussinesq coefficient. Thus similar results are obtained in Figure 11.6 by introducing a polder or by decreasing the roughness coefficient by 25%.

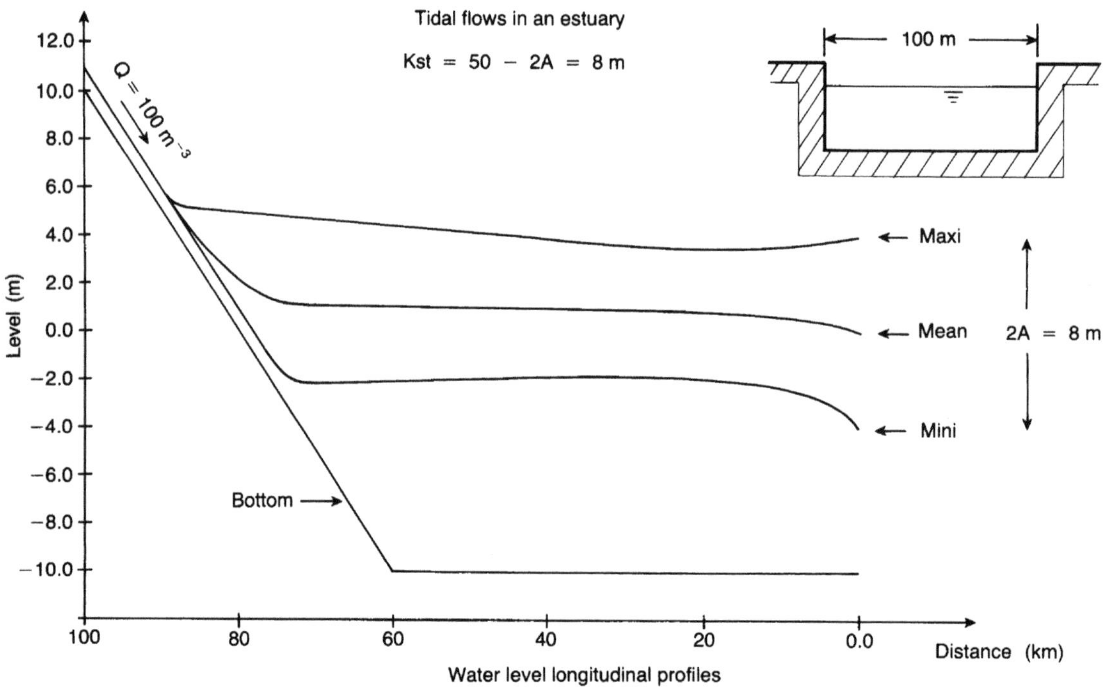

Figure 11.4 Longitudinal profile of schematic estuary. Kst = 50 − 2A = 8 m.

Thus, during calibration, one can obtain the **same degree of coincidence** in various ways. If the methods used correspond to physical reality, the predictive capability of the model increases. If the calibration phase is conducted without a sufficiently deep insight into the physical side, the modeller moves towards a 'black box', deprived of predictive capability. In conclusion, the simulation tool cannot be applied blindly but must be used by an engineer who is fully acquainted with hydraulics and the physical problem involved.

11.8 SPECIFICITY OF COASTAL ONE-DIMENSIONAL MODELS

The tools used for one-dimensional simulation of rivers in coastal areas may be the same as general tools used for long-wave propagation along inland watercourses. Nevertheless there are typical maritime features which should be taken into account and which must be capable of reproduction by such tools. Some of them are described in the sequel.

1. The equations used must comprise inertia terms.
2. Differential equations and their numerical solution must conserve momentum and, when (Δx, $\Delta t \to 0$, Cr = constant) their solution should approach the discontinuous solution (mobile hydraulic jump or bore). This feature is shown in Figure 11.8: the tidal wave steepens and eventually a discontinuity may develop. Such a discontinuity will propagate as a steep front and it is important to know whether or not the tool employed will reproduce or converge to this situation.
3. It must be possible to implement radiating weakly-reflective boundary conditions (see Section 11.4).
4. It must be possible to take properly into account density differences due to the saline intrusion.

One particular feature of models of canals or elongated harbour docks or rivers influenced by tide is the danger of numerical resonance. There is a real danger that the combination of incident-wave frequency, length of modelled watercourse, reflections from boundary conditions and Courant number will 'capture' standing spurious waves of increasing amplitude which eventually reach unrealistic equilibrium with friction forces. A very good description of the phenomenon can be found elsewhere (e.g. Vreugdenhil, 1989). A radiating, weakly reflective, boundary condition again enables spurious waves to be radiated out of the system.

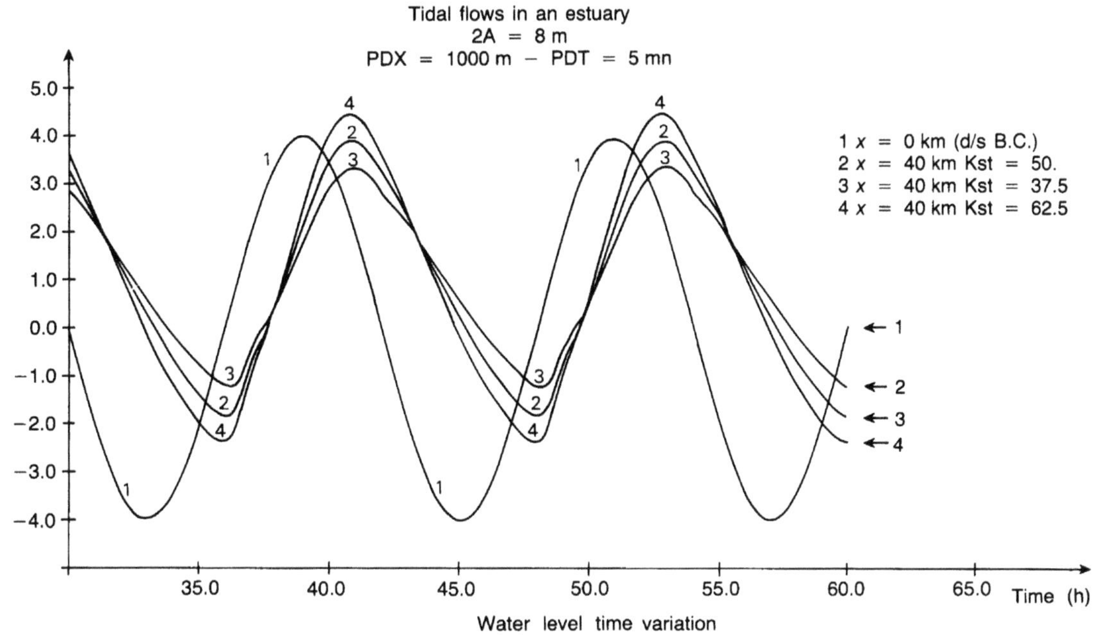

Figure 11.5 Comparison of water-level time variation at km 40 for three different values of Strickler coefficient. Tidal flow in estuary 2A = 8 m; PDX = 1000 m – PDT = 5 mn.

11.9 SCHEMATIC MODEL FOR SAMPLE AND ILLUSTRATIVE COMPUTATIONS

The computation results shown in Figures 11.5–11.8 have been obtained on a schematic model representing a 100 km long estuary having a constant 100 m width rectangular cross-section. As illustrated on the longitudinal profile (Figure 11.4), between km0 (the sea) and km60, the bottom is assumed horizontal, 10 m under the mean sea level. Between km60 and km100, the bottom has a constant 0.5 m km^{-1} slope. All computations have been run with the same boundary conditions: a constant 100 $m^3 s^{-1}$ water discharge upstream and an 8 m amplitude sinusoidal tide downstream.

Figure 11.4 also represents the longitudinal profiles of the minimum, the mean and the maximum water elevation in the estuary for a 50-Strickler roughness coefficient.

Figure 11.5 represents the results of an investigation into the effect on the water elevation of a roughness coefficient variation (±25%). The deviation in the water level at an intermediate location (km40) is not negligible, but still small with respect to the large difference between the roughness coefficients.

In the test represented by Figure 11.6, the effect of a 1 km^2 polder (≈10% of the estuary surface) situated at km60 has been investigated. Again, the water level at km40 is plotted. The polder is seen to have a damping effect comparable to a diminution of the roughness coefficient by 25%.

Between km60 and km80, the tidal wave front notably steepens. Figures 11.7 and 11.8 visualize the arrival of the bore at three locations (km70, 73 and 76). In the test illustrated by Figure 11.7, the effect of the refinement of the computational grid on the steepness of the front has been investigated. In the test the Courant number Cr is kept constant and equal to one. It is evident from the results that the correct representation of the bore necessitates a very fine grid: the results are much better with (Δx = 100 m; Δt = 12 s) than with (Δx = 1000 m; Δt = 120 s), and indeed would be still further improved with a still smaller mesh.

The correct representation of the steepness of the front also necessitates the choice of the Courant number Cr as close as possible to one. The damping effect of a Courant number taken much bigger (Cr = 10) is clearly seen in Figure 11.8.

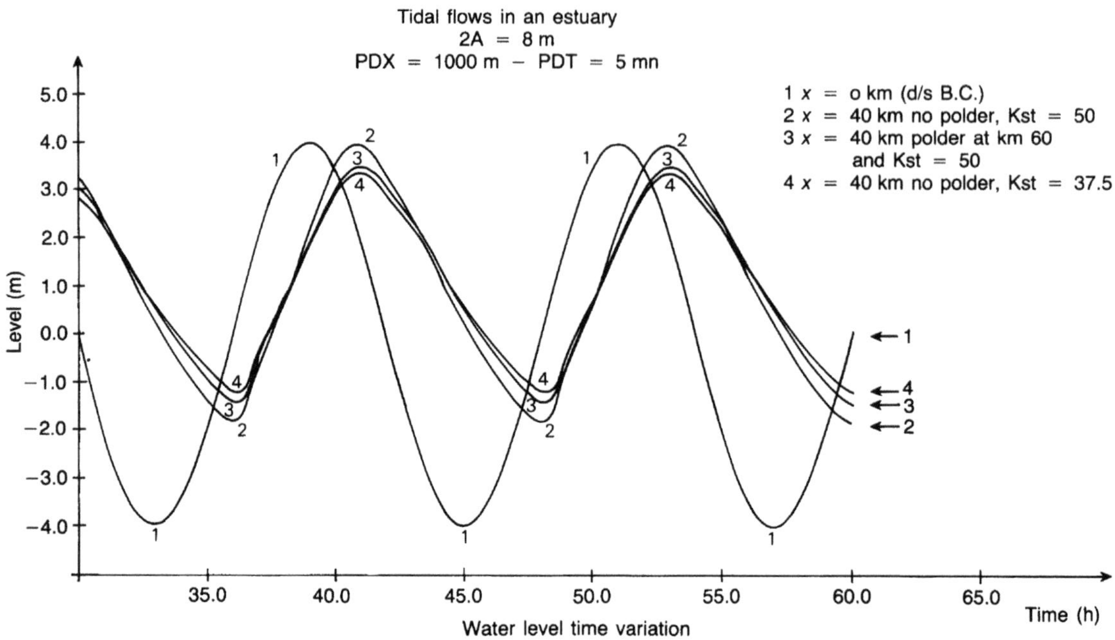

Figure 11.6 Comparison of damping effect on water-level of roughness and of a polder. Tidal flows in estuary 2A = 8 m; PDX = 1000 m − PDT = 5 mn.

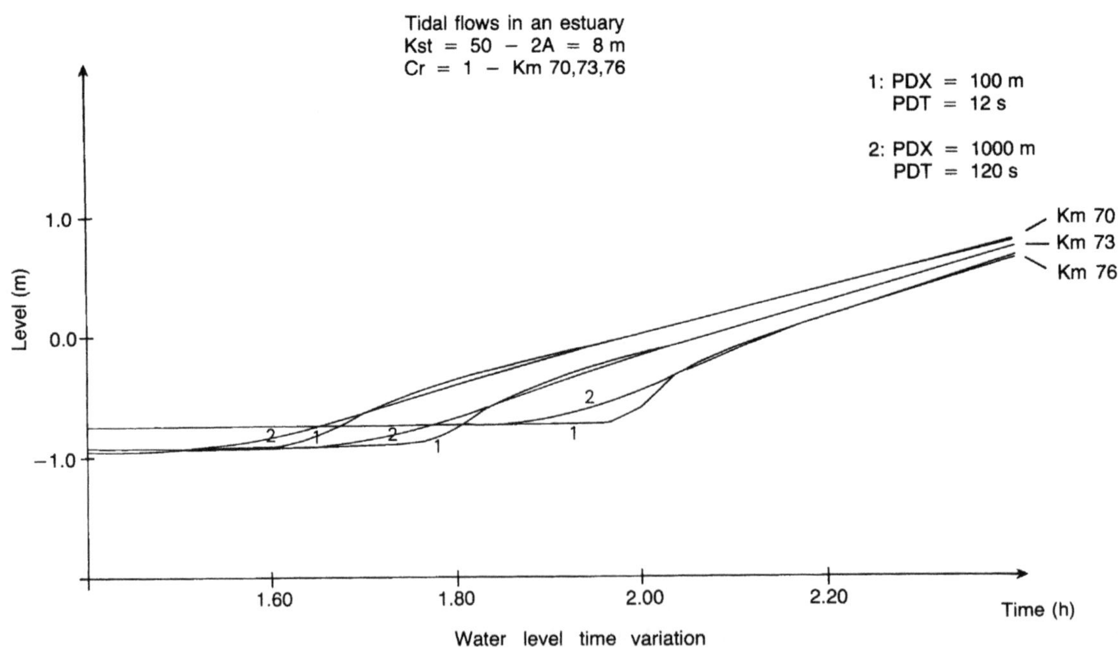

Figure 11.7 Effect of grid refinement on bore representation. Tidal flow in estuary. Kst = 50 − 2A = 8 m; Cr = 1 − km 70, 73, 76.

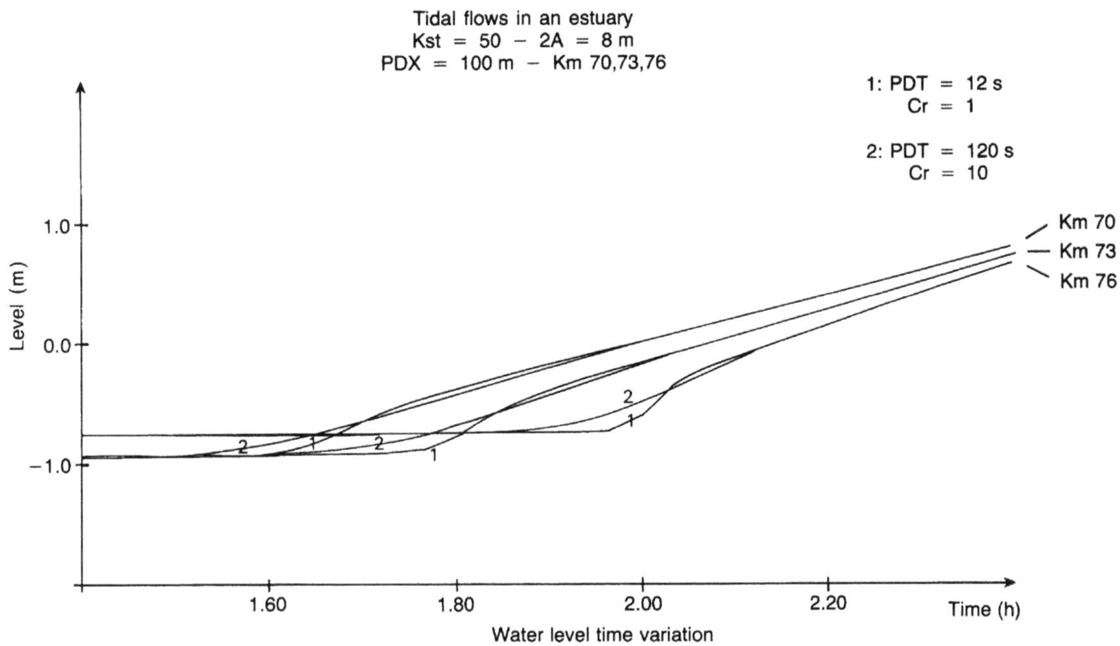

Figure 11.8 Effect of courant number on bore representation. Tidal flow in estuary Kst = 50 – 2A = 8 m; PDX = 100 m – km 70, 73, 76.

11.10 PRESENT AND FUTURE GENERATIONS OF TOOLS, AND SOFTWARE DEVELOPMENT COSTS

The gradual development of improved numerical modelling systems can be described in terms of generations. The first-generation systems were simply computerized empirical formulae. The second-generation systems were one-off numerical models. The third generation are generalized, mainframe-based numerical modelling systems which can generate unsteady flow models for specific situations. Third-generation modelling systems dominated the decade 1975–1985 and are still the basis for most major applications. The advent of the third generation meant that numerical modelling became a practical engineering tool bought or developed by a number of major consulting companies.

The fourth generation are menu-driven modelling systems with integrated database facilities and graphics. The material base for the fourth generation was furnished by the explosion in the development of cheap and powerful microcomputers. From the hydraulics point of view there is little difference between the third and fourth generations, and actually the most efficient and reliable fourth-generation systems are directly derived from their third-generation elder brothers.

The fourth-generation systems are usually **thematically limited**. They cannot normally cover the same range of applications as the mainframe third-generation codes. However, they offer an easy way of handling typical applications difficulties, including the inputting and outputting of data. The fourth-generation systems are very efficient tools for the experienced user (who, however, has no access to the source software) but there is still an important step needed before the systems can be sold to and used by the average engineer without a wider knowledge of hydraulics. The latter observation indicates a need for future systems of a fifth generation which will integrate such knowledge (integration is to be emphasized here, as opposed to a number of rules simply added as an appendix to a system and pompously called an expert system). The fifth-generation will appear within the next few years, but it is not here yet.

Concerning the effort needed to develop software tools, let us consider the sequence of steps which should normally be followed when developing a software system.

1. Feasibility: this step covers the definition of the preferred concept for the software product, its feasibility and superiority over alternative con-

cepts (in other words, choice). The innocent word 'feasibility' actually covers almost the entire span of activities which are often called 'developments': numerical algorithms, stability and convergence problems, etc. all come under this heading. When a software manufacturer starts a job, all these problems are assumed to be solved.
2. Specifications: this step corresponds to the definition of requirements (functions, interfaces, performance) and to the product design.
3. Software development: this step covers: detailed design, coding, verification, validation, integration and implementation, documentation.
4. Maintenance.

A software manufacturer's effort, necessary for carrying out steps 2 and 3 alone, can be estimated as a function of kilo-delivered-source-instructions (KDSI), i.e. a fully checked, verified, validated and documented source (for example FORTRAN) software. A very rough estimate of the number of man-months (MM) needed can be made using the following formula:

$$MM = 2.4*(KDSI**1.05)$$

Typically a river flow modelling system comprises at least 15 000 FORTRAN instructions, i.e. 15 KDSI. To produce such a system, once all numerical problems have been smoothed out and all the algorithms validated, requires an effort of 42 man-months (±25%)!

What is customarily termed development within the university research community corresponds to step 1 and its product is usually a 'personal code', i.e. a pre-competitive software prototype which can only usually be run by its author and serves only to validate the methods and algorithms employed. The real advance in computational river hydraulics, linked to the advent of third- and fourth-generation software is the recognition of this difference. Indeed, such a personal code is not a reliable tool for an engineering design office. Reliability here covers two aspects: not only programming (percentage of bugs still there) but also the range of applicability which enables us to assert that the result (graphical output in report-plan form) will be produced at a well defined cost and within a well defined period. In other words, a reliable engineering tool leaves very little space (if at all) for any risk of trouble in obtaining the results required, at least so far as relatively standard situations are concerned.

11.11 AVAILABLE SOFTWARE

The following list of tools belonging to some of the best known European Institutes which specialize in numerical simulation is by no means exhaustive. The information presented below was supplied in all cases by either the owners or developers of the software, who are solely responsible for its accuracy and value.

Danish Hydraulic Institute (DHI), Agern Allé 5, DK-2970 Horsholm, Denmark

MIKE 11 is a general commercial package for micro-computers which simulates river hydrodynamics, sediment transport and water quality processes. The hydrodynamic description may be based upon complete or reduced forms of the De Saint Venant equations. Special formulations are included for flow over or through structures. Possibilities are also included for simulating operation of structures according to prescribed operation strategies. A pseudo two-dimensional description can be applied for flow over flood plains with a flood cell description (Danish Hydraulic Institute, 1990).

Hydraulics Research Ltd, Wallingford, Oxfordshire OX10 8BA, UK

TIDEFLOW-1D simulates and predicts the interaction of tidal and fluvial flows in branched and looped channel networks of any size. It works with cross-section averaged flows and can handle deltas, varying tidal regimes and multiple river flows. It will predict levels, tidal discharges, residual flow patterns, and area mean velocities. Special features include: velocity variation across the section through modified hydraulic radius; inter-tidal water storage outside the main channel; roughness variable as function of sand transport; bed level and cross-section variable with erosion and deposition; and compatible subroutines simulating weirs, sluices and tidal barrages (Hydraulics Research Ltd., 1990).

Institut für Strömungsmechanik und Elektronische Rechnung im Bauwesen, Universität Hannover, Appelstr. 9a, D-3000 Hannover, Germany

TICAD (Tidal interactive computation and design) simulates one-dimensional branched river and canal systems. It is based on a finite element implicit solution technique and is linked to a digital-terrain model. It allows for structures such as free-flow and drowned weirs and barrages, time- and flow-dependent opening functions, non-linear optimization of opening functions such as maintaining constant level or minimization of sediment transport. The software has been applied for flow and storm-surge simulations in tidal rivers (Holz, 1990).

Laboratoire d'Hydraulique de France (LHF), 6 Rue de Lorraine, 38130 Echirolles, France

The CASCADE modelling system is a third-generation system for simulating unsteady open channel flows in branched or looped networks of canals and rivers. It is based on full momentum and mass-continuity equations written in conservation form and solved numerically by the four-point Preissmann finite difference conservation scheme. It comprises a salt transport (convection-diffusion) module. It is adapted to the simulation of tidal flows including bores and their interaction. Exposed intertidal areas and inundated plains are simulated in full interaction with main channel flows. The CASCADE system allows for insertion of any kind of hydraulic structure (weirs, dams, dikes, tidal gates, orifices, etc.) and comprizes a special regulation interface which permits the user to develop and test both operating rules for manoeuverable structures and the strategies of flow control. The CASCADE system is especially well adapted to and has been already inserted into a user's own informatics environment (comprising databases, graphical pre- and post-processors) (Laboratoire d'Hydraulique de France, 1990).

Laboratoire National d'Hydraulique (LNH) – Electricité de France, B.P. 49-78401, CHATOU Cedex, France

LNH has a number of one-dimensional codes based on the De Saint-Venant equations, each of them adapted to specific problems:

1. LIDO code for a single reach;
2. SARA code for a branched network;
3. REZO code for a looped network;
4. CASIER code for cell-like models which can be coupled with the above-mentioned codes.

These codes can represent most singularities and tributaries as well as overflows (dealt with by CASIER). One of the most important parameters of these codes is bed roughness, which can be automatically and optimally estimated by still another code, CASTOR, (Laboratoire National d'Hydraulique, 1990).

REFERENCES

Cunge, J.A., Holly, F.M. and Verwey, A. (1980) *Practical Aspects of Computational River Hydraulics*, Pitman, London.

Danish Hydraulic Institute (1990) (Private communication).

De Saint-Venant, B. (1871) Théorie du mouvement non-permanent des eaux avec application aux crues des rivières et à l'introduction des marées dans leur lit. *Académie des Sciences, Comptes Rendus*, **73**, 148–54, 237–40.

Holz, K.P. (1990) (Private communication) Universität Hannover.

Hydraulics Research Ltd. (1990) (Private communication).

Isaacson, E., Stoker, J.J. and Troesch, B.A. (1954) Numerical solution of flood prediction and River Regulation Problems (Ohio–Mississippi floods), Rep. II, New York University, Inst. Math. Sci., *Rept. IMM-NYU-205*.

Laboratoire d'Hydraulique de France (1990) (Private communication).

Laboratoire National d'Hydraulique – EDF (1990) (Private communication).

Massau, J. (1889) L'intégration graphique and Appendice au mémoire sur l'intégration graphique, *Association des Ingénieurs sortis des Ecoles Spéciales de Gand, Belgium, Annales*, **12**, 185–444.

Vreugdenhil, C.B. (1989) *Computational Hydraulics; An Introduction*, Springer-Verlag, Berlin.

NUMERICAL MODELLING OF SHORT-PERIOD WAVES 12

M.B. Abbott and P.A. Madsen

Having recently published a full review of this subject (Abbott and Madsen, 1990), the authors here provide an outline of its principal developments. Short-period waves are modelled at a wide range of physical scales, ranging from complete oceans down to small harbours. A range of modelling tools has been developed accordingly.

CONTENTS

12.1 INTRODUCTION

Within this overall area of coastal, estuarial and harbour engineering, that has itself changed so very rapidly over the last two decades, the particular subarea of numerical modelling of short-period waves has probably changed most of all. Whereas in 1980 only a few simple models of this class were in use, by 1990 a considerable range of very sophisticated models were being applied to major projects. During this period, the development of short-period wave models has occurred for all geographical scales, ranging from complete oceans down to small harbour areas. There is space here for only a brief sketch. A more extended view has been given elsewhere by Abbott and Madsen (1990), upon which this outline is based.

12.2 DEEP-WATER WAVES

These are for the most part waves that are generated and subsequently propagate and decay in the oceans and seas. The development of numerical models has here been accompanied by a number of measuring programmes, so that models could be checked against measurements. The further development of remote-sensing techniques using satellites is expected to further augment the testing programmes during the 1990s.

Earlier reference books, as represented by the *Shore Protection Manual*, provided means to make empirical estimates of wave height based on a knowledge of mean wind velocity and duration and fetch. During the 1970s, however, simple spectral methods were developed to provide more reliable estimates. These are based on the radiative-transfer equation of short-period wave energy:

$$\frac{\partial}{\partial t}(E) + \left(\frac{\cos\theta}{C}\right)\frac{\partial}{\partial x}(C\,C_g\,E) + \left(\frac{\sin\theta}{C}\right)\frac{\partial}{\partial y}(C\,C_g\,E) + \frac{1}{C^2}\left[\sin\theta\frac{\partial C}{\partial x} - \cos\theta\frac{\partial C}{\partial y}\right]\frac{\partial}{\partial\theta}(C\,C_g\,E) = S \quad (12.1)$$

where $E(f, \theta)$ is the directional frequency spectrum, C is the phase celerity, and C_g is the group velocity. The source term S represents the net transfer of energy to and from the spectrum and basically comprises three different mechanisms:

1. Atmospheric transfer (Phillips, 1977).
2. Non-linear wave-wave interactions (Hasselmann *et al.*, 1973, 1976).
3. Dissipation (Hasselmann and Collins, 1968).

Two basic approaches have been followed for applying this equation. In the first of these, the discrete spectral approach, a grid is introduced over the ocean or sea area of interest and the directional energy density spectrum is represented by a sum of separate frequency and direction bands at each and every grid point. Typically, 10 frequencies and 16 directions might be used. In the second approach, the

Coastal, Estuarial and Harbour Engineer's Reference Book. Edited by M.B. Abbott and W.A. Price. Published in 1993 by Chapman & Hall, 2–6 Boundary Row, London SE1 8HN. ISBN 0 419 15430 2.

directional energy spectrum is described in terms of its statistical mean and movements, so that typically only five parameters need to be introduced. In a further, intermediate or 'hybrid' class of model, the discrete approach is used for the swell component while only the higher-frequency part of the spectrum is parameterized. Some typical early results (from Brink Kjaer *et al.*, 1984) are shown in Figure 12.1.

12.3 TRANSPORT OF WAVES FROM DEEP WATER INTO THE COASTAL ZONE

The deep-water wave climate having been established along the above lines, and augmented by observations, the effects of this climate have to be routed onto the areas of engineering interest. One of the oldest and best established methods for doing this is by tracing or tracking the propagation paths or 'rays' of wave energy. Commonly only refraction and shoaling effects are considered, with diffraction and reflection effects being neglected.

In the case of regular waves, a popular technique is that of forward wave tracing. This is based, whether explicitly or implicitly, on the equation of conservation of wave action.

$$\frac{\partial}{\partial t}\left(\frac{E}{\omega_r}\right) + \frac{\partial}{\partial x_i}\left\{\frac{E}{\omega_r}(c_{g\alpha})_i\right\} = 0 \qquad (12.2)$$

(a)

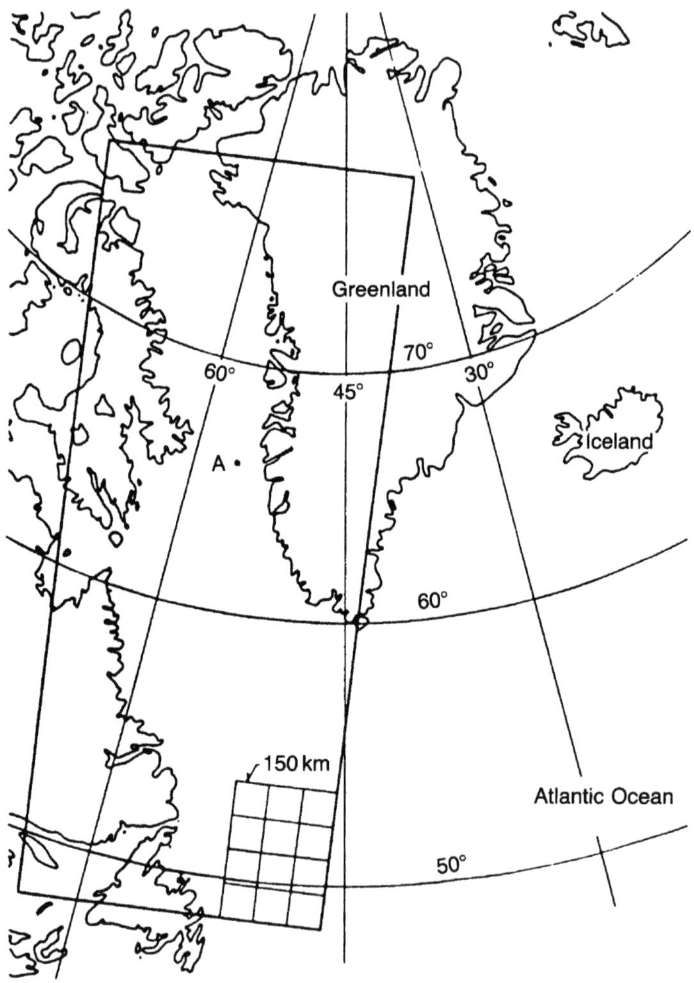

Figure 12.1 For area shown in (a) one of a sequence of barometric pressure fields, (b) is shown together with one of the corresponding sequences of fields of significant wave height (c) as obtained with a second-generation model. A typical directional spectrum is shown in (d).

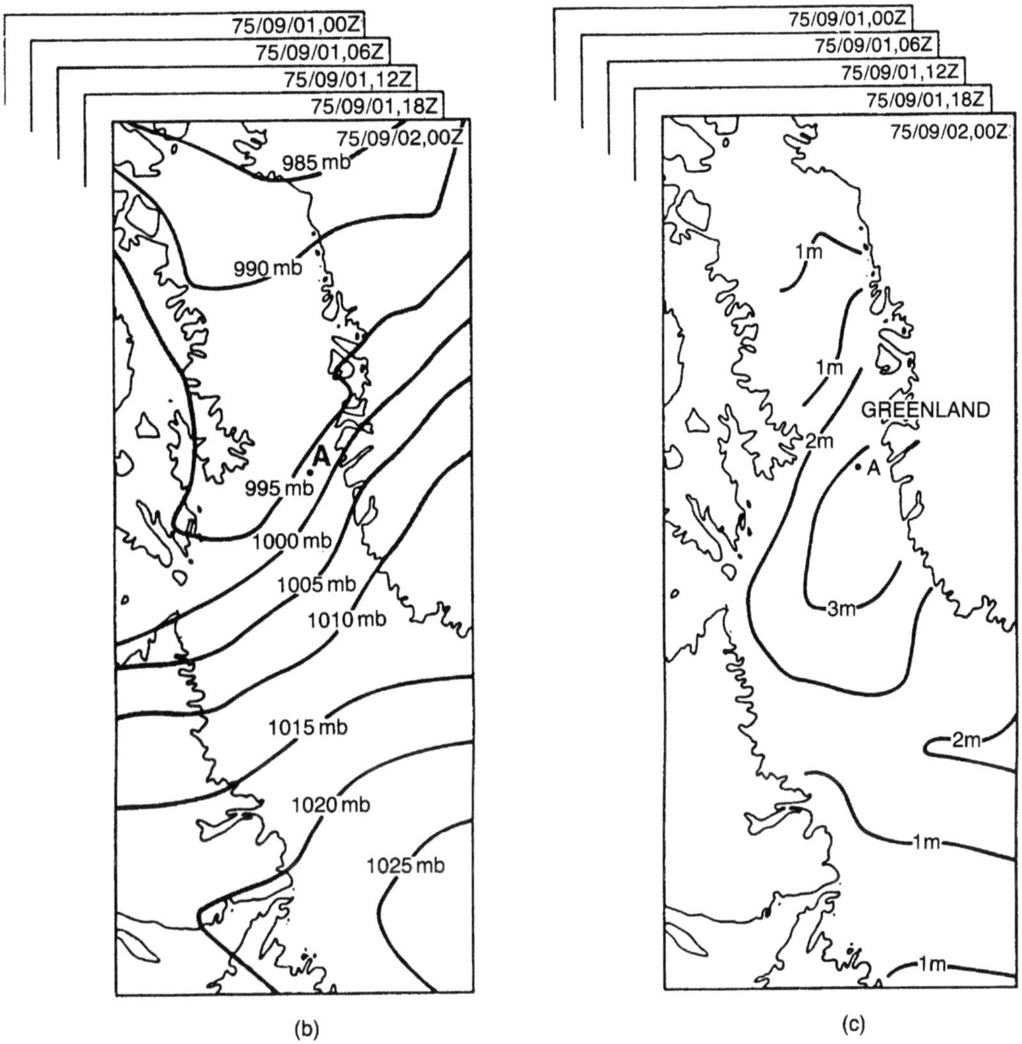

Figure 12.1 *Continued*

A wave ray path is defined as the solution (in time) to the differential equation

$$\frac{dx_i}{dt} = c_{ga} = c_{ga}(\cos\mu, \sin\mu) \qquad (12.3)$$

The wave action density A is defined as

$$A = E/\omega_r \qquad (12.4)$$

where E is the wave energy density

$$E = \tfrac{1}{2}\rho g H^2 \qquad (12.5)$$

and ω_r is the relative cyclic frequency, defined as

$$\omega_r = \omega_\alpha - kU \qquad (12.6)$$

In addition to direct wave tracing procedures, field solutions of Equation (12.2) have also been used, while the effects of refraction and shoaling with wave-current interactions have also been introduced.

These methods have been extended further to irregular waves (Abernethy and Gilbert, 1975), where a backward-tracing method has to be included in order to identify the origins of the different wave com-

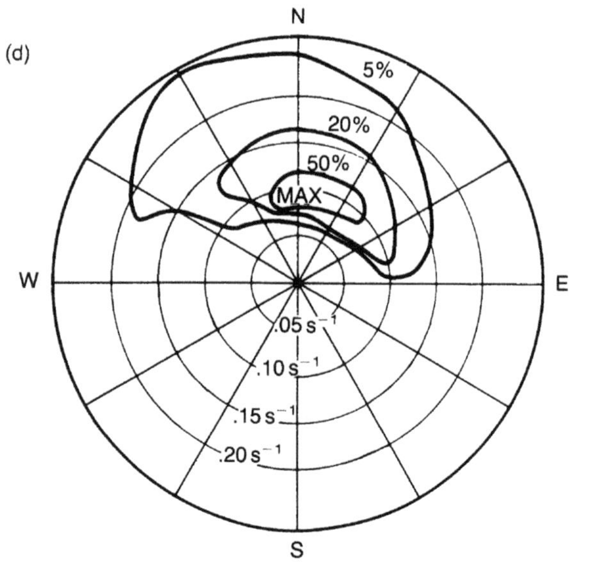

Figure 12.1 *Continued*

ponents in the deeper water. An example of such a tracing procedure, augmented with a back-tracing procedure, is shown in Figure 12.2.

When arriving at this level of resolution, the further step, of using the somewhat more general mild-slope equation (Berkhoff, 1972) which takes account of frequency dispersion but neglects non-linear effects, is justified. It provides a solution for one given frequency at a time, but is applicable in both deep and shallow water and it applies to all frequencies.

The mild-slope equation, valid for steady-state solutions only, can be expressed by

$$\nabla(CC_g\nabla\zeta) - \frac{C_g}{C}\frac{\partial^2\zeta}{\partial t^2} = 0 \qquad (12.7)$$

where ζ is the surface elevation, C is the phase velocity, C_g is the group velocity, and ∇ is the horizontal gradient operator. The phase velocity and the group velocity are related to the local still water depth, $h = h(x, y)$, through the equations

$$C^2 = (g/k)\tan h(kh)$$

and

$$C_g = 1/2\{1 + 2kh/\sin h(2kh)\}C$$

where k is the local wave number.

Very rapid means for solving these equations have been developed in recent years and they are now widely used for obtaining first, low-cost estimates of wave action. By introducing the effects of wave breaking, they can even be used over the whole coastal zone.

Models of this kind can also be used to obtain estimates of those off-diagonal components of the radiation stress tensor of periodic waves that drive longshore currents. An application of this kind is illustrated in Figure 12.3. Great care is, however, required in the use of models of this kind in view of the whole chain of assumption and approximations that are made (Abbott, 1986).

12.4 WAVES IN THE HARBOUR AREAS

Although methods based in the mild-slope equation are often used in these areas also, it is becoming more usual to use more general approaches. These are

Table 12.1

Symbol	*Description*	*Unit*
$\zeta(x, y, t)$	Water surface level above datum	m
$p(x, y, t)$	Flux density in x direction	$m^3 s^{-1} m^{-1}$
$q(x, y, t)$	Flux density in y direction	$m^3 s^{-1} m^{-1}$
$h(x, y, t)$	Water depth	m
$D(x, y)$	Still water depth	m
g	Gravity	$m\,s^{-2}$
$n(x, y)$	Porosity	
α	Resistance coefficient for laminar flow in a porous media	
β	Resistance coefficient for turbulent flow in a porous media	
x, y	Space co-ordinates	m
t	Time	s

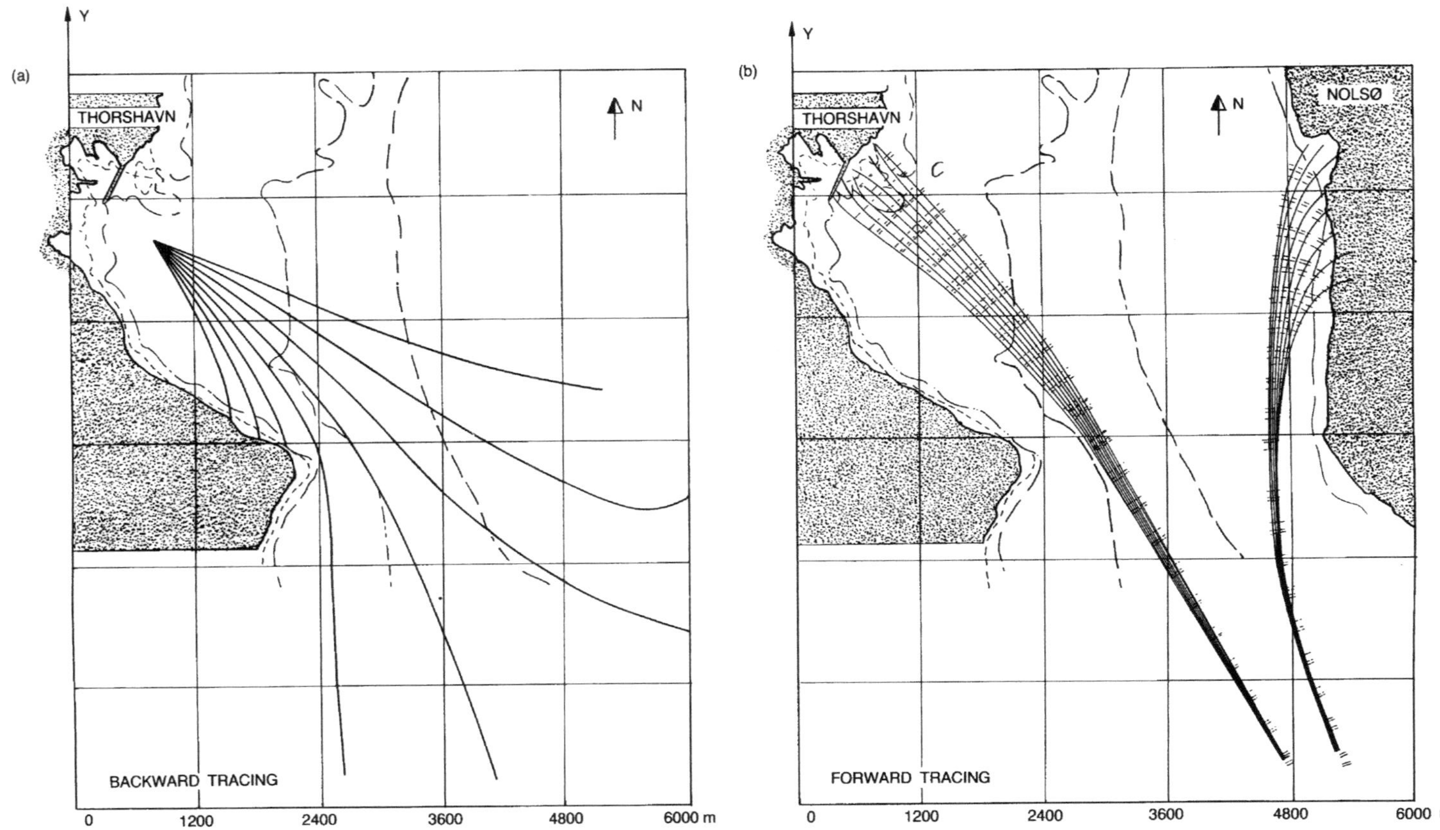

Figure 12.2 Routing of wave rays for site investigations (Faroe Islands). Backtracking from site (a) provides source points for forward tracking of short wave energy (b).

(a)

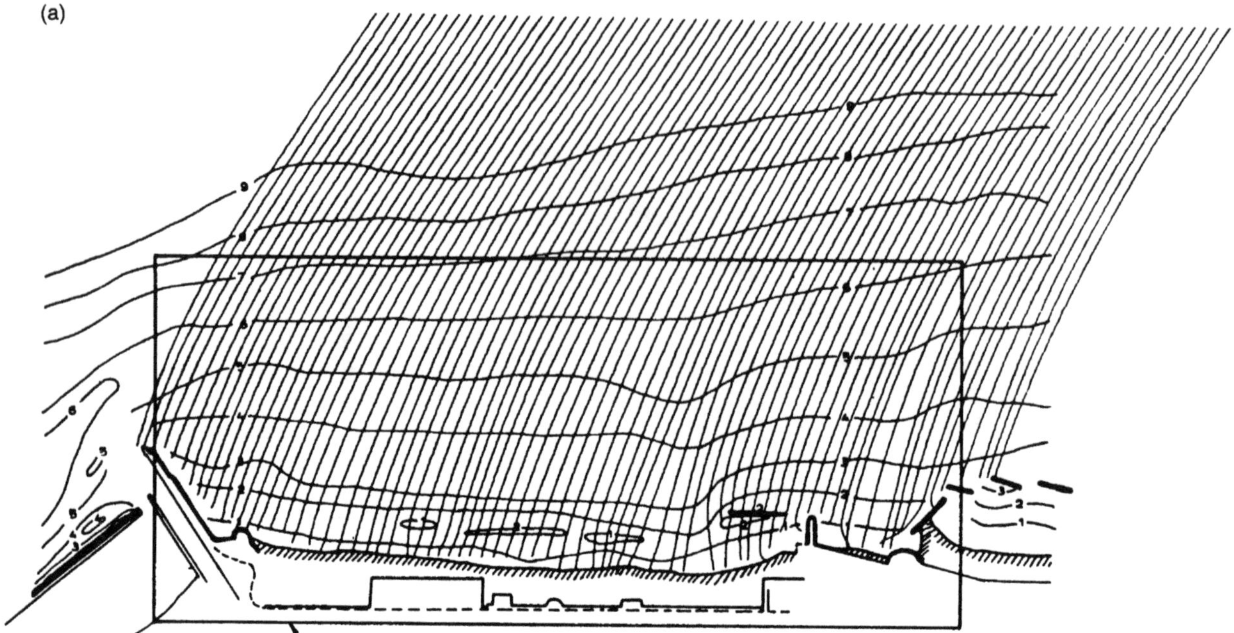

Figure 12.3(a) Model describes how waves approach beach along paths or rays. These paths or rays are shown superimposed upon the bathymetry of the area and beach coastline. It is assumed that all wave energy remaining after cresting and breaking in shallower water is dissipated at the beach, so that no reflection of wave energy occurs. Diffraction effects are not considered.

(b)

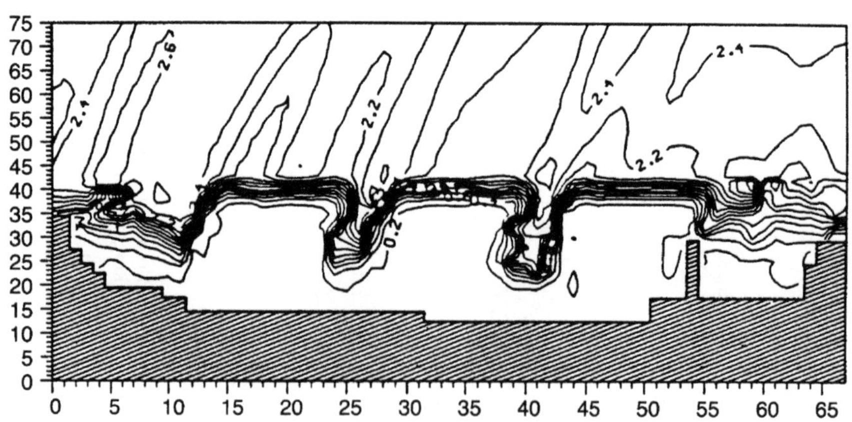

Figure 12.3(b) A contour map of one component of the radiation stress provided by waves corresponding to the rays and bathymetry of (a).

(c)

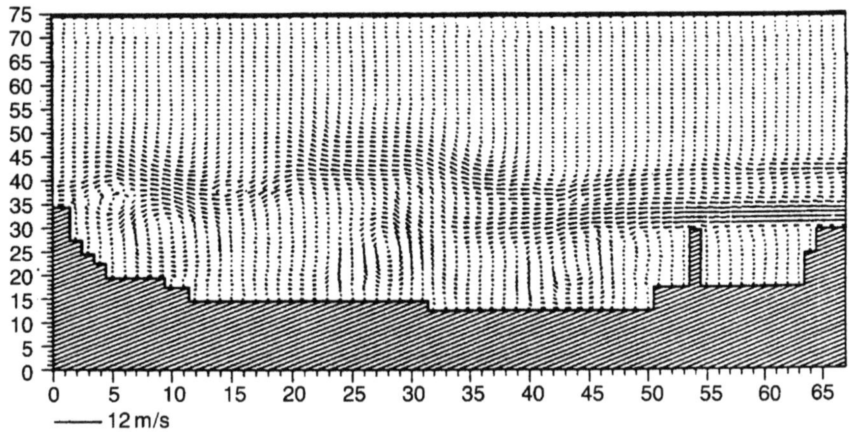

Figure 12.3(c) The flow field produced by the radiation stress of which one component is shown in (b).

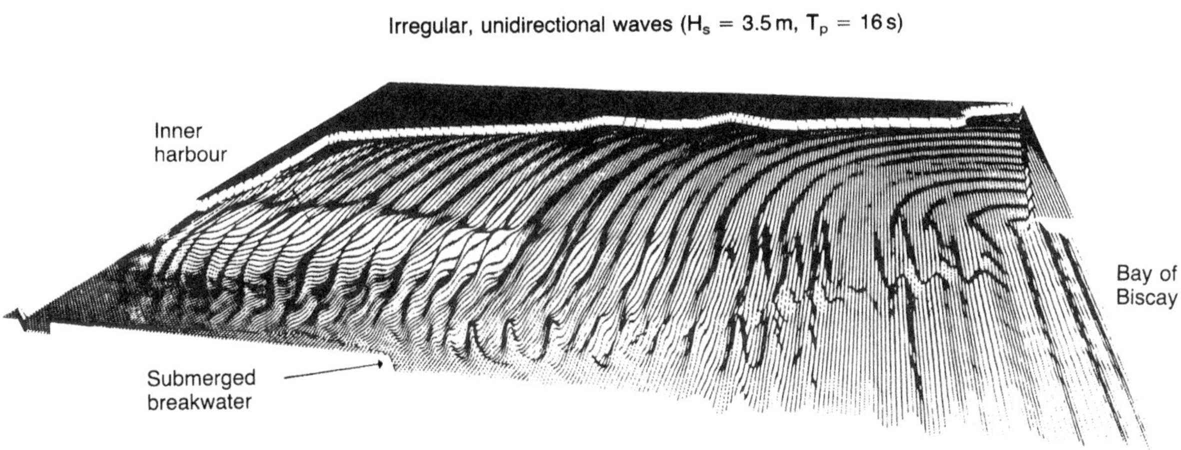

Figure 12.4 A plot of irregular wave action in the outer harbour of Bilbao. The effect of an incomplete, submerged breakwater is apparent, as is also the focusing of wave energy onto the inner harbour.

mainly based on the use of the Boussinesq equations (Abbott *et al.*, 1987; Madsen *et al.*, 1991).

Models of this kind have now been very extensively tested and summaries have been given by Abbott and Madsen (1990) and Murray and Sørensen (1991). It must suffice here to give only a few illustrations of applications. The governing equations can be written as shown in Equations (12.8)–(12.10)

Continuity

$$n\frac{\partial \zeta}{\partial t} + \frac{\partial p}{\partial x} + \frac{\partial q}{\partial y} = 0 \qquad (12.8)$$

x momentum

$$n\frac{\partial p}{\partial t} + \frac{\partial}{\partial x}\left(\frac{p^2}{h}\right) + \frac{\partial}{\partial y}\left(\frac{pq}{h}\right) + n^2 gh\frac{\partial \zeta}{\partial x} + n^2 p\left(\alpha + \beta\left(\frac{p^2}{h^2} + \frac{q^2}{h^2}\right)^{1/2}\right) - \frac{p^2}{nh}\frac{\partial n}{\partial x} - \frac{pq}{nh}\frac{\partial n}{\partial y} = n\frac{D^2}{3}\left(\frac{\partial^3 p}{\partial x^2 \partial t} + \frac{\partial^3 q}{\partial x \partial y \partial t}\right) \qquad (12.9)$$

y momentum

$$n\frac{\partial q}{\partial t} + \frac{\partial}{\partial y}\left(\frac{q^2}{h}\right) + \frac{\partial}{\partial x}\left(\frac{pq}{h}\right) + n^2 gh\frac{\partial \zeta}{\partial y}$$

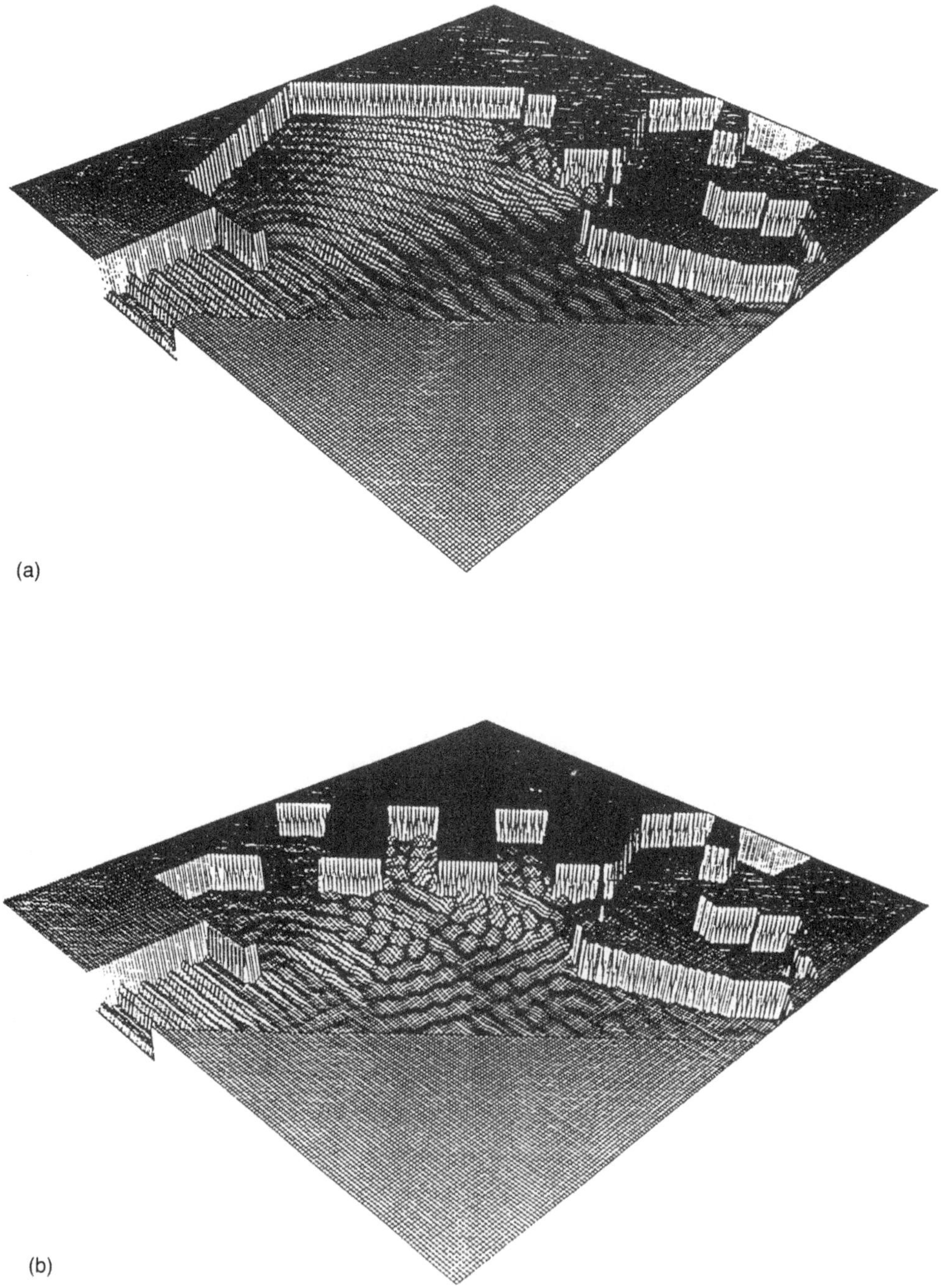

Figure 12.5 Effect of construction modelled in Valencia harbour; (a) typical wave climate in existing situation; (b) corresponding wave climate in proposed new configuration.

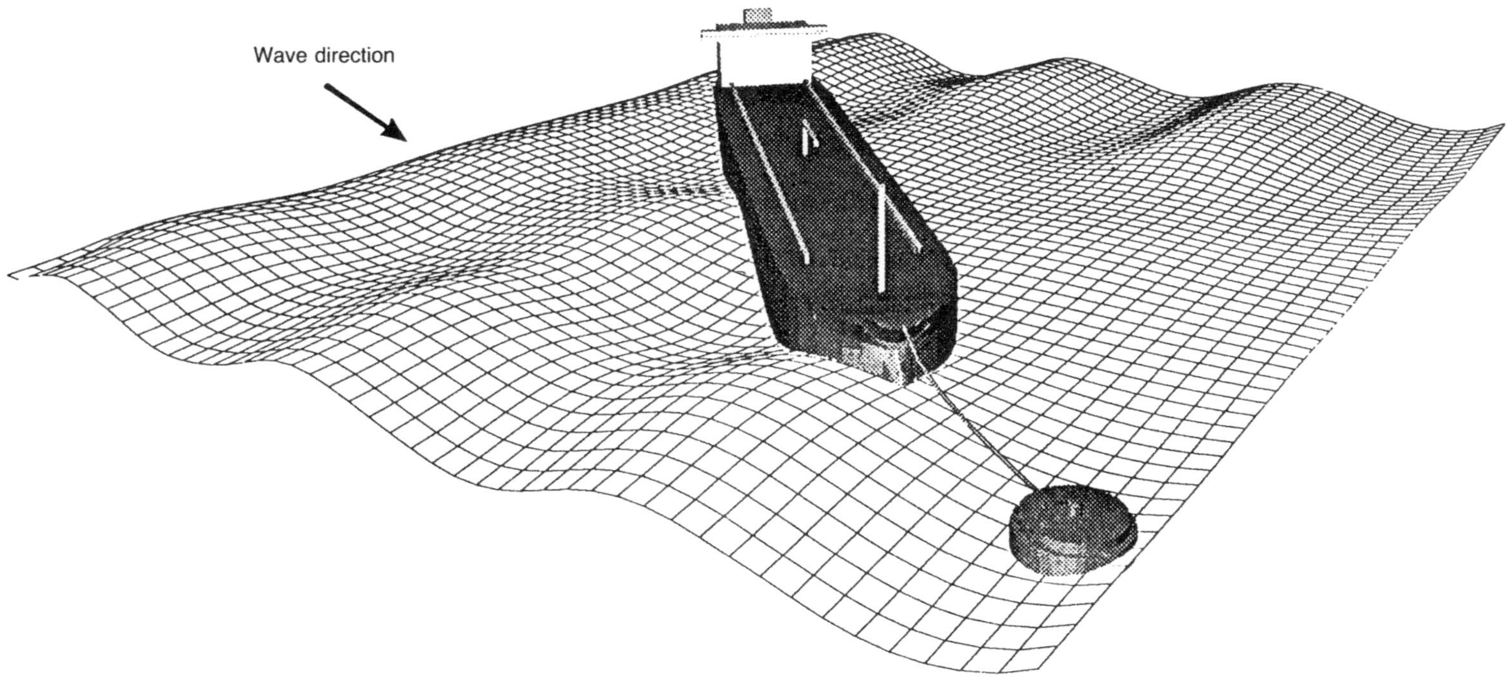

Figure 12.6 Wave action around a moored ship.

$$+ n^2 q\left(\alpha + \beta\left(\frac{p^2}{h^2} + \frac{q^2}{h^2}\right)^{1/2}\right) - \frac{q^2}{nh}\frac{\partial n}{\partial y} - \frac{pq}{nh}\frac{\partial n}{\partial x}$$
$$= n\frac{D^2}{3}\left(\frac{\partial^3 q}{\partial y^2 \partial t} + \frac{\partial^3 p}{\partial x \partial y \partial t}\right) \tag{12.10}$$

where the symbols in Table 12.1 are used.

Figure 12.4 shows an example from the third-generation modelling of the early 1980s, the outer harbour of Bilbao, which was for some years the largest model of its kind. Figure 12.5 shows a comparison between wave fields in the existing and proposed future harbour of Valencia. In the earlier graphic packages used in these systems, land was represented generally by vertical walls, even when beaches that completely absorbed wave energy obtained in practice. A more recent illustration of current fourth-generation modelling – the harbour of Ras Laffan – is subject to irregular wave motion.

This is considerably larger again than the Bilbao harbour (some 300 000 total grid points) but runs on a simple work station. Wave models of this kind have also been extended to allow for the inclusion of ships (Figure 12.6), with complete wave and slip-hull interaction.

REFERENCES

Abbott, M.B. (1987) Computer modelling: a warning notice. *The Dock and Harbour Authority*, March, 251–256.

Abbott, M.B., Petersen, H.M. and Skovgaard, O. (1978) The modelling of short waves in shallow water. *Journal of Hydraulic Research*, **16** (3), 173–203.

Abbott, M.B. and Madsen, P.A. (1990) Modelling of wave agitation in harbours; in Le Méhaute, B. and Hanes, D.M. *The Sea*, Vol. 9, pp. 1067–103.

Abernethy, C.L. and Gilbert, G. (1975) Refraction of wave spectra. *Report INT 117* Hydraulics Research, Wallingford, UK.

Berkhoff, J.C.W. (1972) Computations of combined refraction–diffraction. *Proceedings of the Coastal Engineering Conf.* ASCE, New York, **24**, 471–90.

Brink Kjaer, O., Knudsen, J., Rodenhuis, G.S. *et al.* (1984) Extreme wave conditions in the Central North Sea. *OTL Paper 4809*, Houston, Texas.

Hasselman, K. and Collins, J.I. (1968) Special dissipation of finite-depth gravity waves due to turbulent bottom frictions. *Journal of Marine Research*, **26**, 1–12.

Hasselman, K. (1973) Measurements of wind-growth and swell decay during the North Sea Wave Project (-JONSWAD). *Erganzungsch, Deutsch Hydrogr. Z., Reike A. 12.*

Madsen, P., Murray R. and Sørensen O. (1991) *Coastal Engineering*, **15**, 371–388.

Phillips, O.M. (1977) *Dynamics of the Upper Ocean*, 2nd edn, Cambridge University Press, Cambridge, UK.

3.2 *TRANSPORT–DISPERSION AND WATER QUALITY MODELS*

THREE-DIMENSIONAL ADVECTION–DIFFUSION MODELS 13

E.B. Rassmussen

The hydrodynamic models provide 'platforms' for a considerable number of applications superstructures. Among those superstructures are those for simulating the advection and diffusion of various materials with the fluid. The corresponding advection–diffusion models are most commonly realized using difference methods, which methods have now reached a relatively high level of accuracy and reliability. Special problems occur when it is necessary to follow the consequences of discontinuities in the flow field, such as occur at stratification interfaces and at fronts. This chapter gives an outline of some of the problems and the principal methods used to overcome them.

CONTENTS

13.1 INTRODUCTION

Mathematical models have been widely used in studies analysing the environmental impacts of hydrodynamical processes and associated processes such as water quality and sediment transport. By virtue of these models, our knowledge of the environment and how it responds to both natural and man-made events has increased, and this provides yet another valuable contribution to decision-making processes.

The estuarial and coastal processes are usually conceptually classified separately as hydrodynamical processes and transport processes, and correspondingly they are often numerically treated separately. Some principal aspects of the hydrodynamics have been outlined in Chapter 2, while this chapter will address the aspects of the transport processes.

The transport of scalar quantities is usually viewed as a combination of two physical processes: advection and diffusion. This view has been discussed by Bedford in Chapter 5. The advective process is one in which the scalar quantity propagates unaltered in the flow direction, whereas the diffusive process

Coastal, Estuarial and Harbour Engineer's Reference Book. Edited by M.B. Abbott and W.A. Price. Published in 1993 by Chapman & Hall, 2–6 Boundary Row, London SE1 8HN. ISBN 0 419 15430 2.

tends to smooth out the scalar distribution due to the molecular and turbulent motions. This physical distinction between the two processes is inherently reflected in the mathematical formulations and particularly, in the properties of the numerical models simulating these processes.

13.2 THE ADVECTION–DIFFUSION EQUATION

The advection–diffusion equation for any scalar quantity may be written in conservation form as

$$\frac{\partial(\rho C)}{\partial t} + \frac{\partial}{\partial x_i}\{\rho u_i C\} = \frac{\partial}{\partial x_i}\left\{D_i \rho \frac{\partial C}{\partial x_i}\right\} \qquad (13.1)$$

in which ρ is the density, C is the transported quantity, u_i is the velocity component in the x_i-direction, D_i is the directional diffusional coefficient and t denotes the time. The advection-diffusion equation as given by Equation (13.1) is applicable to both compressible and incompressible flows. The total derivative of the density in Equation (13.2) may be replaced by the space derivatives of the velocity components using the equation of conservation of mass (see Chapter 2 of this volume) whereby Equation (13.1) yields

$$\frac{\partial C}{\partial t} + \frac{\partial}{\partial x_i}\{u_i C\} - C\frac{\partial u_i}{\partial x_i} = \frac{\partial}{\partial x_i}\left\{D_i \frac{\partial C}{\partial x_i}\right\} \qquad (13.2)$$

The first, time-derivative term on the left denotes the local time-rate of change whereas the second, space-derivative term is the contribution to this change provided by the advection. The last term on the left-hand side of Equation (13.2) is due to the compressibility of the fluid and thus is zero for an incompressible fluid. The behaviour of compressible flows will not be considered here. The right-hand side term is the contribution provided by diffusion.

Mathematically, the advection–diffusion equation as given by Equation (13.2) is classified as a mixed-type equation because in the absence of diffusion it is hyperbolic, whereas its properties are parabolic in the absence of advection. This distinction makes it convenient to treat the two processes separately. Hence, the pure advection equation reads

$$\frac{\partial C}{\partial t} + \frac{\partial}{\partial x_i}\{u_i C\} = 0 \qquad (13.3)$$

whereas the pure diffusion equation reads

$$\frac{\partial C}{\partial t} = \frac{\partial}{\partial x_i}\left\{D_i \frac{\partial C}{\partial x_i}\right\} \qquad (13.4)$$

13.3 ADVECTION

The problems of advection range from following the dying-away at smooth variations in the distributions to following quite sharp discontinuities. Standard numerical techniques can be applied to a large class of smooth advective problems in which no discontinuities or fronts appear. Whenever discontinuities of large or small scale are present, however, such techniques may either smooth the discontinuity due to numerical diffusion or may lead to overshooting and undershooting phenomena. Thus, the properties of numerical diffusion and overall accuracy and stability are particularly significant in numerical techniques for the solution of advection problems.

As with all partial differential equations Equation (13.3) needs to be transformed into a description at a finite set of points in a finite-dimensional space. There are many ways of doing this but only the finite difference method will be treated here as this is by far the most commonly used technique in estuarial and coastal engineering. For an outline of other methods and their relation to finite differences, see Abbott and Basco (1989).

The finite difference method transforms the partial differential equation into finite difference approximations which, for any discontinuity in the solution of the partial differential equation, will inherently lead to local errors. Clearly, this is also valid for any distribution which is changing rapidly relative to the grid spacing. Thus, the treatment of discontinuities or, in general, rapidly changing distributions, is essential to the computational algorithm being adopted. Conceptually, the methods for handling such problems are grouped into three classes: discontinuity tracking methods, random choice methods and discontinuity capturing methods. Whereas the first two methods are more directed towards the modelling of discontinuities and are therefore limited in their applications, the latter method is a much more general method. An outline of the three classes has been given by Rizzi and Engquist (1987).

In the capturing method, the sharp resolution of the distribution is simply renounced in favour of accepting larger errors locally. The main advantage of this method is that no *a priori* knowledge of the distribution of the scalar quantity is required. The disadvantage is the lost accuracy and hence the great effort that has to be put into analyses of these errors and how to control them. Despite this disadvantage, it is by far the most frequently used method in advection models.

At the most basic level, there are two main ways of discretizing the partial differential advection equation: central and up-wind differences. The difference in methods is reflected in the truncation errors. As the first-order truncation errors on the space derivatives in Equation (13.3) are equivalent to the diffusion terms in Equation (13.2), so that the effective diffusion coefficient is a summation of the physical diffusion

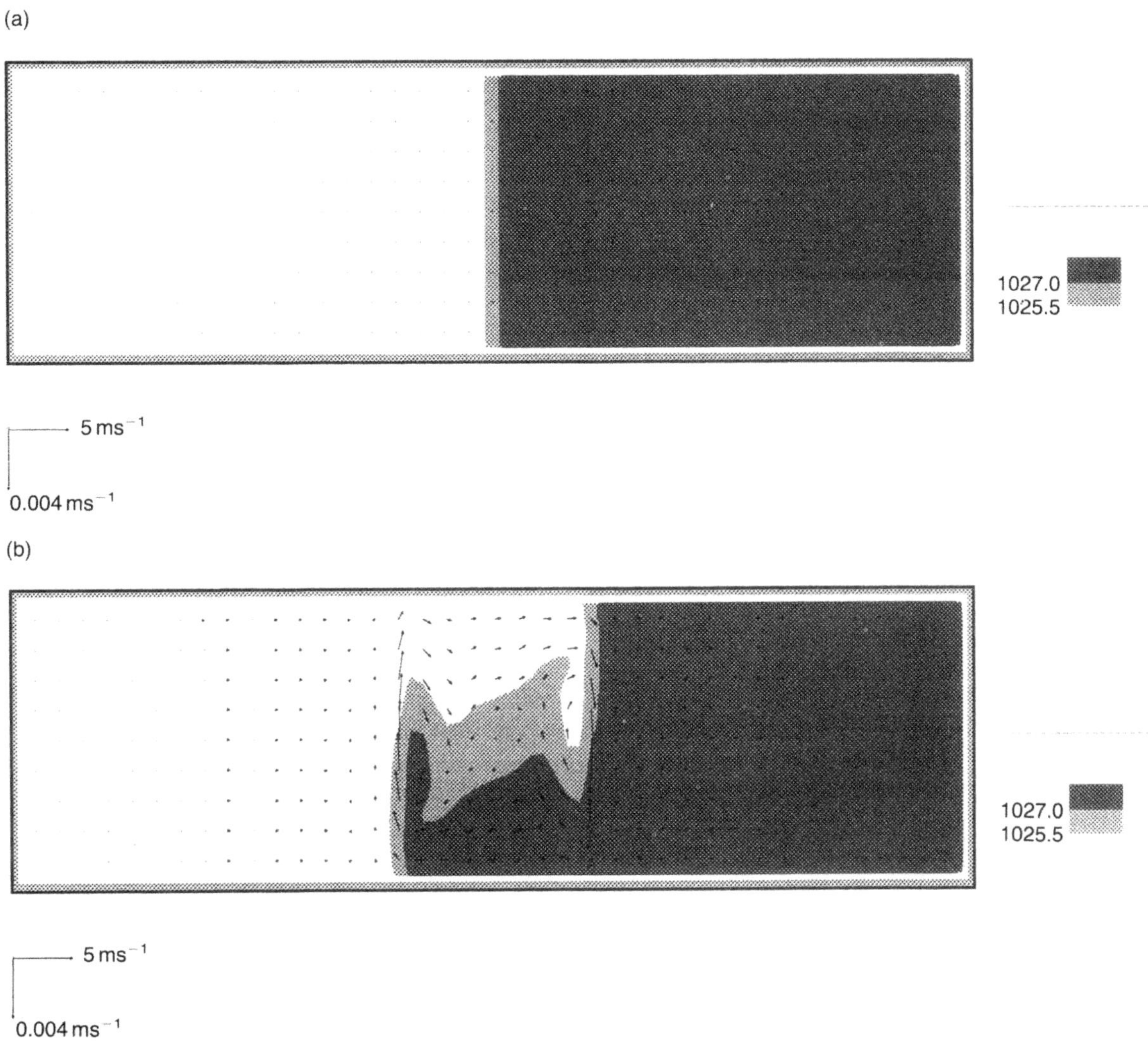

Figure 13.1 Simulation of a lock-exchange with QUICKEST scheme. Initially the two water masses are (a) separated and (b–d) subsequently released. This test is critical to many advection models as it involves a distinct discontinuity – the rotating 'head' at the advancing front, which is also observed in physical model simulations in the laboratory.

and the numerical diffusion, it is evident that higher-order approximations in the advection algorithms are required. By virtue of the central difference, schemes like the Lax–Wendroff and the leap-frog scheme (e.g. Abbott, 1979) have a second-order accuracy in their first space-derivative approximations and so they at first follow the line of higher accuracy. Unfortunately, however, 'the central difference has no inherent numerical stability when applied to a derivative of odd order' as was pointed out by Leonard (1979). This imperative comprehension is central to the discussion, as it will consequently lead to unphysical solutions. Up-wind schemes, on the other hand, have attractive properties regarding stability but are considerably diffusive when applied to first-order approximations. A way out of this dilemma with the central difference has been to introduce an artificial diffusion. The generally accepted legitimacy of this, however, led to a justification of up-wind differences with higher-order approximations, so as to limit the serious numerical diffusivity of up-wind schemes. Examples of such up-wind schemes are the quadratic upstream

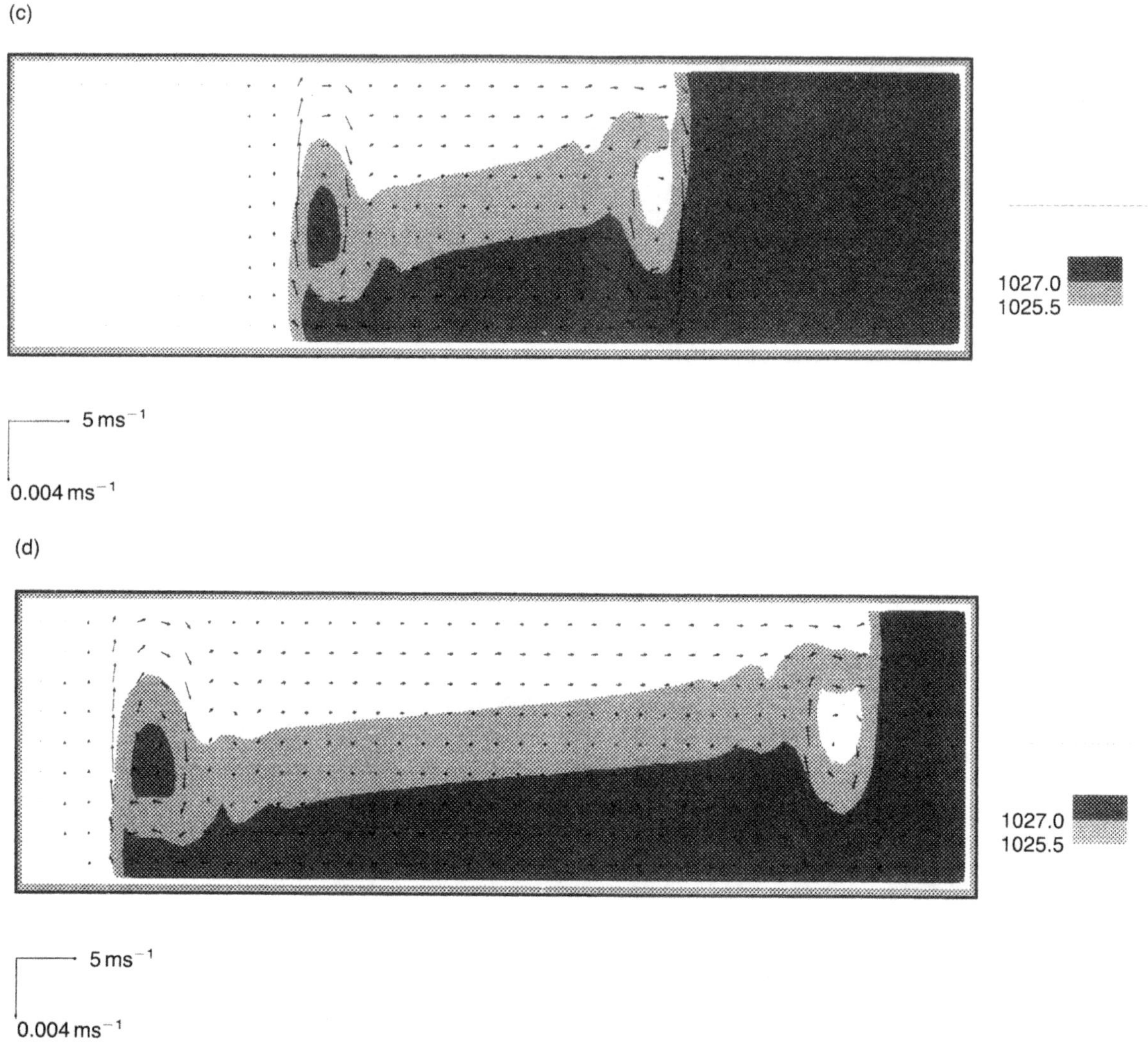

Figure 13.1 *Continued*

interpolation for convective kinematics (QUICK) scheme for steady flow and the corresponding QUICKEST scheme for unsteady flow developed by Leonard (1977). The QUICKEST scheme has since been generalized to two and three dimensions by Justesen *et al.* (1989).

The so-called hybrid schemes seek to overcome the problems through a combination of the central and up-wind techniques so that either a central difference scheme or an up-wind difference scheme is applied dependent on the local grid Peclet number. In three dimensions involving complex geometries, however, this hybrid technique is not straightforward so simpler and highly accurate methods like QUICKEST are preferable (Abbott and Basco, 1989).

13.4 DIFFUSION

The diffusion equation as given by Equation (13.4) is classified mathematically as parabolic. For the most part, it describes a linear problem, whereas the advection equation is commonly non-linear. Physically, the diffusion equation describes the tendency to smooth out variations in the distributions. From a

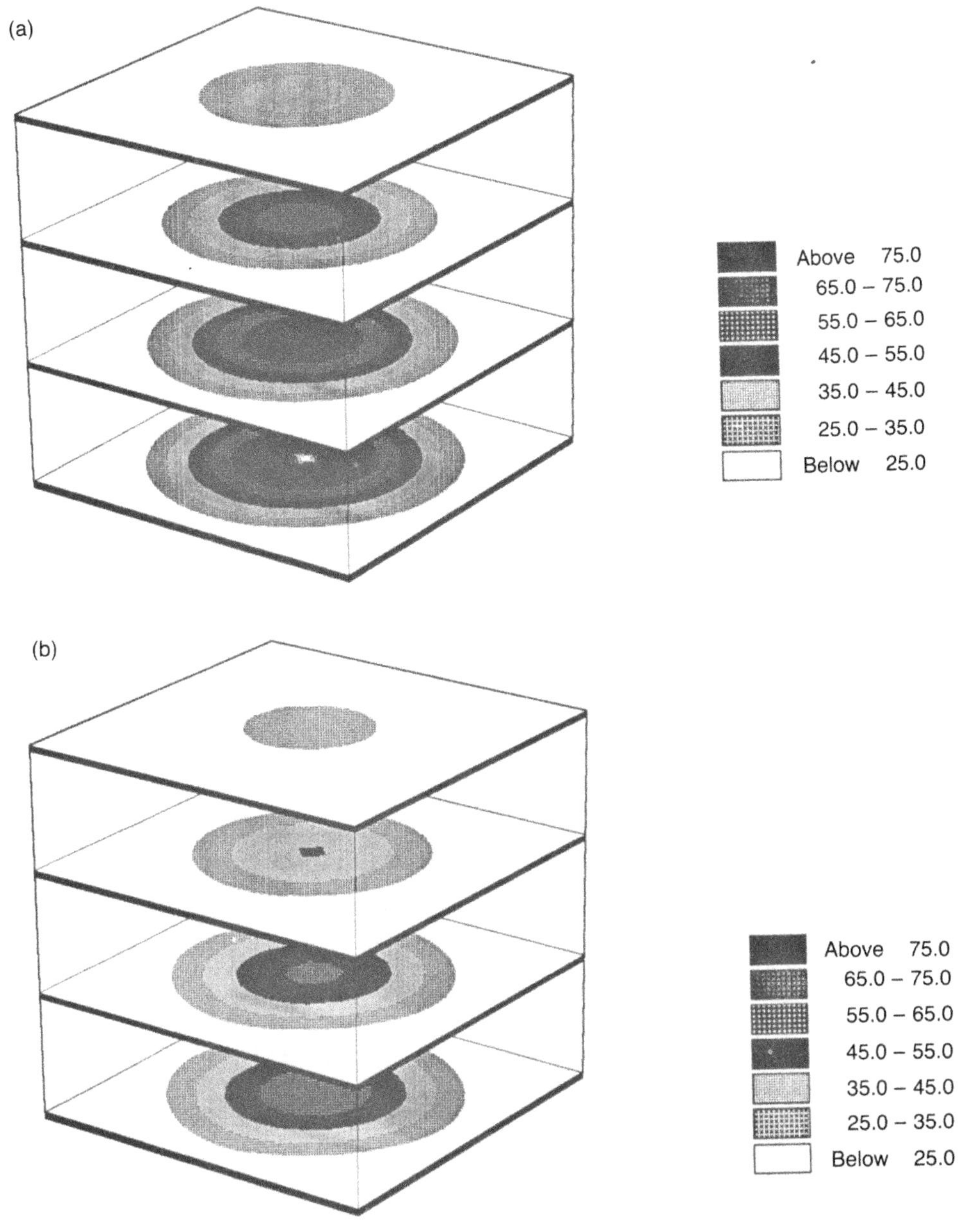

Figure 13.2 Pure diffusion of a hemisphere. (a) initially; (b) after 300 time steps; $D = 0.01\ \text{m}^2\,\text{s}^{-2}$, $\Delta t = 1\ \text{s}$, $\Delta x = \Delta y = \Delta z = 1\ \text{m}$.

numerical point of view, this effect is similar to that provided by numerical filtering processes, and so it has attractive properties so far as numerical stability is concerned.

A great variety of methods has been devised for the solution of both the advection and the diffusion equations. Conceptually, implicit schemes are more complex and require more computations per grid point compared with explicit schemes. But implicit schemes allow for larger time steps, so that, in general,

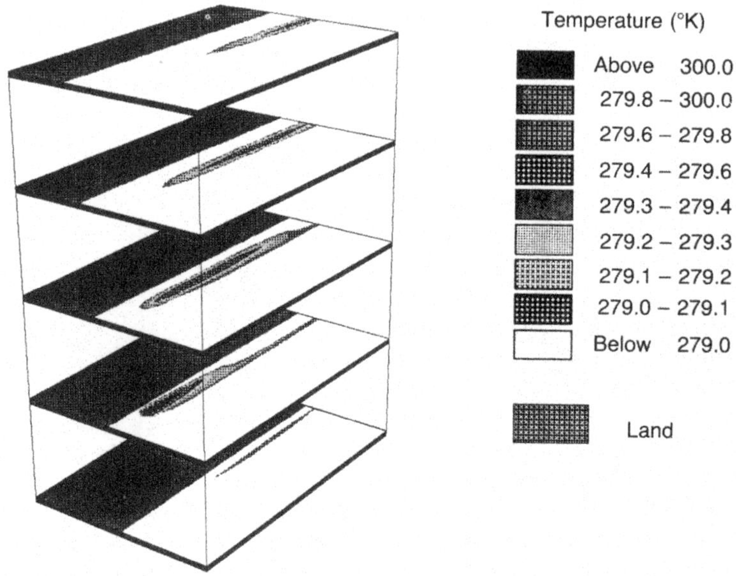

Figure 13.3 Isolines of excess temperature field from outlet of cooling water from a power plant in absence of diffusion.

they are superior to explicit schemes. Imposed on a rectangular space-staggered grid, the diffusion equation forms a sparse and diagonal-dominated matrix suitable for fast solution algorithms like the alternating direction implicit (ADI) technique (Abbott and Basco, 1989).

In this context, it is important that the discretized partial differential equation is expressed in a conservation form. The considerations of conservation and Eulerian (non-conservative) form rest on the fact that finite difference approximations do not always conserve the quantity in Eulerian form, which is the case of conservation form (cf. Roache, 1976; Abbott and Basco, 1989).

13.5 ASPECTS OF THE COMBINATION OF ADVECTION AND DIFFUSION

When using finite difference approximations of the partial differential advection–diffusion equation, spurious overshooting and undershooting phenomena may be seen in the solution. These phenomena, however, can be controlled through the dimensionless grid Peclet number defined by

$$\mathrm{Pe} = \frac{u\Delta s}{D} \qquad (13.5)$$

where u is the velocity vector, D is the diffusion coefficient and Δs is the grid spacing. Small Peclet numbers correspond to diffusion-dominated flows, whereas high Peclet numbers indicate a strong advective influence. The problems associated with the advection are not serious for Peclet numbers less than approximately 2. This condition is, however, not easily achieved in practice, and for this reason precautions have to be taken when selecting the numerical approach.

The advection–diffusion models hold an important position in applications that involve stratification as it is these models that handle the transport of the salinity and temperature determining the density. The modelling of buoyancy effects has no meaning if numerical diffusion will smooth out any physical variation in density.

Also, it is obvious that advection–diffusion simulations very much depend on the velocity fields obtained from the three-dimensional hydrodynamic model. Hence, for practical application it is important to adopt a modelling system that has attractive properties regarding both the advection–diffusion as well as the hydrodynamics. If the hydrodynamic model has been based on a general theory like the Reynolds-averaged Navier–Stokes equations the modelling system will form a powerful tool almost without any limitation in the range of possible applications.

REFERENCES

Abbott, M.B. (1979) *Computational Hydraulics: elements of the theory of free-surface flows*, Pitman/Longman, London.

Abbott, M.B. and Basco D.R. (1989) *Computational Fluid Dynamics: An Introduction for Engineers*, Longman, London, and Wiley, New York.

Justesen, P., Olesen, K.W. and Vested, H.J. (1989) High accuracy modelling of advection in two and three dimensions. *Proceedings of the XXIII Congress of the IAHR*, Ottawa.

Leonard, B.P. (1977) A stable and accurate convective modelling procedure based on quadratic upstream interpolation. *Computer Methods in Applied Mechanics and Engineering*, **19**, 59–98.

Leonard, B.P. (1979) A survey of finite differences of opinion on numerical muddling of the incompressible defective confusion equation, in *Finite Element Methods for Convection Dominated Flows, Rep. AMD, Vol. 34* (ed. T.J.R. Hughes) ASME, New York.

Rizzi, A. and Engquist, B. (1987) Review article, selected topics in the theory and practice of computational fluid dynamics. *Journal of Computational Physics*, **72**, 1–69.

Roache, P.J. (1976) *Computational Fluid Dynamics*, Hermosa Publishers, Albuquerque, NM, USA.

DISPERSIVE PROCESSES IN TWO-DIMENSIONAL MODELS 14

J.R. Vieira

Here the earlier more general introductions by Rodi and Bedford are applied to two-dimensional flows. The problem of representing transport is again treated as one of combining advection and diffusion processes. Confronting the simpler approaches to descriptions of these processes with practice shows marked discrepancies, in particular on the side of diffusion. In this contribution the reasons for these discrepancies are investigated in detail and practical criteria for approximating the processes concerned are obtained.

CONTENTS

14.1 INTRODUCTION

Physical flow processes occur simultaneously with different spatial and temporal scales that span a continuous spectrum ranging from molecular agitation to tidal, barotropic and baroclinic residual circulations. The need to express conservation laws by means of mathematical methods necessitates the treatment of fluid motions as part of continuum mechanics. This implies the use of discretization spacings much larger than the intermolecular spacings associated with molecular agitation in order to define intensive properties such as density or concentration. Furthermore, for the study of practical problems only the circulations with dimensions larger than certain minimum values are of interest; hence the direct modelling of smaller-scale phenomena only incurs unnecessary, and in practice unaffordable, computational costs. Nevertheless, by using convenient filtering procedures it is possible to avoid the explicit inclusion of these phenomena in the dependent variables used in mathematical models. Yet, these non-resolved flow processes cannot be ignored, as, for example, they absorb part of the total energy of the flow. Therefore they have to be empirically related to the dependent variables that are representative of all the resolved scales of motion through the parameterization of specific terms that naturally appear as a consequence of the adopted filtering procedures.

The wide range of time and length scales of flow circulation, as well as the variability of conditions that are found in coastal areas, makes it very difficult, if not impossible, to develop a unified approach with universally applicable expressions for the evaluation of the empirical coefficients associated with terms that represent the non-resolved flow processes. It is thus necessary to resort to calibration values, which have to be based on reliable and sufficient data. However, the values of these coefficients are themselves subjected to large variations, thus raising the need for good first estimates, that can be obtained both by theoretical analysis of schematic representations of the non-resolved physical flow processes, and through the use of experience acquired in similar applications.

The main objective of this chapter is to present and justify the adoption of criteria for the estimation of empirical coefficients of a dispersive nature, with the engineering purpose of applying two-dimensional models of flow circulation and transport to coastal studies. In this context a general revision of fundamental concepts is first presented. Also, as a consequence of the evidence produced by the continued use of numerical models to solve coastal and general environmental problems, a discussion and reassessment of the filtering procedures underlying the establishment of the shallow water equations is presented.

Coastal, Estuarial and Harbour Engineer's Reference Book. Edited by M.B. Abbott and W.A. Price. Published in 1993 by Chapman & Hall, 2–6 Boundary Row, London SE1 8HN. ISBN 0 419 15430 2.

14.2 DISPERSIVE PROCESSES

14.2.1 GENERAL

Fields of instantaneous flow velocities can be decomposed conceptually into two kinds, describing: firstly processes which we can comprehend, either directly from observation or through the resolution of our digital machines; and secondly, processes which we can observe but which we cannot comprehend, even with the help of our digital machines.

This decomposition can be represented by

$$v_i = \hat{v}_i + v_i' \quad (14.1)$$

where v_i is the instantaneous velocity, $\hat{v}_i$ its resolved component and v_i' the unresolved, and so apparently random, fluctuations.

In an Eulerian framework, $\hat{v}_i$ can be obtained by applying Reynolds' original averaging definition (1882), so that

$$\hat{v}_i(x_j, t) = \overline{v_i(x_j, t)} = \lim_{\Delta t \to \infty} \left(\frac{1}{\Delta t} \int_{T-\Delta t}^{T} v_i(x_j, t)\mathrm{d}t \right) \quad (14.2a)$$

which implies

$$\overline{v_i'(x_j, t)} = \lim_{\Delta t \to \infty} \left(\frac{1}{\Delta t} \int_{T-\Delta t}^{T} v_i'(x_j, t)\mathrm{d}t \right) = 0 \quad (14.2b)$$

Thus, to obtain conservation equations in $\hat{v}_i$, the corresponding primitive equations have to be filtered to remove the random processes, and this implies, in practice, the selection of convenient and finite time scales Δt.

When dealing with coastal and environmental flows, the minimum scale of the flow processes that are of interest is still several orders of magnitude above the largest the random motions, and usually even larger than some of the resolvable circulations. This is the case when modelling, for example, tidal waves in shallow water, the flow being nearly horizontal and the depth much smaller than the large-scale circulations induced by the propagation of the waves. It is then convenient to use depth-integrated models, which correspond to a filtering-out of flow processes associated with the distribution of velocities over the vertical.

Models are limited in their ability to simulate environmental flows, because for various reasons some of the physical processes are not explicitly taken into account. Nevertheless, their influence upon the overall phenomena under study cannot be disregarded, due to the important non-linear interactions that occur amongst processes of different scales, as well as due to the existence of a continuous flux of energy towards the smaller scales of circulation in fully three-dimensional flows (Tennekes and Lumley, 1972). As a consequence, the non-resolved flow processes influence the larger-scale resolved circulations in two main ways (Vieira, 1991):

1. Dissipative, which accounts for the amount of energy 'absorbed' by the non-resolved circulations. In depth-integrated hydrodynamic models, bottom friction accounts not only for the energy losses due to shear stresses at the bottom surface, but also for the total amount of energy contained in the non-resolved flow processes over the vertical.
2. Dispersive, which accounts for the transfer of energy and/or flow properties and constituents between resolved scales of circulation due to interactions with non-resolved scales.

In order to take dispersive effects into account, special terms have to be added to the conservation equations, and, as it will be shown, they can be obtained as a result of the application of a generalization of Reynolds' filtering procedures (as expressed in Equations (14.2)). Hence they will be dependent on the fluctuations of velocity, v_i', that are either unknown or of a random nature. The need is then introduced to correlate these to the resolved flow processes, which can only be achieved through the introduction of empirical formulations, the determination of the respective coefficients of dispersion constituting one of the more difficult tasks in modelling practice.

Before proceeding with the analysis of the mathematical formulations used to describe dispersive processes, it is convenient to present the nomenclature adopted. As can be seen in Table 14.1, dispersion is used as a general term to refer to a scattering of fluid particles that depends both on random-type processes (diffusion), and on the effect of velocity gradients (shear). The diffusive processes are never resolved, which feature led to early attempts by Fick (1855) and Taylor (1921) to parameterize the mass fluxes of solutes due to molecular and turbulent motions, respectively. They assumed that the mass fluxes could be set to be proportional to the concentration gradients, the constants of proportionality being

Table 14.1 Transport of fluid particles

Main transport processes	
Advection Movement of centre of gravity of a set of fluid particles	
Dispersion Scattering of fluid particles	**Shear** Differential advection
	Diffusion Molecular motion Turbulence

Table 14.2 Fundamental scales of flow circulation processes

Scale	*Physical processes*	*Filtering procedures*	*Dispersive coefficients*
1	Molecular agitation	Fick's law of diffusion Newton's law of viscosity	Molecular diffusion Viscosity
2	Turbulence	Reynolds' time averaging	Turbulent diffusion Eddy viscosity
3	Vertical velocity distribution	Depth-integration	Dispersion Shear viscosity

called the molecular and turbulent diffusion coefficients, respectively. Taylor (1953, 1954) extended this approximation to shear flows, the combined effect of differential advection and diffusion being then represented by the so-called dispersion coefficients. Elder (1959) applied Taylor's analysis to shallow water flows to describe the shear effects of the vertical velocity gradients. The concept of dispersion of mass of any substance in solution or suspension in a flow can be extended to other properties of the flow. Resorting to Backmeteff's principle of momentum transfer, it follows that the velocity fluctuations act as a mechanism of transfer of momentum between adjacent scales of circulation, i.e. they provide for the dispersion of momentum. The corresponding coefficients are the so-called viscosity coefficients.

14.2.2 FUNDAMENTAL PARAMETERIZATIONS

The fundamental scales and filtering procedures normally used for the development of the momentum and advection–dispersion equations can be summarized as shown in Table 14.2.

Scale 1, which corresponds to filtering out the random molecular motions, is based on Newton's law of viscosity

$$\tau_{zx} = \mu \frac{\partial v_x}{\partial z} \tag{14.3}$$

where the viscosity μ may be regarded as a measure of the resistance of the flow to deformation imposed by tangential stresses τ, which in a broad sense are generated by the transfer of momentum due to velocity fluctuations normal to the corresponding surface.

With respect to the fluctuations due to molecular agitation processes, their effect on the transfer of momentum is independent of the flow conditions; thus the dynamic viscosity μ is a characteristic of the fluid. The corresponding dynamic equilibrium equations are the well known Navier–Stokes equations, where the dispersive effects of the non-resolved random molecular motions are accounted for by the following empirical terms (in the x_i direction)

$$\nu\left(\frac{\partial^2 v_i}{\partial x_j^2}\right) \tag{14.4}$$

$\nu = \mu/\rho$ being the kinematic viscosity (L^2T^{-1}).

The random molecular motions also induce the transfer of mass of any substance dissolved or in suspension in the fluid, which according to Fick's law is given by

$$T_i = D^m \frac{\partial c}{\partial x_i} \tag{14.5}$$

where c represents the concentration of the constituent and D^m the molecular diffusion coefficient. Applying a conservation principle to an arbitrary volume of control, an empirical term similar to Equation (14.4) will appear:

$$D^m\left(\frac{\partial c^2}{\partial x_i^2}\right) \tag{14.6}$$

Although molecular agitation is always present, even for fluids at rest, under flow conditions fluid particles experience additional random motions of much higher magnitude, so that their paths are very irregular and sinuous. It is apparent when observing recorded time series of instantaneous velocities v_i, that turbulent flows can be described by the superposition of a random fluctuating velocity v_i' and a temporal mean velocity $\hat{v}_i$, as defined in Equation (14.1). Taking into account Reynolds' averaging definitions presented in Equation (14.2), it is possible to separate these fluctuations from the mean flow under a selected time scale Δt. As a result the following new terms will be obtained for the momentum equations:

$$\frac{\partial}{\partial x_j}(-\rho\,\overline{v_i'v_j'}) \tag{14.7a}$$

Similarly for the conservation of a property with a concentration given by

$$c = \hat{c} + c' \tag{14.7b}$$

the result is

$$\frac{\partial}{\partial x_j}(-\rho\,\overline{v_i'c'}) \tag{14.7c}$$

In the classical theory of turbulence, as reviewed by Rodi in Chapter 4 of this volume, the terms $-\rho\,\overline{v_i' v_j'}$, called Reynolds stresses, are empirically correlated to the resolved flow through a new coefficient ν^t, usually called the 'eddy viscosity', such that, according to the Boussinesq approximation, the following relationship applies:

$$-\rho\,\overline{v_i' v_j'} = 2\nu^t \hat{S}_{ij} \tag{14.8}$$

where $\hat{S}_{ij}$ is the time-averaged second-order strain tensor. Therefore the new dispersive terms (Equation 14.7a) become for direction x_i for example

$$\frac{\partial}{\partial x_j}\left(\nu^t\left(\frac{\partial \hat{v}_i}{\partial x_j} + \frac{\partial \hat{v}_j}{\partial x_i}\right)\right) \tag{14.9}$$

It is apparent that ν^t is now dependent on the flow conditions. Several models have been developed for its evaluation, each corresponding to a particular kind of turbulence (e.g. isotropy). One of the formulations commonly adopted is based on the so-called Smagorinsky (Smagorinsky, 1963) model for isotropic turbulence given by

$$\nu^t = l^2(x_i)(\hat{S}_{ij}\hat{S}_{ji})^{1/2} \tag{14.10}$$

where $l(x_i)$ is a characteristic mixing length. In general, it is difficult to determine ν^t, but as a reference, for uniform channel flow, it is often assumed to take a value given approximately by

$$\nu^t = \frac{1}{10}hv^* \tag{14.11}$$

where h is the water depth and v^* the shear velocity.

Introducing the Reynolds analogy for turbulent flows, which expresses the fact that the transfer processes of dissolved or suspended matter are equivalent to the transfer processes of momentum, the turbulent diffusion coefficient D^t will be such that $D^t \approx \nu^t$, the new dispersive terms of the conservation equation being

$$\frac{\partial}{\partial x_j}\left(D^t \frac{\partial \hat{c}}{\partial x_j}\right) \tag{14.12}$$

As in many coastal engineering applications the depth is much smaller than the horizontal dimensions of the domain under study, and two-dimensional models are usually adequate to describe the main flow processes. Nevertheless the depth integration will imply the filtering of the vertical velocity profiles, which are responsible for additional spreading in the direction of the flow. As previously mentioned, flows with velocity gradients are often referred to as 'shear flows' and the associated spreading mechanism discussed by Taylor (1954) is currently known as the 'shear effect'. Hence to account for 'shear flows' in a depth-integrated model, additional empirical terms have to be included. Supposing that the distribution of velocities is unknown and using an analogy to the Reynolds time averaging definition, the filtering procedure can now be defined by the following relationships:

$$\hat{v}_i = V_i + V_i' \tag{14.13a}$$

$$V_i = \frac{1}{h}\int_0^h \hat{v}_i \mathrm{d}z \tag{14.13b}$$

$$\frac{1}{h}\int_0^h v_i' \mathrm{d}z = 0 \tag{14.13c}$$

where V_i denotes the mean-temporal depth-integrated velocity and V_i' the deviation from V_i the mean-temporal velocity $\hat{v}_i$ at depth z.

Introducing the above relationships in the Reynolds equations (Navier–Stokes equations after filtering out turbulence under a given time scale Δt) and integrating along the vertical, the convective terms in the x_i direction become

$$\frac{\partial}{\partial x_j}\int_0^h \hat{v}_i\hat{v}_j = \frac{\partial}{\partial x_j}(hV_iV_j) + \frac{\partial}{\partial x_j}\int_0^h (V_i'V_j')\mathrm{d}z \tag{14.14}$$

The last term represents the overall transfer of momentum over the depth due to the shear effect and, in an assumption similar to that applied to the turbulent transfer of momentum (Boussinesq approximation), it is common to accept the following correlations with the resolved variables:

$$\frac{1}{h}\int_0^h (V_i'V_j')\mathrm{d}z = -\nu_{ij}^s \frac{\partial V_i}{\partial x_j} \tag{14.15}$$

The previous assumptions and the Reynolds analogy make it possible finally to write the new terms for the conservation of momentum and constituents, e.g. in the x_i direction:

$$\frac{\partial}{\partial x_j}\left(\nu_{ij}^s h \frac{\partial V_i}{\partial x_j}\right) \tag{14.16a}$$

$$\frac{\partial}{\partial x_j}\left(D_{ij}^s h \frac{\partial C}{\partial x_j}\right) \tag{14.16b}$$

where $\nu_{ij}^s \approx D_{ij}^s$, with ν_{ij}^s and D_{ij}^s representing the horizontal shear stresses and dispersion, respectively, in depth-integrated shear flows. The coefficient ν_{ij}^s will hereafter be called the 'shear viscosity'. C denotes the mean-temporal depth-integrated concentration.

In the case of uniform channel flow (Abbott and Rasmussen, 1977) the shear viscosity can be estimated by

$$\nu^s = \frac{\alpha - (\alpha')^2}{2}C_zVh \tag{14.17a}$$

where

$$\alpha = \int_0^h \hat{v}^3 \, dz / V^3 h \qquad (14.17b)$$

$$\alpha' = \int_0^h \hat{v}^2 \, dz / V^2 h \qquad (14.17c)$$

C_z being the non-dimensional Chézy number.

For example, considering a 10 m deep channel with a depth-averaged velocity of $1 \, m \, s^{-1}$ and a non-dimensional Chézy number of 18, the shear viscosity coefficient that corresponds to a geomagnetic law for velocity distribution along the vertical given by $kz^{0.2}$ is $\nu^s \approx 4 \, m^2 \, s^{-1}$.

As expected, these results suggest that the intensity of transfer of momentum and constituents is related to the scale of the processes considered, the molecular and turbulent effects being in practice negligible in comparison with the shear effect (10^{-6} to $10^{-1} \, m^2 \, s^{-1}$ against $4 \, m^2 \, s^{-1}$).

14.2.3 SUB-GRID SCALE PROCESSES

At this point it is convenient to emphasize that the use of viscosity and dispersion coefficients has been brought about by the need to take non-resolved motions of the fluid particles into account. This concept will be present when the filtering procedures are further extended to higher spatial and temporal scales as a consequence of the practical need for horizontal spatial discretization of the depth-integrated equations. It implies that additional dispersion and viscosity will be required to account for non-resolved large-scale motions; the magnitudes of these will depend on the size of the grid spacing Δx used in the numerical computations. Once again it is only natural to expect that, by increasing Δx, the dispersive coefficients will also increase, with practise confirming that they can be several orders of magnitude larger than the coefficients arising from the depth-integration. In the lower limit, $\Delta x \approx h$, the larger non-resolved circulations will be related to the depth-integrated shear flow and the extensive studies available on this subject provide accurate estimates for the empirical coefficients so long as the velocity profile is reasonably well defined (see section 14.3.1).

One of the main difficulties found when estimating dispersive coefficients, so as to take into account the horizontal spatial discretizations, is the characterization of the new non-resolved circulations, which can be greatly dependent on factors like local bathymetric configurations, density gradients and wind friction and set-up. Although some general formulae can be used for guidance, the accuracy of the final values will always be greatly dependent on the calibration and on the experience of the modeller.

In conclusion, the fact that the numerical resolution of the primitive equations has to resort to additional discretizations in space and time (no analytical solutions being generally available), makes it necessary to extend filtering to larger scales of motion which are dependent on the model resolution Δx. Thus, following the previous schematization, a fourth scale must be considered, such that

Scale 4

Averaging over the model resolution, of order Δx, Δt $\Rightarrow$ $\begin{cases} \text{additional dispersion} \\ \text{additional viscosity} \end{cases}$

In what follows the additional dispersion and viscosity will be referred to as sub-grid dispersion D^G and sub-grid viscosity ν^G, respectively.

14.3 ESTIMATION OF COEFFICIENTS

14.3.1 BASIC FORMULATIONS

Elder's formulation will be used as a basis for the interpretation of dispersive coefficients. Consider the mass conservation equation for a dissolved constituent in an open channel of infinite lateral extent, where the lateral and vertical velocity components are zero and the horizontal diffusion can be neglected, i.e.

$$\frac{\partial \hat{c}}{\partial t} + \frac{\partial}{\partial x}(\hat{v}\hat{c}) = \frac{\partial}{\partial z}\left(D_z^t \frac{\partial \hat{c}}{\partial z}\right) \qquad (14.18)$$

Assuming the vertical profile defined by Equations (14.13) and taking into account the flux continuity for incompressible fluids, Equation (14.18) can be written

$$\frac{\partial}{\partial t}(C + C') + (V + V')\frac{\partial}{\partial x}(C + C') = \frac{\partial}{\partial z}\left(D_z^t \frac{\partial C'}{\partial z}\right) \qquad (14.19)$$

Introducing a non-inertial frame of reference moving at velocity V, C and C' are functions of $x' = x - Vt$ and $t' = t$, which changes Equation (14.19) to the following form:

$$\frac{\partial}{\partial t'}(C + C') + V\frac{\partial}{\partial x'}(C + C') = \frac{\partial}{\partial z}\left(D_z^t \frac{\partial C'}{\partial z}\right) \qquad (14.20)$$

After a sufficiently long time, C and C' will become functions only of x' and z, respectively, which make it possible to write

$$V' \frac{\partial C}{\partial x'} = \frac{\partial}{\partial z}\left(D_z^t \frac{\partial C'}{\partial z}\right), \quad t \to \infty \qquad (14.21)$$

Integrating Equation (14.21), one obtains

$$C' = \left[\int_0^z \frac{1}{D_z^t}\left(\int_0^z V' \, dz\right) dz\right] \frac{\partial C}{\partial x'} \qquad (14.22)$$

Multiplying by V' and integrating again over the vertical, yields

$$\frac{1}{h}\int_0^h (V'C')\mathrm{d}z = \frac{\partial C}{\partial x'}\frac{1}{h}\left\{\int_0^h V'\left[\int_0^z \frac{1}{D_z^t}\left(\int_0^z V'\,\mathrm{d}z\right)\mathrm{d}z\right]\mathrm{d}z\right\} \tag{14.23}$$

It is now obvious that the dispersion coefficient is given by

$$D_x^s = \frac{1}{h}\left\{\int_0^h V'\left[\int_0^z \frac{1}{D_z^t}\left(\int_0^z V'\,\mathrm{d}z\right)\mathrm{d}z\right]\mathrm{d}z\right\} \tag{14.24}$$

Elder evaluated D_x^s using a logarithmic profile for the velocity distribution over the vertical, such that

$$V'(z) = \frac{v^*}{\kappa}\left(1 + \ln\frac{z}{h}\right) \tag{14.25}$$

where v^* is the shear velocity $(\tau_0/\rho)^{1/2}$ and κ is the von Kármán constant ($\approx$ 0.41). The shear stress distribution is considered to be linear:

$$\tau = \tau_0(1 - z/h) \tag{14.26}$$

and the assumption is made that the turbulent transports of mass and momentum are equivalent (Reynolds' analogy). Noting also that the vertical turbulent momentum transfer is

$$\tau = \rho \nu_z^t \frac{\partial V'}{\partial z} \tag{14.27}$$

it follows that

$$D_z^t = \nu_z^t = v^*\kappa z\left(1 - \frac{z}{h}\right) \tag{14.28}$$

Considering finally the independent variable in Equation (14.24) to be $\xi = z/h$ and taking into account Equations (14.25) and (14.28), one obtains:

$$D_x^s = \frac{v^*h}{\kappa^3}\left\{\int_0^\xi (1 + \ln\xi)\left[\int_0^\xi \frac{\int_0^\xi (1 + \ln\xi)\mathrm{d}\xi}{\xi(1 - \xi)}\mathrm{d}\xi\right]\mathrm{d}\xi\right\} \tag{14.29}$$

Performing the integration, Elder finally obtained

$$D_x^s = 0.404\frac{v^*h}{\kappa^3} = K_E v^*h = 5.9v^*h \tag{14.30}$$

Expressions similar to Elder's have been proposed with a wide variety of coefficients, as for example

Krenkel (1962), for open channel flow,	$9.1v^*h$	(14.31a)
Yotsukura and Fiering (1964), for smooth channels,	$13v^*h$	(14.31b)
Thackston (1966), for natural streams,	$7.25\left(\frac{v}{v^*}\right)^{1/4} v^*h$	(14.31c)

to mention but a few, as presented by Bansal (1971).

Jobson and Sayre (1970) have provided some evidence that the Reynolds analogy holds. They also confirmed the parabolic distribution of ν^t and found that the mean-depth value of the dispersion coefficient was about

$$D_z^t = 0.07v^*h \tag{14.32}$$

which justifies neglecting it in comparison with horizontal dispersion due to shear flow processes which occur in the vertical.

In applications for natural streams, Fischer (1968) found that longitudinal dispersion coefficients as predicted by the Elder formula were too small, values as much as 150 times higher having been reported. Bowden (1965) pointed out that the effective coefficients of horizontal dispersion are inversely proportional to the coefficient of vertical turbulent diffusion, which is also apparent in Elder's development (Equation (14.24)). The mixing produced under a shearing current will thus be enhanced if some stability factors are present in the vertical, their effects becoming important for local Richardson numbers of about 0.5 to 1. According to Bowden, the coefficient of vertical turbulent diffusion may then be reduced by a factor of 10 or 20, with a corresponding increase in the horizontal dispersion coefficients, to an order of 10^5 to $10^6\,\mathrm{cm^2\,s^{-1}}$.

The conclusions of these authors clearly show that other mechanisms of spreading may be dominant in comparison with depth-integrated shear effects. Fischer proposed that the main factor contributing to longitudinal dispersion was transversal mixing and not vertical mixing, i.e. the dominant mechanism of spreading should be associated with transversal circulations (this analysis having been done for natural streams). Bowden recognized that shear effects are likely to be most effective in estuaries and near the coastline, and that the existence of large horizontal eddies may dominate among the dispersive processes.

14.3.2 THE EVIDENCE OF EXPERIENCE

It is well demonstrated that dispersive effects due to physical processes at scales 1 and 2 are negligible in relation to those at scale 3. With respect to environmental flows, field measurements and modelling results also show that in most situations the Elder coefficient is several orders of magnitude too small, as mentioned previously. This is illustrated in Table 14.3, in which calibrated coefficients are compared with the Elder formula for four different applications with a two-dimensional circulation model (DHI, 1986).

The magnitude of the calibrated coefficients can only be explained by considering processes at scale 4,

Table 14.3 The Elder coefficient versus calibration

Case	*h* *m*	*u* *m* s^{-1}	*u** *m* s^{-1}	Δx *m*	Δt *s*	*E. calib.* $m^2 s^{-1}$	*6. hu** $m^2 s^{-1}$
A	8	0.7	0.05	50	30	2–5	2.4
B	20	1.0	0.1	500	300	40–50	12
C	30	0.5	0.03	6 000	600	≈500	5.4
D	1000	0.1	0.003	30 000	900	>6000	18

Table 14.4 Sub-grid viscosity coefficients

Case	*h* *m*	*u* *m* s^{-1}	Δx *m*	Δt *s*	ν^s *cal.* $m^2 s^{-1}$	K_1	K_2	K_3
A	8	0.7	50	30	1–5	0.01–0.06	0.03–0.14	0.07–0.34
B	20	1.0	500	300	40–50	0.06	0.10	0.17
C	30	0.5	6 000	600	≈500	0.008	0.17	3.3
D	40	1.0	20	10	1–3	0.025–0.075	0.05–0.15	0.10–0.30
E	1000	0.1	30 000	900	≈6000	0.006	2.0	667

that in general dominate over processes in scale 3 when $\Delta x \gg h$.

As observed by Madsen *et al.* (1988), many modellers attempt to account for the non-resolved processes at scale 4 by increasing K_E in the Elder formulation, which will only be acceptable as long as those processes remain related to flow variations in the vertical. In order to obtain a general interpretation for the sub-grid dispersive processes, the coefficients of viscosity and dispersion have to be related to the length scale Δx and to the time scale Δt, which can be done as follows:

$$K_1 \frac{\Delta x^2}{\Delta t} \tag{14.33a}$$

$$K_2 V \Delta x \tag{14.33b}$$

$$K_3 V^2 \Delta t \tag{14.33c}$$

In Table 14.4 the three different forms of the sub-grid viscosity coefficient for scale 4 have been compared to calibrated values in five different situations. The form $K_2 V \Delta x$ appears to be more convenient since K_2 is almost constant for all the cases.

It is important to emphasize that as Equations (14.33) are not based on a defined flow pattern like Elder's formulation, the estimates can only be expected to be a rough approximation to the true values. Furthermore, as will be analysed in the next section, the values of the dispersive coefficients are mainly dependent on the largest non-resolved circulations and not directly on Δx or Δt. This is very important, and makes it possible to justify that, in situations where $\Delta x \gg h$, the use of the Elder formulation may still be valid although the associated dispersive processes will be irrelevant in comparison with other effects. Nevertheless, the above formulae, when complemented with field observations and experience obtained in similar situations, do provide valuable guidance for the calibration of two-dimensional environmental models.

When Δx and h are of the same order of magnitude, the shear effects along the vertical (processes at scale 3) will be dominant in connection with the transfer of momentum and constituents. In practice, assuming that the friction velocity for coastal waters is about 6% of the mean depth velocity, Elder's formula will give

$$\nu_x^s \approx D^s \approx 5.9 \times 0.06\, Vh \approx 0.4\, Vh \tag{14.34a}$$

Thus, for application in two-dimensional models when $\Delta x \approx h$ the dispersive coefficients can be given as a first approximation by

$$\nu_x^s \approx D^s \approx Vh \tag{14.34b}$$

and possibly varying by an order of magnitude, according to calibration.

It should also be stressed that the transverse shear effects are found to be approximately one order of magnitude smaller than those in the direction of the flow, as indicated by tests reported by Talbot and Talbot (1974). For an arbitrary flow direction in a Cartesian co-ordinate system, cross terms would then be present, complicating the definition of the coefficients. Calling D_L^s and D_T^s the longitudinal and transverse coefficients respectively, a reasonable approach would be to project them on the x and y directions and disregard cross-derivative terms. This can be accomplished by taking the ellipse of semiaxes D_L^s, D_T^s in which D_L^s is aligned with the current vector v_i, and calculate D_i^s accordingly, eventually obtaining:

$$\frac{1}{D_x^s} = \left[\left(\frac{\cos\alpha}{D_L^s}\right)^2 + \left(\frac{\sin\alpha}{D_T^s}\right)^2\right]^{1/2} \quad (14.35a)$$

$$\frac{1}{D_y^s} = \left[\left(\frac{\sin\alpha}{D_L^s}\right)^2 + \left(\frac{\cos\alpha}{D_T^s}\right)^2\right]^{1/2} \quad (14.35b)$$

where $\alpha = \tan^{-1}(v_y/v_x)$.

Although the influence of flow direction is important when evaluating the effects of shear dispersion processes along the vertical in nearly-horizontal flows, in practice isotropy is usually assumed both for flow processes with spatial scales that are larger or smaller than the depth. With respect to small-scale circulations, their structure is mainly dependent on non-linear interactions with larger circulations, and not much dependent on geometric constraints. Isotropic behaviour is then an acceptable hypothesis. The structure of large-scale circulations is more difficult to characterize *a priori*, isotropy being a natural assumption to represent, in an average sense, the associated dispersive processes, so long as they are small enough in comparison with the spatial and temporal scales of the mean flow.

14.3.3 GENERAL INTERPRETATION

Abbott *et al.* (1985) showed that it is possible to realistically simulate secondary large-scale circulations by using two-dimensional models. Madsen *et al.* (1988), continuing on the same subject, referred to the existence of two fundamental mechanisms for the generation of circulations, where the resistance forces (bottom friction) are balanced by: firstly the momentum-transfer resolved at the scale of the spatial discretization adopted (convective accelerations); and secondly, the momentum-transfer non-resolved at the scale of the spatial discretization adopted, represented by horizontal shear stresses (sub-grid viscosity).

The first mechanism is dominant in many situations where the scales of flow circulation and space discretization Δx are much larger than the depth of the flow, and the second mechanism is fundamental to describing secondary circulations when $\Delta x \leqslant h$. Nevertheless, for some situations where $\Delta x \gg h$ it is found in practice that the second mechanism is still necessary to describe the resolved scales of circulation. A possible explanation may be found in the following heuristic hypothesis: the non-resolved transfer of momentum amongst resolved scales of circulation depends upon the relative magnitude of the minimum resolved and of the maximum non-resolved scale. The transfer is only effective when these scales have the same order of magnitude.

To analyse the consistency of this statement the following interpretations are made.

1. When the structure of the flow is well defined by a logarithmic profile of velocities and Δx is of the same order of magnitude of the depth h, the biggest non-resolved flow property is the velocity profile itself, the Elder formula being thus valid for the estimation of the effective viscosity and dispersion coefficients. The second mechanism is therefore fundamental in describing the flow.
2. When $\Delta x \gg h$ different situations may be found.
   (a) The bottom is very regular and three-dimensional effects are negligible. In this case the biggest non-resolved circulations will still be related to the depth, thus being much smaller than the minimum resolved scale. Consequently the transfer of momentum due to the non-resolved circulations (sub-grid viscosity) is negligible. Euler's formula is still valid but the model will not be sensitive to the values of the viscosity coefficient. The circulation is only determined by the first mechanism, i.e. by the dissipative terms, as the non-resolved circulations only contribute as energy absorbers.
   (b) The bottom configurations and the existence of three-dimensional effects originate circulations whose scales are of the same order of magnitude as Δx, i.e. important in relation to the minimum resolved scales. In this case the transfer of momentum due to the non-resolved circulations becomes significant. This can easily be illustrated from the experience of many authors, who for calibration of regional models using large grid spacings ($\geqslant$5000 m) employed dispersion coefficients proportional to Δx and h. For example, one of the formulae commonly used is given by $E = K\Delta x^2/\Delta t$ with $0.01 < K < 0.06$. For a regular mesh with a 30 000 m grid spacing, a 15-minute time step and an average depth of 1000 m, this gives $10\,000 < E < 60\,000\ \mathrm{m^2\,s^{-1}}$, i.e. $10\,h < E < 60\,h$.

   Schwiderski (1981), in his global oceanic tidal model, used an effective viscosity linearly proportional to the depth and Δx that gives values that are in accordance with the previous formula.
   (c) It is logical to expect that by using the Prandtl model for the determination of the dispersive coefficients, the mixing length will be related to the magnitude of the largest non-resolved circulations. Using the Smagorinsky formulation and considering the mixing length given by

$$l = C_s \Delta x \quad (14.36)$$

Madsen *et al.* (1988) recommended values of C_s

in the range $0.4 < C_s < 0.8$. These values are consistent with the hypothesis formulated before, i.e. the transfer processes of momentum are only important when the non-resolved circulations are of the same order of magnitude as Δx.

The importance of the relative magnitude of Δx and h for the simulation of large-scale circulations using two-dimensional models has been emphasized by Abbott *et al.* (1981, 1988) who on the basis of the parameter

$$A_b = \frac{h}{\Delta x} \tag{14.37}$$

associated the conditions described in (1) and (2a) above with $O(A_b) \approx 1$ and $A_b \ll 1$, respectively. The analysis presented in this chapter makes it possible to generalize the information contained in the parameter A_b with respect to the estimation of dispersive coefficients.

Denoting the biggest scale of the sub-grid flow circulations by Δs and considering

$$A_b^* = \frac{\Delta s}{\Delta x} \tag{14.38}$$

together with the hypothesis previously formulated, it follows that the dispersive phenomena due to the non-resolved circulations are only accountable at the scale of the grid resolution adopted, when $O(A_b^*) \approx 1$. In fact, circulations with scales Δs_i smaller than Δx cannot modify the averaging over Δx. But when $\Delta s_i \approx \Delta x$, the non-resolved processes will influence the averaged values in adjacent discretization points, which corresponds physically to a dispersive effect.

The previous hypothesis and conclusions can be extended to the processes of transfer of constituents (dissolved or suspended) on the basis of the Reynolds analogy. In this context another interesting way of analysing the relative importance of the scales of circulation is presented by Abbott and Basco (1989). They considered a superposition of grid-resolved exchanges of fluid determined in terms of a random walk description, such that the property ϕ at instant $n + 1$ is given by

$$\phi_j^{n+1} = \left(\frac{\sigma\Delta V\Delta t}{\sigma\Delta x}\right)\phi_{j+\sigma}^n + \left(1 - \frac{2\sigma\Delta V\Delta t}{\sigma\Delta x}\right)\phi_j^n + \left(\frac{\sigma\Delta V\Delta t}{\sigma\Delta x}\right)\phi_{j-\sigma}^n \tag{14.39a}$$

where σ represents the scale of the considered circulation in terms of Δx. Rearranging the previous equation, it is seen that

$$\frac{\phi_j^{n+1} - \phi_j^n}{\Delta t} = (\sigma\Delta V \cdot \sigma\Delta x)\left(\frac{\phi_{j+\sigma}^n - 2\phi_j^n + \phi_{j-\sigma}^n}{(\sigma\Delta x)^2}\right) \tag{14.39b}$$

which is consistent with the differential equation

$$\frac{\partial\phi}{\partial t} = (\sigma\Delta V \cdot \sigma\Delta x)\frac{\partial^2\phi}{\partial x^2} \tag{14.40}$$

which expresses that any fluid exchange associated with circulations of length scale $\sigma\Delta x$ is represented by an effective dispersion coefficient of $\sigma\Delta V\Delta x$. This schematic model clearly shows that the intensity of dispersion is proportional to the square of the length scale.

Assuming $\sigma = 8$, a measure of the dispersion due to all the possible length scales would be given by the sum:

$$8^2 + 4^2 + 2^2 + 1^2 + (1/2)^2 + (1/4)^2 + (1/8)^2 + \ldots$$

Taking into account the Nyquist limit it is seen that the contribution of circulations smaller than Δx would be irrelevant for the dispersive coefficients, which again is in accordance with the hypothesis previously presented.

Finally it should be noted that the dispersive terms basically account for the effects of interactions between the minimum resolved scales of circulation, which introduce redistributions of energy amongst themselves. The energy involved in these phenomena may result partially from the difference between the gross and net transfers of energy towards smaller scales of circulation. This means that energy is flowing in both directions, in terms of scale, and consequently part of the gross energy transferred to smaller scales is returned to influence the scales where it has been originated, being then subjected to further redistribution. For three-dimensional flows, Leslie and Quarini (1979) estimated the relation between the gross and the net transfers of energy to be about 1.5.

14.4 REVISION OF FORMULATIONS

14.4.1 SHALLOW WATER EQUATIONS

For the integration of the Reynolds equations along the vertical, it is currently assumed that the vertical velocity distribution can be represented in terms of a depth-integrated velocity and of a velocity deviation, as stated in Equations (14.13). It is implicit in this approach that the shape of the velocity profile is unknown, which leads subsequently to the use of a Boussinesq-type analogy to make the closure of the new terms as shown in Equation (14.14). This methodology faces two difficulties of interpretation: firstly, it is not possible to take into account the physical processes associated with the vertical velocity distribution when the horizontal gradients of the depth-integrated velocity are negligible. Secondly, the dispersive terms can only account for shear flow processes along the vertical, i.e. although in modelling practice the dispersive coefficients are assigned values that are orders of magnitude higher than those indicated by the Euler formula, there is no justification in the equations to do so.

As a starting point to overcome these limitations, consider the vertical velocity profile given by

$$v_x(x, y, z, t) = V_x(x, y, t)[1 + \theta_x(z, t)] \quad (14.39a)$$

$$v_y(x, y, z, t) = V_y(x, y, t)[1 + \theta_y(z, t)] \quad (14.39b)$$

where the shape functions $\theta_i(z, t)$ verify the following properties

$$\int_0^h \theta_x(z, t)\mathrm{d}z = \int_0^h \theta_y(z, t)\mathrm{d}z = 0 \quad (14.40)$$

This approximation has been commonly used for the depth-integration of the convective terms, which then become

$$\frac{\partial}{\partial x}(\alpha_{xx}HV_xV_x) + \frac{\partial}{\partial y}(\alpha_{xy}HV_xV_y) \quad (14.41)$$

The correlation coefficient α is given by

$$\alpha_{ij}(t) = \frac{1}{h}\int_0^h [1 + \theta_i(z, t)][1 + \theta_j(z, t)]\mathrm{d}z \quad (14.42)$$

and represents the ratio between the effective momentum of the flow and the momentum calculated using the depth-integrated velocity.

It is apparent from Equation (14.41) that the transfer of momentum is affected by the vertical velocity distribution, as well as by its spatial variation, independently of the horizontal velocity gradients. This is an expected result considering that for the same depth-integrated velocity, the kinematic energy varies with the shape of the vertical velocity distribution, which is therefore enough to induce a transfer of energy.

The use of the previous procedure has obviously to be extended to all the terms of the primitive Reynolds equations. Due to the properties given in Equation (14.40), the depth-integration of the linear terms only implies the substitution of the primitive variables by their depth-integrated values. But, as for the convective terms, the dispersive terms in the Reynolds equations are non-linear, and their depth-integration will provide a formal basis for the inclusion of coefficients that account for the horizontal dispersive effects due to shear flow processes along the vertical (Vieira, 1991). Using again Equations (14.39), one obtains

$$\frac{\partial}{\partial x_j}\int_0^h \nu^t\left(\frac{\partial v_i}{\partial x_j} + \frac{\partial v_j}{\partial x_i}\right)\mathrm{d}z = \frac{\partial}{\partial x_j}\int_0^h \nu^t\left[(1 + \theta_i)\frac{\partial V_i}{\partial x_j} + (1 + \theta_j)\frac{\partial V_j}{\partial x_i} + \left(V_i\frac{\partial \theta_i}{\partial x_j} + V_j\frac{\partial \theta_j}{\partial x_i}\right)\right]\mathrm{d}z \quad (14.43)$$

and supposing that the last term is negligible, it is finally possible to write

$$\frac{\partial}{\partial x_j}\int_0^h \nu^t\left(\frac{\partial v_i}{\partial x_j} + \frac{\partial v_j}{\partial x_i}\right)\mathrm{d}z = \frac{\partial}{\partial x_j}\left(E_i\frac{\partial V_i}{\partial x_j} + E_j\frac{\partial V_j}{\partial x_i}\right) \quad (14.44)$$

where the dispersion coefficients are given by

$$E_i = \int_0^h \nu^t(1 + \theta_i)\mathrm{d}z \quad (14.45a)$$

$$E_j = \int_0^h \nu^t(1 + \theta_j)\mathrm{d}z \quad (14.45b)$$

These coefficients do not account for the shear effects associated with non-resolved sub-grid circulations that result from a further horizontal discretization, and neither do the coefficients obtained through the Boussinesq-type analogy applied to velocity deviations over the vertical, as previously referred. By analogy to what has been considered for the non-resolved, random-velocity fluctuations, the Eulerian representation of the non-resolved sub-grid circulations will be:

$$V_i = \hat{V}_i + V_i'' \quad (14.46)$$

where $\hat{V}_i$ represents the time average of the depth-integrated velocity and V_i'' the corresponding depth-integrated velocity fluctuations obtained over a time interval Δt representative of the sub-grid circulations.

Introducing Equation (14.46) in the depth-integrated equations and integrating over the time interval Δt it is then possible to filter out the non-resolved sub-grid circulations, and it can be shown that by using a Reynolds filter the following additional term will appear

$$\frac{\partial}{\partial x_j}(\rho\,\overline{V_i''V_j''}) = \frac{\partial}{\partial x_j}\left(\frac{\rho}{\Delta t}\int_{T-\Delta t}^{T} V_i''V_j''\mathrm{d}t\right) \quad (14.47)$$

Invoking the Bakhmeteff principle along the same lines used for the interpretation of the Reynolds stresses, it is then possible to speak about 'sub-grid shear stresses' which account for the transference of momentum due to fluctuations of velocity associated with non-resolved sub-grid circulations. The Boussinesq analogy can then be used in its original form, thus yielding

$$\frac{\partial}{\partial x_j}(\rho V_i''V_j'') = \varepsilon_{ij}\left(\frac{\partial \hat{V}_i}{\partial x_j} + \frac{\partial \hat{V}_j}{\partial x_i}\right) \quad (14.48)$$

The new coefficients ε_{ij} have the physical support of the sub-grid-scale processes, their value being naturally dependent on the magnitude of the non-resolved sub-grid circulations.

It is now possible to justify the use of dispersive coefficients of higher orders of magnitude than the values obtained from the Elder formulation.

14.4.2 FILTERING TECHNIQUES

From a practical point of view, when dealing with the very complex nature of coastal environmental flows, the determination of dispersive coefficients can only be achieved through the combined use of general formulae and *in situ* observations which provide for an order of magnitude estimation (Equations 14.31a–c), complemented by calibration and verification simulations for the evaluation of the more probable range of values.

The reliability of the models developed and of the simulation results obtained will therefore depend on the proper understanding of the physical and numerical concepts underlying the use of empirical coefficients.

As for the dispersive coefficients, some of these concepts have been discussed in the present work on the basis of the Reynolds filtering definition (Equation (14.2a)). However, from a theoretical point of view and bearing in mind specific flow conditions in other fields of application, a more general analysis has to take into account the implications of using different filtering procedures for the averaging of non-linear terms. With respect to the terms $v_i v_j$ we can write regardless of the filter employed:

$$\overline{v_i v_j} = \overline{(\hat{v}_i + v_i')(\hat{v}_j + v_j')}$$
$$= \overline{\hat{v}_i \hat{v}_j} + \overline{\hat{v}_i v_j'} + \overline{v_i' \hat{v}_j} + \overline{v_i' v_j'} \qquad (14.49)$$

The term $\overline{\hat{v}_i \hat{v}_j}$ is entirely determined by grid-resolved quantities, so the sub-grid-scale stresses τ_{ij}^G will be given by

$$\tau_{ij}^G = \rho\,(\overline{\hat{v}_i v_j'} + \overline{v_i' \hat{v}_j} + \overline{v_i' v_j'}) \qquad (14.50)$$

In the case of the Reynolds temporal filter one has

$$\overline{\hat{v}_i v_j'} = \overline{v_i' \hat{v}_j} = 0 \qquad (14.51a)$$

and consequently

$$\tau_{ij}^G = \rho\,\overline{\hat{v}_i v_j'} \qquad (14.51b)$$

that are the well-known Reynolds stresses. It should also be observed that for this filter:

$$\overline{\hat{v}_i \hat{v}_j} = \hat{v}_i \hat{v}_j \qquad (14.51c)$$

so that the momentum equations will become

$$\rho\left(\frac{\partial \hat{v}_i}{\partial t} + \frac{\partial(\hat{v}_i \hat{v}_j)}{\partial x_j}\right) = \rho F_i + \frac{\partial \hat{\tau}_{ji}}{\partial x_j} \qquad (14.52)$$

where F_i represents the forces acting per mass unit and $\hat{\tau}_{ji}$ is the total time-averaged stress tensor given by

$$\hat{\tau}_{ji} = -(\hat{p} - \lambda\hat{\Theta})\delta_{ij} + 2\mu\hat{S}_{ij} - \tau_{ij}^G \qquad (14.53)$$

where: $\hat{p}$ is the normal pressure; $\hat{\Theta} = \text{div}\,\vec{v}$; μ is the dynamic viscosity; $\lambda = -2\mu/3$; δ_{ij} is the Kronecker delta function; and $\hat{S}_{ij}$ is the second-order strain tensor.

In general, for other spatial and spatial-temporal filters, the above simplifications do not apply, as appears from the case of the convolution definition of averaging (Leonard, 1974; Bedford, 1987; Abbott and Basco, 1989):

$$\hat{v}_i(x_j, t) = \int_{-\infty}^{\infty} G(x_j - x_j', t - t')v_i(x_j', t')\mathrm{d}x_j'\mathrm{d}t' \qquad (14.54a)$$

where the filter function G satisfies the following fundamental condition

$$\int_{-\infty}^{\infty} G(x_j, t)\mathrm{d}x_j\mathrm{d}t = 1 \qquad (14.54b)$$

Therefore the sub-grid-scale stresses τ_{ij}^G have to be considered in the general form (Equation (14.50)) when making the closure of the equations. Furthermore, as discussed by Leonard (1974),

$$\overline{\hat{v}_i \hat{v}_j} \neq \hat{v}_i \hat{v}_j \qquad (14.55)$$

and thus the advection term in Equation (14.52) becomes

$$\frac{\partial}{\partial x_j}(\hat{v}_i \hat{v}_j + \lambda_{ij}) \qquad (14.56)$$

where the resolvable part of λ_{ij} represents the so-called Leonard stresses. Resorting to a Taylor's series expansion and an isotropic Gaussian filter with effective width Δx_κ, Abbott *et al.* (1981) presented the following approximate method to estimate λ_{ij}:

$$\lambda_{ij} = \frac{(\Delta x_\kappa)^2}{6}\frac{\partial^2(\hat{v}\hat{v})}{\partial x_\kappa \partial x_\kappa} + \begin{array}{l}\text{terms containing}\\ \text{unresolved scales}\end{array} \qquad (14.57)$$

Although the above formulation provides a more complete description of the effects of non-resolved flow processes, it implies the estimation and calibration of additional empirical parameters that are associated with the Leonard stresses. Moreover there is not yet experience concerning the use of these terms in real case applications, and no systematic approach for their estimation is available. On the other hand the principles and concepts discussed in this work are general, and the formulation based on the Reynolds filtering definition has proved to be adequate for coastal and environmental engineering studies, given the degree of discretitation typically used (of the order of magnitude 10^2 to 10^4 m) and the level of accuracy required.

REFERENCES

Abbott, M.B. and Rasmussen, C.H. (1977) On the numerical modelling of rapid expansions and contractions in models that are two-dimensional in plan. International Association for Hydraulic Research, 17th Congress, Baden-Baden, Germany.

Abbott, M.B., McCowan, A. and Warren, I.R. (1981) Numerical

modelling of free-surface flows that are two-dimensional in plan, in *Transport Models for Inland and Coastal Waters*, Academic Press, London.
Abbott, M.B., Larsen, J. and Tao, J. (1985) Modelling circulations in depth-integrated flows. Part 1: The accumulation of the evidence. *Journal of Hydraulic Research*, **23** (4), 309–26.
Abbott, M.B., Shoaling, H. and Warren, I.R. (1988) Auto-dispersion processes in numerical models. International Conference on Computer Modelling in Ocean Engineering, Venice.
Abbott, M.B. and Basco, D.R. (1989) *Computational Fluid Dynamics. An Introduction for Engineers*, Longman, Harlow, UK.
Bansal, M.K. (1971) Dispersion in natural streams. *Journal of Hydraulic Division, ASCE*, **HY11**, 1867–86.
Bedford, K. (1985) Selection of turbulence and mixing parameterization for estuary water quality models. U.S. Army Corps of Engineers, Environmental Research Programme, *Paper EL-85-2*.
Bowden, K.F. (1965) Horizontal mixing in the sea due to a shearing current. *Journal of Fluid Mechanics*, **21** (2), 83–95.
DHI (1986) System 21 HD. Scientific documentation. Danish Hydraulic Institute, Hørsholm, Denmark.
Elder, J.W. (1959) The dispersion of a marked fluid in a turbulent shear flow. *Journal of Fluid Mechanics*, **5**, 544–60.
Fick, A. (1855) On liquid diffusion. *Philosophical Magazine*, **4** (10), 30–9.
Fischer, H.B. (1968) Dispersion prediction in natural streams. *Journal of Sanitary Engineering Divison, ASCE*, **94** (SAS), 927–43.
Fischer, H.B., List, E.J., Koh, R.C.Y. *et al.* (1979) *Mixing in Inland and Coastal Waters*, Academic Press, New York.
Jobson, H.E. and Sayre, W.W. (1970) Vertical transfer in open channel flow. *Journal of the Hydraulics Division, ASCE*, **96**, 703–24.
Justesen, P. (1990) System 3. A three-dimensional hydrodynamic model. Turbulence modelling (Vol. 5). Danish Hydraulic Institute, Hørsholm, Denmark.
Krenkel, P.A. and Orlob, G. (1962) Turbulent diffusion and the reaeration coefficient. *Journal of Sanitary Engineering Divison, ASCE*, **88** (SA2), 53–83.
Leslie, D.C. (1982) Simulation methods for turbulent flows, in *Numerical Methods for Fluid Dynamics* (ed K.W. Morton and H.J. Baines) Academic Press, London, 63–80.
Leslie, D.C. and Quarim, G.L. (1979) The application of turbulence theory to the formulation of subgrid modelling procedures. *Journal of Fluid Mechanics*, **97**, 65–91.
Leonard, A. (1974) Energy cascade in large-eddy simulations of turbulent fluid flows. *Advances in Geophysics, A*, **18**, 237–48.
Madsen, P.A., Rugbjerg, M. and Warren, I.R. (1988) Subgrid modelling in depth integrated flows. 21st International Conference on Coastal Engineering, Malaga.
Rodi, W. (1980) Turbulence models and their application in hydraulics. A state of the art review. International Association for Hydraulic Research, Delft, The Netherlands.
Schumann, V. (1975) Subgrid scale model for finite difference simulations of turbulent flows in plane channels and annuli. *Journal of Computational Physics*, **18**, 376–404.
Schwiderski, E.W. (1981) Global ocean tides. Chief of Naval Material, Department of the Navy, Washington, DC NSWC TR81-122.
Smagorinski, J. (1963) General circulation experiments with the primitive equations, 1. The basic experiment. *Monthly Weather Review*, **91**, 90–164.
Talbot, J.W. and Talbot, G.A. (1974) Diffusion in shallow seas and in English coastal and estuarine waters, in *Physical Processes Responsible for Dispersal of Pollutants in the Sea*, (eds G. Kullenberg and J.W. Talbot) International Council for Sea Exploration, 93–110.
Taylor G.I. (1953) Dispersion of soluble matter in solvent flowing through a tube. *Proceedings of the Royal Society, London, Series A*, **219**, 186–203.
Taylor, G.I. (1954) The dispersion of matter in turbulent flow through a pipe. *Proceedings of the Royal Society of London, Series A*, **223** 446–68.
Tennekes, H. and Lumley, J.L. (1972) *A First Course in Turbulence*, MIT Press, Cambridge, MA.
Thackson, E.L. (1966) Longitudinal mixing and reaeration in natural streams. PhD Thesis, Vanderbilt University, Nashville, Tennessee.
Vieira, J.R. (1990) Contribution for the study of circulation processes in coastal regions. Development and application of a finite element model. PhD thesis, Technical University of Lisbon, Portugal.
Warren, I.R. and Shoaling, H. (1989) Mathematical modelling of turbulence in the coastal zone. International Association for Hydaulic Research, XXIII Congress, Ottawa.
Yotsukura, N. and Fiering, M.B. (1964) Numerical solution to a dispersion equation. *Proceedings of the Royal Society, London, Series A*, **219**, 186–203.

ONE-DIMENSIONAL TRANSPORT–DISPERSION AND WATER QUALITY MODELS 15

F.M. Holly Jr.

We now return to a one-dimensional description, but for the advection–diffusion case. Several of the numerical problems already raised in three- and two-dimensional modelling reappear, but these are now compounded by the exigencies of the higher degree of schematization inherent in the one-dimensional approach. The numerical problems of the previous sections are thus exacerbated even further by the more basic physical–mathematical problems of describing the phenomena of interest that have been raised earlier by Bedford in Chapter 5.

This contribution reviews methods currently used to overcome the numerical difficulties while relating these to the basic problems of describing such complex processes with such simple mathematical means.

CONTENTS

15.1 INTRODUCTION

This chapter assumes that the reader has a basic understanding of contaminant mixing, numerical procedures and the associated vocabulary. It is not a short course in the numerical solution of the one-dimensional contaminant-transport equation. It is, rather, a brief collection of remarks on a few points of practical interest in the creation, selection, or use of contaminant-transport modelling systems.

These remarks are highly subjective, borne of the author's experience, which is devoid of all but formal experience in finite element approaches. Finite difference and Lagrangian techniques have been the author's exclusive practical tools for water quality modelling. This section is not particularly quantitative in nature. It presents the writer's subjective views of some of the more important issues to be considered in obtaining quality simulation of contaminant-transport processes in rivers and canals. These remarks are presented in the hope that they can help bridge the gap between the usual treatises which, in general, are either too theoretical or too practical in the eyes of practitioners and numerical analysts, respectively. Hopefully they bring together the needs and concerns of both groups in a mutually accessible context.

15.2 A SIMPLE PROBLEM – OR IS IT?

It is tempting to view the one-dimensional transport–dispersion problem as somewhat of a poor technical cousin to its two- and three-dimensional analogues as described in earlier chapters. Perhaps this is because the absence of problems of grid generation, split-operator algorithms, generation of complex velocity fields, tensor diffusion terms, sub-grid-scale processes, etc. lull one into a perception that the absence of major computational challenges implies an immediate ability to produce an effective, accurate simulation in one spatial dimension.

Is this in fact the case? The answer is found in the muttered words of many a neophyte in computational hydraulics: 'It is easy to compute one-

Coastal, Estuarial and Harbour Engineer's Reference Book. Edited by M.B. Abbott and W.A. Price. Published in 1993 by Chapman & Hall, 2–6 Boundary Row, London SE1 8HN. ISBN 0 419 15430 2.

dimensional transport, but very difficult to compute it well.' These words of frustration stem from two primary sources. One is the notorious susceptibility of advection computation to the twin evils of artificial diffusion and oscillations. The other is the great difficulty encountered in assigning appropriate mixing coefficients in natural water bodies, an issue already discussed by Bedford in Chapter 5 and explored further below.

15.3 DEALING WITH ARTIFICIAL DIFFUSION AND OSCILLATIONS

It has become something of a ritual in the literature to cite the problems of artificial diffusion and oscillations associated with approximate numerical solution of the advective terms in the transport equation. Repeated here in its one-dimensional (i.e. section-averaged) form, the transport equation is as follows:

$$\frac{\partial C}{\partial t} + u\frac{\partial C}{\partial x} = \frac{1}{A}\frac{\partial}{\partial x}\left(AK_x\frac{\partial C}{\partial x}\right) + S \qquad (15.1)$$

in which $C(x, t)$ is the contaminant concentration, x is the longitudinal spatial co-ordinate, t is the time, $A(x, t)$ is the cross-sectional area, $u(x, t)$ is the cross-sectional average water velocity, $S(x, t, C)$ is the collected source–sink processes, and $K_x(x, t)$ is the longitudinal mixing coefficient. The second term on the left-hand side of Equation (15.1) represents the net advective flux of contaminant into a control volume. The term 'pure advection' is often used to characterize the accumulation of contaminant mass in the control volume due only to net advective transport, as described by setting the left-hand side of the equation equal to zero. We represent the distance and time steps used in the numerical calculation by Δx and Δt, respectively.

It can be shown theoretically, and corroborated by practical experimentation, that unless the so-called 'advection Courant number' C_a, $= u\Delta t/\Delta x$, is strictly unity, many reasonable-looking finite difference and finite element numerical schemes introduce an enormous amount of artificial, or numerical, diffusion into what should be a purely advective process (e.g., Cunge *et al.* (1980), pp. 320–31.) It can also be shown that many so-called 'higher-order' finite difference and finite element schemes which are designed to obviate artificial diffusion do so at the price of introducing artificial oscillations (also called 'dispersion' in a wave propagation sense). Examples can be seen in Holly (1975), Sauvaget (1985), and Gresho and Lee (1979).

Why are artificial diffusion and oscillations a problem? Artificial diffusion is a problem because, in so closely resembling physical diffusion, it can lead one to adopt a calibrated value of K_x which, rather than parameterizing the physical behaviour of the water-course under study is, in fact, heavily influenced by the particular numerical scheme being used and the dimensions of the computational grid on which the simulation is being performed. Such a value of K_x has no validity for use in studying contaminant transport under hydraulic and numerical conditions different from those for which the calibration was performed.

Artificial oscillations are a problem not only because they represent a kind of harmonic behaviour which is completely foreign to diffusive processes, but also because they can easily generate negative concentrations. Of course negative concentrations have no meaning in their own right, and can generate complex computational problems in the context of chemical, biological, or radioactive decay processes whose rates depend on the concentration itself.

Are numerical schemes available which eliminate artificial diffusion and oscillations? The answer is a qualified 'yes' – qualified, because we are dealing with approximate, not exact, solutions to the contaminant-transport equation, and all approximate numerical schemes have some truncation error of more or less importance. There are two classes of numerical scheme which have proved to be fairly effective in minimizing artificial diffusion and oscillations in a one-dimensional computational context. One class adopts the 'anti-diffusion' approach described by Boris and Book (1973), in which a predictable amount of artificial diffusion in a first-order advection computation is neutralized by an equivalent amount of negative diffusion, restoring to near their initial value peak concentrations which would otherwise have been artificially reduced by the numerical error.

Another class of scheme is based on a Lagrangian approach, whereby finite difference or finite element discretization of the advective terms is replaced by integration along a characteristic trajectory. Equation (15.1) is recognized, after the introduction of the water continuity equation, as a total derivative along the trajectory followed by a water particle. Integration of this total derivative along the trajectory in one time step yields an exact solution, namely that the concentration does not change along this trajectory (for pure advection). Thus the problems of artificial diffusion and oscillations in grid-based numerical methods are replaced by problems of calculating space–time characteristic trajectories, obtaining accurate interpolated values of concentration at the foot of the trajectory and, when diffusive and source terms are present, obtaining accurate integration of the right-hand side of Equation (15.1) along that same trajectory. Experience has shown that Hermite cubic interpolation at the foot of the trajectory does an excellent job of minimizing artificial diffusion and oscillations, at the expense of the need to solve a second, auxiliary transport equation for the first derivative of the concentration (Holly and Preissmann, 1977).

Researchers in the finite element domain have gone to great lengths to obtain schemes which minimize artificial diffusion and oscillations (e.g. Morton (1981)), but their efforts have generally focused on multi-dimensional problems considered independently of the considerations that arise in practical engineering implementation.

It is incumbent upon developers and users of numerical contaminant-transport models to be aware of the importance of artificial diffusion and oscillations, to be willing to seek out and implement one of the many schemes which minimize these errors, and to have a quantitative appreciation of the amount of artificial diffusion and oscillations introduced by the adopted scheme. Methodologies for obtaining such a quantification have been described by Cunge *et al.* (1980), Roache (1972), Carnahan *et al.* (1969), Richtmyer and Morton (1967), and others.

15.4 SPLIT-OPERATOR METHODS – PANACEA OR VOODOO COMPUTATIONAL HYDRAULICS?

Implicit in the above discussion of artificial diffusion and oscillations is the notion that one may focus attention on the behaviour of selected (e.g. advection) terms of the governing conservation equation in choosing an acceptable numerical method. But what about the other terms? In fact, the diffusion terms in Equation (15.1) can be approximated with sufficient accuracy with fairly conventional finite difference operators. These terms, when combined with the time derivative on the left-hand side, have no underlying characteristic structure which can be exploited to generate an accurate numerical operator, as is the case for advection. (In fact, a purely diffusive process has characteristic directions of infinite celerity in the x–t plane). Moreover, any source and sink terms of Equation (15.1) are purely algebraic, so the question of an appropriate finite difference operator is replaced by a question of appropriate integration in time.

Split-operator methods have arisen because of the desire to have the best of all possible, worlds, i.e. to use the best numerical method for each of several subsets of terms in the governing equation. As applied to Equation (15.1), for example, this would imply executing three separate operations, each of an equal duration of one computational time step, as follows (e.g. Cunge *et al.*, 1980).

1. Take the concentration field $C(x, t)$ at the end of the previous time step as an initial condition and compute the change in the field due to pure advection only for one time step using, for example, a characteristics method with Hermite cubic interpolation.
2. Take the advected concentration field from step 1 as an initial condition and compute the change in the concentration field due to pure diffusion only for one time step using, for example, a generalized Crank–Nicholson method on a variable spatial grid.
3. Take the advected and diffused concentration field from step 2 as an initial condition and compute the change in the concentration field due to the source/sink terms only for one time step using, for example, trapezoidal- or Simpson-rule integration in time.

The results of this three-step split-operator process would be the new concentration field existing at one time step after the beginning the sequence. The field will have been modified by the combined effects of advection, diffusion, and source and sink, each acting for the same duration of time, but sequentially in the model rather than continuously as in nature. The modeller's self-congratulation arising from use of the 'best' numerical schemes for each process would be tempered by an uneasy feeling that he had tried to fool Mother Nature in treating these simultaneous, continuous physical processes as sequential, discrete ones. In particular, whereas Equation (15.1) requires only two boundary conditions, one upstream and one downstream, the split-operator approach would have required the imposition of five boundary conditions: one upstream for advection, and one upstream and one downstream for both diffusion, and source and sink. Now, all but the downstream diffusion boundary condition may have been rather benign statements of no change in the concentration resulting from the previous process in the split-operator sequence. But these are none the less redundant, overspecified conditions in terms of what the original partial differential equation required.

The mathematician would be appalled at the liberties taken with the original partial differential equation. The computational hydraulics engineer would be less concerned with formalities, and defend the split-operator approach on the basis of the opportunity it provides to use the best numerical method for each process, and on the basis of its apparent satisfactory performance when applied to prototype test cases.

Wherein lies the solution to this apparent dilemma? Probably somewhere between these two points of view. It is indeed true that engineering experience has tended to validate (in the sense that computational catastrophes have not occurred) the use of a split-operator approach not only for one-dimensional, but also for higher-dimensional problems. It is also true that we are in a kind of no-man's land with respect to the formal requirements of the split-operator strategy.

The split-operator dilemma can be mollified to some extent by combining the Lagrangian advective operator and the Eulerian diffusive operator in a hybrid scheme. As described by Toda and Holly (1987), such a scheme is based on recognition of the predominance of advection over diffusion in riverine contaminant transport problems. Equation (15.1) can be written without source–sink processes as

$$\frac{\mathrm{d}C}{\mathrm{d}t} = \frac{1}{A}\frac{\partial}{\partial x}\left(AK_x\frac{\partial C}{\partial x}\right) \tag{15.2}$$

where $\mathrm{d}C/\mathrm{d}t$ denotes the total derivative of concentration along the trajectory

$$\frac{\mathrm{d}x}{\mathrm{d}t} = u \tag{15.3}$$

Formal integration of Equation (15.2) along the trajectory of Equation (15.3) yields the following expression

$$C_i^{n+1} = C_\xi + \int_{t_{n+1}}^{t_n} \frac{1}{A}\frac{\partial}{\partial x}\left(AK_x\frac{\partial C}{\partial x}\right)\mathrm{d}t \tag{15.4}$$

where ξ denotes the foot of the trajectory leading to the computational point i at time t_{n+1} from time t_n.

A hybrid scheme results from using an approximate path integration of the diffusive operator in Equation (15.2), leading to a direct algebraic expression for the unknown concentration C_i^{n+1}. Although logic and intuition would suggest that a trapezoidal integration be adopted, using for example the second derivative of the Hermite interpolating polynomial at the foot of the trajectory and an implicit divided-difference approximation at the head of it, experience has shown that a simple first-order integration using only the implicit approximation at the head is the most robust and accurate.

The predominance of advection is invoked in the treatment of the boundary conditions. At the upstream limit, the known inflow concentration is simply imposed at the first computational point, simultaneously satisfying the boundary condition needs of both the advection and diffusion operators (note that Equation (15.4) is written for only the second and subsequent computational points.) As for the downstream boundary, diffusion is assumed to be negligible in that only pure advection is allowed to operate in the last computational reach.

This hybrid approach has proven to be satisfactory from a practical point of view. It lacks rigour only in its numerical approximations and in its neglect of diffusion in the downstream-most computational reach. Thus the hybrid approach appears to retain most of the advantages of the split-operator philosophy, while replacing the split-operator's blatant over-specification of boundary conditions with the more palatable localized neglect of diffusion in one computational reach.

15.5 COMPUTATIONAL GRIDS: ARE THREE POINTS BETTER THAN TWO?

The basic structure of the computational grid is not only an issue of (alleged) obscure numerical analysis, but also one of practical concern in water-quality modelling. This practical concern arises from the need to accommodate boundary conditions and internal compatibility conditions at the appropriate physical locations in a numerical model.

The basic conservation Equation (15.1) contains two types of differential operators: the first-order time and space derivatives associated with contaminant advection, and the second-order space derivative associated with contaminant diffusion. In a finite difference context, approximation of first-order derivatives by divided differences requires the use of information at at least two computational points, and second-order derivatives require at least three computational points. For the time derivative, this merely formalizes the obvious requirement that the advance of time be represented by passage of the state of the model from one time t_n to the subsequent time t_{n+1} at a time increment Δt later. Actually some numerical schemes work with three time levels in the interest of obtaining improved temporal accuracy, and this poses no particular practical problem.

The first-order space derivative requires only two computational points, although three are used in some schemes of higher-order accuracy. Moreover, the second-order space derivative of the diffusion term would appear always to require three computational points. Thus the most obvious and conceptually simple numerical schemes are based on numerical approximation of the advection and diffusion terms on sets of three computational points in space.

The practical difficulty arises in that, in requiring that one difference equation be associated with the middle of a set of three computational points, these three-point schemes can be written neither for boundary points nor for interior points which represent one side or the other of special hydraulic works or network features such as weirs, retention basins, bifurcations, etc. In such cases one of the three required points would be unavailable, or at least not associated with the other two points by the same transport Equation (15.1). So what is to be done? Most schemes adopt some alternative, generally lower-order, numerical approximation at such special points, or even locally neglect the diffusion operator entirely, accepting the damage that this does to the formal order of accuracy of the scheme.

By contrast, the advantage of a scheme requiring only two computational points for approximation of spatial derivatives is that it is written, not for the centre of three computational points, but for each pair of computational points. And any point, be it on the boundary or adjacent to a bifurcation or weir or whatever, is one of a pair between which a unique conservation law applies. Thus no modification or weakening of the basic scheme is required to accommodate hydraulic and topological features of practical importance.

Now, it is clear that the first derivative of the advection operator requires only two computational points, and indeed the accurate Lagrangian schemes described earlier require interpolation between only two computational points. But what about the second-order diffusion operator, which nominally would appear to always require three points?

The hybrid Lagrangian – finite difference scheme described earlier offers the opportunity to maintain a two-point scheme with the diffusion operator. Since the Hermite cubic advection interpolating polynomial requires that the concentration derivative *CX* be carried as a dependent variable, the second derivative can be expressed as the first-order derivative of *CX*, thus requiring the use of only two computational points. This approach, described in detail by Sauvaget (1985), has been successfully implemented in a simulation code which treats any level of channel branching, and any number of internal hydraulic works, i.e. in which the unavailability of continuous triplets of computational points is routine. The obvious practical advantage is that a user can assemble the physical elements of a water-quality simulation model without any structural constraint on the locations of boundary conditions or internal hydraulic works or junctions, etc.

15.6 IS COMPUTATIONAL COST AN ISSUE?

This question is quickly answered: no. The contaminant-dispersion problem is a linear one (at least insofar as conservative processes are concerned), and as a consequence the required computational resources tend to be a fraction of those required for the equivalent flood-propagation computations. This is true even for accurate Lagrangian advection schemes which must solve an auxiliary problem for the concentration derivative. Simulation of small to medium models on even first-generation personal computers is entirely feasible. Large models can also be processed if the computational code makes efficient use of memory and disk space.

Of course some numerical methods are faster than others, and some codes are faster and more efficient than others. But in the end, the computational time and cost will be insignificant compared to the manpower time and cost, so that computational-time arguments for the use of one numerical method over another are generally of little importance. On the other hand, industrialized simulation codes which provide accurate computation and minimize the manpower time required to construct a model data set and interpret the simulation results can result in significant savings in project cost.

15.7 NON-CONSERVATIVE PROCESSES

Non-conservative processes (e.g. surface heat exchange, biodegradation, re-oxygenation, etc.) appear in the contaminant-conservation equation as strictly scalar source and sink terms, with no direct participation in the differential operators. As a consequence, the numerical treatment of non-conservative processes is to some extent independent of the choice of numerical differential operator. However, nonlinear non-conservative processes may compromise the conditional stability of explicit differential operators, and Lagrangian-based advection operators, especially hybrid ones, require careful path integration of the source and sink terms. The main practical issue is one of being sure, either through formal analysis or practical demonstration, that the presence of the source and sink terms compromises neither the accuracy nor the stability of the basic differential operators.

15.8 LOWER DIMENSIONALITY AND THE GHOST OF TAYLOR – A TRAP FOR THE UNWARY

The title of this section is but the tip of the iceberg which is dispersion theory. This theory is concerned with the description of mixing due to differential advection at various scales as equivalent to gradient diffusion. This is not the place to go into the detailed development and implications of the theory: the interested reader can begin an exploration of this theory by referring to Chapter 5 by Bedford, earlier in this volume and then to Fischer *et al.* (1979) or Holly (1985), for examples. Although many researchers have contributed to the theory, its real godfather is the celebrated British fluid mechanician G.I. Taylor.

So what about lower dimensionality and traps etc? The point is that as one averages turbulent and differential-advective mixing, first over short time periods and then over larger spatial scales, theoretical support for the treatment of mixing as equivalent to diffusive mixing becomes weaker and weaker, and the time which must elapse from the time of contaminant injection to the time during which its mixing can truly be considered to be a gradient process becomes larger and larger.

For example, if one considers instantaneous mixing in three-dimensional turbulent flow, and is prepared to simulate the process at a fine enough spatial and temporal scale to resolve the complete details of the turbulent fluctuations, then the only empirical parameter which is required is the molecular diffusivity, characterizing a well-established gradient-diffusion process at the molecular scale. The mixing due to turbulent velocity fluctuations is fully described by the advective terms in the mass conservation equation, provided again that one is prepared to deal with the equations at the turbulent scale.

Taylor (1921) showed that when one considers the average mixing over time scales which are long compared to turbulent fluctuations, the mixing due to turbulent velocity fluctuations can be described as a gradient-diffusion process with a diffusivity whose quantification is well supported theoretically and experimentally (for homogeneous, isotropic turbulence). But this is true only in a so-called Taylor diffusive period, beginning some time after the introduction of the contaminant into the flow, and the turbulent diffusion coefficient is orders of magnitude larger than the molecular diffusivity.

Taylor (1953) went on to show that as one considers the average mixing over the cross-section of a pipe flow, then once again one can describe the mixing due to differential advection in the cross-section as a gradient-diffusion process, but only after an even longer initial period. Now the appropriate mixing coefficient is again orders of magnitude larger than the turbulent coefficient, but it can be quantified on a sound theoretical basis.

These basic spatial averaging procedures by Taylor have been applied by Elder (1959), Fischer (1967) and others to increasingly complex free-surface flow situations. But a characteristic of these applications is the loss of rigorous theoretical support originally established by Taylor, due primarily to the difficulty of dealing with natural geometries.

At the far end of the scale is the one-dimensional contaminant dispersion under consideration in this chapter. Equation (15.1) embodies the commonly accepted notion that all mixing due to deviations of velocity from the cross-sectional average can be described as a gradient-diffusion process with coefficient K. But Taylor's theory would suggest that even in prismatic canals with uniform flow, the initial period which must elapse before the diffusion process is theoretically justified is so long as to exclude many practical applications. And in natural river geometries, with non-prismatic channels and non-uniform (perhaps unsteady) flow, Taylor's theory would say that a diffusive mixing period simply does not exist, and there is no theoretical basis for quantifying the overall one-dimensional mixing coefficient.

Does this prevent engineers from adopting a one-dimensional advection-diffusion description of mixing in rivers? Not at all, and thus the ghost of Taylor must indeed be unhappy. One can only hope that the ghost is somewhat reassured by engineers' willingness to recognize the limitations and weaknesses of the one-dimensional approach and treat simulation results with the informed scepticism which they deserve.

The coefficient K_x of Equation (15.1) is not in any sense a fundamental physical property of the fluid and the contaminant. Respect for the ghost of Taylor would require that we deny its very existence. Engineers needing nonetheless to solve a problem of contaminant mixing must recognize that to the extent that K_x can describe mixing due to differential advection, it is a function of the particular channel geometry and flow situation prevailing at a particular place and a particular time. K_x can be calibrated for a particular mixing event, for example a dye-dispersion experiment in a river, but the calibrated value of K_x has no universal validity for that river. In particular, it cannot be used to predict mixing for some other flow and channel condition.

Several means of predicting K_x have been developed by Fischer and others (e.g. Fischer *et al.* (1979)), and these predictors provide useful guidance in providing estimates of K_x for the modelling of flow situations for which no mixing observations are available. But responsible use of numerical models of Equation (15.1) requires that the engineer view the adopted value of K_x, and indeed the very validity of Equation (15.1), with much scepticism. In practical terms, this means basing engineering conclusions on careful sensitivity tests, repeating simulations for a range of values of K_x corresponding to the range of uncertainty reflected in the range of values given by the various predictors for K_x.

The only sure solution to this dilemma is to increase the dimensionality of the problem. As one passes from a one-dimensional description to a two-dimensional, then a three-dimensional one, the diffusion coefficients appearing in the appropriate governing equations take on increasingly physical interpretations and are subject to quantification through the authority of Taylor's theory. As available computational resources and automated data processing make it more and more feasible to work with two- and three-dimensional descriptions of contaminant mixing, even for large spatial extents where the higher dimensionality is not strictly necessary from an engineering point of view, engineers should consider abandoning the one-dimensional description so that the ghost of Taylor can rest somewhat more peacefully.

REFERENCES

Boris, J.P. and Book, D.L. (1973) Flux-corrected transport. I. SHASTA, a fluid transport algorithm that works. *Journal of Computational Physics*, **11**, 38–69.

Carnahan, B., Luther H.A. and Wilkes, J.O. (1969) *Applied Numerical Methods*, Wiley, New York.

Cunge, J.A., Holly, F.M. Jr. and Verwey, A. (1980) *Practical Aspects of Computational River Hydraulics*, Pitman, London.

Elder, J.W. (1959) The dispersion of marked fluid in turbulent shear flow. *Journal of Fluid Mechanics*, **5**, 544–60.

Fischer, H.B. (1967) The mechanics of dispersion in natural streams. *Journal of the Hydraulics Division, ASCE*, **93**, 187–216.

Fischer, H.B., List, E.J., Koh, R.C.Y. *et al.* (1979) *Mixing in Inland and Coastal Waters*, Academic Press, New York.

Gresho, P.M. and Lee, R.L. (1979) Don't suppress the wiggles – they're telling you something, in *Finite Element Methods for Convection Dominated Flows*, (ed. T.J.R. Hughes), AMD Vol. 34, American Society of Mechanical Engineers, pp. 37–62.

Holly, F.M. Jr. (1975) Two-dimensional mass dispersion in rivers. Hydrology Paper No. 78, Colorado State University, Fort Collins.

Holly, F.M. Jr. (1985) Dispersion in rivers and coastal waters – 1. Physical principles and dispersion equations, in *Developments in Hydraulic Engineering* (ed. P. Novak), Elsevier Applied Science, London, pp. 1–38.

Holly, F.M. Jr. and Preissmann, A. (1977) Accurate calculation of transport in two dimensions. *Journal of the Hydraulics Division, ASCE*, **103** (HY11), 1259–78.

Morton, K.W. (1981) Finite element methods for non-self-adjoint problems. Department of Mathematics, University of Reading, UK, *Numerical Analysis Rep.* 3/81.

Richtmyer, R.D. and Morton, K.W. (1967) *Difference Methods for Initial Value Problems*, Interscience Publishers, New York.

Roache, P.J. (1972) *Computational Fluid Dynamics*, Hermosa Publishers, Albuquerque, NM.

Sauvaget, P. (1985) Dispersion in rivers and coastal waters – 2. numerical computation of dispersion, in *Developments in Hydraulic Engineering* (ed. P. Novak), Elsevier Applied Science, London, pp. 39–78.

Taylor, G.I. (1921) Diffusion by continuous movements. *Proceedings of the London Mathematical Society, Series A*, **20**, 196–211.

Taylor, G.I. (1953) Dispersion of soluble matter in solvent flowing slowly through a tube *Proceedings of the Royal Society of London, Series A*, **219**, 186–203.

Toda, K. and Holly, F.M. Jr. (1987) Hybrid numerical method for linear advection-diffusion. *Microsoftware for Engineers*, **3** (4), 199–205.

3.3 SEDIMENT TRANSPORT MODELS

THREE-DIMENSIONAL SEDIMENT-TRANSPORT MODELS 16

B.A. O'Connor

Since the fluid flows of interest here are almost always turbulent and turbulence is essentially three-dimensional, the use of a full three-dimensional simulation would appear to be the most desirable. Unfortunately, not only would such simulation be prohibitively expensive in terms of computer costs in most cases, but also it would rarely be justified by the quantity and quality of the field and laboratory data available for calibration and verification (or validation) purposes. None the less, a considerable effort is being expended on advancing the frontiers of three-dimensional simulation, and this is particularly evident in the field of sediment transport. The present contribution provides a short but quite exhaustive overview of the approaches that are now being examined. It is predicted that fully-three-dimensional methodologies will come to be widely applied in practice in the 1990s.

CONTENTS

16.1 INTRODUCTION

The construction of a new port or harbour often requires reclamation of land from the sea, the dredging of deep navigation channels and the building of new breakwaters and quays. Such works can cover many tens of square kilometres and, consequently, can have a major impact on environmental processes. Existing patterns of tides, waves and river flows may be dramatically altered and may lead, in turn, to major changes in the salinity and polluted state of such flows as well as their sediment-carrying capacity. Thus, shallow, muddy areas may be found to deepen rapidly, exposing sandy deposits, while navigation channels through sandy areas may slowly fill in with polluted and flocculated muddy material, which consolidates slowly and is difficult to remove with dredging equipment.

The engineer responsible for the design of such works must attempt, therefore, to produce safe and cost-effective schemes with the minimum of environmental impact. In the past, the engineer made extensive use of integrated programmes of field and

Coastal, Estuarial and Harbour Engineer's Reference Book. Edited by M.B. Abbott and W.A. Price. Published in 1993 by Chapman & Hall, 2–6 Boundary Row, London SE1 8HN. ISBN 0 419 15430 2.

hydraulic model studies and relied heavily on his experience of how similar systems responded to environmental change. More recently, and particularly since the development of powerful digital computers in the early 1970s, the engineer has been able to add an ever more sophisticated array of computer models to his armoury of predictive techniques. Today, the engineer can choose between one-, two- or three-dimensional models. Some of these models, particularly the simpler one- or two-dimensional hydrodynamic versions, have completely replaced the need to conduct expensive hydraulic model studies and it is tempting to conclude that the engineer will soon be able to solve all his problems by the appropriate choice of three-dimensional computer model. While such a state of affairs will undoubtedly occur, the time scale involved is unlikely to be less than ten years. For the present, the engineer must choose an appropriate combination of field, laboratory and computer studies to provide the most cost-effective solution to his particular problem.

As regards models of siltation and erosion processes, it is fair to say that the majority of usage in the 1980s has been confined to the simpler one- or two-dimensional models, mainly because of the prohibitive computer and calibration costs of modelling large areas, which typically may be tens of kilometres in extent. However, attempts have been made to make three-dimensional models more cost-effective, since they have many advantages over field and laboratory-based techniques. For example, the transport of mud, silt and fine sands, which travel predominantly in suspension within the water column and also form the bulk of material contributing to estuary and coastal siltation problems, cannot be properly modelled in the laboratory because of the problem of adequately scaling turbulence levels, while field studies can only provide data on existing conditions and cannot in themselves predict the consequences of change. It is also easier to make changes in bed levels and boundary conditions in any mathematical model since laboratory models require laborious remoulding of flow cross-sections with concrete.

In order to understand recent engineering simplifi-

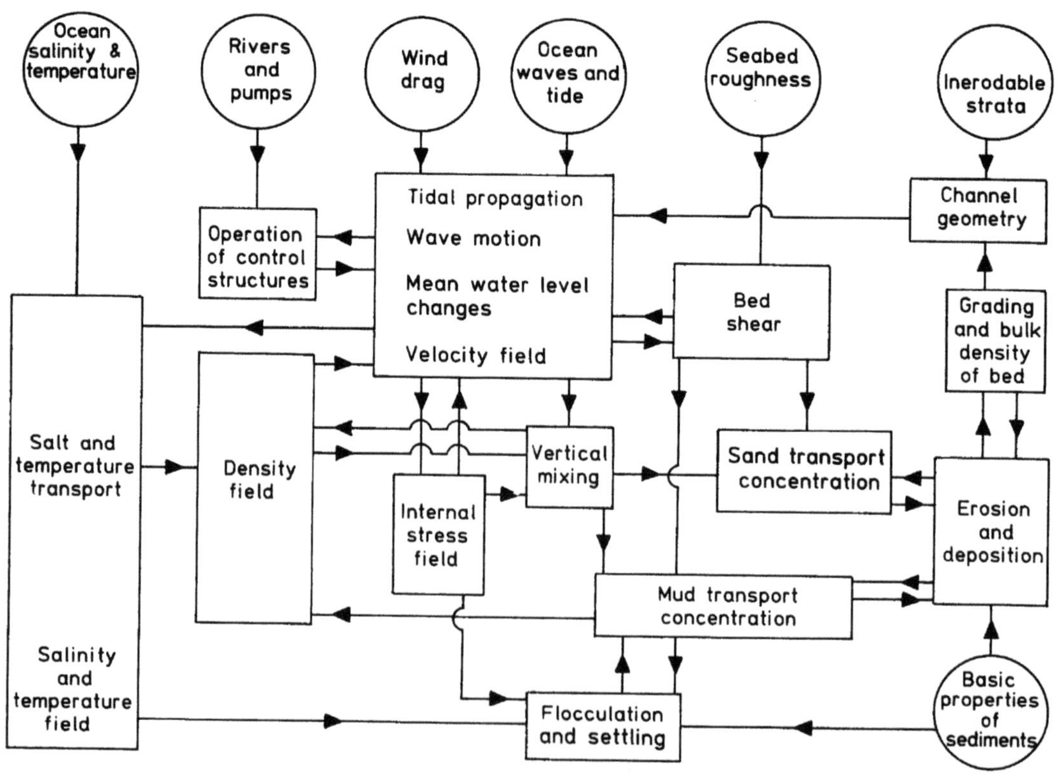

Figure 16.1 Interaction of environmental sediment transport processes (after Hydraulic Research Ltd).

cations, it is appropriate to consider both the interactive nature of the environment, Figure 16.1, and the various elements (sub-models) needed to model it in three dimensions.

16.2 SUB-MODELS

To describe interactive sediment transport processes in three dimensions, it is necessary to build a complete, three-dimensional model, composed of six sub-models, as indicated below.

1. Wind model
   Provides details of near-surface (10 m) wind speeds and, hence, surface shear forces, and directions.
2. Wave model
   Provides details of spectral wave heights (water levels relative to wave-period-mean values) and directions, together with wave-induced velocities and bed shear stresses and directions.
3. Current and mean-water-level model
   Provides details of tide, wave and wind-induced currents over the flow depth as well as wave-period-mean water levels.
4. Turbulence model
   Provides details of internal mixing processes and bed shear stresses as well as mixing coefficients for use in process (transport) models.
5. Process (transport) models
   Provides details of the water temperature, salinity, pollutant levels and suspended-solids concentrations over the water column.
6. Bed level
   Provides details of bed-level changes over the study area.

All the above sub-models now exist in research institutes around the world, although they have not been assembled to provide a single integrated model of the environment. Consequently, only simplified forms of the various sub-models have been used to date.

The most usual, general simplification adopted by engineers has been the assumption that the interactive environment can be represented by a series of simpler interactive or even non-interactive states. Consequently, only dominant physical processes are modelled for particular problems, as indicated in the following three examples.

1. For port and harbour developments within an estuary, such as the deepening of a navigation channel or construction of a new loading terminal, tidal motions rather than waves and winds are likely to be important. Only sub-models 3–6 are needed under such circumstances, although the effect of winds on mean water levels can still be retained by including allowances in the model's seaward boundary conditions. In the case of wider and longer estuaries, surface wind stresses and directions would be retained but values would be provided by the analysis of field observations rather than from a three-dimensional computer model of the atmosphere. The effects of pollutants and temperature changes would also tend to be ignored, although allowances for seasonal temperature changes would be made indirectly by changing particle fall velocities and water viscosities.
2. For offshore sediment problems associated with oil and gas exploration and extraction, such as scour around jacket and caisson structures, or the disposal of drilling muds or, indeed, dredged spoil or sewage sludge, it is usual to ignore the effects of wave- and wind-induced currents and seabed drag on wave motions. The effects of tides would be retained and simpler two-dimensional (depth-average) models would be used for the wave climate. The effects of temperature, salinity, pollutants and winds may also be ignored. Again, the indirect effects of temperature on sediment fall velocities, and thus salinity and wind/wave action on mixing processes could be included but would be based on data from simpler models or field studies.
3. For nearshore coastal situations, such as harbour siltation or erosion of the nearshore seabed and beaches by the construction of long breakwaters, the effects of winds on induced currents would be ignored, although the effect of increased mean water levels would be retained in the determination of wave conditions. The effect of seabed drag on the three-dimensional wave motion would be ignored and wave information would be obtained by the use of simpler depth-average models.

A further, commonly used simplification is that of empirical equations to relate both bed stresses and internal mixing processes to depth-average or local velocities and their gradients, thus eliminating the need for sub-model 4.

A good example of the engineering simplification of environmental models is provided by the TIDEWAY system of computer sub-models produced by Hydraulics Research Ltd., UK (Thorn and Huntington). This system can be used to predict the effect of engineering works on tidal waters, including interactions with marine life. The TIDEWAY system provides a series of hydrodynamic tidal models to give information on flow and water levels, which, in turn, is used by a series of process models, as shown in Figure 16.2 and detailed below.

The various sub-models included in the TIDEWAY system are:

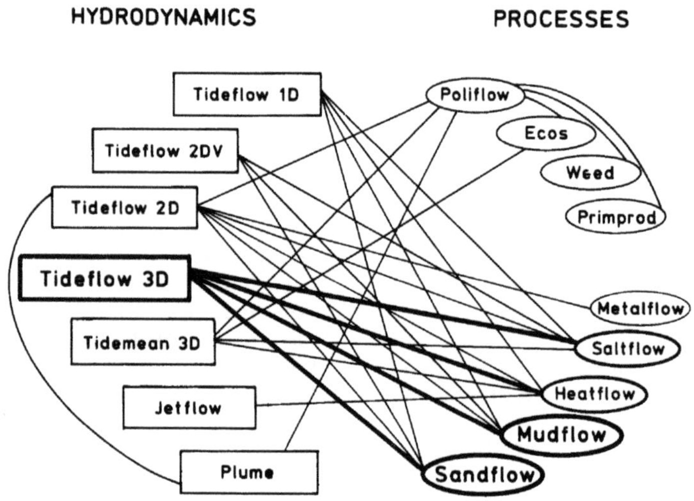

Figure 16.2 The TIDEWAY model system and its submodels (after Hydraulics Research Ltd).

Tideflow-1D
One-dimensional (cross-section-averaged) tidal flow simulation in branched and looped channels;

Tideflow-2D
Two-dimensional (depth-averaged) tidal flow and wind-induced surge simulation;

Tideflow-2DV
Two-dimensional (width-averaged) tidal flow simulation in multiple layers for stratified conditions in channel networks;

Tideflow-3D
Three-dimensional tidal flow simulation in multiple layers for stratified flow conditions;

Tidemean-3D
Three-dimensional tide-averaged simulations of seasonal gravitational circulation patterns generated by salinity and thermal density effects;

Jetflow
Buoyant spread and dilution of momentum jets in tidal flows;

Plume
Transport and diffusion of buoyant effluent plumes in tidal waters;

Saltflow
Transport and mixing of salt, saline intrusion into estuaries;

Heatflow
Recirculation, heat balance, transport and dispersion of heat from power station outfalls and desalination plants;

Mudflow
Scour, transport and deposition of cohesive mud;

Sandflow
Scour, transport and deposition of non-cohesive sediments such as sand and gravel;

Metalflow
Transport of dissolved and adsorbed heavy metals;

Pollflow
Dissolved oxygen balance, nutrient balance, pollution;

Primprod
Photosynthesis by phytoplankton;

Weed
Marine plant life;

Ecos
Complete marine ecosystem model (based on a mathematical model developed by the UK Institute for Marine Environmental Research) including plant and animal life.

For sediment transport problems in tidal conditions, without waves, it would be usual to use Tideflow-3D, together with Mudflow or Sandflow, with density information provided by field observations or simpler models. In some circumstances, involving partially-mixed and stratified estuaries, it may be necessary to include Saltflow and Heatflow to provide more detailed information on density variations, see Figure 16.2. The effect of winds on suspended sediment transport can also be included in Tideflow-3D by providing information on wind drag on the water surface.

While Tideflow-3D provides details of the three-dimensional flow field, it also contains simplifications. For example, internal mixing processes are simulated by the use of empirical equations for effective turbulent viscosities, although allowance is made for reduced mixing caused by vertical density gradients. In addition, a hydrostatic approximation is used when calculating model pressures. This means that vertical forces, other than gravity, are ignored and only lateral and longitudinal momentum equations need to be considered, which considerably reduces modelling costs. The approximation works best where seabed contours show small gradients, but practical tests suggest that it could be used in situations with underwater slopes of up to 1:5 without too much error. A further model of this type for tide and wind-induced currents is the TRISULA system of Delft Hydraulics (Van de Kuur *et al.*, 1989).

Further simplifications have also been tried, including:

1. The use of distribution functions over the water depth to relate local velocities to depth-average values. This approach avoids the use of expensive sub-models, such as Tideflow-3D or more complex models, such as PHOENICS – a three-dimensional hydrodynamic model, including a mixing sub-model ($k-\varepsilon$ turbulence model: see Rodi, Chapter 4 of this volume) for internal processes (CHAM UK Ltd.) – and enables simpler, cheaper sub-models to be used. For example, Tideflow-2D (one or two layers over the flow depth), WAQUA (produced by Delft Hydraulics) or Systems 21 and 22 (produced by the Danish Hydraulic Institute). Such approaches have generally been used for coastal problems with small seabed slopes (<1:10) where density-induced flows are small, since the choice of distribution function becomes difficult under other circumstances.
2. The use of flow data from hydraulic models or field observations. This avoids the use of hydrodynamic computer sub-models entirely. If field data is used, model answers would usually be limited to short-term predictions since large changes in bed level will require changes in the hydrodynamic data used in the predictions. An exception would be a local siltation problem, such as a dock entrance or estuary approach channel, where dredging plant was continuously removing siltation quantities and thus maintaining particular flow depths.
3. The use of a 'steady-state' assumption for sediment concentrations in the water column, which implies that sediment concentrations rapidly adjust to new mixing conditions produced by changes in flow conditions. This 'steady-state' assumption allows the time-dependent terms in process models to be neglected and consequently, leads to the formation of simpler, cheaper models. However, such models are only applicable to problems involving medium to coarse sands or those where tidal flows are weak and the wind and wave field can be regarded as 'steady', at least for a period of several hours. For coastal and offshore sedimentation problems, it is also necessary to add two-dimensional (depth-average) wave models, usually an offshore wave-development model, such as that used by the British Metereological Office or the Royal Netherlands Meteorological Institute (GONO computer model) (Mynett and de Voogt, 1984), and a near-shore model to include wave transformation and wave/current interaction processes, such as the CREDIZ model of Delft Hydraulics. As a first approximation, wave information would be used to increase sediment entrainment rates, to increase vertical mixing within the water column by modifying effective viscosity equations, and to provide extra transport due to wave-induced depth-mean currents; depth-distribution functions being used to convert the depth-mean values to local values. Work is currently in progress to include second-order effects, such as mass-transport and undertow currents, and should lead to improved three-dimensional sediment transport models in due course (Arcilla, 1989).

16.3 BASIC MODEL EQUATIONS

The three-dimensional equations of hydrodynamics and internal mixing have already been discussed along with suitable two-dimensional wave models. The relevant process and bed level change equations for three-dimensional sediment modelling are accordingly only summarized below.

16.3.1 PROCESSES AND SEDIMENTS

For low concentrations of suspended sediment, typical of estuary and coastal situations, the equations can be written in tensor form ($i = 1-3$) as:

$$\frac{\partial c}{\partial t} + u_i \frac{\partial c}{\partial x_i} - w_f \frac{\partial c}{\partial x_3} - \frac{\partial}{\partial x_i}\left[\nu_{si} \frac{\partial c}{\partial x_i}\right] = 0 \quad (16.1)$$

where c is the concentration of sediment in the water column, and w_f is the effective fall velocity of the sediment particles, which is dependent upon water viscosity (temperature), particle density and shape for coarse silts and sands, and on water chemistry (particularly salinity values), turbulence levels and sediment concentrations for muddy sediments composed of fine silts and clays. Empirical equations or standard curves are used to find w_f for the coarse silts and sands, while special field settling tests are needed for muds because of the complexity of the flocculation and deflocculation processes.

ν_{si} are sediment mixing coefficients which are usually taken to be proportional to the effective turbulent viscosity of the fluid (ν_t), that is:

$$\nu_{si} = \beta_i \phi_i \nu_{ti} \quad (16.2)$$

where β_i are coefficients used to distinguish sediment mixing from fluid mixing; ϕ_i are coefficients introduced in order to allow sediment to reduce vertical mixing; and ν_{ti} are the effective turbulent viscosities of the flowing water. Values of the product terms $\phi_i \nu_{ti}$ can be provided by an internal mixing sub-model or from empirical equations. For example: for x_1 and x_2 directions, β_i and ϕ_i are usually taken as unity while the ν_{ti} are related to the flow depth by the equation

$$\nu_{ti}\ (i = 1, 2) = \alpha_i h u_* \quad (16.3)$$

where the α_i are values in the range ($0.3-3\,\mathrm{m^2\,s^{-1}}$) and are effectively calibration factors; and h is the flow depth, as schematized in Figure 16.3.

u_* is the bed shear velocity, which can be related to depth-mean flow velocities (U) for well-mixed flows via a Chezy's coefficient C, that is:

$$u_* = U(g^{0.5})/C \quad (16.4)$$

$$C = 18 \log(12h/k_s) \quad (16.5)$$

where g is the acceleration due to gravity ($9.81\,\mathrm{m\,s^{-2}}$) and k_s is the effective roughness of the seabed.

For vertical mixing, Van Rijn (1984) has suggested that

$$\beta_3 = 1 + 2(w_f/u_*)^2 \quad \text{for } 0.1 \leqslant w_f/u_* \leqslant 1 \quad (16.6)$$

$$\phi_3 = 1 + d^{0.8} - 2d^{0.4} \quad (16.7)$$

$$d = c/c_o \quad (16.8)$$

where c is a volume concentration and c_o is the maximum volume concentration of the seabed (0.65). For many situations, the product $\beta\phi$ is often taken as unity, since increases in β offset reductions in ϕ.

Values of ν_{t3} are often described by a parabolic distribution, appropriate to a logarithmic distribution of u_3 over the flow depth. Thus,

$$\nu_{t3} = 0.24\kappa u_* h[1 - (1 - 2(x_3 - \eta)/h)^a] \quad (16.9)$$

where κ is von Kármán's constant and $a = 2$. Alternatively, Van Rijn (1984) uses a constant viscosity in the upper half of the profile, since this allows larger sediment concentrations at the water surface as found in the field, that is:

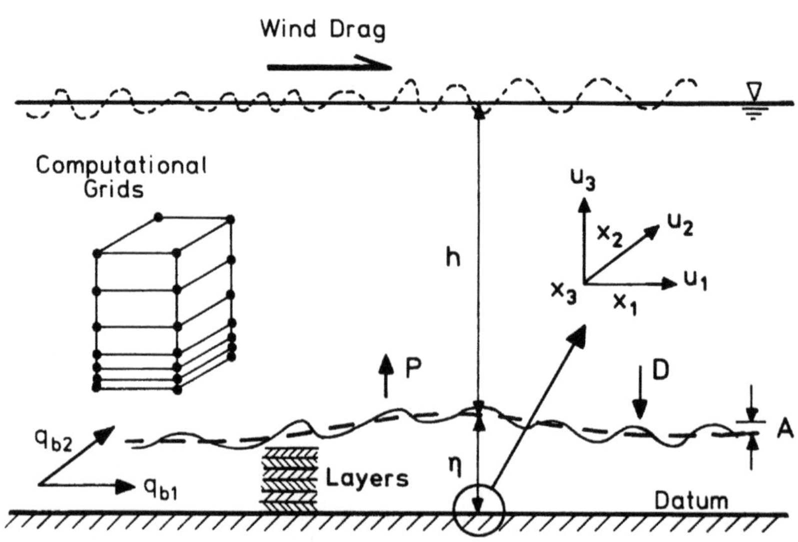

Figure 16.3 Model co-ordinate system.

$$\nu_{t3} = 0.25\kappa u_* h \quad \text{for } (x_3 - \eta)/h \geq 0.5 \quad (16.10)$$

In practice, changes in the upper half of profile have a small effect on sediment transport rates since the bulk of the transport occurs in the lower half of the water column; see also section 16.7.

In coastal situations with waves and currents, it is necessary to modify viscosities to allow for extra entrainment and interaction effects. Van Rijn (1986) has suggested that

$$\nu_{t3} = \nu_{t3}\,(\text{currents}) + \nu_{t3}\,(\text{waves})$$

with ν_{t3} (currents) given by Equation (16.9) with a given as:

$$a = 2 - 0.25\theta \quad \text{for } 0 \leq \theta \leq 4 \quad (16.11)$$

$$a = 1 \quad \text{for } \theta > 4 \quad (16.12)$$

with $$\theta = u_0/|U| \quad (16.13)$$

while u_0, $|U|$ are the maximum orbital near-bed wave velocity, as given by linear wave theory, and the modulus of the depth-mean flow velocity, respectively.

$$\nu_{t3}(\text{waves}) = \nu_{twb} \quad \text{for } (x_3 - \eta) \leq \delta \quad (16.14)$$

$$\nu_{t3}(\text{waves}) = \nu_{twb} + (\nu_{twm} - \nu_{twb})\left[\frac{x_3 - \eta - \delta}{0.5h - \delta}\right]$$

$$\text{for } \delta < (x_3 - \eta) < 0.5h \quad (16.15)$$

$$\nu_{t3}(\text{waves}) = \nu_{twm} \quad \text{for } (x_3 - \eta) \geq 0.5h \quad (16.16)$$

where

$$\nu_{twb} = \alpha'\delta u_o \quad (16.17)$$

$$\nu_{twm} = \alpha''(hH_s/T_s) \quad (16.18)$$

$$\alpha' = 6.5 \times 10^{-4} D_*^2 \quad (16.19)$$

where $\alpha'' = 0.035$; H_s and T_s are the significant wave height and period, respectively, of the wave motion; and δ is the thickness of the wave boundary layer, which is approximately three times the height of wave-induced bed ripples.

16.3.2 PROCESSES, SALINITY AND TEMPERATURE

Identical equations to Equation (16.1) are used, but without the fall velocity term, w_f. The interactive link between processes and hydrodynamics is provided through the density of the flowing mixture, ρ, that is,

$$\rho = \rho_o\,(1 + a_1 s + b_1 c) \quad (16.20)$$

where ρ_o is the density of fresh water at a particular temperature and pressure; s is the salt content (salinity) of the water, and a_1, b_1 are constants depending on water temperature, pressure, and sediment particle density.

16.3.3 BED LEVELS

A sediment continuity equation is used of the form (Figure 16.3):

$$\frac{\partial q_{b1}}{\partial x_1} + \frac{\partial q_{b2}}{\partial x_2} + \frac{\partial}{\partial t}[(1 - n)\eta] - D + P = 0 \quad (16.21)$$

where q_{b1}, q_{b2} are bed load sediment transport rates in volume of solids per unit width of seabed for coarse silt and sand particles, respectively (they are put equal to zero for mud transport problems), n is the porosity of the seabed, η is the elevation of the seabed above a fixed horizontal datum, P is the rate of removal of particles (solids volume) per unit area of seabed; and D is the rate of deposition of sediment (solids volume) from suspension within the water column per unit area of seabed.

16.4 MODEL BOUNDARY CONDITIONS

All sub-models require the specification of appropriate boundary conditions.

16.4.1 PROCESSES, SALT/TEMPERATURE: SUB-MODEL 5

Inflow boundaries

Salinity and temperature values must be provided, either from field data or other simpler models.

Other boundaries

In the absence of outfalls or other sources, it is usual to set 'normal' gradients to zero.

16.4.2 PROCESSES, COARSE SILTS AND SANDS: SUB-MODEL 5

Inflow boundary

The distribution of sediment concentrations would be specified either from field data, or more usually from 'equilibrium' profiles over the flow depth, as appropriate to steady, uniform, open channel flow. For example, for a parabolic distribution of ν_{t3} over the flow depth, Equation (16.9), the equilibrium concentration is given by

$$c = c_a[((h/(x_3 - \eta) - 1)/(h/A - 1)] \quad (16.22)$$

where

$$Z = w_f/(\beta\kappa u_*) \quad (16.23)$$

$$c_a = q_b/(6.35k_s u_*) \quad (16.24)$$

and A is the thickness of a representative layer within which sediment is moving as bed load: usually taken as half the height of any bed forms or k_s in Equation (16.24), Figure 16.3.

If Van Rijn's (1984) v_{t3} profile is used, Equation (16.22) applies in the lower half of the profile, while for the upper half:

$$c - c_a\left[\frac{A}{h - A}\right]^z \exp\left[-4Z\left[\frac{(x_3 - \eta)}{h} - 0.5\right)\right]\right]$$

$$\text{for } (x_3 - \eta)/h > 0.5 \qquad (16.25)$$

The bed load transport, (q_b) can be determined directly from field tracer experiments, or indirectly, by relating bed load to suspended load and then measuring the suspended load in the field with a boundary layer rig placed on the seabed (McDowell and O'Connor, 1979; Thorn, 1981). Alternatively, q_b or c_a can be determined from empirical equations. For example (Van Rijn, 1984):

$$c_a - 0.015\alpha_1 D_{50} T^{1.5}/(AD_*^{0.3}) \qquad (16.26)$$

α_1 is a calibration coefficient for a particular site, D_{50} is the median grain size of the seabed (50% finer), and T is an effective excess bed shear stress for moving sediment grains and is defined by the equation:

$$T - [u'_*/u_{*cr}]^2 - 1 \qquad (16.27)$$

$$u'_* - U(g^{0.5})/C' \qquad (16.28)$$

$$C' - 18\log(12h/(3D_{90})) \qquad (16.29)$$

u_{*cr} is the critical shear velocity for grain motion and can be related to an effective particle grain-size (D_*) defined by the equations:

$$\begin{aligned} B &- 0.24(D_*)^{-1} && \text{for } D_* \leq 4 \\ &- 0.14\,(D_*)^{-0.64} && \text{for } 4 \leq D_* \leq 10 \\ &- 0.04\,(D_*)^{-0.10} && \text{for } 10 \leq D_* \leq 20 \\ &- 0.013\,(D_*)^{0.29} && \text{for } 20 \leq D_* \leq 150 \\ &- 0.055 && \text{for } D_* > 150 \end{aligned} \qquad (16.30)$$

where

$$B - u_{*cr}^2/[(s - 1)gD_{50}] \qquad (16.31)$$

s is the relative density of the sediment (ρ_s/ρ); ρ_s is the sediment density, and D_{90} is the 90% finer grain size of the bed material. D_* is an effective grain size defined by the equation:

$$D_* = D_{50}[(s - 1)g/\nu^2]^{1/3} \qquad (16.32)$$

where ν is the kinematic viscosity of the flowing water.

Outflow and seabed boundaries

The 'normal' derivative of concentration is set to zero.

Seabed (x_3 = A + η)

The bed concentration (c_a) would be specified, for example, by the use of Equation (16.26), with T adjusted to allow for the presence of waves and winds. Alternatively, the quantity of material entrained into the flow can be used ($P = w_f c_a = \varepsilon_{s3}\partial c/\partial x_3$), as shown in Figure 16.3.

Water surface (x_3 = h + η)

It is assumed that there is no loss of sediment through the water surface. Consequently,

$$w_f c + \varepsilon_{s3}\partial c/\partial x_3 = 0 \qquad (16.33)$$

16.4.3 PROCESSES, MUDDY SEDIMENTS: SUB-MODEL 5

Inflow boundaries

Sediment concentrations are specified from field measurements, since empirical equations do not exist.

Solid, non-eroding boundaries

It is assumed that there is no transport through solid boundaries ($u_i = 0$, $u_i c_i = 0$ and $\partial c/\partial x_i = 0$).

Water surface

This is treated in the same way as silt and sandy sediments, Equation (16.33).

Seabed

Unlike sandy sediments, the surface of a muddy deposit is usually plane, eroded material being entrained directly into the water column. Bed load transport is therefore neglected ($q_b = 0$). However, it is necessary to allow for the large changes in porosity that occur as the muddy bed consolidates with time (the porosity of sandy material is usually taken as constant) as described in Chapter 7 of this volume. As explained there, the consolidation process also increases the shear strength of the bed and makes it more difficult to erode. The problem is overcome by dividing the bed into a series of layers of suitable thicknesses; values as small as 25 mm often being used. The density of each layer is assumed to change with time, as determined from laboratory settling tests, and to be related to the layer shear strength, determined by laboratory erosion tests in recirculating flumes (Hydraulics Research). The bed boundary condition uses the laboratory erosion data to directly determine entrainment rates ($P = \varepsilon_{s3}\partial c/\partial x_3$ at $x_3 = \eta$).

16.4.4 BED LEVELS: SUB-MODEL 6

Coarse silt/sands

The bed load (q_b) is determined from field measurements or empirical equations, (16.26). Entrainment rates (P) are determined indirectly from bed concentrations ($w_f c_a$). Deposition rates (D) follow from solution of the sediment process model, that is,

$$D = w_f c_A \qquad (16.34)$$

where c_A is the concentration at level A above the seabed.

Muddy sediments

Entrainment rates (P) are determined from empirical equations with constants (m_e, τ_e) determined from laboratory settling and erosion tests. For example (Thorn, 1981; Hydraulics Research, and Delo, 1988):

$$P = m_e(\tau_{bed} - \tau_e) \qquad (16.35)$$

The values of m_e, τ_e are changed as each layer is eroded to reflect the consolidated state of the bed. Typical values are in the range $2\text{–}4.10^{-2}\,\mathrm{m^{-1}\,s}$ and $0.05\text{–}0.1\,\mathrm{N\,m^{-2}}$, respectively.

Deposition rates (D) are also determined from empirical equations. For example, the following equation (Krone, 1962) is often used:

$$D = w_f c_\eta(1 - \tau_{bed}/\tau_d) \qquad (16.36)$$

where w_f is the effective near-bed fall velocity, a function of concentration and environmental chemistry, which must be determined from field settling tests using suitable equipment, such as an Owen tube (Hydraulics Research); c_η is the near-bed concentration ($x_3 = \eta$) determined from the sediment transport process model, and τ_d is a bed shear stress above which all particles stay in suspension; typical values being $0.05\text{–}0.1\,\mathrm{N\,m^{-2}}$. Its value is again determined from laboratory tests in a recirculating flume (Hydraulics Research).

In some circumstances, where large waves or tidal currents are present with steep underwater banks (side slopes larger than 1:10), or where large concentrations of weakly consolidated mud exist on a gentle slope, it has been found that high concentrations of mud can flow along the seabed as 'fluid mud'. The extra velocities created by these mud flows may lead to further erosion of the underlying strata or to enhanced entrainment rates at down-slope positions. Alternatively, the mud flow may slow down and cause excessive siltation problems, particularly where seabed topography contains over-deepened sections, such as dredged dock entrances, tanker mooring 'boxes' and port navigation channels, or alternatively in low velocity areas created by port and harbour structures.

16.5 NUMERICAL TECHNIQUES

The complexity of field situations, involving time-varying boundary conditions, means that simple, analytical solutions of relevant model equations are impossible and, therefore, numerical techniques must be used. Also the need to obtain results for many tidal cycles means that a computer must be used to speed up the solution process. Unfortunately, computer methods can be expensive and this has led to various attempts to reduce the complex three-dimensional equations into simpler one- or two-dimensional forms (see, for example the splitting technique used by Verboom (1975)), which enables smaller, less expensive computers to be used. These splitting techniques have the additional advantage that the resulting equations can be solved by the best numerical technique available to suit the mathematical characteristics of each equation.

Typically, either finite difference or finite element methods would be employed, although it is possible to use a combination of both (Koutitas and O'Connor, 1980). In the traditional, most widely used versions of these approaches, the finite difference methods divide the water volume into a series of brick-shaped sub-volumes with computation points at the edges of the 'bricks', as shown in Figure 16.3. The boundary is usually represented, therefore, by a series of steps as the water deepens. The finite element method on the other hand, has the advantage that the 'bricks', now called elements, can be more conveniently made into an irregular shape, so that computational points follow the boundary shape more closely. Unfortunately, the finite element method leads to a more complicated computer programme than the finite element approach, and is generally more expensive to use, since it usually requires a main-frame computer to obtain results.

More recently, techniques have been developed (Van de Kuur *et al.*, 1989), whereby the flow depth is divided into a series of elements with vertical thicknesses that progressively increase away from both the seabed and surface and whose lateral and longitudinal dimensions form an orthogonal curvilinear co-ordinate system in plan. These special 'element' shapes are then transformed (sigma (σ) transformation (Blumberg and Mellor, 1985; also Chapter 8 of this volume)), into regular-shaped bricks, which then allows the use of the finite difference approach, (Figure 16.4). This transformation method enables the use of the simpler finite difference approach to model the rapid changes of velocity and sediment concentrations that occur near the seabed.

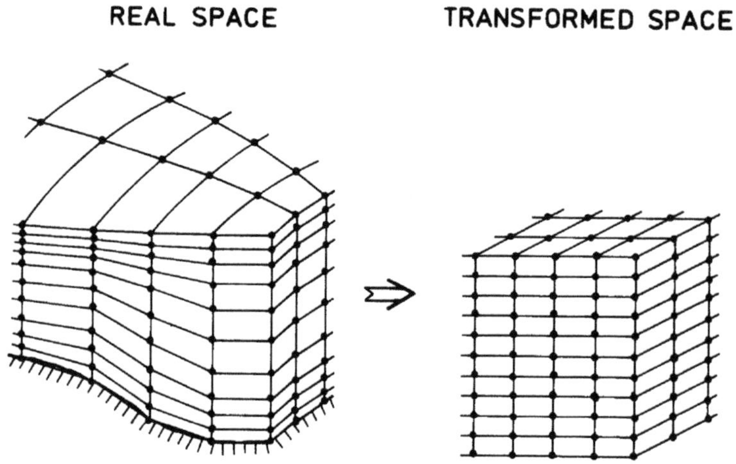

Figure 16.4 Orthogonal curvilinear grid system (after Van de Kuur *et al.*, 1989).

Along with developments in numerical techniques, there have also been developments in computer hardware, with the emergence in the 1980s of powerful personal computers (PCs) and work stations. Given the combination of new numerical techniques and computer hardware, it seems certain that the engineer will be using three-dimensional models, of the type described herein, in the not too distant future, possibly by the turn of the century.

16.6 FIELD DATA REQUIREMENTS

One of the most inhibiting factors in the use of three-dimensional models is the cost of field data for model boundary conditions and calibration/verification purposes. For small-scale local studies covering a few square kilometres in extent, the total cost of an investigation may be the order of £200k, of which 50% may be field data costs. On very large studies covering many hundreds of square kilometres, the cost of acquiring field data alone could exceed £1m, which means that complex three-dimensional model studies are only likely to be used for very large capital schemes costing in excess of £100m, or on small-scale local problems.

For typical estuary studies, where wind and wave action can be neglected, it is desirable to provide field information covering the following items.

1. Recent bed bathymetry, together with any historical information.
2. The distribution of bed sediments, using grabs, gravity corers, vibro-covers, boreholes and seismic acoustic profilers.
3. River and other significant freshwater discharges and their seasonal variation.
4. Variation of tidal water levels from tide gauges located at a number of typical points throughout the estuary. The longer the time series of observations the better.
5. Information over the flow depth (at least five levels) on current velocities and directions, salinity and temperature levels, suspended mud and sand concentrations for typical neap and spring tides (observations at least every 0.5–1 h) during different seasons.

   In addition, a boundary layer rig would be used for suspended sand and velocity information close (<1 m) to the sediment boundary (Thorn, 1981). If possible, the information over the flow depth should be obtained simultaneously at a number (usually 5–10) of stations along the estuary. Often it is too expensive to use simultaneous observations and a single ship will be used to take observations sequentially. Under such circumstances, self-recording current/salinity/temperature and suspended mud observations may be collected at one or two levels over the flow depth for as many tidal cycles as possible (not less than 1 month) during typical seasons at one or two points within the study area.
6. For sandy problems, echo sounding and side-

scan sonar studies would be used to study the occurrence of bed forms and a radioactive or fluorescent tracer study would be done to study bed load movements.
7. For muddy sediments, a 1 m^3 sample of mud would typically be returned to the laboratory for the determination of clay mineralogy, as well as settling, consolidation and erosion tests, using typical site fluid, if possible. In addition, flocculated fall velocities should be determined in the field using suitable equipment.
8. Dredging records which cover as long a period as possible.

For coastal situations, where wave and wind action is significant, it would be desirable to provide additional information.

9. Strength and directions of winds covering the period of any wave observations. Ideally, long-term yearly records should be available.
10. Measurement of offshore and nearshore (outside the surf zone) wave heights, using wave-rider buoys at two or three locations. Information on wave direction would be provided by visual observations, aerial photography or shore-based radar. Observation records should be as long as possible (not less than 1 month for typical sensors).
11. Measurement of surface currents and directions outside the surf zone during calm and stormy conditions using radar techniques, such as OSCR (Prandle, 1989).
12. Measurement of waves, currents, water levels and sediment concentrations in the surface zone using mechanical, electro-magnetic or acoustic current meters, pressure transducers, and acoustic doppler sediment sensors or mechanical traps/gauze bags to trap suspended sediment.
13. Records of beach profile changes and of coastal accretion or retreat. Both hovercraft and photography can be used to speed up the data collection process.

A recent useful development in the acquisition of field data has been the use of instruments capable of sending information ashore via radio telemetry links, often to mobile site offices (BMT Ltd, 1989). Such techniques enable a rapid analysis of data, thereby speeding up planning decisions (Møller, 1989).

16.7 MODEL APPLICATIONS

There have been very few field applications of the sedimentation models described above, although good results have been found in laboratory situations (O'Connor and Nicholson, 1988). In all field cases, simplified versions have been used to both reduce costs and the technical complexity of the problem, and hence enable engineering solutions to be obtained.

16.7.1 SAND MOVEMENTS

Considerable concern has been expressed over the fate of polluted sediment discharged from dredging operations. In 1975, a multi-disciplinary consortium called the Great River Environmental Action Team (GREAT) was formed to determine the environmental effects of dredging and dredge disposal on the Mississippi River. Subsequently, various computer models were developed by the University of Iowa Institute of Hydraulic Research, including a three-dimensional, steady-state sediment model of a dredger plume, the Walden plume model (Wechsler and Cogley, 1977). In 1981, Schnoor *et al.* (1982) modified the Walden plume model and tested its accuracy against field data from dredging plumes at three sites on the Mississippi River between Saverton, Missouri, and Cairo, Illinois.

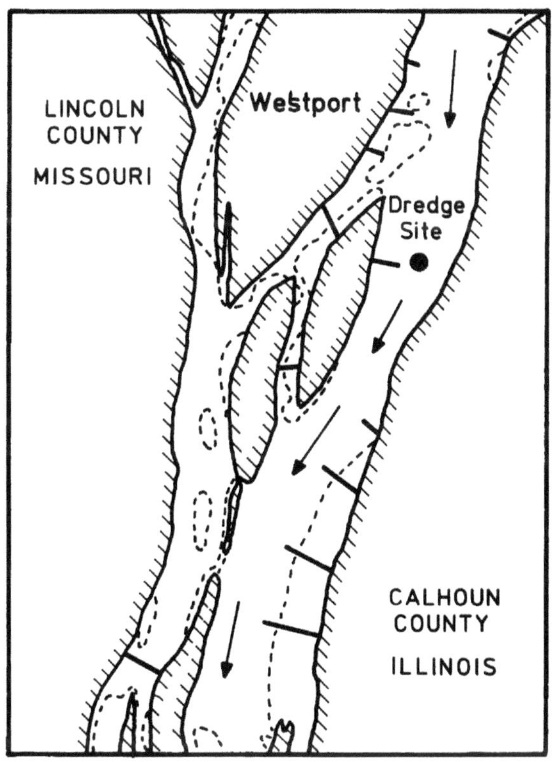

Figure 16.5 Dredge spoil dispersal test site, Westport, Mississippi River (after Schnoor *et al.*).

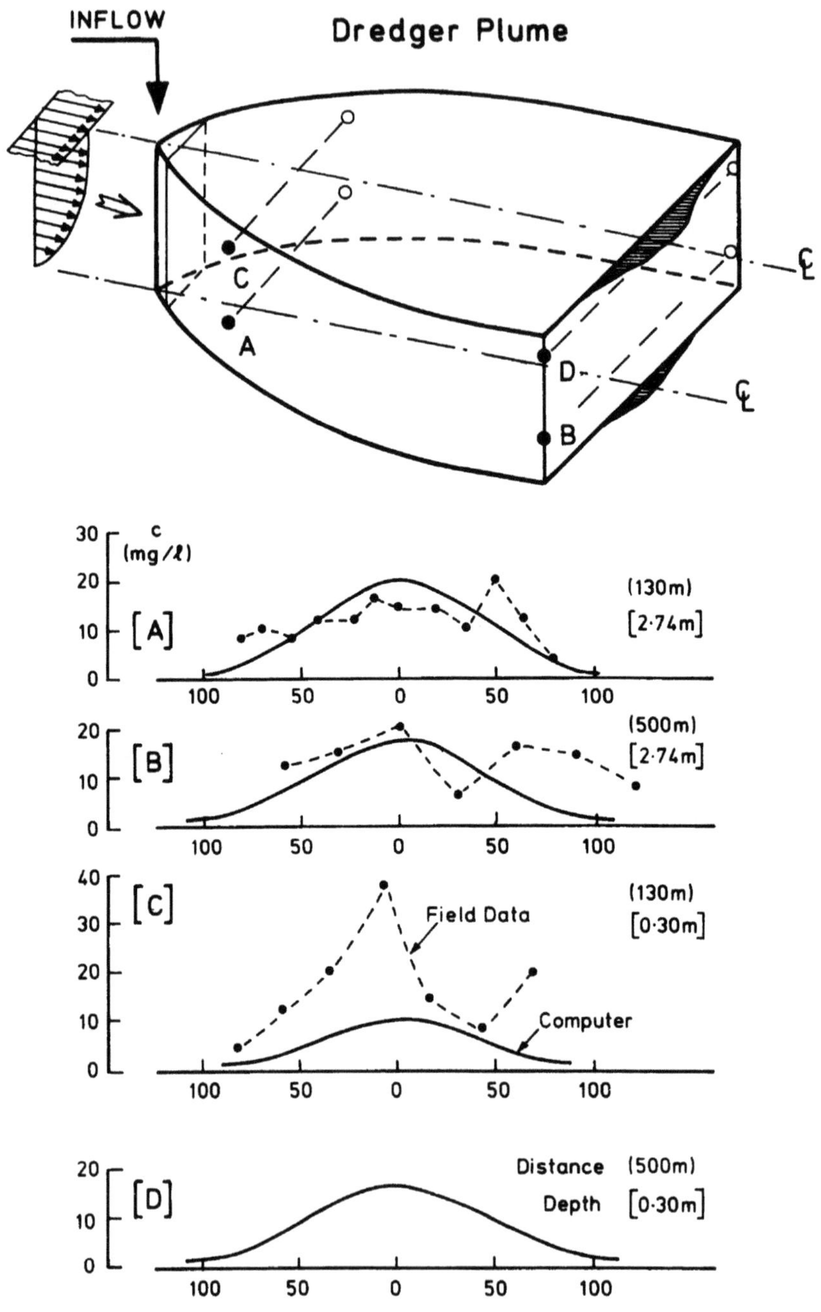

Figure 16.6 Comparison of three-dimensional, steady state, sediment plume model of Schnoor *et al.* with field data.

Water samples were taken at depths of 0.6 m and 2.74 m below the water surface from a boat which traversed each plume at a steady velocity. In addition, velocity profiles were taken over the flow depth, using ten vertical observations, at a number of stations in the study area, together with both bed samples and suspended solids samples at some 6 m, 12 m and 33 m from the disposal points.

Results from the three-dimensional steady state plume model were then compared with the field data. Figures 16.5 and 16.6 show the site location and comparison of field and computer results for the second field site, which involved sediment disposal from the cutter-section dredger *St. Genevieve*, using typical river velocities (fixed) of $0.5\,\mathrm{m\,s^{-1}}$ and depths (fixed) of 4 m, and the assumption of a 50:50 mixture of sand (125 μm) and fines (2 μm) in the plume. The vertical mixing coefficient was given by Equation (16.9), but the lateral coefficient was modified from Equation (16.3), with $\alpha_2 = 5.6–20.0$, by the multiplication of an exponential spreading function with a value of unity at $x_2 = \pm\infty$ and 11 at $x_2 = 0$, the plume centreline. Figure 16.6 shows that the modified coefficients produce reasonable results within the body of the water column, but near-surface results are underestimated somewhat due to the low near-surface mixing produced by Equation (16.9).

Clearly, such plume models have a wide application, including offshore disposal of sewage sludges, and can be improved for tidal conditions by the inclusion of time-varying coefficients and velocities (O'Connor and Zein, 1974). However, they are only applicable to conditions away from the source, once the momentum of the initial injection has disappeared and plume spreading is controlled by the environmental flows.

Van Rijn (1989) reports the use of a sand version of a three-dimensional process model (SUSTRA-3D) to compute transport rates and bed-level changes during a tidal cycle in an estuarine area with velocities in the range $0.6–1.5\,\mathrm{m\,s^{-1}}$ and water depths of 5–15 m. Field observations indicated bed material with D_{50} = 250 μm and a bed roughness k_s = 0.5 m. The three-dimensional velocity field was obtained from a depth-averaged hydrodynamic model (WAQUA) with logarithmic distribution functions to spread the depth-mean values over the flow depth. Mixing coefficients were specified by empirical Equations (16.2–16.19). Changes in bed level during the tidal cycle were calculated from Equation (16.21). Typical erosion/deposition rates of $15\,\mathrm{mm\,hr^{-1}}$ were obtained with good comparisons between observed field data for both depth-mean velocities (within 8%) and sediment transport rates (within 17%), Figure 16.7.

A further application of SUSTRA-3D has been to study inter-tidal erosion of the near-shore seabed due to a proposed extension of the harbour at Great Yarmouth, UK (Delft Hydraulics). Concern was expressed by the planning engineers over possible erosion of the seabed due to the disturbance of the near-shore wave and tidal current climate by the extension works (Figure 16.8). The seabed consists of a medium-sized sand (D_{50} = 285 μm, D_{90} = 425 μm), which is readily moved by the tide and wave-induced currents. An area of some $10\,\mathrm{km^2}$ was modelled assuming both a steady-state tidal current and wave pattern at maximum velocities on an ebb tide. Offshore design-wave conditions (a single, representative wave) were routed inshore using the CREDIZ depth-averaged two-dimensional wave model and tide and wave-induced currents were then determined with the depth-averaged two-dimensional ODYSSEE current model. The three-dimensional velocity field for SUSTRA-3D was then determined from logarithmic depth functions. The model showed realistic erosion rates but, unfortunately, lack of field data prevented detailed calibration of the model.

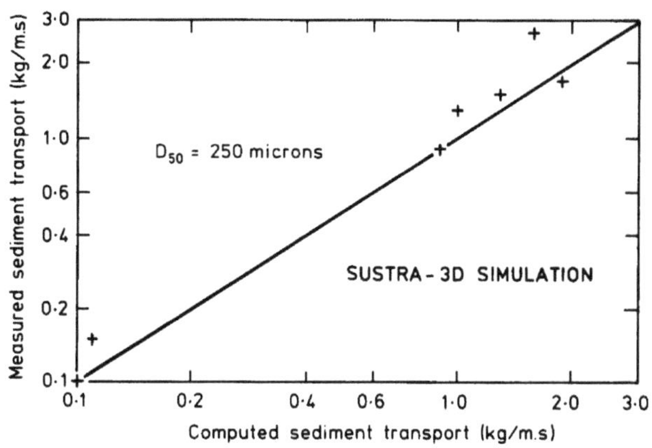

Figure 16.7 Comparison of Delft Hydraulics SUSTRA-3D model with field data.

16.7.2 MUD TRANSPORT

Nicholson and O'Connor (1986) have used a mud version of the three-dimensional model to study siltation in a dock entrance at Grangemouth on the Firth of Forth, Scotland. Redevelopment of the dock in the 1970s necessitated the construction of a new lock and lead-in jetties as well as the marginal deepening of an existing navigation channel to the dock. Subsequently, heavy siltation was experienced in the new lock entrance and needed frequent dredging to maintain navigation depths. In order to seek alleviation of the problem, a combined programme of field

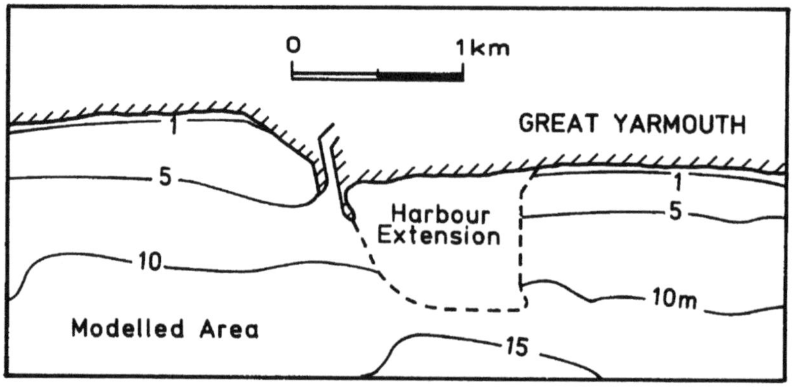

Figure 16.8 Modelled area and depth contours, Great Yarmouth Harbour extension study (after Delft Hydraulics).

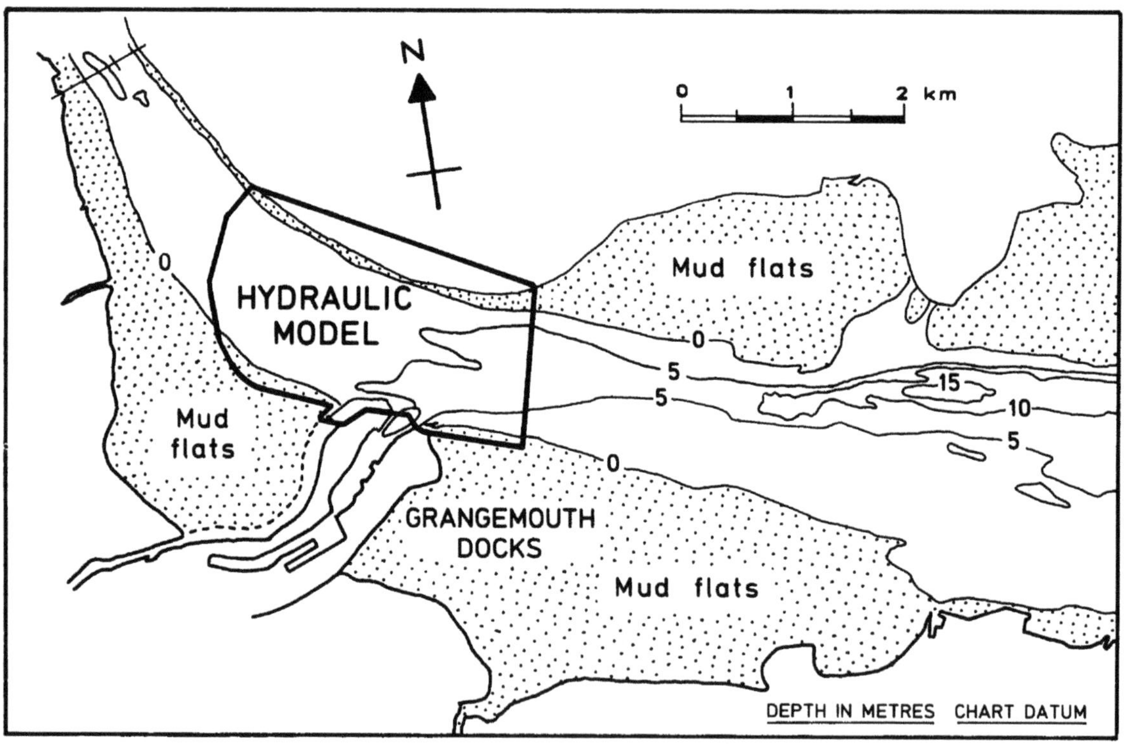

Figure 16.9 Hydraulic model area, Grangemouth Docks siltation study (after Hydraulics Research).

measurements and hydraulic and computer model studies was initiated by the Port Authority; a hydraulic model being chosen because of the complex three-dimensional nature of the recirculating flow in the dock entrance, the presence of piled structures and the lack of a suitable computer model alternative at the time (1978).

The velocity field for the three-dimensional mud model was obtained from depth-average values in the hydraulic model (Thorn and Huntington, 1989), and

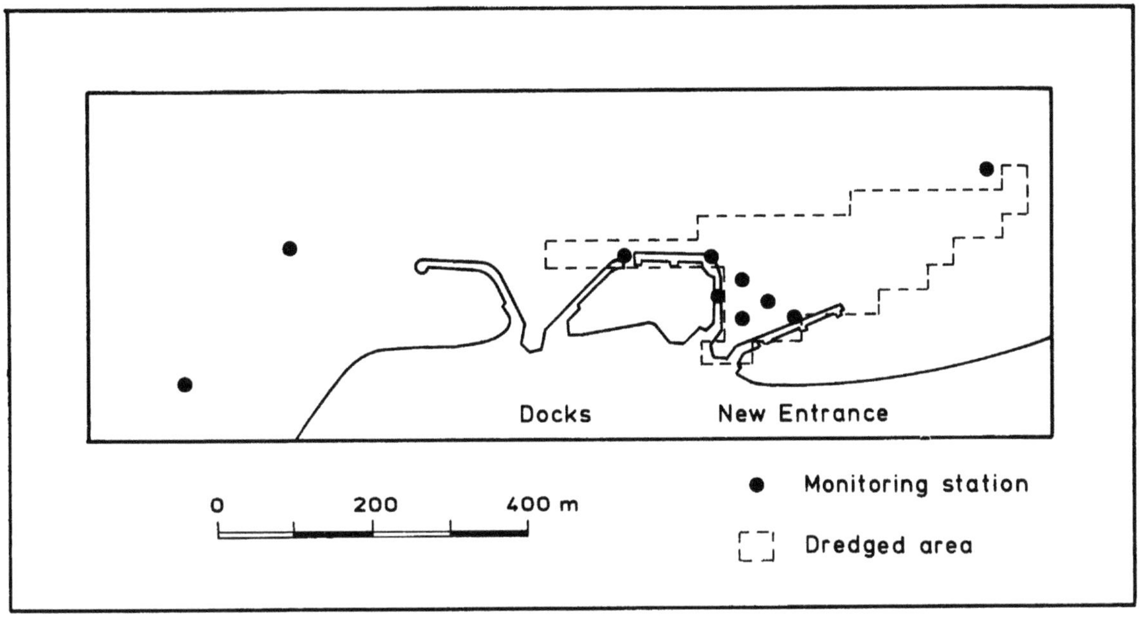

Figure 16.10 Location of field monitoring stations, Grangemouth Docks study. (●), monitoring station; (□), dredged area.

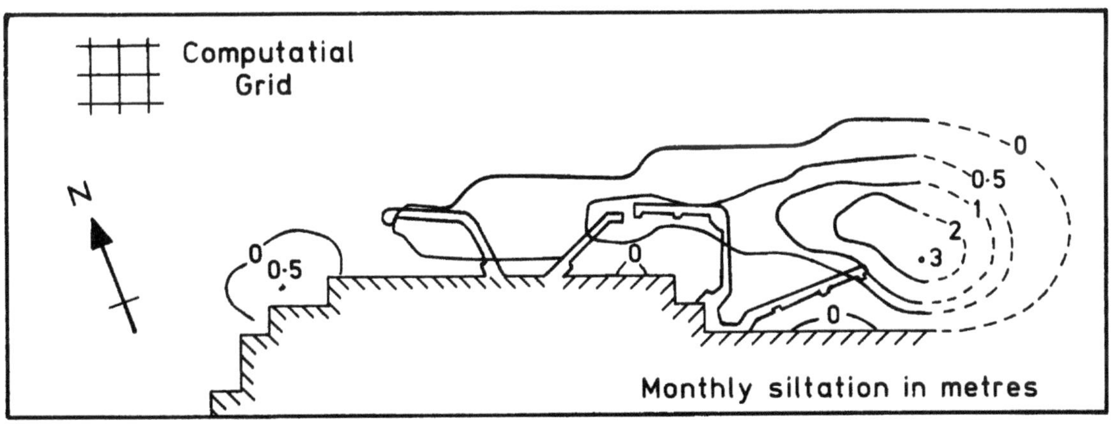

Figure 16.11 Predicted siltation contours for November 1976 tides (after Nicholson and O'Connor, 1986).

the use of logarithmic and other distribution functions, as found from field data, to provide local values within the water column, Figures 16.9 and 16.10.

Model boundary conditions and sediment erosion/deposition parameters were obtained by testing mud in the laboratory (Hydraulics Research, 1987), while mixing coefficients were obtained from Equation (16.3) with $\alpha_1 = \alpha_2 = 0.5$ and $\alpha_3 = 1.0$. The model was operated on a mainframe computer (CDC 7600) and results obtained for a typical spring tide. The resulting sedimentation pattern was then scaled-up to provide dredging predictions for a typical month's tides. Comparison with dredging records showed good results (within 4%) with realistic sedimentation contours, Figure 16.11.

16.8 SUMMARY

Three-dimensional computer models are available to assist the engineer with a wide range of siltation/erosion problems ranging from the safe disposal of dredged spoil and borehole drilling muds to the siltation of estuarine port and harbour navigation channels and port complexes as well as scour of the seabed due to port extensions. Unfortunately, the cost of modelling the environment in three dimensions is prohibitively expensive at present (1990) and the majority of problems have been solved by simplified versions of the models. Both laboratory and field applications suggest that acceptable accuracy can be obtained at reasonable cost, provided sufficient field data is available to determine model parameters. Even so, the high cost of acquiring field data is still likely to limit applications to local problems or large capital schemes for the foreseeable future. Most other problems will be studied using simpler, two-dimensional models.

REFERENCES

Arcilla, A.S. (1989) Hydrodynamic analysis of the surf zone, *International Journal for Numerical Methods in Engineering*, **27**, 87–101.

Blumberg, A.F. and Mellor, G.L. (1985) Modelling vertical and horizontal diffusivities with the sigma co-ordinate system. *Monthly Weather Review*, **113** (8) 1379–83.

CHAM Corporation (1987) The PHOENICS Equations, Report TR99.

BMT Ceemaid Ltd. (1989) Applied Oceanography, Measurement Techniques. Southampton, UK.

Delo, A. (1988) Estuarine muds manual. *Rep. SR164*, Hydraulics Research Ltd., Wallingford, UK.

McDowell, D.M. and O'Connor, B.A. (1979) *Hydraulic behaviour of estuaries*, MacMillan.

Meteorological Office (1985) Annual Report, HMSO, London, UK.

Mynett, A.E. and de Voogt, W.J.P. (1984) Application of wave prediction models in the design of ports and marine structures, Transactions *of the Institute of Marine Engineers* (c), **97**, 33–8.

Hydraulic research Ltd. (1987) *Estuary sedimentation*, Hydraulic Research Ltd., Wallingford, UK.

Koutitas, C. and O'Connor, B.A. (1980) Numerical modelling of suspended sediments. *Advances in Water Resources*, **3**, 51–7.

Krone, R.B. (1962) Flume studies of the transport of sediment in estuarine shoaling processes. Final Report, Hydraulic Engineering Laboratory, University of California, Berkeley, CA.

Møller, J.S. (1989) Denmark's Great Belt Bridge – Environmental and Hydrodynamic Issues. *Proceedings of the ASCE International Symposium on Sediment Transport Modelling* 8–11 October, (ed. S.S.Y. Wang), pp. 1–12.

Nicholson, J. and O'Connor, B.A. (1986) Cohesive sediment transport model. *Journal of Hydraulic Engineering, ASCE*, **112** (7), 621–40.

O'Connor, B.A. and Nicholson, J. (1988) A three-dimensional model of suspended particulate sediment transport. *Coastal Engineering*, **12**, 157–74.

O'Connor, B.A. and Zein, S. (1974) Numerical modelling of suspended sediment. *Proceedings of the 14th ASCE Coastal Engineering Conference*, pp. 1109–28.

Prandle, D. (1989) A review of the use of HF radar (OSCR) for measuring near-shore surface currents, in *Advances in Water Modelling and Measurement*, (ed. M.H. Palmer), BHRA Cranfield, UK, pp. 245–58.

Schnoor, J.L., Giaquinta, A.R., Sato, C. *et al.* (1982) Refinement and verification of predictive models of suspended sediment dispersion and desorption of toxics from dredged sediments. HHR Rep. 249, Iowa Institute of Hydraulic Research.

Thorn, M.F.C. (1981) Physical processes of siltation in tidal channels. *Proceedings of the Conference on Hydraulic Modelling Applied to Maritime Engineering Problems*, Institution of Civil Engineers, London, pp. 47–55.

Thorn, M.F.C. and Huntington, S.W. (1989) *Hydraulic aspects of port and harbour design*, Hydraulics Research Publication, Wallingford, UK.

van der Kuur, P., Roelfzema, A. and Verboom, G.K. (1989) The three-dimensional programme TRISULA with curvilinear orthogonal co-ordinates, in *Advances in Water Modelling and Measurement*, (ed. M.H. Palmer), BHRA, Cranfield, UK, pp. 135–48.

van Rijn, L.C. (1984) Bed forms and alluvial roughness. *Journal of Hydraulic Engineering*, **110** (12), 1733–54.

van Rijn, L.C. (1986) Sedimentation of dredged channels by currents and waves. *Journal of Waterways, Port, Coastal and Offshore Engineering, ASCE*, **112** (5), 541–59.

van Rijn, L.C. (1989) Two and three-dimensional mathematical models for suspended sediment transport by currents and waves. Delft Hydraulics, Delft, The Netherlands.

Verboom, G.K. (1975) The advection-dispersion equation for an anisotropic medium solved by fractional-step methods. Publication 157, Delft Hydraulics, Delft, The Netherlands.

Wechsler, B.A. and Cogley, D.R. (1977) A laboratory study of turbidity generation potential of sediments to be dredged, *Tech. Rep. D-77-14*, US Army Corps of Engineers, Waterways Experimental Station, Vicksburg, MI.

TWO-DIMENSIONAL HORIZONTAL AND WEAKLY THREE-DIMENSIONAL MODELS OF SEDIMENT TRANSPORT DUE TO WAVES AND CURRENTS

17

H.J. de Vriend

At the present time, in practice the use of three-dimensional modelling for sediment transport calculations is still in most cases unjustifiable. Correspondingly, it becomes necessary to use two-dimensional models, either with almost-uniform velocity distributions or with parameterized-variation velocity distributions. This last such class of model provides what is here called 'weakly three-dimensional models'. These two subclasses of model are currently the most widely used for sediment transport modelling in this area.

In effect, the hydrodynamic models of these types 'drive' the sediment transport models with little or no feedback from the sediment model to the hydrodynamic model. As this is one of the most important areas of application in this field, it is treated here in rather more detail, and critically. After an outline of the main mechanisms of sediment transport, the theories and models describing these mechanisms are introduced. A particular significance accrues in the choice of model to the envisaged applications of the model results. The relations to laboratory testing are also introduced. Although, as explained here, this is an area which is still the subject of research and development, much can already be done to solve the problems of coastal, estuarial and harbour engineers.

CONTENTS

17.1 INTRODUCTION

In practical engineering applications, the prediction of sediment transport under the influence of waves and currents involves much more than just selecting an instantaneous relationship between the transport rate and a number of hydrodynamic and sediment parameters, or running a suspended load transport model with a given set of conditions. These short-term transport descriptions are basic, but they are not sufficient to provide a practical model. Engineers are primarily interested in residual transports at time scales of the same order of magnitude as the phenomena that they have to predict (e.g. the impact of a harbour entrance on the adjacent coast).

This means that they have to predict the net effect of a highly variable transport climate, due to random wave fields, varying currents and tides, storms, surges, etc. Possibly, even the effects of a morphodynamic interaction with the sea-bed have to be taken into account: the displacement of morphological features, such as sand banks, can obviously involve significant sediment transports.

In order not to have all complications entering at one time, we will consider only non-cohesive sediment of uniform grain size. Furthermore, we will restrict ourselves to horizontally two-dimensional models, with excursions to one-dimensional and quasi-three-dimensional models wherever necessary.

Coastal, Estuarial and Harbour Engineer's Reference Book. Edited by M.B. Abbott and W.A. Price. Published in 1993 by Chapman & Hall, 2–6 Boundary Row, London SE1 8HN. ISBN 0 419 15430 2.

17.2 SEDIMENT TRANSPORT MECHANISMS

17.2.1 TRANSPORT MODES

Various modes of transport can be found in wave and current environments, depending on the ratio of wave action to current action:

1. Current-induced bed load transport, with the waves causing increased transport rates through their accentuation of the effect on the bottom shear stress.
2. Current-borne suspended-load transport, with the waves stirring up additional sediment and enhancing vertical mixing.
3. Wave-induced 'grab-and-dump' transport in the rippled bed regime, associated with vortex formation due to the bed forms.
4. Wave-induced 'sheetflow' transport, with the top layer of the bed (typically five grain diameters thick) in fluidized motion. This is usually attended by significant suspended load transport higher up in the water column.

There is no unified model formulation covering this entire range of transport modes, nor a reliable predictor of which mode will be present under a given set of conditions. The manner in which this deficiency is overcome in practice, will be described in section 17.2.3.

17.2.2 TRANSPORT MECHANISMS

Apart from the mode in which transport actually takes place, the driving mechanisms have also to be identified before we can select an appropriate model.

A classical, still widely used distinction is that between longshore and cross-shore transport. This seemingly geometrical distinction concerns, in essence, the distinction between current-borne transport (the 'longshore' component) and wave-borne transport (the 'cross-shore' component). This geometrical nomenclature may be appropriate in situations with a straight, almost uniform coast, but it makes no sense in more complex situations. Sometimes it is not even clear what exactly the longshore and cross-shore directions are (Figure 17.1), let alone that they should have anything to do with the distinction between these transport mechanisms. A nomenclature reflecting the actual transport mechanism is therefore preferable.

In broad outline, the principal mechanisms for coastal non-cohesive sediment transport are.

1. Current-borne transport: the wave motion enhances the mobilization of sediment, by increasing the effective bottom shear stress, the sediment entrainment from the bottom and the vertical exchange of suspended sediment, but it is the current that carries the sediment downstream.
2. Wave-asymmetry transport: the sediment in motion is transported by the wave orbital motion, which is asymmetric in time. Due to the non-linear response of the transport to this motion, this can give rise to a residual transport. The effect depends heavily on the transport mode: the same asymmetry can cause an up-wave transport in the grab-and-dump regime, and a down-wave transport in the sheetflow regime.
3. Transport due to wave induced boundary-layer streaming ('mass transport'): boundary layer effects in the wave orbital motion give rise to a net mass flux of fluid and sediment, which is usually down-wave. Note that in a turbulent environment (e.g. in the presence of a tidal current), this boundary layer is not necessarily thin!

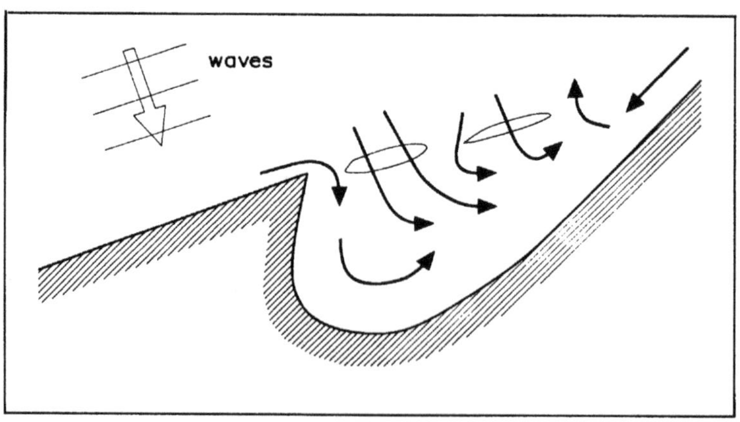

Figure 17.1 Long-shore and cross-shore components.

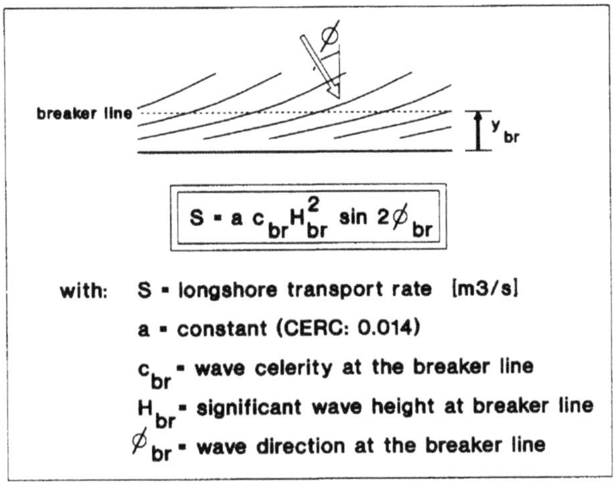

Figure 17.2 The CERC-formula.

4. Wave–wave interaction transport, due to phase coupling effects between sea waves and low frequency (LF) waves. Depending on the phase relationship, and so on the type of LF waves (independent of the short waves, or possibly group-bound), the residual transport direction can be up-wave or down-wave.
5. Down-slope gravitational transport: sediment moving on a slope will tend to deviate down-slope, due to gravitation. Although often a minor effect compared with the other transport components, this turns out to be very important for the morphodynamic behaviour of the coast.

The transport formulation to be chosen in a practical situation depends to a great extent on the transport mechanisms expected to be the most important there.

17.3 SHORT-TERM TRANSPORT MODELS

17.3.1 BULK TRANSPORT

The CERC-formula in Table 1 of the *Shore Protection Manual*, 1984 (Figure 17.2) is the most widely used formula for bulk longshore transport. It concerns the longshore transport rate, integrated over the entire surf zone, at a nearly uniform straight coast.

A good performance of this formula is not always guaranteed for every individual set of input conditions, but when applied over a longer period with naturally varying conditions, it usually gives reasonable indications of the yearly residual longshore transport.

17.3.2 CURRENT-BORNE TRANSPORT FORMULAE

In situations which are more complex than a nearly uniform, straight coast, a bulk transport formula usually makes no sense: we need more spatial resolution and have to allow for more complex combinations of waves and currents.

In the case of not too strong spatial and time gradients in the hydrodynamic and transport conditions, we can utilize a point-by-point relationship between the transport rate and the relevant hydrodynamic parameters: a local transport formula.

Traditionally, these formulae have been derived from a known transport formula for currents alone. Bijker (1971), for instance, started from the Frijlink formula, and there are also coastal versions of the Engelund–Hansen and the Ackers–White formulae. As the incorporation of wave effects is based on an enhancement of the effective bottom shear stress, these formulae typically concern current-borne transport.

As an example of this class of formula, Table 17.1 gives the widely applied Bijker formula (Bijker, 1971). Another formula that performs reasonably well in practice was proposed by Van Rijn (1989). Both are total-load formulae, i.e. they include both bed-load and suspended load transport. More information on current-borne transport models, is given in the comprehensive *Handbook on Sediment Transport by Currents and Waves* (Van Rijn, 1989).

An important factor in these formulae is the bed roughness. This is often introduced as a parameter to be specified by the user, which virtually implies that it is used as a calibration parameter. Only Van Rijn (1989) gives a predictor for the roughness parameter he uses in his formula.

Directionality is not included in the formulae, otherwise than via the shear stress enhancement factor accounting for the wave influence. With the argument that most of the transport is suspended load, its direction is assumed to coincide with the depth-averaged mean current direction.

17.3.3 WAVE-BORNE TRANSPORT FORMULAE (RIPPLED BED)

In the case of wave-borne transport, the previously described bulk and total-load transport formulae fail. The intra-wave transport mechanisms, which are not included in these formulae, predominate here, and the residual transport depends to a large extent on asymmetries in the wave orbital motion.

The interaction of the waves with the bed forms is also a key element in rippled-bed transport (Bijker *et al.*, 1976).

We could attempt to model the processes in detail at the intra-wave scale. Recent publications (Blondeaux

Table 17.1 Bijker's (1971) total-load transport formula

$$s = s_b[1 + 1.83\{I_1(Z_*) + I_2(Z_*)\ln(33h/k_s)\}]$$

$$s_b = bu_{*c}d_{50}\exp[\{-0.27(\rho_s - \rho)gd_{50}\}/\{\rho\mu u_{*cw}^2\}]$$

in which:

s	= total-load transport rate, ($m^2\,s^{-1}$)	h	= water depth
I_1, I_2	= known functions of Z_* (see Van Rijn, 1989)	k_s	= bed roughness height
Z_*	$= \dfrac{w}{\kappa u_{*cw}}$	b	= Bijker's constant (≈ 5 in surf zone)
		d_{50}	= median grain diameter
w	= particle settling velocity	ρ_s	= sediment mass density
κ	= von Kármán's constant	ρ	= fluid mass density
u_{*c}	= bed shear velocity for current only	g	= gravitational acceleration
		μ	= transport efficiency factor
		u_{*cw}	= enhanced bed shear velocity (including waves)

Table 17.2 Nielsen's (1988) 'grab-and-dump' model

$$s_w = 0.0025\frac{\Theta'^3}{(1 - \pi\Delta/\lambda)^{3/2}}\frac{\hat{U}_{\text{off}}^6 - \hat{U}_{\text{on}}^6}{\hat{U}^6}\hat{A}$$

$$\Theta' = 0.5f'_w\frac{\rho\hat{U}^2}{(\rho_s - \rho)gd_{50}}$$

in which:

s_w	= net wave-borne transport rate ($m^2\,s^{-1}$)	g	= gravitational acceleration
w	= particle settling velocity	d_{50}	= median grain diameter
Θ	= mobility parameter	$\hat{U}_{\text{off}}$	= maximum offshore velocity near bed
Δ	= ripple height	$\hat{U}_{\text{on}}$	= maximum onshore velocity near bed
λ	= ripple length	$\hat{U}$	= near-bed velocity amplitude (linear wave theory)
f_w	= grain-related friction factor ($k_s = 2.5\,d_{50}$)	$\hat{A}$	= near-bed orbital excluding (linear wave theory)
ρ_s	= sediment mass density		
ρ	= fluid mass density		

and Vittori, 1989) show that significant progress is being made in this field. Still, intra-wave modelling remains too expensive for many practical applications. There we need simpler models, based upon semi-empirical, or parameterized results of intra-wave models.

Nielsen (1988) has proposed a semi-empirical model type. This 'grab-and-dump' model, derived for rippled beds under regular waves, is outlined in Table 17.2.

Nielsen's model is, by definition, very sensitive to wave asymmetry. This means that it makes rather high demands on the description of the wave orbital motion, including higher-order wave components. The model has to be calibrated with data from natural wave fields, without the parasitic superharmonics that are sometimes found in laboratory experiments. Since the model is derived for regular waves, it has to be reformulated in terms of random waves. Recent research on bed forms under random waves (Ribberink and Al-Salem, 1990) shows that it is not so much the wave orbital motion as, in particular, the bed forms which are different in that situation.

17.3.4 WAVE-BORNE TRANSPORT FORMULAE (SHEETFLOW)

A significant part of the coastal transport takes place in the sheetflow regime. Especially during storms, the top layer of the bed is believed to be intermittently fluidized by wave action.

Though not explicitly derived for sheetflow, an important modelling concept for wave-borne transport is claimed to apply to these conditions. It relates the instantaneous transport rate at the intra-wave scale to the near-bed velocity or the bed shear stress at that instant.

In the flat-bed sheetflow regime, such a local and instantaneous relationship may apply; in the case of a

Table 17.3 Intra-wave transport formulae (horizontal bed)

Madsen and Grant (1976)	*Bailard (1981)*
$S_t(t) = 40wd_{50}\Theta^3 \frac{U}{\lvert U \rvert}$ $\Theta = 0.5f_w \frac{\rho U^2}{(\rho_s - \rho)gd_{50}}$	$s_b(t) = f_w A_b U^3$ $s_s(t) = f_w \frac{A_s}{w} \lvert U^3 \rvert U$

in which:

$s_b(t)$ = instantaneous bed-load transport ($m^2 s^{-1}$)	Θ = mobility parameter
$s_s(t)$ = instantaneous suspended-load transport ($m^2 s^{-1}$)	f_w = friction factor
$s_t(t)$ = instantaneous total transport ($m^2 s^{-1}$)	ρ_s = sediment mass density
$U(t)$ = instantaneous near-bed velocity (wave and current)	ρ = fluid mass density
	g = gravitational acceleration
w = particle settling velocity	A_b = bed-load efficiency factor (Bailard: ≈0.011–0.015)
d_{50} = median grain diameter	A_s = suspended load efficiency factor (Bailard: ≈0.0016–0.0024)

rippled bed this is certainly not the case. There we may at best hope that the wave averaged value of the corresponding calculated transport gets somewhere near reality.

The most important formulae belonging to this class are those proposed by Madsen and Grant (1976) and Bailard (1981), outlined in Table 17.3. If these formulae describe the transport at the intra-wave level, they also provide the possibility of dealing with the residual effects of a variety of wave-related transport mechanisms, such as wave asymmetry. Besides, combinations of waves and currents, even in different directions, can easily be included (De Vriend and Ribberink, 1988).

Models for practical engineering applications are usually formulated in terms of quantities that vary slowly in time, as compared with the waves. Even the wave-field model is often formulated in the frequency domain, without explicitly solving for the time variation. The application of the above transport modelling concept therefore requires:

1. A sub-model describing the time variation of the near-bed velocity or the bed shear stress at the intra-wave scale.
2. An integrator over the waves, if possible analytical, otherwise off-line numerical (utilizing interpolation tables) or on-line numerical (called whenever needed in the computation).

The choice of the intra-wave sub-model (1) should include the mechanisms that are expected to be important in the case to be studied. For example: wave asymmetry can be introduced via non-linear wave theory (e.g. Rienecker and Fenton, 1981); the interaction of short waves with group-bound long waves can be introduced by adding a phase locked, low frequency wave (Roelvink and Stive, 1989); and the interaction of short waves with free long waves (e.g. swell) can be introduced by adding low frequency waves of arbitrary phase.

The Bailard and Madsen–Grant concepts have been tested and verified under a range of conditions. Laboratory experiments at the Delft University of Technology (De Waal, 1987), for instance, showed that the Bailard model works rather well under rough conditions, but much less so in the rippled bed regime. Field applications have sometimes been very successful (e.g. Stive, 1986), but sometimes the model just indicates the right trends and otherwise fails quantitatively. So, although these models are useful tools in orientative computations, much research remains to be done before they become quantitatively reliable transport predictors.

17.3.5 WAVE-INTEGRATED SUSPENDED-LOAD TRANSPORT MODELS

Suspended-load transport, as far as it is not included in a total load formula, is usually described by an advection–diffusion equation for the sediment concentration. Wave effects can be incorporated in such a model via the vertical sediment flux at the bottom and via the vertical mixing coefficient (Figure 17.3). Further details have been given by Van Rijn (1989).

There is a variety of generic models of this type in use: two-dimensional vertical, two-dimensional horizontal, 2.5-dimensional, and three-dimensional. Some of them have been treated in previous chapters (e.g. Chapter 16). We will therefore concentrate on two-dimensional horizontal and 2.5-dimensional versions.

Two-dimensional horizontal models are derived by straightforward depth-integration of the constituting

$$\frac{\partial c}{\partial t} + \frac{\partial}{\partial x_i}(u_i c) - \frac{\partial}{\partial x_3}(wc) - \frac{\partial}{\partial x_i}(\varepsilon_i \frac{\partial c}{\partial x_i} = 0)$$

IC; $c(o,x_i)$ given

Hor. BC: c or $\frac{\partial c}{\partial n}$ given

Surface BC: zero normal flux

Bed BC: c or $\frac{\partial c}{\partial x_3}$ given

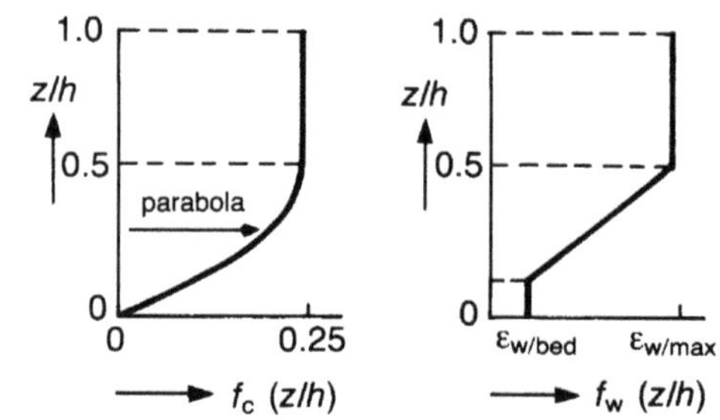

in which:

c = susp. sediment concentration

t = time

x_i = cartesian coordinate (i = 1,2,3)

x_3 = vertical coordinate

u_i = fluid velocity component in the x_i-direction

w = particle settling velocity

ε = sediment mixing coefficient

Van Rijn (1989):

$\varepsilon_1, \varepsilon_2$ = constant (0.1 – 1.0 m²/s)

$\varepsilon_3 = (\varepsilon_c^2 + \varepsilon_w^2)^{1/2}$

$\varepsilon_c = K\, u_{*c}\, h\, f_c\, (\frac{z}{h})$

$\varepsilon_w = \varepsilon_w|_{bed} + (\varepsilon_w|_{max} - \varepsilon_w|_{bed}) f_w (\frac{z}{h})$

$z = x_3 - x_3|_{bed}$

κ = Von Karman's constant

u_{*c} = current-induced shear vel.)

h = water depth

$\varepsilon_w|_{bed}, \varepsilon_w|_{max}$: fnct (wave par.)

Figure 17.3 Wave-averaged suspended-load transport.

equation. Strict vertical similarity, of the velocity profile as well as the concentration profile, is assumed in the advection terms. The shape of these profiles necessarily corresponds with those in uniform equilibrium conditions.

Depth-integration of the settling and vertical mixing terms, in combination with the zero-flux condition at the surface, yields the net vertical sediment flux at the bottom. To describe this flux in terms of the depth-averaged concentration is the key problem in this type of model. For instance, if the entrainment from the bottom is assumed to be the same as under uniform equilibrium conditions (the flow properties are identical, therefore the sediment pick-up rate is likely to be so, too), the net flux is easily shown to be proportional to the difference between the actual concentration and the corresponding equilibrium value (Table 17.4).

The 2.5-dimensional approach (Wang and Ribberink, 1986; Katopodi and Ribberink, 1990, 1992) is more subtle in its treatment of the vertical profiles: it takes account of the response of the profile to non-uniformities in space and time (Table 17.5). One of the constituent models of the SUSTRA-system (Delft Hydraulics) is based on this concept.

The horizontal part of the model outlined in Table 17.5 involves the inherent time and length scales T (adaptation time) and L (adaptation length). Figure 17.4 shows these quantities as a function of the relevant wave and current parameters.

Table 17.4 Depth-integrated suspended-load concentration model

$$\frac{\partial}{\partial t}(h\bar{c}) + \frac{\partial}{\partial x_i}(\alpha \bar{u}_i h\bar{c}) - \frac{\partial}{\partial x_i}\left(\bar{\varepsilon}_i h \frac{\partial \bar{c}}{\partial x_i}\right) = A(\bar{c}_e - \bar{c})w$$

$$\text{or: } (1-\alpha)\frac{\bar{c}}{h}\frac{\partial h}{\partial t} + \frac{\partial c}{\partial t} + \alpha \bar{u}_i \frac{\partial \bar{c}}{\partial x_i} - \frac{\partial}{\partial x_i}\left(\bar{\bar{\varepsilon}}_i \frac{\partial \bar{c}}{\partial x_i}\right) = \frac{A}{h}(\bar{c}_e - \bar{c})w$$

in which:

t = time	$\bar{\varepsilon}_i$, $\bar{\bar{\varepsilon}}_i$ = effective horizontal mixing coefficients
x_i = horizontal Cartesian co-ordinates	A = calibration factor
h = water depth	$\bar{c}_e$ = depth-averaged equilibrium concentration
u_i = depth-averaged velocity component	w = particle settling velocity
c = depth-averaged concentration	
$\alpha \triangleq \overline{uc}/(\bar{u}\bar{c})$	

Table 17.5 2.5-dimensional suspended-load model concept

1DV-part

$$c = \bar{c}\phi_0(\zeta) + \frac{h}{w}\left[\frac{\partial c}{\partial t} - \frac{\partial}{\partial x_i}\left(\bar{\varepsilon}_i \frac{\partial \bar{c}}{\partial x_i}\right)\right]\phi_{11}(\zeta) + \frac{h}{w}\bar{u}_i \frac{\partial \bar{c}}{\partial x_i}\phi_{12}(\zeta)$$

$$\left.\begin{aligned} D[\phi_0] &= 0 \\ D[\phi_{11}] &= \phi_0(\zeta) \\ D[\phi_{12}] &= f(\zeta)\phi_0(\zeta) \end{aligned}\right\} + \text{BC}$$

2DH-part

$$T_A \frac{\partial c}{\partial t} + L_A \frac{\bar{u}_i}{|\bar{u}|}\frac{\partial \bar{c}}{\partial x_i} - T_A \frac{\partial}{\partial x_i}\left(\bar{\varepsilon}_i \frac{\partial \bar{c}}{\partial x_i}\right) = \bar{c}_e - \bar{c}$$

$$\text{with: } T_A = \frac{h}{w}\frac{\phi_{11}(0)}{\phi_0(0)}$$

$$L_A = \frac{h|\bar{u}|}{w}\frac{\phi_{12}(0)}{\phi_0(0)}$$

in which:

$D = \frac{\partial}{\partial \zeta} + \frac{1}{wh}\frac{\partial}{\partial \zeta}\left(\varepsilon_z \frac{\partial}{\partial \zeta}\right)$	t = time
$\zeta = (z - z_b)/h$	x_i = horizontal Cartesian co-ordinates
z = vertical co-ordinate	$\bar{u}_i$ = depth-averaged velocity component
z_b = bed level	$\bar{c}$ = depth-averaged concentration
w = particle settling velocity	$\bar{c}_e$ = depth-averaged equilibrium concentration
ε_z = vertical mixing coefficient	h = water depth
$\phi(\zeta)$ = concentration profile functions	$\bar{\varepsilon}_i$ = horizontal mixing coefficient
$f(\zeta)$ = velocity profile function	T_A = adaptation time
	L_A = adaptation length

In practical applications, it is often possible to estimate the values of T and L before actually applying the model. If we compare these with the 'extrinsic' time and length scales in question (such as the tidal period and the topographical length scale), we can estimate the necessity of applying the advection–diffusion model for the suspended-load transport. If the inherent scales, T and L, are small compared with the extrinsic scales, the suspended-load transport can just as well, or even better, be described with a transport formula.

17.3.6 INTRA-WAVE SUSPENDED-LOAD TRANSPORT MODELS

The suspended-load model described in the previous section concerns current-borne transport only. It may be possible to include a parametric description of wave-borne transport, but this has not been done so far, partly due to lack of insight into how wave-residual suspended-load transport of this type behaves, partly because we assume that the wave-borne transport mainly takes place close to the bed. The latter assump-

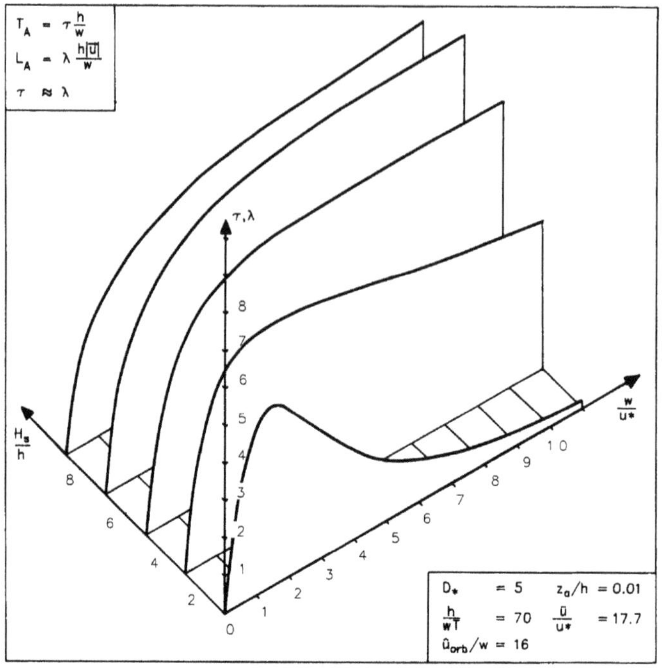

Figure 17.4 Inherent time and length scales of suspended-load transport (Katopodi and Ribberink, 1988).

tion is motivated by the observation that the response time of the suspended sediment concentration increases with the distance to the bed, so that the concentration away from the bed is simply unable to follow the wave motion.

Still we have to be careful here. The above observation originates from models with non-breaking waves. In the surf zone, wave breaking greatly affects the water motion in the top layer of the fluid, including its mixing capacity (see Justesen *et al.*, 1986). Research on intra-wave suspended-load transport processes in this environment is therefore very useful. So far, intra-wave suspended-load modelling concerns two-dimensional vertical cases only. An example of such a model, also in use for practical applications, has been developed in Denmark (Deigaard *et al.*, 1986; Fredsøe, 1993).

Even with the present computer facilities, the incorporation of a 2.5- or three-dimensional intra-wave suspended-load module into horizontally two-dimensional models of longer-term residual transports in practical situations would be too costly. For such problems, a parameterization would be needed.

17.3.7 INCORPORATION OF THE GRAVITATIONAL COMPONENT

Unless the bottom is so steeply sloping that the sediment layers are sliding over each other, gravity is only able to modify the transport, not to generate it. Once the threshold of motion is exceeded and the grains are moving, gravitation tends to move the transport direction downslope.

At present, there are basically two model concepts in use for this effect, one isotropic, the other anisotropic (Figure 17.5). The isotropic concept, used in the Bailard formula, for instance, introduces the gravitational effect as an additional transport component, independent of the direction of the other components.

The anisotropic concept has been developed and tested in river morphology (Struiksma *et al.*, 1985), but there is no reason why it should be less valid in other environments. It introduces the downslope component of gravity as an additional force on the grains and uses the resultant force in a non-linear transport relationship, which leads to a dependency

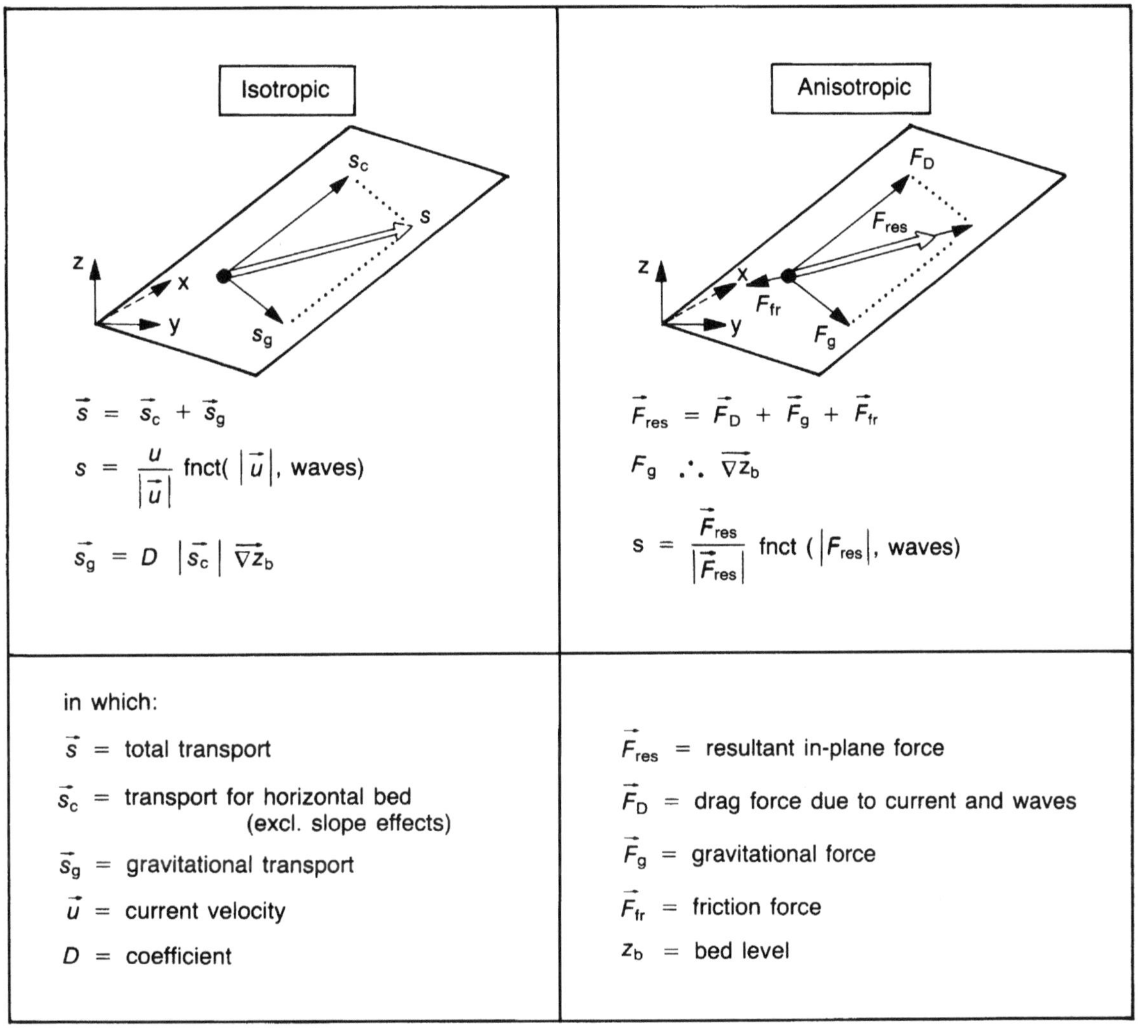

Figure 17.5 Gravitational transport in bed load formulae.

of the gravitational effect on the angle between the flow and the bottom slope. The effect is strongest if these vectors are collinear. The isotropic concept is shown in Table 17.6.

In steady flow, the downslope gravitational transport component is usually rather weak. In oscillatory flow, however, it can be large in comparison with the other residual transport components, its direction being invariant throughout the flow cycle. The gravitational downslope transport is also of utmost importance to the morphology: in its absence, the morphodynamic system would be inherently unstable (De Vriend, 1986). Hence the gravitational transport component can sometimes be left out of consideration in short-term transport computations, but not in models of longer-term residual transport, especially if there is a morphodynamic interaction involved.

The most obvious way for gravitation to exert its influence on the transport is through the direct contact of moving grains with the sloping bed. Clearly, this is associated with bed-load transport. Some suspended-load models also include a downslope effect. Bailard's

Table 17.6 Downslope effect in the Bailard formula

Bed load	*Suspended load*
$s_b = f_w\left[A_b U^3 - B_b \lvert U\rvert^3 \frac{\partial z_b}{\partial x}\right]$	$s_s = \frac{f_w}{w}\left[A_s \lvert U\rvert^3 U - \frac{B_s}{w}\lvert U\rvert^5 \frac{\partial z_b}{\partial x}\right]$
in which: s_b = instantaneous bed-load transport rate [$m^2 s^{-1}$] f_w = friction factor U = near-bed velocity A_b, B_b = coefficients (bed load)	in which: s_s = instantaneous susp.-load transport rate ($m^2 s^{-1}$) w = particle settling velocity $\frac{\partial z_b}{\partial x}$ = bottom slope A_s, B_s = coefficients (suspended load)

(1981) formula, for instance, includes a slope term in its suspended-load part (Table 17.6).

17.4 RESIDUAL TRANSPORT MECHANISMS

17.4.1 WAVE-RESIDUAL TRANSPORT

In the previous sections, we have covered most of the ground concerning the residual effect of regular wave trains. Natural wave fields, however, are random in height and direction, and this has effects on the sediment transport.

The theoretical basis of current-borne transport formulae are usually restricted to regular waves. Through calibration and heuristic arguments, they are extended to random waves (Van Rijn, 1989).

The effects of random directionality have barely been investigated, let alone the combined occurrence of height and direction. The same goes for wave-integrated suspended-load models. The bottom boundary condition in these models is mostly based on a formula for current-borne transport, and hence includes the limitations thereby implied. The vertical mixing coefficient includes wave effects in a semi-empirical manner (Van Rijn, 1989), but without much attention being given to the effects of randomness.

The wave-borne transport models have advanced somewhat further in this respect, primarily because they are easier to handle with random waves. In principle, integration of the instantaneous transport formulae over the waves leads to a number of near-bed velocity moments, which can also be evaluated for random waves (Guza and Thornton, 1986).

Table 17.7, derived from Guza and Thornton (1986), shows that the effects of randomness can be significant, especially in the even moments. They have also included directionality in their analysis, but still this aspect remains weakly developed.

Intra-wave suspended-load models can easily deal with randomness: they work on a real-time basis and can just as well be run with a random wave input signal. Since these models are two-dimensional vertical, they are unsuited to investigate the effects of random directionality. When it comes to their parameterization for more complex and longer-term applications, wave randomness can certainly not be ignored. In summary, wave randomness is claimed to have been calibrated into the models, but remains a topic for further research.

Table 17.7 Effects of wave randomness on velocity moments (after Guza and Thornton, 1986)

Near-bed velocity: $u = \bar{u} + \tilde{u}$, $\langle \tilde{u} \rangle = 0$, $\langle \tilde{u}^2 \rangle = \sigma^2$

in which: $\langle \ldots \rangle = \frac{1}{T}\int_{t_0}^{t_0+T} \ldots dt$, and here $\bar{u} = 0$

Velocity moment	*Wave description*		
	Sinusoidal	*Gaussian*	*Measured*
$\langle \tilde{u} \rangle^3/\sigma^3$	0	0	0.5
$\langle \lvert\tilde{u}\rvert^3 \rangle/\sigma^3$	1.2	1.6	1.7
$\langle \tilde{u} \rangle^4/\sigma^4$	1.5	3.0	3.4
$\langle \tilde{u} \rangle^5/\sigma^5$	0	0	5
$\langle \lvert\tilde{u}\rvert^5 \rangle/\sigma^5$	1.9	6.4	8

17.4.2 TIDE-RESIDUAL TRANSPORT

The morphological time scale in most practical applications exceeds the tidal period by far. Still, the tidal motion is an important agent in large-scale transports and morphodynamics. This means, that we have to know the tide-residual transport accurately if ever we want to model large-scale coastal or estuarial behaviour.

Tide-residual transports can have various causes. Firstly, the 'primary' tidal current, defined here as

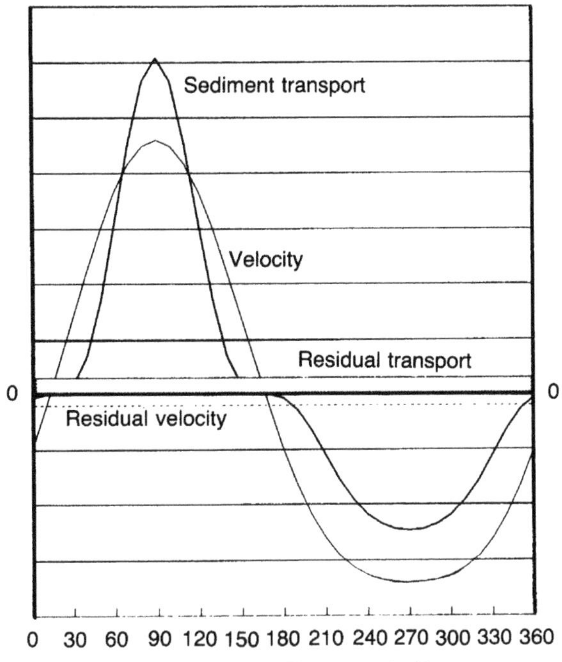

Figure 17.6 'Primary' tide-residual transport.

the current described by the shallow-water equations: possible agents are residual currents and the asymmetry of the tidal oscillation (like wave asymmetry), but note that, in principle, there is no direct relationship between the residual current and the residual transport (Figure 17.6)!

Secondly, the presence of 'secondary' flows and rectification effects, caused by accelerations (inertia, Coriolis, curvature) or depth-varying external forces (wind, waves, density gradients): these give rise to a deviation of the near-bottom velocity and the sediment transport from the 'primary' flow direction, but often in the same direction throughout the tide; consequently, this deviation, though weak as compared with the maximum tidal transport, can be significant as compared with the 'primary' residual transport (Figure 17.7; see also Booij *et al.*, 1985).

And finally, the effect of the tidal water level variation on wave-induced sediment stirring, which is a function of the water depth: this may lead to ebb or flood dominance in the transport (Figure 17.8).

The primary current effect, and also the effect of water-level variations, is incorporated more or less automatically in the present-day coastal and estuarine models (section 17.5).

The secondary flow effects require special model components, which are seldom included. However, along with the increasing interest in large-scale coastal and estuarial behaviour, the understanding and modelling of secondary flows is becoming an increasingly important research topic.

Another effect, particularly important in tidal lagoons, is 'horizontal grain-sorting'. When a sediment-laden flow enters quiet waters, the distance needed for the sediment to settle down will be longest for the finest grains. This means that there will be a horizontal gradient in the grain size of the deposits. If the reverse tidal motion is too weak to exceed the threshold of motion, or to remove all of the deposited sediment, there will be a residual transport and deposition of sediment, which becomes ever finer with the distance from the sea.

No doubt this effect will prove to be quite important when we start building synoptic high-resolution

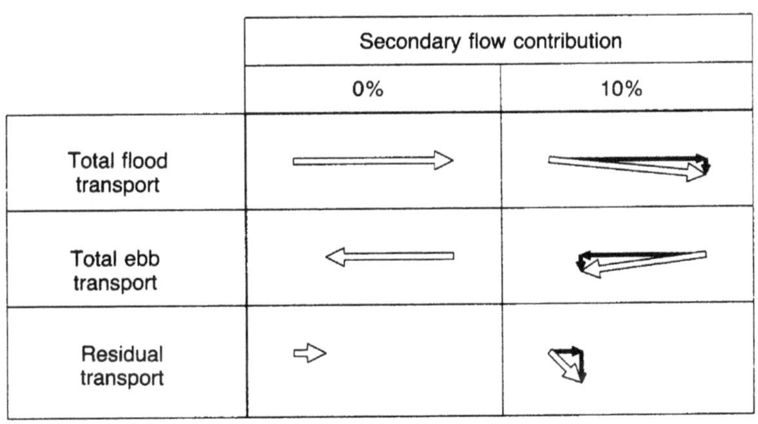

Figure 17.7 'Secondary' tide residual transport.

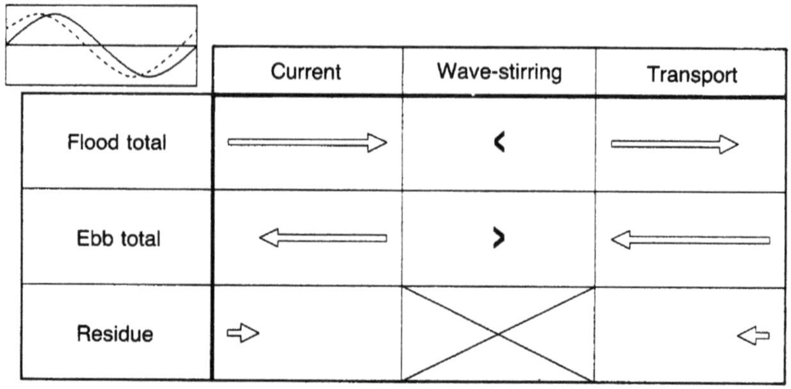

Figure 17.8 Effect of stage-dependent wave-stirring.

models of the morphological processes inside a tidal lagoon.

17.4.3 YEARLY RESIDUAL TRANSPORT

Although the tide can be an important agent in long-term coastal and estuarial behaviour, it is certainly not the only one. Within the yearly cycle, we also have: firstly, the astronomic tidal modulation (neap/spring cycle); for the Dutch coast, for instance, the transport during spring tide is so much higher than during neap tide and mean tide that the large-scale morphology seems to be moulded by spring tides alone. Secondly, storm surges, which can drastically modify the tide during a period of one or two days: since storms also generate high waves, these periods can contribute significantly to the yearly residual transport, although the effect may be mitigated by the low frequency of occurrence of extreme events. Thirdly, the variability of the wave climate (storms, calms), and finally the seasonal variation of conditions, including the water temperature.

The sequence of events may also be of influence, obviously on short-term evolutions (it makes a difference whether a series of storms comes in succession or interrupted by calm periods), but possibly also in the longer run (like the groupiness of waves, perhaps). If this is true, it will have important consequences for the schematization principle in model applications, since linear superposition is no longer adequate. Further research on this topic is therefore extremely relevant.

The relative importance of these effects varies in space: dune erosion is dominated by the combination of storm waves and high tides, including their joint probability, whereas further offshore it is rather the spring tide and the storm surges that count. Besides, the transport and the morphological evolution in an area are not only determined by the local climatic conditions, but also by remote ones, e.g. through swell fields. Apart from this, no year is identical to another, so there is no such thing as an invariant yearly cycle. Every year is a realization of a stochastic process, the parameters of which are gradually changing, in addition to which the climate undergoes long-term changes, and so do extrinsic parameters like the mean sea-level and the tidal amplitude.

In principle, all of these aspects could be taken into account via variation of the input conditions of real-time simulation models. However, especially in horizontally two-dimensional situations, this method of model operation is hardly feasible. Hence we have a schematization problem: how to represent the effects of the actual input time histories by a practicable number of model runs. This problem will be addressed in section 17.5.6.

17.4.4 MORPHODYNAMIC-AVERAGED TRANSPORT

So far, we have implicitly considered the sea-bed topography as an input parameter for the hydrodynamic and sediment transport models. Especially in longer-term evolutions, this idea no longer holds good: the sea-bed evolution can play an essential part in the residual transports, to the effect that the initial sedimentation and erosion patterns bear no relevance whatsoever to the longer-term evolution. Figure 17.9 gives an example.

The only way to model these effects is through morphodynamic simulation, over some time and for some specific classes of problem, possibly followed by parameterization. Such simulations require numerical tools, which at the moment are still limited in their capabilities (see section 17.5).

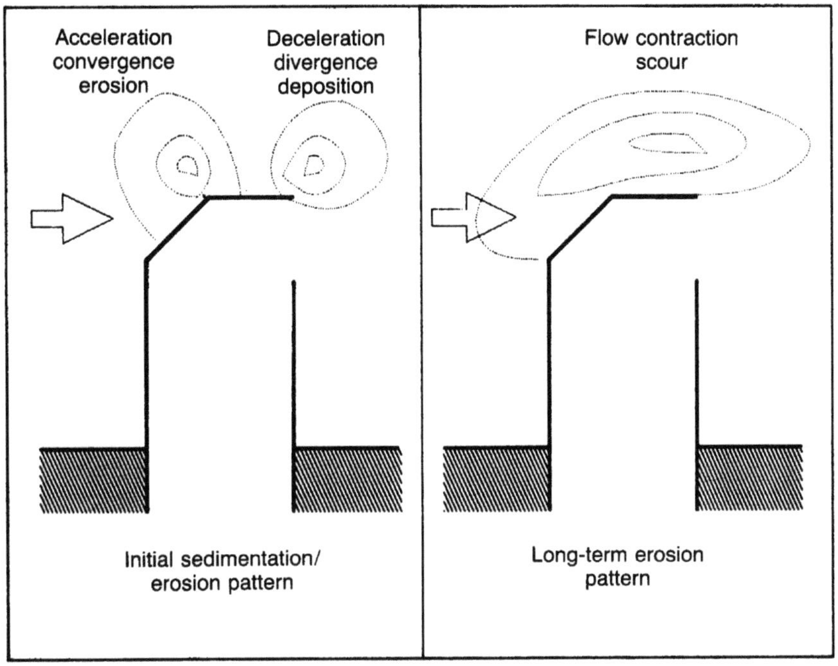

Figure 17.9 Initial sedimentation/erosion versus morphodynamic interaction.

17.5 RESIDUAL TRANSPORT MODELLING

17.5.1 APPLICABILITY OF INITIAL SEDIMENTATION AND EROSION MODELS

Since horizontally two-dimensional morphodynamic simulation models are hardly ever operationally available, we have to base our predictions upon 'initial' sedimentation and erosion models (ISE models), i.e. models using the bottom topography as an input parameter into the hydrodynamic and sediment-transport constituents, with the sediment balance as a post-processing facility to derive the initial sedimentation and erosion rate from the computed sediment-transport field.

Such models cannot claim to predict near-equilibrium situations, were there are subtle interactions between the hydrodynamics, the sediment transport and the bed topography. At best, they describe the first morphological adjustment after a strong distortion of the system, e.g. due to human interference. In such distorted systems, a few 'primary' transport phenomena appear to be predominant, and the morphological adjustment process tends to be monotonous in time. Hence an ISE model can be expected to yield a reliable indication of this process (Figure 17.10).

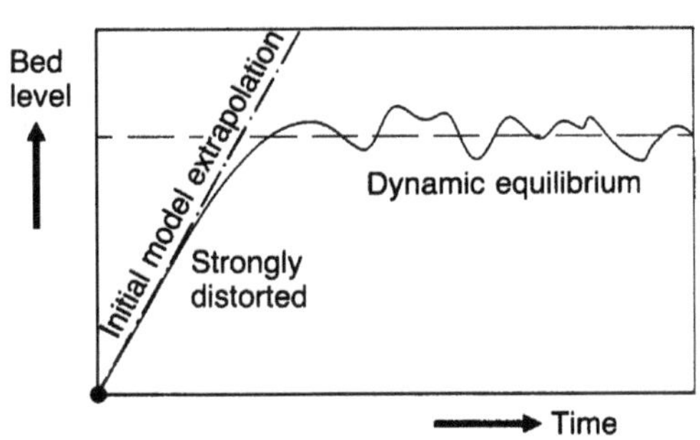

Figure 17.10 Behaviour of a strongly distorted system.

17.5.2 GENERAL SET-UP OF HORIZONTALLY TWO-DIMENSIONAL ISE MODELS

A generalized aggregate flow chart of an ISE model is given in Figure 17.11.

Although their straightforwardness may suggest that ISE models are a sort of panacea, to be utilized blindly wherever this type of model applies, their set-up involves a number of decisions and choices that are far from trivial.

The generic models built together into an ISE model can be rather standard, although combinations of strongly simplified constituents can lead to erroneous

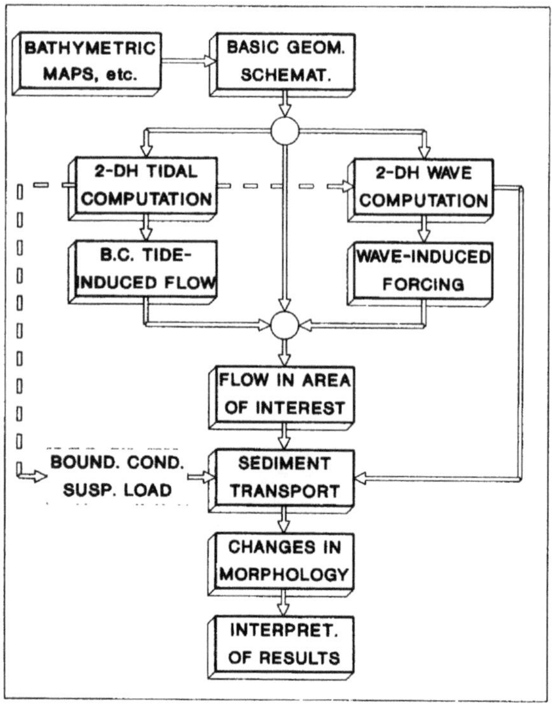

Figure 17.11 Flow chart of an ISE model.

results (Dingemans *et al.*, 1987). Since these generic models are widely distributed, compound ISE models could easily be built by many institutes. The principal bottle-neck at the informatics level is the efficiency of data transfer, while at the physical level it is the skill available to select the right constituents, to schematize the conditions appropriately, and, above all, to interpret the results properly.

17.5.3 PROCESS ANALYSIS AND MODEL COMPOSITION

One of the first and most crucial steps in model application is the analysis of the physical processes in the area of interest, both before and after the envisaged engineering work if this is relevant. This analysis determines which processes are important and which can be ignored, and hence which model components should be incorporated (e.g. a transport formula or a suspended load model?). The principal tools are experience and physical insight, but there are also quantitative aids, such as the so-called 'breaking index,' as a means of indicating where wave breaking can be expected. Other examples are: the long-wave celerity, to estimate the character of the tidal motion in the area of interest; parameters to indicate the importance of secondary-flow effects (e.g. the Ekman-parameter for Coriolis-induced secondary flow, or the depth and curvature product for curvature-induced secondary flow); parameters to indicate which transport regime has to be expected (Van Rijn, 1989); PC-implemented transport formulae (Van Rijn's (1989) handbook, for instance, provides a floppy disk for transport computations), which can be used to estimate the relative importance of the various transport agents; formulae or tables to estimate the inherent time and length scales of suspended load transport, and hence to assess the need for an advection–diffusion model for the suspended sediment concentration (see Figure 17.3). The formulation of such indicative yardsticks is a useful by-product of research and therefore deserves continuous attention.

Apart from physical arguments, the availability of generic models is a factor in the selection of constituent models: not every institute has models of every process available in the context of its sediment-transport modelling environment. For the most part, this implies the inclusion of processes that are expected to be unimportant, but sometimes it implies that important processes cannot be included and have to be taken into account in the interpretation of the results (section 17.5.10).

17.5.4 MODEL DOMAINS AND BOUNDARY CONDITIONS

The model composition in the area of interest is only part of the story about the general model set-up: the constituent models require boundary conditions.

Since we are working in two horizontal dimensions, often with significant parts of the model boundary in open water, the formulation of a consistent set of boundary conditions is not a trivial task. The tidal boundary conditions, for instance, involve the risk of spurious short-circuits in the current near the boundary, spurious reflections, and sometimes even violation of the total mass balance. Similarly, the boundary conditions for the waves involve the problem of disturbances originating from the lateral boundaries (the notorious wedge of unreliability in hyperbolic and parabolic wave models), and in elliptic models they can also produce spurious reflections.

One way to circumvent consistency problems at the boundaries is nesting into larger models, thus bringing the problem away from the area of interest. Sometimes these larger models already exist and have been calibrated and validated in other applications. Otherwise, the costs of construction and validation of such models have to be weighed against their expected benefit. If the outcome is negative, the model boundaries have to be chosen 'sufficiently' far away from the area of interest. An estimate of the

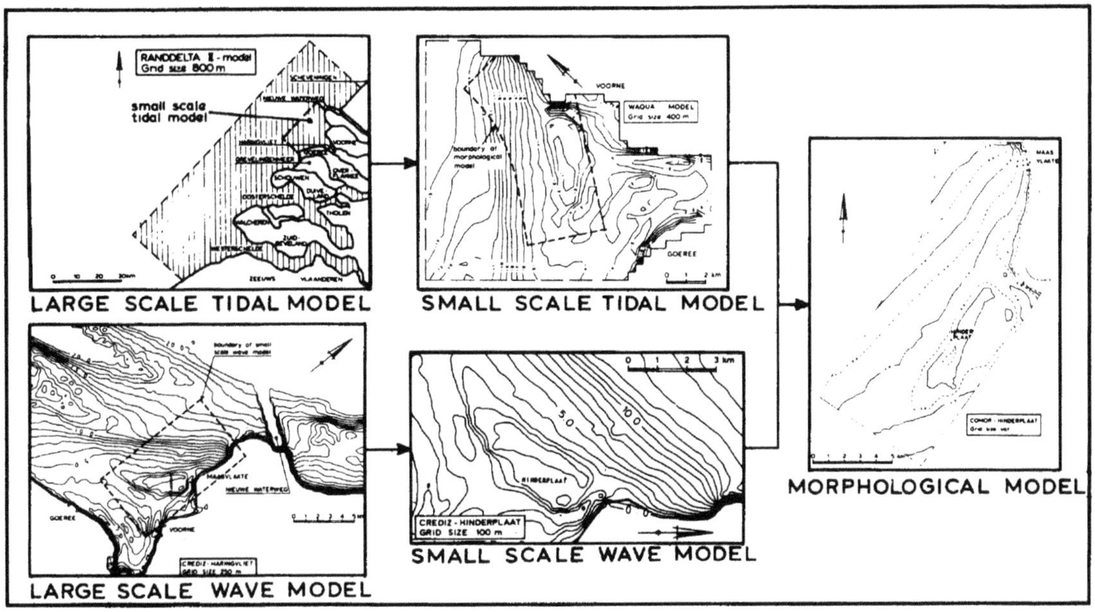

Figure 17.12 Nested model set-up (after Boer *et al.*, 1984).

inherent length scales (see section 17.4.3) may help to decide what is 'sufficient'. Figure 17.12 shows an example of a nested model set-up.

If there are existing models available for nesting, they mostly deal with only one constituent of the ISE model, e.g. tide or waves. This can give rise to trouble with the boundary conditions, especially in areas like the surf zone, where the combined action of tides, waves and currents is important (see section 17.5.7).

17.5.5 GEOMETRICAL SCHEMATIZATION

The geometrical schematization of a two-dimensional horizontal or quasi-three-dimensional model consists of two parts: the grid generation and the discretization of the bottom topography.

Especially in coastal areas, where important transport processes take place in the surf zone, we need either a very fine discretization or a boundary-fitted computational grid (curvilinear finite differences or finite elements). A fine grid usually implies a small time step in order to guarantee numerical stability and accuracy. Consequently, fine-grid models tend to be expensive in terms of computer costs, especially if we are in a tidal environment and have to cover a full tidal cycle.

We will not go into the finite difference versus finite element discussion, since each method has its advantages and the choice between the two uses has to be made at the institute level, not at the application level. In general, there are no indications that one of these methods should perform significantly better than the other in our type of applications.

The discretization of the bottom topography is more crucial than one may expect at first sight and a far from trivial step in the modelling process. Many modellers have found out the hard way that mapping the grid onto a bathymetric map and simply picking out the bottom levels in the grid points can have unexpected and sometimes even disastrous effects. Topographic schematization deserves more attention than this.

Much of the trouble originates from the inherent limitations of the constituent models. In two-dimensional horizontal tidal models of estuaries with pronounced shoals and gullies, for instance, it is sometimes necessary to adjust the bottom topography artificially in order to obtain the correct flow distribution. Combining such a model with wave and sediment transport models can lead to erroneous results (e.g. De Vriend, 1984).

Irregularities in the bottom topography can have strong effects in wave models. This is the most spectacular in ray models (focusing, deflection): 'spaghetti' ray patterns are notorious in this field. But also more sophisticated wave models, especially those for regular waves, are rather sensitive to the topographical

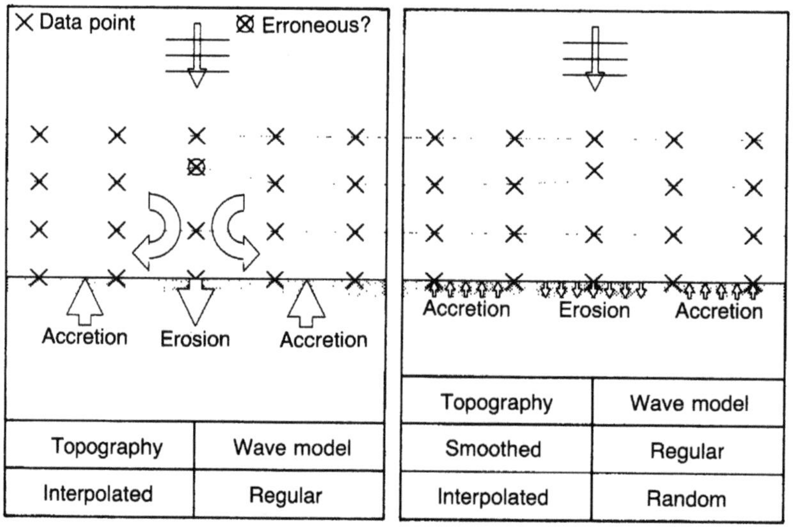

Figure 17.13 Sensitivity to bottom topography schematization due to regular wave propagation model.

schematization. Figure 17.13 illustrates how this can work out on the nearshore current pattern, and hence for the residual transport. There are therefore recommendations to run regular wave propagation models with an artificially smoothed bottom (Dingemans, 1985).

Wave and current models are not always using the same computational grids. Apart from mapping complications, this may give rise to problems if the topographic schematizations are different, especially near the coast. In that case, driving due to wave breaking may occur in the current model at water depths with which it does not comply, with erroneous currents as a consequence. The congruence of the nearshore topographic schematizations in the various constituent models therefore deserves special attention.

Digital terrain models (DTM) form a special class of topographic discretization techniques. In principle, the computational grid is adjusted to the accurate representation of the bottom topography, roughly with the finest grid where the relief is strongest (e.g. Westwood *et al.*, 1984). The criteria for the quality of the representation can be adjusted to the physical processes to be modelled, but they remain focused on the accuracy of the topographic description. In view of the foregoing, this can be in conflict with the objective of having an optimum representation of the physical processes so as to provide accurate residual transports. Therefore, the application of an accurate DTM is by no means a guarantee, and it may not even be necessary, for obtaining an accurate residual transport model.

17.5.6 SCHEMATIZATION OF CONDITIONS

One of the weakest points in residual transport modelling is the schematization of variable conditions, such as tides, storm surges, sea and swell. As was stated in section 17.4.3, the natural variation of these conditions does influence the residual transport, often through a non-linear process. On the other hand, running a compound, two-dimensional, ISE model on a real-time basis with a number of input scenarios is not feasible. Accordingly, we are forced to schematize the conditions, such that the yearly residual transport can be estimated on the basis of a practicable number of model runs.

It is possible that general systems theory will be able to provide an elegant solution for this problem: at the moment, we only have some rather contentious *ad hoc* methods available (Van Banning *et al.*, 1987).

The basic elements of this approach are the separate treatment of tides and waves, and the weighing in terms of contributions to the yearly residual transport in a 'representative' point in the model area.

Van Rijn (1989) describes a method to determine the representative tidal amplitude, such that an invariant tide of this amplitude gives the same yearly transport as the actual tide, which is subject to astronomic modulations and climatologic influences. The ultimate effect is a representative amplitude some 10% above the mean astronomic value. As long as the relationship between the tidal velocity and the transport rate is of the power-law type, with a constant power throughout the model area, this result is

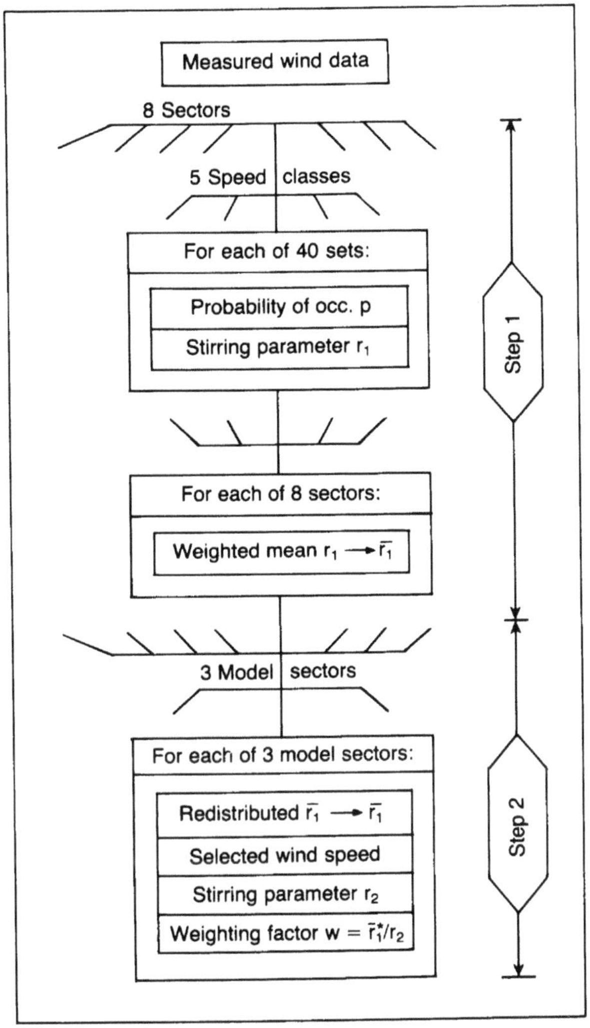

Figure 17.14 Schematization of wave conditions (after De Vriend *et al.*, 1989).

independent of the location of the 'representative' point. The presence of waves, however, tends to violate this condition.

Steijn (1988; Steijn *et al.*, 1989) describes a method to schematize the wave climate into a small number of combinations of wave height and direction. The basic idea is to select a number of wave sectors and to represent the waves from this sector by a single combination of wave height and direction. The model results for each sector, and those for calm weather, are multiplied by a weight factor and composed, to yield the yearly residual transport field.

The key elements in the determination of the representative wave height and the weight factor for each sector are two parameters associated with current-borne sediment transport: the 'stirring factor' (which is the same as Jonsson's (1966) wave-friction factor) and the 'longshore power' (*Shore Protection Manual*, 1984). The latter typically concerns longshore sand drift. It corresponds to a transport rate proportional to the 2.5th power of the wave height. Figure 17.14 outlines how this technique is applied.

The main objections against this approach are the spatial variability of the results (it can make quite a difference where the reference point is taken), and the focus on current-borne transport. Wave-asymmetry transport, for instance, varies with a much higher power of the wave height (≈ 6), and so leads to quite different weighting factors.

When using a transport formula in which the contributions of the various transport mechanisms can be separated (e.g. an elaborated form of the Bailard formula), this problem can be overcome by determining a weight factor for each of them (Steijn *et al.*, 1989).

17.5.7 BOUNDARY CONDITIONS

Boundary conditions derived from an ambient larger model are internally consistent (see section 17.5.4). Still, nesting can give rise to problems, especially if the physical contents of the models are not identical. Such a problem arises when nesting a model, including wave-induced forcing in an ambient tidal model, especially at boundaries intersecting a surf zone. If at such a boundary the velocity has to be prescribed, it is not appropriate to impose the tidal velocity alone. One could simply add a uniform longshore current profile, but that disregards the non-linearity of the flow resistance and leads to locally exaggerated sediment transports. The best option is to assume uniformity perpendicular to that part of the boundary and calculate the velocity distribution as if it were uniform flow with the same longitudinal slope as in the case of no waves at all, or just to assume no wave-induced forcing (which again remains a point of discussion). Such a preliminary calculation can easily be implemented.

If the water level distribution has to be given at the boundary, a similar problem arises. The water level distribution from the ambient tidal model will not comply with the inner part of the model due to the presence of wave-induced forces there. Figure 17.15 shows what can happen if this is ignored.

An option that is available to overcome this is again to assume uniformity perpendicular to the boundary and to calculate the adjustment of the water surface elevation as if it corresponded to wave set-up at a

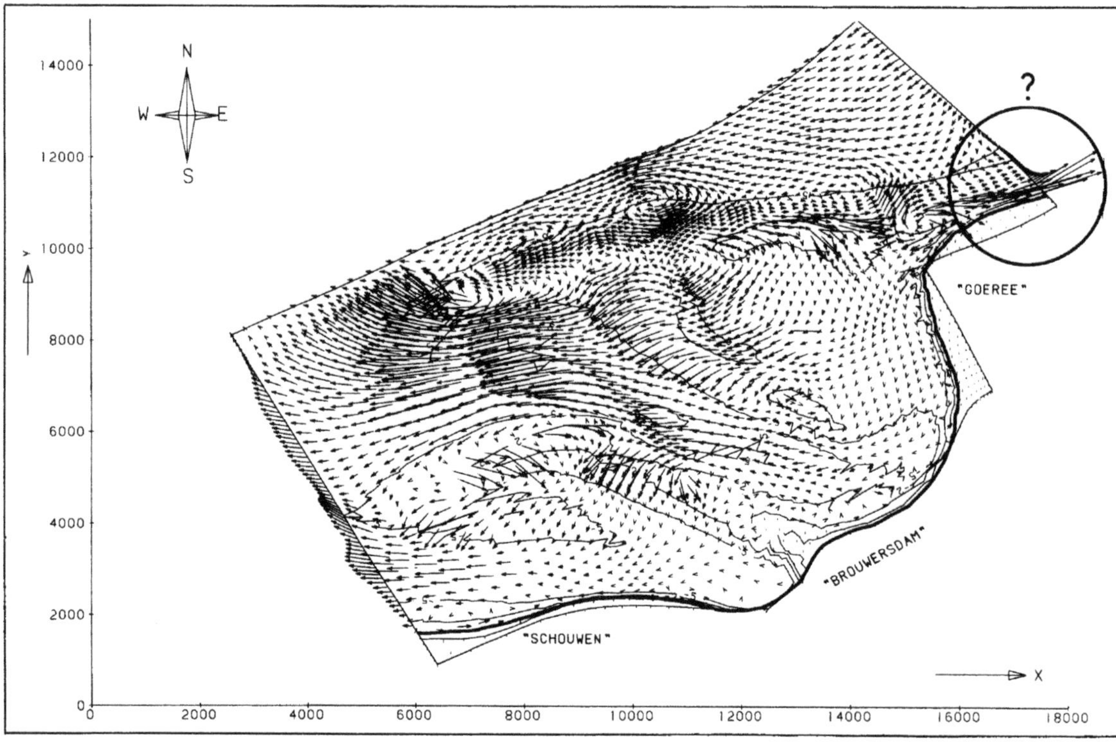

Figure 17.15 Effect of erroneous nesting at a water level boundary.

uniform coast. This set-up computation can easily be implemented on a PC. The result can simply be added to the water level distribution from the ambient model, since non-linearities play a minor part here.

17.5.8 VALIDATION

Validation is a conglomerate of activities, but it is primarily concerned with ascertaining that the model gives an appropriate description of reality in the specific case under consideration.

Part of it is calibration, i.e. the adjustment of model parameters, such that the results agree best with the observations. This used to be done by trial and error, but this becomes a virtually impossible task if there are many degrees of freedom, as is the case when using a spatially variable roughness parameter. Systems theory provides techniques (e.g. inverse scattering) for a more systematic calibration in these cases.

Another part of this process is numerical validation: does the numerical model accurately approximate the solution of the mathematical system it is supposed to solve? For the traditional, uniform grid, tidal models, much experience and generalized insight has now been obtained. In ISE models, and especially in morphodynamic models, we are dealing with much less transparent, highly non-linear, compound systems, in which the description of the tide is only one constituent. The know-how on tidal models is therefore unlikely to suffice.

Introduction of non-uniform grids provides not only new possibilities, but also new sources of inaccuracy, which have been insufficiently investigated (see, for example, the alarming work on internal reflections by Vichnevetsky, 1987). This means, that the accuracy of such models ought to be tested in every new application. The application of systematic test methods is still rare in this field, but perhaps practitioners can learn from other disciplines, such as meteorology (Dee and Da Silva, 1986). Every-day practice, however, is often different. If, in a practical application, this type of validation is given any attention at all, it usually all comes down to performing *ad hoc* numerical experiments, such as grid refinements and other adjustments of the model. Only too often, modellers rely exclusively on previous validation studies, at best in similar situations.

Physical validation is meant to assess the validity of

the model concept and its elaboration in the specific case at issue. This goes for each individual constituent model, as well as for their combination.

ISE models especially rely for their credibility on the physical validation of the constituent models, rather than upon their combined result: residual transports are difficult to calculate, but possibly even more difficult to measure. Still, constituent models should not be validated as if they were applied separately. Their role in the compound model and their interaction with the other constituents are essential aspects, and usually the most time-consuming ones in the validation phase of a model application. For example, tidal models are often validated against water level data. In an ISE model, however, the flow velocities and their time-variation are much more important aspects of the tidal motion. Therefore, these should be the parameters to be validated. Similarly, the principal result of a wave model in nearshore ISE models is not the wave height or direction, but rather the wave-induced effective stress field and the near-bed orbital motion. In general, the results of ISE-model components should be weighed by their contribution to the residual transport.

Physical validation proves to be a time-consuming task in many applications. It is worthwhile to minimize this part of the work, by pursuing maximum transparency of the system and by using systematic validation techniques. Here, again, we may learn something from other disciplines (Murphy and Wiliamson, 1976).

A typical example of how potential sources of errors can be eliminated is illustrated by Figure 17.16. It concerns the wave-induced driving forces due to the same wave field, formulated in terms of radiation-stress gradients and in terms of the dissipation rate (Dingemans *et al.*, 1987). The two force fields should yield identical current patterns. Although theoretically there is nothing wrong with the radiation stress concept, it is not transparent and is error prone, and therefore not to be recommended (De Vriend, 1984).

17.5.9 PRODUCTION AND RECOMBINATION OF RESULTS

Although the objective of schematizing the conditions (see section 17.5.6) is to reduce the number of model runs to an acceptable minimum, the total number of runs can be significant. Table 17.8 gives an example for an application in the Dutch Delta-area (Steijn *et al.*, 1989).

Clearly, such a model makes high demands upon the software environment, as well as on logging and administration facilities. Therefore, it is not surprising that the leading hydraulic laboratories are working with highly structured software systems for ISE models (Danish Hydraulic Institute: System 21; Hydraulics Research Ltd.: Tideway; Delft Hydraulics: AIM/COMOR).

The results of the production runs have to be composed via the weighted summation procedure indicated in section 17.5.6. This, again, requires a well organized and flexible data base. Once this is available, however, it is quite easy to compose results for different future climates on the basis of the same series of model runs: we only have to change the weight factors.

17.5.10 INTERPRETATION OF RESULTS

ISE models describe the residual transport field and the corresponding sedimentation and erosion pattern. The objective of the application, however, is rather the prediction of morphological evolutions. This

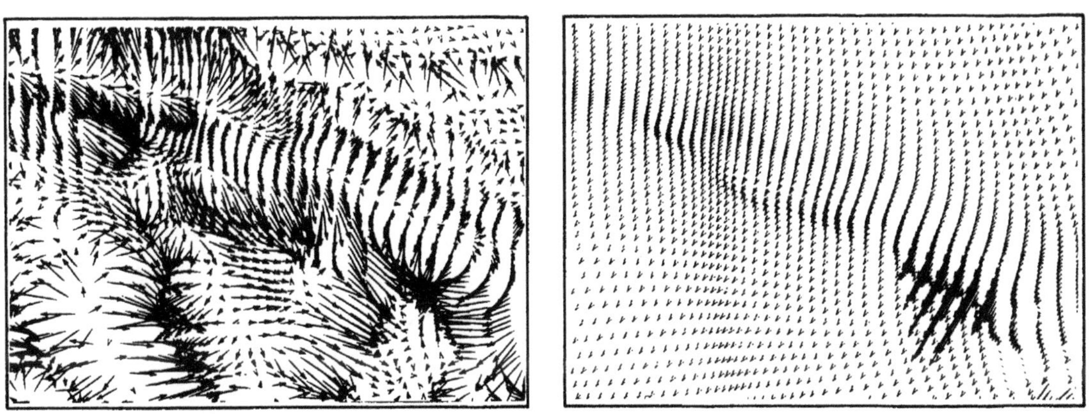

Figure 17.16 Transparency of wave-induced force fields.

Table 17.8 Constituent model runs in a practical ISE model application

			Geometry 1	*Geometry 2*
Overall tidal model			1	1
(Representative tide)		Unsteady		
Wave model	sector 1	stage 1		
	sector 2	·		
	sector 3	·	36	36
quasi-steady	no waves	stage 12		
Current model	sector 1			
(tide + waves)	sector 2			
	sector 3		4	4
unsteady	no waves			
Sediment transport	sector 1	stage 1		
model	sector 2	·		
(waves + current)	sector 3	·	48	48
quasi-steady	no waves	stage 12		
Sediment balance			4	4
		tide-averaged		

means, that we have to interpret the model results in morphological terms.

There is no standard recipe for this interpretation. The starting point is the model with all its inaccuracies, uncertainties and non-transparencies. To illustrate the latter, Figure 17.17 shows a typical result of an ISE model.

Obviously, we have to 'filter' the results, i.e. to eliminate unreliable information and retain what the model 'really' tells us. The resulting picture is doomed to be incomplete, rather than to provide ready-to-use answers to the practical questions at issue. Hence a model of this type is an aid to the engineer, rather than an all-purpose 'problem solver'.

In spite of all this doubt and hesitation, however, the model does provide useful information. It certainly helps to analyse the processes and mechanisms at work, and it often also allows us to identify the onset of a typical morphodynamic behaviour, the evolution of which we know from the past (hindcast), or from elsewhere (theory, models, observations). Besides, it shows the impact of the envisaged works, or whatever changes are studied, on the constituent processes and the residual transport.

On the basis of this information, we can predict the future morphological evolution, qualitatively and to some extent also quantitatively. The model helps us to do this with a much higher spatial resolution than we could ever achieve on the basis of measurements or otherwise.

Figure 17.18 gives an example of how to deal with the model results. Rather than the sedimentation and

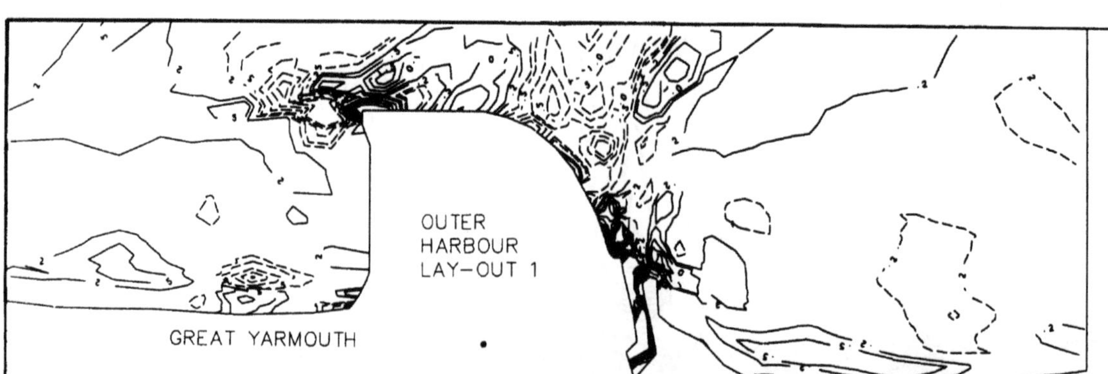

Figure 17.17 Typical sedimentation/erosion pattern resulting from an ISE-model.

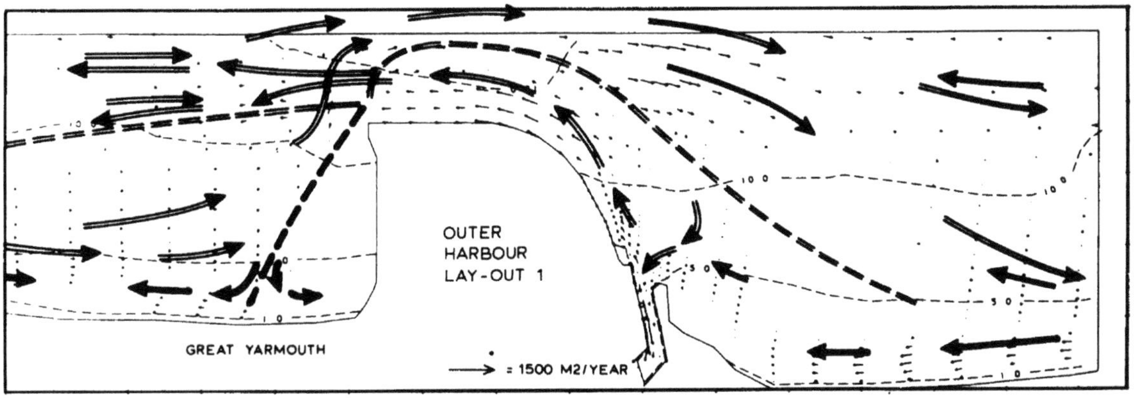

Figure 17.18 Interpretation of model results.

erosion pattern, the residual transport field is the best basis for the interpretation, if only because these transports are easier to interpret in terms of morphodynamic behaviour than the initial sedimentation and erosion rates.

For a quantitative prediction, we recommend that a hindcast phase be included in the study. This provides the ability to calibrate and verify the constituent models in the existing situation, and to validate the interpretation of model results against the past morphological evolution. Thus the role of the model in the prediction becomes rather a comparative one: to assess the impact of the envisaged works, relative to the natural evolution.

17.5.11 HYBRID MODELS

If we have sufficiently long and extensive records of observations in the area of interest and similar situations, we can formulate empirical models. A good example concerns the evolution of tidal-inlet systems, for which empirical models have been formulated on the basis of numerous data sets from all over the world (O'Brien, 1969; Gerritsen and De Jong, 1984). As compared with the models described in the previous sections, empirical models have a narrow physical basis. On the other hand, they provide the maximum agreement with nature, at the study site and during the observation period. Extrapolation into the future, after the works or the changes to be studied have taken place, requires much reasoning and interpretation.

In spite of their limitied predictive capabilities, these models can play a part in engineering predictions. They contain much information, mostly complementary to the contents of a mathematical-physical model. Hence a hybrid approach, combining the two types of model, can be attractive. Figure 17.19 shows how such a hybrid model can function (De Vriend *et al.*, 1989). Note that there is a multiple two-way interaction between the two.

17.5.12 MORPHODYNAMIC SIMULATION MODELS

Because of their inherent limitations, ISE models are likely to be replaced sooner or later by morphodynamic simulation models. In river engineering, where morphodynamic modelling is somewhat further advanced, ISE models are hardly applied, at all. Also in the one-dimensional modelling area, where morphodynamic models are easier to realise, ISE models hardly play a part. Two-dimensional morphodynamic models for coastal, estuarial and marine situations, however, are still rare, especially for coastal applications, and they usually only include depth-averaged constituent models. Besides, they are still lacking a solid theoretical basis for the morphodynamics: too little is known about the inherent behaviour of morphodynamic systems like these (De Vriend, 1987a,b). Taking river dynamics as an example again, this theoretical basis is vital to the development of numerical simulation models (Struiksma *et al.*, 1985; Ikeda and Parker, 1989).

Though still tentatively and with a research approach, various institutes are using morphodynamic models in two horizontal dimensions at the moment. For instance: Laboratoire National d'Hydraulique (France) has been one of the first to develop and investigate two-dimensional models for coastal and tidal areas (Lepetit and Hauguel, 1978; Latteux, 1987; Péchon, 1988). Sir William Halcrow and Partners (UK) have long used a model based on rather simplified constituent models (Fleming and Hunt, 1976). Tokyo University has developed a model (Watanabe,

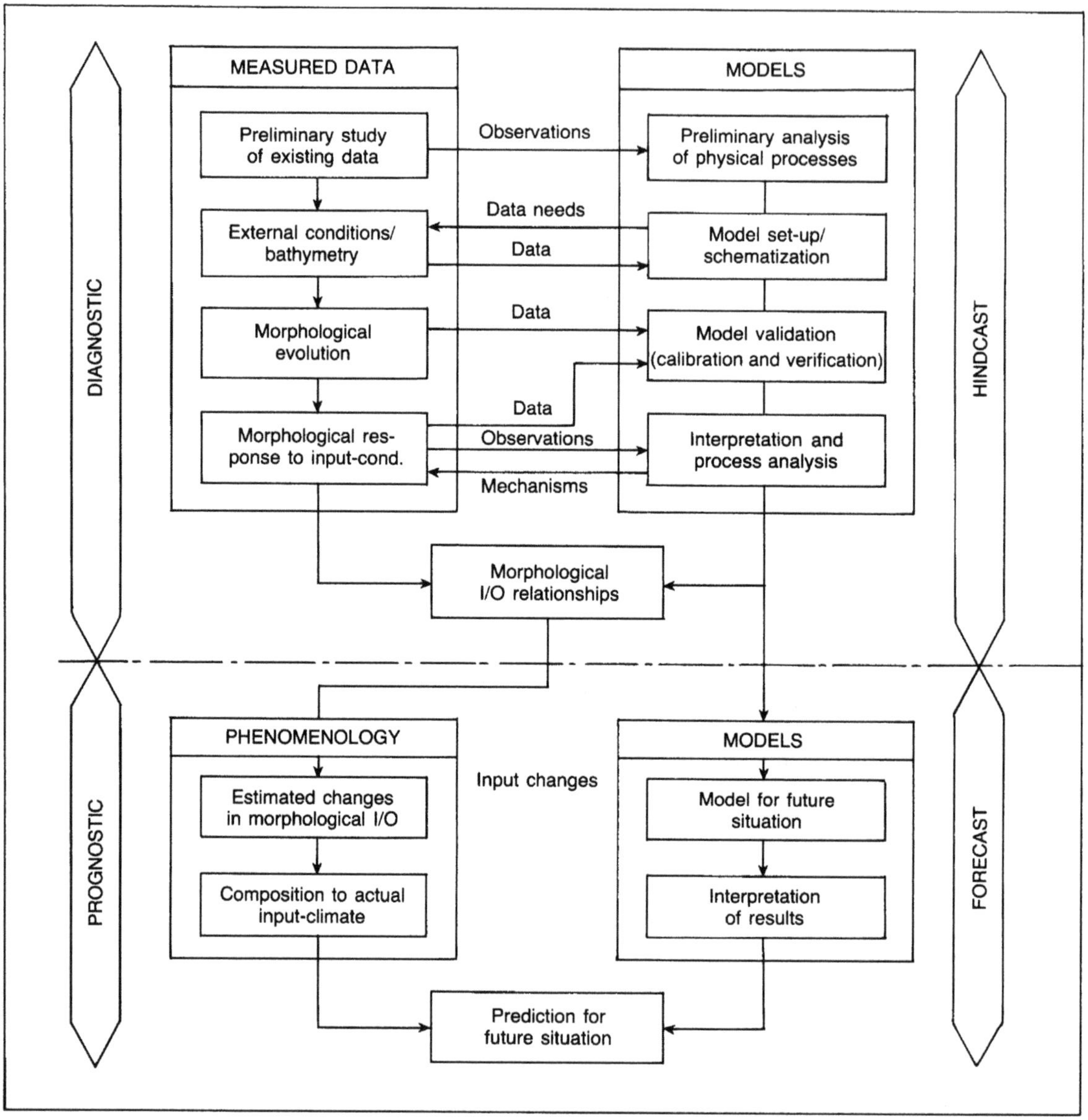

Figure 17.19 Hybrid prediction procedure (after De Vriend *et al.*, 1989).

1985), which is now being used for short-term predictions in practical situations (Maruyama and Takagi, 1988). Danish Hydraulic Institute is developing a coastal model (Andersen *et al.*, 1988), which has been used rather cautiously in practical applications. The Waterways Experiment Station of the US Army Corps of Engineers (Vicksburg, USA) has two systems, one of which concerns estuaries and coastal zones (McAnally *et al.*, 1984, 1986); Delft Hydraulics is developing a 2.5-dimensional system for a wide range of applications, including rivers (Struiksma *et al.*, 1985), estuaries (Wang, 1989) and coasts (De Vriend and Ribberink, 1988) (see also De Vriend *et al.*, 1993); besides, this development involves a first test of a tide-integrated model concept. Various universities are also experimenting with these kinds of model, for example: Liverpool (O'Connor and Nicholson, 1989); Delft (Wang, 1989); Tokyo

(Watanabe, 1985); Osaka (Deguchi and Sawaragi, 1988); Gdansk (Zeidler, 1988); Rome (Gallerano and Rufini, 1989); and Padova (Di Silvio and Gambolati, 1990). In view of the expected role of morphodynamic modelling in the future, these developments deserve to be supported and stimulated.

17.6 CONCLUSIONS

The principal conclusion to be drawn from the foregoing is that two-dimensional and 2.5-dimensional models of sediment transport and morphological changes due to combined waves and currents are useful tools in coastal, estuarial and harbour engineering, provided that they be utilized with care.

Their application requires a thorough insight into the physical background of the constituent models and their interactions, and sufficient experience to avoid the pitfalls. These models also require a certain structuring of software and data transfer. Hence they are not suitable for incidental use: they involve the application of trained personnel and, if necessary, the establishment of a software infrastructure.

There is much more research and development work to be done on morphodynamic models, both on the side of their theoretical basis and on the side of their interpretation, their extrapolation to long-term behaviour, and the periphery of their applications. The worldwide effort in this field indicates that this is a viable development, worth further advancement.

REFERENCES

Andersen, O.H., Hedegaard, I.B., Deigaard, R. *et al.* (1988) Model for morphological changes under waves and current. *Proceedings of the IAHR Symposium on Mathematical Models of Sediment Transportation in the Coastal Zone*, Copenhagen, Denmark, pp. 310–19.

Bailard, J.A. (1981) An energetics total load sediment transport model for a plane sloping beach. *Journal of Geophysical Research*, **86** (C11), 10938–54.

Bijker, E.W. (1971) Longshore transport computations. *Journal of the Waterways, Harbour and Coastal Engineering Divison, HSCE*, **97** (WW4), 687–701.

Bijker, E.W., Van Hijum, E. and Vellinga, P. (1976) Sand transport by waves. *Proceedings of the 15th ICCE*, Honolulu, Hawaii, pp. 1149–67.

Blondeaux, P. and Vittori, G. (1989) On sediment transport over a rippled bed. *Proceedings of the XXIIIrd IAHR Congress*, Ottawa, Canada, pp. C.309–16.

Boer, S., de Vriend, H.J. and Wind, H.G. (1984) A mathematical model for the simulation of morphological processes in the coastal area. *Proceedings of the 19th ICCE*, Houston, Texas, pp. 1437–53.

Booij, R., Pennekamp, J.G.S. and Kalkwijk, J.P.Th. (1985) Secondary flow in the Eastern Scheldt. Delft University of Technology, Fluid Mechanics Laboratory, *Rep. 16–85* (in Dutch).

Dee, D.P. and Da Silva, A.M. (1986) Using Hough harmonics to validate and assess nonlinear shallow-water models. *Monthly Weather Review*, **114** (11), 2191–96.

Deguchi, I. and Sawaragi, T. (1988) Effects of structure on deposition of discharged sediment around rivermouth. *Proceedings of the 21st ICCE*, Malaga, Spain, pp. 1573–87.

Deigaard, R., Fredsoe, J. and Broker Hedegaard, I. (1986) Suspended sediment transport in the surf zone. *Journal of Waterway, Port, Coastal and Ocean Engineering*, **112** (1), 115–28.

De Vriend, H.J. (1984) Compound 2-DH mathematical models in coastal morphology. Delft Hydraulics Laboratory, *Rep. R1747-I.*

De Vriend, H.J. (1986) 2DH computation of transient sea bed evolutions. *Proceedings of the 20th ICCE*, Taipei, Taiwan, pp. 1689–1712.

De Vriend, H.J. (1987a) 2DH mathematical modelling of morphological evolutions in shallow water. *Coastal Engineering*, **11** (1), 1–27.

De Vriend, H.J. (1987b) Analysis of 2DH morphological evolutions in shallow water. *Journal of Geophysical Research*, **92** (C4), 3877–93.

De Vriend, H.J., Louters, T., Berben, F. *et al.* (1989) Hybrid prediction of a sandy shoal in a mesotidal estuary, in *Hydraulic and Environmental Modelling of Coastal, Estuarine and River Waters*, (eds R. Falconer *et al.*) Gower Technical, Aldershot, UK, pp. 145–56.

De Vriend, H.J. and Ribberink, J.S. (1988) A quasi-3D mathematical model of coastal morphology. *Proceedings 21st ICCE*, Malaga, Spain, pp. 1689–1703.

De Vriend, H.J., Zyserman, J., Nicholson, J., Roelvink, J.A., Péchon, Ph. and Southgate, H.N. (1993) Medium-term 2-OH coastal area modelling. Accepted for publication in *Coastal Engineering*.

De Waal, J.C.M. (1987) Investigation on Bailard's sand transport formula, applied to cross-shore transport. *Master's Thesis*, Delft University of Technology, (in Dutch).

Dingemans, M.W. (1985) Surface wave propagation over an uneven bottom; evaluation of two-dimensional wave propagation models. Delft Hydraulics Laboratory, *Rep. W301-5*.

Dingemans, M.W., Radder, A.C. and De Vriend, H.J. (1987) Computation of the driving forces of wave-induced currents. *Coastal Engineering*, **11** (5&6), 539–63.

Di Silvio, G. and Gambolati, G. (1990) Two-dimensional model of the long-term morphological evolution of tidal lagoons. *Proceedings of the International Conference on Computational Methods in Water Research*, Venice, Italy (in press).

Fleming, C.A. and Hunt, J.N. (1976) Application of a sediment transport model. *Proceedings of the 15th ICCE*, Honolulu, Hawaii, pp. 1184–1202.

Fredsøe, J. (1993) Modelling of non-cohesive sediment transport processes in the marine environment. Accepted for publication in *Coastal Engineering*.

Gallerano, P. and Rufini, P. (1989) Sediment erosion and deposition around a groin perpendicular to coastline. *Proceedings of the XXIIIrd IAHR Congress*, Ottawa, Canada, pp. C.177–184.

Gerritsen, F. and De Jong, H. (1984) Stability parameters of Western Scheldt estuary. *Proceedings of the 19th ICCE*, Houston, Texas, pp. 3078–93.

Guza, R.T. and Thornton, E.B. (1986) Velocity moments in the nearshore. *Journal of Waterway, Port, Coastal and Ocean Engineering*, **111** (2), 235–56.

Ikeda, S. and Parker, G. (eds) (1989) *River Meandering*, Am. Geoph. Union, Water Research Monograph 12.

Jonsson, I.G. (1966) Wave boundary layers and friction factors. *Proceedings of the 10th ICCE*, Tokyo, Japan, pp. 127–48.

Justesen, P., Fredsoe, J. and Deigaard, R. (1986) The bottleneck problem for turbulence in relation to suspended sediment transport in the surf zone. *Proceedings of the 20th ICCE*, Taipei, Taiwan, pp. 1225–39.

Katopodi, I. and Ribberink, J.S. (1988) Quasi-3D modelling of suspended sediment transport by currents and waves. Delft Hydraulics, *Rep. H460*.

Katopodi, I. and Ribberink, J.S. (1990) A quasi-3D model for suspended sediment transport by currents and waves. *Proceedings of the 22nd ICCE*, Delft, The Netherlands (in preparation).

Katopodi, I. and Ribberink, J.S. (1992) Quasi-3D modelling of suspended sediment transport by currents and waves. *Coastal Engineering*, **18**, 83–110.

Latteux, B. (1987) Transport modelling of particulate matter – methodology of long-term simulation of bed evolutions. Lab. Nat.

d'Hydr., Chatou, France, *Rep. HE-42/87.25* (in French).

Lepetit, J.P. and Hauguel, A. (1978) A numerical model of sediment transport. *Proceedings of the 16th ICCE*, Hamburg, Germany, pp. 1715–24.

Madsen, O.S. and Grant, W.D. (1976) Sediment transport in the coastal environment. MIT, R.M. Parsons Laboratory, *Rep. 209*.

McAnally, W.H., Letter, J.V., Stewart, J.P. *et al.* (1984) Application of Columbia hybrid modelling system. *Journal of Hydraulic Engineering* **110** (5), 627–42.

McAnally, W.H., Letter, J.V. and Thomas, W.A. (1986) Two and three-dimensional modelling systems for sedimentation. *Proceedings of the 3rd International Symposium on River Sedimentation*, Jackson, MS., pp. 400–11.

Maruyama, K. and Takagi, T. (1988) A simulation system of near-shore sediment transport for the coupling of the sea-bottom topography, waves and currents. *Proceedings of the IAHR Symposium or Mathematical Modelling on Sediment Transport Coastal Zone*, Copenhagen, Denmark, pp. 300–9.

Murphy, A. and Williamson, D. (eds) (1976) *Weather forecasting and weather forecasts: models, systems and users* – Vol. 2. National Center for Atmospheric Research, Boulder, Colorado, 450 pp.

Nielsen, P. (1988) Three simple models of wave sediment transport. *Coastal Engineering*, **12** (1), 43–62.

O'Brien, M.P. (1969) Equilibrium flow areas of inlets on sandy coasts. *Journal of Waterways, Harbours and Coastal Engineering*, **15** (WW1), 43–52.

O'Connor, B.A. and Nicholson, J. (1989) Modelling changes in coastal morphology. *Proceedings of the International Symposium on Sediment Transport Modelling*, New Orleans, LA, pp. 160–5.

Péchon, P. (1988) Comparison of a numerical model for longshore sediment transport with measurements, in *Computer Modelling in Ocean Engineering*, (eds B.A. Schreffler and O.C. Zienkiewicz) Balkema, Rotterdam, pp. 451–60.

Ribberink, J.S. and Al-Salem, A. (1990) Bedforms, sediment concentrations and sediment transport in simulated wave conditions. *Proceedings of the 22nd ICCE*, Delft, The Netherlands (in preparation).

Rienecker, M.M. and Fenton, J.D. (1981) A Fourier approximation method for steady waves. *Journal of Fluid Mechanics*, **104**, 119–37.

Roelvink, J.A. and Stive, M.J.F. (1989) Bar-generating cross-shore flow mechanisms on a beach. *Journal of Geophysical Research*, **94** (C4), 4785–800.

Shore Protection Manual (1984) Waterways Experiment Station, US Army Corps of Engineers, Vicksburg, MS.

Steijn, R.C. (1988) Schematization of the natural conditions in multi-dimensional numerical models of coastal morphology. Delft Hydraulics, *Rep. H526-1*.

Steijn, R.C., Louters, T., Van der Spek, A.J.F. *et al.* (1989) Numerical model hindcast of the ebb-tidal delta evolution in front of the Deltaworks, in *Hydraulic and Environmental Modelling of Coastal, Estuarine and River Waters*, (eds R. Falconer, *et al.*) Gower Technical, Aldershot, UK, pp. 255–64.

Stive, M.J.F. (1986) A model for cross-shore sediment transport. *Proceedings of the 20th ICCE*, Taipei, Taiwan, pp. 1551–64.

Struiksma, N., Olesen, K.W., Flokstra, C. *et al.* (1985) Bed deformation in curved alluvial channels. *Journal of Hydraulic Research*, **23** (1), 57–79.

Van Banning, G.K.F.M., de Vriend, H.J. and Boer, S. (1987) Schematization and validation of 2DH mathematical models in coastal morphology. *Proceedings of the Coastal Sediments '87 Conference*, New Orleans, LA., pp. 600–15.

Van Rijn, L.C. (1989) Handbook of Sediment Transport by Currents and Waves. Delft Hydraulics, *Rep. H461*.

Vichnevetsky, R. (1987) Wave propagation analysis of difference schemes for hyperbolic equations: a review. *International Journal for Numerical Methods in Fluids*, **7**, 409–52.

Wang, Z.B. (1989) Mathematical modelling of morphological processes in estuaries. *Doctoral thesis*, Delft University of Technology.

Wang, Z.B. and Ribberink, J.S. (1986) The validity of a depth-integrated model for suspended sediment transport. *Journal of Hydraulic Research*, **24** (1), 53–67.

Watanabe, A. (1985) Three-dimensional predictive model of beach evolution around a structure. *Proceedings of the Symposium on Water Wave Research*, Hannover, Germany.

Westwood, I., Holz, K.P. and Ratke, R. (1984) On the application of digital surface representations in hydrodynamic modelling. *Proceedings of the 5th International Conference on Finite Elements in Water Resources*, Burlington, Vermont.

Zeidler, R.B. (1988) Modelling of coastal processes and shore evolution. *Proceedings of the IAHR Symposium on Mathematical Modelling of Sediment Transport in the Coastal Zone*, Copenhagen, Denmark, pp. 240–49.

ONE-DIMENSIONAL MODELS OF SEDIMENT TRANSPORT UNDER THE INFLUENCE OF CURRENTS 18

J.A. Cunge and J.L. Rahuel

Once again we arrive at the most schematized of all dynamic modelling tools, the one-dimensional variety, but now at the level of sediment transport. As the influence of wave action is usually very small in such models, this is not considered here. An outline is provided of the basic problems that arise in this area and some typical tools that are now available to solve these problems are introduced.

CONTENTS

18.1 INTRODUCTION

A universal tool for numerical simulation of one-dimensional unsteady morphological processes in the coastal zone is not, as yet, available. In what follows, the reader will find more constraints and limitations upon using various existing simulation codes than cases of their applicability. Indeed, the essential phenomena in movable-bed river hydraulics have not yet been synthesized in mathematical formulation.

A modelling tool in this context (Cunge and Holly, 1986) should have the following features:

1. Simulate unsteady processes in multiply-connected networks of rivers and channels, i.e. furnish the time-dependent variations of water discharge, water stage, mean velocity, bed-load and suspended-load discharges, bed elevation and cross-sectional shape at every point.
2. Be sufficiently general for use on flat tidal deltas as well as on steeper inland reaches, taking into account graded sediments and capable of simulating both short-term and long-term changes.
3. Be 'engineer friendly', relieving the user of topological constraints and the need for specialised knowledge of the software while setting-up, extending or running the model.
4. Be reliable in terms of robustness, furnishing results which are qualified as to their numerical accuracy.
5. Take into account specific coastal-zone problems such as mixed cohesive and sand beds, flocculation of suspended sediments as a function of salinity, etc.

All morphological changes result from non-uniform sediment transport and water flow as well as changes in sediment properties. These processes are not sufficiently well understood for a reliable and complete tool to be developed. Consequently, simplifications are unavoidable, and non-industrialized prototype-like codes have to be used which cannot be applied without a proper understanding of their limitations. Most often, tools developed for rivers unaffected by

Coastal, Estuarial and Harbour Engineer's Reference Book. Edited by M.B. Abbott and W.A. Price. Published in 1993 by Chapman & Hall, 2–6 Boundary Row, London SE1 8HN. ISBN 0 419 15430 2.

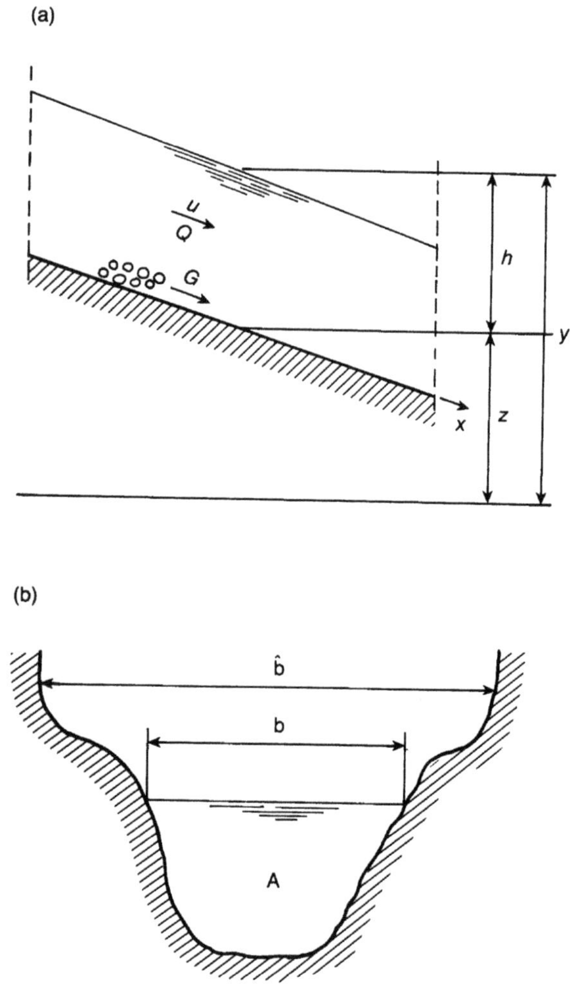

Figure 18.1 Definition sketch for one-dimensional flow in movable bed river. (a) longitudinal profile; (b) cross-section.

the sea are used in coastal areas, because the physical complexity of these areas is so high that specific maritime problems cannot be taken into account. In most of the numerical codes developed so far, the dependent variables describing the process are the flow variables, the sediment transport and the bed level. These variables are interdependent and should normally be considered as a whole.

But a mathematical model can be regarded as an assembly with three components (Nordin, 1988):

1. Flow component, for which the main unresolved factor is the roughness formulation, which is of far greater importance for sediment and morphological processes than for water flow simulation.
2. Transport component: since there are almost as many transport functions (formulae) as there are modellers, it follows that, depending on the opinion of which seems to be the most reasonable, one or several of these are better than all others for a given river. Hence the role of the modeller, who should be allowed a freedom of choice of the transport function to be applied in any particular application of the modelling tool.
3. Procedures for taking into account bed composition which are the basis for the graded granulometry approach and which are still very crude. For many, if not most maritime cases, the granulometry of the sediment is relatively uniform (except for cohesive sediments and flocculations), and this makes things simpler in the present chapter. However, if sediment size cannot be considered uniform, then it may have a considerable influence and severe simulation errors may arise if size uniformity is assumed when it does not in fact obtain.

Summarizing, numerical simulation of morphological processes, even limited to one-dimensional situations, is still in its infancy and, as described below, for each specific problem several approaches (can we speak of tools in such situations?) are available. Engineering judgement is needed every time in order to decide which approach should be chosen.

18.2 BASIC HYPOTHESES AND EQUATIONS

The simplest acceptable mathematical description of a one-dimensional situation is summarized by the following system of equations (see Figure 18.1):

1. Continuity equation for water flow:

$$\frac{\partial h}{\partial t} + u\frac{\partial h}{\partial x} + h\frac{\partial u}{\partial x} = 0 \tag{18.1}$$

2. Dynamic equation for liquid flow:

$$\frac{\partial u}{\partial t} + u\frac{\partial u}{\partial x} + h\frac{\partial h}{\partial x} + g\frac{\partial z}{\partial x} + gS = 0 \tag{18.2}$$

3. Continuity equation for sediment transport

$$(1 - n)\bar{b}\frac{\partial z}{\partial t} + \frac{\partial G}{\partial x} = 0 \tag{18.3}$$

together with the sediment transport formula

$$G = G(u, h, d, \ldots) \tag{18.4}$$

where: (x, t) are space and time independent variables, u and h are, respectively, the velocity and depth of water, z is the bed level elevation above datum, S is the steady-state energy line slope (friction

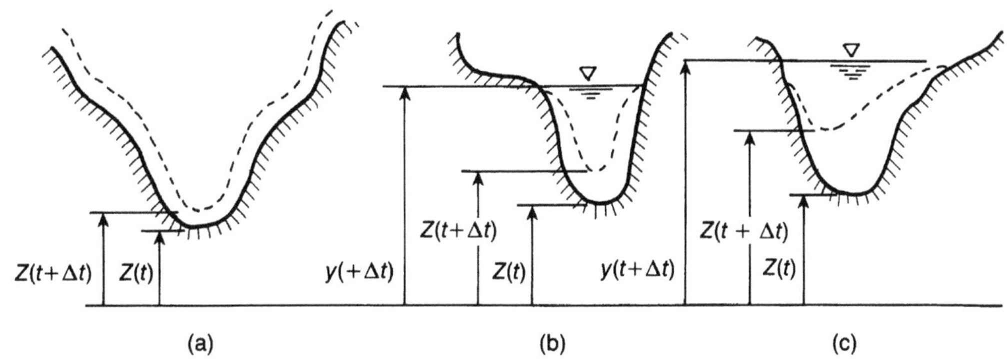

Figure 18.2 Hypotheses relating cross-sectional and bed elevation changes. (a) entire cross-section rises; (b) cross-section affected only below water surface; (c) lateral distribution of deposition assumed.

slope), $\bar{b}$ is the width affected by bed load transport, d represents the river bed and sediment characteristics, and n is the porosity. Equation (18.4) is usually a 'total-load formula' which encompasses both suspended-load and bed-load and, usually, does not take into account the grading of the sediments (i.e. a uniform, constant-diameter material is assumed).

Equations (18.1)–(18.4) do not take into account the influence that a suspended solid phase can have on the flow. They are based on the de Saint-Venant hypotheses (see Chapter 11 of this volume). Any conceptual model based on these equations should be supplemented by laws relating the variations of cross-section to deposition or erosion, e.g. such as shown in Figure 18.2.

The friction slope S can be an explicit function of the flow and river bed characteristics, such as is expressed by the Manning–Strickler formula. Such a representation is very crude when the bed is composed of fine sand. Indeed, for alluvial streams, the hydraulic relations should allow for the existence of bed forms, and resistance to flow is variable according to the discharge, depth and recent history of bed evolution (due to the presence of ripple and dunes). In this case there is no explicit expression for the friction slope; instead an implicit non-linear equation should be considered, of the form:

$$\varphi(S, u, h, d) = 0 \tag{18.5}$$

The interrelations between various groups of variables gives rise to a general scheme of a computational procedure in a numerical simulation, as shown in Figure 18.3a (Wang, 1989).

Until recently, this procedure was rarely adopted for numerical codes because of its complexity and the equations were often uncoupled, i.e., during certain time intervals, Equations (18.1) and (18.2) are considered to be independent of Equations (18.3) and (18.4) and *vice versa*. In this approach the bed level is supposed to be invariable during the flow computation and, conversely, during the bed level computation the flow is supposed to be invariable, as shown conceptually (Wang, 1989) in Figure 18.3b.

The next simplification in the formulation concept was the quasi-steadiness of the flow (de Vries, 1959, 1965). Neglecting h- and Q- derivatives with respect to time leads to a much simpler system, such as Equation (18.6), composed of steady-state liquid flow and unsteady sediment transport equations:

$$u\frac{\partial u}{\partial x} + g\frac{\partial h}{\partial x} + g\frac{\partial z}{\partial x} + g\frac{1}{C^2}\frac{u|u|}{h} = 0$$

$$\frac{\partial z}{\partial t} + \frac{\partial G}{\partial x} = 0$$

$$uh = q = \text{const}$$

$$G = G(u, h, d, \ldots) \tag{18.6}$$

In general, the bed level varies much more slowly than the water flow and this observation justifies such an approach.

Most of the existing tools are based on these hypotheses and equations, and there are situations in which codes based on such simplified concepts of mobile-bed processes can give reliable results. However, such simplified approaches must be used carefully and only in appropriate situations. While in most riverine studies the one-dimensional approach can be used, the assumption that the sediment geometry can be represented by a single particle diameter may produce misleading predictions. A similar remark can be made regarding the assumption of quasi-steady water flow when inertial forces are significant with respect to gravity forces. The total-load approach, which makes no distinction between the fundamentally different processes of bedload and

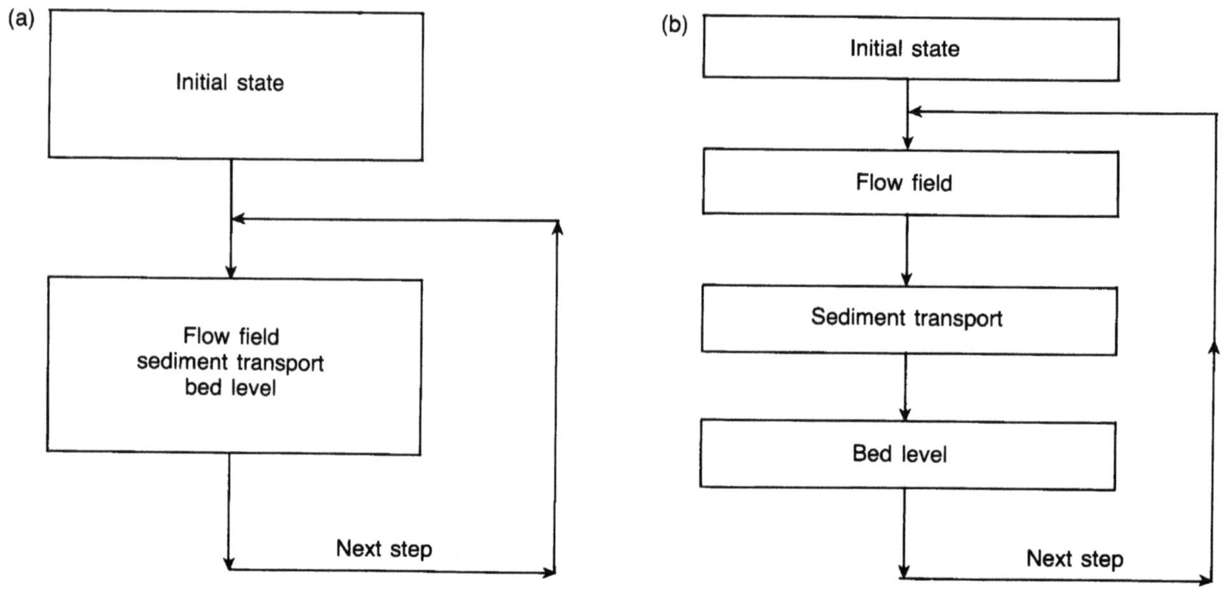

Figure 18.3 Computation procedure; (a) morphological model without simplifications; (b) uncoupled morphological model.

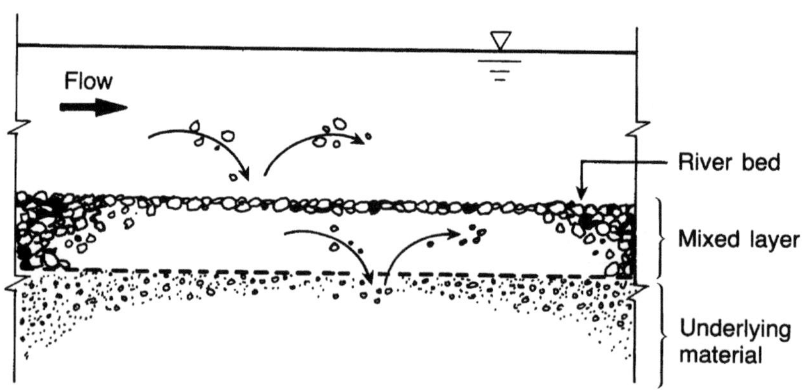

Figure 18.4 Schematic representation of mixed layer.

suspended load transport, may give very distorted predictions when applied to areas of current deceleration and deposition of sediment. Most of the existing codes are based on the assumption that in each section of the river the sediment transport is in equilibrium with respect to the local hydraulic conditions; this view of the process may also distort the predictions and is notoriously inaccurate in estuaries. Recent developments have led to more ambitious formulations based on the following imperatives:

1. Recognition of the fact that bed-load moves at a relatively slow 'kinematic wave'-type propagation velocity, while suspended-load moves at the order-of-magnitude-greater water velocity.
2. Accommodation of the possibility that sediment of any particular size class may be transported as wash-load (with no interaction with bed material), suspended-load or bed-load or, on the other hand, that it may be immobile, depending on the local hydraulic conditions.

3. Incorporation of recent research on non-equilibrium transport effects.
4. Representation of the bed-material mixed-layer as the transfer layer between the transported material and the lower bed strata (see Figure 18.4).
5. Adoption of the full momentum- and mass-conservation equations for water flow, allowing treatment of inertia-dominated unsteady situations, and, in particular, reversing flow in estuaries.

The equations resulting from this approach are as follows (Holly, 1988; Holly and Rahuel, 1989a):

Water mass conservation

$$\frac{\partial(\rho A)}{\partial t} + \frac{\partial(\rho Q)}{\partial t} = \rho_0 q \tag{18.7}$$

Water momentum conservation

$$\frac{\partial(\rho Q)}{\partial t} + \frac{\partial}{\partial x}\left(\rho\frac{Q^2}{A}\right) + gA\frac{\partial}{\partial x}(\rho y) + \rho g A\frac{Q|Q|}{D^2} = 0 \tag{18.8}$$

Suspended-load transport

$$\frac{\partial}{\partial t}(C_jA) + \frac{\partial}{\partial x}(C_jQ) = \frac{\partial}{\partial x}(A\psi\frac{\partial C_j}{\partial x}) + S_j \; ; \; j = 1,2, \ldots, J \tag{18.9}$$

Bed-load loading law

$$\frac{\partial G_j}{\partial t} = K_j(G_j^* - G_j) + (G_j/G_j^*)\frac{\partial G_j^*}{\partial x} \; ; \; j = 1,2, \ldots, J \tag{18.10}$$

Global bed-material conservation

$$(1-p)\frac{\partial \Gamma}{\partial t} + \sum_{j=1}^{J}\frac{\partial G_j}{\partial x} + \sum_{j=1}^{J} S_j = 0 \tag{18.11}$$

or

$$\sum_{j=1}^{J} \beta_j = 1 \tag{18.12}$$

Bed-material sorting

$$(1-p)\frac{\partial}{\partial t}(\beta_j BE) + \frac{\partial G_j}{\partial x} + S_j + (1-p)\beta_j^*\frac{\partial}{\partial t}[(z - z_0 - E)\beta] = 0$$

$$j = 1,2, \ldots, J \tag{18.13}$$

In these equations, x = streamwise co-ordinate; t = time: $\rho(C)$ = density of water-sediment mixture: $C_j(x, t)$ = suspended-load concentration of size class j; $C = \Sigma C_j$; $q(x, t)$ = lateral inflow; $Q(x, t)$ = water discharge; $A[y(x, t), z(x, t)]$ = flow area; $y(x, t)$ = water surface elevation; $D[z(x, t), y(x, t)]$ = channel conveyance; $\Psi(x, t)$ = longitudinal dispersion coefficient for suspended sediment; $S_j(C_j(x, t), Q(x, t), A(x, t), \ldots)$ = suspended-sediment source term for size-class j; $G_j(x, t)$ = bed-load transport for size-class j; $K_j(x, t)$ = loading coefficient for size class j; p = sediment porosity; $\Gamma(x, t)$ = bed-material area above a certain datum; $\beta_{0,j}$ = fractional representation of size-class j in the bed material underlying the active layer; $\beta_j^*(x, t)$ = fractional representation of size-class j in the active bed layer; $\beta_j^* = \beta_j$ if the floor of the active layer is rising, or β_{0j} otherwise; $B[y(x, t), z(x, t)]$ = active bed-evolution channel width, $z[\Gamma(x, t)]$ = bed elevation; $E[Q(c, t), A(x, t), \ldots]$ = active-layer depth; J = total number of discrete sediment size classes used to represent the non-homogeneous material.

Equations (18.7) and (18.8) are the conventional de Saint-Venant equations for unsteady one-dimensional flow. The significant departure from their usual fixed-bed context is the dependence of the area A and conveyance D on the variable thalweg elevation, as well as the conveyance dependence on a variable alluvial friction factor as evaluated by an appropriate empirical expression. The density dependence on suspended-sediment concentration is normally very weak, but is included for generality.

Equation (18.9) represents the advective/diffusive transport of suspended sediment for each size-class j. It is implicitly assumed in Equation (18.9) that the grains in suspension are advected downstream at the cross-sectional water velocity, although this assumption could be made less rigid by introducing an appropriate functional relation between the mean water velocity and mean grain transport velocity. The source-term S_j, representing suspended-sediment exchange with the bed surface, comprises two independent processes: deposition and entrainment, which may or may not be in equilibrium.

Equation (18.10) reflects recent attempts to quantify the asympototic approach of actual bed-load transport to the equilibrium capacity. The loading coefficient K_j is recognized by researchers to be inversely proportional to the particle size d_j and, to some extent, of the excess bed shear stress over that required for initiation of motion, although only tentative empirical expressions have been suggested for the quantification of K_j. Equation (18.10) suggests that for large values of K_j (i.e., small particle diameter and/or conditions just above incipient motion), the actual bed-load G_j approaches the equilibrium capacity G_j^* rapidly. As K_j decreases (large particle diameter and/or high particle mobility), G_j may differ significantly from G_j^*.

Equation (18.11) is the conventional transport equation for conservation of bed material, but with separate recognition of the roles of bed-load gradient and net suspended-load acquisition of material through the source term. Equation (18.12) is, at least in principle, an equivalent statement of global bed-sediment mass conservation at any location.

Equation (18.13), the so-called 'sorting equation', can be recognized as another form of Equation (18.11), but written for each size-class j and for an 'active layer' control volume of vertical thickness E below the bed surface. The second and third terms of Equation (18.13) represent sediment fluxes through the active-layer ends and ceiling (i.e. the bed surface). The fourth term represents the flux of material across the floor of the active layer, with β_j^* representing either the active-layer fraction b_j for a rising floor, or the underlying material β_{0j} for a descending floor.

18.3 NUMERICAL SOLUTIONS

The numerical solutions on which various tools are built depend on the equations and on the choice of algorithms. Only the main lines and principles of the methods encountered will be given below.

The system of Equations (18.1)–(18.4) is usually uncoupled (see Figure 18.3b). The water flow equations can then be solved by any method (see Chapter 11 of this volume used for one-dimensional models), while the first order partial differential transport equation is usually solved by a finite difference scheme. An uncoupled numerical algorithm of the solution to these equations is described elsewhere (Cunge *et al.*, 1980).

There are two well-defined finite difference methods for solving the system of Equation (18.6) numerically.

One method is based on the coupled solution of the first two of Equations (18.6) using the implicit finite difference, numerically stable four-point Preissmann scheme. In this method, time steps can be chosen as compatible with the time scale of the physical phenomena, and its accuracy is unaffected by variable spacing of computational points. The solution fails, however, if critical flow is encountered along the simulated river reach. The method and algorithm are described by Cunge and Perdreau (1973) and they are the heart of the code called CHAR2, used as industrial tools for rivers by SOGREAH and Laboratoire d'Hydraulique de France.

The second method was developed by de Vries (1965, 1973) and is used mostly by the Delft Hydraulics Laboratory. It is based on the uncoupled concept of the solution: as the first step, the first equation of Equations (18.6) is solved as a steady-state backwater equation considering a constant bed level; as the second step the transport equation is solved using a diffusive explicit finite difference scheme. Being explicit, the method does not allow for arbitrary time steps (transport equation Courant number must be less than unity) and its accuracy depends upon the variability of the space step. On the other hand, it is as flexible as steady-state backwater computational method can be.

The coupled system of Equations (18.7 to 18.13) is solved by a hybrid finite difference/characteristics method. Rahuel (1988) was the first to conduct an analysis of the system of equations for sediment-mixture transport and to define the number of initial and boundary conditions needed to ensure the existence of solutions. Essentially the numerical solution follows two algorithms:

1. Water mass and momentum conservation (Equations (18.7) and (18.8)), bed-material conservation (Equation (18.11)) and the loading law (Equation (18.10)) are solved by a generalized Preissmann implicit finite difference scheme as given by Equations (18.14)–(18.16) and Figure 18.4.

$$f = \Psi[(1-\theta)f_{i+1}^n + \theta f_{i+1}^{n+1}] + (1-\Psi)[(1-\theta)f_i^n + \theta f_i^{n+1}] \quad (18.14)$$

$$\frac{\partial f}{\partial x} = \frac{\theta(f_{i+1}^{n+1} - f_i^{n+1}) + (1-\theta)(f_{i+1}^n - f_i^n)}{\Delta x} \quad (18.15)$$

$$\frac{\partial f}{\partial t} = \frac{\Psi(f_{i+1}^{n+1} - f_{i+1}^n) + (1-\Psi)(f_i^{n+1} - f_i^n)}{\Delta t} \quad (18.16)$$

The Preissmann scheme is of first-order truncation error in space and time if θ and Ψ differ from 0.5, whereas it becomes of second-order when θ and Ψ are equal to 0.5. The scheme is inherently conservative, being equivalent to an approximation of the integral conservation relations, and thus preserves this property when applied to conservative partial differential equations. Moreover, the scheme lends itself well to the convenient imposition of boundary conditions, both internal and external, since it is based on the use of only two computational points in space.

While Ψ is normally taken as 0.5 for the de Saint-Venant equations, it may be advantageous to off-centre the scheme for the sediment conservation equations.

The advection process incorporated in Equation (18.9) is computed by a high-order characteristic-based algorithm which avoids numerical dispersion effects and allows accurate simulation of steep fronts of suspended-load concentration.

The primary dependent variables are A, Q, Γ, Cj, Gj and βj, and the independent variables are the spatial co-ordinate x and the time t. The system is closed by addition of two hydrodynamic boundary conditions, $J = \Sigma j$ upstream boundary conditions (one per sediment class), J additional conditions for βj and one condition for Γ. The result of discretization is a set of $3 + 3J$ non-linear algebraic equations for each computational reach between two computa-

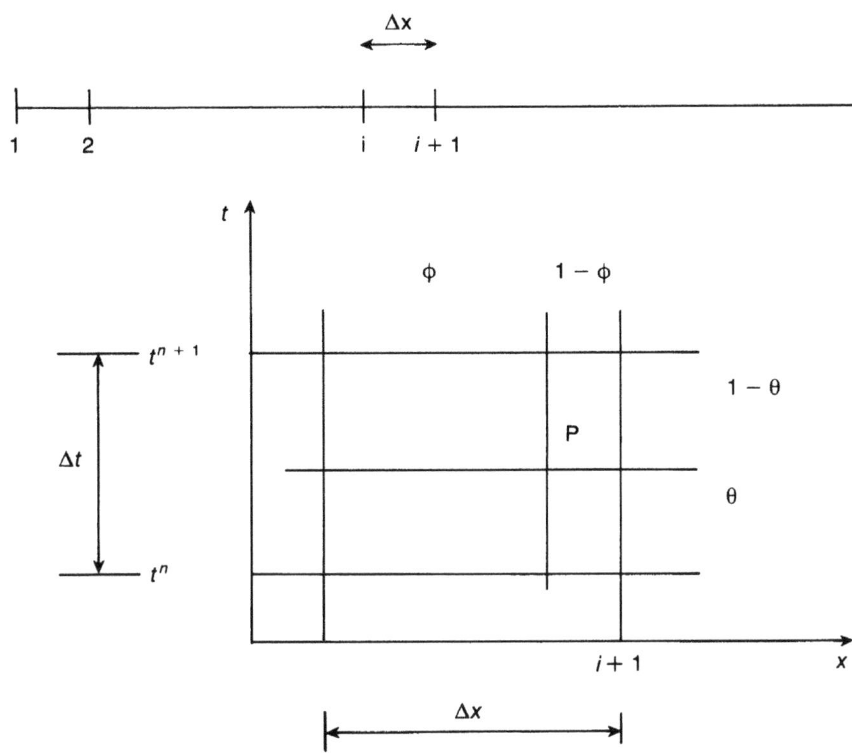

Figure 18.5 Description of generalized Preissmann four-point numerical scheme.

tional points. The system is solved by the Newton–Raphson iteration in which a matrix generalization of double-sweep algorithm is used to limit the size of matrices inverted. Numerical tests showed the method to be robust, convergent and stable for hydrodynamic Courant numbers of the order of 800 and suspended load advective Courant numbers 100. This approach is used in the SEDICOUP code developed by IIHR (Iowa, USA) and LHF (Grenoble-Echirolles, France).

18.4 INITIAL AND BOUNDARY CONDITIONS

The number and kind of initial and boundary conditions needed to use a modelling tool depend on the physical laws (equations) on which the tool is founded. Consider the system of Equations (18.1)–(18.6). Such a system will need the same liquid-phase conditions as one-dimensional fixed bed models, as described in Chapter 11 of this volume (i.e. two conditions at every computational point at initial time, plus two boundary conditions at every instant at the limits). Moreover, it will require the specific conditions of the solid phase. It can be shown that a coupled system of Equations (18.1)–(18.6) corresponds to a system of three characteristic curves in plan (x, t) while Equation (18.3) alone (considered such when the system is uncoupled) has only one family of characteristics. Hence, the user of a tool based on Equations (18.1)–(18.6) must furnish one initial sediment-linked condition at every point and one boundary time-variable upstream sediment condition.

The most natural initial condition is the bed-level initial profile. The most natural upstream boundary condition is the sediment discharge hydrograph at this point. The latter, apparently simple, is in practice very difficult to define. It is difficult to accept that the upstream boundary (physically speaking, whether on the land side or on the seaward side of a river in a coastal area) cross-section is 'in equilibrium', i.e. that the uniform flow transport formula is applicable all the time. It can also be affected by deposition or erosion from the inside of the model. Non-equilibrium suspended-load is very difficult to estimate as a function of time. It is the user's obligation to make sure that the proper conditions are both defined and implemented. The uncoupled algorithms, such as Equation (18.6), need only two boundary conditions (e.g. downstream free surface elevation and upstream sediment discharge), and not three. A SEDICOUP-type system of Equations (18.9)–(18.13) requires the time-variable sieve curve as upstream boundary condition and the initial sieve curves at all computational points! Since it is already difficult to define total sediment discharge at the boundary, this requirement will most likely give rise to some hostility! On the other hand, when an engineer plans protection works or diversions, is he not thinking in terms of 'propagation and deformation of a sieve curve'?

18.5 VALIDATION OF SIMULATION TOOLS

As for fixed-bed models, the engineering judgement concerning the physical situation is essential for the optimum choice of tool, but even more so since the tools are all either extremely crude or not yet extensively tested. The reader can find in the corresponding paragraph of Chapter 11 of this volume all essential remarks concerning the validation problem and requirements. It should be stressed that proper formal validation is even more important for sediment transport models since the uncertainty as to the accuracy of results is even higher here. Correct validation should minimise all sources of error other than specific uncertainties concerning sediment transport, erosion or deposition.

18.6 REPRESENTATION OF PHYSICAL FEATURES; NECESSARY DATA

Very little can be added here to what is said on the subject in Chapter 11 of this volume. Representation of topographic, hydraulic and topological features in one-dimensional sediment transport modelling involves the same problems, difficulties and requirements as in one-dimensional fixed bed modelling. There are, however, specific points to be considered:

1. Topology of the modelled network of channels may be a cause of major difficulties. Except for rather crude uncoupled explicit algorithms, it is difficult to find a sediment transport code which can reliably simulate dendritic (not to mention multi-connected) networks. It is important to stress that it is more a physical problem (how to describe the junctions of canals transporting sediment) than a data processing or computational problem. Hence, the user should neither ask for nor accept the impossible. No doubt, this problem will find a solution in the near future, but for the time being (1990), it is still here.
2. Structures and transitions, such as dams, weirs gates, etc.: the problem is almost exactly the same as that mentioned above, and the same caution should apply. Whatever available tools produce as simulation, it would be an approximation to be interpreted. The best solution for a user is to be able to insert one's own formulation for transition. Unfortunately, most of the existing codes would not permit such an approach, unless the code developer is deeply involved himself in the task.
3. Sedimentation data needed for initial and boundary conditions and for calibration purposes are extremely costly and difficult to obtain. Moreover, they are generally unreliable, at least not as accurate as liquid flow information.

18.7 CALIBRATION AND EVALUATION OF RESULTS

Here again, the general modelling philosophy is exactly the same as that applied to fixed-bed situations (see Chapter 11 of this volume). Specific problems concern movable bed and sediment transport characteristics. In theory, it should be possible to calibrate a numerical model of a movable bed river by comparing observed and computed results and modifying the empirical coefficients included in the equations. Thus, by a trial and error procedure, a reliable predictive tool is again obtained. While this procedure is the only one available to a user, its application is extremely difficult because of the scarcity of data.

As compared with the fixed-bed problem, assuming the geometric data correctly measured and introduced, three main physical phenomena may be subjected to 'calibration': resistance to flow, sediment transport formula (see Equation (18.4)) and loading laws expressing a non-equilibrium situation (or adaptation time/space-lag) of transported sediments. Corresponding formulae and equations contain a number of empirical coefficients and relationships which depend, in turn, on hydraulic or geometric, local or general conditions.

The only available calibration criteria consist of measured longitudinal bed profile change and sediment discharge, the latter being very difficult, and often impossible to determine in larger rivers. While it is known that uniform-flow equilibrium formulae may give, in real life, a sediment discharge estimate with some 100% error, *in situ* measurements of bed-load or suspended load may well be no more accurate. Such measured data are to be compared with the computational results and, from the comparison, the user should deduce the corrections to be applied: depth-variable bed-form resistance laws, critical shear stresses assumed by applied formula, an exponent in the transport formula, hiding factor for graded sediments, etc.

The calibration factors, as mentioned at the very beginning of the chapter, are not independent among themselves. Thus the flow-resistance law, especially for fine-sand covered beds, should include a possibility of taking account of variable bed-forms (dune and ripples as well as a smooth bed) as a function of the flow depth and velocity. Indeed, the transport of sand in such water courses, either in suspension or as bed-load, is strongly related to the magnitude and forms of the resistance. It has been shown in Chapter 11 of this volume that the free-surface elevation is calibrated by modifications in roughness coefficient (a correct way of proceeding) but that the same results can be obtained by arbitrary modifications of the geometry (which must be avoided). In movable-bed situations the same danger is present, except that the consequences of the mistake might be transmitted to transport formulae and might well falsify all the simulations.

In this context, engineering knowledge and judgement stemming from experience in evaluating the results of both calibration and predictive simulation are essential. In the present state of sediment transport modelling, one should not use available codes without knowledge of their inner constitution, and without hydraulic-sedimentology expertise.

18.8 SPECIFICITY OF COASTAL ONE-DIMENSIONAL SEDIMENT TRANSPORT MODELS

When considering a modelling approach to a project, one must know if the capability of a tool corresponds to the real-life situation. Most existing codes were developed for inland waterways and there are several differences between the modelling of coastal and inland river flows.

To begin with, there is the problem of salt intrusion and density differences. In the present chapter, this problem is excluded since only one-dimensional modelling is considered and stratification is not taken into account, and also because the specific influence of saline water on sediment transport introduces difficulties of an even higher order. Although there have been attempts to model such situations, corresponding codes definitely cannot as yet be called 'tools'. Thus the reasonable way currently available to take salinity into account is to include it in the hydrodynamics modelling component, as described in Chapter 11.

Another problem is the flow direction reversibility during the tidal cycle. The sediment transport is governed by the first-order partial differential Equation (18.3) or, in the SEDICOUP approach, by Equations (18.9), (18.11) and (18.13). These equations describe convective transport phenomena and the sediments are transported in the flow direction from physical upstream towards physical downstream. Not all existing codes can deal with the situation where the simulated water course is flowing to the right along one part of its length, to the left along another part, and the zero velocity section continuously changes its locus.

The third difficulty is the time-scale problem. When long-term (say 20 to 50 years) evolution of a river is considered, the flow and sediment equations are usually uncoupled and steady-state water flows are simulated over long time-steps, compatible with slow river-bed evolution (cf. Section 18.2). It is, however, impossible to apply a steady state liquid flow concept to a tidal river in a straightforward way. In many situations, the succession of irregular tides is as important for sediment simulation as any other external influence and one has to face very long simulation periods and corresponding computational efforts.

On the other hand, it is costly to simulate tides over several years with a time step of few minutes, knowing that the bed elevation changes would be microscopic during one tidal cycle. Hence, for such situations, it is assumed that flow and sediment transport are periodic over several tidal cycles. In practice, this means that the time step for bed-level computation corresponds to several tidal periods. Such an approach is based on continuity correction methods as shown in Figure 18.6 (Wang, 1989).

In this method the distribution of the water discharge and the water level is assumed to be the same as in the previous tidal period. The depth and velocity change, however, due to the bed-level change, and so does the sediment transport. Continuity correction (which gives a new velocity field) is carried out with frequency depending upon the rate of bed-level change. When this change exceeds a given limit, the complete flow computation is carried out again.

Finally, the problem of the sediment boundary condition on the seaward side of models should be mentioned. When the flow enters an estuary or a tidal inlet or channel the sediment load carried by it must be imposed in the model as a boundary condition. It is extremely difficult to estimate this solid discharge and measurements are not often feasible over long periods, yet its importance for the simulation results (such as creation of sand bars) may be critical.

18.9 SOFTWARE TOOLS

The following list of tools developed by some of the best-known European Institutes specializing in numerical simulation is by no means exhaustive. The information presented below was supplied in all cases

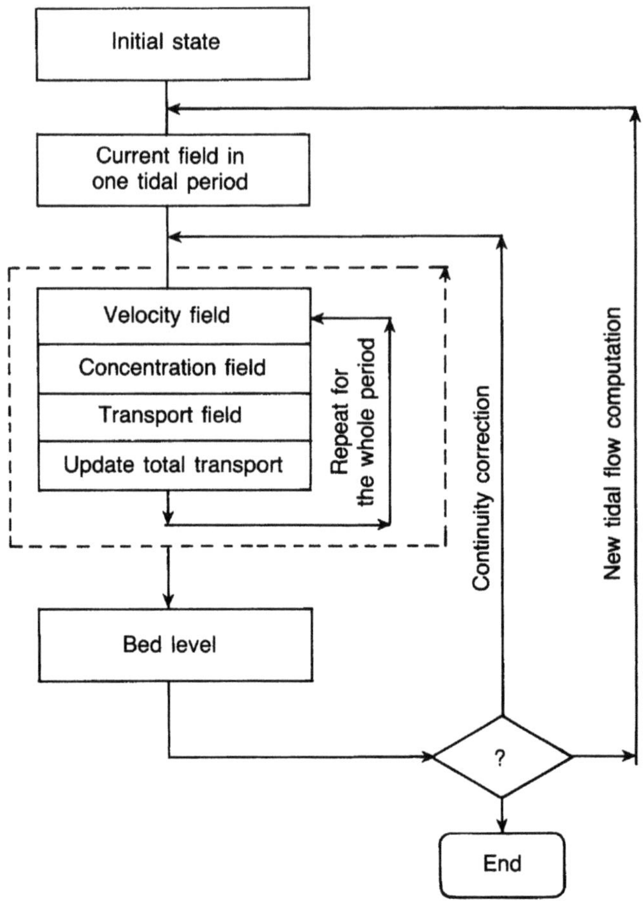

Figure 18.6 Computation procedure for a morphological model in tidal regions.

by either the owners or developers of the software, who are solely responsible for its accuracy and value.

Danish Hydraulic Institute (DHI), Agern Allé 5, DK-2970 Horsholm, Denmark

MIKE 11 is a general commercial package for micro-computers which simulates river hydrodynamics, sediment transport and water quality processes. A hydrodynamic-module description has been given earlier in this volume (Chapter 11, section 11.11). The sediment transport module includes three alternative sediment transport formulations: Engelund–Hansen, Ackers–White and Engelund–Fredsoe. With the last one it is possible to compute dune dimensions and associated resistance coefficients. The system can also account for sediment transport between the river and its flood plains. It is also possible to make morphological (movable bed) simulations. The effect of the change in cross-section can be taken into account in the hydrodynamic simulation. Cohesive sediments are treated separately with a special formulation of the erosion and deposition processes. A separate system (MIKE 12) is available for stratified flows, but this does not currently (1990) incorporate sediment transport components within itself. (Danish Hydraulic Institute, 1990).

Delft Hydraulics, P.O. Box 177, 2600 Delft, The Netherlands

The WENDY (Water and ENvironmental DYnamics) package is based on a one-dimensional dynamic flow-routing model. Additional to this core, several modules are available: water flow including density effects, e.g. salt intrusion in an estuary (SAFLOW);

sediment transport and bed-level changes (SEFLOW); suspended sediment transport and bed-level changes (SUSFLOW). In sediment transport option distinction is made between total and suspended loads. In total load mode Engelund–Hansen, Ackers–White, Rottner, Van Rijn and Meyer–Peter–Mueller formulae are available. (Delft Hydraulics, 1990).

Hydraulics Research Ltd, Wallingford, Oxfordshire OX10 8BA, UK

TIDEFLOW-1D simulates and predicts the interaction of tidal and fluvial flows in branched and looped channel networks. A description of the hydrodynamic module has been given in Chapter 11, section 11.11. It can be run interactively with other process models to take into account siltation and erosion. Special features: roughness variable as a function of sand transport; bed-level and cross-section variable with erosion and deposition. (Hydraulics Research Ltd., 1990).

Institut für Strömungsmechanik und Elektron. Rechnen im Bauwesen, Universität Hannover, Appelstr. 9a, D-3000 Hannover, Germany

The TICAD system (Tidal Interactive Computation and Design) simulates one-dimensional branched river and canal systems. It is based on a finite element implicit solution technique and is linked to a digital-terrain model. A description of the hydrodynamic module has been given in Chapter 11, section 11.11. Sediment transport is calculated in terms of transport capacity from empirical formulae. The resulting morphological changes are determined by balancing transport capacities between neighbouring discrete elements. (Holz, 1990).

Laboratoire d'Hydraulique de France (LHF), 6 Rue de Lorraine, 38130, Echirolles, France

The SEDICOUP modelling system is a one-dimensional code based on coupled unsteady state sediment and water flow taking into account extended graded sediments, simulating suspended load and bed-load transport, deposition and erosion (including sorting and armouring), bank erosion and deposition in bank zones (Holly, 1988; Holly and Rahuel, 1989a,b; Belleudy, 1992; Belleudy *et al.*, 1992).

Laboratoire National d'Hydraulique (LNH), Electricité de France, B.P. 49-78401, Chatou Cedex, France

LNH has a number of one-dimensional codes available based on the de Saint-Venant equations, each of them adapted to specific problems, (see Chapter 11, section 11.11). Water flow computations can be coupled with a sediment transport component. Thus the COURLIS code allows the treatment of erosion, deposition, consolidation and transport of cohesive materials. The TSAR code, under development in cooperation with the LHF, will allow simulation of bedload and suspension. (Laboratoire National d'Hydraulique, 1990).

REFERENCES

Armanini, A. and Di Silvio, G. (1988) A one-dimensional model for the transport of a sediment mixture in non-equilibrium conditions. *Journal of Hydraulic Research*, **26** (3), 275–92.

Belleudy, Ph. (1992) Modeling Danube and Isar Rivers morphological evolution (2nd part): Comparison of field data with modeling results, *5th International Symposium of River Sedimentation*, Karlsruhe, Germany, 6–10 April 1992.

Belleudy, Ph., Loy, G., Kellermann, J. and Sohngen, B. (1992) Modeling Danube and Isar Rivers morphological evolution (3rd part): Range of uncertainty caused by scattered field data with special respect to grain sorting, *5th International Symposium of River Sedimentation*, Karlsruhe, Germany, 6–10 April 1992.

Cunge, J.A. and Perdreau, N. (1973) Mobile bed fluvial mathematical models, *La Houille Blanche*, **7**, 561–80.

Cunge, J.A. and Holly, F.M. Jr. (1986) Development Strategy for One-Dimensional Mobile-Bed Modeling System, *3rd Intern. Symp. on River Sedimentation, 31 March–4 April*, Jackson, MS, USA.

Danish Hydraulic Institute (1990) (Private communication).

Delf Hydraulics (1990) (Private Communication).

Holly, F.M. Jr. (1988) CHARIMA and SEDICOUP Codes of Riverine Mobile-Bed Simulation, *Interagency 2nd Seminar on Stream Sedimentation Models, 19–20 October*, Denver, CO, USA.

Holly, F.M. Jr. and Rahuel J.-L. (1989a) Riverine Mobile-Bed Modeling Under Non-Equilibrium Conditions. *Proc. HYDROCOMP'89, June*, Dubrovnik, Yugoslavia.

Holly, F.M. Jr. and Rahuel J.-L. (1989b) Advances in Numerical Simulation of Alluvial River Response to Disturbances. *Proceedings of the IAHR 23rd Congress, August*, Ottawa, Canada.

Holz, K.P. (1990) (Private communication).

Hydraulics Research Ltd. (1990) (Private communication).

Laboratoire National d'Hydraulique (1990) (Private communication).

Nordin, C.F. Jr. (1988) Closing comments. *Interagency 2nd Seminar on Stream Sedimentation Models, 19–20 October*, Denver, CO, USA.

Rahuel, J.-L. (1988) Modélisation de l'évolution du lit des rivières alluvionnaires à granulométrie étendue. *PhD de Grenoble thesis*, Institut National Polytechnique de Grenoble, Université, France.

Vries, M. de (1959) Transients in bed-load transport. *Delft Hydraulics Laboratory Report N° R3*, Delft, The Netherlands.

Vries, M. de (1965) Considerations about non-steady bed load transport in open channels. *Delft Hydraulics Laboratory Publication 36*, Delft, The Netherlands.

Vries, M. de (1973) River bed variations – aggradation and degradation. *Delt Hydraulics Laboratory Publication 107*, Delft, The Netherlands.

Wang, Z.B. (1989) Mathematical modelling of morphological processes in estuaries. *Communications on Hydraulic and Geotechnical Engineering Rep. 89-1, March*, Delft Technical University, The Netherlands.

PART FOUR

PHYSICAL MODELS AND THEIR APPLICATIONS

PHYSICAL SHORT-PERIOD WAVE MODELS 19

O.J. Jensen

Although numerical models of short-period waves are now very general and reliable tools for engineering practise, the more traditional physical models continue to be used in this area. In fact the two types of model complement one another, the numerical being better suited to investigating a number of configurations and incident wave fields and the experimental to investigating the long-term effects of a few chosen incident wave fields on a few selected configurations. This complementarity even goes so far as using the same input data for numerical and physical models, so that the computer codes that are used to drive the models in the two cases also have much in common. Some of the most significant of the recent developments within this interactive environment are introduced in this chapter, which also includes a contribution on ship manoeuvring simulations by A. Hasle Nielsen and S.E. Sand.

CONTENTS

19.1 INTRODUCTION

Short-period wave modelling, like other branches of engineering science, is a subject that has been constantly developing since the beginning of this century. Physical wave models are now an important tool for assessing complex wave phenomena and have considerable economic importance for the planning and design of harbours and marine terminals, as well as for the testing of structures (wave impact and stability, etc.). For some applications, numerical models are equally reliable (see Chapter 12).

The choice of model, type and configuration, for a specific investigation, depends on a large number of factors but the most important is the reliability of the results to be obtained. Consistently reliable results are obviously obtained only with models which adequately describe the characteristics of the physical processes involved. Another pre-condition for obtaining reliable results is that test cases are properly selected to represent the natural conditions of importance to the problem in question. This problem of representativeness requires, in the case of sea waves, site investigations of the wave climate, including statistical analysis and analysis of joint occurrence probabilities of wave directions and wave heights and periods (H_s, T_p) and water levels (WL).

19.2 RECENT HISTORY OF WAVE MODELLING

Regular waves were used for physical modelling of marine and maritime subjects until about 15 years ago, and many researchers have investigated how to use the results from such tests to predict representative conditions for the prototype. A review of the literature shows that there is a great variety in the results obtained and in the conclusions derived, depending on the type of physical process or structure in question.

The state-of-the-art underwent a great change after the introduction of irregular waves in laboratory tests during the 1970s, and since then the use of irregular waves has become standard practice in hydraulic modelling. The focus in research has shifted from how to manipulate the results to be representative, to

Coastal, Estuarial and Harbour Engineer's Reference Book. Edited by M.B. Abbott and W.A. Price. Published in 1993 by Chapman & Hall, 2–6 Boundary Row, London SE1 8HN. ISBN 0 419 15430 2.

how to obtain the most reliable natural wave conditions in the laboratory. However, two lines of reproduction philosophy were followed among the leading laboratories, i.e. either reproduction based on an artificial wave-spectrum description (random/irregular waves) or reproduction based on measured wave records (natural/irregular waves). Later, this discussion became more refined and included the importance of second-order phenomena associated with wave groups and bounded long waves etc. and how to reproduce these phenomena correctly in hydraulic models.

In the 1980s, the development went further and the deterministic understanding and description of natural waves has now reached a level where it has become possible to construct hydraulic machinery for the reproduction of three-dimensional natural waves in the laboratory. The reproduction comprises all natural phenomena such as correct wave energy distribution on both frequencies and directions (directional wave spectra) and thereby includes the correct reproduction of short-crested waves, with all the characteristics of a real sea-state occurring in nature.

Further, as the technique for the reproduction of correct natural waves was developed, the introduction of scale models of vessels has made hydraulic investigations a reliable and in many cases indispensable tool in the planning and design process of ports, harbours and marine terminals.

19.3 MODEL LAW

Physical wave modelling of marine structures, harbours, breakwaters, etc. is based on reproduction in scale of the properties of gravity waves. Froude's model law is applied because the essential forces involved are inertia, pressure and gravity; viscous and surface tension forces are ignored.

The Froude model law implies the following scales:

Length scale	$1:\lambda$
Time scale	$1:\lambda^{1/2}$
Acceleration scale	$1:1$
Velocity scale	$1:\lambda^{1/2}$
Volume scale	$1:\lambda^3$
Force scale	$1:\lambda^3$

All other scales in Froudian modelling can be derived from the physical units. For example, the scale for pressure can be found from the rule that the scale for pressure is the force scale divided by the length scale squared, i.e.

$$1:\lambda^3/\lambda^2 = 1:\lambda$$

In certain cases where wave motion in porous media is involved, serious scale effects may occur and special means need to be introduced to reduce their influence (Juul Jensen *et al.*, 1983). Models further need to be so large as to eliminate the influence of surface tension. This is normally supposed to be achieved when the wave length is significantly longer than the lengths of surface tension waves. ('Surface tension waves' are waves where the surface tension is the restoring force.) However, for wave lengths above, say, 0.05 m, the influence of surface tension can generally be neglected (Wiegel, 1964).

19.4 WAVE FLUME STUDIES

Wave flumes are used for studying many hydraulic problems where the configuration can be reproduced two-dimensionally with acceptable accuracy. In a wave flume, it is nowadays possible to reproduce very accurate irregular waves with prescribed wave conditions, i.e. correct reproductions of wave spectrum, wave height and period distribution are attainable.

This has been possible for about the last ten years, under the condition that a wave absorbing structure is present in the end of the flume. Figure 19.1 shows an example from Mansard (1984) of reproduction of waves in a wave flume. Besides the first-order waves, the so-called SIWEH function is shown. This function describes the 'smoothed instantaneous wave energy history' and is thus a signal representing the wave groups. An important feature of irregular waves is the long wave bounded to the wave groups and travelling with the group velocity. The third signal shows, for the same wave signal, the bounded long wave. This long wave, which is the 'instantaneous mean water level' under the wave motion, has a trough under the 'high waves' in the wave group and a crest under the 'small waves'. It should be noted that there is an almost inverse proportionality between the SIWEH function and the bounded long wave.

This bounded long wave plays a very important role in physical modelling as it may become a free wave if the short waves are subject to a rapid decrease in wave height, such as occurs during diffraction around a breakwater and in the wave breaking zone. These phenomena are not yet completely understood. The liberation of long waves is very important for harbours as they are the primary cause of seiching (long periodic oscillations) in harbours.

In wave flumes, it has until recently been difficult to perform hydraulic model tests with well-defined wave conditions in cases where the structure in the flume has a high wave-energy reflection. This is, for example, the case for vertical face breakwaters and other structures with steep slopes. In such cases, extended analysis has been made to determine the true incident waves, i.e. to take into consideration the reflection of the incoming waves produced by the wave generator. When this reflected energy reaches

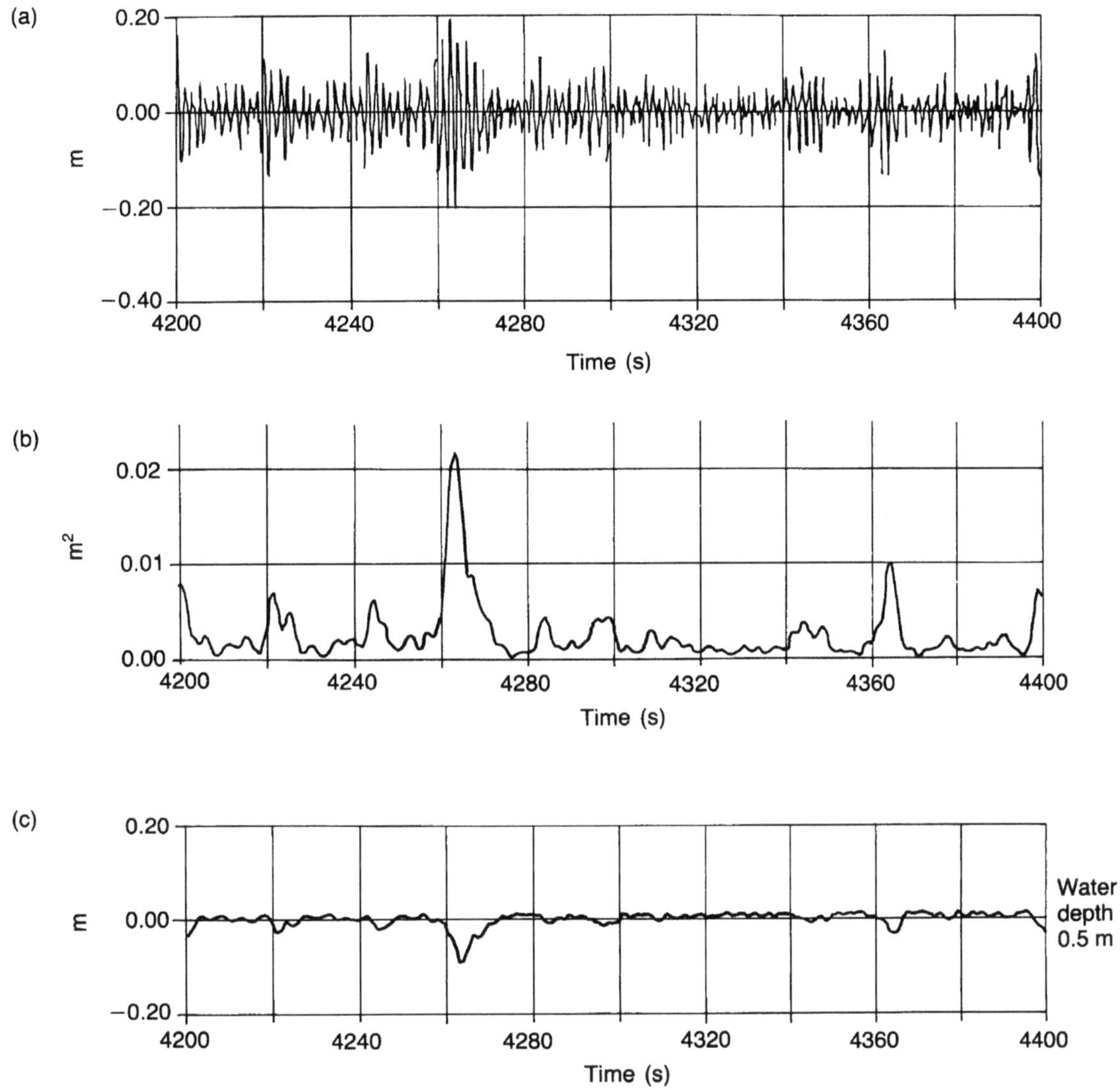

Figure 19.1 Example of a wave train (Mansard, 1984). (a) first-order time series; (b) SIWEH first-order time series; (c) bounded long wave (water depth 0.5 m).

the wave paddle it is reflected once more (re-reflection), and thereby adds to the generated incoming wave energy.

To overcome this problem, systems which automatically absorb the reflected waves have recently been developed. The principle used is to measure the reflected waves on the wave paddle, and, by means of analogue filters and a servo-system, to isolate the reflected part and then impose an opposite movement of the generator, which results in the desired absorption. Figure 19.2 shows as an example the spectra of the specified and the measured incoming wave series without and with an active absorption system for a structure with a reflection coefficient of $R = 0.7$.

19.5 WAVE BASIN MODELS

Almost twenty years ago irregular natural waves were introduced in harbour models with moored vessels

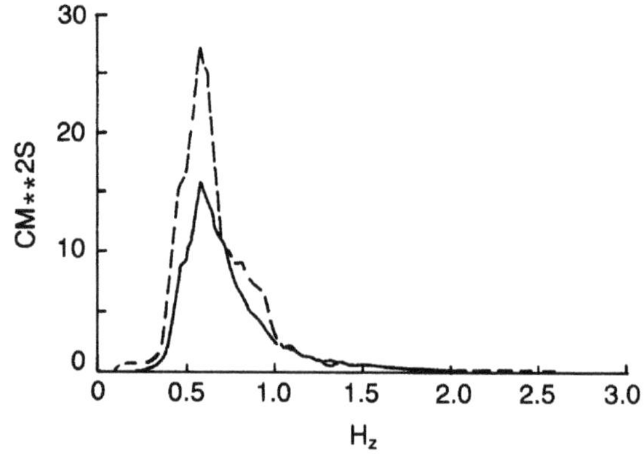

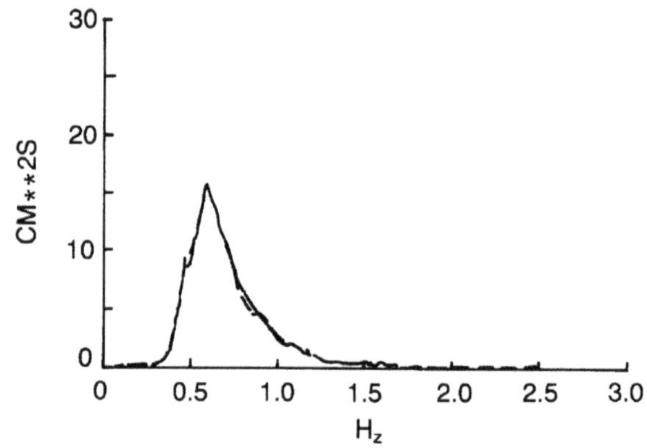

Figure 19.2 Specified and measured spectra with and without DH1 active absorption system. Tests in a wave flume with vertical partly permeable plate as structure with reflection coefficient, $R = 0.7$; (——), specified; (---), measured.

(Gravesen *et al.*, 1974). This large step forward meant a great improvement towards reliable simulation of real natural wave conditions. In recent years, however, it has been shown that this method also has problems. The work by Bowers (1977, 1980), Sand (1980, 1982) and Ottesen-Hansen (1980) showed that the model results are highly dependent on the correct reproduction of wave groups and the attached long waves.

19.6 SPURIOUS LONG WAVES IN HARBOUR MODELS

Spurious free long waves may appear in laboratory models if the applied wave generation technique only considers the first-order waves. Spurious or parasitic waves appear because of incorrect boundary conditions at the wave generator board.

One of the effects is due to the fact that the wave board is impermeable and hence the water surface at the board is always horizontal while the bounded long wave should, to get the physics correct, have a varying slope at the surface. For this reason, the wave generator board will produce undesirable parasitic long waves if the wave generation technique only considers the first-order waves (the normal short waves).

Therefore, in order to compensate for incorrect reproduction, a method has been developed to compensate the parasitic long waves by introducing low frequency components in the wave board motion which correspond to free long waves, but travel in the opposite phase and consequently eliminate the free parasitic waves.

19.7 OTHER IMPORTANT ASPECTS OF WAVE GENERATION IN MODELS

Another important aspect of wave generation in harbours is the use of irregular waves with indefinite wave crests, whereas the real waves in the sea are short crested and exhibit directional spread.

Sand *et al.* (1983) showed that the wave energy arriving at a specific point in a harbour after diffraction around the entrance breakwater(s) is different in the cases of two-dimensional (uni-directional) and three-dimensional (directional) waves.

Figure 19.3 shows results from numerical and physical simulations of the diffraction pattern of irregular three-dimensional waves around a single breakwater. For comparison with theory, results for the Sommerfeld solution are also presented. It can be seen that more wave energy reaches in behind the breakwater for three-dimensional waves than for two-dimensional waves. It can thus be foreseen that three-dimensional testing of harbours and marine terminals will eventually become common practice. Kirkegaard *et al.* (1980) showed that for an open terminal (freely moored vessel), the motions of a moored vessel are different in this case from motions with two-dimensional waves.

Figure 19.4 shows the movements of a 20000 DWT bulk carrier in a beam sea, the difference between two-dimensional and three-dimensional waves is clear. This example emphasizes the importance of using three-dimensional waves in cases where the first-order waves are the main factors responsible for vessel motions. However, for many ports the short wave agitation is not the most important. Long

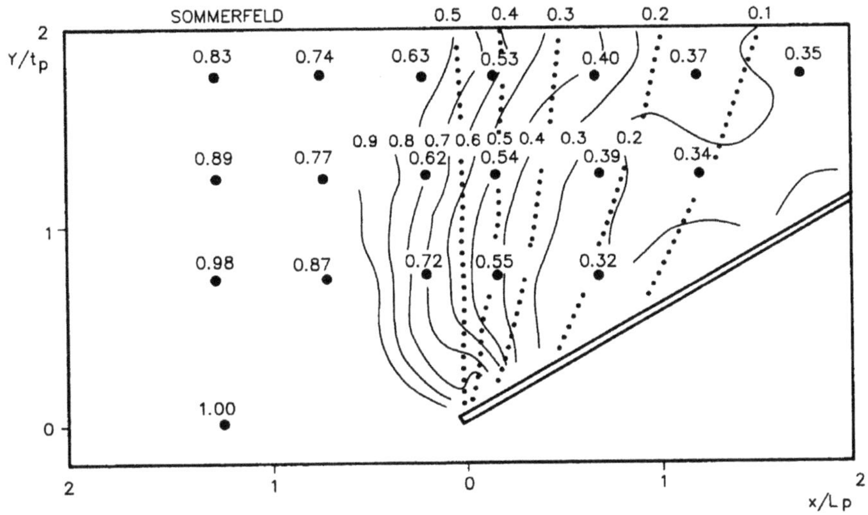

Figure 19.3 Diffraction patterns of three-dimensional waves. Contour lines are numerical results; single values are measured from a physical model. L_p = wavelength corresponding to wave period T_p.

periodic planar motions of relatively large vessels in large harbour basins are mainly a result of the wave-group-induced long waves (seiching).

The future introduction of three-dimensional waves as a normal procedure in the testing of ports and marine terminals would, in the ultimately developed stage, have to consider compensating for the three-dimensional long periodic parasitic waves in basically the same way as for the two-dimensional waves. Sand (1982) showed that the directional spreading is much larger for the long waves than for the first-order short periodic waves. This implies that testing with uni-directional compensated wave trains (two-dimensional) should give rise to larger wave amplification in shallow water and thus to more severe long wave action than a true, natural, three-dimensional wave simulation. However, as is well known, the refraction in shallow water tends to make the waves more long crested and more uni-directional than in deep water. Accordingly, from this point of view, it is desirable to have a relatively limited depth at the wave generators (provided that the wave front is almost straight at the boundary of wave generation). The above reasoning shows that a number of points need to be carefully considered in the design of a physical model and in the input waves and boundary conditions.

If free long waves (seiching) are present in a harbour basin, long waves will in principle radiate offshore from the harbour entrance. In nature, these waves will disappear from the area, while in a physical model they will be reflected and possibly re-reflected from non-dissipating boundaries such as the wave generator board and guiding walls.

The above considerations may seem pessimistic in relation to the reliability of simulating long waves (seiching) in scale models of harbours. However, examples such as those given below, demonstrate that such models can be reliable and in almost perfect agreement with the prototype. The main demand made upon the model seems at present to be that the input waves are representative for prototype conditions, that the bathymetry in the model area is correctly reproduced and that the depth at the outer boundary is chosen in an optimal way, i.e. taking due consideration to minimize the effect of parasitic long waves.

19.8 RELIABILITY OF PHYSICAL WAVE MODELS

19.8.1 LONG WAVE OSCILLATIONS (MODEL AND PROTOTYPE MEASUREMENTS)

With regard to the reliability of physical models for wave disturbance investigations of harbours, only limited data is available for comparisons between the model and prototype. In most cases, the comparisons show good agreement with the prototype measurements. A few case studies will be presented below. Beaudelaire *et al.* (1956) described how seiching is inconvenient for moored vessels in the Port d'Alger and presented prototype measurements in the po-

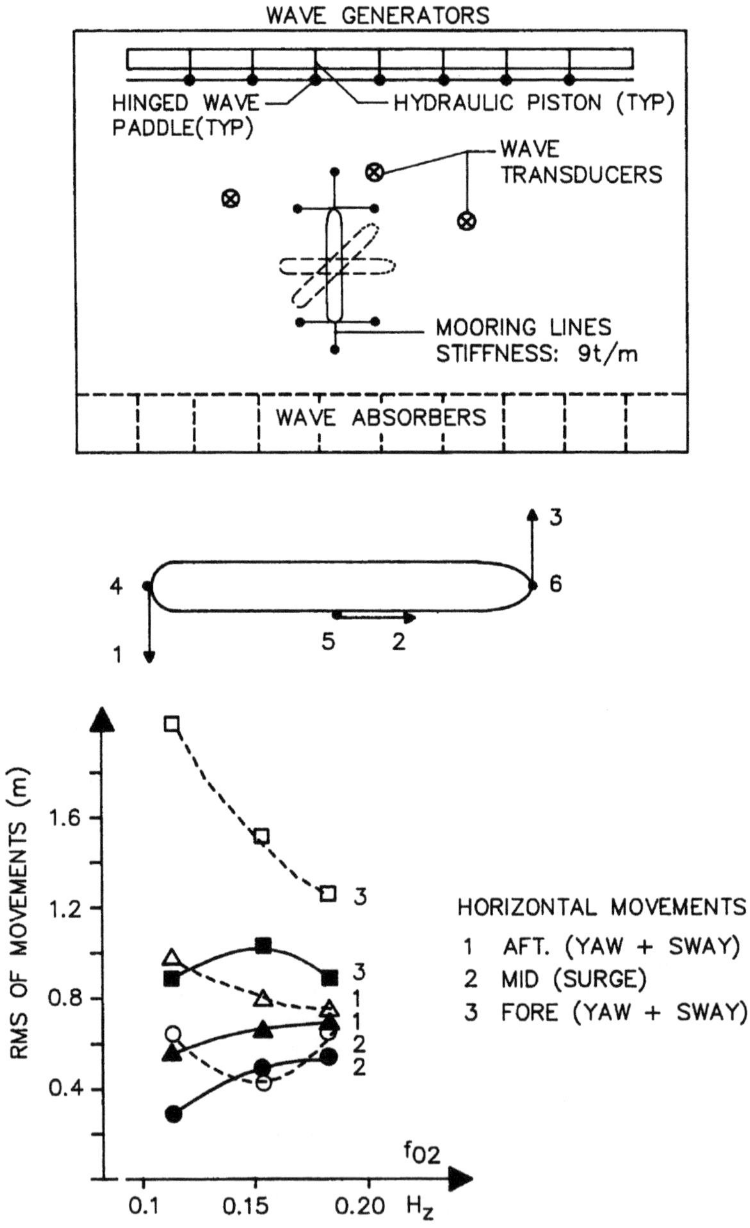

Figure 19.4 Test set-up and horizontal movements of 20 000 DWT bulk carrier in beam sea. (——), three-dimensional; (---), two-dimensional; $H_s = 2.0$ m.

sitions (A–E) as shown in Figure 19.5.

In Figure 19.6, the measurements are shown together with measurements from a physical model (scale 1:100). $H_{max,L}$ is the maximum long wave measured in the harbour during test runs of 3 h prototype duration. The long waves were determined by filtering of the signal as seen in Figure 19.7. H_s refers to the waves outside the harbour in 35 m water depth.

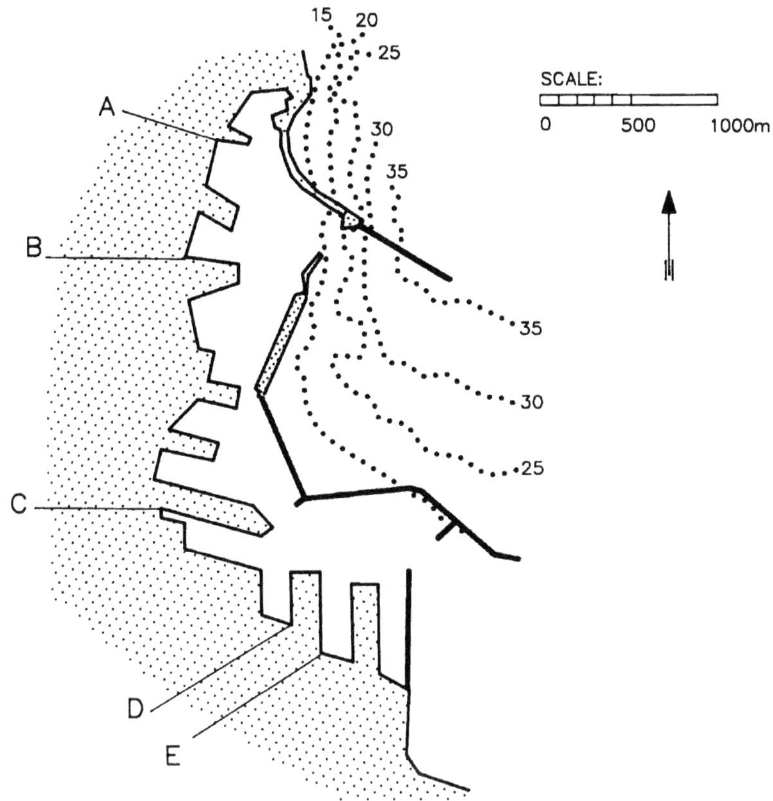

Figure 19.5 Plan of Port d'Alger.

Since only the maximum waves measured in nature are reported, the maximum long waves in the model are used in this comparison. It can be seen that there is good agreement between model and prototype. The model tests were run with irregular natural waves reproduced directly from wave records from near the site, and the equivalent depths at the wave board were equal to 35 m.

Figure 19.7 shows prototype and model records from Station C and demonstrates that the long waves in the model are not only representative with respect to wave heights but also with respect to wave periods. The long waves with periods of about 100 s on which short periodic first-order waves are superimposed can be clearly seen in both cases. This case demonstrates that by applying the techniques currently available, it is possible to realistically simulate long wave effects in physical scale models.

19.8.2 SHORT PERIODIC WAVE AGITATION (MODEL AND PROTOTYPE MEASUREMENTS)

In Rønne Harbour in Denmark, wave measurements were made both outside the harbour and in the basins. A comparison of the measurements with model test results from a hydraulic model (scale 1:80) with reproduction of irregular natural wave records is given in Figure 19.8. The harbour layout and the position of the wave measuring instruments are also shown. There is again, good agreement between model and prototype.

This example also shows that for smaller harbours where the short periodic first-order storm waves are the main source of harbour wave agitation, physical scale models are reliable provided that the input wave conditions accurately simulate the wave conditions which occur in nature.

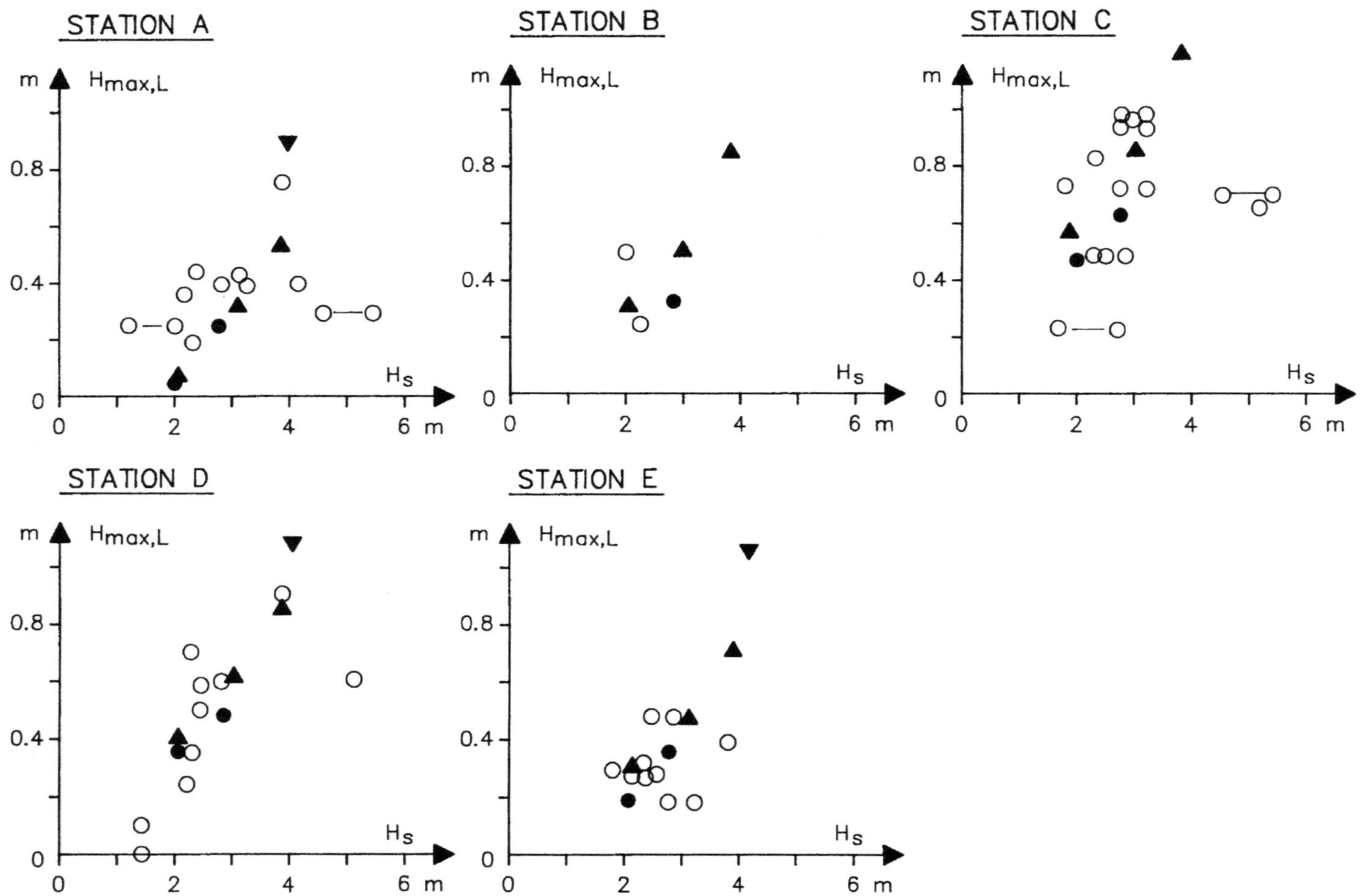

Figure 19.6 Comparison of seiching in Port d'Alger in model and prototype (Ottesen–Hansen *et al.*, 1980). (○), prototype, Beaudelaire *et al.* (1956). (●), model; $T_p = 8$ s; (▲), $T_p = 10$ s; (▼), $T_p = 12$ s.

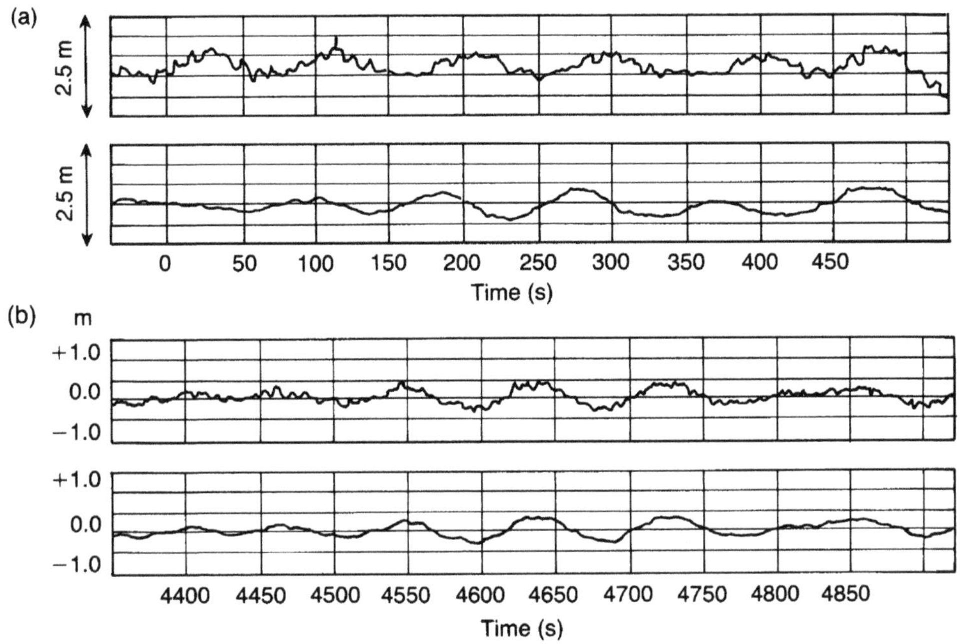

Figure 19.7 Seiche measurements from station C in (a) prototype (Beaudelaire 1956) and (b) model. Lower prototype and model signals have been filtered to eliminate short waves.

19.9 NAVIGATION

So far this chapter has dealt with physical model studies aimed at reducing wave activity and hence ship movements in coastal harbours. However, ease of navigation is also a prime requirement. Ports should be designed so that pilotage and tug assistance is unnecessary or at least reduced to a minimum. Harbour configurations, especially the harbour entrances, should be planned with emphasis upon the manoeuvring characteristics of the types of ship likely to use the facility. This applies also to changes and modifications to existing harbours and marine terminals. Many ports built in the past are more difficult to negotiate than would have been necessary if timely consideration had been paid to navigational aspects.

The development in recent years of accurate and realistic ship manoeuvring simulators has now provided the harbour designer with a unique and powerful tool, which allows all navigational aspects, including the human element, to be taken into account in harbour planning and design in a rational and realistic manner.

19.10 OVERVIEW OF MARINE SIMULATORS

The heart of marine simulators is a mathematical model that describes the real world's physics with emphasis on the manoeuvring performance of vessels. Graphical representation of the ship movements, ranging from simple track plots to three-dimensional colour vision on large movie-type screens may add realism to the simulations. Ship-handling simulators make it possible to include the human factor in risk assessments of harbour manoeuvres, and to evaluate and train pilots and captains.

Several levels of simulation are available.

1. Macro level, dealing with traffic pattern simulation, including simulation of vessel traffic surveyance (VTS) systems.
2. Micro level, where detailed manoeuvring characteristics of the individual ship or ship type are reproduced, including shallow water effects, ship–ship interaction etc., as well as environmental effects (water levels, currents, wind and waves), tug boats, anchor and fender forces, etc. Wide screen three-dimensional colour graphics of the landscape with daylight, dusk, night, rain and fog effects as well as sound effects may also be

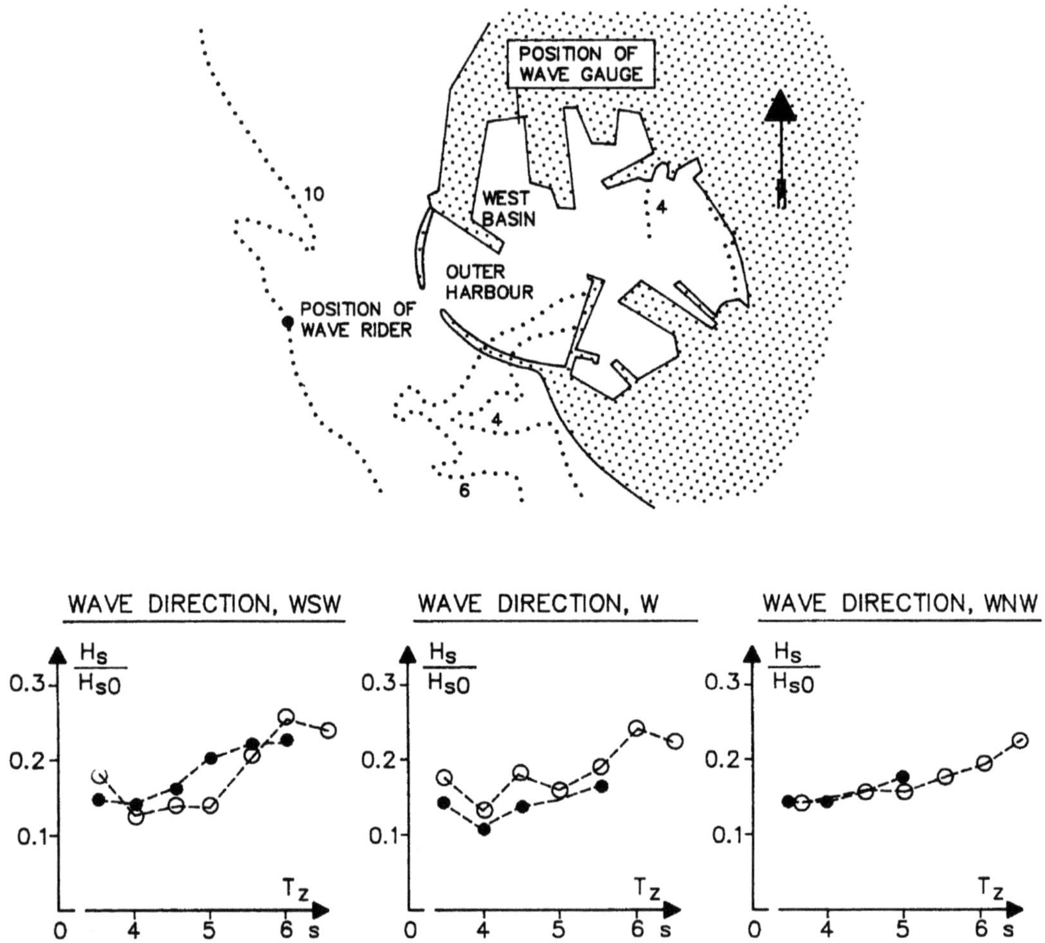

Figure 19.8 Measurements inside and outside Rønne Harbour, Denmark. (○), model; (●), prototype; H_s = significant wave height in harbour; H_{so} significant wave height outside harbour. Wave directions: (a), WSW; (b), W; (c), WNW.

provided, as may moving platforms reproducing the physical movements of the vessels in up to 6 degrees of freedom.

Marine simulators are primarily used for three purposes.

(a) Education and training, where realistic real-time audio-visual representation of the real world is of primary importance, and where real ship handles and instruments are available in the simulator to provide a realistic environment for the trainee.
(b) Planning and design of vessels, harbours, navigational channels, bridges, marine operations, etc. where reliable and accurate representation of the physical phenomena is of predominant importance.
(c) Determination of limiting conditions for safe arrivals and departures, with respect to environmental conditions and ship types and sizes.

Design simulations are performed at the micro level (2) in fast-time and/or real-time.

(a) Fast-time, where the ship is 'navigated' by a mathematical model of the human navigator (non-interactive simulation). By this method a number of different environmental situations can be quickly tested. However, the most important mode for harbour design purposes is:

Figure 19.9 Ship simulators.

(b) Real-time, where the simulations are performed by professional navigators, preferably pilots, captains and other navigators associated with or engaged in the project under study (man–machine interactive simulation). By this method psychological effect, ability to handle unexpected situations, etc. can be evaluated and included in an overall risk assessment.

19.11 HARBOUR DESIGN SIMULATION

For design purposes it is essential to use a simulator, that realistically and accurately simulates the real world phenomena, e.g. simulators based upon recognized and well proven mathematical models. It is of course also very important that the input data are of high quality.

The ship manoeuvring characteristics may be obtained from ship research institutes on the basis of model tests (planar motion test, free sailing model tests), and/or analysis of trial trip data. Wind effects may be obtained from wind tunnel tests. Environmental data input may be based upon existing data, supplemented with field measurements and numerical calculations, notably of the current patterns in the approach areas. Water depths may be obtained from sea charts and sounding plans. Land maps and colour photographs provide information for simulation of the landscape with emphasis on high buildings and other characteristic features of importance for the navigation. Exact locations of navigational marks, buoys etc. are also of importance for the simulations.

If possible, the actual simulations should be carried out by local pilots and captains. The local experience ensures that all relevant aspects are included, and that the selected solutions are acceptable from a navigational point of view. On the simulator the navigators are able to demonstrate the navigational problems to the harbour designer, and in front of the simulator consensus between conflicting interests may be reached.

The tracks of the ships are normally printed out during the simulations and form the basis for the important debriefing after each run. Relevant results are stored in the computer for later statistical and risk analysis.

It should be noted that the established simulation model may be used for the subsequent training of pilots, ferry operators, captains etc. as well as for analysis of the feasibility of future changes in the ports.

Based upon experience from more than 50 harbour design simulations in many countries, some of the benefits to be obtained may be listed as follows:

- Optimization of alignment and dimension of navigational channels;
- Optimization of harbour entrances (breakwater alignment, entrance orientation and width);
- Optimization of berth locations and orientations, including ferry terminals;

- Evaluation of mooring/fender systems;
- Optimization of dredging limits;
- Determination of suitable location of navigational aids;
- Simpler and safer ship manoeuvring in the harbour approaches as well as inside the ports;
- Evaluation of the possibility of allowing larger vessels into existing ports;
- Definition of limiting environmental conditions for safe entrance (wind-, current- and wave 'windows');
- Evaluation of tug requirements;
- Consensus may be obtained for new port project at the planning stage often resulting in improved working relations between different categories of personnel, e.g. between port engineers/administators and pilots/captains;
- Training of pilots, ferry operators, captains, etc. enhances confidence and skill, possibly resulting in safer and more efficient ship handling.

It should be noted, that marine simulators are also used for many other purposes, such as ship design (evaluation of power/truster requirement, propellers, rudders etc.), crew training including familiarization with new advanced ship types (one-man bridge, joystick steering etc.), planning of difficult offshore operations including first- and second-order wave effects, deep sea mining etc. Marine simulators with two or more vessels may also be used to evaluate the required free span of bridges spanning over navigable waters. As a result of simulations with more than 20 local pilots it was thus decided to increase the free span of the Great Belt bridge from the tentatively planned 780 m to 1624 m.

REFERENCES

Beaudelaire, J.G., Gabriel, J. and Michel, G. (1956) Agitation à longue période dans le Port d'Alger, Premiers Résultats IV *Journées de l'Hydraulique*, Paris, (Jane).

Bowers, E.C. (1977) Harbour resonance due to set-down beneath wave groups. *Journal of Fluid Mechanics*, **79** (1), 71–92.

Bowers, E.D. (1980) Long period disturbances due to wave groups. *Proceedings of the 17th Coastal Engineering Conference, ASCE*, Sydney.

Cellier, F.E. (1991) *Continuous System Modeling*, Springer-Verlag, New York.

Gravesen, H., Frederiksen, E.W., and Kirkegaard, J. (1974) Model tests with directly reproduced natural wave trains. *Proceedings of the 14th Coastal Engineering Conference*, pp. 372–85.

Jensen, O.J. and Klinting, P. (1983) Evaluation of scale effects in hydraulic models by analysis of laminar and turbulent flows. *Coastal Engineering*, **7**, 319–29.

Kirkegaard, J., Sand, S.E., Ottesen-Hansen, N.-E. *et al.* (1980) Effects of directional sea in model testing. *Ports '80 Conference, ASCE*, Norfolk, Virginia.

Mansard, E.P.D. (1984) Variabilité Statistique des Paramètres de Vagues. *Proceedings of the International Symposium on Maritime Structures in the Mediterranean Sea*, Athens.

Ottesen-Hansen, N.-E., Sand, S.E., Lundgren, H. *et al.* (1980) Correct reproduction of group-induced long waves. *Proceedings of the 17th Coastal Engineering Conference*, Sydney.

Roberts, N., Anderson, D.F., Deal, R.M. *et al.* (1983) *Introduction to Computer Simulation: The Systems Dynamics Approach*, Addison-Wesley, Reading, MA.

Sand, S.E. (1980) Detailed calculations of the second-order paddle control signal for reproduction of group-induced long waves (Appendices to Ottesen-Hansen, *et al.*, 1980). Institute of Hydrodynamics and Hydraulic Engineering, Technical University of Denmark (internal note), 10 pp.

Sand, S.E. (1982) Long wave problems in laboratory models. *Journal of Waterways, Port, Coastal Engineering Division, ASCE*, **198** (WW4), 492–503.

Sand, S.E. (1982) Long waves in directional seas. *Coastal Engineering*, **6**, 195–208.

Sand, S.E., Kirkegaard, J., Larsen, J. *et al.* (1983) Numerical and physical modelling of directional diffraction of waves. *Proceedings of the Conference on Coastal and Port Engineering in Developing Countries*, Colombo, March 1983.

Wiegel, R.L. (1964) *Oceanographical Engineering*, Prentice-Hall, Englewood Cliffs, NJ.

Zeigler, B.P. (1976) *Theory of Modelling and Simulation*, Wiley, New York.

HYDRAULIC–STRUCTURE INTERACTION 20

J.C. Huis in t'Veld and J.L.M. Konter

Given the waves, currents and average free-surface levels, the loads on structures may be calculated. In many cases the load calculation derives directly from the given hydrodynamics, but in other cases the structure may itself influence the hydrodynamics, and the extent of the influence may depend in turn upon the loadings. In this section an overview is given of the range of such hydraulic–structure interactions.

CONTENTS

20.1 INTRODUCTION

The development of coastal and estuarial engineering in The Netherlands, as a science, started with the closure of tidal basins: the former Zuiderzee and the estuaries in the south-western part of The Netherlands.

The design and construction of the Eastern Scheldt Barrier demanded an approach which was scientifically based, because prior experience with this type of structure was not available. For large hydraulic structures, such as closure dams, high water barriers and overflow dams, a combined approach using physical models and numerical models will ensure the optimal use of hydraulic results for design, construction and operational purposes.

In general, numerical models can be used to predict the boundary conditions for hydraulic structures. Physical models can make the translation from boundary conditions to the different design aspects of structures or elements of a structure (stability of bottom protection, local erosion, flow patterns, loads on structural elements etc.). Especially for structures in tidal conditions, both in operation and under construction, there are numerous possible geometries, each of which each has its own complex flow pattern. In such cases the connection between the numerical models and physical models can be made by choosing a selected number of hydraulic parameters, which, as they govern the hydraulic process, can be predicted with a reasonable accuracy.

A physical scale model can also give designers insights into the possibilities inherent in new kinds of structures and indications for the improvement of preliminary designs. Moreover, critical construction phases can be determined. Based on this insight, relatively simple mathematical models can be developed for the loads-to-strength relations. After a few weeks of such 'playing' with a physical model the experienced designer will have sufficient insight and inspiration to develop basic designs for new structures under the given hydraulic load conditions.

Any structure placed in moving water will experience loads, but the hydrodynamic processes in the vicinity of the object are also effected by the presence of the object. The same applies, in reverse order, when a structure is moved in stagnant water. The moving structure will generate pressures and currents in the fluid, which, in turn, will cause loads on the structure. These processes will occur simultaneously when, as a result of the hydrodynamic loads, the structure tends to move in the water. These water–structure interactions may result in a strong amplification of linitial movements, so a careful study of this phenomenon is important. The most reliable way to study this hydraulic–structure interaction is to use physical scale models. Most of the phenomena are dominated by gravity forces; and it is important to reproduce correctly the water movement induced by the moving structure and the boundary conditions. The scale models for water–structure interaction are mainly based on the Froude law on similarity. However, scaling down means a reduction of the Reynolds number. Some of the forces are a function of the Reynolds number, especially at structures

Coastal, Estuarial and Harbour Engineer's Reference Book. Edited by M.B. Abbott and W.A. Price. Published in 1993 by Chapman & Hall, 2–6 Boundary Row, London SE1 8HN. ISBN 0 419 15430 2.

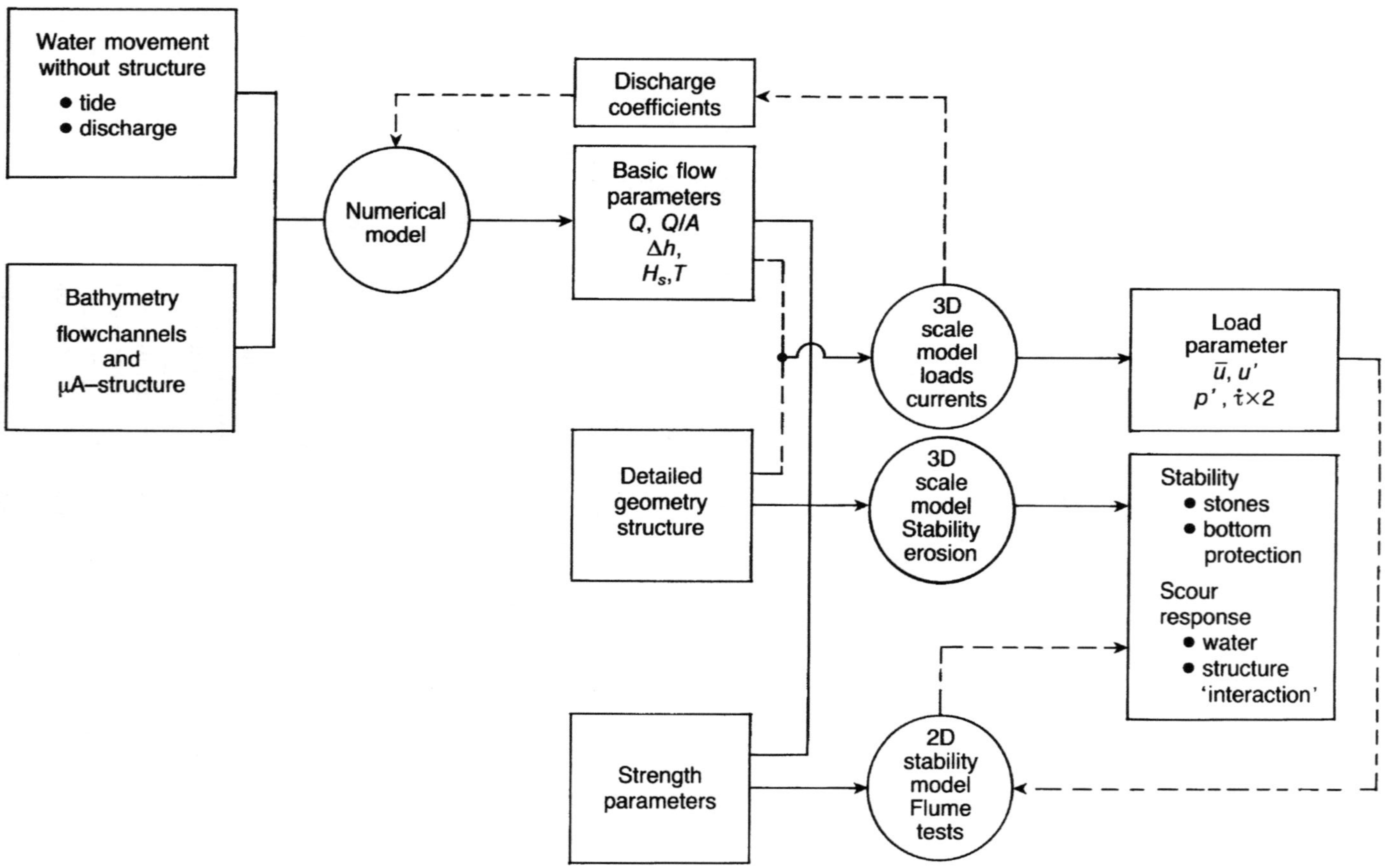

Figure 20.1 Hydraulic investigation.

with round edges. This problem can be solved by experience or, if this is not available, by varying the scale factor of the model. Also investigations into the effects of some changes in the geometry of the structure can give much insight into possible scale effects. In some cases the structure is not completely rigid, and therefore the deflection of the structure will initiate further water–structure interactions. These aspects can be studied in an elastic-similarity model, in which the elasticity of the structure is correctly reproduced. It is not always necessary to reproduce the exact elasticity of the structure. In many cases a study of two alternatives, one in which the elasticity is overestimated and the other in which the elasticity is underestimated, gives more insight into the response of the structure (including the stability of the bed protection).

Many attempts have been made to study the water–structure interaction by mathematical models. A reliable mathematical model for complex three-dimensional interaction phenomena needs much effort and time before the same accuracy can be reached as in a physical model. The mathematical model can give more insight than the 'black-box' of the physical scale model, but for specific design purposes scale models remain necessary. A combination of relatively simple mathematical models provided with coefficients obtained from scale models will often be the optimum tool.

For the impact of the hydraulic structure on the large-scale flow patterns, discharge characteristics of the structure are the governing parameters. These are expressed by the value of μA, the effective opening of the structure. The total effective flow opening of the entire structure determines, for example, the tidal range in a tidal basin which has to be closed, or protected against high water. To represent the flow conditions in the vicinity of the hydraulic structure, the following hydraulic parameters are usually selected: Q is the total discharge through the structure; Q/A is the average velocity in the opening of the structure; and Δh is the head difference over the structure.

With this approach, the flow parameters that have to be predicted (Q, Q/A and Δh) depend only on the global geometry of the hydraulic structure (expressed as μA). Details of the flow pattern can be incorporated in the 'translation' from basic flow parameters to hydrodynamic loads or stability criteria. For example, the stability of stones in a bottom protection is expressed as a 'load-strength' relation between q/A (or Δh) and the dimensions of the stones.

20.2 APPROACH OF HYDRAULIC INVESTIGATIONS

The set up of investigations for hydraulic structures as presented in Figure 20.1 indicates that physical models can be used for calibration of numerical models (measuring discharge coefficients), finding the 'loads' (flow pattern, pressures and process), finding the load–strength relations (stability and erosion models) and studying the interaction between water and structure. The 'load-strength' relation for the several aspects of a hydraulic structure can be found with two strategies. The first of these is a strategy in which loads and strength are determined separately.

The local boundary conditions, estimated by the numerical models (see Figure 20.1), are expressed as the governing flow parameters Q, Q/A and Δh and the wave parameters h_s and T at the structure. The detailed geometry of the structure transforms these boundary conditions into a flow pattern, and this causes the loads on the structure elements, expressed

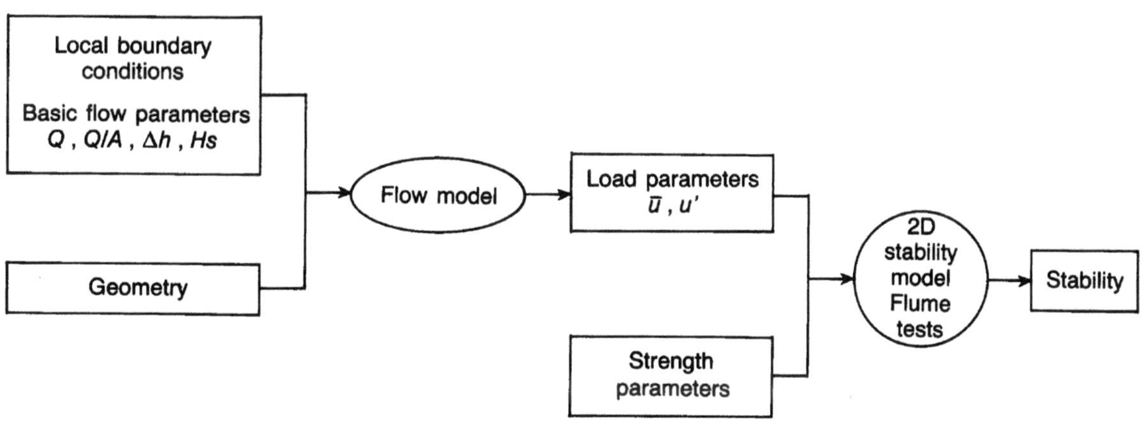

Figure 20.2 First strategy.

(a) (b)

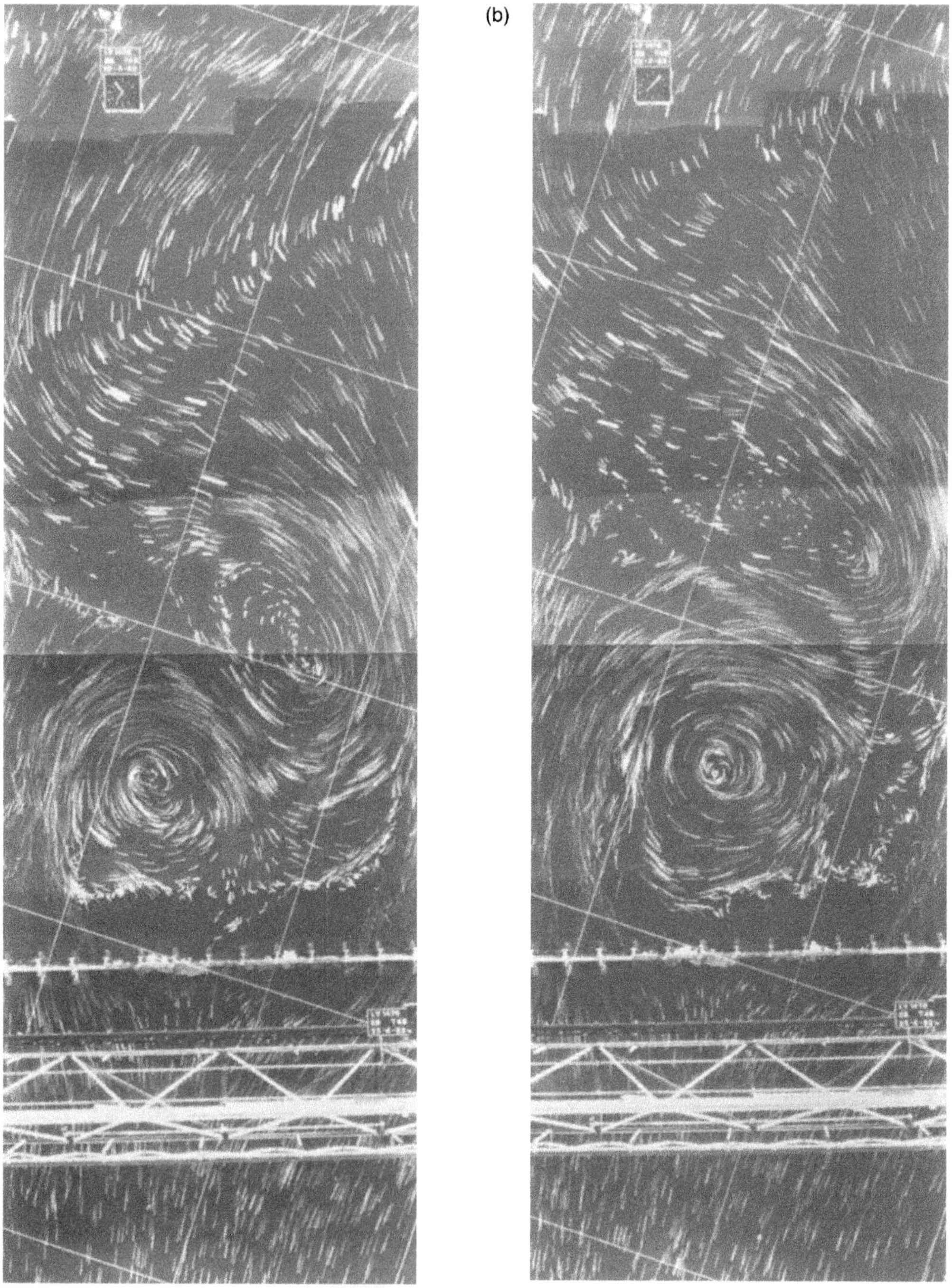

Figure 20.3 Current pattern behind closed gates (top views).

in terms of a mean velocity $\tilde{u}$ and a turbulent velocity u'. These 'loads' can be measured in a scale model from the flow pattern.

These loads provide the input for other models, for example: erosion; sedimentation models, in which the flow velocity is the most important input parameter; models for the stability of some new types of bottom protection structures in full scale; and models to calculate the forces on elements, ($F = (C_D)^{1/2} \rho v^2$).

The flow model gives the value of $\tilde{u}$, so F can be calculated if C_D is known, such as for instance from previous investigations, or measured in flume tests. This strategy is often applied to avoid scale effects. Flow problems can be studied on a small scale (e.g. n_L = 80) but, especially when the strength parameter is not only dominated by gravity forces, the relation between load and strength has to be found with scale factors, varying between 1 and 30. This strategy is illustrated in Figure 20.2.

However, flume tests in full scale (for example, to investigate the influence of water permeability and stiffness on the stability of several new types of bottom-protection structures) introduce schematization effects. The bottom protection can be reproduced at full scale, but not the water depth and the flow pattern.

Especially for very complex flow patterns (Figures 20.3 and 20.4), it is difficult to give one unique critical

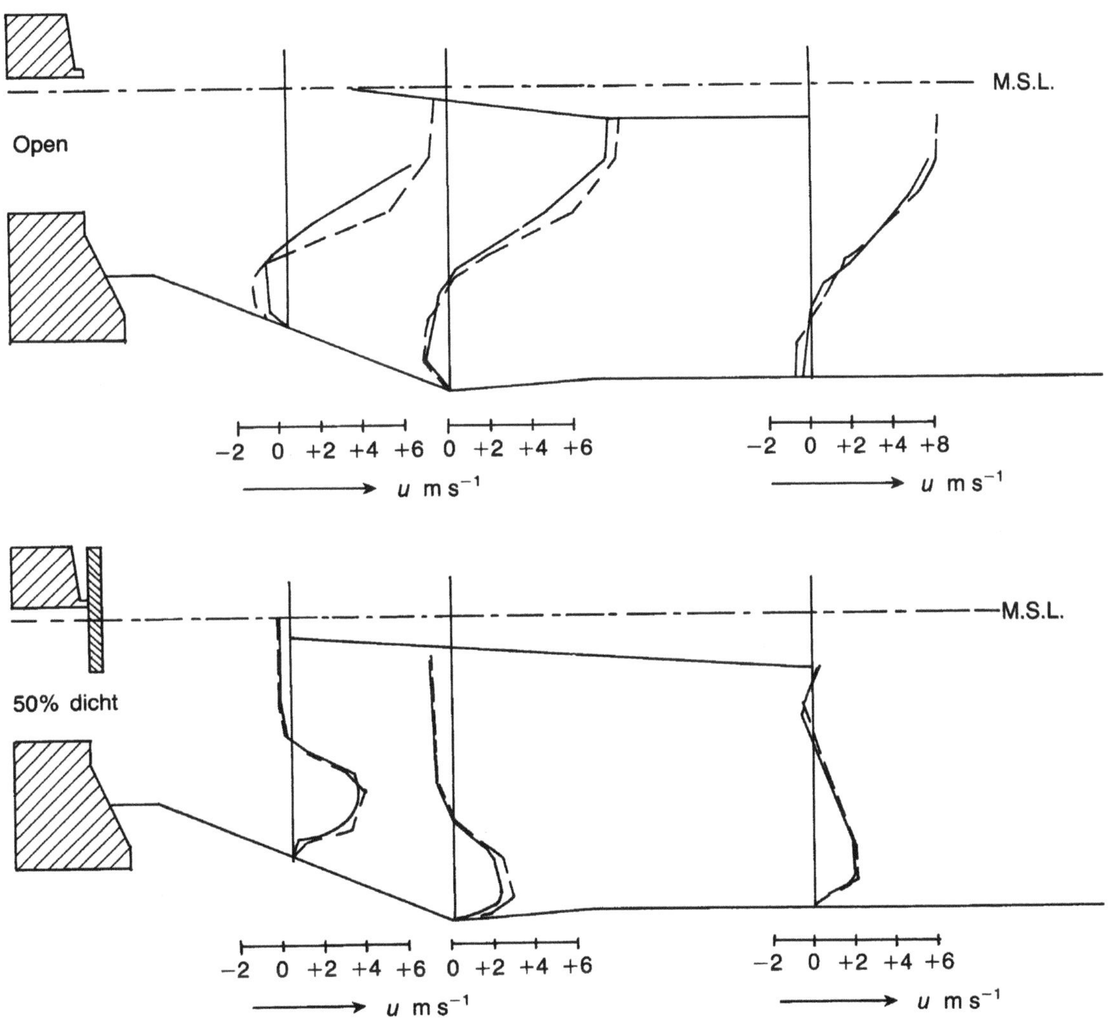

Figure 20.4 Vertical current pattern downstream of Eastern Scheldt barrier.

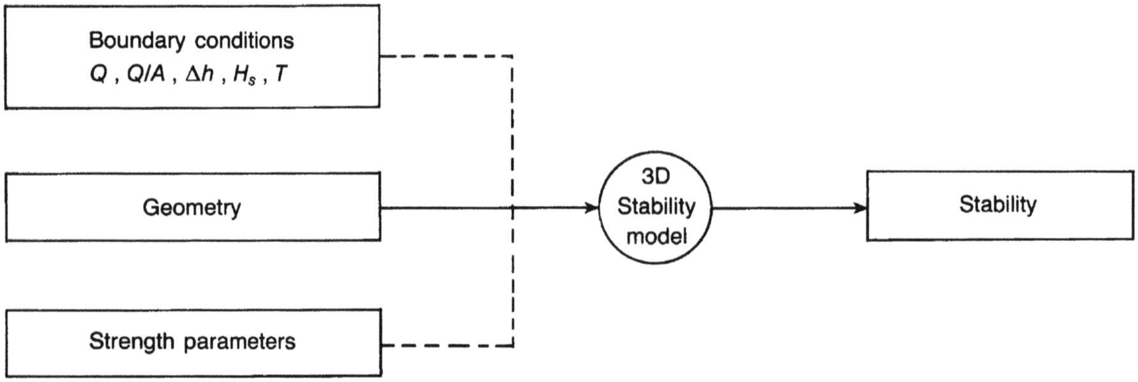

Figure 20.5 Second strategy.

velocity which represents both the current pattern and the strength. Figure 20.3 shows vortices with vertical axes, which are moving in time. Beside this there are also vortices with a horizontal axis (see Figure 20.3b). The vertical velocity distribution depends on the detailed geometry of the structure. Another problem is that an initial movement of one element of the structure may have a great influence on the current pattern and therefore on the load-strength relation. Thus, in a probabilistic design, not only is the initial movement important, but also the behaviour of the structure up until complete failure. In such cases, it is often difficult to make a connection between the flow model and the two-dimensional stability model.

The problems of defining (and measuring) the critical velocity can be avoided in the second strategy by using one three-dimensional model, in which the loads and the strengths are reproduced according to the same scale. The stability of the structure element and the structure–water interaction can then be tested directly.

In a physical, three-dimensional stability model, the criterion governing the selection of the model scales is that the forces acting on the structure element due to the various types of fluid motion and the forces resisting them should all be reproduced to the same scale. As the flow through and around the hydraulic structure and the surface waves are gravity phenomena, the general scales of the model should be in accordance with the Froude similarity rule. In order to ensure that the flow forces will be on the Froude scale, it is necessary that the structure of the turbulence, and in particular the amplitude and frequency of the vortices, is correctly reproduced. This can be achieved by a more or less detailed reproduction of the geometry of the structure. Moreover, the Reynolds number (Re) of the model should exceed a certain minimum value. This minimum value corresponds to the condition that macro-turbulence overwhelms the micro-turbulence at which energy dissipation takes place too rapidly by viscous forces.

20.3 EXAMPLES

20.3.1 LOCAL SCOUR AROUND STRUCTURES

The occurrence of local scour plays an important role in the design of hydraulic structures. At both sides of structures under tidal conditions, bed protection is necessary to prevent excessive scour. For economic reasons this bed protection will have limited dimensions, so that a scour hole will develop at the downstream part. This scour hole can be characterized by the maximum scour depth h_{max} and the upstream slope of the scour hole β (see Figure 20.6).

If the maximum scour depth becomes too large and the upstream slope becomes too steep, a slide or even a flow slide at the end of the bed protection occurs. The geotechnical instability causes the failure of a part of the bed protection, after which the scouring of the non-protected part of the bed will start, now closer to the structure. If the scouring process cannot be stopped in time, a new instability may occur, which may lead to the failure of the structure. Two kinds of measure can be applied: either, lengthening of the bed protection by which less steep slopes and smaller scour holes will develop; or scour hole protection by means of dumping stones before β reaches the critical gradient β_{cr}. In both cases a proper prediction of the time development of scour is necessary.

The prediction of the scour development is commonly solved by means of physical model investigations. However, for the translation of the model results to prototype predictions, a proper knowledge of the time scale is needed. For this reason, a sys-

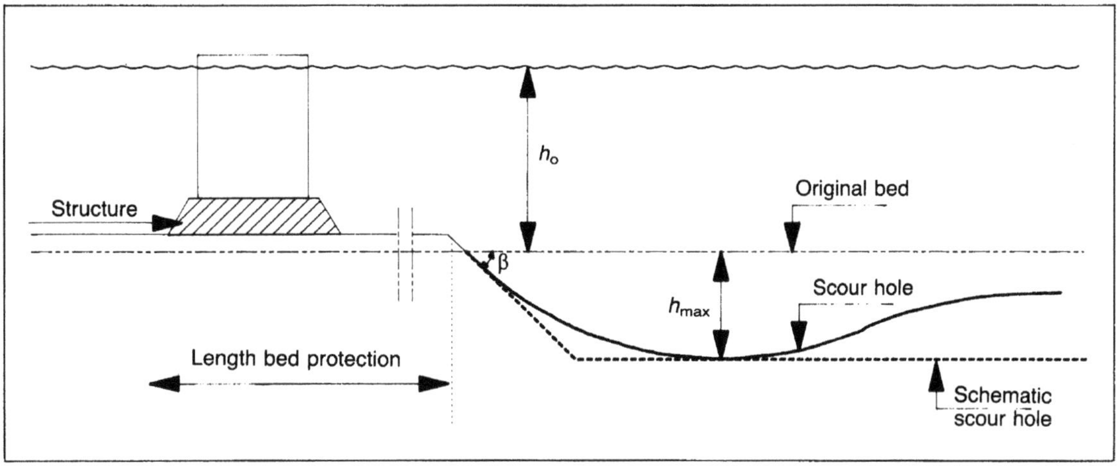

Figure 20.6 Definition sketch.

tematic investigation on the time scale for two- and three-dimensional local scour in loose sediments is often conducted.

Systematic research on local scours has resulted in the following empirical relations for the time development of the maximum scour depth h_{max} for steady flow:

$$\frac{h_{max}(t)}{h_0} = \left(\frac{t}{t_1}\right)^p \tag{20.1}$$

$$t_1 = \frac{K \cdot \Delta^{1.7} h_0^2}{(\alpha U - U_{cr})^{4.3}} \quad \text{for steady flow} \tag{20.2}$$

or after substitution of Equation (20.2) in (20.1):

$$h_{max}(t) = \frac{(\alpha U - U_{cr})^{4.3p}(h^{(1-2p)})t^p}{K^P(\Delta^{1.7}P)} \tag{20.3}$$

in which: $h_{max}(t)$ is the maximum scour depth (m); h_0 is the original water depth (m); t is the time (h); t_1 is the time in which $h_{max} = h_0$ under steady flow conditions (h); K is a dimensionless coefficient, ≈ 330; U is the current velocity (m s^{-1}); U_{cr} is the critical velocity for initiation of the bed material (m s^{-1}); Δ is the relative density of the bed material under water (dimensionless); α is the dimensionless geometry parameter, dependent on the geometry of the structure and the length of the bed protection; P is the coefficient (in two-dimensional flow conditions p = 0.4 and in three-dimensional flow conditions the value of p varies between 0.2 and 0.6 depending on the water depth). The value of α is the same in the model and in nature. The time scale of the scouring phenomenon can be deduced from Equation (20.2):

$$n_t = n_1^2 n^{1.7} n(\alpha U - U_{cr})^{-4.3} \tag{20.4}$$

In nature the values of α, U, U_{cr} and Δ are known, so the time scale is known and the development of the scour depth in time can be determined. The influence of unsteady flow (tidal movement) and upstream sand transport can be introduced (Konter and van der Meulen, 1986).

Equation (20.3) indicates that the scouring rate depends mainly on the scour rate parameter $\alpha_1 U_1 = \alpha U$ and the parameter p. In this case: the value of α_1 is constant for a specific geometry; the velocity U_1 is defined, as the local depth averaged velocity at the downstream side of the bed protection; and the value of p is a parameter, constant for a specific geometry but depending on the scour depth.

The values of αU and p can be determined for each specific geometry by means of physical model tests. It is recommended that the value of αU for each specific construction stage be investigated. The value of αU gives a direct indication of the critical construction stages. A model test also gives insights into the steepness of the upstream slopes for a specific geometry of the structure. It can be shown that vortices with vertical axes cause very steep slopes and are therefore relatively dangerous for the geotechnical stability of the bottom protection. The above method can also be applied to predict the local scour around structures without a bottom protection, and gives a very accurate prediction of the time development of local scour.

20.3.2 BOTTOM PROTECTIONS

For a Reynolds number related to flow depth and flow velocity in the axis of the structure, the model value should exceed the critical value of 1500 to 3000

(Konter and van der Meulen, 1986). The bottom protection has to be scaled according to the length scale. In this way, both the roughness factor (important for the flow pattern) and the main, motion resisting force of the bottom protection, i.e. its weight, are reproduced correctly. A minimum Reynolds number in the model (now related to the size of the elements and shear stress velocity) is then required ($Re^* > 500$). In principle, a third minimum Reynolds number in the model (now related to the flow through the subsoil and in the case of mattresses the permeable geotextile), is required to ensure the reproduction of percolation forces to the same scale as the other forces. A minimum Reynolds number in the model, now related to the size of the element and filter velocity, is then required ($Re_{uf} > 576$). Generally this third requirement is not a problem. In situations where the bottom protection is attacked by overflow and percolation forces, the forces due to the currents and waves in the channel are dominant and the percolation forces can be neglected.

In order to protect the bottom of the Eastern Scheldt barrier against scouring, about $4.5 \times 10^6\,m^2$ of bottom protection were required. Prefabricated mattresses have been developed, composed of permeable geotextiles with ballast attached to them. In this way, compared to the application of other types of bottom protection, a considerable reduction in materials and costs could be achieved. Several types of mattress have been studied. One of the most appropriate is the so-called 'block-mattress'.

The block-mattresses were prefabricated in sizes of 30 m × 200 m and consisted of a polypropylene fabric with concrete blocks attached with hard plastic pins, shot through the geotextile prior to the pouring of the concrete. In the factory (onshore) the mattress is wound onto a floating drum (measuring 10 m in diameter and about 42.5 m in length). This drum with the block mattress is then towed to an accurately positioned pontoon and locked in a kind of trap. During the turning of the tide, the mattress is unwound from the cylinder in a direction parallel to the current and sunk into position on the sea-bed. Each mattress has been laid down on the sea-bed with a small overlap over the previously-laid mattress. Under extreme circumstances, high and turbulent current velocities with a large component perpendicular to the overlapping mattress edge may endanger its stability (Figure 20.8).

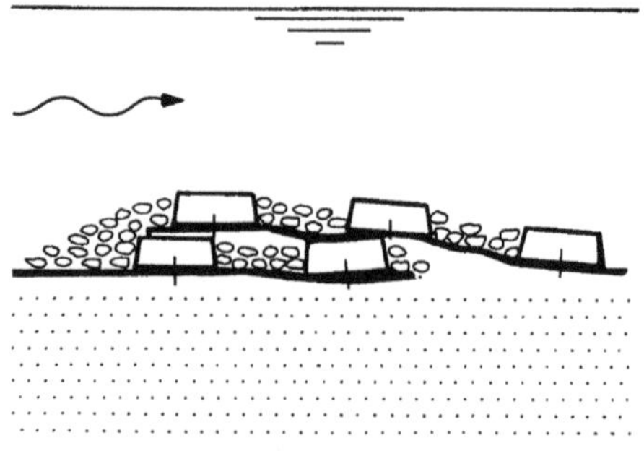

Figure 20.7 Block mattress.

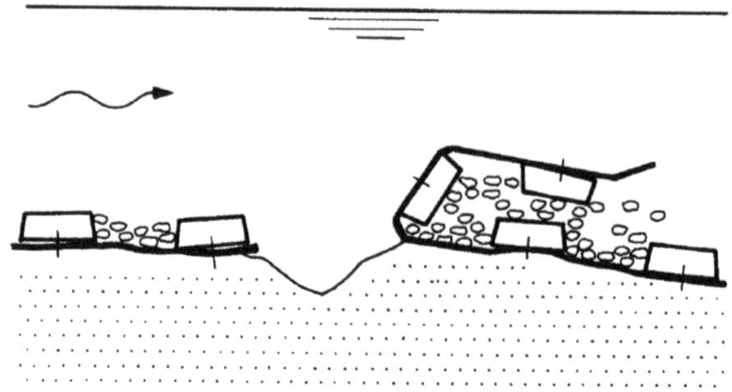

Figure 20.8 Instability of mattress edge.

Physical model tests had to be carried out to predict the stability. The investigation was started with the first strategy (flume tests on full scale) because it seemed not to be possible to fulfill the Froude law of similarity for such mattress properties as stiffness and water permeability. A systematic investigation has given insights into the physical phenomena determining the stability. With regard to the water permeability, it has been found that the ratio between the water permeability of the geotextile and the water permeability of the surrounding elements (stones or concrete blocks) should be the same in model and prototype (Huis in t'Veld *et al.*, 1986).

From Table 20.1 it follows that the geotextile is impermeable as compared to the blocks of the block mattress. However, in a scale model, the difference in water permeability between the geotextile and the blocks must be large enough to represent a correct pressure distribution. However, a better look at the stiffness characteristics of a certain type of mattress can solve this problem. For example, the concrete block mattress could be considered as a row of completely rigid parts (the blocks) connected by flexible

Table 20.1

Material	*Water permeability coefficient K (m s^{-1})*
Sand	10^{-4}
Geotextile	10^{-3}
Stones 0.1 m or blocks	10^{-1}
Channel flow	10

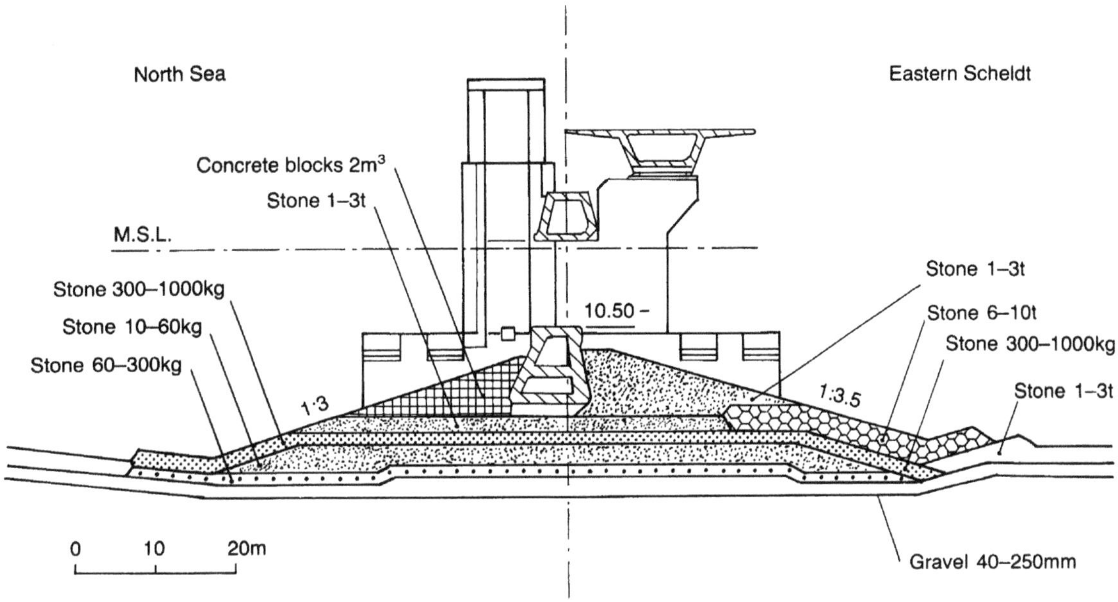

Figure 20.9 Rubble sill construction at Roompot 12 pier location.

hinges (the geotextile). The most important function of the geotextile is connecting the individual blocks in a horizontal direction. The stiffness of the geotextile in the vertical direction is negligible, so only the weight of the concrete blocks provides a force which resists uplifting. The shape of the blocks is also important for the stability, and especially their length in the current direction. If the blocks both in model and in prototype are completely rigid and have the same shape (reproduced on a length scale), the stability of the block-mattress can be investigated in a Froude model. In the model, the stiffness of the geotextile in the vertical direction has to be as small as possible.

20.3.3 STABILITY OF STONES

Stone stability has to be predicted with the second strategy. One three-dimensional scale model gives a direct translation from basic flow parameter q/A or Δh and required stone size. Scale tests for the sill of the Eastern Scheldt Barrier (see Figure 20.9) have shown that it is impossible to use the first strategy, because: firstly, the flow pattern is very complex and many unexpected three-dimensional vortices, which have considerable effect on the stone stability, occurred; and secondly, the relation between current velocities and stone stability for this flow pattern is not known. Maximum damage occurred at locations with a very low, mean velocity, but relatively high pressure fluctuations. Accordingly, first a systematic investigation of stone stability in turbulent flow conditions is necessary to determine local load parameters (e.g. turbulent shear stress, turbulent pressure fluctuation etc.) which can describe the stone stability. Afterwards it can be decided whether this load parameter can best be measured in a physical model or best calculated with a mathematical model. But even if it seems to be possible to predict the loads, a three-dimensional stability model has some advantages because it can also give the behaviour of the rubble-mound structure from initial movement until failure.

20.3.4 NEW TYPE OF BARRIER

A good example of a water–structure interaction study is the feasibility of a 'spinaker barrier'. This type of barrier can be used to close a flow channel for protection against high waters or pollution. The 'spinaker barrier' is composed of a synthetic fabric strung between two very strong synthetic cables. The ends of the cables are attached to abutments. In these abutments the ends of the cables can be manoeuvred in the correct vertical position. In a Froude-scale model, the whole closing procedure has been tested, including transport of the 'floating' barrier from embankment to embankment, closing the barrier by bringing down the under cable, the behaviour of the closed barrier, and opening the barrier. The model has yielded many insights into the possibilities and impossibilities of such a type of barrier. Ideas for improvement of the preliminary designs were developed and, after the physical model tests, mathematical models could be developed for the load–strength relations.

REFERENCES

Delft Hydraulics and Rijkswaterstaat (1989) Storm Surge Barrier Eastern Scheldt: Evaluation of Water Movement Studies for Design and Construction of the Barrier.

Huis in t'Veld, J.C., Stuip, J, Walther, A.W. *et al.* (1986) *The Closure of Tidal Basins: Closing of Estuaries, Tidal Inlets and Dyke Breaches*, Delft University Press, Delft, The Netherlands.

Konter, J.L.M. and van der Meulen, T. (1986) Influence of upstream sand transport on local scour. *IAHR Symposium on Scale Effects in Modelling Sediment Transport Phenomena*, Toronto.

PHYSICAL MODELS OF ESTUARIES 21

W.A. Price and M.F.C. Thorn

CONTENTS

21.1 INTRODUCTION

Professor Osborne Reynolds constructed what is generally acknowledged to be the first scientifically designed model of an estuary in 1885. It reproduced the Mersey estuary between the Liverpool Narrows and Runcorn at a horizontal scale of 1:31 800 and a vertical scale of 1:960 i.e. it had a vertical distortion of about 33:1. The bed of the model was formed in fine sand with a median particle diameter of about 165 μm, and the tide was created by a hinged dust-pan-like device that alternately emptied and filled at the seaward end of the model. The model bed was initially flat, and it is reported that after 2000 tides had passed up and down the model the configuration of the banks and channels in the model estuary reproduced the natural features in both character and scale. 'And, what is as important, the causes of these as well as all minor features could be distinctly seen in the model.'

Broadly speaking, during the history of the development of estuary models three distinct ranges of scales have been used. Reynolds and Vernon Harcourt at the end of the last century built to scales of about 1:30 000 to 1:40 000 although they did use others. Gibson and Allen in the 1920s built to scales of about 1:6000 to 1:8000 and today physical models of estuaries are built to scales of 1:500 to 1:1000. Also, alongside the development of mathematical models more part models are being built for the study of local problems where the flows are complex. These have scales of 1:100 to 1:200 and are often undistorted. These will be described later in the chapter.

The early models would now be called 'mini models'. Some engineers, but not many, still build such models (Beckett, 1992) but in the authors' view they are of limited use. However, to assist in the debate on their usefulness, Hydraulics Research, Wallingford are at present engaged in a research project where a mini model will be built and checked against a larger-scale model and field data. We do not have photographs of the early models built at the end of the last century but we do have one of the Gibson–Allen model of the Mersey estuary (Figure 21.1). It had a horizontal scale of 1:7040 and a vertical scale of 1:190. Readers who would like to know more of the history of the development of physical models should see Reynolds (1887), Gibson (1934), British Association (1889), and Allen (1946). Some thoughts on the philosophy of scale modelling can be found in Price (1978).

The success of Reynolds' model in simulating the hydraulic and sedimentary regime of an estuary is one of the cornerstones on which the science of engineering hydraulics has been founded. It was, however, a very small model that sought only to reproduce the major features of the estuary. As the commercial development of estuaries has increased in the twentieth century, as the size and draught of vessels has demanded greater and more reliable navigable depths, and as society has become more aware of the environmental consequences of estuary developments, so increasingly more detailed studies of estuary hydraulics and sedimentation have been demanded. Reynolds' model criteria have been extended to increasingly large-scale models until the quantum jump has become too great, particularly in respect of mobile-bed models, and difficulty has been encountered in simulating all the natural major and minor features, as Reynolds claimed to have done.

Coastal, Estuarial and Harbour Engineer's Reference Book. Edited by M.B. Abbott and W.A. Price. Published in 1993 by Chapman & Hall, 2–6 Boundary Row, London SE1 8HN. ISBN 0 419 15430 2.

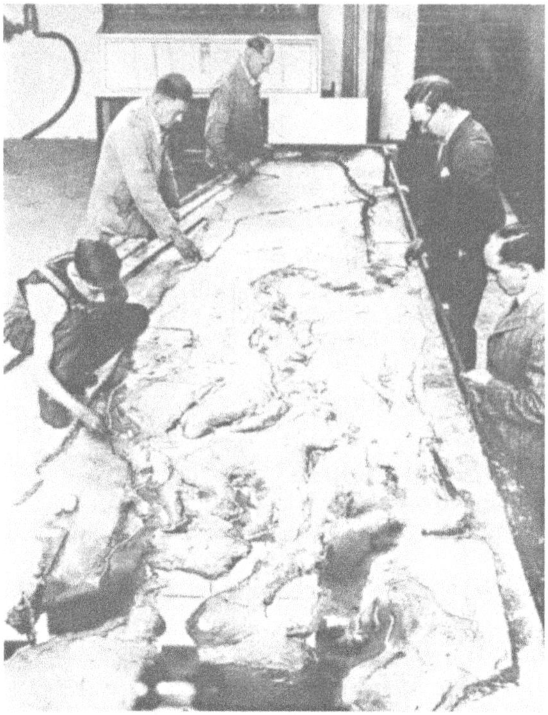

Figure 21.1 Gibson/Allen model of the Mersey estuary (1920).

This eventual over-stretching of the application of a technology is a common occurrence in many branches of applied science, and it says much for the robustness of Reynolds' original concepts that the break point was not reached for nearly a century.

21.2 NUMERICAL MODELS

These difficulties were being encountered at a time when numerical modelling techniques were in their ascendency, and held out the prospect of a new technology that would overcome the obstacles encountered in physical modelling. In particular, it seemed as though the problem of representing estuarine sediments in large-scale models so that the model bed topography properly represented the natural features might be overcome by incorporating sediment properties in a numerical model. Thus, in the early days of numerical modelling in the 1960s and early 1970s, there was a sharp swing away from physical modelling of estuaries towards numerical models that seemed to offer the prospect of a universal answer to all the problems that had been encountered.

Experience of numerical estuary models has since shown that they also have limitations which restrict their applications in some engineering problems and predictions. The physical model has regained some of its former status as an engineering design tool, and there is now a much more realistic balance in the choice and use of physical and numerical models in hydraulic studies. The strengths and weaknesses of these two alternatives are different, and largely complementary, so that a well informed, judicious choice of technique can be made to achieve the best possible result. However, this chapter will concentrate on the physical modelling of estuaries.

21.3 PHYSICAL MODELS OF ESTUARIES

The physical hydraulic model is effectively an analogue computer in which the physical processes are reproduced at a reduced scale. Properly designed, it reproduces tidal propagation and tidal currents, including secondary flow characteristics, accurately. Density stratification can be simulated by using saline water at the seaward boundary and fresh water to represent the flow from rivers. The great strength of the physical model is that it provides, in addition to the scientific data obtained by measurement, a synoptic view of the whole estuary which gives the experienced scientist or engineer an insight into its flow structure and its sensitivities. With some empirical experimentation this often leads to the development of engineering ideas and solutions that do not readily suggest themselves by study of the measured data alone.

The principal disadvantages of the physical model are its cost and its relative inflexibility. There are few estuarine engineering schemes that can be studied on models as small, or as vertically distorted, as those used by Reynolds. The objective of the model designer is to achieve minimum vertical distortion while maintaining an adequate overall constraint of a manageable physical size. If the model is too large it becomes expensive to construct, difficult to control, and demands a large staff to operate it. For these reasons large general-purpose physical models of estuaries are being supplanted by general numerical models which can be used to supply boundary input data to more localized physical models of specific parts of an estuary that require closer study. The general adaptability of the physical model is constrained simply by the effort required to re-model any area in which changes are to be investigated. Large changes, or a substantial range of options, might be more economically investigated, in principle, in a relatively simple numerical model, and the more promising schemes then studied in detail and refined in a physical model.

Physical estuary models may be simplified by reproducing a limited reach of the estuary in a steady-state or 'frozen tide' condition. In such a model, tides are not propagated through the model by means of a tide generator; instead the flow conditions appertaining to particular moments during the tidal cycle are created and held steady. This saves the cost of a model tide generator, solves the problem of creating the right total tidal storage volume for the estuary, and has the positive advantage that critical tidal states in the performance of an engineering scheme can be studied at leisure. This type of model will be discussed in more detail later and an example of its use will be described. In a fully tidal model with a typical model tidal cycle of perhaps only ten minutes, the critical moment may last for only a few seconds in each cycle. More sophisticated types of part models can be designed to reproduce tidal propagation and discharge. These are described later in the chapter.

21.4 SEDIMENT TRANSPORT TO SCALE

Simulation of sediment movement in a physical model is a problematical subject in which there are diverse opinions. Clearly the cohesive characteristics of estuarine muds cannot be correctly scaled, and siltation predictions are best made by computer-assisted desk calculations based on velocity measurements from a model. Non-cohesive, sandy sediments have been simulated in tidal models. The most successful simulation seems to have been in relatively small-scale models in which the reproduction of gross sedimentary features, and not fine details, has been the objective. Difficulties have been encountered in large-scale models, where the necessarily limited simulation of estuary sediment properties by a model bed material has become crucial to the performance. The science of simulating sediment characteristics and movement in a physical model has thus come up against a limitation, which is as yet unresolved. The erosion and accretion of bed sediments can be computed from physical data, but if they lead to substantial changes in bed topography the model has to be altered and re-run to provide a new set of data for the changed topography, and so on until a new equilibrium state is reached. It is at this point that the numerical model, continuously calculating erosion and accretion and adjusting the topography accordingly, becomes a more efficient and attractive option.

The calculation and prediction of sediment processes in estuaries, whether using physical or numerical models, depends ultimately on an adequate understanding of the physical properties of the sediments in an estuarine environment. The properties of cohesive sediments depend on chemical, mineralogical and environmental parameters as well as on particle size. They can be determined by properly designed field and laboratory experiments; this field of technology has expanded over three decades of effort and much expertise has been gained (see Chapters 38–41).

The dynamics of non-cohesive sediments have been extensively studied in the context of steady flows in rivers or laboratory flumes, but the semi-empirical relationships deduced for transport are not truly applicable to estuaries where the tidal flow is inherently unsteady and fully-developed conditions of sediment transport are rarely attained. The effect of wave action in enhancing the tidal transport of both cohesive and non-cohesive sediments in estuaries is a further complication that is also the subject of current field and laboratory research. It is research of these kinds to improve the understanding of estuarine sediment processes that will now lead to improved model predictions, rather than further sophistication of the intrinsic model techniques themselves.

21.5 CHOICE: MATHEMATICAL OR PHYSICAL

The final presentation of results from physical and numerical models is remarkably similar. In the case of the physical model, tidal current patterns might be shown as trajectories determined by photographic measurement of floats in the model: in the case of the numerical model they might be presented as a plotted matrix of arrows in the model grid squares. In both cases, tide curves and tidal variations of velocity magnitude and direction at specified points in the model can be produced in tabular and graphical form. The non-specialist client more readily assimilates the results of a physical model, as they can be demonstrated to him. He is therefore more likely to have an active and constructive input to the evolution of a physical model study, than if he is required to evaluate, and make comments on, the printed output of a numerical model. The recent development of computer animation techniques is providing similar accessibility to the results of numerical models, but as yet without the opportunity for interactive involvement by the inexpert user.

The choice between a numerical or a physical model is a matter for careful consideration and informed judgement. In some applications the choice is clear. For example, a numerical model will almost always be the choice when the area to be modelled is large and is also the most likely choice when tidal dispersion of pollutants or transport of sediments is being studied. However, when flow obstruction of a jetty structure, for example, is of importance, only a physical model will suffice. A physical model is also usually necessary when fine detail and secondary flow effects form an integral part of the study. An example of such a model is described later.

Between these clear-cut decisions there lies a large mid-band of estuary engineering projects which can be investigated by either numerical or physical model techniques, or by a combination of the two. The eventual choice is governed by non-technical as well as technical considerations. Cost and time are clearly major influences, but credibility and presentation of the results to both technical and non-technical clients also plays a part. The relative importance and complexity of tidal flow patterns and sediment transport also influence the decision. It is not a decision to be taken lightly without full appreciation of the capabilities, limitations and suitability of the available model techniques to produce the best possible result. Techniques and knowledge continually advance, balances of advantage change, and the client with a need to carry out an investigation of a proposed estuary development is well advised to discuss his proposals with competent specialists at an early stage, without preconceptions, so that he receives the optimum evaluations and predictions for his needs.

21.6 REPRODUCTION OF PHENOMENA TO SCALE

In considering the modelling of an estuary there are essentially three forces that need to be considered: gravitational, viscous and inertial. For dynamic similarity where gravitational forces are important it is necessary to maintain the same ratio of inertial forces to gravitational forces in model and prototype. The magnitude of the inertial force is the product of mass times acceleration. If we take a typical length L then volume is proportional to L^3 and mass proportional to ρL^3 (where ρ is the fluid density). If V and T represent a particular velocity and time interval, respectively, then acceleration is proportional to the rate of change of velocity i.e. V/T. Time can also be represented by a typical length divided by a typical velocity so that the inertial force F_i may be written as:

$$F_i \propto \frac{\rho L^3 V}{L/V} \quad \text{or} \quad \propto \rho L^2 V^2$$

The gravitational force, F_g, adopting the same procedure as above is:

$$F_g \propto \rho L^3 g$$

where g is the acceleration due to gravity. Thus if it is necessary to maintain the same ratio of inertial and gravitational forces and hence satisfy dynamic similarity,

$$\frac{F_i}{F_g} = \frac{\rho L^2 V^2}{\rho L^3 g} = \text{constant}$$

or

$$\frac{V^2}{Lg} = \text{constant}$$

The square root of this ratio is known as the Froude number (F_r) and it is necessary that this is the same in model and prototype.

The drag force on the bed of a channel may be derived from the basic equation for fluid shear stress:

$$\tau = \mu \frac{dV}{dy}$$

where the differential represents the rate of change of velocity near the bed and μ is the dynamic viscosity of the fluid in motion.

Thus the viscous force F_v may be determined as follows. Shear stress is the force per unit area. Therefore

$$\frac{F_v}{L^2} \propto \mu \frac{V}{L}$$

and thus

$$F_v \propto \mu V L$$

and so the requirement for dynamic similarity between inertial and viscous forces is then

$$\frac{F_i}{F_v} = \frac{\rho L^2 V^2}{\mu V L} = \frac{\rho V}{\mu L} = \text{constant}$$

or

$$\frac{VL}{\nu} = \text{constant}$$

Alternatively VL/ν = constant where ν is defined as the kinematic viscosity of the fluid (μ/ρ). This ratio is known as the Reynolds number.

The essential problem is the difficulty of fulfilling the Froude and Reynolds numbers to a scale of unity. Yalin, in the preface to his book *Theory of hydraulic models* (1971), points out that as the prototype and its model are on the same planet this implies that they are subject to the same force of gravity, or that the scale of acceleration due to gravity is necessarily equal to unity. Similarly, the use of the fluid (water) in the model means that the scales of fluid density and viscosity are also equal to unity. It is often very difficult, and sometimes impossible, to satisfy the criteria of similarity by using such 'unadjustable' values of the characteristic properties. In such cases the model cannot be designed according to the exact principles of the theory of models. It is up to the knowledge, experience, and even intuition of the experimenter to decide which compromises are needed to design a reliable model. Even with estuary models such compromises are not always the same. The engineer has to ask himself which characteristics of the model he is most interested in, and from the answer make the choices of scale etc. These are not easy decisions; thus most models are built in, or with the guidance of, a major hydraulic laboratory. Physical models are not exact tools but designed and used properly they are powerful aids to the solution of problems.

As intimated earlier a much greater difficulty arises in attempting to scale sediment transport. For correct simulation there must be an identical ratio of sediment transport rate between model and prototype at all corresponding places and times. This is impossible to achieve precisely, and the aim has been to minimize deviations. For example, this might be achieved by concentrating attention on specific areas of interest and scaling the bed material in these areas according to the criteria established by Yalin (1971). It is stressed, however, that these criteria apply only to granular and not to cohesive beds.

21.7 MATHEMATICAL AND PHYSICAL MODELS IN PARALLEL: MAPLIN PROJECT

On major projects complementary studies are often carried out on physical and mathematical models. Such an approach can add a great deal of confidence to the forecasts that have to be made when large changes to an estuarine system are contemplated. In 1971 the British Government proposed to site a major airport on reclaimed land at Maplin in the outer Thames estuary (Figure 21.2). The interaction between the physical and mathematical models built for this study is of some interest. The purpose of the models was to assist in the design of the shape of a large reclamation, to predict changes in the regime of estuaries close to the new works, and to determine the areas from which the large amounts of fill required for reclamation could be taken without causing serious effects on the estuary.

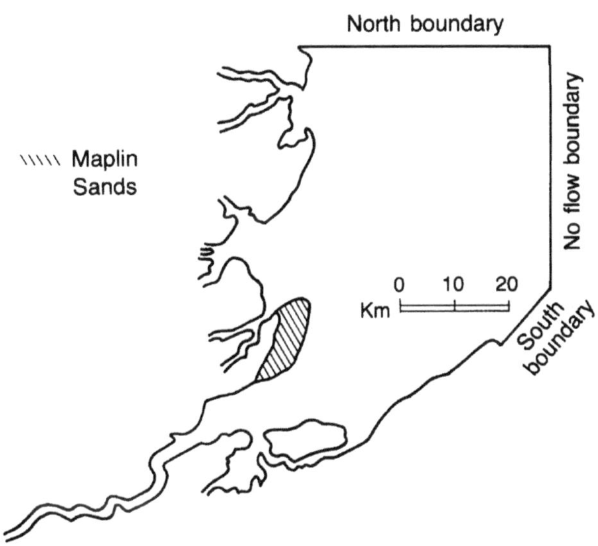

Figure 21.2 Model of Thames and its outer estuary (1:1000, 1:100).

The physical model of the outer Thames estuary, horizontal scale 1:1000 and vertical scale 1:100, is one of the largest ever built and covered an area bounded by the coasts of Essex and Kent and extended some 50 km offshore (Figure 21.2). There are two main components of the tide into the Thames estuary, from the north and through the Straits of Dover to the south. One of the problems in designing a model of such a large area with relatively high tidal currents is to decide whether or not the effect of the earth's rotation should be included. It was felt that the multiplicity of channels, separated by extensive intertidal banks, which are the main features of the topography of the outer Thames, would tend to minimize its significance. For this reason no attempt was made to reproduce the Coriolis forces in the physical model. However, to establish the importance of the earth's rotation, a mathematical model was built in which the Coriolis force could be included or excluded as desired. Both models could be checked against the results from an extensive field data collection programme. Examination of the data obtained during the proving of the physical model proceeded with two questions in mind. Firstly, how well does the model reproduce tide levels, velocity distributions and flow patterns? And secondly, does the model suffer significantly due to the absence of the Coriolis force?

As with most vertically exaggerated models, additional roughness had to be introduced. Desk studies had shown that friction factors in the outer Thames were very high due to the presence of large sand dunes. Initial attempts to prove the model with tidal height control along the northern limit and a tidal discharge control along the southern boundary calculated from spring tide conditions given in the *Admiralty Tidal Stream Atlas* met with some success in that the phasing of high and low water were reasonably well reproduced over most of the model. However, the shape of the tide curves was poor. This meant that tidal discharges would be in considerable error. At this stage it was difficult to decide whether the poor shape of the tide curves was due to the absence of the Coriolis effect, or to difficulties associated with the boundary conditions. Consequently it was decided to use the numerical model to assist in diagnosing the problem, and also, if possible, to suggest a remedy.

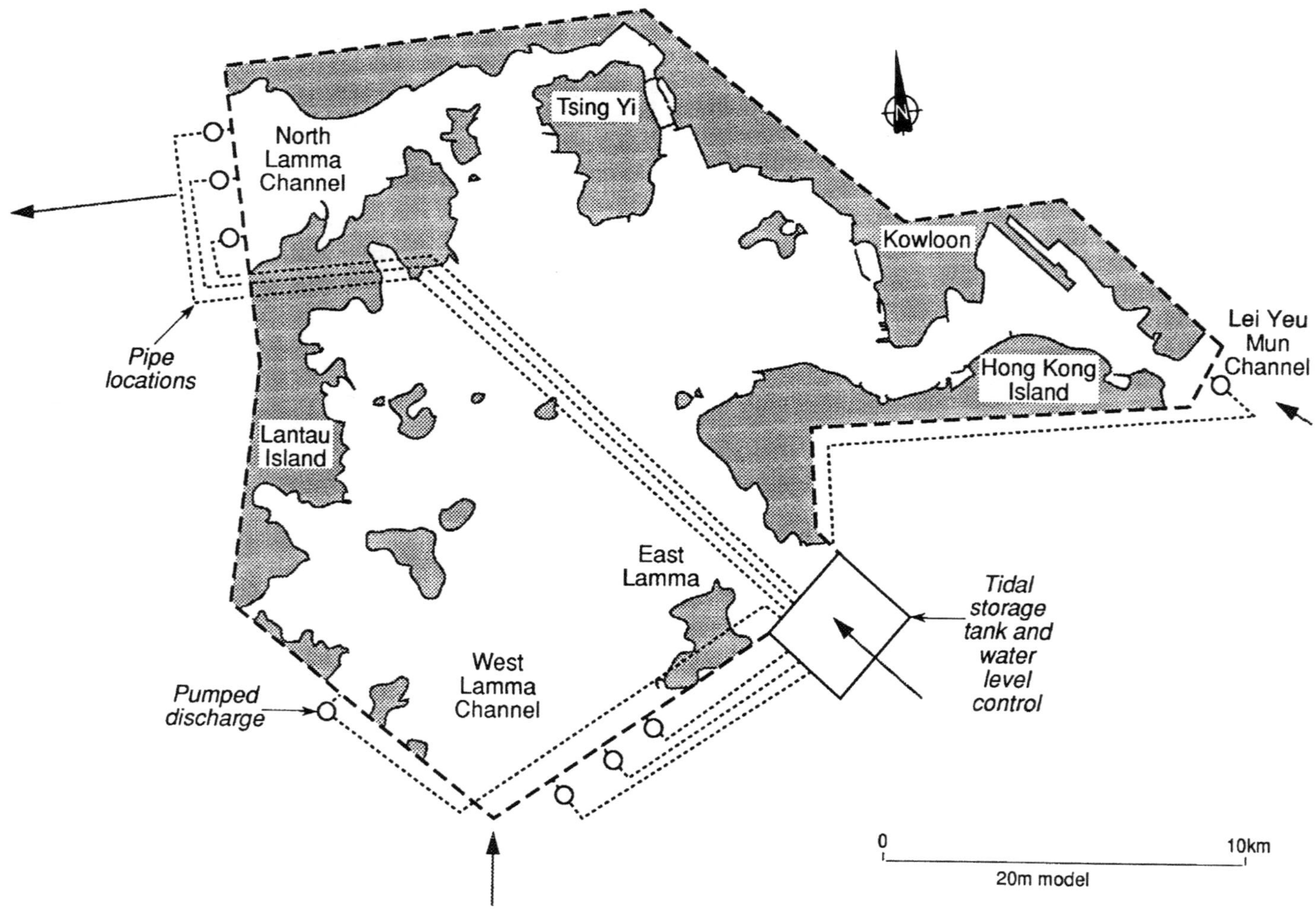

Figure 21.3 Physical model of Hong Kong. Arrows show relative discharges at approximately 5.6 hours after high water at North Point.

21.8 NUMERICAL MODEL ASSISTANCE

The numerical model was run first with completely open seaward limits, where the only conditions imposed were the tidal heights around all three boundaries: discharges were free to adjust themselves during the calculations to give the correct tidal propagation. This test confirmed that the tidal discharge input at the southern boundary was an accurate interpretation of the *Tidal Stream Atlas*. However, it also showed that flow existed across the eastern boundary which, although small in comparison to the flows along the boundary, was nevertheless significant compared with the flow across the southern boundary. The next stage, therefore, was to run the mathematical model with a no-flow eastern boundary, as in the physical model. As before, tidal heights were specified around all three boundaries, but discharges were only free to adjust themselves at the northern and southern boundaries. The results showed that correct tidal propagation was achieved by greater discharges at the southern boundary which compensated for the lack of flow across the eastern limit of the model. With these new input conditions to the physical model a dramatic improvement in tidal propagation was achieved. It was concluded that in this model of the outer Thames, with its system of deep channels bounded by extensive sand banks, the effect of omitting the earth's rotation is such that it can be largely counteracted by minor adjustments to the model input conditions and to the distribution of roughness.

This study has been described to illustrate the way in which a mathematical model can help to determine the boundary conditions for a physical model and correct malfunctions.

21.9 PHYSICAL MODEL: HONG KONG

A physical tidal model has been built to study the effect of proposed large-scale reclamations for urban and port development in Victoria Harbour, Hong Kong. Hydraulic models are usually built in established laboratories, but in this case there was a need for local engineers to have immediate access and input to the studies. Consequently it was set up in a specially constructed building in the Hong Kong New Territories.

The model (Figure 21.3) only simulates repeated dry season tidal flows and discharges. This represents conditions in the winter period in Hong Kong when discharges from the Pearl River are low and flow conditions in Hong Kong harbour are well mixed i.e. there is little or no vertical saline stratification in the water column. The input conditions in terms of discharges and tidal levels are complex, and were obtained from a two-dimensional depth-averaged numerical model of Hong Kong and the surrounding waters at its five open-sea boundaries. This meant that to test large schemes within the physical model area, which might themselves affect boundary conditions, they had first to be tested in the numerical model to provide new input conditions. Of course the numerical model itself had to be proved and this was done by comparing its output at many places where field observations had been carried out during two intensive data collection exercises.

The physical model has five open water boundaries and four of these have computer-controlled, reversible pumps to reproduce the discharges predicted by the numerical model. The flows into and out of the model are continuously monitored by electromagnetic flow meters and a feed back system ensures accurate reproduction of the discharges. In addition to these pumped flows the fifth boundary in the East Lamma channel has a pneumatic tide generator which accurately controls the rise and fall of the tide: Perhaps the most important function of such a model, apart from predicting the effect of proposed developments on the overall regime of the harbour, is to provide a detailed picture of flow conditions close to the proposed works. Specially made miniature current meters were used to measure current speeds at particular points of interest, such as proposed ferry terminals and container ship berths. Tidal flow patterns were obtained from cameras mounted near the roof of the building using a control system and miniature floats to produce photographs which show current velocities and directions as lines on the water surface. The length of the line corresponds to the current speed (Figure 21.4).

It is worth stressing that such a model enables the resolution of flow patterns, particularly near shorelines and structures to a greater degree of accuracy and detail than could be obtained from the numerical models. It is thus a most useful tool in optimizing the detailed alignment of the proposed reclamations and breakwaters and minimizing the adverse effect of eddies and slack water zones on navigation and siltation.

The model scales of 1:500 horizontally and 1:70 vertically were chosen taking account of model size, minimum water depths, tidal range, velocities and turbulent flow reproduction and the experience gained from a previous tidal model of Hong Kong eastern harbour built at Wallingford in the 1960s. These scales gave rise to diurnal and semi-diurnal tidal periods of about 25 min and 12.5 min, respectively, on the model.

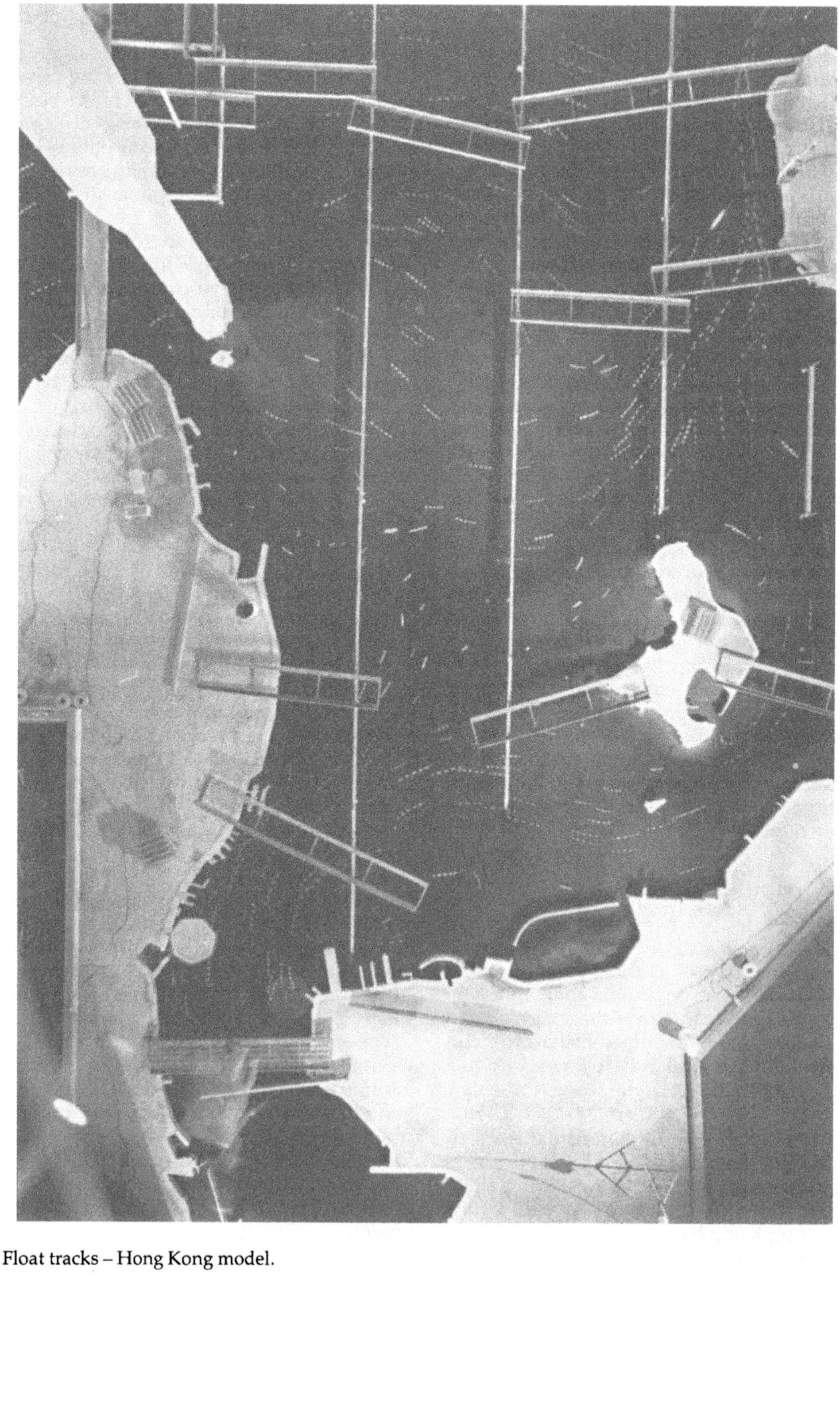

Figure 21.4 Float tracks – Hong Kong model.

(a)

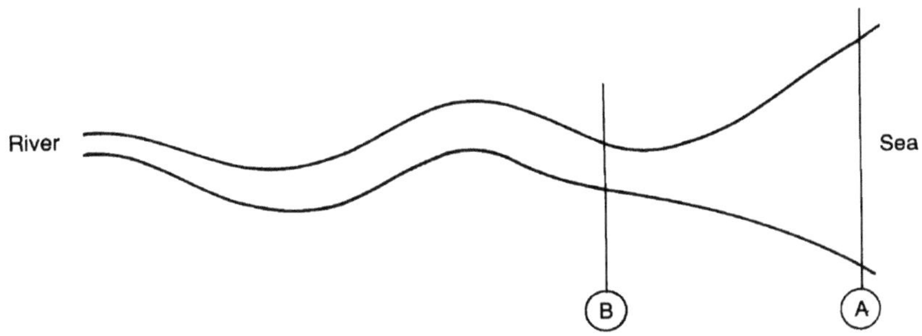

Position A

1. Ideal for overall studies especially when large scale works in one part of the estuary are likely to affect the regime elsewhere.
2. Adjust model - usually by adding frictional elements - to produce correct tidal propagation and enhance mixing.
3. Salinity/density currents can be reproduced.

Position B

1. Refer to 1 and 2 above.
2. Salinity/density currents can be generated but it requires a special tide generator where tidal storage can be compartmented to cope with salinity change at the boundary between HW and LW.

(b)

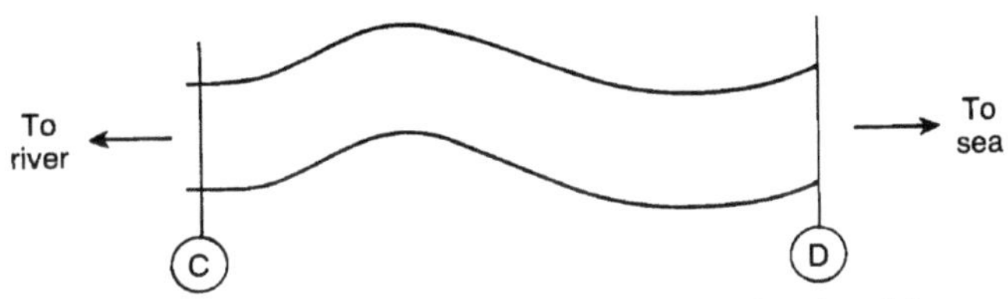

There are variations in the types of control that can be applied at C and D

1. Tidal level control at D and tidal discharge control at C either programmed or servo controlled.
2. Fixed state models where say a weir controls a fixed tide level at either C or D and a discharge is introduced corresponding to a particular tide level on flood or ebb.

(c)

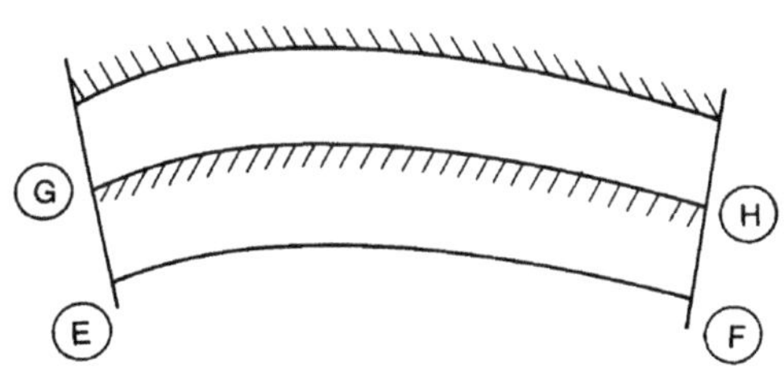

1. The same comments apply as to (b) above.
2. GH is situated along a flow line as far as this is possible.
3. Advantage – large scale and sometimes undistorted.

Figure 21.5 Types of estuary model (a) estuary models for general studies; (b) estuary models for detailed studies: part estuary – full width; (c) estuary models for detailed studies: part estuary – part width.

21.10 THE PART MODEL OPTION

Models of estuaries of the type illustrated in Figure 21.5a were very common 20 to 30 years ago. At that time they were the only option where studies of large-scale works likely to have an impact on the regime of the system as a whole were necessary. Nowadays there is the alternative of numerical modelling. Even so physical models of complete estuarial systems still have their place but their role in many cases has changed. Their value in presenting a global moving picture to an observer should not be underestimated. Often this can greatly assist in an overall appreciation of a problem. However, today they are more likely to be built to provide input data to numerical models or to part models.

Rather than build a physical model of an estuary from its source to its mouth, part models of the system are now often built which are either operated as steady-state models run at constant discharge or where tides are reproduced (see Figure 21.5). Steady-state models are particularly useful for investigating certain parts of the tidal cycle, which are of particular interest to the proposals under study, on a large scale.

These types of model are built when it is necessary to study very local conditions in detail. Because they cover a small part of the estuary they can be built to large scales (1:100 to 1:200) and are often undistorted. They frequently represent the full width of the estuary but in some cases they are designed to be part width (Figures 21.5b,c). They can be controlled in many different ways and some of these are illustrated in Figure 21.5. If, for example, it is necessary to study problems near jetties, bankside reclamation or lock entrances this is the type of model that is best suited to the task. Mathematical models as yet cannot reproduce the fine detail of the complex flow patterns that can occur around and through structures (jetties, for example) and it is in this area that the physical model scores. However, boundary conditions at the model limits still need to be defined and this can be done by either a mathematical or a physical model covering a larger part of the system, or in some cases by field studies concentrated on or near the proposed model boundaries.

Some words of warning are needed. One option, already described, is to run such models at steady states. This means that for various fixed tide levels, flood or ebb, appropriate discharges can be applied to the model. Such a discharge can only be correct across one section of the model because in this case tidal storage is absent. This means that the model should not represent an estuarial compartment that is too long and the 'correct' discharge should be arranged across the system near the locations of most interest. It should also be remembered that eddy shedding is a product of current velocity interacting with an obstruction. In a tidal estuary, where the current velocity varies with time, eddies grow and decay. In a fixed-state model this variation will not occur, so interpretation of the results by an experienced engineer is particularly important. Even with these provisos such models can be powerful tools. They are illustrated in Figures 21.5(b) and 21.5(c). Note that Figure 21.5(c) extends the technique to part-width models where the longitudinal 'wet' boundary is aligned as far as possible with a flow line. Again such a model enables scales to be kept large, sometimes with no distortion (1:50 to 1:100). A more sophisticated model of this type is illustrated where controls can be designed to reproduce correct tidal heights and discharges. There are variations to these types of control:

1. Control tidal levels along one of the boundaries and tidal discharges at the others.
2. A less sophisticated system to (1) is to programme the discharge, rather than control it, through a tidal cycle.

21.11 TILBURY GRAIN TERMINAL

One such example of the type of model study just described was carried out for the Port of London Authority to assist in the design of a riverside container jetty to be sited between Tilbury Lock Entrance and the existing Grain Terminal (See Figure 21.6).

The terms of reference were, to test and report on the following:

1. The effect of the proposed container jetty on;
   (a) The existing grain terminal;
   (b) The existing impounding pump intakes;
   (c) The existing lock entrance;
2. Navigation into an improved lock entrance.

It became apparent during the proving stage of the model that existing problems associated with the berthing and mooring of ships at the Grain Terminal jetty required solution prior to the testing of the container jetty designs. Consequently the terms of reference were amended and priority was given to solving the problems at the Grain Terminal. What follows confines itself to this particular aspect of the study to illustrate where physical models of parts of the system have their special use. Also, where model radio-controlled ships can be used to advantage in undistorted large-scale models.

21.11.1 THE MODEL

A 1:150 natural-scale model which covered the full width of the river from Broadness to just seawards of

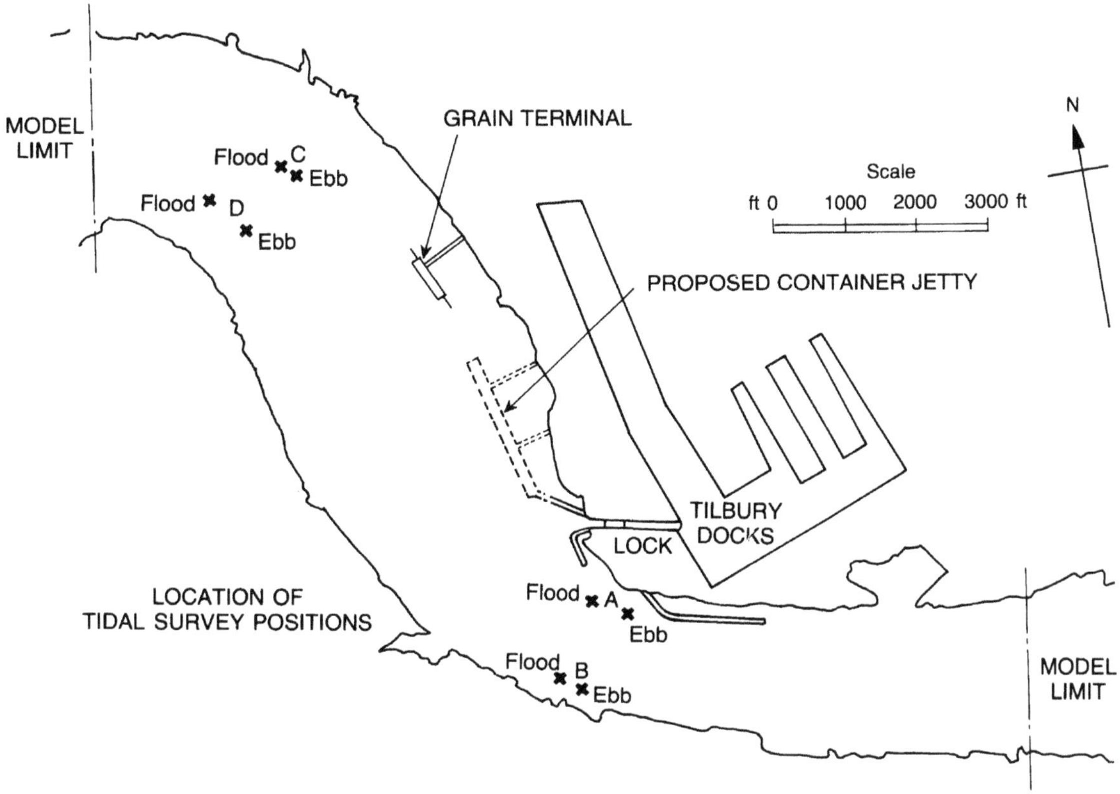

Figure 21.6 Tilbury Grain Terminal model.

the Tilbury tidal basin was used (see Figure 21.6). The Froude relationship gives time and velocity scales of 1:12.25 and a discharge scale of 1:275 500. As it was intended to use a radio control ship to illustrate the handling techniques, the upper part of Gravesend Reach was included to provide a sufficient length of estuary for the model ship to attain its correct speed before docking or berthing. The model covered the full width of the estuary. The flow control apparatus was designed to provide any selected state of the tidal cycle and to maintain a fixed level and discharge. In this case the parts of the tidal cycle that were of special interest could also be considered quasi-steady from the point of view of current velocity. It was also argued that as all the tests would be judged on a comparative basis it was reasonable to operate the model at fixed states of the tide. This simplified form of model operation reduced the testing time of the investigation.

All schemes were investigated at six states of tide, namely, 3, 2 and 1 h before high water, then 1, 2 and 3 h after high water. Most problems occurred at these times. The range of the tide was 20.0 ft and the data for proving the model was obtained by the Port of London Authority Hydrographic Department. A survey of tide and tidal currents was carried out to check the performance of the model. Boats were stationed at four positions in Northfleet Hope (see Figure 21.6) and recorded tidal velocities for flood and ebb near the bed, at mid-depth and just below the water surface.

The correct simulation of the water levels and associated velocities was obtained by adjusting the discharge through the model. Tidal velocities in the vicinity of the impounding pump intakes and also in the lock entrance were recorded in the model and were compared with the prototype survey data. The permeability of the upper and lower lead-in jetties was adjusted to correctly simulate flow across the lock entrance. Piles were grouped together and due allowance was made for their dimensions and shape so that the correct drag coefficient applied at the maximum velocity. Flow patterns were recorded photographically. Model ship manoeuvres were also recorded by a camera and were analysed. Diagrams depicted the ship in successive stages of docking together with details of the ship's engine speed, position of the rudder and the tug forces.

21.11.2 SHIP MODEL

The model container ship was constructed to line plans supplied by the Technical Services Division of the P&O Steam Navigation Co. The length and breadth on the water line were 775 ft and 105 ft, respectively, the length overall being 815 ft. The draught was kept uniform at 30 ft. The prototype ship had a forecast draught of between 26 ft and 30.5 ft, depending on its cargo, with a maximum trim of 6 ft at the stern.

The model ship (Figure 21.7) when fully instrumented was weighted and balanced to float at a level 30 ft draught. It was adjusted to have the correct moments of inertia in all planes. Speed trials were carried out in a large tank and the motors were adjusted to give the ship the correct through-the-water speeds at model scale for six settings. Trials were also conducted to match the turning circle of the model ship with that of the proposed container ship. Due allowance was made during operation of the model to simulate correctly the response time of the rudder to orders issued by the pilot. The simulation of tug forces was obtained by using small fans mounted on the deck of the ship. Three variable speed fans were used, each calibrated to produce a maximum force of 30 t, equivalent to two tugs. The fans at the bow and stern were placed across the ship whilst the midship fan was mounted fore and aft. A tug pulling at say 45° on the starboard bow could be represented by using the bow fan and the midship fan with equal strength. At the same time a force could be applied at the stern. The apportioning of the force applied by the midship fan to either bow or stern would give a varying angle of pull. The direction of the forces was resolved along the axes of the fans. The fans simulated tug forces but they could also represent forces applied by mooring lines heaved by winches. It was assumed that when a ship was close by a jetty then forces applied by the fans were in fact, mooring line forces or a combination of tug and mooring line forces. During the berthing manoeuvres every effort was made by the model operator to apply actions and forces realistically. Where it was considered that only one tug would have room to manoeuvre, then the model ship was restricted to a maximum tug pull of 15 t. There were occasions, however, when due to the speed of operation it was not possible to control the ship through intricate manoeuvres without inserting an unrealistic force. Where this occurred it was noted and due allowance made in interpreting the final result.

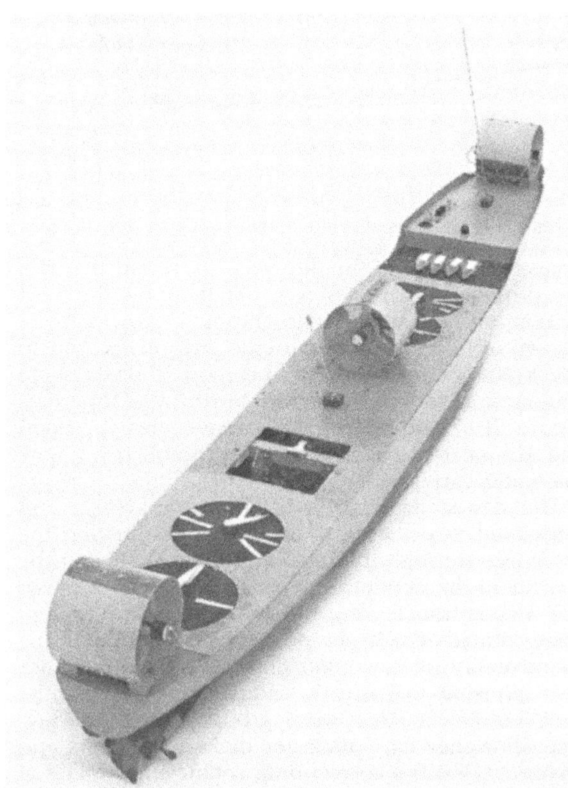

Figure 21.7 Radio controlled ship.

21.11.3 THE GRAIN TERMINAL

It is worth stressing that the object of this section of the chapter is not to describe all the studies but only to demonstrate the value of large-scale part models where complex flow patterns need to be altered to improve an existing situation. The physics of such flow patterns is still little understood so recourse to physical models is, at present, the only alternative. Mathematical models will one day be able to handle such problems but this is still some way off.

The problems at the Grain Terminal were fourfold:

1. Difficult berthing during the ebb tide;
2. Excessive mooring forces;
3. The development of a scour hole on the river bed, close to the structure and associated siltation at the lower end of the jetty;
4. The tendency for moored vessels to sheer away from the jetty.

The cause of all four problems was thought to be the same: misalignment of the jetty to the ebb tide flow and also the presence of a bluff upstream face. The flow during the ebb impinged upon the upstream end of the Grain Terminal at an angle of about 10° and separated from the face of the structure. This angle of flow imparted a force tending to move a ship away from the jetty. The eddies cast from the structure made the berthing of ships almost impossible shortly after high water and as the run of the tide increased towards mid ebb the berthing of large ships could not be completed (see Figure 43.9a).

A solution to the problems was to provide an additional structure which allowed the tidal flow to negotiate the obstruction (the Grain Terminal) with little disturbance and no separation. Three modifications were considered:

1. An asymmetrical bull nose shape.
2. A curved guide wall 220 ft long.
3. A curved guide wall 370 ft long.

Assessment of the modifications to the jetty was by visual observation of the flow patterns. Surface floats (polystyrene chips) were released onto the water surface and observed and photographed as they passed the Grain Terminal (see Figure 43.9b). At the same time potassium permanganate crystals were sprinkled on the bed of the model and the dye traces observed by eye. Velocity measurement were taken inshore of the river berth along the line of the access road. These were compared with velocities obtained for existing conditions.

Tests were carried out with modifications (1) and (2), and it was observed that they both created parallel flow along the Grain Terminal jetty. However, it was later specified that the parallel flow should be obtained at least 150 ft upstream of the jetty and in order to achieve this a 370 ft long guide wall was tested. It was formed of a 206 ft straight section and continued on the projected line of the river side face of the jetty, progressing into a curved section some 164 ft in length.

Tests were carried out for three states of the tide during the flood and showed that there were no adverse effects created by this 370 ft guide wall. For most of the flood tide a large reverse eddy existed near the Grain Terminal and flow lines had similar directions but the velocities were lower than during the ebb tide.

Visual observation and the evidence of the float tracks recorded on film indicated that the worst flow conditions occurred at 2 and 3 h after high water near the Grain Terminal. Assessment of the guide-wall design was made for 2 h after high water with general notes on behaviour for other states of tide.

Observation of the float tracks during the ebb tide showed an angled approach by the flow to the north-western end of the Grain Terminal. With the existing jetty, separation took place and considerable turbulence occurred. The angle of the flow was deflected even further towards the centre of the river and this produced extremely difficult conditions for the berthing of a ship on the ebb tide. The turbulence added to the difficulties in that a ship's head moved first one way and then the other, indicating that the angle of approach was critical. Current velocities in this area could be as high as 5 knots on a 20 ft range spring tide. The float tracks also indicated a confused pattern of flow lines below the Grain Terminal; this was due to turbulence created at the upstream end of the jetty. The turbulent eddies could be seen as far downstream as the impounding pump intakes. The scour hole on the river bed near the north-western end of the Grain Terminal, revealed by recent surveys, was a direct result of the turbulence shed from the Terminal structure.

With a 370 ft guide wall in position the flow negotiated the upstream end of the Grain Terminal without separation and very little turbulence. If anything there was a tendency for the surface flow to provide an on-berth component towards the centre and down river half of the berth. Dye crystals placed on the bed showed that parallel flow lines were established at least 150 ft upstream of the jetty.

Ship manoeuvres were carried out with the Grain Terminal in its existing condition. This was done to provide a basis for comparison and to a certain extent illustrate the ability of the radio-controlled ship to behave in a realistic manner. The addition of a 370 ft guide wall considerably improved berthing conditions.

Subsequent to these studies and after demolition of the model the Port of London Authority required answers to some further questions. Although they accepted the conclusions of the studies just described they debated whether less ambitious schemes could provide the necessary solution to the problem at the Grain Terminal. An existing model of the full estuary was used to extend the studies, which is briefly described in Chapter 43. The photographs that are reproduced in this subsequent chapter as Figures 43.9a and b illustrate by float tracks the eddy shedding at the existing structure and a scheme for its elimination.

21.12 CONCLUSIONS

This chapter has aimed to show the many ways physical models of estuaries, of various types, can be used to solve a variety of problems. Although the physics of similitude is an exact science the chapter demonstrates that its laws cannot be entirely satisfied in scale models so that compromises have to be made. Such decisions do not always lead to the same conclusions because they depend on the type of problem to be studied and at what scale. Experience is vital in this respect and this is why the majority of physical models are built in international laboratories or at least under their control.

When considering the type of studies required to solve estuarial problems there are a number of options available to the hydraulic engineer. Sometimes, but not often, a solution can be found by carefully planned field studies followed by an analysis of the results (see Chapter 34). In the majority of cases, however, some type of modelling is necessary. There are still entrenched views in some quarters that numerical models are the ultimate panacea. Others hold similar opinions about physical modelling. Fortunately such polarization is now not so widespread with the result that the strengths and weaknesses are now fully considered before embarking on a preferred methodology.

The hydraulic engineer is invariably faced with the problem of establishing confidence in his predictions and this is often achieved by using the strengths of both techniques. He might decide to run a numerical model in parallel with a physical model. This can be the best of both worlds.

REFERENCES

Allen, J.A. (1946/7) Model experiments in relation to harbours and waterways. *Journal of the Institution of Civil Engineers*, **25**, 378–409.

Allen, J.A. (1947) *Scale Models in Hydraulic Engineering*, Longmans Green & Co. London.

Beckett, T.K.H. (1992) Hydraulic mini model studies. Beckett Rankine Partnership, London, UK, in *Marina Technology*. Proceedings of the 2nd International Conference, 31st March–2nd April 1992, Computational Mechanics Publications.

British Association (1889) Report of the committee appointed to investigate the action of waves and currents on the beds and foreshores of estuaries by means of working models. *British Association Report*.

Fourth International Congress on Inland Navigation, Brussels, Belgium, 1890.

Gibson, A.H. (1934) The problems of tidal estuaries: forecasting by working models. Royal Institution of Great Britain. *Report on a weekly evening meeting*, Friday 13 April 1934, pp. 1–23.

Price, W.A. (1978) Can we learn from the past? *Proceedings of the 16th Coastal Engineering Conference*, Hamburg. ASCE Vol **I**, pp. 25–36.

Reynolds, O. (1887) On certain laws relating to the regime of rivers and on the possibility of experiments at small scale. *British Association Report*.

Yalin, M.S. (1971) *Theory of Hydraulic Models*, Macmillan, London.

Yalin, M.S. (1973) *Mechanics of Sediment Transport*, Pergamon Press, Oxford.

THE PHYSICAL ENVIRONMENT

THE PHYSICAL ENVIRONMENT: AN INTRODUCTION

W.A. Price

The introduction by M.B. Abbott to the first part of this book stresses the importance of mathematical modelling of hydraulic systems as a design tool and the way in which such modelling techniques have advanced rapidly in the last two decades. In a physical model the problems of scale can be considerable, but provided the physics of a phenomenon are well known then the mathematical model can be considered to represent the prototype to a scale of 1:1.

Whether one chooses to use a mathematical or physical model approach or a combination of both to solve a particular problem is not an arbitrary decision as two examples will illustrate. In a physical model of an estuary it is impossible to represent the movement of cohesive sediments accurately even though the water movements can be reasonably well reproduced: in this case one would usually choose a mathematical approach. However, to reproduce the movements of a ship at a berth in a harbour subject to wave action is a difficult mathematical problem complicated by the characteristics of its moorings and fenders, and a physical model approach would often be preferred. Sometimes the best of both worlds can be achieved by using both approaches in parallel.

At the risk of labouring the argument I would stress that physical models have not had their day, and more research money could sensibly be spent on their improvement.

The future improvement of mathematical models lies mainly in the better description of physical processes whereas physical modelling depends on practical experience, from which adjustments can be made which improve the model as a tool for forecasting. In recent years, the improvement of field measurements has considerably improved our understanding of the basic physics of our subject. The marine environment is often hostile and we owe a great deal to those who work in the field and have developed such an array of sophisticated instrumentation.

Turning to the real problems with which coastal engineers are faced, the best solution to a problem is frequently a compromise between the many interrelated aspects that make up the whole job. For example, is the best hydraulic solution necessarily the most economic? What is the best type of plant to carry out the work, and is it available, especially in remote areas? How well documented is the hydraulic environment, especially under extreme conditions? These few examples demonstrate the iterative nature of the design process and the broad spread of expertise that the coastal engineer must draw on. Some further comment on the practical nature of the subject is called for, and I will first deal with the coastal and shoreline aspects and provide an introduction to the individual chapters which follow.

COASTAL AND SHORELINE ENGINEERING

I see beaches and their reaction to man's interference as follows:

> An eroded beach is normally crying out:
> 'Give me some material, I am starving. If you feed me, make my diet coarser, if possible, than I have been used to. This way I will survive much longer. If you feed me I will promise not to starve my neighbours.'
>
> At the same time groynes are saying:
> 'We are not magicians – we cannot trap sand that isn't there.'

With few exceptions coastal management used to be a fairly haphazard affair. It is now generally accepted that it cannot be done successfully on a local or piecemeal basis. Cooperation between interested parties is essential. Chapter 22 by Townend reviews the many frameworks that can be adopted to achieve good coastal management for specific situations. It involves long-term monitoring which can, for example, help to identify the first signs of a problem and provide data which can be used as an aid in its solution.

The book stresses the need in dealing with coastal problems of first establishing what is going on at the site in question and what has happened in the past. The need for well planned beach monitoring, even if the beach at the present time is satisfactory cannot be stressed too forcibly.

At one time groynes were invariably the chosen solution to a beach erosion problem whereas nowadays the coastal engineer has many alternatives to consider: beach nourishment, offshore breakwaters, artificial headlands etc. A major contribution to the solution of such problems was the establishment of long stretches of coastline (geographical cells) under one authority. These important administrative changes started to take place in the 1970s. At about the same time beach nourishment was first tried successfully in the UK at Bournemouth, a town on the south coast of England.

Fleming reviews the way beaches react to such works as groynes, detached breakwaters and beach nourishment in Chapter 23.

This is followed by a very practical chapter by Roberts on the way in which the coastal engineer should approach coastal problems. It emphasizes the importance of an understanding of fundamental beach processes as the foundation of sound problem solving. Once again, as elsewhere in this book, the importance of beach monitoring is stressed and the many activities within this field are described. The author illustrates the interplay that has to take place between the fundamental thinking and the practical options available to solve a particular problem within the design process. The chapter represents an overview of the subject paying particular regard to beach nourishment, offshore dredging and sand bypassing.

A study was carried out in the early 1960s on improving entrance depths to a harbour in Australia. With a harbour bar subject to waves and currents, dredged depths can disappear overnight in a storm and this happened in this case. The chosen solution was to increase current velocities over the bar by a training wall. However, one of the schemes considered involved laying pipes under the bar and pumping water through holes drilled in the pipes, with the aim of decreasing the shear stress of the indigenous material and hence making it more easily erodible. At that time we were not able to make sufficiently reliable calculations of the discharge required to obtain a satisfactory result and the idea of fluidization was shelved. The idea, which has now been extended to other applications, has now been resurrected and full-scale trials are being carried out in the United States on channel improvement, beach build up, and pipeline burial. I would like to thank Dr Jim Parks of DYNEQS Ltd for his contribution to this subject which forms an addition to the chapter by Professor Roberts.

In Chapter 25 Fleming demonstrates that although the movement of sediment in the near shore zone is extremely complex considerable advances have been made in recent years to the point where engineers can now make reasonably reliable forecasts of the effect of structures on beaches by using mathematical models.

If sea levels rise as a result of the greenhouse effect this will have a significant effect on how we manage our coastlines and design coastal structures. One authoritative prediction is that global mean sea levels will rise by about 6 cm per decade over the next century with an uncertainty range of 3–10 centimetres per decade. Some countries are particularly vulnerable. For example, eight to ten million people live within 1 m of high tide level in each of the unprotected river deltas of Bangladesh, Egypt and Vietnam and many of the 1200 islands in the Maldive archipelago are barely 3 m above sea level. Many coastal structures are subject to waves which are limited in height by shallow offshore depths. As sea level rises they will therefore be exposed to increasing wave heights. Wave run-up, overtopping and wave loading will increase. Changes in storm tracks could alter incident wave conditions and beaches will have a greater tendency to erode. These are only some of the consequences of an increase in mean sea level, and coastal engineers will need to plan their work with a growing regard for the phenomenon.

Between the time I wrote these words and the date of publication many conflicting opinions have been expressed on the greenhouse effect. Some forecast larger changes than quoted while others are more optimistic of the consequences. Consequently this chapter (26) should be read as an introduction to the subject and the figures quoted for mean sea level rise should not be used in design because these will invariably change with time and more research.

In an environmentally friendly world one would like to see beaches allowed to erode and accrete naturally, but this is not an acceptable solution in many places. More often than not we are involved in maintaining or improving the *status quo*.

Protection of the environment is now taken very seriously in Europe, the United States and many other countries, and in more and more countries, environmental protection measures are being established by legislation. Clean seas, beaches and estuaries are becoming political issues, which demand ways of forecasting changes within a given framework. It is an area which means more work for coastal, estuarial and harbour engineers while at the same time demanding of them that they improve their forecasting techniques.

COASTAL STRUCTURES

I now move on to the hardware of our business – the design and construction of coastal, estuarial and harbour structures. They are subject to much more variation in loads than land-based structures and have to withstand extremes of sea level and wave height. Even in a moderate storm the breakwater at Sines, in Portugal was subject to a wave 21 m high with a period of 23 s. The energy associated with such a wave is awesome and as designers we have to be able to calculate the forces which are coupled with such events. We have to know the wave climate at the site in question and calculate return periods often from inadequate data. The calculation of wave forces is perhaps less of a step in the dark and original work on this subject is presented.

Forces imposed by waves on impermeable slopes are the main subject of Chapter 27 by the late Professor Führboter. Unlike rubble mound or similar structures no percolation takes place in the energy transformation. The wave loads act perpendicular to the slope as pressure forces and tangential to the slope as shear forces. An important design parameter is run-up due to wave up-rush. Although vertical walls are not included the design methods apply to the slopes of dikes, sea-walls, revetments and similar shore protection structures. The main emphasis of the chapter is on uniform slopes. Composite slopes are considered towards the end.

The main difference between a land-based structure and one subject to wave action is that the latter has to take large horizontal intermittent forces. In Chapter 28 Tickell concentrates on the loading on slender cylindrical members. The essential prerequisite to the calculation of these forces, as mentioned many times elsewhere in this book, is the estimation of extreme events – waves, tides, surges and often their combination. Central to the discussion is Morrison's equation which is derived and applied to a number of different systems. A review of particular methods that coastal engineers have developed to estimate the wave loads on vertical walls closes the chapter.

In the 1950s the maximum size of tankers was no more than 50 000 tonnes. The 1960s and 1970s saw this figure rise five-fold to the order of 250 000 tonnes. Part of the reason for this escalation in size was the closure of the Suez Canal following hostilities in the Middle East. This meant a much longer sea route from the oil fields to Europe and encouraged tanker owners to seek economies of size. To accept these larger ships very large coastal harbours were built some with breakwaters in deepish water. There was also a general increase in the number of coastal harbours being built partly to cope with the increased consumption of oil and partly to cater for a new type of vessel, the container ship. The demand for deeper water meant a move out of estuaries and calm water to the coast and its associated hostile environment. The main problem to emerge was the stability of breakwaters subject to large storm waves.

As a direct result of costly failures much effort has been devoted to research on rubble mound structures worldwide. On the whole this has been well directed to the point where it is now possible to design and construct such structures with confidence. Burcharth deals with this subject in Chapter 29.

Coastal engineers all recognize the pioneering work of Irribarren and Hudson on the stability of rubble mound structures. However, in the 1970s our confidence was shaken by the serious damage which occurred to a number of recently built breakwaters. A PIANC report concluded that one of the main causes was an underestimation of the wave climate. As a result of a great deal of subsequent research it is now possible to feel confident in our design and construction of such structures. At the time of writing a second PIANC report is soon to be published which will make a major contribution to this subject.

This confidence in design is reflected in parallel advances that have been made in associated fields like geophysics. Geophysical techniques are complex even when applied to land-based structures but the problems are compounded when they are subject to wave loading. Chapter 30 by Barends illustrates how recent advances have made it possible to study by simulation techniques hydromechanical processes relevant to the design of works in the marine environment. This is a major step forward bearing in mind the many types of soil and different types of failure that have to be considered, for example, on projects like the Delta scheme and the stability of major breakwaters.

Marinas are a prime example of an expanding leisure industry and Lacey reviews the history of their development in Chapter 31. Some of the most important aspects governing the permission to build a marina are concerned with the likely effects on the existing regime. Usually results from field studies and mathematical or physical models are necessary to allay the concern of planners and to ensure safe moorings. The chapter reviews many of the engineering aspects that need consideration in working towards a final design. It is appropriate for an industry that is expanding so rapidly that the chapter concludes with a look into the future.

Attention to the method of construction is essential to the successful design of maritime works. Even as early as the conceptual stage of a project thought given to construction will contribute to savings of time and money. Knowledge of the environmental conditions, especially their extremes and seasonality, are vital constituents of the design and construction processes. Chapter 32 by O'Hara describes the many types of contractors' plant now available to carry out such projects as rubble mound breakwaters, caissons, block walls, piling and reclamation.

Too often the maintenance of a coastal structure is neglected sometimes with dire consequences. Good design necessarily includes considerations of ease of maintenance and this will pay dividends in the long run. Chapter 33 by Simm stresses the need for monitoring and suggests how this can best be planned. Design for failure is just as important as design for success and will highlight the best form monitoring and maintenance should take.

The marine environment is often hostile and it requires special skills and careful planning to collect reliable field data. Often the results are used to provide the input to the boundaries of mathematical and physical models and to check their reliability. Chapter 34 by Rayner describes how such observations are carried out and the types of instruments that are used. This is an invaluable chapter to any engineer planning field work. However, it illustrates that more often than not field studies should be left in the hands of experts.

MATERIALS FOR COASTAL STRUCTURES

As a result of recent research the subject of rock mechanics has been put within a sound scientific framework. Chapter 35 by Latham and his colleagues demonstrates that by integration of information of the known output from a quarry and appropriate geological considerations the most economical design of a marine structure can be achieved. For example, it is now possible to predict the range of block sizes which will be produced by a particular quarry and such information is invaluable within the design process. The way in which further evaluation of the properties of rock should be carried out on site and in the laboratory in order to achieve good quality control is described.

Maritime structures are built in hostile physical and chemical environments, so the choice of structural materials has an important bearing on their design and construction. Concrete, steel and timber are often used and in Chapter 36, Spragg concentrates on their properties to ensure minimum maintenance during the lifetime of the structure.

It is interesting that coastal engineers were the first to see the value of using synthetic fabrics to solve geotechnical problems. By normal civil engineering standards the subject has a short history but even so considerable advances have been made in the last 25 years. New materials have been developed and design methods are now based on sound scientific principles. The subject is highly specialized and Heerten in Chapter 37 reviews the background to the successful use of geotextiles in maritime engineering.

COHESIVE SEDIMENTS, ESTUARIES AND DREDGING

The ability to predict the movement of cohesive sediments within coastal and estuarial waters has great significance in the development of engineering projects both for their ecological implications and for the maintenance of existing installations. Thirty years ago forecasts were made from field studies and physical models. Nowadays such problems can be handled by mathematical models because the processes of deposition, hindered settling, consolidation and erosion are better understood. Even the transport of granular sediments is difficult to quantify but the problems associated with fine sediments are further complicated by the fact that their hydraulic properties are controlled by a complex array of physical and chemical factors which even now are only partly understood. Chapters 38–42 under the general editorship of Parker, provide a thorough explanation of the behaviour of cohesive sediments and associated problems.

The first chapter of the series written by Parker sets the scene by reviewing the scientific background of the behaviour of cohesive sediments. It describes the many factors that control the behaviour of fine sediments and is an excellent starting point to this complex subject. The following chapter by Mehta describes how the science can be used in terms of practical engineering to handle such subjects as free settling, flocculation, hindered settling, rates of erosion and deposition, entrainment and erosion by waves. Towner then describes the important role that these fine materials have on the question of estuarial pollution. They are of critical importance due to their interactions with contaminants which are mainly of industrial origin like lead, copper, zinc, mercury. In estuaries fine sediments are more often than not associated with high concentrations of organic matter and this causes a demand for dissolved oxygen from the overlying waters.

It is understandable why in most cases coastal and estuarial engineers necessarily adopt the well-tried approach to the solution of a particular problem even when they suspect that a new idea could prove a better prospect. Civil engineering hydraulics is a high risk business and this is the main reason why new ideas take many years to become acceptable. There are many examples. Beach nourishment was new some 30 years ago but is now a well-tried technique. Fluidization is a newish idea which the profession will wait to see well proved. In his chapter 'Depth determination and navigability in fine sediment areas' Parker describes the relatively new concept of nautical depth. Put simply the thesis is that it should be possible to navigate through soft, silty deposits and hence accept a greater depth 'on the chart' than that established by normal methods. Success has been obtained using this method. It can lead to huge savings in cost because less maintenance dredging is necessary.

One tends to associate silts with estuaries but there are cases where they create problems in the coastal environment. The main thrust of the chapter by Mehta is an illustration of the importance of the many factors which can affect erosion and siltation in the complexity of the coastal environment.

Apart from deepening by dredgers there are a number of complementary activities which could reduce or even eliminate the need to dredge. The chapter by de Vlieger considers these alternatives for cohesive sediment areas. Among others there is the ability to improve natural depths by training walls, fixed dredging installations such as silt pumping stations, bypass plants and bed-levellers.

An estuary is a highly complex hydraulic environment. The general circulation patterns are controlled by tides, river flows and the penetration of salt water from the sea. Often the sediments in movement are cohesive which adds another level of complication. It is perhaps in this area more than any other that mathematical models have represented a major breakthrough. Chapter 43 by Kendrick reviews the various types of problem that occur in the development of an estuary and describes the thinking behind their solution often using specific examples as illustrations.

Hydraulic engineering is mainly involved with controlling the natural environment to the benefit of mankind. Barriers and barrages built in estuaries and at their mouths are structures which can produce larger changes than most others. They are built to control flooding, to improve amenity, to control saline intrusion and more recently to harness tidal energy. In Chapter 44 Reilly deals with the many types of structure that have been built for such specific purposes and highlight the problems that have to be addressed in their design. Two of the most impressive engineering projects of recent decades, the Thames Barrier and the Oosterschalde Storm Surge Barrier, are described in the chapter.

Major port and harbour projects invariably involve dredging and the disposal of material. Dredging is a specialized subject and Bates in Chapter 45 discusses the various types of dredger to be used for specific purposes and describes their operation in different materials and situations. In a section on capital dredging, attention is drawn to the problems of working in difficult ground. Maintenance dredging poses its own problems, especially difficulties of measurement, and various survey methods are discussed. Disposal of dredged material at sea is subject to increasing environmental pressures especially for materials containing pollutants and alternative methods of disposal and legislation constraints are described. The economic evaluation of a dredging, dumping or land reclamation project is obviously complex and in the final section of the chapter the author points to the many factors that have to the considered.

Until quite recently few restrictions were placed on the disposal of dredged material at sea but, increasingly, more countries are passing legislation to control such operations. The problem materials are fine grained cohesive silts and clays which usually arise from maintenance dredging. They are often highly polluted. In Chapter 47 Sergent reviews the complex scientific, engineering and legislative aspects associated with this relatively new topic and indicates the direction in which the subject is heading.

In many cases outfalls are not adequate to achieve satisfactory dispersion with the result that inshore areas become heavily polluted. Chapter 48 by Neville-Jones and Thompson describes the contaminants that can enter the sea from these sources. Better outfalls are now being designed especially with the aid of advanced computer modelling and one such scheme is described. An intermittent pollutant is oil and one of the important factors in planning a clean-up operation is to predict its movement. Methods available to do this are reviewed.

ECONOMIC AND CONTRACTUAL CONSIDERATIONS

One of the consequences of our greater understanding of physical processes is that we can now, with reasonable confidence, carry out an economic appraisal of maritime engineering works and quantify with a reasonable degree of accuracy the effect of various alternative schemes proposed to solve a particular problem on a benefit/cost basis. Using examples from coastal engineering, Barber describes in Chapter 49 the calculations that should be made in carrying out such a benefit/cost exercise.

Chapter 50 by Brinson describes the key aspects of tendering, contract management and site supervision. The first requirement of major projects is generally a feasibility study which usually includes an environmental impact assessment. This involves the study of existing site data and the specification of additional information required for detailed design, and the preparation of tender documents. The realistic specification of Conditions of Contract is particularly important and although these vary from country to country they form a framework within which the client can insert and amend clauses to meet his own particular requirements. Control of the works on site requires tight supervision and the chapter discusses how this can be achieved and ends with a section on Insurance.

When things go wrong the parties to a dispute sometimes go to arbitration rather than to court. There are many advantages to this procedure. It is normally quicker and less expensive than the courts, mainly because there is not so great a need for lawyer participation. Although this subject is not a mainstream engineering one engineers likely to be involved in the settlement of disputes will find that Chapter 51 by Miles is a welcome review of the subject not only as it is applied in the UK but also when litigation is conducted under the rules of the International Chamber of Commerce.

CONCLUDING REMARKS

Looking back over 40 years I am amazed at the advances that have been achieved. Up to the early 1960s Hydraulics Research, Wallingford employed one mathematician; now over 30 are employed. In 1955 I built a pseudo-mathematical model of the Thames estuary to determine the effect of overspill on the tide levels in London. Working with hand calculators solidly for nine months an answer was achieved. That same programme would probably take five minutes to run on modern computers. But the size and speed of modern day computers are only of use because advances have been made at the same time in our understanding of basic processes. There is still a lot to learn, many areas still rely on empiricism, but the advances are undeniable.

I have attempted in my introduction to the second half of this volume to offer a concise view on every chapter that I have edited. I am conscious, however, of the considerable advances that have been made in topics outside my direct field of interest. For example, rock mechanics has now been put on a highly scientific basis, so too has the choice of other materials we use in our structures. Too often the maintenance of a coastal structure is neglected with dire consequences. Good design, as this book stresses, necessarily includes considerations of ease of maintenance, and maintenance involves regular monitoring. Even as early as the conceptual stage of a project thought given to methods of construction will invariably contribute to savings of time and money. The coastal engineer works in a hostile environment which puts special responsibilities on him so far as contract management and site supervision are concerned.

Gone are the days when research was done for its own sake. Certainly in civil engineering hydraulics the object of the majority of research programmes is to improve design and construction. This means that the results of research need to be disseminated to industry. In the UK, the Construction Industry Research and Information Association (CIRIA), an independent non-profit-distributing body, initiates, organizes, and manages research and provides information on all topics of concern to professionals in the design and construction industries.

To act as a focus for research in the UK the Coastal Engineering Research Advisory Committee (CERAC) was set up in the early 1980s by the Maritime Engineering Board of the Institution of Civil Engineers, London. An on-line data base is now established which contains information on individuals and organizations working in and funding coastal engineering research, together with details of those responsible for the design, construction and maintenance of coastal works. Every opportunity has been taken to make contacts and form links with other committees and organizations – for example, the Commission of the European Community.

The main reason for mentioning CIRIA and CERAC in this introduction is to emphasise that research needs to be co-ordinated. Too often in the past the jam has been spread too thinly.

Finally, as engineers and scientists, we naturally put value to the contributions we make to society. It would be as well, however, if we admit that on major issues the politicians have the last word. In 1954 I investigated the idea of a Thames Barrier. It was not until 28 years later that it became operational. The delay was not due to engineers – the politicians had other things to spend their money on. The event that finally prompted action was a brilliant, well reasoned article by two newspaper reporters who spelled out the dire consequences of doing nothing. This story illustrates the fact that as engineers and scientists we have a duty to make our concerns easily understood by the general public who in turn elect our politicians. Only then will they sit up and take notice. Some of the long hours spent at meetings and in conferences talking about the 'nuts and bolts' of our particular scientific interests might be better spent educating the public. Only then are applications for adequate funds for research likely to receive the necessary support. The present growing concern for our natural environment presents a timely focus for further efforts to solve the coastal engineering problems that confront us through cooperation, research and the dissemination of information.

ACKNOWLEDGEMENTS

I would especially like to thank Dr R. Parker for editing the following chapters, which are mainly concerned with the complex behaviour of cohesive sediments (Chapters 38–42 and 46).

PART FIVE

COAST, SHORELINE AND BEACH MANAGEMENT

COASTAL MANAGEMENT 22

I.H. Townend

Coastlines are subject to increasing pressures on many fronts yet their resources are finite. It is now recognized that shorelines cannot be managed successfully on a local or piecemeal basis. Problems have to be anticipated and when they arise it is important that information is available which can be used to produce the best solution. The collection of such data in its most useable form, whether it be from historical charts or measurements in the field, is central to good coastal management. This chapter reviews the many frameworks which can be adopted to achieve the best results (see also Chapters 23, 25, 33 and 34).

CONTENTS

22.1 INTRODUCTION

The coast has always played a central role in our affairs. Initial settlements formed to take advantage of the food and living area available. As they developed, so travel by sea provided a link between settlements on the coast. At the same time this mobility led to the need for defences against invasion. With the growth of trade, large ports and cities developed, taking advantage of the ready access to the sea. The arrival of rapid travel, provided by the steam train, gave rise to seaside holidays and the advent of tourism. Consequently, pressure for coastal space and coastal resources has steadily increased over the centuries.

There is now wide recognition that these resources are finite and, particularly in the developed countries, are in danger of being over-exploited. The problem is exacerbated by the very sensitive nature of the interface between land and sea.

The best way forward would seem to be 'to work with rather than against the natural system' (Carter, 1988). This demands a 'process' based approach. As we shall see this is at variance with the more usual 'issue' based approach, which tends to dominate current practice. In this chapter we will review the frameworks for coastal management and consider how these can be aided by suitable coastal mapping techniques and information systems. When talking of 'management' we mean 'a deliberate effort to direct or control conditions and actions, without suggesting that such efforts are necessarily successful' (Armstrong and Ryner, 1981). Some further definitions of the terminology used are provided in the Glossary at the end of the chapter.

22.2 COASTAL MANAGEMENT FRAMEWORKS

One can argue, that without our desire to make use of the coast and its vast range of resources, there is no need for coastal management. This being the case, the following relatively simple definition of coastal management seems appropriate:

> Our use of the coast must be organized so as to allow the reconciliation of the demands of DEVELOPMENT and the requirements of PROTECTION
> (based on the European Coastal Charter: CPMR, 1980)

The same sentiment has been more elaborately stated in the many other definitions which abound (Gub-

Coastal, Estuarial and Harbour Engineer's Reference Book. Edited by M.B. Abbott and W.A. Price. Published in 1993 by Chapman & Hall, 2–6 Boundary Row, London SE1 8HN. ISBN 0 419 15430 2.

bay, 1990; CAMPNET, 1989; 4th World Wilderness Congress, 1987; US Coastal Zone Management Act, 1972).

Some of the definitions cited have a leaning towards conservation, reflecting the authors' prime concern. This must, however, co-exist alongside rational development, if there is to be economic benefit so making protection of the coast sustainable in the long term. There must therefore be a recognition that the system is both multi-dimensional and dynamic. The coastal resource base comprises bio-physical, socio-cultural and economic elements all of which interact. Organization of these elements implies planning, which must be sufficiently detailed to provide a national, regional and local framework, but with the flexibility to be adopted to the particular physical, economic and legal circumstances of the country or region in question. There is also a need to recognize that the 'region' may need to be a marine or coastal region, which extends beyond the confines of national boundaries.

22.2.1 BOUNDARIES AND REGIONS

The question of how to define the boundaries of the coastal zone has been extensively discussed (Burbridge *et al.*, 1989; Vallega in IGU, 1985; UNOETB, 1982). Unfortunately no single criterion is universally acceptable or meets the requirements of an effective definition, as summarized in Table 22.1 The combination used must reflect the characteristics of the country or region in question. An alternative view (Smith, 1990) is that the coastal land/sea environment can be divided up as urban, rural and wilderness. With this sub-division the question of boundaries to the coastal zone are of secondary importance, although the importance of considering both land and sea is stressed. As Salm and Clark (1984) noted, however, 'to define a coastal management unit requires more pragmatism than theory, because the unit must be a realistic entity'. Whatever means of delineating the area to be managed is adopted, the coastal margin must be regarded as mobile and therefore changing with time. Only when this is an integral part of the thinking behind planning and management, as well as an accepted public perception, can appropriate response strategies be adopted.

The coastal zone is the most widely recognized entity for management purposes. There is, however, an alternative, which derives from a systems approach to the problem. This is based on a coastal region, which exhibits a degree of independence from external elements and has the potential to be organized within an integrated framework. Vallega (in IGU, 1985) develops the concept of such regions and outlines how they must be delimited to respect existing legal frameworks, minimize differences due to administrative jurisdictions and recognize the spatial extent of the coastal environment. This is, of course, wholly consistent with the much discussed holistic approach, seeking to optimize the use of resources working from the whole to the part (Green, 1990; Vallega in IGU, 1985). The concept was critically reviewed by Morgan (1989), who noted that although regions are constructs, either homogeneous or functional, they can also be natural and therefore really exist. This is a crucial observation and brings us back to Carter's first rule of working with rather than

Table 22.1 Criteria for management boundaries

Criteria	*Examples*	*Advantages*	*Disadvantages*
Physical	Area defined by such things as: coastal plain, watershed and major roads on land; high water, low water and continental shelf edge to seaward	Simple to describe Easy to understand	Seaward extent difficult to establish Not necessarily consistent with natural processes
Administrative	National, county and district boundaries	Easy to understand Legally viable	May be too restrictive May artificially divide resources
Arbitrary	Set back lines, territorial waters	Ease of application	Arbitrary distance bears no relation to coastal topography, the natural system or economic activity
Environmental	Selected units such as inlets, bays, reefs, mangrove etc.	Sound morphological and/or ecological basis	Not easily understood Difficult to chart and legislate for

Table 22.2 Institutional frameworks

	Government framework	*Voluntary framework*
Centralized decision making	I. All embracing legislation with a statutory coastal zone and specific decision making agencies	
	II. Lead agency established to co-ordinate individual agencies who have a responsibility for activities or resources in the coastal zone	A. Voluntary organizations and pressure groups monitor and, in some cases, police coastal management practices
	III. Coastal zone recognized but not established as a statutory zone with agencies directed to co-ordinate activities through the planning process, in order to take account of coastal zone needs	
	IV. Inter-agency co-ordination at national and local level, often through co-ordinating councils or commissions. In many cases these have no statutory or executive decision making powers	B. Voluntary groups actively involved in practical coastal management (often supported by pressure groups)
Devolved decision making	V. Coastal zone not recognized in legislation and individual agencies make decisions without any formal framework for co-operation	

against nature (Carter, 1988). A fundamental requirement for a sustainable coastal region is therefore that it must recognize the true regional extent of the natural processes. This favours environmental boundaries (Table 22.1) and merits the additional effort needed to define them.

22.2.2 ACTIVITIES AND CONFLICTS

The wide ranging nature of coastal activity has been extensively reported (Gubbay, 1990; Clark, 1987; Townend, 1986; Smith and Lalwani, 1984). Similarly the potential for conflict is well documented (Johnson and Pollnac, 1989; Clark, 1987; Sorensen, 1971). In particular Clark (1987) has outlined how conflicts can be internal, where the practices of an industry can be self destructive (e.g. fisheries and some aspects of tourism); external, when conflict arises between activities (e.g. offshore exploration and shipping, industry and conservation); and finally regional, where national or international interests have to be taken into account. Interestingly he concluded, 'In some instances it may be possible to regulate activities in a particular area so that all have tolerable access to the marine resource. Often however, conflict is too great for compromise and particular areas have to be dedicated to particular functions'. This inevitably leads to the need for planning, similar to land-use planning.

A further dilemma arises because of the conflicts which exist within the natural environment. For example morphological evolution is often forcing change upon wildlife habitats (e.g. a spit may provide protection for a particular habitat, saltings or wading grounds, which is lost when the spit breaches and a more exposed and energetic environment results). This raises a dichotomy for conservationists, who must choose between allowing natural change, or trying to preserve the *status quo* by imposing artificial conditions as in a museum. In either case, the problem is exacerbated by our limited knowledge of how particular ecosystems work, so making it difficult to establish sound management guidance.

In addition to acknowledging that some conflicts cannot be overcome by compromise, it is important to recognize that conflicts will always be present. This is due to the dynamic nature of the problem, where the physical form, the processes and the uses are all changing. Consequently, conflicts are also likely to change with time. This means that a strategy cannot be set on a tablet of stone, but must recognize change. Any framework must therefore be responsive; capable of continuous update and renewal.

22.2.3 LEGAL, INSTITUTIONAL AND MANAGEMENT FRAMEWORKS

The discussion so far provides four tenets:

1. The coast is dynamic and any management philosophy must reflect this;

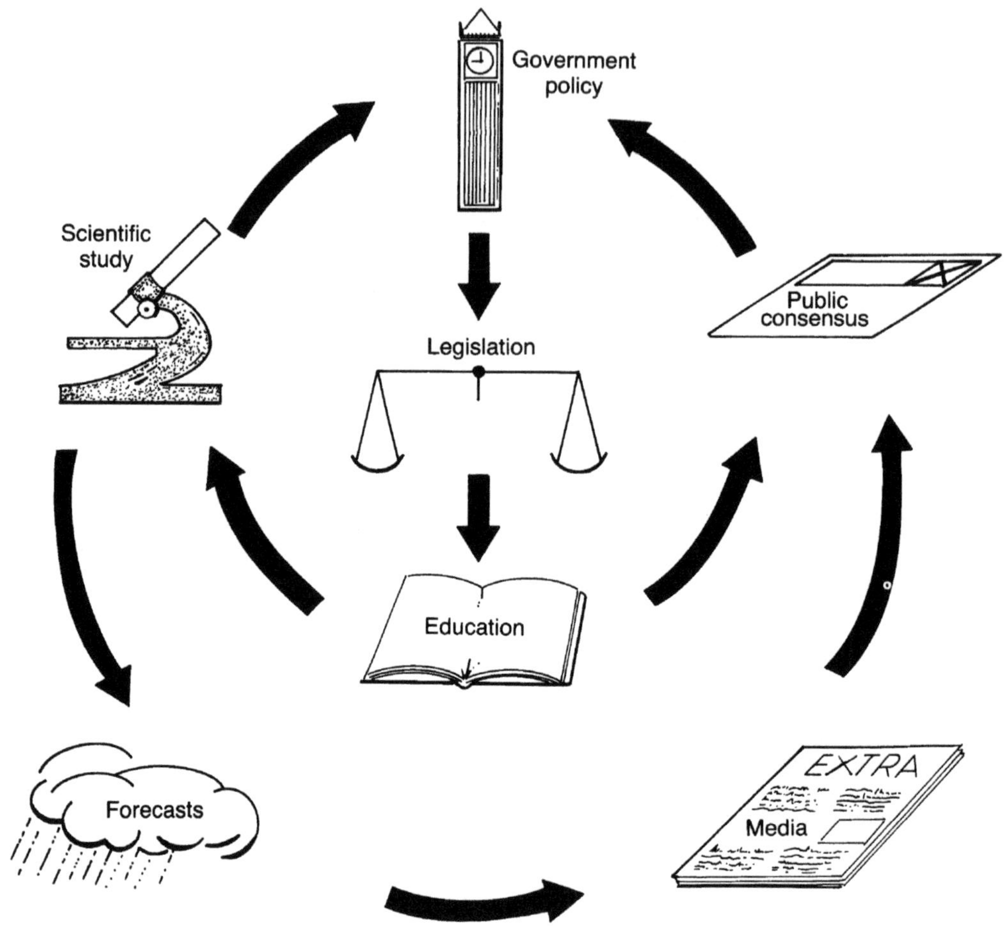

Figure 22.1 Decision making process.

2. The use of management boundaries to define a coastal zone or region must be pragmatic and yet recognize the regional extent of natural processes;
3. Conflicts cannot always be resolved, so requiring planning and regulation;
4. Conflicts change with time, so that a responsive management framework is essential.

These must then be related to the decision making process. The information flow for a democratic society is summarized in Figure 22.1. In this context the law provides the instrument for implementing government policy. It defines the framework and obligations for the institutional/organizational framework. These tend to take a variety of forms (Table 22.2), ranging from almost complete state control as in the Netherlands, to non-existent or speculation-led management as in some Third World countries (Carter, 1990). There have been a number of reviews of the institutional frameworks in existence around the world (Carter, 1988; UNOETB, 1982), some detailed comparisons of the US and UK systems (Carter, 1988; Ricketts, 1986), and a wide range of presentations on specific national frameworks (Smith, 1990; Chua, 1989; UNOETB, 1982). The particular legal and institutional framework(s) which apply to a given region, determine the scope of any management strategy and ensuing plans or programmes.

The tasks involved in establishing a management strategy have been set out with particular reference to tropical islands by Bacon *et al.* (1988) – although the exercises presented are relevant to coastal zone plans in general. A broader discussion of the needs is given by Cendrero (1989) and in the European Coastal

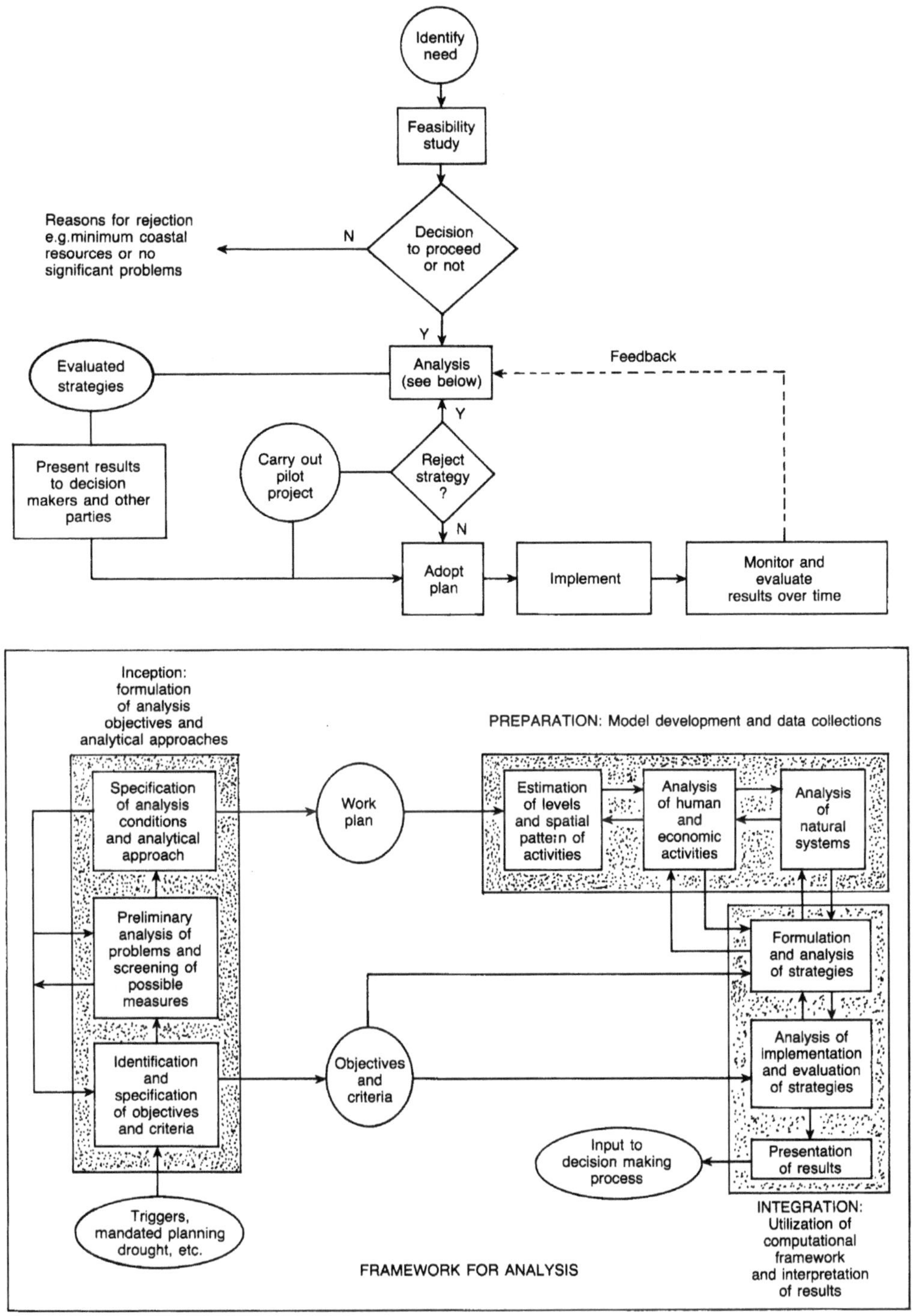

Figure 22.2 Stages in development of a management strategy (after CAMPNET, 1989; Rijskerman and Koudstaal, 1989).

Charter (1980). The overall plan should embrace: a regional strategy to protect and exploit resources; planning for specific activities allocated to management units; and management and control to ensure successful implementation.

The steps to be taken are summarized in Figure 22.2. It is important that the goals and objectives set reflect an accurate interpretation of government policy. Management strategies and plans must be adaptive and have appropriate provision for feedback. The planning process must establish an understanding of the natural regime, the criticality of habitat, the impacts of human usage, and present and potential threats. Whilst the strategy may have provision for long term aims, the management plan should not set unattainable objectives and should include a mechanism for review and revision (Salm and Clark, 1984; Amir, 1983).

There are many other issues within the realms of coastal management which are worthy of more detailed discussion such as the conceptual and computational frameworks available, the concept of zoning, the need for regulation and enforcement, the need to develop public awareness and promote public consultation, and the legal, economic and social considerations. In addition, there are a wide range of disciplines and organizations involved, often each with their own particular interests (e.g. conservation, planning, tourism, industrial development, marine protected areas, etc.). In the limited space available we will only consider a discipline of particular interest to engineers, that of shoreline management. This relates to control of the shoreline interface, with the emphasis on coastal hazards and the need to protect against either flooding or erosion.

22.2.4 SHORELINE MANAGEMENT

Many shorelines around the world are prone to flooding or subject to erosion. Sea-level rise and, in particular, any increases which may arise as a consequence of global warming, will only serve to exacerbate the situation. Over recent years there have been some fundamental changes to the provision of flood protection and measures to limit erosion. In practical terms this has resulted in new types of structure, with the traditional 'hard' approach, using sea-walls and groynes, being replaced by a 'soft' approach based on techniques of beach nourishment and subsequent management. Implicit in such changes is a greater understanding of coastal processes, which has come about as a consequence of international research efforts over the last few decades.

As such techniques are based on understanding the dominant coastal processes of an area, this has led to the study of whole coastal regimes, rather than just the frontage of interest. It soon became clear from these more regional studies, that there were great benefits to be gained from adopting a strategic approach to the provision of coastal works and this has provided the incentive to develop suitable shoreline management techniques.

One approach is to carry out a wide range of detailed studies to look at particular aspects of the natural system. These can then be used as input to a coastal policy model. The model is set up to compute the morphological changes which will take place for given climate and management scenarios. The cost of implementing a particular scheme can then be evaluated in simple benefit-cost terms (Rijkswaterstaat, 1989). This approach is particularly useful as a catalyst to establish appropriate policy guidelines.

A more fundamental approach is to develop a knowledge base as an integral part of the shoreline manager's working practice. This can be in the form of a database or, more usefully given the spatial nature of the data, make use of a geographic information system (GIS). By including routine monitoring as a part of the overall management framework, this ensures that information is kept up-to-date. This leads to the sort of responsive management framework illustrated in Figure 22.3 (Townend, 1990b). The management strategy derives from the information contained within the management system and the legal and institutional frameworks which prevail. It will inevitably comprise policy guidelines to reflect both the constraints imposed from without and strategic decisions which endeavour to make the best use of available resources. Whilst the policy options will guide future planning, the response options define how it can be implemented.

This framework is consistent with the stages of shoreline management identified by Fischer (1985) and the requirements proposed by Ricketts (1986), who suggested that a shoreline management programme should include:

1. Information and analysis by independent professionals;
2. National and regional policy as well as local;
3. The ability to consider alternative strategies;
4. Public involvement should be an integral component;
5. The programme must be flexible with continual re-evaluation.

A number of authors have extolled the need for data to provide a well-defined baseline (Salm and Clark, 1984; UNOETB, 1982), and as a basis for long term planning (Charlier, 1989; Townend, 1986). This is at the core of the framework outlined. In addition it can

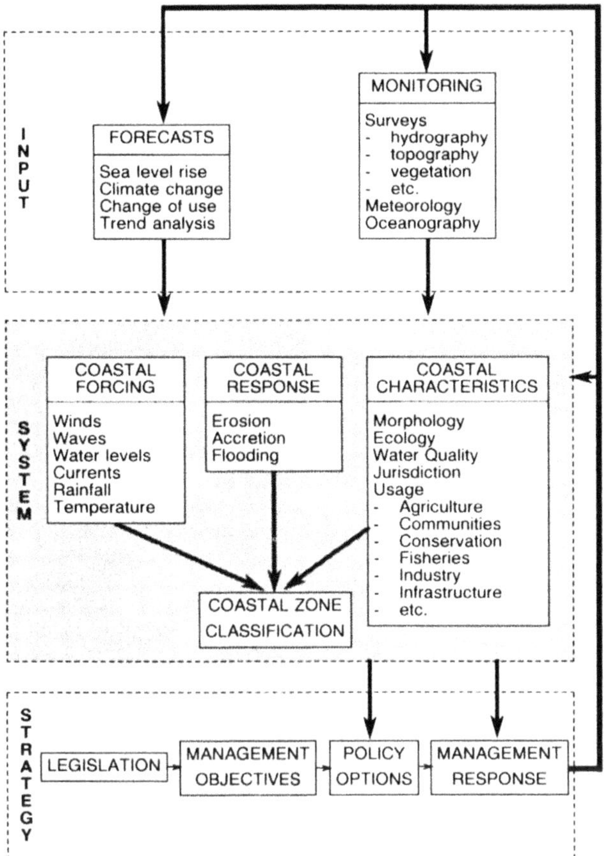

Figure 22.3 Responsive management framework.

readily be applied at a regional scale and can meet many of the above requirements, subject to the constraints of the existing legal framework.

22.3 COASTAL MAPPING AND CLASSIFICATION

Although a coastal region will contain a number of different and distinct environmental units, these will be interconnected. It is these interconnections which lead to the complexity of the coastal zone, particularly in respect of multiple resource planning (UNOETB, 1982). In order to identify and plan around these various units, they need to be clearly identified. The use of mapping and in particular the preparation of data atlases is a well-established tool for strategic planning (Bacon *et al.*, 1988).

In a comprehensive review of mapping for coastal planning, Cendrero (1989) identifies three levels of planning, macro, meso and micro, which can be aligned with international/national, regional and local levels of planning, respectively. The map scales will of course vary according to the area of study and the following are suggested as common ranges:

1:1 000 000	international/national
1: 500 000	\
1: 250 000	regional
1: 100 000	\
1: 50 000	local
1: 10 000	\
1: 5000	detailed
1: 1000	\

Some examples at the macro level are the CORINE project covering Europe (Quélennec, 1989), the US coastal hazards database (Gornitz and Kanciruk, 1989), the estuaries of New Zealand (Hume and Herendorf, 1988), the mapping of the Italian coast (Fierro, 1985), and the mapping of the UK coast to define Heritage coast sites (Countryside Commission, 1970). On a regional scale there is the FMG study for the Bay of Fundy in Canada (Ricketts *et al.*, 1989), the Anglian study in the UK (Townend and McLaren, 1988) and the atlas prepared for the submerged lands of Texas (Wermund, 1989). The micro level is the one usually adopted by planning authorities and mapping at this local scale has been widely reported (e.g. Menanteau and Vanney, 1990; Roberts and McGown, 1987).

The majority of maps prepared are descriptive, presenting measurable or observable features, such as the physical setting, coastal usage, coastal processes and coastal response (Townend and McLaren, 1988). Cendrero (1989) identified a further three types of map, namely:

1. Interpretive maps. These are obtained by combining or deriving new information from the descriptive maps.
2. Evaluation maps. These are used to illustrate how coastal activities impact and use measures such as capacity, suitability, vulnerability, etc.
3. Prescriptive maps. These set out the land/sea-use options or the zoning plan. As such they are one of the end products of the planning task.

A 'top down' approach to mapping first sets out a set of consistent spatial units and then maps descriptive elements onto these units. This is a form of *a priori* integration of the problem and has the advantage that the units can be rapidly (and manually) combined as interpretive, diagnostic or prescriptive maps. The alternative is a 'bottom-up' approach where maps of individual elements are mapped independently, without any preconceived notion of fixed reaches or units. Individually these provide thematic maps of each variable (e.g. sediment type, vegetation, tidal range). By superimposing these maps, or manipulating them based on the rules of some form of model, they can be combined to give a better understanding and to develop classifications as the basis for defining units or zones. The former approach is well suited to manual mapping with the ability to provide clear and concise information, whereas the latter can be a particularly powerful approach when used in conjunction with a GIS and the associated spatial analysis capabilities.

The objective of the mapping exercise is to present spatial information in a way that can be readily understood. To obtain some form of order within the map, it is usually necessary to remove excessive detail by generalization or classification. In some cases the data may be split into classes as it is collected, or standard classification systems may exist for particular variables. Whenever possible, however, the original values should be retained and classification only used for interpretive and presentation purposes (Burrough, 1986). Many coastal attributes are of a descriptive nature and for these some form of classification is inevitable. Some examples of classification systems which have been proposed for different aspects of the coast include, marine habitats (Salm and Clark, 1984), coastal habitats (NCC/NERC, 1979), aquaculture technology (Dutrieux and Guélorget, 1988), estuaries (Hume and Herendorf, 1988), coastal morphology (Cendrero, 1989), coastal works (Fleming and Townend, 1989) and dune systems (Ranwell and Boar, 1986).

The interpretive phase of planning can make use of a number of other tools to support and direct map production. In order to try and identify patterns or appropriate groupings within a number of variables, it may be appropriate to use correlation and regression techniques or some form of cluster or factor analysis. The latter, in particular, have been successfully applied to identify coastal units by Cin and Simeoni (1989) and Price (1990). The evaluation stage is often supported by the preparation of a compatibility matrix as a means of cross-relating different uses, or resources and uses (Cendrero, 1989; Carter, 1988; Salm and Clark, 1984). This concept is taken a stage further by Sorensen (1971) and Amir (1983) who link a series of matrices in a network in order to relate use-cause-condition-effect. This enables the potential impact of specific events or resource uses to be simulated.

22.4 INFORMATION SYSTEMS

At its most basic an information system provides access to data through some form of referencing. Traditionally, this is through an index system to an archive of records and documents. The alternative now available is to use some form of computer database to store and retrieve information. For coastal studies a convenient and relatively quick approach is to define a reference coastline and chainages along this line from some arbitrary start point. All attribute entries into the database are then referenced to this fixed chainage system. This imposes a reduction on the data to either point or line type on the fixed reference line. Its advantage is that it readily allows spatial correlations in the alongshore dimension and can be used to develop both maps and x–y plots showing values of variables against chainage along the coast.

The processes, the resources, and the uses we make of the coast all have a clear spatial dimension. By using geographic information systems (GIS), it is possible to describe real world objects in terms of their spatial description (point, line, area), their attributes (e.g. name, value, classification), and their relationship with other objects (i.e. topological relationship). The fundamentals of GIS and the concepts involved in the related issues of data structures, storage techniques, modelling (e.g. terrain models), data analysis, classification and most importantly data quality issues have been comprehensively set out in the book by Burrough (1986). The specific application of GIS to coastal and shoreline management has been considered by Law (1990), Townend (1990a), Mitchener *et al.* (1989), and McCue (1989), among others.

A GIS offers the potential for a geographic and spatially holistic framework (Roberts, 1989). This ideal must, however, be treated with some caution. Firstly, within coastal management applications one is usually trying to satisfy the requirements for analysis and the requirements for management. Secondly, the capabilities of a GIS are dependent on the scale at which data is captured because detail cannot readily be removed or added. This may require compromise to meet the needs of different users, or alternatively multiple sets of graphics linked to a common database to meet differing presentational and analytic needs. In either case, the user requirements must be clearly specified and the data model designed to meet these needs. This has always been recognized when building large relational databases but is often overlooked when setting-up a GIS because of the apparent definition provided by real world objects. This, however, is something of an illusion because within a GIS they are represented schematically as points, lines or areas (e.g. when mapping a river one could use the water surface area, the two banks, a centre-line, or even a point when working at very large scale). The right choice depends on how the data is to be manipulated and presented.

As a basic data retrieval tool a GIS can be used to find where items are, what an item is, or to obtain a summary of all occurrences of an item. The real advantages are gained from the analytic capabilities. The tool can be used heuristically for classification, by comparing different data sets to seek patterns or combinations which suggest some form of relationship. By starting with known problems or phenomena one can adopt a more inductive approach, to seek more general relationships. Finally it is possible to work in a deductive manner, starting with an accepted principle and seeing if this can be substantiated by the available data. In all such analysis, it is important to seek the underlying physical explanation for any relationships observed (usually by means of carefully focused modelling), and to be aware of spatial and temporal limitations of the data within the GIS.

Roberts (1989) noted that 'the thematic map is only a summary of the spatial dimensions of phenomena', to which one might add that it is also only a snapshot in time. Nonetheless it is a powerful tool for portraying large and complex data sets. For a coastal manager to obtain maximum benefit, however, there is a need for continual revision as new information becomes available from routine monitoring. In consequence the process of updating the system is of particular importance, if the GIS is to have a sufficient life to justify the initial costs of establishing the system. Once again this is an aspect which requires careful consideration at the design stage (Townend, 1990a).

Although GIS is still a relatively new tool in coastal management applications, it has been widely applied in a range of studies, ranging from macro scale (Quélennec, 1989; Gornitz and Kanciruk, 1989), through meso scale (Ricketts *et al.*, 1989; Fleming and Townend, 1989; Law, 1990), to micro scale (Cendrero, 1989; McCue, 1989). These studies, however, only serve to illustrate the potential. For the proposed approach to succeed and justify both the initial set-up and subsequent maintenance costs, such tools must form an integral part of routine management. They must therefore be available for use by those responsible for the day-to-day management of the coast. Success will depend on well-designed systems, appropriate training and a commitment to regional understanding as the premise for local planning and management as already discussed.

REFERENCES

Amir, S. (1983) Tools for a coastal resources management framework with multiple objectives. *Journal of Environmental Management*, **17**, 121–32.

Armstrong, J.M. and Ryner, P.C. (1981) *Ocean Management, A New Perspective*, Ann Arbor Science, Michigan.

Bacon, P.C., Deane, C.A. and Putney, A.D. (1988) *A Workbook of Practical Exercises in Coastal Zone Management for Tropical Islands*, Commonwealth Science Council, London, UK.

Burbridge, P.R., Dankers, N. and Clark, J.R. (1988) Multiple-use assessment for coastal management. *Coastal Zone '89*, ASCE, Proceedings of the 6th Symposium on Coastal and Ocean Management, pp. 33–45.

Burrough, P.A. (1986) *Principles of Geographical Information Systems for Land Resource Assessment*, Clarendon Press, Oxford.

CAMPNET (1989) The status of integrated coastal zone management: a global assessment. Report of Charleston Workshop, Rosenstiel School of Marine Sciences, University of Miami, USA.

Carter, R.W.G. (1990) Coastal zone management: Comparisons and conflicts, planning and management of the coastal heritage. Sefton Metropolitan Borough Council, Merseyside.

Carter, R.W.G. (1988) *Coastal Environments*, Academic Press, London.

Cendrero A (ed.) (1989) Land-use problems, planning and management in the coastal zone. (special issue) *Ocean and Shoreline Management*, **12**, 365–589.

Charlier, R.H. (1989) in Cendro, A. (1989) pp. 383–402.

Chua, T.E. (1989) Developing coastal area management plans in the South East Asian region. *Coastal Zone '89*, ASCE, Proceedings of the 6th Symposium on Coastal and Ocean Management. 11–14 July 1989, South Cardina (ed. O.T. Magoon *et al.*). ASCE, New York, pp. 2192–201.

Cin, R.D. and Simeoni, U. (1989) Coastal zoning and vulnerability: application to the Middle Adriatic (Italy), *Coastal Zone '89*, ASCE, Proceedings of the 6th Symposium on Coastal and Ocean Management, pp. 98–110.

Clark, R.B. (1987) *The Waters Around the British Isles: Their Conflicting Uses*, Clarendon Press, Oxford.

Countryside Commission (1970) *The Coastal Heritage*, Her Majesty's Stationery Office, UK.

Dutrieux, E. and Guélorget, O. (1988) Ecological planning: a possible method for the choice of agricultural sites. *Ocean and Shoreline Management*, **11**, 427–47.

European Coastal Charter (1989) Conference of Peripheral Maritime Regions of the EEC.

Fischer, D.W. (1985) Shoreline erosion: a management framework. *Journal of Shoreline Management*, **1**, 37–50.

Fleming, C.A. and Townend, I.H. (1989) A coastal management database for East Anglia. *Coastal Zone '89*, ASCE, Proceedings of the 6th Symposium on Coastal and Ocean Management, pp. 4092–107.

Gornitz, V. and Kancituk, P. (1989) Assessment of global coastal hazards from sea level rise. *Coastal Zone '89*, ASCE, Proceedings of the 6th Symposium on Coastal and Ocean Management, pp. 1345–1359.

Green, B. (1990) Changing realities and perceptions in coastal management, planning and management of the coastal heritage. Sefton Metropolitan Borough Council, Merseyside, UK.

Gubbay, S. (1990) A future for the coast? Proposals for a UK coastal zone management plan. Marine Conservation Society, Ross-on-Wye, UK.

Hume, T.M. and Herendorf, C.E. (1988) A geomorphic classification of estuaries and its application to coastal resource management – a New Zealand example. *Ocean and Shoreline Management*, **11**, 249–74.

IGU (1985) *Coastal Planning: Realities and Perspectives*. Proceedings of an International Seminar, Geneva, International Geographical Union, 24 October, Geneva.

Johnson, J.C. and Pollnac, R.B. (1989) Managing marine conflicts, (special issue). *Ocean and Shoreline Management*, **12**, 191–284.

Law, M.N. (1990) Application of GIS to local shoreline management issues. Ontario Ministry of Natural Resources, Toronto, Canada.
McCue, J.W. (1989) The use and potential use of geographical information systems (GIS) in coastal zone management. MSc Thesis, Department of Biology, University of Newcastle-upon-Tyne, UK.
Menanteau, L. and Vanney, J.R. (1990) La carte physiographique du littoral Atlantique de L'Andalousie, *Littoral 1990*, Eurocoast, Marseille, France.
Michener, W.K., Cowen, D.J. and Shirley, W.L. (1989) Geograhic information systems for coastal research. *Coastal Zone '89*, ASCE, Proceedings of the 6th Symposium on Coastal and Ocean Management, pp. 4791–805.
Morgan, J.R. (1989) Marine regions: myth or reality. *Ocean and Shoreline Management*, **12**, 1–21.
NCC/NERC (1979) Nature Conservation in the Marine Environmet. Nature Conservancy Council, Peterborough, UK.
Price, A.R.G. (1990) Rapid assessment of coastal zone management requirements: a case study from the Arabian Gulf. *Ocean and Shoreline Management*, **13**, 1–19.
Quélennec, R.-E. (1989) The Corrine coastal erosion project: identification of coastal erosion problems and data base on the littoral environment of eleven European countries. *Coastal Zone '89*, ASCE, Proceeding of the 6th Symposium on Coastal and Ocean Management, pp. 4594–601.
Ranwell, D.S. and Boar, R. (1986) Coast dune management guide. Institute of Terrestrial Ecology, NERC, UK.
Ricketts, P.J., Mclver, A.R. and Butler, M.J.A. (1989) Integrated information systems, the key to coastal zone management. *Coastal Zone '89*, ASCE, Proceedings of the 6th Symposium on Coastal and Ocean Management, pp. 4138–50.
Ricketts, P.J. (1986) National policy and management responses to the hazard of coastal erosion in Britain and the United States. *Applied Geography*, **6**, 197–221.
Rijkswaterstaat (1989) Kustverdediging na 1990, Discussienota, The Hague, The Netherlands.
Rijsberman, F.R. and Koudstaal, R. (1989) Development and application of a framework for analysis for coastal zone management in different national and international contexts, *Coastal Zone '89*, ASCE Proceedings of the 6th Symposium on Coastal and Ocean Management, pp. 58–71.
Roberts, A.G. and McGown, A. (1987) Coastal area management system as developed for Sealsalter-Reculver, North Kent. *Proceedings of the Institution of Civil Engineers, Part 1*, **82**, 777–97.
Roberts, S.I. (1989) Problems of developing a GIS data base for the FMG. *Coastal Zone '89*, ASCE, Proceedings of the 6th Symposium on Coastal and Ocean Management.
Salm, R.V. and Clark, J.R. (1984) *Marine and Coastal Protected Areas: A Guide for Planners and Managers*, International Union for Conservation of Nature, Switzerland.
Smith, H. (1990) Coastal zone management and planning: making it work, planning and management of the coastal heritage. Sefton Metropolitan Borough Council.
Smith, H.D. and Lalwani, C.S. (1984) The North Sea: sea use management and planning. University of Wales Institute of Science and Technology Cardiff, UK.
Sorensen, J.C.A. (1971) Framework for identification and control of resource degradation and conflict in the multiple use of the coastal zone. University of California, Berkley, Department of Landscape Architecture.
Townend, I.H. (1990a) The application of GIS to coastal zone management. EGIS '90 First European Conference on Geographic Information Systems, Amsterdam, 10–13 April 1990 (eds J. Harts *et al.*). EGIS Foundations, Utrrecht.
Townend, I.H. (1990b) Frameworks for shoreline management, *PIANC Bulletin* 71, pp. 72–80.
Townend, I.H. and McLaren, P. (1988) Anglian coastal management atlas. Sir William Halcrow & Partners, Swindon, UK.
Townend, I.H. (1986) Coastal studies to establish suitable coastal management procedures. *Journal of Shoreline Management*, **2**, 131–54.
UNOETB (1982) *Coastal area management and development*, United Nations Ocean Economics and Technology Board, Pergamon Press, Oxford.
Vallega, A. (1985) pp. 23–57.
Wermund, E.G. (1985) in IGU (1985) pp. 411–26.

GLOSSARY

(Items are grouped rather than placed in alphabetic order.)

Coastal characteristic: a distinguishing feature of the coast, either natural or human (e.g. morphology, ecology, usage, etc).

Coastal forcing: the natural processes acting on the coast (e.g. winds, waves, tides, etc).

Coastal processes: the interaction of coastal forcing with coastal characteristics (e.g. sediment transport, pollutant dispersion, etc).

Coastal response: change resulting from coastal processes (e.g. erosion, flooding, etc).

Coastal resources: the coastal characteristics which can be used for human benefit (life support, economic gain, recreation, etc).

Coastal elements: any physical or human component on the coast, this being the aggregate of characteristics, forcing and response.

Shoreline: one characteristic of the coast. Poorly defined but essentially the interface between land and sea.

Shoreline management: relates to efforts to control the shoreline interface. As such, this usually focuses on coastal hazards and the need to provide protection against either flooding or erosion.

Coastal zone: some combination of land and sea area, delimited by taking account of one or more elements.

Coastal zone management: the planning and allocation of resources within a given area.

Coastal region: a spatial region where the elements have a degree of independence from external elements and the potential to be organized within an integrated framework.

Coastal region management: the planning and implementation of a structure plan, embracing all elements, for the development of a coherent spatial region.

Management framework: the complete structure, which includes all aspects of the management process (i.e. data inputs, analysis and understanding, planning and allocation, dissemination and consultation, regulation and enforcement, etc).

Management system: the tools used to develop and implement a management strategy. These can range from paper-based archives to a computer database with facilities to present information spatially and undertake various forms of analysis.

Management strategy: sets out the requirements for a management plan. The aims and objectives set out what is to be achieved and will be judged against defined criteria. Implementation will make use of policy options or guidelines and an appropriate range of response options.

Management plan: defines how the strategy will actually be implemented.

Management aims or goals: focuses on long term aims and may be somewhat open ended.

Management objectives: shorter term aims which are measurable and provide a means of fulfilling the goals.

Assessment criteria: the measures or standards used to develop the management plan and subsequently assess progress towards implementation.

Policy options/guidelines: a definition of resource use and constraints (e.g. promotion of tourism development in an area, limits on mineral extraction, protection of a particular habitat, etc).

Management response options: means by which policy options can be implemented (i.e. regulation, planning, education, engineering, eco-management, etc).

Legal framework: defined by Government policy and ensuing statutes.

Institutional/Organizational framework: devolve from the legal framework as the basis for implementing the statutes.

Conservation: the protection of an area, or a particular element within an area, whilst accepting the dynamic nature of the environment and therefore allowing change.

Preservation: static protection of an area or element, attempting to perpetuate the very existence of a given 'state'. This in essence leads to a natural museum, where artificial conditions have to be maintained to avoid degradation of the artefact.

Systems analysis: a study which includes any or all of the parts of a system; notably the level of interest, the levels within (the components), and the levels without (the environment).

GROYNES, OFFSHORE BREAKWATERS AND ARTIFICIAL HEADLANDS 23

C.A. Fleming

The best form of sea defense is a good beach, and groynes, offshore breakwaters, and artificial headlands are some of the methods used by coastal engineers to encourage the build up of beach material in the back shore and near shore zones. The chapter discusses the various factors that have to be considered in arriving at the best solution to a beach erosion problem. Very similar considerations would apply, for example, if works were required to prevent siltation of a navigation channel or a coastal harbour. In dealing with problems in the near shore zone the decisions that have to be made are not always straightforward and often call for a coast engineer who has experience of hindcasting or predictive models (see also Chapters 25, 33, 34 and 49).

CONTENTS

23.1 INTRODUCTION

Groynes, offshore breakwaters and artificial headlands are all classes of structure that are used to attempt to control, or at least modify, beach response to the dominant wave and tidal conditions. Whilst a clear distinction can be made between groynes and offshore breakwaters, artificial headlands may take a number of forms which include features of the first two.

A thin depth of beach provides an ineffectual dissipator of wave energy. The deeper the beach, up to a certain critical depth related to the mobility of the beach and percolation from breaking waves, the less will be the rate of drift. Differences in drift related to depth of beach may be considerable on a shingle foreshore, but veneer sand beaches on impermeable sub-strata such as clay are also inherently much more volatile than deep beaches.

Thus, whatever type of structure is used the objective is usually to increase the volume of beach material in the backshore and nearshore regions in recognition of the fact that a natural beach of sand or shingle is the most satisfactory and economic means of absorbing the energy of breaking waves and thus protecting an erodible coastline. However, other applications include the arrest of alongshore drift to prevent siltation of a navigation access channel or the creation of artificial beaches for recreational purposes.

Both regional sediment transport processes and local sediment transport mechanisms need to be considered when assessing the stability of the foreshore and the most appropriate beach control structures. The foreshore at the site must be considered within the framework of the regional unit of which it forms part. An appreciation of the dominant physical processes of sediment movement is the most important component of the design of such structures, whether for new works or for the modification or extension of existing works. A regional unit may be defined as a geographical cell in which all sediment sources and sinks are inter-related with the sediment travelling along the beach and in the nearshore zone. A regional unit may be a closed cell such as a beach contained between two major headlands, past which there is negligible bypassing of sediment.

23.2 PLANNING CONSIDERATIONS

When planning the application of beach control structures it is necessary at the outset to define a number

Coastal, Estuarial and Harbour Engineer's Reference Book. Edited by M.B. Abbott and W.A. Price. Published in 1993 by Chapman & Hall, 2–6 Boundary Row, London SE1 8HN. ISBN 0 419 15430 2.

of fundamental quantities and facts relating to the littoral drift (which is taken to include both alongshore and onshore–offshore sediment transport, see Chapters 16, 17 and 18):

1. The prevailing direction of alongshore drift, its seasonal variation and the ratio of net to gross movement;
2. The extent of onshore–offshore sediment transport and its seasonal variation;
3. The variability of the magnitude of littoral drift along the length of coastline under study and adjacent lengths of coastline;
4. The present rate of change of the line of the foreshore along the section under review and what this represents as an annual volumetric loss or gain of material;
5. Identification of all sources and sinks of beach material contributing to the sediment budget;
6. Identification of the possible existence of a sediment circulation system within the geographical cell;
7. Any trend in time of the annual rate of change of any of the factors and whether they undergo any cyclic variation;
8. The need to maintain the shoreline at a particular position or whether retreat is a viable option. If retreat is viable the extent allowable in relation to the rate of retreat;
9. Additional factors that may affect the future trend, such as sea-level rise, climatic change, change in sediment supply, etc.

In many situations definition of the above factors will not be straightforward and requires the combination of analysis of historical maps and charts, careful observation by a trained eye and the judicious use of hindcasting and predictive models.

Whatever type of control structure is to be employed, the definition of the nearshore regime in terms of waves, currents and sediment movement is a common requirement.

On exposed coastlines waves are the main agency for the transport of material in the nearshore zones and hence control natural beach geometry. However, in areas of high tidal range the possibility of strong tidal currents and their impact on wave propagation and sediment movement should not be ignored.

The deepwater wave climate close to any site can often be effectively determined by the application of parametric wave hindcasting techniques or by synthesizing visually observed wind and wave data from ships in passage. To transform the deepwater wave climate into a shallow-water wave climate, it is necessary to consider the effects of refraction, diffraction and shoaling as the waves approach shallow water.

Determination of the nearshore wave climate is essential to the design of any type of beach control structure. In particular, wave direction plays a dominant role in the generation of alongshore currents, the corresponding alongshore sediment transport and the change in orientation of the beach when the control structures are present. The definition of design criteria for wave heights in conjunction

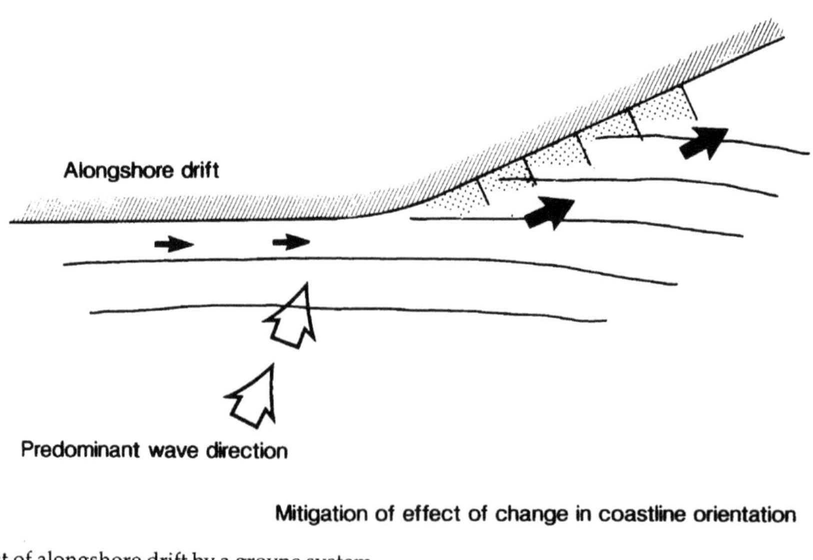

Figure 23.1 Arrest of alongshore drift by a groyne system.

with water level is important for both the structural integrity and design of the beach profile. Whilst considerable emphasis is generally given to establishing extreme event statistics and joint probability, the derivation of the lesser wave climate which represents the 'average' or 'normal' conditions on a seasonal or annual basis is equally important where sediment transport processes and beach response need to be predicted.

The following sections describe the basic outline for designing the geometry of groynes, offshore breakwaters and artificial headlands.

23.3 GROYNES

Groynes are shore protection structures which are used to control the alongshore movement of beach material principally by altering the natural orientation of the beach line and intercepting wave-induced alongshore currents. Groynes rarely extend beyond the zone in which wave-induced alongshore currents occur naturally, but may also be designed to deflect tidal currents. They can have little direct effect on onshore/offshore movement of beach material which can, in some situations, be the dominant mechanism for loss of beach material. Groynes are generally straight structures constructed normal or near normal to the coastline although zig-zag shapes, or Y-, T- and L-shaped heads have been used.

A well-designed groyne system can:

1. Arrest or slow down the alongshore drift of material and, by building up material in the groyne bays, stabilize the foreshore and protect the coastline (see Figure 23.1);

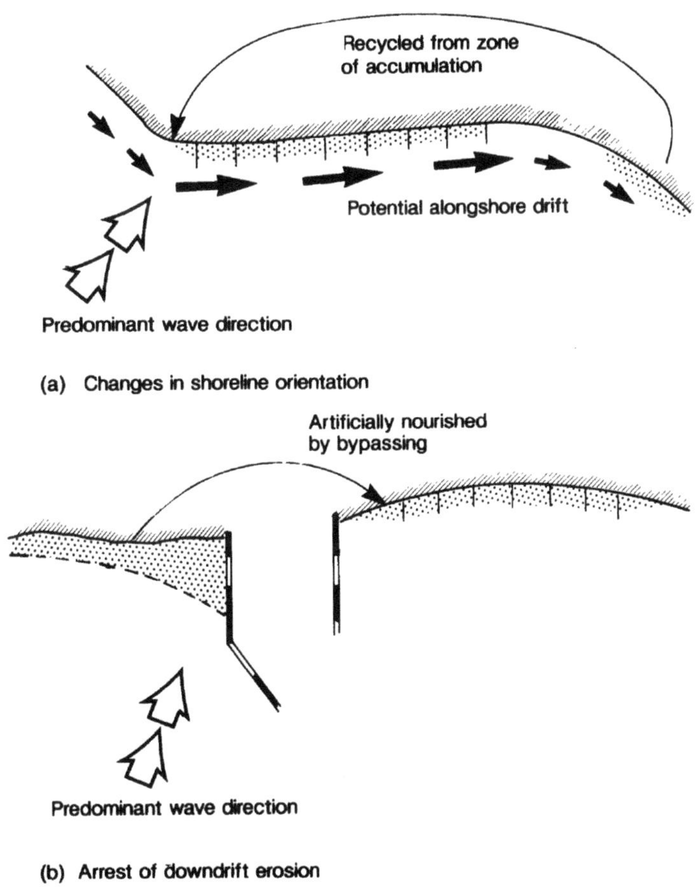

Figure 23.2 Use of groyne system on nourished beach. (a) changes in shoreline orientation; (b) arrest of downdrift erosion.

2. Deflect strong tidal currents away from the shoreline;
3. Assist in holding material on a beach that has no natural supply and has been artificially nourished (see Figure 23.2);
4. Control the seasonal shifts of material alongshore and hence movements of a shoreline within a bay (see Figure 23.3);
5. Reduce the long-term erosive effect of wave activity in an area of coastal defence by accumulating beach material in front of prominent hard beach heads such as sea-walls, revetments or cliffs as long as there is an adequate supply of material moving alongshore;
6. Improve the extent and quality of an amenity beach;
7. Increase the depth of beach material cover to an otherwise erodible sea-bed.

Groynes can be either permeable or impermeable and there are several methods of construction for both types (CIRIA 1990b); results of physical model studies and corresponding prototype experience can be found in the literature, e.g. Bakker *et al.* (1984).

Simple mathematical models, if properly calibrated, can be used for engineering purposes to predict the development of the beach plan shape between widely spaced groynes. A discussion of different types of beach response modelling can be found in Chapter 25. A physical model may be used to investigate and optimize the effect of beach shape, groyne length and groyne spacing (Kemp, 1963; Barcelo, 1968; Hulsbergen *et al.*, 1976; Bakker *et al.*, 1984 and CIRIA, 1990b). However, there are difficulties in creating a littoral environment of alongshore and onshore–offshore sediment transport in a model as well as the problems of interpretation to full scale. It does seem that some success can be achieved for modelling of shingle-sized material at a suitable scale.

Design criteria for a groyne system should define the approximate proportional reduction of alongshore drift that is required. If it is assumed that the volumetric transport rate of alongshore drift is directly proportional to the sine of the angle between the wave crest and the beach contour at the breaker line, then Figure 23.4 shows how the ratio of drift, with and without groynes (Q_g and Q_o, respectively) are associated by the ratio:

$$\frac{Q_g}{Q_o} = \left[\frac{\sin 2(\alpha_o - \alpha_g)}{\sin 2\alpha_o}\right] \approx \frac{(\alpha_o - \alpha_g)}{\alpha_o} \qquad (23.1)$$

where α_o represents the angle of incidence of waves to the ungroyned beach and α_g represents the change in angle due to groyning. The approximation relates to small angles of incidence. The progressive filling of the upstream boundary area and the groyne bays can be determined by reference to the continuity equation described in Chapter 23.

The top level of a groyne will determine the maximum potential beach depth updrift of the groyne. The groyne structure should be designed for any

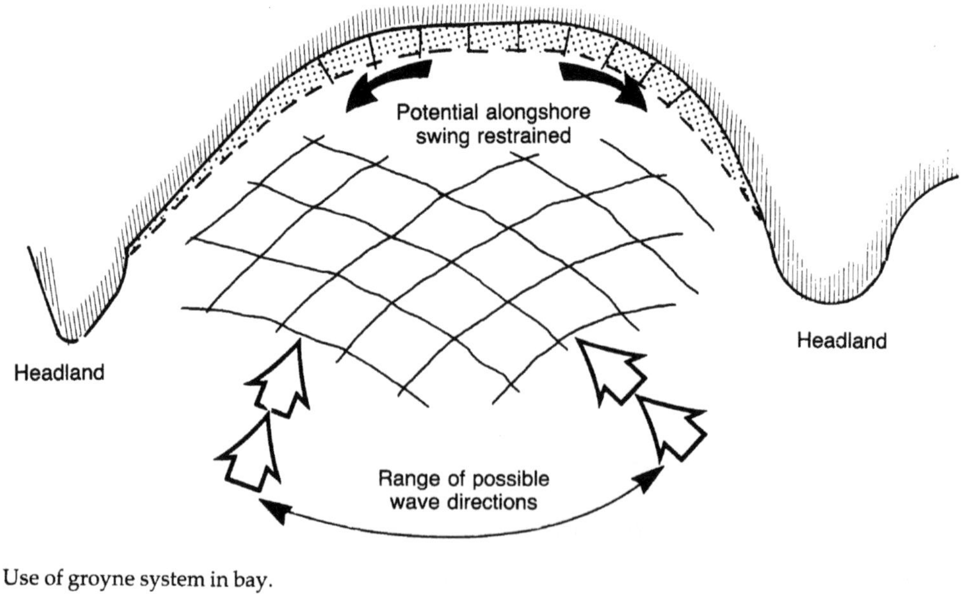

Figure 23.3 Use of groyne system in bay.

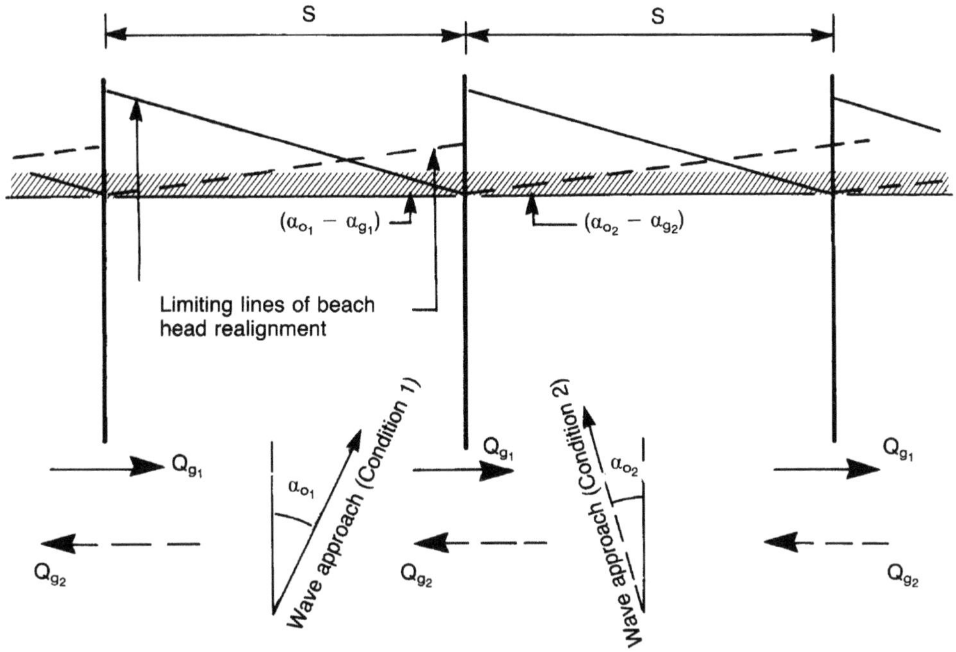

Figure 23.4 Change in beach orientation due to groynes.

combination of beach levels on either side of it between the local scour level and the desired maximum beach depth. The extremes will usually be determined by the natural limiting winter and summer beach profiles.

Variations in concentration of wave energy and, hence, in rate of alongshore drift along a coastline, can cause local variation in the transport rate, and hence produce local scour. This is particularly likely to occur during a period in which the direction of drift has been reversed due to a change in wave direction. Some time is required for beach material within a groyne bay to adjust to a new form. Localized erosion of beach material may thus occur before a new regime of littoral drift is established. This means that the groyne height may need adjustment from time to time to ensure that sufficient alongshore drift is trapped to prevent scour when there is a temporary change in wave conditions. This is, of course, only possible with groynes that are designed to allow their crest height to be adjusted. The significance of groyne height on the effectiveness of the system differs according to different beach types because the extent of scour also depends on the size of beach material (CIRIA (1990b)), on shingle beaches greater groyne heights are permissible and practical. Also where beach recharge is contemplated, groyne heights must be determined to suit the nature of the recharge material.

The cost of building conventional groynes is greatly influenced by the period of accessibility of the foreshore between tides. Without special provision in construction, an economic limit on length is reached inland of the mean low water mark of spring tides (MLWS). In practice the length is often determined by tidal range and beach slope. The required length is, however, also related to the desired trapping effectiveness of the groyne system. In order to control sufficient alongshore drift, it may be necessary to lengthen the groyne, but this cannot be dissociated from the spacing and height of the groyne. When long groynes are built at close spacing, a proportion of the alongshore drift will not be captured by each intermediate groyne bay and may be directed offshore and lost from the nearshore system (see Figure 23.5).

For a given groyne length, maximum groyne spacing should take into account the resulting variation in beach level each side of the groyne. Thus, in a bay where the direction of wave attack is confined, groynes may be more widely spaced than on an exposed promontory. It also follows that steeper beaches require more closely spaced groynes.

The rational determination of groyne spacing in-

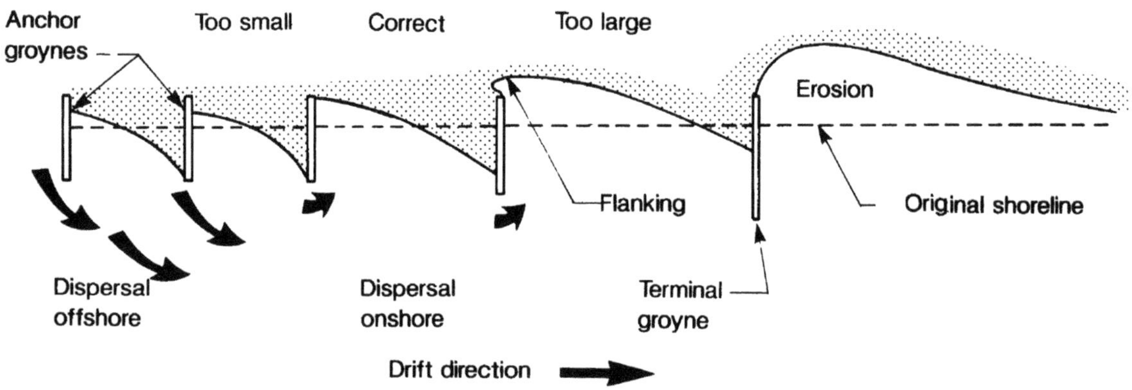

Figure 23.5 Impact of groyne spacing for given groyne length (Carter, 1988).

volves estimating the possible variation on beach shape that may take place within each groyne bay, while at the same time ensuring that an adequate reservoir of beach material is allowed to accumulate. The spacing should be selected so that, at periods of maximum drift, the angle of the beach head allows the requisite length of groyne to be constructed economically and avoids excessive loading on the groyne due to variations in beach level on either side. It can be seen from Figure 23.4 that the horizontal distance between the beach crest on either side of the groyne is $s\tan(\alpha_o - \alpha_g)$. For a beach slope of β a conservative estimate of the difference in level across the groyne will be $[s\tan(\alpha_o - \alpha_g)]\tan\beta$.

A simple representation of the effect of relative spacing to length ratios is shown in Figure 23.5 from Carter (1988). In addition, the beach crest must be sufficiently far seaward to ensure that any sea wall or revetment is provided with reasonable protection by the beach at all times. The latter design consideration requires a good estimate of the equilibrium beach profile geometry under storm conditions.

From the practical viewpoint, groynes are generally constructed transverse to the general direction of the coastline. In order to minimize structural damage during storms, groynes should ideally be aligned directly into the direction of the maximum storm waves. In practice this is not usually possible. At many sites there is, in fact, a substantial drift in both directions due to the multi-directional nature of the wave climate.

Special considerations are required when dealing with the last downdrift groyne in a system. The importance of considering a length of coastline as a geographical cell has already been mentioned. Often the beach will terminate at an inlet to a tidal estuary or creek. Terminal groynes or training walls are sometimes constructed in these cases to perform two functions. Firstly, they arrest as much of the alongshore drift as possible to prevent siltation of the inlet; and secondly, they preserve and improve a beach on the updrift side.

It can be appreciated that a terminal groyne in a system might deliberately be made longer and higher in order to create a reservoir of material which can be mechanically transported updrift to nourish depleted beaches. Alternatively, in order to reduce the immediate impact of downdrift erosion, the groynes may be made progressively shorter in the downdrift direction.

A nationwide study of groyne systems in the UK, initiated by a review of available information (CIRIA, 1983) has resulted in the compilation of a large volume of data covering many different beach types (CIRIA, 1990a) as well as a guide to the uses of groynes in coastal engineering (CIRIA, 1990b). For the purpose of data analysis beaches were classified as:

1. Shingle;
2. Shingle upper/sand lower where a distinct boundary between the upper shingle beach and the lower sand beach exists;
3. Mixed shingle/sand where no such division exists but the material sizes span the full range;
4. Sand, which was further subdivided by mean grain diameter as coarse (0.6–2.0 mm), medium (0.2–0.6 mm) and fine (0.06–0.2 mm).

The CIRIA guidelines (1990b) provide design notes related to the above specific beach types and groyne types. CIRIA Technical Note 135 (1990a) provides table of many different facets of the data. The average of the primary dimensions of interest are shown in Table 23.1.

The CIRIA groyne study also included collection of

Table 23.1 Average dimensions of groyne systems (CIRIA 1990b)

Beach type	*Average lengths and ratios*		
	Length (m)	*Spacing (m)*	*Range of spacing length ratio*
Shingle	60	60	0.5–1.7
Shingle upper/sand lower	50	50	0.5–1.5
Shingle/sand mixed	70	85	0.6–2.4
Sand	95	130	0.8–2.7

field data and a comprehensive series of physical model tests. The latter showed that broadening the head of the groyne as with a 'T' or 'Y' shaped groyne made significant differences to the current flow patterns within the groyne bays indicating beneficial effects. This shape of structure is that generally used for artificial headlands, as described in section 23.5.

Groyne types tend to vary from country to country depending on practice often developed over decades of experience and also on the construction materials economically available. In the UK the most common type of groyne has been the timber-fence type of groyne where horizontal planks are fixed to the king piles driven into the beach. Whilst this type of groyne can be designed to allow the vertical height to be varied by adding or removing planks, the vertical nature of the structure results in wave reflections that increase turbulence within the groyne bays. With the availability of imported rock at competitive prices there has been a significant increase in the use of rock groynes which are inherently more satisfactory due to their ability to dissipate incident wave energy.

In The Netherlands, much of the coastline is protected by groynes. Along the southern coast the groynes are relatively high and subject to strong tidal currents. The groynes are made of natural stone at the base with a single or double row of circular piles protruding through the water surface. A number of parameters have been used to investigate the effectiveness of these groynes and this has been summarized by Fleming (1990).

The conclusion drawn from these investigations are applicable elsewhere, in that while well-designed groynes can be effective in controlling beach movements, the degree of success achieved will, to a large extent, be dependent on the rate of sediment supply. Where macro-scale movements such as sand waves or onshore–offshore movement are taking place groynes can have little influence in the long term. A number of references are provided, which although not directly cited are worthwhile background reading.

23.4 OFFSHORE BREAKWATERS

Detached or offshore breakwaters have been used extensively for coast protection or the creation of crescentic beaches with some considerable success, particularly on coastlines where the tidal range is negligible or small.

Detached breakwaters create a zone of reduced wave energy behind the breakwater as well as local patterns of wave-induced currents that, in turn, create a zone of sand deposition in the lee of the structure. In the absence of other influences beach material will be transported into the area to form a tombolo or salient as shown in Figure 23.6. Detached breakwaters can be used in much the same way as groynes to build up the volume of material held on the beach that is the capable of accommodating the drawdown that occurs under storm conditions. The shape of the beach that forms between adjacent breakwaters is that of a crenulate bay which is inherently more stable and less volatile than the abrupt discontinuities caused by groynes. However, if attachment of tombolos (see below) is permitted downdrift erosion problems will still occur. Detached breakwaters may also be used to deliberately create an area of deposition, for example, updrift of a harbour entrance, so that it can conveniently be dredged and deposited on the downdrift beach.

An offshore breakwater is generally sited with its major axis parallel to the coastline thus intercepting incident wave energy so that both alongshore and onshore–offshore sediment transport is reduced. The primary influence is that of wave diffraction which not only reduces wave heights in the lee of the structure, but creates wave height gradients that drive wave-induced currents towards the sheltered area. Thus, in both instances the trend is towards the transportation of sediment to the centre of the shoreward side of the structure. The net result is the growth of a salient (an advance of the water line behind the structure) or a tombolo (the attachment of the shoreline to the structure).

A number of rules of thumb exist for predicting whether either a tombolo or a sediment might form. These are independent of the incident wave height and beach material properties and relate to the geometry of the structure in terms of its length, its distance from the shoreline and the gap between adjacent structures, relative to the wavelengths of the incident waves. Examples are:

1. If the ratio of the length of the breakwater to its distance offshore is less than 1.0 and the breakwater length is of the order of dimension of at least one wavelength in lengths, then attachment of the tombolo will take place.
2. If the isoline of the wave diffraction coefficients are

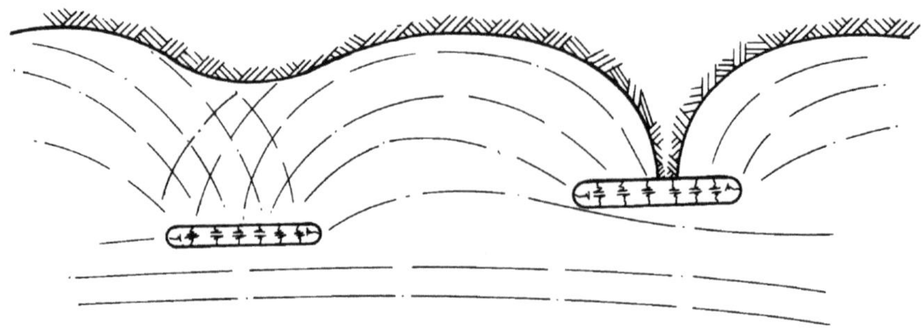

Figure 23.6 Formation of salient or tombolo.

plotted for the various predominant wave conditions, and if the value of 0.3 remains seaward of the undisturbed shoreline then the salient will not form into a tombolo.

A comprehensive summary of other such rules based on prototype experience and physical model tests has been given by Dally and Pope (1986) and Herbich (1989). Hanson and Kraus (1990) have compared numerical simulations from GENESIS (Hanson and Kraus, 1989) with field measurements.

Hsu and Silvester (1990) have developed an empirical method of defining the curvature and apex position of the salient that will form in terms of the length of the breakwater relative to its distance from the original shoreline. The data they have analysed does indeed suggest that a universal relationship exists that is independent of either incident wave conditions represented by wave height or the sediment characteristics represented by grain size. However, such relationships describe the 'final' form of the shoreline and the rate of development must depend on these other parameters. It is also apparent that detached breakwaters are most effective in situations where the tidal range is small. Where this is not the case, typically in UK waters, a more rigorous analysis is required to determine the optimum length, spacing, and distance offshore of the breakwaters.

Design conditions will also need to consider the impact of storm surges on the top of the tide and for both aesthetic and economic reasons it is necessary to allow the structure to become partially submerged during the highest levels of the tidal cycle. Such structures therefore become reef breakwaters for which design rules have not been well developed and physical model tests are recommended for determining the stability of the breakwater armour, whether rock or concrete armour units. A review of design data has been provided by Ahreus and Cox (1990) and a typical cross-section of such a structure is shown in Figure 23.7.

23.5 ARTIFICIAL HEADLANDS

The fundamental difference between a groyne and an artificial headland is that the latter is a more massive structure designed to eliminate problems of downdrift erosion and promote the formation of beaches. Whilst these structures may take a number of different forms, their geometry is such that, as with the offshore breakwater, wave diffraction is used to assist in holding the beach in the lee of the structure.

Hence, artificial headlands are structures which bridge the gap between groynes and detached breakwaters and in some instances the differences might be viewed as subtle. However, the fundamental mode of application of artificial headlands is to create stable bays between adjacent structures. This implies less passage of alongshore drift of material that might be accommodated in a groyne or detached breakwater system, but does not rule it out completely,

Artificial headlands attempt to emulate the stable bay formations that occur in nature. The rate of erosion of a shoreline can be reduced by dividing an existing, developing bay into a number of sub-bays as there is a geometric relationship between the distance between headlands and the maximum recession measured from the line facing the headlands.

The formation of crenulate bays (sometimes called ζ – curves or spirals) as the equilibrium form between headlands has long been recognized. The form arises where oblique prevailing waves both diffract around the headland and refract into the bay itself. Silvester and others (Silvester, 1976; Silvester *et al.*, 1980) have derived empirical expressions for the equilibrium bay shape based on the angle of wave approach,

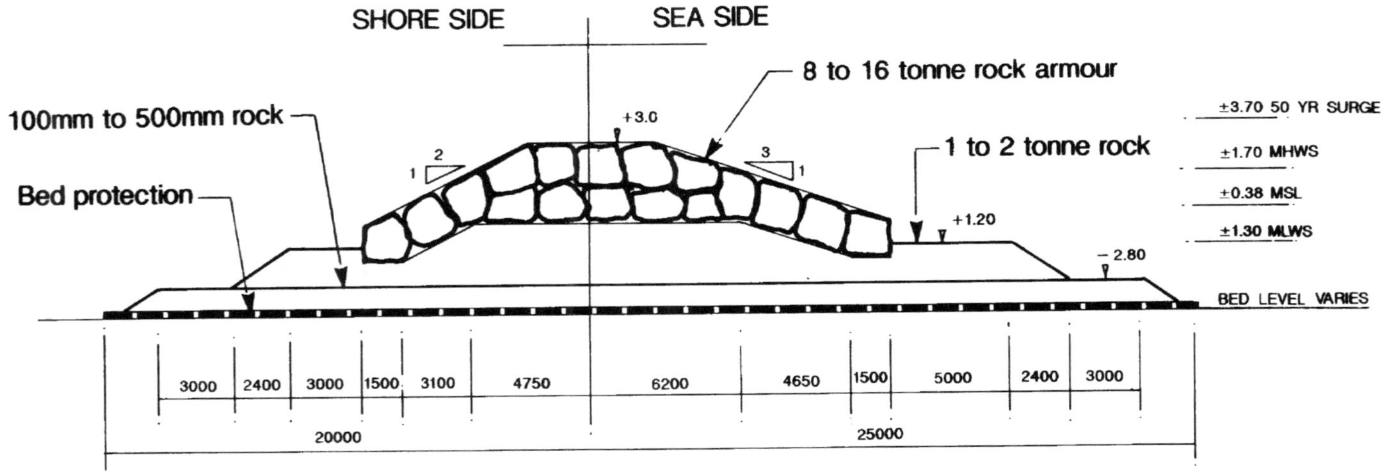

Figure 23.7 Cross-section of typical offshore reef structure.

the indentation ratio and the radius of a logarithmic spiral. The rate of development of such equilibrium bays depends on the level of wave energy, but can be determined through the use of a suitable beach response model (see Chapter 25).

More recent studies have led Silvester and his co-workers to revise the static equilibrium bay description in recognition of the fact that, whilst the spiral shape applies mainly to the shadow zone behind the upcoast headland and a new method has been proposed (Hsu *et al.*, 1989) using arcs from the point of wave diffraction to the shoreline, whose lengths and angles to the wave crest line are related to those of a control line. Some of the expressions have also been simplified by Hsu and Evans (1989) whereas Chiew and Tan (1991) have proposed slightly different expressions based on the finding that the alignment of the beaches at the down-coast headland are not necessarily parallel to the incident wave crests.

Clearly the expressions derived assume that a single dominant wave direction can be defined and that the tidal range is limited in magnitude. In practice the designer will have to consider a range of incident wave directions and conditions to evaluate the possible variability of the equilibrium bay shape. Nevertheless, these expressions have been found to be very useful for planning and design purposes particularly when combined with other methods of determining beach response.

A particular type of artificial headland is the 'fishtail breakwater' which has been largely developed by P.C. Barber in the UK. The following is drawn from material supplied by him.

The concept of the fishtail breakwater is to combine the beneficial effects of the groyne, offshore breakwater and tombolo and reduce the undesirable effects of the separate structures. The basic plan shape of a fishtail breakwater is shown in Figure 23.8. The breakwater arms OA and OB act as wave energy dissipators whilst the arm ACE provides the interceptor to alongshore drift. Thus the updrift beach is formed by the normal accretion processes associated with a groyne whilst the downdrift beach is formed by processes associated with an offshore breakwater.

The arm AC is curved in plan so that the axial alignment at A is normal to the streamline of the diverted alongshore and tidal currents and the axial alignment of C is generally normal to the shoreline. The curvature of COA is designed to minimize wave reflection effects on the concave side of the breakwater. The arm OB is located in plan to allow the waves sufficient distance to transform out of the current field and its length is dependent on the length of OC, but generally dependent on achieving the desired wave diffraction effects. The overall breakwater dimensions are thus interdependent and depend on wave height, direction and period, tidal range, beach morphology and the extent of required influence. The distance of A offshore depends on the length of coast the breakwater is intended to influence, but should be greater than three inshore

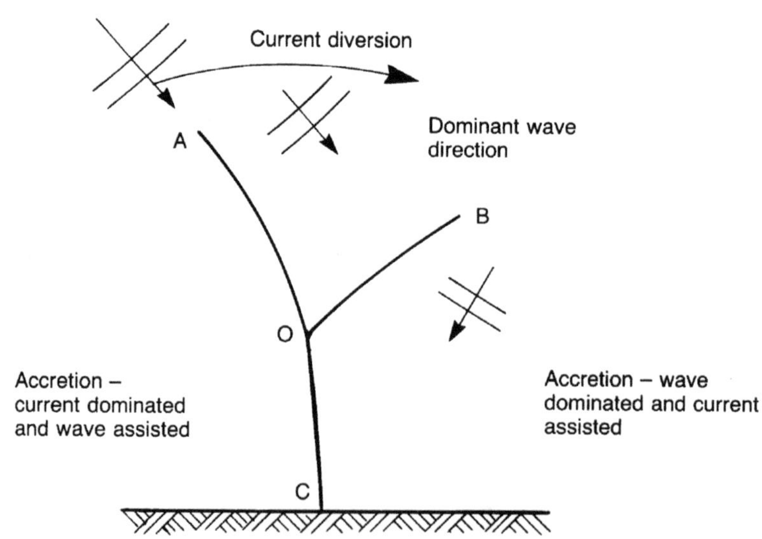

Figure 23.8 Definition sketch for fishtail breakwater (Barber).

wave lengths as well as less than half of the width of the active littoral zone.

The crest levels vary throughout a fishtail breakwater and are dependent on the frequency of the water levels and wave exposure considerations along the length AOB. The crest between O and C should follow the equilibrium beach profile.

The fishtail breakwater can be expected to influence the beach in a number of ways. There is usually a small steepening of the beach gradient due to the current reductions caused by the breakwater. The beach may form a crenulate bay form if the wave conditions are dominantly from an oblique wave direction. However, more often multi-directional conditions exist and more complex geometries will evolve. Where there is a high tidal range and a varied wave climate the beach will be constantly changing in plan level and gradient.

Artificial headlands are powerful coastal control structures in that the creation of stable bay formations is one of the most satisfactory forms of sea defence. However, in situations where it is desirable to allow significant volumes of alongshore drift to pass through the system or where shore connected structures are not environmentally acceptable, detached breakwaters or reef structures may be preferable.

REFERENCES

Ahrens, J.P. and Cox, J. (1990) Design and performance of reef breakwaters. *Journal of Coastal Research* (special issue No 7) 61–75.

Bakker, W.T., Hulsbergen, G.H. Roelse, P. *et al.* (1984) Permeable groynes: experiments and practice in the Netherlands. *Proceedings of the 19th Coastal Engineering Conference, ASCE*, pp. 2026–41.

Barber, P.C. (personal communication).

Barcelo, J.P. (1968) Experimental study of the hydraulic behaviour of inclined groyne systems. *Proceedings of the 11th Coastal Engineering Conference, ASCE*, pp. 526–48.

Baslsillie, J.H. and Berg, D.W. (1972) State of groyne design and effectiveness. *Proceedings of the 13th Coastal Engineering Conference, ASCE*, pp. 1367–83.

Biss, M.A. and Craig, C.W. (1987) Clacton sea defences. Ministry of Agriculture Fisheries and Food, Conference for Coastal and River Engineers, Loughborough, UK.

Brampton, A.H. and Motyka, J.M. (1983) The Effectiveness of Groynes. *Proceedings of Conference on Shoreline Protection, Southampton, ICE*, Thomas Telford, London.

Carter, R.W.G. (1988) *Coastal Environments*, Academic Press, London.

Chiew and Tan (1991) Discussion of 'Static Equilibrium Bays: New Relationships' (J.R.C. Hsu, R. Silvester and Y.-M. Xia) *Journal of Waterway, Port, Coastal and Ocean Engineering*, ASCE May/June 1989, 285–98. In *Journal of Waterway, Port, Coastal and Ocean Engineering*, ASCE, March/April 1991, 194–6.

CIRIA (1983) Groynes in coastal engineering: a review. *Tech. Note III*, Summers, L. and Fleming, C.A., CIRIA, London.

CIRIA (1990a) Database of existing groyne systems in britain. *Tech. Note 135*, CIRIA, London.

CIRIA (1990b) Guide on the uses of groynes in coastal engineering, *Rep. 119*, Fleming, C.A. CIRIA, London.

Dally, W.R. and Pope, J. (1986) Detached breakwaters for shore protection. US Army Corps of Engineers, *Tech. Rep. CERC-86-1*, Coastal Engineering Research Centre, Vicksburg.

Everts, C.H. (1979) Beach behaviour in the vicinity of groynes – two New Jersey field examples. *Coastal Structures '79, ASCE.*

Fleming, C.A. (1990) Principles and effectiveness of groynes. *Coastal Protection*, Proceedings of the Short Course on Coastal Protection, Delft University of Technology, July 1990, A.A. Balkema, Rotterdam, pp. 121–56.

Fleming, C.A. and Lunt, H. (1988) The Lincolnshire groyne study. *Proceedings of the 1st Coastal Engineering Conference, ASCE.*

Hanson, H. and Kraus, N.C. (1990) Shoreline change behind detached breakwaters. *Proc 22nd Coastal Engineering Conference, ASCE.*

Hanson, H. and Kraus, N.C. (1989) GENESIS: generalised model for simulating shoreline change rep. 1: technical reference, *Tech. Rep. CERC-89-19*, US Army Waterways Exp Station, CERC, Vicksburg.

Herbich, J.B. (1989) Shoreline changes due to offshore breakwaters. *XXIII Congress of IAHR*, pp. C317–27.

Hsu, J.R.C. and Silvester, R. (1990) Accretion behind single offshore breakwater. *Journal of Waterway, Port, Coastal and Ocean Engineering, ASCE*, **116**, (3) 362–80.

Hsu, J.R.C. and Evans, C. (1989) Parabolic bay shapes and applications. *Proceedings of the Institution of Civil Engineers, Part 2*, **87**, 557–70.

Hulsbergen, C.H., van Bochove, G. and Bakker, W.T. (1976) Experimental verification of groyne theory. *Proceedings 15th Coastal Engineering Conference, ASCE.*

Kemp, P.H. (1963) A model study of the behaviour of beaches using groynes. *Proceedings of the Institution of Civil Engineers*, **22**, 191–210.

Price, W.A. and Tomlinson, K.W. (1968) The effect of groynes on stable beaches. *Proceedings of the 11th Coastal Engineering Conference, ASCE*, pp. 518–25.

Price, W.A. and Tomlinson, K.W. (1970) The effect of groynes on eroded beaches. *Proc 12th Coastal Engineering Conference, ASCE*, pp. 1053–58.

Silvester, R. (1976) Headland defense of coasts. *Proc 15th Coastal Engineering Conference, ASCE*, pp. 1394–1406.

Silvester, R., Tsuchiya, Y. and Shibano, T. (1980) Zeta bays, pocket beaches and headland control. *Proc 17th Coastal Engineering Conference, ASCE*, pp. 1306–19.

Thorn, R.B. and Roberts, A.G. (1981) *Sea Defence and Coast Protection Works*, Thomas Telford, London.

Tomlinson, J.H. (1980) Groynes in coastal engineering – a literature survey and summary of recommended practice, *Hydraulics Research Rep. IT199*, Hydraulics Research, Wallingford (reproduced in full in *CIRIA Tech. Note 111* (1983)).

BEACH NOURISHMENT, OFFSHORE DREDGING, SAND BYPASSING AND FLUIDIZATION — 24

Professor A.G. Roberts

This is a very practical chapter on the way in which the coastal engineer should approach coastal problems. It emphasizes the importance of an understanding of fundamental beach processes as the foundation of sound problem solving. Once again, as elsewhere in this book, the importance of beach monitoring is stressed and the many activities within this field are described. The author illustrates the interplay that has to take place between the fundamental thinking and the practical options available to solve a particular problem within the design process. The chapter represents an overview of the subject, paying particular regard to beach nourishment, offshore dredging and sand bypassing (see also Chapters by 22, 23, 25 and 45), and also includes a contribution on beach stabilization, sand bypassing and channel maintenance by fluidization, by J. Parks.

CONTENTS

24.1 INTRODUCTION

The important role, now recognized world-wide, of sand and shingle as integral parts of a sea defence has emphasized the need for an improved understanding of the behaviour of beaches. The construction of hard sea defences such as sea-walls and the armouring of eroding coastal cliffs have reduced the traditional source of supply for longshore transport. Further, the coastal engineer himself can all too often cause disruption to longshore drift by building obstructions in the form of groynes and breakwaters, thus increasing the need to carry out artificial nourishment of beaches. This chapter reviews the techniques available to create such schemes.

24.2 MONITORING

The characteristic behaviour of an existing beach must be well understood by the coastal engineer before embarking on any works. He must draw up a comprehensive monitoring plan which, when implemented will allow him to draw conclusions as to the likely behaviour of the proposed beach both during and after construction. Normally a period of some two years would be required to ascertain the beach's behaviour, however, if time is short then the beach should be measured frequently after exposure to storms.

Beach behaviour cannot be assessed on a narrow geographical basis. Roberts and McGown (1987) describe the concept of the coastal process unit or cell; as an area or location which does not significantly interfere with, nor is subject to significant interference from, an adjacent area. And the engineer

Coastal, Estuarial and Harbour Engineer's Reference Book. Edited by M.B. Abbott and W.A. Price. Published in 1993 by Chapman & Hall, 2–6 Boundary Row, London SE1 8HN. ISBN 0 419 15430 2.

must firstly identify the coastal cell or cells in which he intends to construct a beach (quite often a simple inspection of the site can suggest the likely process unit boundaries) and then collect data pertaining to that cell. It is necessary to establish all hydraulic criteria for the process unit, including the maximum high, minimum low and highest annual maximum sea-levels, maximum and significant wave heights and their associated periods. A comprehensive knowledge of the inshore currents must be gained together with an accurate topographical survey of the site between the high- and low-water lines. In addition a survey (which need not be too detailed) of the offshore area should be carried out.

If there are existing beaches in the area under consideration, they should be comprehensively surveyed to establish their shape and extent. The survey should also include data on the beaches' geological formation and particle size distribution. All these surveys should then be repeated at fixed time intervals. Extra surveys should also be carried out immediately after any severe storms.

There are many techniques available to ascertain the topographical shape of a beach, the principal ones (which will be outlined in this section) being:

1. Aerial survey and contour modelling;
2. Physical ground surveys and contour modelling;
3. Cross-sectioning;
4. Photographic records.

All rely on the application of a consistent approach to data collection. It is necessary to ensure that the data is always taken at the same tide state, at regular time intervals and to the same datum with all calculations carried out in the same way.

The advent of the micro-computer has revolutionized the handling of data, but as yet no standard software exists for the storage of beach profiles and behaviour. The engineer will have to tailor off-the-shelf systems to store, recover and display data in the manner which seems best to him.

24.2.1 AERIAL SURVEYS

Aerial surveying and contour modelling for the study of beach behaviour is a well-established technique. It is particularly useful for the monitoring of long lengths of coastline and is very cost effective. Some reservations have to be made, however, with regard to the tolerances that exist in surveys of this nature, where ±75 mm in level datum is not uncommon. Such tolerances can, in the case of large areas of beach exposed between high- and low-water marks, give rise to substantial apparent changes in beach volume where only minimal changes exist in reality. In such cases aerial surveys should be calibrated by physical measurement on the ground. The Southern Unit of the National Rivers Authority uses this technique to monitor its coastline. It has led to the drawing up of a coastal management strategy which has shifted their emphasis from sea-walls as sea defences to beaches.

24.2.2 PHYSICAL SURVEYS

Conventional land surveying techniques have the advantage of simplicity. They are, however, labour intensive but bring the benefit, especially in the case of small-scale studies of small areas, of speed of response coupled with a high degree of provable accuracy. The use of computer programmes permits easy computation and data comparison.

24.2.3 CROSS-SECTIONING

Either aerial or conventional techniques can be used as the basic survey method. This approach has the advantage of simplicity allied to flexibility as more cross-sections may be added as and when it is found necessary. Sections should be sited to reflect and monitor changes within the beach; slavish adherence to setting cross-sections at a regular chainage is not recommended. Again the use of micro-computers permits the handling and analysis of very large data sets; data being normally displayed in conventional form, sometimes showing previous records by way of comparison.

24.2.4 PHOTOGRAPHS

In some cases the need for physical measurement is not of paramount importance, and here photographic records may permit a sufficiently detailed study of beach behaviour to be carried out. Copies of the vertical photographs used in aerial surveying can be obtained and will repay study not only as stereoscopic pairs but also as photographic records of any changes over the years. An alternative and less expensive method is to have oblique photographs taken from a lower altitude (about 1000 ft) with subsequent sets taken from the same height and at the same state of the tide. Ideally they should be taken when the sun is comparatively low in the sky as this gives better contrast to the photographs. Careful study of such photographs can provide much valuable information on the bahaviour of beaches and, used in combination with cross-section measurements, can form the bulk of the predictive data base.

Some members of the public, who have lived near a beach for years, take a genuine interest in the shoreline. Provided they are reliable observers (and this has to be judged) they can provide valuable information on beach changes together with opinions on causes and effects which might assist with the understanding of a particular problem under study.

24.3 THE DESIGN PROCESS

When providing beach sea defences it must be recognized that the presence of high, longshore, drift rates may soon cause the transportation of material away from the chosen site. The steps that should be taken to provide the necessary control structures, either by offshore breakwaters or by provision of a groyne system within the beach itself are described elsewhere.

Owen (1989) has identified the following design procedures:

1. Identification of the minimum beach profile necessary to give the required level of protection at all times.
2. Identification of the plan shape necessary to produce the required beach profile at all important locations.
3. Evaluation of any beach structures or management which are necessary to maintain the desired beach shape.

Owen then goes on to describe the use of mathematical modelling to define the anticipated beach movement and beach shape. This technique has been used in connection with the evaluation of Hengistbury Head at Christchurch, on the south coast of England.

Section 24.2 outlined the need for a comprehensive data base of beach behaviour. Using that knowledge it is possible to calculate the volume of beach material necessary to provide the desired level of protection. Van Hijum and Pilarczyk (1982) produced a model for

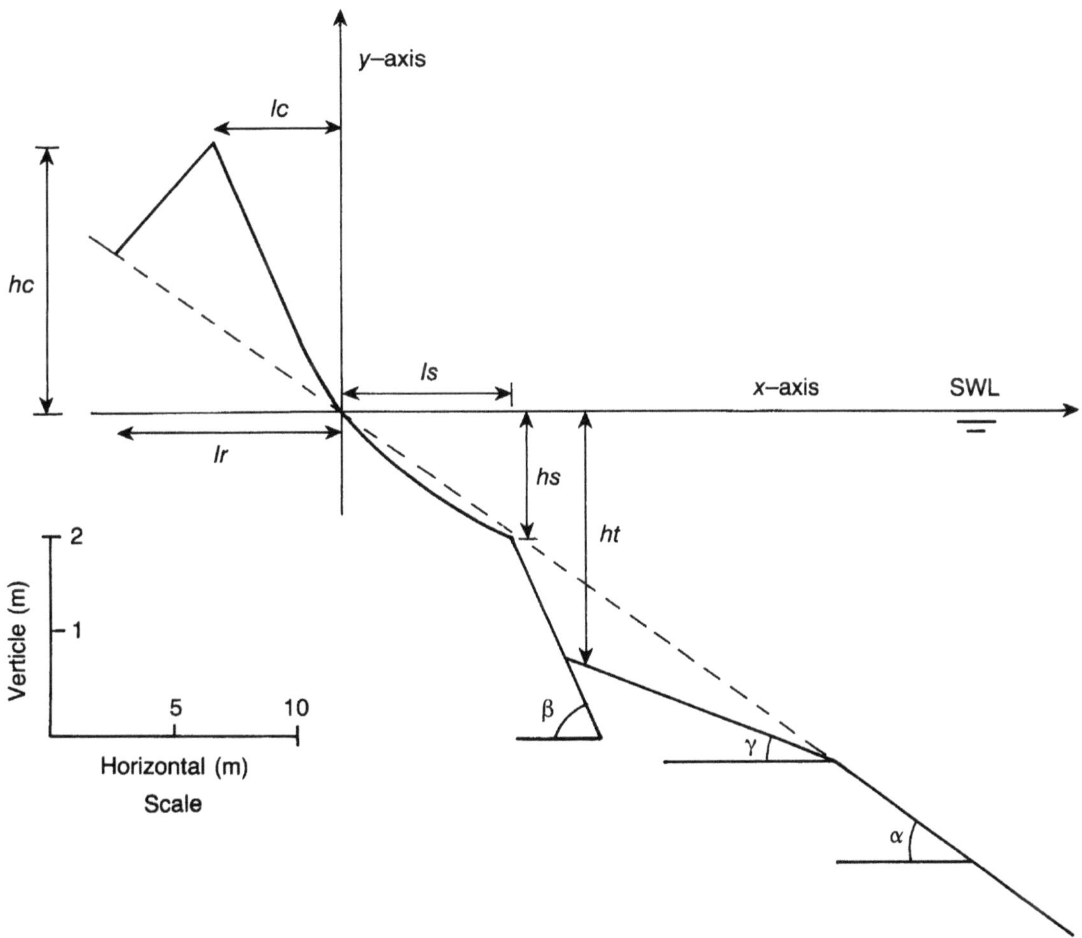

Figure 24.1 Schematized beach profile (van der Meer, 1988).

a dynamically stable beach profile. A study should be made of the existing beach profiles to determine the characteristic shape and dimensions of the beach under both normal and storm conditions. If the beach is so badly eroded as to prevent the identification of the normal beach profile, it is suggested that a nearby beach of similar characteristics be studied. Van der Meer (1988) explores, for gravel beaches, the set-up and analysis of dynamically stable structures and the development of a calculative model which permits the evolution of a theoretical beach shape. A worked example is given in Figure 24.1 and Appendix A. Van der Meer used the storm wave for his design. The same approach is equally valid in the design of a shingle beach shape by using the significant wave height.

These theories can also be used to study existing beach profiles to determine their likely resistance to storm attack and the maximum hydraulic conditions that the beach can resist before it is subjected to catastrophic erosion.

24.4 BEACH NOURISHMENT

Once the required beach dimensions have been determined, consideration has to be given to the provision of beach nourishment. The method selected will depend on the scale of the nourishment scheme proposed, as techniques for small-scale projects are different from that needed for major works.

Access to the site is of paramount importance, whether by land or by sea. Small replenishment contracts (up to 20 000 t) often require the delivery of material by lorry or off-road earth-moving plant. Both require a clear indication of the routes that are available to the contractor together with any restrictions that may exist. Delivery from the sea, which is normally confined to medium or large schemes, demands the provision of all necessary information to allow the maritime specialist contractor to formulate his bid. This must include tidal range, bed contours and navigational obstructions (permitting the assessment of approach channels), together with enough weather records to enable the contractor to determine the likely exposure to inclement weather and hence downtime and to identify the best weather window for the execution of his deliveries.

For small-scale works material will normally be obtained either direct from a quarry or, in the case of sea-dredged material, from the nearest port, and will be transported to the site by lorry. In either case the adoption of a contract based on payment by weight, together with a knowledge of the placed density of the specified recharge material, will allow the necessary volume to be accurately assesssed. The delivered material is normally spread roughly to the desired profile and then left for sea action to trim it to its final shape.

The grading of the material which will go to make up the proposed beach must also be defined. The first step here is to analyse the grading of the existing beach and then decide if an exact match should be specified or whether improvements could be gained by changing the proportions of the coarse and fine constituents. Finer material is probably better for amenity purposes but the placing of the material with a larger grain size, even in the sand range, will produce a steeper beach and reduce the losses due to offshore transport in storms. Even so, amenity can, more often than not, be maintained. Whatever the chosen grading, an unequivocal specification will allow comparison to be made of the possible sources of supply.

Gravel is an important sea-defence material because when it is placed on a beach it has less tendency to move offshore than sand. However, the availablility of quarry-won gravel in the UK is diminishing for a number of reasons, not least those relating to the protection of the environment and this reduced availability brings with it an inevitable decrease in the types and grading of material economically available to the engineer. He must decide, therefore, whether to modify his ideal grading to reflect the availability of material from a single source, or to obtain a range of materials from different sources, each with its own characteristics, and allow his final grading to be achieved by mixing. The use of vehicular delivery permits a close mixing of these various materials to be carried out at the site. This method of replenishment has the advantage of great flexibility and is ideally suited to small contracts of less than 20 000 t and to annual maintenance contracts involving small quantities of recharge material being delivered to a number of sites in an area. It is often the only practical method in areas of very shallow water or where the risk to maritime plant is too great.

The restrictions on land extraction of suitable material for both beach replenishment and the general construction industry has led to a considerable expansion of the offshore dredging industry. However, great care has to be taken in the employment of offshore dredging to ensure that it neither damages the ecology of the dredged area nor causes other adverse effects within the immediate locality, or indeed, elsewhere. For example, the presence of a borrow area dredged too close to the shore can so effect the wave climate on adjacent beaches in such a way as to change the pattern of littoral drift. Such changes can lead to serious erosion.

These problems have been identified in the UK by the management agency, the Crown Estate Commissioners, in the operation of the licensing system.

Parrish (1989) sets out the parameters under which licences are granted. The Crown Estate is of the view that sea-bed materials are neither valueless nor limitless resources which may be released only after all the effects of extraction have been taken into account. The time scale for the release of the material must also be taken into account. If large-scale replenishment is planned utilizing material from coastal and United Kingdom continental shelf waters it is sensible to involve the Crown Estates at an early stage of scheme planning.

In addition, European practice requires licences to be obtained for the placement of material below the high-tide mark. In the UK the licensing authority is the Ministry of Agriculture, Fisheries and Food acting under the provisions of the Food and Environment Protection Act, 1985.

Occasionally supplies of recharge material can be obtained as a by-product of the dredging of navigation channels or from the deepening of harbour basins. Again, investigations have to be undertaken to ensure that its removal does not have any undesirable effects on the total coastal system.

Before considering the methods used for beach recharge from the sea it is appropriate to review the types of dredger available. In general they can be assigned to one of the following types:

1. Mechanical dredgers;
2. Static suction dredgers;
3. Trailer suction dredgers.

The reader is referred to Chapter 45 for a detailed description of dredger types. However, for completeness a brief description of the plant most commonly used in beach nourishment schemes will now be given.

Mechanical dredgers consist of a backhoe, grab or endless bucket chain mounted on a vessel which may be a ship with a hopper for holding the dredged material for transport to the discharge site or a simple pontoon which loads the material into barges. Hopper vessels are normally fitted with grabs whilst bucket and backhoe dredgers normally discharge into barges. Such vessels are seldom employed on beach recharge projects.

Static suction dredgers are essentially pontoon

Figure 24.2 Suction dredger.

mounted pumps which suck up material from the sea-bed and discharge it either into barges or through a pipeline. Pumping distance is limited for any particular horse power, though booster pumps may be used to increase the pumping power.

Trailer suction dredgers (Figure 24.2) are normally used for beach recharge contracts. They are hopper vessels with a suction pipe which is trailed behind the ship, sucking material into its hold. The laden ship is then sailed to the discharge point. Discharge can be by bottom dumping, grab or conveyor belt (dry discharge), or by pumping through a jet or a pipeline.

Some vessels are fitted with screening devices so that only selected parts of the dredged material are loaded into the hopper, the remainder being returned over the side at the dredging ground.

The methods of transferring the material from the dredger to the beach are classified as follows:

1. bottom dumping;
2. spraying;
3. pumping;
4. a combination and other methods.

Bottom-dumping vessels have doors set in their keel below the hopper through which the hopper contents are discharged when the ship is correctly positioned. Sufficient water is needed at the discharge point to allow the vessel to sail and have clearance over the material it has dropped beneath itself. Split-hopper vessels (Figure 24.3) can operate in slightly shallower water. Their hulls are hinged at either end to allow the vessel to split open over its whole length. Because of draught considerations bottom dumping is normally restricted to small, short haul, projects where such lesser vessels may be economically used. Unless it can be guaranteed that the dumped material will be carried to its final position by sea action it will be necessary to employ draglines or the like to rehandle it into its required position. Theoretically the coarser the material the greater the chance of moving onshore. Even so the losses associated with this method are normally very high.

Where spraying is employed as the discharge technique the dredger is placed bow on to the beach and the material pumped out through a nozzle fitted to its bow. The distance the material can be sprayed will be

Figure 24.3 Split hopper dredger.

about 40 m on average, with a spread at this distance of some 20 m. Losses are fairly high as the material tends to form a pancake rather than a heap with a lot of run-back to the sea which can even force the dredger to back off. The advantage of the method is in its cost. No mooring or pipelines are required and the discharge is quite fast, with the dredger cycle time being thus reduced. The operation must of necessity be tidal and tidal constraints can negate part of these advantages. Other drawbacks are the losses (which can also be affected by wind) and the dangers inherent in a large vessel approaching so close to the shore. The amount of 'dry earthwork' is higher than when direct pumping is employed.

Pumped supply systems minimize both losses and the dangers from materials ending up where they are not wanted. They also minimize the length of beach rendered unfit for everyday use. They may be subclassified as follows:

1. Direct pumping from the borrow area;
2. Direct pumping from the dredger;
3. Method involving a booster station.

Direct pumping from the borrow area requires the borrow area to be within reasonable distance of the beach to be recharged. The prime tool is the static cutter suction dredger which operates in the borrow area and discharges into a floating pipeline. Floating pipeline is expensive, at risk during heavy weather and a hazard to navigation, hence the main pipe run is normally a submerged, plain steel, pipeline laid on the sea-bed. The connection between the floating pipeline, usually 250–300 m long, and the submerged section is normally made at a floating pontoon anchored in position. The submerged pipe is fabricated on the beach, towed out and sunk into position. It can be raised again by blowing it full of air. At the beach end the submerged pipe has further sections added to it as necessary to bring its open end to the required position on the beach. Recharge material is pumped mixed with water at a concentration of around 15% solids with suspension maintained by sufficiently high velocities in the pipe. A bulldozer is normally employed to prevent material build-up directly beneath the pipe. The operation can be continued on a round-the-clock basis, extra pipes being added to the line as the material builds up. To prevent delays the pipe is often split into two outlets, with one being extended while the other is used for pumping. There is no segregation of the coarse and fine materials and fill placed in this way has an even grading. The available pumping power and material grading are the determining factors for both pumping distance and production rates. Static suction dredgers vary in power from just a few hundred up to 20 000 horse power, with production varying from 200 to 2000 $m^3 h^{-1}$.

Pumping from a hopper dredger differs from pumping direct from the borrow area in that the floating pipeline has a free end held on the sea surface. There are several methods of doing this. When the hopper dredger arrives at the site the free end of the floating pipeline is attached to the discharge pipe of the hopper dredger and its contents are pumped to the beach using the dredger's discharge pumps. The connection between the dredger and the pipeline is made using either a single buoy mooring incorporating the pipeline or by having a fixed pontoon to which the dredger can moor and which can hold the pipeline while the dredger is away. The position for the mooring point is governed by navigational requirements, particularly the depth of water available. The dredger must have sufficient pumping power to achieve the distance from the mooring point to the beach. Hopper dredgers used by the dredging industry have relatively high pumping capacities whilst those of the aggregate dredging industry do not. However, aggregate dredging industry vessels often have screening facilities, a rare feature on dredging industry vessels.

Where the pumping distance is too great for the capacity of the onboard pumps, booster pumps can be incorporated into the pipeline. It is often the case, however, that the optimum position for the booster is still offshore and so much of the advantage is lost. This may be overcome by using the booster station as the main power point. For instance a static suction dredger could be set up as though to execute direct pumping at a location which enabled the hopper dredger to come alongside it. When the hopper dredger arrives, its discharge pipe is connected directly to the inlet side of the static suction dredger. By this means pumping distances of 1800 m from the shore combined with a further 1000 m along the beach can be achieved with a material having a mean grain diameter of 5 mm. Some caution must be expressed as to the suitability of pumping material with larger grain diameters which can cause substantial wear problems to both pump and line.

Combinations of these methods can be used when access to the shore is difficult. For instance the material may be discharged from a hopper dredger by bottom dumping and then redredged using a static suction dredger. The major drawback is the loss of material involved both during dumping and while laying in the temporary sea-bed stockpile, in addition to the normal losses associated with suction dredging. The advantage over the method described in the previous paragraph using the two connected dredgers is in the improved cycle time, shortened by virtue of the hopper dredger not having to moor and by the greater speed of discharge when dumping. Dry discharge, hopper dredgers have also been used on beach recharge projects with the material being transshipped to another vessel for ferrying to the beach. Here small hopper barges are used in the final placing of the material by the usual dump and recover method. Alternatively the material may be discharged onto a flat-topped shallow draught pontoon which can then make a close approach to the shore where the material is bulldozed off.

A side-push stone dumper (Figure 24.4) is a specialized craft designed for placing stone rather than sand or small shingle. The cargo is carried on the deck and off-loaded over the sides by the controlled application of bulldozer blades permanently installed on the vessel. The craft itself is made to be highly manoeuvrable with a shallow draught allowing close approach to the shore and even entry between groynes. As the material is dumped over each side, rather than beneath the vessel, there is no problem of the craft being grounded on its own cargo.

Lelliott (1989) has described the work carried out since 1971 at Bournemouth on the south coast of England, which culminated in the adoption of a revised policy of coast protection. Over the years the sea-wall had become damaged and it was decided that it would be more economical in the long term and a better engineering solution to nourish the beach in front of the wall rather than spend large sums on repairs. Beach nourishment accompanied by a comprehensive groyne building programme was adopted. The contracts were of the pump ashore type involving the initial placing of 764 000 m^3 of sand material in the period 1974/75. It is interesting that this initial quantity of material was obtained from the offshore borrow area and had a mean diameter of just over 0.2 mm, almost twice the size of the natural beach sand. It consequently formed a slightly steeper beach but had the advantage that less offshore transport took place during storms. The amenity of the beach was preserved. A contract to provide an improved access channel to nearby Poole Harbour gave Bournemouth the opportunity to place a further 632 000 m^3 in 1989 by using the material won from the deepened channel as a mutually beneficial alternative to transporting the dredged material to a dis-

Figure 24.4 Side-dump barge.

tant offshore dumping ground. As this operation was a joint one with the Harbour Commissioners, substantial economic savings accrued for both parties.

Once the material has been delivered to site it must be spread to the desired profile. This can be done either by allowing natural littoral processes to take place or by employing conventional earth-moving plant to place and grade the material to the necessary profile. This latter method has the advantage that accurate measurements can be taken to determine the quantity actually placed on the beach. Spreading and trimming need to be more precise than described earlier in this section, as in this case it is an aid to measurement and payment.

It has to be recognized that dredging contractors require to be properly reimbursed for the execution of the works. Major problems in measuring beach recharge projects arise from losses and movement of materials. The losses occurring during the actual placing process by the contractor may generally be assumed to be part of his risk, and to a very large degree under his control by virtue of his expertise and the method he chooses. However, beach recharges are inevitably carried out where material losses have occurred and are almost certainly still occurring. It is equally certain that, within a limited regime, continuous movements of the beach materials are occurring. These losses and movements are totally beyond the contractor's control and it would be expected that the engineer, due to his pre-contract studies will have a far better knowledge of these.

Generally one of two methods of measurement are used: either measurement in the hopper; or measurement on the beach. Measurement in the hopper is only applicable if the material is transported in a hopper at some stage of the contract. The method results in payment for all material at the time it is in the hopper regardless of where it ends up after that. Furthermore, if measured by volume then this volume is at the bulk density in the hopper rather than on the beach, and this is variable, depending on the means of loading the hopper. The measurement can be made in a number of ways. Probably the most accurate is to calculate the volume of the empty hopper and then for each load to calculate the difference between this and the actual load carried by reference to a theoretical top line to the hopper. The alternative is to establish the average change in vessel draught before and after discharge and thus calculate the weight of material discharged. This method works best in calm water. When direct pumping by static dredger is employed quantities may be established by sounding the borrow area to calculate the volume removed. As the area is quite likely to be backfilling itself by tidal action during the dredging operation, this introduces yet another unknown factor. It may be, however, that this is the only practical solution if measurement on the beach is for some reason impossible.

Measurement on the beach is a satisfactory method provided that it is carried out as a regular operation to minimize naturally occurring changes on the beach. Practically the operation is carried out by taking cross-sections and calculating volumes. To enable proper comparison of tender prices it is imperative that the chosen method of measurement is clearly defined in the enquiry document.

24.5 SAND BYPASSING

Coastal inlets commonly become obstructed with beach material as a result of natural littoral drift. If the inlet has to be kept open for navigation then dredging will be necessary. The quantity of material to be removed may not justify the mobilization of expensive specialist plant, so the problem may best be solved by the use of a small grab dredger. The material recovered can then be placed on the foreshore at the updrift end of the coastal cell. To avoid the leeside scour problem the recovered material may alternatively be placed downdrift of the inlet, thus ensuring a source for continued transport and the avoidance of starvation problems downdrift.

An alternative is the use of fixed or semi-mobile plant feeding pipelines which pass round or under the inlet to a deposition point downdrift; a method often referred to as 'bypassing'. The major difference between this and dredging is that the former acts on material before it reaches the inlet, while the latter attacks it only after it has been deposited in the channel.

Before the technique of bypassing is adopted it is essential to determine the amount of sediment transfer to be expected. Provided that the shoreline is in equilibrium with the wave climate, then one assumption that can be made is that the quantity of material to be bypassed is equal to the net drift on the open shoreline. Equations exist which can be used to calculate the drift in terms of the wave height and direction at the shoreline. If necessary such an approach can be extended to the form of a mathematical model.

The use of bypassing is restricted by the current state of pump technology which limits its use to littoral materials that fall within the sand/silt spectrum. The operational costs of bypassing have to be properly evaluated and allowance made not only for the repayment charges on the fixed installation but also the maintenance costs, in particular the pumping plant, where the replacement costs of the pump impellers can be high. That said, the technique can provide an attractive solution for harbour inlets where the need to guarantee an approach channel is often of overriding importance and where any obstruction to navigation incurs a high financial penalty.

The deposition point for beach material won from dredging or bypassing has to be carefully chosen to ensure that it is spread by wave action, thus avoiding the further use of costly mechanical plant. The cross-referencing of the disposal point to a good hydrographic survey is essential. This will indicate the presence and formation of any offshore banks and current velocities. In this way it will be possible to allow the continuous distribution of the material at a rate which will cause little effect on the natural beach processes (see Appendix B).

24.6 THE 'DO NOTHING' APPROACH

All the foregoing descriptions have relied on the interference of man in the natural process. In certain situations the best form of beach replenishment is to allow the natural processes to continue unchecked. Doing nothing still requires a good understanding of the physical processes involved at the site. Roberts and McGown (1987) describe such an approach in the Reculver area on the north Kent coast. Here coastal erosion of low-grade agricultural land is allowed to continue to provide a source of beach material which is transported by longshore processes to the adjacent coastal cell.

Interference by man with no clear understanding of the effect of such interference can have far reaching and sometimes catastrophic effects on adjacent coastal cells. Great care has to be taken to ensure that works which may solve a problem at one point do not create other, even worse problems elsewhere.

ACKNOWLEDGMENTS

The author is deeply indebted to his colleagues at Canterbury City Council, especially P. Brooks for editing the text, M. Kent, for provision of illustrations, and S. McFarland for background research and many pages of calculations.

REFERENCES

Lelliot, R.E.L. (1989) Evolution of the Bournemouth defences. *Proceedings of the Institution of Civil Engineers Coastal Management Conference* 9–11, May 1989, pp. 263–77.

Owen, M.W. (1989) Research on beaches and coastal structures. *Proceedings of the Institution of Civil Engineers Coastal Management Conference*, 9–11 May 1989, pp. 77–92.

Parrish, F. (1989) Marine resources. *Proceedings of the Institution of Civil Engineers Coastal Management Conference*, 9–11 May 1989, pp. 187–96.

Roberts, A.G. and McGown, A. (1987) A coastal area management system as developed for Seasalter-Reculver, north Kent. *Proceedings of the Institution of Civil Engineers, Part 1*, **82**, 777–97.

Van der Meer, J.W. (1988) Rock slopes and gravel beaches under wave attack. *Delft Hydraulics, Publ. 396.*

Van Hijum, E. and Pilarczyk, K.W. (1982) Equilibrium profile and longshore transport of coarse material under regular and irregular wave attack. *Delft Hydraulics, Publ. 274.*

APPENDIX A: CALCULATION OF A BEACH PROFILE USING VAN DER MEER (1988)

The profile is described by six points, two power functions and two angles. Van der Meer introduces two dimensionless parameters which depend on the wave climate and on the physical nature of the beach. These are the dimensionless wave parameter, H_0T_0, and the fictitious wave steepness, sm. Van der Meer's calculations as used below, apply when H_0T_0 exceeds 1000.

An illustration of the profile calculation is given below, using the following design parameters:

Significant wave height, H_s	1.5 m
Wave period, T_m	6.5 s
Water depth at the toe of beach, h	2.5 m
Median grain size of shingle, D_{n50}	0.01 m
Relative density of stone, (flint), Δ	1.65
Initial foreshore angle, α	10°
Duration of storm conditions, h	4
Angle of wave attack, ϕ	20°

The number of waves, N, is calculated from the wave period and the storm duration. The number of waves during storm conditions = (4 × 3600)/6.5 = 2215

(1) *Calculation of dimensionless wave parameters*

$$H_0T_0 = (H_s/\Delta)D_{n50}(g/D_{n50})^{1/2}T_m = 18\,517$$

(2) *Calculation of the fictitious wave height parameter*

$$sm = 2\pi H_s/(gT_m^2) = 0.023$$

(3) *Calculation of profile points*

Run-up length	$H_0T_0 = 2.9(l_r/D_{n50}N^{0.05})^{1.3}$ $l_r = 12.42$ m
Crest height	$(h_c/H_s)N^{0.15} = 0.089sm^{-0.5}$ $h_c = 2.81$ m
Crest length	$H_0T_0 = 22(l_c/D_{n50}N^{0.12})^{1.2}$ $l_c = 6.90$ m
Step height	$(h_s/H_s)N^{0.07} = 0.22sm^{-0.3}$ $h_s = 1.76$ m
Step length	$H_0T_0 = 3.8(l_s/D_{n50}N^{0.07})^{1.3}$ $l_r = 11.77$ m

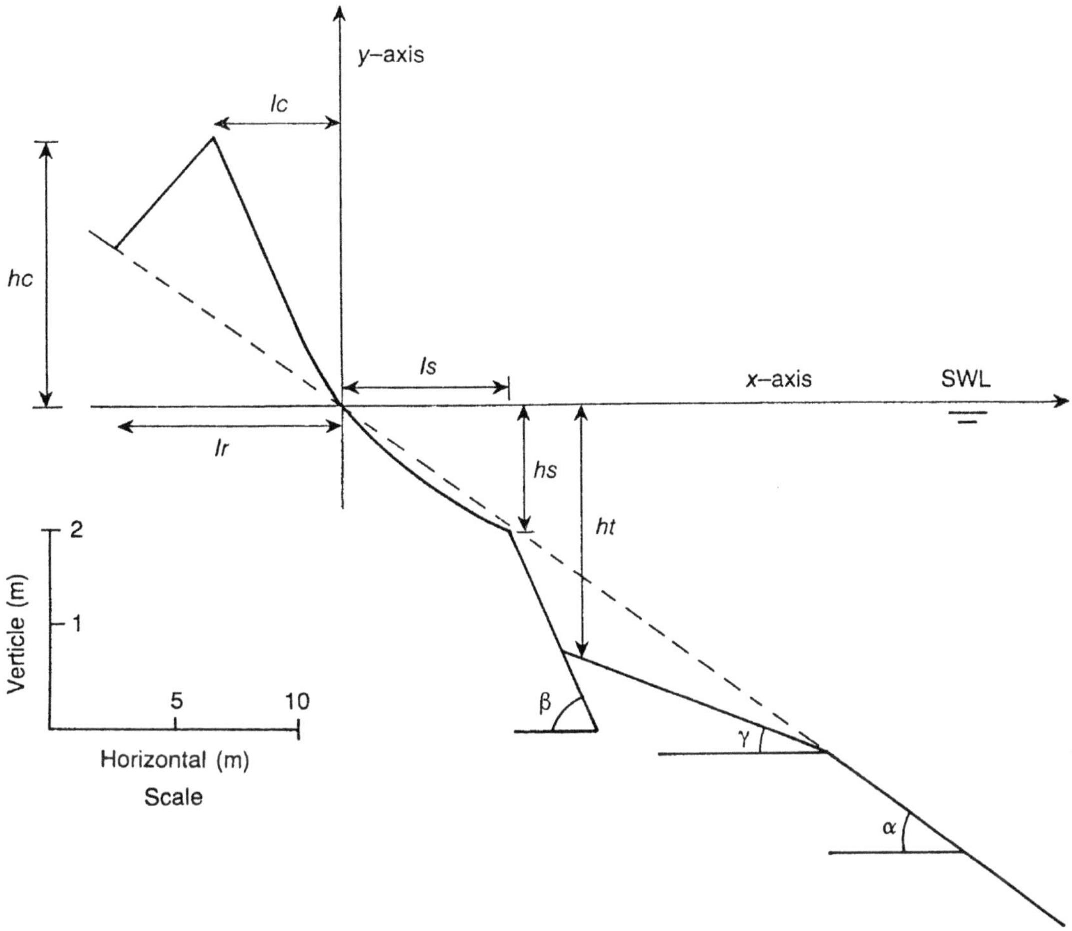

Figure 24.5 Schematized beach profile (after van der Meer, 1988).

Transition height $(h_t/H_s)N^{0.04} = 0.73sm^{-0.2}$
$h_t = 3.18\,\text{m}$

(4) Correction for shallow foreshore, reduction factor r

For $h/H_s < 2.2$, $r = 1 - 0.75(2.2 - h/H_s)^2$

For $h/H_s > 2.2$, $r = 1$

so for $h/H_s = 2.5/1.5$, $r = 0.787$

The reduction factor applies to h_s and l_s only.

(5) Correction for oblique wave attack, cos ϕ

For a wave approach 20°, $\cos\phi = \cos 20° = 0.94$.

The correction is applied to all length and height parameters except l_c

(6) Final parameters including corrections

Run-up length $(l_r) = 12.42(0.94)$
$= 11.67\,\text{m}$

Crest height $(h_c) = 2.81(0.94)$
$= 2.64\,\text{m}$

Crest length $(l_c) = 6.90$
$= 6.90\,\text{m}$

Step height $(h_s) = 1.76 \times (0.94) \times (0.787)$
$= 1.30\,\text{m}$

Step length $(l_s) = 11.77 \times (0.94) \times (0.787)$
$= 8.71\,\text{m}$

Transition height $(h_t) = 3.18 \times (0.94)$
$= 2.99\,\text{m}$

(7) Profile curves

These are described by the following equations:

Below SWL $y = a_4 x^{0.83}$ $a_4 = 0.22$

Above SWL $y = a_t \cdot (-x)^{1.15}$ $a_5 = 0.29$

where a_4 and a_5 are calculated by substituting h_c, l_c and h_s, l_s for x and y, respectively.

(8) Slope angles

Two angles are calculated using the initial slope angle

β: $\tan\beta = 1.1 \tan(\alpha)^A$ $\beta = 19.9°$

where $A = 1 - 0.45\exp(-500/N)$

γ: $\tan\gamma = 0.5 \tan\alpha$ $\gamma = 5.04°$

(9) The calculated profile (Figure 24.5)

The final stage in the construction of the profile involves moving along the still water level until mass balance is achieved, that is, that the deposition of material at the crest equals the erosion of the profile below still water level. A full explanation of the derivation of equations and their limitations has been given by van der Meer (1988).

APPENDIX B: BEACH STABILIZATION, SAND BYPASSING AND CHANNEL MAINTENANCE BY FLUIDIZATION

Jim Parks

One of the fundamental problems in the coastal zone is that of sand management. Often sand is eroded from where it is wanted (recreational and storm-protection beaches), and accumulates where it is not wanted (shoals in harbours, entrances, passes and inlets). Conventional approaches to these problems – dredging, beach nourishment, jetties, groins, seawalls and breakwaters – are not always adequate for the task and can be expensive and environmentally damaging.

The new technologies of fluidization (for channel maintenance and for effective bypassing of sand around inlets) and of dewatering (for beach stabilization and accretion) are believed to offer improved methods of sand management in appropriate circumstances. They are intended to be less expensive in the long run, more effective and less environmentally damaging than conventional methods of dredging and beach nourishment.

For channel fluidization and for fluidized sand bypassing, the new methods of pipe burial underwater allows the use of cheaper, longer lasting, plastic pipe and more efficient configurations. For beachface dewatering, the drain-pipe can now be placed seaward of the low-tide line, thus increasing its effectiveness and increasing efficiency by reducing pumping costs.

FLUIDIZED SAND BYPASSING

Fluidization is defined as the pumping in of additional water beyond the quicksand point to make a 50:50 sand–water slurry that will flow down a slight grade and is pumpable for significant distances. In the fluidized sand bypassing version (Figure 24.6), the perforated fluidizer pipes are buried lengthwise along the axis of the channel, sloped from each distal end down to the mid-length position. A jet-eductor pump located immediately above the fluidizing pipes moves the sand through a discharge pipe to the downdrift beach. After fluidization is initiated in one direction with the full flow capacity of the clear-water pump, a part of that clear-water stream is diverted to the fixed

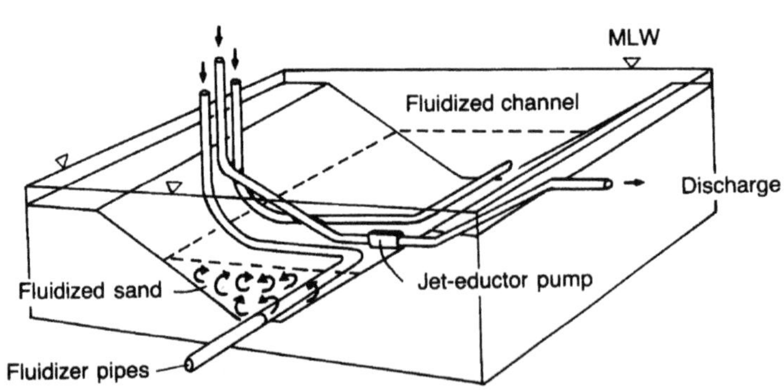

Figure 24.6 Fluidized sand bypassing.

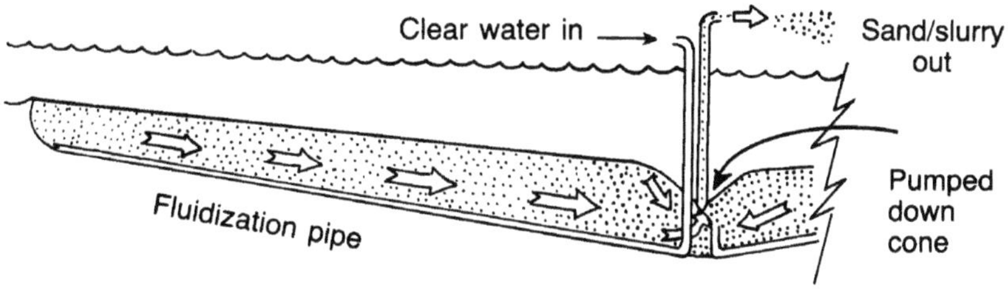

Figure 24.7 Induction of lateral flow in a fluidized channel.

jet-eductor pump, and as the level of fluidized sand is pumped down in a local 'cone of depression', lateral flow along the length of the trench 'feeds' the jet pump.

Beachface dewatering works by lowering the top of the water saturated zone (water table) immediately beneath the active beach face (see Figure 24.7). Part of the wave uprush soaks in, causing reduced backwash and thus part of the suspended sand brought onto the beach with each wave uprush is left on the beach. Secondary effects include local 'hardening' or densification of the sand by withdrawing supporting, interstitial pore water, and reducing seepage outflow at the toe of the beach.

Testing of similar systems in Denmark over several years has shown that a widened beach could be maintained 20–30 m (65–100 ft) seaward of the buried drain-pipes. The system at Sailfish Point, Stuart, Florida, was installed near the limit of its potential effectiveness.

COASTAL SAND DUNES

Coastal sand dunes are recognized as an important line of defence against storm-driven shoreline recession. Much effort has gone into artificially vegetating existing dune areas, and 'snow-fence' methods to accrete additional dune build-ups from wind-driven sand.

The natural source of sand for dunes is the 'dry' beach width above normal high tide. Only from this 'permanently dry' area is it possible for wind to pick up appreciable quantities of sand and carry it inland a few tens/hundreds of feet/metres and deposit it on dunes.

On an eroding beach, there is commonly only a very narrow width of permanently dry sand, if any. On an accreting beach there is usually an appreciable width of dry sand. A restored or nourished beach starts off, at least, with a wide dry beach zone that accomplishes some dune nourishment from wind action.

Stabilization of a beach (reducing seasonal beach width changes) would also help stabilize and accrete beach dunes by providing year-round dry beach zones for the wind to act on.

SAND BYPASSING

Some form of sand bypassing is currently being recommended in many inlet sand management plans now being prepared for most Florida inlets. One particular approach is to use the boat navigation channel itself as the sand trap, and to empty that trap by pumping out fluidized sand on a frequent periodic basis. The frequency of the pumping schedule will be site-specific; some inlets may require weekly pumping while others may only require it monthly or every other month. Pumping can also be done during and immediately after each storm event. Thus adequate depths of water will be maintained at all times, and the location and alignment of the boat channel will be fixed in position, allowing permanent channel markers for increased boating safety. The typically mid-length fixed jet-eductor pumps down a cone of depression in the fluidized sand and thus induces lateral flow along the length of the fluidized channel (Figure 24.8).

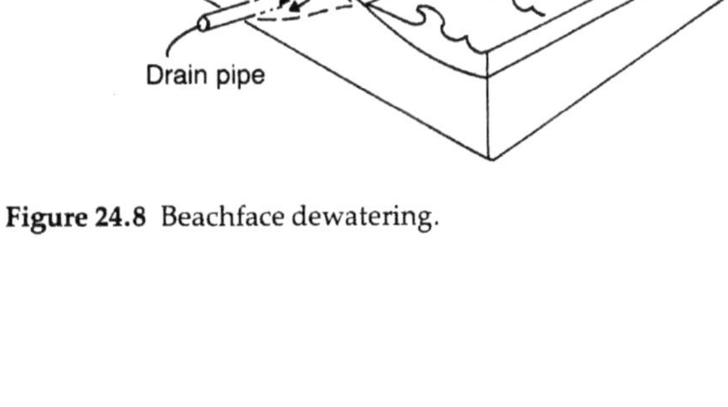

Figure 24.8 Beachface dewatering.

TRANSFER OF TECHNOLOGY

Several full-scale demonstrations in a variety of conditions are planned (Summer 1991), which will be monitored in detail for at least a year. These include Flag Harbour Yacht Marina (sand bypassing), Long Beach, Calvert County, Maryland (beach build-up), St Petersburg Beach, Penellas County, Florida (beach build-up), Vero Beach, India River County, Florida (beach build-up in a different wave climate) and Long Beach, Brunswick County, North Carolina, USA.

REFERENCES

FLUIDIZATION

Bailard, J.A. and Inman, D.L. (1975) Analytical model of duct-flow fluidization. *Proceedings of the Symposium on Modeling Techniques, ASCE*, 1402–21.

Bruun, P. and Adams, J.W. (1988) Stability of tidal inlets: use of hydraulic pressure for channel and bypassing stability. *Journal of Coastal Research*, **4** (4), 687–701.

Collins, A.G., Weisman, R.N., Parks, J.M. *et al.* (1987) Anna Maria, Florida: case study of sand fluidization for channel maintenance. *Shore & Beach*, April, 42–8.

Inman, D.L. and Harris, R.W. (1970) Cratersink sand transfer system, *Proceedings of 12th Coastal of Engineering Conference, ASCE*, 919–33.

Jenkins, S.A., Wasyl, J., and Skelly, D.W. (1992) Tackling trapped sediments, *Civil Engineering*, February pp 61–3.

Lennon, G.P., Chang, T., and Weisman, R.N. (1990) Predicting incipient fluidization of fine sands in unbounded domains. *Journal of Hydraulic Engineering*, **116** (12), 1454–67.

Parks, J. US Patents 5 052 857; 5 061 117; 4 645 377

Weisman, R.N., Roberts, E.W., Lennon, G.P. *et al.* (1987) Update: fluidization for tidal channel maintenance. *Proceedings of the 1987 National Conference on Hydraulic Engineering, ASCE*, August 3–7, 1987, Williamsburg, VA. ASCE, New York, pp. 339–44.

BEACHFACE DEWATERING

Bruun, P. (1989) Coastal drain, what can it do or not do?. *Journal of Coastal Research*, **5** (1), 123–5.

Machemehl, J.E., French, T.J., and Huang, N.E. (1975) New method for beach erosion control. *Civil Engineering in the Oceans, ASCE*, pp. 142–60.

Ogden, M.R. and Weisman, R.N. (1991) 'Beach stabilization using drains: an experimental model study', *Proceedings Coastal Sediments 91, ASCE*, 25–27 June 1991, pp. 1955–69.

Terchunian, A.V. (1990) Performance of beachface dewatering: the STABEACH system at Sailfish Point (Stuart), Florida. *Proceedings of the 1990 National Conference, on Beach Preserv. Techn., Beaches: Lessons of Huricane Hugo*, St. Petersburg, FL, February 16–18, 1990, pp. 185–201.

Vesterby, H. (1991) Coastal drain system: a new approach to coastal restoration. *Proc. 306 Sept 1991, Yokohama, Geo-Coast '91*, pp. 651–4.

BEACH RESPONSE MODELLING 25

C.A. Fleming

The prediction of shoreline evolution following the construction of works that might interfere with natural beach processes is of considerable interest to the coastal engineer. The simplest mathematical model operates on the principle that there is a relationship between the alongshore flux of wave energy and the bulk transport of sediment transport along the beach and in the nearshore zone. Even though such models do not attempt to simulate sediment transport in detail it has been found that the bulk transport approach can produce reliable forecasts of plan shape and beach levels following the construction of works in the nearshore zone. If a problem involves strong components of onshore/offshore transport then this has to be handled empirically and separately. The chapter, however, discusses the development of mathematical models that truly embrace the complex physics of three-dimensional sediment transport. These will be used increasingly by engineers as our knowledge of the basic physics of sediment transport improves (see also Chapters 22, 23, 25 and 33).

CONTENTS

25.1 INTRODUCTION

Beach response modelling is distinct from one-, two- or three-dimensional sediment transport modelling as it is more concerned with simulating the changes in line and level of the beach than the detailed sediment transport processes as described in previous chapters. Prediction of shoreline evolution is a subject of considerable interest to the coastal engineer who is involved with the planning of any coastal works that might interfere with natural beach processes. This would include costal defences of any type, jetties, marinas, ports and any activity in the nearshore zone that might have an influence on beach response (i.e. sand and gravel extraction).

There are a number of different approaches to shoreline evolution problems which include analytical solutions, 'empirical' solutions, numerical modelling and physical modelling. The last of these is dealt with in Chapter 21. There are a limited number of analytical solutions, whilst numerical solutions vary considerably in their complexity ranging from the simplest one-dimensional models which include representation of a number of complex physical processes.

25.2 SHORELINE PROCESSES

Sediment transport on beaches is schematized in two models; alongshore and onshore-offshore (see Figure 25.1). Various models may include one or both of these modes of transport, but to date they have been treated in a decoupled way, i.e. there is no interaction in process between the two modes of transport. Theories relating to these two components are described in the following sections.

25.2.1 ALONGSHORE SEDIMENT TRANSPORT

Alongshore sediment transport is driven by alongshore currents that are generated when waves break obliquely to the shoreline. At the simplest level of a straight parallel contained shoreline it can be shown that the bulk rate of transport can be related to the alongshore wave energy flux leading to the derivation of the well known CERC formula as presented in the *Shore Protection Manual* (1984). The bulk volumetric rate of alongshore transport is given in a number of forms such as:

Coastal, Estuarial and Harbour Engineer's Reference Book. Edited by M.B. Abbott and W.A. Price. Published in 1993 by Chapman & Hall, 2–6 Boundary Row, London SE1 8HN. ISBN 0 419 15430 2.

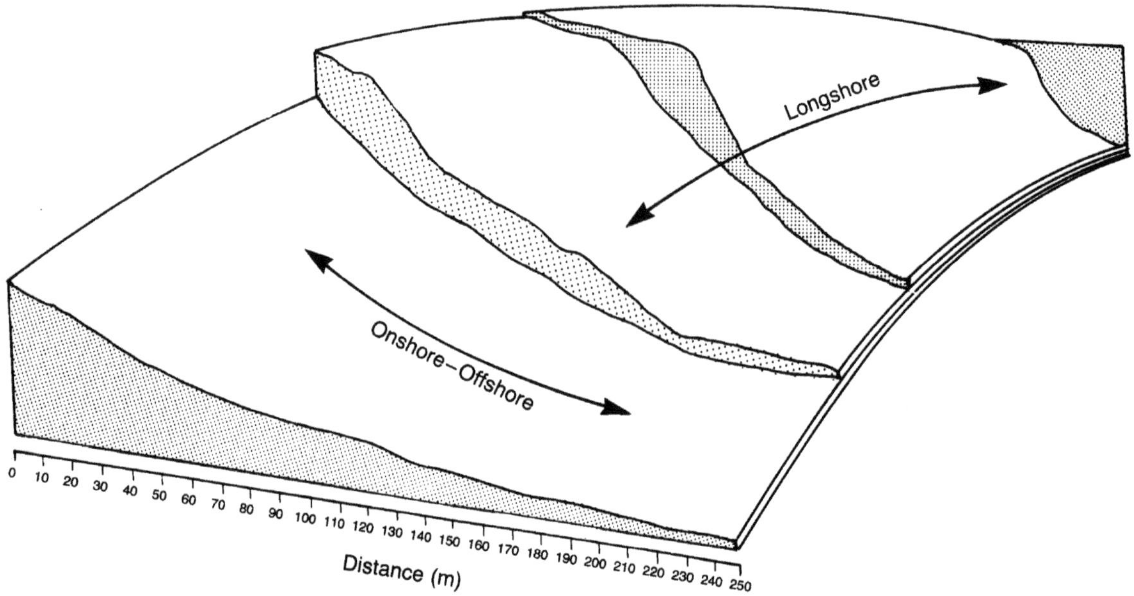

Figure 25.1 Basic modes of beach movement.

$$Q = K_c(H^2C_g)_b \sin(2\alpha_b) \qquad (25.1)$$

where: K_c is an empirical constant, whose value depends on the units employed; H_b is the breaking wave height; α_b is the breaking wave angle relative to the shoreline; C_o; and C_g.

The limitation of the CERC formula is that it is only valid for straight shorelines, and does not include the effects of sediment grain size, beach slope, wave period, breaker type etc. Many of these parameters are interdependent (see section 25.2.2). The value of the 'constant' K_c is also based on relatively limited data which exhibits a significant degree of scatter around the theoretical relationship.

Swart (1982) proposed the introduction of a variable factor related to the sediment grain size:

$$Q_o = K_s(H_oK_r)^2_{rms}T \sin\alpha_b \cos\alpha_b \qquad (25.2)$$

where H_o is the deepwater wave height, K_r the refraction coefficient, T the wave period and

$$K_s = 0.116 \log\left(\frac{1.46}{D50}\right)$$

within the limits $0.1\,\text{m} < D50 < 1.0\,\text{m}$. Q_o is given in $\text{m}^3\,\text{s}^{-1}$ and $D50$ is the median grain diameter.

A recent variant on this type of formulation is known as the QUEST formula. This includes the influence of grain size and beach slope:

$$Q_o = \frac{0.0024mH_b^{7/2}\sin 2\alpha_b}{D50} \qquad (25.3)$$

An inherent advantage of the QUEST formula is that not only does it include more meaningful physical parameters, but it also allows variations in time (or surge) levels to be included in the calculation of time varying transport rates by way of varying the slope for the active part of the beach profile. The value of doing this in improving estimates has been demonstrated as part of the Canadian coastal sediment study (Kamphus and Kooistra, 1983).

Ozasa and Brampton (1980) proposed a predictive expression for alongshore sediment transport which includes a contribution from both oblique wave incidence and the alongshore varieties in breaking wave height. This can take the form:

$$Q_0 = (H^2Cg)_b\left(K_1 \sin 2\alpha_b - \frac{K_2}{M}\cos\alpha_b \frac{\partial H_b}{\partial x}\right) \qquad (25.4)$$

The first term in this expression corresponds to the CERC formula Equation (25.1), whilst the second represents the increase or decrease in sediment transport that occurs due to a breaking wave height gradient which will be greatest in areas of significant diffraction due to structures or natural features.

The foregoing expressions are applicable to sand-size sediments. When dealing with shingle sizes it is necessary to introduce a threshold of motion parameter below which no sediment transport takes place.

A second class of alongshore sediment transport modelling encompasses the detailed predictors. These are models which link theoretical concentration profiles as a function of waves and currents with theoretical models of alongshore current distributions across the surf zone. In most instances the sediment transport model has been based on non-breaking wave conditions and is applied to surf zone wave and current variations. Bijker (1967) developed the first approach of this type and many other similar models have been developed since. These based on work by a number of researchers were evaluated as one of the Canadian Coastal Sediment Study sub-projects (Fleming *et al.*, 1986). Models of this type that have been incorporated in shoreline response models are described in the appropriate sections of this chapter.

25.2.2 ONSHORE–OFFSHORE SEDIMENT TRANSPORT

Onshore–offshore sediment transport models are most commonly based on changes in beach profile response, thus inferring sediment transport on each part of the profile. Both Bruun (1954) and Dean (1977) proposed that the nearshore equilibrium profile could be represented by

$$h = Ay^r \tag{25.5}$$

where y is the distance offshore and h is the water depth. The exponent r is theoretically equal to ⅔ and has been found to be very close to this value when fitted to field data. Clearly such a formulation is monotomically varying and cannot accommodate important features such as nearshore bars. Neither can it provide any indication of the time dependency of profile development.

The Brunn rule of erosion by sea level rise was first published in 1962 (Bruun, 1962) and is based on beach profile response due to offshore sediment transport. As can be seen from Figure 25.2 for a given profile $y = f(y)$ it can be shown on the basis of equilibrium profile development and material balance that the shoreline recession of the waterline is:

$$\Delta y = l_a \frac{\Delta s}{h_a} \tag{25.6}$$

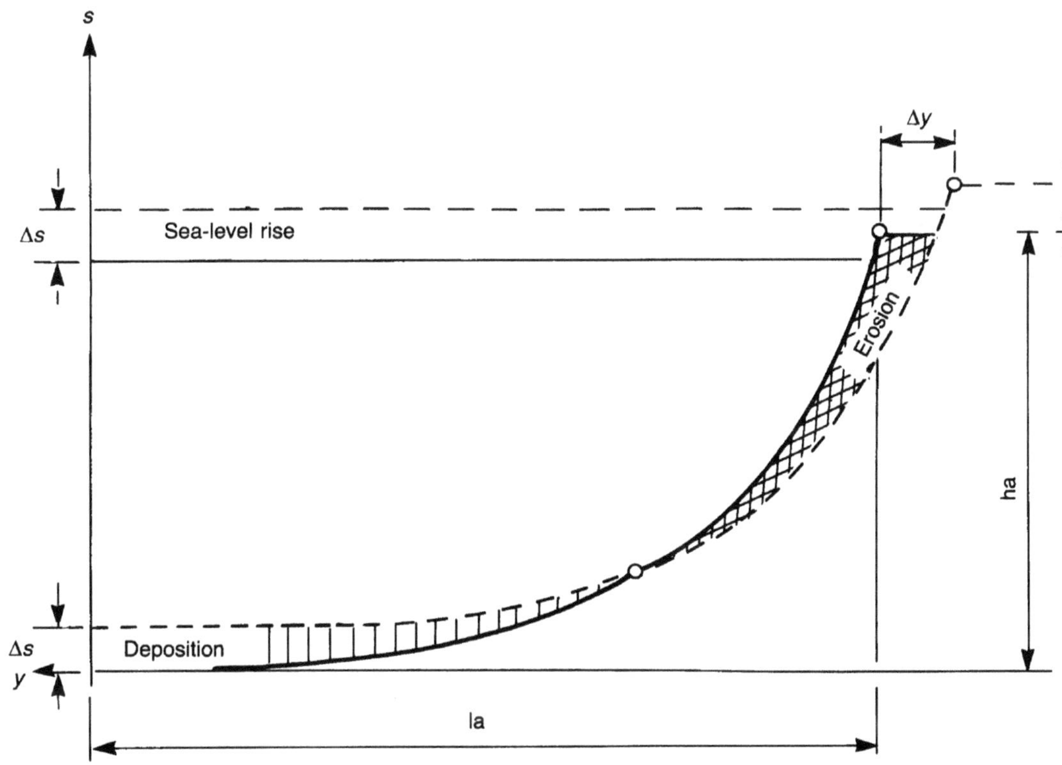

Figure 25.2 Definition sketch for Bruun's rule.

where Δs is the rise in sea level and l_a and h_a are the length and depth, respectively, corresponding to the point of maximum exchange of material between the nearshore and the offshore. A method of evaluating this limit has been given by Hallermeier (1981). Neither the slope of the profile, the point of intersection of the profiles or the seaward slope of the profile has been used in the derivation of the formula. A review of the limitation of the use of Equation (25.6) has been given by Brunn (1983).

Swart (1974) developed a semi-empirical time-dependent equilibrium profile model based on an extensive series of model tests. The onshore/offshore sediment transport takes the general form:

$$Sy_t = sy[W - (L_2 - L_1)_t] \qquad (25.7)$$

where W is an equilibrium distance, L_1 and L_2 are the schematized onshore and offshore profile lengths respectively, t is time and sy is a function of a number of variables including the deep water wave height, wave steepness, grain size and the still water depth at the seaward extremity of the profile. The relationships developed are largely empirical within a framework of physical parameters. A semi-empirical time-dependent profile response has also been proposed by Kriebel and Dean (1985), whilst Vellinga (1984) has proposed an equilibrium profile relationship applicable in the range 0.0001 m to 1.0 m given in metric units by:

$$y = 0.7(H_0/L_0)^{0.17}W^{0.44}x^{0.78} \qquad (25.8)$$

where w is 4.36 D_{50} for a relative density of 2.6.

Beach profile response relationships have also been proposed by van Hijm and Pilarczyk (1982) and Powell (1990) based largely on fitting data measured in a large series of physical model tests. These relate specifically to shingle beaches with particles typically in the size range of 0.01 to 0.04 m, for which it is necessary to include a threshold of motion parameter (see Chapter 21).

25.3 ANALYTICAL SOLUTIONS

Analytical or closed form analytical solutions to beach response generally apply to very simple situations in terms of initial shoreline positions, structure position, structure geometry and boundary conditions. The first fundamental model of this type was developed by Pelnard–Considère (1956). His underlying assumptions were that:

1. The relative geometry of the beach face remains similar at all times, so that erosion and accretion are in the form of parallel slices.
2. Angles of wave approach to the shoreline are small (20°) and thus the alongshore transport may be linearized with respect to the angle of incidence.
3. The active beach is a constant vertical distance between the beach crest and a closure depth.

As can be seen from Figure 25.3 the basic continuity equation is:

$$\frac{\partial Q_x}{\partial x} + D\frac{\partial y}{\partial t} = q \qquad (25.9)$$

where q is the net input of sediment (e.g. beach nourishment). Putting this to zero and linearizing

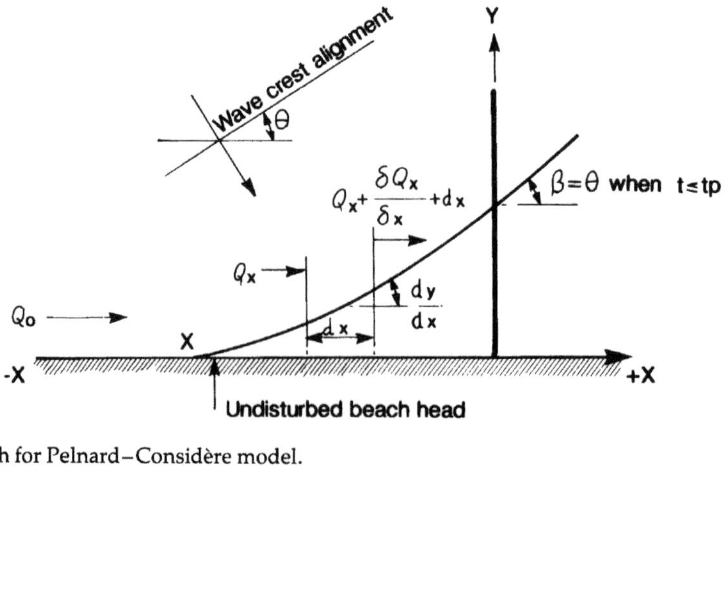

Figure 25.3 Definition sketch for Pelnard–Considère model.

the transport equation (Equation 25.1) and assuming constant conditions

$$Q_x = K_c\left(\theta - \frac{\partial y}{\partial x}\right) \qquad (25.10)$$

The full solution to Equations (25.9) and (25.10) is given for the example of a barrier such as a groyne in Muir Wood and Fleming (1981) or Le Méhauté and Soldate (1977). The most useful quantitative result is the beach line at the groyne

$$(y)_{x=0} = 2\left(\frac{Q_0 t\theta}{\pi D}\right)^{1/2} \qquad (25.11)$$

where t denotes time.

The point on the updrift side (x) at which the shoreline is modified by accreting material is approximately

$$X = \frac{3}{\theta}(y)_{x-0} \qquad (25.12)$$

When the beach accretes so that it starts to bypass the groyne, the boundary conditions change so that the angle of the beach contour at the breakwater is no longer equal to that of the angle of wave attack. In this case the rate of bypassing is:

$$Q_p = Q_o\left[1 - \frac{2}{\pi\left(\frac{t}{t_p} - 0.38\right)^{1/2}}\right] \qquad (25.13)$$

where t_p is the time at which bypassing commences. A number of 'one-line' analytical solutions have been developed. These have been described by Le Méhauté and Soldate (1977). However, such theories become generally inadequate when the alongshore sand transport is only partially interrupted by a structure or in any situation where onshore–offshore sediment transport becomes significant.

Bakker (1968) developed a two-line analytical solution which included cross-shore transport as described in section 25.2.2 so that the concept of an 'equilibrium distance' between the two contours out of the influence of any cross-shore structures was introduced. Thus the profile is divided into two parts; the upper part extending to a depth which may be effected by a groyne structure and the lower part extending to the practical seaward limit of sediment movement. It was assumed that the alongshore transport at each contour is linearly dependent on wave attack and that the onshore–offshore sediment transport is dependent on the steepness of the profile in relation to the assumed equilibrium profile. The first solution did not include wave diffraction, but later developments (Bakker *et al.*, 1970) did.

Bakker's two-line model was the forerunner to equilibrium profile models such as Swart (1974) (see section 25.2.2) and the N-line models described in section 25.5. Larson *et al.* (1987) have compiled a large number of analytical solutions to the generalized shoreline change equation by using a number of simplifying assumptions.

25.4 BEACH PLAN SHAPE MODELS

25.4.1 EMPIRICAL MODELS

Empirical beach plan shape models have been developed mostly in relation to the formation of equilibrium bay formations between headlands. The most recent summary of these developments has been given by Hsu *et al.* (1989) and Hsu and Evans (1989). Similar principles have also been used to deal with the change in the beach plan shape behind an offshore breakwater. These approaches and their use in design is discussed in Chapter 23 of this volume.

25.4.2 NUMERICAL MODELS

As can be seen from Figure 25.4, and Equation (25.9) the simplest form of the continuity equation may be written in finite difference form as:

$$Y_{n+1,t+1} = \frac{1}{ha}\left(\frac{Q_{n,t} - Q_{n+1,t}}{\Delta x} - q_t\right)\Delta t + y_{n+1,t} \qquad (25.14)$$

which gives the advance or recession of the shoreline. The angle between the breaking wave front and the shoreline is given by:

$$\alpha_n = \theta_n - \tan^{-1}\left(\frac{y_{n,t} - y_{n-1,t}}{\Delta x}\right) \qquad (25.15)$$

so that using a suitable time step successive beach changes may be calculated.

This type of model can accommodate arbitrary initial plan shapes, varying input parameters alongshore, various boundary conditions and generally allows a large amount of flexibility in the way it is applied. For example, such models are frequently combined interactively with wave refraction models so that the changing contour orientations are accounted for. Similarly the type and complexity of the sediment transport process may be varied to allow for other phenomena. For example, beach nourishment (or extraction) may be included in the continuity equation, short groynes may be included internally by applying bypassing rules, sea-walls may be included by restraining plan shape development and stable points may be imposed.

There have been numerous models of this type and a good summary has been given by Horikawa (1988). Price *et al.* (1972) compared their model, which excluded refraction but included diffraction effects, with laboratory experiments of beach response be-

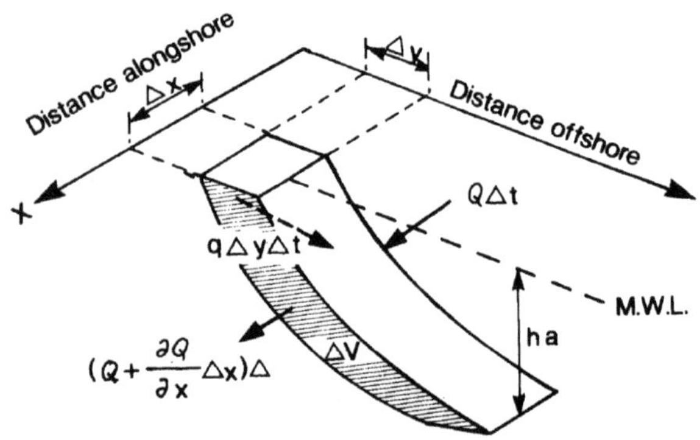

Figure 25.4 Definition sketch for one-line beach response model.

hind an offshore breakwater. They found that the comparison was good up for to three hours, but deteriorated thereafter as a result of wave refraction not being represented in the numerical model.

Many other examples of applications can be found in the literature e.g. Ozasa and Brampton (1980), Larson and Kraus (1989). The latter describes the GENESIS model developed for the US Army Corps of Engineers which has the capability to deal with arbitrary numbers of groynes, sea-walls, detached breakwaters etc., beach nourishment, wave transmission through detached breakwaters and wave diffraction at structures.

25.5 THREE-DIMENSIONAL BEACH RESPONSE

25.5.1 *N*-LINE MODELS

The first presentation of a generalized *N*-line model was by Perlin and Dean (1983). It is called an *N*-line model simply because the beach profile is discretized into an arbitrary number (N) of depth contours along which alongshore sediment transport is calculated and across which onshore–offshore sediment transport takes place as shown schematically in Figure 25.5. The CERC formula for alongshore transport was used and distributed across the profile whilst the simple exponential form of equilibrium beach profile as described by Equation (25.5) was used. It should be noted that this profile as wave height, period or wave steepness so that the equilibrium profile formulation is independent cannot pretend to represent reaction of the beach to single wave events or a short term wave climate. Rather its use should be restricted to investigating long term trends when sediment grain size which is accounted for dominates the supposed reaction.

A number of variants of Perlin and Dean's original *N*-line model can be found in the literature which consider different alongshore transport models and equilibrium profile formulations. Their use has been found to be effective especially when coastal structures, which are bypassed by sediment transport, are involved. However, the underlying methodology requires that the profile is monotomically varying so that offshore bars, so often an important feature of storm profile response, cannot be accommodated.

25.5.2 PROFILE MODELS

The latter half of the past decade has seen the development of onshore–offshore profile development models based on the physical processes taking place in the surf zone rather than fitting data. Processes represented include wave height transformation across the surf zone, wave set-up, mean return flow and sediment dynamics. These models have been successful in reproducing beach face erosion and bar formation under storm events, but at present appear to be less reliable in handling the accretion phase of the upper beach. Both Stive (1986) and Nairn (1988) have investigated the application of the energetic sediment transport model proposed by Bailard (1981). Developments by de Vriend (1987) and Deigaard *et al.* (1988) have also resulted in realistic cross-shore morphological models. Space does not permit any detailed description of such models.

25.5.3 MULTI-PROFILE MODEL

The multi-profile model attempts to simulate three-dimensional transport by linking a profile model with an alongshore transport model much in the same way as do the *N*-line models described in section 25.5.1, so that the two fundamental modes of transport still remain decoupled.

Southgate (1987) and others have developed an alongshore current model which allows an arbitrary cross-shore profile including bars and a tidal current boundary condition. This has been coupled with the Bailard (1981) sediment transport model and Nairn's (1988) cross-shore hydrodynamic and sediment transport model thus providing predictions of both modes of transport for a given profile during its development (Nairn and Southgate (1989)). This has been further developed by Dales (1991) to link a series of profiles along the beach to determine the three-dimensional response of a beach with slowly varying contours.

REFERENCES

Bailard, J.A. (1981) An energetics total load sediment transport model for a plane sloping beach. *Journal of Geophysical Research*, **86** (C11) 10938–54.

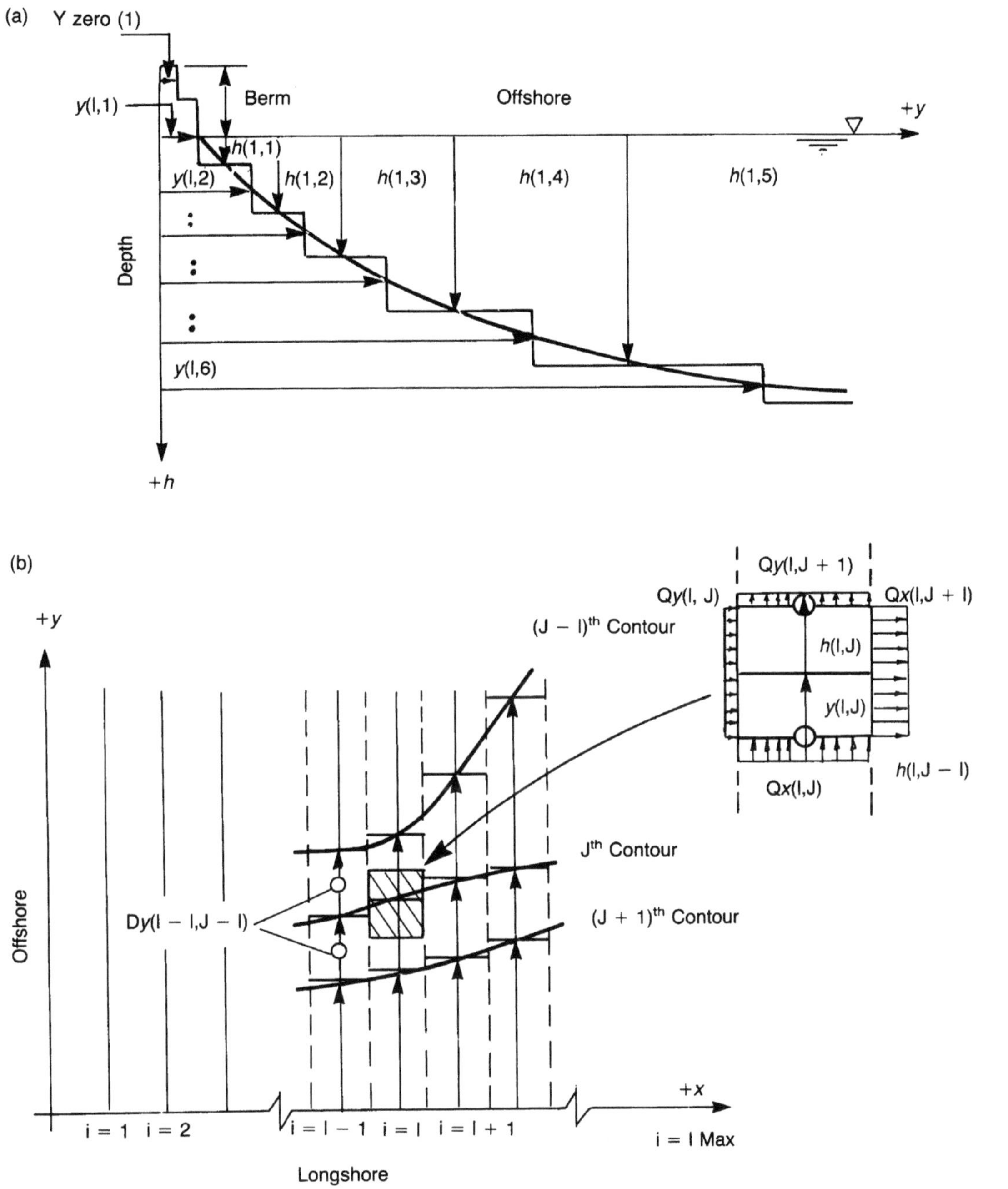

Figure 25.5 Definition sketch for *N*-line beach response model. (a) beach profile representation; (b) beach planform representation.

Bakker, W.T., Klein Breteler, E.J.H. and Roos, A. (1970) The dynamics of a coast with a groyne system. *Proceedings of the 12th Conference on Coastal Engineering, ASCE*, pp. 492–517.

Bakker, W.T. (1968) The dynamics of a coast with a groyne system. *Proceedings of the 11th Coastal Engineering Conference, ASCE*, pp. 492–517.

Bijker, E.W. (1971) Longshore transport calculations. *Proceedings of ASCE*, **97** (WW4), 687–701.

Bruun, P. (1954) Coast erosion and development of beach profiles. US Army Corps of Engineers, *BEB Tech Memo 44*.

Brunn, P. (1962) Sea-level rise as a course of beach erosion. *Journal of the Waterways and Harbours Division, ASCE*, **88** (WWI), 117–30.

Brunn, P. (1983) Review of conditions for uses of the Brunn rule of erosion. *Coastal Engineering*, **7**, 77–89.

Dales, D.C. and Al-Mahouk, M. (1991) Application of a 3-D beach response model to the Norfolk coast. *Proc MAFF Conference for River and Coastal Engineers*, Loughborough, UK.

Dean, R.G. and Maurmeyer, E.M. (1983) Models for beach profile response, in, *Handbook for Coastal Processes and Erosion* (ed. P.D. Kanar) CRC Press, Boca Raton, Fl, p151.

Dean, R.G. (1977) Equilibrium beach profiles: US Atlantic and Gulf Coasts. *Ocean Engineering Rep. 12*, University of Delaware Press. Delaware, USA.

Deigaard, R., Hedegaard, I.B. and Fredsoe, J. (1988) A model for off/onshore sediment transport. *IAHR Symposium on Mathematical Modelling of Sediment Transport in the Coastal Zone*, Copenhagen.

De Vriend, H.J. (1987) 2DH mathematical modelling of morphological evolutions in shallow water. *Coastal Engineering*, II (1), 1–27.

Fleming, C.A., Pinchin, B.M. and Nairn, R.B. (1986) Evaluation of coastal sediment transport estimation techniques, phase 2: comparison with measured data. Canadian Coastal Sediment Study, National Research Council, Canada *Rep. C252–19*.

Hallermeier, R.J. (1983) Sand transport limits in coastal structure design, *Proceedings of Coastal Structures 1983, ASCE*, pp. 703–16.

Hallermeier, R.J. (1981) Seaward limit of significant sand transport by waves. US Army Corps of Engineers, Coastal Engineering Research Centre, Fort Belvoir (now Vicksberg), *Rep. CETA 81–2*.

Hanson, H. and Kraus, N.C. (1989) Gensis: generalised model for simulating shoreline change, *Rep. 1, Ref. Manual, Tech. Rep. LERC-89-00*, US Army Waterways Experimental Station, Vicksburg.

Horikawa, K. (ed.) (1988) *Nearshore Dynamics and Coastal Processes*, University of Tokyo Press, Tokyo.

Hsu, J.R.C. and Evans, C. (1989) Parabolic bay shapes and applications. *Proceedings of the Institution of Civil Engineers, Part 2, 87*, 557–70.

Hsu, J.R.C. and Silvester, R. (1990) Accretion behind single offshore breakwater. *Journal of Waterway, Port, Coastal and Ocean Engineering, ASCE*, **116** (3), 362–80.

Hsu, J.R.C., Silvester, R. and Xia, Y.-M. (1989) Static equilibrium bays: new relationships. *Journal of Waterway, Port, Coastal and Ocean Engineering, ASCE*, **115** (3).

Kamphuis, J.W., Davies, M.H., Nairn, R.B. *et al.* (1986) Calculation of littoral sand transport rate. *Coastal Engineering*, **10**.

Kooistra, H. and Kamphuis, J.W. (1984) Scale effects in alongshore sediment transport rates. National Research Council, Canadian Coastal Sediment Study, *Rep. C2S2–13*.

Kraus, N.C. (1989) Beach change modelling and the coastal planning process. *Proc. Coastal Zone 1989*, **1**, p. 553.

Kriebel and Dean, R.G. (1985) Numerical simulation of time dependent beach and dune erosion. *Coastal Engineering*, 9.

Larson, M., Hanson, H. and Kraus, N.C. (1990) Schematised numerical model of three-dimensional beach change. *Proceedings of the 22nd Coastal Engineering Conference, ASCE*.

Larson, M. and Kraus, N.C. (1989) Prediction of beach fill response to varying waves and water level. *Proc. Coastal Zone, 1989, ASCE*, **1**, pp 607–21.

Larson, M., Hanson, H. and Kraus, N.C. (1987) Analytic solutions of the one-line model of shoreline changes. US Army Corps of Engineers, *CERC Tech. Rep.*

Le Méhauté, B. and Soldate, M. (1977) Mathematical modelling of shoreline evolution. US Army Corps of Engineers, Coastal Engineering Research Centre. Fort Belvoir (Now Vicksburg), *Misc. Rep. 77–10*.

Muir Wood, A.M. and Fleming, C.A. (1981) *Coastal Hydraulics*, Macmillan, London.

Nairn, R.B. (1988) Prediction of wave height and mean return flow in cross shore sediment transport modelling, *IAHR Symposium on Mathematical Modelling of Sediment Transport in the Coastal Zone*, Copenhagen.

Ozasa, H. and Brampton, A.H. (1980) Mathematical modelling of beaches backed by sea walls. *Coastal Engineering*, **4** (1), 47–64.

Perlin, M. and Dean, R.G. (1983) A numerical model to Simulate sediment transport in the vicinity of coastal structures. US Army Corps of Engineers, Coastal Engineering Research Centre, Fort Belvoir (now Vicksburg) *Misc. Rep. MR-83-10*.

Powell, K.A. (1990) Predicting short term profile response for shingle beaches. Hydraulics Research, Wallingford, *Rep. SR219*.

Pelnard–Considère, R. (1956) Essai de théorie de l'evolution des formes de riverages en plages de sable et de galets. *Quatrième Journées de l'Hydraulique, Les Energies de la Mer*, Question III, Rapport No 1, pp. 298–298.

Southgate, H.N. (1987) A one-dimensional model of wave-current interaction. *ASCE Speciality Conference on Coastal Hydraulics*, University of Delaware, USA.

Stive, M.J.F. (1986) A model for cross-shore sediment transport. *Proceedings of the 20th Coastal Engineering Conference, ASCE*, pp. 1550–64.

Swart, D.H. (1974) A schematisation of onshore-offshore transport. *Proceedings of the 14th Coastal Engineering Conference, ASCE*, pp. 884–900.

Swart, D.H. (1972) Predictive equations regarding coastal transports. *Proceedings of the 15th Coastal Engineering Conference, ASCE*.

van Hijum, E. and Pilanczyk, K.W. (1982) Equilibrium profile and longshore transport of coarse material under regular and irregular wave attack. Delft Hydraulics Laboratory, *Pub. 274*.

Vellinga, P. (1984) A tentative description of a universal erosion profile for sandy beaches and rock beaches. *Coastal Engineering*, **8**.

THE GREENHOUSE EFFECT 26

W.A. Price

CONTENTS

26.1 INTRODUCTION

Fluctuations in climate are nothing new. The hot Mediterranean period that sustained Britain's thriving vineyards eight hundred years ago was not caused by the greenhouse effect but was part of a natural, long term, weather cycle. During a later phase of this cycle the River Thames froze over regularly during the eighteenth and nineteenth centuries and fairs held on the ice were almost annual events. This fact is often quoted, but it has to be realized that since that time, many power stations have been built on this estuary which take cooling water, which raises the temperature on being returned to the system. This is one of many examples where the argument could be confused by what is taken in error as cause and effect. However, since the industrial revolution the concentration of carbon dioxide in the earth's atmosphere has increased by about 30%. At the same time other 'greenhouse' gases have increased, namely methane, chlorofluorocarbons and nitrous oxide. All have the effect in the upper atmosphere of increasing the reflection of heat back to the earth's surface.

Today, fossil fuels are releasing 5.5 billion tonnes of carbon into the atmosphere each year. Forests are being cleared at such a rate that over the next 30 years an area of woodland equal in size to India will be lost.

In looking to the future, authorities differ in their assessments of the rise in the surface temperature of our planet but it could be of the order of 0.3°C per decade into the next century. Such a warming could produce an average rate of global rise in mean sea level of 6 cm per decade. Such rises are large compared with the variations that have occurred over the last 10 000 years and need to be taken into account especially by coastal, estuarial and harbour engineers.

It is difficult enough to reproduce large circulation patterns in mathematical models of an almost incompressible fluid like water, but meteorologists have to handle a compressible fluid – air. Their atmospheric models have also to be coupled to oceanic models in order to predict ocean-system temperature changes. This is a difficult but necessary task because large changes of energy take place on the interface between land and sea.

The general warming of the planet will lead to a rise in sea levels, because of thermal expansion of the oceanic waters, and to melting of glacial ice. Precipitation will be affected and hence run-off to rivers will change and this, in turn, will eventually alter the balance between salt intrusion in our estuaries and fresh water discharge. Apart from the effect on ecosystems this could alter flood levels. These examples demonstrate that changes are likely to be significant and difficult to forecast.

The Report of the Coastal Zone Management Sub-Group of the Intergovernmental Panel on Climate Change – Response Studies Working Group published an authoritative review of the greenhouse effect in 1990. A summary of their findings will be found later in this chapter. Also, the paper by Sorensen (1991), *Impact of expected climate change effects on coastal structure design*, is an excellent addition to the reading on this subject.

Hundreds of millions of people live in coastal areas and have, over the centuries, traded with other nations. Travel by sea was, and is, the most convenient method for bulk shipment of goods. Such people have always, however, had to live with the threat of storms, hurricanes, tsunamis, cyclones and typhoons. The concern now is that these threats might increase.

Some nations are particularly vulnerable. The twelve

Coastal, Estuarial and Harbour Engineer's Reference Book. Edited by M.B. Abbott and W.A. Price. Published in 1993 by Chapman & Hall, 2–6 Boundary Row, London SE1 8HN. ISBN 0 419 15430 2.

hundred islands of the Maldive archipelego are barely 3 m above sea-level (Delft Hydraulics, 1989) and similar coral atolls like the Marshall Islands find themselves in the same position. Almost ten million people live within normal tide level in each of the river deltas of Bangladesh, Egypt and Vietnam. Even coastlines protected by dunes or sea-walls will suffer because waves will be able to travel further inshore and cause beach erosion. Overtopping of existing sea defenses will increase as sea level rises. Some of the conclusions of the Executive Summary in the Report of the Coastal Zone Management Sub-Group (1990) are worth quoting here:

> Global climate change may raise sea level as much as one metre over the next century and, in some areas, increase the frequency and severity of storms. Hundreds of thousands of square kilometres of coastal wetlands and other lowlands could be inundated. Beaches could retreat as much as a few hundred metres and protective structures may be breached. Flooding would threaten lives, agriculture, livestock, buildings and infrastructures. Salt water would advance landward into aquifers and up estuaries, threatening water supplies, ecosystems and agriculture in some areas.
>
> Even in nations that are not, on the whole, particularly vulnerable to sea level rise, some areas could be seriously threatened. Examples include Sydney, Shanghai, coastal Louisiana and other areas economically dependent on fisheries or sensitive to changes in estuarine habitats.
>
> The responses required to protect human life and property fall broadly into three categories: retreat, accommodation and protection.
>
> *Retreat* involves no effort to protect the land from the sea. The coastal zone is abandoned and ecosystems shift landward. This choice can be motivated by excessive economic or environmental impacts of protection. In the extreme case, an entire area may be abandoned.
>
> *Accommodation* implies that people continue to use the land at risk but do not attempt to prevent the land from being flooded. This option includes erecting emergency flood shelters, elevating buildings on piles, converting agriculture to fish farming, or growing flood or salt tolerant crops.
>
> *Protection* involves hard structures such as sea walls and dikes, as well as soft solutions such as dunes and vegetation, to protect the land from the sea so that existing land uses can continue.

Environmental implications are discussed in the report, and significant changes are predicted in coastal wetlands, upon which two thirds of the world's fish catch and marine species depend for their survival.

The conclusions of the Report are that in some cases, the impact of global warming could be disastrous; that in a few cases impacts would be trivial; but that for most coastal nations, at least for the foreseeable future, the impacts of sea-level rise would be serious but manageable if appropriate actions are taken.

Coastal nations are urged to begin the process of adapting to sea-level rise not because there is an impending catastrophe, but because there are opportunities to avoid adverse impacts by timely action. The following actions are recommended by the Report:

Coastal planning

1. Coastal nations should implement comprehensive coastal zone management plans.
2. Coastal areas at risk should be identified.
3. Nations should ensure that coastal development does not increase vulnerability to sea-level rise.
4. Emergency preparedness and coastal zone response mechanisms need to be reviewed and strengthened.

International co-operation

5. A continuing international focus on the impacts of sea-level rise needs to be maintained.
6. Technical assistance for developing nations should be provided and co-operation stimulated.
7. International organizations should support national efforts to limit population growth in coastal areas.

Research data and information

8. Research on the impacts of global climate change on sea-level rise should be strengthened.
9. A global, ocean observing, network should be developed and implemented.
10. Data and information on sea-level change and adaptive options should be made widely available.

The seriousness of the problem is perhaps best highlighted by putting cost estimates to a particular problem. Some time ago the UK Ministry of Agriculture, Fisheries and Food put the present level of marine flood protection at £5000 M to £8000 M based on the assumption that the sea-level will rise 1650 mm by 2050. Even if these rises have to be adjusted downwards it still means large expenditures to cope with the greenhouse effect.

To some extent we have created more of a problem by allowing development right up to the edge of the coastline. Professor Keith Clayton of the University of East Anglia maintains that to cope with all the coastal changes that are predicted engineers and planners will have to change their approach to coastal management. He points out that coastal engineers have sought in the past to stabilize coastal features but in future we need to give these features room to move as they adjust to changing conditions. If they are allowed to adjust they will remain effective natural barriers to the 'destructive' effects of the sea. Such a philosophy is difficult to implement when the coastline is already highly developed but even so there are still many places where it makes sound sense.

The effects of a rise in sea-level on the design of coastal structures have been discussed by Sorensen (1991) and are summarized below.

Typical design lives for most substantial coastal structures are fifty or more years. The actual service lives of these structures may often exceed one hundred years. A sea-level rise of a metre or more over a century will have a significant impact on most of these structures.

With the strong likelihood of a significant sea-level rise during the next century, the designer should incorporate ease of retrofitting into a structure's design in order to respond to a raised sea-level as it occurs. If a structure cannot easily be designed for this retrofitting, then it should be designed for the ultimate sea-levels likely to occur during its anticipated service life.

Many coastal structures with significant ocean exposure have a design wave height limited by the wave breaking depth just seaward of the structure. Irrespective of the wave amplitude spectrum generated by the design storm, the maximum wave height at the structure will be about equal to the water depth at the structure. Consequently, as sea-level rises the structure will be exposed to increasing wave heights. This in turn will lead to increased wave loadings on the structure. Wave run-up, overtopping and energy transmission to the lee of the structure will also increase owing both to the raised sea level and to the increased incident wave height.

A rise in sea-level not only moves the MSL line landward due to the increased water level flooding further up the beach; but there is also an adjustment of the equilibrium shore profile resulting in beach erosion owing to the movement of sand from the beach face to offshore. Bruun (1962) explains this phenomenon and provides a method for calculating its effect.

When a structure such as a sea-wall or bulkhead is in place at the water's edge, it restricts the offshore movement of sand from the portion of the beach face located landward of the structure. Beach profile adjustment in response to sea-level rise then requires that an additional volume of sand be removed from the beach face just seaward of the structure.

The probabilities of given storm plus tide water level at a site are increased (return periods decreased), usually significantly (Pugh, 1990).

A half-metre in sea-level could result in say a three-quarter metre increase in breaking wave height and, the required weight of armour stone for a stone mound structure is a function of approximately the cube of the wave height. So, if there was a three-quarter metre increase in the total incident wave height, the structure would incur increased damage with time and might ultimately require placement of a larger stone size armour layer. Also, the toe erosion caused by shore profile adjustment would require continued maintenance and strengthening of the structure's toe protection.

Groyns, jetties and shore-connected breakwaters would suffer similar consequences at their seaward ends. As sea-level rises, the shoreward retreat of the MSL line might make the structure susceptible to flanking during storms. Shore-parallel offshore breakwaters would decrease in effectiveness as the distance from the structure to the MSL line increases. Also, any significant changes in the alongshore sediment transport regime might affect the functional behaviour of some of these structures and require structure modification.

26.2 CONCLUSIONS

This chapter is not intended to be a comprehensive review of the greenhouse effect, but it is hoped that it draws engineers' attention to the far-reaching consequences that are predicted for the future, especially for coastal engineering works. However, even since drafting this chapter, which has mainly relied on two sources (Report of the Coastal Zone Management Sub-Group, 1990; Sorensen, 1991) disagreement as to the extent of the changes still persists. It is an area in which there has been an explosion in research interest. Consequently engineers who need to incorporate these effects in design need to seek the best information that exists at the time.

REFERENCES

Bruun, P. (1962) Sea level rise as a cause of beach erosion, *Journal of Waterways and Harbours Division, ASCE*, **88** (WWI).

Delft Hydraulics (1989) Republic of Maldives, implications of sea-level change, *Report*.

Pugh, D.T. (1990) Is there a sea level problem?, *Proceedings of the Institution of Civil Engineers, Part 1*, **88**.

Report of The Coastal Zone Management Sub-Group of the Intergovernmental Panel on Climate Change (1990) Strategies for Adoption to Sea Level Change. Copies available through Ministry of Transport and Public Works, Rijkswaterstaat, Tidal Waters Division, PO Box 20907, 2500 Ex The Hague, The Netherlands.

Sorensen, R.M. (1991) Impact of expected climate change effects on coastal structure design. *Conference on coastal structures and breakwaters, November 1991*, Institution of Civil Engineers, London.

PART SIX

DESIGN, CONSTRUCTION AND MAINTENANCE OF COASTAL STRUCTURES

WAVE LOADS ON SEA DIKES AND SEA-WALLS

27

I.A. Führböter

Forces imposed by waves on impermeable slopes are the main subject of this chapter. Unlike rubble-mound or similar structures no percolation takes place in the energy transformation. The wave loads act perpendicular to the slope as pressure forces and tangential to the slope as shear forces. An important design parameter is run-up due to wave up-rush. Although vertical walls are not included the design methods apply to the slopes of dikes, sea-walls, revetments and similar shore protection structures. Although the main emphasis of the chapter is on uniform slopes composite slopes are considered towards the end (see also Chapters 28 and 29).

CONTENTS

27.1 INTRODUCTION

The following sections deal with the forces imposed by waves on impermeable slopes where, unlike rubble mound breakwaters or similar constructions, percolation effects do not participate in the energy transformations. The wave loads act perpendicular to the slope in the form of pressure forces (impacts) and tangential to the slope as shear forces due to the velocities in the wave motion. An additional important parameter for the design is the run-up height caused by the wave up-rush.

The slope can be those of dikes or sea-walls but also of revetments and similar constructions for shore protection. Vertical walls are not included in this treatment. The outer slopes of dikes are rarely steeper than $\tan\alpha = 1:n = 1:3$, where α is the angle to the horizontal.

Mostly uniform plane slopes are considered here, but composite slopes are mentioned in section 27.7. The wave loading is not related to any special dike cover like grass, concrete etc. The loading applies generally to all sorts of slopes and is only dependent on the parameters of the incoming waves and the slope. The influence of roughness will be treated separately.

27.2 WAVES IN FRONT OF A DIKE (INCIDENT WAVES)

The computations of wave loads on dikes are based on the incident waves directly before the dike. Mostly it can be assumed that the foreland of the dike is a nearly horizontal plane with a constant water depth d determined by the still water level at the location during a storm. The waves from the open sea are mostly transformed by diffraction (islands etc.), by refraction (due to underwater morphology), by shoaling effects and by breaking processes in the offshore and nearshore areas. It is generally difficult to compute reliably the incident wave parameters from the deep water values. Therefore, it will be necessary to have measured values or to estimate the incident wave parameters for the worst conditions. If necessary, the parameters of the incident waves according to the water depth d can be referred to the deep water conditions, i.e. the wave height H_0 and the length L_0.

A further very important parameter for design is the highest storm surge water level to be considered. Its determination, usually by statistical methods, is

Coastal, Estuarial and Harbour Engineer's Reference Book. Edited by M.B. Abbott and W.A. Price. Published in 1993 by Chapman & Hall, 2–6 Boundary Row, London SE1 8HN. ISBN 0 419 15430 2.

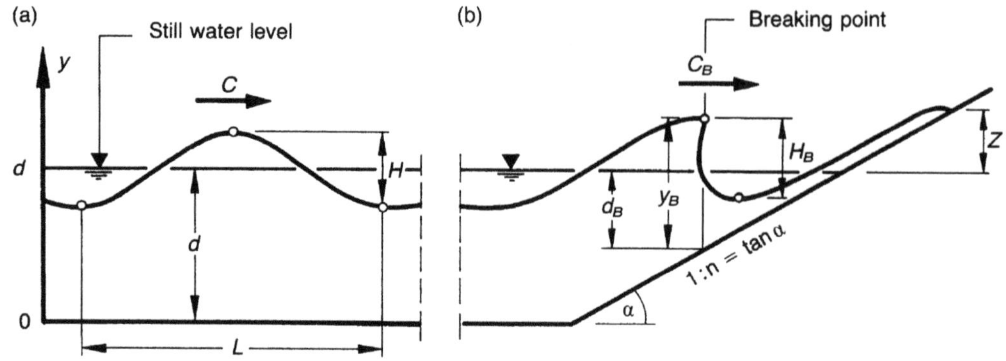

Figure 27.1 Definition sketch (a) incident wave; (b) breaking wave.

not included here. This design water level is assumed to be a given input like the wave climate mentioned above. Together with the elevation of the foreland this highest still water level determines the water depth d before the dike. This water depth d limits the maximum wave height H_{max} before the dike. According to the solitary wave theory this height is

$$H_{max} = 0.78d \qquad (27.1)$$

For periodic waves, however, a better approximation is (Kishi, 1959)

$$H_{max} \leqslant d \qquad (27.2)$$

which has also been verified by intensive field studies on areas of the North Sea by Siefert (1974). Because the wave energy increases with the square of the wave height the wave loads on a dike depend strongly on the elevation of the foreland (Hensen, 1954), i.e. on maximum water depth.

The incident waves perpendicular to the dike axis usually produce the highest loads, since then for given H_o, L_o the wave energy approaching the unit length of a dike is a maximum. In the following, therefore, perpendicular waves are assumed. This also leads to a two-dimensional treatment similar to theory and experiments in wave flumes.

27.3 SLOPE AND BREAKER TYPE

The definitions of parameters of the incident and the breaking wave are shown in Figure 27.1. The incident wave on the foreland with the uniform depth of d below the still water level (= highest water level) is completely described by its height H and length L which is related to the wave period T through the equation $L = cT$ where c is the wave celerity in depth d. On the uniform slope with $\tan\alpha = 1{:}n$, at the breaking point the wave height increases to the breaker height H_B with the crest elevation y_B and the breaking depth d_B.

The physical process of breaking is that the velocity c_B of the water particles of the crest (Figure 27.1) becomes higher than the wave celerity itself. During breaking, a large amount of wave energy is dissipated by very complicated turbulent mixing processes connected with intensive air entrainment (Führböter, 1970).

Three main types of breakers can be defined; the spilling, the plunging and the surging breaker (Figure 27.2). For the spilling breaker (predominant on very flat slopes of natural beaches) the breaking begins at the crest with the formation of a foam sheet running gradually upon the front slope of the wave into the trough. With this type of breaker, the wave energy in a surf zone is dissipated gradually over a distance of several wave lengths. The air entrainment is only near the surface and very little wave energy is reflected.

The plunging breaker forms the characteristic breaker tongue like a waterfall. The path of the falling water is described by a free fall parabola. Where it is intercepted by the slope of the dike high normal loads occur (impact pressures). The energy dissipation takes place over a very short distance, of the order of one wave height, with strong air entrainment through the depth. A part of the wave energy is reflected (Figure 27.2).

The surging breaker (predominant on steep slopes) is similar to the plunging breaker but does not form the free falling waterfall from the crest. At the breaking point the water mass forms a nearly vertical front which flattens producing a strong up-rush swell

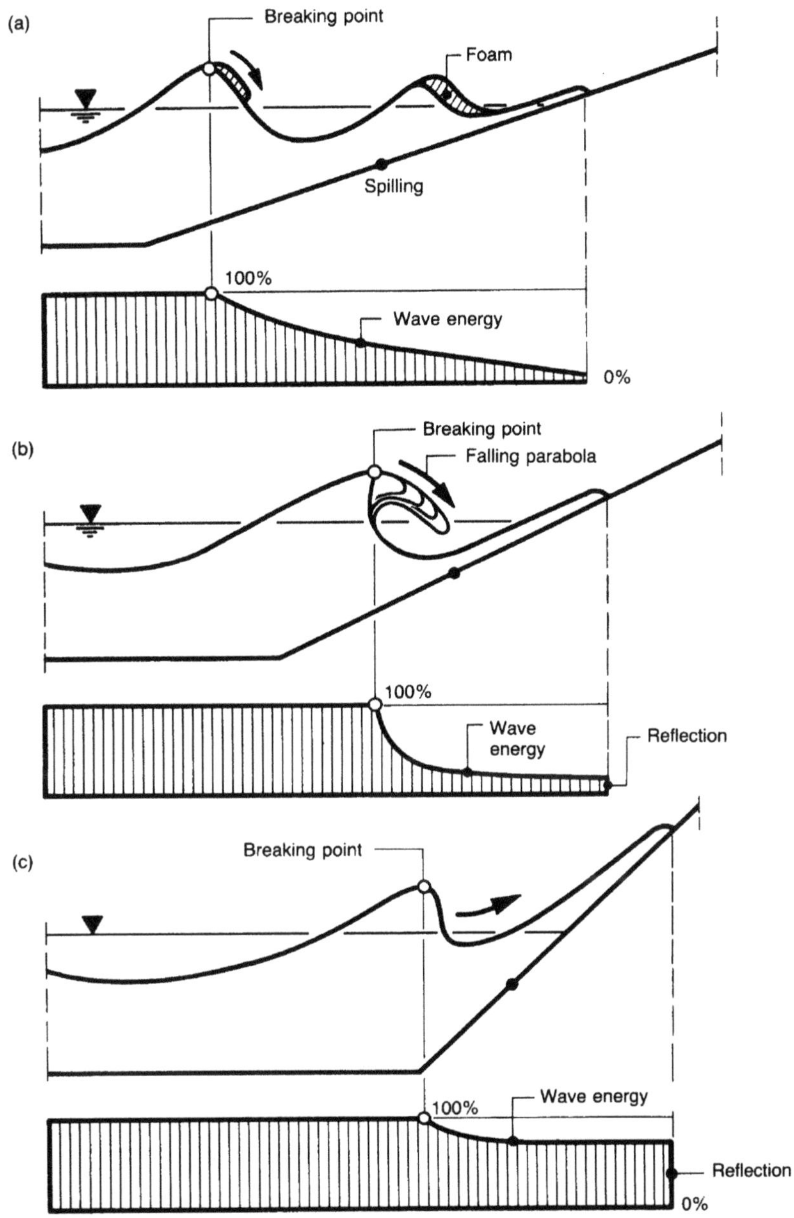

Figure 27.2 Breaker types (schematic) (a) spilling; (b) plunging; (c) surging.

with a high reflection of the wave energy (Figure 27.2). The high up-rush and down-rush velocities on the slope, cause high tangential forces (shear stresses) over the large up-rush height z (Figure 27.1).

The type of breaker is characterized by the dimensionless breaker number ξ, first introduced by Iribarren and Nogales (1949, see Battjes, 1974), which combines the geometry of the slope ($\tan \alpha$) with the

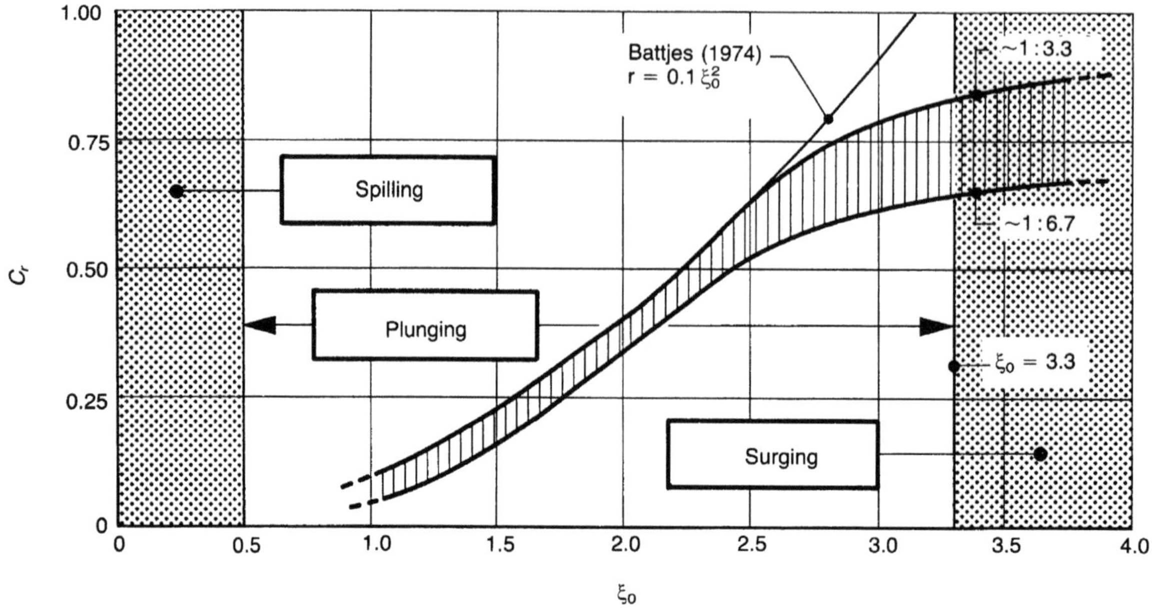

Figure 27.3 Reflection coefficient c_r versus breaker number ξ (after Battjes, 1974).

geometry of the wave H_o and L_o (for deep water conditions) in the form of:

$$\xi_o = \frac{\tan \alpha}{\sqrt{H_o/L_o}} \tag{27.3}$$

which can be written as:

$$L_o = \frac{g}{2\pi} T^2 \tag{27.4}$$

and also in the form of

$$\xi_o = \sqrt{\frac{g}{2\pi}} \frac{1}{\sqrt{H_o}} T \tan \alpha \tag{27.5}$$

The limits for the breaker types are (Battjes, 1974):

$\xi_o < 0.5$ spilling breaker
$0.5 < \xi_o < 3.3$ plunging breaker
$\xi_o > 3.3$ surging breaker

The transition between breaker types is gradual. The transition between spilling and plunging types is at times also described as the transition or the intermediate breaker (Mason, 1951), and between the plunging and surging types as the collapsing breaker (Galvin, 1968).

The energy reflection affects energy dissipation. The reflection coefficient c_r as a ratio between the heights of the reflected and the incident wave is as shown in Figure 27.3 as a function of the breaker number ξ_o.

Very low reflection (c_r less than 10%) is found for the spilling breaker ($\xi_o < 0.5$). With increasing ξ_o, in the range of the plunging breaker, the reflection coefficient increases up to 60%–80% at the limit for surging breakers ($\xi_o \sim 4$). On steeper slopes (1:3.3 in Figure 27.3) the reflection coefficient is higher than on flatter slopes (1:6.7 in Figure 27.3). For breaker numbers up to $\xi_o = 2.5$, the upper limit of the reflection coefficient can be approximated (Battjes, 1974) from $c_r = 0.1(\xi_o^2)$.

On conventional dike profiles usually only the forms of plunging and surging breakers together with their transition type, the collapsing breaker, are observed. Only on very flat dike slopes (flatter than ~1:10) will the spilling type occur.

27.4 WAVE IMPACT

The wave impact is associated with the occurrence of the plunging breakers where the free falling tongue impinges upon the slope of the dike. At the moment of impact, very high pressures occur for a very short time on relative small areas. For full-scale experiments ($H = 1.25$ m) with a slope of 1:4, Figure 27.4a

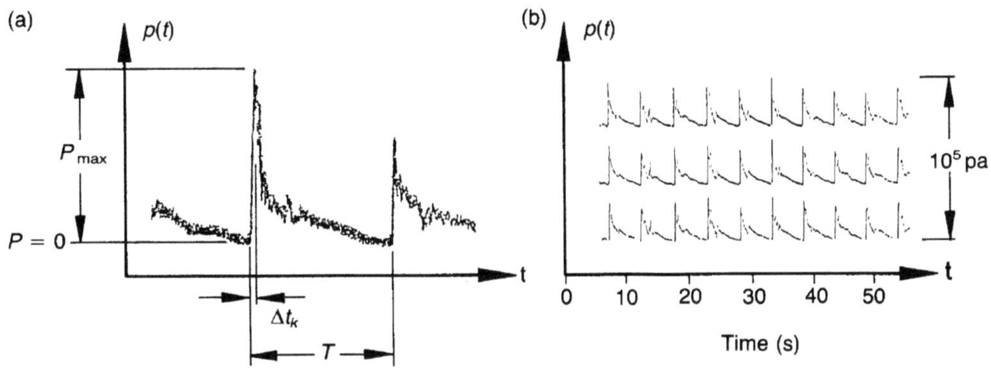

Figure 27.4 (a) time history of pressure produced by an individual breaker with impact pressure p_{max}; (b) series of impacts at different measuring points on slope.

shows the short rise or compression time t_k of the maximum impact pressure p_{max}. The average observed value of t_k was about 0.1 s, with large scatter, and the lowest recorded value was 0.003 s (Witte, 1988). As Bagnold (1939) has shown for the special case of a vertical wall, the shortest values of t_k are connected with the highest pressures p_{max}. In Figure 27.4b a series of ten impacts at three different measuring points on the slope are shown. Although produced by exactly regular waves, a stochastic pattern of the pressure peaks is obvious. On the area of impact, the maximum pressures p_{max} do not occur simultaneously; the time lag of p_{max} at the different locations, however, is so short that in the first approximation the spatial envelope of the time maxima p_{max} can be assumed to produce the total load on the slope. As shown in Figure 27.5, a new peak pressure p_{max} can be defined as the overall maximum pressure during one individual impact on the total area. From Figure 27.5 (tests with $H = 1.25$ m on an 1:4 slope) it can be seen that not only the peak pressures (Figure 27.4b) but also the spatial pressure distributions vary on the impact areas from impact to impact even with regular waves.

Due to the geometry of the free falling water from the crest at the breaking point (Figure 27.2), the location of maximum pressure p_{max} (Figure 27.5) on slopes flatter than about 1:3 is found near the level of about $0.5H$ below the still water level. The length of the impact area (which cannot be defined exactly) is of the order of about $1H$ (Figure 27.5).

The main reason for this stochastic pattern of impact pressures and forces is the air content in the area between the front of the impinging tongue or jet and the plane surface as was first considered by Bagnold (1939) for a vertical wall. Denny (1951) first described the stochastic pattern of wave impact on a vertical wall. The ratio between the elasticity of water and air is about 15:500. This strongly influences the velocity of sound in the water-air mixture (Lundgren, 1969). If the influence of air contents is neglected the impact pressure would rise to

$$p_{max} = \rho v c \tag{27.6}$$

as was shown by von Karman (1929); ρ is the density and c the sound velocity of water and v the impact velocity perpendicular to the impact plane. With a physical model considering the time history of an impact as a sequence of compression of air and water and following expansion at the free borders of a jet, a differential equation can be found with special solutions for p_{max} and t_k (Führböter, 1966):

$$p_{max} = \delta_i \sqrt[3]{\frac{c}{v}}\, \rho v c \tag{27.7}$$

$$t_k = \frac{R}{\sqrt[3]{vc^2}\sqrt{\delta_i}} \tag{27.8}$$

with

$$\delta_i = \left(\left(\frac{E_a}{E_w}\right)\left(\frac{R}{D_i}\right)\right)^{2/3} \tag{27.9}$$

with E_a and E_w = elasticity of air and water, R = the hydraulic radius of the impact area and D = the thickness of a sheet of air representating the enclosed air content in the impact area. The term ρ_i is a dimensionless impact number which strongly depends on the air content D_i. The reciprocal relation between the peak pressure p_{max} and the compression time t_k, as follows from Equation (27.8), was also found by Bagnold (1939).

Through the dimensionless impact number ρ_i, the peak pressure $p_{max\,i}$ is connected with the air content

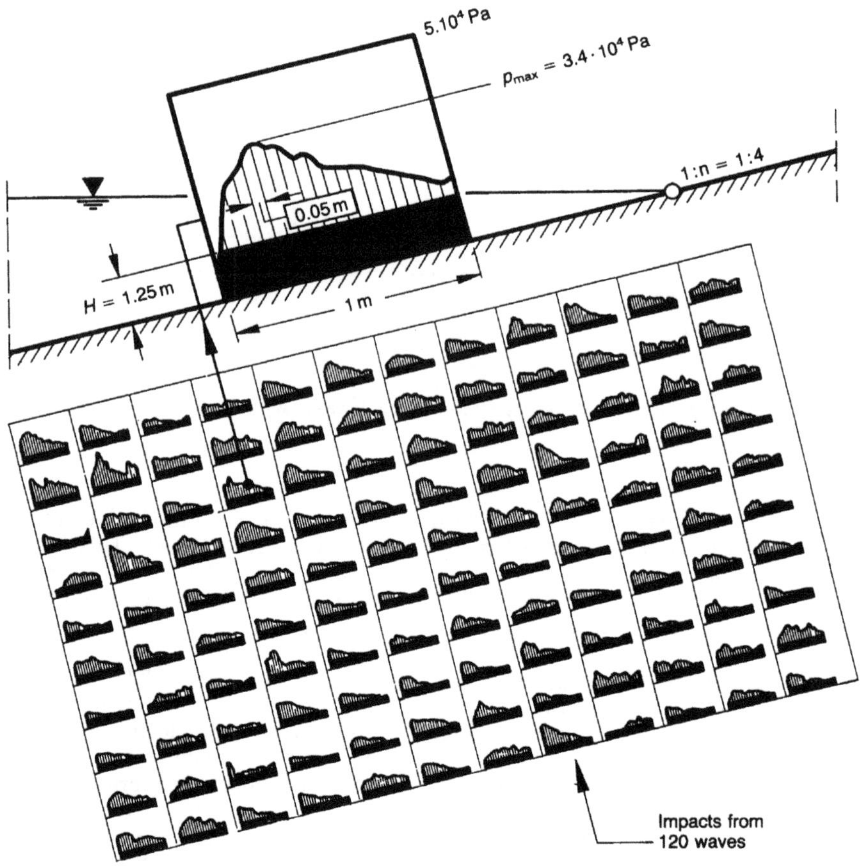

Figure 27.5 Spatial pressure distributions (envelope of impact pressures p_{max} in Figure 27.4) from series of 120 waves.

D_i through a power of −2/3; only when these equations are written in logarithmic form, a linear function is found between log $p_{max\,i}$ and log D_i. The air content D_i is produced by very complex conditions during the impact of the front of the impact jet on the plane. In first approximation, a random process can be assumed with a normal (Gaussian) distribution of the air content D_i. The consequence of this for the peak pressures p_{max} is that their probability distribution must follow the log-normal distribution. This was verified by experiments with impinging jets (Führböter, 1966).

If Equations (27.7) and (27.9) are combined with the impact velocity v perpendicular to a slope with $n = \cot\alpha$

$$v = (gy_B)^{1/2}\left(\frac{1 + 2n^2}{1 + n^2}\right)^{1/2} \qquad (27.10)$$

where y_B is the crest height at the breaking point (Figure 27.1) and g is the gravitational acceleration. If y_B and the hydraulic radius R of the impact area are inserted as ratios of the incident wave height H, it follows that:

$$p_{max\,i} = \left(\underbrace{\frac{E_a}{E_w}\left(\frac{c^2/g}{D_i}\right)}_{\text{stochastic with } D_i}\underbrace{\left(\frac{R}{H}\right)\left(\frac{y_B}{H}\right)^{1/2}}_{\sim\text{ const.}}\underbrace{\left(\frac{1 + 2n^2}{1 + n^2}\right)^{1/2}}_{=\text{ const.}}\right)^{2/3}\rho g H \qquad (27.11)$$

or

$$p_{max\,i} = \text{const}_i(\rho g H) \qquad (27.12)$$

so a linear relationship can be stated between the peak pressure p_{max} and the incident wave height H.

From Equations (27.11) and (27.12), the same normal-log distribution of impact forces for wave

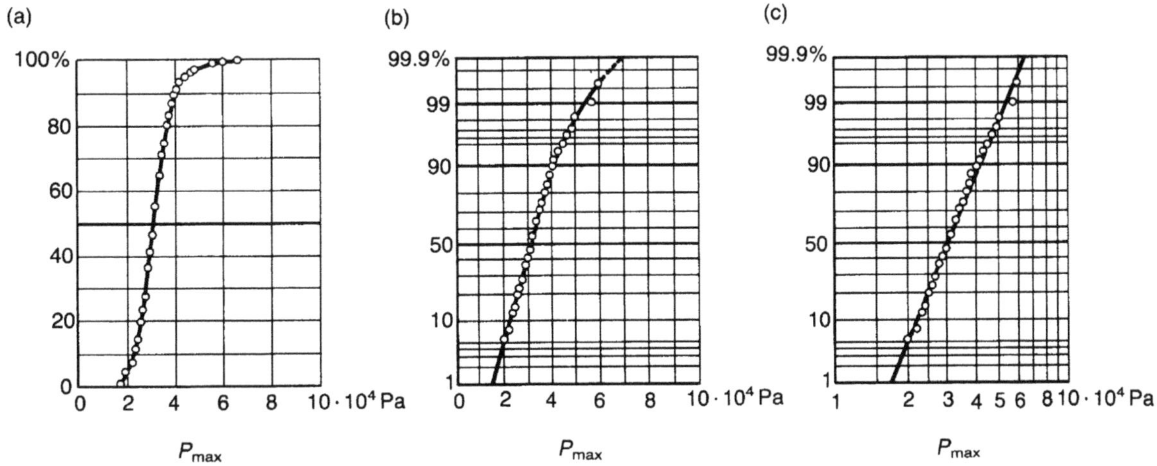

Figure 27.6 Probability distributions of impact pressures on smooth 1:4 slope (full-scale tests with regular waves $H = 1.3$ m and $T = 4.6$ s). (a) linear; (b) Gauss; (c) normal-log.

impacts is to be expected as for jets. This was demonstrated by experiments for wave impacts on slopes (Führböter, 1986) and on vertical walls (Witte, 1988). As an example, Figure 27.6 shows the good agreement of the measured values with the log-normal-function, whereas this is not the case with the simple linear function and the Gauss-distribution. With exceedences of $i = 50$, 90, 99 and 99.9% in full scale tests on an 1:4 slope (smooth) the following values to the constant in Equation (27.12) were found (Führböter, 1986):

slope $1{:}n = 1{:}4$:

$$i = 50 \quad p_{max\,50} = (2.2 \pm 0.4)\rho g H$$
$$i = 90 \quad p_{max\,90} = (3.0 \pm 0.5)\rho g H$$
$$i = 99 \quad p_{max\,99} = (3.9 \pm 0.6)\rho g H$$
$$i = 99.9 \quad p_{max\,99.9} = (4.8 \pm 0.7)\rho g H$$

On the 1:4 slope, these highest pressures occur at breaker numbers ξ between 1.0 and 1.5 in the regime of fully developed plunging breakers. At breaker numbers lower than 1.0 and higher than 2.0 the maximum pressures decrease appreciably until they approximate to hydrostatic pressures of the order of the wave height. On steeper slopes, however, the critical breaker number for the highest peak pressures is shifted towards the higher values of ξ; with model tests on a 1:3 slope, Bruun and Günbak (1977) found the highest pressures at a breaker number ξ near 3. Here negative pressures (up to 40% of p_{max}) were also recorded. However, for flatter slopes this phenomenon does not seem to predominate (Witte, 1988).

With jet impact, it can be shown that even a thin layer of water upon the impact area hat a high damping effect on the peak pressures (Führböter, 1966). This is due to the high reduction of sound velocity in a water-air mixture (Lundgren, 1969). On flat slopes ($n > 3$) because the point of impact usually lies below the still water level, the impinging jet of the plunging breaker mostly falls into the down-rush water of the foregoing wave (Figure 27.2). The thickness of this back-rush water increases with decreasing angle α, therefore the peak pressures will also decrease with decreasing α or $1{:}n$. It also follows that an increase of peak pressure is to be expected with steeper slopes. Model tests by Greslou and Montaz (1962) showed an increase of peak pressures of 2.3 for a 1:2 slope and of 4.4 for a 1:1.3 slope compared with pressures on slopes between 1:3.3 and 1:5. Führböter and Sparboom (1988) stated that impact pressures on slopes between 1:4 and 1:6 are related to the factor $1{:}n$. Further research is necessary for very flat ($n > 6$) and very steep ($n > 2$) slopes. Since for $1{:}n = 1{:}0$ (vertical wall) this relationship is clearly not valid, application of the results with jet impacts on a water-free surface gives $p_{max\,99.9} \sim 45\rho g H$ (Führböter, 1966). For very flat slopes on the other side ($\tan\alpha = 1{:}n$ less than about 1:10, at the start of the range of spilling breaker) only hydrostatic pressures due to the wave height are to be expected without the characteristic peak pressure shown in Figure 27.4.

For wave spectra, the probability of wave impacts is closely combined with the number of waves breaking in the form of plunging breakers. Here it is possible, with an unfavourable sequence of waves, that the plunging breaker falls into a thinner layer of the back-rush water than with regular waves and thus causes higher peak pressures.

For waves with a significant height $H_s \sim 1\,\text{m}$, Grüne (1988) from field investigations on smooth (asphalt-concrete) 1:4 slopes found an overall highest pressure with $p_{max}/\rho g H_s \sim 7$ which is not very far from $p_{max\,99.9}$ for regular waves ($p_{max} = (4.8 \pm 0.7)\rho g H$ mentioned before. For rough surfaces, however, lower pressures are to be expected due to the higher air content in the impact area (Stive, 1984; van Vledder, 1990). The danger of impact pressures lies in the high and short pressure peak on a small area (see section 27.8). Reviews (in English) on impact forces have been compiled by Ramkema (1978) and, especially for slopes, by van Vledder (1990).

27.5 WAVE RUN-UP HEIGHT

The wave run-up height z is one of the most important parameters of wave load because it determines; together with the highest still water level; the necessary crest height of a dike. It is defined as the vertical distance between the still water level and the highest point of the wave up-rush swell (Figure 27.1). In the run-up height z the wave set-up is included. It is especially significant at low breaker number near to the type of spilling breaker (Battjes, 1974).

The physical process of wave run-up lies in the transformation of potential to kinetic energy at the breaking point. Part of the wave energy is dissipated here by turbulent mixing, the remaining energy forming a swell or a bore running up the slope to the point where due to the influence of gravity a return motion occurs (wave down-rush or back-rush). The reflection increases with the strength of the down-rush. For the same wave heights, the lowest run-up height is caused by the spilling and the highest by the surging breakers (Figure 27.2).

A good summary of the relations between the slope angle, the wave parameters and the run-up height z is given by the results of the model investigations (for regular waves and smooth slopes) by Saville (1958). For ratios $d/H_o \geqslant 3$ the water depth d on the foreland was found to not influence the run-up. Figure 27.7 shows for $d/H_o \geqslant 3$ the function z/H_o versus the

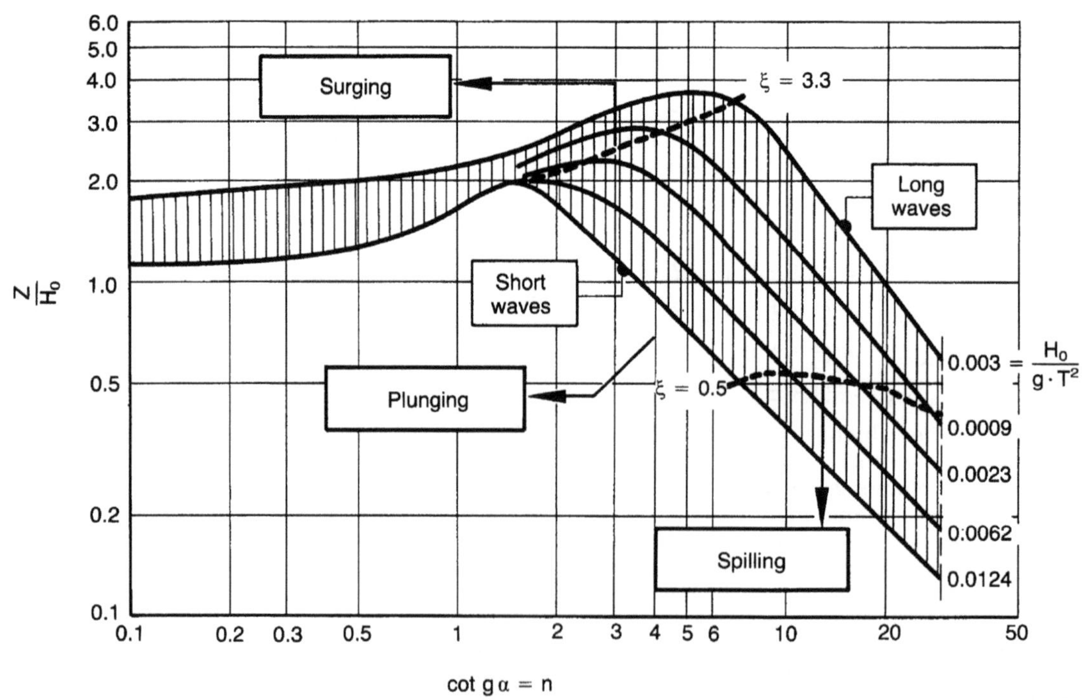

Figure 27.7 Dimensionless run-up height z/H_0 as a function of $\cot\alpha = n$ and steepness factor $(H_0/g)T^2$ for regular waves, smooth slopes, $d/H_0 \geqslant 3$ (after Saville, 1958).

slope angle $\cot\alpha = n$ with the steepness factor H_o/gT^2 as the parameter. For very steep slopes, the influence of this parameter is small (Figure 27.7). In the broad area between the surging, plunging and spilling breaker, however, the run-up characteristics (straight lines in Figure 27.7) can be approximated by the well-known formula (Hunt, 1959):

$$\frac{z}{H} = \xi = \frac{\tan\alpha}{(H/L_o)^{1/2}} \tag{27.13}$$

Here H_o is substituted by the incident wave height H (Figure 27.1). Equation (27.13) can be written in the form:

$$z = (HL_o)^{1/2}\tan\alpha \tag{27.14}$$

or with L_o, after Equation (27.4) as

$$z = \left(\frac{g}{2\pi}\right)^{1/2}\sqrt{H'}\,T\tan\alpha \tag{27.15}$$

Here z is a linear function of the wave period T and a function with the power 1/2 of the wave height H. Although empirically, Hunt's formula is justified by dimensional considerations, the dimensionless ratio z/H is seen to be the wave number ξ (Equation 27.3), i.e. a function of the breaker type.

Even for exactly regular waves, a scatter of the values of z is observed in all experiments. It is nearly independent of the experimental facilities and mostly due to the stochastic processes from turbulent mixing during breaking. It has therefore become usual to define a value of z_{98} as a significant run-up height with an excedence of 2%, i.e. if the crest of the dike lies exactly on the level of z_{98}, one in 50 waves will overtop. From experience, this amount of overtopping is not dangerous for the inner slopes, even for grass-covered earth dikes. The definition of z_{98} has the advantage that it can also be used for the run-up of waves which are spectrally distributed.

After the earlier investigations by Saville (1955, 1956, 1958), Wassing (1957), Savage (1958, 1959) and Hunt (1959) further research on wave run-up on smooth slopes has been carried out, e.g. Schijf (1974), Battjes and Roos (1975), Bruun and Günbak (1977), CERC (1984) and Bruun (1985). Mostly model tests were used where scale effects cannot be excluded (CERC, 1984). In Figures 27.8 and 27.9, results of full scale tests (wave heights up to 2 m, wave periods up to 15 s) are presented (Führböter *et al.*, 1989). These are for a smooth (asphalt concrete) 1:6 slope and regular waves. In Figures 27.8 and 27.9, as in Equations (27.13)–(27.15) the breaker number ξ is based on the incident wave height H taken from Figure 27.1:

$$\xi = \frac{\tan\alpha}{(H/L_o)^{1/2}} \tag{27.3}$$

In Figure 27.8, the experimental results of the ratio z_{98}/z_{Hunt} are compared by the values from the formula by Hunt. The measured points are fitted by a correction function $k(\xi)$ where $k = 1$ would express an exact agreement with Hunt's formula. As can be seen from Figure 27.8, for breaker numbers lower than ~3.3 the experimental results give higher values and the

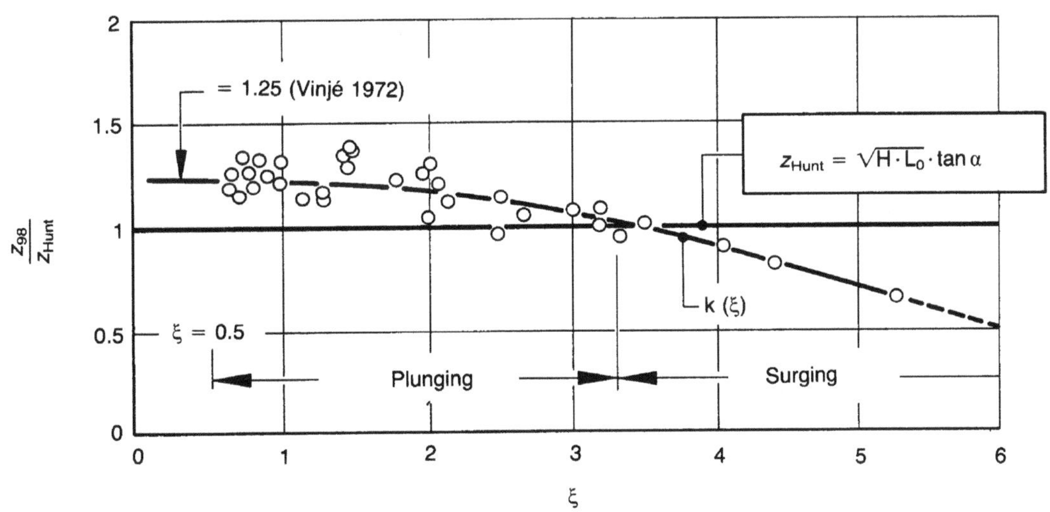

Figure 27.8 Ratio z_{98}/z_{Hunt} versus breaker number ξ for regular full-scale waves on smooth (1:6) slope.

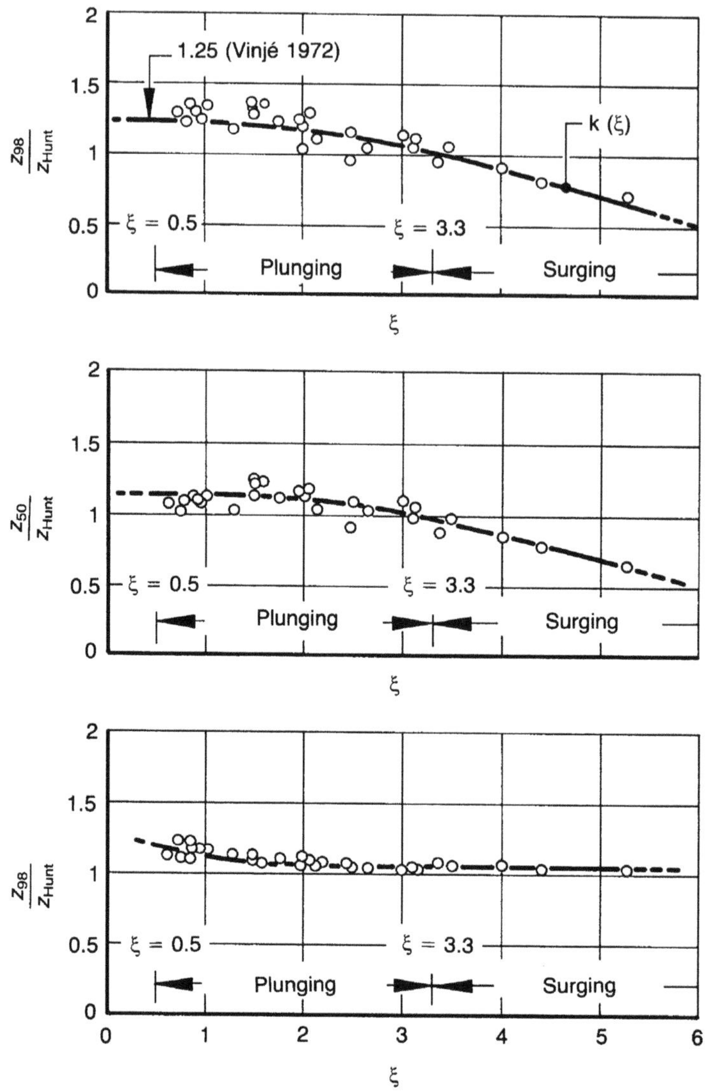

Figure 27.9 z_{98}/z_{Hunt}, z_{50}/z_{Hunt}, and scattering factor z_{98}/z_{50} versus breaker number ξ for regular full-scale waves on smooth (1:6) slope.

function $k(\xi)$ rises asymptotically to 1.25 for $\xi < 1$. This correction factor, as a constant, was introduced by Vinjé (1972). In the range of the surging breakers only ($\xi > 3.3$), the function $k(\xi)$ becomes lower than 1 and the formula by Hunt lies on the side of safety.

The full-scale tests also showed that the formula by Wassing (1957)

$$z_{98} = 8H \tan \alpha \quad (27.16)$$

fits the data well over the range of fully developed plunging breakers ($1 < \xi < 1.5$). For slopes in the range between 1:4 and 1:6 and wave periods ≤ 6 s, the Wassing's formula represents the upper limit of the experimental results after Figure 27.8 and lies on the side of safety (Führböter *et al.*, 1989).

It is very important that, for slopes flatter than about 1:3, the run-up height z decreases linearly with $\tan \alpha = 1{:}n$, and with flatter slopes the wave run-up can be reduced significantly.

In Figure 27.9, the scattering of data is shown by the ratio z_{98}/z_{Hunt} compared with z_{50}/z_{Hunt}, where z_{50} can be seen as a mean run-up height. The scattering

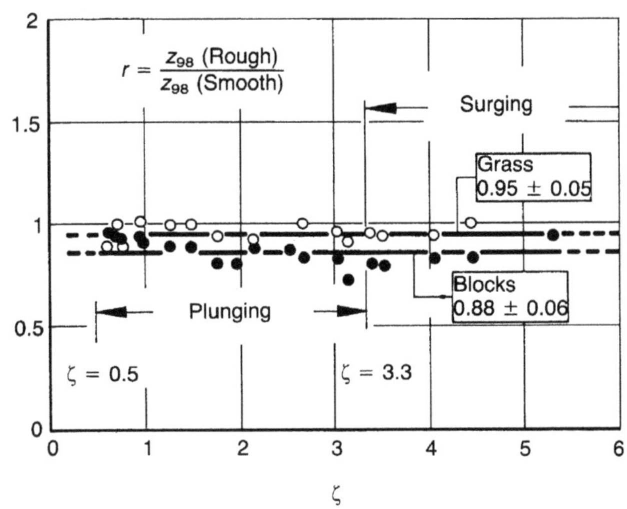

Figure 27.10 Reduction factor r for two types of roughness versus breaker number ξ for regular full-scale waves on 1:6 slope.

factor formed by the ratio z_{98}/z_{50} is nearly constant ~1.1 for breaker numbers $\xi > 3$. In the range of high energy dissipation and low reflections (see Figure 27.3) it rises to 1.2 at breaker numbers near $\xi = 0.5$. Some differences between the results of different authors may be explained not only by scale effects due to very small models, but also by the definition of z either as z_{98} or z_{50}.

On rough slopes additional wave energy will be dissipated by the friction effects, especially during the up-rush, causing a reduction of the wave run-up height z. The influence of a rough slope can be expressed by a dimensionless reduction factor r formed by the ratio

$$r = \frac{z_{98}(\text{rough slope})}{z_{98}(\text{smooth slope})} \tag{27.17}$$

Additional to the tests with a smooth (asphalt-concrete) 1:6 slope (shown in Figures 27.8 and 27.9) tests with two types of roughness were carried out at full scale. One roughness used was formed by an artificial grass cover with a grass height of 2.7 cm similar to a natural vegetation on a grassed dike and the other consisted of concrete blocks (16 cm × 16 cm, 14 cm height) with central distances of 80 cm horizontal and along the slope. These blocks were mounted on the surface of the asphalt-concrete.

Figure 27.10 shows the reduction factor r for these two roughnesses as a function of the breaker number ξ. A weakly decreasing trend of r with an increasing breaker number ξ is indicated. A similar observation was made by Savage (1958, 1959) who found that the reduction factor r decreases with decreasing wave steepness and slope angle α. The latter is due to the longer up-rush paths on flat slopes. In Figure 27.10 for a 1:6 slope this trend, however, is only weak and for practical purposes an average value of r = 0.95 for grass and 0.88 for the block roughness can be used. It should be noted that grass dikes with a short-cut vegetation show a roughness very near to that of a smooth surface.

Reduction factors for different types of roughness are given by, for instance, Schijf (1974) or CERC (1984). For impermeable slopes, the reduction factor r can fall to 0.5 for a cover of very large and rough stones.

It is very difficult to state general rules or laws for the significant run-up height z_{98} for wave spectra. This is because there is a very high variability of spectral types. The wave spectra before a dike are also strongly affected by the special conditions of the foreland.

One method is to use a theoretically constructed transfer function between the spectra parameters and the wave run-up. Using a Rayleigh distribution for both H_o and L_o as spectrum parameters and the formula by Hunt (Equation 27.13) for regular waves of the indivial spectrum components, Battjes (1971) derived correction factors for Hunt's formula between 1.78 and 2.33, dependent on a special correlation factor ρ between the individual spectra of H_o and L_o. Similar investigations by van Oorshoot and D'Angremond (1968) led to correction factors between 1.5

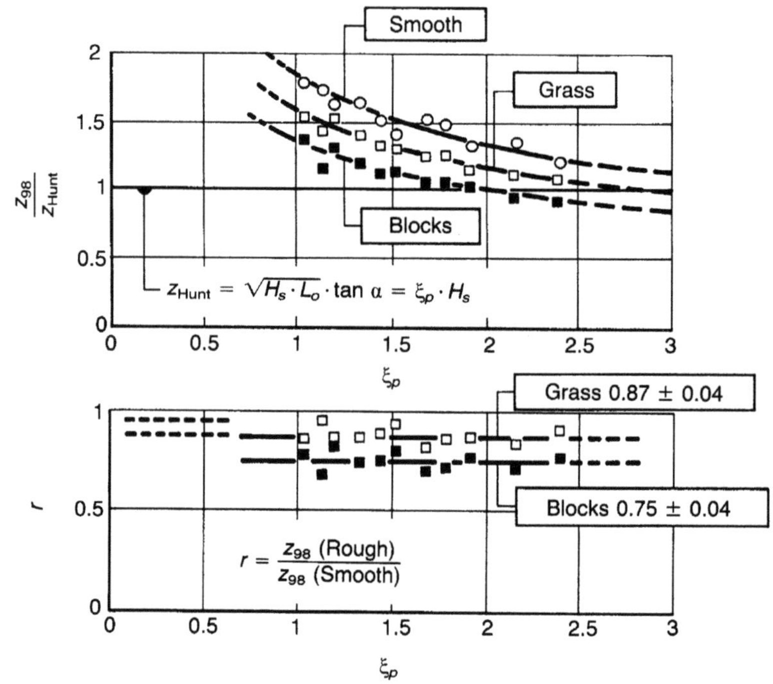

Figure 27.11 z_{98}/z_{Hunt} versus breaker number ξ_p for different roughnesses (see Figure 27.10) using Pierson–Moskowitz spectra on 1:6 slope.

and 1.93. As an example, the results of full scale experiments with spectra of the Pierson–Moskowitz type with the same experiments arrangement as used for the results shown in Figures 27.8–27.10 are shown in Figure 27.11. If ξ_p is used to signify the breaker number with significant wave height H_s instead of H and the peak period T_p instead of T then for the smooth 1:6 slope, the correction factors are between 1.3 and 1.7 over the range of $1 < \xi_p < 2.5$. This correction factor obviously decreases with increasing breaker number ξ_p because spectra with long waves show a similar pattern to regular waves with similar average periods. The decrease of the correction factor with an increasing breaker number ξ is also to be seen for the rough slopes with grass and blockcovers. It may be of interest that here the reduction factors r (0.85 for grass and 0.75 for blocks) are lower than those for regular waves (0.95 for grass and 0.88 for blocks) (Figure 27.10). An explanation is that high values of z_{98} are combined with long up-rush paths on the slope where friction effects are stronger than for short up-rush ways.

Another possibility for determining z_{98} from the parameters of wave spectra is to define a most unfavourable design wave (with H_{max} and T_{max}) combined with a formula for regular waves. The exceedance of this unfavourable design wave in the spectrum can then be interpreted as the exceedance of the corresponding run-up height z.

It must always be borne in mind that in wave spectra unfavourable sequences of waves can occur. When a very long wave follows a short one, the smaller down-rush of the short wave will cause a stronger up-rush for the following long wave (wave grouping). Since in most cases, even the determination of the wave parameters in front of a dike is only a matter of estimation, the calculation of z_{98} for design work should err on the side of safety.

Unlike breakwaters, overtopping should not be considered for dikes as a steady-state condition. If the relative number of waves causing overtopping increases, the amount of water passing over the dike crest will increase rapidly (Schijf, 1974; CERC, 1984; Waterloopkundig Laboratory, 1987). It is a question for the construction of a dike whether or not strong overtopping is dangerous for the inner slope (see section 27.8).

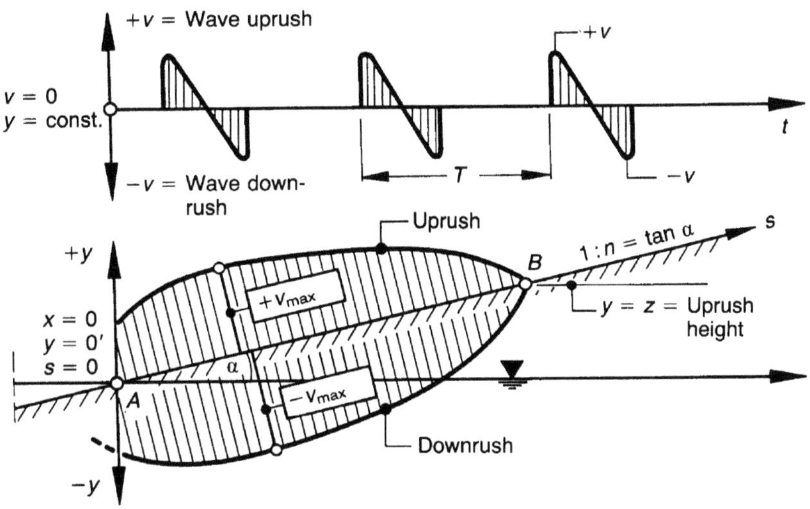

Figure 27.12 Up-rush and down-rush velocities on 1 : 1 slope.

27.6 UP-RUSH AND DOWN-RUSH VELOCITIES

By wave up-rush and down-rush, high velocities are produced on the slope causing high tangential forces (= shear stresses). Because of the complicated transformations of energy from potential over-pressure (impact!) into kinetic energy during breaking; combined with high turbulence and aeration effects; it is not possible to give a general and exact mathematical solution for up-rush and down-rush velocities. After the formation of the up-rush bore some mathematical treatment is possible (Shen and Meyer, 1963; Freeman and Le Méhauté, 1964). Most investigations on this topic have been carried out experimentally using scale models (Saville 1955, 1956; Battjes and Roos, 1975).

In Figure 27.12 above, the time histories of up-rush (+) and down-rush (−) velocities at one point of the slope are shown schematically. The highest velocities appear near to the time when the up-rush or down-rush front is passing (front velocities). For all points along the slope, envelopes for $+v$ (up-rush) and $-v$ (down-rush) can be constructed. These functions follow approximately the square root law as for the mechanics on an inclined plane. Near the still water level ($y = 0$), two maxima exist with $+v_{\max}$ for up-rush and $-v_{\max}$ for down-rush; these maxima represent the highest velocities on the total slope during one individual wave period (Figure 27.12). They are to be found between the still water level ($y = 0$) and a level up to $1H$ above it for plunging breakers with $\xi < 2$ and about the still water level for breaker number $\xi > 2$.

As for wave impact and wave run-up, and for the same reasons mentioned in discussing them, even for exactly regular waves, stochastic effects are present causing a scatter of experimental results. Similarly then velocities v_{98} and v_{50} can be defined for the wave run-up height z, where v_{98} denotes a velocity with an exceedance of 2% like z_{98} and v_{50} an expression for an average velocity. In Figure 27.13 (Führböter and Witte, 1989), the characteristic distributions of up-rush and down-rush velocities v_{98} and v_{50} are shown for a typical plunging ($\xi = 0.96$) and surging ($\xi = 4.03$) breaker as functions of the vertical distances y to the still water level ($y = 0$); they are measured as front velocities by step gauges.

For up-rush and down-rush, the surging breaker shows a more symmetric pattern like that of a standing wave or clapotis. The plunging breaker, however, shows a stronger difference between the velocities of up-rush and down-rush together with higher differences between v_{98} and v_{50}. The main reason for this is that, near the still water level especially, the down-rush water layer is thinner than that of the up-rush.

Similar to Battjes and Roos (1975), the measured velocities can be presented in a dimensionless form. With Hunt's formula for z a run-up length on the slope of $s = z/\sin\alpha$ is given; assuming the half wave period $T/2$ to be the same time for up-rush and down-rush on the slope (which is not exactly correct) a fictitious velocity can be defined called v_{Hunt}, which may be interpreted as a mean velocity along the slope. It is:

$$v_{\text{Hunt}} = \frac{z_{\text{Hunt}}/\sin\alpha}{T/2} \qquad (27.18)$$

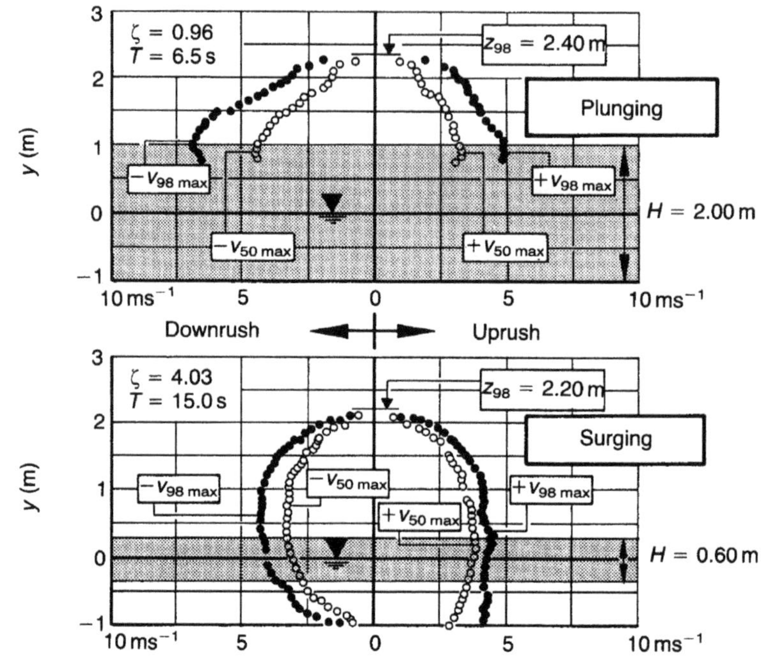

Figure 27.13 Velocities v_{98} and v_{50} of up-rush (+) and down-rush (−) versus vertical distance y from still water level ($y = 0$) for plunging and surging breaker with regular waves on 1:6 smooth slope.

With z_{Hunt} after Equation (27.15):

$$v_{Hunt} = \left(\frac{2}{\pi}\right)^{1/2} \frac{1}{\cos\alpha}(gH)^{1/2} \qquad (27.19)$$

For flat slopes cos α becomes ~1 and v_{Hunt} is a function of $(gH)^{1/2}$ only as a swell velocity with the incident wave height H. Similar to Battjes and Roos (1975), the measured velocities, here v_{max} after Figures 27.12 and 27.13, can be written as a dimensionless ratio and as a function of the breaker number ξ:

$$\frac{v_{max}}{v_{Hunt}} = f_1(\xi) \qquad (27.20)$$

With a factor 1/π added to Equations (27.19) and (27.15), a second dimensionless ratio can be formed containing the circular orbital velocity $H(\pi/T)$ of the incident wave:

$$\frac{v_{max}}{H(\pi/T)} = f_2(\xi) \qquad (27.21)$$

with

$$f_2(\xi) = \frac{2}{\pi \sin\alpha}\xi(f_1(\xi)) \qquad (27.22)$$

In Figure 27.14 (Führböter and Witte, 1989) the velocities $+v_{98max}$ (for up-rush) and $-v_{98max}$ (for down-rush) (see Figure 27.13) are evaluated from Equation (27.20) with $f_1(\xi)$ and Equation (27.21) with $f_2(\xi)$ from full scale tests with regular waves on a smooth 1:6 slope (see Figures 27.8 and 27.9). The functions $f_1(\xi)$ and $f_2(\xi)$ are different for up-rush and down-rush according to Figure 27.13.

From the functions $f_1(\xi)$ both for up-rush and down-rush, a clear maximum is to be seen near ξ = 3.3. It can be explained by a special interaction between up-rush and down-rush, first described as a resonance phenomenon by Bruun and Johannesson (1977). Here for up-rush motion $f_1(\xi)$ rises up to ~3 and for down-rush ~4. For the function $f_2(\xi)$, however, the transition between the types of plunging and surging breaker is clearly expressed by $f_2(\xi)$ ~ constant for $\xi > 3.3$. For up-rush $f_2(\xi)$ is constant at about ~40 and for down-rush at about ~45 (Figure 27.14).

From model tests, with a similar evaluation of the dimensionless parameters for $f_1(\xi)$, Battjes and Roos (1975) found nearly identical functions for slopes between 1:3 and 1:7. Thus, as a first approximation the function $f_1(\xi)$ in Figure 27.14 can be used also for slopes flatter than 1:3. As shown for the wave run-up height z, the up-rush and down-rush velocities decrease as a function of $\tan\alpha = 1{:}n$. This means, these velocities can be reduced by decreasing the slope. As

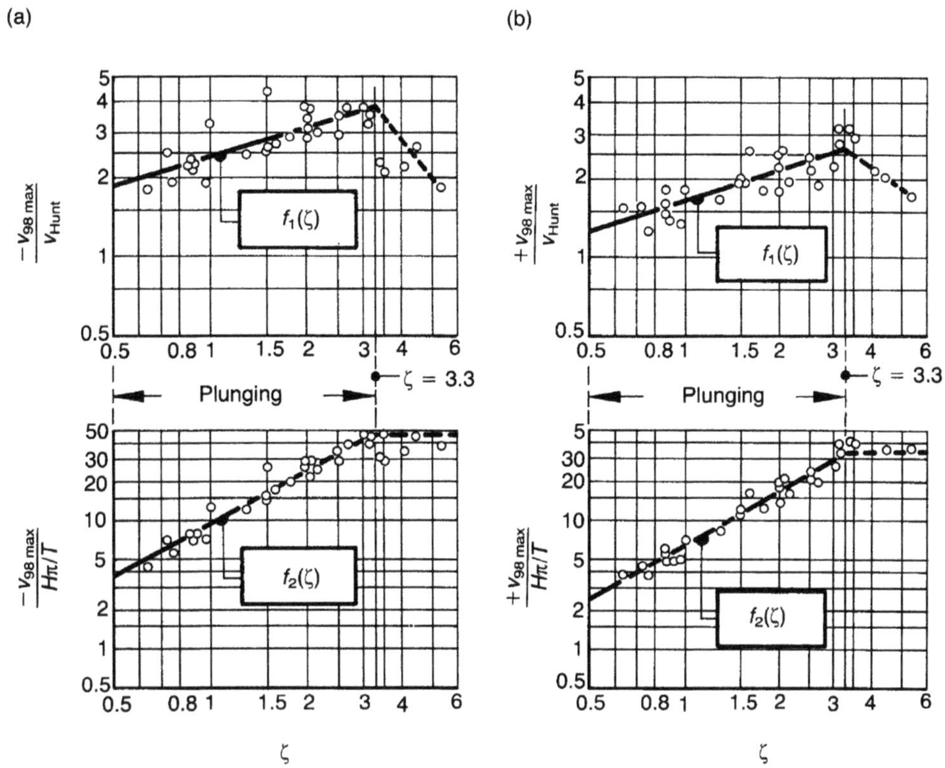

Figure 27.14 Regular waves in full-scale, 1:6 slope (smooth). Dimensionless descriptions of (a) up-rush ($+v_{98\max}$) and (b) down-rush ($-v_{98\max}$) velocities versus breaker number ξ.

with reduction of the wave run-up heights z, a reduction of velocities is also produced by increasing the roughness of the slope. For grass and block roughnesses, (Figures 27.10 and 27.11) a higher reduction was observed for the downrush than for the up-rush velocities both for regular waves and for wave spectra (Führböter and Witte, 1989).

27.7 COMPOSITE SLOPES

Composite slopes consist of two or more slope angles, sometimes combined with berm as well as with vertical walls. Because of the numerous possible combinations of the slope angles and their transition levels it is difficult to describe the wave attack on such structures in a simple generalized form. A comprehensive summary of results from research relating to forces on composite slopes can be found in CERC (1984).

However, for most cases of composite slopes the slope angle at the level of the highest storm surge is of special interest because here the highest wave impact loads, wave run-up heights and velocities are to be expected (see section 27.2). For a composite slope with two angles α corresponding to $\tan\alpha = 1{:}m$ for the lower and $1{:}n$ for the upper slope (Figure 27.15), experience shows that all wave loads approach asymptotical towards the values on a uniform slope with $\tan\alpha = 1{:}n$ with increasing vertical distance, a, between the still water level (point A in Figure 27.15) and the transition of the slope angles (point B in Figure 27.15). As an approximation the asymptotical conditions are reached when the still water level is more than the incident wave height, H, above the slope change. For the design of complex composite slopes, the use of model tests is recommendated.

27.8 DAMAGE BY WAVE LOADS

Independent of the type of dike cover (concrete or asphalt-concrete, artificial stones or pavements of natural stones or grass) the damage usually starts very near to the highest still-water level of the storm

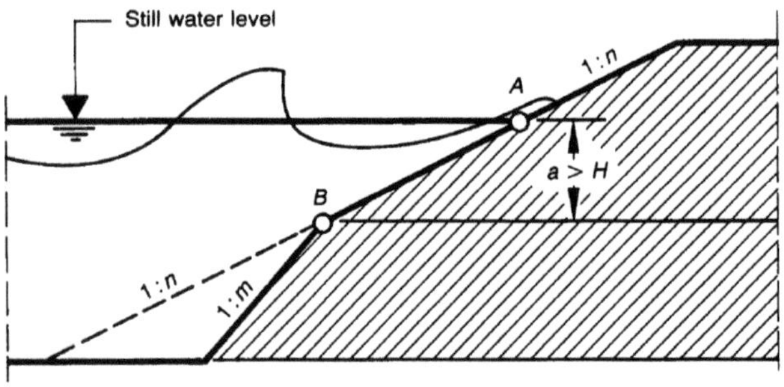

Figure 27.15 Wave loads on composite slope.

surge. One reason is that the highest waves are combined with the maximum height of a storm surge. Another reason is that the duration of stage (= the time the water level is between two given levels) is also higher at the maximum of a surge. The main reason for the damage near the still-water level, however, is that wave impacts and wave velocities are concentrated near the still-water level. It was shown in section 27.3, that the location of highest impact pressure lies about $0.5H$ below the still-water level and the highest up-rush and down-rush velocities are present at about the still-water level (section 27.6). Thus, a concentration of wave loading occurs over a relatively small area of the slope around the still water level.

Damage is caused by the wave impact pressures which penetrate the water-filled fissures, cracks and joints. The peak pressure propagates with the velocity of sound into the deeper layers of the cover causing effects like those from explosives. The high uplift forces can eject individual stones out of a stone cover. Water-filled fissures due for instance to dry seasons on grass dikes can be the source of destruction. On roughness elements, such as blocks, high concentration of impact energy is possible. Therefore, such elements should not be mounted below the maximum still-water level.

Even small discontinuties can lead to high pressure forces, together with impact and inertia effects during the passing of the up-rush and down-rush. For grass covered earth dikes, the high shear stress of up-rush and down-rush can cause erosion, again especially around the still water level where the maxima of these velocities are to be found. The maximum danger zone on a slope is around the highest still-water level because of the interaction of impact and velocities.

Closed and completely impermeable dike covers, such as concrete or asphalt-concrete, can be endangered by the presence of a high ground water level area below the cover, causing uplift. On the other hand cavities under such covers, caused by wash-outs or settlement, may reduce support and lead to damage by wave impacts.

The interaction of impact forces and high velocities progressively transforms the damage on the outer slope into a cliff which moves towards the crest of the dike.

The destruction by overtopping, however, begins at the inner slope by the accelerated flow down the slope which may be weakened by seepage. With special surface protection, overtopping may not be dangerous, although for earth dikes with grass cover the danger by erosion effects is very high. It was found that grass dikes with inner slopes of 1:3 or flatter can resist overtopping for some time. The best solution, however, is to avoid overtopping if possible.

Storm surges and wave loads are governed by the laws of probability. This, the question of the safety of a dike is a problem of probability (Losada and Gimenez-Curto, 1981). From the physical processes of wave loads on dikes, a general observation is that all forces and loads – impacts, wave run-up heights or up-rush and down-rush velocities – are strongly decreasing with decreasing slope angle or $\tan\alpha = 1{:}n$. Therefore, independent of dike cover, a flatter slope will always be safer.

REFERENCES

Bagnold, R.A. (1939) Interim report on wave pressure research. *Journal of the Institution of Civil Engineers*, **12**.

Battjes, J.A. (1971) Run-up distributions of waves breaking on slopes. *Proceedings ASCE*, **97** (WW1).

Battjes, J.A. (1974) Surf similarity. *Proceedings of the 14th Coastal Engineering Conference*, Copenhagen, Denmark.

Battjes, J.A. and Roos, A. (1975) Characteristics of flow in run-up of periodic waves, Delft University of Technology. *Rep.* 75-3.

Bruun, P. (1985) *Design and construction of mounds for breakwaters and coastal protection*, Elsevier, Amsterdam.

Bruun, P. and Grünbak, A.R. (1977) Stability of sloping structures in relation to $\xi = \tan\alpha / \sqrt{H/L_0}$ risk criteria in design, *Coastal Engineering*, **1** (4).

Bruun, P. and Johannesson, P. (1977) Parameters affecting stability of rubble mounds, Closure, *Journal of Waterways, Port, Coastal and Ocean Division, ASCE*, **103** (WW4).

CERC (1984) *Shore Protection Manual*, Vol. I–III, (1984 edn) Coastal Engineering Research Center, Vicksburg.

Denny, D.F. (1951) Further experiments on waves pressures. *Journal of the Institution of Civil Engineers*, **35**.

Freemann, J.C. and le Méhauté, B. (1964) Wave breakers on a beach and surges on a dry bed. *Proceedings of ASCE*, **90** (HY2).

Führböter, A. (1966) Der Druckschlag durch Brecher auf Deichböschungen, Mitt. des Franzius-Institutes der TU Hannover, Heft 28.

Führböter, A. (1970) Air entrainment and energy dissipation in breakers. *Proceedings of the 12th Coastal Engineering Conference*, Vol. I, Chapter 24, Washington.

Führböter, A. (1986) Model and prototype tests for wave impact and run-up an a uniform 1:4 slope. *Coastal Engineering*, **10**.

Führböter, A. and Sparboom, U. (1988) Full scale wave attack on uniformly sloping sea dykes, *Proceedings of the 21st Conference on Coastal Engineering*, Malaga, Spain.

Führböter, A., Sparboom, U. and Witte, H.-H. (1989) Großer Wellenkanal Hannover: Versuchsergebnisse über den Wellenauflauf auf glatten und rauhen Deichböschungen mit der Neigung 1:6, Die Küste, Heft 50.

Führböter, A. and Witte, H.-H. (1989) Strömungsgeschwindigkeiten beim Wellenauflauf auf einer Böschnung der Neigung 1:n = 1:6, *Jahrbuch der Hafenbautechnischen Gesellschaft, 44. Band*, Springer Verlag, Berlin.

Galvin, C.J. (1968) Breaker type classification on three laboratory beaches. *Journal of Geophsical Research*, **73** (12).

Grüne, J. (1988) Anatomy of shock pressures induced by real sea state breaking waves. International Symposium on Modelling Soil Water Interactions (SOWAS), Delft.

Greslou, L. and Montaz, J.-P. (1962) Mesure des pressions sur une dique á talus. *Bulletin de l'Association Internationale de Navigation*, Vol. III.

Hensen, W. (1954) Modellversuche über den Wellenauflauf an Seedeichen im Wattengebiet, Mitt. des Franzius-Institutes der TU Hannover, Heft 5.

Hunt, I.A. (1959) Design of seawalls and breakwaters, *Proceedings of ASCE*, **85** (WW3).

Irribarren, C.R. and Nogales, C. (1949) Protection des ports. Section II, Comm., 4, XVIIth Int. Nav. Congress, Lisbon.

v.Karman, Th. (1929) The impact on seaplanes during landing. National Advisory Comitee for Aeronautics, TN 321, Washington.

Kishi, T. (1959) The possible highest gravity waves in shallow water. *Coastal Engineering in Japan*, **2**.

Losada, M.A. and Gimenez-Curto, L.A. (1981) An approximation to the failure probability of maritime structures under a sea state. *Coastal Engineering*, **5**.

Lundgren, H. (1969) Wave shock forces. Proceedings of Research on Wave Actions, Delft.

Mason, M.A. (1951) Some observations of breaking waves, in: Gravity waves. Proceedings of the NBS Semicentennial Symposium on Gravity Waves.

v.Oorschoot, J.H. and d'Angremond, K. (1968) The effect of wave energy spectra on run-up. *Proceedings of the 11th Conference on Coastal Engineering*, London, UK.

Ramkema, C. (1978) A model law for wave impacts on coastal structures, *Proceedings of the 16th Conference on Coastal Engineering*, Hamburg, Germany.

Savage, R.P. (1958) Wave run-up on roughened and permeable slopes, *Proceedings of ASCE*, **84** (WW3).

Savage, R.P. (1959) Laboratory data on wave run-up on rough and inpermeable slopes, B.E.B., *Techn. Mem. 109*.

Saville, T. (1955) Laboratory data on wave run-up and overtopping on shore structures. Beach Erosion Board, *Technical Memorandum 64*.

Saville, T. (1956) Wave run-up on shore structures. *Proceedings of ASCE*, **82** (WW2).

Saville, T. (1958) Wave run-up on composite slopes. *Proceedings of the 6th Conference on Coastal Engineering, ASCE*, Council of wave research.

Schijf, J.B. (Chairman) (1974) Wave run-up and overtopping, Technical Advisory Committee on Protection against Inundation, Government Publishing Office, The Hague.

Shen, M.C. and Meyer, R.E. (1963) Climb of a bore on a beach, Part 3. *Journal of Fluid Mechanics*, **16**.

Siefert, W. (1974) Über den Seegang in Flachwassergebieten, Mitt. des Leichtweiß-Institutes der TU Braunschweig, Heft 40.

Stive, R.J.H. (1984) Wave impact on uniform steep slopes at approximately prototype scale. IAHR Symposium on scale effects in modelling hydraulic structures, Esslingen.

Vinjé, J.J. (1972) Investigations for coastal protection with concrete mats. North Sea Coastal Engineering Group.

van Vledder, G. Ph. (1990) Literature survey to wave impacts on dike slopes, Technische Adviescommissie voor de Waterkeringen, A4, Delft Hydraulics, *Report H 976*.

Wassing, F. (1957) Model investigations of wave run-up carried out in the Netherlands during the last twenty years. *Proceedings of the 6th Conference on Coastal Engineering*, Gainesville, Florida, USA.

Waterloopkundig Laboratorium, Delft Hydraulics Laboratory (1987) Golfoverslag bij dijken, verslag modelonderzoek, H 181/M 1258, Juli 1987.

Witte, H.-H. (1988) Druckschlagbelastung durch Wellen in deterministischer und stochastischer Betrachtung, Mitt. des Leichtweiß-Institutes der TU Braunschweig, Heft 102.

WAVE FORCES ON STRUCTURES 28

R.G. Tickell

One of the most important aspects in the design of a coastal structure is the determination of the wave forces to which it is to be subjected during its lifetime. The establishment of extreme events for particular return periods is central to the design process and must include such environmental parameters as winds, waves, tides and surges. However, this chapter concentrates on the determination of wave forces, including the derivation and application of Morison's equation, and also outlines the principal characteristics of wave loading on a number of different systems. It ends with a review of particular methods that coastal engineers have developed to estimate the wave loads on vertical walls (see also Chapters 27, 29 and 34).

CONTENTS

28.1 INTRODUCTION

28.1.1 STRUCTURAL SYSTEMS AND DESIGN APPROACH

One of the first features to strike an engineer when designing a marine structure is the great importance of the horizontal components of environmental forces when compared with a land-based structure. Waves, in combination with appropriate currents, usually contribute the major part of such forces. This is a large topic and the present discussion only provides a description of the principal load mechanisms, together with a route map for the engineer to find the necessary further detail in each case.

Table 28.1 summarizes the principal types of structure and forces to be considered. Rubble-mound structures are given detailed treatment in Chapter 29 and are not considered here. The many aspects of dike and sea-wall design are covered in Chapters 27 and 28. The present discussion will only comment on the wave forces acting on vertical walls. Most weight is given to loading on slender cylindrical members since other topics are addressed elsewhere in this volume.

Usually the design process will be governed by some code of practice or standard but no particular code is adopted here. However, it is important to appreciate that load factors or margins of safety may be applied at different stages of the design by the various code drafting committees. It is stressed that mixing codes can be dangerous.

28.1.2 DEVELOPMENT OF THE ENVIRONMENTAL CONDITIONS

Although a design method may use the concept of a set of long-crested regular (deterministic) waves, in reality storm waves are random and short-crested. The intensity, frequency content and directional distribution of the wave field varies over the lifetime of the structure (Ochi, 1982), and these variations must be combined with tidal, surge and set-up effects on water depths and with current and wind fields to predict the extreme design conditions and the fatigue effects of the many, less severe conditions acting on the structure. In addition, the transition from the relatively simple deep water conditions to the site of a coastal structure requires the modification of the wave field to allow for shoaling, depth and current refraction, diffraction, reflection, frictional losses or secondary wave generation and wave breaking. These matters are addressed at the appropriate points elsewhere in this text, while Sarpkaya and Isaacson (1981)

Coastal, Estuarial and Harbour Engineer's Reference Book. Edited by M.B. Abbott and W.A. Price. Published in 1993 by Chapman & Hall, 2–6 Boundary Row, London SE1 8HN. ISBN 0 419 15430 2.

Table 28.1 Structural types and appropriate force mechanisms

Structural type	*Load mechanisms*	*Additional comments*
1. Piled quays, dolphins and oil/gas platforms composed on slender cylindrical members	Morison's equation for force in predominant wave direction. Possible lift mechanism transverse to predominant flow direction. Vertical forces due to buoyancy and wave impact (slam) forces in the free surface zone.	Morison and lift forces are affected by inclination of members, dynamic response, surface roughness and the presence of currents. Special treatment needed for sea-bed pipelines.
2. Breakwaters and sea-walls, ranging from vertical walls to rubble-mounds via composite structures	Expressions for wave pressures, forces and moments for breaking and non-breaking wave action on walls. Special methods for designing rubble structures. Also buoyancy and uplift forces to be considered.	Extensions for oblique wave attack and overtopping of structures. Special treatment needed for broken wave bores on dikes or revetments.
3. Large structures other than in 2, above, such as gravity-type oil platforms and floating systems	First order wave loading from wave diffraction analysis, leading to second order, low frequency excitation of floating systems. Also buoyancy and uplift forces to be considered.	Extensions to deal with local loading due to wave impact, etc. Special treatment for nonlinear moorings on floating vessels.

is a useful reference for both deterministic and random wave theories.

For the present purposes of wave load estimation, it will be assumed that the variation in surface elevation (η) is given by a series of appropriate regular waves with heights, H, periods, T and directions θ, together with a range of water depths (d) and current strengths and directions (U, ϕ). Alternatively, if a random wave approach is adopted, some form of spectral density function of surface elevation ($S_\eta(\omega)$ or $S_\eta(\omega, \theta)$) is given where, by definition, the variance of η is given by

$$\sigma_\eta^2 = \int_0^\infty S_\eta(\omega)d\omega \tag{28.1}$$

or

$$\sigma_\eta^2 = \int_0^\infty \int_0^{2\pi} S_\eta(\omega, \theta)d\theta\, d\omega \tag{28.2}$$

and the use of linear random wave theory results in η and the water particle velocities and accelerations being random processes with appropriate Gaussian probability distributions.

28.2 MORISON'S EQUATION

Now consider the problem of the force acting on the slender, surface-piercing, vertical cylinder shown in Figure 28.1, when subject to a long-crested regular wave. In the direction of wave travel, the horizontal force per unit length (F) is given by Morison's equation as

$$F = C_d(\rho/2)A\, u|u| + C_m \rho V\, du/dt \tag{28.3}$$

$$= K_d\, u|u| + K_i\, du/dt \tag{28.4}$$

where u and du/dt are the horizontal components of water particle velocity and acceleration. The projected surface area, A, and displaced volume V per unit length become D and $\pi D^2/4$, respectively, for the particular case of a circular cylinder, of diameter D, considered by Morison *et al.* (1950). The density of sea-water is ρ. The first term in equations (28.3) or (28.4) is the drag component of F and is a non-linear function of u while the second term is the inertia component of F and is a linear function of du/dt. Finally C_d and C_m are the much discussed Morison drag and inertia (mass) coefficients, respectively.

The term slender cylinder implies that the incident wave field is not strongly disturbed by the presence of such members and the usual criterion for this is that the diameter to wavelength ratio, D/L, should be less than 0.2, otherwise diffraction effects become important, see section 28.4.

28.2.1 DERIVATION OF MORISON'S EQUATION

A simplistic explanation of Morison's equation starts with the well known form of mean drag force, F_d, in a steady uniform velocity, u, such that,

$$F_d = C_d(\rho/2)Du^2 \tag{28.5}$$

where it may be shown that the mean drag coefficient C_d is a function of the Reynolds number Re and the cylinder surface roughness ratio (k^*/D) with:

$$\mathrm{Re} = uD/\nu \tag{28.6}$$

and ν is the kinematic viscosity of sea-water. In the present context k^* is some average measure of the

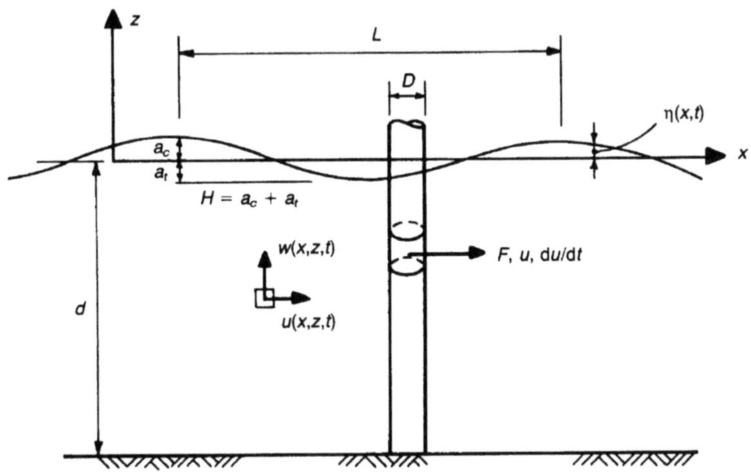

Figure 28.1 Vertical cylinder under wave loading.

roughness of the cylinder, e.g. root mean square height of roughness.

The simple form of Equation (28.6) belies the complex hydrodynamics of flow around the cylinder in which boundary layers develop and may eventually separate from the cylinder wall, forming areas of concentrated vorticity (vortices) which are shed into the wake of the cylinder, see Figure 28.2a. As individual vortices are shed, the fluid pressure distribution around the cylinder varies, producing what is usually a small dynamic component in the drag force and also a significant dynamic force (F_ℓ) in the direction transverse to the flow. The problem of lift forces is discussed in the next section.

The inertia term (F_i) in Morison's equation arises from the wave-induced fluid acceleration field. To resist the uniform acceleration of the body of ideal fluid (no viscosity/vorticity), shown in Figure 28.2b, a force would be needed of $\rho(\pi D^2/4) \cdot du/dt$ per unit length. If the body of fluid is now replaced by the cylinder, then the streamlines shown in Figure 28.2c imply an added set of pressures acting on the cylinder, due to the disturbance of the flow, which would have a resultant force represented in the form $C_a\rho$ $(\pi D^2/4) \cdot du/dt$. So the total inertia force is,

$$F_i = (1 + C_a)\rho(\pi D^2/4) \cdot du/dt \qquad (28.7)$$

$$= C_m\rho(\pi D^2/4) \cdot du/dt \qquad (28.8)$$

In the general case, the first term is called the Froude–Krylov force and the second term is often called the added mass component, where C_a is the added mass coefficient.

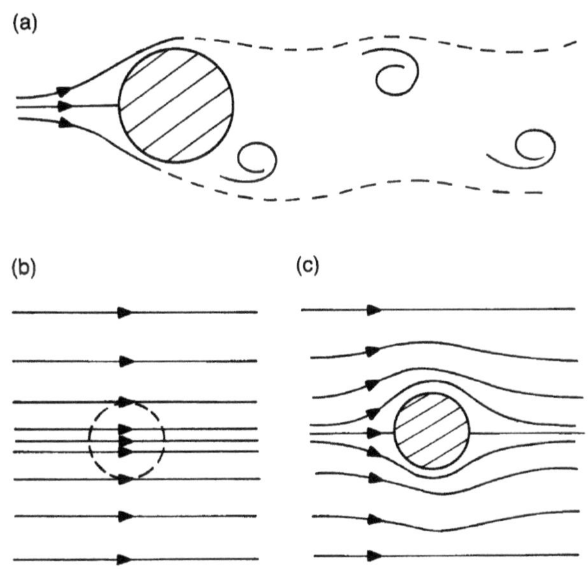

Figure 28.2 Development of Morison's equation. (a) generation of vortices in cylinder wake; (b) uniformly accelerating flow; (c) disturbance due to cylinder.

The drag and inertia terms can be combined to give Morison's equation for a circular cylinder in waves as

$$F = C_d(\rho/2)D\,u|u| + C_m\,\rho(\pi D^2/4) \cdot du/dt \quad (28.9)$$

where u^2 in the steady drag expression is replaced by

$u|u|$ to deal with flow reversals. While $C_a = 1$ and $C_m = 2$ for an ideal fluid, in reality the presence of the boundary layers and wake for the cylinder complicate the situation. It has been shown that the choice of C_d, C_a or C_m is now dependent on the Keulegan–Carpenter number (K_c) as well as Re and k^*/D. K_c reflects the movement (orbital path) of fluid particles relative to the characteristic size of the cylinder and for a regular wave with period (T), it is commonly taken as

$$K_c = u_m T/D \tag{28.10}$$

where u_m is the amplitude of the particle velocity. Low values of K_c, say $K_c < 5$, correspond to the dominance of the inertia component in equation (28.9) while $K_c > 30$ indicates drag dominance. A significant proportion of the applications for Morison's equation lie in the intermediate range $5 < K_c < 30$.

28.2.2 SELECTION OF COEFFICIENTS AND APPLICATION OF MORISON'S EQUATION

Many laboratory and full-scale monitoring exercises have been carried out to identify suitable ranges of Morison coefficients, e.g. Bearman *et al.* (1985) in large-scale wave tank tests and Tickell and Bishop (1985) and Bishop (1987) at the Christchurch Bay Tower test facility in the UK. Coefficient selection is reviewed in detail by Sarpkaya and Isaacson (1981) and more recently by Barltrop *et al.* (1990), where the characteristics and quality of the various major monitoring exercises are discussed. Many of the data sets show significant scatter in the values of C_m and C_d. However, at prototype scales and for the so-called post-critical conditions (i.e. $\text{Re} > 5 \times 10^5$), the range of coefficients used in design for smooth circular cylinders ($k^*/D < 1/5000$) is typically $1.7 < C_m < 2.0$ and $0.6 < C_d < 0.8$. While for heavily marine-fouled cylinders, with $k^*/D > 1/50$, the drag coefficient range increases to $1.0 < C_d < 1.5$ and C_m tends to reduce to about 1.5 for $K_c > 10$. Baltrop also gives guidance on coefficients for non-circular cross-section members, which may be met in some piled structures etc.

Study of the data referenced above will frequently show larger or smaller values of the coefficients derived from measurements over a short length of force transducer sleeve. However, the values used in design take into account, often in an unspecified way, the spatial correlation of loads on the multitude of members, at various orientations, which constitute the total structure.

Where simple structures, involving a few vertical members are concerned, use may be made of the design charts given in the *Shore Protection Manual* (1984); which are based on Dean's stream function wave theory and give the maximum overturning moments and shears acting at the sea-bed. Otherwise numerical integration of the Morison force over each member or 'lumped' equivalent areas and volumes must be used to develop the required nodal forces for a standard, finite element, stiffness analysis of the structure. In carrying out these lengthy computations it is necessary to consider various extensions to the basic Morison force developed in this section.

28.3 EXTENSIONS TO MORISON'S EQUATION

28.3.1 COMBINED WAVES AND CURRENTS

The presence of a current requires, as a minimum, that the combined wave and current velocities are used in estimating the drag force since it is a non-linear function of u and linear superposition of the separate drag forces due to waves and currents would be incorrect. In fact, the problem is more complex as consideration must be given to treating waves and currents which are not acting in the same horizontal direction and to the way in which the two velocity components are combined in the free surface zone. The hydrodynamics of wave–current interaction has been reviewed by Hedges (1987) and the estimation of combined forces is discussed by Baltrop *et al.* (1990), including the changes necessary to the Morison C_m and C_d coefficients in the presence of a current.

28.3.2 MORISON FORCES ON INCLINED MEMBERS

So far attention has been concentrated on the horizontal loads on vertical cylinders. In the more general case horizontal and vertical components of velocity and acceleration must be resolved into the plane normal to the cylinder axis and the inertia and drag forces calculated along each member. For slender members it is usual to neglect the forces acting parallel to the cylinder axis. Borgman (1958) or Sarpkaya and Isaacson (1981) give procedures to generalize Morison's equation for an arbitrary orientation of the member in three-dimensional flow. There is less data for the selection of C_m and C_d appropriate to inclined members and the so-called cross-flow principle is doubtful for narrow angles of incidence between the particle kinematic vectors and the member axis.

28.3.3 DYNAMICALLY RESPONSIVE CYLINDERS

Equation (28.3) refers to force on a rigid cylinder and requires modification when the cylinder responds dynamically such that the structural velocities and accelerations are significant in comparison with the particle kinematics. If the structural displacement in the direction of u is r and the relative velocity between the cylinder and fluid is denoted by b where $b = u\text{-}dr/dt$ then equation (28.3) is rewritten as

$$F = C_d(\rho/2)A\,b|b| + \rho V\,du/dt + C_a\,\rho V\,db/dt \quad (28.11)$$

where the Froude–Krylov term is unaltered and only the added inertia component depends on the relative acceleration. The first term in equation (28.11) implies the presence of non-linear hydrodynamic damping which can be particularly important when the structure is excited at or near to its natural frequencies since dynamic amplification of stresses may be severe in these resonant situations, see Eatock Taylor and Rajagopalan (1982). Further discussion of wave loading and dynamic structural analysis will be found in Sarpkaya and Isaacson (1981) or Blevins (1990).

28.3.4 MORISON FORCES DUE TO RANDOM WAVES

So far in this chapter it has been implied that the particle kinematics are given by some regular wave theory. However, if it is necessary to specify random waves with spectra such as those described in Equations (28.1) or (28.2) then linear random wave theory must be used to describe the probabilistic properties of the particle kinematics. The probability density function of the Morison force given in equation (28.4), for a long-crested random sea, becomes

$$p(F) = B \cdot \int_{-\infty}^{\infty} \exp[-0.5[a^2/\sigma_a^2 + (u - \bar{u})^2/\sigma_u^2]]du \quad (28.12)$$

where $a = du/dt = (F - K_d\,u|u|)/K_i$, $\bar{u}$ is the steady current in the direction of F and $B = 1/(2\pi\,\sigma_a\,\sigma_u\,K_i)$. As discussed by Tickell (1977), $p(F)$ is not a simple Gaussian distribution. It can be shown that the Morison forces and resultant structural stresses need careful treatment when the non-linear drag forces are a significant proportion of the total load. Nevertheless knowledge of the wave spectrum ($S_\eta(\omega)$) and the application of linear random wave theory enables the necessary parameters, such as the standard deviations of particle kinematics (σ_u, σ_a), to be found as part of the probabilistic analysis of wave loading and response, see Barltrop *et al.* (1990).

28.3.5 WAVE SLAM

While not strictly an extension of Morison's equation, it is necessary to recognize that slender members can be subject to high loads due to wave breaking or, in the case of members in the free surface zone, waves slamming the members as they first become submerged. These load mechanisms can cause high local stresses or additional fatigue in member or joint welds. Barltrop *et al.* (1990) review these problems and possible models such as:

$$F = C_s(\rho/2)A\,v^2 \quad (28.13)$$

where v is a characteristic velocity of impact and C_s is an empirical slam or slap coefficient with a wide range of values ($2 < C_s < 20$), depending on conditions.

At this point the designer is reminded that the simple case of buoyancy forces can be particularly important for members which are intermittently submerged and the cycling of such vertical forces can make a significant contribution to fatigue damage of the structure.

28.3.6 LIFT FORCES

The movement of the wake around a circular cylinder due to vortex shedding was described in section 28.2. The designer must always consider the possible significance of fluctuating lift forces (F_ℓ), generated transverse to the flow, which may arise from the redistribution of normal pressures on the cylinder as vortices form and are shed. This is particularly important in the case of slender and dynamically responsive members.

The frequency with which pairs of vortices (eddies) are shed in steady flow (a current) is given by

$$f_\ell = Su/D \quad (28.14)$$

where S is the dimensionless Strouhal number, D is the member diameter and u is the fluid velocity.

In steady flow S lies in the range $0.2 < S < 0.3$ depending on Re and to a lesser extent, the surface roughness of the cylinder and the free stream turbulence (Blevins, 1990). Even in steady flow the cycles of force are irregular and lift coefficients (C_ℓ) are defined in terms of some average, e.g. the root mean square value,

$$C_{\ell\text{rms}} = 2F_{\text{rms}}/(\rho D u^2) \quad (28.15)$$

for a circular cylinder. The coefficient is typically in the range $0.2 < C_{\ell\text{rms}} < 1.2$, depending on the conditions.

The case of lift forces in waves or waves and currents is somewhat more difficult than steady flow. For example, there must be time during one half of a wave period for vortices to form and be shed, before the flow reverses, if lift forces are to be generated (i.e. $1/f_\ell < T/2$, where T is the wave period). Using steady flow results this would indicate that the Keulegan–Carpenter number ($K_c = uT/D$) should be greater than 5–10, depending on S. As K_c increases, the time available for shedding vortices also increases and more than one cycle of F_ℓ may occur. In practice, it is found that lift frequencies tend to increase as integer multiples of the wave frequency, as K_c increases. The situation is complex since interaction may occur between vortices shed in different wave cycles (Sarpkaya and Isaacson, 1981).

Various oscillatory forms of Equation (28.15) have been investigated, (Barltrop *et al.*, 1990) and one of the more successful models, from the analysis of planar oscillatory flow experiments, is discussed by Bearman *et al.* (1984).

Work on a rigid vertical cylinder at realistic Re values (Bearman *et al.*, 1985) indicates that significant lift forces are not likely for $K_c < 5$ and the strength of lift forces reaches a maximum for $10 < K_c < 15$ and then decreases for higher K_c conditions. These measurements gave $C_{\ell rms}$ values peaking to 1.4 at $K_c \simeq 10$ with values around 0.8 for $15 < K_c < 20$.

It is commonly found that the spanwise correlation of F_ℓ is weak for rigid members (2–3D), particularly in wave action and the total member loading may not be significant. However, this can change radically when the member responds dynamically to lift excitation. Transverse vibration at or near the vortex shedding frequency tends to increase the strength of the vortices shed and to organize the shedding so that the spanwise correlation is significantly increased. Amplitudes of motion can become large (0.2–2.0D) and f_ℓ can be captured by the natural frequency (f_n) of the member (lock-on). Several studies (Blevins, 1990) have shown that large transverse motions may occur when $3.5 < (u/f_nD) < 8$ and smaller in-line oscillations may occur when $1.5 < (u/f_nD) < 3.5$. The mass and damping properties of the member are important in dictating the magnitude of response.

Many studies correlate the dynamic response with the dimensionless damping parameter $2\bar{m}\delta/(\rho D^2)$, where $\bar{m}$ is the effective mass per unit length (dependent on vibrational mode shape as well as actual mass distribution) and δ is the logarithmic decrement of damping. Thus care must be taken with the design of slender cylindrical members of frames, individual piles, risers and conductors or moorings. The range of reduced velocity (u/f_nD), K_c and the damping parameter ($2\bar{m}\delta/(\rho D^2)$) should be checked (see Barltrop (1990)), and specialist advice taken where dynamic response due to vortex excitation is suspected.

28.3.7 INTERFERENCE EFFECTS IN WAVE LOADING ON CYLINDERS

The complexity of forces on just a single cylinder, discussed in previous sections, indicates that loading on groups of cylinders or members near free or fixed boundaries can be even more difficult to estimate accurately. When two or more cylinders are within say five diameters of each other significant changes can occur to the individual loads and the total load experienced in waves and current compared with the corresponding forces for an isolated cylinder. The pattern of cylinders and the relative angle of approach of the waves can have a large effect on the forces experienced (Zdravkovich, 1977) and the reviews of other experimental studies given by Sarpkaya and Isaacson (1981) and Barltrop *et al.* (1990). Where spacings are not too close, it may be possible to justify a simplified approach by designing each individual member of the group to withstand say 1.5 of the isolated cylinder load while taking say 1.2 of total force for the group of members. Special care is needed to account for the interference or group effects when dealing with dynamically responsive members.

The presence of sacrificial anodes, service ducts or other fixtures on the structural members can also lead to enhanced wave forces and it is not only important to design the main structural members accordingly but also to account for the increased fluid loading on the fixtures and fixing brackets produced by the presence of the main member.

Finally an important interference problem arises for pipelines at or near to the sea-bed. An asymmetry of flows is caused (see Figure 28.3) and the situation is complicated by the effect of the high velocity jet which occurs beneath the cylinder for non-zero clearances ($e \neq 0$) and the effect of the boundary layer existing at the sea-bed even in the absence of the pipeline. Drag, inertia and transverse forces may be estimated using Morison's equation and some of the other concepts described in this and the previous section. The transverse or lift force is not only caused by the vortex shedding mechanism described earlier but would be present even in ideal fluid flow since the asymmetry of the streamlines implies an asymmetry in the normal pressures acting on the pipeline cross-section.

The drag, inertia and lift coefficients are functions of the clearance parameter (e/D) and the bed boundary layer in addition to Re, K_c and k^*/D. Field measurements of force coefficients have been reported by Grace *et al.* (1976, 1979a,b) and Sarpkaya (1977) gives

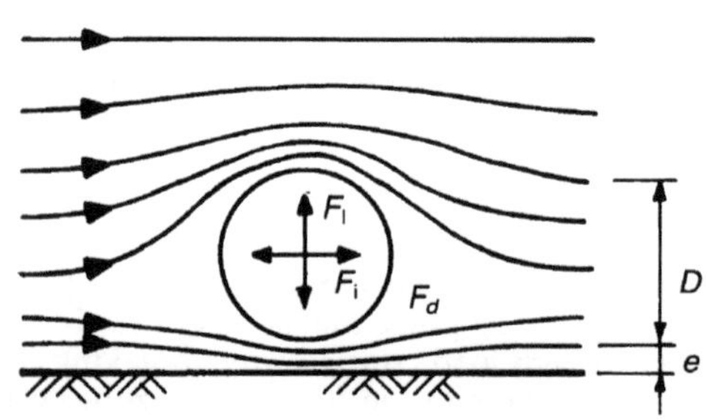

Figure 28.3 Loading on cylinder near a wall or sea-bed.

results at appropriate K_c and Re values from laboratory tests of a cylinder near a plane wall. In common with comments made at the close of the previous section, pipeline design and stability checks are a specialist topic.

28.4 DIFFRACTION ANALYSIS OF WAVE LOADING

Morison's equation and its various extensions deals with wave loading on slender members which do not generate a significant scattered or diffracted wave field. The theory of diffraction wave loading is used when this is not a valid assumption. It is generally considered that diffraction effects become important when the ratio of the structure size to wavelength D/L exceeds 0.2. The theory is based on irrotational flow such that the incident and diffracted wave fields can be described by velocity potential functions and it is widely applied to concrete or masonry gravity structures and to ship forms or other floating systems with large D/L ratios. While simple problems can be solved analytically (see Sarpkaya and Isaacson (1981)), arbitrary geometry usually requires a wave source or finite element numerical approach. Computer packages for linear diffraction analysis are available commercially. The present discussion can only cover the principles of the approach and details can be found in the references given.

28.4.1 DIFFRACTION ANALYSIS APPLIED TO FIXED STRUCTURES

Consider the body shown in Figure 28.4 subject to a specified single train of linear waves with frequency ω, where the velocity potential of the waves is,

$$\phi_w'(x, y, z, t) = \phi_w(x, y, z) \cdot \exp(-i\omega t) \quad (28.16)$$

and let the body generate a scattered or diffracted wave field described by,

$$\phi_s'(x, y, z, t) = \phi_s(x, y, z) \exp(-i\omega t) \quad (28.17)$$

where i is $\sqrt{-1}$. The temporal term $\exp(-i\omega t)$ is common to Equations (28.16) and (28.17) and only the spatial variation of the scattered wave field ($\phi_s(x, y, z)$) is unknown.

Now the total potential ($\phi = \phi_w + \phi_s$) must satisfy Laplace's equation subject to all the usual boundary conditions for wave motion. For example, that the horizontal sea-bed is impermeable ($w = \partial\phi/\partial z = 0$) and that the dynamic and kinematic free surface boundary conditions can be combined and linearized at MWL ($\partial^2\phi/\partial t^2 + g\,\partial\phi/\partial z = 0$ at $z = 0$). In addition, the scattered wave field must satisfy a radiation condition which ensures that the potential decays as the waves propagate from the body (see Garrison (1978)). Finally the body is assumed impermeable so that, at any point on the structure, the component of velocity normal to the surface due to the incident wave is

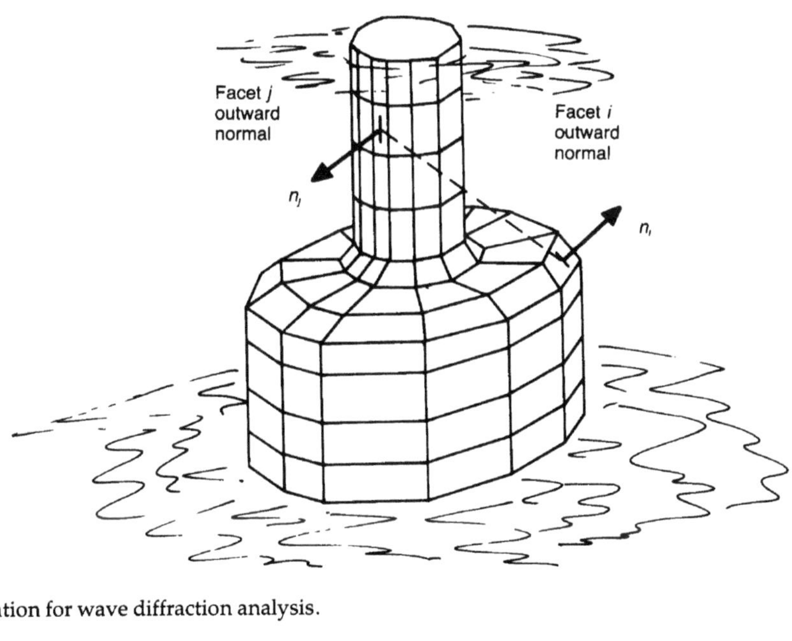

Figure 28.4 Idealization for wave diffraction analysis.

balanced by the corresponding normal velocity component of the diffracted wave (see Figure 28.4):

$$\partial\phi_w/\partial n = -\partial\phi_s/\partial n \qquad (28.18)$$

The numerical solution of the diffraction problem is obtained via the wave source method by dividing the surface of the structure into a series of facets and supposing that there is a distribution of wave sources of an unknown strength over each facet. The contribution to the outward normal, diffracted wave velocity at each facet due to the wave source at every facet may be calculated from the geometry of the structure and an appropriate influence or Green's function. Using the boundary condition given by equation (28.18), a matrix equation may be formulated and solved for the unknown wave source strengths on each facet (see Hogben and Standing (1974)). Once the facet strengths are known, the value of ϕ_s may be calculated at any point within the fluid, including on the body surface. Since the normal pressures acting on the body surface can be written as

$$p(x, y, z) = -\rho g z - \rho\partial(\phi_s + \phi_w)/\partial t \qquad (28.19)$$

the pressures and integrated forces and moments acting on the structure may be evaluated.

Clearly there are many practical considerations in the application of such a numerical method e.g. choice of facet size and geometry, numerical accuracy and stability of the algorithms etc. and the reader should consult the references given for details of these. However, once the solution of the linear diffraction problem is known for a set of unit amplitude waves of given frequencies and directions, it is a straightforward matter to calculate the random wave forces corresponding to a given surface elevation spectrum.

28.4.2 DIFFRACTION ANALYSIS APPLIED TO FLOATING STRUCTURES

The analysis of wave action on floating systems is a large topic in its own right involving the consideration of hydrostatic stability of the floating body, the response of the system at the frequencies of the waves and certain important non-linear effects which give rise to low-frequency or so-called slow-drift motions. Again, only the concepts of diffraction wave loading are introduced here.

Consider a ship form, such as shown in Figure 28.5, subject to the six rigid-body motions (r_j) where:

$$r_j = a_j \exp(-i\omega t) \qquad (28.20)$$

Each component of motion generates a wave field described by a velocity potential (ϕ_{rj}), which must satisfy Laplace's equation and all the boundary conditions discussed in the previous section. Now consider the system subject to an incident linear wave. To a first-order the diffracted potential (ϕ_s) arising from the incident wave potential (ϕ_w) may be found by treating the vessel as a fixed body divided into facets and solving for ϕ_s as described previously. In addition the six body motion potentials may be found

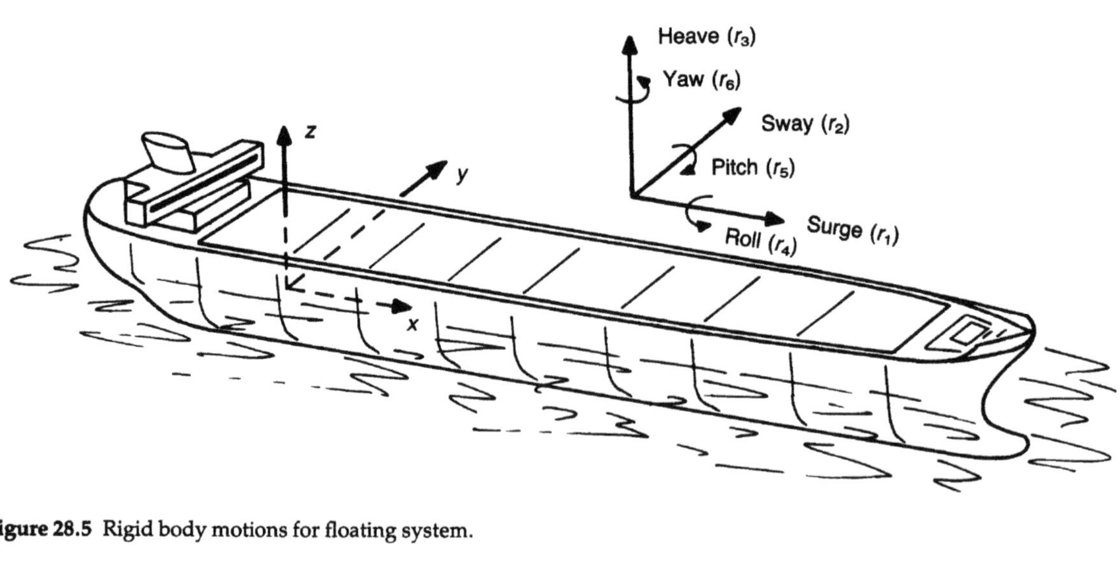

Figure 28.5 Rigid body motions for floating system.

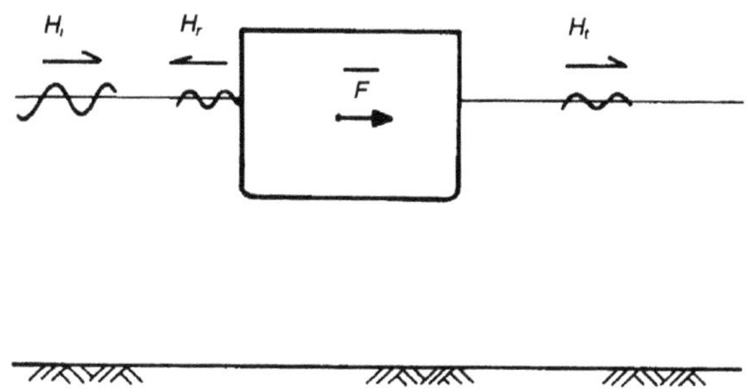

Figure 28.6 Second-order mean drift force.

by performing the diffraction analysis in still water, in which the surface boundary condition given by Equation (28.18) becomes

$$\partial\phi_{rj}/\partial n = v_j \tag{28.21}$$

where v_j is the outward normal velocity of the surface at each facet due to a unit amplitude motion of the form given in Equation (28.20). Integration of the pressures found from each ϕ_{rj} give the forces or moments corresponding to the rigid body motions and those components in phase with the body accelerations constitute the added mass terms for the vessel **A**, while those in phase with the velocities correspond to the wave or linear potential damping terms **C**, in matrix form. Thus the dynamics of the vessel at the wave frequency may be described by

$$(\mathbf{M} + \mathbf{A})\{d^2r/dt^2\} + \mathbf{C}\{dr/dt\} + \mathbf{K}\{r\} = \{F\} \tag{28.22}$$

where **M** is the mass matrix of the vessel in air, **K** is the restoring force matrix including hydrostatic effects and linear mooring springs and $\{F\}$ is the wave loading from the first stage of the diffraction analysis. It should be noted that **A** and **C** are functions of the frequency of excitation.

The solution of equation (28.22) enables frequency response functions or response amplitude operators (RAO) to be developed for the vessel motions and, as with the static diffraction analysis, random motion spectra may be simply calculated from the surface elevation spectra.

It has long been recognized that floating systems under wave action are subject to low frequency forces and motions with non-zero mean values (drift offsets). Longuet–Higgins (1977) demonstrates that, for the simple two-dimensional problem of a regular wave acting on the body shown in Figure 28.6, a mean force ($\bar{F}$) is generated such that

$$\bar{F} = \rho g[1 + 2kd/\sin h(2kd)](H_i^2 + H_r^2 - H_t^2)/16 \tag{28.23}$$

where H_i, H_r and H_t are the incident, reflected and transmitted waveheights, respectively, in a water depth of d, for a wave with a wave number of k. Equation (28.23) indicates the quadratic non-linearity of the drift force. When random waves containing many frequencies act on the vessel the quadratic force not only contains a steady component ($\bar{F}$) but also components at the sums and differences in frequency between the incident wave components. For example, if F is a function of η^2 then for a bichromatic wave field

$$\eta = a_1 \cos\omega_1 t + a_2 \cos\omega_2 t \tag{28.24}$$

and

$$\eta^2 = \underline{0.5(a_1^2 + a_2^2)} + 0.5a_1^2\cos 2\omega_1 t + 0.5a_2^2\cos 2\omega_2 t + 0.5a_1 a_2\cos(\omega_1+\omega_2)t + \underline{0.5a_1 a_2\cos(\omega_1-\omega_2)t} \tag{28.25}$$

The underlined terms correspond to the mean and low frequency excitations. There are a number of sources of such quadratic behaviour such as the force resulting from the first-order wave forces acting on a moving body and the force acting between MWL and the instantaneous free surface. These components have been fully discussed by Pinkster (1979) and almost all of them may be calculated using the linear potentials (ϕ_s, ϕ_{rj}, etc.) found earlier. The practical importance of such mean and slow drift forces lies in the design of vessel mooring or dynamic positioning systems since large responses at the natural frequencies of floating systems in surge, sway and yaw can be generated due to the low damping in these modes (Standing, 1988). There is a plethora of papers dealing with these problems in *Applied Ocean Research*, *Ocean Engineering* and recent preprints from the Offshore

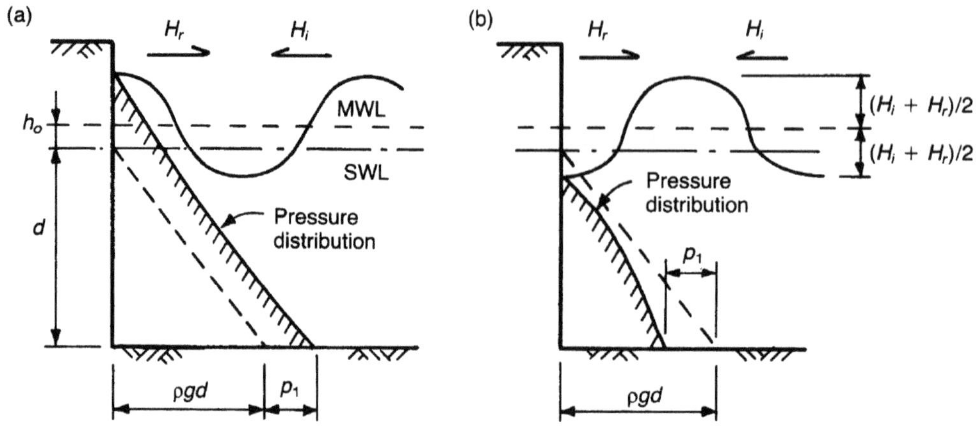

Figure 28.7 Non-breaking wave forces on a wall (a) crest condition; (b) trough condition (after *Shore Protection Manual*).

Technology Conferences (OTC) and the Offshore Mechanics and Arctic Engineering Symposium (OMAE), to which the reader is referred for detailed discussions.

28.5 WAVE LOADING ON WALLS

The load mechanisms described in previous sections have general applications to offshore and coastal structures. Particular methods have been developed by coastal engineers for the estimation of wave loads on long walls typified by caisson or sheet-pile breakwaters. The theory used depends on whether the walls are subject to non-breaking, breaking or broken wave action. For completeness, an outline of the available theories is given here with further discussion available in the references quoted.

28.5.1 NON-BREAKING WAVE FORCES

A train of non-breaking waves approaching a vertical wall at 90° will be reflected to some degree and form a standing wave pattern. The pressures acting on the vertical face of the wall can be calculated from the velocity potential of the standing wave in the same way as in the diffraction analyses. However, to improve the estimated forces it is necessary to consider the finite amplitude effects in the wave field. The *Shore Protection Manual* (1984), gives a series of design charts, based on original work by Sainflou and modified by Rundgren, which provide the total force and overturning moments per unit length of the wall during the crest and trough conditions of the standing wave, see Figure 28.7. Due to the finite amplitude of the waves, it will be seen that MWL is enhanced by h_o above SWL and this must be taken into account in setting the limits for crest and trough elevations as part of the construction of the pressure distributions. Appropriate values of h_o are also given in the *Shore Protection Manual* and the incremental pressure (p_1) at the sea-bed is given by

$$p_1 = 0.5(1 + H_r/H_i)\rho g H_i/\cosh(kd) \qquad (28.26)$$

where H_i and H_r are the incident and reflected waveheights and k is the wavenumber.

The Shore Protection Manual also deals with a design approach for walls which are overtopped or have rubble bases. At this point it is worth noting that the Japanese approach is different and a set of design equations have been developed to estimate breaking and non-breaking wave forces (Goda, 1985). This approach also deals with the dynamic pressure distribution across the base of the wall where it is seated on rubble or other permeable material.

28.5.2 BREAKING WAVE FORCES

The largest pressures experienced by vertical walls will result from the direct impact of breaking waves. The duration of the impact pressure is very short and its magnitude (p_m) is highly sensitive to the geometry of the wave front and wall, the deepwater steepness of the incident waves, the beach slope and the degree of air entrainment in the water. This makes the estimation of breaking wave forces one of the more difficult topics in the area of wave loading.

For many years the *Shore Protection Manual* has given Minikin's method for the estimation of forces and overturning moments arising from the assumed pressure distribution shown in Figure 28.8. It has been recognized that this approach may be conserva-

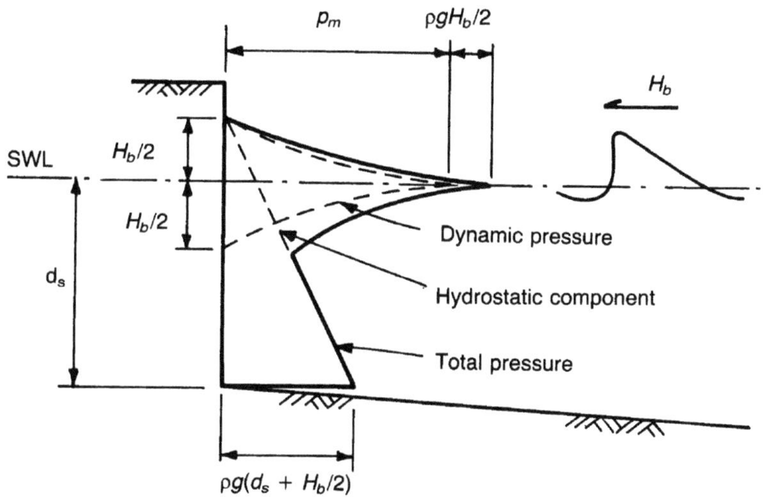

Figure 28.8 Breaking wave force on a wall (after *Shore Protection Manual*).

tive in some conditions and that the relatively weak correlation of impact pressures along the wall, coupled with the inertia of massive walls, will reduce the overall dynamic effect of such forces. Some further light on the problem has been shed by reports of laboratory studies (Kirkgoz, 1982; Whillock, 1987) and full-scale measurements (Blackmore and Hewson, 1984). The dynamic pressure component does appear to reach the peak value (p_m), at or near SWL, in the form:

$$p_m = 0.5 C_i \rho u_b^2 \qquad (28.27)$$

where u_b is some characteristic velocity for the breaking wave (e.g. the breaker celerity) and C_i is an impact coefficient. These studies, together with a numerical comparison of the Minikin and Goda methods carried out by Chu (1989), indicate that conflicts remain in accurately estimating the breaking wave pressure distributions for realistic conditions and further improvements will require extensive and expensive field studies coupled with an acceptance of a probabilistic approach to the variation of the initial conditions and resultant pressures, forces and moments.

28.5.3 BROKEN WAVE FORCES

The previous sections have outlined the approach to the estimation of non-breaking and breaking wave forces on walls. It is also necessary to consider the forces generated by broken waves which tend to produce a bore of water. There is not the same volume of data or theoretical studies for this case. The *Shore Protection Manual* tentatively suggests a relatively simple procedure which follows one of two possible routes.

Where the wall is seaward of the still water line, the dynamic pressure on the wall in resisting the bore is taken as

$$p_m = 0.5\,\rho c^2 \qquad (28.28)$$

$$= 0.5 \rho g d_b \qquad (28.29)$$

where c is the wave celerity; which is then approximated by $c = (g d_b)^{1/2}$, in which d_b is the water depth at the breaking point. The constant dynamic pressure (p_m) is taken to act from SWL up to a height $h_c \simeq 0.78 H_b$ above SWL, where H_b is the breaker height. To this rectangular pressure block is added a hydrostatic pressure distribution from zero at h_c above SWL down to $\rho g(h_c + d_s)$ at the wall base, where d_s is the water depth at the wall.

If the wall is shoreward, at a distance x_1, from the still water line and it is estimated that the unrestricted uprush would travel to a point x_2 from the still water line (i.e. if the wall were absent) then the velocity in Equation (28.28) would be linearly reduced to $c(1 - x_1/x_2)$ while h_c is replaced by $h_c(1 - x_1/x_2)$. A modified dynamic pressure is then combined with the hydrostatic component. Since the *Shore Protection Manual* method was originally put forward, other studies have attempted to improve the load model and some of this work is reviewed by Ramsden and Raichlen (1990) in discussing the forces on vertical walls subject to broken solitary waves, as measured in a series of laboratory studies.

28.5.4 UPLIFT FORCES

Clearly any member which is either immersed or exposed by the passage of a wave train will experience a vertical hydrostatic force of appropriate sign. However, it is also important to consider possible dynamic uplift loads arising from waves with crest heights in excess of the soffit or underside level of the platform or suspended pier structure. Lee and Lai (1986) discuss the numerical analysis of these forces via a finite element procedure and the results indicate that the pressures can be significantly greater than the corresponding hydrostatic values for the immersed depths, depending on the wave conditions and the seabed and wall geometries.

28.6 CONCLUSIONS

The previous sections have outlined the principal characteristics of wave loading on a number of different systems. Slender structural members are subject to inertia, drag and lift terms, which are largely formulated using empirical coefficients to modify the basic terms in the force equation to account for the complex hydrodynamics of the situation, including wave impacts.

Wave diffraction analysis is applicable to the situations where the fixed or floating structure is large relative to the incident wave lengths and a diffracted wave field is generated. Special care is needed with floating systems in order to deal with important second-order effects (the slow drift forces).

Finally wave forces on vertical walls are dependent on whether the waves are non-breaking, breaking or broken. In addition to the pressure on the vertical face it is necessary to consider the uplift pressures across the base of the wall.

REFERENCES

Barltrop, N.D.P., Mitchell, G.M. and Attkins, J.B. (1990) *Fluid Loading on Fixed Offshore Structures*, UK Department of Energy (OTH 90 322) (2 Volumes) HMSO Books, London.

Bearman, P.W., Graham, J.M.R. and Obasaju, E.D. (1984) *Applied Ocean Research*, **6**, 166–72.

Bearman, P.W., Chaplin, J.R., Graham, J.M.R., *et al.* (1985) *Behaviour of Offshore Structures*, Elsevier, Amsterdam, pp. 213–25.

Blackmore, P.A. and Hewson, P.J. (1984) *Coastal Engng.*, **8**, 331–46.

Blevins, R.D. (1990) *Flow-Induced Vibrations*, Van Nostrand Reinhold, New York.

Bishop, J.R. (1987) *Wave Force Coefficients for Horizontal and Vertical Cylinders with Kelp Fouling*, UK Department of Energy (OTH 87 268), HMSO Books, London.

Borgman, L.E. (1958) *Transactions of the American Geophysical Union*, **39**, 885–8.

Coastal Engineering Research Center, (1984) *Shore Protection Manual* (3 Volumes), US Goverment Printing Office, Washington.

Chu, Y.-H. (1989) *Journal of Waterway, Port, Coastal and Ocean Engineering, ASCE* **115**, 58–65.

Eatock Taylor, R. and Rajagopalan, A. (1982) *Journal of Sound and Vibration*, **83**, 401–16.

Garrison, C.J. (1978) in, *Numerical Methods in Offshore Engineering*, (ed. Zienkiewicz, O.C. *et al.*) Wiley, Chichester. pp. 87–140.

Grace, R.A. and Nicinski, S.A. (1976) *Offshore Technology Conf.*, OTC 2676, 681–94.

Grace, R.A. (1979) *Marine Outfall Systems – Planning Design and Construction*, Prentice-Hall, Englewood Cliffs, NJ.

Grace, R.A., Castiel, J., Shak, A.T. *et al.* (1979) *Civil Engineering in the Oceans IV*, ASCE, **1**, 99–110.

Goda, Y. (1985) *Random Seas and Design of Maritime Structures*, University of Tokyo Press, Tokyo.

Hedges, T.S. (1987) *Proceedings of the Institution of Civil Engineers*, **82**, 567–85.

Hogben, N. and Standing, R.G. (1974) *Proceedings of the International Symposium on Dynamics of Marine Vehicles and Structures in Waves*, London, pp. 258–77.

Kirkgoz, M.S. (1982) *Journal of Waterway, Port, Coastal and Ocean Engineering, ASCE*, **108**, 81–95.

Lee, J.-J. and Lai, C.-P. (1986) ASCE *Proceedings of the 20th Coastal Engng. Conf.*, Taiwan, pp. 2023–34.

Longuet-Higgins, M.S. (1977) *Proceedings of the Royal Society, Series A*, **352**, 463–80.

Morison, J.R., O'Brien, M.P., Johnson, J.W. *et al.* (1950) *Petroleum Transactions, AIME*, **189**, 149–57.

Ochi, M.K. (1982) *Advances in Hydroscience*, (ed. Ven-Te-Chow) Academic Press, New York, 13, pp. 217–375.

Pinkster, J.A. (1979) *Ocean Engineering*, **6**, 593–615.

Ramsden, J.D. and Raichlen, F. (1990) *Journal of Waterway, Port Coastal and Ocean Engineering, ASCE*, **116**, 592–613.

Sarpkaya, T. (1977) *Offshore Technology Conference*, OTC 2898, 161–6.

Sarpkaya, T. and Isaacson, M. (1981) *Mechanics of Wave Forces on Offshore Structures*, Van Nostrand Reinhold, New York.

Standing, R.G. (1988) *Journal of the Society of Underwater Technology*, **14** (4), 2–12.

Tickell, R.G. (1977) *The Structural Engineer*, **55**, 209–22.

Tickell, R.G. and Bishop, J.R. (1985) ASME Proceedings of the 4th International Offshore Mechanics and Arctic Engineering Symposium, Dallas, **1**, 142–50.

Whillock, A.F. (1987) *Coastal Engineering*, **11**, 297–308.

Zdravkovich, M.M. (1977) *ASME Journal of Fluid Engineering*, **99**, 618–33.

THE DESIGN OF BREAKWATERS 29

H.F. Burcharth

CONTENTS

29.1 INTRODUCTION

The main purpose of a breakwater is to reduce the waves in an area. More often than not the objectives are to obtain acceptable mooring and manoeuvring conditions for ships, but breakwaters are also used for beach protection. Moreover, many of the principles related to the design of breakwaters are valid for groynes and sea-walls.

Breakwaters are generally expensive structures the cost of which increases dramatically with water depth. Another important aspect which must be considered in the overall layout is the influence on currents and sediment transport. This can cause significant accumulation and erosion, not only close to the structures but also at a distance. The various functions and typical layout of breakwaters are illustrated in Figure 29.1.

The final layout in plan of the breakwaters is determined by an optimization with respect to construction and maintenance costs, and down-time costs related to non-accessibility of the port and cargo-handling operations. For the evaluation of the last mentioned points the most powerful tool available is real-time numerical simulation of manoeuvring of ships and hydraulic wave-disturbance model tests with moored ships (see Chapter 19). The layout of breakwaters will not be discussed any further in this chapter which deals with the design and the construction of the structures. Definition of the incident wave climate is treated in one section because the waves constitute the dominating loads on breakwaters.

29.2 TYPES AND PRINCIPLES OF BREAKWATERS

Wave energy is roughly proportional to the wave height squared. Thus the sheltering effect of a breakwater is established through a reduction or a cut-off of the incident wave energy. Physically this is done by reflecting the waves back into the sea and by turbulent dissipation of the wave energy. A vertical impermeable wall causes almost 100% wave reflection. An extremely effective dissipation mechanism is wave breaking. A sloping structure will cause most wind generated waves to break because of the decreasing water depth which reduces the wave celerity but not the velocities of the crest particles which eventually overtake the main form. Another effective dissipation mechanism is the turbulent flow in a porous structure. The three principles of sheltering against wave energy are illustrated in Figure 29.2.

In many cases it is too expensive and also unnecessarily restrictive to design for no-energy transmission across the structure. Usually some **overtopping** and, in the case of a porous structure, also some **wave penetration** is acceptable; see Figure 29.3.

The way breakwaters function is illustrated in Figures 29.2 and 29.3 together with considerations of the cost and availability of materials and methods of construction. Combining Figure 29.3b and c leads to the most simple **rubble mound breakwater**, consisting of a mound of stones (Figure 29.4a). However, a homogeneous structure built of stones large enough to resist displacements due to wave forces is very permeable and will cause too much penetration not only of waves but also of sediments if present in the area. Moreover, large stones are expensive because

Coastal, Estuarial and Harbour Engineer's Reference Book. Edited by M.B. Abbott and W.A. Price. Published in 1993 by Chapman & Hall, 2–6 Boundary Row, London SE1 8HN. ISBN 0 419 15430 2.

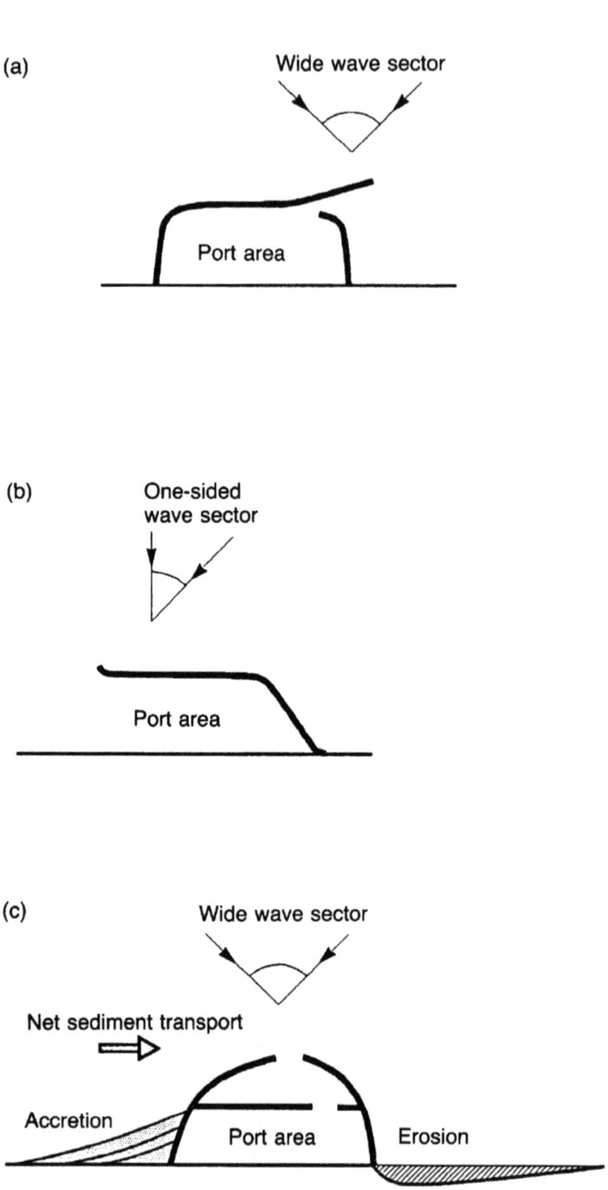

Figure 29.1 Typical functions and layouts of breakwaters (a) breakwater layout for the case of wide wave selector and no significant sediment transport. Entrance protected by extended outer breakwater to improve manoeuvering positions; (b) breakwater layout in the case of one-sided sector for severe waves; (c) breakwater layout in the case of a wide wave sector and significant sediment transport. Alignment of outer breakwaters designed to minimize eddies and related sedimentation in outer basin. Inner breakwater necessary for protection of port area. Potential areas of accretion and erosion; (d) harbour protected by detached breakwater; (e) protection of inlet to inland port, lagoon or estuary in order to maintain access channel water depths and improve manoeuvering conditions; (f) detached breakwaters used for beach protection (tombolo effect); (g) groynes.

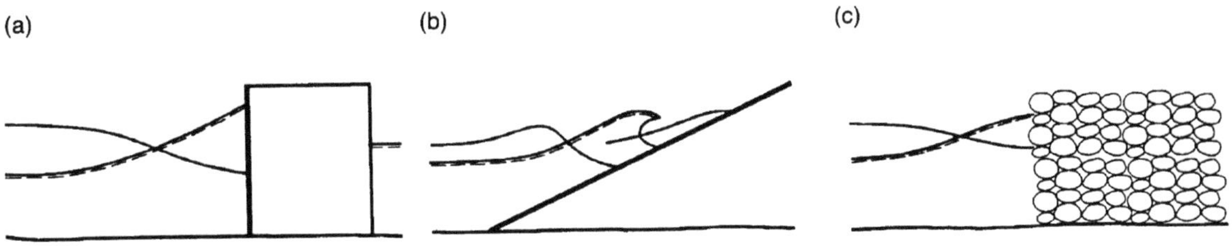

Figure 29.2 Reflection and dissipation of wave energy. (a) wave reflection; (b) dissipation by wave breaking (some reflection); (c) dissipation in porous structure (some reflection).

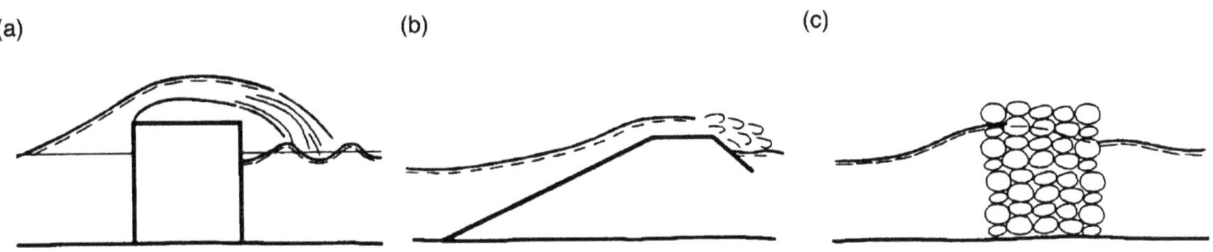

Figure 29.3 Illustration of energy transmission by overtopping and wave penetration.

most quarries yield a lot of finer material and only relatively few large stones. Consequently a real structure will consist of a core of fine material (quarry run) covered by big stones forming the so-called **armour layer**. To prevent the finer material being washed out, filter layers must be provided, Figure 29.4b. The first filter layer is also called the **underlayer**. If the structure is designed for significant overtopping it is called **low crested**.

In areas with rough wave climates or at sites where only small or too few large quarry stones can be produced, concrete armour units such as cubes, tetrapods and Dolosse are used (Figure 29.5).

Generally the volume of the structure and the surface area to be armoured will decrease as the steepness of the slopes increases. Consequently the steepest possible slope, which satisfies stability under design wave conditions, is used. For a given rate of overtopping the amount of material can be reduced by the use of a **wave-wall superstructure** (Figure 29.4c). Superstructures can have several purposes, e.g. providing access for vehicles (and cranes for repair) and accommodation of pipelines.

The armour units in conventional multi-layer structures (Figures 29.4b and c) are designed to stay in place as built, i.e. the profile remains unchanged with only minor displacements of armour units. In the case of rock armour a design can also be based on some natural reshaping of the seaward profile during wave action (self-adjusted profile). In this way relatively smaller rocks can be used because nature will optimize the profile to be the most resistant to wave action. If the mound is large enough to prevent complete erosion of the crest, then an S-shaped profile will develop as indicated in Figure 29.4d. This type of structure is often called a berm breakwater due to the large berm of armour stones placed during construction, and is designed for no overtopping. Multi-layer rubble mound structures might be given an S-shaped or bermed profile as shown in Figure 29.4e in order to reduce the wave forces, run-up and overtopping.

Completely submerged breakwaters are called **reef breakwaters**. They are constructed either as conventional multi-layer structures or as homogeneous structures as shown in Figure 29.4f.

Another category of breakwater is the **vertical wall** type which as the phrase suggests has a vertical or almost vertical seaward face. With reference to Figure 29.6 this category might be divided into fully reflective and dissipative types. The first is the conventional type with an impermeable seaward face and is most often constructed of reinforced concrete caissons filled with sand or quarry rock and capped with *in situ* cast concrete, Figure 29.5a. Many old structures of this type were constructed of stacked precast mass concrete blocks, Figure 29.5b. The upper part of the front face might be given a slope to ensure a more favourable aspect and a reduction of the wave forces which would contribute to an improvement of the stability, Figure 29.6c. Because the first rational

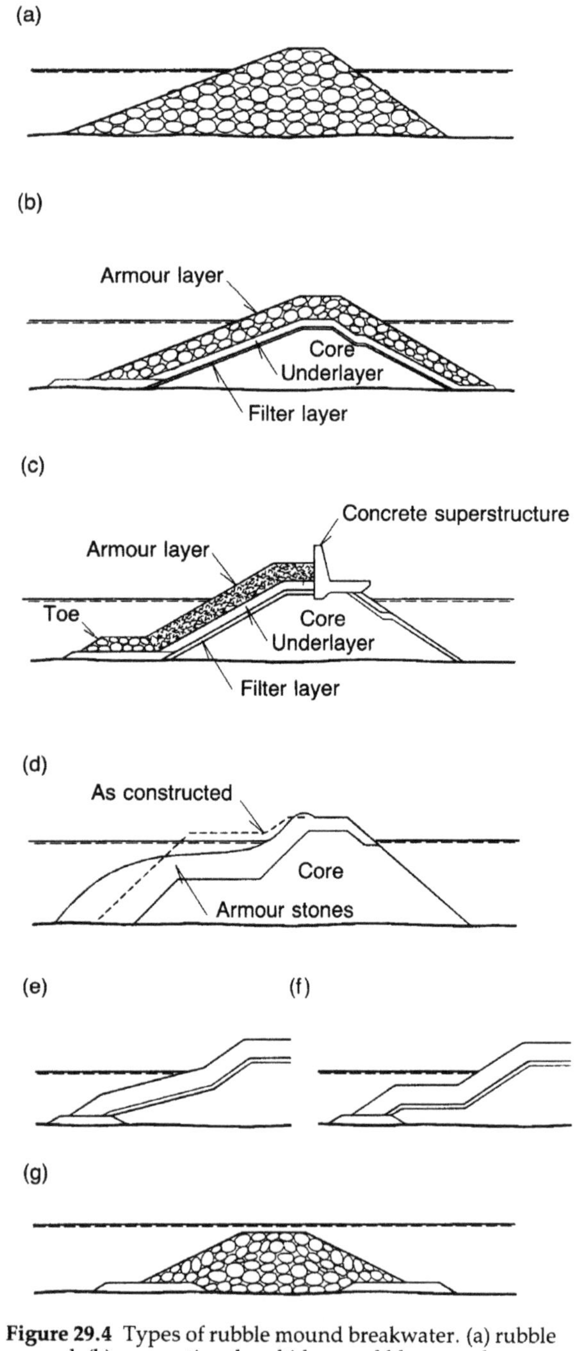

Figure 29.4 Types of rubble mound breakwater. (a) rubble mound; (b) conventional multi-layer rubble mound breakwater; (c) multi-layer rubble mound breakwater with superstructure; (d) reshaping rubble mound breakwater (berm breakwater); (e) S-shaped front (composite slope); (f) bermed front; (g) reef breakwater.

consideration of this shape is related to the Hanstholm breakwater in Denmark this type is sometimes referred to in the literature as the Hanstholm type.

The dissipative type consists of caissons with holes or slots in the front face and a wave chamber behind, as originally proposed by Jarlan. Wave energy is dissipated by the turbulence created during in and outflow, Figure 29.6d. The advantages obtained are the reduction of the overtopping and of the wave forces, especially of the large forces caused by waves breaking on the structure. Another advantage is reduced reflection, often desirable from a navigational point of view.

The same positive effects can be obtained by placing a mound of armour units in front of the wall (Figure 29.6e). This solution is suitable mainly in shallow water wave situations for combatting frequent high loads from breaking waves. It is a common structure in Japan where the armour units are called wave dissipation blocks. It is sometimes referred to as the Japanese type breakwater.

In the case of deep water an economical design often consists of caissons placed on a rubble mound foundation (Figure 29.6f). This type is now referred to as a composite breakwater although this name was originally related to the functional circumstances where waves are reflected from the vertical face at high water and break on the rubble slope at low water. However, this idealized functional description is not realistic because the slope will increase the wave breaking and thereby the loads on the vertical wall. In Japan, the composite breakwater is called the **mixed type**. Smaller caissons might be designed without a floor in which case they cannot float and must be transported and lowered by a crane travelling on the already placed caissons or by a floating crane. After placement a bottom of *in situ* underwater concrete is cast. Most caissons have rectangular cross-sections but circular sections are also used.

The breakwaters shown in Figure 29.6 are the most commonly used vertical wall types for larger structures. Smaller and less exposed structures are sometimes built as series of sand-filled steel sheet pile cells with *in situ* cast concrete capping. Timber cribs filled with rocks and stones are also used. Non-gravity structures might be used for rather protected locations and sites with poor soil conditions. Such structures basically consist of a vertical or inclined front wall made of piles or prefabricated plates supported by spaced batter piles.

29.3 MODES OF DAMAGE

All possible modes of damage must be identified and considered in the design process. The term **failure modes** is commonly used. In this context failure means

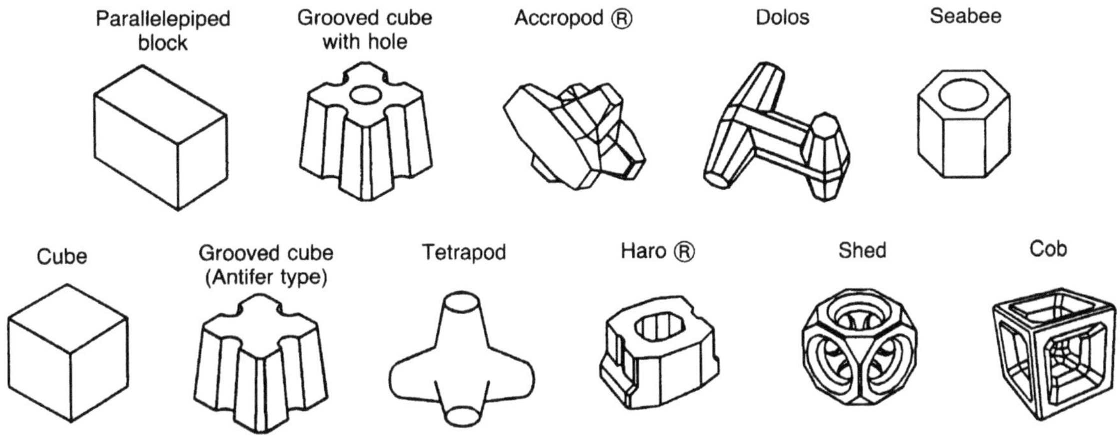

Figure 29.5 Examples of concrete armour units.

exceedance of a design damage criterion, e.g. excess in displacement of armour units or excess overtopping. The failure modes of a rubble mound breakwater are indicated in Figure 29.7. A more detailed discussion is given in section 29.6. The failure modes of a vertical wall breakwater are indicated in Figure 29.8. A distinction is made between those related to the overall stability of the caissons and failure of the structural elements themselves. Loads and failure modes due to ice and ship collision will not be discussed.

29.4 HYDRAULIC BOUNDARY CONDITIONS

29.4.1 TYPES OF WAVE BREAKING

The kinematics of waves breaking on smooth, impermeable slopes can be qualitatively described by the so-called surf-similarity parameter or Irribarren number, ξ, which characterizes the type of wave breaking (Battjes, 1974). If viscous forces are neglected dimensional analysis shows that the flow characteristics are a function of the parameters $(gT^2)/H$ and α, where α is the slope angle. Using the expression for the deep water wave length $L_o = (g/2\pi)T^2$ the parameters can be combined into the surf-similarity parameter originally defined for regular waves as $\xi_o = \tan\alpha(H_o/L_o)^{-0.5}$, where H_o/L_o is the deep water wave steepness. The wave height H_b at the breaking point is sometimes substituted for H_o in which case the parameter is denoted by ξ_b. Breaker types and related ranges of ξ-values are given for impermeable slopes in Table 29.9 (Günbak, 1977).

For irregular waves some characteristic values of the wave steepness must be used, e.g.

$$\xi_m = \tan\alpha(s_{om})^{-0.5}$$

or

$$\xi_p = \tan\alpha(s_{op})^{-0.5}$$

29.4.2 DEPTH LIMITED WAVE HEIGHT AND ITS DISTRIBUTION

It is important to know whether the structure will be exposed to non-breaking, breaking or broken waves. As indicated through the definition of ξ both the wave steepness and the slope angle determine the breaker height, H_b, and the water depth at wave breaking, h_b. Limiting values of H_b/h_b for regular waves can be found in Goda (1985) and in the *Shore Protection Manual* (1984).

In the case of irregular waves the wave breaking normally takes place within a fairly wide surf zone and, consequently, some characteristic value of the breaker height must be used. Goda (1985) provides diagrams and formulae for the estimation of the variation of H_s and H_{max} (defined as $H_{1/250}$, i.e. the 1 in 250 exceedence probability value) with water depth for various deep-water wave steepnesses and sea-bed slopes.

Experience from various site specific model tests indicate that the following typical ranges are valid for sea-bed steepnesses flatter than 1:30.

$$0.5 \leqslant \max[H_s/h_b] \leqslant 0.7,$$
$$0.7 \leqslant \max[H_{max}/h_b] \leqslant 1.2$$

For design wave conditions where wave breaking takes place due to limited water depths the wave height distribution will differ from the Rayleigh distribution in that relatively fewer large waves heights will exist. With respect to the distribution of extreme

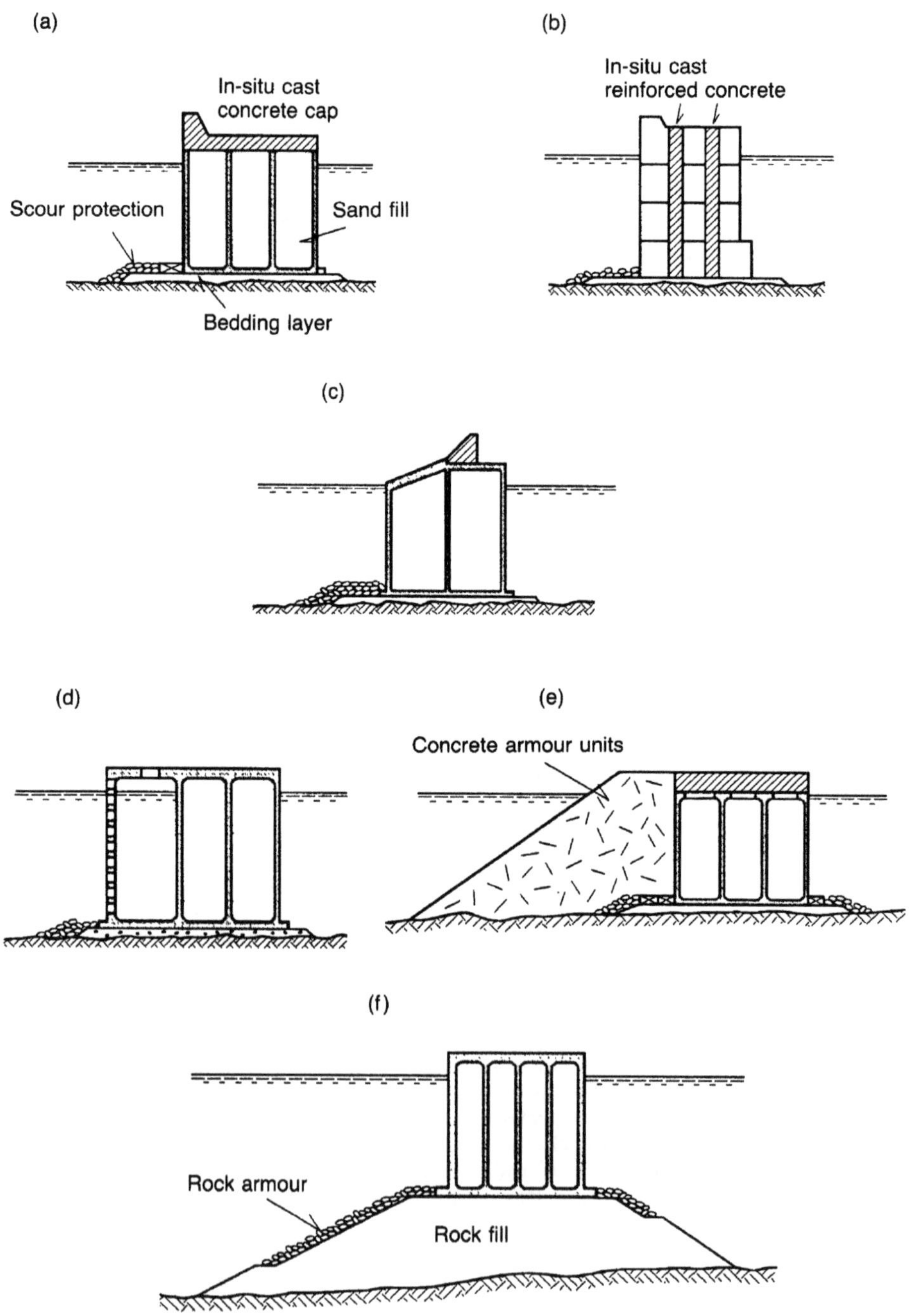

Figure 29.6 Types of vertical wall breakwater. (a) conventional caisson type; (b) conventional block type; (c) sloping-top type; (d) perforated wall type; (e) Japanese type; (f) composite type.

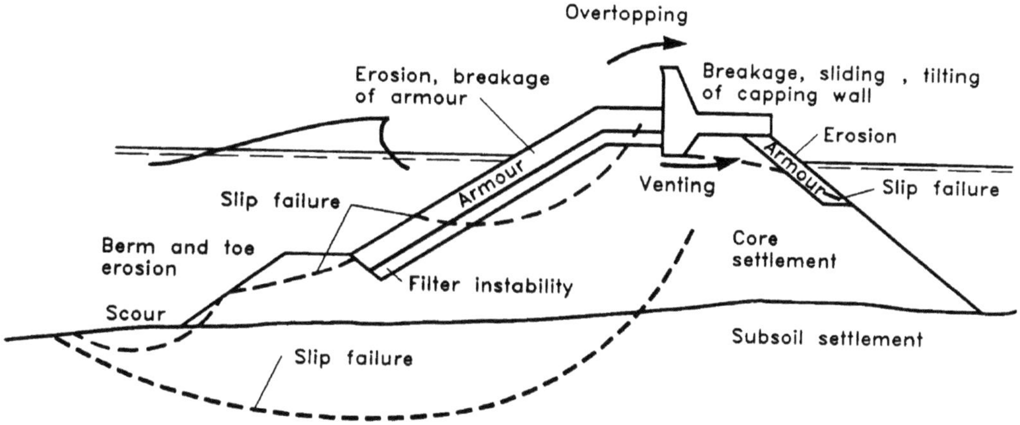

Figure 29.7 Failure modes of a rubble mound breakwater.

wave and crest heights there is a lack of reliable model formulations. Thornton and Guza (1983) show that at least up to $H_{10\%}$, i.e. the wave height exceeded by 10% of the waves, the Rayleigh distribution is valid. For higher waves in water depth d, Stive (1986) proposed the following empirical correction to the Rayleigh distribution based on model tests but roughly checked against some prototype data:

$$\begin{aligned} \hat{H}_{1\%} &= H_{mo}\left(\frac{\ln 100}{2}\right)^{1/2}\left(1+\frac{H_{mo}}{d}\right)^{-1/3} \\ \hat{H}_{0.1\%} &= H_{mo}\left(\frac{\ln 1000}{2}\right)^{1/2}\left(1+\frac{H_{mo}}{d}\right)^{-1/2} \end{aligned} \quad (29.1)$$

where $\hat{H}_{1\%}$ means the 1% exceedence value of the wave height determined by zero crossing analysis, whereas the significant wave height H_{mo} is determined from the spectral moments. Stive also discusses the reliability of the corrections. The correction formulae are very useful for checking the wave height distribution in small-scale physical model tests, cf. Figure 29.10.

29.4.3 WAVE SPECTRUM IN SHALLOW WATER

For shallow-water wave conditions Hughes (1984) proposed an empirical spectrum, the so-called TMA spectrum. It is a one-dimensional spectrum (two-dimensional wave) for fetch limited sea where shallow-water effects are included. The analytical expression is given by the JONSWAP spectrum expression multiplied by a shallow-water wave effect function $\Phi(f, h)$:

$$\Phi = \begin{cases} 2\pi^2 f^2 h/g & \text{for } f \leq 1/(2\pi(h/g)^{0.5}) \\ 1 - 1/2(2 - 2\pi f(h/g)^{0.5})^2 & \text{for } f > 1(2\pi(h/g)^{0.5}) \end{cases} \quad (29.2)$$

where f is the frequency and h the water depth. At the same time the α-factor in the JONSWAP expression should be replaced by $\alpha = 0.076(gF/U^2)^{-0.22}$, where F is the fetch and U the wind velocity.

It is important to notice that the information given above for shallow-water waves is valid only for plane or almost plane sea-beds. A complicated strongly varying sea-bed topography will influence the asymmetry and kinematics of the waves and thereby also the impact on a structure. For example, a hump on the sea-bed might cause waves to break. This kind of wave instability cannot yet be described by simple formulae. Only in physical models, where the sea-bed topography can be modelled, is it possible to take this important effect into account. Consequently, the available failure mode formulae related to shallow-water conditions, which are presented in what follows, are valid only for gently sloping and almost horizontal sea-beds if no other information is given.

29.5 DESIGN WAVE CLIMATE

29.5.1 RETURN PERIOD AND ENCOUNTER PROBABILITY

An event, e.g. H_s, with a return period of T years will, on average, be exceeded only once every T years. In statistical analysis this time interval is calculated as

$$T(H_s) = \frac{\tau}{1 - F(H_s)} \quad (29.3)$$

if we assume H_s to be a continuous random variable. τ is the time interval between the observations of H_s and $F(H_s)$ is the cumulative distribution function of H_s expressing the probability of a value less than H_s.

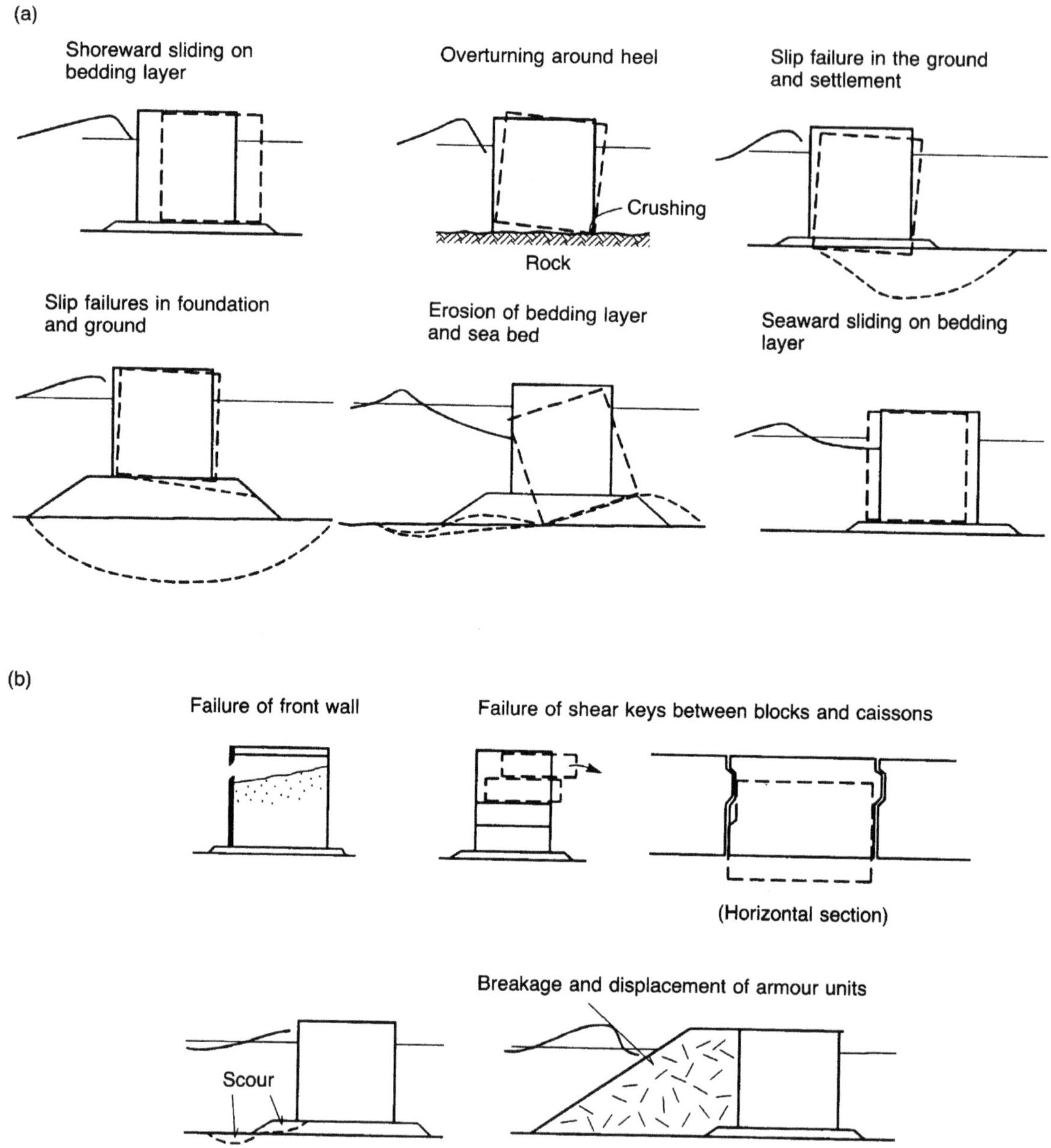

Figure 29.8 Examples of failure modes of vertical wall breakwaters (a) related to overall stability of caissons; (b) related to structural elements.

The **encounter probability** is the probability, p, that the T-year return period event will be exceeded during an L-year structural life, given by

$$p = 1 - \left(1 - \frac{1}{T}\right)^L \quad (29.4)$$

29.5.2 METHODS OF SINGLE VARIABLE EXTREME ANALYSES

The wave height is the most important environmental parameter in breakwater design. The significant wave height H_s, is the most commonly used parameter for

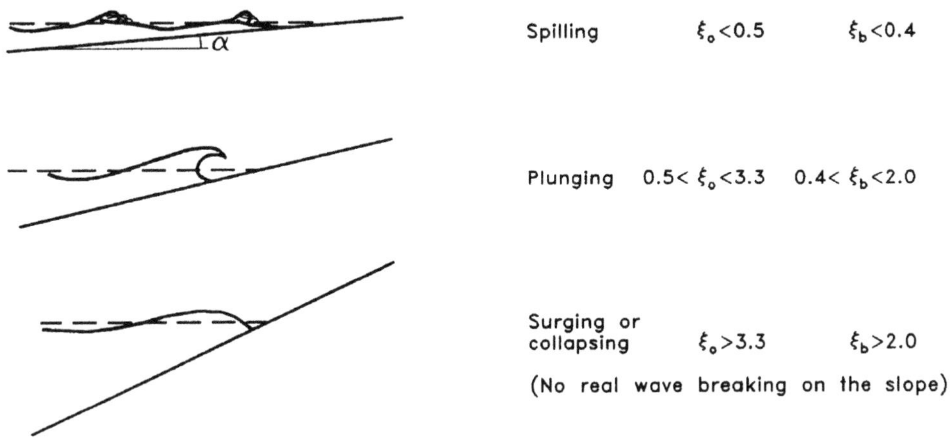

Figure 29.9 Types of wave breaking and ξ-values.

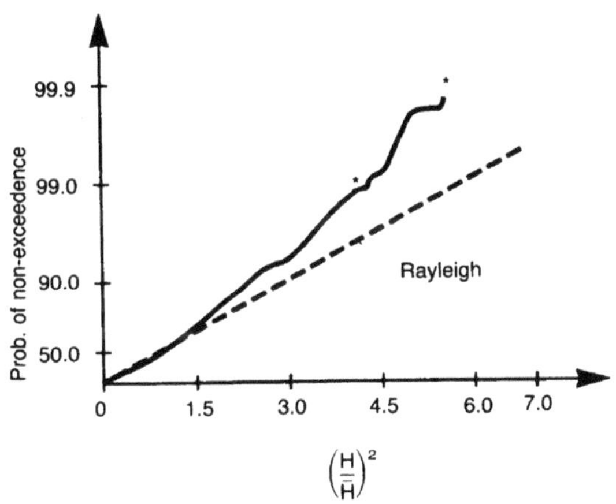

Figure 29.10 Comparison of expression by Stive (1986) for shallow water wave height distribution with model test results. (Aalborg University Hydraulics Laboratory, 1990).*, formulae by Stive (1986).

single-parameter characterization of sea states and is generally applied in statistical analysis of extreme events. Traditionally the H_s value corresponding to a certain return period (e.g. T = 50 or 100 years) has been used as the design value.

Estimates on events of low probabilities are often performed in the following three different ways:

1. Extrapolation of data from frequent measurements or observations of H_s equally spaced in time. The data are often compiled at intervals Δt = 3, 6 or 24 h, which gives a large sample, N events, even in the case of a short time of observation or record length Y in years. The order of magnitude of N is often 1000–10 000.
2. Extrapolation of data representing the largest significant wave height H_S in each year (annual series data), see Figure 29.10.
3. Extrapolation of relatively few data sets representing the maximum significant wave height a certain level, H_s' (partial duration series). This is called the peak-over-threshold (POT) method. The data are often determined from hindcasts and the sample size N is typically within the range 10–30 (see Figure 29.11). As will be discussed later a sample size of 10 is too small to produce a reliable estimate of extreme waves. Samples larger than 30 are used in some studies.

It is often debated whether it is best to use annual series analysis or partial series analysis (POT). The most frequent objection encountered with respect to the use of annual series is that only one event per year is used. Quite often the second largest event in a year, which the annual series neglects, may outrank annual events of other years.

In general, an extreme analysis based on partial series should be preferred, but the POT method is conditioned by the validity of series representing independent events.

Methods of fitting distribution functions to data sets

There are four commonly used methods of fitting the parameters of a chosen theoretical distribution to a data sample, namely the maximum likelihood method,

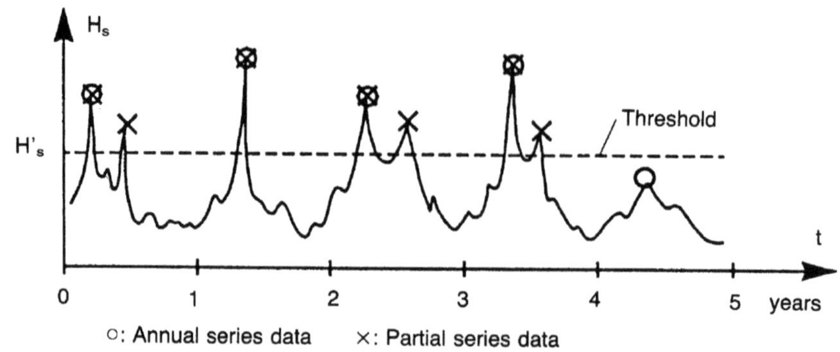

Figure 29.11 Representation of data reduction to constitute annual and partial series. (○), annual series data; (×), partial series data.

the method of moments, the least square method and the graphical method.

It is generally accepted that the **method of maximum likelihood** uses the maximum amount of the information inherent in the sample of the parameters. The maximum likelihood estimators are consistent and asymptotically unbiased. Furthermore, the maximum likelihood estimates, adjusted for bias, will be efficient, i.e. with minimum variance. The method is a 'large sample' method, which for small samples may give biased estimates.

By the **method of moments** the parameters in a chosen distribution are estimated by equating moments of the distribution to those of the data. The minimum number of moments equals the number of parameters to be estimated. The moment method is unbiased and consistent and requires fewer calculations than the maximum likelihood method and the least square method. However, the method is not efficient, i.e. generally the estimates have larger variance than in the case of the maximum likelihood method.

The **least square method** operates on plotting positions, i.e. the probabilities attached to the ranked events. A proper selection of the plotting position formulae is therefore very important.

The **graphical method**, which also operates on plotting positions, is a subjective visual judgement of the best fit of the data points to a chosen distribution function.

Examples of theoretical extreme distributions and related uncertainties due to sample size

When evaluating the uncertainties due to sample size the assumption is made that the chosen distribution is the true one. This means that the uncertainties are related exclusively to the quality of the estimates on the parameters of the distributions, which again are affected by the sample size.

What follows lists the formulae for fitting data to the Weibull distribution, including estimates of distribution parameters, formula for computing the T-year event and the uncertainty on the T-year event in terms of the standard deviation. With respect to other distributions reference is made to Burcharth (1989).

The following notation is used:

f	probability density function
F	cumulative distribution function
T	return period
N	total number of data (sample size)
λ	average number of data per year
$H_{s,i}$	events included in the data sample
H_s'	threshold value
H_s^T	estimate on T-year event
σ_T	standard deviation related to H_s^T
$\hat{\alpha}, \hat{\beta}$	estimate of distribution parameters α and β which are assumed normally distributed
$\sigma_\alpha, \sigma_\beta$	standard deviation related to α and β

Weibull distribution

$$f(H_s; \alpha, \beta) = \frac{\alpha}{\beta}\left(\frac{H_s - H_s'}{\beta}\right)^{\alpha-1} \exp\left[-\left(\frac{H_s - H_s'}{\beta}\right)^{\alpha}\right],$$

$$H_s > H_s', \quad H_s' \text{ fixed}$$

$$F(H_s; \alpha, \beta) = 1 - \exp\left[-\left(\frac{H_s - H_s'}{\beta}\right)^{\alpha}\right]$$

The maximum likelihood estimate $\hat{\alpha}$ is obtained by solving the following equation by an iterative procedure:

$$N + \hat{\alpha}\sum_{i=1}^{N} \ln B_i = N\hat{\alpha}\sum_{i=1}^{N} (B_i^{\hat{\alpha}} \ln B_i) / \sum_{i=1}^{N} B_i^{\hat{\alpha}}$$

Table 29.1 Typical coefficient of variation $\sigma' = \sigma/\mu$ (standard deviation over mean value) for measured and calculated sea-state parameters (Burcharth, 1989)

Parameter	*Methods of determination*	*Estimated typical values*		*Comments*
		σ'	*Bias*	
Significant wave height, offshore	Accelerometer buoy, pressure cell, vertical radar	0.05–0.1	~0	
	Horizontal radar	0.15	~0	
	Hindcast, num. models	0.1–0.2	0–0.1	Very dependent on quality of weather maps.
	Hindcast, SMB method	0.15–0.2	?	Valid only for storm conditions in restricted sea basins.
	Visual observations from ships	0.2	0.05	
Significant wave height nearshore determined from offshore significant wave height taking into account typical shallow-water effects (refraction, diffraction, shoaling)	Numerical models	0.1–0.20	0.1	σ' can be much larger in some cases
	Manual calculations	0.15–0.35		

where $B_i = H_{s,i} - H'_s$.

The maximum likelihood estimate of $\hat{\beta}$ is

$$\hat{\beta} = \left[\frac{1}{N}\sum_{i=1}^{N}(H_{s,i} - H'_s)^{\hat{\alpha}}\right]^{1/\hat{\alpha}}$$

$$H_s^T = (H'_s + \hat{\beta})(\ln(\lambda T))^{1/\hat{\alpha}}$$

$$\sigma_T = (\ln\lambda T)^{1/\hat{\alpha}-1}\hat{\beta}^2\left[\frac{1}{\hat{\alpha}^2 N} + \frac{(\ln\lambda T)^2}{N}\left(\frac{\Gamma\left(1+\frac{2}{\hat{\alpha}}\right)}{\Gamma^2\left(1+\frac{1}{\hat{\alpha}}\right)} - 1\right) + \frac{1}{\hat{\alpha}^4}((\ln\lambda T)^{1/\hat{\alpha}}\ln(\ln\lambda T))^2\,\mathrm{Var}[\hat{\alpha}]\right]$$

(Nielsen and Burcharth, 1985)

$$\mathrm{Var}[\alpha] \simeq \frac{1}{N}$$

$$\sigma_\alpha \simeq \frac{1}{N^{0.5}}$$

$$\sigma_\beta = \left(\frac{\beta}{N^2}\left[\frac{\Gamma\left(1+\frac{2}{\alpha}\right)}{\Gamma^2\left(1+\frac{1}{\alpha}\right)} - 1\right]\right)^{0.5}$$

Uncertainties related to extreme value estimates

The sources of uncertainties related to an extreme value estimate of H_s are: firstly, errors in measurement, visual observation or hindcasting of the H_s data set values, on which the extreme statistics are based, and short-term randomness (variability) of the wave records applied for calculation of the H_s values. Table 29.1 gives typical values of the coefficient of variation related to H_s and other sea-state parameters.

Secondly, uncertainty related to extrapolation from short samples (e.g. limited numbers of H_s-values) to events of low probability of occurrence. Monte Carlo simulations can also be used to provide a reliable measure of the standard deviation. Several such studies have shown that the T-years estimate closely follows a normal distribution (Knudsen, 1983). This means that confidence limits on H_s^T can be obtained, e.g.

Interval	*Confidence*	*Probability of exceedence of upper bound*
$H_s^T \pm 1.28\,\sigma_T$	80%	10%
$H_s^T \pm 1.65\,\sigma_T$	90%	5%
$H_s^T \pm 1.96\,\sigma_T$	95%	2.5%

Finally, uncertainty due to limited number of years of observations. The statistical uncertainty depends not only on the total number of data, N, as discussed under B but also on the length of the period of

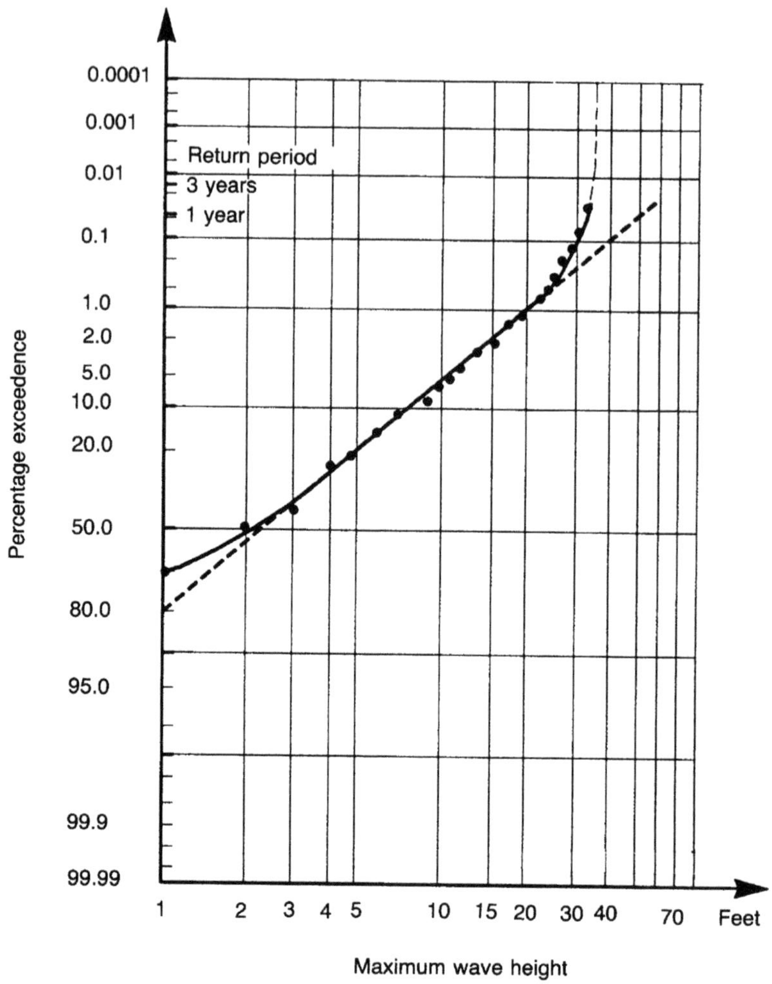

Figure 29.12 Example of wave data fit for a shallow water location.

observation, as might be expected. However, the last source of uncertainty is not and cannot be implemented in the formulae for σ_T, because in the statistical analysis it is assumed that the data samples are equally representative of the true distribution. This puts restrictions on the data set in the sense that data must be sampled with a frequency and over a length of time which ensures that periodic variations (e.g. seasonal) are not biasing the sample. A minimum sampling period of 15 years is recommended.

A discussion on the uncertainty of climatological changes and the influence the choice of distribution, threshold value on the extreme value (in P.O.T. analyses) and plotting position formulae has been given by Burcharth (1989) (see also Chapter 26 of this volume).

Combined errors

The various sources of uncertainty discussed above might be assumed to be mutually independent except for an unknown but probably weak correlation between the climatological variation and the data samples. The total normalized standard deviation might then be estimated as the square root of the sum of the squared normalized standard deviations.

Physical limitations in extrapolation to events of low probability

Any extrapolation must respect possible physical limitations such as those in wave heights due to limited water depths and fetch restrictions (Figure 29.12).

The effect of limited water depth for wave heights larger than 25 ft is clearly seen as a deviation from the linear fit. The deviation seen for waves smaller than 2 ft is due to swell and ship generated waves in the area. Such waves do not belong to the same statistical population as the wind generated waves.

29.5.3 MULTI-PARAMETER EXTREME ANALYSIS

For the general case where several variables are of importance but the correlation coefficients are not known the best joint probability approach would be to establish long-term statistics for the response in question, e.g. for the run-up, the armour unit stability, the wave force on a parapet wall, etc. If we assume that the variables of importance are H_s, T_m, α (wave direction) and z then by hindcasting and/or measurements several data sets covering some years can be established $(H_{s,1}, T_{m,1}, \alpha_1, z_1)$, $(H_{s,2}, T_{m,2}, \alpha_2, z_2) \ldots (H_{s,n}, T_{m,n}, \alpha_n, z_n)$. For each data set the response in question is either calculated from formulae or determined by model tests. If for example run-up, R_u, is in question a single variable data set $R_{u,1}, R_{u,2}, R_{u,n}$ is obtained and the related long-term statistics can be established by fitting to a theoretical extreme distribution as explained above. It is important to note also that for joint probability cases physical limitations might impose restriction to the extrapolation to low probability situations.

29.5.4 INFLUENCE OF THE STRUCTURAL RESPONSE CHARACTERISTICS ON THE CHOICE OF DESIGN SEA-STATE LEVEL

The choice of design sea-state depends not only on the structure's lifetime and the uncertainty related to the estimation of the extreme sea-state, but also on the character of the structural response.

Figure 29.13 indicates as an example the differences in armour layer damage development for various types of rubble structure. The figure illustrates the importance of evaluation of prediction and confidence limits related to the estimated design sea-state, especially in the case of structures with brittle failure

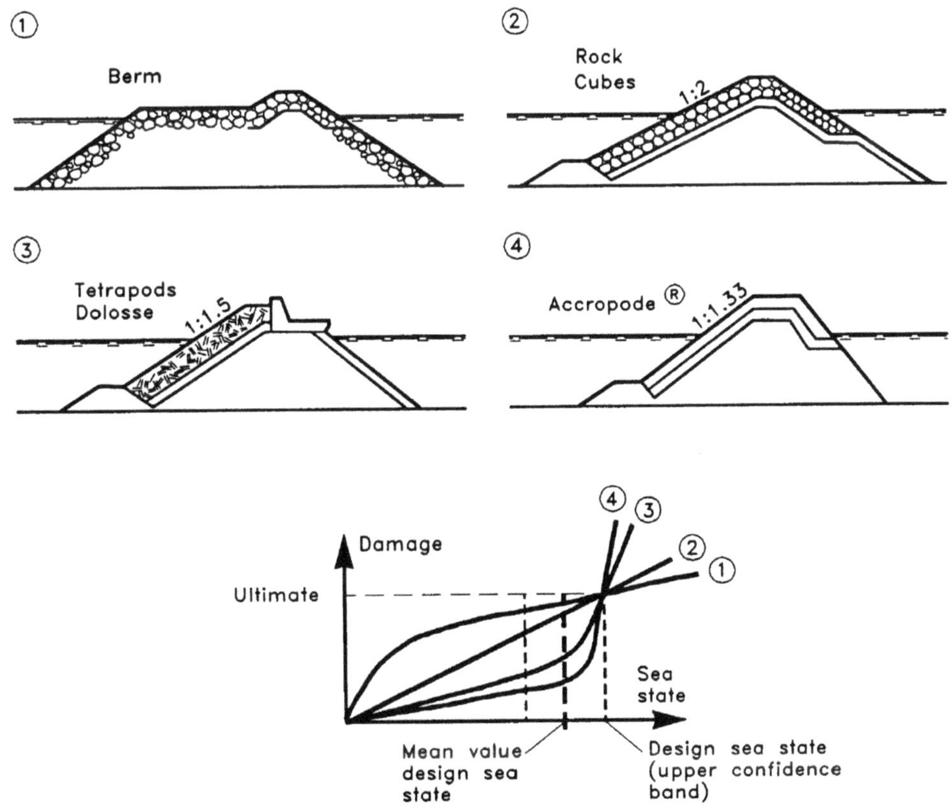

Figure 29.13 Typical armour layer failure characteristics for various types of rubble mound structure.

characteristics. For such cases a lower damage level must be chosen for the mean value design sea-state. The figure is illustrative. In reality the confidence bands for the damage curves should also be considered. Such bands can be calculated from the information given on standard deviations related to the various failure mode formulae in section 29.2.

29.6 DESIGN OF RUBBLE MOUND BREAKWATERS

29.6.1 HYDRAULIC ASPECTS OF WAVE–STRUCTURE INTERACTION AND FAILURE MODE FORMULAE

This section discusses the wave–structure interaction and the related modes of damage. Functional relationships between the main environmental parameters, structural parameters and the structural responses are given in terms of formulae, where they exist. The formulae are derived mainly from physical considerations and scale model tests and are only valid for partial parameter ranges. These ranges are given together with information about the uncertainty of the formulae, when known.

Because of the limitations in available failure mode formulae and their validities, it is in most cases necessary to test and optimize designs by performing laboratory model tests. Preliminary designs will, however, always be based on formulae and the experience of the designer.

Run-up, run-down and internal flow

Wave breaking on a slope causes up-rush and down-rush. The maximum and minimum elevation of the water surface measured vertically from the still-water level are denoted by R_u, run-up, and R_d, run-down, respectively, see Figure 29.14a.

R_u and R_d depend on the height and steepness of the incident wave and its interaction with the preceeding reflected wave, as well as the slope angle, the surface roughness and the permeability and porosity of the slope. Maximum values of flow velocities and of R_u and R_d for a given sea-state and slope angle are reached on smooth impermeable slopes.

Figure 29.14a illustrates the variation of the flow velocity vectors along an impermeable slope. Figure 29.14b illustrates the variation in the case of a permeable slope. Both the size and the direction of the velocity vectors are important for the stability of the armour units. The most critical flow field occurs in a zone around still water level and down-rush normally produces the largest destabilizing forces, except for slopes flatter than approximately 1:3.5 in which case up-rush is more detrimental. The velocity vectors shown in Figure 29.14b explain why breakwaters which are designed to reshape attain S-profiles as indicated in Figure 29.4d.

Increasing the permeability of the slope reduces the flow velocities along the slope surface because a larger proportion of the flow takes place within the slope. The internal water table is also affected as indicated in Figure 29.14c.

The wave action will cause a rise of the internal water level leading to an increase in the mean pore pressures. This internal set-up is due to a higher water level being attained during wave up-rush than the water surface reached during down-rush. The mean flow path for inflow is also shorter than that for outflow. The rise of the phreatic line will continue until the outflow balances the inflow. The lower the permeability of the structure the higher the set-up, as is also indicated in the figure.

Barends (1988) has suggested practical formulae for the calculation of the penetration length and the maximum average set-up which occurs after several cycles. Two cases are considered: a conventional breakwater structure with open (permeable) rear side, and a structure with a closed (impermeable) rear side. The latter case causes the largest set-up.

Increasing porosity also reduces the overflow velocities because a larger portion of the incoming water volume can be stored in the pores which then act as reservoirs. This positive reservoir effect is reduced in the case of large internal set-ups of the water table.

Breakwaters with crest levels lower than the run-up level are called low-crested breakwaters. While the up-rush velocities are almost unchanged the down-rush velocities are reduced due to the overtopping of some part of the incoming wave (Figure 29.14c). The larger the overtopping the greater is the reduction of the down-rush and of the destabilizing flow forces on the armour units.

Parapet walls which cut off the hypothetical run-up wedge shown in Figure 29.14e will increase the down-rush velocities and consequently also the destabilizing flow forces on the armour units.

From dimensional analysis it is to be expected that R_u and R_d expressed as the dimensionless variables, R_u/H and R_d/H, respectively, are functions of ξ as well as of the permeability, porosity and surface roughness of the slope. In this case it is the slope angle which enters into ξ and not the slope angle of the sea-bed. The influence of the water depth in front of the slope on R_u and R_d and on the reflection is fairly weak and is often neglected. Also the degree of instability of the waves approaching the structure (non-breaking, breaking or broken waves) is normally neglected and only a characteristic steepness of the waves enters in ξ through the use of s_{om} or s_{op}. However, although a characteristic wave height as H_s in front of the slope is used, it is still traditional to use the deep-water wave

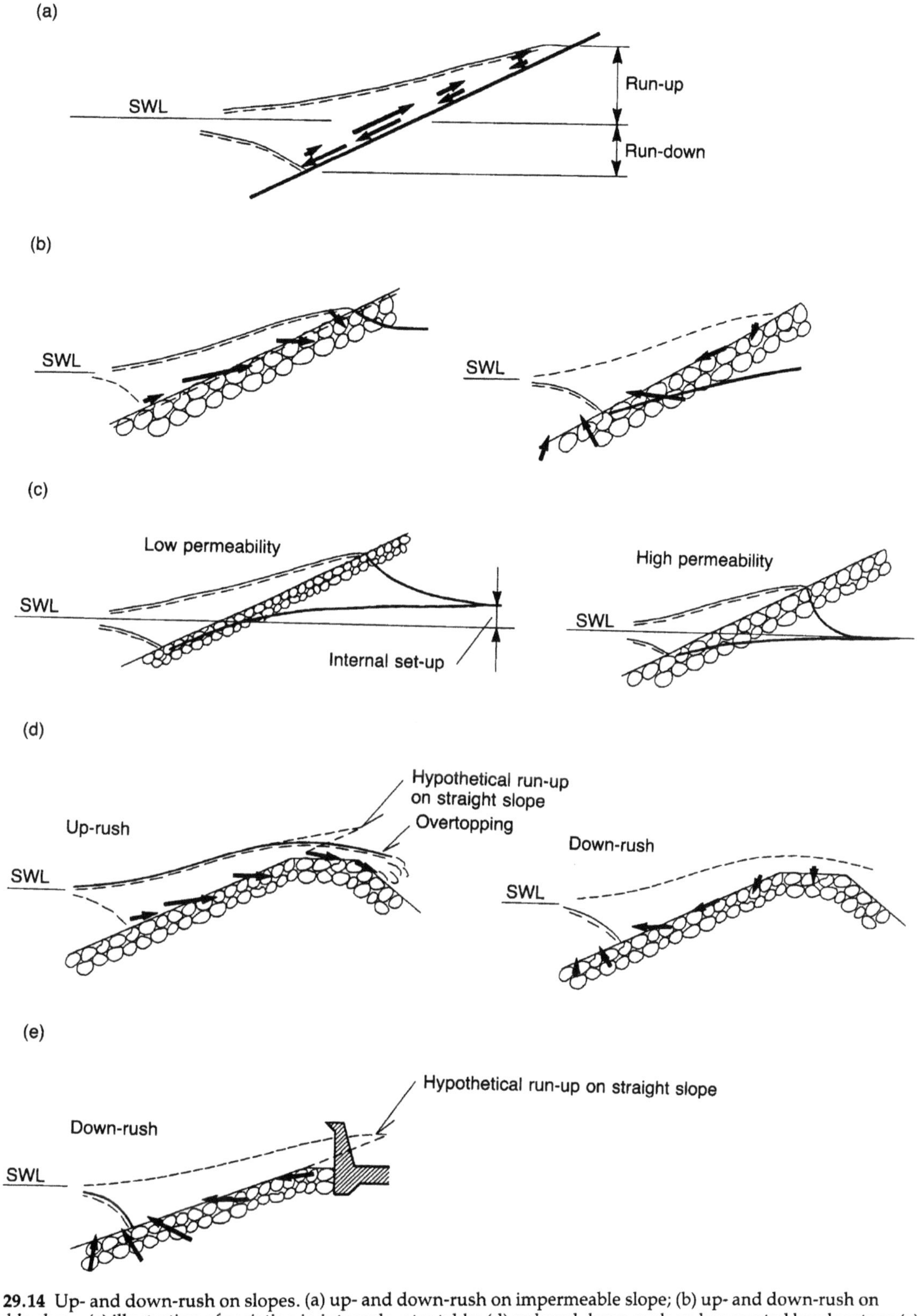

Figure 29.14 Up- and down-rush on slopes. (a) up- and down-rush on impermeable slope; (b) up- and down-rush on permeable slope; (c) illustration of variation in internal water table; (d) reduced down-rush on low crested breakwaters; (e) increase in down-rush velocities due to reflection from parapet wall.

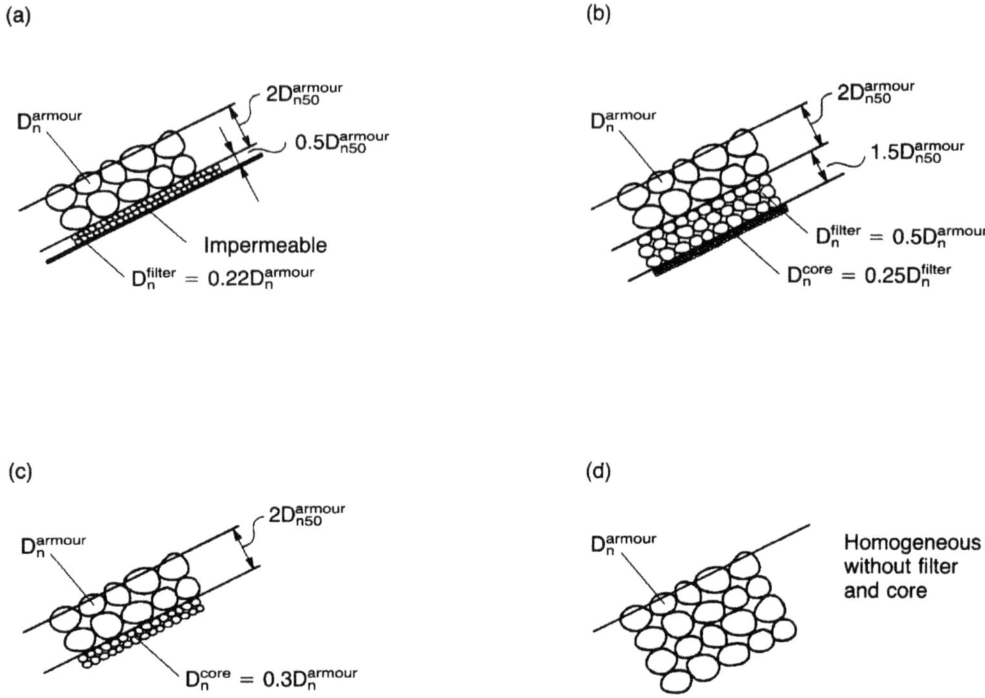

Figure 29.15 National permeability factor P for various structures. (a) $P = 0.1$; (b) $P = 0.4$; (c) $P = 0.5$; (d) $P = 0.6$.

length, L_o, to obtain the steepness, s_{om} or s_{op}. It would be better to use the wave length in front of the structure, but this wave length is more difficult to determine.

One of the few well documented formulae for run-up on rubble mound breakwaters is based on tests with rock armoured straight slopes by van der Meer (1988), who also introduced a notional permeability factor P for the structure, see Figure 29.15. This factor also includes the effect of the pore volume.

Defining R_{up} at the run-up level of exceedence probability p the formula valid for $0.1 \leqslant P \leqslant 0.6$ and head-on waves reads

$$R_{up}/H_s = \begin{cases} a\xi_m & \text{for } \xi_m < 1.5 \\ b\xi_m^c & \text{for } \xi_m > 1.5 \end{cases} \qquad (29.5)$$

R_{up}/H_s reaches a constant maximum value equal to d in the case of permeable structures, i.e. $P > 0.4$. This corresponds to the region of surging waves (no real wave breaking) where wave steepness and slope angle do not influence the run-up.

The coefficients a, b, c and d are given in Table 29.2. The coefficient values reflect the fact that the run-up level varies considerably from wave to wave in irregular seas.

With respect to run-up levels on smooth slopes reference is made to Ahrens (1981), Allsop *et al.* (1985), and TAW (1974). The latter also contains information on the reduction in run-up caused by various types of surface roughness.

In real seas with short-crested waves the highest run-up is caused by head-on waves (angle of incidence, $\beta = 0$) and is equal to the run-up for long-crested waves. For smooth slopes the run-up decreases very gradually with β.

The run-down is usually between 1/3 and 1/2 of the run-up. The maximum run-down level might be used to indicate the minimum downward extention of the

Table 29.2 Coefficients of Equation (29.5) for rock armoured slope

Run-up exceedence probability level, p %	*a*	*b*	*c*	*d*
0.1	1.12	1.34	0.55	2.58
2	0.96	1.17	0.46	1.97
5	0.86	1.05	0.44	1.68
10	0.77	0.94	0.42	1.45
50	0.47	0.60	0.34	0.82

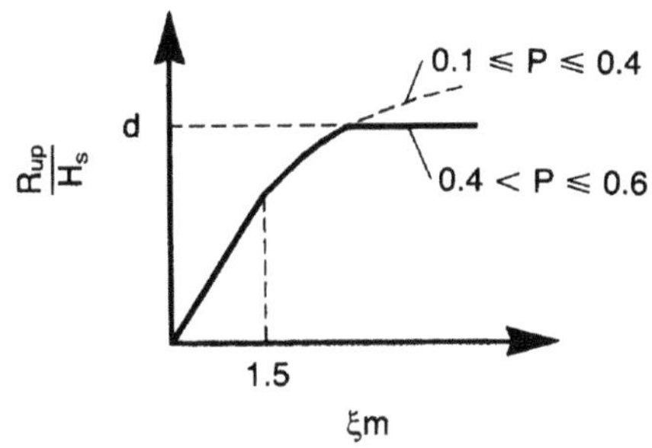

Figure 29.16 Run-up formula (Equation (29.5)).

main armour and a possible upper level for introducing a berm with reduced armour size.

Overtopping

Overtopping occurs only if the run-up level exceeds the freeboard, R, of the structure; see Figure 29.17, which shows the notation used to describe the profile geometry.

The relative freeboard, R/H_s, is the simplest possible but most important dimensionless parameter for the prediction of overtopping. However, the wave period or steepness is also of significant importance as is the geometry including permeability, porosity and surface roughness of the structure. A recurved wave wall as shown in Figure 29.16d is under certain conditions effective in reducing overtopping. For small values of R/H_s (<0.3) when the overtopping is excessive, the detailed geometry of the crest part of the structure becomes less important because the waves just travel over the structure. Formulae for overtopping are empirical as they are fitted to hydraulic model test results for specific breakwater geometries.

Owen (1980, 1982) studied the overtopping of straight and bermed, smooth and rough slopes without crown walls (see Figure 29.17a,b). He defined a dimensionless freeboard which includes the wave steepness:

$$R_m^* = \frac{R}{H_s}\left(\frac{s_m}{2\pi}\right)^{0.5} \qquad (29.6)$$

and arrived at the following equation for the mean overtopping discharge:

$$\bar{Q}/(g H_s T_m) = a \exp(-b R_m^*/r) \qquad (29.7)$$

where a and b are empirical coefficients given in Tables 29.3 and 29.5 and r is a reduction coefficient to include the effect of surface roughness relative to the smooth surface. Owen suggests the use of the TAW (1974) run-up reduction factors some of which are given in Table 29.6.

The overtopping of breakwaters with crown walls has been studied by Bradbury *et al.* (1988a,b). Jensen (1984) has reported the model test results of several case studies of overtopping for various breakwater configurations. He also discusses the horizontal distribution of the average overspill intensity. Fukuda *et*

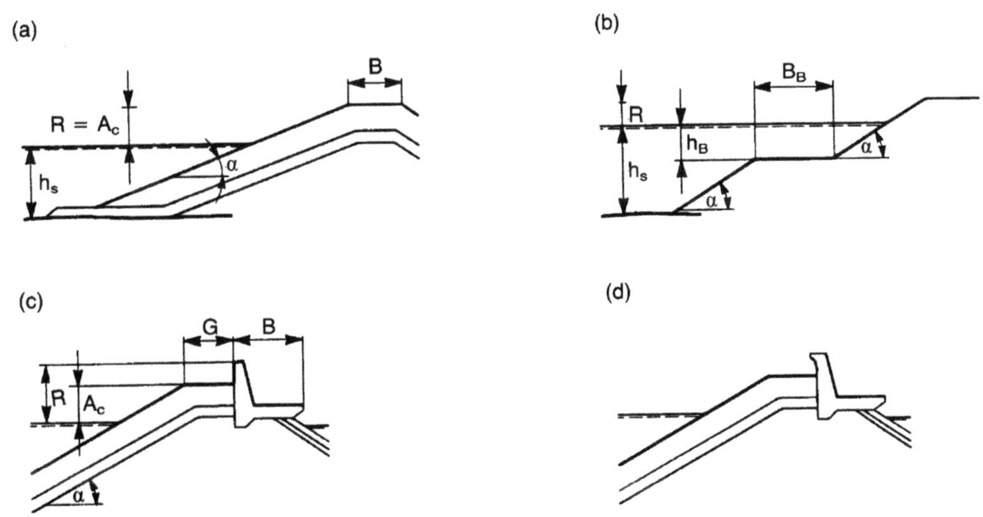

Figure 29.17 Geometrical parameters related to overtopping. (a) straight slope; (b) bermed slope; (c) sloping crown wall; (d) recurred crown wall.

Table 29.3 Coefficients in Equation (29.7) for overtopping discharges for bermed smooth slopes. Non-depth limited waves

Slope	h_B *(m)*	B_B *(m)*	$a \cdot 10^4$	*b*
1:1	−4.0	10	64	20
1:2			91	22
1:4			145	41
1:1	−2.0	5	34	17
1:2			98	24
1:4			159	47
1:1	−2.0	10	48	19
1:2			68	24
1:4			86	46
1:1	−2.0	20	8.8	15
1:2			20	25
1:4			85	50
1:1	−2.0	40	3.8	23
1:2			5.0	26
1:4			47	51
1:1	−1.0	5	155	33
1:2			190	37
1:4			500	70
1:1	−1.0	10	93	39
1:2			340	53
1:4			300	80
1:1	−1.0	20	75	46
1:2			34	50
1:4			39	62
1:1	−1.0	40	12	49
1:2			24	56
1:4			1.5	63
1:1	0	10	97	42
1:2			290	57
1:4			300	80

Table 29.4 Coefficients in Equation (29.7) for overtopping discharges for straight smooth slopes. Non-depth limited waves

Slope	*a*	*b*
1:1	0.008	20
1:1.5	0.010	20
1:2	0.013	22
1:3	0.016	32
1:4	0.019	47

al. (1974) presented, based on field observation, the extent of the influence of various overtopping intensity on persons, cars and buildings situated a few meters behind the breakwater crest. As mentioned earlier for run-up, overtopping on impermeable smooth slopes is hardly affected by the angle of incidence, especially in the case of short-crested seas.

Table 29.5 Run-up reduction factors *r* (TAW, 1974)

Smooth impermeable (including smooth concrete and asphalt)	1.0
One layer of stone rubble on impermeable base	0.8
Gravel, gabion mattresses	0.7
Rock rip-rap with thickness greater than $2D_{n50}$	0.5–0.6

Wave transmission

Waves behind a structure can be caused by overtopping, and for permeable structures also by wave penetration. Wave transmission can be characterized by a transmission coefficient, C_t, defined either as the ratio between transmitted and incident characteristic wave heights (e.g. H_{st} and H_s) or as the square root of the ratio between transmitted and incident time averaged wave energy (e.g. E_t and E),

$$C_t^{H_s} = H_{st}/H_s \tag{29.8}$$

$$C_t^E = (E_t/E)^{0.5} = (m_{0t}/m_0)^{0.5} \tag{29.9}$$

where m_0 is the first moment (area) of the wave amplitude density spectrum. The relationship between H_s and m_0 depends on the wave height distribution, e.g. $H_s = 4.00\sqrt{m_0}$ for wave trains with Rayleigh distributed wave heights.

The wave height distribution of the incident and the transmitted waves are not generally Rayleigh distributed but if we assume as a reasonable assumption the same distribution for incident and transmitted waves then $C_t^{H_s} = C_t^E$, and we can define a common transmission coefficient

$$C_t = H_{st}/H_s = (E_t/E)^{0.5} = (m_{0t}/m_0)^{0.5} \tag{29.10}$$

Specific transmission coefficients for wave overtopping (sub-index *o*) and wave penetration (sub-index *p*) can be defined as:

$$C_{to} = H_{so}/H_s \tag{29.11}$$

$$C_{tp} = H_{sp}/H_s \tag{29.12}$$

If again the same wave height distribution is assumed for incident waves and those caused by penetration and overtopping we can relate the various coefficients:

$$C_t = (C_{to}^2 + C_{tp}^2)^{0.5} \tag{29.13}$$

It should be noted that the wave height distribution of waves caused by overtopping can depart significantly from the Rayleigh distribution as can incident waves in very shallow waters.

Values of C_t given in the literature are almost all from laboratory experiments, many of these per-

Table 29.6 K_D-values for a two layer quarry rock armour. $H = H_s$. Slope angles $1.5 < \cot\alpha \leqslant 3.0$ (*Shore Protection Manual, 1973*)

Stone shape	*Placement*	*Damage, D*			
		0–5%		*5–10%*	*10–15%*
		*Breaking waves**	*Non-breaking waves†*	*Non-breaking waves*	*Non-breaking waves*
Smooth, rounded	Random	2.1	2.4	3.0	3.6
Rough angular	Random	3.5	4.0	4.9	6.6
Rough angular	Special§	4.8	5.5		

* Breaking waves means depth limited waves, i.e. wave breaking takes place in front of the slope.
† No wave breaking, due to depth limitation, takes place in front of the slope.
§ Special placement with long axis of stone placed perpendicular to the slope face.

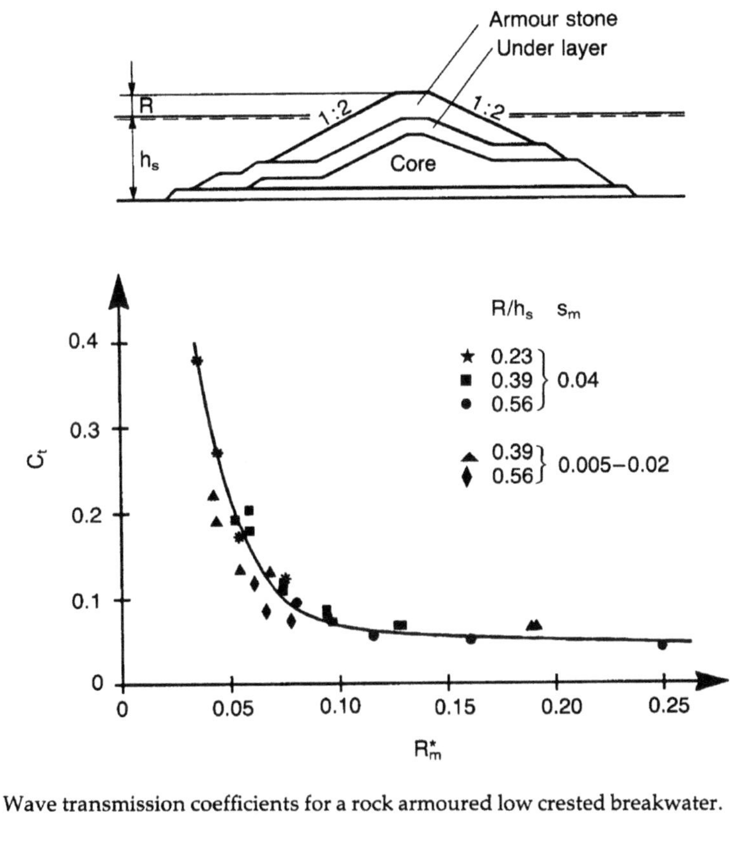

Figure 29.18 Wave transmission coefficients for a rock armoured low crested breakwater.

formed to rather small scales. Significant scale effects might be present especially for the proportion of C_t stemming from wave penetration.

Allsop (1983) tested a rock armoured low-crested rubble mound breakwater under random wave attack. The test specifications and the related total coefficient of wave transmission, C_t, are given in Figure 29.18. A clear trend between C_t and R_m^* (defined by Equation (29.6)) is seen.

Powell and Allsop (1985) tested the cross-section shown in Figure 29.18 under submerged-crest conditions. Besides this they also tested homogenous structures. The results are given in Figure 29.19.

Breakwaters with complex types of concrete armour

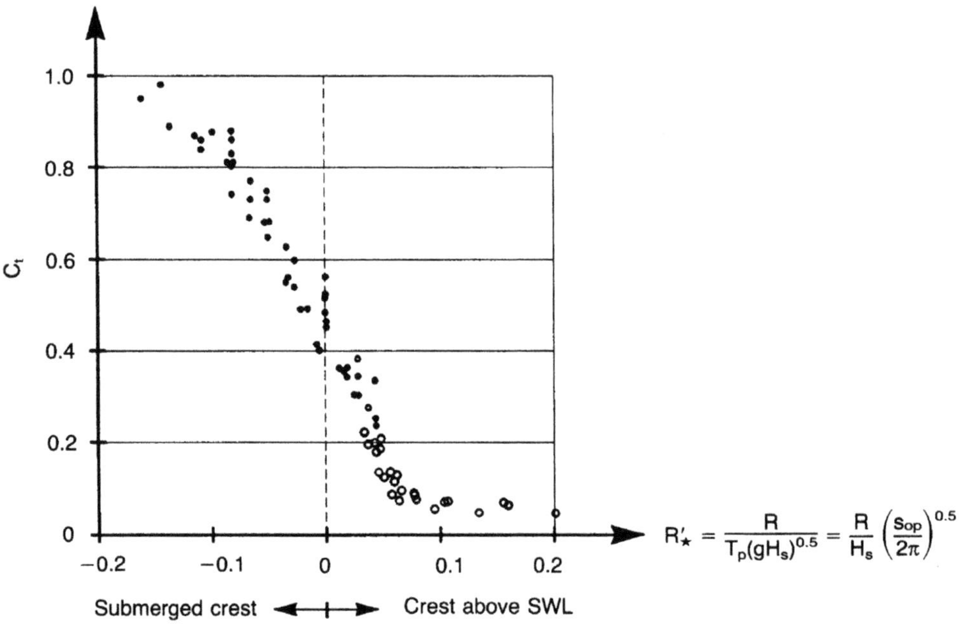

Figure 29.19 Wave transmission coefficients, C_t, for submerged and surface piercing rock armoured breakwater; (○), multi-layer structure; (●), homogeneous structure; crest width/wave length = 0.03 − 0.07.

units, such as tetrapods, have generally a more permeable crest than rock armoured breakwaters, and consequently also larger transmission coefficients.

Detached breakwaters for coastal protection are invariably sited in very shallow water (Figure 29.2f) and are often built entirely of armour blocks, i.e. without underlayer and core. Such breakwaters are very permeable and C_{tp} can reach 0.8 in the case of complex armour units and small wave steepnesses.

Wave reflection

Breakwaters reflect some proportion of the incident wave energy. If significant, the interaction of incident and reflected waves can create a very confused sea with very steep and often breaking waves. It is a well known problem in many harbour entrance areas where it can cause considerable manoeuvring problems to small vessels. A strong reflection also increases the sea-bed erosion potential in front of the structure. Moreover, waves reflected from breakwaters can in some cases create or increase erosion of neighbouring beaches.

Non-overtopped impermeable smooth steep slopes will reflect almost all the incident wave energy while a permeable mild slope rubble mound will absorb a significant portion of the energy and is therefore well suited as a wave absorber, for example, in harbour basins.

The reflection can be quantified by the reflection coefficient:

$$C_r = H_{sr}/H_s = (E_r/E)^{0.5} \qquad (29.14)$$

where H_s and H_{sr} are the significant wave heights of incident and reflected waves, respectively. E and E_r are the related energies. The last equality in (29.14) holds only if the same wave height distribution is present in the incident and the reflected waves (compare with the section on wave transmission).

Values of C_r given in the literature are almost without exception obtained from rather small-scale laboratory model studies and for this reason some bias due to scale effects may be present.

Non-overtopped slope

For non-overtopped slopes the wave reflection is related to the wave dissipation, which for smooth impermeable slopes depends strongly on the type of wave breaking, which again can be characterized by ξ, see section 29.4.

Based on the reflection criterion by Miche, Battjes (1974) suggested for regular waves on smooth impermeable slopes

$$C_r = 0.1\xi^2 \qquad (29.15)$$

which fits well the experimental data for mild slopes ($\cot\alpha \simeq 3 - 6$) and $\xi < 2.5$. Seeling and Ahrens (1981) and Seeling (1983) proposed:

$$C_r = a\xi_p^2/(b + \xi_p^2) \qquad (29.16)$$

where $a = 1.0$ and $b = 5.5$ for smooth impermeable slopes; $a = 0.6$ and $b = 6.6$ for rubble mound breakwaters (conservative estimate). Other references on wave reflection are van der Meer (1991) and Allsop and Hettiarachchi (1988).

Overtopped slopes

Transmission of waves by overtopping and penetration will reduce the reflection coefficient. For such cases estimates of C_r can in principle be obtained from the equation

$$E = E_r + E_t + E_d \qquad (29.17)$$

where E, E_r, E_t and E_d are the incident, reflected, transmitted, and dissipated energies, respectively. The equation can be reformulated as

$$1 = \frac{E_r}{E} + \frac{E_t}{E} + \frac{E_d}{E} = C_r^2 + C_t^2 + C_d^2 \qquad (29.18)$$

C_t can be estimated, but only if C_d can be quantified will it be possible to calculate C_r.

Hydrodynamics related to armour layer stability

The movement of armour units due to flow forces is called 'hydraulic instability'. Breakage of the units is another type of instability, which will be discussed later in this chapter.

The forces acting on an armour unit are schematized in Figure 29.20.

The resulting force F_F, which tries to move the unit from its original position in the mound, can be interpreted as the vectorial sum of a drag force F_D, a lift (drag) force F_L, and an inertia force F_I, i.e. $\bar{F}_F = \bar{F}_D + \bar{F}_L + \bar{F}_I$.

Applying the Morison equation the flow forces on a unit at rest can be expressed as follows:

$$\begin{aligned} F_D &\approx C_D \rho_w A \bar{v}\,|v| \\ F_L &\approx C_L \rho_w A \bar{v}\,|v| \\ F_I &\approx C_I \rho_w V \frac{d\bar{v}}{dt} \end{aligned} \qquad (29.19)$$

where ρ_w is the mass density of water, A is the cross-sectional area at right angles to $\bar{v}$ and V is the volume of the unit. C_D, C_L and C_I are time dependent empirical coefficients, which depend on the shape of the unit, on the ratio between inertia and viscous forces (i.e. a Reynolds number, Re) and on the ratio between the amplitude of the water particle oscillations and the armour unit diameter (i.e. Keulegan–Carpenter number, KC).

The stabilizing force is the gravitational force F_G. The forces acting at the contact points with neighbouring units might be stabilizing or destabilizing depending on the position of the contact point and the direction of the contact point force.

The flow, characterized by the velocity v, is highly non-stationary with respect to both velocity and direction. Thus all forces except the gravity force vary in size and direction with time.

The forces at the contact points are static when no wave action and settlements are present, but will be quasi-static (pulsating) under wave action. Breaking waves can cause wave slamming which causes impact-like contact forces. Solid body impact forces occur when the units are moving.

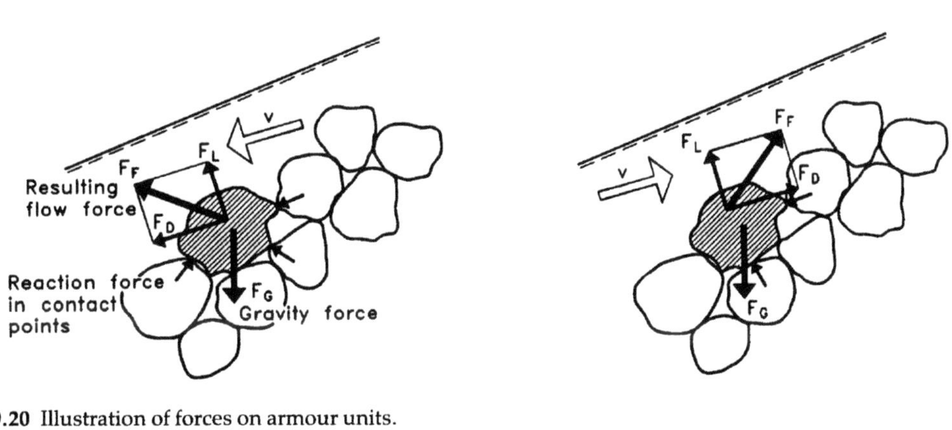

Figure 29.20 Illustration of forces on armour units.

v depends on the properties of the incoming wave and its action on the slope. The latter includes interaction with the preceding wave, i.e. the succession of the waves in irregular seas plays a role (wave grouping). v is also affected by the permeability, the porosity and the surface roughness of the structure (see Figure 29.13).

Also when considering the complicated shape of armour units, the random placement and the very complex flow field it is clear that a deterministic calculation of the instantaneous forces and the stability conditions of the armour units are impossible.

Simple assumptions about the geometry of the units and the flow make it possible to formulate expressions for the armour stability. The armour unit geometry might be characterized by an equivalent cube length, $D_n = (M/\rho)^{1/3}$, where M is the mass and ρ the mass density. If the flow field is assumed quasi-stationary, i.e. the inertia forces are neglected, then a qualitative stability ratio can be formulated as the gravity force divided by the drag plus the lift force:

$$\frac{F_G}{F_D + F_L} \approx \frac{g(\rho - \rho_w)D_n}{\rho_w v^2} K \qquad (29.20)$$

where the dimensionless K is a function of the structural parameters, a Reynolds number, Re, and the amount of armour unit movements, i.e. the damage level. v might be substituted by the characteristic particle velocity, $v \approx (gH)^{1/2}$, in which case K also depends on T_m (or s_m), ξ and the number of waves N_z. If $\Delta = (\rho/\rho_w) - 1$ is introduced we obtain

$$\frac{H}{\Delta D_n} \leqslant K \qquad (29.21)$$

for non-exceedence of a certain degree of damage (instability). $H/(\Delta D_n)$ can be regarded as a stability parameter. It is often denoted by N_s in the literature.

The term signifying the influence of the slope might be derived under very simplified conditions as is exemplified in Figure 29.21, where the gravity force and the resultant flow force are assumed to act at the same point of the unit of 'diameter', D_n.

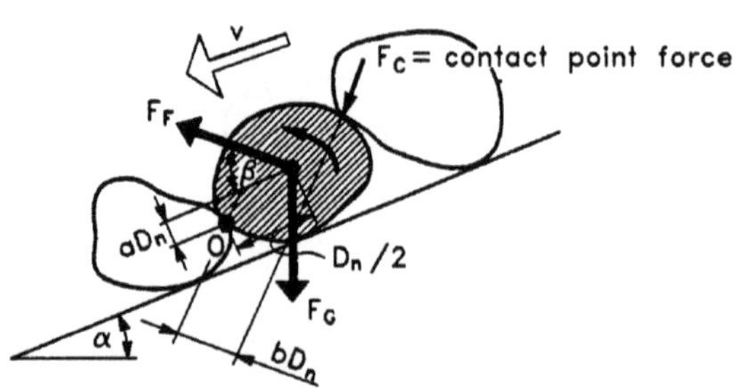

Figure 29.21 Displacement by rotation during down-rush.

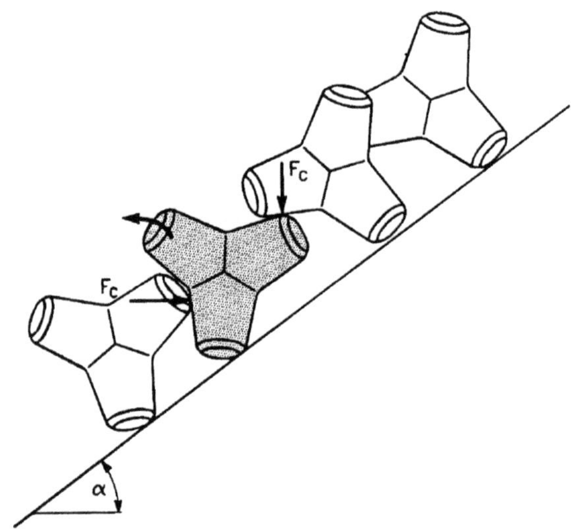

Figure 29.22 Illustration of contact forces for interlocking types of armour units.

The contact point force F_C is a function of the gravity forces and flow forces on the up-slope units. The direction of F_C depends on the surface friction and shape of the units. F_C depends, in the case of fairly mild slopes, mainly on flow forces, i.e. $F_C \approx nF_F$, whereas in the case of steep slopes F_C depends mainly on gravity forces, i.e. $F_C \approx nF_G$. The factor n might typically be 0–0.5 for mild slopes, but up to and even larger than 1 for very steep slopes. Figure 29.22 illustrates the large stabilizing, interlocking effect of the contact forces which are one of the features of the complex types of concrete armour units. The maximum interlocking effect in terms of pull-out resistance is obtained for fairly steep slopes, the size of which depends on the type of units. Price (1979) found by pull-out tests in the dry maximum resistance at slopes of $\cot\alpha = 2$ for Dolosse. The effect of interlocking and surface friction on stability is schematized in Figure 29.23.

For rock armour on the basis of model tests with regular waves Hudson (1958, 1959) found that:

$$\frac{H}{\Delta D_n} \leqslant K(\cot\alpha)^{1/3} \qquad (29.22)$$

The empirical coefficient K is determined from model test results but will be different because the slope terms are different. The values of K given in the literature are related to the type of armour and the degree of damage.

Formulae of this type will never be very precise over the whole realistic parameter ranges because important parameters, as for example wave steepness,

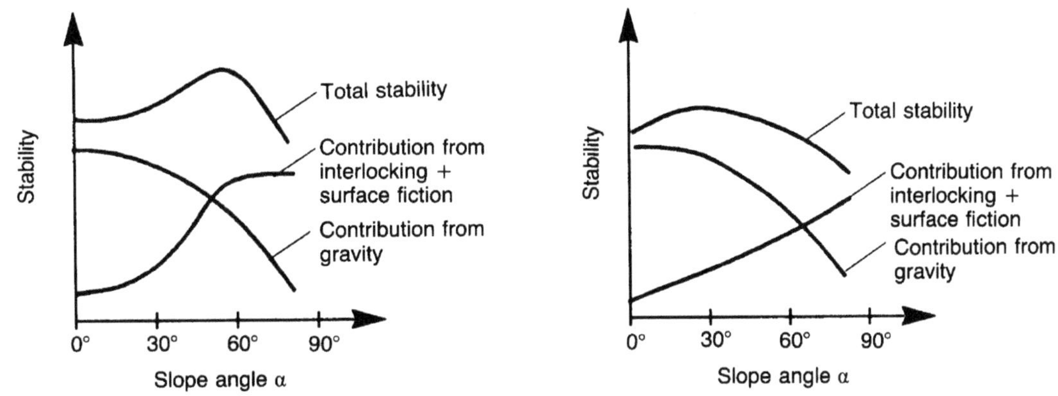

Figure 29.23 Illustration of effect of interlocking and interblock surface friction on stability. (a) complex interlocking types of armour units like Dolos; (b) bulky types of armour units like cubes and quarry rock.

do not enter explicitly into the formulae. It is therefore important to check the range of the model test parameters on which the fitted coefficient K is based.

Some stability tests given in the literature are presented by graphs where wave steepness or wave period or wave length are taken into account. Günbak (1976) related $H/\Delta D_n$ to ξ for various values of α.

Figure 29.24 Definition of eroded cross-section area.

Definition of armour layer damage

Each of the functional relationships between the environmental parameters and the structural parameters contains a parameter which signifies the amount of damage. The damage related to hydraulic stability can be characterized either by counting the number of moving units or by measurement of the surface profile of the slope. They are called relative damage (D) and damage number (S), respectively.

Relative damage D is defined as

$$D = \frac{\text{number of displaced units}}{\text{number of units in test area}} \qquad (29.23)$$

Broderick (1984) defined for rip-rap and rock armour a dimensionless damage number S

$$S = A_e/D_{n50}^2 \qquad (29.24)$$

where A_e is the eroded area and $D_{n50} = (W_{50}/\rho g)^{1/3}$, W_{50} is the median weight of rocks (see Figure 29.24).

If the porosity of the armour is considered then the actual number of units displaced within a strip width D_{n50} can be related to S if settlements are disregarded. Assuming the eroded profile to be two-dimensional, i.e. A_e is constant and represents an average along the length of the breakwater, then the actual number of eroded units is $N_{od} = G(1 - p)S$, where p is the porosity of the armour layer and G is a factor dependent on the gradation. A very significant influence of slope angle on the relationship between S and N_{od} was found by van der Meer (1988) and Holtzhausen and Zwamborn (1990).

Practical formulae for hydraulic stability of armour layers and toe berms in trunks, exposed to head-on waves

Rock armour, non-overtopped slopes

Hudson formula Hudson considered all relevant parameters for rubble mound stability but arrived at the following simple equation containing only the most important parameters.

Medium mass of rock

$$M_{50} = \frac{\rho H^3}{K_D\left(\dfrac{\rho}{\rho_w} - 1\right)^3 \cot\alpha} \qquad (29.25)$$

or written in terms of Δ and D_{n50}

$$N_s = \frac{H}{\Delta D_{n50}} = (K_D \cot\alpha)^{1/3} \qquad (29.26)$$

The *Shore Protection Manual* (1973) recommends for quarry stones placed in two layers the K_D-values given in Table 29.6.

The definition of the coefficient, D, is given in the previous section. The damage level $D = 0$–5% is denoted 'no-damage' in the *Shore Protection Manual* and signifies a very moderate degree of damage. The wave height to be used in Equation (29.25) is recommended to be H_s in the *Shore Protection Manual* (1973). The 1984 edition of the *Shore Protection Manual* recommends a much more conservative approach by using $H = H_{1/10}$ together with generally smaller K_D-values.

The uncertainty inherent in the use of the Hudson formula for rock armour has been estimated by van der Meer to a coefficient of variation of 18%.

Van der Meer formula – Deep water conditions

Plunging waves, $\xi_m < \xi_{mc}$

$$\frac{H_s}{\Delta D_{n50}} = 6.2\, S^{0.2} P^{0.18} N_z^{-0.1} \xi_m^{-0.5} \qquad (29.27)$$

Surging waves, $\xi_m > \xi_{mc}$

$$\frac{H_s}{\Delta D_{n50}} = 1.0(S^{0.2} P^{0.13} N_z^{-0.1}) \cot\alpha^{0.5} \xi_m^P \qquad (29.28)$$

where

$$\xi_m = \tan\alpha\, s_m^{-0.5} \quad \text{and}$$
$$\xi_{mc} = (6.2\, P^{0.31} \tan\alpha^{0.5})^{1/(P+0.5)} \qquad (29.29)$$

For $\cot\alpha \geq 4.0$ only Equation (29.27) should be used.

P and S have been defined in the sections on run-up, run-down and internal flow, and on the definition of armour layer damage. N_z is the number of waves.

The limits of Equations (29.27) and (29.28) are given by

$$0.1 \leq P \leq 0.6, \quad 0.005 \leq s_m \leq 0.06,$$
$$2.0\, t/m^3 \leq \rho \leq 3.1\, t/m^3$$

$N_z \leq 7500$ after which number equilibrium damage is more or less reached. The formulae are based on tests with rock material gradings of $D_{85}/D_{15} = 1.25$ and 2.25. The coefficient of variation in Equations (29.27) and (29.28) are estimated at 6.5% and 8%, respectively.

Van der Meer formula – Shallow water conditions (depth limited waves)

Plunging waves

$$\frac{H_{2\%}}{\Delta D_{n50}} = 8.7\, S^{0.2} P^{0.18} N_z^{-0.1} \xi_m^{-0.5} \qquad (29.30)$$

Surging waves

$$\frac{H_{1\%}}{\Delta D_{n50}} = 1.4\, S^{0.2} P^{-0.13} N_z^{-0.1} \cot\alpha^{0.5} \xi_m^P \qquad (29.31)$$

The wave height distribution for shallow water conditions is truncated compared to the Rayleigh distribution for which $H_{2\%} = 1.4 H_s$. Equations (29.27) and (29.28) are equal to (29.30) and (29.31) for Rayleigh distributed wave heights.

Rock armour, overtopped, but not submerged, breakwaters

The stability formulae (Equations (29.27)–(29.31)) are used with a reduction factor f_i multiplied by D_{n50}, suggested by van der Meer (1991) to be

$$f_i = \left(1.25 - 4.8\frac{R}{H_s}\sqrt{\frac{s_m}{2\pi}}\right)^{-1} \qquad (29.32)$$

where $R \geq 0$ is the freeboard.

Limits of the formula are given by

$$0 < \frac{R}{H_s}\sqrt{\frac{s_m}{2\pi}} < 0.052$$

Other references are Ahrens *et al.* (1982), Ahrens (1984), Allsop (1983), Jensen (1984).

Rock armour, submerged (reef type) breakwaters

The reader should consult Ahrens *et al.* (1982), Ahrens (1984), Givler and Sørensen (1986) and Ahrens (1987) for information on this topic.

Rock armour, rear slope of breakwaters without superstructure

Jensen (1984) reports results from two case studies of conventional rock armoured rubble mound breakwaters with the main armour carried over the crests and the upper part of the rear slope and crest width of approximately 3–4 stone diameters.

Cubes, non-overtopped slopes

The Hudson formula

$$M = \frac{\rho H_s^3}{K_D\left(\frac{\rho}{\rho_w} - 1\right)^3 \cot\alpha}$$

or

$$N_s = \frac{H_s}{\Delta D_n} = (K_D \cot\alpha)^{1/3} \qquad (29.33)$$

is often used for concrete cubes.

For a two-layer armour of randomly placed cubes K_D is typically in the range 4–8 for zero and moderate damage.

Table 29.7 Average N_s and corresponding K_D-values for a two-layer concrete cube armour. Random placement. Slope angles $1.5 \leqslant \cot\alpha \leqslant 2.0$. Non-depth limited waves (Brorsen *et al.*, 1974)

Damage level	$N_s = \frac{H_s}{\Delta D_n}$	K_D	
		slope 1:1.5	*slope 1:2*
Onset, D = 0%	1.8–2.0	3.9–5.3	2.9–4.0
Moderate, D = 4%	2.3–2.6	8.1–12	6.1–8.8

Brorsen *et al.* (1974) performed comparative stability tests with rock, cube and Dolos armoured slopes and found for $1.5 < \cot\alpha < 2.0$, non-depth limited waves and $0.01 < s_m < 0.04$ the stability coefficients given in Table 29.7 apply.

Hudson (1974) reports a K_D-value of 7.8, corresponding to a damage level of $D = 0–5\%$, for a two-layer random placed cube armour exposed to non-depth limited waves.

Van der Meer (1988b) tested a two-layer cube armour with one steep slope angle, $\cot\alpha = 1.5$, and found for $3 < \xi_m < 6$ (corresponding to $0.05 > s_m > 0.01$), and non-depth limited waves that:

$$N_s = \frac{H_s}{\Delta D_n} = (6.7\, N_{od}^{0.4}/N_z^{0.3} + 1.0)\, s_m^{-0.1} \quad (29.34)$$

The reliability of Equation (6.30) is estimated to correspond to a coefficient of variation of 0.1.

Analysis of the van der Meer test results gives the following relationship between the damage level definitions as $D = N_{od}/33.5$ where 33.5 is the number of units in the model within a strip of width D_n = cube length.

If Equation (29.34) is converted to K_D-values we obtain for $N_{od} = D = 0$, i.e. for the start of damage, $1.6 < K_D < 2.7$, corresponding to the wave steepness range $0.01 < s_m < 0.05$. These K_D-values are only half of those given in Table 29.13. For $D = 4\%$ corresponding to $N_{od} = 1.34$ Equation (29.34) gives for the same wave steepness range $9.7 < K_D < 15.8$ which compares well to the figures in Table 29.13. Special placement in patterns is sometimes used instead of random placement. In this way the stability can be increased but at the same time the overtopping will generally increase.

Parallelepiped blocks, with superstructure

Aalborg University Hydraulics Laboratory tested the hydraulic stability of parallelepiped blocks randomly placed in two layers on a slope of 1:2. The stability number is given by

$$N_s = \frac{H_s}{\Delta D_n} = 1 + 2.72 D^{0.25} \quad (29.35)$$

The reliability of the formula is estimated to correspond to a coefficient of variation 0.16.

Tetrapods, non-overtopped slopes

SOGREAH generally recommends for two-layer tetrapod armour slope angles in the range $1.25 < \cot\alpha < 1.5$. SOGREAH has no recent recommendations for the K_D-values, but has some recommendations as to the method of placement of the units.

The K_D-values given by Hudson (1974) and the *Shore Protection Manual* (1984) are shown in Table 29.8.

Van der Meer (1988b) tested a two-layer tetrapod armour on a slope with one slope angle, $\cot\alpha = 1.5$, and found that for $3.5 < \xi_m < 6$ and non-depth limited waves:

$$N_s = \frac{H_s}{\Delta D_n} = (3.75\, N_{od}^{0.5}/N_z^{0.25} + 0.85) s_m^{-0.2} \quad (29.36)$$

where $N_{od} = 0$ and $N_{od} = 1–2$ corresponds to the start of damage and severe damage, respectively. The reliability of Equation (29.36) may be given by a coefficient of variation of 0.10. Note that breakage of tetrapods and not hydraulic instability might be the

Table 29.8 K_D-values for a two-layer random placed tetrapod armour for damage level, $D = 0 - 5\%$

	Non-breaking waves	*Breaking waves*
Hudson (1974)	8.3*	7.2*
Shore Protection Manual (1984)	8.0†	7.0†

* Values to be used with H_s in equation (29.33)
† Values recommended in *Shore Protection Manual* (1984) to be used with $H_{1/10}$ instead of H_s in Equation (29.33) in order to introduce a safety factor. For comparison with other formulae (like (29. 36)) expressing central estimates on the stability, H_s should be used.

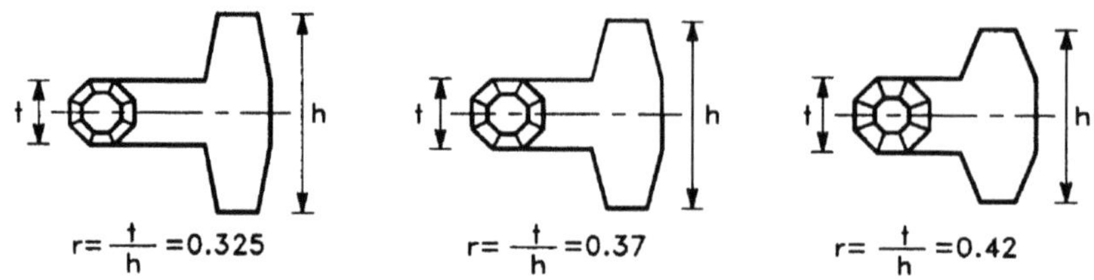

Figure 29.25 Examples of Dolosse with different waist ratios but with equal mass.

limiting design condition (see section on the structural integrity of concrete armour units, later in this chapter).

Accropodes (R), non-overtopped slopes

SOGREAH recommends that the units be placed in one layer on steep slopes corresponding to $\cot\alpha = 1.33$. The units are recommended to be placed in a regular pattern with a down-slope spacing of 0.6 block height and along slope (horizontal) spacing of 0.64 block height. For an initial choice of size SOGREAH recommends the use of the Hudson formula with $K_D = 15$ for non-breaking wave conditions and $K_D = 12$ for breaking wave conditions. SOGREAH states that these K_D-values correspond to a safety factor of 1.2 on the design wave height. No information about the related damage level is given. K_D-values of 15 and 12 correspond to N_s-values of 2.71 and 2.52, respectively, for $\cot\alpha = 1.33$.

Van der Meer (1988b) at Delft Hydraulics studied accropode stability under non-overtopped conditions for one slope angle, $\cot\alpha = 1.33$. Holtzhausen and Zwamborn (1991) at CSIR studied accropode stability under non-overtopped conditions for slopes, $\cot\alpha = 1.33$, 1.5 and 2.0, and compared the results with those of van der Meer. Stability tests with accropodes have also been presented by the First Design Institute of Navigation Engineering, Tianjin (1987). Generally the agreement between the results is not very good.

Delft Hydraulics found that 'no damage' corresponds to $N_s = 3.7$ and failure corresponds to $N_s = 4.1$. This means that an increase in H_s of only 10% may turn Accropode armour from a no-damage stage to failure. The CSIR tests also showed that Accropode armour on steep slopes fail, as a sudden collapse in which most of the units are dislocated as they slide towards the bottom of the slope.

Consequently, the design values of N_s must be chosen well below the limit of the onset of damage in order to accommodate the uncertainties related to design wave climate and model tests.

Dolos, non-overtopped slopes

The Dolos is a multi-shaped unit, the strength of which can be adjusted to prevent breakage by changing the waist ratio, r = t/h, Figure 29.25. This section deals only with the hydraulic stability while structural integrity is discussed in a later section where design diagrams are also presented.

The hydraulic stability of Dolos armour is dependent on the waist ratio and the packing density. Based on the results of Brorsen *et al.* (1974), Burcharth and Brejnegaard-Nielsen (1986), Holtzhausen and Zwamborn (1990) and Burcharth and Liu (1992), the following formula for the hydraulic stability of Dolos armour on a slope 1:1.5 can be established (Burcharth and Liu, 1992).

$$N_s = \frac{H_{mo}}{\Delta D_n} = (47 - 72r)\varphi_{n=2} D^{1/3} N_z^{-0.1} \qquad (29.37)$$

where: H_{mo} is the significant wave height in front of breakwater; r is the Dolos waist ratio; $\varphi_{n=2}$ is the packing density for two layer armour; D is the relative number of units within levels SWL $\pm$ 6.5 Dn displaced one Dolos height h, or more (e.g. for 2% displacement $D = 0.02$); and N_z is the number of waves. For $N_z \geqslant 3000$ use $N_z = 3000$.

Damage is expressed only in terms of relative displacement. The amount of rocking units is not included. This is because the rocking mode is relevant only to breakage aspects which are dealt with specifically in the structural integrity diagrams presented later in this chapter.

Equation (29.37) covers both breaking and non-breaking wave conditions, the limits being given by the following parameter ranges:

$$0.32 < r < 0.42$$
$$0.61 < \varphi_{n=2} < 1$$
$$1\% < D < 15\%$$

The uncertainty of the formula is estimated to correspond to a coefficient of variation of approximately 0.22.

Hydraulic stability of armour in roundheads

The hydraulic stability of armour in cone shaped roundheads is reduced compared to the stability in trunks. One reason is the very high cone-overflow velocities, sometimes enhanced in certain areas by

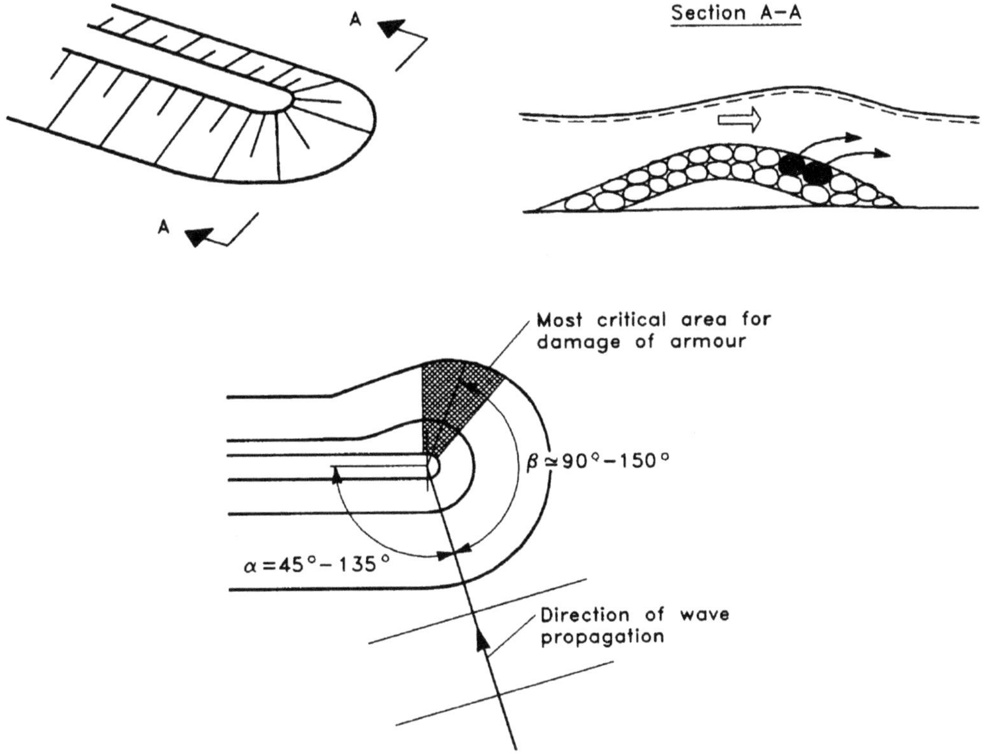

Figure 29.26 Illustration of critical areas for damage to armour layers in cone shape roundheads.

refraction. Another reason is the reduced support from neighbouring units in the direction of wave overflow on the lee side of the cone, Figure 29.26. The figure also illustrates the position of the most critical area for armour layer instability. The toe within the same area is also very vulnerable to damage in the case of shallow water situations, and a toe failure will often trigger failure of the armour layer. As can be seen from Figure 29.26, section A-A, a flatter slope means a better lateral support of the individual armour units against pullout. However, at the same time the stabilizing friction forces and interlocking forces between the units reduce. Consequently, it is very difficult to find the slope which produces optimum hydraulic stability. For complex types of armour blocks a steeper slope than for bulky units might be advantageous, also because a steeper slope reduces rocking movements which can cause breakage of slender units.

Table 29.9 shows ratios of K_D-values for roundheads and trunks extracted from SPM (1984) for two layer randomly placed armour. The coefficients given in Table 29.9 have not been verified by tests.

Hydraulic stability of berm breakwaters (reshaping breakwaters)

Figure 29.4d shows a typical cross-section of a berm breakwater. The efficiency of the S-profile has been known for many years and used for dikes and breakwaters with conventional armour layers. Within recent years (starting with papers by Baird and Hall (1984)), more attention has been given to the concept of natural reshaping of the front slope by wave action. The natural reshaping produces the most efficient profile, in terms of resistance, for given conditions. Development of a front face towards an S-profile takes place when $H_s/\Delta D_{n50} > \Omega 2.5$.

One of the basic ideas of the berm breakwater concept is to utilize all classes of stones produced in the quarry. Another aspect is that smaller armour stones can be applied compared to the stones in the armour of conventional multi-layer structures. Also construction is less demanding, although experience has shown that berm breakwater armour layer stones should also be carefully controlled with respect to size and be kept free of larger quantities of fine

Table 29.9 Ratios of K_D-values for roundhead and trunk for two layer randomly placed armour. Extracted from *Shore Protection Manual* (1984)

Armour	*cot a*	$K_D^{roundhead}/K_D^{trunk}$ *for damage* $D = 0\text{–}5\%$. *Breaking waves*	*Non-breaking waves*
Quarry rock rough, angular	1.5	0.95	0.80
	2.0	0.80	0.70
	3.0	0.65	0.58
Tetrapod	1.5	0.71	0.75
	2.0	0.64	0.69
	3.0	0.50	0.50
Dolos	2.0	0.51	0.50
	3.0	0.44	0.44

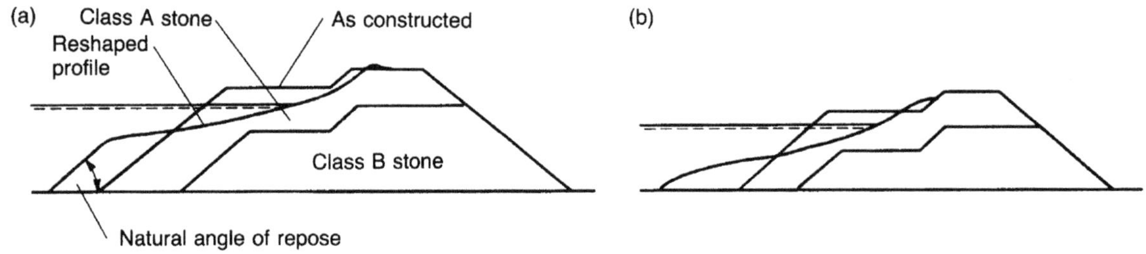

Figure 29.27 Typical berm breakwater profiles before and after reshaping in: (a) deep water; (b) shallow-water conditions.

materials in order to ensure wave dissipation in the otherwise blocked pores. This means that construction roadways made of fine materials placed directly on already constructed parts of the berm should be avoided. If they exist as part of the construction then the contaminated part of the armour should be removed at a later stage. Fournier *et al.* (1990) studied the influence of contamination and found that contamination due to a narrow roadway close to the core crest did not significantly influence the reshaped front slope profile. However, the contamination had a significant effect on the stability of the crest and the rear slope due to increased overtopping.

Stability of trunks

Figure 29.27 shows a typical cross-section built of two classes of stone gradations. The profile as constructed and after reshaping due to wave attack is indicated. In the case of deep water conditions the lower part of the profile attains a slope corresponding to the natural angle of repose for the stone material.

Van der Meer (1988a) presented a parametric study of the deformation of the seaward profile for non-overtopping conditions in head-on waves, and produced on this basis an empirical computational model BREAKWAT, which predicts the profile under given conditions.

Based on model tests the following stability factor values were proposed by Burcharth and Frigaard (1988) as the critical conditions for the start of significant transport of stones along the structure after the reshaped profile has formed. The trunk was exposed to oblique waves with angle of incidence ≤30°.

$$\frac{H_s}{\Delta D_{n50}} = \begin{cases} 4.5 \text{ steep waves} \\ 3.5 \text{ long waves} \end{cases} \tag{29.38}$$

Stability of roundheads

Parametric studies of the stability of berm breakwater roundheads have not yet been performed. Based on model test results Burcharth and Frigaard (1987, 1988) found that $H_s/(\Delta D_{n50})$ should be smaller than 3 if significant continuous erosion is to be prevented. Examples of berm breakwaters can be found in the Proceedings of the ASCE Conference on Berm Breakwaters held at Ottawa, Canada in 1987.

Structural integrity of concrete armour units

Figure 29.5 shows examples of the wide variety of existing concrete armour units. These might be divided into the following categories related to the structural strength:

Massive or blocky	(e.g. cubes including antifer type, parallelepiped block)
Bulky	(e.g. grooved cube with hole, Seabee, Accropode, Haro, Dolos with large waist ratios)
Slender	(e.g. tetrapod, Dolos with smaller waist ratios)
Multi-hole cubes	(e.g. shed, cob)

The units are generally made of conventional unreinforced concrete except the multi-hole cubes where fibre reinforcement is used.

For slender units such as Dolos with small waist ratios various types of high-strength concrete and reinforcement (conventional bars, prestressing, fibres, scrap iron, steel profiles) have been considered, but only used in a few cases as it generally seems to be less cost effective.

The hydraulic stability of armour layers is hampered if the armour units disintegrate because this causes reduction of the stabilizing gravitational forces and possible interlocking effects. Moreover, broken armour unit pieces can be thrown around by wave action and thereby trigger accelerated breakage. To prevent this it is necessary to ensure structural integrity of the armour units.

Unreinforced concrete is a brittle material with a low tensile strength, S_T, in the order of 1.5–3 MPa ($N\,mm^{-2}$) and a compressive strength, S_C, which is one order of magnitude larger than S_T. Consequently, the reason for crack formation and breakage is nearly always that the load induced tensile stresses, σ_T, exceeds S_T. The magnitude of S_T is therefore of much more interest than S_C, a fact that should be reflected in the specifications for armour unit concretes. It is important to notice that S_T decreases with repeated loading due to the effects of fatigue.

The different categories of units are not equally sensitive to breakage. Slender units are the most vulnerable because the limited cross-sectional areas give rise to relatively large tensile stresses. Many recent failures of breakwaters armoured with Tetrapods and Dolosse were caused by breakage of the units before the hydraulic stability of intact units expired. These accidents could have been avoided if design diagrams for structural integrity had been available at the time of design. The situation at present is that structural integrity diagrams are available only for Dolos (Burcharth and Liu, 1990, 1992). Dutch research (Ligteringen *et al.* 1990) and research carried out by the Franzius Institute, University of Hannover (Bürger *et al.* 1990) might result in design diagrams for tetrapods.

Massive units will generally have the smallest tensile stresses due to the large cross-sectional areas. However, breakage can take place if the units are

Table 29.10 Types and origins of loads on armour units. Burcharth (1983)

Types		*Origin of loads*
Static		Weight of units Prestressing of units due to wedge effect and arching caused by movement under dynamic loads
Dynamic	Pulsating	Gradually varying wave forces Earthquake
	Impact	Collisions between units when rocking or rolling, collision with underlayers or other structural parts Missiles of broken units Collisions during handling, transport and placing High-frequency wave slamming
Abrasion		Impacts of sand, shingle etc. in suspension
Thermal		Temperature differences during the hardening (setting) process after casting Freeze – thaw
Chemical		Alkali-silica and sulphate reactions, etc. Corrosion of steel reinforcement

impacting due to the application of less restrictive hydraulic stability criteria and if the concrete quality is poor in terms of a low S_T. The last point is related mainly to larger units where temperature differences during the hardening process can create tensile stresses which exceed the strength of the weak young concrete, thus resulting in micro-cracking of the material (thermal cracking). If massive units are made of good quality concrete and not damaged during handling, and designed for marginal displacements, there will be no breakage problems. With the same precautions this statement holds also for the bulky units.

No structural integrity design diagrams exist for the massive units. Some information is available on maximum impact speeds which is useful for the assessment of hydraulic stability design criteria, and handling and construction methods.

Types of loads

The different types of loads on armour units and their origins are listed in Table 29.10. Static, pulsating and impact loads are critical for slender units whereas impact loads and thermal loads are more important for massive and bulky units. However, very few of the loads and related stresses or deteriorational effects can be quantified.

Scaling stresses from static and dynamic loads

The various methods for the calculation or measurement of stresses from static and dynamic loads have been discussed by Burcharth and Liu 1990. Generally, stress determination is a very difficult task because of the stochastic nature of wave loads, the complex shape of the units and their random placement. One of the main problems is that the two main failure modes for armour, namely displacements (hydraulic instability) and breakage (structural instability) are interrelated and must be studied together. However, while the first can be studied conveniently in small Froude-scale models the second cannot because the stress levels are too small to cause any breakage of the model armour units when made of mortar or other conventional model materials. In order to understand the problem the model scales for stresses, σ, caused by static, pulsating and impact loads in a Froude model are summarized as follows:

$$\lambda_{\sigma_{\text{Static}}} = \lambda_{\sigma_{\text{Pulsating}}} = \lambda_{\rho_w}\lambda_L \qquad (29.39)$$

$$\lambda_{\sigma_{\text{Impact}}} = (\lambda_{\rho_A}\lambda_{E_A}\lambda_L)^{0.5} \qquad (29.40)$$

E_A is the modulus of elasticity of the armour unit concrete and L is a characteristic length. ρ_W and ρ_A are the mass densities of water and armour units, respectively.

It can be seen from Equations (29.39) and (29.40) that the scaling laws are different which means that, in general, the two categories of stresses must be identified and measured separately in the model in order to produce a correct up-scaling of the total stresses to prototype conditions.

Moreover, the static and pulsating stresses scale linearly with the length scale (e.g. D_n) whereas the impact stresses scale with the square root of the length scale. This means that the static stresses grow much more rapidly with the size of the armour than the impact stresses. Consequently, the static stresses are often dominant in the very large slender armour units.

Determination of stresses from static and dynamic loads

Scaled material properties Because breakage of a larger proportion (say > 10%) of the armour units will accelerate the hydraulic instability it would be ideal to test a design in a hydraulic model with armour units with correctly scaled material properties. Such a model will exhibit a true picture of prototype damage.

It is extremely difficult to produce such a material when the density has to be kept almost as high as for normal concrete. Timco (1981) had some success in producing a material which almost fulfilled the criteria except for the elasticity of the material, which was much too high. A practical problem was the low surface resistance of the material which caused a rapid rounding (abrasion) of the armour units. Timco and Mansard (1983) describe model tests with the use of the scaled material.

A drawback of the method of scaled material is that a new material must be produced for each length scale. Another is, that no general information on stresses is obtained except the information on the exceedence of the strength level when the units break.

Recording of armour unit movements Impact stresses might be estimated on the basis of impact velocities determined in a hydraulic model by cine/video technique or by accelerometers installed inside the units. The methods involve many problems. Cine/video techniques generally fail to give information in the splash zone. Moreover, because it is almost impossible to calculate stresses from information of the impact speed of the impinging body alone, it is necessary to calibrate the stress calculations method against known behaviour of prototype armour. van der Meer and Angremond (1991) and van der Meer and Heydra (1991) describe a Dutch CUR project where the probability density functions of the impact speeds of cubes and Tetrapods were studied. Pulsating and static stresses, which are of great importance for large slender units, cannot be determined by the method described.

Surface or bar mounted strain gauges A direct way of determining stresses is to use strain gauges mounted directly on the concrete surface or on bars cast in the concrete close to the surface where the largest strains occur. The limitation of the method is the small strain values in the case of small units exposed only to static and pulsating loads. The method cannot be applied even to large model armour units, say of 50 kg, unless a very sophisticated strain gauge technique is used. Strain gauge mounted bars have been used successfully in the 38-ton prototype Dolosse in the CERC Crescent City research project, (Howell, 1985, 1988).

Load cell technique Load cells have been used to study the resulting wave induced flow forces on armour units for quite some years: the stresses in slender/bulky types of armour units (Delft Hydraulics, 1980; Scott *et al.*, 1986; Aalborg University, 1987; Markle, 1990; Bürger *et al.*, 1990). The method consists of inserting a load cell able to record the component forces and moments in the critical sections of the armour units. The critical sections are those where the largest stresses and, consequently, the fractures generally occur.

Small-scale load cell technique is of great importance because design diagrams must be based on a very large number of parametric hydraulic model tests which, for economic reasons, cannot be performed at larger scales.

Ultimate impact velocities for concrete armour units

An indication of the relative strength of the various types of armour units can be given by the maximum impact velocity which a unit can resist without serious breakage. No standard method exists for the impact testing of geometrically different units and consequently no directly comparable results are available. The values of ultimate impacting velocities given in Table 29.11 are rough estimates corresponding to solid body impact against a heavy rigid concrete base, causing breakage in terms of a mass loss of 20% or more.

If the armour units are not dropped on a hard rigid surface, but on soil or a rock underlayer the ultimate impact velocities are significantly higher than those given in Table 29.11. For an evaluation of the placing technique during construction it is important to consider the ultimate impact velocities. The lowering speed of the crane at the moment of positioning of the units must be much lower than the figures given in Table 29.11.

When placing units underwater a heavy swell might impose rather large horizontal velocities on the unit when hanging from the crane. It is obvious from the figures in Table 29.12 that free fall dropping of concrete armour units by quick release from a crane or similar device should be avoided because even small drop heights cause breakage. This is true also in the case of underwater placement because the maximum free fall velocity underwater exceeds the limiting values given in the table except for very small massive types of units.

Fatigue

The strength of concrete reduces with the number of stress cycles. Each stress cycle larger than a certain range will cause partial fracture in some parts of the material matrix resulting in a low strength. Repeated loading causes an accumulative effect which might result in macro-cracks and, eventually, breakage of the structural element.

The number of stress cycles caused by wave action will be of the order of 200 million during 50 years' structural life in the North Atlantic area. About 10 million will be caused by larger storm waves. In subtropical and tropical areas the number is generally one or two orders of magnitude less.

Table 29.11 Approximate values of ultimate rigid body impact velocities for concrete armour units

Armour unit	*Impact velocity of centre of body* ($m\,s^{-1}$)	*Equivalent drop height of centre of drop height* (*m*)
Cube < 5 t	5–6	1.2–1.8
20 t	4–5	0.8–1.2
50 t	3–4	0.4–0.8
Tetrapod	2	0.2
Dolos, waist ratio 0.42	2	0.2
0.32	1–1.5	0.05–0.12

Since 1903 it has been known that concrete shows significant fatigue. Considering the high stress levels in some of the slender types of armour units it is important to evaluate the fatigue effect. A more detailed discussion of this has been given by Burcharth (1984).

Thermal stresses

During the curing of the concrete the heat of hydration will increase the temperature. Because of the fairly low conductivity of concrete and because of the poor insulation of conventional formwork (e.g. steel shutters) a higher temperature will be reached in the centre of the body than at the surface. The temperature difference will create different thermal expansion, but because of the coherence, the various parts of the body are not free to move and thus stresses are created. The larger the distance from the centre to the surface, the larger will be the temperature difference and the stresses. The tensile stresses can easily exceed the very limited strength of the fresh young concrete thus causing the formation of micro-cracks. Unfortunately, it is not possible to see thermal cracks because they will be close at the surface due to the thermal contraction of the concrete when cooling off. For a full discussion on thermal stress, see Burcharth *et al.* (1991).

Design diagrams for Dolos of different waist ratios

The design diagrams presented in this section are based on model tests at Aalborg University (Burcharth and Liu 1992) with instrumented Dolos exposed to irregular waves. The diagrams contain design curves both for stress and displacement levels. The diagrams make it possible to choose a combination of mass and waist ratio which ensures both structural integrity and hydraulic stability (see Figure 29.8). The hydraulic stability in terms of displacements has also been given by formula in an earlier section. The amount of rocking is not given because this model is relevant only to aspects of breakage which are dealt with specifically in the stress curves.

The test ranges are as follows:

Waves	irregular, breaking and non-breaking, head-on
Slope	1:1.5
Dolos	waist ratios, r = 0.32–0.42. Random placement in two layers with packing density, $\vartheta_{n=2}$ = 0.74.

The hydraulic instability level, D, is given in terms of the relative number of units within levels SWL ± 6.5 D_n which are displaced one Dolos height or more. The design diagrams have been checked against observed

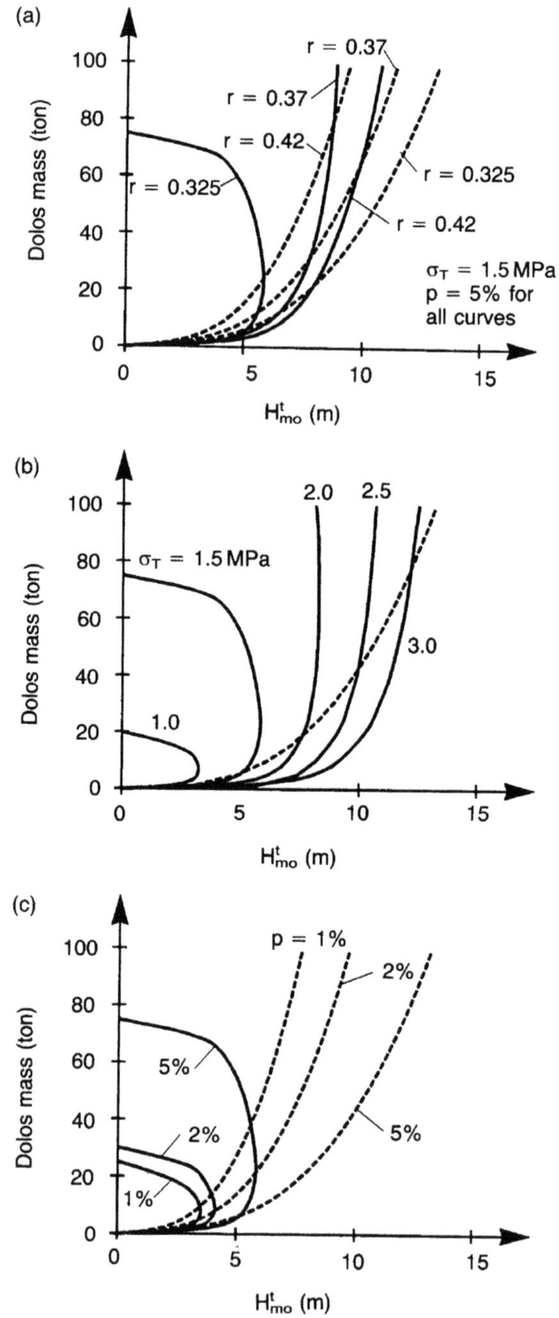

Figure 29.28 Examples of design diagram for Dolos on slope 1:1.5 (–––), hydraulic stability limit (N_z = 3000; $\psi_{n=2}$ = 0.74; Δ = 1.29; (——), teasile strength limit; H^t_{mo} is the significant wave height at toe of breakwater; r is the Dolos waist ratio; σ_T = concrete tensile strength (MPa); p is the strength exceedance probability and relative number of displaced units. For all curves r = 6.325; σ_T = 1.5 MPa; p = 5%.

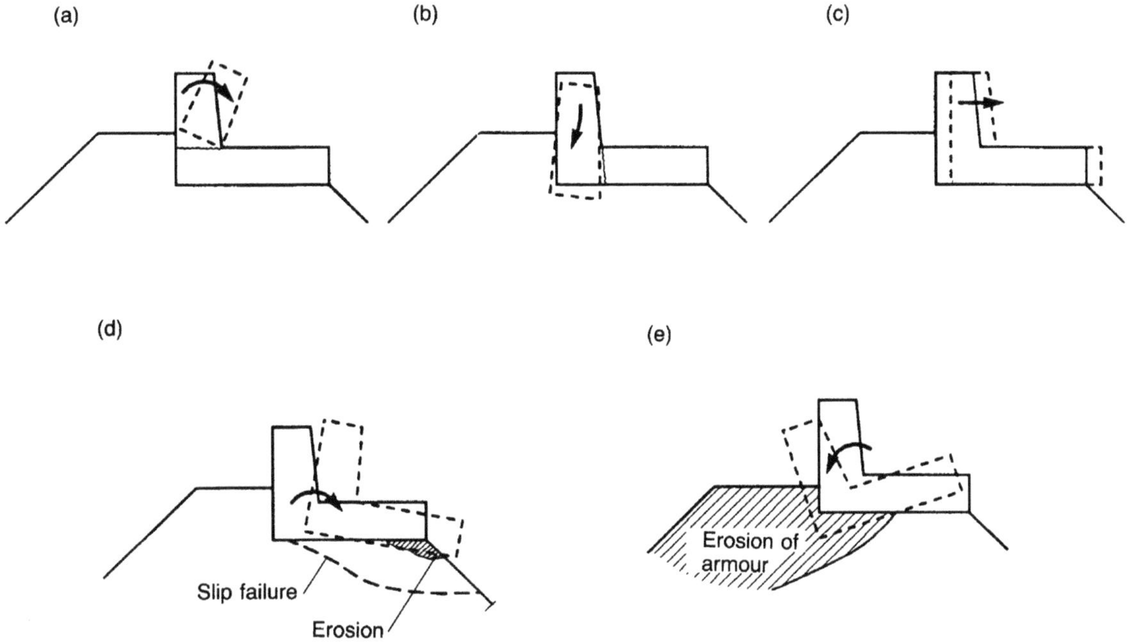

Figure 29.29 Common failure modes of concrete crown walls. (a) breakage of wall; (b) breakage of base plate; (c) sliding on foundation; (d) backward sliding/tilting due to slip failure in foundation or rear slope erosion; (e) foreward tilt due to undermining by erosion of front armour.

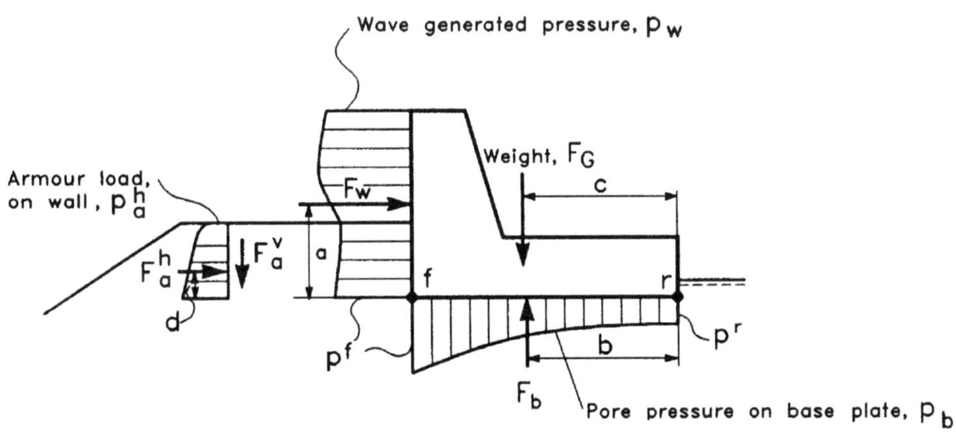

Figure 29.30 Forces on a superstructure.

behaviour of prototype Dolos breakwaters and good agreement was found.

Stability of concrete crown walls

The most common failure modes of concrete superstructures are shown in Figure 29.29. Quite often combinations of the four types of failure are experienced.

The wave forces on a monolithic superstructure exposed to irregular waves are of a stochastic nature. The pressure distributions and the related resultant forces at a given instant are schematized in Figure 29.30.

The wave generated pressure, p_w, acting perpen-

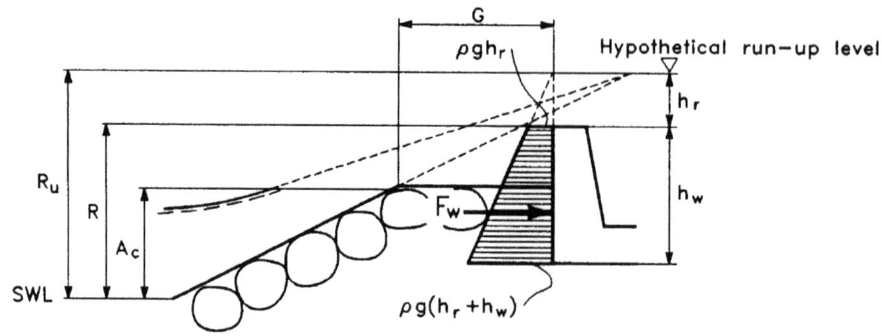

Figure 29.31 Crude estimation of wave force on wave wall.

dicular to the front of the wall is the one, which would be recorded by pressure transducers mounted on the front face. The distribution of p_w is greatly affected by very large vertical velocities and accelerations which often occur. F_w is the instantaneous resultant of the wave generated pressures.

The instantaneous uplift pressure, p_b, acting perpendicular to the base plate is equal to the pore pressure in the soil immediately under the plate. The resultant is F_b. At the front corner, f, the uplift pressure p_b^f, equals the pressure on the front wall. At the rear corner, r, the uplift pressure, p_b^r, equals the hydrostatic pressure at r. The actual distribution of p_b between p_b and p_r^r depends on the wave-generated boundary pressure field and on the permeability and homogeneity of the soil. The distribution cannot be determined in normal wave flume scale tests because of the strong scale effects in modelling the porous flow. However, the corner pressures p_b^f and p_b^r can be measured or estimated, and in the case of homogeneous and rather permeable soils and quasi-static conditions, a safe estimate of the most dangerous uplift can be found by assuming a linear pressure distribution between a maximum value of p_b^f and a minimum value of p_b^r.

A very crude method of estimating the wave load on a wall would be to consider the hypothetical run-up, R_u, on an imaginary elongation of the slope and assume a part of a related hydrostatic pressure distribution to act on the wall (see Figure 29.31). The run-up level might be estimated from the formulae given in the section on run-up, run-down and internal flow, earlier in this chapter.

Calculation of the total force in this way gives:

$$F'_w = \rho g(h_r + 0.5h_w)h_w \tag{29.41}$$

which, however, is a very conservative estimate. A central estimate has been found, by the author, to be

$$F_w = \left(0.304 + 0.054\left(1 + \frac{R - Ac}{G}\right)^{Hs/Ac}\right)F'_w \tag{29.42}$$

where R and G are given in Figure 29.36. The uncertainty of the formula corresponds to a coefficient of variation of 0.31. Equation (29.42) is developed on the basis of results from model tests with five different geometries ($0 < (R - A_c)/G < 1.7$), all except one with conventional rock armoured slopes (1:2) and G corresponding to approximately three stone diameters.

The total stability of the superstructure must be investigated by considering both the resistance to sliding and overturning and the bearing capacity of the foundation; see the failure modes depicted in Figure 29.28.

Stability against sliding exists when the ratio of resultant horizontal forces to vertical forces is equal to or less than the coefficient of friction, μ, between the superstructure and the underlayer, i.e.

$$F_w + F_a^h \leq (F_G - F_b)\mu \tag{29.43}$$

(see Figure 29.29). The friction will be rather low corresponding to $\mu \leq 0.5$ if the foundation of the superstructure is plane and precast. Higher values of the order of $\mu \simeq 0.5–0.7$ are obtained if a plane base plate is cast *in situ* directly on the underlayer. If the base plate is not flat-bottommed but is well keyed into the underlayer, then larger horizontal resistance can be mobilized because the sliding plane is forced to pass through the stone material. The resistance should then be determined by considering various slip failure planes. Apparent values of μ in the order of 0.8–1.0 might be obtained in the case of good stone material. Due to the uncertainty related both to wave loads, uplift pressures and coefficients of friction it is important to apply a safety coefficient in Equation (29.39).

With reference to Figure 29.29 the equation expressing stability against overturning around point r reads

$$F_w a + F_b b + F_a^h d \leq F_G c \qquad (29.44)$$

The vertical load, F_a^v, from the armour has been disregarded because its stabilizing effect is very uncertain. A safety factor should be applied to Equation (29.44).

The first step in the assessment of the bearing capacity of the foundation is to calculate the magnitude, position and direction of the resultant force, R, on the foundation. The soil mechanics methods of analysing the bearing capacity of a foundation when exposed to eccentric inclined loads are then applied, i.e. slip failure analyses or bearing capacity diagrams.

29.6.2 THICKNESS, POROSITY AND PACKING DENSITY OF ARMOUR LAYERS

These interrelated quantities are of importance for run-up and armour layer stability. They must be quantified in the design and respected in the construction.

The layer thickness is usually given by

$$t_a = ncD_n \qquad (29.45)$$

where n is the number of armour unit layers making up the total thickness of the layer; c is an empirical layer thickness coefficient which depends on n, the shape of the blocks, the method of placement and the packing density see Table 29.13; and $D_n = (M/\rho)^{1/3} = (V)^{1/3}$, where M, ρ, and V are the mass, the mass density, and the volume of the block, respectively. For graded materials D_{n50} is used.

The number of blocks placed within a unit area is, as a consequence of Equation (29.45) given by

$$N_a = nc(1-p)D_n^{-2} = \vartheta_n D_n^{-2} = \vartheta_n V^{-2/3} \qquad (29.46)$$

where p is the porosity = (pore volume/bulk volume). The coefficient $\vartheta_n = nc(1-p)$ is independent of the block size and is called 'packing density'.

29.6.3 DESIGN OF UNDERLAYERS AND FILTER LAYERS

The purpose of underlayers and filter layers is to prevent the finer material in the core and the seabed from being pulled out through the cover layers. Significant migration of finer material into the coarser material will lead to settlements and might cause reduced porosity and permeability of the underlayers resulting in reduced hydraulic stability of the cover layers.

It is important to notice that in some parts of a breakwater the internal flow is changing both with respect to velocity and flow direction and large hydraulic gradients (maybe up to 1) might occur in some regions of the front face (Figure 29.14). The

Table 29.12 Examples of applied layer thickness coefficients, porosities and packing densities for various types of rock materials and concrete armour units

Type of armour	*Number of layers* n	*Placement*	*Layer thickness coefficient* c	*Porosity* p	*Packing density* ψn	*Reference*
Quarry stones, narrow graded						
smooth	2	random	1.02	0.38	1.26	*Shore Protection Manual* (1984)
rough	2	–	1.00	0.37	1.26	–
rough	>3	–	1.00	0.40	1.20	–
Cube	2	random	1.00	0.33	1.34	van der Meer (1986)
Modified cube	2	–	1.10	0.47	1.17	*Shore Protection Manual* (1984)
Grooved cube (Antifer)	2	–	1.10	0.46	1.18	De Rouck *et al.* 1987
Parallelepiped*	2	–	1.05–1.1	0.37–0.40	1.37–1.31	CEPYC (1992)
Tetrapod	2	–	1.02	0.5	1.02	SOGREAH
Dolos	2	–	$\frac{\vartheta n}{n(1-p)}$	0.81–0.80,$r^{\dagger}$	0.74	Burcharth and Christiansen (1991)
Accropode	1	special§	1.29	0.50	0.64	SOGREAH
Haro	2	special**	1.10	0.52	1.06	De Rouck *et al.* (1987)

* Ratio between length of sides, app. 1:1:1.25.
† r is Dolos waist ratio. Also fits data from SPM (1984) and Zwamborn (1978)
§ Regular placement with spacing 0.6 block height in down-slope direction and 0.64 block height in horizontal along-slope direction.
** Holes orientated normal to slope, otherwise fairly random placement.

Table 29.13 Filter rules for stone materials in order to prevent migration

Reference		D_{15f}/D_{85b}	D_{50f}/D_{50b}	D_{15f}/D_{15b}	D_{20f}/D_{20b}
		≤4–5		>4*, <20–25	
		≤5			
Lee (1972)		<5	<25	>4*, <20	
Thompson and Shuttler (1976)		≤5	≤7	≤7	
van Oorschot (1983)			≤3		
CIRIA/CUR (1991)	Narrow graded materials		<5		>5*
	Wide graded materials	<5	>5, <20–60		>5*
Shore Protection Manual (1984)†	Very narrow graded rock		≃2.2–2.5		
–	Narrow graded rock underlayer and filter layer		≃2.7		
–	Narrow graded rock filter layers		≃2.7–3.1		
–	Graded rip-rap	≤5			

* Values to ensure adequate permeability.
† The recommendations in the *Shore Protection Manual* (1984) are given in terms of ratios of average mass of the materials but is here converted to diameter ratios.

Table 29.14 Filter rules for concrete armour unit underlayers

Reference	*Type of unit*	*Mass of concrete armour unit* / *Average mass of underlayer stones*	*Volume of armour unit* / *Average volume of underlayer stones*
Shore Protection Manual (1984)	Massive (cubes etc.)	10	
	Slender (Tetrapods etc.)	5	
van Oorschot (1983)	Massive (cubes etc.)		≤30
	Slender (Tetrapods etc.)		≤15

conditions are generally very different from those of conventional soil mechanics filters where, in most cases, the hydraulic gradients are small and the flow unidirectional.

It is easy to formulate filter rules which assure no or marginal migration. However, such restrictive rules necessitate more filter layers to overcome a certain difference in the size of core and cover layer materials. A medium gradation corresponding to D_{85}/D_{15} = 1.5–2.5 is frequently used for underlayers and filter layers.

Table 29.13 gives an overview of some filter rules for stone materials. The subindex b and f relate to the base (finer) and the filter (coarser) materials, respectively. The base material is the one to be protected against being washed out.

The *Shore Protection Manual* (1973, 1984) recommendations are rather conservative as they are based on narrow gradations, i.e. $_{max}\text{mass}/_{min}\text{mass} = 1.67–3$ or $_{max}D_n/_{min}D_n = 1.10–1.44$. It is general experience that less restrictive criteria are applicable. In the case of uni-size concrete armour units some recommendations from the literature are referred to in Table 29.14.

The above filter criteria are rather safe as they ensure geometrical stability, i.e. almost no penetration of the base material into the filter material. However, in some cases less restrictive criteria might save a filter layer or might allow the use of cheaper local materials. More advanced criteria which allow some migration must take into account the magnitude of the hydraulic gradient which then has to be estimated. The influence of a hydraulic gradient on filter stability has been studied by Graauw *et al.* (1984) and Bezuijen *et al.* (1987) but only for rather fine materials.

The thickness of underlayers and narrow graded large diameter filters should never be less than a double layer for functional reasons. The layer thickness as a function of the number of layers can be estimated from the formulae given in section 29.6.2. For graded filters of finer material laid underwater a minimum thickness of four times D_{85} is recommended. However, the difficulties in constructing a thin layer under water and the poor possibility for effective monitoring determines in fact the minimum thickness. The absolute minimum thickness of a layer of fine material (say $D_{50} \simeq 5$ cm) placed under water in favourable shallow water conditions is 0.3 m and double or more than that under more difficult conditions.

Sufficient frictional resistance between the armour and the underlayer is important in the case of steep slopes. The larger the underlayer stone the larger the resistance against sliding of the armour. The selection of the underlayer stone size and the evenness of the underlayer surface profile will also depend on the type of armour and the prescribed placement pattern. For example random placement of cubes will be difficult on a very smooth underlayer surface. On the other hand placement of tetrapods in a regular pattern demands a smooth underlayer surface.

During the construction phases it is inevitable that some parts of the filter and underlayers will be exposed to wave action. Depending on the wave climate it might be economically advantageous to use coarser materials than those deduced from the filter rules in order to reduce the risk of erosion and ensure a more steady construction procedure.

29.6.4 DESIGN OF CORE

The core material is generally quarry run. Exceptions are cases where suitable glacial deposits, such as gravel, are readily available as for example in Denmark. Quarry run usually has a wide gradation.

The core should not be constructed of materials that are too coarse in order to avoid undesirable transmission of waves and sediment transport through the breakwater. Coarse core material might also facilitate venting underneath superstructures. On the other hand there should also be a lower limit for the size of material to prevent wash-out of the finer material. This is due to the consequent risk of larger settlements and inconvenient deposition of the materials. Also, if very fine material is used for the core it might be necessary to use several filter layers to avoid outwash. Finer materials are also less suitable for construction due to the smaller natural angle of repose (very gentle slopes) and the vulnerability to erosion in waves and currents. Unfortunately, it can be very expensive to remove finer material from quarry run because an extra operation, such as screening, is necessary. This again makes the optimum choice of quarry and blasting technique very important (see Chapter 32 of this volume).

29.7 DESIGN OF VERTICAL WALL BREAKWATERS

29.7.1 INTRODUCTION

Today larger vertical wall breakwaters are almost without exception built as caisson structures, see Figure 29.7. Some typical failure modes of caisson breakwaters are shown in Figure 29.9. The figure indicates a distinction between failure modes related to the overall stability of the (monolithic) structure and failure modes related to the integrity of the structural parts. This distinction is important because local wave loads for which structural parts should be designed are often much higher but of shorter duration than the space and time averaged wave loads which are relevant for overall stability assessments.

The stability of caisson and block-type structures relies heavily on the net gravitational force, F'_G, which is the weight of the structure, F_G, reduced for the buoyancy, $F_{H,b}$, (Figure 29.32). The figure illustrates the definition of the wave induced pressure as the difference between the total pressure (as recorded on the caisson surface) and the hydrostatic pressure at SWL. The resultant forces of the various pressures are indicated together with the gravitational force. The magnitude and direction of the resultant, R, can be calculated either from total pressures and the weight F_G, or from the wave induced pressures and the buoyancy reduced weight F'_G.

Figure 29.32a illustrates a situation with a wave crest at the front face and a wave through at the rear face. This condition gives the largest shoreward net force and shoreward tilting moment and is usually responsible for all the types of failures shown in Figure 29.9 with the exception of the failure modes involving a seaward sliding of the structure. The condition which would create seaward sliding is a deep wave trough at the front face combined with a wave crest at the rear face, as illustrated in Figure 29.32b. Such conditions can also trigger a seaward tilt failure mode when erosion of the front face bedding layer and/or the sea-bed takes place. The wave load on the top part of the structure caused by overtopping should be considered in the design. This includes the impact forces from the water which is falling down.

29.7.2 WAVE LOADS ON FRONT FACE

The wave generated pressures along the front face are very complicated functions of the wave conditions and the geometry of the structure. It is therefore recommended that model tests be performed, at least for the final designs of important structures. For preliminary designs of structures with straight vertical faces the formula by Goda (1974) can be applied under conditions where extensive wave breaking does not take place directly on the front (see Figure 29.33).

When the crest of a wave approaches a vertical wall the water particles will be forced upwards in a rather sharp curve. The sharper the curvature, i.e. the smaller the radius of the particle path, the larger the pressure as it is balanced by the centrifugal forces.

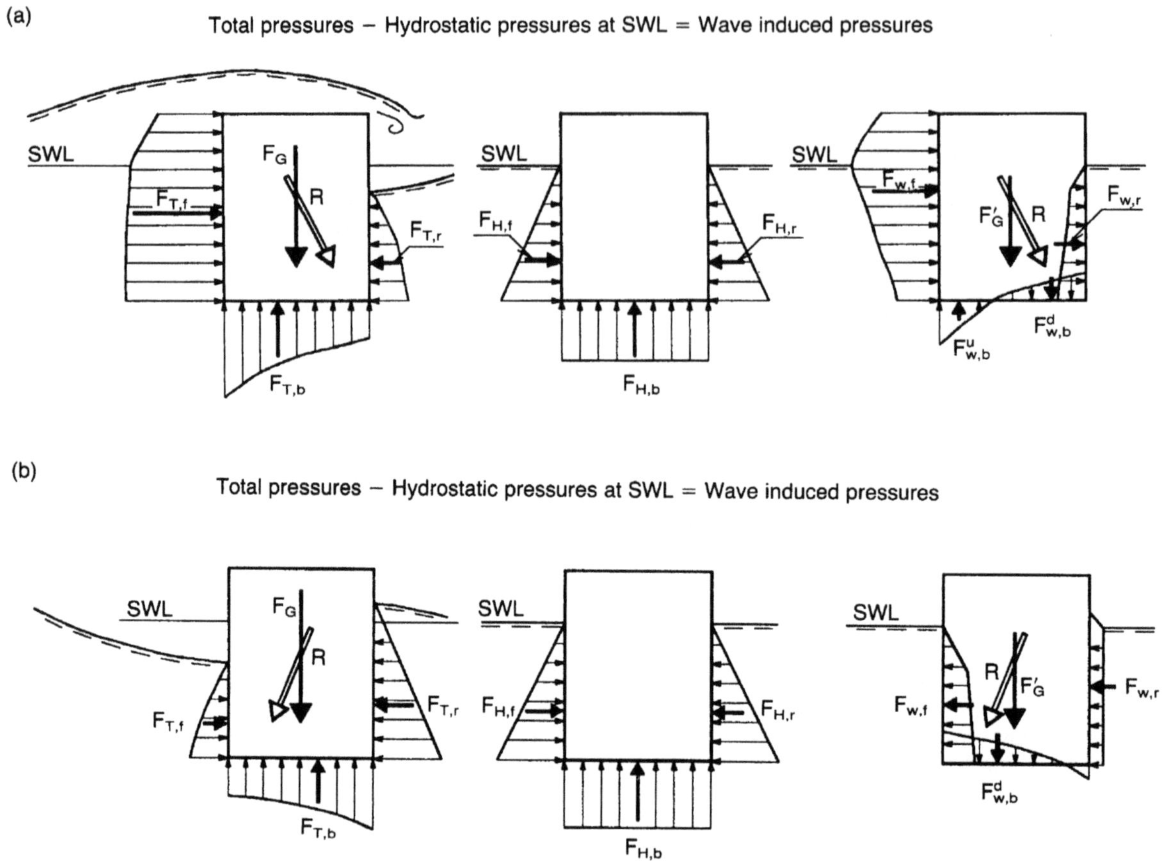

Figure 29.32 Wave induced loads on a vertical wall caisson; (a) wave crest at front face, wave at rear face; (b) wave trough at front face, wave trough at rear.

Figure 29.33 illustrates two different situations of wave–structure interaction. In the case of a non-breaking wave no airpocket will be trapped on the wall and the pressure at the wall will have a relatively gentle variation in time in accordance with the wave period. These wave loads might be called quasistatic loads because the period is much larger than the natural period of caisson breakwaters (approximately ten to a hundred times larger). Consequently, the wave load can be treated like a static load in the stability calculations, except that special considerations are required if the caisson is placed directly on fine soils like sand where pore pressure build-up might take place.

The wave load from non-breaking waves does not always vary in accordance with the wave period. In irregular seas some waves can cause a very fast rise in the wave pressure, say within tenths of a second. Lundgren (1969) terms this a ventilated shock.

Breaking waves of the plunging type develop an almost vertical front before they curl up. If this, almost vertical, front happens to develop just before the interaction with a vertical wall then very high, but extremely short duration, pressures occur. The highest pressures occur at point A where the crest of the wave hits the wall. Slightly lower pressures are recorded at point B due to the larger compressibility of the entrapped air pocket. Figure 29.33b illustrates the variation in time of the pressures integrated over wall height, h. The duration of the pressure peaks are typically of the order of hundredths of a second. The shorter the duration, the larger the peak pressure for constant momentum of the impinging water.

Lundgren (1969) denotes the forces recorded at points A and B in the case of wave-breaking on the wall hammer shock and compression shock, respec-

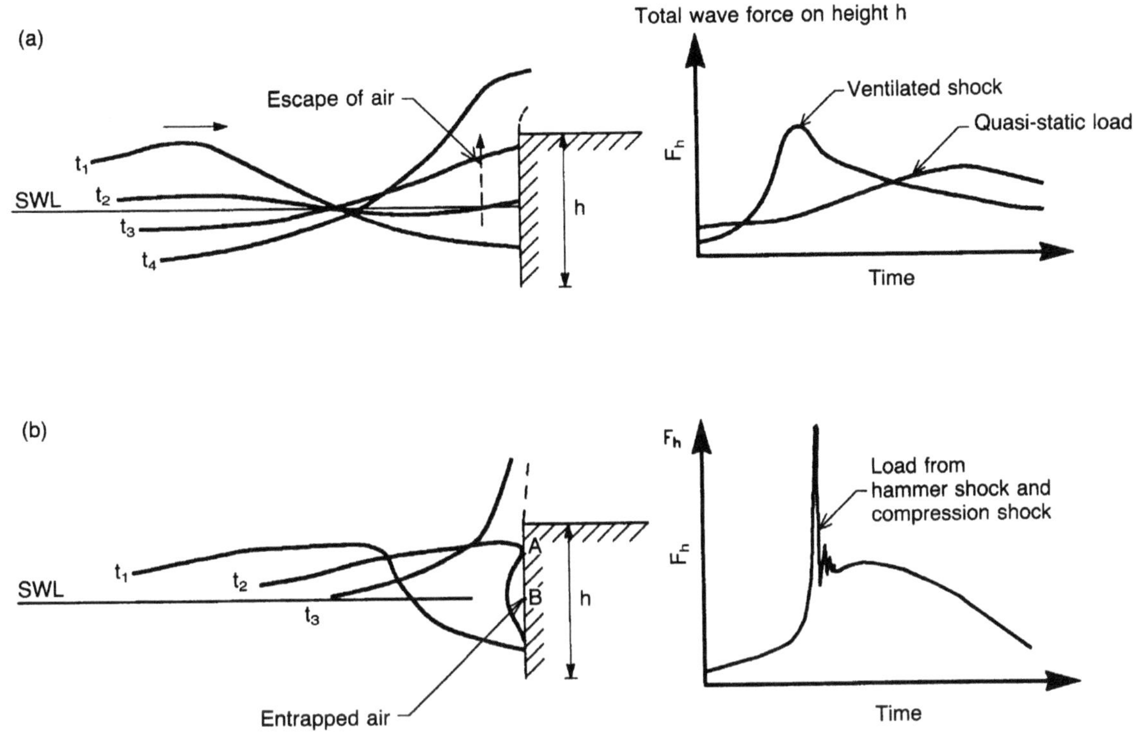

Figure 29.33 Wave forces from: (a) non-breaking wave; (b) breaking (plunging) wave.

tively. In the literature all wave generated pressures with fast rising time are generally called either impact pressures or impulsive pressures.

There are no reliable formulae for the prediction of impulsive pressures caused by breaking waves. Furthermore, the determination of such pressures in model tests is difficult because of scale effects related to the amount and size of the air bubbles and the size and shape of the air pockets. Also the instrumentation, data sampling and analyses need special considerations in order to avoid bias by dynamic amplification and misinterpretations at prototype scales. For the latter Lundgren (1969) has formulated a model law for the transformation of compression shock loads. Another problem related to model tests is the sensitivity of the shock loads to the shape and kinematics of the breaking waves. This calls for a very realistic and statistically correct reproduction of natural waves.

Shock loads from breaking waves are very large. Moreover, they are of a stochastic nature which means that the extreme values increase with the number of loads. For this reason frequent wave breaking at the structure should be avoided and if not possible a rubble mound structure might be the best solution. Alternatively, a mound of armour units might be placed in front of the vertical wall structure, see Figure 29.6e.

Frequent wave breaking at the structure will not take place in oblique waves with angles of incidence larger than 20°. Nor will it take place if the sea-bed in front of the structure has a mild slope, say 1:50 or less, over a distance of at least some wave lengths.

The use of a sloping top from about SWL to the crest, Figure 29.6c, is very effective in almost eliminating impact pressures from breaking waves and in reducing the load from non-breaking waves. This solution, however, is not feasible where large water-level variations occur because the sloping front must start approximately at SWL. The efficiency of the sloping top is illustrated in Figures 29.34 and 29.35 which show model test results performed by Waterloopkundig Laboratorium, Holland, and Laboratoriet for Havnebygning, Technical University of

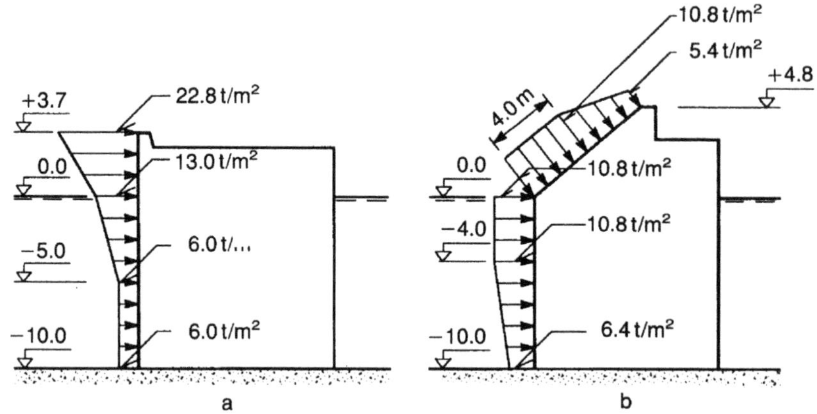

Figure 29.34 Comparative maximum values of local pressures recorded in model tests with wave height corresponding to 8 m at prototype scale. Levels are in meters.

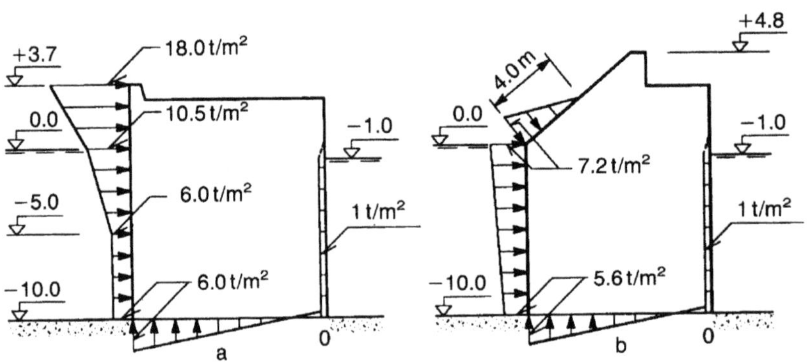

Figure 29.35 Comparative distributions of simultaneously recorded pressures in model tests with wave height corresponding to 8 m at prototype scale. Levels in meters.

Denmark, related to the early 1960s Hanstholm harbour project. The figures show a comparison between the pressures on a conventional vertical front face caisson and a sloping-top caisson when exposed to head-on waves 8 m high. The crest level is 1.1 m higher on the sloping-top caisson in order to compensate for the otherwise larger overtopping. Figure 29.34 shows the maximum recorded pressures at each location of the fronts during the test, i.e. the pressures are not simultaneous, but might, on the safe side, be used for the design of the structural parts of the caisson. Figure 29.35 shows the pressure distributions to be used for the calculation of the stability of the caissons. These pressures at the fronts are simultaneous values recorded in the models. The pressure at the rear faces is calculated corresponding to a wave trough at level −1.0 m. The pressure distributions on the bottom of the caissons are assumed to be linear.

The total horizontal force and the overturning moment around the heel (point 0) are for the sloping-top caisson approximately 60% and 40%, respectively, of the values for the vertical face caisson.

It is important to notice that a semi-high rubble slope protection should be avoided because it triggers wave breaking and thereby frequent impact loads on the vertical wall (Figure 29.36a). A relatively high rubble foundation can have the same effect (Figure 29.36b) dependent on the ratios between wave length, width of berm below SWL and water depth. Tanimoto (1976), (also reproduced in Goda, 1985) and Tanimoto *et al.* (1987) give some guidelines based on model test results.

Wave pressure formulae have been developed by

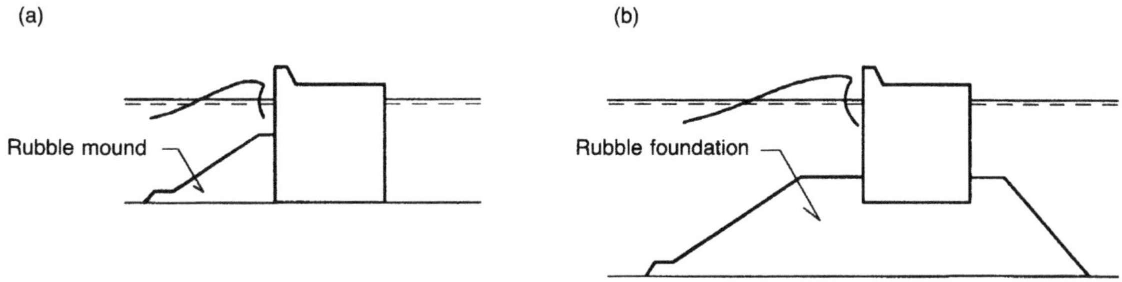

Figure 29.36 Enhanced frequency of wave breaking on caisson front walls due to: (a) semi-high rubble slope protection; (b) high rubble foundation.

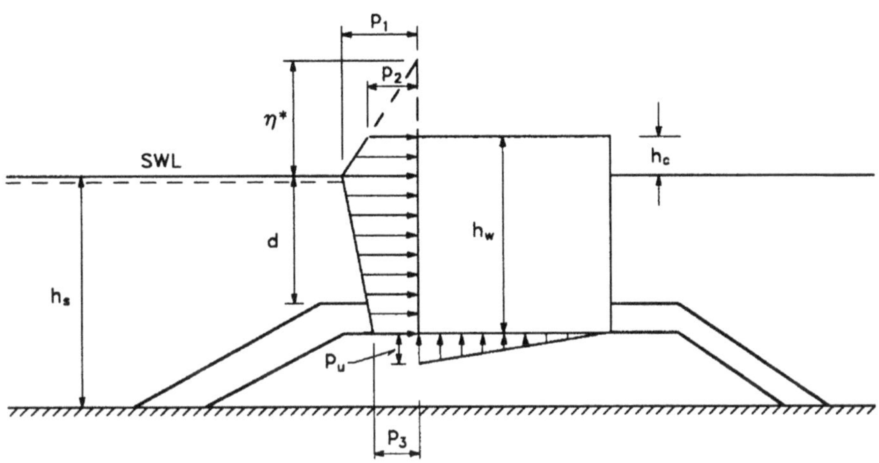

Figure 29.37 Definition sketch for Goda formula for wave induced pressure under wave crest.

Goda *et al.* (1972) and Goda (1974) for the case of a conventional vertical front where wave breaking on the wall is not enhanced by a steep sea-bed or structural configurations, see Figure 29.35. The formulae are based on model tests in head-on waves but were modified to also cover oblique waves on the basis of work by Tanimoto (1976). The formulae include the effect of breaking waves to the extent of normal accidental (non-provoked) wave breaking. The formulae are currently used in the Japanese standards. Figure 29.37 shows the related definitive sketch for the wave induced pressure under a wave crest.

$$\eta^* = 0.75(1 + \cos\beta)H_{design} \quad (29.47)$$

$$p_1 = 0.5(1 + \cos\beta)(\alpha_1 + \alpha_2\cos^2\beta)\rho_w g H_{design} \quad (29.48)$$

$$p_2 = \begin{cases} \left(1 - \dfrac{h_c}{\eta^*}\right)p_1 & \text{for } \eta^* > h_c \\ 0 & \text{for } \eta^* \leq h_c \end{cases} \quad (29.49)$$

$$p_3 = \alpha_3 p_1 \quad (29.50)$$

where: β is the angle of incidence of waves (angle between wave crest and front of structure); H_{design} is the design wave height defined as the highest wave in the design sea state at a location just in front of the breakwater. Seaward of the surf zone a value of 1.8 H_s might be used corresponding to the 0.1% exceedence value for Rayleigh distributed wave heights. If within a surf zone H_{design} is taken as the highest of the random breaking waves at a distance $5H_s$ seaward of the structure. L is the wave length corresponding to the significant wave period $T_s \simeq 1.1T_m$, where T_m is the average period.

$$\alpha_1 = 0.6 + \frac{1}{2}\left[\frac{4\pi h_s/L}{\sin h(4\pi h_s/L)}\right]^2 \quad (29.51)$$

$$\alpha_2 = \text{the smaller of } \frac{h_b - d}{3h_b}\left(\frac{H_{design}}{d}\right)^2 \text{ and } \frac{2d}{H_{design}} \quad (29.52)$$

$$\alpha_3 = 1 - \frac{h_w - h_c}{h_s}\left[1 - \frac{1}{\cosh(2\pi h_s/L)}\right] \quad (29.53)$$

h_b is the water depth at a distance of $5H_s$ seaward of the breakwater front wall.

Although the wave induced uplift pressure, p_u, at the front edge of the base plate is equal to p_3 Goda has suggested a somewhat reduced value:

$$p_u = \frac{1}{2}(1 + \cos\beta)\alpha_1\alpha_3\rho_w g H_{\text{design}} \quad (29.54)$$

This is because analyses of the behaviour of Japanese breakwaters revealed that the use of $p_u = p_3$ together with an assumed triangular distribution of the uplift pressure gave results that were too conservative.

When a wave trough reaches the front wall the pressure becomes less than the hydrostatic pressure corresponding to SWL and the resultant wave pressure will be directed seawards (see Figure 29.32b). In the case of shallow water vertical faced caissons sliding if it occurs will be shorewards. However, in the case of deep water structures Tanimoto *et al.* (1991) have also shown the potential for seaward sliding, due to the very deep troughs which can occur in reflected deep water seas at the wall. Goda (1985) has presented diagrams for the wave pressure under troughs, based on the theory for finite amplitude standing waves.

29.7.3 STABILITY OF CAISSONS

The various failure modes related to the overall stability of caissons are shown in Figure 29.8. The stability against sliding exists when the ratio of resultant horizontal forces to vertical forces is equal or less than the coefficient of friction, μ, between the caisson and the underlayer, i.e.

$$|F_{T,f} - F_{T,b}| \leq (F_G - F_{T,b})\mu \quad (29.55)$$

(see Figure 29.32).

The value of μ has been discussed in section 29.6.1. In Japan the value of μ is usually taken as 0.6. The stability against overturning around the heel of the caisson is expressed in an equation similar to Equation (29.44). It is recommended that a safety factor against sliding and overturning be applied. The Japanese standard prescribes the value to be not less than 1.2.

The overturning of a caisson implies very high stresses at the point of rotation. The bearing capacity of the stone underlayer will be exceeded and the crushing of the stones and the caisson heel will take place. In reality the bearing capacity of the underlayer and the sea-bed sets the limiting conditions. The soil mechanics methods of analysing the bearing capacity of a foundation when exposed to eccentric inclined loads should be applied, i.e. slip failure analysis or the use of bearing capacity diagrams. As a first estimate of the allowable pressure under the heel values in the range 0.4–0.6 MN m^{-2} might be used, according to Japanese pratice.

29.7.4 HYDRAULIC STABILITY OF RUBBLE MOUND FOUNDATION

This has been discussed by Tanimoto *et al.* (1987).

29.7.5 OVERTOPPING AND REFLECTION, CAISSON BREAKWATERS

This topic has been considered by Goda (1985).

REFERENCES

Ahrens, J.P. (1981) Irregular wave run-up on smooth slopes. Technical paper. US Army Corps of Engineers, CERC, Fort Belvoir, *Tech. Paper 81-17*.
Ahrens, J.P. (1984) Reef type breakwaters. Proceedings of the 19th Coastal Engineering Conference, Houston, Texas, USA.
Ahrens, J.P. (1987) Characteristics of reef breakwaters. US Army Corps of Engineers, Vicksburg, *Tech. Report CERC, pp. 87-17*.
Ahrens, J.P. and Heimbaugh, M.S. (1986) Irregular wave overtopping of sea walls. Proceedings of Conference, Oceans '86. IEEE, Washington, USA.
Ahrens, J.P., Viggoson, G. and Zirkle, K.P. (1982) Stability and wave transmission characteristics of reef breakwaters. CERC Interim Report, Fort Belvoir, USA.
Allsop, N.W.H. (1983) Low-crest breakwaters, studies in random waves. Proceedings of Coastal Structures '83. ASCE.
Allsop, N.W.H. (1990) Reflection performance of rock armoured slopes in random waves. Proceedings of the 22nd Coastal Engineering Conference, Delft, The Netherlands.
Allsop, N.W.H., Hawkes, P.I., Jackson, F.A. *et al.* (1985) Waves run-up on steep slopes – model tests under random waves. Hydraulics Research, Wallingford. *Rep. SR 2*.
Allsop, N.W.H. and Hettiarachchi, S.S.L. (1988) Reflection from coastal structures. Proceedings of the of the 21st Coastal Engineering Conference, Malaga, Spain.
Aminti, P. and Franco, L. (1988) Wave overtopping on rubble mound breakwaters. Proceedings of the 21st ICCE, Malaga, Spain.
Baird, W.F. and Hall, K.R. (1984) The design of breakwaters using quarried stones. Proceedings of the 19th Coastal Engineering, Houston, USA.
Barends, F.B.J. (1988) Discussion on paper 5, Proceedings of the Conference on Breakwaters 1988, Eastbourne, UK, Institution of Civil Engineers, London.
Battjes, J.A. (1974) Surf similarity. Proceedings of the 14th Conference on Coastal Engineering, Copenhagen, Denmark, 1974.
Bezuijen, A., Klein Bretteler, M. and Bakken, K.I. (1987) Design criteria for placed block revetments and granular filters. Proceedings of the International Conference on Coastal and Port Engineering in Developing Countries, Beijing, China.
Broderick, L.L. (1984) Riprap stability versus monochromatic and irregular waves. MSc thesis, George Washington University, USA.
Brorsen, M., Burcharth, H.F. and Larsen, T. (1974) Stability of Dolos Slopes. Proceedings of the 14th Coastal Engineering Conference, Copenhagen, Denmark.
Burcharth, H.F. (1983) Material, structural design of armour units. Seminaral rubble mound breakwaters, Royal Institute of Technology, Sweden, *Bull. TRITA-VBI-120*.
Burcharth, H.F. (1984) Fatigue in breakwater armour units. Proceedings of the 19th ICCE, Houston, USA.

Burcharth, H.F. (1989) Uncertainty related to environmental data and estimated extreme event. Report of sub-group B, working group 12, PTC II. PIANC, Brussels.

Burcharth, H.F. and Brejnegaard-Nielsen, T. (1986) The influence of waist thickness of Dolosse on the hydraulic stability of Dolos armour. Proceedings of the 20th ICCE, Taipei, Taiwan.

Burcharth, H.F. and Frigaard, P. (1987) Stability of roundheads and trunk erosion. Proceedings of ASCE Conference on Berm Breakwaters, Ottawa, Canada.

Burcharth, H.F. and Frigaard, P. (1988) On 3-dimensional stability of reshaping breakwaters. Proceedings of the 21st Coastal Engineering Conference, Malaga, Spain.

Burcharth, H.F. and Liu, Z. (1990) A general discussion of problems related to the determination of concrete armour unit stress including specific results related to static and dynamic stresses in Dolosse. Proceedings of the Seminaral Stresses in Concrete Armour Units, ASCE, Vicksburg, USA.

Burcharth, H.F. and Christensen, C. (1991) On stationary and non-stationary porous flow in coarse granular materials. Report of the MAST G6-8 research programmes. Aalborg University, Hydraulics Laboratory.

Burcharth, H.F. and Liu, Z. (1992) Design formulae for hydraulic and structural stability of Dolos armour. Proceedings of the 23th Coastal Engineering Conference, Venice.

Burger, W.W., Oumeraci, H. and Partenscky, H.W. (1990) Impact strain investigations on Tetrapods: Results of dry and hydraulic tests. Proceedings of the Seminar Stresses in Concrete Armour Units, ASCE, Vicksburg, USA.

CEPYC (1992) Personal communication.

CIRIA CUR (1990) *Manual on the use of rock in coastal and shoreline engineering*, CIRIA, London, UK.

First Design Institute of Navigation Engineering (1987) The stability of Accropode (R) and the comparison with Dolosse. First Design Institute of Navigation Engineering, Tianjin, China.

de Graauw, A.F.F., van der Meulen, T. and de Bye, D. (1984) Granular filters: Design criteria. *Journal of Waterways, Port, Coastal and Ocean Division, ASCE*, **110** (1).

De Rouok, J., Wens, F., Van Damme, L. and Lemmes, J. (1987) Investigation into the merits of the HARO breakwater armour units. Proceedings of Coastal and Port Engineering in Developing Countries, Beijing, China.

Fournier, C.P., Sayao, O.J. and Caldas, F. (1990) Berm breakwater contamination study Sergipe Marine Terminal, Brazil. Proceedings of the 22nd Coastal Engineering Conference, Delft, The Netherlands.

Givler, L.D. and Sørensen, R.M. (1986) An investigation of the stability of submerged homogeneous rubble mound structures under wave attack. H.R. IMBT Hydraulics, Lehigh University, Lehigh, USA. *Rep. IHL 110-86*.

Goda, Y.C. (1972) Experiments on the transition from nonbreaking to post breaking wave pressures. *Coastal Engineering in Japan*, **15**, 81–90.

Goda, Y. (1974) A new method of wave pressure calculation for the design of composite breakwater. Proceedings of the 14th Coastal Engineering, Conference, Copenhagen, Denmark.

Goda, Y. (1985) *Random seas and design of maritime structures*, University of Tokyo Press, Japan.

Günbak, A.R. (1976) The stability of rubble mound breakwaters in relation to wave breaking and run-down characteristics and to the $\zeta \sim \tan \alpha T/\sqrt{H}$ number. Division of Port and Ocean Engineering, The Norwegian Institute of Technology, Trondheim, Norway.

Günbak, A.R. (1977) Rubble Mound Breakwaters. Division of Port and Ocean Engineering, Technical University of Norway, Trondheim, *Rep. 12/77*.

Holtzhausen, A.H. and Zwamborn, J.A. (1990) Stability of Dolosse with different waist thickness for irregular waves. Proceedings of the 22nd ICCE, Delft, Holland.

Holtzhausen, A.H. and Zwamborn, J.A. (1991) The stability of Accropode (R) and the comparison with Dolosse. *Coastal Engineering*, **15** (1–2).

Howell, G.L. (ed.) (1985) Proceedings of workshop on measurement and analysis of structural response in concrete armor units, 23–24 January Waterways Experiment Station, Coastal Engineering Research Center, Vicksburg, Mississippi, USA.

Howell, G.L. (1988) Measurement of forces on dolos armor units at prototype scale. Proceedings of the 21st Conference on Coastal Engineering, Malaga, Spain.

Hudson, R.Y. (1958) Design of quarry stone cover layers for rubble mound breakwaters. WES, Vicksburg, *Res. Rep. 2.2*.

Hudson, R.Y. (1959) Laboratory investigation of rubble mound breakwaters. *Journal of Waterways*, Harbors Division, *ASCE*, **85** (WW3).

Hughes, S.A. (1984) The T.M.A. shallow water spectrum description and applications. U.S. Army Corps of Engineers, *Tech. Rep.* 84-7.

Jensen, O.J. (1984) *A monograph on rubble mound breakwaters*, Danish Hydraulic Institute, Horsholm, Denmark.

Knudsen, J. (1983) Description of EVA – A programme for extreme value analyses. Danish Hydraulic Institute, Horsholm, Denmark.

Ligteringen, H., Altink, H. and van Orchot, J.H. (1990) Strength of concrete armour units, joint industry research. Proceedings of a Seminaral stresses in concrete armour units, ASCE, Vicksburg, USA.

Lundgren, H. (1969) Wave shock forces: An analysis of deformations and forces in the wave and in foundation. Proceedings of a Symposium on Research on Wave Action II, Paper 4, Delft, The Netherlands.

Markle, R.G. (1990) Grescent City instrumental model Dolos study. Proceedings of a seminaral stresses in concrete armour units, ASCE, Vicksburg, USA.

Nielsen, S.R.K. and Burcharth, H.F. (1985) On the uncertainties related to estimates on Weibull distributed parameters. Hydraulics and Coastal Laboratory, Department of Civil Engineering, University of Aalborg, Denmark (in Danish).

Owen, M.W. (1980) Design of seawalls allowing for wave overtopping. 294, Hydraulics Research, Wallingford. *Rep. Ex 294*.

Owen, M.W. (1982) The hydraulic design of seawall profiles. Proceedings of the Conference on Shoreline Protection, ICE, Southampton, UK.

Powell, K.A. and Allsop, N.W.H. (1985) Low-crest breakwaters, hydraulic performance and stability. Hydraulics Research, Wallingford, UK. *Rep. SR 57*.

Price, A.W. (1979) Static stability of rubble mound breakwaters. Report of Hydraulics Research Station, Wallingford, UK.

Seeling, W.N. (1983) Wave reflection from coastal structures. Proceedings of Conference on Coastal Structures '83. Arlington, Virginia.

Seeling, W.N. and Ahrens, J. (1981) Estimation of wave reflection and energy dissipation coefficients for beaches, revetments and breakwaters. US Army Corps of Engineers, Coastal Engineering Research Center, Fort Belvoir, Va.

Shore Protection Manual (1973) USA Army, Coastal Engineering Research Center. (4th edn) I and II.

Shore Protection Manual (1984) USA Army, Coastal Engineering Research Center I and II.

Stive, M.J.F. (1986) Extreme shallow water wave conditions. Delft Hydraulics. *Int. rep. H533*.

Scott *et al.* (1986) Delft Hydraulic 1980 Aalborg University (1987), p. 53.

Tanimoto, K. (1976) Wave forces on a composite-type breakwater. Proceedings of the 1976 Annual Research Presentation of Port and Harbour Research Institute, Japan (in Japanese).

Tanimoto, K., Takahashi, S. and Kimura, K. (1987) Structures and hydraulic characteristics of breakwaters – The state of the art of breakwater design in Japan. Report of the Port and Harbour Research Institute, Japan. Vol. 26, No. 5.

TAW (1974) Wave run-up and overtopping. Technical Advisory Committee on Protection against Inundation.

Thompson, D.M. and Shuttler, R.M. (1976) Design of riprap slope protection against wind waves. CIRIA report 61.

Thornton, E.B. and Guza, R.T. (1983) Transformation of wave height

distribution. *Journal of Geophysical Research* **88**, 5925–38.
Timco, G.W. (1981) The development, properties and production of strength – reduced model armour units. Hydraulics Lab., Tech. rep. Nov. 1981, NRC, Ottawa, Canada.
Timco, G.W. and Mansard, E.P.D. (1983) On the interpretation of rubble-mound breakwater tests. Proceedings of Coastal Structures '83, Arlington, Virginia.
van der Meer, J.W. (1988a) Rock slopes and gravel beaches under wave attack. Doctoral thesis. Delft University of Technology.
van der Meer, J.W. (1988b) Stability of cubes, Tetrapodes and Accropode. Proceedings of Breakwater '88, Eastbourne, UK.
van der Meer, J.W. and Angremond, K. (1991) Wave transmission at low crested structures. Proceedings of Conference on Coastal Structures and Breakwaters, ICE, London.
van der Meer, J.W. and Heydra, G. (1991) Rocking armour units: number, location and impact velocity. *Coastal Engineering* **15**, 21–39.
van der Meer, J.W. (1986) Stability of rubble mound breakwaters. Stability formula for breakwaters armoured with Cubes. Delft Hydraulics, Report on Basic Research S467, **VI**.
van der Meer, J.W. (1991) Stability and transmission at low-crested structures. Delft Hydraulics Report H-453, **May**.
Van Oorschot, J.II. (1983) Breakwater design and the integration of practical construction techniques. Proceedings of Coastal Structures '83, Arlington, Virginia, USA.

HYDROMECHANICS OF POROUS MEDIA IN THE MARITIME ENVIRONMENT

30

F.B.J. Barends

The subject refers to the mechanical behaviour of porous structures, like breakwaters, when exposed to hydraulic loading. This means that the methods developed to handle such problems are different from those used in standard geomechanics. Simulation techniques are described to assess, by the use of models, the hydro-mechanical processes. The importance of the many soil parameters that need to be established, like cohesion, porosity, permeability and many others is discussed. Fracture mechanics and the way they can occur by excessive deformation and erosion of materials are characterized differently and some general aspects of modelling and simulation are mentioned. These vary from simple empirical formulae to sophisticated computer simulations and examples of these are described. An interesting section ends the chapter where the author suggests where engineering intuition has to be applied and where certain processes like the hydraulic loading imposed on individual particles by waves, internal friction and interlocking cannot properly be described (see also Chapters 29, 34 and 37).

CONTENTS

30.1 INTRODUCTION

The term hydro-mechanics of porous media makes a distinction with regard to standard geomechanics. It refers to the mechanical behaviour of porous structures when exposed to hydraulic loading, such as waves. Two facts make a difference: firstly, multiphase mechanics, in which air, water and rock/soil play a role; and secondly, dynamics, where critical situations occur during dynamic loading. Both aspects render common geotechnical methods applicable with special considerations, which will be explained in this chapter. For common geotechnical methods reference is made to the geotechnical handbooks (Lambe and Whitman, 1969).

Wave loading generates normal stresses (pressures) and shear stresses in the porous structure. The water phase in the porous structure can sustain pressures well, but not shear stresses. For that part of the phenomenon the granular skeleton of the porous medium is involved. When shear stresses in the granular skeleton surpass a limit strength, the skeleton will slide and fail. This limit strength depends on the actual stress in the granular skeleton. For the evaluation of hydro-mechanics in porous media the distinction between stresses in the granular skeleton and pore pressures is essential. The inter-granular stresses determine the functioning of the structure. In hydraulic engineering the granular response is often disregarded resulting in an incomplete understanding of the real process and in over-estimation of the functionality.

The procedure to assess the hydro-mechanical behaviour is given in Table 30.1 and the checklist. The table contains the most essential processes and some characteristic parameters. The checklist refers to the most essential steps in an evaluation procedure.

Checklist

Determination of:

- Critical cross-sections, geometry, material parameters;

Coastal, Estuarial and Harbour Engineer's Reference Book. Edited by M.B. Abbott and W.A. Price. Published in 1993 by Chapman & Hall, 2–6 Boundary Row, London SE1 8HN. ISBN 0 419 15430 2.

Table 30.1 Procedure to assess hydromechanical behaviour of porous media

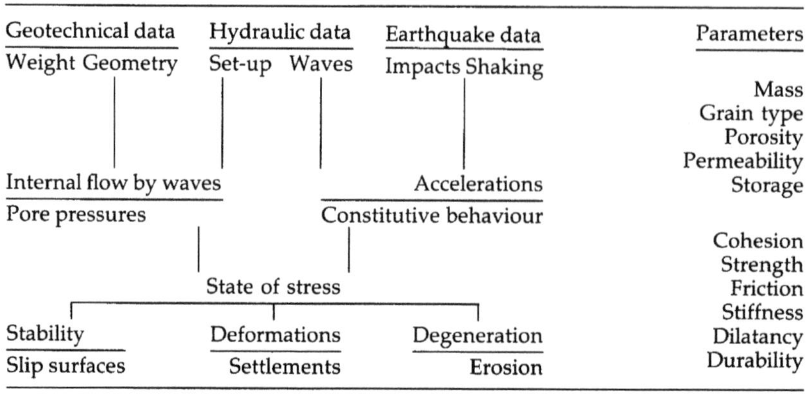

- Design wave loading/design earthquake loading;
- Pore pressures (seepage pressures), transmission, internal set-up;
- Hydraulic stability of filters;
- Excess pore pressures;
- Stability of elements/structure;
- Settlements;
- Dynamic stability;
- Residual deformations.

The above items are addressed in this chapter.

30.2 PARAMETERS

The hydro-mechanical processes are assessed by simulation techniques, by using models. Information is required to operate the models and formulae. The selection of critical cross-sections and the geometry of material layers is one part. The other parts are material parameters, a provisional list of which is presented below.

Symbol	*Name*	*Remark*
D_i	Grain size (m)	i (%) fraction smaller than D_i
$\tilde{\rho}$	Specific density (kg m^{-3})	Density of bulk granular medium
ρ'	Specific density (kg m^{-3})	Density of granular material
ρ	Specific density (kg m^{-3})	Density of water
γ	Specific weight (N m^{-3})	$\gamma = \rho g$ (also γ' and $\tilde{\gamma}$)
Sh	Grain shape	Ratio maximum/minimum diameter
Rf	Surface smoothness (mm)	Irregularities grain surface
S	Strength (N m^{-2})	Breaking strength grain
c'	Cohesion (N m^{-2})	Interlocking, cementation
n	Porosity	Volumetric density granular skeleton
k	Permeability (m s^{-1})	$k(D, n, Sh, Rf,$ pressure gradient)
ϕ	Friction angle	$\phi(R, n, S, \sigma', D)$
ψ	Liquefaction potential	Rate of excess pore-pressure
n_0	Critical density	Porosity at zero dilatancy
G	Shear modulus (N m^{-2})	$G(n, \sigma', S)$
K	Bulk modulus (N m^{-2})	$K(n, \sigma', S)$

Some additional specific information in relation to the above mentioned parameters are important for rubble mound marine structures and are discussed next.

Cohesion c'

Cohesion is related to the fact that removal of a particle at zero effective stress requires some effort. For rock and artificial armour units cohesion is related to the interlocking. For sand, cohesion is negligible. For clay it is related to electromechanical forces at the microscale. Kobayashi *et al.* (1987) suggests that for rock the use of an apparent cohesion to account for the dependence of friction to isotropic pressure or for the overburden pressure (sand) should be assumed, particularly when applying a Bishop stability analysis. An average value for this apparent c' is 20 kN m^{-2}.

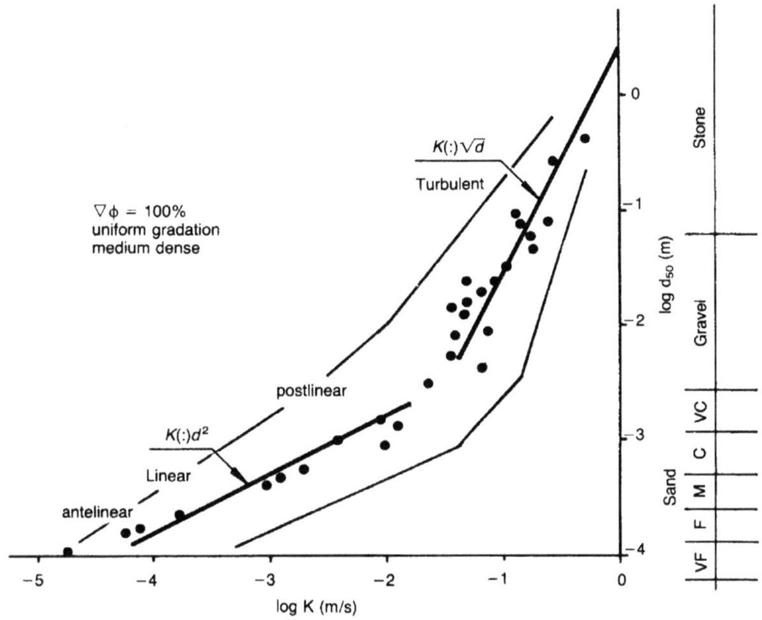

Figure 30.1 Permeability versus grain size.

Porosity *n*

The porosity depends on the packing of the particles (skeleton structure). An average porosity value for large well-placed normally graded stones is 42%. Randomly placed it may vary between 30% and 50%. In wide-graded mixtures this range is even larger. It is not practically possible to determine the porosity of existing coarse granular marine structures in a direct manner.

Permeability *k*

The permeability depends on the flow regime (linear, turbulent). An approximate formula for linear flow is $k = 0.006\,n^3D^2g/(1 - n)^2\nu$, which is valid for uniform gradation with $D_{50} < 0.01$ m. Here, ν is the kinematic viscosity of water, and g the gravitational acceleration. Hannoura and Barends (1981) describe various formulae. The apparent permeability is related to D^2 for linear flow and to $\sqrt{D}$ for turbulent flow (Figure 30.1). The regime of flow depends also on the actual pressure gradient, which may vary in space and time (waves). Scaling problems are unavoidable in physical modelling.

Friction angle ϕ

The granular friction angle is a material constant for sands in the range of 35° to 45°. For stone/rock media the actual friction angle depends not only on various material characteristics, but also on the actual stress. Barton and Kjaernsli (1981) suggest $\phi = \phi_0 + R\log(S/\sigma')$, where ϕ_0 is a basic friction coefficient between polished surfaces, R a factor dependent on type, shape and packing, S the material strength and σ' the actual effective stress. ϕ_0 is in the range of 27° to 33°. The value for R is given in Figure 30.2. The effective stress can be determined by standard methods (stability or deformation models).

Dilatancy

The sensitivity of a granular material to becoming liquefied is expressed by the so-called liquefaction potential. The liquefaction potential can be obtained by special laboratory tests (cyclic shear and cyclic triaxial tests) on samples from the site at various (manufactured) densities. Results should be calibrated against the *in situ* density. Excess pore pressures arise depending on the drainage capacity. Particularly fine loose sands ($D < 300\mu$) are sensitive to excess pore pressure generation.

30.3 SCOPE OF HYDRO-MECHANICAL ASPECTS

Hydro-mechanical aspects are important in those circumstances where the construction will not function properly, referred to as failure states. The man-

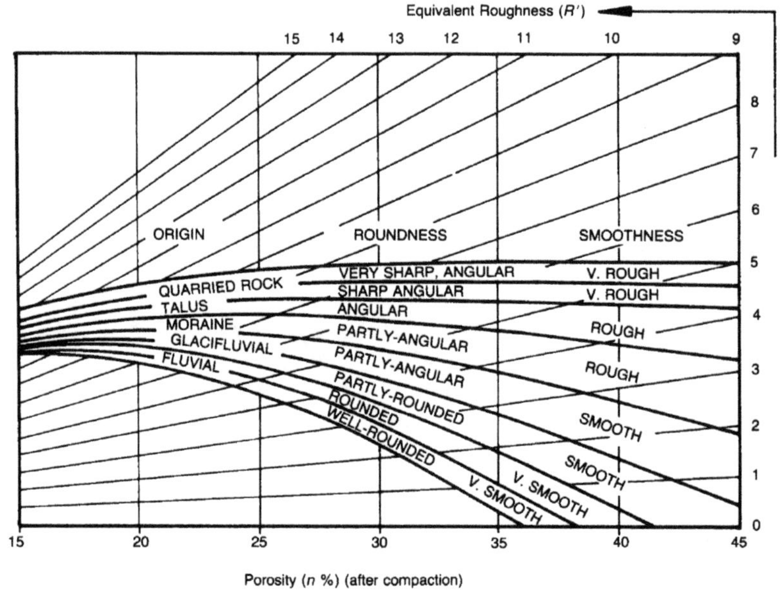

Figure 30.2 Equivalent roughness and porosity (Barton, 1981).

ner in which such a state can occur is called a failure mechanism. Two major geotechnical mechanisms of a porous structure under dynamic loading are distinguished: excessive deformation and erosion of materials. Each mechanism is characterized by different aspects. The aspects are evaluated by applying models and formulae.

Mechanism	Aspect
Excessive deformation	Porous flow
	Shear strength
	Liquefaction
	Accelerations
	Excess pore pressures
	Settlement of the fill
	Residual deformation
Erosion of materials	Filter gradation
	Change of material properties
	Dynamic gradients
	Local stability of elements

30.4 MODELLING AND SIMULATION

Models can be distinguished with respect to applicability, reliability, cost, sophistication, user-friendliness and accessibility. It is not possible to give a complete list of models, nor to give a detailed picture of a particular model. Some general aspects are discussed and references are mentioned.

Four types of models (methods) are distinguished: sophisticated computer models, which provide a complete solution; uncoupled computer models, which solve a porous flow problem or a deformation/stability problem; simple models (hand-formulae) for particular phenomena; and engineering experience (intuition).

30.4.1 LOADING

Geotechnical aspects of rubble mound marine structures concern the mechanical behaviour of rock/soil and the foundation being subjected to weight, to seepage forces, to pore pressures and to accelerations caused by waves and earthquakes. Wave pressures on the structure are usually measured in hydraulic models. The design loading essential for geotechnical failure is usually the maximum loading occurring during the lifetime of the structure. To release this condition probabilistic methods are suggested.

30.4.2 GEOMECHANICAL PRINCIPLES

The mechanical behaviour of a two-phase material (fluid and grains) is determined by loading (action) and by reaction, in terms of forces. In a porous medium the inter-granular forces acting at the grain contacts are represented by the so-called inter-granular or effective stress. This stress is equal to the average of all inter-granular forces. It has normal and tangential components. Fluid pressures in the pores present another force, with only normal components.

The reaction to a loading, expressed in terms of total stresses, comprises effective stresses and pore pressures. Because pore-pressure gradients cause flow, which in turn change the pore pressures with time, the part of the reaction by the pore pressures changes in time. Effective stresses will vary to compensate for the change in pore pressures. Deformations corresponding to the effective stresses are therefore time-dependent.

If the individual grains are relatively rigid, Terzaghi's effective stress rule applies:

for normal stress components: $\sigma = \sigma' - p$

for tangential stress components: $\tau = \tau'$

where (σ, τ) is the total stress; (σ', τ') is the effective stress; and p is the pore pressure. For example, the critical stress state along a slip failure surface is expressed by: $\tau = c' + (\sigma - p)\tan\phi$. The fact that p appears in the formula clarifies the importance of the flow field for the actual state of stress.

30.4.3 MODELLING FACILITIES

The state-of-the-art assessment of the geotechnical aspects mainly consists of numerical methods varying from simple empirical formulae to sophisticated computer simulations. Physical model testing of geotechnical aspects is rarely performed because of difficulties in coping with non-linear behaviour and scale effects.

The description of physical processes with formulae, physical or numerical models, graphs or engineering intuition can provide realistic approximations. The purpose of using a model is to optimize the design or a particular element of the structure.

For the assessment of an optimum, most reliable and economic solution one may use probabilistic methods. The methods that have been discussed fit this purpose. For a probabilistic approach the probabilistic distribution of loading and the stochastic variation of material and geometrical parameters is required (CIAD, 1985).

It is difficult to make observations of hydrodynamic processes in a porous medium. Such a medium is not transparent. In addition to the measurement of the pore pressures the response of the granular body should be recorded. However, the behaviour of one particular pore or one particular grain is not necessarily characteristic of the overall behaviour. In model studies not all scale effects can be accounted for simultaneously. An important geotechnical scale effect which is difficult to match is the dependence of strength and stiffness to the absolute effective stress. A promising facility which could correct this deficiency is geocentrifuge testing.

For the simulation of hydro-mechanical aspects empirical and numerical methods are common. The empirical methods are simple formulae taken from geotechnical practice and adjusted for the maritime environment. Numerical methods comprise computational models, suited for a schematic structure. Numerical methods have become widely used in the last decade. In numerical models there are no scale effects, but a proper mathematical formulation of the material behaviour and an accurate solution procedure are a problem.

30.5 MATHEMATICAL BACKGROUND

The essence of numerical modelling is the mathematical description of the physical processes. A survey is presented in the form of basic equations and rules. Three media are involved: air, water and rock/soil. The influence of air is important because its entrainment affects compressibility and hence impact pressures. Practical information about air effects is limited.

30.5.1 PORE-WATER MECHANICS

For water and rock/soil two inter-related equations of state (mechanical equilibrium) are involved. The equation of state of the pore water yields (action = reaction):

$$\underbrace{-np_{,i}}_{\text{Pressure gradient}} - \underbrace{n\rho g z_{,i}}_{\text{Gravity}} = \underbrace{n\rho\dot{v}_i}_{\text{Inertia}} + \underbrace{n^2\rho g(v_i - u_i)/k}_{\text{Interaction}} \tag{30.1}$$

Here, v and u are the velocity vectors for the pore fluid and the granular matrix, respectively. The subscript i represents a co-ordinate direction. If preceded by a ',' it is a partial derivative. The '$\cdot$' represents a time-derivative:

$$\dot{v}_i = \underbrace{cv_{,t}}_{\text{Local}} + \underbrace{bv_jv_{i,j}}_{\text{Convective}} + \underbrace{a\dot{u}_i}_{\text{Interaction}} \tag{30.2}$$

where c includes virtual mass effects ($c > 1$) related to the unsteadiness of flow, b is the momentum distribution coefficient, and the last term represents mass effects in the flow field because of the presence of the granular phase. The last term in Equation (30.1) includes viscous effects at pore size dimensions.

30.5.2 GRANULAR MATRIX MECHANICS

The pore pressure p and the porous matrix velocity u are important to the deformation behaviour of the granular skeleton, which is described by an equation of motion for the porous matrix. This equilibrium is governed by the second equation of state (action = reaction):

$$\underbrace{-(1-n)p_{,i}}_{\text{Pressure gradient}} \underbrace{- (1-n)\rho' z_{,i}}_{\text{Gravity}} = \underbrace{(1-n)\rho' \dot{u}_i}_{\text{Inertia}} - \underbrace{\sigma'_{ij,i}}_{\text{Effective stress}} \underbrace{- n^2 \rho g (v_i - u_i)/k}_{\text{Interaction}} \tag{30.3}$$

For the inertia term an expression like Equation (30.2) can be formulated.

Beside the two equilibrium equations the conservation of mass and material behaviour are required to complete the mechanical description. For a rock granular media the individual grains are considered incompressible. For high stresses this assumption has to be reconsidered because it leads to other relations. The pore-water mass-conservation equation yields:

$$\underbrace{(\rho v_i)_{,i}}_{\text{In/out flow}} + \underbrace{(n\rho)_t}_{\text{Storage}} = 0 \tag{30.4}$$

A compressibility law for the pore fluid can be introduced accounting for the presence of entrained air.

A stress-strain constitutive law describes the relation between the porous matrix stresses and the deformations (elastic, plastic, creep). For practical application rigorous concessions are made with regard to the stress-strain behaviour, for example linear elastic or elasto-pure-plastic.

Equations (30.1)–(30.4) represent a complete description of the mechanical process in a saturated rock/soil porous medium. Sophisticated numerical models exist covering these equations. In most applications much simpler approaches will do. Simplification is possible when interaction is disregarded. Then the porous flow is independent from the porous matrix behaviour. The opposite, however, is not the case: the effective stress then depends on the pore pressure.

30.5.3 CONSOLIDATION, LIQUEFACTION AND POROUS FLOW

Introduction of the filter velocity: $q_i = n(v_i - u_i)$ and a potential function: $\rho g \Phi_{,i} = p_{,i} + \rho g z_{,i}$ renders Equation (30.1) into a familiar flow law:

$$\underbrace{\Phi_{,i}}_{\text{Gradient}} = \underbrace{-cq_{,t}}_{\text{Unsteadiness}} - \underbrace{q_i/k(|q|)}_{\text{Forchheimer/Darcy}} \tag{30.5}$$

Substitution into the conservation Equation (30.4) disregarding the unsteadiness yields after some elaboration the so-called storage equation

$$(K/\rho g)\Phi_{,ii} = n\beta\Phi_{,t} + \varepsilon_{,t} \tag{30.6}$$

The term ε represents volumetric strain. It is important for soft slowly-draining layers (consolidation), for example in the foundation, and for loose sand deposits, which may show dilatancy. Then the last term of (30.6) is replaced by $-\gamma\psi$, where ψ is the liquefaction potential. If ε is assumed constant and the compressibility of the pore-water is disregarded ($\beta = 0$), the familiar equation for potential flow remains.

30.5.4 WAVE TRANSMISSION

In the case of a phreatic surface (internal water table) the motion of this surface under time-variant loading has to be considered. Local flow and storage determine the position, mathematically described by the so-called moving boundary condition. Saturation, air flow, and capillarity complicate the model simulation of phreatic motion. In coarse porous media the unsteadiness and the inertia effects become significant. A simple way to address this phenomena is by the averaging and integration of Equations (30.1) and (30.4) over the vertical co-ordinate (like the shallow-wave approach), and disregarding the interaction with the skeleton deformation. The following equations are obtained:

$$nc_m(\bar{v})_{,t} + \bar{v}(\bar{v})_{,x} = -n^2 g(c_p D)_{,x} - n^2 g\bar{v}/k \tag{30.7a}$$

$$nD_{,t} + (D\bar{v})_{,x} = 0 \tag{30.7b}$$

where D is the local water depth and $\bar{v}$ the mean velocity, c_m and c_p are correction factors for vertical mass and pressure distribution, respectively.

30.5.5 SIMPLE APPROACH

Simplification of the equation of state of the granular matrix is obtained by adding Equations (30.1) and (30.3) while disregarding inertial terms. The result is:

$$-p_{,i} - \tilde{\rho} g z_{,i} + \sigma'_{ij,i} = 0$$

$$\text{and} \quad \tilde{\rho} = n\rho + (1+n)\rho' \tag{30.8}$$

or with Terzaghi's rule:

$$\sigma_{ij,i} - \tilde{\rho} g z_{,i} = 0$$

In a practical approach the stress field is simplified rigorously by only considering vertical stresses: a direct relation between local stress and overburden:

$$\sigma_{zz} = \tilde{\rho} g z \tag{30.9a}$$

and by adopting a simple stress-strain law, assuming negligible deformations until a failure state given by Coulomb's law

$$\tau \leqslant c' + \sigma'_{zz} \tan\phi = c' + (\sigma_{zz} - p)\tan\phi \tag{30.9b}$$

where c' is the (apparent) cohesion and ϕ the internal friction angle.

30.6 MODELS

30.6.1 SOPHISTICATED MODELS

Sophisticated models deal with non-linear multiphase deformation and dynamics. The core of the model is the solution of Equations (30.1)–(30.4) simultaneously with appropriate constitutive relations and initial/boundary conditions. Complicating features are the formulation of non-linear behaviour and dynamic interaction, the determination of the initial state and the dynamic effect of air.

Sophisticated models are applied in special situations: for unique or large structures, when a high degree of safety is required, and for the qualitative understanding of certain phenomena.

The application of sophisticated models is relatively costly. It requires an extensive parameter collection (site investigation and laboratory testing). The results of numerical models should be validated by physical modelling. A sophisticated model is the SATURN code. It is suited to simulate non-linear dynamic behaviour of two-phase soil structure interaction utilizing the finite element technique. A Darcy–Biot approach is used for the water–soil interaction. The soil–structure interaction includes slip and splitting. A general friction law can be applied at the interface. Equilibrium equations are solved using explicit time integration for dynamic wave propagation, and implicit techniques for unsteady motion. By updating spatial positions large strain effects are taken into account. The initial state problem for non-linear behaviour is generated implicitly to provide compatibility to the imposed non-linear dynamic problem. Available soil models are: von Mises, Drucker–Prager, Mohr–Coulomb, critical state, double hardening. The pore-water is compressible. The code is fully tested and applied to various comprehensive dynamic problems (Barends, *et al.*, 1983).

30.6.2 SIMPLIFIED (UNCOUPLED) MODELS

The behaviour of the water and the rock matrix can sometimes be separated, which simplifies the modelling significantly. For some phenomena a static geotechnical approach is sufficient, such as for settlements of the subsoil. For structures in a maritime environment the dynamic character of wave loading or earthquakes is essential. Uncoupled models simulate the porous flow without consideration of the influence of the rock/soil deformation, or they simulate the stability with a simplified pore-pressure field.

Porous flow

Wave loading generates internally a pore-pressure field. Differences in pore pressures generate porous flow. For the determination of effective stresses the internal flow field must be known. The flow in coarse porous media is non-linear (turbulent). A moving boundary represents the fluctuating water table inside the structure. For mainly horizontal flow, Equation (30.7) can be used. Specific features are the precise boundary conditions: the moving water table, overtopping, air entrainment, and the formulation of the wave action on the slope. Few models suited for wave-generated dynamic porous flow are known, one of which is the HADEER code (Hölscher and Barends, 1990). An application is shown in Figure 30.3.

Deformation

The deformation behaviour of rock and soil is typically non-linear and irreversible. It is not easy to find a simple proper description of this behaviour. Field and laboratory tests are indispensible for the characterization of the soil/rock behaviour.

The deformation can be obtained in a semi-uncoupled fashion. The pore-pressure field must be known. Simulation of deformations of a rubble mound structure itself is seldom performed. Most deformation takes place during construction. For large structures deformation analysis is important. Settlements of compressible layers in or under the structure due to the overburden weight is a common feature in soil mechanics practice (consolidation). Various methods and models are available. Retarded settlements can be expected. Non-linear behaviour must be used to model squeezing (plasticity).

Stability

Effective stresses are responsible for deformations of the granular matrix. Significant deformations may occur in a narrow band, a slip surface conditioned by shear strength and kinematics, or deformations may occur throughout the structure (plastic strain). Various numerical models dealing with elasto-plastic deformations can handle these phenomena. Models suited to simulate two-dimensional elasto-plastic consolidation are few in number. An example of the occurrence of slip zones and plastic zones is presented in Figure 30.4, showing two situations in a state of failure, modelled at the Technical University, Delft, in centrifuge tests. A different failure model is observed in this case which is dependent on the geometry.

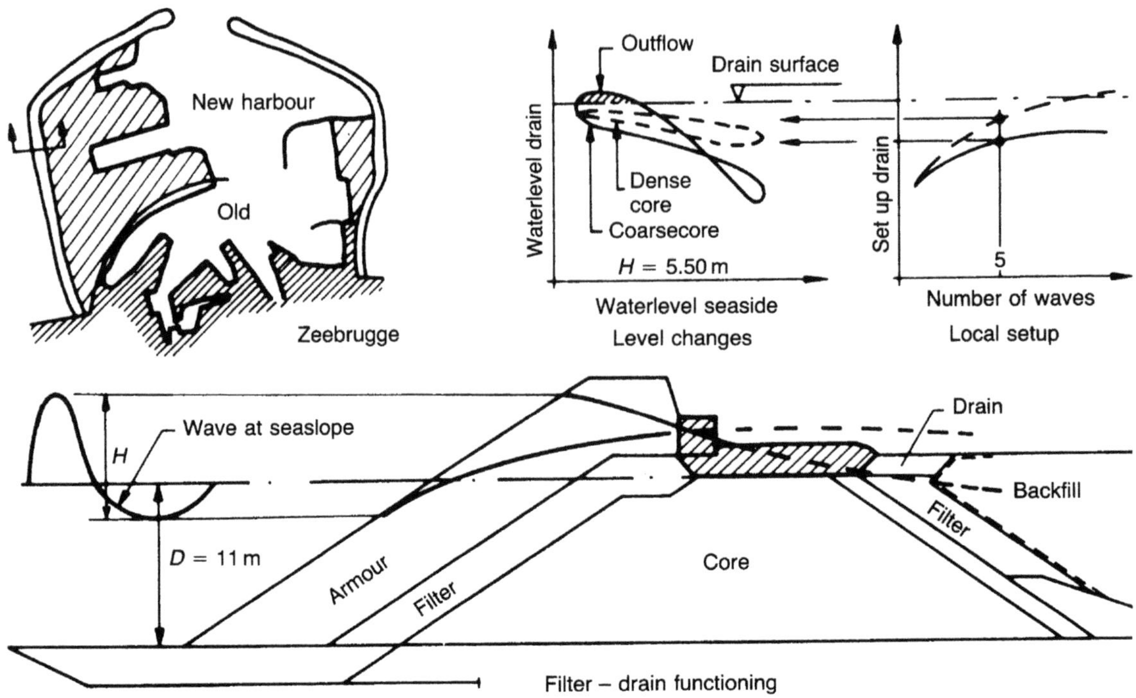

Figure 30.3 Simulation of porous flow field by HADEER.

Liquefaction

Dilatancy expresses a volume change of granular material under shear deformation; dilatancy (increase) and contractancy (decrease). This volume change is related to a change in the granular structure, by sliding and rolling of the particles in a different composition. Because water which is trapped in the pores cannot be expelled immediately, excess pore pressures are generated, negative for dilatancy and positive for contractancy. At the so-called critical density the material is volume constrained. The sensitivity to dilatancy–contractancy depends on the actual *in situ* density. In an ultimate state the skeleton structure collapses. Particles are free. A state of liquefaction – a thick fluid without significant shear resistance – is produced. The character of this phenomenon also depends on the type of loading. A sudden loading (earthquake, solitary wave) may cause a rapid decrease of pore volume and a corresponding decrease of shear strength due to excess pore pressures, in less than a second. The dissipation of the pore pressures may take some minutes or more, and in the intermediate period catastrophic consequences may have occurred. Constant cyclic loading may cause slow build-up of excess pore pressures with similar consequences. Both situations may occur simultaneously.

The suggested assessment of wave induced liquefaction of sea-bed sand is, according to the state-of-the-art (Rahman and Jaber, 1978), referred to as the uncoupled approach. In the uncoupled approach the phenomenon is divided into two inter-related problems: firstly, the wave induced state of stress without cyclic excess pore pressures; and secondly, with excess pore pressures by the cyclic stress ratio including dissipation. The actual liquefaction potential can be determined when the actual porosity and the critical porosity (at zero dilatancy) are known. The first one has to be determined at the site the second in the laboratory. A practical method has been described by Barends and Calle (1985). It is particularly useful at the toe of a rubble mound construction, where local failure subsequently jeopardizes the entire structure.

Dynamic effects

During wave impacts and earthquakes the changes in rapid loading changes induce inertial effects in the pore fluid and the porous matrix. In some cases accelerations have to be taken into consideration. For

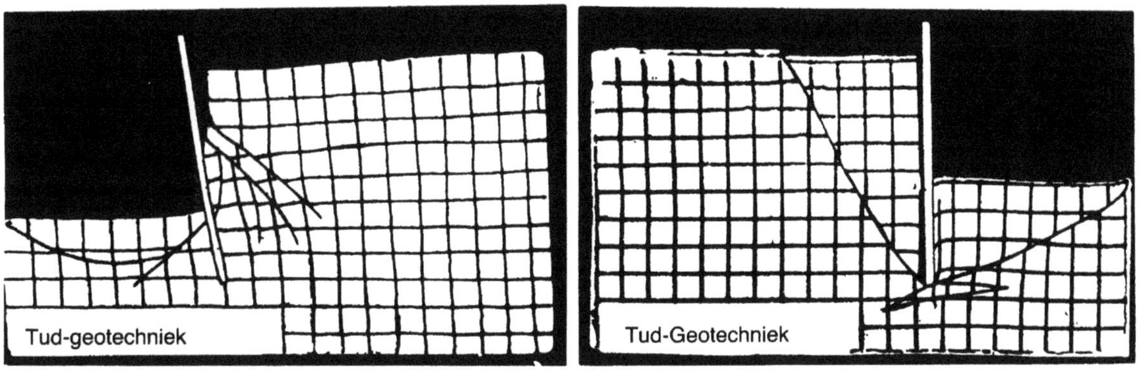

Figure 30.4 Centrifugal modelling of failure.

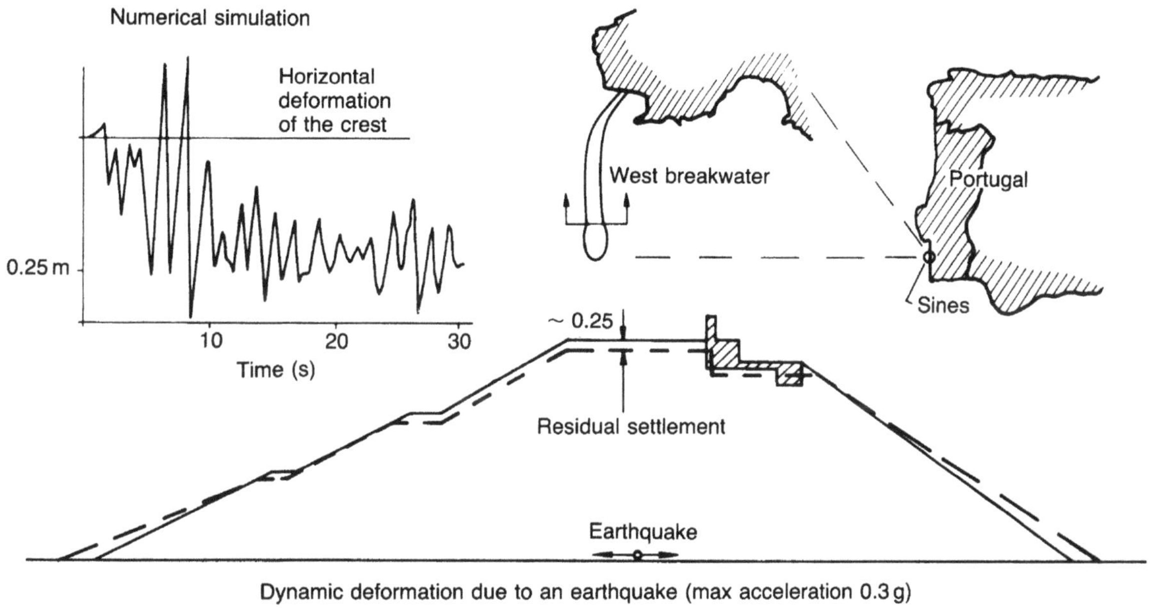

Figure 30.5 Simulation of earthquake response.

the dynamic reaction, whether in the pore-water or in the granular skeleton, the permeability plays a significant role (Barends, 1985).

A realistic behaviour under earthquake loading can be obtained by a dynamic model based on a consistent stress and deformation state. Figure 30.5 shows the result of such an analysis where the residual deformation has been calculated as the result of a complete earthquake loading.

For coarse rubble media the earthquake stability analysis can be employed without taking the pore water weight into consideration. This is permitted, because the sliding soil mass will not convey the pore-water in its motion. This approach yields a reduction of the driving momentum.

30.6.3 SIMPLE MODELS (FORMULAE)

Wave penetration/transmission

Wave loading causes time-dependent porous flow. The time dependency is due to storage. One kind of storage occurring in coarse granular media is phreatic storage. The storage acts as a damping mechanism which attenuates the wave effect inside the structure. A formula is derived based on conservation of mass and on a simplified flow field. The volume of water penetrating equals the volume of water stored:

$$Q\Delta t = k\Delta t A(H/\lambda) = nH(\lambda(t + \Delta t) - \lambda(t))/2 \quad (30.10a)$$

which gives

$$(n/2)(d\lambda/dt) = kA/\lambda$$

and integrated

$$\lambda = \eta(kAt/n)^{1/2} \quad (30.10b)$$

where λ is the penetration length (m), η is a parameter dependent on the loading ($\eta = 2.0$ for a sudden constant water level change, $\eta \approx 0.5$ for cyclic water level changes), A is the drainage surface through which water seeps (m), and t is the elapse time for a sudden constant water lever change or period of cyclic loading. The significance of rapid or slow varying wave loading for the geotechnical stability can be evaluated. For coarse sand with $n = 0.20$, $k = 0.0001\,\text{m}\,\text{s}^{-1}$ and $A = 10\,\text{m}$, the following is found:

Water waves	$t = 10\,\text{s}$	$\lambda = 0.11\,\text{m}$
Tides	$t = 12\,\text{h}$	$\lambda = 7.35\,\text{m}$
Rain river (constant)	$t = 3$ day	$\lambda = 72.00\,\text{m}$

Internal phreatic set-up

The slope of a rubble marine structure subjected to waves, swell or tides represents a geometric non-linear effect. The inflow surface along the slope at the moment of a high water-level is larger than the outflow surface at the moment of a low water-level. Moreover, the average path for inflow is shorter than the outflow path. Hence, during cyclic water-level changes more water will enter the structure than will leave. Consequently, there will be a state, in which the outflow of the surplus of water is realized by an average internal set-up of the water level inside the structure (Barends and Hölscher, 1988).

The slope steepness causes an extra storage of water volume every cycle in comparison to a linear situation with a vertical slope. This extra volume can be expressed as an average infiltration, according to:

$$I = I_0 \exp(-x/\lambda)$$

with

$$I_0 = cH^2/(2\lambda T \tan\alpha) \quad (30.11a)$$

using the simplified formula for the penetration length, λ. Solving a schematic flow problem gives the expression for the average water height h as a function of x. The solution is

$$h = Ax + B - \xi D^2 \exp(-x/\lambda)$$

with

$$\xi \approx 0.1\, cH^2/(n\lambda D \tan\alpha) \quad (30.11b)$$

where A and B depend on boundary conditions. The resulting formula is

$$S/D = (1 + \xi F)^{1/2} \quad (30.11c)$$

where S is the maximum set-up, D the mean water depth, c is a constant depending on the effects of air entrainment and rush-up/run-down ($c > 1$), H is the water level amplitude at the slope, λ is the penetration length, and α the slope angle. The function F is presented in Figure 30.6 for two cases: a closed; and an open lee side. In the first case the maximum set-up occurs at the backside. In the second case the maximum set-up occurs at bL with L the length of the inner water-table. For the real water-table inside the structure the cyclic motion must be superimposed on the set-up.

Slip-circle analysis

The evaluation of the stability of rock marine structures is a standard procedure. Usually, a failure state is simulated by the choice of a critical kinematic system assuming continuous deformation surfaces (slip surfaces), and comparing the mobilized force with the exposed one. Key factors are internal friction and cohesion. The stress state along the slip surfaces is determined by wave induced pore-pressures or earthquake-induced accelerations and a simplified effective stress field according to Equation (30.9) (Bishop's method). Local excess pore-pressures (liquefaction) can be accounted for by adjusted pore-pressures or a reduced friction angle. Several studies have shown that this method is sufficiently accurate for rubble slopes when proper values for friction and cohesion are applied.

Bishop's method is most commonly used because of its convenience and is widespread in simple computer codes. The most critical circle of sliding is calculated by considering the moment of weight and the shear resistance along the sliding surface. The soil mass is divided into several slices. The mechanical equilibrium of each of them is considered. Horizontal forces between adjacent slices are disregarded. The error in the safety factor is 2–6%. The stability of the soil mass is expressed as a safety factor F. If F is larger than 1 the structure is safe. Usually an extra margin is taken into account to allow for uncertainties, i.e. $F > 1.25$. The slip-circle analysis is based on effective stresses. Hence, pore pressures are important, even more when they vary in time. The typical rock friction angle dependent on the actual stress can be incorporated easily. Wave-induced porous flow has

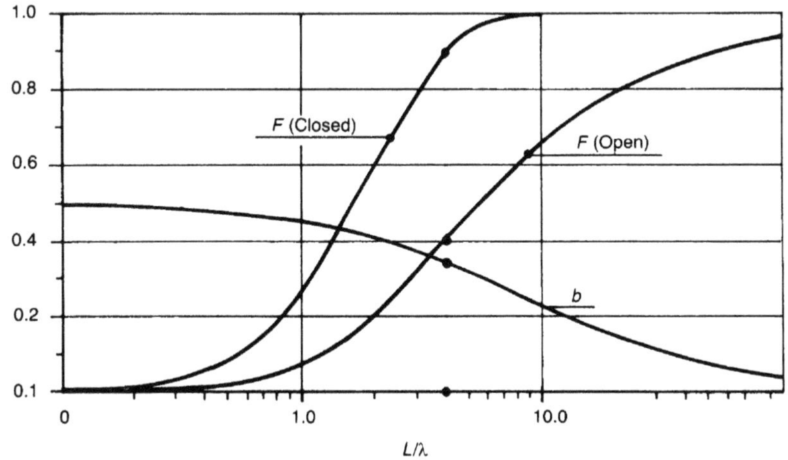

Figure 30.6 Internal set-up.

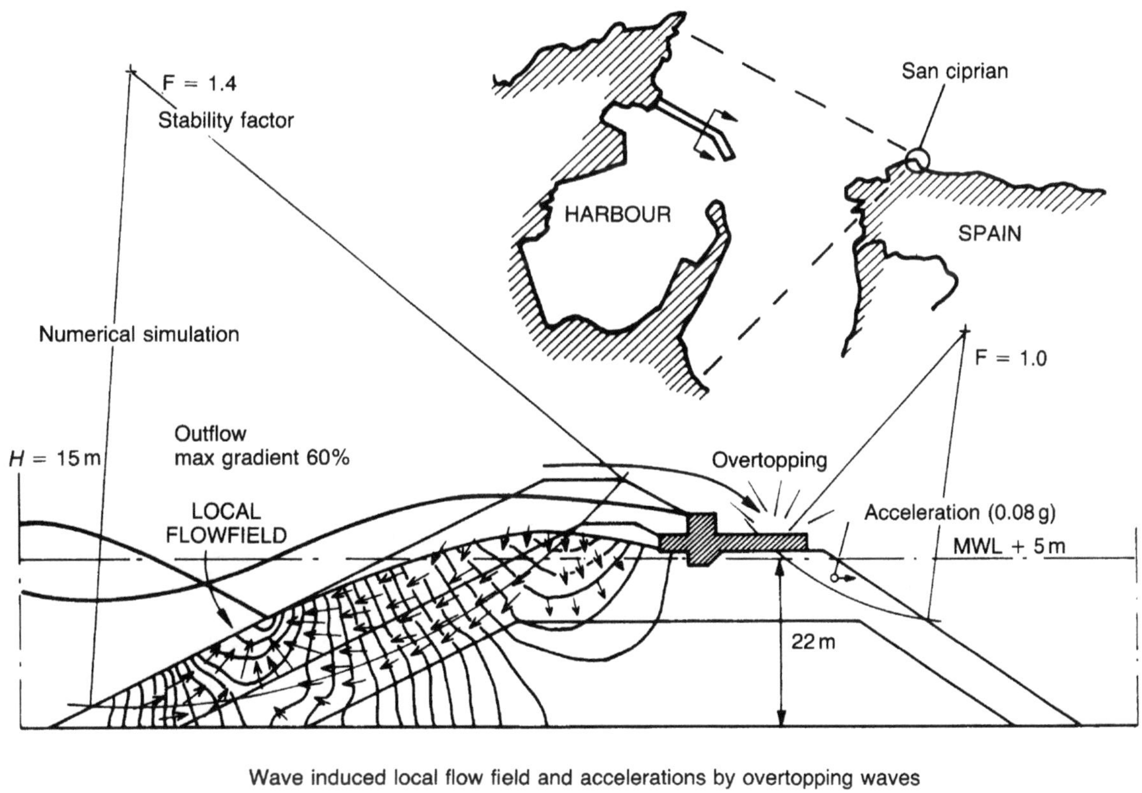

Figure 30.7 Stability under wave attack.

shown an effect up to 25% of the safety factor (Figure 30.7).

The stability of slopes subjected to earthquakes is commonly simplified by the introduction of an additional inertial force, whose magnitude is equal to the product of the mass of the slice in failure and its horizontal acceleration. Vertical acceleration is not usually considered. The horizontal acceleration can be allowed for in a Bishop stability analysis.

Morgenstern has presented stability charts to facilitate the computation of the factor of safety of earth slopes during rapid draw-down based on Bishop's stability analysis. A homogeneous structure with uniform c' and ϕ is considered on a firm impervious bed. After draw-down no dissipation has yet occurred. Two graphs are shown in Figure 30.8 (left) for a slope α of 2:1 and (right) 3:1 for a draw-down L of original height H. Heavy lines refer to $c'/\gamma H$ = 0.025 and dotted lines 0.0125.

In some cases the most likely sliding face is not circular, but rather along weak strata or along strong rock surfaces. The interface between different rock layers may reveal lower internal friction because of different stone sizes. A reduction of the internal friction up to 30% may be expected. In this case one divides the sliding soil mass in an active wedge, a central block and a passive wedge. The factor of safety is the ratio between the active driving force and the passive resistance force.

In the classical soil mechanics approach the conventional stability analysis assumes that the peak shear strength of the soil is fully mobilized simultaneously along the total length of the potential sliding plane. This is not always the case, as progressive failure can occur. Particularly under certain conditions, slides in over-consolidated plastic clays and clay shales are preceded by the development of a continuous failure surface by a mechanism of progressive failure. Such situations have to be analysed with special care.

Filter stability

Granular filters are applied to protect against erosion by scour and migration. The filter stability is characterized by filter rules. The classic filter rules are

Migration	Uniform material	$D_{50f}/D_{50b} < 5$
	Wide graded	1) $D_{15f}/D_{85b} < 5$
		2) $5 < D_{50f}/D_{50b} < 20$ to 60
Permeability		$D_{20f}/D_{20b} > 5$

The suffixes f and b relate to the filter and basic (fine) material, respectively.

Less strict filter rules are available which take into consideration the gradation:

$$D_{10f}/D_{10b} < 2.50\ D_{60b}/D_{10b} + 5.00$$
$$D_{60f}/D_{10f} < 0.94\ D_{10f}/D_{10b} - 5.65$$
$$D_{50f}/D_{50b} < 2.41\ D_{60f}/D_{10f} + 8.00$$

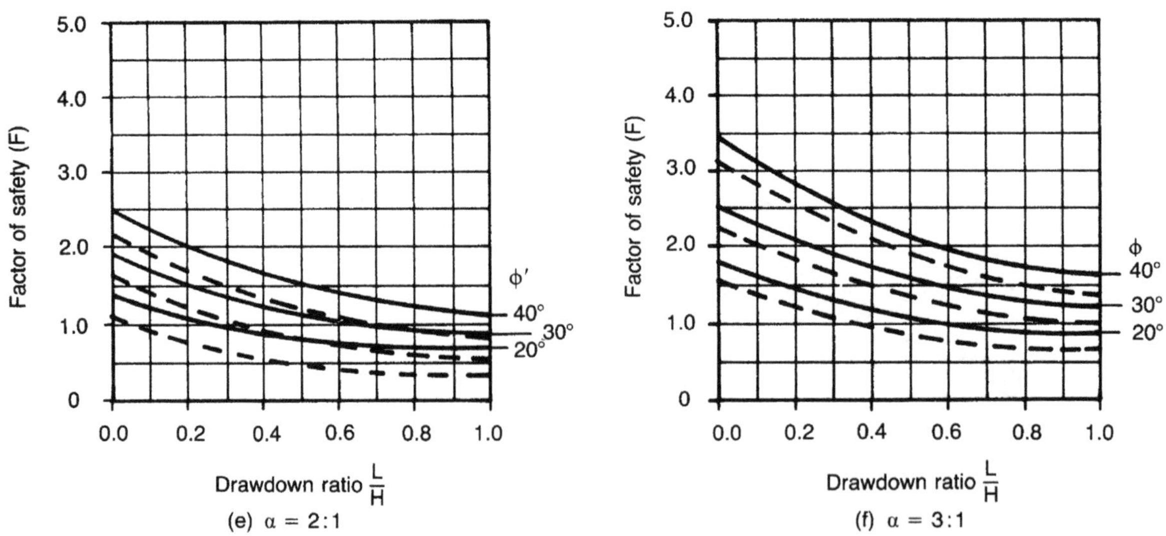

Figure 30.8 Stability factors during rapid draw-down.

The internal stability of a wide graded material covering more orders of grain size must be treated in a different way. Kenney *et al.* (1985) refers to a normalized steepness of the grain size distribution curve to express the internal stability. A minimum of 1.5 is advised. The application of this approach is validated for sand, gravel, and mine stone.

A geotextile on a fine granular bed must prevent migration through the textile, which may be caused by 'driving gradients'. The following criteria are determined for the evaluation of the tightness of its weave:

Cohesive beds

$O_{90}/D_{50b} < 10$; $O_{90}/D_{90b} < 1$; $O_{90} < 0.1\,\mathrm{mm}$

Non-cohesive beds

Stationary loading	$O_{90}/D_{90b} < 1$	regular woven mattresses
	$O_{90}/D_{90b} < 18$	non-regular non-woven
Cyclic loading	$O_{98}/D_{15b} < 1$	no-clogging not widely graded
	$O_{98}/D_{85b} < 1$	clogging, widely graded

In some situations the fines from the bed may clog in the fabric and decrease the permeability significantly, causing high pressure gradients. An acceptable criterion is that the geotextile pressure gradient equals the gradient in the bed under design load conditions. Since the flow through the textile can be turbulent, the precise definition of the permeability needs careful attention.

The strength of geotextiles is important when sliding of soil masses may occur through the textile, or when settlements are opposed by the geotextile. The required strength of a fabric can be determined by a modified Bishop stability analysis. For more information, see Geotextiles (1986).

Dynamic excess pore-pressure

A pore-pressure build-up in a rubble marine structure is possible due to external time-dependent loading and local deformation of the granular matrix. Pore-pressure build-up dissipates by porous flow and by unloading. For the evaluation of the possibility of a local cyclic pore-pressure build-up the order of magnitude of the dissipation period has to be considered. If the dissipation period is small compared to the relevant loading period, a cyclic pore-pressure build-up will not occur. Three phenomena need to be considered;

1. Phreatic storage;
2. Propagation of dynamic disturbance along the inner water table;
3. Field storage (elastic storage).

For the phreatic storage effect see the section on wave penetration. A perturbation in the pore-pressure will decay due to dissipation according to: $p(t) = p_0 \exp(-Bt)$ where p_0 is sudden pressure excitation and B is a decay factor depending on the phenomenon considered: for (2) $B = ng/K$ and for (3) $B = 3c/rl$.

An example clarifies the use for the other cases. A structure has the following properties: porosity $n = 0.4$, gravity acceleration $g = 10\,\mathrm{m\,s^{-2}}$, permeability $k = 0.05\,\mathrm{m\,s}$ (coarse gravel), elastic storage coefficient $c = kK/\gamma$ with bulk modulus $K = 3\,\mathrm{MPa}$ and $\gamma = 15\,\mathrm{kN\,m^{-3}}$ giving $c = 10\,\mathrm{m^2\,s^{-1}}$, average drainage path $l = 4\,\mathrm{m}$, hydraulic radius $r = V/A$ with volume $V = 140\,\mathrm{m^3}$ and drainage surface $A = 15.6\,\mathrm{m^2}$ giving $r = 9\,\mathrm{m}$. Applying the formula yields: for (2), $B = 40$; and for (3), $B = 0.8\,\mathrm{s^{-1}}$. A local pore-pressure build-up will not occur, if the relevant loading return period exceeds $1/3B$, which yields for (2) 0.008 s and for (3) 0.4 s. Excess pore-pressures are expected (breaking waves and wave impacts), but cyclic accumulation is unlikely.

30.6.4 ENGINEERING EXPERIENCE (INTUITION)

Settlement of rock fill

The natural settlement (densification) of rock fill subjected to gravity, hydraulic and earthquake loading is sometimes important. Little is known about this aspect. For earth reservoir dams usually an increase of height of 0.2 to 0.5% is applied, but during the construction densification activities are executed and controlled carefully. In marine structures controlled densification is rarely performed.

Loose dumped gravel and rock fill can be increased in density by special methods, such as the falling weight and shaking plate method. The Menard method is popular and densification effects to a depth of 10 m have been reported. For gravel a 10–15% densification can be achieved and for rock fill 10/60 kg 15%. For large stones heavy mechanical densification causes crushing.

As a first estimate one may compare the energy of natural causes to that of machines. Densification equipment produces an acceleration of $10g$ maximum. Earthquakes may produce up to $2g$ in the entire structure, which will, therefore, lead to a densification of 2% to 4%. The result of failure by internal sliding is not considered.

Large wave impacts may generate in the structure, on average an acceleration of the order of $1g$. The average densification will be of the order of 1%. Slamming waves may, however, produce locally higher accelerations and higher densifications. This can be observed along coastal defence structures around the storm water-line.

Internal erosion/fatigue

Local pore fluid velocities may convey fine particles through the pores of the coarse. This may lead to internal erosion and become the onset of deterioration of the structure. The occurrence is controlled by so-called filter rules, which define limits of adjacent coarse and fine layer grain sizes.

Artificial armour and rock fill degenerates with time, breaks at the contacts, and the fines are washed out. If not washed out, the skeleton fills up and behaves differently. This process can be characterized as material fatigue (Fookes and Poole, 1981).

Water may convey free sediments which settle in the pores and change the hydraulic permeability. It may become a serious problem under critical loading conditions (suffocation). Growth of marine organisms in pores may have a similar effect.

Local stability of rock fill

The hydraulic loading imposed on individual particles by waves, internal friction and the interlocking of randomly placed stones cannot be described properly. Physical tests provide some general understanding of the stability of a stone at the border subjected to hydraulic gradients. The experiments show that for a compacted granular medium porous outflow first activates a group of coherent particles to expand, creating a loose packing, until one favourable surface particle starts moving (boiling), giving way to outflow. This surface particle seems to possess a self-healing stability potential due to the inhomogeneity which is induced by the motion. The space generated by the lifting decreases the extrusion pressure. This behaviour has been observed in tests on sand and gravel (turbulent porous flow). The lift can, however, be a catalyst for the erosion, since a lifted extruding particle can be conveyed more easily by the parallel free surface flow.

REFERENCES

Barends, F.B.J. *et al.* (1983) Westbreakwater – sines – dynamic – geotechnical stability of breakwaters. Proceedings of the Conference Coastal Structures, Arlington, USA.

Barends, F.B.J. (1985) Geotechnical aspects in breakwater design, 6 in, *Developments in Breakwaters*, Thomas Telford, London.

Barends, F.B.J. and Calle, E.O.F. (1985) A method to evaluate the geotechnical stability of off-shore structures founded on a loosely packed seabed sand unde wave loading. BOSS '85, Delft.

Barends, F.B. and Hölscher, P. (1988) Modelling interior process in a breakwater, 4, in *Design of Breakwaters*, Thomas Telford, London.

Barton, N. and Kjaernsli, B. (1981) Shear strength of rock fill, *Proceedings of ASCE*. Geotechnical Engineers, **107** (GT7) 873–91.

CIAD (1985) Computer aided evaluation of the reliability of a breakwater design (ed. F.B.J. Barends), CIAD Assoc., Zoetermeer, Holland.

Fookes, P.G. and Poole, A.B. (1981) Some preliminary consideration on the selection and durability of rock and concrete materials for breakwaters and coastal protection works, *Quarterly Journal of Engineering Geology* **14**.

Geotextiles (1986) IIIrd International Conference on Geotextiles, Vienna.

Hannoura, A. and Barends, F.B.J. (1981) Non-Darcy flow; a state of the art (Euromech-143) *Flow and Transport in Porous Media*, A.H. Balkema, Rotterdam.

Hölscher, P. and Barends, F.B.J. (1990) Finite difference schemes for wave transmission in rubble mounds. *International Journal for Numerical Methods in Engineering*, **30**, 1129–45.

Kenney, T.C. *et al.* (1985) Internal stability of granular filters, *Canadian Geotechnical Journal* **22**, 215–55.

Kobayashi, M. *et al.* (1987) Bearing capacity of a rubble mound supporting gravity structure. Port and Harbour Research Institute, Ministry of Transport, Nagase, Yokosuka. *Rep.* 26(5).

Lambe, T.W. and Whitman, R.V. (1969) *Soil mechanics*, Wiley, New York.

Rahman, M.S. and Jaber, W.Y. (1986) A simplified analysis for wave-induced liquefaction in ocean floor sands. *Journal of Soils and Foundations*, **26**, 57–68.

MARINAS 31

P. Lacey

Marinas are a part of an expanding leisure industry and a brief history of their development and the many factors that have to be considered when selecting a suitable site constitute the beginning of this chapter. In dealing with engineering considerations the author points out that statutory legislation governing environmental impact assessments means that marina development is becoming more difficult. Most coastal marinas require studies involving field work and either physical or model investigation. The complexity of designing a marina is well illustrated in a section which considers the many issues that have to be considered in their design. The marina scene is changing rapidly so that it is relevant that the chapter closes with a look into the future (see also Chapters 19 and 29).

CONTENTS

31.1 INTRODUCTION

The word 'marina' was coined in 1928 by the US National Association of Engine and Boat Manufacturers to describe a recreational boat facility, where a boat owner can launch, repair, fuel and victual his boat. Craft can be made ready and food and shelter are available. The owner can walk aboard at any state of the tide and does not need to row to a yacht moored at a buoy or anchor by dinghy. Safety is an essential aspect for both boats and people as the journey from shore to a boat on moorings is a dangerous part of the whole activity.

The needs of the yachting fraternity have changed considerably in recent decades. A great increase in demand has resulted in the need for more moorings and, with increasing affluence, the desire for bigger and better amenities. There is still a place, however, for types of moorings to suit all pockets. The balance between the types of mooring amenities is changing and this may not be the case in the future. The types may be summarized as follows.

1. Moorings in rivers, estuaries and small harbours, mostly on buoys or chains (Figures 31.1 and 31.2).
2. Small hard berthings, floating pontoons in sheltered locations.
3. Large purpose-built marinas originally catering for the upper end of the yachting fraternity (Figure 31.3).
4. The modern 'village harbour marinas' which are really housing developments situated around water. Some of these use old and disused dock basins (Figure 31.4). They sometimes also incorporate retail and commercial developments.

These developments have meant greater complexity in the planning and construction of marinas. Marinas generate traffic and the more marinas there are in a selected area the more boats will be attracted. The papers from the international conference at Southampton in 1989 form a very useful state-of-the-art (Blain and Webber, 1989), and material for this chapter has been drawn from them.

31.2 SITE SELECTION AND INVESTIGATION

The search for suitable sites for marinas is governed by the following criteria:

1. Sheltered locations;
2. Sufficient depth of water – natural or impounded;
3. Minimum maintenance dredging;

Coastal, Estuarial and Harbour Engineer's Reference Book. Edited by M.B. Abbott and W.A. Price. Published in 1993 by Chapman & Hall, 2–6 Boundary Row, London SE1 8HN. ISBN 0 419 15430 2.

Figure 31.1 Traditional tidal harbour, Megavissey, Cornwall, UK.

Figure 31.2 Traditional marina and moorings, River Hamble, UK.

Figure 31.3 Residential marina, Hythe, UK.

4. Road access;
5. Closeness to an established population centre;
6. Closeness to good sailing waters and places to sail to within half a day;
7. Environmental factors. This issue has become vital due to the requirement for higher standards brought about by national and international legislation.

The following information is needed by planners and engineers to assess any proposed marina site especially in coastal, river or estuarine situations, and to develop the best layout having consideration for the long term effects on the existing coastal regime:

- River/sea-bed levels;
- Wave, wind and currents;
- Tidal ranges;
- Sediment transport rates;
- Longshore drift (littoral drift);
- Soil investigations;
- Sources or potential sources of pollution.

From feasibility or pre-planning studies, decisions on cost-effective designs for the necessary marine structures can be made, based on this information.

In old docks and harbours the importance of good site investigations cannot be overstated as the multiplicity of buried obstructions and services makes planning very difficult. The closure of existing basins and the need to produce new promenades or edges poses many design problems, some of which are noted below.

Soft material (silt, mud, etc.) overlying the dock bottom could cause problems when filling to provide developable land. The composition of the soft material or any other dumped material must be assessed

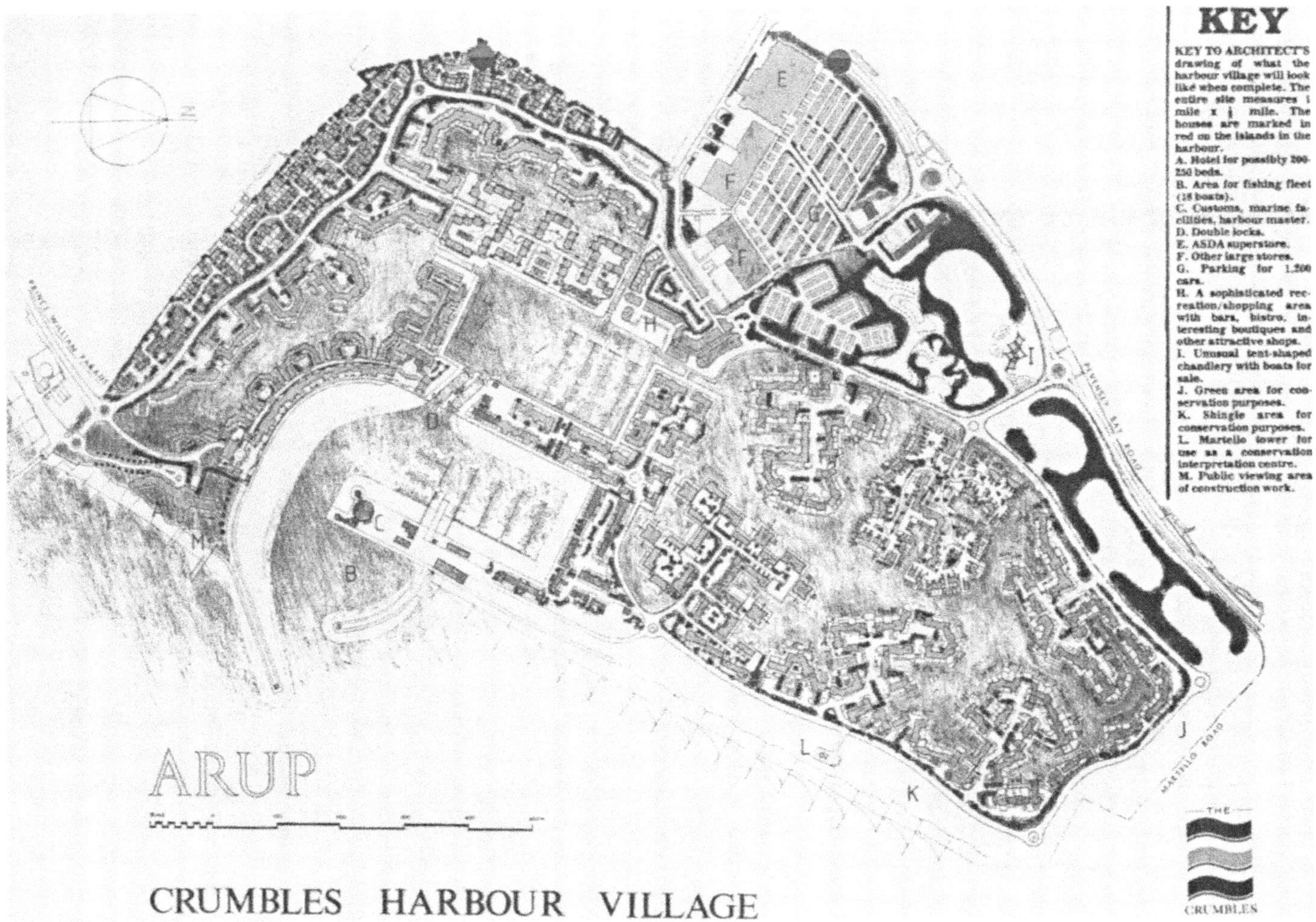

Figure 31.4 Village harbour marina, Crumbles Harbour Village, Eastbourne, UK.

to detect any toxic substances or the possibility of methane being generated. Poor quality fill in existing basins or graving docks can cause problems with foundations.

The condition of existing dock or river walls must be assessed with respect to the loading by the new development. The position of buried obstructions, ducts and services which are very common in urban/dockland sites, and the position of buildings or loads near existing walls, must be thoroughly investigated.

The water levels, which may be either retained at a nearly static level or some intermediate levels between the tides, need to be established. Proposed usage of the dock basin for leisure or other uses and any requirements for navigable access into the basin must be considered with the information on water levels. Existing and future water quality must be assessed.

31.3 MARINA DEVELOPMENT

The best natural harbour sites in the United Kingdom and many other countries have, in many cases, already been developed, leaving sites between where development is possible but expensive. The village harbour concept either on a greenfield site or in an existing dock relies on property development to fund the upfront engineering costs. The re-use of existing old docks has become attractive in the UK and elsewhere due to government funding possibilities and the fact that services are already available from the previous port working, whereas providing a marina on a greenfield site requires special study of the area and its needs (see PIANC, 1979).

Planning and legal aspects of marina development must be considered in depth at an early stage. The boundaries affecting land, foreshore, harbours, seabed, rivers and estuaries is a difficult subject and produces many problems for would-be developers. Dromgoole *et al.* (1989) have provided very full guidance on the topic.

Ownership of foreshores and the sea-bed in adjacent harbours, rivers and estuaries needs to be identified together with the leaseholders of land falling within a proposed development. Third parties may have a license to use, or gain access to land or water. Existing landowners are entitled to be protected from nuisances that might be generated by the development.

There are many commercial and environmental interests that need to be considered during the development and operation of a marina. These could include existing pipelines, cables, sewage outlets and fisheries. Environmental interests might include minimum interference with nature conservation and coast protection. Control of pollution and land drainage are equally important.

The costs of investigations, feasibility studies, environmental impact studies and the provision of infrastructure services can be large and can affect the initial cash flow of new developments.

31.4 DEVELOPING FACILITIES FOR MARINAS

The basic requirements for a marina are shelter for the boats, and an acceptable water depth, which is determined by the tidal range, wave and current conditions and the wind. Three types of marina can be identified:

- The locked harbour;
- The tidal basin;
- The haul-out marina.

The locked harbour has a special appeal to developers, as the water vista is pleasingly constant and has no areas of mud drying out or large walls covered with slime exposed at every tidal cycle. It can, however, have pollution problems.

Yachtsmen require (besides shelter and sufficient water depth) navigational room, bunkering and mooring facilities, services, access to the shore and good car parking space. It is a strange aspect of design that in many schemes the most valuable piece of land, along a waterfront, is dedicated to cars. Also, security must be provided, and amenities for washing, showering and toilets, and club facilities. Apart from the important aspects of finding and developing the site on its landward side, its location and accessibility, both from the land and sea, is paramount.

The main features of a modern marina development are:

1. *Harbour infrastructure*: Reclamation, dredged basin and entrance channel, protective breakwaters, water impounding structures to ensure a good water depth at low water and floating pontoons to moor against.
2. *Direct services*: Car and trailer parking, supplies of fresh water and electricity, toilets, showers, refuse collection, telephones and administration with services and security.
3. *Indirect services*: Slipping and lifting-out, boat storage, hull cleaning, repairs and refitting, chandlery, bunkering and boat sales.
4. *Ancillary facilities*: Internal road access, club-house with restaurant or bar, hotel, shops, leisure facilities and residential units.

Recently planned marinas are often incorporated in complexes of residential and office blocks, conference and exhibition centres, hotels, restaurants, industrial facilities, leisure complexes (cinemas, shops, sports and historical centres) and so on. By comparison with the 1928 definition of a marina, the sailing enthusiast's attitudes and requirements have changed.

31.4.1 DEMAND

Boating as a consumer leisure market is poorly documented. Figures that are available usually refer only to the yachting fraternity. The following are the observations of the author using rule-of-thumb methods to establish demand. Potential demand can be estimated by desk studies, telephone interviews and visits with members of yacht clubs who sail in the area, general assessment of demand in the country, and examination of local or regional factors.

Taking an example from the UK, The General Household Survey (1983) found that 0.9% of the British population had been sailing (i.e. excluding windsurfing and powered craft) in the four weeks before being interviewed in the summer. This represents some 390 000 adults.

The size of the boating market can be judged very approximately from the number of craft in each sector if such data are available from national industry sources, such as the *Digest of Sports Statistics* (Sports Council, 1986).

31.5 ENGINEERING CONSIDERATIONS

Marina development is a combination of assessing inland restrictions, weather elements affecting orientation and design of structures, seaward restrictions and any other factors likely to cause damage or loss of amenity at berth or entering, leaving or manoeuvring.

Village harbour marinas are being developed by people whose main interest is in the housing development by the water. The aesthetics, ambience and engineering often have to take a back seat.

Some marinas in the past have been created without sound engineering and operational judgement and with little care for the environment. Statutory instruments governing the requirements for environment impact assessments have now made marina development much more difficult.

The following guidelines cover the principal engineering aspects of marina development:

1. Access time for boats and the necessity for a full locking system, flap gates or a permanent sill. This gives a range from full access to a limited access window depending on tidal range, retained water depth and entrance channel depths.

   The sizing of the lock barrel should be consistent with boat usage and economy and requires careful consideration. Some older marina locks are inadequate for the passage of peak traffic. Guidance on basic calculations for locks is given at the end of this section.
2. The type of breakwater protective structure will vary according to the quality of protection required, impact on local beach or river regime and materials available. Structures could be semi-submerged, overtopped, permeable, impermeable, rubble mound, vertical walled or a variation of any of these. Each will be selected to suit the design conditions and protection required or which can be afforded. They may also be required to protect an approach channel from littoral drift or to stabilize or train the alignment of a tidal entrance.
3. Most coastal marinas and some river and estuarial marinas will need mathematical or physical hydraulic modelling to check the water movements outside and inside protective breakwaters. Ship simulators can also be valuable aids to the planning of the harbour entrance and breakwater layout. Models are also helpful in forecasting the effects of coastal structures on coastal regimes. The interruption of littoral drift can be disastrous if no account has been taken of drift rate, build up time or arrangements for bypassing if this is found to be necessary.
4. The presence of breakwaters can produce special navigation conditions at the harbour entrance. Current may be generated across a harbour entrance as a result of the deflection of currents and by wave diffraction around the head of the breakwater. Wave reflections may occur from the breakwaters, and as a vessel moves from the open sea to sheltered water there are significant changes in environmental conditions affecting the vessel over a short distance.

   A wide harbour entrance, to ease navigation, conflicts with the objective of limiting wave penetration, and some compromise is needed. Navigation may not always be possible in exceptional wind and wave conditions. The extent of these interruptions should be assessed from a knowledge of the wave and current climates and the output from mathematical or physical models.

   The advice of experienced mariners is essential in determining the optimum layout of breakwaters at the harbour entrance, taking into account the economic aspects of cost and any limits on navigation and port operation.
5. The most important determinant of harbour response is wave penetration through the entrance. It is first necessary to establish wave conditions just outside the entrance, then to determine the effect of the entrance in permitting waves to enter the harbour and finally to determine the response at critical positions within the harbour.

   Long-period waves, over say 15 s, are difficult to

Table 31.1

Length	*Sail and motor sail*		*Motor cruiser*		*Trimaran and catamaran*		*Minimum clearance*
(m)	*Depth (m)*	*Width (m)*	*Depth (m)*	*Width (m)*	*Depth (m)*	*Width (m)*	*(m)*
8	1.5	2.8	1.0	3.3	0.6	4.5	26
8–10	1.6	3.0	1.0	3.5	0.8	5.0	32
10–12	1.7	3.2	1.0	4.0	1.0	5.5	40
12–15	2.0	3.8	1.3	4.6	1.2	7.0	52
15–18	2.5	4.5	1.5	5.0			64
18–25	3.0	5.5	2.0	5.5			90
25							

For depth of water required add a minimum of 1.0 m.
For width required add an allowance for fendering, etc., plus say 1.0 m.
Based on earlier ASCE and PIANC work.

Table 31.2

L Boat LOA *(m)*	*L2* Berth length including deck *(m)*	*W* Berth width including deck *(m)*	*F* Share of Fairway $\frac{1.5L}{2}$ *(m)*	*A* Total area *(F + L2) × W* *(m²)*	$\frac{A}{L}$
8	9	2.74	6.0	56.2	7.0
10	11	4.25	7.5	78.6	7.9
12	13	5.00	9.0	110.0	9.2
15	16	6.50	11.25	177.1	11.8
20	21	9.0	15.00	324.0	16.2

exclude from a harbour and may cause ranging of moored ships. They may also cause harbour resonance, for which guidance is also given in BS 6349:Part 1 (29.4). This undesirable phenomenon is aggravated by the use of reflecting faces within the harbour and such faces should be avoided if possible. Absorbing faces may need to be provided.

The most useful parameters for assessing the acceptability of conditions within a marina basin are:

(a) Maximum wave heights likely to occur within a period ranging from one to ten years.
(b) The duration and frequency of the occurrence of waves exceeding 150 mm maximum height in any one year.

31.6 PLANNING AND ENGINEERING GUIDE (RULES OF THUMB)

Planning density

Berths per hectare 75 to 120

Caution must be used in estimating berth numbers. How the marina is to be marketed is important in taking a decision regarding width of berth and other dimensions. If berths are to be sold on long leases, then each berth must be capable of taking most designs at a specific length. If berths are to be let on annual leases then density can be increased by fitting in broad and thin boats as required.

There is a difference in the amount of water certain boats command as can be seen in Tables 31.1 and 31.2. Comparison shows that ratio of occupied sea area to boat length increases. Charging for berths on a purely linear boat length clearly favours the larger boat.

Water to land ratio 1.25–1.0

31.6.1 CRITICAL DIMENSIONS

Entrance width	12–18 m
Entrance channel width	20–30 m
Channel width inside marinas	20 m minimum

Areas within breakwater arms, etc., must have sufficient room for turning, etc. This will depend on boat size, traffic density and quality of protection provided from environmental forces.

Walkway dimensions and loadings

	Dimensions	*Loading (UDL)* (kN/m^2)
Main walkways width	2–2.5m	0.75–2.0 (Depending on use)
Finger piers width (9–15 m boats)	0.75–1.25 m	0.75
Length of finger piers		
Boats up to 10 m length (*L*)	0.75*L*	
Boats from 10–15 m length (*L*)	0.875*L*	
Boats in excess of 15 m length (*L*)	*L*	
Areas varying from heavily used restricted access to public areas		2–4
Access and support structures		4
Minimum freeboard (unloaded)		
Boats not exceeding 15 m		500 mm
Boats exceeding 15 m		750 mm

Access bridges

A maximum slope (tidal conditions) is suggested of 1 in 4.

Car parking

Car parking space which can also double up as winter boat storage is always a problem and depends on the local judgements made by planning authorities.

Parking ratio guide

A ratio of 24–27 m^2 per car and 0.75 to 1.0 space per leisure craft is suggested. This ratio may increase if commercial use is also intended in the marina.

A car-to-berth distance of 250–350 m and a Berth-to-amenity-block distance of 200–300 m are suggested. The length of the main berthing spines should not be too long and anything over 190 m should be considered carefully.

31.6.2 ESTIMATING LOCK CAPACITY

This example demonstrates the type of calculations that have to be made.

Assume 35–40% of boat population would use the lock over a five-hour peak period. This period may occur say two or three weekends per year. A figure of 25–30% may be considered for say 6–8 weekends a year. A fine spell of weather would mean the demand figures being reached consistently.

Although times will vary from area to area most exits occur between 0900 and 1300 hours. Return to base occurs mainly between 1500 and 1900 hours. When considering locking hours, to be consistent with outer tidal range and inner basin controlled water level, the use of 'free flow', i.e. all gates open to allow uncontrolled access, is possible to control near the top of the tidal ranges.

Consideration of currents caused by free flow must be made to avoid unacceptable turbulence or even erosion. If free flow currents are very strong then only experienced sailors will be able to navigate safely.

Consideration must also be made of the locking cycle. Naturally this will vary with the lock size but much will depend on the efficiency of the lock operators. For estimation purposes comparative throughputs are shown below.

Lock size	*Boat numbers per locking*	*Cycles per hour*	*Throughput per hour*
24.0 × 9.0 m	4–6	4	16–20
45.0 × 10.0 m	9–10	3	28–30
75.0 × 12.0 m	20	2	35–40

These figures are dependent on tidal patterns, water levels, distance to arrive at lock and peak time delays.

31.7 MARINA PROBLEMS

Marina operators and developers will wish to obtain a site which is, or can be, sheltered with a minimum need for breakwaters, requires little or no dredging, especially maintenance dredging, and is close to a large population centre and is easy to get into. These sites are now almost impossible to find.

Large tidal ranges mean either the provision of a lock or tidal flap gates for the marina basin or restrictions sailing in and out, using the available tidal window either side of the high water level.

To cope with variations in water level during a tide, floating pontoons, if not within a locked basin, require bridges or stepped access.

Regular maintenance dredging must be avoided if possible as it is costly, and moving or dismantling the pontoons is time consuming, requiring yachts to be removed and stored elsewhere.

The biggest problem is the viability of a marina project. Marina projects are expensive to construct and do not earn large revenues without the back-up of property development or grants.

The correct choice of pontoon and pontoon material is important. Much depends on the exposure to waves and current at each particular site and the strength of a proprietary pontoon. Pontoons are made from concrete, glass fibre, reinforced plastic and polystyrene. The prime requirement is to prevent finger piers from twisting and breaking under adverse conditions.

Services and their provision often cause problems. The supply and protection of electricity for yacht owners is fraught with problems. The number of yachts to a fuse, type of fuse and the resetting fuses when yachts have faulty equipment must be carefully considered.

Protection of water supplies throughout the year in cold climates, and providing sufficient pressure when large numbers of yachts are using water are important considerations.

Rubbish containers have to be numerous and strategically placed. However, they should be onshore and large, as the rubbish produced by yachtsmen is bulky.

Animals can also be a hazard in a marina, causing mess and nuisance and always providing the fear of rabies if foreign visitors use the marina.

People in marinas like to feel secure, but as marinas encourage visitors, it is difficult to ensure security apart from extremely good lighting especially on pontoons and minimum access points to the berths. Twenty four hour security using permanent staff is not uncommon.

Boats need servicing and therefore facilities for lifting out of the water are needed. Such a provision is not unduly costly and earns revenue, but it does need an experienced operator.

Winter storage is also a concern, now that developable land near water is attracting large sums of money. The balance of providing car park/boat storage against possible profit by building becomes an important cash sheet item in the initial planning. It is, however, clear that yachtsmen want to drive near their berths, park and also take the boats out of the water during the winter months. A problem which was formerly not considered no more than a nuisance is floating debris which is forced into corners by winds or currents. Owners of expensive boats and houses beside marinas do not want to find the water obscured by oil slicks, polystyrene, weed or other flotsam. This problem is particularly prevalent in locked basins and requires special measures by the marina operators to remove the debris regularly, usually by hand.

Lastly the problem of persuading people to pay their rental fees, etc, is a constant worry which cannot be ignored if a marina is to be run successfully.

Marina management is an important and highly specialized subject and guidance is given in the paper by Hirst and Lacey (1989).

31.8 THE FUTURE

Subject to the fluctuation of interest rates, and therefore the role of residential building, the construction of village-style marinas will continue either on greenfield sites or in existing disused docks or harbours at a rate which will be dictated by the availability of investment funds. Schemes that do not need extended planning or complicated permissions will be more attractive than sites which are at the feasibility stage. Planning and feasibility studies will still be carried out even in recessionary times until large residential schemes become viable again.

Internationally, there is a great interest in leisure engineering which encompasses village harbour marinas, marinas and waterside properties with limited mooring facilities. However, there is also evidence that certain areas are being overcrowded. An example is the Solent in southern England where moorings have increased from 10 000 in the mid 1970s to approximately 28 000 in the late 1980s. This has led to the County Planning Office scrutinizing new planning requirements very carefully, and restricting development.

Boating and interest in boating is increasing and at the same time boat sizes are increasing. Mooring layouts must be flexible to accommodate these changes. The length and possibly the width of finger piers might well increase to cope with the larger boats. Fairway and entrance dimensions will also need to be adjusted.

The interest in boating has brought its own problems in that, whereas before the yachting community was small and well experienced in sailing, the community now consists of less experienced sailors and motor cruiser enthusiasts. Amateur sailors probably prefer to go not longer than a day without slipping into a well set-up marina with good landside facilities.

Potential mooring sites, even those which are difficult or expensive to develop, will now receive attention. Sites requiring lower civil engineering costs with good existing access and lower established infrastructure such as docks, canal or river sites, will be eagerly contested. The large village harbour developments which have received great publicity may run into problems.

Land acquisition will become more difficult as national legislation inquiries and protest groups will slow the advance of projects. Naturally developers will need to assess their financial risks more carefully. There is already a case where, after spending considerable time and money, a marina scheme has been stopped by an efficiently mounted, local, protest group.

The amount of services required, access and car parking together with the ever increasing costs of Section 52 requirements (in the UK) adds to the developer's problems and is a far cry from dropping some buoys and trots in the water or providing a few hard moorings, some ground for car parking and a shed.

The attention quite rightly paid to the architecture and aesthetics of large schemes is often a surprise to a developer and there will be much more care taken with approvals for developments in the next few years on these grounds. Above all the ambience of large residential marinas must be established at a high level of quality.

Proposals for coastal village harbour marinas will also receive much more scrutiny in relation to environmental issues. This is seen as becoming a major hurdle for developers as authorities struggle with implementing new legislation. The implications for sea defences, littoral drift, erosion, pollution and aesthetics will inevitably mean very close scrutiny by investors before developments commence.

Developments not on the coast but which retain a body of water by means of locks or flap gates will have to pay particular attention to water strategy. Pollution by the boats or the development itself which would effect water quality cannot be ignored. Many large schemes promote the water leisure activities available to house or flat dwellers.

Land, near water, which will become part of the development may contain contaminated fill. Removal or containment of such material is subject to strict legislation in the UK and this will have an increasing effect on the prefeasibility study of marinas, especially those in urban areas.

The use of a suitably experienced professional team which can foresee and deal with these potential problems will become even more important. In summary, the following trends are foreseen in marina developments in the 1990s.

- Modernization and reassessment of existing marinas wherever possible.
- Conversion of swinging and piled moorings into marinas wherever possible.
- Continued regeneration of old docks and harbours using marinas as a catalyst to attract public grants and redevelopment funding.
- A few, carefully planned, large village harbour marina projects comprising housing, commercial and leisure uses.
- Larger involvement of the boating industry to encourage provision of affordable berths.
- Much more emphasis by coastal towns encouraging and promoting marinas to attract tourists and income.

REFERENCES

Adie, D.A. (1984) *Marinas*, (3rd edn.) Architectural Press, London.

Blain, W.C. and Webber, N.B. (1989) *Marinas '89* Proceedings of an International Conference, Southampton. Computational Mechanics Publications, Southampton, UK.

BS 6349 (1984) Code of practice for maritime structures: Part 1; General Criteria.

BS 6349 (1985) Part 4, Design of fendering and mooring systems. British Standards Institution, London.

BS 6349 (1988) Part 2, Design of quay walls, jetties and dolphins. British Standards Institution, London.

BS 6349 (1988) Part 3, Design of dry docks, locks, slipways and ship building berths, ship lifts and dock and lock gates. British Standards Institution, London.

BS 6349 (1989) Part 6, Design of inshore moorings and floating structures. British Standards Institution, London.

BS 6349 (1990) Part 5, Recommendations for dredging and land reclamation. British Standards Institution, London.

BS 6349 (1990) Part 7, Guidelines for design and construction of breakwaters. British Standards Institution, London.

Dromgoole, S., Gaskell, N. and Grant, M. (1989) Planning and legal aspects of marina development. International Conference on Marinas, Southampton.

Fleming, R. (1987) Village schemes dominate in general expansion in UK marina development, *Dredging and Port Construction*. issue? page?

Hirst, M. and Lacey, P. (1989) Village marina developments, United Kingdom. PIANC Congress, Osaka.

Lacey, P. (1987) Changing face of marina development in the United Kingdom. *PIANC Bull. 59*.

National Yacht Harbour Association Ltd (1980) A Code of Practice for the Construction and Operation of Marinas and Yacht Harbours. *Leisure Harbours in the 80's Conference*, Canterbury, September 1980.

PIANC (1976) Final report of the International Commission for Sport and Pleasure Navigation. *PIANC Bull. 25*, Brussels.

PIANC (1979) Standards for the construction, equipment and operation of yacht harbours and marinas.

PIANC (1987) Report on dry docks, PIANC, Brussels.

PIANC (1979) Final report of the International Commission for the Study of Locks. *Suppl. Bull. 55* (published 1986).

UK Sports Council (1986) Digest of sports statistics.

Wheeler, A.J. (1988) Marina development seminar. Hydraulics Research Ltd, Wallingford, October 1988.

The General Household Survey (1983).

CONSTRUCTION OF MARITIME WORKS – CONTRACTORS' METHODS AND EQUIPMENT 32

J.F. O'Hara

Good design cannot be achieved without a knowledge of the practicalities of construction. A knowledge of the output from a quarry can, for example, affect the final design of a breakwater. So too can the environmental conditions to which it might be subjected during its construction. These and other factors will in turn affect the plant chosen for construction. This chapter reviews the construction of maritime works in a hostile environment and deals specifically with quarries, rubble mound breakwaters, caissons, block walls, piling and reclamation (see also Chapters 24, 30, 35, 36 and 45).

CONTENTS

32.1 INTRODUCTION

In the early conceptual stage of a project thought given to construction will result in worthwhile savings in money and/or time. Two simple illustrations are:

1. A steel structure design using light sections produces a lower steel tonnage than an alternative with heavier sections. However, per tonne fabrication costs and painting costs (surface area per tonne) of the lighter structure will be greater and can produce an overall larger bill.
2. A textbook breakwater with a no fines core and with strictly regimented armour layers meeting precisely specified grading curves and layer thicknesses may well be achievable. However, if it results in 25% of the quarried material remaining in the quarry as waste and takes 50% more crane hours than a more relaxed specification, the designer needs to question his requirements. A greater recognition is needed by designers of the fact that stone/rock is a non-'manufactured' product, with consequences on product control, which is frequently placed blind in far from ideal circumstances.

32.2 ENVIRONMENTAL EFFECTS

For most marine structures the greatest destabilizing forces are environmental; waves, currents, tides, wind; and much recording and analytical effort is put into establishing the design maxima for the chosen return period. Similar attention needs to be paid to the production of the short-term environmental data, which control the feasibility and accuracy of the construction operations.

Thus a breakwater whose primary armour is sized to cope with a design wave of 9 m on a 1 in 50 year basis may have a total construction shutdown each winter, when sustained wave action a fraction of that height would cause unacceptable erosion of an exposed core. In this particular case both the normal seasonal environment and the effects of this on the partially completed structures need evaluation, e.g. model testing of the scour end of a breakwater or of a lowered construction crest level.

In addition to the potential damage to the partially completed structure there is the inhibition on plant operations caused by environmental conditions. Cranes cannot operate during periods of high wind. Construction plant on a breakwater may have to retreat

Coastal, Estuarial and Harbour Engineer's Reference Book. Edited by M.B. Abbott and W.A. Price. Published in 1993 by Chapman & Hall, 2–6 Boundary Row, London SE1 8HN. ISBN 0 419 15430 2.

to shore during periods of high waves. Floating operations are weather sensitive, both because of the potential damage to plant and also because of the deterioration in accuracy of construction.

Cranes are affected in two ways. Firstly, by the dynamic increase in loading caused by the crane barge and/or the load carrying barge pitching and rolling, and secondly, by the limitation on permitted tilt of an operating crane. The latter is frequently governed by the shear capacity of the pinion on the slewing ring which has to act as a brake preventing uncontrolled down-hill rotation of the slewing ring under load. A typical derrick can tolerate a maximum 3° tilt while Manitowoc recommend a permitted maximum 2.5° tilt on the cranes they manufacture, when barge mounted. Dynamic 'snatch' loadings have caused damage or loss overboard of cranes on fixed offshore platforms, when picking loads off pitching decks. That particular problem has now been alleviated by snatch compensating devices incorporated in the lifting tackle.

It is not possible to suggest hard and fast limiting wave conditions on marine operations. The variables are the height and the period of the wave, associated with the size and natural frequency of the barge.

While unstabilized natural roll periods for ocean going vessels may range from 8 s for long and lean warships to 22 s for a more comfortable liner, a small, flat bottomed, low draught, barge with a large metacentric height could have a period approaching 1 s at light draft. It will therefore be adversely affected by much lower and shorter period waves than ocean going vessels. As a rough guide, floating plant operations are curtailed by wave heights between 1 m and 3 m high (approximately 3–7 s period), depending on plant type. Dredgers which are likely to have to operate in exposed conditions are more wave tolerant, but even the most resistant trailing suction hopper dredger is limited to about 2.5 m wave height, while a cutter suction dredger would have an efficient operational limit of about 1 m. The 1 m figure is also the limiting criterion for safe and efficient operation for most crane barges used in civil engineering, smaller bottom dump barges (500 Te) being limited to 0.5 m.

Alleviating measures can achieve limited improvements, e.g. ballasting a pontoon can increase the natural rolling period and extend the potential working conditions. Similarly altering a cutter suction dredger from a conventional spud/floating type to a semi-submersible or self-elevating type results in the improved work availability illustrated in Figure 32.1.

A graph of the following type, included with tender documents, would be of great benefit to contractors, in assessing probable weather downtime on their operations (Figure 32.2).

32.3 QUARRYING/QUARRY PRODUCTS

Most coastal structures incorporate stone and rock. Normally design is tailored to the nearest economic source of suitable material. This source could be a

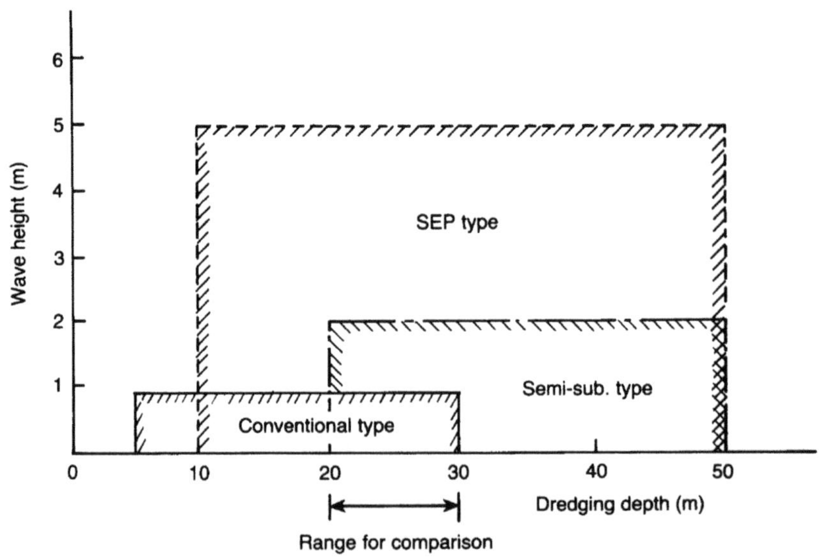

Figure 32.1 Cutter suction dredger, operational limits of different types (taken from WODA paper).

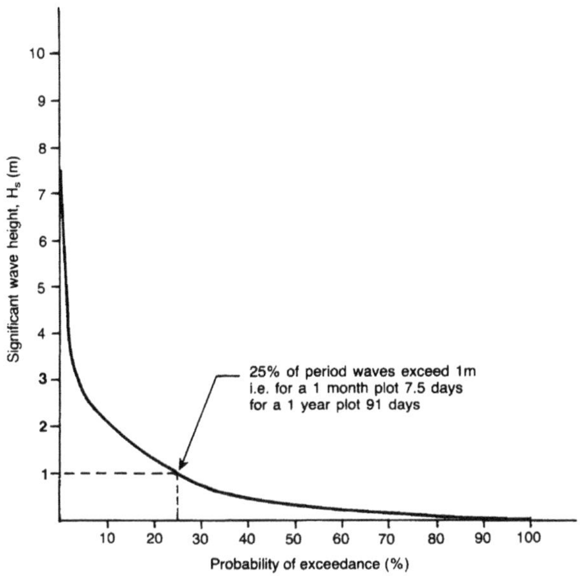

Figure 32.2 Construction wave climate.

quarry in which rock is won by drilling and blasting with faces up to 15 m or higher. It could be a thin layer of duricrust found in desert regions, in which small grains are cemented by salts which have been borne upwards by capillary rise. Stone may also be available in river headwaters or in alluvial outwash in desert areas caused by flash floods.

In the final case, quoted above, winning is achieved by normal earthmoving equipment using dozers and loading shovels. Stone is processed by passing through a grizzly. The grizzly can be either fixed or vibrating, the latter being suited to smaller size material. A fixed grizzly may be formed from structural steel sections or railway rails set in diverging lines to give a screen of varying opening size. Wadi outwashes can on occasion be kind to contractors in providing materials naturally sorted by size along relatively short lengths. An alluvial source usually produces smaller sizes suitable for underlayers, particularly for more lightly armoured shore protection schemes.

The construction of the Ras Lanuf breakwaters in Libya is an example of the successful application of a duricrust source. Core and armour rock up to 6 Te in size was 'quarried' at a location 25 km inland from the harbour site. The quarry was a limestone caprock layer varying from 0.5 m to 1.5 m in thickness and with specific gravity of 2.2 to 2.6. The armour requirement was mainly 2–4 Te, used as secondary armour under tetrapods or inside face armour. Relatively light explosive charges with millisecond delays were placed in shotholes, just under the base of the caprock, at 2 m centres. A total of about 9 million tonnes was won in a production period of approximately 18 months. (A very high output rate which required a proportionately large plant fleet.)

The most common source by far is, however, a quarry in which a large rock mass is worked. Such a quarry can be opened and operated specifically for the project as is usual in developing countries, i.e. a dedicated quarry. Alternatively, it may be an established facility developed to produce concrete aggregates, road stone or limestone for cement, or 'dimension stone' for cladding buildings or providing monuments.

The normal working methods of the non-dedicated quarry need some adaptation to suit the needs of a rubble mound breakwater. Drilling and blasting patterns in such a quarry are usually geared to the production of material small enough for the primary crusher with the minimum of secondary breaking. Handling, loading and hauling plant is also tailored to this objective and size sorting equipment is usually limited to 75 mm down or smaller sizes. In some, larger size rock stockpiles may accumulate over the years as a byproduct of the avoidance of secondary breaking. As the primary objective is usually the production of small crushed stone, jointing and bedding patterns of the rock mass may either not lend themselves to production of larger armour size pieces, or if they do, only as a small percentage of the total product. However, even if the armourstone capability is small, it can still be of value. The Port Talbot breakwater in Wales, built in the late 1960s obtained most of its stone from a dedicated quarry about 9 miles from the site. Armour rock (2–8 Te in size and 43% of the total 3.4 million Te requirement) was, however, also obtained from 30 different sources in Wales with substantial tonnages of the larger grades hauled by rail from Derbyshire (one man's waste may be another man's cheap armour).

Quarrying specifically for rock for breakwaters or revetments is normally carried out nowadays in benches of 10 to 15 m height, loading explosives into vertical shotholes fired with delay detonators. This produces the most controllable fragmentation, permitting adjustment of explosives ratios, hole spacing and burden to optimize the product size range. However, large breakwaters (Tema in Ghana in the 1960s which absorbed about 8 m t of gneiss, Escravos in Nigeria with 1.75 m t of granite and Aden Oil Harbour) have been served by quarries operating a heading blast system with faces up to 60 m in height. The advantages of the latter are much reduced equipment needs, the ability to quarry where ground surface is highly irregular and the production of large quantities in a single blast, its disadvantages being

Figure 32.3 Multi-deck static grizzly.

less control of fragmentation and a more irregular and potentially more dangerous face. The Tema quarry and that operated by Costain for Escravos had blasting ratios of 4.6 Te kg^{-1} and 3.1 Te kg^{-1}, respectively.

Conventionally operated limestone quarries at Dubai and igneous rock at Dibba in the UAE had blasting ratios of 4.9 Te kg^{-1} and 11 Te kg^{-1}. The latter had faces up to an unusually high 45 m, vertical drilling being carried out by down the hole drills associated with some toe drilling.

Ammonium nitrate/fuel oil mixture (ANFO) combined with high explosive (HE) is now commonly used to reduce explosives costs. ANFO can be site-manufactured, further reducing costs, while the transport and storage requirements of the ANFO constituents are not subject to the same restrictions as HE.

The maximum size of rock and the gradation will of course vary enormously. An indication of what may be expected is as follows:

Maximum size	15 Te
Stone up to 200 kg	50% of quarry product
200 kg to 1 Te	15% of quarry product
Greater than 5 Te	10% of quarry product

For a breakwater at Bangor in Wales needing 6 to 10 Te rocks in olivine basalt only a lower than expected 5% of the quarry product fell in this category necessitating a change to artificial (concrete) primary armour. An Icelandic breakwater quarry, also in basalt, yielded about 20% of its product in the 1.7–7 Te range. Of the 1.7 Te down-core, constituting the balance of the 1.85 million tonnes, 25% was less than 2 kg with about 5% being sand-sized grains. Many specifications nowadays would have placed a lower limit of 2 kg or even 10 kg on the acceptable core material with this fraction remaining on the quarry floor as very expensive waste.

32.4 HAUL

Most projects involve the movement of large quantities of materials, mainly rock, aggregates and sand, to the site. The main options are road vehicle, rail or water-borne transport and occasionally long conveyor systems. The factors influencing choice are both economic and social. The social considerations arise mainly when public roads are to be used as haul routes, perhaps on a 24 h basis, by heavy trucks. Restrictions on operating hours, load size or routeing may result.

In an industrialized country, assuming origin and destination are convenient to established transport modes, the cost per tonne/mile of bulk haul is cheapest if water-borne, and most expensive if hauled by road. For a particular project, however, the options have to be evaluated in detail, including the costs of rehandling, the provision of transfer facilities where a number of different transport modes are involved, the flexibility of many 20 Te truck deliveries versus the logistic problems of 20 000 Te barge consigments.

Where a rock source is relatively close to the site, road transport is the normal solution, although in developing countries where no link exists between quarry and site, a dedicated rail system may be more economical (as was used to haul 8 m tonnes of rock on a 3′ 6″ gauge track for Tema Harbour in Ghana and about 2 m tonnes of rock for Escravos in Nigeria), particularly if the link has no foreseeable subsequent use.

Road haul can use on/off highway trucks or off highway dump trucks if public roads are not involved. In the latter case up to 150 Te capacity trucks, as used in open cast mining operations, are available, while in the former the capacity is usually in the range 20–50 Te. In both cases trucks need to be fitted with heavy duty rock bodies when carrying loose material. Alternatively, a flat bed configuration, on which rock trays are placed, may be adopted.

Where dedicated roads are provided a high quality road is desirable to minimize truck downtime, very expensive tyre wear and driver errors, while increasing haul speeds with a consequent reduction in

Figure 32.4 Marine transfer point for rock and armour units.

Figure 32.5 Flat-top barge with 'greedy boards', under tow.

number of trucks. A dedicated quarry road on a Costain project in the UAE had two surfaced lanes with a heavier road structure on the loaded haul lane.

Marine haul of rock, particularly that containing larger sizes, is mainly by flat top barge rather than in ships' holds (except for bottom dump barges which are also a placing tool). Loading and offloading is carried out, or assisted by trucks, loaders or cranes running on the barge deck. Barge capacity ranges from about 250 Te (30 × 9 × 2.4 moulded depth) through 2500 Te (65 × 18 × 4.2) 10 000 Te (90 × 27 × 6) up to 20 000 Te. Haul speeds are in the 7 to 10 km h^{-1} range with a 720 hp tug hauling a 600 Te capacity barge and a 7000 hp tug hauling a 20 000 Te barge. Smaller barges, not specifically designed for construction work, may have permissible deck loadings as low as 2 Te/m^2 and require strengthening to take wheel or truck loads over the whole of the deck area. Large, purpose designed flat-top barges, used for hauling small particle sizes (aggregates) are discharged by underdeck conveyors fed by rubber-tyred loaders moving about the deck.

As with trucks, the contact faces of marine transport, when hauling rock, also need to be reinforced. The reinforcement in this instance is usually renewable. It can take the form of timber linings if timber is cheap and readily available, or, more commonly, sheet piling. Thus, the inner faces of the hopper on a split barge would be lined with tack welded sheet piling.

A 50 km marine haul in the Red Sea involved movement of 500 000 Te of material on 500 Te flat-top barges. To protect the deck from rock impact and to allow a Cat. 966 wheeled loader or a 32.5 Te GVW rock truck to move on the deck of the barges, which were originally built for the offshore oil industry needs, required an overdecking of Larssen 3 sheet piles laid side by side and tack welded to the deck. This, of course, reduced the carrying capacity of the barges. For the Port of Ramsgate breakwater in the

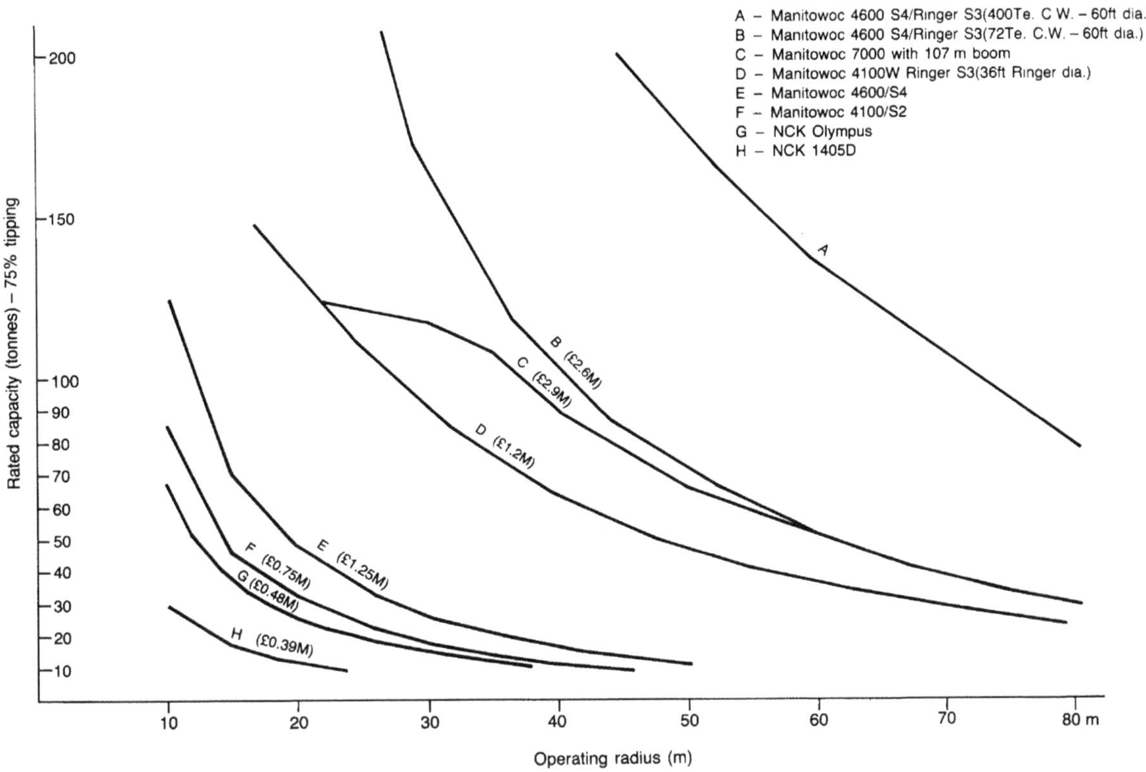

Figure 32.6 Crane capacities. Prices are indicative – 1989 new machine.

south-east of England 240 000 Te of 2 to 8 Te granite rock was hauled from Sweden on 20 000 Te barges, flat top barges of this size being commonly used in the Middle East for long rock hauls.

32.5 BREAKWATER CONSTRUCTION

Building of breakwaters can involve either land or marine haul of materials, land or marine placing of these materials and construction from shore outwards or towards the shore. One or all of the above combinations may be required on a particular project for reasons of economy, programme, access or other physical constraints.

End-tipping from trucks is probably the most common method of core placing, particularly when the quarry source is not too distant from the site. It can also be used for placing secondary armour. End-tipped core will form slopes at the natural angle of repose in the order of 1 to 1.25 horizontal. To achieve flatter slopes the wedge outside the tipped profile will have to be placed or marine dumped. The breakwater surface requires a relatively smooth, fine grained, surface to allow trucks to run, and this may require the temporary blinding of open textured rock. The running surface should normally be about 1.5 m above sea-level, although 1.0 m can be tolerated in sheltered conditions. A minimum of 7 m width is needed where two-way traffic is operating, 4 m for one-way operation. Turning bays are needed at intervals to allow trucks to reverse to the dumping point at the construction face. Where breakwater-mounted cranes operate on the haul route the working platform width needs to allow for the overall track width plus a shoulder of about 0.75–1.5, depending on crane size, in addition to a single truck haul lane. Crane tail swing will obstruct truck passage so some interference between truck and crane operation is inevitable. A dozer or tracked loader (the latter being more versatile) is used at the construction face to push and level material. A typical unhampered truck cycle for the final 250 m length of the breakwater is about 3 minutes, i.e. the 500 m final portion of its round trip including turning (typically 30 s) and

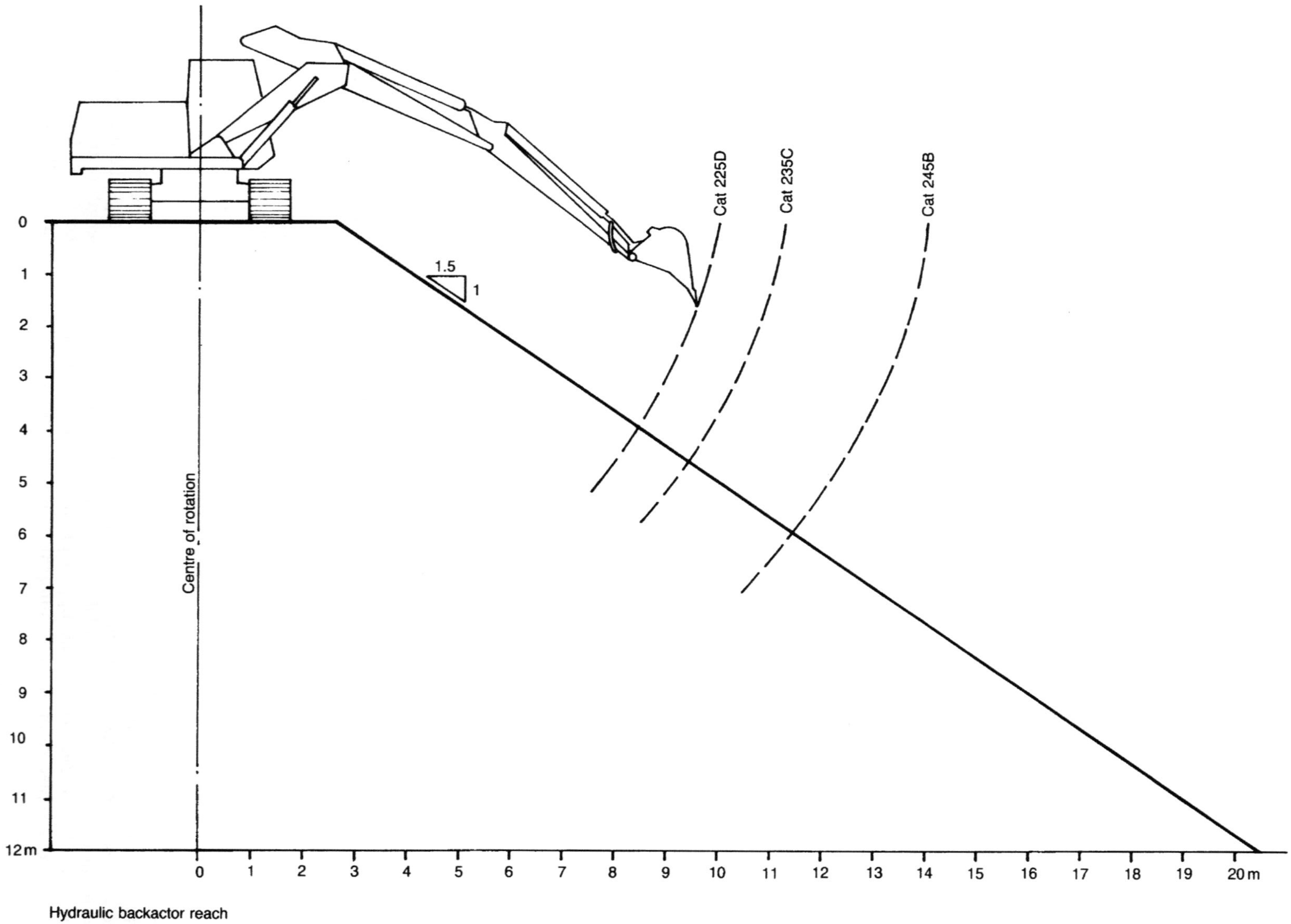

Figure 32.7 Hydraulic backactor reach.

tipping. Queuing, waiting to pass an obstructing crane, or other causes could easily double that cycle time.

Crane placing of core material can be used for the wedge outside a dumped profile, for placing toe blankets or where rock is supplied by marine feed. Whatever the reason, it is normally both convenient and economic to place core or secondary armour in containers (rock trays or skips) to reduce crane time. Typical rock tray weights range from 15 Te (2 trays to a 30 Te haul truck) to 50 Te, with the self-weight of tray being about 15% of the total. Trays are purpose designed to permit easy tipping. Control of underwater placing on the Dubai breakwater project was achieved by producing detailed tipping tables which specified the horizontal orientation of the crane and the boom angle for each tray load. The crane tracks ran on a defined alignment and level on the breakwater. The underside of the jib had a welded pointer which followed a circle divided in degrees painted on the crane's slueing ring, thus defining the jib alignment relative to the breakwater centreline. The vertical boom angle was read on an indicator within the cab. The lowered crane jib was also used as a survey tool, soundings being taken by steel basket suspended from a walkway on the underside of the crane jib. The same Manitowoc 4600 crane also placed secondary and primary armour.

Primary armour units are normally placed individually to a pattern, a density or an interlock. Where the reach is short and the units not too heavy (e.g. small breakwaters and revetments) a hydraulic backactor fitted with hydraulic grapple is a suitable and precise placing tool. For most applications cranes are necessary, rock being mainly handled with grapples and concrete armour units by wire slings. The latter has the advantage of adding little to the crane payload, while the former has a self-weight of about half that of the rock lifted. Capacity problems encountered at extreme radii; e.g. toe units on a roundhead with flattened slopes; are sometimes overcome by immersing the load at short radius and jibbing out with the load underwater (a 30% to 40% reduction in apparent unit weight). Another solution is to add buoyancy collars to those units otherwise outside the load radius.

Grab equipped cranes or backhoes are also used to trim armour which has been bulk placed, while bucket equipped backhoes are used to fill in or trim shorter core slopes.

The productive capacity of cranes and backhoes is a function of the cycle time. Cycle time is in turn governed by organizational efficiency (minimizing hook movement, crane rotation, jib luffing, etc.), handling devices used and accuracy required. Thus rock tray handling will be faster than individual armour placing which itself could vary by a factor of 2 to 3, depending on the precision required on location, interlock, etc. While a cycle of 1 min or less may be attainable for one crane load the long term norm can be expected to lie in the 5–10 minute cycle range. Mechanical availability, weather and the reliability of supply will reduce this further.

Marine placing can be used for the construction of the total breakwater profile or to supplement land-based construction. The former would apply to submerged breakwaters or to those whose core is at, or just below, sea-level. The latter can speed overall construction by placing the underwater core and toe armour ahead of land-based construction, allowing two independent construction fronts, or it may be used only for work out of reach of land-based plant.

Bulk placing of core materials can use split barges, bottom door, tilting or side unloading barges, the last being the most accurate with the load being discharged gradually while the others dump their contents as one large mass. All but the bottom-door type can place material to within 3 m below water level, the bottom-door barge needing additional clearance equal to the below keel door swing. Commonly available barges have capacities in the 400 to 1000 tonne range, but sizes can range up to 20 000 Te or more.

Marine placing of individual units or of trayfulls of material by floating plant also has a place. The crane barge and material haul are usually separate, allowing the crane barge to remain on station while a material shuttle operates. The placement accuracy is unlikely to be as good as that achieved by a land-based crane, and the risk of damage to concrete armour units during placement is greater.

32.6 CAISSON-TYPE STRUCTURES

The largest marine 'building blocks' are caissons. Placed contiguously they can form breakwaters, quay walls or dry docks while berthing dolphins, mooring dolphins, breakwater roundheads, lighthouses, bridge piers, tunnel ventilation buildings and offshore platforms may use a single caisson. Some are towed a few hundred metres from construction site to emplacement, while others may travel hundreds of miles (the World War II Mulberry harbour units had a 'day-trip' across the English Channel, North Sea platforms from the East Coast of Scotland or Norway to the North Sea, Dartford Bridge Pier from Holland to the UK, etc.).

Construction facilities for units of this size are both large and expensive and final caisson design is frequently tailored to suit economic construction restraints, e.g. if one or two caissons only are needed, the first reaction would be to check the dimensions of existing dry docks within range of the site and if possible, tailor the caisson proportions to fit existing docks.

As a general rule it is desirable to produce as much as possible of the structure in the initial casting area – normally to a height which permits safe towing and sinking without further addition afloat. This general rule is frequently breached, however, when only a small number of units is required and the cost of the moving and launch facility cannot be spread over a large number of units.

A reinforced concrete tunnel ventilation building, produced as a caisson for the first cross-harbour tunnel in Hong Kong, provides an illustration of the one-off approach. The desired end result was a caisson measuring 28 m × 16.5 m × 18 m to high water level with a sinking displacement of 9000 Te. The economic approach here was an initial steel caisson, with a reinforced concrete floor slab lowering the centre of gravity, and weighing only 480 Te at launch. This was constructed in an existing shipyard in Hong Kong and launched over an existing slipway, the caisson being 12.5 m high at this stage. It was then towed to a fitting-out area where construction in reinforced concrete continued while it was afloat. The addition of water ballast at the permanent building location caused the structure to settle onto its screeded granular bed, following which completion of the upper portion of the structure was effected. When adequate structural deadweight was in place the caisson was dewatered. A method such as this required very careful planning. Floating construction stages have to consider stability, trim, freeboard, wave action, moorings, the ability of recently cast concrete to sustain hydrostatic pressure as further pours cause the caisson to sink. External walls were cast to a thickness adequate to withstand hydrostatic pressures but thin enough to 'keep the caisson's head above water'. Final wall thickness was added after sinking. Other than the provision of moorings, no special construction facilities were built for this caisson.

More spectacular one-offs include some Scandinavian lighthouses built on temporary piled jetties and launched by the demolition of the piling. Floating dry docks, if available provide another construction option.

Most caissons will, however, require or merit the provision of special temporary construction facilities. This may take the form of a dock, a construction area with a slipway, or a construction area with a shiplift. Each has its advantages and disadvantages and choice is dependent on the particular project requirements.

The dock option is normally most suited to the production of one or a small number of units. At its simplest it is an unlined excavation on the foreshore with slope stability maintained by wellpoints. Flooding the dock and floating out the structure is accomplished by dredging a channel from dock to open sea. Within the dock one or a number of units may be built. The advantages are minimum facility investment and specialized plant requirement, the disadvantages being the launch time (dredging for a large structure could take weeks or even months) and the reinstatement time for subsequent uses. The flooding and reinstatement times can be much reduced by the provision of a floating gate as at the oil platform sites at Graythorp and Nigg Bay, but this is usually considered only for more permanent facilities.

An example of a single-use dock approach is the Royal Sovereign Lighthouse which now lies 8 miles off the coast at Eastbourne in the south-east of England. The structure is supported on a 31 m^2 × 6 m high reinforced concrete foundation caisson, resting on the sea bed. This supports a reinforced concrete tower approximately 27 m high, surmounted by the accommodation cabin and light. The structure started life in a basin excavated above the high water mark on a shingle beach near Newhaven. Initial float out, in this case, needed the assistance of large polystyrene blocks strapped to the sides to reduce the launch draft by 1 m. Final sink needed temporary supplements in the form of five large Admiralty buoys, to control the rate of sink of the first corner to touch down.

A new quay wall at Devonport in the south-west of England illustrates the construction yard and slipway approach. Here a 483 m long quay wall was formed from 29 2000 Te caissons, each measuring 19 m × 100 m × 13 m high. The caissons were compartmentalized into eight cells. In the production line the caisson base slab was cast at the back of the yard and then moved seaward to the wall casting area. Here 225 mm thick double reinforced walls were cast to full height by slipforming. The slipform was a conventional, climbing-on-jack-rods type, which was dismantled and re-set for each caisson. Tower cranes handled the concrete and rebar supplies during the slide, an on-site batcher being located at the caisson yard. Following wall casting the caisson moved to the fitting out area, from which it progressed to the slipway head.

Each caisson was moved on to launch wedges at the head of a temporary 1 in 9 slipway. The underside of the wedges housed deflated water skates. At launch, high pressure pumps mounted on an offshore pontoon forced water into the skates, causing them to inflate and elevate the wedges (and caisson) by 50 mm. Water continuously escapes via the flexible skirts to the skate. A single 100 Te winch with wire passing through a fixed multi-reeved block at the slipway head provided launch restraint. The water skate is similar in principle to the air cushion (hovercraft) but capable of much higher pressure and hence smaller load footprints, 24 × 100 Te capacity units were used, each measuring about 2400 mm × 1200 mm × 160 mm high deflated or 210 mm high inflated. The inflation pressure was 6 bar. Approximately 1% coefficient of friction was produced, a very low figure. (Twenty years earlier, launching a 6000 Te × 100 m long tunnel unit on a 1 in 8 slipway in Hong Kong, using a petroleum based grease to lubricate the sliding surface, had produced a friction coefficient of 7% associated with a similar contact pressure of 5 bar).

The Dubai dry dock complex walls (including a 1 million tonne ship dock) were formed by approximately 160 31 m × 17.5 m × 18 m high caissons each containing about 3000 Te of reinforced concrete. Launch from the construction area here was via a 4000 Te capacity shiplift, 1000 Te of sand ballast being conveyed on board the caisson while on the shiplift in order to convert a potential negative metacentre into a positive 300 mm. The high capital outlay was justified by the large number of units handled.

The slipform operation on the Dubai caissons used a top hung slipform, the higher slipform investment being justified by the large number of uses and the rapid turnaround on the slipform which did not require dismantling and re-erection after each use. Two parallel construction lines flanked a central launch line which terminated in the shiplift.

Slipforming has a number of advantages for caisson walls: concrete placing into a shallow shutter about 1 m high allows well compacted concrete even when heavily reinforced, continuous construction eliminates cold joints and need for water bars and a typical slide rate of 4 m per day produces a 20 m high caisson in only five days.

The movement of massive, partially complete or complete, structures about the construction yard is a feature of all launched caissons. At Dubai double flanged rail bogeys were used. Each four-wheel bogey incorporated coil springs and had a capacity of 80 Te. Caissons were jacked forward from their casting location to the launch line. Subsequent caisson movement at right angles to their original path involved lowering the structure from the first set of bogeys to a new set (using synchronized hydraulic jacks) on the launchway tracks.

Devonport used lubricated ways for the about yard movements, four 500 Te capacity greased metal sledges running on two 300 mm wide steel plate standing ways. In each case accurately lined and levelled standing ways were required as irregularities greatly increase apparent friction and winch or jack loads. Standing way movements induced by settlements under load increased apparent friction at Devonport from an initial 4% to a final figure of more than 8%.

32.7 BLOCKWORK

Large precast mass concrete blocks are frequently used to form quay walls and more rarely breakwaters. The block size chosen has a considerable influence on plant, method and therefore cost. Some blockwork examples are.

Port of Tema, Ghana	16 Te solid, 3 m deep × 1.5 m high × 1.5 m long/and horizontally jointed and coursed.
Jebel Ali, UAE	45 Te solid, 5.4 m deep × 2.0 m high × 1.8 m long.
Jubail, Jeddah, Saudi Arabia	About 40 Te hollow with handling reinforcement, 6 m to 7 m deep × 1 m high placed in vertical columns.
Buchanan Harbour, Liberia	80 Te blocks
Abijan, Ivory Coast	130 Te blocks
Dublin, Ireland	350 Te blocks (built in 1871)

Because relatively tight tolerances are required, steel shutters are normally used in block casting. Fresh water and sulphate-resisting cement are frequently specified, although many examples of mass blocks exist which used OPC and salt water.

Numbers of blocks required is in the hundreds (small) through 3000 up to 60 Te blocks in a slicework quay wall in Turkey to 23 000 × 18 Te slicework blocks in a Barbados breakwater.

The large numbers involved call for production line techniques. Blocks normally remain in the as-cast location for a few days until sufficient strength is attained to allow movement to a curing area, from which they are taken to be laid, usually after 28-day strength has been achieved. As space is frequently limited in the construction yard a commonly adopted configuration is a long rectangular blockyard covered by a gantry of appropriate span and capacity. Blocks are stacked at least three or four high to minimize space demand. All blocks have type number and casting date painted on after demoulding. Careful organization of storage is essential to allow easy location and retrieval, bearing in mind that in a stack of four blocks the oldest block is at the base of the stack and youngest on top. A gantry has the advantage of imposing least dynamic loading on newly cast blocks during the movement process. The gantry will ideally continue on to a pair of finger jetties within which a block transport barge can be moored for direct load out.

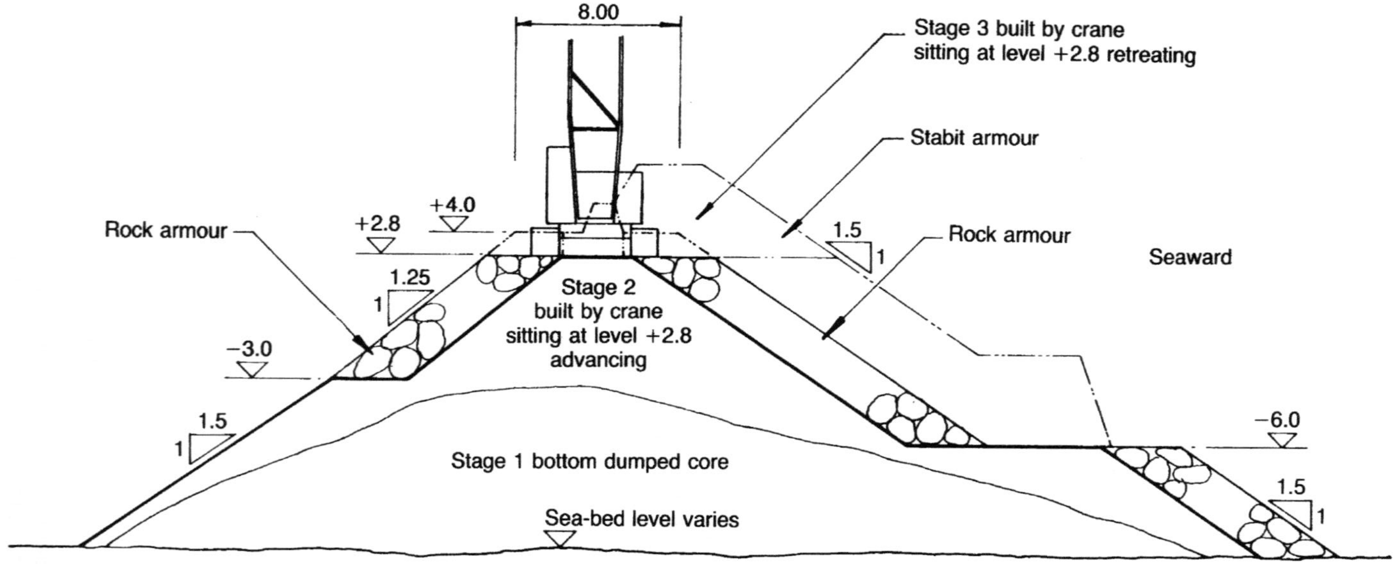

Figure 32.8 Breakwater construction with marine material feed.

If block numbers are small a dedicated gantry is uneconomic. A 421 block contract (50 to 60 Te blocks) in the Red Sea, producing a 240 m long quay, used a 110 Te NCK Eiger tracked crane and low loader to move blocks from casting bed to store and store to barge load out.

Handling of units is normally achieved without cast-in inserts (which could become placing obstructions and/or corrode in the future). Solid blocks usually have a pair of holes which wholly or partly penetrate the block. Inverted T-shaped (Lewis bars) high strength steel rods are inserted into these holes and then rotated through 90° in plan, the process being reversed after the block is laid. Slicework blocks require a suspension system which allows the block to hang at the correct placing angle ('Fidlers gear'). An alternative to Lewis bolts with lower load capacity is the re-usable RMD anchor screw, which is cast into the concrete and unscrewed after use. The largest available has a 12 Te safe working load in 21 N mm^{-2} concrete. Blocks are also handled with external tongs which rely on friction. Tongs tend to be heavier than the equivalent Lewis bar assemblies but the cost of additional crane capacity may be set against the expense of void forming in the block.

Blocklaying plant is frequently the most expensive element of the construction cost. The simplest solution is to lay blocks with a shore based crane but this is usually not economic because of the reach required. Quay walls normally front reclamation whose toe must be behind the walls' foot and whose head is at least 1.5 × wall height away from the back of the wall. At Dibba, in the UAE, where circumstances were favourable, a 180 Te capacity Manitowoc 4100 sitting on shore placed all 400 quay blocks in less than four weeks of continuous operation (blocks up to 25 Te size).

A sliced block breakwater in Barbados was wide enough (10 m) to allow a rail mounted crane to progress along its top laying the wall blocks in front of it.

The more usual approach, however, is the waterborne crane which can work at close radius. Greatest flexibility and speed is provided by a large crane temporarily mounted on a flat top or jack up barge.

Figure 32.9 Construction of breakwater fed by rail.

Figure 32.10 Breakwater construction, fed by trucks.

The job size has to justify the mobilization cost of such equipment. Where it does not, purpose-made laying devices provide the solution. The 421 block contract, mentioned previously, used a purpose designed barge mounted gantry which loaded seven 60 Te blocks at a temporary jetty, travelled to the laying location and placed the blocks. This type of gantry together with another common solution, sheerlegs, although purpose designed, contain components with high residual values – winches etc. They suffer the inability to slew, this movement requiring the barge to be manoeuvred on its moorings (an imprecise adjustment) although gantries usually have a fore and aft movement capability, while sheerlegs have a luffing ability.

32.8 PILING

Bearing piles or short piles installed in a marine environment use similar equipment to that used

Figure 32.11 Slipway, launch wedges and water skates. Note also caisson base-slab below caisson.

on land, and are subject to the same installation problems. The extra problems in marine work are getting the equipment to the installation point and providing the ability to work.

A basic initial choice is between fixed and floating platforms. A floating platform is not suitable for drilled piles except in very calm conditions, or for oscillators which require a platform with ability to resist torque. A floating platform can be used for driving with drop hammers and diesel, hydraulic or air hammers, although again relatively calm conditions are needed unless piles are driven through a sea-bed supported template with a sit-on-the-pile rope suspended hammer. For driven piles the pile guide (leader) may be a frame fixed to the barge, into which a pile is pitched and driven – any damage caused by barge movement during the driving process being sustained by the barge and/or pile. Alternatively the leader may be attached to a crane on the barge, the crane/leader then being at risk if significant barge movement takes place during driving. The problem can be reduced with a spud leg pontoon. A bigger investment is represented by purpose-designed piling frames. These can take the form of relatively standard pile frames, as produced by Menck, or special frames resembling floating sheer-legs with ability to rake fore and aft only or, additionally, to rake laterally ('moonbeam').

Where for reasons of site exposure or other causes, a fixed piling platform is needed two options are open – a jack-up platform or hand-over-hand construction. In the latter, it may be possible to programme construction to allow the piling equipment to travel over an already completed permanent deck; on temporary waybeams supported on already driven piles; or on a temporary launch gantry, progressing

Figure 32.12 Barge-mounted crane placing sliced blockwork. Note 'Fidler's gear' suspension of angled blocks.

over the piles as installed. It may thus be possible to build an along-shore jetty with no major marine plant, suffering no programme impediment. The hand-over-hand method applied to finger jetty work is subject to much greater programming restraint, each type of operation being limited to one working face and, unless marine fed, with a narrow restricted material feed path. A snag in one part of the chain causes all parts to grind to a halt.

Jack-ups are more mobile than hand-over-hand construction. They are, however, expensive pieces of equipment and take time to move about on the site and re-set-up again. When used in estuarial environments with very poor sea-bed material they need enlarged elephant's feet to minimize bed penetration of the jack-up legs. Subsequent extraction of the legs can be a problem exacerbated in bored piling by the additional vertical loads caused by casing extraction. Jack-ups used in civil engineering are frequently modular in form made up from flotation units, each of which is small enough for road transport, locked together with shear connections to form a large barge.

32.9 RECLAMATION – ACCELERATED CONSOLIDATION

An aspect of earthworks which recurs frequently in coastal or estuarial work is that of settlements associated with very soft clay layers and the need to accelerate this to avoid long-term problems to structures and services. To facilitate rapid settlement two things are usually needed: the creation of water conduits to allow easy dissipation of excess pore pressure and surcharge loading to increase total stress. The former is achieved by the installation of vertical drains at close centres (usually on a grid of about 1 m to 3 m). These drains were once invariably formed from vertical columns of sand, but recently wick drains have gained a very large market share because of their ease of installation. Wick drains comprise a band of PVC, grooved on both sides to provide small continuous channels, and sheathed in a permeable geotextile. The drain can be spooled on the base of the installation machine and installed by driving a mandrel to the foot of which the drainage band is attached, usually by a sacrificial metal or

Figure 32.13 Tube piling from floating barge.

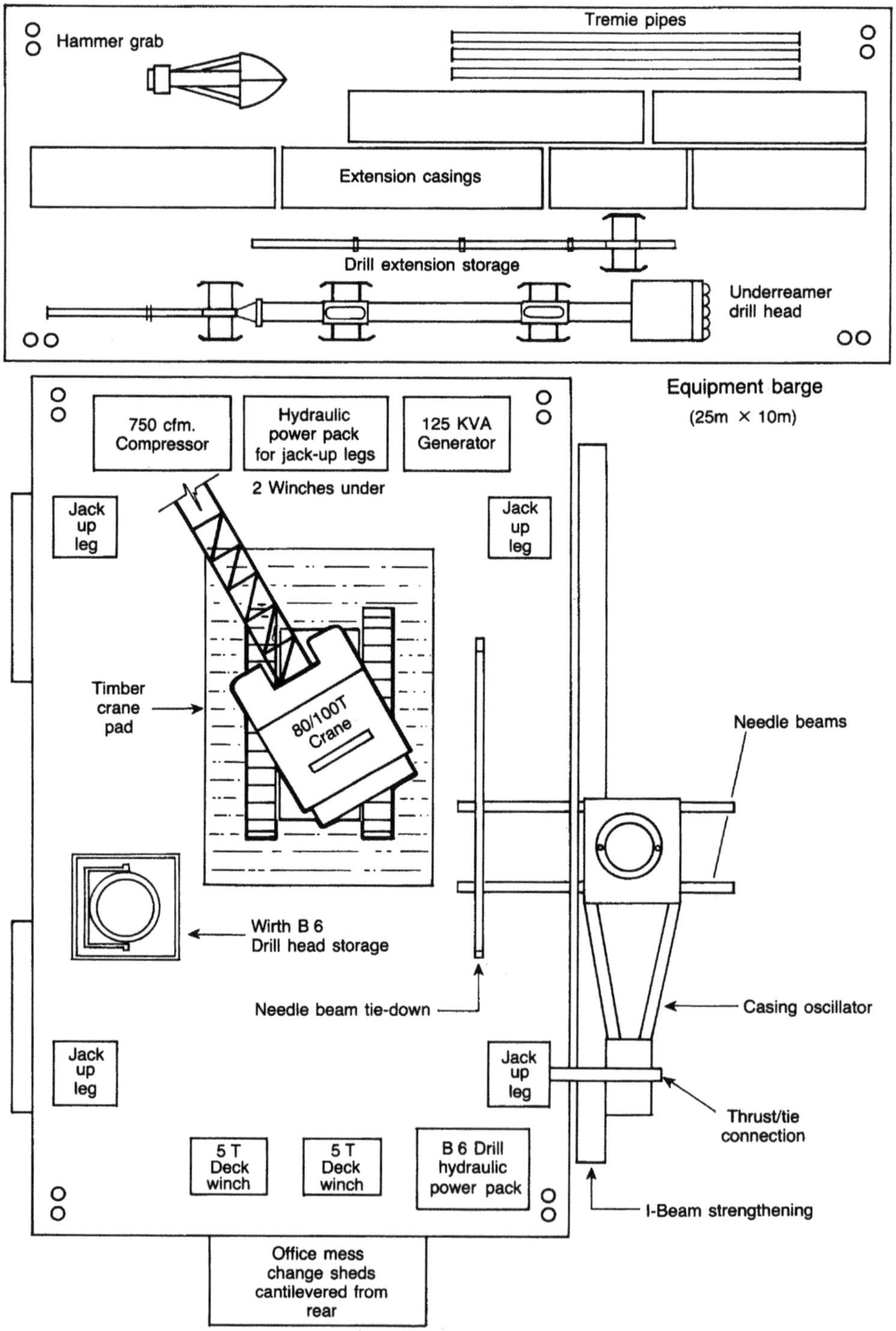

Figure 32.14 Bored piling from jack-up. Plan of 'Humber' and 'Thames' (self-elevating pontoon 24.3 m × 15.2 m).

Figure 32.15 Vacuum surcharge. Note discharge at bottom right-hand corner and adjacent suction pump. Evacuated water is lying on the impermeable surface membrane.

concrete shoe. Sand drains are installed by driving or vibrating a casing down through the layer to be consolidated, removing the soft clay from within and filling the hole with sand as the casing is withdrawn. Great care has to be exercised to ensure a continuous sand column.

As installation time is very short, as little as one or two minutes for a 10 m long wick, and the numbers of drains are large, rapid movement of the rig is essential or, if the rig is marine, simultaneous installation of many drains at each set-up.

Surcharge is usually provided by placing a temporary additional depth of soil over the area to be consolidated, perhaps 4–5 m high representing a 6–8 Te m^{-2} surcharge. To avoid importing vast quantities of additional material, which has to be subsequently disposed of, a rolling surcharge is adopted, re-using the same material on successive areas. Where the original ground has low shear strength (e.g. tidal estuary flats) a very flat slope to reclamation/surcharge is needed to avoid edge instability.

An alternative to solid surcharge is a vacuum surcharge. As with conventional surcharge vertical drains are installed and a drainage blanket of sand placed over the area to be consolidated. The whole is then covered by an impermeable plastic sheet. Water is evacuated by pump from under this sealing membrane, creating a vacuum and mobilising atmospheric pressure (full atmospheric pressure = 9.81 Te m^{-2}). Vacuum surcharge, in addition to removing the need to import and rehandle solid surcharge has two other desirable features:

- Quick release of surcharge (vacuum) if problems arise.
- While the same total stresses may be induced by the two surcharge methods effective stresses differ. In the vacuum method a reduction in pore pressure at the top of the soft layer takes place creating a hydraulic gradient which induces water to flow upwards. Edge instability of the surcharge area is therefore not as great as it is with solid surcharge.

A vacuum surcharge of 0.8 to 0.9 bar is achievable.

MAINTENANCE OF COASTAL STRUCTURES 33

J.D. Simm

It is essential at the design stage of any maritime structure to pay particular attention to maintenance. The ease with which maintenance can be carried out might well determine the choice of a final design. Attention needs to be paid, for example, to the availability of plant to carry out repairs. Regular inspections of the structure, especially after storms, might well save considerable expenditure in the long run. Failure modes of such structures are discussed in this chapter and it is pointed out that the most likely mode of failure will determine the best monitoring methods to adopt (see also Chapters 22 and 29).

CONTENTS

33.1 INTRODUCTION

Maintenance of a coastal structure should be viewed as comprising all those activities that need to be carried out on a periodic basis after construction to ensure that the structure performs to an acceptable standard during its lifetime. A maintenance programme will, therefore, include the following essential elements discussed in this chapter.

1. Inspection and monitoring of environmental conditions and structural response.
2. Appraisal of monitoring data to assess compliance of performance with predetermined standards, which may themselves vary through the lifetime of the structure (e.g. due to secular trends in water level or wave climate, or developments in the understanding of hydraulic or geotechnical processes and interactions).
3. The repair or replacement of components of a structure whose life is assessed to be less than the overall structure, or of a localized area which is assessed to have failed. This last item is sometimes viewed as 'maintenance' in a narrower sense and, therefore, will be termed for the purposes of this chapter 'repair' or 'replacement'.

The activities to be carried out will be based on a maintenance policy developed at the design stage in a management view of the whole-life performance and costs of the structure in question. This design overview of maintenance will have taken account of likely requirements and costs for plant, labour and materials. The policy developed will also have considered access for monitoring and maintenance and likely engineering and financial resources of the owner for carrying out both practical monitoring and repair activities and also the interpretation of the structure's performance.

The development of a maintenance policy should be followed at the design/construction stage by production of a maintenance manual. This is particularly important for rock or rubble mound structures in view of their flexible, and sometimes even dynamic, nature. In many situations, particularly where the structure has been designed to minimize maintenance, the manual may be very simple. However, it must contain the minimum necessary guidance on techniques and criteria for the three basic elements of the maintenance programme listed above. It will also set out the inter-relations of the various activities involved, a guide to which is given in the flow chart in Figure 33.1. This chapter sets out the principles for establishing such a maintenance programme and deals with the three basic elements.

33.2 MONITORING

Flexible structures comprised of rock and/or a flexible revetment respond to the potentially damaging ef-

Coastal, Estuarial and Harbour Engineer's Reference Book. Edited by M.B. Abbott and W.A. Price. Published in 1993 by Chapman & Hall, 2–6 Boundary Row, London SE1 8HN. ISBN 0 419 15430 2.

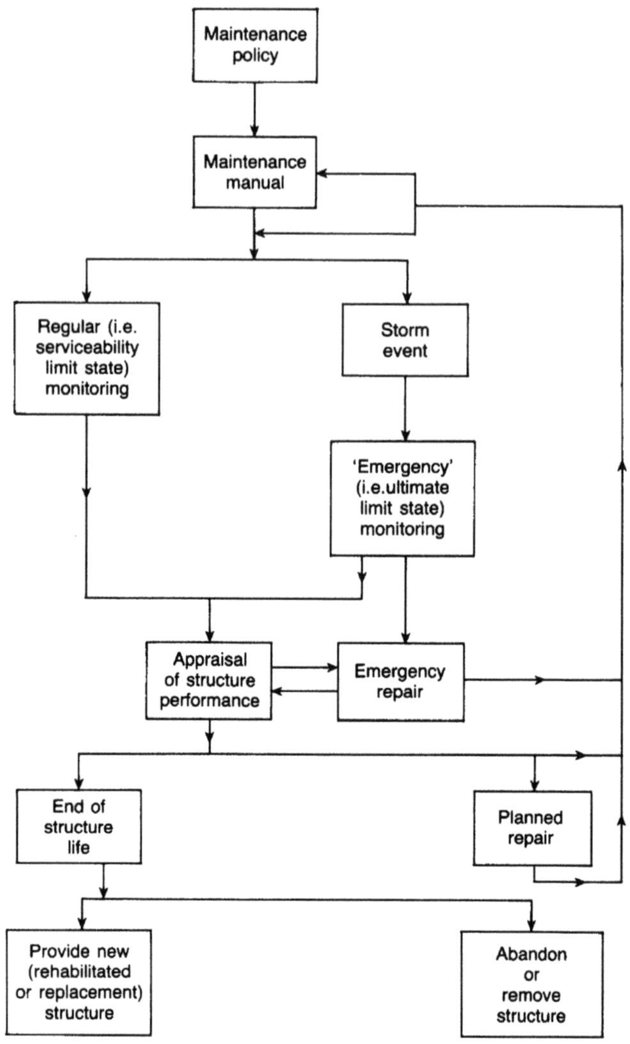

Figure 33.1 Maintenance programme flow chart.

fects of wave and tidal action by changes in their structural profiles and in the size and shape of their component parts. The changes in shape can also be due to loss or changes to foundation or core material. Typically, rock may be displaced, abraded, fractured or even dissolved.

Rigid structures may respond rather differently. Damage or change in the structure itself may occur in the form of abrasion, fracture or dissolution arising from water and sediment action. Often of more concern are the overall processes of settlement, tilting or sliding and possibly major structural failure. Such failures are related as much to wave and current induced settlement, or to scour of the beach beneath or in front of the structure, or to erosion of backfill, as to the direct action of waves and tides.

Any quantitative description of the state of such structures must be able to identify these different responses, which may either take place gradually, or catastrophically after a major storm. It must also be able to identify the environmental forces driving the responses.

A regular monitoring programme for both structure and environment allows the designer or owner to plan repair and replacement activities with a good understanding of the failure mechanisms and damage trends. Gradual deterioration of components such as armour layers, settlement or scour of the foundations, may go unnoticed without the aid of a monitoring programme, and may ultimately result in the failure of the armour layers or unacceptably large settlements. By permitting a comparison of measures of the state of a structure at a number of points in time, a structural monitoring programme allows these changes to be identified at an early stage, thus enabling the appropriate maintenance action to be carried out.

33.2.1 TYPES OF MONITORING

From the above, it follows that there are two principal areas of monitoring: measurement of structural state and measurement of environmental loading conditions.

Structural state monitoring concerns the performance of the fabric of the structure and its foundations. Details of the measures of structural state which may be monitored, together with appropriate survey techniques, are illustrated for rock structures in Table 33.1. Fuller details of the survey techniques mentioned in Table 33.1 are given in CIRIA/CUR (1991).

Formal structural state monitoring is worthwhile supplementing with walk-over surveys carried out to record, with the aid of photographs, the overall condition of the structure including any obvious movements, changes in profile, fractures, etc. This type of survey must remain very subjective and has only limited use in a detailed quantitative assessment of a structure. However, if fixed reference points can be established on the structure and checked at the time of the survey, walk-over surveys carried out by experienced personnel can form the basis of a suitable monitoring programme.

Environmental monitoring concerns the external loading on the structure and the effect the structure has on the total environment. A brief summary of environmental conditions or loadings which may be monitored, together with appropriate monitoring

Table 33.1 Measures of state of a rock structure

Aspect of structure state	*Survey technique (full details in CIRIA/CUR (1991))*
Level I: Location	
2 to 10 points on the structure measured in relation to a well established grid and datum levels. Redundant survey points needed if major movement possible.	Conventional survey techniques (any settlement markers should be installed during construction)
Level II: Geometry	
Outer surface description, related to level I survey points.	Conventional survey, using profiling techniques recommended in CIRIA/CUR (1991), but with profiles at wider spacings say 20–30 m. For underwater surveying, bathymetric techniques may give useful information.
Level III: Composition	
Position and attitude of each piece of rock armour, including unstable pieces. Position and size of major voids and exposures of core or underlayers.	Armour degradation inspection techniques Comparative photography Photogrammetry For underwater surveying, side scan sonar techniques may be used in co-ordination with diver surveys to identify features on sonar traces
Level IV: Element composition	
Shape and size of armour rock, including any fractures	Armour degradation inspection techniques

Note: Level III and IV armour degradation inspection techniques are difficult to implement on wide gradings and/or gradings with D_{15} less than about 0.3 m. The techniques, however, work extremely well on large 'single sized' rock armour ($D_{15} > 0.3$ m; D_{85}/D_{15} in range 1.2 to 1.4).

Table 33.2 Measures of environmental conditions or loadings

Environmental condition or loading	*Monitoring survey technique*
Water level	Tide board, visually inspected. Data from nearest local tide recording stations. Use of surface elevation monitor (step gauge or resistivity gauge) recordings, if available.
Wave climate	Sea-bed pressure meter (robust and cheap). Surface elevation monitor mounted on robust support (e.g. pile, or triangulated scaffold tube arrangement). Wave rider buoy or similar (will be expensive to maintain for long periods). Hindcasting analysis for storm events using wind records.
Wind climate	Standard anemograph device (depending on correlation between wind and wave direction may be a useful way of assessing directionality of wave climate).
Bathymetry and beach topography	Below high water, standard bathymetric techniques possible. Above low water, conventional land survey techniques may be used, or photogrammetry from aerial photography. Land based photography of waterline from fixed positions gives useful assessment of low to high water beach form.
Stress in foundation	Pressure pads
Pore pressure in foundation	Piezometers (simple standpipe or, for continuous measurements, vibrating wire electronic recording devices may be used).

techniques, is given in Table 33.2. For a fuller discussion of environmental data collection, the reader is referred to other chapters of this volume.

In addition, for critical structures, it may be worth monitoring the hydraulic/geotechnic response to wave loading. As an example, brief suggestions on measurements and techniques appropriate to rock structures are given in Table 33.3. The monitoring methods selected should relate to the potential failure modes for the structure in question and, in particular, to those which have been identified as the most likely.

33.2.2 FREQUENCY OF MONITORING

As with the methods of monitoring, the frequency of monitoring should be predetermined in relation to the risk associated with particular failure mechanisms, structural elements, foundation conditions, exposure conditions and design criteria. This frequency may well be different for different types of monitoring. It could also increase during the life of the structure reflecting the structure's reducing resistance to failure as it degrades with time and approaching need to carry out repair (Figure 33.2).

The typical behaviour of structures may commence with some initial 'bedding down' movement, such as settlement and packing of individual pieces of rock within a rock structure. Otherwise structures are often stable for some years and changes are often minor. Later, the degradation of individual elements of a structure may become significant and/or environmental conditions (including loadings due to settlement or scour) may become more severe. Changes in coastal structures may then become more rapid. This pattern of overall behaviour under what could be described in limit state terms as serviceability conditions, may be overridden by a very severe storm event and the behaviour of the structure in response to such an ultimate limit state loading.

It is important to commence the monitoring of new structures by establishing a 'baseline' set of information regarding the structure and its environment at the time of construction and during the construction guarantee period. This information should

Table 33.3 Measures of hydraulic/geotechnic response of rock structure to wave loading

Aspect of hydraulic/geotechnic response	*Monitoring survey technique*
Wave run up	Parallel steel wire resistivity gauge (survival a problem)
Wave transmission (for breakwaters)	Appropriate wave gauge at rear of breakwater
Mound pore pressures	Piezometers installed within mound with automatic recording facility

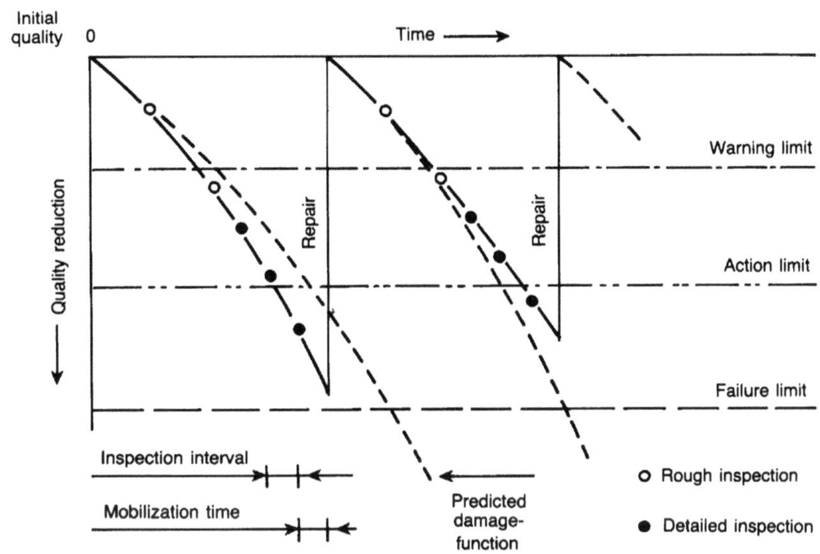

Figure 33.2 Preventative condition-based maintenance (de Quelerij and van Hijum, 1990).

be stored in a form which allows for its retrieval for purposes of comparison in future years. The recording of details about the structure during the construction stage is also required as a check against the assumptions and details established at the design stage. Information recorded should not only include basic geometric survey data of profiles etc., but also records of any failures of the structural elements during the construction period. The basic monitoring information should be documented by the contractor and handed over to the owner or designer by the end of the guarantee period.

The frequency of structure state monitoring adopted thereafter during the life of the structure depends on a number of factors, principally the following:

1. Location on structure;
2. Type of construction;
3. Design risk levels;
4. Exposure conditions;
5. Foundation conditions.

Two basic frequencies of monitoring will be involved, one related to normal conditions or the serviceability limit state and the other related to extreme conditions or the ultimate state. Monitoring related to the serviceability state should take place on a planned basis at frequencies identified at the design stage. Ultimate state monitoring will only take place during the following severe storm events, the minimum wave conditions for which should be stated at the design stage.

Monitoring frequency above low tide level

As an example of the typical frequency of planned monitoring that should be adopted for normal (serviceability limit state) conditions, Table 33.4 indicates the recommended maximum period between planned monitoring inspections for a rock structure above low tide level.

After the first few years of life of a structure, satisfactory performance may indicate that adequate

Table 33.4 Frequency of planned monitoring above low tide level for a rock structure

Location of whole structure	6 months visual – 12 months survey
Geometry of the structure	12 months
Position of individual rocks	12 months
Shape, size, condition of individual rocks	12 months
Foundations, scour, etc.	6 months visual – 12 months survey

Table 33.5 Outputs from comparison of measures of state of rock structures over a period of time

Aspect of structure state measured (see Table 33.1 for details)	*Output from comparison of structure state at a number of points in time*
Level I: Location	Settlement of foundation. Change in alignment.
Level II: Geometry	Consolidation of structure. Comparison of slope profiles enables overall armour damage parameter, *S*, (see CIRIA/CUR, 1991) to be determined. Scour damage.
Level III: Composition	Loss or movement of armour rocks. Overall sliding of armour layers if this has occurred. Voids requiring emergency/planned repair.
Level IV: Element composition	Rounding of rocks and loss of material, enables revised assessment of D_{n50}. With the design wave climate, or measured wave climate, or revised design wave climate from wave measurements, this enables re-assessment of armour stability parameter $H_s/\Delta D_{n50}$ using equations in CIRIA/CUR (1991). Comparison with design and measured damage parameter, *S*, is also possible.

monitoring will still be achieved if the detailed surveys are reduced in frequency from 12 to 24 months or even longer. However, the experience of authorities responsible for monitoring is that there is a danger of surveys being forgotten if they are not carried out every year. There is also the danger with such infrequent surveys that rapid deterioration near the end of the life of a structure will not be detected before a failure occurs.

For structures which have elements below the lowest tide level, it should be recognized that the effects of wave exposure will not be as great. However, the effects of local currents will have to be considered when developing a monitoring frequency for elements of a structure which are permanently submerged.

Monitoring frequency below low tide level

Unless special circumstances exist it is recommended that the submerged elements of a flexible structure are fully inspected at least every five years and that rigid structures are inspected at least every ten years. Both structure types should be fully inspected after extreme storm events. In addition, the regular annual monitoring of the upper sections may indicate possible problems on the submerged section of a structure worthy of further investigation. Instrumentation can also be introduced at the construction stage, into structures which are partially or permanently submerged, to allow certain performance aspects (e.g. foundation settlement) to be monitored during the more frequent inspections of the upper structure.

33.3 APPRAISAL OF STRUCTURE PERFORMANCE

Having instigated a comprehensive monitoring programme a designer or owner needs to document all the monitoring information and carry out a review and/or analysis of the data which will form the basis for decisions regarding the need for and extent of maintenance works.

Each structure state monitoring report should be assessed in relation to the environmental conditions recorded during the inspection and compared with previous structure state monitoring reports. As an example, typical assessments which may be made for rock structures are listed in Table 33.5. Such assessments enable a continual assessment of design and performance criteria.

It is also useful after each monitoring survey to make an assessment of the remaining life of the structure in relation to each potential damage pattern in order that the overall safety of the structure may be monitored. The time scale of the deterioration effect can usefully be presented by constructing a kind of 'measuring watch'. On this watch (see Figure 33.3) both the actual damage level and the damage limits (warning limit, action limit and failure limit) can be presented on a time scale (de Quelerij and van Hijum, 1990). The watch provides the managing organization with a simple overview of the relative safety condition of the structure in relation to the relevant damage pattern, together with an idea of the remaining safe lifetime.

To provide a good presentation, the 'measuring watches' of all relevant damage patterns for the rock structure can be collected together into one control panel as illustrated for a sea-wall in Figure 33.4, thereby providing a complete overview of its overall actual and near-future safety condition.

Having completed the assessment, the owner or designer may decide to opt for one or more of the maintenance actions indicated in the flow chart in Figure 33.1. The options may be briefly listed as:

1. Do no repair/replacement work and await next planned monitoring report;
2. Do no repair/replacement work but instigate additional future monitoring of structure state and/or environmental conditions;
3. Carry out further detailed inspection before making decision;
4. Undertake temporary or 'emergency' repair/replacement works;
5. Undertake permanent repair/replacement works;
6. Instigate development of a new (rehabilitated or replacement) structure;
7. Instigate abandonment/removal of the structure.

The decisions made in response to a monitoring report should be set against performance and failure criteria which have been established at the design stage. It must be recognized, however, that performance and failure criteria can change as technical understanding develops and as the requirements for the function of a structure change due to changes in use, safety standards, or standards for environmental assessment.

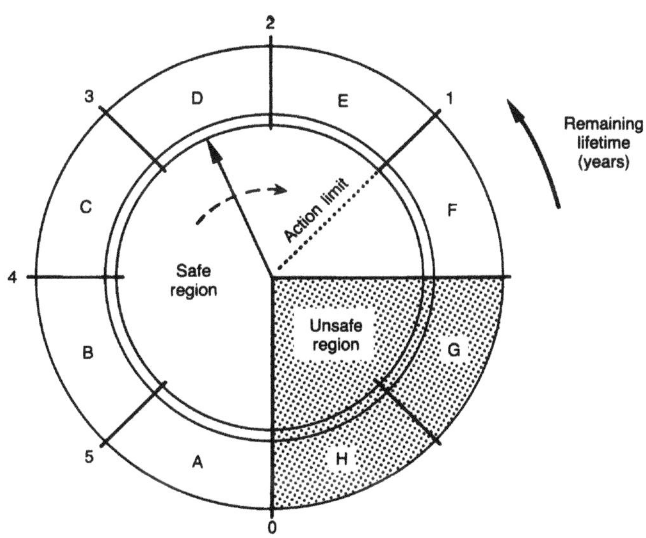

Figure 33.3 Measuring watch for a selected damage pattern (de Quelerij and van Hijum, 1990).

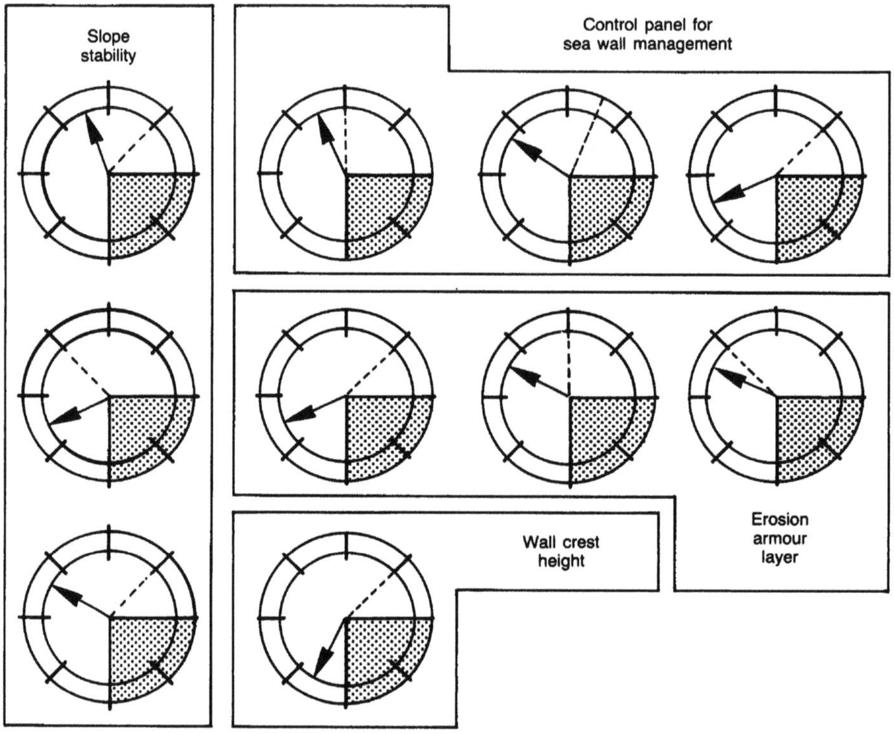

Figure 33.4 Safety control panel for sea-wall management (de Quelerij and van Hijum, 1990).

33.4 REPAIR/REPLACEMENT CONSTRUCTION METHODS

The repair construction methods adopted for any particular structure will be strongly influenced by decisions taken during the design and project appraisal phase prior to the initial creation of the structure. Financial considerations will be one aspect, the maintenance budget set obviously having a strong influence. Generally, it will be uneconomic to so underdesign the structure that when damage occurs it affects more than the surface of the structure (e.g. the armour layer of a rock structure). When damage is more than superficial, repairs become more in the nature of a major rehabilitation and are correspondingly costly. Economic maintenance procedures will therefore generally be confined to repairs to the surface of the structure such as re-profiling rock armour layers or reinstating joint seals in rigid structures. Having set the financial boundary conditions, the main considerations then become availability of, and access for, materials and plant.

The adoption of re-usable initial structural materials such as armour stone may well be possible and worthwhile in many cases. In such cases, if suitable access to the coastal structure is possible, repair works will only require dislodged materials to be retrieved and re-positioned/re-fixed into the structure.

In other cases fresh materials will be required for repairs. In this case, if provision for suitable access has been made at the design stage and there are no financial constraints, new materials can be imported as required. However, importation of small quantities of additional materials may be expensive, particularly if the source is remote. In addition, access after construction is complete may be difficult or impossible for large or bulky materials such as armour stone. In this case, stockpiling of spare material at the site as part of the main initial construction operation should be considered.

Access for plant to, along and around the structure for handling and placing or fixing repair materials is of prime importance and should be considered carefully when detailing or dimensioning new structures. Access for initial construction is very often gained along the underlayers of a partly completed structure or by purpose-built and expensive temporary works, but these forms of access are no longer available for repair works. Even purpose-built roads are only of value if it can be guaranteed they will not be damaged by storms.

Equipment used for repair activities is very specific to the type of structure being repaired and may be very different from the equipment used for initial construction. An important general consideration is that the most useful pieces of equipment tend to be those that can both retrieve/remove damaged materials for re-use or disposal as well as placing repair materials back in the structure.

ACKNOWLEDGMENTS

The assistance of Messrs. C. Davies (Metropolitan Borough of Wirral/Bullen & Partners), N.W.H. Allsop (Hydraulics Research, Wallingford) and K. Pilarczyk and W. Leeuwestein (Rijkswaterstaat, The Netherlands) in developing the concepts for this paper as part of the preparation of the CIRIA/CUR (1991) *Manual on the Use of Rock in Coastal and Shoreline Engineering* (for which the author was general editor) are gratefully acknowledged.

REFERENCES

de Quelerij, L. and van Hijum, E. (1990) Maintenance and monitoring of water retaining structures, in, *Coastal Protection*, (ed. K.W. Pilarczyk), Balkama, Rotterdam, pp. 369–401.

CIRIA/CUR (1991) *Manual on the Use of Rock in Coastal and Shoreline Engineering*. CIRIA Special Publication 83, CIRIA, London.

FIELD STUDIES AND DATA ANALYSIS 34

R.F. Rayner

This chapter describes position fixing and the measurement of currents, dispersion coefficients, tides, waves, meteorological conditions and water column structure. It also provides an introduction to bathymetric and geophysical survey techniques. Normally field data is collected as part of a study to support the design and construction of marine works. Field studies are necessarily expensive so that good planning is a prerequisite of such excursions. The environment is often hostile and the field equipment sophisticated and complex which means that specialized firms are usually entrusted with this type of work. (See also Chapters 23, 24, 25, 29, 30, 31, 38, 42, 43, 47 and 48).

CONTENTS

34.1 POSITION FIXING

An essential prerequisite of all measurement programmes is the need for positional data. Positional data may be obtained using a variety of different techniques. The selection of an appropriate technique depends on the accuracy required and the nature of the study area. Navigation systems may be broadly divided into two main categories: electronic position fixing devices and non-electronic position fixing methods.

34.1.1 ELECTRONIC DEVICES

A wide range of systems is available using a variety of operational principles and methods. As a general rule the greater the cost, the greater the accuracy. A number of available devices are described although an exhaustive list is not intended. The systems have been chosen as representative of particular types of system currently in common use in coastal investigations.

Decca mainchain

A permanent chain is established in a number of areas world-wide, and position is derived from the transmitted signals. Receiving equipment may be hired or purchased. Earlier receivers output the distance from the master beacon as a lane number. This may be subsequently converted by computer or plotted onto special charts. More recent devices have incorporated microprocessors and latitude and longitude is output directly. Accuracy is highly site dependent but it is estimated to be in the range 50–200 metres. Reception of Decca mainchain signals can be seriously affected at dawn, dusk and during atmospheric electrical disturbances or heavy rain.

The use of Decca mainchain has largely been superseded by the satellite Global Positionary System (GPS).

Radar

In particular situations a ship's radar may be used to estimate range and bearing to a known position. Accuracy depends on the quality of the radar and distance to the reference object. Alternatively radar may be used from a fixed location to track the move-

Coastal, Estuarial and Harbour Engineer's Reference Book. Edited by M.B. Abbott and W.A. Price. Published in 1993 by Chapman & Hall, 2–6 Boundary Row, London SE1 8HN. ISBN 0 419 15430 2.

ment of individual reflectors (e.g. surface floats) by triangulation.

Microwave

The most commonly used short range (up to 80 km) systems are Miniranger, Trisponder and Mircrofix. Microwave systems are generally used in range-range mode with several beacons (remotes) deployed at co-ordinated points onshore. The master installation or installations on board the survey vessel (mobile) transmit a signal to several remotes and calculate the distance from the time delay in the returned signal. A minimum of two remotes is required to calculate position, although three are recommended in order to provide a measure of accuracy. The accuracy of the system is dependent on the configuration of the beacons. The most accurate position for a two range system is when the angle of cut between the two lines of position is 90°. Three range operation is less critical although as the angle of cut between ranges from any two mobiles diminishes, the accuracy decreases. The accuracy of a single range is typically ±1 m. Overall accuracy for a well configured chain is of the order of 1.5 to 5 m.

Other microwave systems, such as Artemis, use a single fixed master station and derive position from range and bearing from the master station. Accuracy decreases with distance up to the maximum range of 30 km. Expected accuracies are ±1.5 m in distance and 2 min of arc in bearing. This is equivalent to an accuracy of ±3.0 m at maximum range.

Microwave systems are generally very reliable and can operate in poor weather conditions. However, null (phase cancellation) effects can occur when the systems are used in the marine environment and there are differences in height between mobile and remote units antennae. Dual antenna systems can reduce this risk, but increase the cost.

Laser systems

The range of laser based systems varies from simple hand-held devices which output distance directly to complex range and bearing systems which are integrated with computers. The range of most laser based systems is less than 5 km. At extreme ranges a substantial array of reflective prisms is required to ensure sufficient reflection. For the range and bearing systems the main station is located at a co-ordinated position onshore with line of sight to the whole survey area. Operation is controlled onshore with position telemetered to the survey vessel offshore, or controlled from offshore using automatic tracking techniques. Range accuracies are of the order of ±0.5 m with directional accuracy of ±0.01°. This is equivalent to an overall accuracy of ±0.5 m at a maximum range of 5 km.

The accuracy and maximum range of laser based systems is highly dependent on the prevailing conditions. In particular, vessel motion may compromise the ability of the reflector to return an adequate strength signal. The degree to which this is the case will vary with the type of system and type of reflector array.

Satellite positioning

Global coverage positioning is provided by the global positioning system (GPS) operated by the United States Military. Although capable of high accuracy (better than ±10 m) signals are coded so that civilian users can only access a degraded positional accuracy. This can be overcome using so-called differential techniques. A GPS signal receiver is established at a static point with an accurately known position and then positional corrections are transmitted to the vessel being positioned. Here they are used to correct the degraded GPS signal such that a greatly improved positional accuracy is achieved.

GPS receivers may be purchased or hired. In some regions permanent differential systems are installed with access to the signals being provided based on a daily charge.

Use of electronic navigation information

The use of navigation computers has greatly changed the techniques used to carry out surveys at sea. Navigation computers allow the surveyor to pre-program the course to be followed by the vessel and to provide real-time plotting and displays.

Calibration is a particularly important part of any survey and it is recommended that all navigation systems are calibrated over an accurately known baseline prior to the start of the survey. Should this not be possible, calibration may be carried out at sea by comparison with another more accurate means of position fixing. Calibration checks may be performed at sea by crossing the line connecting two shore beacons and checking that the ranges measured correspond with the length of this baseline. This technique is known as 'crossing the baseline'.

34.1.2 NON-ELECTRONIC DEVICES

Non-electronic devices are generally restricted to close inshore applications. In particular circumstances they can provide a cost-effective and accurate method of fixing a survey vessel position. Three main methods are used:

Horizontal sextant resection

In a dynamic situation this requires two surveyors; each observes the angle between two co-ordinated shore marks. One, but not both of the marks must be common to both observers. The simultaneous sextant angles are logged and converted to a position fix using a microcomputer or conventional graphical techniques. The shore marks would usually be co-ordinated prior to the start of the survey and marked clearly, although naturally occurring marks which are easily identifiable may be used. The main limitations of this method are visibility and range. A well practiced pair of surveyors would be able to record position fixes at 30 s intervals to an accuracy of ±5 m with well conditioned geometry and independent plotting.

Theodolite

The use of theodolites is sometimes a convenient method of obtaining a fix offshore. A pair of observers may simultaneously observe bearings from co-ordinated positions onshore to a point on the survey vessel. A single observer may be used in conjunction with distance measuring equipment or with the survey vessel steaming along a transit line.

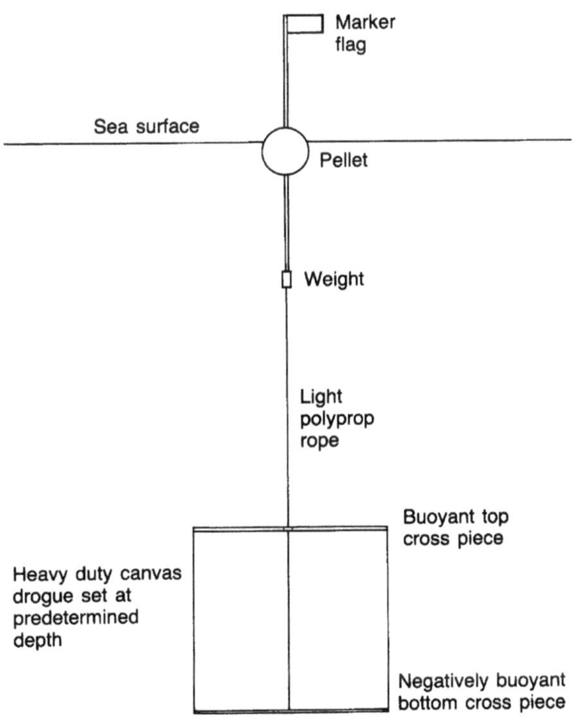

Figure 34.1 Current following drogue design.

Transits

For surveys very close inshore it is often convenient to co-ordinate a series of transits onshore. Transits usually consist of poles or markers. These are set up such that pairs of transit markers may be lined up and used as a means of steering a straight line towards the shore. The survey vessel steams along these transits. Distances along each transit line are measured electronically or by means of observing the change in angle from a known position using a sextant on board or a theodolite or sextant onshore.

Selection of a positioning method

The selection of a method appropriate to a particular survey area is best left to a qualified surveyor since there are many criteria to be balanced in the selection process. Points to be considered include:

- Configuration of the coastline;
- Range requirements;
- Accuracy requirements;
- Likelihood of environmental conditions which affect navigation (e.g. fog);
- Obtaining necessary radio frequency licenses;
- Number of vessels requiring positioning.

In many cases it may prove necessary to employ more than one means of positioning (e.g. in areas of cliffs sextants may be the only satisfactory way of obtaining good navigation close inshore due to shadowing effects).

34.2 CURRENT MEASUREMENT

34.2.1 DROGUES AND DRIFTERS

Drogues

Drogue tracking is the most frequently used water path or Lagrangian measurement technique. Drogue tracking provides data on the path of currents. Such data is well suited to verifying flows predicted by numerical models.

A variety of techniques are available for tracking drogues. The simplest and most commonly employed is to use subsurface drogues with surface markers (such as a small buoy or a pole marked with a flag). These are tracked by a vessel. The depth of the drogue may be adjusted to track water movement at different levels in the water column. Drogues of this type must be carefully designed to ensure that wind effects on the surface float and marker are small in relation to the drag on the sub-surface drogue. The size of drogue necessary to ensure that this is the case will vary with location. In areas of very weak tidal streams where low speed wind induced currents dominate it may prove necessary to use very large surface area drogues. However, in the majority of UK waters drogues of the type illustrated in Figure 34.1 provide a reasonable performance. Many alternative designs are available including the use of parachute shaped drogues although these are more often employed for tracking deep water masses.

The number of drogues that may be tracked by a single vessel depends on the desired frequency of position fixing (which in turn depends on the use for which the drogue track data is being collected), the speed of the currents, the speed of the survey vessel, the distance between floats at release and the weather conditions under which the operation is to take place. As a general rule it is wise to use no more than ten drogues per vessel. Tracking during the hours of darkness can be particularly difficult and great care must be taken in selecting the means of lighting the drogues and in providing a means of identifying individual drogues. A variety of small flashing light assemblies are available which may easily be attached to a drogue marker. Useful techniques for identification at night are the use of different coloured lights or attachment of different coloured chemical light sticks to each drogue.

The procedure for tracking drogues is to fix a position as the survey vessel comes alongside the drogue. Positions are plotted onboard to allow prediction of locations and relocation of the drogues at the appropriate fix interval. It should be remembered that as the drogues separate the minimum interval between fixes will increase. This will particularly be the case if the drogues are tracking water at different depths.

Radio and radar-tracked drogues

As well as the conventional technique described above there are a variety of other methods used to track water movement.

Drogues may be equipped with small radio transmitters to allow them to be tracked from the shore using simple triangulation techniques. This method, whilst attractive from the point of view of minimizing vessel use, has a number of disadvantages. Firstly, the transmitter greatly increases the cost of the drogue assembly (a number of which are inevitably lost during any investigation). Secondly, there are the problems of obtaining radio frequency licenses for such devices particularly if a number of drogues are to be tracked simultaneously since each requires a different frequency. The additional wind drag imposed by the transmitter and antenna can also cause problems.

Drogues may also be tracked using radar techniques. This allows rapid position fixing of a number of drogues by triangulation using two radar sites. It also offers the added advantage that an observer fixing drogue positions onshore can instruct a vessel to modify drogue release patterns to maximize data collection in areas of particular interest. Because radar reflectors greatly increase the wind drag on the drogue assembly it is necessary to calibrate the 'windage' of the drogue assembly so that observed tracks may be corrected for wind influences (Charlton, 1985).

Drifters

Other water movement tracking techniques make use of a range of designs of drifter. Drifters are designed to follow either surface or near sea-bed water movement and are usually released in large numbers. Each is individually identifiable and the drifter carries a message requesting that anyone finding it returns it with a note of the location in which it was found and the time and date. These techniques have a number of obvious limitations and are little used in coastal investigations.

One drifter technique that can be very useful in determining details of regional circulation is aerial tracking of surface drifters. In this technique a large number of highly reflective surface drift cards are released and the resulting advection of these is monitored by means of time lapse photography from an aircraft or in some cases a kite or balloon. Logistics for this technique are sometimes complex and tracking at night is, of course, not possible!

Limiting conditions

Each of the techniques described above suffers from weather limitations. Under most circumstances it is unwise to attempt any sort of drogue or drifter tracking in conditions above the Beaufort wind strength 4–5. Under certain circumstances the limiting sea state may be less than this, particularly in areas of weak currents where wind influences on drogue floats and markers are most pronounced.

Presentation of drogue tracks

Drogue tracks are usually presented on charts showing the path followed by a series of drogue releases. It is usual to indicate the individual position fixes and time of position fix relative to high water. The speed of travel between position fixes may also be indicated. A typical example of such a presentation is given in Figure 34.2.

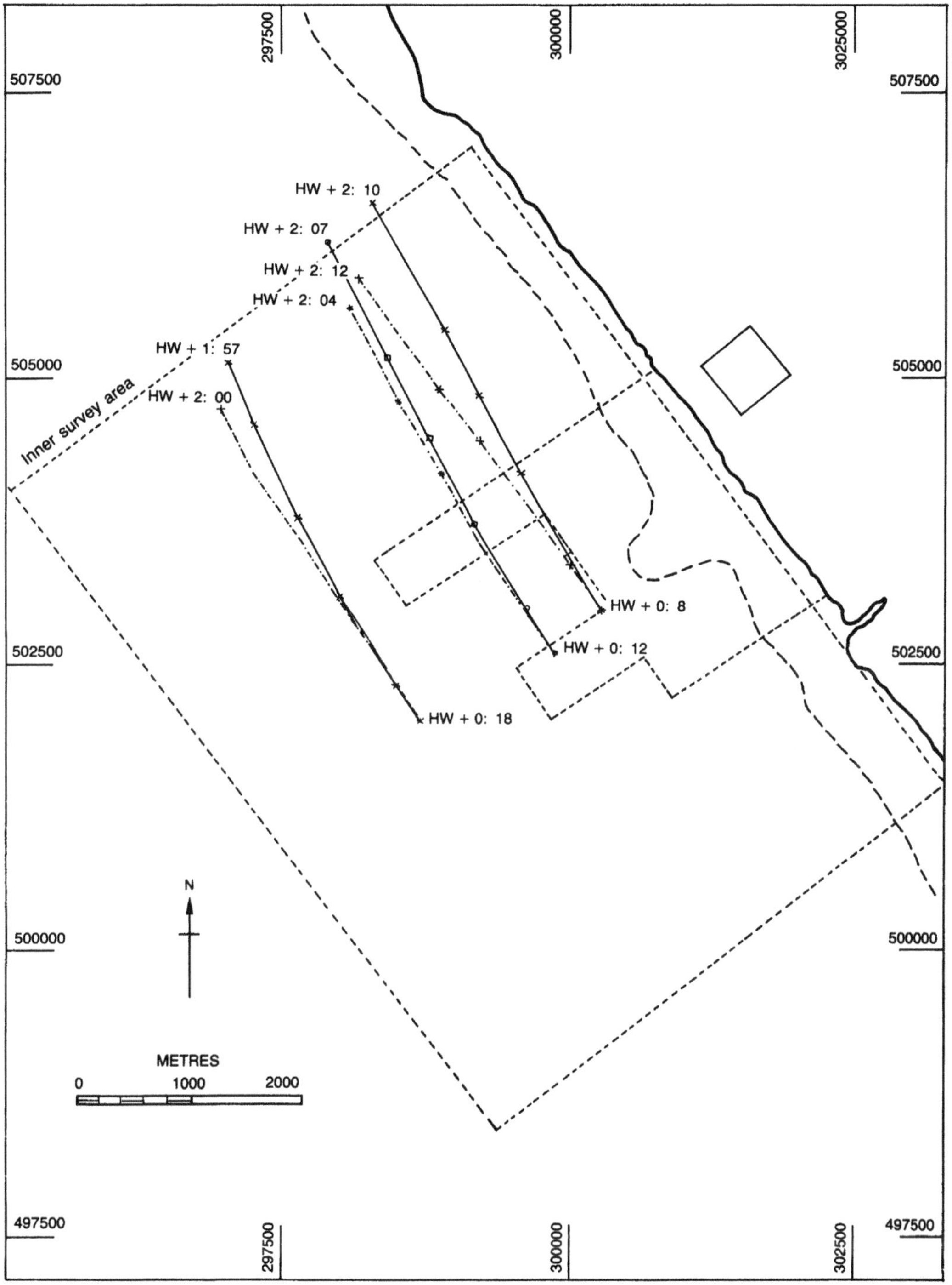

Figure 34.2 Float streamline plot. Float velocities are in cm s^{-1}. ——, surface float; –·–·–, subsurface float.

34.2.2 CURRENT METERS

There is probably a wider variety of instruments available for measuring currents at a fixed point (Eulerian measurement) than for any other single oceanographic parameter. There are two basic principles of current measurement upon which instrument design is based: firstly, scalar averaging of readings. Speed is integrated over a pre-set sampling period and direction is either averaged separately or derived from a single measurement during the sampling period. This principle is appropriate to reasonably steady flows but produces significant errors when used in turbulent regimes (e.g. the near surface zone where waves introduce oscillatory flows). Secondly, vector averaging of readings; rapidly sampled individual measurements of speed and direction are vectorially averaged over a pre-set sampling period. These instruments are suited to use in both steady and turbulent flows. In turbulent flows (e.g. in the wave zone) high frequency oscillatory flows are filtered out by the vector averaging process.

A number of methods are used to sense flow speed.

1. Propeller/impellor system utilizing a multi-bladed arrangement set about the axis of revolution so that the water flow imparts a torque to the blades.
2. Rotor system (often a Savonius rotor) causing a volume of water to be trapped thus extracting the maximum kinetic energy from the current stream and thereby imparting a high rotational velocity.
3. Acoustic methods relying on sensing the differential velocity of sound in opposing directions or measurement of Doppler backscatter from particulate matter carried in the water column.
4. Electromagnetic system sensing water flow with a transducer that has no moving parts. A magnetic field generated by an electromagnet with the transducer induces an electromotive force in the moving water which is proportional to flow speed. This voltage is detected by the use of orthogonally oriented pairs of electrodes.

In the case of propeller and rotor based instruments, current direction may be determined either through orientation of the instrument itself pivoting on an axis using large fins aligning to the mean flow, or by separate fins pivoting on the instrument case. In the case of acoustic or electromagnetic instruments, two perpendicular components of velocity are generally measured, allowing current direction to be determined.

Vector averaging instruments generally use either acoustic or electromagnetic methods to measure water flow components in two or three axes, although orthogonally mounted ducted propellers are also used. The measured components are generally resolved along two principal axes before recording, or averaging and recording. One of the principal advantages of acoustic and electromagnetic sensors is that they have no moving parts and are therefore less likely to become fouled or damaged by weed or other debris in the water column.

In each type of instrument a compass is generally included within the instrument case to allow directions measured relative to the current meter to be converted to directions relative to magnetic north. In instances where a sensor is rigidly mounted with known orientation, the provision of a compass may not be necessary.

It should be noted that the convention for water flow direction is the opposite to the meteorological term for air flow direction. Currents are stated as the direction towards which they are flowing and winds the direction from which they are blowing.

Two types of data recording in common use are: manual recording from a direct or real time display of data; and automatic recording of data on to tape reel, tape cassette or solid state memory. In some instances, both of these can be provided by one instrument, with the further possibility of telemetering data in real time to a remote point for display or recording.

Choice of instrument

The most appropriate type of instrument in any situation will depend on a number of factors;

- Purpose of measurements;
- Suitable deployment methods;
- Length of measurement period;
- Appropriate sampling scheme;
- Data processing requirements and facilities;
- Deployment position (geographic location, depth below surface);
- Probable flow characteristics;
- Level of accuracy required.

Additional sensors

It is common when measuring currents to measure other parameters using the same instrument. The most obvious addition is depth determination by means of a pressure transducer. Other common sensors are temperature and conductivity (salinity). Less commonly turbidity, light intensity, pH, dissolved oxygen and other chemical parameters may also be recorded.

Use of current meters

Current meters may be used in two principal ways. Firstly, one or more instruments is moored in the

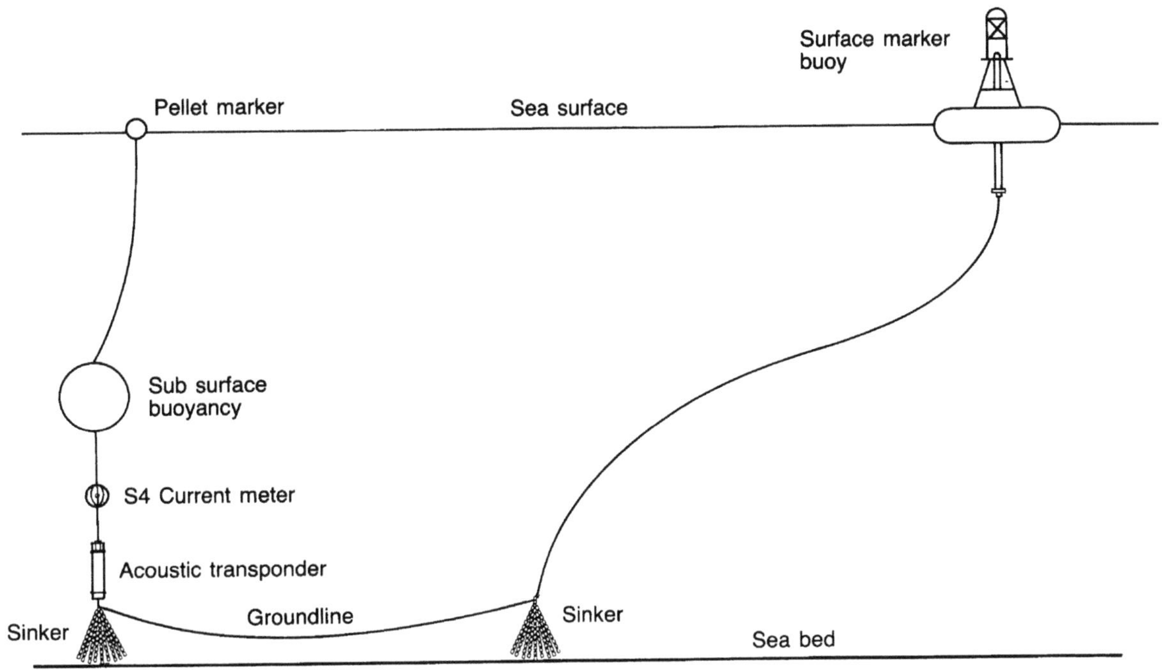

Figure 34.3 S4 current meter mooring, 'U' shaped mooring configuration.

water column to monitor currents at a fixed height above the sea-bed usually over long periods (typically one to two months). Secondly, the use of a single instrument lowered on a wire (often fitted with sensors to measure other parameters, e.g. water temperature, conductivity (salinity), depth) to obtain profiles through the water column. Measurements of this type are typically conducted over much shorter time periods (typically one tidal cycle, approximately 13 h).

34.2.3 MOORED CURRENT METERS

If data collection over a long period is required, it is usual to moor the instrument(s) at a specific level or levels. For near sea-bed measurements the instruments are often installed in a cage. When a 'profile' through the water column is required, a series or string of instruments may be deployed to record at a series of levels.

Two common mooring configurations are 'U' shaped and single point. The 'U' shaped mooring consists of a subsurface buoy, with a taut line to a sea-bed sinker. The instrument(s) are attached to this line. A ground line connects the sinker to a second sinker some distance away, with a line connecting this to a surface marker buoy. The subsurface buoy may also have a small surface marker buoy attached to it. A diagram illustrating this type of mooring is presented in Figure 34.3. The single point mooring consists only of the subsurface buoy, taut line with instrument(s), and single sea-bed sinker. An example is shown in Figure 34.4. In both cases the subsurface buoy must be moored below the lowest possible tide level.

The design of a mooring to be used at a given site should take into account the following factors:

- Whether acoustic releases should be used;
- Means of recovery of lost or abandoned mooring components;
- Adequate surface marking;
- Minimizing vertical displacement of instrument due to current drag on the moorings;
- Corrosion and cathodic protection of mooring components;
- Level of fishing and other activity;
- Specifications of the deployment/recovery vessel.

Sampling periods and locations

For most investigations it is usual to obtain fixed point current measurements over a period of at least a full lunar month (28 days). Instruments are usually set up to log data at ten minute intervals throughout this period. The number of instruments employed very much depends on the nature of the site.

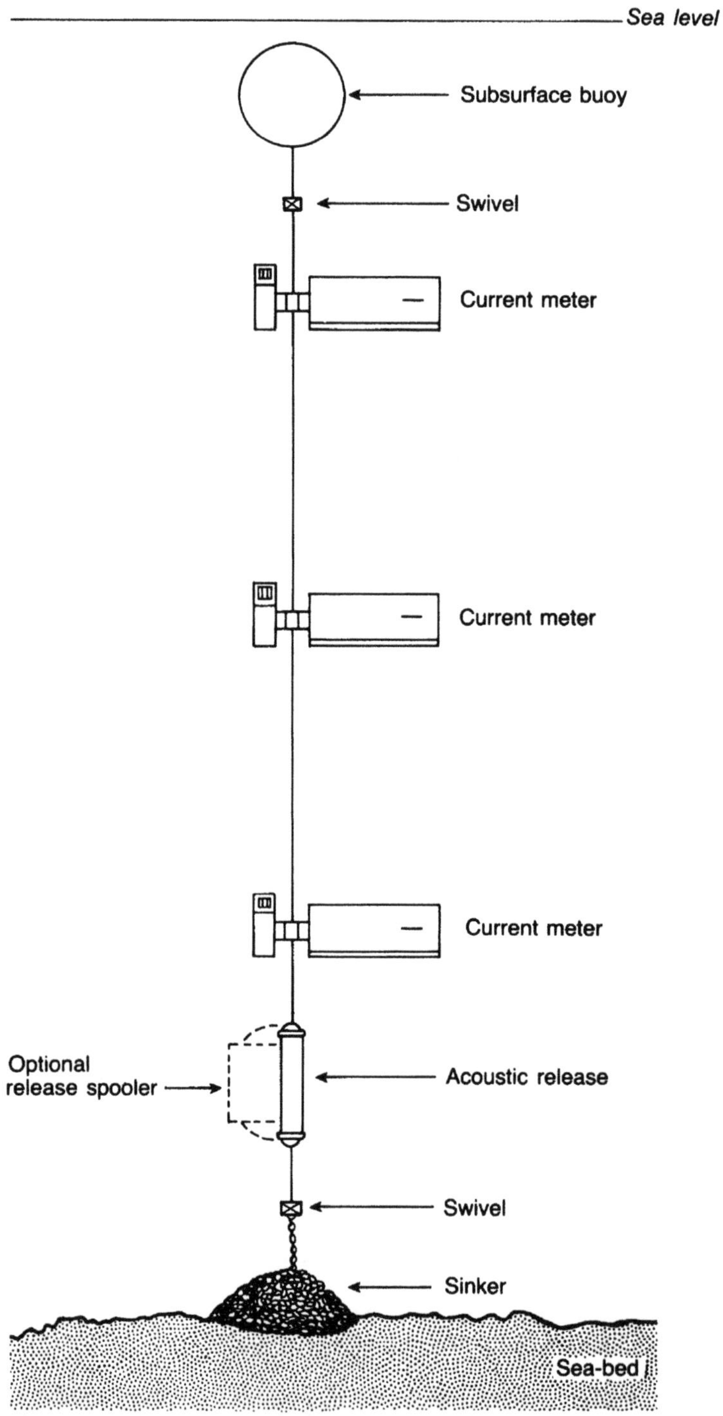

Figure 34.4 Recording current meter mooring, single point mooring configuration.

Quality control

In order to maximize data quality it is essential that instruments are calibrated before deployment with a check calibration upon recovery. Careful maintenance of the instruments is essential to prevent failure of moving parts, flooding of electronics due to 'O' ring seal failure, etc.

Service visits

The frequency of service visits to moored instruments depends upon the capacity of the data recording system, battery life, site access and the likelihood of outside interference, e.g. by fishing or shipping activities, storms etc. Normally a one month period of unattended operation is possible. In areas of frequent vessel activity or where there is a high risk of fouling of the instruments then weekly service visits may prove necessary.

Data analysis and presentation

The simplest current data presentation is a listing and time series of speed and direction. Percentage exceedence, rose diagrams, progressive vector plots and persistence diagrams are also commonly produced. Examples of these presentations are given in Figures 34.5–34.12. Other analyses include the calculation of orthogonal components normal and parallel to a shoreline or proposed pipeline, harmonic analyses for tidal current prediction, calculation of highest and lowest astronomical currents, and estimates of residual currents. The computation of return periods for predicted maxima may also be required. The accuracies achievable depend upon the quality and length of the data set.

34.2.4 CURRENT, TEMPERATURE AND SALINITY PROFILING

Detailed profiles of current speed, current direction, water temperature and conductivity (salinity) are usually obtained from a moored vessel using a single direct reading current meter fitted with appropriate additional sensors. Measurements are usually obtained over a typical spring and a typical neap tidal cycle. In UK waters where tides and therefore tidal currents are strongly semi-diurnal (approximately 2 high waters each day) each exercise usually comprises measurements over a 13 h period (thus covering one ebb and one flood tide).

Profiles are usually taken at half hourly intervals. The instrument is lowered through the water column and samples of the current speed, current direction, temperature, conductivity (salinity) and other parameters (e.g. suspended solids or dissolved oxygen) are taken at regular depth increments. The number of depth intervals depends on the depth of water. A minimum of three are required (near surface, mid depth and near sea-bed), but in deeper waters up to ten may be taken. Readings closer together than one metre are difficult to justify.

Limiting conditions

When planning a profiling exercise it is important to ensure that measurements are taken from a stable platform, and that vessel motion is minimized. Motion may be caused by a number of factors. Firstly, the vessel swinging on its anchor; a multi-point anchoring system is recommended to reduce errors. Secondly, wave effects, the most successful profiling exercises are carried out during calm conditions. Waves induce errors both by movement of the current meter resulting from vessel motion and by the direct influence of wave orbital currents. These errors may be reduced by using vector averaging instruments. The maximum wind conditions that are workable depend on both the nature of the site and the type of vessel used. As a guide profiling should not be attempted in stronger winds than Beaufort force 3–4. Finally, the effects of noise on the data are most noticeable during the slack water period and are usually manifested by unstable direction readings and greater than expected current speeds.

Data analysis and presentation

The data from a profiling study are used to determine the extent of density stratification. The density is derived from the temperature, salinity and depth information. The current speed data are used to determine vertical shear and to estimate vertical mixing. In conjunction with the suspended sediment data, current data may also be used to estimate the sediment transport. The results are usually tabulated although graphical presentations showing the distribution of an individual parameter through depth and time or a time series plot at a particular depth may also be prepared. Examples of typical presentations are shown as Figures 34.13 and 34.14.

34.2.5 REMOTE SENSING OF SURFACE CURRENTS

This section describes two recent technological advances that have resulted in dramatic improvements in our ability to understand both the horizontal and the vertical variability of water movement in coastal areas.

Although the two techniques differ in application

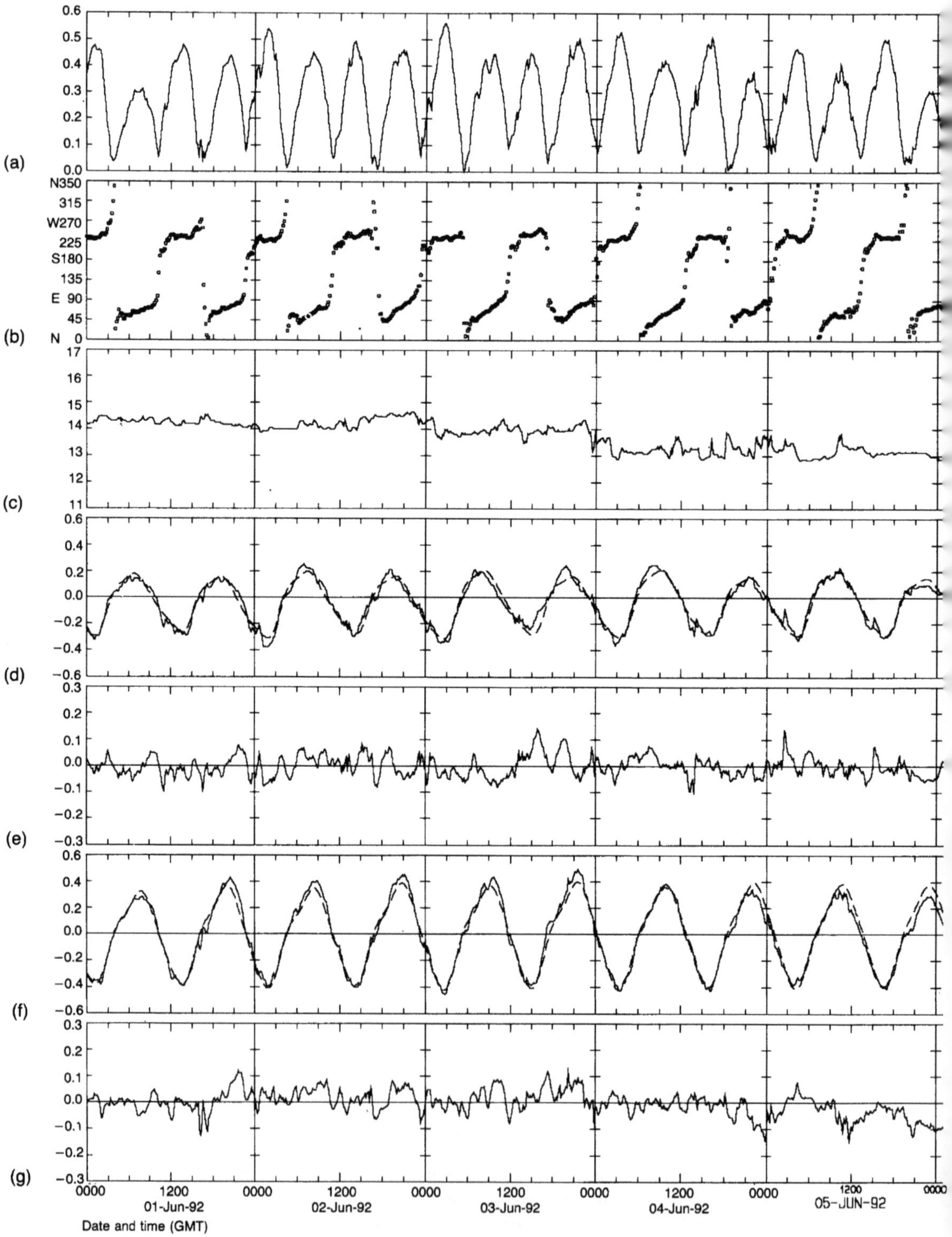

Figure 34.5 Current data times series (a) observed current speed ($m\,s^{-1}$); (b) observed current direction (°true); (c) observed sea temperature (°C); (d) north component of current velocity; (—), observed; (---), predicted ($m\,s^{-1}$); (e) east component of residual current velocity (observed – predicted) ($m\,s^{-1}$); (f) east component of current velocity; (—), observed; (---), predicted $m\,s^{-1}$; (g) west component of residual velocity (observed – predicted) ($m\,s^{-1}$). Current direction is that towards which current is flowing.

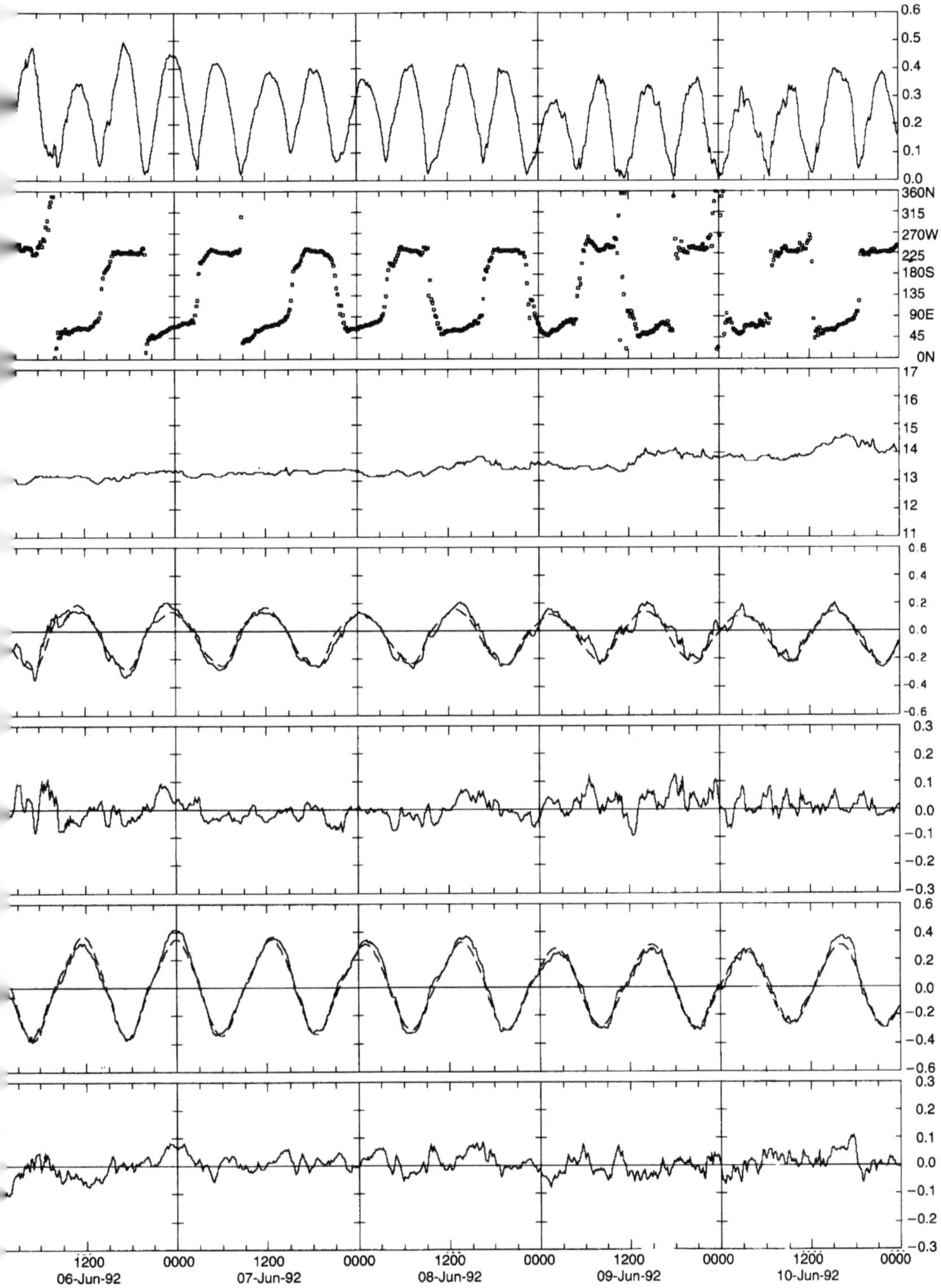
0.6
0.5
0.4
0.3
0.2
0.1
0.0
360N
315
270W
225
180S
135
90E
45
0N
17
16
15
14
13
12
11
0.6
0.4
0.2
0.0
-0.2
-0.4
-0.6
0.3
0.2
0.1
0.0
−0.1
−0.2
−0.3
0.6
0.4
0.2
0.0
−0.2
−0.4
−0.6
0.3
0.2
0.1
0.0
−0.1
−0.2
−0.3
1200
06-Jun-92
0000
1200
07-Jun-92
0000
1200
08-Jun-92
0000
1200
09-Jun-92
0000
1200
10-Jun-92
0000

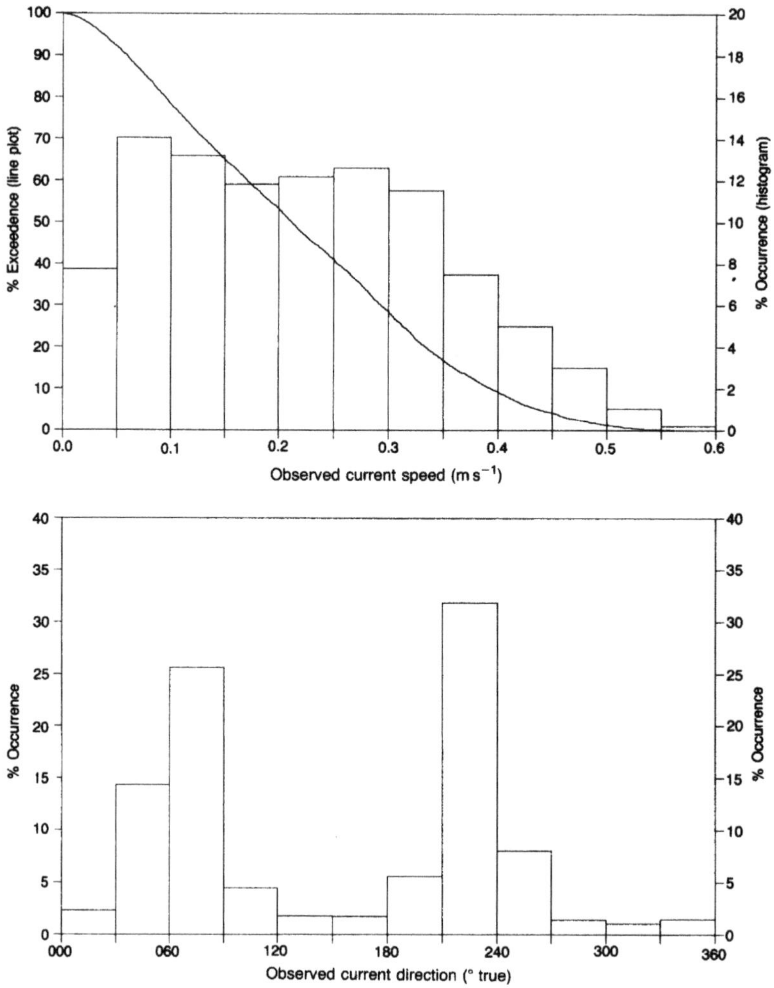

Figure 34.6 Current occurrence/exceedence graph.

they both have central reliance on a very well known physical phenomenon, the Doppler shift.

The first example of a recent innovation is the use of radar techniques to measure surface currents.

When radio waves are reflected from seawaves, having exactly half the radio wavelength and travelling either directly away from or directly towards the radar site, the radio waves constructively interfere to produce a large return signal. Two 'Bragg lines' appear in the Doppler frequency spectrum of the radio echoes which, in the absence of any surface current, are symmetrical about zero Doppler shift. The magnitude of any surface current present in the area of sea from which the radar signal has been reflected causes a displacement of the Bragg lines from their symmetrical position. This displacement is directly proportional to the radial component of the surface current towards or away from the direction of the radar beam (see Figure 34.15 which illustrates the principle of operation of Ocean Surface Current Radar (OSCR)). If the reflected signal from the sea surface is divided into 'time slices' then the radial component can be measured at each of a number of cells out to the limit of the range of the radar.

If two radar sites are positioned on the coast so that transmitted radar beams intersect then, by combining the radial components measured at the intersection of the two beams, it is possible to determine current speed and direction at this point. By use of receiving antennae which allow the signals back-scattered from

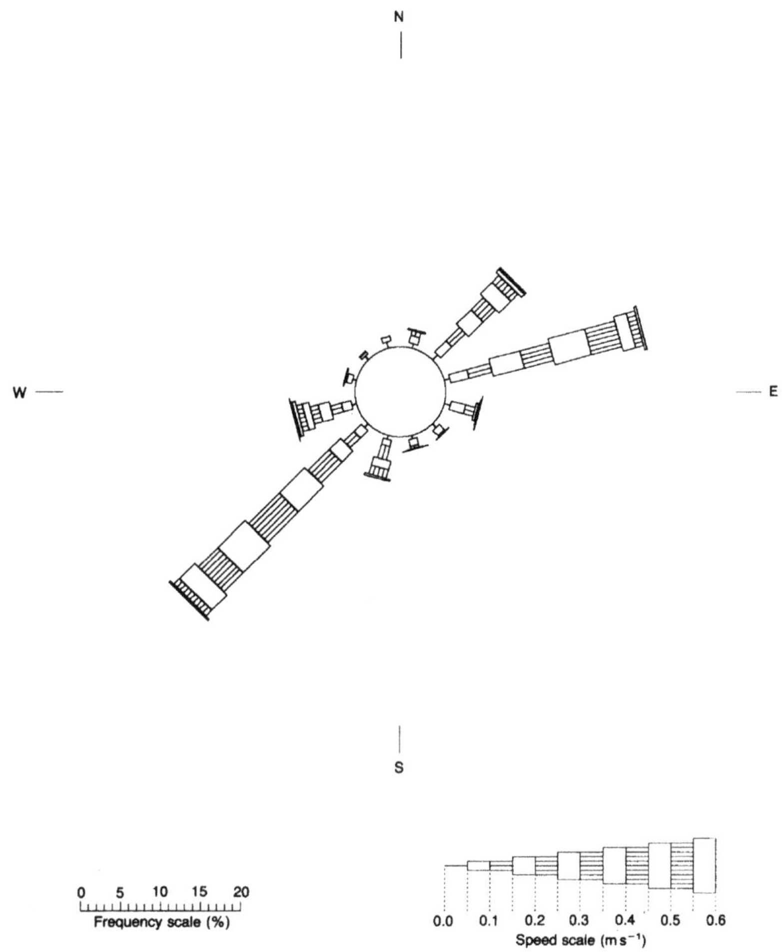

Figure 34.7 Observed current rose.

the sea surface to be acquired from different directions, it is possible to measure current speed and direction over a grid of points on the sea surface.

Figure 34.16 shows the typical configuration and coverage of an OSCR system. Surface currents can be defined at fixed intervals over a regular network of up to 700 cells, giving high resolution coverage over a wide area.

Surface current radars of this type are now in regular use for mapping the spatial variability of currents in coastal areas. Similar principles can be used in measurements of vertical current structure, only in this case the Doppler shift in an acoustic signal is measured.

If an acoustic pulse is directed through seawater the sound reflected by particles and plankton in the water is Doppler shifted by the speed at which the particles are moving. By 'slicing' the reflected signal into time segments a component of a current velocity can be computed for different distances along the beam. If pulses are transmitted from an instrument located on the seabed or mounted on a vessel such that they propagate at angles as shown in Figure 34.17 then an analysis of Doppler shifts in the reflected sound will provide information on current speed and directions at different depths in the water column.

Acoustic Doppler Current Meters (ADCPs) are able to provide a high resolution and accurate profile of current throughout the water column. A moored upward or downward looking ADCP provides information directly comparable with an array of conventional current meters in which an instrument is located within each depth cell.

ADCPs can be operated from a moving vessel and

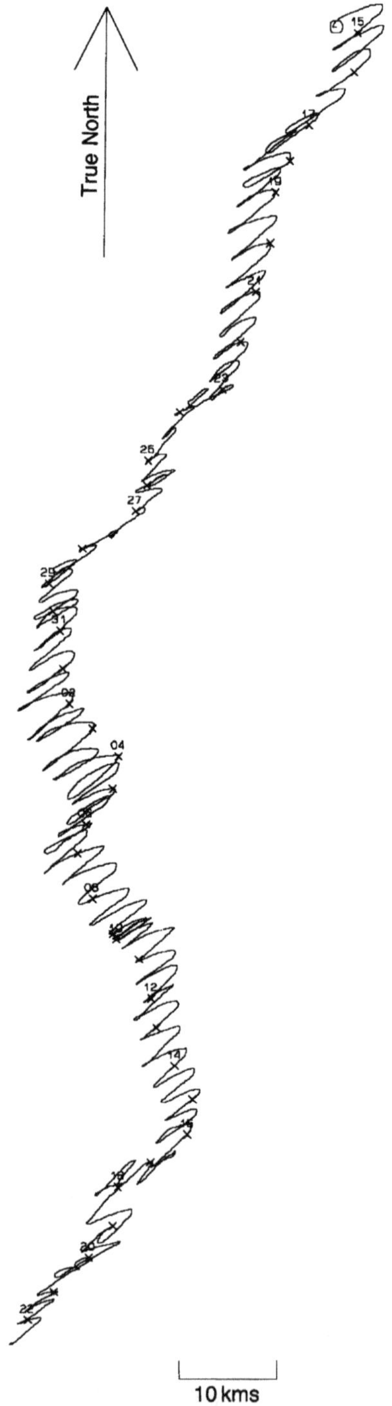

Figure 34.8 Observed current progressive plot. ○, start of analysis period; ×, start of day.

so can be used to measure current structure in vertical sections through the water column. These provide an invaluable insight into variability in pollutant fluxes.

Figure 34.18 shows a typical graphical output of the current velocity field measured over an estuarine transect using an ADCP in the vessel mounted mode. This technique effectively provides 3D cross-sectional current profiles enabling detailed definition of current discontinuities and layering within the water column in addition to direct measurement of total flux through the measurement section.

34.3 DISPERSION MEASUREMENT

Monitoring the dilution and movement of a tracer substance introduced into the water column provides a means of determining the path that may be followed by a coastal discharge and the rate at which dilution is likely to take place. The most frequently used tracers for such investigations are fluorescent dyes. Radioactive and bacteriological tracers may also be employed.

Tracers

Two distinct types of tracer experiment are in common use for the measurement of dispersion: batch release of a discrete slug of dye which is tracked as a distinctly identifiable 'patch'; and continuous release of dye to generate a plume.

Batch release techniques are used to determine the dispersion characteristics of the waters into which a discharge is to be introduced. Dispersion is usually quantified in terms of the dispersion coefficients K_x, K_y and K_z which represent the longitudinal, transverse and vertical components of dispersion, respectively. The values of these coefficients will vary with the size of the dye patch and variation in tide, current, waves and water structure. Because of this variability it is usual to combine the results of a number of releases carried out at different states of the tide and under varying meteorological conditions in order to determine appropriate coefficients for an area.

Continuous releases are used to model the increase in size and decrease in concentration of a discharge plume as it is advected away from a discharge point.

The most commonly used dyes are Fluorescene and Rhodamine. The latter is more widely used and available in a number of variants, including Rhodamine B and Rhodamine WT. Rhodamine B is considerably cheaper than Rhodamine WT, but has the disadvantage that it has a greater tendency to absorb onto sediment and other water-borne particles. It is also considerably more difficult to introduce into the sea. Rhodamine WT was specifically

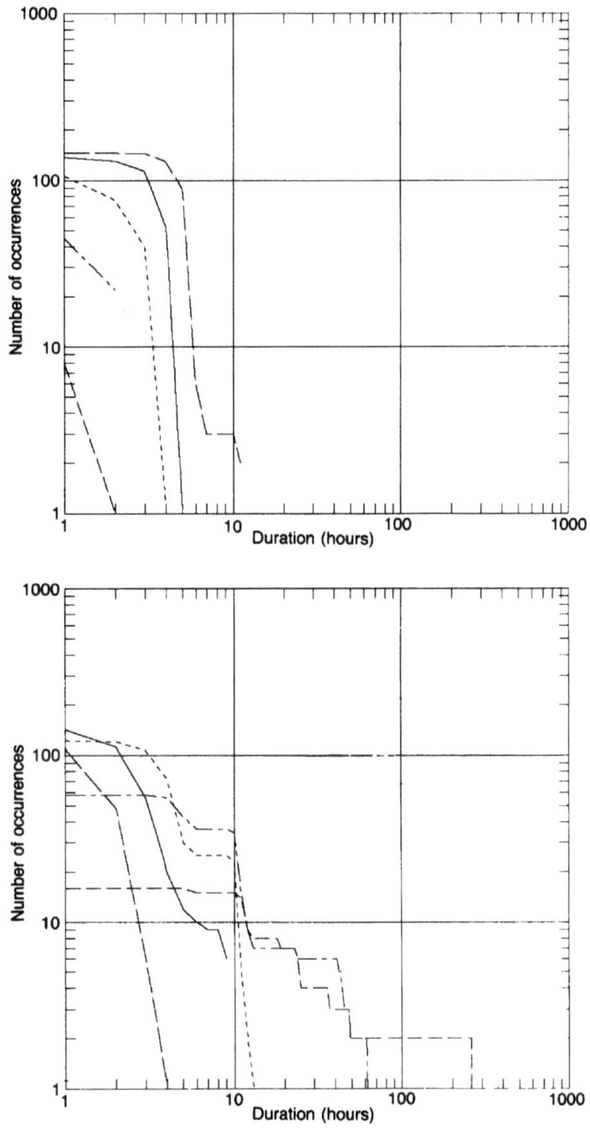

Figure 34.9 Persistence of observed current speed; (———), >0.1 m s^{-1}; (———), >0.2 m s^{-1}; (------), >0.3 m s^{-1}; (–·–·–) >0.4 m s^{-1}; (–––-), >0.5 m s^{-1}.

produced as a water tracing dye, although it still has a tendency to be absorbed. The Rhodamine dyes are red and Florescein green. It should be noted that Rhodamine is toxic at high concentrations and if not used correctly can cause long lasting stains on the skin. Protective clothing should thus always be worn and recognized safety procedures followed.

Fluorescene and Rhodamine fluoresce when exposed to ultraviolet light, emitting light in characteristic bands. It is this property that is used when measuring their concentration by means of a fluorimeter fitted with appropriate filters for the dye in use. Fluorimeters can be classified into two main types.

Firstly, pumped sampling e.g. the Turner Designs Model 10. This is the most widely used type of fluorimeter and has been designed either to analyse individual water samples or the concentrations of

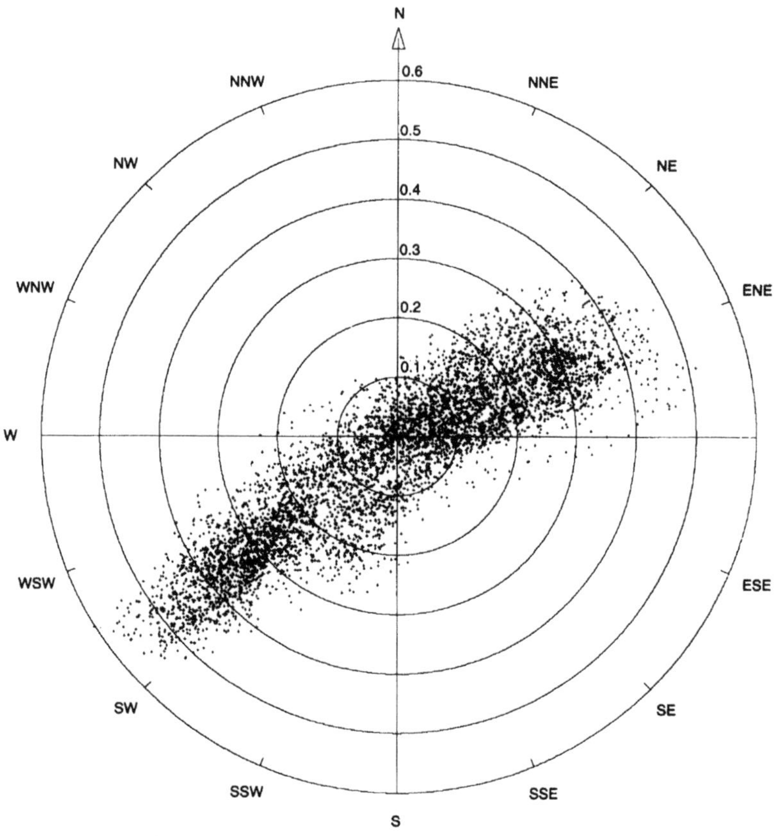

Figure 34.10 Observed current velocity scatter graph.

water pumped through a flow cell. It is this continuous monitoring mode that is employed in measurements of dye dilution. Sea water is continuously pumped from a depth of approximately one metre and passed through the flow cell.

Secondly, non-pumped e.g. the Chelsea Instrument Aquatraka. These instruments tend to be more expensive and have been designed for use in deep water. The electronics and light source are all contained in a pressure housing which is mounted at the required sensing depth.

Fluorimeters require regular calibration using standard solutions covering the range of concentrations to be monitored. When calibrating a pumped sampling system it is essential to calibrate in the pumped sample mode as employed on the survey vessel since calibration solutions will give quite different results if they are not pumped through the flow cell of the instrument under the same conditions as when operationally sampling dye.

When carrying out a dye release it is important to ensure that the dye has the same density as the ambient seawater into which it is to be injected. The density of the dye solution is adjusted to approximately $1.025\,kg\,l^{-1}$ by dilution with methanol. For batch releases this mixture is either poured or pumped into the surface water at the injection site. For continuous releases prepared dye is released into the seawater at a controlled rate prior to injection into the surface water of the site. This ensures that the dye is well mixed with ambient seawater when it is released.

The quantity of dye required for a given release depends on the period over which measurements are required and the local current conditions. Typically 5–10 l of dye concentrate are required to produce a batch release which is traceable for up to eight hours. Continuous releases are carried out with quantities in the range of 1–2 l of dye concentrate per hour.

With both batch and continuous releases dye concentrations are monitored by traversing the dye patch or plume at appropriate time intervals and continuously monitoring the concentration. During con-

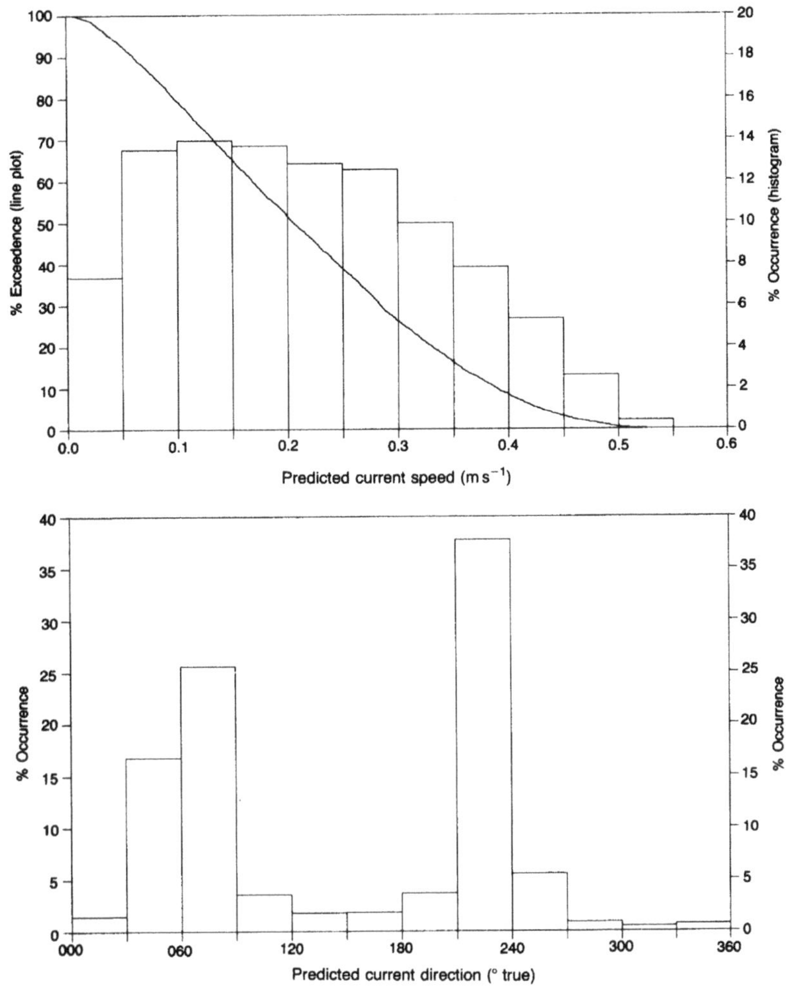

Figure 34.11 Predicted current frequency analysis.

centration measurement regular position fixes are taken and these are correlated with the measured concentrations.

Batch release tracing

Following release of the dye a surface current drogue is usually deployed in the centre of the patch to assist in the tracking process. The patch is then allowed to stabilize and to dilute to a level where continuous concentration measurements are possible (at high concentrations the colour of the dye masks the fluorescence).

Each measurement run commences well away from the patch and is continued until the vessel is some distance past the patch to ensure that background fluorescence values are accurately recorded. Following a sampling run the position of the drogue deployed in the patch is fixed and survey operations are halted for 15–30 min. This break between runs minimizes the disturbance to the patch and simplifies the final data presentations. When the patch is small two orthogonal lines of concentration measurement are sufficient. One of these lines should be parallel to the ambient current direction. As the patch increases in size a greater number of lines are required.

A convenient method of tracking is for the survey vessel to carry out a series of zig-zags at a slight angle to the minor axis of the patch, making sure that background concentrations are recorded at the end of

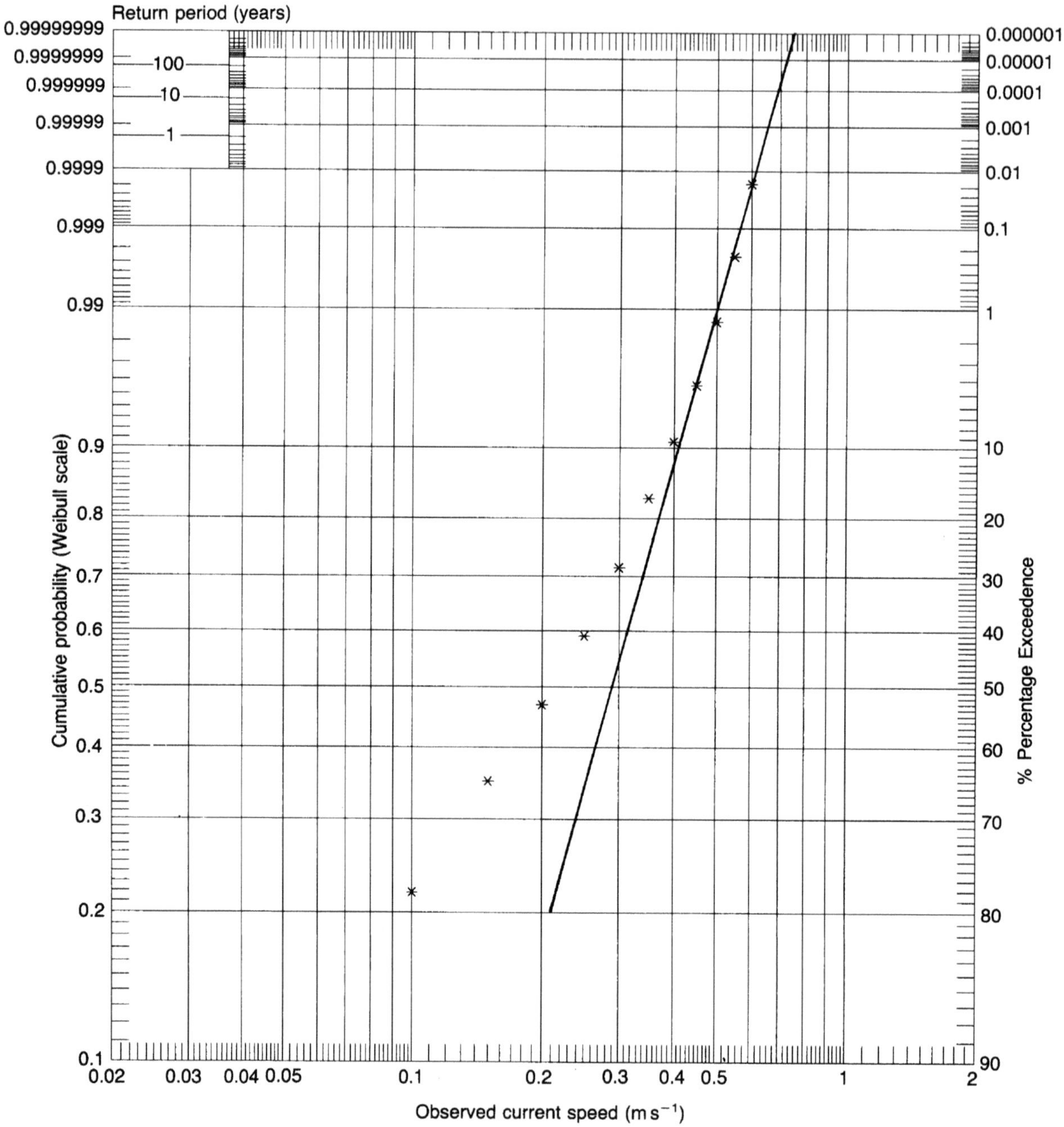

Figure 34.12 Weibull extrapolation of observed current speed.

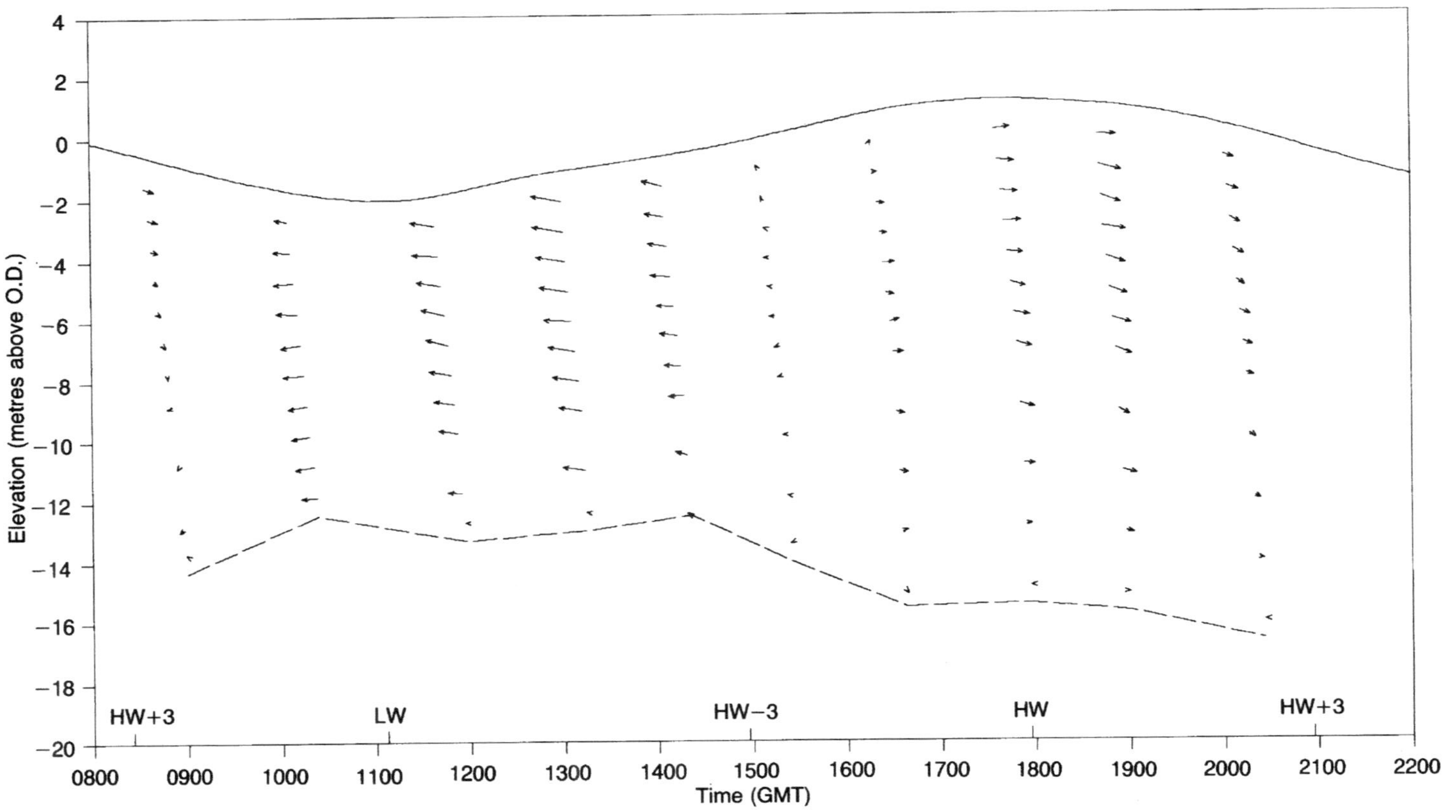

Figure 34.13 Depth/time plot, isometric vector.

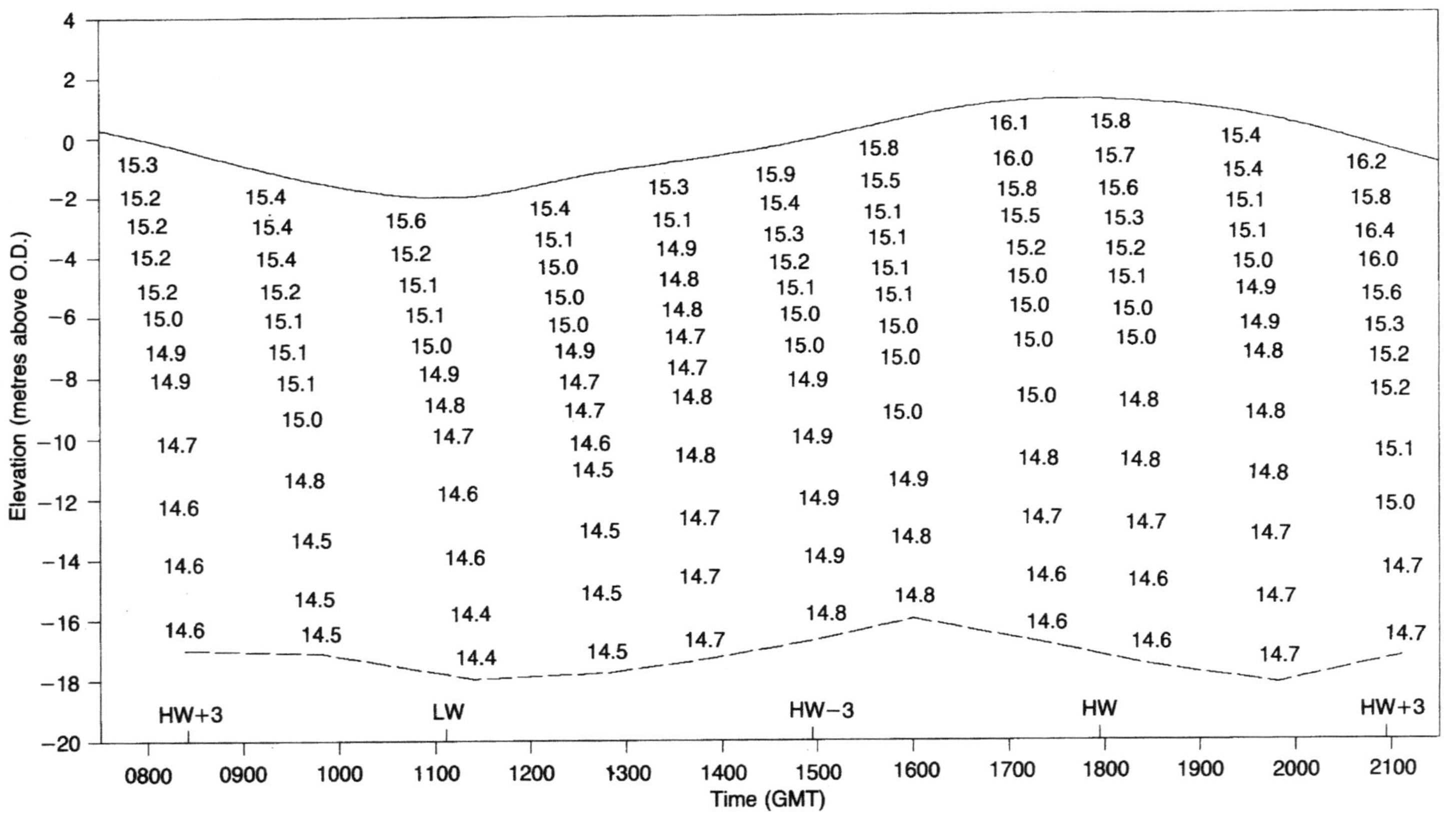

Figure 34.14 Depth/time plot, temperature.

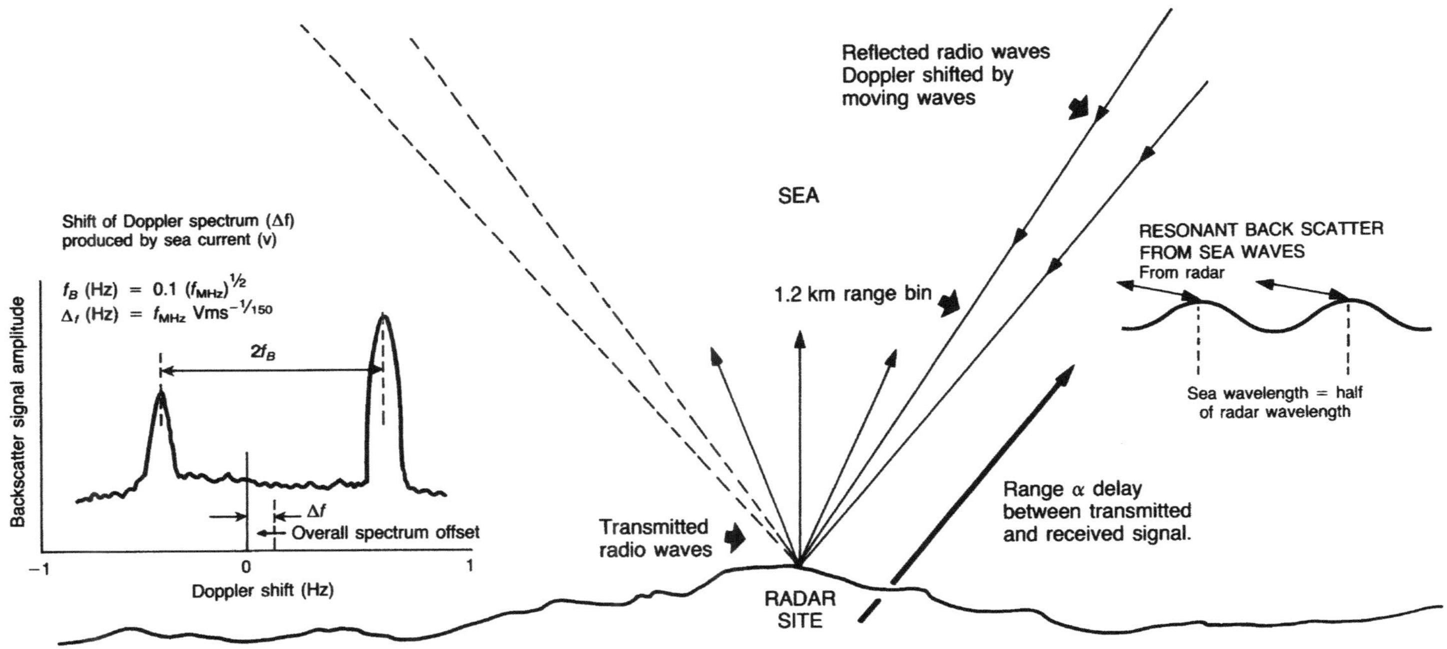

Figure 34.15 Principle of operation for Ocean Surface Current Radar (OSCR).

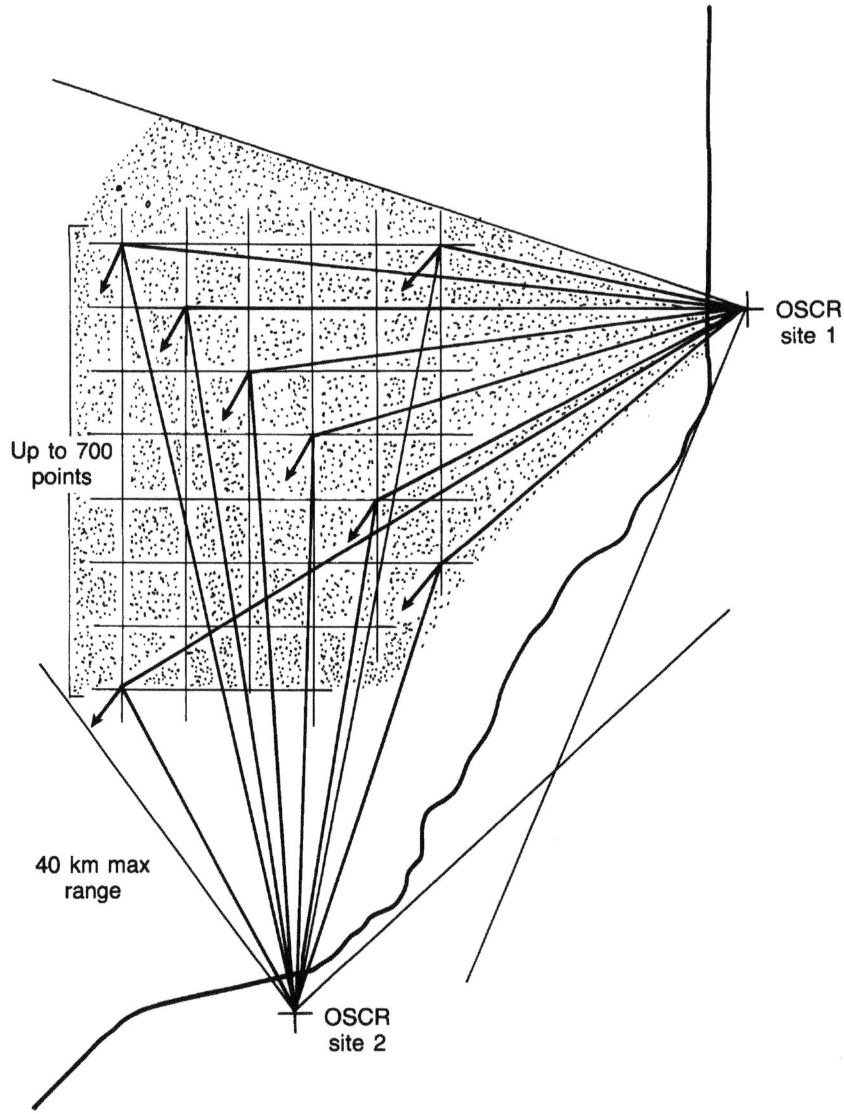

Figure 34.16 Typical configuration and coverage of OSCR system ↙, vector current/measurement points (not all shown); ▨, measurements area; /, receive paths; >, radar coverage.

each line. When the whole patch has been sampled, a final line is run along the major axis through the area of maximum concentration. The total time to survey an individual patch is not normally longer than 30 min. At the patch increases in size the rate of decrease in concentration falls away, and longer periods between successive runs should be allowed.

Tracking continues until the patch cannot be reliably de-lineated in a 30-minute period, or the concentrations have fallen to near background levels. Night time tracking is not recommended since the oceanographer relies to a large extent on visual observation when planning a track through the patch. The maximum wind speed suitable for survey work is dependent on the type of sensor being used and in the case of the pumped system the depth of the inflow pipe. This would usually be about 30 cm below the surface, but may need to be deeper to avoid

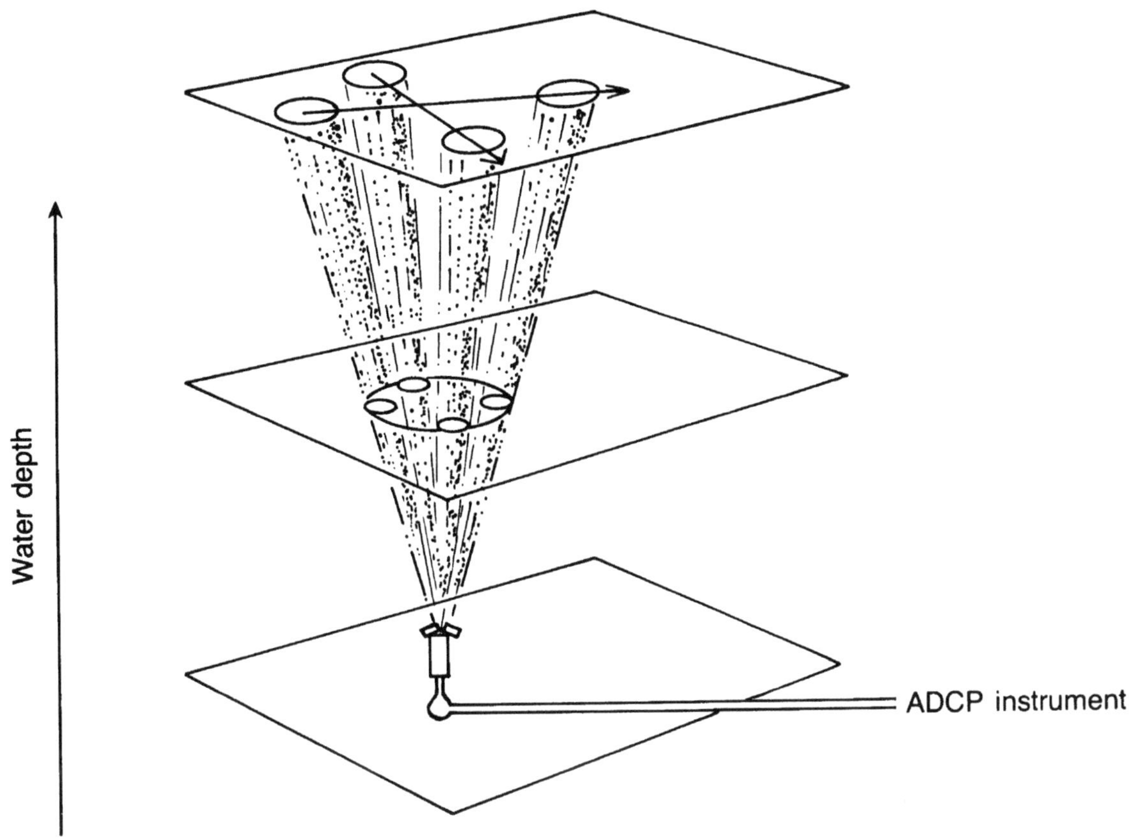

Figure 34.17 Principle of operation for acoustic Doppler current profiling.

aeration in marginal weather conditions. As a general rule dye tracing should not be attempted in conditions exceeding Beaufort force 4–5.

In order to derive the vertical dispersion coefficient (K_z) it is useful to have information on the vertical structure of a dye patch. Such measurements are particularly desirable in areas where there is significant stratification in the water column. Use of several sensors at different depths may be considered but this is very expensive and usually unnecessary. Commonly used methods of collecting vertical data include carrying out a profile using a single sensor at the point of maximum concentration, or collecting a series of small water samples using a string of bottles or syringes.

Continuous release tracing

The methods of measurement used are similar to those described for batch releases. The dye plume is released from a craft at anchor. Dye injection is achieved by introducing concentrated dye into a pumped sea-water flow at a precisely controlled rate (by means of an accurately metered pump). The pumped sea-water flow containing dye injected at a known rate is then allowed to return to the surface waters. The resulting plume of dye is monitored at suitable time intervals (usually 30 min) by carrying out a series of traverses of concentration measurements across the developing plume.

Aerial tracing

More recently work has been carried out on tracking dye patches and plumes from the air. In the simplest form the extent and concentration of the patch is estimated from a photographic print. Dye patches have been photographed from helicopters and aeroplanes and experiments have been carried out using balloons and model aircraft. A more quantitative

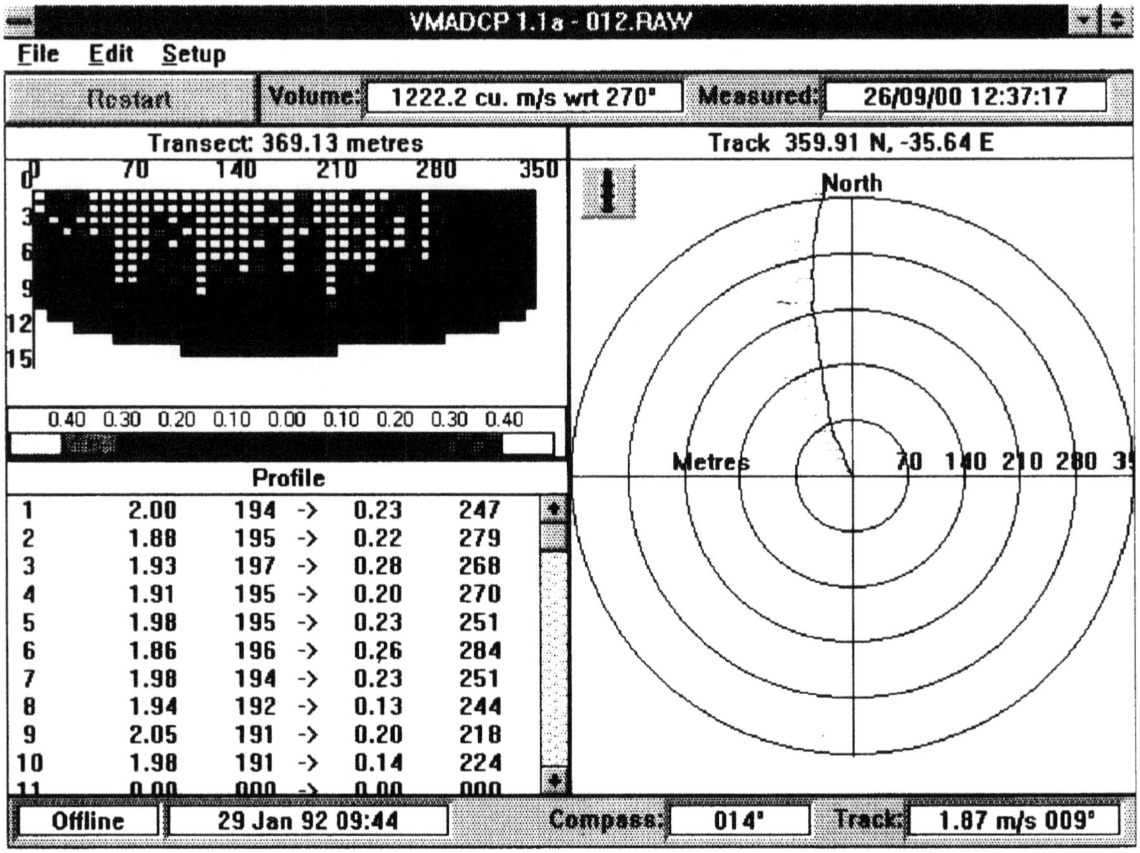

Figure 34.18 Typical presentation of ADCP profiles to show current structure along an estuarine transect.

technique involves the use of a multispectral scanner (Wilkinson and Curran, 1985, Curren, 1987). This technique involves sensing light reflected from the sea surface in a number of discrete spectral bands. Measurements made using this technique must be calibrated by direct measurements of concentration.

34.3.1 ANALYSIS OF DYE CONCENTRATION DATA

Batch releases

Measured raw concentrations are adjusted by reference to the appropriate calibration curve and are plotted at the appropriate positions. Since a finite period has elapsed in sampling concentrations it is necessary to recalculate all the positions to produce a synoptic picture. The position data are adjusted with reference to the observed current speed and direction. Figure 34.19 shows a typical example of the presentation of dye plume information. A number of methods are available to determine dispersion coefficients. Perhaps the simplest is to fit a Gaussian curve to the major and minor axes of each patch. The standard deviations of these curves are used to determine the dispersion coefficients K_x and K_y. Other methods make allowance for the irregular shape of the patches with calculations based on areas inside each concentration contour (Crofts, 1979; Bowden, 1983).

Continuous releases

Raw dye concentrations are calibrated and plotted similarly to the batch release method. If it can be assumed that current speed is constant and in a straight line, the plume can be assumed to be in a steady-state and no positional corrections to observed concentrations are necessary. Further, if dispersion in the vertical can be ignored, dispersion can be considered to occur in one direction only, i.e. transverse

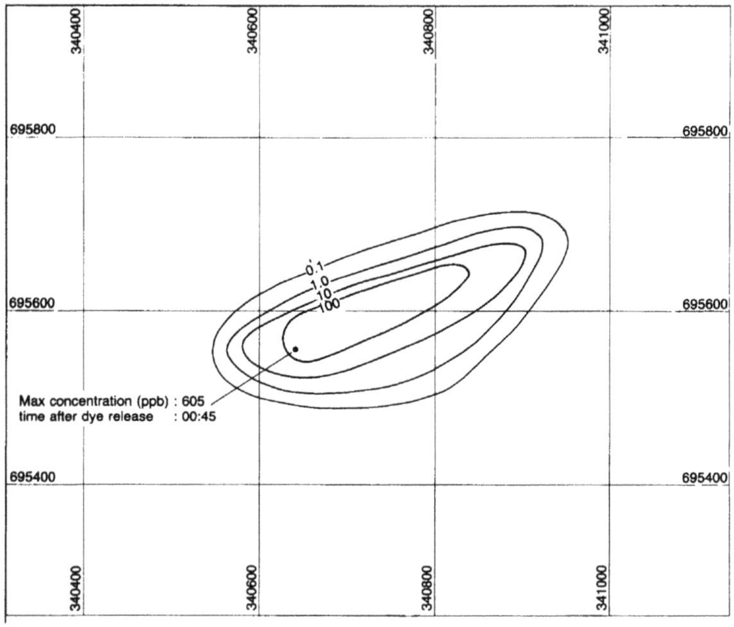

Figure 34.19 Dye track plot. Positions are corrected for current.

to the current (Bowden, 1983). The transverse dispersion coefficient (K_y) can be determined as a function of the cross plume standard deviation, y, which itself is a function of time (or, in steady state, distance from source).

In meandering plumes or when flow reverses, the analysis becomes more complex. Absolute dispersion about the fixed axis in this case is determined by adding the relative dispersion about the plume axis to the standard deviation of the varying centre of the plume (Bowden, 1983).

Radioactive tracers

An alternative to the use of fluorescent dyes is the use of radioactive tracers. Tracking and analysis of the data from both techniques are broadly equivalent, although for radioactive tracers the sampling equipment is smaller and has a greater operating range of concentrations.

The disadvantages of using radioactive tracers include the requirement to employ licenced operators to handle the isotopes and the possible public concern with the introduction of radioactive isotopes into the sea. Isotope tracers are also costly.

The advantages of this technique are that tracking can be continued for considerably longer than is possible with dyes, and that monitoring can begin as soon as the isotope is injected. The small size and cost of the sensor also means that monitoring at 3 or 4 depths simultaneously is feasible.

Microbiological tracers

Micro-organisms can be used to trace water movement since they are not capable of independent movement and therefore only reflect the dispersive capacity of the receiving waters. This technique is commonly employed in feasibility studies for long sea outfalls.

The most commonly used micro-organisms are bacteria. Two bacterial states are commonly used depending upon the specific objectives of the study and the local water and geographical conditions.

Bacterial vegetative cells

A limited number of bacterial species have been used for tracer studies. It is important to use species not naturally occurring in seawater. The antibiotic resistant mutant of *Serratia marcescens* is one of the more popular strains. Sampling is carried out to establish background levels of cells prior to release. Typically, large volumes of cells (of the order of 200 l) are cultured on-site and released as a batch at a designated tidal state. A small amount of dye and a float are used to mark the bacterial patch to facilitate relocation for subsequent water sampling runs.

Before release, samples of the batch are taken and cultured in order to derive a value for the initial numbers of bacteria in the release.

Two approaches to offshore sampling have been used. Firstly, for each release four to five hours of sampling at hourly intervals and at random positions over the projected area of the patch may be conducted. Secondly, sampling for a minimum of eight hours at known positions may be carried out on day 1 followed by sampling over two large grids (to define the patch at high and low water) parallel to the prevailing current direction on day 2. For both approaches onshore samples are collected from beaches of interest during the 48 h following the release.

Continuous release rather than batch release of bacteria can be adopted. However, since there is then a period of time elapsing between the beginning and end of the release, when the bacterial cells may be dying off at some unknown rate. Data derived from this release method can only be used for dilution and dispersion studies and can give no real indication of die-off. The water samples are tested by standard techniques (Anon, 1969), involving filtration incubation colony counting.

Evaluation of the decay rate from the numbers of bacterial colonies requires ancillary data on the impinging solar radiation, water temperature, turbidity and the wind speed and direction during the bacterial release and subsequent sampling programme.

The advantages of using bacterial vegetative cells are the provision of direct counts of viable cells reaching beaches and the indication (assuming the availability of ancillary data) of the rate of die-off of the cells in the form of T90 values. The application and derivation of T90 values has been the subject of much study and discussion (Gameson, 1986).

The disadvantages of this tracing technique are that once the culturing process has begun then there is approximately a three-day period from the time when suitable cell numbers have been cultivated until the time of release, and there are certain potential health risks involved in the handling of these cells. The samples must be processed as soon after collection as is possible, necessitating good logistic planning.

Bacterial spores are used in a similar way as the vegetative cells. *Bacillus globigii* is a commonly used species. Whereas vegetative cells can be released as a batch or as a continuous discharge, spores are typically released as a continuous discharge since they are not subject to decay or die-off. They can be collected and recovered from sea-water using similar methods as used for vegetative cells.

The main advantage of using spores is the almost indefinite storage times that can be adopted before processing, thus enabling very large numbers of samples to be collected without overloading the laboratory facilities. The spores of different bacterial species have different, and often discrete, tolerances to heat. This variable thermodurability of spores enables the spores of interest to be selectively isolated from indigenous spores relatively easily. The cost of culturing large numbers of bacteria on site is avoided. The number of spores per sample provide a direct measure of the rate of dispersion and requires no correction for mortality.

The main disadvantage associated with the use of spores is their conservative nature and the relatively long periods of time that are required between consecutive tracing exercises in order to allow the spores to disperse and thus not affect any subsequent release. Of course, this is not a problem if a continuous release method is adopted. Since the spores are quiescent no die-off rates can be calculated.

34.4 TIDAL MEASUREMENTS

There are a number of field proven instruments available and these fall into four categories.

1. Electrical pressure sensors. There are two types, those that measure absolute pressure (i.e. including atmospheric pressure) and those that measure differential pressure between the head of water and atmospheric pressure.
2. 'Bubbler type' sensors. Air, from a constant supply, is introduced to a small sub-sea reservoir where it escapes through a small hole. The variation in pressure at this reservoir due to the variation in water level is then recorded.
3. 'Stilling well' type. A float is suspended within a tube open to the atmosphere and to the sea. The up and down movement of the float is then transferred to a recorder.
4. The tide board. A graduated board which is read by an observer.

Associated measurements

Accurate tidal analysis depends on a number of factors. Atmospheric pressure measurements are essential for absolute pressure measuring devices and desirable for all tidal measurements. The atmospheric pressure measurements should be made as geographically close as possible to the tidal measurements. Water density is also an important factor. The density of sea-water is affected primarily by temperature and salinity. The degree of the effect depends on a number of factors but its magnitude can be as much as several centimetres. Wind records are desirable to assist in interpreting the records, particularly for examining surge effects in long-term tidal records. Tidal readings should always be related to a datum, preferably a national datum such as Ordnance Datum (Newlyn).

Timing

Accurate timing is essential for any tidal work. This is particularly vital if the data are to be used for any predictive work where timing should be checked to the nearest second.

Installation methods

Tide gauges are often attached to fixed structures such as jetties or piers. If a structure is not available then the tide gauge has to be deployed in a frame which can sit on the sea-bed. If the instrument is deployed on a mooring supported by a sub-surface float then the tide gauge should be as near the sea-bed as possible and the mooring as short as possible. Care must be taken to select a site at which minimal settlement of the mooring will occur.

Sampling periods and locations

Various sampling regimes are possible depending on the use to be made of the information. A minimum requirement would be a two minute average once every hour. More usual is the recording of a two minute average every ten minutes. Data should be gathered for at least one lunar month (29.5 days) in order to establish the principal tidal components.

Quality control

To maximize the data quality the points noted above must be followed. In addition, calibrations and checks on instrument performance should be carried out on a regular basis. During field operations, checks should be made against other instruments and account taken of the other factors affecting the data. Pre- and post-deployment calibrations should be performed in accordance with the manufacturer's recommendations.

Service visits

Analogue chart recorders should be checked as frequently as possible, normally at least daily. The frequency of service visits for other types of instruments depends on site conditions. Usually for self-recording instruments a one month unattended period is quite reasonable.

Data analysis and presentation

Tidal data is usually presented as a time series plot together with the recorded mean and extreme values. A harmonic analysis is often carried out to derive the various astronomic constituents and the residual which is the sum of all the non-astronomic effects such as wind and atmospheric pressure variation. Predictions for the harmonic portion of the tide are reasonably straightforward and give reliable results. Predictions of residual extremes can also be made providing at least a year of observed data are available. An example of the presentation of tidal data is given in Figure 34.20. Detailed information on the collection, analysis and presentation of tidal data has been given by Pugh (1987).

34.5 WAVE MEASUREMENT

Offshore surface moored buoys

These are of two types. The first makes use of an accelerometer to measure the acceleration of the buoy as it follows the wave profile. The output from this is double integrated to determine elevation which is continuously transmitted as a frequency modulated signal to a shore station. The range of this system is of the order 30–50 km and in the UK the buoys would usually operate in the 35 MHz waveband. Data may also be internally recorded. The second category of moored buoy system is capable of determining directional wave spectra. These instruments again make use of a vertical accelerometer to sense heave. Pitch and roll are also measured in order to derive directional sea-state data. Data may be telemetered to shore, logged on board or transmitted as computed spectra by satellite telemetry.

Pressure bases systems

The use of a high precision pressure transducer may be appropriate in particular situations. The instrument may be deployed at sea, recording data internally or cabled to shore. In the former it is necessary to replace the data tapes at regular intervals, usually monthly. In the latter the long lengths of cable often required are expensive and subject to damage. The use of a pressure transducer is only recommended in shallow water, since the pressure fluctuations due to waves are attenuated with depth. This effect is particularly severe for short period waves, and it is necessary to apply a transfer function to the derived spectra before calculating the wave parameters.

Wave staffs

A capacitive or resistance sensor may be installed on a suitable structure and the observed wave heights recorded. These instruments measure the change in resistance or capacitance between two conductors

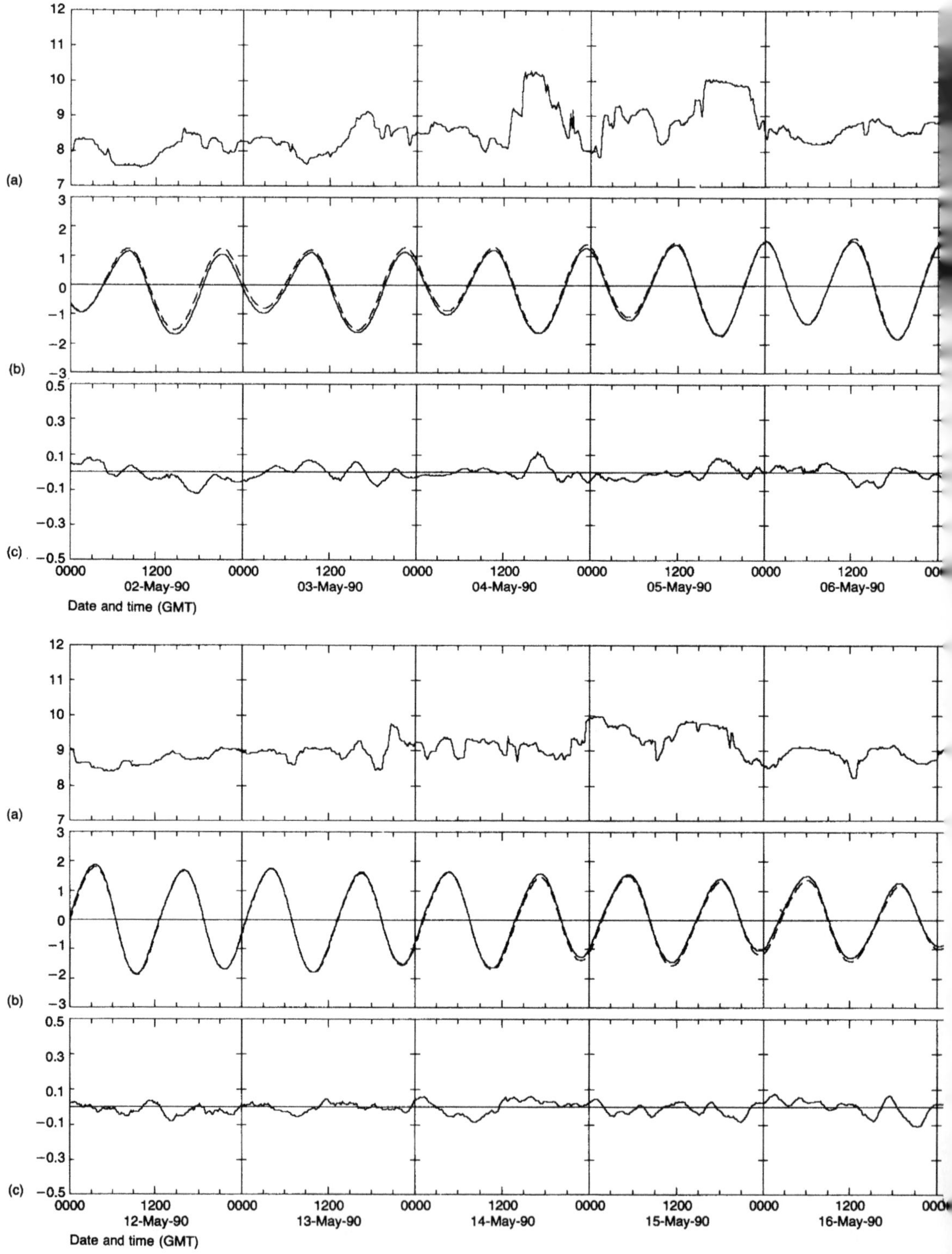

Figure 34.20 Tidal data time series over 20 days (a) observed sea temperature (°C); (b) tidal elevation (metres above MSL; —, observed; ----, filtered predicted); (c) filtered residual tidal elevation (metres above MSL).

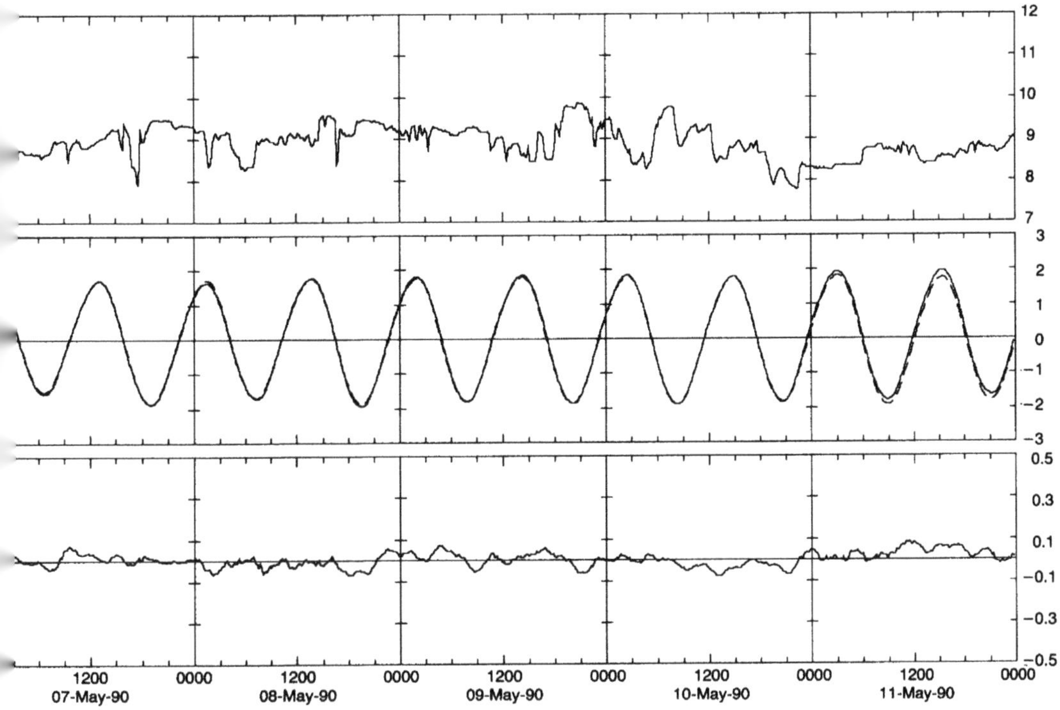
12
11
10
9
8
7
3
2
1
0
−1
−2
−3
0.5
0.3
0.1
−0.1
−0.3
−0.5
1200
07-May-90
0000
1200
08-May-90
0000
1200
09-May-90
0000
1200
10-May-90
0000
1200
11-May-90
0000

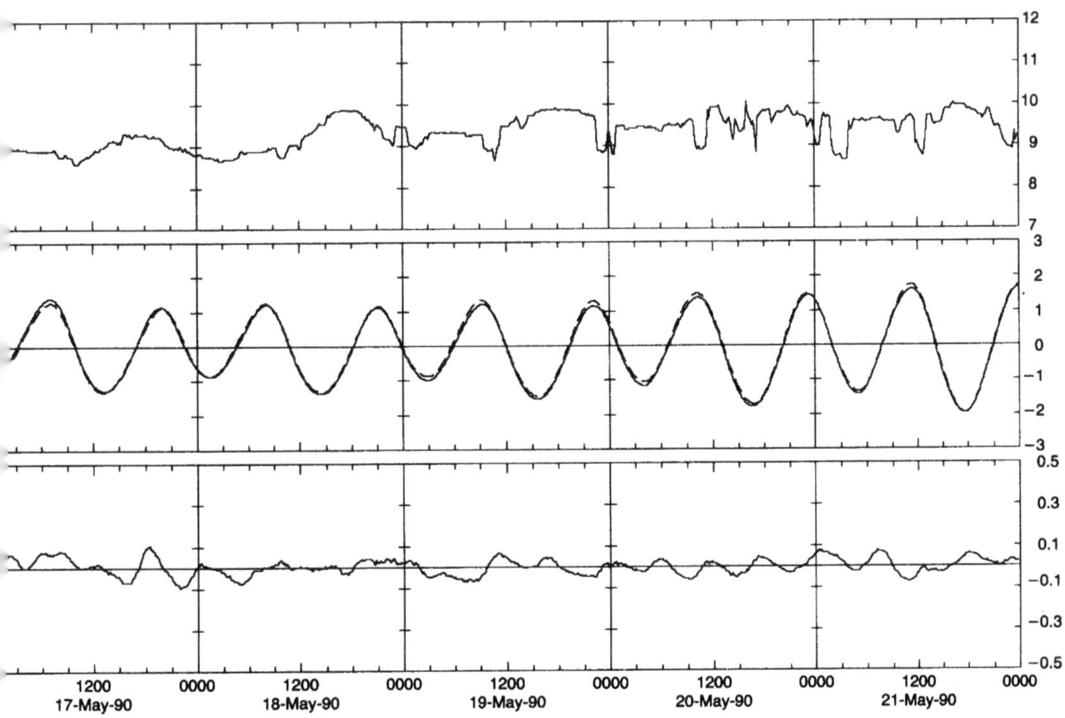
12
11
10
9
8
7
3
2
1
0
−1
−2
−3
0.5
0.3
0.1
−0.1
−0.3
−0.5
1200
17-May-90
0000
1200
18-May-90
0000
1200
19-May-90
0000
1200
20-May-90
0000
1200
21-May-90
0000

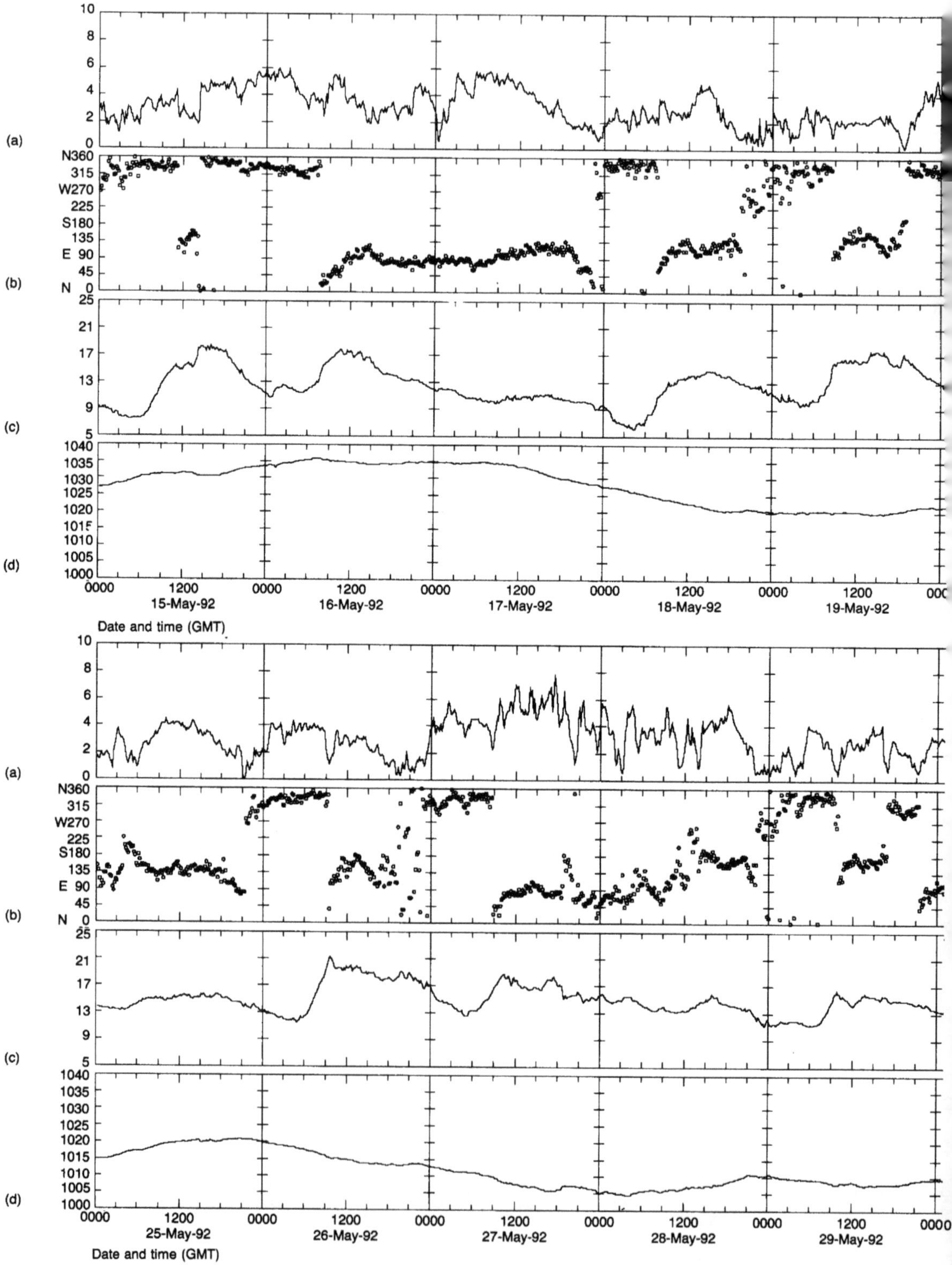

Figure 34.21 Meteorological data time series over 20 days (a) wind speed (m s^{-1}); (b) wind direction (°true); (c) air temperature (°C); (d) atmospheric pressure (millibars). Wind direction is that from which wind is blowing.

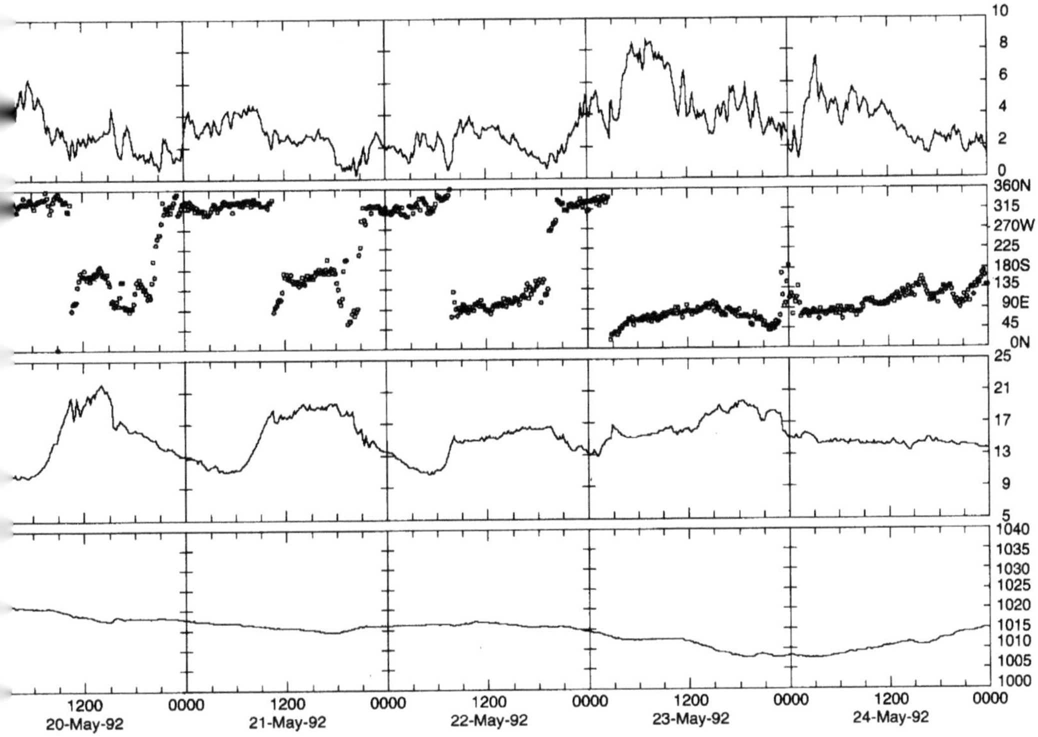
10
8
6
4
2
0
360N
315
270W
225
180S
135
90E
45
0N
25
21
17
13
9
5
1040
1035
1030
1025
1020
1015
1010
1005
1000
1200 0000 1200 0000 1200 0000 1200 0000 1200 0000
20-May-92
21-May-92
22-May-92
23-May-92
24-May-92

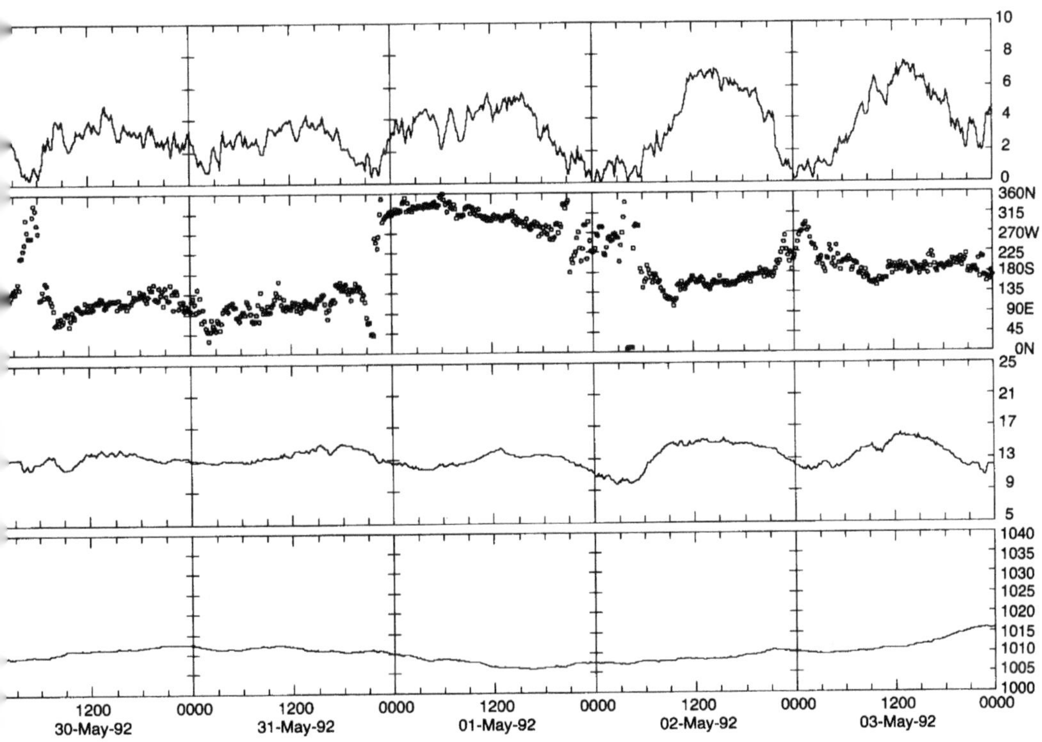
10
8
6
4
2
0
360N
315
270W
225
180S
135
90E
45
0N
25
21
17
13
9
5
1040
1035
1030
1025
1020
1015
1010
1005
1000
1200 0000 1200 0000 1200 0000 1200 0000 1200 0000
30-May-92
31-May-92
01-May-92
02-May-92
03-May-92

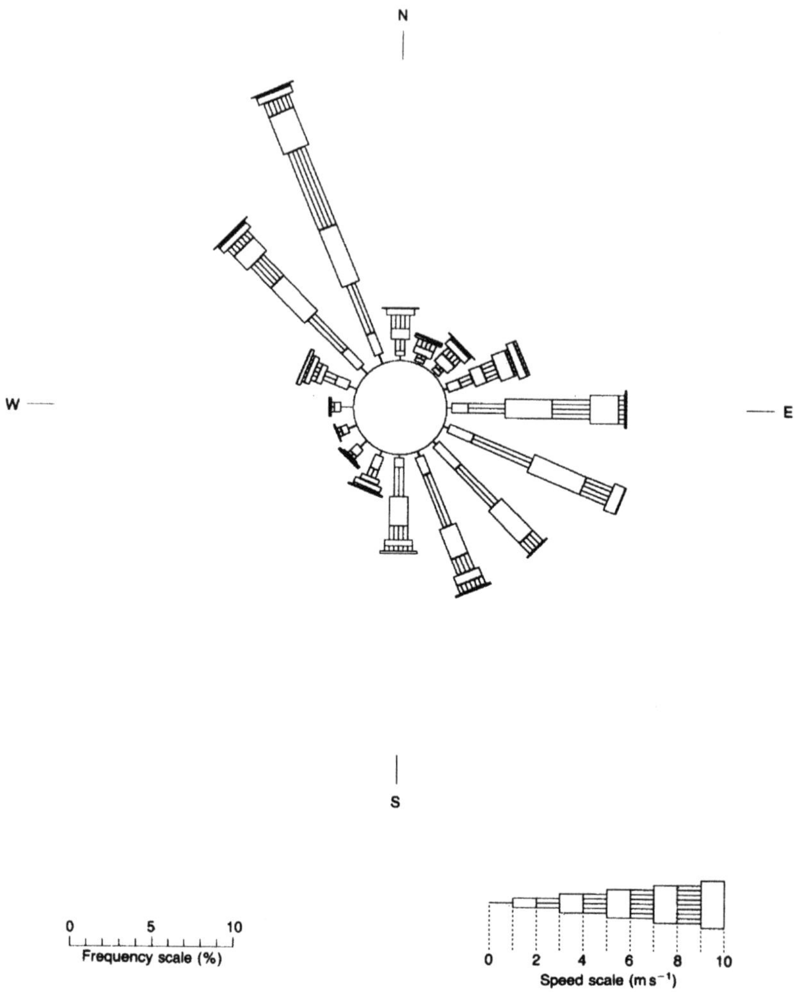

Figure 34.22 Wind rose.

mounted across the air–sea interface. The main disadvantage of this system is the requirement to mount the sensor on an existing structure. In many areas suitable structures do not exist and in areas where a suitable structure does exist, it is usually the case that the structure will influence the wave climate.

Radar

Recent innovations have made it possible to accurately measure the wave climate remotely using radar. The method is expensive to operate over a long period but provides the capability of rapidly assessing regional variations in wave climate, and in particular variations between offshore and close to shore conditions. Systems currently being developed offer the facility to determine directional wave spectra (Wyatt, 1985).

Data analysis and presentation

Data are usually logged in blocks of typically 20 minutes data every three hours. Analysis is conducted using spectral techniques. Wave data are usually presented as time histories of wave statistics (e.g. significant wave height, zero crossing period, etc.) or as spectra for particular time periods. Further more detailed information on wave data analysis are given in the *Shore Protection Manual* (US Army Corps of Engineers, 1984).

Wind direction (°true)											% Occurence
000	12	39	93	91	125	63	18	12	1		9.5
	36	72	157	199	203	93	35	22	8	3	17.3
315	23	55	91	62	39	44	21	20	5		7.5
	30	27	22	5	6	3		2			2.0
270	38	24	8	4	1						1.6
	24	23	10	3	1	2	1				1.3
225	20	37	17	6	3	9	5				2.0
	14	33	42	36	16	15	16	8			3.8
180	8	32	122	112	48	27	14	3			7.7
	20	79	187	92	50	13	18	5			9.7
135	17	89	150	150	60	12					10.0
	10	59	166	161	93	47	10				11.4
090	15	44	68	89	95	90	19	17	4		9.2
	10	24	20	30	13	22	8	11	6		3.0
045	7	12	6	25	33	12	4	1			2.1
	12	9	13	24	20	11	2				1.9
000											
% Occurrence	6.2	13.8	24.5	22.8	16.9	9.7	3.6	2.1	0.5	0.1	Number of samples
% Exceedence	100.0	93.8	80.1	55.6	32.8	15.9	6.3	2.7	0.6	0.1	4783
Wind speed ($m\ s^{-1}$)	0		2		4		6		8		10

Figure 34.23 Wind occurrence table. Table entries are actual number of occurrences.

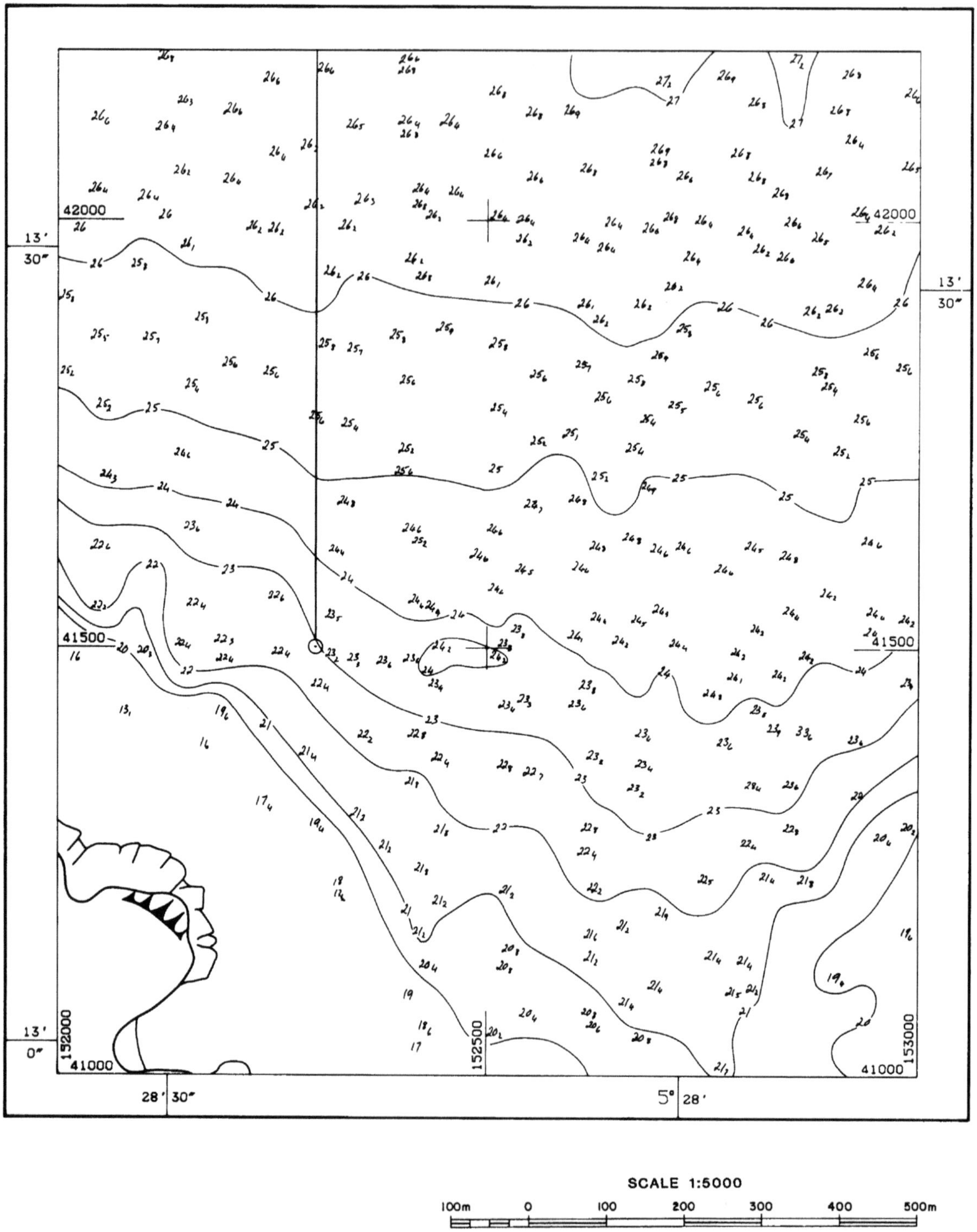

Figure 34.24 Bathymetry. Contours shown at 1 m intervals.

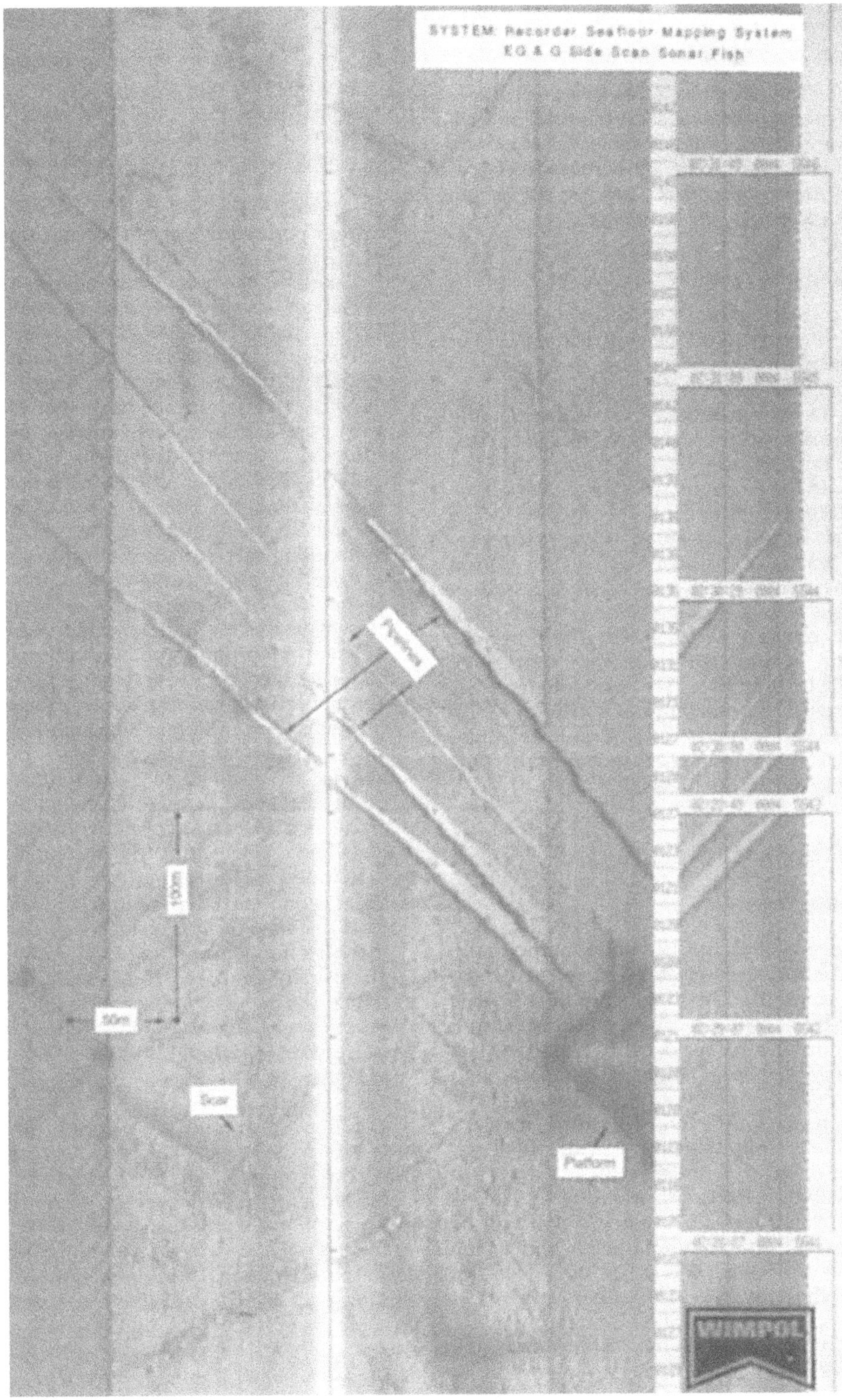

Figure 34.25 A side scan sonar record.

34.6 METEOROLOGICAL MEASUREMENT

Local meteorological measurements are particularly useful for interpreting other measurements. The location for a station must be carefully selected to ensure that a representative wind climate is measured (e.g. care should be taken to ensure that the site is not sheltered from particular wind directions).

Data analysis and presentation

All wind and pressure data must be corrected to be representative of a standard vertical elevation. For winds this is normally 10 m above mean sea-level and for pressure mean sea-level.

Analysis of wind speed and direction data is mainly statistical. Information is presented by average month, season and year. This may be in the form of wind roses or presented numerically as percentage occurrence by direction for each wind speed band. More detailed analysis includes calculation of persistence. The complete time series is analysed and the period of time that the wind speed exceeds or is less than certain threshold values is determined on each occasion the speed crosses that threshold. The durations are ranked and cumulative frequency curves are plotted for all thresholds. This plot enables the engineer to assess the probability of a calm or storm period lasting for a particular time period. Examples of the presentation of wind speed and direction data are shown in Figures 34.21 to 34.23.

34.7 BATHYMETRY AND GEOPHYSICS

34.7.1 ECHO SOUNDING

Echo sounding is used to produce bathymetric charts which show variation in sea-bed topography as soundings or depth contours. The echo sounder operates by transmitting an acoustic pulse and measuring the time delay in the received echo. This delay is termed the two-way travel time. The variation in sophistication of the available range of instruments is substantial, from simple non-recording displays to multi-frequency equipment fitted with digitizers. A dual frequency echo sounder has the advantage that echoes will be received from different levels in the sea-bed. The higher frequency signal will be reflected at the sediment/water interface, whereas the lower frequency will penetrate soft muds. This has the advantage that the type of sea-bed and thicknesses of very soft muds can often be determined.

The echo sounder measures the distance from the transducer, which is attached to the survey vessel, to the sea-bed. A number of corrections are applied before the depth can be accurately determined.

1. Speed of sound in water. In deeper water, this is usually determined directly using a purpose built instrument or would be calculated from temperature and salinity data. In shallow water (typically less than 30 m) a more direct calibration using a bar or disk check is used. This must be carried out at least daily and during calm conditions in a water depth equivalent to the maximum expected in the proposed sounding area.

   A metallic bar/disk is lowered from the stationary vessel on a calibrated wire or chain, until it is a short distance below the transducer and at a convenient depth below the surface. The instrument is adjusted so that the output is equivalent to the observed depth. The bar is then lowered to the maximum depth expected and the speed of sound adjusted until the correct depth is output. A series of check readings are then carried out as the bar is raised. The bar is often covered with plastic foam to increase the strength of the echo.
2. Meteorological conditions. Sounding is best carried out during calm conditions, since any movement of the vessel will affect the water depth recorded. However, it is rarely practical to restrict surveying to ideal conditions and it is often necessary to make allowance for vessel movement. This may be automatic, using complex and expensive heave compensation systems, although for inshore surveys it is normal practice to make the corrections graphically on the echo sounder trace. The maximum sea state during which sounding can be carried out successfully depends on the site, the depth of water and the accuracy requirement demanded by the survey specification.
3. Tidal reductions. The echo sounder measures the total distance from the surface to the seabed; this includes the height of the tide. Bathymetric charts are usually prepared to a standard datum (Ordnance Datum Newlyn, or local Chart Datum), so when preparing the final chart it is necessary to remove the tidal influence by reference to measured local tides.

To reduce costs and vessel time, the bathymetric survey is usually carried out concurrently with the side scan sonar survey and sub-bottom profiling, making use of the same vessel. Typical survey line spacing for a coastal survey would be between 50 and 100 m. The results of the sounding are usually prepared following the completion of the survey, although the use of on board computers makes it possible to prepare preliminary charts in real time. An example of a bathymetric chart is given in Figure 34.24.

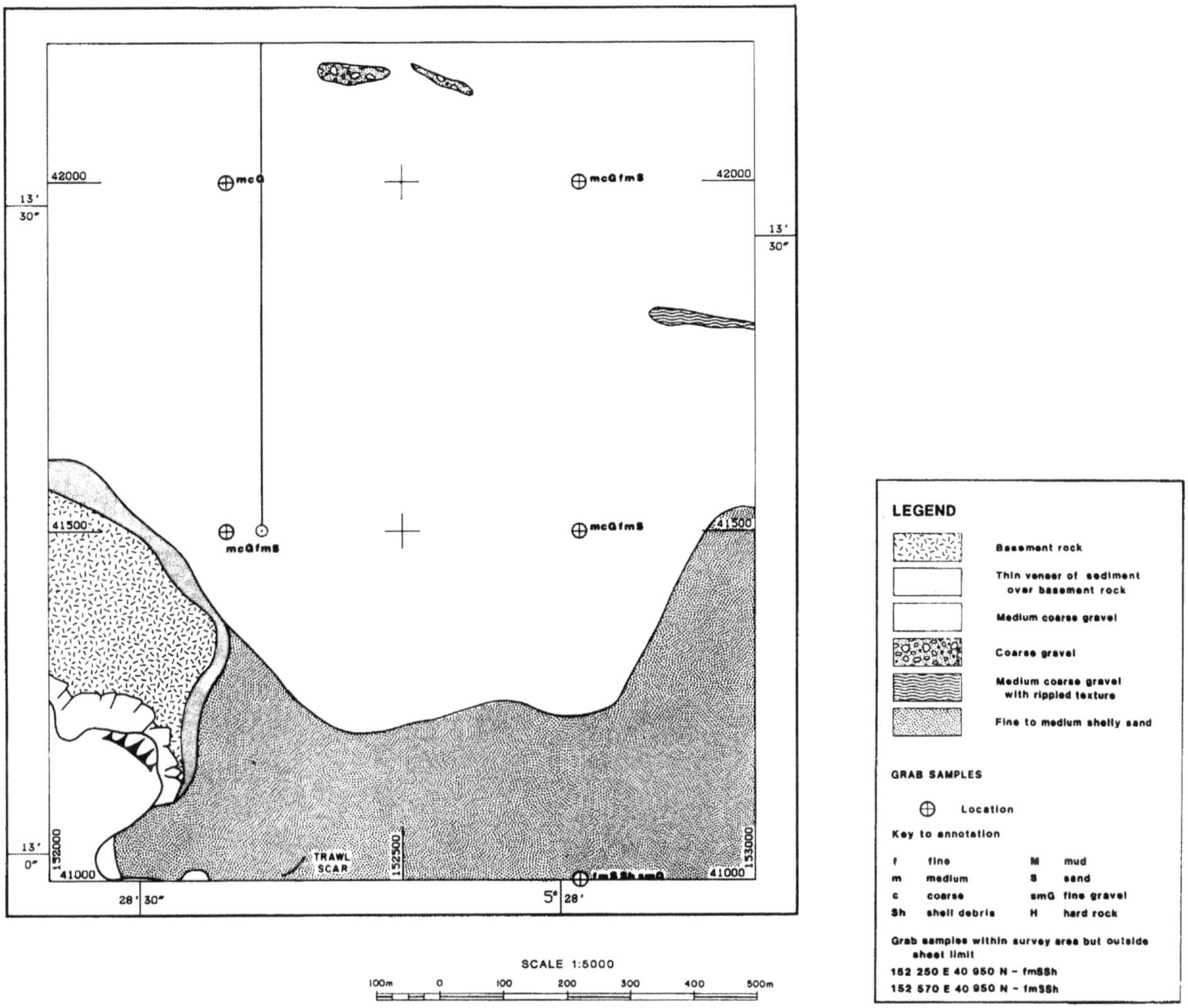

Figure 34.26 Sea-bed features.

34.7.2 SIDE-SCAN SONAR

Side-scan sonar is a powerful tool for inspecting the sea-bed for obstructions and mapping the distribution of sediments and features. Sonographs may also be used as a coarse tool to estimate the sediment transport regime of an area. The side-scan sonar may be hull mounted or, more usually, towed behind the survey vessel. Towed fish are usually dual beam, collecting data both sides of the fish. Each beam typically has a narrow horizontal beam angle of less than 2° and a wide vertical beam angle, usually greater than 20°.

Sonographs are collected by presenting the echoes detected from a transmitted pulse on a recorder as the survey vessel moves along a chosen survey line. The greater the delay between the transmitted pulse and the received echo, the greater the distance to the reflector. These distances, from the transducer to the reflector are called slant ranges and it is these that are presented on the recorder. Good reflectors, for example rough surfaces or obstructions, appear as dark patches on the trace and poor reflectors such as smooth surfaces, as white patches. The trace is thus similar to a negative image of the picture that would be seen by the human eye if a light was shone across the sea-bed. An example is presented as Figure 34.25.

Preliminary interpretation of the data is usually

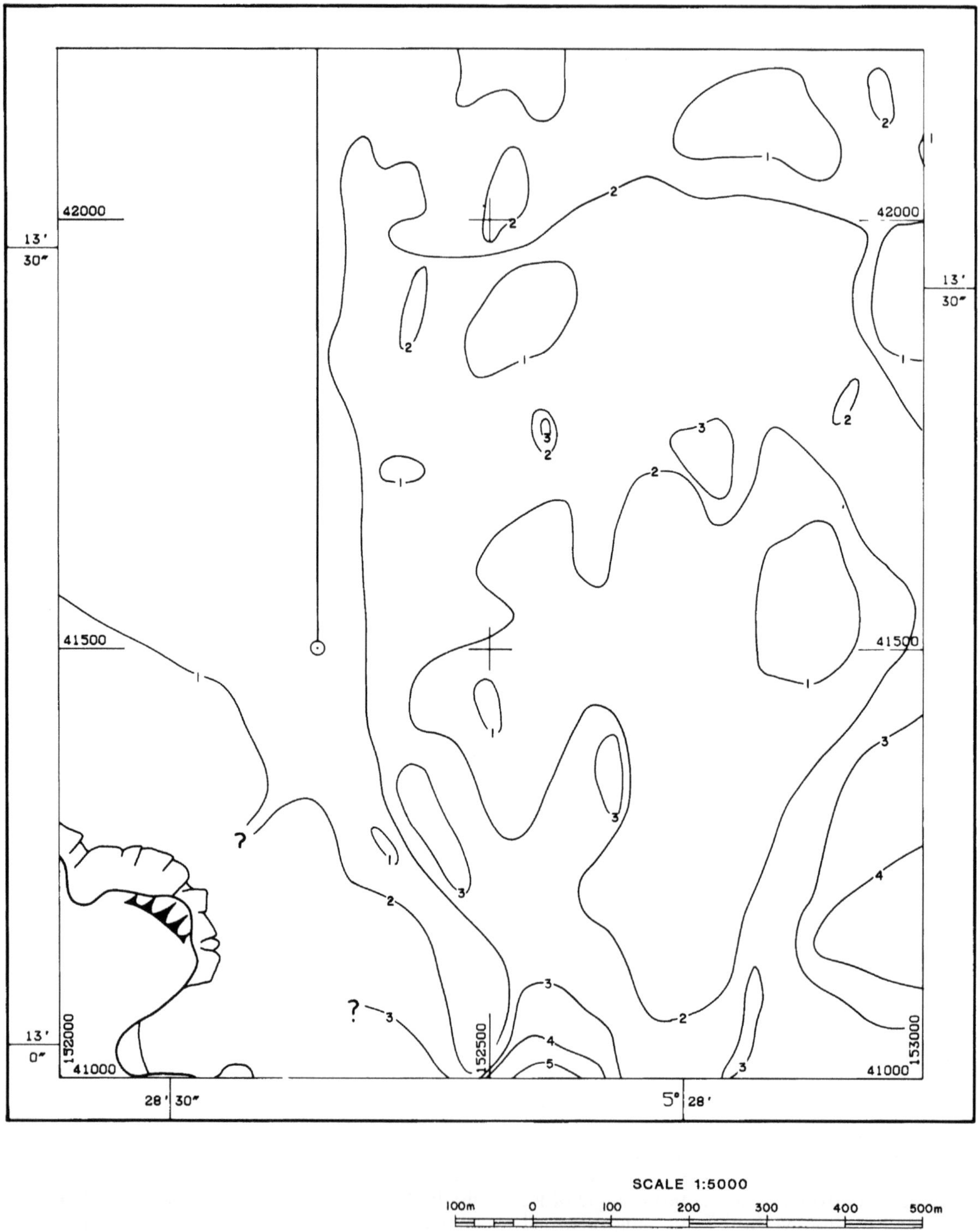

Figure 34.27 Isopachyte chart showing thickness of sediment overlying basement rock Isopachs shown at 1 m intervals.

carried out by the geophysicist on site. This consists of de-lineating areas of similar acoustic properties. Representative sea-bed samples from each of these areas can then be taken using conventional grab sampling equipment or by the use of divers. The use of a diver is particularly valuable when inspecting obstructions. Following the demobilization of the survey the side-scan sonar data is correlated to the navigation data and the appropriate lay back applied (the distance behind the vessel that the fish was towed). Appropriate obstructions or features are picked off the record, corrected for slant range and plotted as a plan view of the sea-bed (see Figure 34.26).

34.7.3 SUB-BOTTOM PROFILING

Sub-bottom profiling may be used to map the depth of sediment below the sea-bed. The technique provides a relatively cheap and convenient tool to investigate superficial sediments and shallow geology. Sub-bottom profiling entails creating an acoustic energy pulse in the water column which propagates through the water and sediment. A percentage of the energy is absorbed or reflected at each density change. The reflected signals are sensed by hydrophones and recorded on board the survey vessel. The depth of penetration depends on the frequency of the source and the type of strata.

The choice of energy source allows the geophysicist to plot shallow sediment with high resolution or deeper sediment with lower resolution. The available equipment may be conveniently divided into three categories:

1. Sparkers: an electric discharge is induced between several electrodes which ionizes the water creating a bubble which oscillates and then collapses after a few milliseconds, creating an acoustic pulse. Typically, sparkers return data from depths of up to 70 m below the sea-bed, although resolution is typically limited to ±2 m.
2. Boomers: the rapid movement of a metal plate by an electrical pulse through a coil induces a broad band acoustic pressure pulse. Boomers typically have a penetration of between 30–50 m with a resolution of ±0.5 m.
3. Pingers: a piezoelectric or magnetostrictive transducer similar to an echo sounder is used to produce a narrow frequency band, usually in the range 3–10 kHz.

Many other sources are available, but would not commonly be used for inshore shallow seismic studies. For all methods, the energy source and the hydrophones are towed behind the survey vessel.

The interpretation of the results of the survey is usually carried out by the geophysicist following completion of the survey. The analogue records are linked to the navigation chart and appropriate laybacks applied to correct for the distance behind the vessel of the hydrophones. Generally interpretation consists of identifying discontinuities which mark the junction between materials of different density. The thickness of individual layers is determined by assuming an appropriate speed of sound. Results are usually presented as an isopachyte chart. This is a contoured drawing showing thickness above the chosen reflector. The isopachyte may be used to draw profiles of the strata along chosen lines (see Figures 34.26 and 34.27).

Seismic refraction

For determining the nature of sub-surface material very near to shore, seismic refraction techniques may sometimes be used. These techniques again use a seismic energy source but the refracted rather than the reflected signals are detected using an array of moored sonobuoys. Sub-bottom geophysical data may be validated using drop-cores, jet probing, vibrocores or boreholes. At many sites only the upper 2–4 m are of primary interest and provided that bedrock does not outcrop vibrocores provide sufficient validation data. Detailed information on geophysical techniques has been given by McQuillin and Ardus (1977).

REFERENCES

Anon (1969) The bacteriological examination of water supplies. Reports on Public Health and medical subjects, 71, HMSO, London 1–52.

Bowden, K.F. (1983) *Physical Oceanography of Coastal Waters*, Ellis Horwood, Chichesher, UK.

Charlton, J.A. (1985) *Sea Outfalls. Development in Hydraulic Engineering*, Elsevier, Amsterdam.

Crofts, I.P. (1979) Practical guidelines for the evolution of outfall lengths. Symposium on Long Sea outfalls for Sewage Disposal. Institute of Public Health Engineers.

Gameson, A.L.H. (1986) Investigations of sewage discharges to some British coastal waters, in, Bacterial Mortality. *Water Research Centre Tech*. Rep. 239.

McQuillin, R and Ardus, D.A. (1977) *Exploring the Geology of Shelf Seas*, Graham and Trotman, London.

Pugh, D.T. (1987) *Tides, Surges and Mean Sea-Level. A Handbook for Engineers and Scientists.*, John Wiley, Chichester.

Shore Protection Manual (1984) US Army Corps of Engineers, Vicksburg, MS.

Wilkinson, W.B., Curren, P.J. (1985) The use of serial photography and image processing to measure dye dispersion from a long sea outfall. *Proceedings of the Institute of Civil Engineers Part 2*, **79**, 327–49.

PART SEVEN

MATERIALS FOR COASTAL STRUCTURES

ROCK FOR MARITIME ENGINEERING 35

J.-P. Latham, H. Wang and A.B. Poole

A considerable amount of research has been carried out in recent years to the point where it is now possible to put the subject of rock mechanics within a sound scientific framework. This means, as the chapter demonstrates, that by the skilful integration of the known output from a quarry and appropriate geological considerations the most economical design of a marine structure can be achieved. For example, it is now possible to predict the range of block sizes from a particular quarry and such information is invaluable within the design process. To achieve good quality control the authors describe the way in which further evaluation of the properties of rock should be carried out on site and in the laboratory (See also Chapters 29, 32 and 33).

CONTENTS

35.1 INTRODUCTION

Loose rock is much used in marine structures for core, filter, underlayers and armour layers. The design and specification of rock for so-called rubble mound structures such as groynes, protection for sea-walls, revetments and conventional or berm breakwaters is not a discipline full of accurate models and tools; but nor is it the unsophisticated task sometimes imagined when picturing a mound of rocks in the sea. Rock is not an off-the-shelf commodity with fixed properties. For this reason the engineer must refer to the available expertise in order to fully realize the potential economies of using rock with its inherent strength and massiveness and the flexibility of design that it allows.

Two recent books provide a wealth of geological and scientific background as well as practical engineering guidance. The *Manual on the use of rock in coastal and shoreline engineering* (CIRIA/CUR 1991) has been prepared jointly by CIRIA in the UK and CUR in The Netherlands and is referred to in this chapter as the *Rock Manual*. Its sections dealing with materials take several steps towards a much needed standardized approach to rock specification. These proposals have been summarized by Simm *et al.* (1991).

Les Enrochments (LCPC 1989) ('armourstone'), prepared in France for the Laboratoire Central des Ponts et Chaussées does not include the geotechnical and hydraulic applications of rock, which are extensively treated in the *Rock Manual*. Rather than present design procedures, it gives references with many diagrams, tables and pro formas, providing a practical, self-contained guide to rock armour. The report represents an almost entirely French overview of rock exploration, testing and specification procedures, and naturally gives a slightly different emphasis from the *Rock Manual*. Just as marine structures are unique, so too are quarries and their products. The skilful integration of quarry and geological considerations into a project's evolution should include a study of these two documents, though site visits are also essential.

In addition to engineering geology and quarrying text books and papers in learned journals, there are numerous national standards and codes identifying tests for quarried materials, many of which are geared to aggregates for use in highways and concrete. Such test results have often been quoted or have misguidedly been presented as evidence of the likely durability of armourstone. They may allow a rudi-

Coastal, Estuarial and Harbour Engineer's Reference Book. Edited by M.B. Abbott and W.A. Price. Published in 1993 by Chapman & Hall, 2–6 Boundary Row, London SE1 8HN. ISBN 0 419 15430 2.

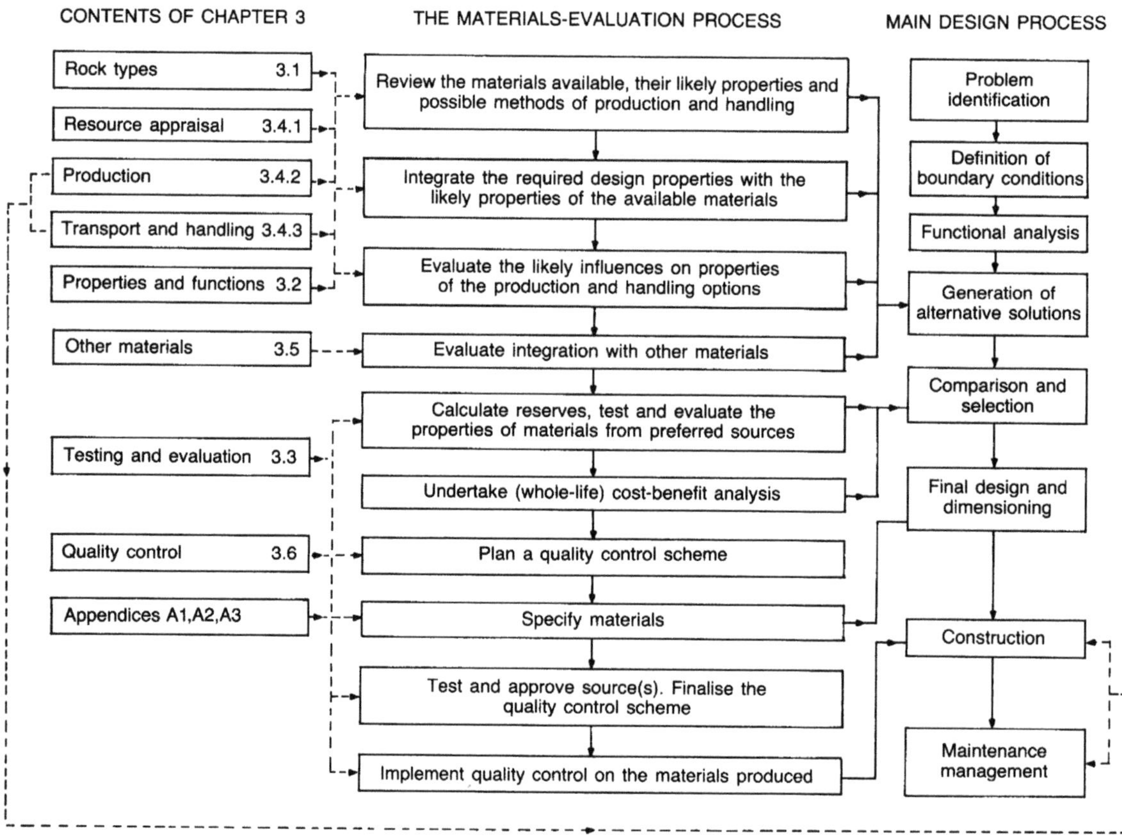

Figure 35.1 The materials evaluation process in sequential stages shown in relation to the main design process and sources of information as given in the CIRIA/CUR *Rock Manual*. →, action flow; →, data flow.

mentary assessment of sources of rock but will be superseded by new recommendations such as those in the *Rock Manual* and *Les Enrochments*. Engineers need to keep up with progress in the standardization of materials testing (Fickelson, 1990) and new Euronorms (EN) and ASTM publications on armourstone which have been prompted by concern over rising sea-levels and more frequent storms.

35.2 GENERAL PRINCIPLES

The cost of production and transport of the very large quantities of rock often required is an important consideration in design. A large breakwater in deep water could require two million tonnes of rock and several hundred thousand tonnes of armour. It is therefore necessary to establish at an early stage which materials are available and their properties, especially the density and block size gradings. Alternatives to rock (concrete, or industrial byproducts such as slags), rock with stability improvements (by asphalt or colloidal cement grouts), and composite rock systems (gabions and mattresses) which are summarized in the *Rock Manual* and described by Pilarczyk (1990) should also be considered.

An existing quarry near the construction site may be the cheapest source of rock, but more distant quarries can also prove economic sources, even after considering transport and handling costs. In many countries, particularly in the third world, opening a temporary quarry needs to be considered, especially if vast quantities of rock are required. Such an option requires a more thorough appreciation of geological factors, some of which are noted in this chapter.

The flow diagram in Figure 35.1 from the *Rock Manual*, indicates how the materials evaluation process is introduced into an evolving design. In a simple low-cost scheme, or where sources of suitable armourstone already exist (a typical European situation), much of the responsibility for evaluating materials

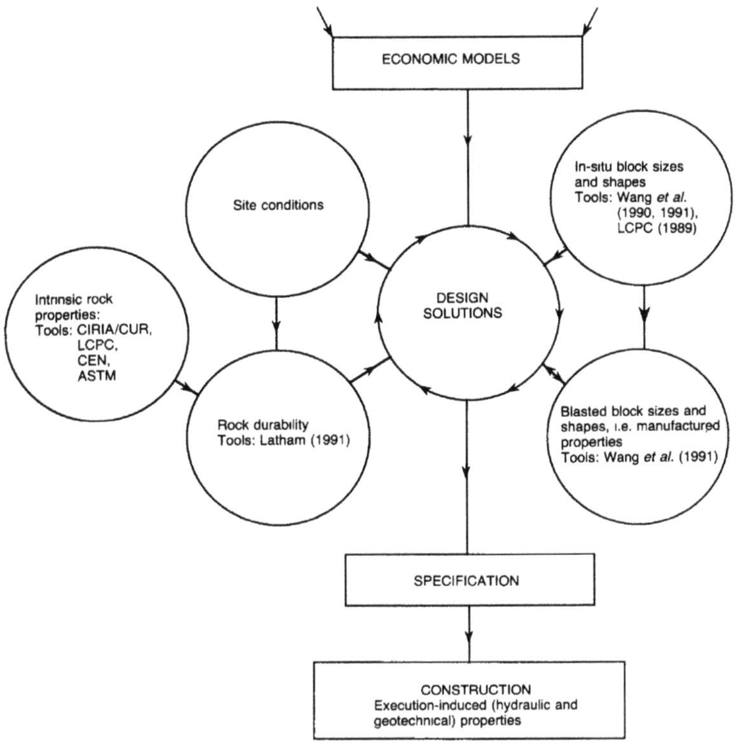

Figure 35.2 Illustration of the possible integration of quarrying considerations into the design process.

will fall on the contractors. Several stages in the designer's actions may then be omitted. The designer should, however, be familiar with the subjects on the left-hand side of the figure to ensure that the final specification is written with a realistic view of economically attainable and relevant rock properties and quality control procedures to be employed. Reference to the model specification in Appendix A1 of the *Rock Manual* will help with this task and in ensuring that profile levels and tolerances on drawings are compatible with the material geometries specified, a common cause of disputes.

Opening a temporary quarry is most economic if all the quarry products can be used. For a major rock project such as a berm breakwater supplied from a temporary quarry, the aim should be to integrate quarry considerations into the design process, as indicated schematically in Figure 35.2.

The right-hand side of Figure 35.2 deals with production. The rock terrain in which a new or existing quarry is excavated must have the necessary width of spacing between joints and weakness planes in order that *in situ* blocks of suitable size and shape are plentiful for economic production of armourstone (An *in situ* block is a piece of rock bounded by discontinuities located in the rock mass prior to excavation.). There are now survey methods and analysis tools (though relatively crude) to confirm that the rock is capable of producing blocks of a given size, and also for designing a blast which will minimize or optimize further breakage. Accurate prediction of the range of block size in the blasted pile can be of great value in the design process, for reasons explained more fully later. One of the arrows in Figure 35.2 also indicates how a preferred design solution with certain size requirements can in turn influence the blast design.

The left-hand side of Figure 35.2 indicates how intrinsic properties such as density, water absorption and resistance to weathering, abrasion and impacts have important implications for durability. (Rock weathering is defined as physical and mineralogical decay processes in rock brought about by exposure to climatic conditions either at the present time or in the geological past. Durability is defined as the ability of the rock to retain its physical and mechanical properties (i.e. resist degradation) in engineering service.) Important though often neglected influences

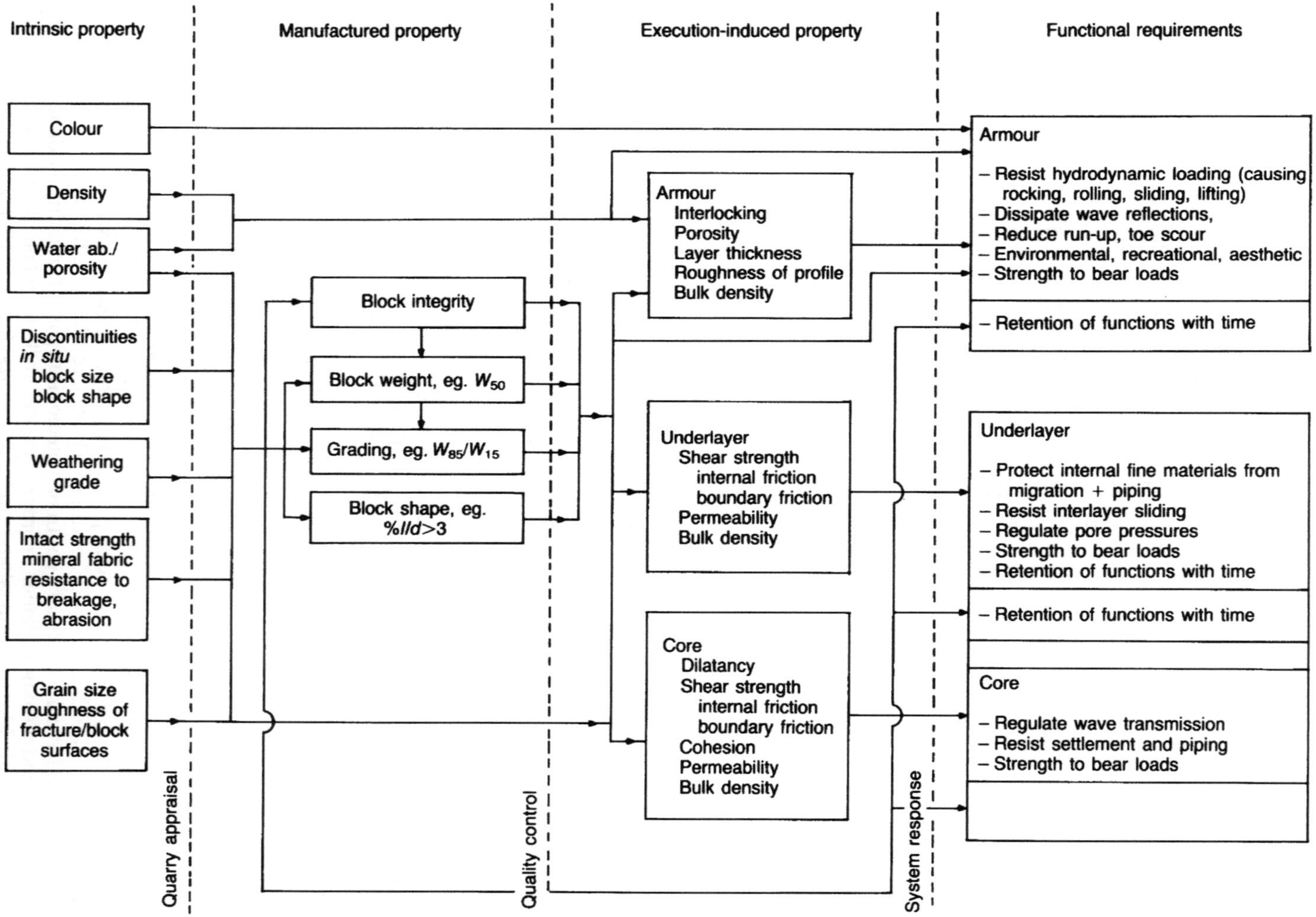

Figure 35.3 Flow diagram for rock properties and functions (taken from CIRIA/CUR *Rock Manual*).

on how the rock will last or degrade in service is the aggressiveness of site conditions such as storm frequency and wave heights, the abrasive load of water-borne attrition agents, and the metereological climate which affects rates of freezing and thawing, salt precipitation, cracking and spalling etc. A first attempt to give quantitative guidance on the interplay between rock quality and site aggressiveness is given in the QMW rock degradation model (Latham, 1991), which also allows for the fact that dynamic stability design options with 'mobile' or reprofiling armour can lead to greater rates of wear.

Successful design is of course a question of matching properties to requirements. Figure 35.3 indicates how these are linked.

35.3 BASIC CHARACTERISTICS OF ROCKS

Geologists divide rock into three primary groups depending on their mode of formation.

1. Igneous rocks: formed by crystallization and solidification of a molten silicate magma, giving strong rocks with interlocking crystals. Usually divided into seven groups: granite, diorite, gabbro, rhyolite, andesite, basalt and serpentinite.
2. Sedimentary rocks: formed by the sedimentation and subsequent lithification of mineral grains, usually underwater, giving bedded rocks with grains cemented by interstitial material of varying strengths. Usually divided into six groups: quartzite, sandstone, siltstone, shale, limestone and chalk.
3. Metamorphic rocks: formed by the effect of heat and pressure on igneous or sedimentary rocks for geological periods of time with the consequent development of new minerals and textures within the pre-existing rock, giving crystals which usually interlock but may have planar or linear grain alignments that weaken the rock. Usually divided into five groups: slate, phyllite, schist, gneiss and marble.

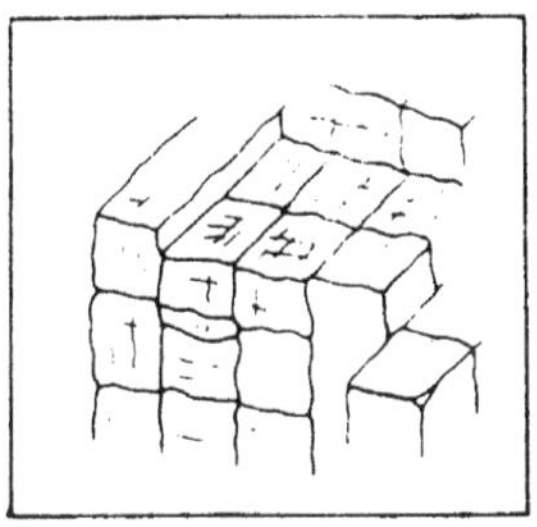

BLOCKY
Typical of igneous rocks such as granites; also massive sandstones and limestones

IRREGULAR
Typical of igneous rocks and some hard sandstones and limestones

FLAGGY
Typical of bedded sedimentary rocks such as siltstones and sandstones

BLADED
Typical of many metamorphic rocks such as phyllites and schists

SLATY
Slates and some shales

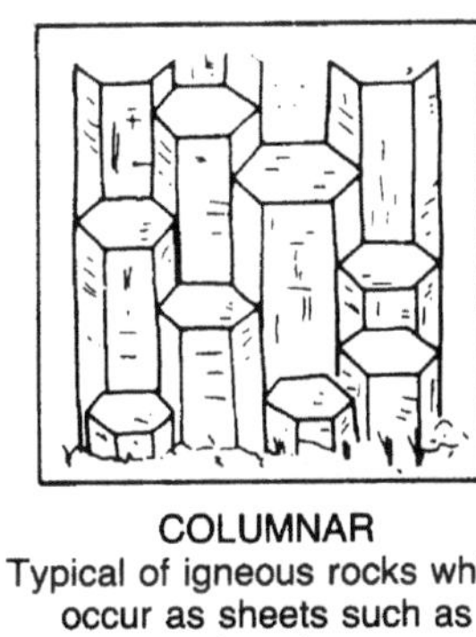

COLUMNAR
Typical of igneous rocks which occur as sheets such as basalt

Figure 35.4 Idealized sketches of common natural outcrop forms (taken from CIRIA/CUR *Rock Manual*).

These modes of origin give rise to a small number of typical discontinuity patterns in the rock mass, as illustrated in Figure 35.4. They arise from the lamination or orientation of mineral grains, systematic variation in grain size, or shrinkage due to drying of sediments or cooling of igneous rocks. These patterns are often overprinted with additional discontinuities resulting from tectonic folding, faulting and jointing of the earth's crust.

Rocks within each of the 18 groups have essentially similar characteristics, so that generalizations about their engineering properties are often made. Reference should be made to books on engineering geomaterials for information on aggregates, intact rock and rock mass properties such as Collis and Fox (1985), Bell (1983), Farmer (1983), Johnson and Degraff (1988) and Prentice (1990), as well as the *Rock Manual* and LCPC (1989). Lama and Vutukuri (1978) and Carmichael (1982) contain extensive data on mechanical and physical rock properties listed under rock types and may be useful as specialist reference books.

Geological weathering can often seriously reduce the quality of the rock concerned, though it is not always identifiable by simple visual inspection. Test results outside the typical range for the group (e.g. an oven dry density of 2.65 is unusual for fresh basalt) will usually indicate suspect rock quality. Rock weathering is brought about by the exposure of the rock during long periods of geological time to the climatic conditions at the earth's surface and involves mechanical disintegration and chemical decomposition acting together (Fookes *et al.*, 1988). The effects are most marked in hot humid climates. Numerous classifications, including some based on field assessment, are available for grading rocks in terms of degree of weathering (e.g. Fookes *et al.*, 1971; Geological Society, 1977; Lee and de Freitas, 1989) but petrographic thin-section examination is always advisable with potentially weathered or altered rock.

Certain rock types, such as siltstone, shale, chalk, slate and phyllite, are generally unable to provide suitable blocks for all but core materials. The required mechanical characteristics and durability of rock in a structure depend on where they will be used (armour, facings, underlayers and filters generally require high, durability; core and fill – generally low) and the site aggressiveness. Once the general expectations of the various sources have been considered on the basis of rock type from a desk study, proper evaluation procedures with quarry site investigation and laboratory testing should follow (section 35.8).

35.4 BLOCK SIZES AND PRODUCTION

Armourstone can be produced from three types of quarry.

1. Dedicated armourstone quarries: opened temporarily for a large project where there is a need to maximize usable fractions in order to minimize waste. There are a few permanant specialist quarries with deep-water access; these often have mobile crushing plant to process the undersize material as aggregates.
2. Dimension stone quarries: large blocks are produced, often without blasting, for masonry and monumental works, leaving many offcuts which may be ideal for rock armour.
3. Aggregate quarries: the largest group. Blasting and processing are geared to give 40 mm and finer sizes. In most, the production of some oversize blocks greater than three tonnes is inevitable but unlikely to exceed 5%. These may be stockpiled if there is room and if standardized size ranges become established. Collecting small stocks from several sources is practised (e.g. in North Wales) but is inefficient and rarely provides sources which can respond rapidly to large orders over 30 000 t.

In any design the desired proportion of large blocks for armourstone to that of underlayer, filter and core materials is always greater than that proportion in the rock pile after blasting. This means that the time taken to produce and bring to site a certain amount of armourstone is often the controlling factor in the rate of construction. Greater awareness of production constraints would assist the contractors in assessing the producer's supply rate forecasts when sufficient quantities of rock are not already on the quarry floor for inspection.

35.4.1 DISCONTINUITIES

As discussed in Figure 35.1, armourstone cannot be produced from a quarry in which the *in situ* block sizes formed by the dissecting discontinuities are too small and the block shapes are not favourable. A virgin site for an armourstone quarry will be viable only if the discontinuities in the rock mass are sufficiently widely spaced to contain enough blocks of suitable size. The *in situ* block size distribution in the rock mass can be most accurately predicted where good exposure allows for detailed manual survey techniques. The location and orientation data may then be input to powerful computer modelling methods (Wang *et al.*, 1991a) or assessed with the most recent mean discontinuity spacing–block size equations (Wang *et al.*, 1990), which supersede the volumetric joint counting methods (ISRM, 1978).

However, down-hole techniques such as closed circuit television (CCTV) and sonic velocity profiling (as developed in France: LCPC, 1989) combined with the less reliable rock quality designation (RQD)

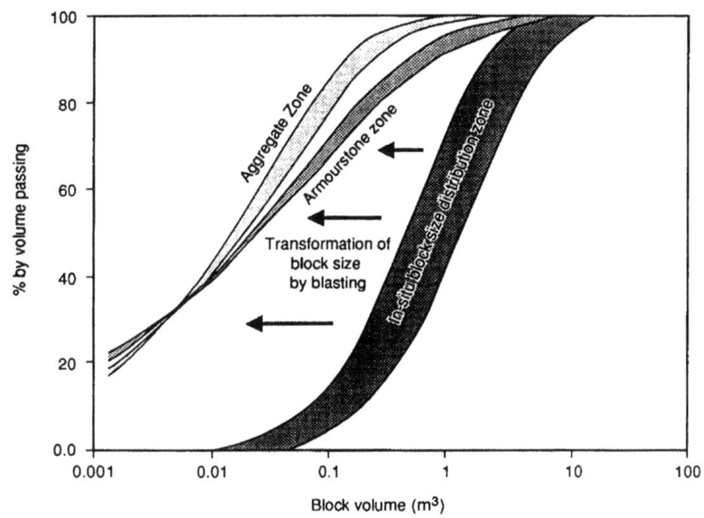

Figure 35.5 Transformation of the block-size distribution resulting from blasting (taken from Wang *et al.*, 1991b).

methods (see, Hudson and Priest, 1979) may have to be used. Such methods can indicate *in situ* block sizes in poorly exposed sites and will generally increase confidence in estimation.

The predicted fragmentation curve for the blast products of a given blast design, though it is often extremely inaccurate in practice (with serious consequences, see Van Oorschot, 1983), is potentially useful for planning production and thus establishing the supply rates of the required gradings of blocks. Furthermore, in a temporary dedicated quarry, the curve may be used to design a structure which matches the expected as-blasted block sizes and aims for a 100% utilization of blasted rock (Baird and Woodrow, 1988). Matching procedures which help determine the most economic block sizes for specification are illustrated by Vrijling and Nooy van der Kolff (1990), using examples of conventional and berm breakwater designs. Standardized gradings should therefore be avoided for such temporary quarry sources, which need to be more flexible. Two methods for predicting fragmentation are outlined in the *Rock Manual*.

35.4.2 BLAST DESIGN

Blasting of rock is a highly specialized skill. Advice in many books is aimed at producing high fragmentation (Gustaffson, 1973; Morehard *et al.*, 1987). A helpful model for understanding how the transformation by blasting from the *in situ* block size distribution can give different fragmentation curves is given in Figure 35.5. Blast design for low fragmentation and a maximum yield of armour-sized blocks follows a procedure which attempts to dislodge blocks by opening existing discontinuites and destabilizing the rock mass. Blast patterns will have to be adjusted to conditions in the actual blast face for the best armour production. Blocks are also liable to have fractures induced by the blast itself. For this reason non-explosive extraction techniques may receive future consideration.

Wang *et al.* (1991b) reviews armourstone blast design, listing ten measures to increase armourstone yields (which are not simply the reverse of well-known measures designed to improve fragmentation). Table 35.1 (from this reference) indicates in a very practical way with the aid of new formulae, how scan-line discontinuity data can be used to provide three sets of principal mean spacings (PMS) which can also be related to RQD. These data can be used to predict the *in situ* block sizes. Taking the 80% passing size (V_{i80}) as the most meaningful *in situ* size parameter, depending on whether a high- or low-fragmentation blast design is used, a certain reduction factor for blasting gives the blasted sizes V_{b80}. For armourstone blasting where a factor of six has been assumed, the percentage of armourstone-sized blocks, arbitrarily taken as $0.2\,m^3$ or about 0.5 t, is also given in Table 35.1. Thus a useful quarry classification, based not on rock types but on discontinuity data is proposed. The three quarry classes correspond to three ranges of the principal mean spacings. Range I cannot supply armourstone. Range II can supply armourstone but all measures designed to maximize

Table 35.1 *In situ* and blastpile block sizes (V_{i80} and V_{b80}) relating to principal mean spacings (PMS) (Source: Wang *et al.*, 1991b)

PMS (m)	*min. MS (m)*	*max. RQD (%)*	*min. RQD (%)*	V_{i80} *in-situ* (m^3)	V_{b80} *armour.* (m^3)	V_{b80} *aggreg.* (m^3)	% ($V_i > 0.2\,m^3$)	*Range*
0.1	0.06	73.58	48.37	–	–	–	–	I
0.2	0.12	90.98	78.49	0.07	0.01	–	–	
0.3	0.17	95.54	88.55	0.24	0.04	0.01	–	
0.4	0.23	97.35	92.94	0.57	0.10	0.03	7	II
0.5	0.29	98.25	95.22	1.12	0.19	0.06	18	
0.6	0.35	98.76	96.56	1.93	0.32	0.10	30	
0.7	0.40	99.07	97.40	3.07	0.51	0.15	40	
0.8	0.46	99.28	97.97	4.58	0.76	0.23	52	
0.9	0.52	99.43	98.37	6.52	1.09	0.33	62	III
1.0	0.58	99.53	98.66	8.95	1.49	0.45	70	
1.1	0.64	99.61	98.88	11.91	1.99	0.60	75	
1.2	0.69	99.67	99.05	15.46	2.58	0.77	>80	

large block yields, e.g. those listed in Wang *et al.* (1991b) should be applied. Range III runs the risk of excessive production of unusable mammoth blocks, and in this case the blast design must be tailored to the requirements.

In addition to spatial information, the persistence, aperture, filling and water seepage of discontinuities can give a strong indication of rock quality. The state and distribution of weathered zones is particularly important in assessing durability. A geologist's visual inspection, based on discontinuity spacings and weathered zones, may designate certain faces in a quarry as unsuitable for armourstone production.

35.5 INTRINSIC PROPERTIES

Figure 35.3 gives a scheme which identifies relevant intrinsic (i.e. geological) properties of rock. One factor which is often overlooked is the importance to coastal authorities of a 'natural' look and colour for rock structures in recreational areas.

35.5.1 DENSITY AND WATER ABSORPTION

Determination of density is critical and it is important to be precise with terms and symbols for rock density, particularly when dealing with a rock with significant water absorption for the following reasons.

1. The same terms used in different countries have different meanings, which can lead to serious misunderstandings, particularly about water absorption.
2. Hydraulic stability of armour is highly dependent on density in the governing Van der Meer or Hudson equations.
3. Density is used in size–weight relations for dimensioning designs.
4. Density and durability are generally well correlated.

For armour, the most important requirement of the rock blocks is stability against waves. For a given storm condition, the necessary block weight for stability can be reduced if a rock of greater density is used. This is because instability or the amount of block displacement damage in a storm after a given number of waves is inversely proportional to ΔD_{n50} raised to the fifth power (Van der Meer, 1987) where Δ is the relative buoyant density, which for rocks with significant water absorption is given by

$$\Delta = (\rho_r/\rho_w)(1 + 0.01w_{ab}) - 1 \qquad (35.1)$$

or

$$\Delta = (\rho_{ssd}/\rho_w) - 1 \qquad (35.2)$$

where ρ_r is the oven dry rock density, ρ_{ssd} is the saturated surface dry density, ρ_w is the density of seawater (~1.03), w_{ab} is the atmospheric water absorption, D_{n50} is the median block size given by $(W_{50}/\rho_r)^{1/3}$, and W_{50} is the median block weight. The significance can be illustrated by contrasting rocks with extreme densities (ρ_r) of 3.1 and 2.6 t m^{-3} and assuming w_{ab} = 0.5%. For an identical but typical packing of armourstones for each rock type, the same stability is given by W_{50} = 5 t for the high density rock, as would be achieved with W_{50} of 8.3 t for the low density rock. Also there is an accompanying increase in total weight of placed armour of about 5% (although fewer blocks would need placing) for the lower density rock. There are therefore clear, quantifiable advantages in using higher density rocks.

It should be noted that for two rocks having the same oven dry density, Δ will be greater for the more absorbent rock. (These and other density relations are explored fully in the *Rock Manual*.) Confusion arises because ρ_r is more meaningful for the durability correlation whereas ρ_{ssd} is the correct density term to give Δ directly.

35.5.2 RESISTANCE TO BREAKAGE BY IMPACTS

The strength of a mineral fabric can be divided according to the main types of loading in hydraulic structures, into firstly, resistance to breakage, and secondly, resistance to abrasion. From observations of armourstone, underlayer and filter-sized materials it is apparent that two extreme types of breakage can be identified:

Type 1: when blocks break into major pieces along flaws such as cracks, bedding planes or poorly cemented joints. For quality control of armourstone, a block integrity test is needed.

Type 2: when cracks develop along new fractures within the mineral fabric usually resulting in edges and corners being knocked or sheared off. For quality control, fracture toughness (ISRM, 1988), which for most rocks is well correlated with dynamic (impact) strength for small rock pieces (Grady and Kipp, 1987), or a quasi-static test such as point-load (ISRM, 1985), may be used for classification.

In extreme cases the grading of a material after transporting to site may differ significantly from that loaded from the quarry floor, and this has implications for quality control.

Size effects

Rocks small enough to be handled in bulk and armourstones which must be handled individually are both subject to severe loading, but of different kinds. Consequently the degradation that may occur during handling in the quarry, transport and completion of the structure, and in service, tends to differ. The smaller gradings, such as filters, are sometimes compacted, and the significant static and dynamic compressive loads cause abrasion and rounding and some breakage often occurs. The higher ratio of surface area to volume of small fragments results in rapid degradation. In contrast, high impact loads (inevitable during handling of large blocks or during rocking in storm conditions) will be most likely to cause damage of armourstones, for example by whole-block failure if there are any planes of weakness through the blocks.

The strength of rock cannot be assessed straightforwardly because of the increasingly important role of possible cracks and planes of weakness (i.e. flaws) as pieces become larger. With no flaws, the strength of a rock depends on its minerals, matrix cement, grain boundaries, pore structure and general grain texture, so that measurement on laboratory-size aggregate or core specimens is meaningful; this measurement will usually also detect chemical weathering, which tends to be pervasive. With pieces larger than, say, 50 to 300 kg – the approximate value depending on the natural discontinuities and weaknesses in the quarry and the way it has been blasted or crushed – natural flaws and blast-induced cracks become more significant in determining the impact strength of a rock piece. For large blocks from typical armourstone gradings, much research in quality control measures is directed at determining the integrity of the block so that blocks weakened by flaws can be identified. (The integrity of the block is defined as the degree of wholeness of a block as reflected by the degree to which its strength against impacts is reduced by the presence of flaws.).

Block integrity

Once blasted, the rock can be examined in stockpiles. Often, a large number of armour-sized blocks will be found to have very poor integrity or 'wholeness', so that they would clearly break up along visible cracks on handling. Other blocks may be equally weakened by flaws propagated during the blasting process but not easily visible, while others showing banding, bedding-plane partings and veins can be surprisingly impact-resistant. This is clearly a major problem for suppliers and users of armourstone, and assessing block integrity is now an active research topic. In France and Holland, sonic velocity measurements are taken across rock blocks; poorer acoustic propagation across 'dangerously' open cracks or weakness planes indicates the possibility of flaws.

The French method (LCPC, 1989) compares three mutually perpendicular measurements of the velocity of sound V_m with the rock's theoretical velocity V_t based on the composition and proportions of the constituent minerals obtained from thin-section analysis, thus providing an index of continuity I_c (= $100{*}V_m/V_t$). Some corrections for porosity and conditions of water saturation are needed to obtain an indication of the integrity of the block. One clear example of how I_c can be used together with a drop test series to set the threshold value of acceptable I_c from a particular source was presented in the French report (LCPC,

1989). However, such a histogram of results is unlikely to be typical as drop tests (from 3 m onto a base of similar armour blocks) are notoriously difficult to standardize. The Dutch approach (Niese *et al.*, 1991) involves taking 10 to 20 velocity values at various points round the block and presenting these as parameters of a cumulative curve to see if the number of low velocity values is exceptionally large when compared with the median velocity (indicating fissuring). Obtaining correlations with drop test results, though also promising, is very slow as a method to obtain a statistically valuable appraisal of a stockpile's block integrity.

The *Rock Manual* recommends a drop test for armourstone which examines the change in median weight (resulting from breakages) of a large sample of armour blocks dropped individually from a fixed height onto a carefully prepared substrate of similar armour blocks. There is little experience with either this or sonic velocity tests but the importance of developing a quantitative method is widely recognized.

Sometimes the loss of integrity is delayed and cracking occurs while rocks are left 'curing' after the blast. According to Lienhart and Stransky (1981), residual stresses are released after blasting which can propagate blast induced cracks called 'popping', reportedly for up to 30 days or more after blasting. This phenomenon may be enhanced when excavation is done in freezing conditions, for example in the northern USA and Canada. It is not widely acknowledged in the UK, perhaps owing both to the lack of monitoring for this effect and to the warmer climate.

35.5.3 RESISTANCE TO ABRASION

Materials handled in bulk (typically core and underlayer materials of less than 300 kg) and/or loaded many times will undergo considerable mutual shearing, resulting in abrasive degradation. Weight losses are proportionately greater for finer materials. Resistance to abrasion in service is most important for sites where foreshore attrition agents such as shingle or sand can attack the armour. Also, for structures using dynamic design concepts, the increased risk of rocking, sliding and rolling of stones will require that only highly abrasion resistant materials be used. The preferred tests are those which tumble coarse aggregate-size pieces of rock in wet conditions.

35.6 MANUFACTURED PROPERTIES

Apart from block integrity, which is often only marginally affected by production methods and was therefore treated above as an intrinsic property, these manufactured properties (see Figure 35.3) are concerned with geometry and have been summarized by Simm *et al.* (1991). The recommended standardization of terms in the *Rock Manual* has been long overdue. The size of a piece of rock universally accepted by coastal engineers in considering armour stability is the side length of the cube with equivalent weight i.e., $D_n = (W/\gamma_r)^{1/3}$, where γ_r is the unit weight of rock. Conversion factors are still needed for relating actual dimensions to D_n and these are clearly explained in the *Rock Manual*. For example, taking D_{n50} to be 84% of the equivalent square sieve size for the 50% passing value gives the single best conversion factor for the range of angular block shapes encountered in practice, but exact factors for individual sources will vary somewhat with rock type and sizes of stones, since smaller blocks tend to be more tabular and elongate.

35.6.1 BLOCK SHAPE

Rock type, in association with discontinuity patterns, jointing, bedding and schistosity exerts an obvious control on the block shapes extracted (Figure 35.4). The amount of slabby or elongate blocks is most conveniently and most unambiguously controlled during block selection by using a threshold value of the aspect ratio l/d of 3 where l is the maximum distance between two points on the block and d is the minimum distance between two parallel bars through which the block will just pass. Other shape defining systems, such as those based on side lengths of the enclosing cuboid, cannot be so generally used. In future, the influence of block surface texture and roughness on permeability of bulk rock media in hydraulic applications may prove sufficient to warrant measurement of particle roughness by Fourier and/or fractal methods (e.g. Latham and Poole, 1987; Bradbury *et al.*, 1990).

35.6.2 GRADING

Grading requirements and their verification methods have recently been rationalized by the recommendations given in the *Rock Manual*. There is support from hydraulic model tests that gradings specified for armour are often far narrower than are needed for it to perform satisfactorily. For bulk-handled materials which are both smaller and efficiently produced in wider gradings, fears over segregation affecting performance may be justified, but for armourstone where segregation can be controlled it has been suggested (Van der Meer, 1987) that W_{85}/W_{15} as high as 16 ($D_{85}/D_{15} = 2.5$) may perform adequately in a double armour or thicker cover layer.

A set of standardized gradings has been recommended in the *Rock Manual* in order to encourage producers to stockpile routinely in sizes that are most often specified by designers. In the USA, standard gradings for rip rap have been in operation for many years, but generally these have been unsuccessful because the system had too many separate and overlapping classes and an unrealistically imposed identical grading width (e.g. as expressed by W_{85}/W_{15}) in each class. A considerable amount of Dutch research at the Rijkswaterstaat has produced a system which attempts to combine a method of imposing suitable grading limits with realistically producible oversize and undersize tail requirements. An additional requirement is set for the effective mean weight which is intended to promote the use of a rapid control method which uses bulk weighing and counting.

In the *Rock Manual*, three types of standard gradings are proposed. 'Heavy' gradings are for blocks handled individually, i.e. heavier than about 0.3 t, and are specified in terms of weight by defining the class limits ELCL, LCL, UCL and EUCL (Figure. 35.6) chosen from a table of standard gradings where:

- ELCL = less than 2% by wt. lighter than this weight
- LCL = between 0 and 10% by wt. lighter than this weight
- UCL = between 70 and 100% by wt. lighter than this weight
- EUCL = greater than 97% by wt. lighter than this weight

For example, in what is termed a 3–6 tonne grading, the class limits are 2, 3, 6 and 9 tonnes, respectively, and the effective mean weight is required to be in the range 4.2 to 4.8 tonnes.

For materials that are small enough to be tested by using sieves, i.e. up to 250 mm, a series of standard 'fine' gradings has been introduced that are neatly defined by sieve size.

There is, however, a third group of gradings called 'light' gradings, with extreme pieces weighing roughly between 2 and 450 kg. These do not naturally fall into a size or a weight system and so, after much consultation, an option has been added to the *Rock Manual* gradings to specify by weight (with greater control being exerted but with a possible greater expense in production and verification), or by sieve size and average weight (for more rapid testing with perhaps less stringent control).

For rock products that are produced by mechanical selection using screens or grizzly grids (parallel bars in one direction only), it is difficult to predict the weight distribution that will result, because of many factors such as efficiency (whether screens or grids are vibrating, sloping, of non-standard aperture shape, worn etc.), intrinsic shape of the rock pieces and density. However, empirical rules of thumb can be devised, and some guidance is given in the *Rock Manual* when visual selection is to be used to limit upper sizes, and grizzly grids can be used to remove undersized pieces.

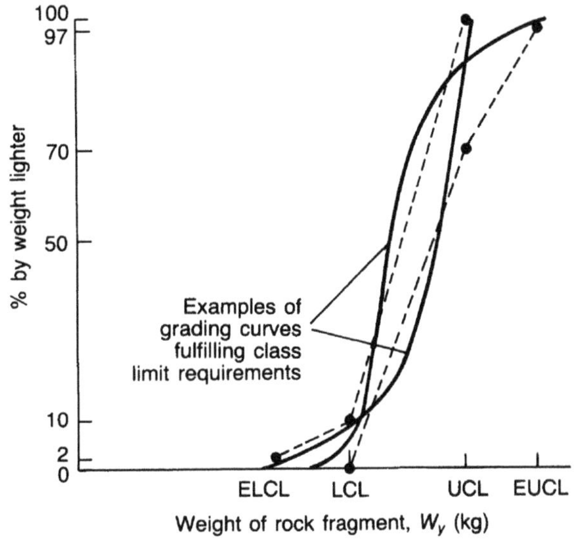

Figure 35.6 Explanation of the grading class limits proposed in the CIRIA/CUR *Rock Manual*.

These approaches to grading specification should be studied, as grading is the most critical and awkward of the rock properties to control. For example, it is rarely noted that for 'light' gradings, as the grading width increases, production methods will invariably yield a grading curve which is concave upwards and not straight when plotted on a log-linear plot. If in a design a target W_{50} is calculated, and then upper and lower limits are set to define maximum and minimum weights, they should reflect this curve or else W_{50} that is supplied will be well down on that assumed in the design.

35.7 EXECUTION-INDUCED PROPERTIES

35.7.1 GENERAL

As shown in Figure 35.3, these are the properties of the bulk materials in place which directly affect the hydraulic and structural performance of the main structural components. For example, to avoid blocks being plucked out and rolling down (or up) a slope, a combination of, firstly, sufficient block weight and density; secondly, sufficient layer porosity, and thirdly a sufficient degree of interlocking is generally necessary for armourstone.

The term 'interlock' often applied to blocks in armour layers, is generally understood to apply to a pack of blocks and to reflect the degree to which individual particles are constrained by their neighbours from rotating or sliding relative to each other. It depends mainly on placement methods. However, improved interlocking of armour does not accompany increased porosity. For example, an instruction to produce a tightly packed arrangement of rectangular sectioned blocks can lead to high interlocking but at the expense of a lower layer porosity and poorer energy dissipation.

The surface roughness and the porosity of the armour layer will influence the wave reflections and run-up/run-down on a given slope. Careful placement methods may achieve specific results, for example, in structures within high value recreational areas where safer outer slopes for people intent on climbing can be built. The topographic roughness also has implications for the accuracy and interpretation of profile survey methods. Inter-layer friction between, say, armour and under-layer is normally considered sufficient if the size and grading relations between layers have been set by filter rules. However, it has been suggested that there are considerable stability advantages in using a W_{50}(above)/W_{50}(below) ratio of 10 to 15.

Payment of the contractor on completion of works involving placed armourstone can be by volume or by weight and the arguments for either system have come down in favour of payment by weight for armour-layer materials. The system for billing is outlined in Simm *et al.* (1991) and explained in full in the *Rock Manual*. The survey accuracy and certification procedures for the weight of supplied armour in large barge deliveries may be a justified cause of concern with this system.

35.7.2 LAYER THICKNESS

Porosity and layer thickness of armour layers and underlayers need to be calculated in order to estimate the volume or weight requirements of rock to cover a given area of a structure. The perpendicular layer thickness and porosity of a layered system of armouring will depend on the following factors:

1. Weight and grading of rock (W_{50} and W_{85}/W_{15});
2. Weight density of rock (γ_a);
3. Method of placement;
4. Block shapes;
5. Definition of the surface and the survey method of measurement.

The last three factors provide the greatest areas of uncertainty for design and construction. For narrow gradings, good approximations of armour layer thickness (t_a) and layer porosity (n_v) are given by the equations

$$n_v = 1 - (\gamma_b/\gamma_a) \tag{35.3}$$

$$t_a = nk_tD_{n50} \tag{35.4}$$

$$N_a = nAk_t(1 - n_v)D_{n50}{}^{-2} \tag{35.5}$$

where γ_b is the bulk weight density as laid; n is the number of layers making up the total thickness of armour; k_t is the layer thickness coefficient; N_a is the total number of blocks in area considered; and A is the area considered.

The porosity for double layers surveyed by a method similar to that recommended in the *Rock Manual* ranges from 35 to 42% in narrow-graded armour layers. Recent experience with numerous contracts in the UK indicates that the selection of k_t and n_v from values tabulated in the *Shore Protection Manual* (1984), together with lack of clarity caused by profile survey methods, has often caused later contractual disputes between designer and contractor. This is the motivation for the standardized survey method and tolerances given in the *Rock Manual*. Usually the *Shore Protection Manual* values of k_t are higher than those given by the thickness of double armour layers constructed with normal satisfactory methods.

Reported values for layer thickness coefficients vary considerably and it is acknowledged that further guidance in the selection of appropriate values for k_t and n_v is required, although some suggested values have been tabulated in the *Rock Manual*. Another procedure designed to help with uncertainties over layer thickness and porosity is the requirement for an initial 'test panel' to be built so as to establish placement methods that will give acceptable results and with which the remaining sections can be built.

35.8 EVALUATION OF DURABILITY IN SERVICE AND ROCK SPECIFICATION

Ideally, the suitability of a rock source for given applications should be fully evaluated prior to specification.

35.8.1 EVALUATION

The scheme proposed in the *Rock Manual* gives a table (Table 35.2) of preferred tests which may be used to assess rock quality. Each property may then be graded into four quality categories: excellent, good, marginal or poor, according to suggested ranges of results. This is an improvement on the rigid acceptance test criteria given in previous papers (Wakeling, 1977; Fookes and Poole, 1981; Allsop *et al.*, 1985).

The overall appraisal of a quarry and the evaluation of its rock for marine applications will normally follow the procedure given in Figure 35.7. This flow diagram

Table 35.2 Guide to rock durability from test results (taken from CIRIA/CUR *Rock Manual*, 1991)

Test	*Excellent*	*Good*	*Marginal*	*Poor*	*Comments*
Rock density (oven dry) Ap. A2.6 ρ_r (t/m^3)	>2.9	2.6–2.9	2.3–2.6	<2.3	Physical property affecting hydraulic stability. Good indicator of durability except dense but weathered basic rocks.
Water absorption Ap. A2.7 W_{ab} (%)	<0.5	0.5–2.0	2.0–6.0	>6.0	Single most important indicator of resistance to degradation. Good indicator of weathering resistance. Often misleading for porous limestones with large free draining pores.
Magnesium sulphate soundness BS 812/BS 6349 MSS (%)	<2	2–12	12–30	>30	Indicates resistance to weathering. Important test for porous sedimentary rocks for use in hot dry climate. Good correlation with water absorption.
Freeze/thaw Ap. A2.4 FT (%)	<0.1	0.1–0.5	0.5–2.0	>2.0	Important test for freezing winter climates (especially reservoir dams). Good correlation with water absorption.
Methylene blue absorption Ap. A2.10 MBA (g/100 g)	<0.4	0.4–0.7	0.7–1.0	>1.0	Indicates presence of deleterious clay minerals.
Fracture toughness ISRM (1988) K_{IC} (MPa m$^{1/2}$)	>2.2	1.4–2.2	0.8–1.4	<0.8	Indicates resistance to Type 2 breakage, degree of weathering (geological). Good correlation with abrasion resistance. Can be misleading for impact resistance of large blocks.
Point load index ISRM (1985) $I_{s(50)}$(MP_a)	>8.0	4.0–8.0	1.5–4.0	<1.5	Indicates resistance to Type 2 breakage. Quick test with high test variability. Can be misleading for impact resistance of large blocks.
Wet dynamic crushing value Ap. A2.5 WDCV (%)	<12.0	12–20	20–30	>30	Indicates resistance to Type 2 breakage. Quick test.
Mill abrasion resistance index Ap. A2.9 k_s (loss per 1000 revs)	<0.002	0.002–0.004	0.004–0.015	>0.015	Indicates resistance to abrasion by mutual grinding of saturated rock surfaces. Test result may be used with rock degradation model (*Rock Manual*).
Block integrity drop test	<2	2–5	5–15	>15	Indicates resistance to Type 1 breakage of large blocks.

has several important features. Petrographic analysis by an experienced geologist is a critical stage, as discussed in Latham *et al.* (1990). It is used to establish whether quick tests are likely to be adequate for assessment and, if not, which kinds of tests should be performed. For example, if clay minerals are suspected, then a test to examine whether breakdown due to swelling clay is likely, such as methylene blue absorption, and a wet attrition test are recommended. If, however, the microstructure is suspect, with arrays or networks of microcracks and/or small voids present and it is likely to encourage rapid in-service degradation by freeze-thaw or salt action, an appropriate weathering resistance test can be chosen.

In most circumstances a test for abrasion resistance will be necessary. Since the wet coarse aggregate type of tumbling tests are particularly good simulators of wear, these are considered the best indicators of general resistance to all wear processes. Laboratories in France, the UK and elsewhere have equipment for the Deval test (AFNOR, 1990) and attrition test (BS 812, 1951). An improvement in this type of test has resulted in the QMW abrasion mill test (see Latham, 1991 and *Rock Manual*), for which there is currently a short but encouraging history of experience with interpreting results.

Whether by rapid or more detailed lengthy tests, the quality category with respect to each type of

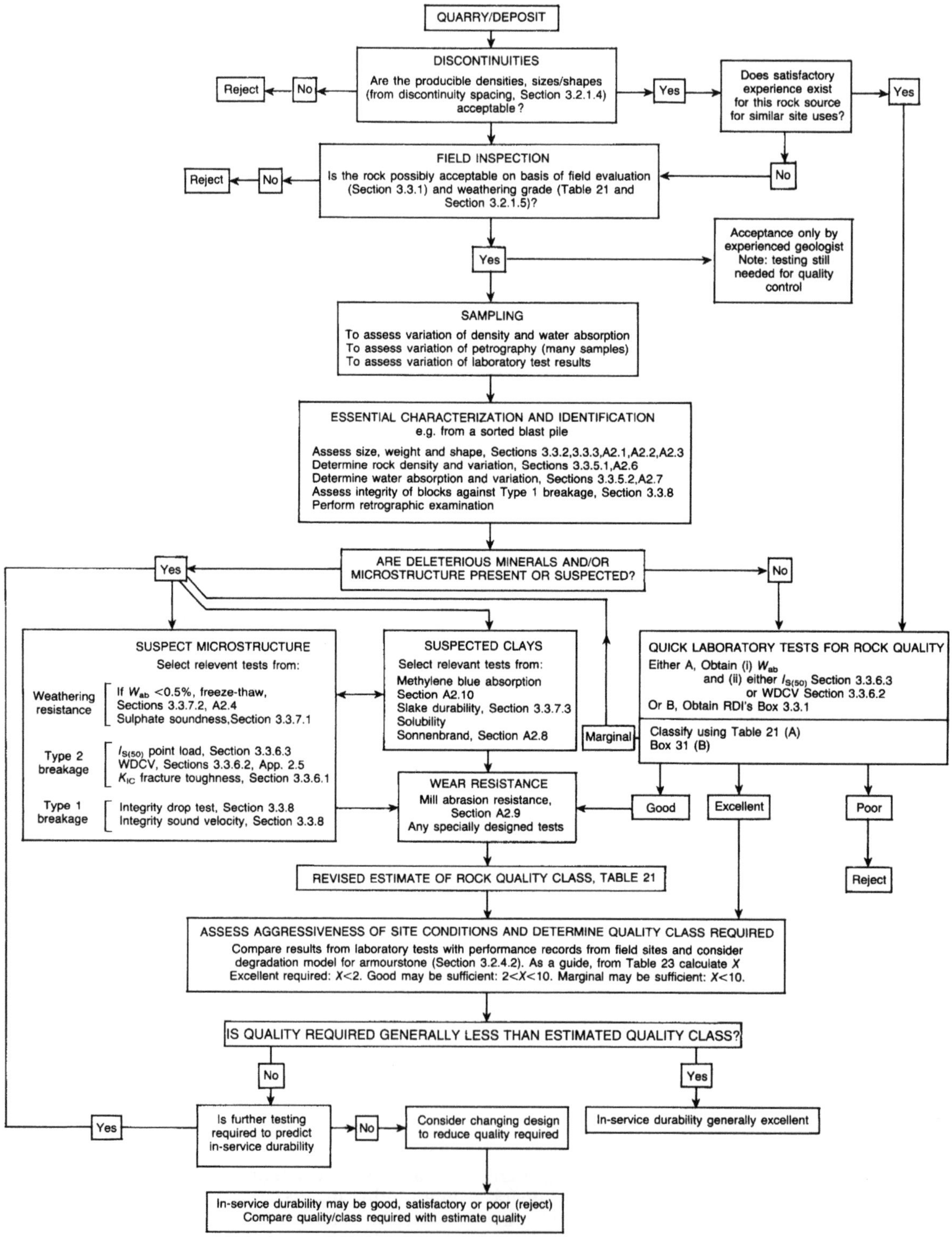

Figure 35.7 Flow diagram for in-service rock durability evaluation (taken from CIRIA/CUR *Rock Manual*).

intrinsic property can be assessed. It is then necessary to consider the aggressiveness of the site conditions under specific headings or overall. The QMW degradation model for armourstone (Latham, 1991) was proposed as a tool for the prediction of rates of degradation by weight loss from armour layers. First, an assessment of the overall aggressiveness of the site is made and second, the QMW mill abrasion test results are considered in relation to wear time on the structure.

These procedures should, for example, allow for the consideration of local sources of marginal quality rocks for armour where nothing else is available in sufficient sizes except for costly imports, and also, where wave conditions are very mild, as in the sheltered seas of the Middle East. (Where sizeable rock is scarce, concrete might be preferred.)

35.8.2 SPECIFICATION

The required rock quality class based on the site conditions should be compared with the intrinsic rock quality class available from different sources or parts of the quarry. This information, which should also draw on any service records or evidence from weathering of outcrops of the source rock, can provide considerably more information than an uncritical acceptance or rejection of the rock source.

If the rock sources are not known at the design stage, the design should generally impose no more severe a requirement than that the rock be better than the lower limit of 'good' but, if site conditions are extreme, 'marginal' or 'excellent' quality may be specified for all or some of the intrinsic properties. In following the flow diagram procedure (Figure 35.7) in parallel with the design process, specification and quality control of rock properties from known source(s) can be realistically defined in the contract. Changes in the design may also be considered which could allow for the use of 'marginal' or 'good' rather than 'excellent' quality rock.

35.9 QUALITY CONTROL

The objective of quality control is to ensure that the client obtains the product required. The quality control of supply and placement of rock materials to at least the minimum requirements specified is often the most important part of the quality systems of both the contractor and the client. It will include control of equipment, training and procedures, with a clear understanding of responsibilities including safety, for example during blasting.

Essentially quality control divides into three parts: firstly, quarry control, which covers selection of faces and beds, blasting techniques, handling, sorting, selection and loading for delivery; secondly, construction control, which starts with checking deliveries, and progresses via stockpiles through detailed placement and control of layer thicknesses and profiles to the finished product; and finally, post-construction monitoring and maintenance (see *Rock Manual* and LCPC, 1989).

The effort needed for quality control strongly depends on the variability of the relevant rock properties and the margins between the requirements and the available or producible quality. It is essential that the work is concentrated at the quarry to minimize handling and avoid excessive and expensive on-site rejection, although a degree of further official control is always necessary on site to account for possible transport effects and since strictly the specification applies to materials in the works.

The homogeneity of the rock formation may vary greatly. This is particularly important for the required intrinsic material properties. Compared with quality control of industrial production, the human factor is important in the production and selection process of armourstone, particularly for the blasting process and the visual selection of gradings, shape and block integrity.

Many aspects of quality control are discussed in the *Rock Manual*, which emphasizes practical procedures during quarry inspections that can help producers and contractors meet their quality plan targets. These include unofficial methods of control that do not lead to certification by testing. It is particularly important that stockpiles and production rates are realistically assessed. Selection of armourstone with the appropriate frequency of individual and bulk check-weighing with feedback to the machine operators is also vital. This is best carried out with a grading plan, as explained in the *Rock Manual*. The increased use of block handling machinery with automated weight output in the driver's cabin is an important advance towards controlling gradings. However, certain designs of weigh-loaders will be very inaccurate for armour blocks, and machines with grabs attached through axially mounted load cells (e.g. as described in LCPC, 1989) will give the most reliable results.

The frequency at which the different test requirements given in the specification will be subject to official testing is the primary concern of the quality plans between client and contractor, and contractor and producer. Certification of most intrinsic property test results may sometimes be required only at the outset of the project, although from a knowledge of the source quarry and production methods the designer or client engineer may be quick to order testing of any suspect loads perhaps containing unsound material, whether or not that is stated in the quality plan. Usually a change in colour, fall in density or increase in water absorption will indicate a reduction in durability caused by weathering. The weathering, breakage and abrasion resistance tests may then have to be repeated. The manufactured properties of shape and grading will require many tests for a large contract with a greater frequency at the start. An example of a test frequency schedule has been given by Reed (1988).

ACKNOWLEDGEMENTS

The writing of this chapter owes a great deal to the knowledge gained during numerous research projects funded by the Science and Engineering Research Council and the CIRIA/CUR Anglo-Dutch project to write a manual on the use of rock in coastal and shoreline engineering. The authors wish to thank Mr Gerard Laan of the Rijkswaterstaat in Delft, Holland for the many free exchanges of views and research results that took place during the preparation of the CIRIA/CUR Rock Manual. Thanks are also due to Jonathan Simm of Robert West and Partners.

REFERENCES

AFNOR (1990) Determination of the index of continuity, L'Association Francaise de Normalisation, p. 18–556.

Allsop, N.W.H., Bradbury, A.P., Poole, A.B. *et al.* (1985) Rock durability in the marine environment. Hydraulic Research Ltd, Wallingford, UK *Rep. SR11.*

Baird, W.F. and Woodrow, K.W. (1988) The development of a design for a breakwater at Keflavik, Iceland, in, *'Design of breakwaters'* Thomas Telford, London. pp. 225–34.

Bell, F.G. (1983) *Engineering Properties of Soils and Rocks,* (2nd edn), Butterworths, Borough Green, UK.

Bradbury, A.B., Latham, J.-P. and Allsop, N.W. (1990) Rock armour stability formulae – influence of stone shape and layer thickness. *Proceedings of the 22nd International Conference on Coastal Engineering,* Delft, 2–6 July, 1990.

BS 812 (1951, 1975), Methods of sampling and testing of mineral aggregate sands and fillers. British Standards Institution, London.

Carmichael, R.S. (1982) *Handbook of Physical Properties of Rock,* Volume 2. CRC Press, Baca Raton, USA.

CIRIA/CUR (1991) *Manual for the Use of Rock in Coastal and Shoreline Engineering.* Published jointly by the Construction Industry Research and Information Association in the UK and the Centre for Civil Engineering Research, Codes and Specifications in The Netherlands.

Collis, L. and Fox, R.A. (eds) (1985) *Aggregates: Sand, Gravel and Crushed Aggregates for Construction Purposes,* Geological Society of London.

Farmer, I.W. (1983) *Engineering Behaviour of Rocks* (2nd edn), Chapman and Hall, London.

Fickelson, M. (ed.) (1990) *Testing Quality and Quality Assurance in Testing Houses.* Chapman and Hall, London.

Fookes, P.G., Dearman, W.R. and Franklin, J.A. (1971) Some engineering aspects of rock weathering with field examples from Dartmoor and elsewhere. *Quarterly Journal of Engineering Geology,* **4**, 139–85.

Fookes, P.G., Gourley C.S. and Ohikere, C. (1988) Rock weathering in engineering time. *Quarterly Journal of Engineering Geology,* **21**, 33–57.

Fookes, P.G. and Poole, A.B. (1981) Some preliminary considerations on the selection and durability of rock and concrete materials for breakwaters and sea defence works. *Quarterly Journal of Engineering Geology,* **14**, 97–128.

Geological Society, (1977) The description of rock masses for engineering purposes: Geol Soc. (London) Eng. Group Working Party, *Quarterly Journal of Engineering Geology,* **10**, 355–88.

Grady, D.E. and Kipp, M.E. (1987) Dynamic rock fragmentation, in *Fracture Mechanics of Rock,* (ed B.K. Atkinson) Academic Press, London, pp. 429–76.

Gustafsson, R. (1973) *Swedish Blasting Technique,* SPI, Gothenburg, Sweden.

Hoek, E. and Bray, J.W. (1981) *Rock Slope Engineering,* (revised 3rd edn) Institution of Mining and Metallurgy, London.

Hudson, J.A. and Priest, S.D. (1979) Discontinuities and rock mass geometry. *International Journal of Rock Mechanics and Mining Sciences and Geomechanics Abstr.,* **16**, pp. 339–62.

Johnson, R.B. and De Graff, J.V. (1988) *Principles of Engineering Geology,* Wiley, New York.

ISRM (1978) Suggested methods for the quantitative description of discontinuities in rock masses: International Society for Rock Mechanics Communcation on Standardization of Laboratory and Field Tests. *International Journal of Rock Mechanics and Mining Sciences and Geomechanics Abstr.* **15**, 319–68.

ISRM (1985) Suggested method for determining point load strength: International Society for Rock Mechanics Commission on Testing Methods. *International Journal of Rock Mechanics and Mining Sciences and Geomechanics Abstr.* **22**, 51–60.

ISRM (1988) Suggested methods for determining the fracture toughness of rock. F. Ouchterlony (co-ordinator): International Society for Rock Mechanics Commission on Testing Methods. *International Journal of Rock Mechanics and Mining Sciences and Geomechanics Abstr.* **25**, 71–96.

Lama, R.D. and Vutukuri, V.S. (1978) *Handbook on Mechanical Properties of Rocks Vol 2: Testing Techniques and Results* Trans Tech Publications, Clausthal, Germany.

Latham, J.-P. (1991) Rock degradation model for armourstone in coastal engineering. London, *Quarterly Journal of Engineering Geology,* **24**, 101–18.

Latham, J.-P. and Poole, A.B. (1987) The application of shape descriptor analysis to the study of aggregate wear. *Quarterly Journal of Engineering Geology,* **20**, 297–310.

Latham, J.-P., Poole, A.B., Laan, G.J. *et al.* (1990) Geological constraints on quarried rock for use in coastal structures, in *Proceedings of the 6th International Association of Engineering Geology,* Amsterdam, 6–10 August 1990; (ed D.G. Price), A.A. Balkema, Rotterdam, pp. 217–25.

LCPC (1989) *Les Enrochments,* Ministere de L'Equipement, du Logement, des Transports et de la Mer, Laboratoires Central des Ponts et Chaussees.

Lee, S.E. and de Freitas, M.H. (1989) A revision of the description and classification of weathered granite and its application to granite in Korea. *Quarterly Journal of Engineering Geology,* **22**, 31–48.

Lienhart, D.A. and Stransky, T.E. (1981) Evaluation of potential sources of riprap and armourstone – methods and considerations. *Bulletin of the Association of Engineering Geologists,* **18**, 322–32.

Morhard, R.C., Chiappetta, R.F. and Borg, D.G. (1987) *Explosives and Rock Blasting,* Altas Powder Company, USA.

Neise, M.S.J., van Eijk, F.C.A.A., Laan, G.J. *et al.* (1990) Quality assessment of large armourstone using an acoustic velocity analysis method. *Bulletin of the International Association of Engineering Geologists,* **42**, 11–21.

Pilarczyk, K.W. (1990) Design of seawalls and dykes – Including overview of revetments, in *Coastal Protection* (ed K.W. Pilarczyk), A.A. Balkema, Rotterdam, pp. 197–289.

Prentice, J.E. (1990) *Geology of Construction Materials,* Chapman and Hall, London.

Reed, J. (1988) The control of rubble-mound construction, with particular reference to Helguvik Breakwater in Iceland, in *Design of breakwaters,* Thomas Telford, London, pp. 235–250.

Simm, J.D., Latham, J.-P. and Orbell-Durrant, C. (1991) Standardizing the approach to specifying quarried rock in coastal structures and breakwaters. Conference on Coastal Structures and Breakwaters '91, London, 6–8 November.

Shore Protection Manual (1984) (4th edn) US Army Corps of Engineers, Coastal Engineering Research Center. US Government Printing Office, Washington.

Van der Meer, J.W. (1987) Stability of breakwater armour layers – design formulae. *Coastal Engineering*, **11**, 219–39.

Van Oorschot, J.H. (1983) Breakwater design and the integration of practical construction techniques. *Proceedings of ASCE conference*, Coastal Structures '83. pp. 108–121.

Vrijling, J.K. and Nooy van der Kolff, A.H. (1990) Quarry yield and breakwater demand, in, *Proceedings of the 6th International Association of Engineering Geology*, Amsterdam, 6–10 August 1990; (ed D.G. Price) A.A. Balkema, Rotterdam, Volume 4, pp. 2927–36.

Wakeling, H.L. (1977) The design of rubble breakwaters. Symposium on the design of rubble-mound breakwaters. Experimental and Electronic Laboratories. British Hovercraft Corporation, Paper no. 5.

Wang, H., Latham, J.-P. and Poole, A.B. (1990) In-situ block size assessment from discontinuity spacing data, in, *Proceedings of the 6th International Association of Engineering Geology*, Amsterdam, 6–10 August 1990; (edited D.G. Price) A.A. Balkema, Rotterdam, pp. 117–127.

Wang, H., Latham, J.-P. and Poole, A.B. (1991a) In-situ block size distribution for quarrying. *Quarterly Journal of Engineering Geology*, **24**, 91–9.

Wang, H., Latham, J.-P. and Poole, A.B. (1991b) Blast design for armourstone production. *Journal of Quarry Management*, Part I (July), 17–21, and Part II (August), 19–22.

CONCRETE, STEEL AND TIMBER IN THE MARINE ENVIRONMENT

36

P.G. Spragg

Maritime structures are built in hostile physical and chemical environments. Consequently the choice of structural materials has an important bearing on the design and construction of such structures. Two of the materials often used are concrete and steel and the author concentrates on the properties, specification and use of these materials to ensure minimum maintenance during the lifetime of the structure. The use of timber is also described.

Other important considerations are the forces to which the structures are to be subjected and these are dealt with in Chapters 27 and 28 (see also Chapters 29 and 35).

CONTENTS

36.1 INTRODUCTION

Generally the structures that maritime engineers are concerned with involve heavy civil engineering and are capital intensive. They can be expected to have effective lives of 40 to 100 years. Although this period may well exceed their useful economic life because of the pace of technological change and market forces, it is unrealistic to design for shorter lives.

The main reasons for this are the significant magnitude and variability of the forces that such structures have to resist. (Effects of wind, waves, ground pressures, seismic effects, ship forces, crane loadings etc.) Because they normally involve large capital investment programmes and maintenance costs must be kept to a minimum the maritime engineer needs to consider how structures can be made durable in the aggressive marine environment. The physical and chemical forces on structures are summarized in Table 36.1.

The most aggressive zones are the tidal zone between high and low water and the splash zone, immediately above high water. Estuaries are often polluted with waste products from manufacturing processes and are tidal.

Modern construction techniques rely largely on the use of concrete and steel, either separately such as blockwork to quay walls, and steel piling to jetties, or combined, as in reinforced or prestressed concrete decks.

Table 36.2 lists the principal materials in typical coastal and harbour structures. It illustrates the overwhelming use of steel and concrete, generally in combination, to form the main structural elements. There is, however, a fundamental difference in approach to their manufacture.

Steel is essentially a factory-produced material so close and continuous control can be exercised on the quality of the completed article. It can be delivered to a site with known and predetermined properties which can be verified by reference to a manufacturer's certificate, and in its finished condition, only requiring incorporation in the structure. Concrete is either delivered in a fluid state from a remote ready-mix plant or produced by the construction contractor on site. Although maritime structures often incorporate precast concrete units manufactured off-site in near-factory conditions the often massive or non-standard shape of maritime precast units favours the contractor making them on site himself.

Coastal, Estuarial and Harbour Engineer's Reference Book. Edited by M.B. Abbott and W.A. Price. Published in 1993 by Chapman & Hall, 2–6 Boundary Row, London SE1 8HN. ISBN 0 419 15430 2.

Table 36.1 Environmental forces

Atmospheric	Carbon dioxide (carbonation of concrete)
Polluted air:	Sulphur dioxide, chlorides
Sun:	expansion – contraction
Rain:	Acid rain, freeze – thaw
Splash zone	Water and air: chlorides, sulphates, alkalis
Immediately above high water	Condensation: salt laden
	Deposited salts
	Galvanic action
	Impact: waves, gravel, rock, ice
	River pollution: acids, alkalis.
Tidal zone	Alternate wetting and drying
Between high water and low water	Chlorides
	Sulphates
	Alkalis
	Impact: waves, gravel, rock, ice
	Abrasion: gravel, sand
	Marine growth
Submerged zone	Abrasion: gravel, sand
Below low water	Scour
	Marine growth
	Marine borers, timber
	Sulphate-reducing bacteria
	Stray electric currents

Whilst a ready-mix plant may apply strict quality control procedures during the production of fresh concrete the final product relies on the contractor to complete the work. On remote sites the contractor often needs to set up his own production capability, and to source the ingredients locally.

Continuous control has to be exercised to achieve satisfactory concrete, and much of the expertise gained by the contractor on one project has to be re-learnt on the next using different concrete constituents and labour force. For the design/specifying engineer to be satisfied that the required concrete will be produced on site a full and detailed specification needs to be written which is relevant to the prevailing conditions of the site covering the main topics of mix ingredients, performance, workmanship and quality control.

36.2 CONCRETE

Concrete is a composite material of water, cement, aggregate and admixtures and usually contains embedded reinforcement. The aggregates and water are usually sourced locally and differ from one site to another. To achieve concrete of satisfactory strength and durability each constituent requires careful specification together with suitable mix design, minimum mix performance requirements, quality control parameters and satisfactory workmanship.

For good durability, the completed concrete should be impermeable and be of the greatest possible density, when cured, compatible with the provisions of adequate workability for placing. This latter requirement is most important to ensure that the concrete is well compacted particularly around reinforcement or embedded metal.

Specification requirements are usually arrived at by a combination of reference to appropriate national standards, particularly for the quality of the constituents, and instructions on performance and workmanship levels based on practical experience. Some of the topics the engineer should consider particularly in relation to the needs of the structure under design and its siting when writing the specifications are discussed further below.

36.2.1 CONCRETE CONSTITUENTS

Water

Water should be clean and free from toxic compounds. Generally water that is safe for drinking and meets the requirements of BS 3148 (1980) is satisfactory for concrete making.

The water should be analysed chemically to establish the levels of the major ions present, and BS 3148 gives guidance on total maximum levels that can be tolerated. Of particular concern are the levels of chlorides, sulphates and alkalis since they can have deleterious effects on the concrete. The tolerance levels in BS 3148 may be difficult to achieve for

Table 36.2 Principal materials in typical maritime structures

Structure	*Precast Concrete*	*In situ*	*Steel*	*Other*
1 Quay walls				
1.1 Gravity	(a) Blockwork	Cope slab	Reinforcement	Rock fill
	(b) –	Diaphragm wall and cope slab	Reinforcement	
	(c) Caisson	Cope slab	Reinforcement	Sand fill
	(d) Monoliths	Cope slab, Plug	Reinforcement	Sand fill
	(e) –	Cope slab	Circular cell	Sand fill
1.2 Bulkhead	–	Cope beam	Main sheet pile and Anchor Ties Reinforcement (also rock anchors and tie backs)	Granular Backfill
2 Open jetties				
2.1 Suspended deck	(a) Large diameter prestressed piles (Concrete beams and slabs)	Deck slab Beams	Reinforcement Prestressing Wire	Granular fill and stone armouring
	(b) Prestressed slab – deck soffit	Deck slab – topping	Tubular piles Reinforcement Prestressing Wire	
	(c) Pile caps/beam planks	Deck slab – topping	Tubular piles Reinforcement	
	(d) Retaining wall	Beam and deck abutment	Retaining wall Reinforcement	Granular fill Stone armouring
2.2 Approachway	Pile cap	Infill	Reinforcement Tubular piles	
	Roadway slab		Support beams Cross beams	
2.3 Dolphins Fixed	(a) Outer shell to pile cap	Infilling	Reinforcement Tubular piles	Timber and rubber Fendering
Flexible	(b) –	–	High tensile Tubes of varying grade/thickness	Timber and rubber Fendering
These structures have many miscellaneous fixtures and fittings in common, such as bollards, access ladders, handrails, lighting standards, walkway bridges, sliphooks, which are either structural or cast steel. In addition quay walls and open jetties are usually fitted with fendering of rubber with steel frames or timber.				
3. Shore protection				
Groynes			Piling (sometimes)	Timber piles and planking
Breakwaters	Armour units	Top slab	Reinforcement	Armour stone Core and filter
Flood protection	Armour units	Top slab	Reinforcement	Core material Filters Reclamation

particular waters, for example water in desert regions is often saline. Therefore the amount of each ion type that can be accepted in the mix water should be considered in relation to the total in the concrete from all sources.

Under no circumstances should chloride-rich sea-water be used for making or curing reinforced or prestressed concrete or concrete with embedded metal. Although it has been used for making plain or mass concrete it will normally cause the structure to weaken and disintegrate within the design life.

Cement

A variety of cements is available ranging from ordinary Portland cement, to sulphate resisting, to cements in which part is replaced by ground granulated blastfurnace slag, pulverized fuel ash or silca fume.

Structural members in maritime structures are usually large which can cause the effects of heat of hydration to increase the risk of cracking; the use of cement replacement materials has much to commend it, to assist in reducing this risk. Such concretes often have a slower rate of gain of strength and may therefore require longer curing times and formwork to be kept in place for longer. Although sulphate-resisting cements were once used in maritime situations to resist sulphate attack, which was seen as the main cause of deterioration, and to provide a low heat of hydration, their use is not normally recommended now. To counteract possible sulphate attack on the tricalcium aluminate (Ca_3A) phase of ordinary Portland cement by sea-water its content is limited to not more than 10% by weight of cement. However, at least 4% Ca_3A by weight of cement is required to resist chloride attack on the steel reinforcement.

For durability in maritime situations the British Standard Code of Practice for Maritime Structures (BS 6349) restricts the cements which can be used to OPC to BS12, low-heat Portland cement to BS 1370, both with the Ca_3A limits noted above, and cements containing blast furnace slag or pulverized fuel ash (which are covered by BS 146, BS 4246 and BS 6588).

Aggregates

Coarse and fine aggregates are the major constituent of concrete, forming about 80% of the total mass. They must be physically durable – strong, dense, hard, chemically sound and well graded. Increasing awareness during the 1970s and 1980s of chemical attack on concrete by different agents, and the problems arising therefrom, have led engineers increasingly to specify chemical as well as physical tests on aggregates. Once an aggregate source has been identified, and accepted, testing should be continued to ensure that satisfactory quality is maintained.

Physical testing

Aggregates are normally tested in accordance with the requirements of the appropriate national standards, such as BS 812 and BS 882 in the UK. Very often the specifying engineer's requirements for marine concrete are more severe particularly for properties that may affect the properties of the concrete. For example elongated and flaky aggregates lower the workability of the concrete mix. The tests that are normally required are given in Table 36.3.

Chemical testing

Sea-water is rich in chlorides and sulphates and they can give rise to corrosion of the reinforcement and concrete. Recognized forms of deterioration are associated with chloride and sulphate levels found in the concrete matrix and limits on the contribution from the aggregate are usually specified. Deterioration can also occur from the disruptive effects of alkali aggregate reactivity.

Various theories on how chlorides initiate corrosion in reinforced concrete have been advanced. Once started, however, the process of corrosion of the reinforcement is well documented (Simms and Fookes, 1989). The expansion of corroding steel (fivefold or more increase in volume) causes the concrete cover to split and drop away exposing the reinforcement to further attack. Tests for chlorides should be undertaken and limits which the engineer considers acceptable be specified.

Some constituents of the cement matrix, particularly tricalcium aluminate, Ca_3A, are vulnerable to attack from sulphates, and expansive products of this reaction can give rise to cracking in the concrete. Although limits are normally recommended on the Ca_3A content of cements, (for example BS 6349 sets the range between 4% and 10% by weight of cement for OPC to BS 12 and BS 1370), limits are also placed on the acid soluble content of the aggregates, particularly of man-made aggregates. In the UK, BS 882 does not specify levels for natural aggregates. From untried sources testing should be undertaken and limits specified by the engineer.

Three types of deleterious alkali-aggregate reaction are understood to occur: alkali – silica reaction, alkali – carbonate reaction, and alkali – silicate reaction.

They are all very rare in relation to the volume of concrete placed and the number of structures built (Concrete Society, 1987, Building Research Establishment). Alkali – silica reaction (ASR) is the most common in structures in the UK, but the others are

Table 36.3 Physical tests on aggregates. The following list is of recommended tests based on those defined in BS 812. Other countries have similar relevant national standards which give similar requirements. Maximum or minimum limits which the aggregate must achieve should be specified by the engineer relevant to the prevailing site conditions

Recommended tests to BS 812

1. Description and classification.
2. Visual examination of fine and coarse aggregates.
3. Particle size distribution by sieve analysis. Fine aggregate to BS 882 continuously graded. Coarse aggregate to BS 882 single sizes.
4. Flakiness index.
5. Elongation index.
6. Angularity number.
7. Aggregate crushing value.
8. Aggregate impact value.
9. '10% fines' value.
10. Relative density of coarse and fine aggregates.
11. Water absorption of coarse and fine aggregates.
12. Bulk density and voids of aggregate.
13. Organic impurities in fine aggregates, (may also be tested to ASTM C-40).
14. Clay, fine silt and fine dust material (finer than 75 micron sieve).
15. Voided shell content.
16. Field settling test.
17. Drying shrinkage BS 812: Part 120 in conjunction with BRE Digest 357.

Additional tests to other Standards

18. Clay lumps ASTM C-142

more significant in some other parts of the world, depending on the local materials and conditions.

For ASR to occur, high levels of moisture must be present within the concrete matrix together with a source of alkalis and an alkali reactive constituent in the aggregate. The reaction is between hydroxides of sodium and potassium and certain forms of silica in the aggregate which form a gel. This gel takes up pore-water; it then swells exerting a pressure which can cause the concrete to crack.

Sea-water provides a continuing source of both moisture and alkalis and if an aggregate is used which contains potentially reactive substances the possibility of reaction is present. Testing of aggregates for potential reactivity, particularly from untried sources, can be undertaken. At present there is no British Standard specifying tests and reference is made to the American ASTM standards C-227 and C-289 for guidance used in conjunction with petrographic examination (C-295).

Test results need careful interpretation. For example some UK aggregates tested to the mortar bar method in ASTM C-227 have been indicated to be unreactive but ASR has occurred in structures built with them. Research continues on this subject.

The most appropriate measure to avoid ASR is to use aggregates containing no potentially deleterious substances. Records over at least the previous ten years evidencing satisfactory performance of the same aggregate/cement combinations in the same exposure location as that proposed will give guidance, but may be difficult to establish particularly in remote locations. Other precautions are listed in BS 5328: Part 1 (1990).

Admixtures

Controlled doses of chemical admixtures are often added to concrete mixes to alter their behaviour. The engineer must be certain that these admixtures do not adversely affect the performance of the hardened concrete when used singly or in combination. Typically, admixtures are used to influence:

1. Setting times: retarders decrease the initial reaction rate between cement and water;
2. Rate of hardening: accelerators increase rate of gain of strength;
3. Water content: water reducers reduce the water content while still providing the necessary workability;
4. Workability: plasticizers/superplasticizers increase workability, e.g. when the water/cement ratio

specified is low, and improve pumpability of the mix;

5. Freeze-thaw resistance: air entraining admixtures provide a microscopic network of air bubbles in the concrete.

To determine the effect that an admixture will have trial mixes with and without varying dosages of admixture are essential. Admixtures should comply with the relevant national standard (BS 5075 in the UK) and when used correctly can enhance the properties of the fresh or hardened concrete. For example, maritime concrete usually needs a high cement/concrete and a low water/cement ratio, but this can give rise to a harsh and unworkable mix. Water reducing agents/plasticisers can enable a dense concrete to be produced which can be compacted well around the reinforcement and still comply with the specified minimum mix requirements and performance. This aspect is described next.

36.2.2 MINIMUM MIX AND PERFORMANCE REQUIREMENTS

In specifying minimum mix and performance requirements the engineer needs to consider the mix constituents, strength grade and any special requirements.

The preferred route is to specify certain criteria and leave the contractor or concrete supplier free to use their expertise to design the mix. The information normally required is (as a minimum):

	Typical requirements depending on exposure
i) Minimum characteristic strength at 28 days	40–50 N mm^{-2}
ii) Maximum aggregate size	Normally 20 mm
iii) Minimum cement content	350–400 kg m^{-3}
iv) Maximum free water/cement ratio	0.42 or less
v) Special requirements	e.g. air entrainment
vi) Workability – slump, or compacting factor	25–50 mm

Guidance on these values is given in the relevant Standards, such as BS 6349, BS 8110, BS 5328 and BS 5400. A common feature of these codes is the importance placed on using high strength, cement rich mixes of low water/cement ratio combined with large reinforcement cover for conditions of severe exposure. The Code of Practice for Maritime Structures (BS 6349) specifies a preferred minimum cover of 75 mm. This is in line with the trend of increasing cover (see Simms and Fookes, 1989) but may still be considered low for concrete in the splash zone where cover of the order of 90 mm is often considered more appropriate. Increasing cover can, however, give rise to problems of increasing crack widths, and can be counter productive.

The minimum cement contents are relatively high and may give rise to a high heat of hydration, a further possible source of cracking. This risk may be reduced by specifying cements with a low heat of hydration and using cements containing ground granulated blast-furnace slag (GGBFS) or pulverized fuel ash (PFA).

36.2.3 CONTROL OF QUALITY

Having set the minimum mix and performance requirements for the concrete it is most important that compliance is demonstrated. Regular checks should be made on the weights of the batched ingredients, workability, placing, compaction and curing. The main measure normally adopted for hardened concrete is to check the strength compliance of concrete cubes. Preliminary testing before the main works start allow mixes to be developed for particular applications, e.g. reinforced, prestressed or underwater concrete. Records of ingredients of acceptable mixes should be kept together with mixing times, water temperature and other relevant factors.

Such records can be of use if tested cubes are found not to comply with the specification. Non-compliance can be a serious matter since completed permanent works are at risk. Further testing may be required such as testing cores cut from the structure and non-destructive testing.

36.2.4 WORKMANSHIP

One of the most important areas that the contractor must address in order to ensure satisfactory placement of concrete is workmanship. Guidance must be given in the specification on the expected levels of competence and supervision required. Topics to be covered include the following.

Transportation	methods, time between mixing and delivery, avoidance of contamination, segregation, loss of constituents, pumping, spraying.
Placing	correct position and sequence, layer thickness, means of spreading, free fall height, chutes, no damage to form-work and reinforcement, tidal pours.

Compaction	complete against surfaces, ensure greatest density and impermeability, use of vibrators essential.
Screeding	ensure that the top surface is closed and dense, state finish required.
Curing	methods, minimum times required, curing membranes. (Curing prevents moisture loss, avoids large temperature gradients and reduces the risk of cracking).
Construction joints	types, waterstops if required, spacing and positions specified or guidance given, treatment after striking of formwork.
Weather precautions	additional precautions during inclement, cold or hot weather (ACI, 1989; FIP, 1986).
Formwork	design requirements – strength, no movement, supports, chamfers at edges, grout tight to ensure no loss of grout, finishes required, moulds for precast elements.

Any of these aspects can give rise to various forms of defects and non-structural cracks, which may adversely affect the durability of the hardened concrete (Concrete Society, 1982, rev. 1992).

36.2.5 REINFORCEMENT

When concrete is required to resist tensile stresses it is normally reinforced or prestressed. In both cases the steel is factory produced under controlled conditions and its quality is witnessed by test certificates. In UK practice reinforcement is specified as mild steel or high tensile steel to comply with BS 4449 and is cut and bent to required dimensions to BS 4466.

Site welding of reinforcement is not recommended although with considerable care it can be satisfactory. Steels complying with BS 4449 are normally considered to be weldable or readily weldable and welding should be carried out in accordance with the requirements of BS 7123.

Prestressing strand should meet the requirements of BS 5896 and mesh to BS 4483.

The major site requirements are to ensure that the delivered reinforcement is kept clean and salt free and when fixed is in a rigid cage set within the formwork to give the specified cover. Binding wire can be mild steel or stainless steel and loose ends should be turned inwards to the cage centre.

With the problems that have occurred over the last two decades with corrosion of reinforcement, often the result of chloride ingress, specifying engineers have shown increasing interest in alternative means of providing durability to the reinforcement.

1. Coatings on conventional reinforcement;

(a) Galvanizing	a sacrificial zinc coating. BS 729 (ASTM A123 or ISO 1461) (BRE, IP 14/88)
(b) Fusion bonded epoxy coating	an organic coating to BS 7295 (previously to American Standard ASTM A775–81) (BRE, IP 14/88)

2. Cathodic protection — the impressed current makes the reinforcement cathodic. (Pithouse, 1986)
3. Alternative steels — austinitic grades 304 S31 and 316 S33, and bars complying with BS 6744:1986. (BRE, IP 14/88)
4. Alternative reinforcing materials, such as
   - glass fibre (alkali resistant)
   - aramid – fibres
   - carbon fibres
   - polymers –
     - polypropylene
     - high density polyethylene

   (See Miesseler and Preis, 1988, Veldhuijzen Van Zenten, 1989, Gerritse and Werner, 1990)

The use of these materials does not permit any reduction in the necessary concrete quality or any reduction in cover but they enable the durability of the structure to be enhanced.

36.2.6 ADDITIONAL MEASURES TO PROMOTE DURABILITY

Durability can be enhanced by attention to detail at the design stage, such as: good drainage; ensuring no ledges or areas for corrosive material to accumulate; use of simple, robust units; specifying pre-cast units wherever possible, including precast reinforced and prestressed concrete piles. Also: reinforcement design, to limit flexural cracking and to limit shrinkage/thermal cracking; and reinforcement detailing which is rigid and simple to fix, to promote ease of achieving correct cover.

Coatings applied to the concrete surface are being investigated and increasingly used. Such materials include bitumastics, epoxy resins and silanes.

Silanes combine chemically with the concrete surface and prevent the ingress of moisture while permitting its egress and have particularly increased in popularity. The UK Department of Transport specifies the material for use in bridge structures for areas subject to attack from de-icing salts (Department of Transport, 1990a,b). Further guidance on surface coatings can be found in CIRIA (1987).

Microsilica is an ultra fine pozzolana formed by condensing the fumes arising from the smelting of silicon alloys in an electric arc furnace. Its chemical composition contains over 90% silicon dioxide and in concrete it forms part of the cementitious components either as a concrete replacement or additive. Levels of up to 10% by weight of cement are used. It is considered to promote the durability of concrete by reducing permeability and heat of hydration increasing the strength and sulphate resistance, and suppressing ASR (Bradley-Williams, 1986; Parker, 1985).

36.3 STEEL

Steel is used extensively in maritime structures. It can be fabricated into complete items for delivery to site, such as fender frames, in a partially completed state, such as long piles for site jointing or used as a reinforcement to concrete, as previously described.

Steel has high strength and stiffness and good ductility and is relatively easily worked. However, like many metals it readily corrodes in moist air and hence requires protection (see section 36.3.3).

Typical applications are listed in Table 36.2 and usage can be broadly classified as follows:

1. Structural steel beams, joists, columns and angles hot rolled in shapes (BS 4).
2. Structural hollow sections (BS 4848)
3. Piles –
   tubular (from plate circumferentially and longitudinally welded, also helically welded)
   box (Larssen)
   H (universal bearing piles)
   sheet piling (various shapes U, Z, H straight web)
4. Tie bars and couplings for anchorage and post tensioning applications.
5. Rails (for cranes, ship unloaders BS 11, 105)
6. Miscellaneous fixtures and fittings – fastenings, holding down bolts, ladders, handrails, light standards, chains, shackles, bollards, mooring rings, sliphooks.

36.3.1 STRUCTURAL STEEL, PILES AND FABRICATED MISCELLANEOUS FITTINGS

For most structural applications steel sections and plate to the equivalent grades specified in BS 4360 (1986)* give satisfactory performance. Although four main grades are specified, (40, 43, 50 and 55), grade 40 is rarely specified, and most steel used is grade 43. Apart from strength these steels give a satisfactory structural performance from the point of view of weldability and toughness.

Weldability

Steel is essentially iron alloyed with carbon. Small quantities of other elements are also present, some intentionally and some as impurities, which affect the physical properties of the steel. Carbon has the greatest effect on weldability and steels to BS 4360 normally have the percentage limited to no more than 0.25%. A measure of weldability is the carbon equivalent (CE) value which is defined in BS 4360 as:

$$CE = C + \frac{Mn}{6} + \frac{Cr + Mo + V}{5} + \frac{Ni + Cu}{15}$$

when the elements are measured by weight. Values of CE for a particular steel may be calculated using data given with the mill certificates. BS 4360 gives values of CE varying between 0.39% and 0.54% and generally the lower the value the more readily weldable the steel is. Each steel grade is subdivided into grades A to F, weldability increasing from grade A to F. The operation of welding, type of weld and methods of testing quality are referred to later in the chapter.

Toughness

Toughness is a measure of the energy required to break the steel and is determined under test in an Izod or Charpy impact machine, the latter being the more popular. Standard square section samples with a V-notch cut in one face are broken under impact at temperatures between +20°C and −60°C and the energy recorded. BS 4360 specifies a minimum energy level that each steel grade must attain at a particular temperature. The engineer's main concern is that a steel is tough enough to avoid brittle fracture that is failure arising from the rapid propagation of a small crack.

BS 5950 *Structural use of steelwork in building* gives guidance on this mainly through specifying a 'safe' maximum material thickness for a given grade of material for normal applications and thus avoiding the likelihood of cracking. A similar but more detailed approach is given in BS 5400, the bridge code. Generally steels specified in BS 4360 will have adequate toughness for normal maritime structures. However, in certain applications, for example at low temperatures, or thick plates the engineer may need to consider fracture toughness in more depth through the application of fracture mechanics (See BS 5762) and is advised to seek specialist advice.

* Currently individual European national standards are being unified under a single series of standards of which BS EN 10.025 (1990) is now available for use in conjunction with BS 4360 (1990). Equivalent steel grades are grade 40 – Fe 360, grade 43 – Fe 430, grade 50 – Fe 510.

36.3.2 WORKMANSHIP

The engineer, having selected and specified the material grades for a structure's design, needs to give guidance to the contractor on workmanship requirements for fabrication, erection procedure and corrosion protection.

Fabrication

The engineer will show on his drawings fabrication requirements such as bolting and weld details. Bolt types should be specified, e.g. black, precision or high strength friction grip and for the maritime environment be of generous diameter. Weld types such as fillet and butt should be detailed showing the thickness required. Welding is normally specified to be carried out to the requirements of BS 5135 and careful consideration needs to be given by the fabricator to his welding methods to avoid distortion in the welded parts. All fillet welds should be continuous to avoid corrosion traps and all connections details kept as simple as possible to assist in erection and maintenance.

Welders are normally required to undertake tests to demonstrate their competence. This is of particular importance on projects in remote parts of the world where welders of many different nationalities and backgrounds may be employed. BS 4871 and 4872 give procedures to be adopted and BS 709 is on the testing of the completed welds. With important welds such as butt welds in main members, weld procedures need to be established and both non-destructive and destructive testing undertaken to prove that the procedure is satisfactory. Acceptance levels for defects must be stated. Normally once a procedure is accepted, welds in the project are tested on a random percentage basis using non-destructive methods (NDT) to confirm that welders are continuously achieving the required standard.

Typical non-destructive tests are:

Visual	surface defects	no permanent record
Dye-penetrant	surface cracks	no permanent record
Magnetic particle	surface cracks	no permanent record
Radiographic	porosity, slag inclusions cavities, lack of penetration	permanent record
Ultrasonic	all sub-surface defects and laminations	difficult to obtain record

Typical destructive tests are:

Tensile	across the weld	to test weld strength
Bending	face, root, side	to test weld ductility
Impact	Charpy V notch	to test weld toughness
Hardness	Brinell, Rockwell, Vickers	to check weldability, (also abrasion resistance)

(other tests for fatigue and cracking are also possible)

(the above listing is based on Table 1 of Gourd (1986). Also see Gourd (1986) for a detailed description of welding).

Erection procedures

Once the individual items of steelwork have been completed they need to be installed in the works. Requirements for erection and installation need to be specified. This could range from the simple grouting up of base plates to a complete specification for the installation of important structural elements such as piles. The details of required pile performance should be stated, as should methods of handling piles, piling plant suitability, positional tolerances and final testing (ICE, 1988).

36.3.3 CORROSION PROTECTION

In the marine environment corrosion can be higher than on land for several reasons:

1. Cathodic/anodic dissolution in sea-water (galvanic action);
2. Continual loss of corrosion products by sea or river action;
3. Abrasion of the surface from floating fenders or sand or gravel moved by sea action or river currents;
4. Sulphate reducing bacteria;
5. Difficulty in maintaining anti-corrosion measures.

Because of the aggressive nature of the environment and the difficulty of maintenance the engineer needs to consider and specify appropriate anti-corrosion measures. These could include;

1. Providing a sacrificial corrosion allowance (see BS 6349 Table 22 for typical rates of loss);
2. Adding wear plates;
3. Designing to grade 43 stresses and specifying grade 50 material;
4. Ensuring fabrication details are simple, robust and do not promote the entrapment of corrosion elements;
5. Specifying a cathodic protection system;
6. Applying a coating system;
7. Applying a thick polypropylene heat shrink sleeve;
8. Installing a water repelling wrapping.

Common practice for maritime structures has been to apply a thick coating of, for example, coal tar epoxy

Table 36.4 Typical uses of timber

1. Jetties
   load bearing piles, jetty and mooring dolphins
   fendering
   decking
   miscellaneous, kerbing
2. Sea defences
   Groynes, piles and planking
3. Marinas
   piling for moorings
   sheeting to walls
   decking to fixed and floating walkways
4. Lock gates
   the structure of small span mitre gates
   large spans, fendering, rubbing strips and closers to steel gates (heel and mitre posts and seals to the cill)

paint and specify an impressed current or sacrificial anode cathodic protection system.

When specifying coating systems the engineer needs to ensure the specification includes adequate requirements for;

1. Surface preparation;
2. Coating method;
3. Ambient conditions under which coatings may be applied;
4. Wet and dry film thicknesses to be achieved;
5. Testing and inspection of completed work;
6. Repair of damaged surfaces.

It is most important that the advice of the coating supplier is sought and the workmanship in application accords with his requirements. BS 5493 gives very helpful and relevant advice on the specification and application of coating systems.

36.4 TIMBER

Timber is used in maritime structures in both structural and non-structural applications. It possesses a high strength to weight ratio, is relatively easy to handle, can readily be cut to length and can have a long useful life span. Typical uses are shown in Table 36.4. In all cases, careful selection of the timber species for the particular requirements is essential in order to meet the demands of strength and durability.

In this respect BS 5268, *Structural use of timber*, is able to give guidance. Timber can be strong; for example, Greenheart and Ekki have permissible compression and bending stresses parallel to the grain approaching $20\,\mathrm{N\,mm^{-2}}$, a tensile strength of about $10\,\mathrm{N\,mm^{-2}}$ and good resilience. It is therefore structurally ideal for load-bearing piles and piles subjected to bending such as fender piles.

However, by its very method of creation, which results in relatively small section sizes and lengths, timber is of only limited structural application where applied loads are low. Thus modern port and jetty design and construction requiring increasingly heavy loads to be supported in deeper waters together with the decreasing availability of suitable timbers, and hence increasing coat, has seen a decline in its use. Its use has thus often been limited to fixed fendering where its property of resilience comes to the fore or kerbing where its durability is exploited.

However, in shallower coastal waters, such as marinas, and inland waterways, timber is often used, particularly for small-span lock gates. Typical species specified for use in the maritime environment include both hardwoods and softwoods such as:

Hardwoods

Balau	Jarrah
Basralocus	Meranti
Ekki	Oak
Elm	Okan
Greenheart	Opepe

Softwoods
Pitch Pine
Douglas Fir
Western Hemlock
Baltic Redwood

Many species can be used, however, and the engineer will need to make his final choice and specification based on all relevant factors including availability in the area where the works are to be constructed.

As previously noted, durability is an important requirement. Timber can suffer attack by physical damage, fungal decay or rot and marine borers.

Physical damage can be caused by mechanical impact and abrasion by vessels or floating objects or by the simple wearing away with time by road or pedestrian traffic or mechanical handling equipment. Certain sea-beds subjected to fast currents may also cause an abrasive scouring action when brought into motion locally where a pile enters the sea bed. This can give rise to a loss of section and will need special measures such as protection fixed to the pile or stabilizing of the sea-bed with a heavy, much coarser, material like stone.

Biological damage may occur through rot or attack by marine borers. Normally air-dried or kiln-dried timber which achieves a moisture content of less than 20% will, if this level is maintained, not decay through fungal attack or rot. However, in the marine environment, its moisture content will naturally rise although where oxygen is precluded the timbers will remain rot-free. A typical example of this is timber piles driven into clay to form foundations to old masonry quay walls which, upon extraction, are often found to be as sound as the day they were installed. The main area of attack is in the structure above sea- or river-level where poor detailing can give rise to water ponding permitting penetration of the joints.

Horizontal surfaces are the most susceptible. Some timber species can be treated with preservatives although the preservative may not penetrate very deeply because of the density of the timber. Thus for a maritime situation, a careful selection of the timber species is important both for natural resistance to attack and the ability to accept a preservative.

The second form of biological attack is by marine borer. The more prevalent types are the ship worm (*Teredo*) and gribble or sea louse (*Limnoria*) although other types exist. Depending on the environment conditions prevailing these borers can cause extensive damage quite quickly. The *Teredo* is a mollusc and enters the timber through small holes or incisions destroying the timber as it grows. The gribble is a form of crustacean and attacks the surface of the timber to create an interlocking network of holes. Their level of activity depends on water salinity, temperature, amount of pollution, current velocity and oxygen in the water. No timber is known to completely resist attack from borers but by careful selection resistance for up to a period of 20 years is possible. The Building Research Establishment gives guidance on timbers resistant to borers and for marine work such as piling these would typically be Greenheart, Ekki and Basralocus. (BRE, 1972).

Care should be exercised in the handling of timber to ensure that its surface, particularly if treated, is not penetrated. Where metal fixings are used, as in groyne construction and fendering, which penetrate the timber after pre-boring, the fixing should be treated by hot-dip galvanizing and covered with a generous coating of bituminous paint immediately before fixing and erection. For fendering, timber holes should be countersunk and capped with timber plugs set in mastic. Several British Standards give recommendations on the design and construction of fastenings and are listed in the reference section at the end of this chapter.

REFERENCES

ACI (1989) Committee 305: Hot weather concreting. American Concrete Institute.

Bradley – Williams, T.S. (1986) The production of microsilica. *Concrete*, 17–18.

BRE (1972) Marine borers and methods of preserving timber against their attack. Building Research Establishment, Garston, Watford, (UK) *Technical Note* 59.

BRE (1988) Corrosion-protected and corrosion-resistant reinforcement in concrete. Building Research Establishment, Garston, Watford, UK BRE Information Paper IP14/88.

BRE (1988) Alkali aggregate reactions in concrete. Digest 330. Building Research Establishment, Garston, Watford, UK.

CIRIA (1987) Protection of reinforced concrete by surface treatments CIRIA – Technical Note 130. Construction Industry Research and Information Association, London, UK.

Concrete Society (1982) Non Structural Cracks in Concrete. *Tech. Rep.* 22.

Concrete Society (1987) Alkali – silica reaction – minimising the risk of damage to Concrete. *Tech. Rep.* 30.

Department of Transport (1990a) Impregnation of concrete highway structures. Departmental Advice Note BA 33/90. Highways and Traffic. Department of Transport, London, UK.

Department of Transport (1990b) Criteria and material for the impregnation of concrete highway structures. Departmental Standard BD 43/90. Highways and Traffic. Department of Transport, London, UK.

Federation Internationale de la Précontrainste (1986) *Concrete Construction in Hot Weather. FIP Guide to Good Practice*, Thomas Telford, London, UK.

Gerritse, A. and Worner, J. (1990) Arapree – the aramid prestressing tendon. Properties and recent applications A. Gerritse & J. Werner. TRB 69th Annual Meeting.

Gourd, L.M. (1986) *Principles of Welding Technology* – (2nd edition) Edward Arnold, London.

ICE (1988) *ICE Specification for Piling*, Thomas Telford, London, UK.

Miessler, H.J. and Preis, L. (1988) Heavy Duty Composite Bars made of Glass Fibre as reinforcement in concrete and Foundation Constructions. H.J. Miesseler & L. Preis *Bauen Mit Kunstoffen*. **2**. 4–14.

Pithouse, K.B. (1986) Cathodic protection of steel reinforcement. *Civil Engineering*.

Parker, D.G. (1985) Microsilica concrete. Concrete Society Current Practice Sheet 104 'Concrete', 21–22.

Simms, J.D. and Fookes, P.G. (1989) Improving reinforced concrete durability in the Middle East during the period 1960–1985; an analytical review. *Proceedings of the Institution of Civil Engineers, Part* 1 **23** (5) 333–58.

Veldhuijzen Van Zenten, R. (1989) Aramid Fibres as a construction material in semi-permanent structures. 28th International Manmade Fibres Congress, Austria, 1989.

APPENDIX: BRITISH AND AMERICAN STANDARDS

BRITISH STANDARDS

General

BS 6349 Code of practice for maritime structures
BS 8110 Structural use of concrete
BS 5400 Steel, concrete and composite bridges
BS 5950 Structural use of steelwork in building
BS 8004 Code of practice for foundations
BS 8007 Code of practice for design of concrete structures for retaining aqueous fluids

Concrete

BS 5328 Methods of specifying concrete including ready mixed concrete

Water

BS 3148 Methods of test for water for making concrete.

Cement

BS 12 Specification for Portland cements
BS 146 Specification for Portland-blast furnace cement
BS 1370 Specification for low heat Portland cement
BS 4246 Specification for low heat Portland blast furnace cement
BS 6588 Specification for Portland pulverized – fuel ash cement

Aggregates

BS 812 Testing aggregates
BS 882 Specification for aggregates from natural sources for concrete

Tests

BS 1881 Testing concrete
BS 1377 Methods of test for soils for civil engineering purposes.

Admixtures

BS 5075 Concrete admixtures

Reinforcement

BS 4449 Specification for carbon steel bars for the reinforcement of concrete
BS 4466 Specification for scheduling, dimensioning, bending and cutting of steel reinforcement for concrete
BS 7123 Specification for metal arc welding of steel for concrete reinforcement
BS 5896 Specification for high tensile steel wire and strand for the prestressing of concrete
BS 4483 Specification for steel fabric for the reinforcement of concrete
BS 4486 Specification for high tensile alloy steel bars for the prestressing of concrete
BS 7295 Fusion bonded epoxy coated carbon steel bars for the reinforcement of concrete
BS 729 Specification for hot dip galvanized coatings on iron and steel articles
BS 6744 Specification for austenitic stainless steel bars for the reinforcement of concrete

Steel

BS 4 Structural steel sections
BS 4848 Specification for hot rolled structural steel sections
BS 11 Specification for railway rails
BS 105 Light and heavy bridge type railway rails
BS 4360 Specification for weldable structural steels
BS 5135 Specification for arc welding of carbon and carbon manganese steels
BS 4870 Specification for approval testing of welding procedures
BS 4871 Specification for approval testing of welders working to approved welding procedures
BS 4872 Specification for approval testing of welders when welding procedure approval is not required
BS 709 Methods of destructive testing fusion welded joints and weld metal in steel
BS 5762 Methods for crack opening displacement testing
BS 5493 Code of practice for protective coating of iron and steel structures against corrosion.

Timber

BS 5268 Structural use of timber
BS 5756 Tropical hardwoods graded for structural use
BS 5589 Code of practice for preservation of timber

AMERICAN STANDARDS

ASTM C-33 Standard specifications for concrete aggregates
ASTM C-88 Soundness of aggregates by use of sodium sulphate or magnesium sulphate

ASTM C142 Clay lumps and friable particles in aggregates

ASTM C227 Potential alkali reactivity of cement aggregate combinations (mortar bar method)

ASTM C289 Potential reactivity of aggregates (chemical method)

ASTM C295 Petrographic examination of aggregates for concrete

ASTM C342 Potential volume change of cement aggregate combinations

ASTM 586 Potential alkali reactivity of carbonate rocks for concrete aggregates (rock cylinder method)

UK DOCUMENTS

National Buildings Studies 1958 Research Paper 25. Reactions between aggregates and cement. Building Research Station.

Shrinkage of natural aggregates in concrete. Building Research Establishment Digest 357, January 1991.

GEOTEXTILES IN COASTAL AND HARBOUR ENGINEERING

37

G. Heerten

The profession is seeing an increasing use of geotextiles in coastal, estuarial and harbour engineering. This chapter reviews the many applications of these materials and describes the type of geotextile best suited to solve a particular problem. Field trials are described which illustrate the long-term performance of the various types of material including among others multi-filament woven fabrics of polyester or polyamide and needle-punched non-woven fabrics. Filter efficiency is dealt with at length to illustrate the significance of this important design parameter. A number of case studies are described as further illustrations of design principles (see also Chapter 30).

CONTENTS

37.1 INTRODUCTION

Coastal engineering was the starting point for the use of synthetic fabrics in geotechnics in the 1950s. Pioneers in coastal engineering first used synthetic fabrics to form huge sand bags for building groins and closing dikes. Based on the advantages of these fabrics (low weight, high strength, long-term resistance) woven and, more recently, non-woven fabrics have been increasingly used in coastal engineering and, latterly, in waterway engineering, dam construction, road construction, railroad construction, tunnelling – all fields of geotechnical engineering.

In the 25 years of developments in geotextile applications, many tests and investigations have been carried out and many construction and design methods have changed due to increased experience. Today we have sufficient expertise to use geotextiles in design and construction in coastal engineering applications. This chapter considers the fundamental principles for the successful use of geotextiles in coastal and harbour engineering.

37.2 APPLICATIONS OF GEOTEXTILES

The following examples give a rough idea of geotextile applications in coastal and harbour engineering today;

1. Filters in erosion control structures as revetments of sea-dikes (Figure 37.1), sea-walls or bottom protection structures, e.g. of sluices or storm surge barriers (Figure 37.2);
2. Separators in the foundations of groins and breakwaters (Figure 37.3);
3. Fabric forms for sand-filled bags or tubes as construction elements of groins and dikes (Figure 37.4);
4. Flexible scour protection mats at different offshore and coastal structures.

Due to the importance of geotextiles for the safety and longevity of many coastal protection structures, selection of the most suitable product has to be made carefully. Unfortunately, however, difficulties arise for the planner because he is sometimes faced with data on various products that cannot be compared. Moreover, important loads can often only be calculated by means of very complex methods and thus experience and a feel for the subject are essential in considering the following functions;

Coastal, Estuarial and Harbour Engineer's Reference Book. Edited by M.B. Abbott and W.A. Price. Published in 1993 by Chapman & Hall, 2–6 Boundary Row, London SE1 8HN. ISBN 0 419 15430 2.

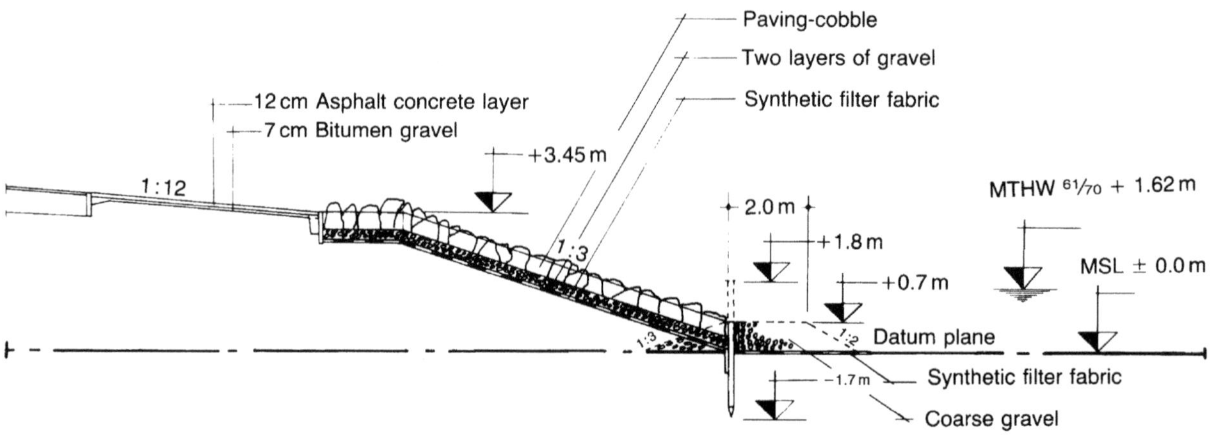

Figure 37.1 Cross-section of revetment of sea dike without foreland (North Frisian Coast).

Figure 37.2 Willow-fascine mattress for bottom protection.

1. Wave attack and current loads;
2. Loads by seepage flow;
3. Appearance of erosions and sedimentation.

The application of geotextiles represents in reality a true challenge for the engineer in choosing the right product. Possible failures during construction or execution very often become evident after the first big storm when deformation of the structure or even total destruction occurs. The following thesis can be applied for offshore techniques as well as coastal protection: think big and do it right the first time! To achieve

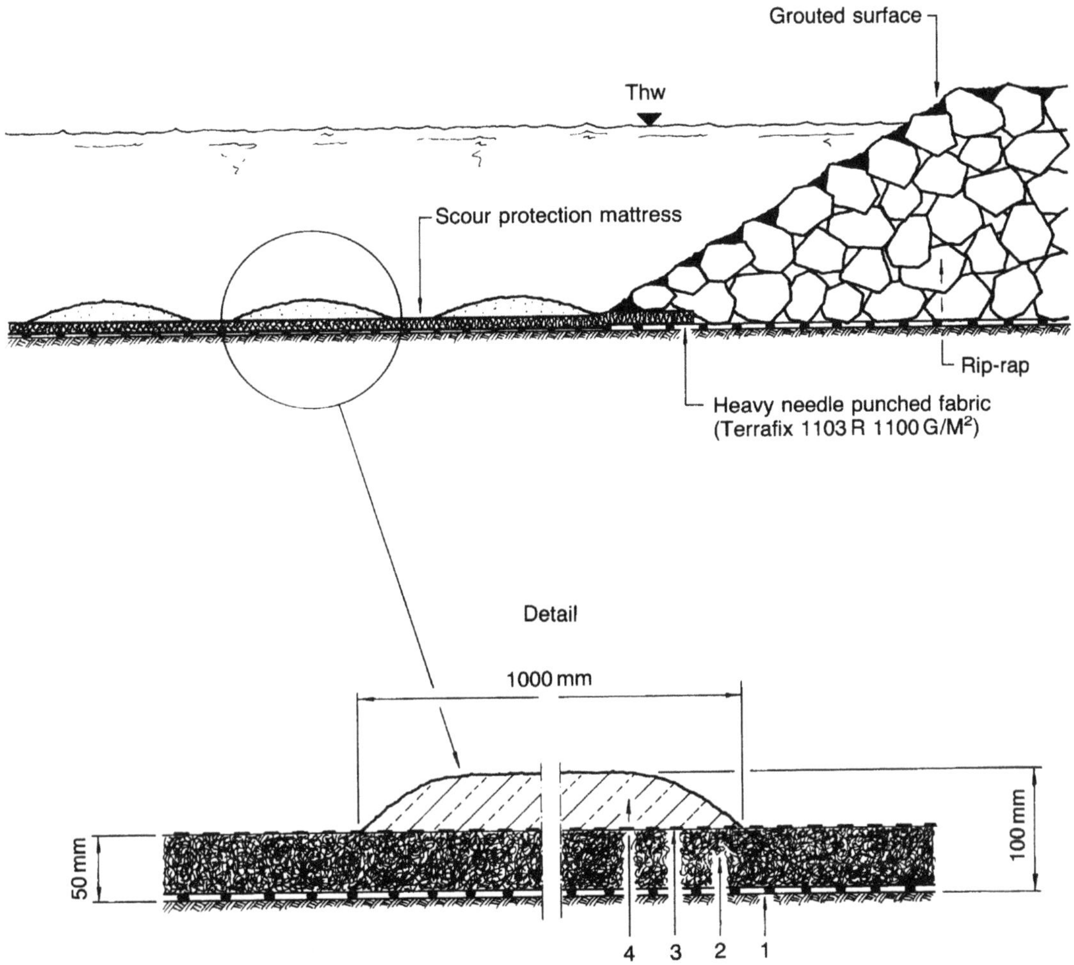

Figure 37.3 Foundation of a groyne on top of a geotextile and scour protection at toe. (1) filter layer (woven or non-woven); (2) sedimentation layer; (3) high strength reinforcement woven fabric; (4) concrete ballast elements.

this the following properties of geotextiles in coastal engineering are of importance;

1. High filtering efficiency (soil retention function and water permeability);
2. Sufficient resistance to all installation and structural loads (e.g. puncture resistance, abrasion resistance, heat resistance to bituminous construction materials installed when hot);
3. Sufficient long-term resistance to increased UV-degradation in coastal areas and in salt water regimes;
4. Good frictional behaviour on existing subsoil slopes.

Thus only comparatively heavyweight geotextiles can be recommended for coastal protection; these geotextiles being characterized by a high resistance and filtering efficiency. But above all, the construction method selected has to be suitable for synthetic materials. This means that the 'geotextile' has to be installed so that its particular strengths – especially the filtering and soil retention properties – are guaranteed against all possible influences. It is essential to make sure that the properties of the material, like puncture and abrasion resistance are not exceeded during the construction process.

Figure 37.4 Closing a dike with huge sand bags.

37.3 LONG-TERM RESISTANCE OF GEOTEXTILES

37.3.1 FIELD INVESTIGATIONS

In order to provide safeguards and expand the knowledge of the long-term resistance of geotextiles used in coastal protection, the Franzius Institute for Hydraulic Research and Coastal Engineering of the University of Hanover carried out an extensive research programme on the German North Sea Coast from 1978 to 1980 (Heerten, 1981).

The most important part of the investigation concerned 39 samples of fabrics which had been installed for between 5 and 21 years at 13 locations on the North Sea Coast of the German Bight. 16 samples were excavated from the revetments and 23 samples were taken from sand bags and sand-filled tubes. Figure 37.5 shows a needle-punched non-woven fabric after removal of the filter and cover layers. An example of the application of geotextiles to sand-filled tubes as stabilizing elements for beach feeding areas likely to erode is shown in Figure 37.6. Two differently loaded samples were taken from these tubes or sand bags: samples from the top side (weathered) and from beneath (protected against UV-radiation).

37.3.2 RESULTS FOR LONG-TERM RESISTANCE

The long-term resistance of a geotextile is normally judged by its tensile strength, namely a comparison of the original and final strengths. All samples mentioned above have been tested in accordance with the 'conventional strip tensile strength tests for textile materials'. Various products such as multi-filament woven fabrics of polyester or polyamide, needle-punched non-woven fabrics of polyester, polyamide or polypropylene and tape fabrics of polyethylene and polypropylene have been tested. The average values of residual strengths of the raw materials for both the weathered and unweathered (UV-protected) samples were tested. It could be shown that the residual strengths of the weathered tapefabrics of polyethylene or polypropylene are higher than those of the weathered multi-filament woven fabrics of polyester, polyamide or polypropylene. Comparable results can also be seen for the unweathered samples. However, the losses in strength are considerably lower (≙50%).

An interpretation and evaluation of these results has to take into consideration the considerable influence of fibre fineness on the probability of damage.

Figure 37.5 Needle-punched non-woven fabric after removing revetment cover layers.

This explains the relatively good results obtained with the heavy tape fabrics, whose fibre fineness is many times that of the multi-filament woven fabrics. Furthermore, there are many influences which are difficult to evaluate, e.g. the suspended load of the sea-water (turbidity), the duration of the daily tidal flow, the temporary changes of the geotextile covered with mud, micro-organisms or rip-rap, drifting wood, ice and shipping, which influence the unprotected weathered fabrics. The following, generally valid, statements can be derived from the investigations, described in detail by Heerten (1981).

1. An effect on the resistance of geotextiles, caused by biological or chemical influences, could not be established for the salt-water environment.
2. In the case of free weathering, considerable losses in strength do occur by UV-radiation. This influence can be mitigated by using coarse fibres (protection of the core by damaged margin zone) or thick, heavy products (protection of internal fibres). All tested products were intact from the point of view of technical construction because the strength of the tubes and bags, designed for the loads at installation, included considerable reserves for the load at application in the structure.
3. Proven damage to geotextiles (only woven fabrics in revetment construction) was clearly caused by mechanical influences during installation (dumping procedure) or by external abrasion, pushing or cutting loads (Kniess and List, 1982). This emphasises the need for a construction method suited to the use of synthetic materials and robust geotextiles in coastal protection.

On the basis of these investigations and existing experience, minimum weights per unit area – comparable to the requirement for minimum thicknesses for synthetic sealing membranes – should be fixed in order to guarantee a sufficient robustness. This should be a necessary requirement as it is impossible to derive any information concerning robustness from the measurement of the tensile strength of the product. But the robustness normally increases with higher elongation. The following weights per unit area may serve as a guide.

1. For the dumping of mineral grain filter layers (crushed material/gravel) or beneath feeble loaded

Figure 37.6 Sand-filled tube in beach feeding area.

concrete block revetments without an additional protection layer (e.g. reed mattress)

$$F_G \geqslant 500\,\mathrm{g\,m^2}$$

2. For the dumping of rip-rap layers of grade II to III (approx. 15 to 50 cm stone size) with a single weight of the stones ≤50 kg or beneath highly loaded concrete block revetments without an additional protection layer (e.g. reed mattress)

$$F_G \geqslant 750\,\mathrm{g\,m^2}$$

3. For the dumping of rip-rap layers with a single weight of stones >50 kg (stone size >50 cm)

$$F_G \geqslant 1000\,\mathrm{g\,m^2}$$

Evidently higher single weights of the stones or heavy artificial breakwater elements (e.g. tetrapodes) necessitate field trials.

The following characteristics should always be taken into account: mechanically bonded (needle-punched) non-woven fabrics have the most favourable behaviour because of their high elongation which is due to the displacement of single fibres at dumping and pushing loads and that the high tensile strength of a woven fabric necessitates its application e.g. as a sinking element in cases where it has to fulfill corresponding functions in the structure. If details are needed for the tensile strength, the values of ≥1200 N for dry installation (strip tensile strength DIN 53857 or ISO 5081) and ≥1800 N for underwater installation can be taken as a basis under normal installation loads, according to experience in hydraulic construction. Furthermore, geogrids (integrally connected extruded grids) in combination with needle-punched non-woven filter layers are increasingly used in coastal engineering.

37.3.3 DESIGN AND TESTING OF FILTER EFFICIENCY

After careful installation of the geotextiles and covering layers the longevity of a bank or bed revetment depends mainly on the efficiency of the filter layer. Therefore filter layers have to fulfil the following two basic requirements: sufficient soil retention function (soil tightness); and sufficient water permeability (hydraulic efficiency).

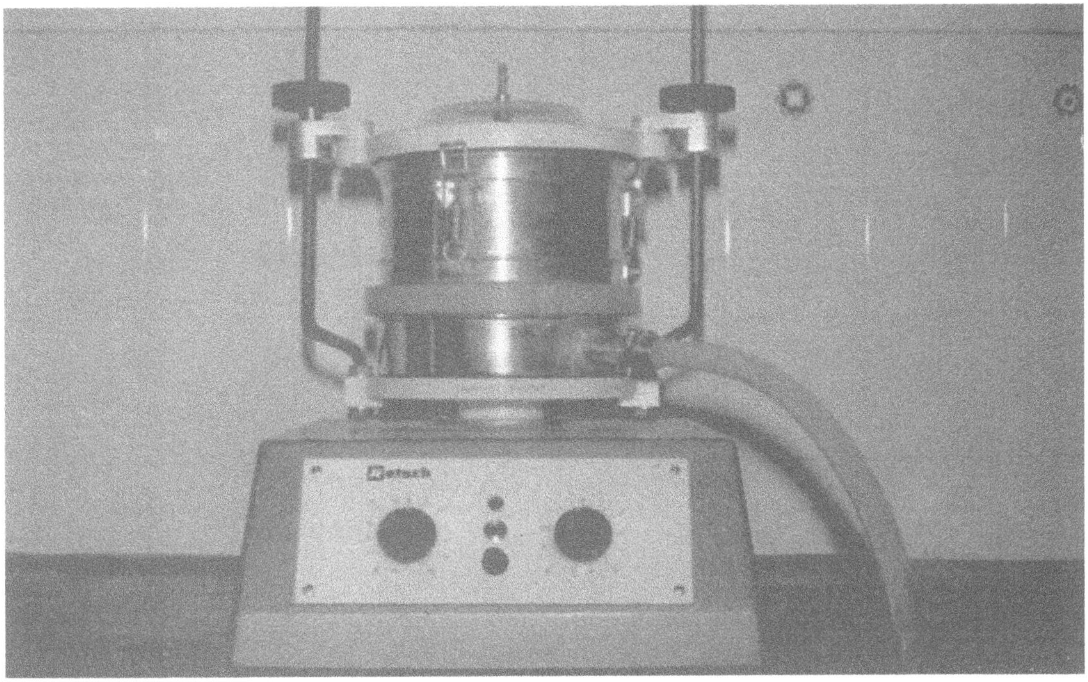

Figure 37.7 Testing equipment for estimation of effective opening size, d_w of geotextiles.

The properties by which the filtering behaviour can be described, are the 'effective opening size D_w' and the water permeability coefficients k_v (vertical to the plane of the geotextile) and k_H (in the plane of the geotextile for drainage functions). Test equipment for the determination of these product data can be found e.g. at the Franzius Institute for Hydraulic Research and Coastal Engineering, University of Hanover. The testing methods can briefly be described as follows (Kohlhase and Saathoff, 1986).

To estimate the effective opening size a wet sieving with a defined testing sand is used. In the test, the fabric has to operate as a sieve. The complete testing equipment for the wet sieving is shown in Figure 37.7.

A grain-size analysis of the retained material and the material that passes through leads to the effective opening size as an essential filtration parameter by a fixed evaluation method (Kohlhase and Saathoff, 1986).

For testing the permeability of the filter fabrics, a permeability test with constant hydrostatic head, generated by two overflow reservoirs, is used.

Figure 37.8 shows the test layout. A sample of several layers of the fabric is placed in the permeability cell and after measuring the flow, the difference in piezometric level, the water temperature and the settlement of the sample, a DARCY coefficient can be determined. Repeating this procedure for various superimposed loads the permeability as another essential filtration parameter can be obtained as a function of load conditions. In addition this method of testing allows the determination of the permeability of aged geotextiles which have been in operation in the structure for several years and have been carefully excavated. In order to ensure that there is no change in the clogging rate (the amount of incorporated soil material) during transport and storage of the samples, it must be ensured that there is no variation in the moisture content and that the samples are carefully handled.

The re-examination of filtering properties of the geotextile samples excavated from the sea-dikes and revetments led to the following important results.

1. Mineral grain filter layers (crushed stone material/gravel) installed above the geotextile were clearly filled up with sand and mud particles. This means

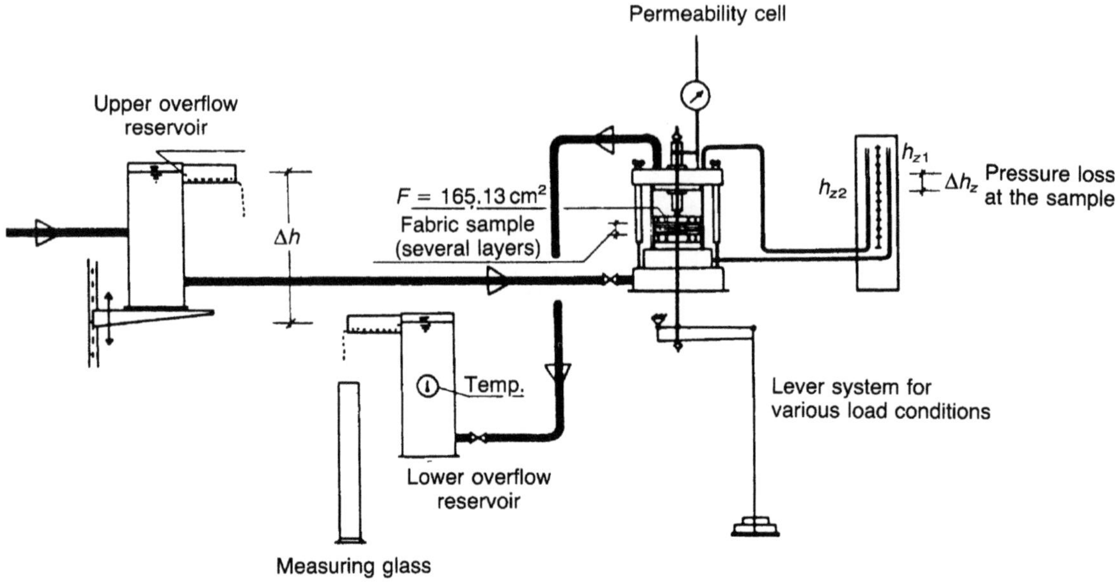

Figure 37.8 Equipment for water permeability test.

that even those revetment sections which remained stable contained geotextiles with opening sizes, which were two large according to today's knowledge.

2. The excavated mechanically bonded non-woven fabrics did not show a compression in correspondance with the revetment loads, but they showed thicknesses greater than or equal to the production thickness. This unexpected behaviour by spontaneous grain/fibre/grain contacts could, in the meantime, also be demonstrated in the case of high road embankments and railway tracks.
3. The pores of the mechanically bonded non-woven fabrics were partly filled with soil particles so that the weight of the filter mats had increased tenfold by soil incorporation. As the open pore structure of, for example, needle-punched non-woven fabrics cannot attain the same density as the existing soil, the observed good hydraulic filtering efficiency of these products is explained by the fact that they are 5 to 12 times more permeable than the existing soil with k-values of 1.0 to $5.0 \times 10^{-5}\,m\,s^{-1}$.
4. The portion of pores of the non-woven fabrics still approaches 35% to 75% after many years of installation (Figure 37.9).
5. The water permeability of the woven fabrics, in contrast to the non-woven fabrics, partly fell below that of the existing soil. Permeability ratios of only 0.16 to 1.8 were established. This behaviour can be explained by the blocking of single mesh openings by soil particles (sieve blockage).
6. Due to the sieve effect woven fabrics react more sensitively to fluctuations in the grain distribution of the existing soil than non-woven fabrics with a wider pore spectrum.

In accordance with the recommendations of the German Committee Küstenschutzwerke (Coastal Structures) of the Deutsche Gesellschaft für Erd- und Grundbau e.V. (German Society for Soil Mechanics and Foundation Engineering) and the Hafenbautechnische Gesellschaft e.V. (Committee for Waterfront Structures) the following basic dimensioning formulation for the soil tightness should be considered for existing fine and medium sand in coastal areas ($0.09\,mm \leqslant d_{50} \leqslant 0.34\,mm$), d_{50} = grain size at 50% sieve passage (EAK, 1981):

$$D_w \cong d_{50}$$

In addition an exact and stable installation of the geotextile has to be ensured. An extensive, fixed placement of thin geotextiles can, in general, only be achieved by installing an intermediate layer of, for example, crushed stone material or gravel. Especially on steep embankments, on which thin woven and non-woven fabrics are directly layed on rip-rap layers of grade II, water running off (down-rush of a wave) into the hollow spaces of the rip-rap layer can act

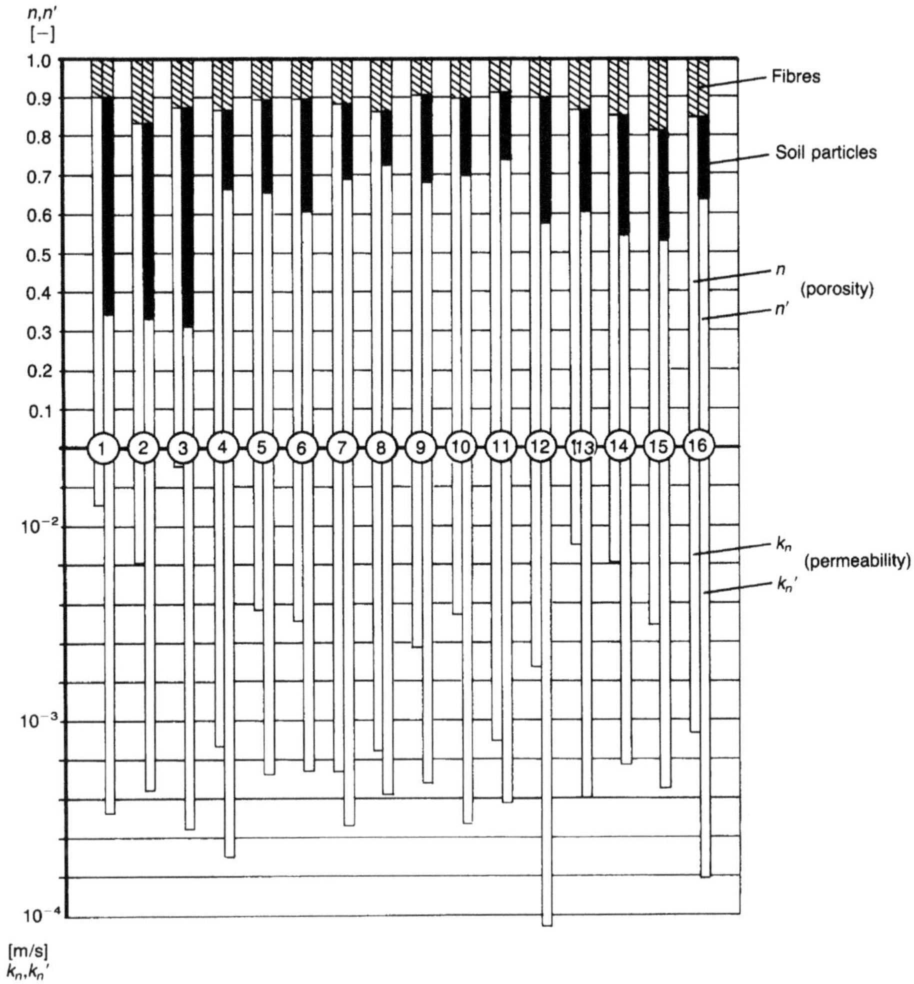

Figure 37.9 Clogging of voids volume and permeability decrease of 16 dug-up needle-punched non-woven geotextiles.

through the pore spaces of a geotextile that is not thoroughly fixed and can cause an inadmissible soil migration of the relatively mobile sands beneath the geotextile. On corresponding problem soils the application of filter mats with a stabilization layer as is common in waterway engineering, is recommended (Figure 37.10).

The following criteria indicate, independently of each other, a soil endangered to soil migration beneath the geotextile;

1. Grain fraction <0.06 mm contained in soil and more: coefficient of uniformity $C_u = d_{60}/d_{10} < 15$;
2. Quantity part of soil in the section $0.02\,\text{mm} < d < 0.1\,\text{mm} > 50\%$;
3. For fine grained soils (Ø 0.063 mm > 40%); plasticity index $I_p < 0.15 = 15\%$ or alternatively the relationship quantity of clay/quantity of silt <0.5.

A typical revetment damage caused by soil migration beneath the geotextile and alternative developments of filter layers for the prevention of soil migration beneath the geotextile on steep embankments are shown in Figure 37.11.

According to current knowledge $0.5\,\text{mm} \leqslant D_w \leqslant 2.0\,\text{mm}$ can be taken as an approximate value for the

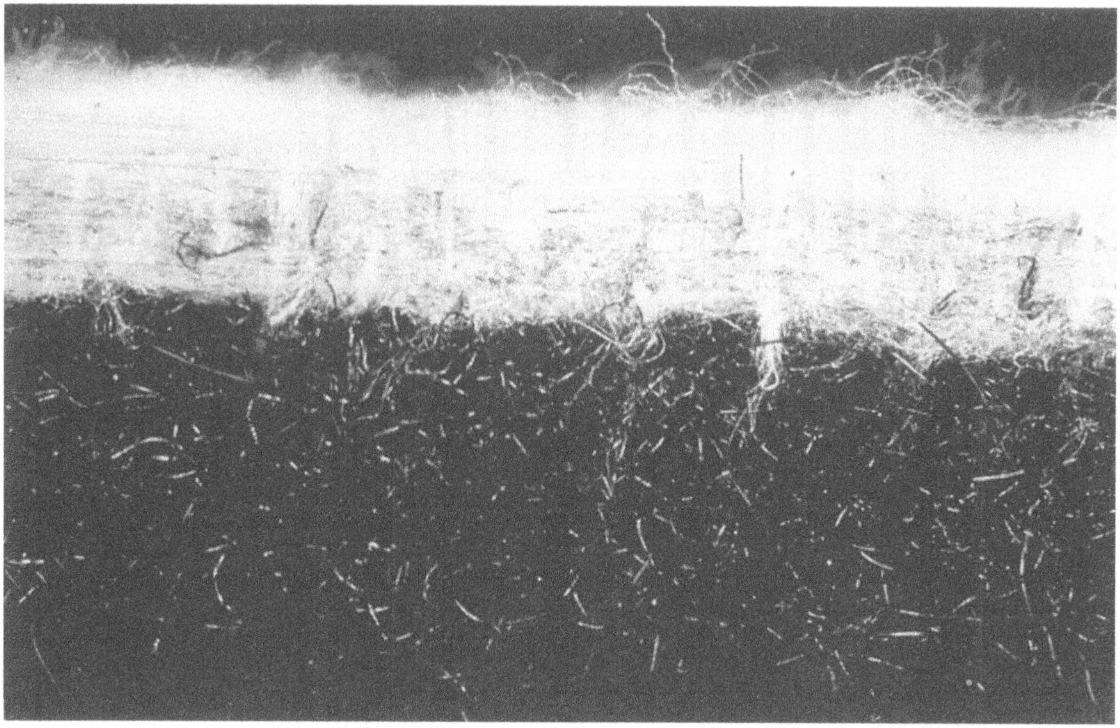

Figure 37.10 Cross-section of three-layered composite material with stabilizing layer.

pore structure of a corresponding stabilization layer for the sands of coastal areas (Heerten, 1986). The thickness of a stabilization layer should be $5 \geqslant d \geqslant 15$ mm. Corresponding filter mats as shown on excavated samples (Kniess and List, 1982), form a fibre stabilized interface filter mat/soil, which means a stable bridging of the hollow spaces beneath a rip-rap layer.

In general, proof of the hydraulic filtering efficiency is essential for the application of geotextiles in coastal engineering as stability and the correct functioning of the structure are significantly influenced by it.

For relatively thin geotextiles (wovens and thin non-wovens with a thickness <2.0 mm), whose filtering behaviour is comparable, a reduction factor η_G can for a certain valid section ($0.01\,\text{mm} \leqslant d_{10} \leqslant 0.5\,\text{mm}$) be indicated depending on the water permeability coefficient k_v and the grain diameter d_{10} of the soil to be filtered (Heerten, 1981).

It is known from investigations on needle-punched non-woven fabrics installed for many years that the hydraulic filtering efficiency of these geotextiles (thickness of filtering layer >2.0 mm) for silts and sands is guaranteed if the water permeability k_v measured on the virgin product at a load of $2\,\text{kN}\,\text{m}^{-2}$ does exceed the permeability coefficient of the soil by approximately 50 times ($\eta \leqslant 1/50 = 0.02$)

$$k_v \geqslant 50\,k_{\text{soil}}$$

37.4 GEOTEXTILES AT PORT KEMBLA

In June 1977 the government of New South Wales, Australia decided to construct a new coal loader at Port Kembla, approximately 90 km south of Sydney. To provide additional land for coal stockpiles, it was necessary to construct a 1200 m long sea-wall, fronting the Pacific Ocean, in water depths up to 2 m below low water and to reclaim the stockpile area behind the wall. The sea-wall is part of an embankment containing a roadway and rail mounted stackers servicing half of the total stockpile area. The integrity of the sea-wall is important in the sea-wall design and during construction as damage to it or settlement of it will affect the operation of the stackers. The sea-wall

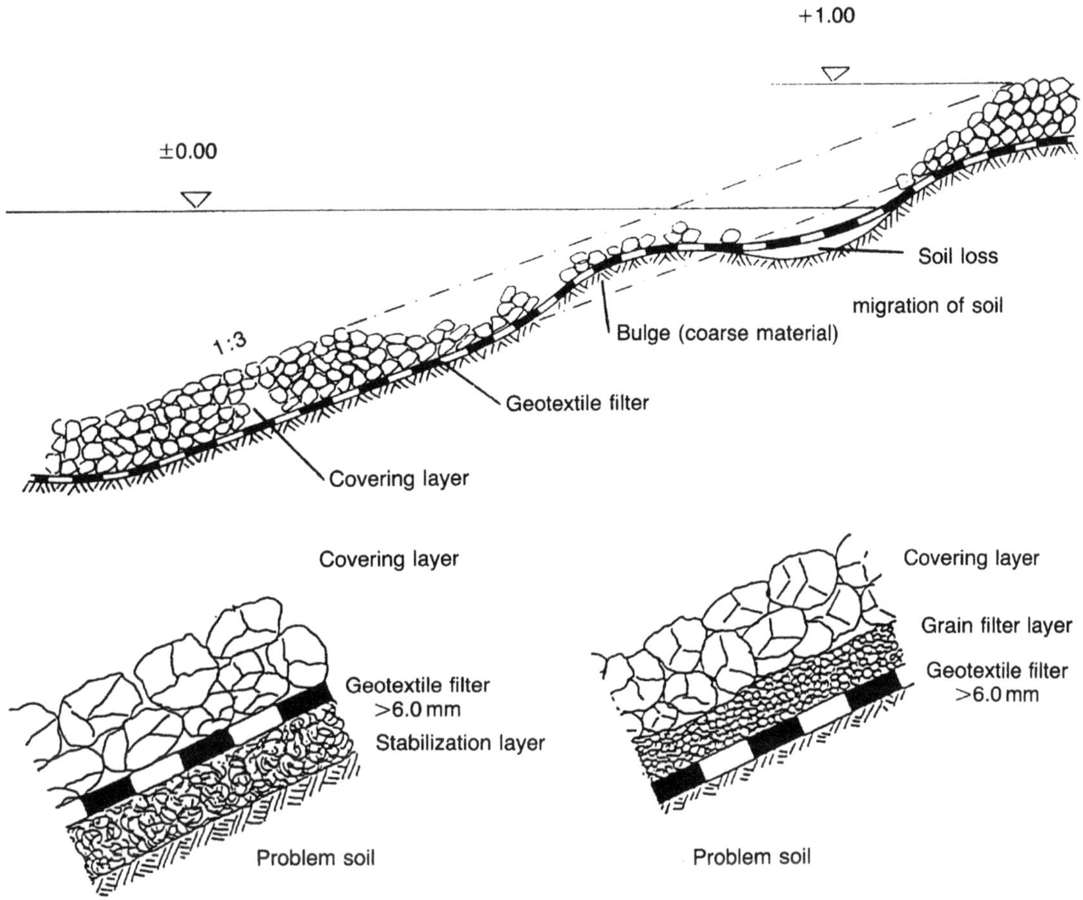

Figure 37.11 Revetment damage caused by migration of soil particles beneath geotextile and alternative formation of filter layers to prevent migration of soil particles beneath geotextile filter or steep embankments (not to scale).

design must take account of a water level change of 4.0 m, including wave set-up, a design breaking wave height of 5.5 m and a scouring depth at the toe of 2.0 m. Two preliminary sea-wall designs were prepared with 12 t Hanbars and 7 t Dolosse as alternate primary armour layers. Both designs included heavy needle-punched non-woven geotextiles replacing standard filter material. Some construction details are given in Figure 37.12.

The stability of the sea-wall depended on the integrity of the toe of the structure. The difficult construction requirements of excavating a toe within the surf zone where there were large quantities of mobile sand and the impossibility of satisfactorily placing conventional filter material, led to the adoption of a synthetic filter cloth material. Materials were subject to extensive field tests to ensure that a suitably robust cloth was available to withstand placement and movement of sub-armour (1.5 to 3.0 t rocks) and primary armour (15 t Hanbars) without damage to the cloth. On the loose fill slope under the 1.5 to 3.0 t rocks a needle-punched non-woven fabric produced from polyester fibres with a weight of 1200 g m^{-2} has proved its robustness against this heavy stone dumping operation. Lighter geotextiles were punctured and destroyed during field tests.

On the slope of the toe excavation directly under the 15 t Hanbars a needle-punched non-woven geotextile with a weight of 2000 g m^{-2} was selected as a suitable fabric for this special application. Potential problems associated with the placement of the non-woven fabric in the toe excavation and securement before overlying material was placed, were recognized. A concrete block filter mat was designed as a method of holding

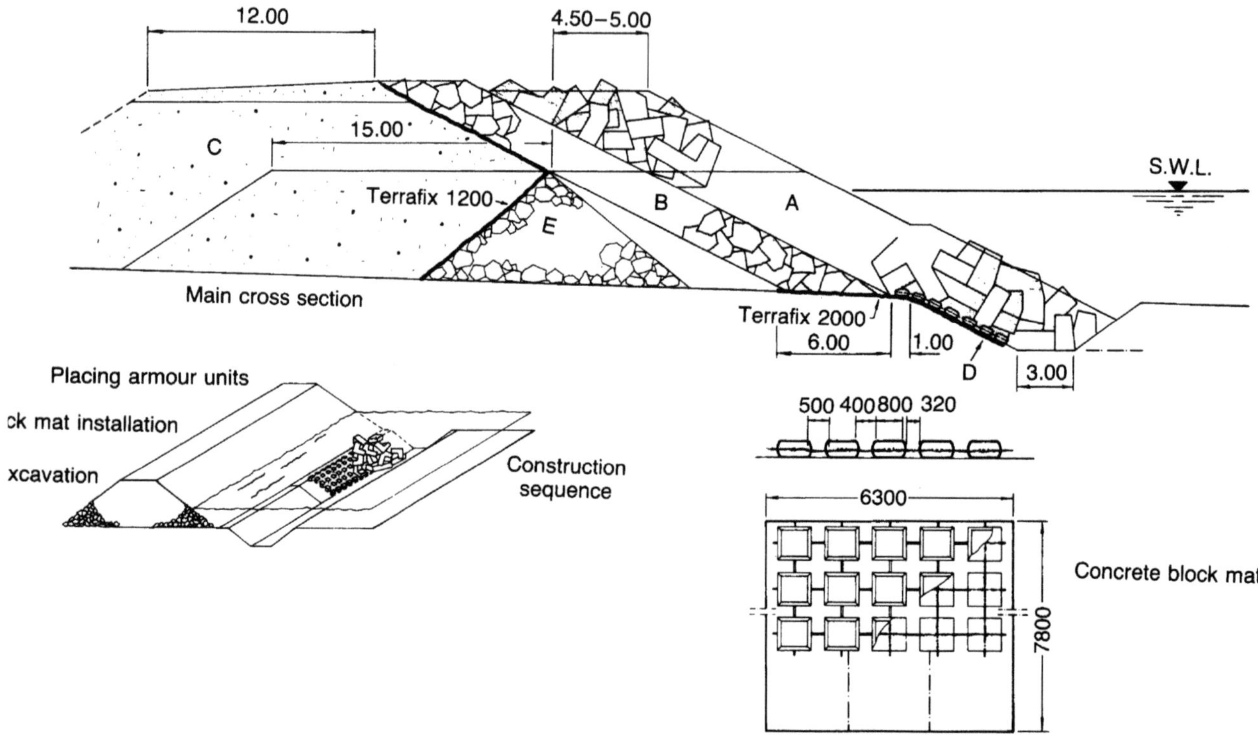

Figure 37.12 Port Kembla sea-wall, Australia, Construction details, S.W.L. = sea-water level. Layer A, 15 t Hanbars; layer B, 1.5–3.0 t rocks; layer C, slag fill; layer D, concrete block mat plus terrafix 2000; layer E, 100 kg to 4 t quarry run.

Figure 37.13 Port Kembla sea-wall, Australia. Fabric and concrete block mat installation.

the geotextile in position until rock and armour units were placed. The prefabricated block mat consisted of 25 × 670 kg concrete blocks laced together with polyester ropes (see Figure 37.12).

On achieving the design section for the toe trench, the dragline laid the concrete block filter mat from the toe of the trench back to the working platform. The geotextile was tied back to prevent excessive movement in wave action. Rock sub-armour was placed on the landward end of the filter cloth, and Hanbars were placed in the toe trench. This operation had to be carried out in one construction sequence so that sand did not move onto the filter cloth before the placement of the armour and the sub-armour. Each concrete block filter mat was overlaid on the previous mat with a 1.5 m overlap.

Although the concrete block filter mats weighed approximately 18 t in all, and acted as huge sails in strong winds they could be positioned in winds up to 25 knots (Figure 37.13). The adoption of the concrete block mats with heavy non-woven geotextiles in lieu of rock resulted in a total saving of approximately $600 000 over the 1200 m length of wall (Southwell, 1982).

37.5 CONCLUSIONS

Some fundamentals of geotextile applications in coastal and harbour engineering have been given in this chapter. The case studies of geotextile applications in coastal engineering works show the necessity of knowing the geotextile performance for successful design and construction. In many cases the geotextile, its suitable selection, its safe installation and its adequate long-term behaviour are of utmost importance for the life-time and safety of the whole structure. In coastal engineering there is a demand for heavy fabrics of high robustness.

REFERENCES

Heerten, G. (1981) Geotextilien im Wasserbau – Prüfung, Anwendung, Bewährung. Mitteilungen des Franzius-Instituts der Universität Hannover, Heft 52.

Heerten, G. (1986) Functional design of filters using geotextiles. Proceedings of the 3rd International Conference on Geotextiles, Vienna, Austria.

Kohlhase, S. and Saathoff, F. (1986) Research at the Franzius-Institut on geotextile filters in hydraulic engineering. Fifth Congress of the Asian and Pacific Division, International Association for Hydraulic Research, Seoul, Republic of Korea.

Kniess, H.-G. and List, H.-J. (1982) Langzeitbeständigkeit geotextiler Filter im Verkehrswasserbau, Forschungsbericht der Bundesanstalt für Wasserbau (BAW), Karlsruhe, (unpublished).

Southwell, B.S. (1982) Design and construction of a seawall at Port Kembla, N.S.W. Fifth Australian Conference on Coastal and Ocean Engineering, Perth.

EAK (1981) *Empfehlungen für die Ausführung von Küstenschutzwerken*, Ausschuß 'Küstenschutzwerke' der DGEG, Die Küste, Heft 36, 1981.

PART EIGHT

COHESIVE SEDIMENTS, ESTUARIES AND DREDGING

COHESIVE SEDIMENTS – SCIENTIFIC BACKGROUND

38

W.R. Parker

Twenty or so years ago, we were only able to describe the science of cohesive sediments in the most general terms. It was known that their geological origin was important and could significantly effect the aggregation or flocculation of particles. Salt water could also affect the rate at which flocs were formed but the complex structure of the settled particles could be imagined but not quantified so far as re-entrainment in the flow was concerned. This chapter and the ones that follow introduce the fundamental science relevant to cohesive sediments and illustrate its application in practical engineering solutions. They also illustrate the explosion in research that has taken place in the last 10–20 years. There is much more that needs to be known but at least the engineer now has quantitative tools at his disposal. Even so if he is not well versed in the subject he would be wise to call in the specialist. Even they disagree to some extent (see Migniot (Chapter 7) and Parker (Chapter 42)). Both approaches are presented but it should not be too long before the differences are resolved.

This chapter is the first of six on cohesive sediments which have been co-ordinated by Dr Parker. The others are Hydraulic behaviour of fine sediment (Chapter 39), Contaminants of estuarine sediments (Chapter 40), Cohesive sediments in coastal engineering (Chapter 41), Determining depth and navigability (Chapter 42) and Dredging in cohesive sediment areas (Chapter 46).

CONTENTS

38.1 SCIENTIFIC BACKGROUND

Natural cohesive sediment consists principally of mineral particles and organic debris as well as micro-organisms and their secretions. It is characterized by the tendency of the particles to stick together.

The bulk of the sediment is made up of small mineral grains which are usually fragments of much larger crystals. The surfaces of the particles have a net surface charge arising from the interaction of Van der Waals (attractive) forces and electrostatic (repulsion) forces (Figure 38.1). In estuarial water a layer of counter-charged ions gathers at the particle surface. The nature of this layer varies with the mineralogy of the particles and the pH, ionic strength or salinity and composition of the water. It modifies the distribution, amplitude and sign of the net surface charge of the particles.

When fine particles are brought together in estuarial water by, for example settling, turbulent shear or Brownian motion their surface charges interact such that at very small separations (e.g. $<10^{-8}$ cm) there is very strong net positive interaction (repulsion). This interaction decays through a negative (attraction) minimum, sometimes more than one, to an insignificant value at larger ($>10^{-4}$ cm) distances (Figure 38.1). The depth of the minimum and thus the strength of the attraction between particles has been shown to be principally a function of the size and mineralogy of the particles and the pH and ionic strength of the water (Edzwald *et al.*, 1974). This leads to a tendency for the particles, when brought together, to position themselves at the minima in the interaction curve giv-

Coastal, Estuarial and Harbour Engineer's Reference Book. Edited by M.B. Abbott and W.A. Price. Published in 1993 by Chapman & Hall, 2–6 Boundary Row, London SE1 8HN. ISBN 0 419 15430 2.

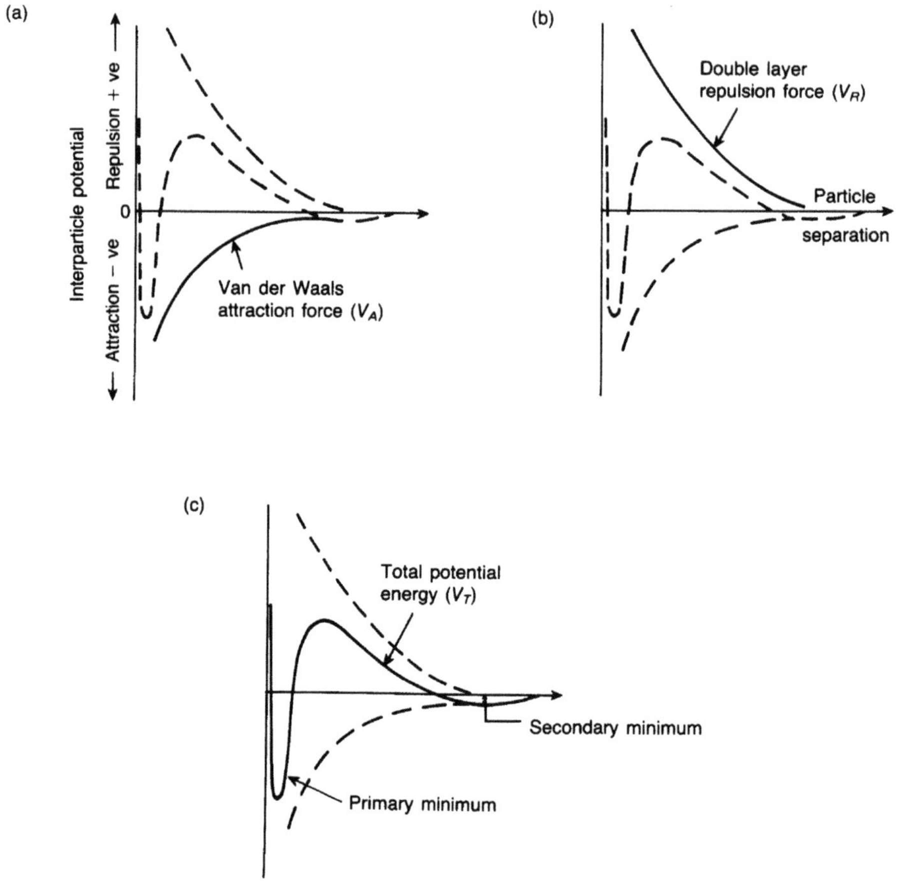

Figure 38.1 Schematic illustration of distribution of inter-particle potentials due to Van der Waals forces and double layer repulsion, giving a net total inter-particle potential with primary and secondary minimum; (a) distribution of Van der Waals force (V_A) with inter-particle distance; (b) distribution of double layer repulsion V_R with inter-particle distance; (c) distribution of net inter-particle potential V_T with inter-particle distance.

ing rise to their sticky (cohesive) behaviour. Although these forces also occur with larger particles the magnitude of the net attraction in the potential energy minimum and the mass of the particles means that the forces are insufficient to hold the particles together. Thus sediments coarser than 20–30 μm, for example sands, are non-cohesive. This charge-influenced behaviour is further modified by any organic macro-molecules which are sorbed on the surface of the mineral grains or the mucous produced by bacterial colonies living on the mineral surfaces. The organic mucous or other macro-molecules can themselves assist in the cohesion of the mineral grains. Where large populations of mucous-secreting organisms occur, for example in intertidal muds, the micro-organisms can exert a dominant influence on sediment behaviour. This can extend to sandy sediments as well. Most natural estuarine sediments are a mixture of sizes so their cohesiveness is most strongly influenced by the proportion of fine particles (less than 20 or 30 μm) that they contain. The tendency of fine particles to stick together has a number of important effects on the properties of cohesive sediment and how its behaviour must be considered.

38.1.1 SIZE

Analyses by a variety of methods gives a description of the size of individual mineral grains. In nature many hundreds or thousands of mineral particles may be gathered into an aggregate. The behaviour and properties of the aggregate may be quite different from the individual grain and it is the behaviour of the aggregate that is important. The size of the aggregate is limited by local shear stresses and its density (including water occluded between the particles) by the conditions of shear it has grown under.

38.1.2 SETTLING VELOCITY

The settling velocity of an aggregate is determined largely by its size, its density and the proximity of other aggregates. It has long been appreciated that settling speeds determined in the laboratory are very much lower than those derived from field experiment (Burt, 1986). With variations in tidal currents and associated shear, the settling properties of the aggregates can change with time, with position in the water column and with position in the estuary.

38.1.3 CONCENTRATION

For practical purposes, e.g. calculation of fluxes, the concentration information required is the dry mass per unit volume of water. A wide variety of methods are used to measure concentration and various types of 'concentration' are concerned in understanding sediment behaviour.

Mass concentration

This is the mass of dry solid per unit volume of water and is generally determined by gravimetric analysis. For continuous observations it can be measured indirectly by empirically calibrated instruments such as turbidity meters and nuclear or acoustic concentration meters.

Volumetric concentration

This is the proportion of any volume element occupied by sediment. It can be considered in terms of:

Solid volume concentration: Volume of solid per unit volume of suspension which transforms with mass assuming a known particle density.

Effective volume concentration: The particles are aggregated into flocs which include a lot of water. Thus a given mass of solids can occupy a much greater volume than the solid volume concentration. This has important consequences for settling at high concentration and the development of mud beds by self weight consolidation.

Hydrodynamic volume concentration: As the flocs move relative to the fluid the hydrodynamic effects of their presence extend well beyond their physical limits. This affects their interaction with other flocs and thereby the rheology and flow of high concentration fine particle suspensions.

In considering the rheological behaviour of cohesive sediments, volume concentration is of particular importance. When the aggregates form a continious three-dimensional network, flow behaviour changes fundamentally. The volume concentration at which this change occurs is referred to as the **critical volume concentration** or the **space filling volume concentration**.

Number concentration

The number of particles per unit volume influences various behavioural characteristics and metrological aspects of the study of fine sediment. For example, depending on local turbulence, a given mass of sediment may gather into fewer larger flocs or be broken into more smaller flocs. This changes the transparency of the water affecting the response of optical or acoustic concentration meters or the light available to local fauna or flora.

The small particle size of muds results in their having a large surface area per unit mass and gives rise to their considerable capacity to adsorb pollutants. Concentration factors (K_D), the ratio of the concentration of pollutant per unit mass of sediment and unit volume of water, may vary from 10^3 to 10^7 for metals for example. Thus fine sediments play a major role in the transport of pollutants.

38.1.4 MATERIAL PROPERTIES

Fine particles in suspension can change the behaviour of the estuarial or coastal water (Gust, 1976). At low concentrations it can induce changes in turbulent energy dissipation, and at high concentrations it can lead to fundamental changes in the stress/strain behaviour of the medium giving non-Newtonian characteristics. These may vary from pseudo-plastic and shear-thinning to elasticoviscous or even dilatent, depending on the solids content, mineralogy and particle size spectrum (Figures 38.2 and 38.3).

Study of the rheology of cohesive sediments has a long history. Pioneering efforts have been made by Krone (1963), Migniot (1968) and others. The application of modern rheological and rheometric theory and technology (Williams and Williams, 1989) reveals that the extension of methodologies used for the study of relatively dilute suspensions into the study of materials above the critical space filling volume concentration is inappropriate. The most notable example of this is the concept of 'initial rigidity' (Migniot, 1968) and its relationship to 'yield stress' in Bingham models. This is a particular area of interest because of the attempts made to relate rheological properties to erosion function. The most up-to-date studies of cohesive sediment rheology are those from the Swansea school. Other evaluations of the concept and determination of yield stress have been given by Barns

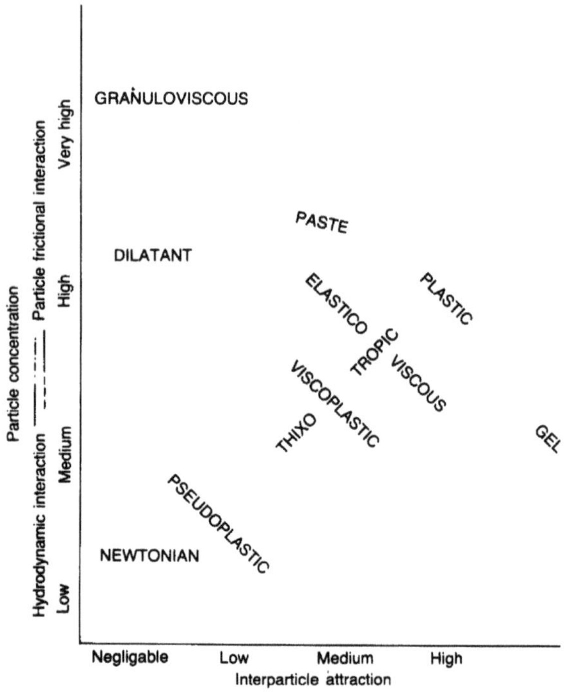

Figure 38.2 Influence of particle concentration and interparticle attraction on flow behaviour.

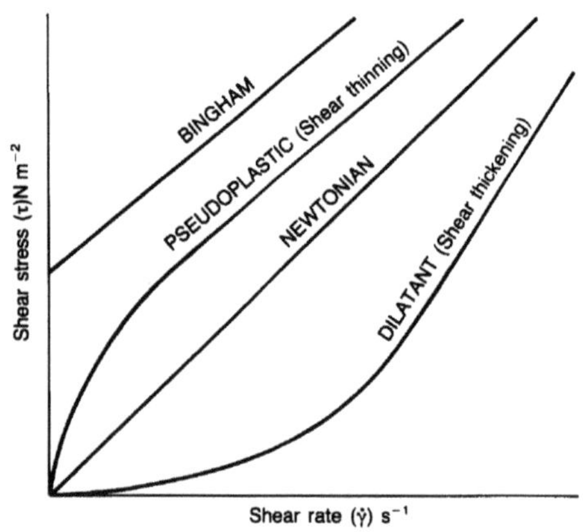

Figure 38.3 Idealized equilibrium flow curves.

and Walters (1985) and Cheng (1986). These studies indicate that the properties of cohesive sediments of most interest to engineers are both difficult to measure and sensitive to the stress history of the material. In particular they draw attention to the inadequacy of measurements using concentric cylinder viscometers and imposed constant shear rate, as opposed to controlled stress methods, when determining yield at volume concentrations above the critical space filling volume concentration (James *et al.*, 1987, 1988).

Another key aspect of fine sediment behaviour is the time dependence of cohesive sediment properties. This is particularly important in considering the process of deposition, the development of the bed and the response of the bed to fluid shear.

All materials tend to deform under an applied stress. Relaxation of this stress by molecular or particle diffusion will take place over a timescale characteristic of the material. The effects are related by the ratio of the relaxation time to the period of the stress, known as the Deborah number (D_e). Where $D_e \ll 1$ the material behaves as a viscous fluid and where $D_e \gg 1$ the material behaves as an elastic solid. In the natural environment the period of the stress may vary from 10^{-1} s for some types of turbulent stress to 10^5 s for tidal or other long period variations in stress. Thus the same mud may respond as a viscous fluid or elasticoviscous 'solid' depending on its situation or the nature of the stress field applied to it.

Another behavioural area where time is important is in the consolidation of very soft and liquid muds. The self-weight consolidation of these materials leading to the expulsion of pore-water and upward growth of higher density layers means that the structure and properties of the bed change with time. As well as problems in detecting the bed for navigation or dredging purposes, these time-dependent changes in bed structure lead to problems in predicting the erodability of the bed at a particular time (Parker, 1987, 1989). The fundamental properties of fine sediments and their response to hydrodynamic stresses mean that their overall behaviour needs to be viewed in a completely different context from that of non-cohesive sediments. The size of the mineral particle may be a time and energy invariant characteristic but the size of the aggregate is not and thus particle size has quite a different connotation from that associated with, for example, sands. The operational transport unit is not the individual grain but the aggregate. As a consequence settling velocity can vary spatially and temporally, bed properties are time and space dependent and the response of the bed is both stress and strain amplitude as well as strain rate dependent.

38.1.5 CONSOLIDATION OF SOFT MUD

Individual aggregates in suspension are supported by buoyancy and drag forces. At high volume concentration, settling of these aggregates is limited by the displacement and expulsion of water between them in the condition known as 'hindered settling'. When the volume concentration is sufficient for a continuous network structure to form a transitional condition develops between a fluid-supported suspension phase and a particle-framework-supported bed phase (Migniot, 1968; Sills *et al.*, 1986). In the suspension phase the pore-water pressure equals the total stress. Once structure forms the pore-water pressure reduces below the total stress as more of the sediment mass is transferred to and supported by the particle framework. With continued settlement the aggregates compact together as pore-water is expelled, pore pressures remaining above hydrostatic pressure. At equilibrium the sediment forms a self-supporting particle framework and the pore pressures are at hydrostatic.

This self-weight consolidation is accompanied by large strains of the soil skeleton resulting in increases in sediment density. Layers of greater density build up from the base of the settling suspension accompanying the transfer of mass from the suspension phase to the structured bed phase. Thick layers consolidate more slowly but reach higher basal densities than shallow layers which consolidate more rapidly. The evolution of this structure is central to considerations of erosion resistance, definition of depth for calibration or verification of numerical models or quantifying dredging need or dredging achieved (Parker, 1989).

During consolidation the biogeochemical changes taking place within the mud release sorbed pollutants which may be expelled along with pore-water (Parker and Lee, 1981; Sills and Been, 1984). Other microbial processes lead to the formation of gas micro-bubbles which affect consolidation, the response of the mud to tidal or wave stresses and sound speeds used in calibration of acoustic survey systems (Sills *et al.*, 1991).

38.1.6 FIELD BEHAVIOUR

Cohesive sediment behaviour can be viewed in the context of high and low concentration regimes and locations which are wave affected or not subject to wave action. The classical models of estuarine fine sediment behaviour describe a conceptual view of the transport of solids within an estuary (Postma and Kalle, 1955).

The general view derived from estuarine circulation models and fine sediment behaviour models is that fine sediment tends to gather in the areas of strongest salinity gradients, or in areas determined by local mixing, as regional or local turbidity maxima. Within these turbidity maxima the sediment is subjected to cyclical transfer over spring-neap-spring, seasonal or annual periods between various relatively ephemeral states of occurrence. These states and various processes linking them together are shown in Figure 38.4. Similar models have been proposed by Allen *et al.* (1984), Parker and Kirby (1977) and Mehta *et al.* (1982). The states of occurrence are the host for a spectrum of other physical or chemical processes, as the sediment passes from the settled bed via the mobile suspended sediment population back to the

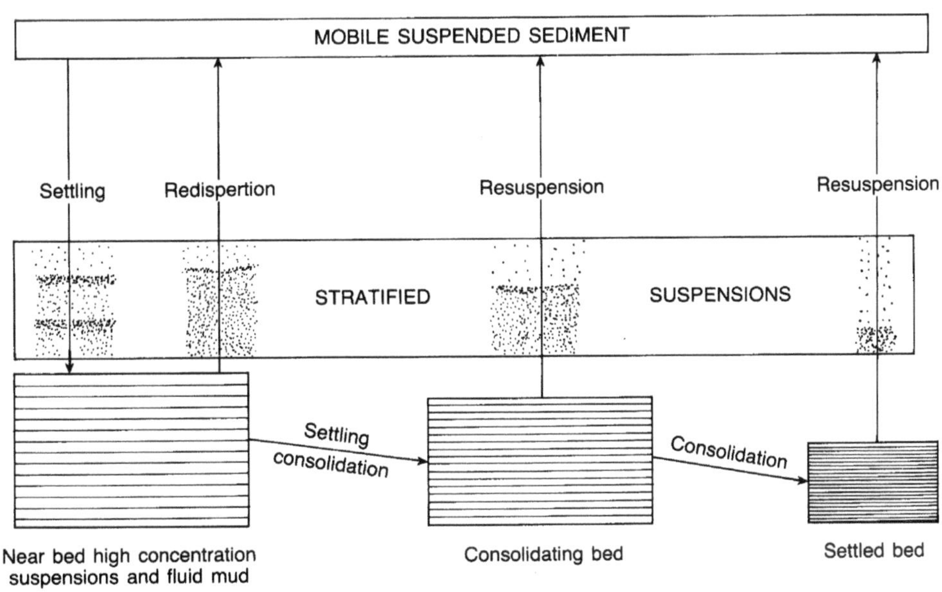

Figure 38.4 Network of phenomena and processes illustrating cycling of fine sediment within turbidity maximum.

settled bed. In terms of sedimentation engineering, three phenomena may be highlighted as being of principal importance. These are stratified suspensions, high-concentration benthic layers with an abrupt concentration gradient (lutocline) at the upper surface and 'fluid mud'.

Stratified suspensions can arise when available current energy or time is insufficient to mix the erodable fine sediment throughout the water column or whilst the sediment is settling. The layering is thought to affect mixing and encourage settling (Ross and Mehta, 1989; Smith and Kirby, 1989).

High-concentration benthic layers develop locally by settling, often evolving from stratified suspensions. Flowing below otherwise low concentration waters they may transport significant quantities of mud (Parker and Kirby, 1977). The occurrence of high-concentration bed layers allows a high mass settling flux to develop in suitable conditions (e.g. docks, locks, basins). They may be the precursor of the widely reported, acoustically detected, 'fluid mud'.

Fluid mud has been shown to be a high concentration mud slurry. Its detection by survey echosounders has been extensively discussed (Parker and Kirby, 1977). Detection has been shown to be dependent upon the characteristics of the echosounder being used and the state of evolution of the fluid mud structure (Parker and Kirby, 1982). Within fluid mud deposits biochemical reactions occur as the sediment consolidates and porewater is expelled (Sills and Been, 1984). Erosion of this mud or its disturbance by dredging can release nutrients, oxygen deficient water, reduced chemical species and oxygen-consuming biological species into the water column with serious environmental consequences.

In areas with relatively low suspended solids populations the development of layered or stratified suspension is not usually reported and the development of high concentration near-bed layers is much more restricted. However, the effects of waves in coastal areas can generate near-bed high concentration layers which may flow into otherwise low turbidity situations (Kendrick, 1980; Parker and Kirby, 1977).

REFERENCES

Allen, G.P., Bonnefitte, R., Courtois, G. *et al.* (1974) *La Houille Blanche*, **1/2**, 129–36.

Barnes, H.A. and Walters, K. (1985) *Rheologica Acta*, **24**, 323–26.

Burt, T.N. (1986) in, *Estuarine Cohesive Sediment Dynamics*, (ed. A.J. Mehta), Springer-Verlag, Berlin, pp. 126–50.

Cheng, D. and Cheng, C-H, (1986) *Rheologica Acta*, **25**, 542–54.

Edzwald, J.K., Upchurch, J.B.O. and Melia, C.R. (1974) *Environment Science and Technology*, **8**, 58–63.

Gust, G. (1976) *Journal of Fluid Mechanics*, **75**, 29–47.

James, A.E., Williams, D.J.A. and Williams, P.R. (1988) in Dronkers, J. and van Leussen, W. (eds.), *Physical Processes in Estuaries*, Springer-Verlag, 488–500.

James, A.E., Williams, D.J.A. and Williams, P.R. (1987b) *Rheologica Acta*, **26**, 437–46.

Kendrick, M. (1980) *Report EX 879 (1&2)*, Hydraulics Research Ltd., Wallingford, UK, 103.

Krone, R.B. (1963) *U. Calf. Berkeley Report SERL*, **63–81**, 91.

Mehta, A.J., Parchure, T., Dixit, J.G. *et al.* (1982) in, *Estuarine Comparisons*, (ed. V.S. Kennedy), Academic Press, New York, 591–609.

Migniot, C. (1968) *La Houille Blanche*, **23** (7), 591–620.

Parker, W.R. (1989) *Journal of Coastal Research*, (Special Issue 5), 175–84.

Parker, W.R. (1987) *Continental Shelf Research*, **7** (11/12), 1285–93.

Parker, W.R. and Kirby, R. (1982) in, *Estuarine Comparisons*, (ed. V.S. Kennedy), Academic Press, New York, 15–26.

Parker, W.R. and Kirby, R. (1977) in, *Proceedings of the 2nd International Symposium on Dredging Technology*, (eds H.S. Stephens, N.G. Coles and J.A. Clarke). BHRA, Cranfield, UK, pp. 15–26.

Parker, W.R. and Lee, K. (1981) *Rapp. P-V. Reun. Con. Int. Explor. Mer.*, **181**, 28–34.

Postma, H. and Kalle, K. (1955) *Deutsche Hydrographische Zeitschrift*, **8**, 137–44.

Ross, M.A. and Mehta, A.J. (1989) *Journal of Coastal Research, (Special Issue 5), 51–62.*

Sills, G.C. and Been, K. (1984) in, Transfer Processes in Cohesive Sediment Systems, (eds W.R. Parker and D.J.J. Kinsman), Plenum Press, pp. 109–25.

Sills, G.C., Elder, D. and McG, (1986) in, *Estuarine Cohesive Sediment Dynamics*, (ed. A.J. Mehta), Springer-Verlag, Berlin, pp. 192–206.

Sills, G.C., Wheeler, S.J., Thomas, S.D. and Garner, T.N. (1991) *Geotechnique*, **41** (2), 227–41.

Smith, T.J. and Kirby, R. (1989) *Journal of Coastal Research*, (Special Issue 5), 63–74.

Williams, D.J.A. and Williams, P.R. (1989) *Journal of Coastal Research*, (Special Issue 5), 165–74.

Williams, P.R. and Williams, D.J.A. (1989) *Journal of Coastal Research*, (Special Issue 5), 151–64.

HYDRAULIC BEHAVIOUR OF FINE SEDIMENT 39

A.J. Mehta

The physics of the transport of granular materials like sands and gravels is quite complex but not as difficult to handle as the transport of fine sediments. The main reason for this difference in behaviour is that fine sediments including silts and especially clays are cohesive, and this means that their transport, settlement and erodibility are to an extent controlled by electrochemical effects. The geological origin of the sediments is therefore important and whether they exist in fresh or brackish water also affects their hydraulic behaviour. For example, they can aggregate as flocs whose properties, especially settlement, are quite different from those of the individual particles. This chapter presents a perspective of the problems and establishes a physical framework to enable an engineer to deal with the properties of fine sediments. The nature of the settling velocity of cohesive sediment flocs or aggregates is discussed together with simple formulations for the rates of deposition and erosion.

CONTENTS

39.1 INTRODUCTION

In order to examine the hydraulic behaviour of fine, cohesive sediment it is essential to establish a physical framework for a perspective on the problem of dealing with fine sediment properties and their transport. In this chapter the matter of transport is considered by reviewing the sediment mass fluxes associated with fine sediment movement as they govern the bahaviour of the suspension concentration profile.

Mass fluxes that characterize the dynamic response of the concentration profile to forcing by currents and waves are those that ultimately determine the nature and behaviour of sediment layering in the water column. The horizontal velocity (u) profile and the associated concentration (C) profile shown in Figure 39.1 inherently reflect the strong interaction between the two profiles, particularly near the bed where high concentrations generally prevail. It is convenient to conceive of the mass fluxes in terms of five vertical processes that are critical in determining the thicknesses of the mobile suspension layer, the mobile fluid mud layer and the stationary mud layer, as well as the level of the cohesive bed. Of these five processes two, i.e. entrainment of the lutocline and fluidization of the bed, may be construed as erosion related processes, while two more, i.e. settling and bed formation, are deposition related processes. Consolidation and gelling of the fresh deposit with aging increase the strength of the bed, thereby strongly influencing erosion.

In engineering practice it is found to be convenient to treat both deposition- and erosion-related processes as being controlled by the hydrodynamic bed shear stress acting at the level of the cohesive bed. In reality, bed fluidization and entrainment of fluidized mud are distinct processes which must be treated as such. Bed formation and settling likewise are governed by distinct physical principles. Furthermore, the presence of a dynamic fluid mud layer over the bed and below the lutocline complicates the manner in which stress is transmitted from the turbulent water column through the lutocline (Mehta, 1989). In spite of these difficulties, however, simple formulations for deposition and erosion seem to perform well in many coastal, estuarial and harbour engineering applications of common interest (Mehta *et al.*, 1989b).

In the following sections we will first consider the nature of the settling velocity of cohesive sediment

Coastal, Estuarial and Harbour Engineer's Reference Book. Edited by M.B. Abbott and W.A. Price. Published in 1993 by Chapman & Hall, 2–6 Boundary Row, London SE1 8HN. ISBN 0 419 15430 2.

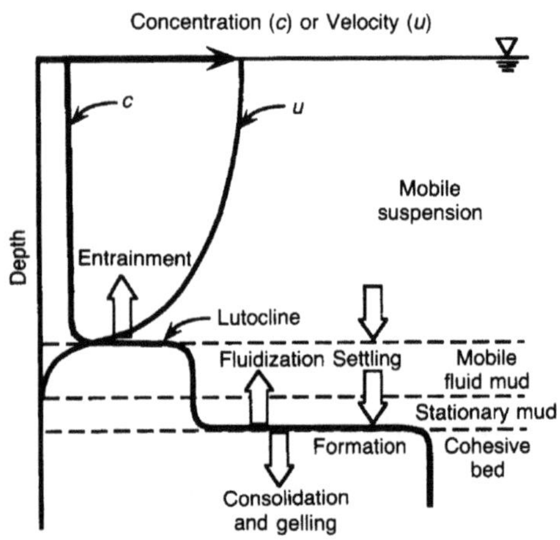

Figure 39.1 Schematic description of velocity and concentration profiles and vertical sediment mass fluxes (adapted from Mehta, 1989).

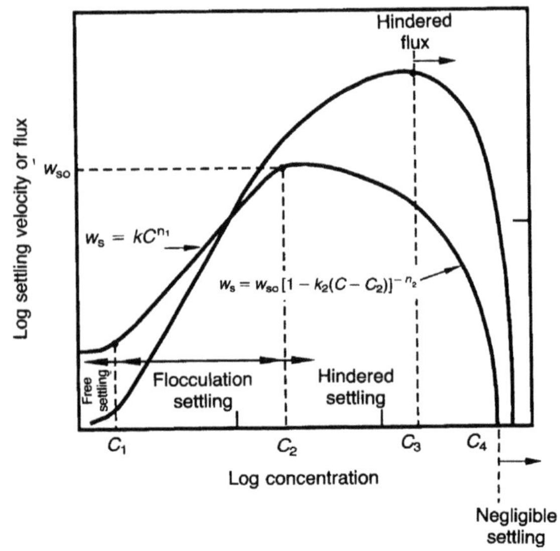

Figure 39.2 Variation of settling velocity and settling flux with suspension concentration (after Costa, 1989).

flocs, or aggregates, followed by simple formulations for the rates of deposition and erosion.

39.2 SETTLING VELOCITY

As in the case of cohesionless sediment, the settling velocity of the floc is a fundamental property characterizing cohesive sediment transport. As a result of floc aggregation due to inter-particle collision and mutual particle interference, however, the settling velocity of the flocs varies strongly with the volume fraction of the suspension concentration. Consequently the settling flux, i.e. the product of the settling velocity and the concentration, also varies nonlinearly with concentration.

Aggregation of fine, clayey particles occurs by virtue of net attractive forces between particles brought close enough by Brownian motion, differential settling or flow shearing (Hunt, 1986). Although the relative importance of the collision frequency due to each of these mechanisms depends on the floc diameter, shearing under turbulent flows seems to be the most important factor contributing to the formation of strong flocs, with the exception of slack water times when differential settling becomes dominant (Krone, 1962). Flocs differ from the constituent individual, or primary, particles in four main aspects:

1. Their size is orders of magnitude larger than that of the primary particles; for example, macroflocs can attain a size on the order of millimeters (Dyer, 1989);
2. Their density is considerably less than that of the primary particles due to the large volume fraction of interstitial water;
3. Their shape is much more spherical than the plate-like shape of the primary particles, which in turn leads to reduced drag;
4. They are extremely weak, tending to break up easily.

Among the above factors, the increase in fall diameter and, to an extent, drag reduction relative to a primary particle, are far more significant than the decrease in density, with the result that, in general, the settling velocities of flocs are substantially higher than those of individual particles. The magnitude of the floc diameter and settling velocity are, moreover, only weakly dependent on the primary particle diameter (Migniot, 1968; Mehta *et al.*, 1989a).

The dependence of the settling velocity on concentration, neglecting the usually (but not always) secondary effects such as those of temperature and salinity, falls within three ranges as shown in Figure 39.2, in which the variation of the logarithm of settling velocity with logarithm concentration is depicted schematically.

39.2.1 FREE SETTLING

Free settling occurs at low concentrations, typically lower than concentration $C_1 = 0.1–0.3\,\mathrm{g\,l^{-1}}$. Here, concentration has been conveniently expressed as dry sediment mass per unit volume of suspension, rather than as volume fraction. In this concentration range the particles settle without mutual interference, and their terminal settling velocity is a result of a force balance between drag and submerged weight. The velocity, w_s, is given by

$$w_s = \left[\frac{2gd}{C_d}\left(\frac{\rho_f}{\rho_w} - 1\right)\right]^{1/2} \qquad (39.1)$$

where d is the floc diameter, ρ_f and ρ_w are the floc and fluid densities, respectively, g is the acceleration due to gravity and C_d is the drag coefficient. As the flocs become very small, of the order of 1 μm in diameter, the settling velocity is modified from that given by Equation (39.1) by a multiplier which is smaller than one (Lick, 1982). Measurement in the laboratory can be carried out by accumulation or pipette analysis (Vanoni, 1975).

39.2.2 FLOCCULATION SETTLING

When the suspension concentration becomes higher than the free settling limit C_1, increased concentration and associated inter-particle collision frequency cause an increase in aggregation, or flocculation, leading to the formation of larger flocs and correspondingly higher settling velocities. The general expression for w_s in the flocculation settling range is

$$w_s = k_1 C_1^n \qquad (39.2)$$

The coefficients in Equation (39.2) may be determined from laboratory settling columns or in field tests. Ross (1988) used a laboratory procedure which was adapted from one developed by McLaughlin (Vanoni, 1975) involving sample withdrawal from a settling column at different elevations and times. It has been found that settling velocities in this range determined by laboratory procedures can be lower by as much as an order of magnitude than those in the field, given the same sediment. This is mainly due to the effect of the local rate of flow shearing on k_1, since high turbulence in the field produces strong, tightly packed aggregates which settle out rapidly. Likewise n_1, which is typically equal to 4/3, may change as well depending upon the level of turbulence. *In situ* measurement using an 'Owen tube' is one way to obviate the problem of scaling laboratory data for field application (Burt, 1986). This is essentially a settling tube which is designed so as to enable the collection of a requisite sample of the suspension from any desired depth, followed by immediate analysis on board the vessel at the site.

39.2.3 HINDERED SETTLING

At concentrations higher than a value C_2 of about $2–5\,\mathrm{g\,l^{-1}}$, the settling velocity decreases with increasing concentration. This is the result of hindered sedimentation, a phenomenon in which the flocs become so closely packed that the interstitial fluid is forced to flow upward between them through increasingly smaller pores (Kynch, 1952). The general expression for the settling velocity in this range is

$$w_s = w_{so}[1 - k_2(C - C_2)]^{n_2} \qquad (39.3)$$

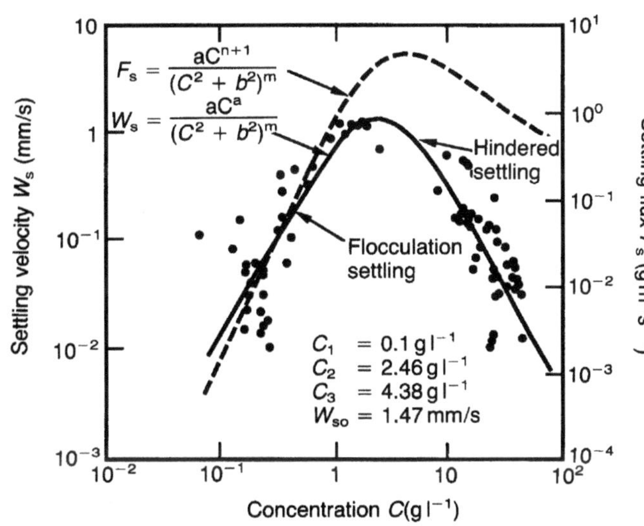

Figure 39.3 Settling velocity data from a laboratory column using as natural sediment (after Hwang, 1989).

where w_{so} is the value of w_s at concentration C_2, k_2 is the inverse of the concentration at which $w_s = 0$, and theoretically for falling spheres $n_2 = 4.66$ at low Reynolds numbers (Richardson and Jeronimo, 1979). In general, w_{so}, k_2 and n_2 are required to be determined in settling columns. Note that in the range $C_2 < C < C_3$, while the settling velocity decreases the flux continues to increase. The value of C_3 can typically vary from 4 to 20 g l^{-1}. In the range $C_3 < C < C_4$ the flux is also hindered, and leads to the formation of near-bed lutoclines (Mehta, 1989). Beyond C_4, which can be of the order of 80–100 g l^{-1}, settling is practically nil.

Figure 39.3 shows an example of settling velocity data using a natural mud in a 2 m tall settling column (Hwang, 1989). The relationship

$$w_s = \frac{aC^n}{(C^2 + b^2)^m} \tag{39.4}$$

which is modified from Wolanski *et al.* (1989), has been used to represent both flocculation settling and hindered settling, with $a = 33.38$, $b = 2.537$, $n = 1.83$ and $m = 1.89$. Note that the corresponding curve for the settling mass flux, $F_s = w_s C$, is obtained by replacing n by $n + 1$ in Equation (39.4). Note further that the peak value of w_s is defined by

$$w_{so} = ab^{n-2m} \frac{\left(\frac{2m}{n} - 1\right)^{m-n/2}}{\left(\frac{2m}{n}\right)^{1/2}} \tag{39.4a}$$

and furthermore

$$C_2 = \frac{b}{\left(\frac{2m}{n} - 1\right)^{1/2}} \tag{39.4b}$$

Values of concentrations C_1, C_2 and C_3 as well as w_{so} for the particular sediment used are given in the figure.

39.3 DEPOSITION UNDER TURBULENT FLOW

39.3.1 RATE OF DEPOSITION

The time-rate of decrease of sediment mass per unit bed area, m, under steady or quasi-steady (as in the case of tide), turbulent flow is given by

$$\frac{dm}{dt} = -pW_1 C \tag{39.5}$$

where $p[0, 1]$ is defined as the probability of deposition, w_s is the settling velocity and C in this case is the depth-averaged suspended sediment concentration. The product, $pw_s = w_s'$, is the apparent settling velocity, and $w_s'C$ is the flux of depositing sediment. Whenever p is less than one, the rate deposition is less than the rate of settling, since not all settling flocs are successful in withstanding the high stresses that prevail near the bed; strong ones deposit while weak ones tend to break up and re-entrain. Since w_s and p depend on the inter-particle collision frequency, and therefore on concentration, it is useful to examine deposition in various ranges of concentration in order to provide simplified descriptions for the time-rate of decrease of concentration in the water column as represented by the depth-average concentration.

Three concentration ranges defined by two characteristic concentrations, C_1 and C_2, can be realistically identified. In the range $C < C_1$, which is compatible with free settling (w_s independent of C), C_1 typically ranges from 0.1 to 0.3 g l^{-1}, with the precise value depending upon the type of sediment–fluid mixture. The probability of deposition, p, is approximated by

$$p = 1 - \frac{\tau_b}{\tau_{cd}} \tag{39.6}$$

where τ_b is the time-mean (over the scale of turbulence) value of the bed shear stress and τ_{cd} is referred to as the critical shear stress for deposition. When $\tau_b < \tau_{cd}$, all initially suspended sediment will eventually deposit, and when $\tau_b \geqslant \tau_{cd}$ there is no deposition, since $p = 0$ (Krone, 1962). Given the initial suspension concentration, C_0 (Equation (39.5)) can be integrated under these conditions to yield

$$\frac{C}{C_0} = \exp\left[-\left(1 - \frac{\tau_b}{\tau_{cd}}\right)\frac{w_s}{h}\right]t \tag{39.7}$$

where h is the flow depth. Equation (39.7) is a simple exponential decay law implying suspension dilution by deposition.

In the concentration range $C_2 > C > C_1$, where C_2 ranges from 0.3 to about 10 g l^{-1}, the concentration is high enough for inter-particle collision and associated kinetics of particle aggregation to become quite important. Note that C_2, in the case of deposition under turbulent flows, is higher than that for quiescent settling, since turbulence tends to retard the onset of hindered settling. The time–concentration relationship is described experimentally by

$$\log C = -\frac{pk'}{h} \log t + k'' \tag{39.8}$$

where p is given by Equation (39.6), and k' and k'' are empirical coefficients whose magnitudes reflect the influence of aggregation on settling. The critical shear stress is typically different in the two concentration ranges. In flume studies using mud from the San Francisco Bay in salt water, Krone (1962) found $\tau_{cd} = 0.060$ N m^{-2} in the lower concentration range ($<C_1$)

and $0.078\,N\,m^{-2}$ in the higher range (C_1 to C_2). This difference implies the formation of stronger flocs in the latter case, under continuing aggregation due to inter-particle collision.

Finally, in the high concentration range ($>C_2$), the time–concentration relationship is found to be similar to Equation (39.8), with pk'/h and k'' replaced by empirical coefficients, which particularly reflect the hindered nature of the settling process (Krone, 1962).

When the sediment is graded, as is commonly the case, the decay law for concentration becomes complicated, since under given flow conditions the probability of deposition varies with particle size. The outcome is a sorting of the initially suspended material (Mehta, 1988).

39.4 EROSION

The terms erosion and resuspension are typically considered to be synonymous when dealing with erosion of mud. Resistance to erosion is contingent upon a number of factors including sediment composition, pore and eroding fluid compositions, and the degree of consolidation of the deposit (Parchure and Mehta, 1985). The deposit itself may be in the form of a stationary, high concentration suspension, or a cohesive bed. The bed may be soft, partially consolidated, with a high water content (>100%), or it may be a more dense, settled bed of lower water content.

The mode of erosion varies both with the magnitude of the bed shear stress and the stress history of the deposit. Three modes have been identified:

1. Floc-by-floc, or surface erosion, of a bed;
2. Mass erosion of a bed;
3. Entrainment of high concentration fluid mud suspension.

The hydrodynamic regime can be conveniently divided into that determined by steady or quasi-steady (e.g., tidal) current and that due to waves. These two regimes will be discussed separately.

39.4.1 EROSION BY CURRENTS

The time-rate of increase of suspended sediment mass per unit bed area, m, is given in a functional form by:

$$\frac{dm}{dt} = f(\tau_b - \tau_s, \nu_1, \nu_2, \ldots, \nu_n) \qquad (39.9)$$

where $\tau_b - \tau_s$ is the bed shear stress in excess of the bed shear strength with respect to erosion, τ_s, and $\nu_1 \ldots \nu_n$ are bed resistance defining parameters. Expressions of the form of Equation (39.9) obtained in laboratory flume studies have been based on surface erosion studies on soft beds with non-uniform properties, and on dense, uniform beds.

Soft beds

These beds, which are usually composed of freshly deposited mud undergoing consolidation, exhibit non-uniform property variation with depth, and are therefore stratified. Typically, the bed density and the shear strength, τ_s, increase with depth, z, in the top few centimeters. The erosion rate, $\varepsilon(= dm/dt = h(dC/dt))$ is given by (Parchure and Mehta, 1985; Kuijper *et al.*, 1989):

$$\frac{\varepsilon}{\varepsilon_f} = \exp\{\alpha[\tau_b - \tau_s(z)]^{1/2}\} \qquad (39.10)$$

where ε_f is defined as the floc erosion rate and α is a rate coefficient. Values of ε_f and α depend the type of sediment-fluid mixture. Thus for instance, ε_f may vary from 10^{-4} to $10^{-7}\,g\,cm^{-2}\,min^{-1}$, and α from 5 to $20\,m\,N^{-1/2}$ (Mehta, 1988).

As a consequence of the increase of τ_s with depth in a stratified bed, at a given τ_b, as bed scour proceeds the rate of erosion decreases with time because of a corresponding decrease in the excess shear stress, $\tau_b - \tau_s$. When τ_b equals τ_s at some depth of scour, a small amount of floc erosion, represented by ε_f, continues to occur because of the probabilistic nature of the bed shear stress and the variability of the shear strength over any horizontal plane.

Two examples of $\tau_s(z)$ profiles are shown in Figure 39.4, which presents results from two tests, both with beds of kaolinite in tap water, but with consolidation periods of 1 and 8 days (Parchure and Mehta, 1985). The bed, formed initially by deposition, was eroded by increasing the bed shear stress, τ_b, in steps of con-

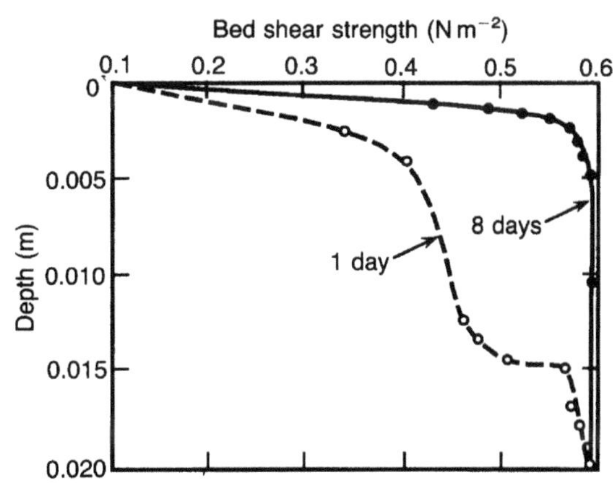

Figure 39.4 Erosion shear strength profiles for kaolinite beds (after Parchure and Mehta, 1985).

stant duration. In each step the suspension concentration first increased rapidly and eventually attained a near-steady value. At this stage therefore, $\tau_b = \tau_s$ at a known depth of bed scour, z. In this manner, τ_s was evaluated at different depths z and the $\tau_s(z)$ profile constructed.

Dense beds

Consolidated cohesive beds, in which the bed properties are uniform over depth, i.e., τ_s is independent of z, comprise this category. In Equation (39.10), the erosion rate consequently becomes constant at a given τ_b. A first-order approximation of this equation can be interpreted as

$$\frac{\varepsilon}{\varepsilon_M} = \frac{\tau_b - \tau_s}{\tau_s} \tag{39.11}$$

where ε_M is a rate constant. The magnitudes of ε_M and τ_s can vary widely depending upon the properties of the sediment–fluid mixture and those of the bed. For example, the range of ε_M can be from 10^{-4} to $10^{-2}\,\mathrm{g\,cm^{-2}\,min^{-1}}$, while τ_s can vary from nil for an organic-rich floc layer to as much as $10\,\mathrm{N\,M^{-2}}$ for hard soils (Lavelle and Mofjeld, 1985; Mehta, 1988).

Given a sediment, the engineer typically faces the difficult problem of selecting specific values of the rate constant, ε_M, and the erosion shear strength, τ_s, for the wide ranging conditions of the bed. It has been found that these two coefficients can be related very crudely to the bed-bulk density (Mehta, 1988; Hwang, 1989). This relationship must, in general, be established by testing the sediment at different bed densities in flume experiments. Some results are referenced in the following paragraphs.

Mass erosion

Equation (39.11), although derived from surface erosion studies, has also been used for simulating mass erosion in an approximate way. In mass erosion, which is actually described by a Mohr–Coulomb-type formulation, the bed fails at some level beneath the surface where the soil shear strength is unable to withstand the induced stress. Erosion occurs sometimes by dislodgement of large pieces of the soil. The rate coefficient, ε_M, is typically larger than that for surface erosion under comparable conditions, and must be evaluated either experimentally or by calibration against available data for specific eroding conditions.

Figures 39.5a and b show variations of τ_s and ε_M with bed-bulk density for a natural mud with a very high (40% by weight) fraction of organic matter. During the flume tests it was found that for a bed of given density, while surface erosion occurred at low stresses, the bed failed at higher stresses. By testing beds of different densities it was feasible to construct these plots which characterize the two different erosion regimes. It should be noted that at densities less than $1.065\,\mathrm{g\,cm^{-3}}$ the mud occurred as a fluid supported slurry, which was assumed to erode at a fixed, pre-set rate (Hwang, 1989).

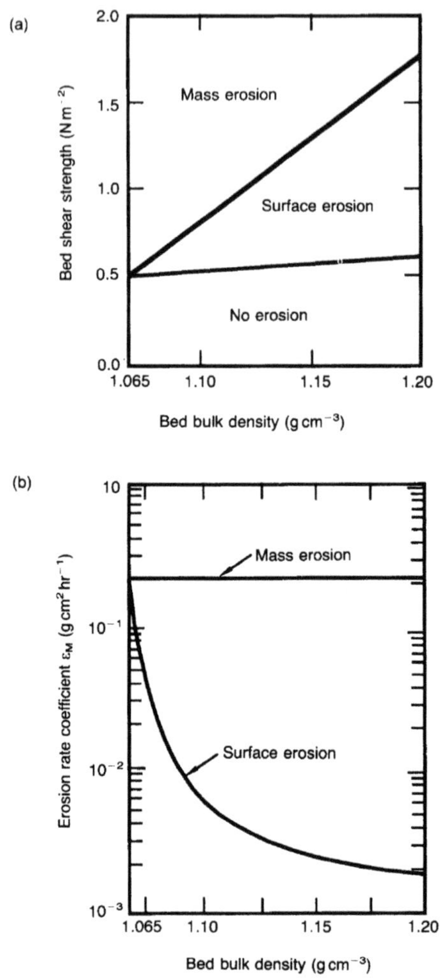

Figure 39.5 (a) Variation of bed shear strength with bed bulk density; definition of domains of surface erosion and mass erosion of a natural mud; (b) variation (lines) of corresponding erosion rate constants with bed bulk density (after Hwang, 1989).

Fluid mud entrainment

Entrainment of fluid mud is an erosion-related process which, however, must be considered from a somewhat different physical perspective than bed erosion, since entrainment essentially occurs at the interface between two fluids – water above and heavier mud below. Under shear driven, upper layer, flow the interface becomes unstable and entrainment occurs by virtue of mixing of the two fluids brought about by breaking interfacial waves and associated vortices. Laboratory flume experiments show that the entrainment process is characteristically dependent on the negative buoyancy associated with the heavier fluid and the kinetic energy associated with the flowing upper fluid. It can be shown therefore that the rate of entrainment varies with the product of the mean flow velocity of the upper layer multiplied by the shear stress acting over the mud. Furthermore, the rate relationship is found to be somewhat different from that, for example, for the entrainment of salt water into fresh water (Srinivas, 1989).

In the absence of entrainment data on given sediment under consideration for a particular problem, common practice is to calibrate Equation (39.11) to account for the mass brought into suspension by this process.

39.4.2 TIME-CONCENTRATION RELATIONSHIP

Since the nature of the bed erosion process manifests itself in the corresponding variation of the suspension concentration, it is worth examining the time-variation of the water column depth-average concentration, C, resulting from erosion at a constant applied stress. In Figure 39.6, a typically observed variation of C with time is sketched together with what is believed to occur at and near a stratified bed. Considering the situation in which initially the bed shear stress τ_b is greater than the bed shear strength τ_s, C will increase relatively rapidly at first. Eventually, as bed scour proceeds, τ_b will equal τ_s at some final depth of scour at which C will attain a practically constant value, with a very small rate of further increase due to floc erosion. Looking at the bed (initially at $z = 0$) in consonance with the variation of C, the erosion flux (arrow up from the bed) will eventually becomes zero as well, ignoring the small floc erosion flux.

The mass fluxes can also be viewed from another, compatible perspective. This perspective is based on what occurs at $z = \delta$, a small height above the bed. We may conceive of a comparatively thin near-bed layer whose upper and lower boundaries straddle the $z = \delta$ level as shown in Figure 39.6. This layer can be construed as a storage volume for the eroding bed sediment. Initially there would be no sediment in the layer, so that the mean concentration in the layer, C_b, would be zero. With the commencement of erosion, however, C_b will rise and, per unit time, more sediment will go into the layer from the bottom by erosion than will leave through the upper level by virtue of upward entrainment by turbulent diffusion. Once, however, there is sediment in suspension, settling will occur and a convective circulation cell will be set up, with sediment moving upward as well

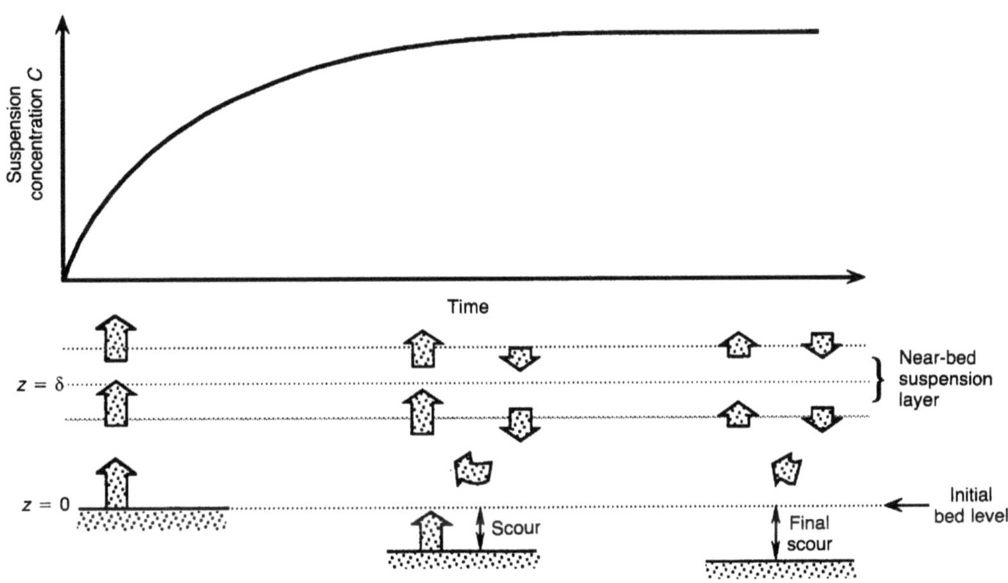

Figure 39.6 Entrainment and settling during erosion of a stratified bed at a constant applied bed shear stress.

as downward across the $z = \delta$ level. Finally, as bed erosion stops, the upward and downward fluxes (arrows) will all become equal in magnitude, implying a condition of equilibrium. Thus the mechanism of erosion can be explored either as one of bed scour at $z = 0$, or as one of simultaneous exchange involving entrainment and settling across $z = \delta$ (Parchure and Mehta, 1985; Lick, 1982). Exchange of flocs between those in the layer practically bounded by $z = 0$ and $z = \delta$ levels, and those in the water column above, has been demonstrated by tagging some flocs in suspension by radioactive gold atoms and measuring the rate of change of their concentration during deposition (Krone, 1962).

A manifestation of the sediment storing role of the near-bed layer surrounding $z = \delta$ level is that the ratio C_b/C takes on a value greater than one in the early phases of erosion, then levels off to a value just over one at equilibrium; the excess material having been entrained upward (Cervantes, 1987).

39.4.3 EROSION BY WAVES

A characteristic feature of the action of waves over mud is the gradual weakening of the mechanical and erosion strengths of the bed. This type of response is the outcome of the transmission of both pressure and shear forces into the bed, as a result of which the bed initially undergoes elastic deformation. However, with the build-up of excess pore pressure the bed matrix loses its integrity due to the break-up of cohesive bonds, and fluidization occurs (Mehta, 1989).

The expression for the rate of surface erosion is found to be of the same form as Equation (39.9), but the shear strength changes with time as the waves work their way through the bed, hence a representative average value of the shear strength must be used. The coefficients ε_M and τ_s are found to be somewhat smaller than in the case of erosion by current under comparable bed shear stresses (Mehta, 1988; 1989). Note that for waves the bed shear stress is characterized by its amplitude.

Useful formulations for the rate at which material that may become potentially available for re-suspension as a result of fluidization have not yet been fully developed. It is worth noting, however, that this amount can be overwhelmingly large compared with the mass that is re-suspended via surface erosion alone. By carefully measuring the time-rate of change of the near-bed suspension *in situ*, the erosion rate coefficients in Equation (39.9) can be calibrated to account for the contribution from fluidization in a very approximate way.

REFERENCES

Burt, T.N. (1986) Field settling velocities of estuary muds, in, *Estuarine Cohesive Sediment Dynamics*, (ed. A.J. Mehta), Springer-Verlag, New York, pp. 126–50.

Cervantes, E.E. (1987) A laboratory study of fine sediment resuspension by waves. M.S. Thesis, University of Florida, Gainesville.

Costa, R.C.F. (1989) Flow-fine sediment hysteresis in sediment-stratified coastal waters. M.S. Thesis, University of Florida, Gainesville.

Dyer, K.R. (1989) Sediment processes in estuaries. *Journal of Geophysical Research*, **94** (C10), 14327–39.

Hunt, J.R. (1986) Particle aggregate breakup by fluid shear, in, *Estuarine Cohesive Sediment Dynamics*, (ed. A.J. Mehta), Springer-Verlag, New York, pp. 85–109.

Hwang, K.-N. (1989) Erodibility of fine sediment in wave-dominated environments. M.S. Thesis, University of Florida, Gainesville.

Krone, R.B. (1962) Flume studies of the transport of sediment in estuarial processes. *Final Report*, Hydraulic Engineering Laboratory and Sanitary Engineering Research Laboratory, University of California, Berkeley.

Kuijper, C., Cornelisse, M. and Winterwerp, J.C. (1989) Research on erosive properties of cohesive sediments. *Journal of Geophysical Research*, **94** (C10), 14341–50.

Kynch, G.J. (1952) A theory of sedimentation. *Transactions of the Faraday Society*, **48**, 166–76.

Lavelle, W.J. and Mofjeld, H.O. (1985) Do critical stresses for incipient motion and erosion really exist?. *Journal of Hydraulic Engineering*, **113** (3), 370–85.

Lick, W. (1982) Entrainment, deposition and transport of fine-grained sediment in lakes. *Hydrobiologia*, **91**, 31–40.

Mehta, A.J. (1988) Laboratory studies on cohesive sediment deposition and erosion, in, *Physical Processes in Estuaries*, (eds J. Dronkers and W. van Leussen), Springer-Verlag, New York, pp. 427–45.

Mehta, A.J. (1989) On cohesive sediment suspension behavior. *Journal of Geophysical Research*, **94** (C10), 14303–14.

Mehta, A.J., Hayter, E.J., Parker, W.R., *et al.* (1989a) Cohesive sediment transport I: process description. *Journal of Hydraulic Engineering*, **115** (8), 1076–93.

Mehta, A.J., McAnally, W.H., Hayter, E.J., *et al.* (1989b) Cohesive sediment transport II: application. *Journal of Hydraulic Engineering*, **115** (8), 1094–112.

Migniot C. (1968) A study of the physical properties of different very fine sediments and their behavior under hydrodynamic action. *La Houille Blanche*, **7**, 591–620 (In French with English abstract).

Parchure, T.M. and Mehta, A.J. (1985) Erosion of soft cohesive sediment deposits. *Journal of Hydraulic Engineering*, **111** (10), 1308–26.

Richardson J.F. and Jeronimo, M.A. da S. (1979) Velocity-voidage relations for sedimentation and fluidization. *Chemical Engineering Science*, **34**, 1419–22.

Ross, M.A. (1988) Vertical structure of estuarine fine sediment suspensions. Ph.D. Dissertation, University of Florida, Gainesville.

Srinivas, R. (1989) Response of fine sediment-water interface to shear flow. M.S. Thesis, University of Florida, Gainesville.

Vanoni, V.A. (1975) *Sedimentation Engineering*, American Society of Civil Engineers, New York.

Wolanski, E., Asaeda T. and Imberger, J. (1989) Mixing across a lutocline. *Limnology and Oceanography*, **34** (5), 931–8.

CONTAMINANTS IN ESTUARINE SEDIMENTS 40

J.V. Towner

Especially since the Industrial Revolution man has used estuaries and the coastal zone to dispose of his waste. This has been the case since man first lived near rivers and estuaries but then pollutants were mainly organic, were in small quantities and hence easily biodegradable. As populations expanded and mainly for geopolitical reasons, people organized themselves into larger and larger groups and the capitals of countries more often than not were sited on rivers and estuaries for many obvious reasons. The pollution load even 200 years ago was mainly organic but now the quantities were large so that long stretches of rivers and estuaries became anaerobic and hence foul smelling. This was the case on the Thames estuary even up to the 1950s. The logical consequence was disease. With the Industrial Revolution came the advent of obnoxious and dangerous chemicals as byproducts and the 'obvious' place for these was in rivers and estuaries also.

This undesirable practice was only seriously tackled as late as 30–40 years ago but has received increasing attention in the last ten years. Whereas in earlier times the larger groups created problems they are now co-operating in almost a worldwide clean-up act. Perhaps what is even more significant, and the European Economic Community is a good example, these groups are producing legal standards which specify the decreases in the levels of pollution that have to be achieved by a particular date. This means that hydraulicians will increasingly be called on to predict the consequences of schemes to decrease or even eradicate the problem. They now have the help of mathematical models to do this but still need more knowledge of the chemistry. This chapter will help to achieve this.

CONTENTS

40.1 INTRODUCTION

Sediments have a major influence on the behaviour of chemical constituents in estuaries. The high organic matter content in estuaries and the high biological productivity of sediments results in sediment exerting a demand for dissolved oxygen from the overlying waters. The deposition of nutrient rich organic detritus and its subsequent biological decomposition means that estuarine sediments act as both a sink and source of nutrients such as phosphorus and nitrogen; sediments act as a reservoir for nutrients.

In addition to their role in the nutrient and oxygen balance of estuaries, sediments are of critical importance due to their interactions with contaminants introduced to estuaries by industrial activities. These contaminants may include trace metals and metalloids such as lead, zinc, copper, mercury, arsenic or selenium, and organic micropollutants, such as pesticides, petroleum hydrocarbons, polycyclic aromatic hydrocarbons (PAH) and polychlorinated biphenyls (PCBs) amongst others. With large chemically reactive surface areas, sediments adsorb and concentrate many of these pollutants from the water column, and as a result estuarine sediments act as a major sink for pollutants introduced to estuaries (Salomans and Forstner, 1984). A consequence is that the analysis of sediments can often provide a useful means of determining the degree of contamination of an estuary by pollutants (Forstaner and Wittman, 1983). Pollution histories of estuaries and other waters can be reconstructed by examining the vertical variation of contaminant concentrations in undisturbed, dated sediment cores (e.g. Goldberg *et al.*, 1977, 1979; Christenson and Lo, 1986).

Coastal, Estuarial and Harbour Engineer's Reference Book. Edited by M.B. Abbott and W.A. Price. Published in 1993 by Chapman & Hall, 2–6 Boundary Row, London SE1 8HN. ISBN 0 419 15430 2.

40.2 BASIC ASPECTS OF SEDIMENT BIOGEOCHEMISTRY

40.2.1 EARLY DIAGENESIS

In order to understand the various chemical interactions that are affected by sediments it is necessary to briefly examine some basic biogeochemical reactions (biologically mediated geochemical reactions) that are known as early diagenic processes. Diagenesis refers to the various post-depositional processes that transform the nature of sediments, where early diagenesis refers to processes in the upper sediment column. Diagenetic processes may be physical, chemical or biological. Of particular interest here are those processes whose driving force is the heterotrophic decay of organic matter in sediments. An excellent overall review of these processes has been given by Berner (1980). In considering these processes it is necessary to note that the sediment pore waters must be included as a component part of the sediments. They provide an extremely sensitive indicator of diagenetic change since they act as both a transport medium for chemical reactants as well as the reaction medium.

In steadily accreting sediments, a number of important chemical reactions affect particles as they are buried. The driving force of these reactions is the microbial degradation of organic matter. In the upper 1–2 cm of the sediments the oxidative microbial decomposition of organic matter exhausts dissolved oxygen in the sediment pore-waters. The rate of oxygen diffusion into the sediments from the overlying waters is insufficient to replace the oxygen used, thus microbes must use other compounds to continue the decomposition of organic matter. At first they may use nitrate, but this too is also rapidly exhausted; manganese and iron oxides may then be used as substitutes. Subsequently, microbes switch to sulphate which is in abundant supply in seawater. The microbial decomposition of organic matter by sulphate-reducing bacteria and concomitant production of hydrogen sulphide is an extremely important process in estuarine sediments. Following exhaustion of sulphate, the predominant microbial process is methane production (fermentation). The succession of microbial decomposition reactions leads to a vertical stratification of sediments giving a thin brown or yellow oxic layer, where oxygen is present, underlain by black anoxic sediment containing hydrogen sulphide and iron sulphides which are responsible for the colour. An intermediate grey layer is often not particularly noticeable.

Berner (1981) has proposed a simple, schematic vertical classification of the sediments into four main functional zones:

1. Oxic zone;
2. Post-oxic zone;
3. Sulphidic zone;
4. Methanic zone.

Each zone is characterized by a dominating biogeochemical process related to organic matter decay and can normally be differentiated by colour changes or through insertion of a redox (reduction-oxidation) electrode into the sediments; the 'post-oxic' zone in estuaries is normally too thin to be differentiated. Much work has been carried out on modelling diagenetic processes and fluxes of solutes using one-dimensional vertical models of the following general form:

$$\underset{\text{Chemical flux}}{\frac{SC}{St}} = \underset{\text{Diffusion term}}{\frac{DS^2C}{SX^2}} - \underset{\text{Advection term}}{\frac{WSC}{Sx}} + \underset{\text{Reaction term}}{R(C)} \qquad (40.1)$$

where C is the concentration of constituent e.g. sulphate, nitrate or ammonia, x is depth, t is time (thus sc/st is the flux of c at any depth), D is the diffusion coefficient, W is the rate of burial and $R(C)$ is a net rate constant for reactions affecting C (Berner, 1980).

Whilst this type of simple, steady-state model has been successful in examining distributions of certain components (particularly sulphate) in steadily accreting sediments, 'real' sediments do not behave in this way. Of particular importance are the effects of the following processes:

1. Bioturbation: the surface layers of sediments are subject to mixing by burrowing organisms which increases the depth of oxygen penetration. While some of this effect takes the form a general mixing of surface material, the burrows of organisms and associated irrigation profoundly affect rates of exchange of pore waters with overlying water and thus also chemical fluxes (Aller, 1984; Aller and Yingst, 1985). The sediment zonation discussed above spreads radially around the oxygenated burrows so sediment zonation becomes three-dimensional. Approaches to examining such complex effects are discussed and described by Aller (1980) and Berner (1980).
2. The presence of microzones in sediments, which are formed by localized concentrations of organic matter and have geochemical characteristics distinct from the bulk sediments (Jorgensen, 1977), can again impose a three-dimensional effect on chemical zonation and fluxes.
3. The physical mixing and re-suspension of sediments, due to wave and current stirring, results in enhanced fluxes of chemical substances from sediments (Vanderborght *et al.*, 1977).

4. Seasonal variations in microbial reaction rates due to changes in temperature and macroinvertebrate activity may lead to corresponding variations in the zonation of sediments and rates of diagenetic processes (Krantzberg, 1985).

A further aspect of considerable potential interest is the nature of biogeochemical processes in fluid muds although as yet much work on this problem is required.

In the following sections, a brief review is given of the effects of sediment diagenesis and associated processes with regard to sediment oxygen demand (SOD) and nutrients. These issues are examined on the basic that they are likely to be of greatest interest to engineers concerned with water quality problems in relation to coastal or estuarial engineering.

40.2.2 OXYGEN CONCENTRATION

It is generally acknowledged that SOD resulting from the aerobic breakdown of organic matter has a major influence on the oxygen balance in estuaries (Streeter and Phelps, 1925). Most state-of-the art water quality models incorporate a term to represent SOD, usually based on an empirical determination. Reviews of appropriate techniques for determining SOD have been given by NCASI (1981, 1982).

In general, the techniques used for determining SOD are based on the measurement of oxygen consumption rates (the flux of oxygen into sediments from overlying waters) normally expressed in units of $gO_2\,m^{-2}\,day^{-1}$. In terms of examining total benthic respiration (i.e. including nitrate and sulphate reduction) these techniques are open to criticism, since they do not take account of respiration by organisms using compounds other than oxygen (Patmatmat, 1975; Hargrave and Phillips, 1981). Anderson *et al.* (1986) have proposed a benthic respiration method based on total carbonate production in order to account for nitrate and sulphate reduction. Although water quality modellers are not normally interested in benthic respiration, other than that related to oxygen consumption), problems may arise due to the effects of methane formation. Di Toro *et al.* (1988) have noted that SOD does not necessarily increase proportionally as rates of diagenetic organic matter decay increase and have attributed this to methane gas bubbles escaping from the methanic zone and reducing oxygen in overlying waters The ability of methane to reduce oxygen is well established (Berner, 1980). Accounting for reactions between oxygen and methane is therefore critical in SOD measurement. De Rooij and de Vries (1988) have presented a diagenetic model for estimating SOD for the proposed Cardiff Bay Barrage in the UK.

In addition to the influence of sediments on oxygen due to diagenesis and benthic respiration, the erosion of bottom sediments or fluid mud by tidal currents or waves can result in re-suspension SOD. This phenomenon is often apparent as a variation in the magnitude of the oxygen sag in an estuary between spring and neap tides with oxygen concentrations lower on rising spring tides than on neaps or high springs. The possible causes for this additional oxygen demand may be:

1. The re-suspension of sediment from anaerobic zones, which is made available to decay by aerobic micro-organisms (i.e. there is an effective increase in the suspended BOD load),
2. The release of reduced chemical species such as iron (II), manganese (II), sulphides and ammonia from sediment pore water. This effect may be of considerable significance. Apart from re-suspension SOD related to natural physical factors, dredging, reclamation and spoil disposal give rise to substantial impacts on oxygen concentrations in overlying waters due to sediment re-suspension (Jeane and Pine, 1975; Khalid *et al.*, 1978; Maurer *et al.*, 1985; Rosenberg, 1977).

40.2.3 NUTRIENTS

The role of sediments as a reservoir of nutrients in estuaries is of interest principally because of the nutritional requirements of estuarine primary producers such as algae. Of particular significance are the various forms of the micronutrients, nitrogen and phosphorus, associated with sediments and to a lesser extent silicon which is an essential nutrient for diatoms. The particulate nutrients in sediments may be derived from a variety of natural sources including debris from plant and animal life, both from marine or terrestrial sources, as well as sewage and other contaminants. In general, the sedimentary nitrogen is predominantly associated with organic matter with a small proportion of inorganic nitrogen while phosphorus tends to be present mainly as inorganic forms. These latter forms include phosphate bound to carbonates, iron oxides or clay surfaces, vivianite (an iron (II) phosphate mineral) and apatite (a calcium phosphate mineral), which are often precipitated during diagenesis (Berner, 1980). However, in estuarine sediments a large part of the decomposition of organic matter is mediated by organisms using sulphate rather than oxygen (see section 40.2.1).

In addition to nutrient regeneration associated with the decay of organic matter, phosphate may be released to pore waters during the reduction and dissolution of iron oxides, which occurs in the post-oxic zone of sediments. Adsorbed phosphate is released as the oxides dissolve. Some of the released phosphate is subsequently precipitated as vivianite and apatite (Aller, 1980, Berner, 1980).

Transfer of micronutrients from pore waters to the overlying waters may occur by molecular diffusion or, more rapidly, by bioturbation, current induced advection and consolidation (Parker and Lee, 1981; Sills and Been, 1984). Dreging, spoil disposal and related activities may also lead to nutrient release (Maurer *et al.*, 1985). The reader is referred to Berner (1980) and Aller (1980) for detailed examinations of the fundamental approach to the diagenesis of nutrients, and Day *et al.* (1989) for a discussion of micronutrient cycling and its ecological significance in estuaries.

40.3 TRACE METALS IN ESTUARINE SEDIMENTS

40.3.1 INTRODUCTION

Many trace metals (e.g. copper, zinc, cobalt, etc.) are essential elements for living systems but above threshold concentrations may become toxic. Others, such as lead, cadmium and mercury are not essential and are potentially highly toxic.

Rivers transport trace metals to the ocean in dissolved, colloidal and particulate forms. The mixing of river water with saline marine waters in estuaries produces marked gradients in chemical and physicochemical properties (Boyle *et al.*, 1974; Morris, 1986; Morris, *et al.*, 1978; Sholkovitz, 1976). As a result, the relative contributions of various forms of trace metals can be modified considerably. Consideration should be given to these processes since they affect the overall chemical behaviour of each element, the rate of transport of the element through the estuary and the availability of the element to organisms (bioavailability).

The major chemical interactions affecting the distribution of trace metals between the dissolved and particulate forms are adsorption, desorption and flocculation (Duinker, 1980; Olsen *et al.*, 1982). In some cases biological processes can also affect trace metal behaviour. The adsorption of dissolved trace metals by particulate matter leads to their accumulation and retention in estuarine sediments (Morris *et al.*, 1986). Because of the importance of this 'scavenging' effect, it is essential to consider trace metals in both the water column and sediments in any assessment of their impact on water quality. The concentrations of dissolved trace metals are subject to considerable temporal variability in most estuaries. For this reason, the monitoring of the trace metal contents of sediments, which are subject to less temporal variability, often provides a more stable, integrated indication of the pollution status of estuarine waters (Forstner and Wittman, 1983).

40.3.2 REACTIONS AFFECTING TRACE METALS IN ESTUARINE SEDIMENTS

Adsorption reactions

Adsorption reactions entail the removal of trace metals from the dissolved state onto particles which contain a variety of metal binding sites. The principal binding sites in oxic sediments are associated with iron oxides, manganese oxides, organic matter, calcium carbonate and ion-exchange sites on clays and related minerals (Jenne, 1976; Gibbs, 1973; Forstner and Wittman, 1983). All these binding phases may exist in a wide variety of structural forms but as a first approximation classification into these broad classes of binding site is useful. The binding phases in sediments provide a variety of strengths of metal binding and the fate of a metal bound to particulate material is determined by its distribution among the available binding sites.

In addition to being bound to the phases outlined above, all of the metals are associated with the crystalline structures of detrital minerals (e.g. clays, feldspars, etc.) that are present in the sediments and particulate matter. However, metals associated with these crystalline structures are essentially unavailable for desorption or biological uptake (Luoma, 1983).

Flocculation

Many substances in estuaries exist in the form of stabilized colloids which, although not truly dissolved, do not behave as particles. Substantial quantities of certain metals, particularly iron and aluminium, exist in colloidal form in estuaries (Aston and Chester, 1973; Eckert and Sholkovitz, 1976; Fox and Wofsy, 1983). For example, much of the iron in estuaries exists as 'micro-particles' of iron hydroxide stabilized by a coating of natural organic matter (Moore *et al.*, 1979; Crerar *et al.*, 1981; Mayer, 1982; Tipping, 1981; Tipping and Heaton, 1983; Hunter, 1983; but also see Fox (1984) for a dissenting view). As these colloids enter the very low salinity estuarine environment the major ions of sea-water and hydrogen ions interact chemically and electrostatically with the surfaces of the colloids (Mantoura, 1987; Morris, 1978; Morris *et al.*, 1978). These interactions destabilize the colloids and result in their flocculation. The flocculated colloids, and their associated trace metals, may then accumulate in the sediments.

Metal release from sediments and desorption reactions

The sequence of microbial decomposition reactions outlined in section 40.2.1 has important implications for the behaviour of metals associated with the sediments as conditions in the sediments progress from oxic to anoxic. The important processes may include:

1. Metals bound to iron and manganese oxides are released to the sediment pore waters as the oxides dissolve under anoxic conditions (Burdige and Gieskes, 1983).
2. Metals bound to sedimentary organic matter may be released to the sediment pore waters as the organic matter undergoes microbial degradation (Presley *et al.*, 1972; Ridgway and Price, 1987).
3. Metals released into the sediment pore waters may diffuse upwards into the oxic sediment layer, and thence to the overlying waters, or downwards into the anoxic sediments (Corignan and Tessier, 1985; Corignan and Nriagn, 1985; Davies–Colley *et al.*, 1985; Elderfield *et al.*, 1981b).
4. Dissolved metals in anoxic sediment pore water containing hydrogen sulphide are precipitated as highly insoluble metal sulphides (Stumm and Morgan, 1981; Boulegue *et al.*, 1982; Kremling, 1983; Dyrssen, 1985; Jacobs *et al.*, 1985; Giblin *et al.*, 1986) or in association with iron sulphides (Jean and Bancroft, 1986).
5. Dissolved metals diffusing into the oxygen containing sediment pore waters of the upper oxic layer are removed from solution by re-adsorption onto iron oxides, manganese oxides and organic matter present in this layer (Heggie and Lewis, 1984; Gendron *et al.*, 1986; Giblin *et al.*, 1986; Sundby *et al.*, 1986).
6. A small proportion of the metals diffusing upwards may escape to the overlying waters (Gaillard *et al.*, 1986; Emerson *et al.*, 1984; Westerlund *et al.*, 1986).
7. Dissolved metals in porewaters are advected upwards during consolidation and can escape into the overlying waters (Parker and Lee, 1981; Sills and Been, 1984).

The overall balance of these processes results in the transformation of the chemical forms of metals in anoxic sediments so that most are present as either pure metal sulphides, in association with iron sulphides or organically-bound. Chromium is present as Cr(III) hydroxide. Evidence for the actual release of metals to estuarial waters due to the above processes is sparse, although the partial release of manganese (Elderfield *et al.*, 1981a; McCaffrey *et al.*, 1981; Sundby *et al.*, 1981; Sundby and Silverberg, 1985) and arsenic (in the +3 and +5 forms) (Knox *et al.*, 1984) from sediments has been established for some estuaries. Release from sediments to anoxic overlying waters of both arsenic (Peterson and Carpenter, 1983) and mercury (Lu *et al.*, 1986) has been established in intermittently anoxic fjords, while the benthic regeneration and release of copper has been established for deep sea sediments (Fischer *et al.*, 1986). As a general point, it is apparent that the release of metals from sediment pore waters to the overlying waters requires physical mixing or pore-water advection processes that allow more rapid solute exchange than molecular diffusion. These processes include the bioturbation and irrigation of sediments by benthic fauna, the scour and resuspension of sediments by strong tidal, wind or wave-induced currents, and pore-water transport arising from self-weight consolidation. Nevertheless, at present the role of sediments as a long-term sink for dissolved trace metals in estuaries is believed to outweigh their importance as a potential short-term source of metals to the overlying waters (Forstner and Wittman, 1983).

Other processes which may lead to the desorption of trace metals from sediments and particulate matter to the dissolved state are also relevant. They include, firstly, weakly adsorbed metals associated with, for example, ion-exchange sites which may be displaced from particulate matter by the major cations of seawater (sodium, magnesium, potassium and calcium) (De Groot *et al.*, 1976; Van der Weijden *et al.*, 1977). Some evidence suggests that this process may cause the partial release of some metals, particularly cadmium (up to 25–40% of total) from sedimentary material in estuaries (Rohatgi and Chen, 1976; Van der Weijden *et al.*, 1977).

Secondly, when anoxic sediments are re-suspended into oxygenated waters the associated metal sulphides may be oxidized and metals released to the dissolved state (Salomans, 1985; Ackroyd *et al.*, 1986). Desorption of metals from anoxic dredge spoils due to re-suspension in oxygenated waters has been observed (Holmes, 1977; Khalid *et al.*, 1978, Delaine *et al.*, 1984). Experiments have been conducted where anoxic sediments have been re-suspended in oxygenated seawater. These results showed no desorption of copper or zinc and, in fact, the re-suspended sediments adsorbed these metals from the sea-water (Hirst and Aston, 1983). Similar results have been obtained for lead (WRC, unpublished). In contrast, short-term releases of mercury followed by re-adsorption by sediments have been noted (Lindberg *et al.*, 1975).

Methylation of trace metals

Finally, mention should be made of microbial methylation (biomethylation) processes, whereby a metal is transformed from the inorganic state to an organometallic compound. Metals and metalloids such as mercury, arsenic, selenium, thallium and tin are all readily biomethylated in the aquatic environment (Forstner and Wittman, 1983). The bacteria *Chlostridia spp.* are particularly effective in methylating metals (Yamada and Tonomura, 1972). The major environmental methylating agent is methylcobalmin [vitamin

B12] (Wood, 1972). A number of other biological co-enzymes can also achieve methylation (Beijer and Jornelov, 1979). This process is particularly important regarding mercury since methyl-mercury is more toxic and more readily taken up by organisms than the inorganic form. By contrast, although methylated forms of arsenic are more readily taken up by organisms than the inorganic forms, they are also more readily excreted and not particularly more toxic (Langston and Bryan, 1984).

It has been suggested that inorganic lead can be biomethylated (Wong *et al.*, 1975; Schmidt and Huber, 1976). The evidence for this process is ambiguous. By contrast, the methylation of alkyl-lead has been established. There is evidence, however, that alkyl-lead species can alkylate mercury by a chemical mechanism and so render mercury more toxic (Beijer and Jorelov, 1979).

In sulphide-rich sediments there is evidence that methyl-mercury interacts with sulphide ions to form bismethyl mercury sulphide $(CH_3Hg)2S$. This species then disproprionates to form mercuric sulphide and dimethyl mercury, which being volatile, is lost by evaporation from the sediment environment (Bartlett and Craig, 1981; Craig and Moreton, 1984).

There is evidence that microbes, in particular *Pseudomonas sp*, are also able to demethylate methyl-mercury and so detoxify it. Microbes can also reduce oxidized mercury (Hg^{2+}) to elemental mercury (Hg), which due to its volatility may be removed to the atmosphere from the aquatic environment (Billen *et al.*, 1975; Beijer and Jornelov, 1979).

40.3.3 BIOLOGICAL UPTAKE OF TRACE METALS

Organisms in estuaries are able to take up dissolved or particulate trace metals from pore-water or sediments. Dissolved metals are generally more readily assimilated than are particulate metals.

The chemical form of dissolved trace metals and biological uptake

Dissolved trace metals in pore-waters can exist in a number of different chemical forms (species). This is because trace metals interact with the major dissolved inorganic anions in sea-water (chloride, sulphate, bicarbonate, etc.) or sulphide and polysulphide complexes and natural dissolved organic matter to form dissolved complex compounds. (Boulengne *et al.*, 1982; Kremling, 1983).

The balance (speciation) of a dissolved metal among the various complex species is critical in determining the biological activity, toxicity and uptake availability of the metal to organisms (Luoma, 1983; Langston and Bryan, 1984; Rainbow, 1990). In particular, free metal ions display high biological availability (Luoma, 1983). For example, it has been demonstrated that although total dissolved copper concentrations in estuaries may exceed those in clean, open ocean, waters, the level of free Cu^{2+} ions may approach the toxic threshold level for phytoplankton more closely in the open ocean waters (Sunda and Gillespie, 1979). Organometallic compounds such as methyl-mercury and methylated arsenic compounds show extremely high biological availability due to their high solubility in the lipids of organisms (Luoma, 1983; Langston and Bryan, 1984). Uptake of manganese from pore-waters by sediment fauna may be important (Bryan and Hummerstone, 1973).

The chemical form of particulate trace metals and biological uptake

The variety of chemical forms of particulate (suspended or sedimentary) trace metals has been reviewed in section 3.2. The biological uptake of trace metals from particulate matter and sediments depends on the chemical form of the metals (Luoma, 1983; Langston and Bryan, 1984; Langston, 1986; Forstner and Wittman, 1983; Marquenie, 1985).

Uptake of metals from particulate matter and sediments may occur following the ingestion of particles through feeding or through uptake of particles over the body surface; this occurs in invertebrates and is termed pinocytosis (Louma, 1983). Uptake may also occur following surface contact between sediments and organisms. Organisms also take up metals from sediment pore-waters (Bryan and Hummerstone, 1973); this could be important when sediments are undergoing processes of microbial decomposition that occur during early burial (see sections 40.2.1 and 40.3.2). Uptake of metals need not take place through ingestion and digestion in the alimentary tract. The seaweed *Fucus vesiculosus* can scavenge copper and lead from particles, possibly through an interaction with natural complex substances (polyphenols) present in Fucus (Luoma and Bryan, 1982).

Generally, fine-grained oxidized surface sediments are the most important sources of available metals (Luoma and Bryan, 1980). The uptake of trace metals from surface sediments is dependent on their chemical form (Luoma and Bryan, 1980). For example, lead is strongly associated with iron oxides, and the concentrations of lead in clams (*Scrobicularia plana* and *Macoma balthica*) in many estuaries are related not to total sediment lead contents but rather to the ratios of Pb/Fe in acid extracts (1 M HCl) of sediments (Luoma and Bryan, 1978). Similarly, the concentrations of mercury in *S. plana* and *M. balthica* are related to the ratio Hg/organic matter in surface sediments. This is due to the strong interaction of mercury with organic matter which confers a 'protective' effect on mercury uptake and toxicity (Langston, 1982, 1986).

Food chain magnification and availability of metals in food

Biomagnification of trace metals along food chains in natural marine communities is less common than is popularly believed and has only been established for mercury (Langston and Bryan, 1984). This is because of the selective retention of methyl-mercury at each level of the food chain, due to its high assimilation efficiency and long biological half-life, combined with the longevity of most top predators. Nevertheless, uptake of metals from food can be an important route for the accumulation of metals in organisms (Fowler *et al.*, 1981; Windom *et al.*, 1982).

Biochemical control of trace metals

When an organism is confronted with trace metals in excess of its normal metabolic requirement it will attempt to eliminate, metabolize or detoxify the excess concentration in order to prevent the inhibition of sensitive enzyme systems that could lead to serious physiological injury and ultimately death. Organisms may employ the following biochemical mechanisms to prevent toxic effects:

1. Organisms can eliminate metals as rapidly as they are accumulated, particularly when exposed to dissolved inorganic metal species (Fowler, 1982). Elimination of organometallic species such as methyl-mercury may be slow due to their high lipid solubility and strong affinity for certain complexes (sulphahydryl groups) present in organisms.
2. An alternative detoxifying mechanism employed by organisms is the storage and isolation of metals (Luoma, 1983). Of major importance are:
   (a) The formation of strong metallothionein metal-binding protein complexes that sequester metals and increase the tolerance of organisms to high concentrations of metals, although if the binding capacity of these compounds is exceeded toxic effects may occur (Williams, 1981).
   (b) The storage of metals in association with lysosomal membranes and other membrane limited vesicules, often in association with metal-rich granules (Fowler *et al.*, 1981).
   (c) The formation of metal-rich granules containing sulphur, calcium (carbonate, orthophosphate or pyrophosphate) (Mason and Nott, 1981), iron (often as ferritin) (George *et al.*, 1977) or pure mercuric selenide (Martoja and Viale, 1977).

40.3.4 TRACE METAL CONTENTS OF ESTUARINE SEDIMENTS

The importance of grain size

An important elemental partitioning effect is brought about by the textural control of the mineralogy of sediments, whereby some elements may be preferentially associated with certain grain-size fractions of the sediments. For example, zirconium, is often found to be preferentially associated with silts because zircon, the principal mineral in which zirconium is found, is more abundant in that size fraction of sediments (Calvert, 1976). In contrast, fine-grained clays contain high proportions of octahedral lattice sites which may concentrate many transition metals (Mayer and Fink, 1980). A second textural partitioning effect is brought about by the dependence of the specific surface area (SSA) upon grain size, whereby the SSA of sediments increases with decreasing grain size. Thus, smaller particles present a greater SSA for adsorptive reactions with metal ions, whether through direct adsorption onto clay particle surfaces or through adsorption by oxide or organic matter coatings on the surfaces of the clays. In this context, Jenne (1976) has expressed the view that the role of clay minerals as hosts for trace metals in sediments is relatively unimportant, but that their main role is as mechanical substrates for the precipitation and flocculation of hydrous iron oxide and organic matter, both of which are powerful metal adsorbents.

A variety of methods have been proposed to account for and reduce the effect of grain size on metal contents. Commonly used techniques include.

1. The normalization of metal contents to standardized granular composition.
2. Relating metal content to SSA where SSA is determined geometrically, analytically or using an algorithm.
3. Normalization of metal contents to an element that is mainly contained in fine particles (e.g. caesium).
4. Separation of sediments into a single class of fine material.

Baseline trace metal contents of sediments

In order to assess the extent of trace metal contamination in sediments it is necessary to establish a baseline for uncontaminated sediments, which allows excess metal contents to be derived. The baseline may be defined by examining sediment samples collected from a known depth in dated sediments (see Kemp *et al.*, 1978; Goldberg *et al.*, 1977, 1979). Table 40.1 shows the average trace element compositions of nearshore muds and shales which have been adopted by some geochemists as useful standards for base-

Table 40.1 Average trace element compositions of nearshore muds and shales (expressed as ppm)

Trace element	*Nearshore muds**	*Shales†*
Al	80 000	80 000
V	130	130
Cr	100	90
Mn	850	850
Fe	69 000	47 200
Co	13	19
Ni	55	68
Cu	48	45
Zn	95	95
Pb	20	20

* Wedenpohl (1960)
† Turekian and Wedenpohl (1961)

line metal levels (Wedenpohl, 1960; Turekian and Wedenpohl, 1961; Riley and Chester, 1969). For any one study area use of locally derived historical sediments is preferred for pollutant flux estimates but the data given in Table 40.1 is useful for a preliminary assessment of pollution.

The comparison between elemental analyses and baseline standard compositions becomes extremely difficult if the mineralogical compositions of the environmental sample and baseline standard are different. A very simple method for avoiding this problem is to compare the element/aluminium ratios of the samples to that of the baseline standard. This comparison is achieved by calculating the element enrichment factors (Chester and Aston, 1981), or the sediment enrichment factors (Kemp *et al.*, 1976). Enrichment factors may be computed using elements other than aluminium as the reference element, e.g. iron, potassium, titanium, caesium (Ackermann, 1980).

40.4 ORGANIC MICROPOLLUTANTS

40.4.1 INTRODUCTION

The organic micropollutants comprise a wide range of compounds, many of which, such as PCBs and PAH are ubiquitous in the estuarine environment. Given the extremely wide range of chemical structure, behaviour and toxicity of individual compounds (the PCBs alone comprise 209 individual compounds), it is beyond the scope of this review to give more than a brief overview of these compounds in estuarine sediments.

In estuaries a number of processes may lead to the removal of organic micropollutants from the water column. These include evaporation, adsorption by suspended sediments (particularly the organic fraction), photodegradation, 'salting out' due to increasing salinity and microbial degradation. However, there is evidence that dissolved organic matter can bind to and enhance the solubilities of a range of organic micropollutants (Chiou *et al.*, 1986; Madhun *et al.*, 1986), and may affect uptake and toxicity to organisms (McCarthy *et al.*, 1985).

The various organic micropollutants found in estuaries have varying degrees of affinity for adsorption into particles, particularly organically enriched particles. The adsorption affinity of organic pollutants onto particles is largely dependent on their lipophilicity (affinity with lipids), which is often described by an octanol partition coefficient (Xie *et al.*, 1986; Lodge and Cook, 1989). This is an experimentally derived coefficient which expresses the distribution of a substance within an octanol–water test system. In general, a linear relationship between logarithms of the octanol–water and sediment organic matter–water partition coefficients is observed (McElroy *et al.*, 1989). However, measured water–sediment coefficients show a strong dependence on solids concentration, which becomes stronger with increasing octanol–water partition coefficient (Di Toro 1985; Elzerman and Coates, 1987). Furthermore, for substances such as phenolics or chlorophenolics, which have lower lipophilicities and are influenced by acid-base characteristics, the partioning towards sediments is influenced by pH (Xie *et al.*, 1986).

As with trace metals, organic micropollutants are subject to speciation effects such that compounds show varying affinities for natural or anthropogenic substances (Means and Wijayraine, 1982; Madhun *et al.*, 1986; Readman *et al.*, 1982) or different size fractions of sediments (Readman *et al.*, 1982; Preston and Al-Omran, 1989). Tidal variations in the concentration of kepone in suspended matter have been documented by Huggett and Bender (1980), and were correlated with tidal variations in the organic content of the suspended sediments.

Accumulation of organic compounds, such as PAH and PCBs, by organisms has been shown to be related to the octanol–water partition coefficient.

40.4.2 ORGANOCHLORINE COMPOUNDS (e.g. PCBs)

Origins

PCBs have been widely used in industry since 1929. They have been used as capacitor fluids, fire retardants, additives to paints, hydraulic fluids and as plasticizers. The organochlorine pesticides include DDT and its metabolites (DDD and DDE), dieldrin, aldrin, and lindane w,4-D toxaphene, hexacholorbenzene, the pyrethroids, mirex and its degradation product kepone among others. These compounds have found widespread past use in agriculture and forestry. Other industrially derived, environmentally significant chlorophenols, include the chlorinated dibenzofurans and dibenzo-dioxins (known as PCDFs, PCDDs, TCDFs, TCDD) which are highly toxic and are derived from a wide variety of sources, including paper bleaching, waste inceneraton and sewage treatment.

Environmental interactions in estuaries

The adsorption of most chlorinated compounds is strongly related to the organic matter content of sediments (Brownawell and Farrington, 1986), and sorption affinity increases with increasing octanol –water partition coefficients for individual compounds. Octanol–water partition coefficients generally increase with an increasing degree of chlorination within homologous series of compounds (Hansch, 1979). During early diagenesis, there is evidence that PCBs are associated with both sedimentary organic matter and dissolved and colloidal organic matter in the pore-waters (Brownawell and Farrington, 1986), the remainder being truly dissolved. The equilibria between these three phases is critical in terms of desorption and remobilization of PCBs to the water column from sediments and bioavailability to organisms. An important consequence of the three-phase equilibria is that the less soluble, highly chlorinated PCBs tend to be enriched in the pore-waters relative to the more soluble less chlorinated PCBs, due to the former compounds association with pore-water organic matter.

PCBs and organochlorine compounds in sediments are subject to slow processes of microbial and chemical degredation. For example, DDT is degraded to DDE and DDD, whereas PCBs may be oxygenated and the benzene rings cleaved to form p-chlorobenzoic acid. The nature of these degradation reactions varies depending on whether they occur in the oxic, sulphide, or methanic zones of the sediments. Under oxidized conditions susceptibility to, and rates of, degradation depend on the number and substitution of chlorine atoms (Bailey *et al.*, 1983). For example, monochlorobiphyl is readily microbially degraded while higher chlorinated compounds are not, and the rates of degradation decrease with greater degree of chlorine substitution. Under reducing conditions it has been suggested that degradation of organochlorine compounds proceeds by reductive dehalogenation (Zoro *et al.*, 1974). There is dispute as to whether these reactions are chemical (Zoro *et al.*, 1974), involving Fe(II) compounds as catalysts, or microbial (Brown *et al.*, 1984), possible involving methanogenic bacteria (Sulfita *et al.*, 1983). Under reducing conditions, the ease of reduction appears to increase with an increasing degree of chlorine substitution (Brownawell and Farrington, 1986). Nevertheless, despite these processes many PCBs and organochlorine compounds have considerable persistence in the environment. Bioturbation of sediments may play an important role in the diagenesis of organochlorine compounds by two mechanisms: firstly, by increasing pore-water exchange and fluxes of compounds to overlying waters and secondly, by introducing oxygen into the lower anoxic zones of the sediments thereby changing the nature of degradation reactions. Accumulation of organo-chlorine compounds by organisms tends to corelate with octanol–water partition coefficients and storage occurs in lipid fractions.

40.4.3 POLYCYCLICAROMATIC HYDROCARBONS (PAHs)

Origins

These compounds are natural constituents of petroleum hydrocarbons. Although they rarely exceed 40% of the compounds present in crude oil, they may comprise 60% of the content of some refined products. They can also be introduced to the environment via fossil fuel combustion. The dominance of unsubstituted parent PAH in environmental samples is indicative of combustive origins of the PAH. A number of the PAHs are highly carcinogenic, in particular, 3,4-benzo(a)pyrene, 1,2-benzanthracenes, 1,2-benzphenanthrene (chrysene), dipenylmethane, florene, phenanthrene and dibenzthiophene. There is some indication that fish can detoxify some PAH by metabolism in the liver, followed by transfer of hydrocarbons and metabolites to the bile, and finally excretion (Neff, 1979).

Environmental interactions in estuaries

On the basis of environmental behaviour, physical and chemical properties, and microbial degradation rates the PAH compounds may be divided into two major groups. The first group consists of low molecular weight compounds that are characterized by relatively high water solubility, high vapour pressure and low particle affinity. Naphthalene, phenanthrene and anthracene are incorporated into this group (Readman *et al.*, 1984). Napthalene as a typical member of this group is subject to rapid rates of microbial decay by free-living microbes. This group of compounds is characteristically volatile and subject to high degradation rates which results in their having short residence times in estuaries. Due to their low particle affinity they are not readily adsorbed ('scavenged') by particles and thus tend to be enriched in the water column.

The second group consists of high molecular weight PAH compounds such as fluoranthrene, pyrene, chrysene, benz(a)-anthracene, benzo(b)-fluoranthrene, benzo(k)fluorantherene and benzo(a)pyrene. These compounds are characteristically of low water solubility, low vapour pressure and exhibit high particle affinity. A typical PAH of this group benzo(a)pyrene is only slowly degraded by microbes, which are probably associated with particles. The high particle affinity and microbial refractivity of this high molecular wight group of PAH leads to the accumulation of these compounds in estuarine sediments. Interestingly, PAH compounds do not show the expected adsorption affinity for fine grain-size particles that is shown by trace metals and some organic contaminants (Readman *et al.*, 1982, 1984). Rather these compounds show a marked preference for coarse, sand-sized particles; this could be due to the presence of charcoal and combusted organic material in this fraction, which would be expected to show a high affinity for the adsorption of PAH.

REFERENCES

Ackermann, F. (1980), *Environ. Technol.*, Letts. **1**, 518–527.

Ackroyd, D.R., Bale, A.J., Howland, R.J.M., Knox, S., Millward, G.E. and Morris, A.W. (1986), *Estuar. Coast. Shelf Sci.*, **23**, 621–640.

Aller, R.C. (1980), *In Salzman, B. (Ed): Advances in Geochemistry; Estuarine Physics and Chemistry: Studies in Long Island Sound*, Vol 22, Academic Press.

Aller, R.C. (1984), *Geochim. Cosmochim. Acta*, **48**, 1929–1934.

Aller, R.C. and Yingst, J.Y. (1985), *J. Mar. Res.*, **43**, 615–645.

Anderson, L., Hall, P.O.J., Iverfeldt, A., Rutgers, van der Loeff, M.M., Sundby, B. and Westerlund, S.F.G. (1986), *Limnol. Oceanogr.*, **31**, 319–329.

Aston, S.R. and Chester, R. (1973), *Estuar. Coast Mar. Sci.*, **1**, 225–231.

Bailey, R.E., Gonslor, S.J. and Rhinehart, W.L. (1983), *Environ. Sci. Technol.*, **17**, 617–621.

Bartlett, P.D. and Craig, P.J. (1981), *Water Research*, **15**, 37–47.

Beijer, K. and Jernelov, (1979), *In Nriagu, J.O. (Ed), The Biochemistry of mercury in the environment*, Elsevier/North-Holland Biomedical Press, Amsterdam, 203–210.

Benjamin, M.M. and Leckie, J.O. (1981), *J. Colloid Interface Sci.*, **79**, 207–217.

Berner, R. (1980), *Early Diagenesis – A Theoretical Approach*, Princeton University Press, New Jersey, 241.

Berner, R. (1981), *J. Sediment. Petrol.*, **51**, 359–366.

Billen, G., Joins, C. and Wollast, R. (1979), *Water Res.*, **8**, 219–225.

Boulegue, J., Lord, C.J. and Church, T.M. (1982), *Geochim. Cosmochim. Acta*, **46**, 453–464.

Boyle, E.A., Collier, R., Dengler, A.T., Edmond, J.M., Ng, A.C. and Stallard, R.F. (1974), *Geochim. Cosmochim. Acta*, **38**, 1719–1738.

Brownawell, B.J. and Farrington, J.W. (1986), *Geochim. Cosmochim. Acta*, **50**, 157–169.

Brown, J.F., Wagner, R.C., Bedard, D.L., Brennan, M.J., Carnahan, J.C., May, R.F. and Tofflemire, T.J. (1984), *Environ. Sci.*, **3**, 167–179.

Bryan, G.W. and Hummerstone, L.G. (1973), *J. Mar. Biol. Assoc., U.K.*, **53**, 858–872.

Burdige, D.J. and Gieskes, J.M. (1983), *Am. J. Sci.*, **283**, 29–47.

Calvert, S.E. (1976), *In: Chemical Oceanography*, Eds: Riley, J.P. and Chester, R. Vol 6. Academic Press, New York, 187–280.

Carignan, R. and Nriagu, J.O. (1985), *Geochim. Cosmochim. Acta*, **49**, 1753–1764.

Carignan, R. and Tessier, A. (1985), *Science*, *228*, 1524–1526.

Chester, R. and Aston, S.R. (1981), In: *Techniques for Identifying Transuranics in Aquatic Environments*, IAEA, Vienna, pp. 173–183.

Chiou, C.T., Malcolm, R.L., Brinton, T.I. and Kile, D.E. (1986), *Environ. Sci. Technol.*, **20**, 502–508.

Christensen, E.R. and Lo, C.-K. (1986), *Environ. Pollut.*, Ser B, **12**, 217–232.

Craig, R.J. and Moreton, P.A. (1984), *Mar. Pollut. Bull*, **15**, 406–408.

Crerar, D.A., Means, J.L., Yuretich, R.F., Borcsik, M.P., Amster, J.L., Hastings, D.W., Knox, G.W., Lyon, K.E. and Quiett, R.F. (1981), *Chem. Geol.*, **33**, 23–44.

Davies-Colley, R.J., Nelson, P.O. and Williamson, K.J. (1985), *Mar. Chem.*, **16**, 173–186.

Day, J.W., Jr, Hall, C.A.S., Kemp, W.M. *et al.* (1989), *Estuarine Ecology*, Wiley Interscience, New York, p. 558.

de Rooiy, N. and de Vries, I. (1988), *Report to Wallace Evans and Partners by Delft Hydraulics Institute*, **34**, plus figures.

Delaine *et al.* 1984, p. 13.

Di Toro, D.M. (1985), *Chemosphere*, **15**, 1503–1538.

Duinker, J.C. (1980), *In Olausson, E. and Cato, I. (Eds): Chemistry and Biogeochemistry of Estuaries*, Wiley, New York, 121–151.

Dyrssen, D. (1985), *Mar. Chem.*, **15**, 285–293.

Eckert, J.M. and Sholkovitz, E.R. (1976), *Geochim. Cosmochim. Acta*, **40**, 847–848.

Elderfield, H. (1981), *Am. J. Sci.*, **281**, 1184–1196.

Elderfield, H., Bender, M., McCaffrey, R.J. and Luedtke, N. (1981a), *Am. J. Sci.*, **281**, 768–789.

Elderfield, H., McCaffrey, R.J., Luedtke, N., Bender, M.L. and Truesdale, V.W. (1981b), *Am. J. Sci.*, **281**, 1021–1055.

Elzerman, A.W. and Coates, J.T. (1987), *In: Sources and Fates of Aquatic Pollutants*, Eds: Hites, R.A. and Eisenreich, S.J. Adv. Chem. Ser 216, ACS Washington, D.C., 263–317.

Emerson, S., Jahnke, R. and Heggie, D. (1984), *J. Mar. Res.*, **42**, 709–730.

Fischer, K., Dymond, J., Lyle, M., Soutar, A. and Rau, S. (1986), *Geochim. Cosmochim. Acta*, **50**, 1535–1543.

Forstner, U. and Wittman, G.T.W. (1983), *Metal Pollution in the Aquatic Environment*, Second Edition, Springer-Verlag, Berlin.

Fowler, B.A., Carmichael, N.G., Squibbs, K.S. and Engel, D.W. (1981), *In: Biological Monitoring of Marine Pollutants*, Eds: Vernberg, F.J., Calabrese, A., Thurberg, F.P. and Vernberg, W.B. Academic Press, London, pp. 127–144.

Fowler, S.W. (1982), *In: Kullenberg, G (Ed). Pollutant Transfer and Transport Processes*, Vol Ii, CRC Press Inc, Boca Raton, Florida, pp. 1–65.

Fox, L.E. (1984), *Geochim. Cosmochim. Acta*, **48**, 879–884.

Fox, L.E. and Wofsy, S.C. (1983), *Geochim. Cosmochim. Acta*, **47**, 211–216.

Gaillard, J.-F., Jeandel, C., Michard, G., Nicolas, E. and Renard, D. (1986), *Mar. Chem.*, **18**, 233–247.

Gendron, A., Silverberg, N., Sundby, B. and Lebel, J. (1986), *Geochim. Cosmochim. Acta*, **50**, 741–747.

George, S.G., Pirie, B.J.S. and Coombs, T.L. (1977), *Biochem. Soc. Trans.*, **5**, 136–137.

Gibbs, R.J. (1973), *Science*, **180**, 71–73.

Giblin, A.E., Luther, G.W. and Valieba, I. (1986), *Estuar. Coast. Shelf Sci.*, **23**, 477–498.

Goldberg, E.D., Gamble, E., Griffin, J.J. and Koide, M. (1977), *Estuar. Coast. Mar. Sci.*, **5**, 549–561.

Goldberg, E.D., Griffin, J.J., Hodge, V., Koide, M. and Windom, (1979), *Environ, Sci. Technol.*, **11**, 297.

Groot, A.J., de, Salomans, W. and Allersma, E. (1976), *In Burton, J.D. and Liss, P.S. (Eds), Estuarine Chemistry*, Academic Press, London, pp. 131–157.

Hansch, C. (1979), *Substituent Constants for Correlation Analysis in Chemistry and Biology*, Wiley Interscience, New York.

Hargrave, B.T. and Phillips, G.A. (1981), *Estuar. Coast. Shelf Sci.*, **12**, 725–737.

Heggie, D. and Lewis, T. (1984), *Nature, London*, **311**, 453–455.

Hirst, J.M. and Aston, S.R. (1983), *Estuar. Coast. Shelf Sci.*, **16**, 549–558.

Holmes, C.W. (1977), *J. Res. U.S. Geol. Surv.*, **5**, 243–251.

Huggett, R.J. and Bender, M.E. (1980), *Environ. Sci. Technol.*, **14**, 918–923.

Hunter, K.A. (1983), *Geochim. Cosmochim. Acta*, **47**, 467–473.

Jacobs, L., Emerson, S. and Skei, J. (1985), *Geochim. Cosmochim. Acta*, **49**, 1433–1444.

Jeans, C.S. and Pine, R.E. (1975), *J. Water. Pollut. Contr. Fed.*, **47**, 553–561.

Jean, G.E. and Bancroft, G.M. (1986), *Geochim. Cosmochim. Acta*, **50**, 1455–1463.

Jenne, E.A. (1976), *In Chappell, W. and Peterson, K. (Eds): Symposium on Molybdenum*, Vol 2, Marcel Dekker, New York, 425–553.

Jorgensen, B.B. (1977), *Mar. Biol.*, **41**, 7–17.

Kemp, A.L.W., Thomas, R.L., Dell, C.I. and Jaquet, J.M. (1976), *J. Fish. Res. Board Can.*, **33**, 440–462.

Kemp, A.L.W., Thomas, R.L., Dell, C.I. and Jaquet, J.M. (1978), *J. Fish. Res. Board Cam.*, **33**, 440–462.

Khalid, R.A., Patrick, W.M., Jr, and Cambrell, R.P. (1978), *Estuar. Coast, Mar. Sci.*, **6**, 21–35.

Knox, S., Langston, W.J., Whitfield, M., Turner, D.R. and Liddicoat, M.I. (1984), *Estuar. Coast. Shelf Sci*, **18**, 623–638.

Krantzberg, G. (1985), *Environ. Pollut., Ser A.*, **39**, 99–122.

Kremling, K. (1983), *Mar. Chem.*, **13**, 87–108.

Langston, W.J. (1982), *J. Mar. Biol. Assoc., U.K.*, **62**, 667–684.

Langston, W.J. (1986), *Estuar. Coast. Shelf Sci.*, **23**, 239–261.

Langston, W.J. and Bryan, G.W. (1984), *In: Kramer, C.J.M. and Duinker, J.C. (Eds). Complexation of Trace Metals in Natural Waters.*, Martinus Nijhoff/Dr W Junk, The Hague, pp. 375–392.

Lindberg, S.E., Andren, A.W. and Harriss, R.C. (1975), *In Cronin, L.E. (Ed), Estuarine Research*, Vol. I, Academic Press, New York, 64–107.

Lodge, K.B. and Cook, P.M. (1989), *Chemosphere*, **19**, 439–444.

Luoma, S.N. (1983), *Sci. Total Environ.*, **28**, 1–22.

Luoma, S.N. and Bryan, G.W. (1978), *J. Mar. Biol. Assoc., U.K.*, **58**, 793–802.

Luoma, S.N. and Bryan, G.W. (1982), *Estuar. Coast. Shelf Sci.*, **15**, 95–108.

Lu, X., Johnson, W.K. and Wong, C.S. (1986), *Mar. Pollut. Bull.*, **17**, 263–267.

Madhun, Y.A., Freed, V.H. and Young, J.L. (1986), *Soil Sci. Soc. Am. J.*, **50**, 319–322.

Mantoura, R.F.C. (1987), *Nature, London*, **328**, 579–580.

Marquenie, J.M. (1985), *Environ. Technol.*, Letts **6**, 351–358.

Martoja, R. and Viale, D. (1977), *C.R. Hebd. Seances Acad. Sci. Paris Ser. D*, **285**, 109–112.

Mason, A.Z. and Nott, J.A. (1981), *Aquat. Toxicol.*, **1**, 239–256.

Maurer, D., Church, T.M., Lord, C. and Wethe, C. (1985), *Int. Revue. ges. Hydrobiol.*, **70**, 369–377.

Mayer, L.M. (1982), *Geochim. Cosmochim. Acta*, **46**, 2527–2535.

Mayer, L.M., Fink, L.K. Jr (1980), *Estuar. Coast. Mar. Sci.*, **11**, 491–503.

McCaffrey, R.J., Myers, A.C., Davey, E., Morrison, G., Bender, M., Luedtke, N., Cullen, D., Froelich, P. and Klinkhammer, G. (1980), *Limnol. Oceanogr.*, **25**, 31–44.

McCarthy, J.F., Jimenez, B.D. and Barbee, T. (1985), *Aquat. Toxicol.*, **7**, 15–24.

McElroy, A.E., Farrington, J.W. and Teal, J.M. (1989), In: *Metabolism of PAH in the Aquatic Environment*, 1–39.

Means, J.C. and Wijayratune, R.D. (1982), *Science*, **215**, 968–970.

Moore, R.M., Burton, J.D., William, P.J., Le, B. and Young, M.L. (1979), *Geochim. Cosmochim. Acta*, **43**, 919–926.

Morris, A.W. (1978), *In Krumbein, W.E. (Ed): Environmental Biogeochemistry and Geomicrobiology*, Vol 1 Ann Arbor Science, Ann Arbor, 179–187.

Morris, A.W. (1986), *Sci. Total Environ.*, **49**, 297–304.

Morris, A.W., Bale, A.J., Howland, R.J.M., Millward, G.E., Ackroyd, D.R., Loring, D.H. and Rantala, R.T.T. (1986), *Wat. Sci. Tech.*, **18**, 111–119.

Morris, A.W., Mantoura, R.F.C., Bale, A.J. and Howland, R.J.M. (1978), *Nature, London*, **274**, 678–680.

NCASI, (1981), *National Council of the Paper Industry for Air and Stream Improvement Inc., Technical Bulletin 351, July 1981, 18 plus Appendix.*

NCASI, (1982), *National Council of the Paper Industry for Air and Stream Improvement Inc.*, Technical Bulletin 386, Nov 1982, pp. 9.

Neff, J.M. (1979), *Polycyclic Aromatic Hydrocarbons in the Aquatic Environment: Sources, Fates and Biological Effects*, Applied Science Publishers, London.

Olsen, C.R., Cutshall, N.H. and Larsen, L.L. (1982), *Mar. Chem.*, **11**, 501–533.

Parker, W.R. and Lee, K. (1981), *Rapp. P-V. Reun. Con. int, Explor. Mer.*, **181**, 28–34.

Patmatmat, M.M. (1975), *Cah. Biol. Mar.*, **16**, 613–633.

Peterson, M.L. and Carpenter, R. (1983), *Mar. Chem.*, **12**, 295–321.

Presley, B.J., Kolodny, Y., Nissenbaum, A. and Kaplan, I.R. (1972), *Geochim. Cosmochim. Acta*, **36**, 1073–1099.

Presley, B.J. and Trefry, J.H. (1980), *In Olauson, E. and Cato, I. (Eds): Chemistry and Biogeochemistry of Estuaries*, Wiley – Interscience, New York, pp. 187–232.

Preston, M.R. and Al-Omran, L.A. (1989), *Environ. Pollut.*, **62**, 183–193.

Rainbow, P.S. (1990), *In: Furness, R.W. and Rainboaw, P.S.: Heavy Metals in the Marine Environment*, CRC, Boca Raton, Florida, pp. 67–79.

Readman, J.W., Mantoura, R.F.C. and Rhead, M.M. (1984), *Fresnius Z. Anal. Chem.*, **319**, 126–131.

Readman, J.W., Mantoura, R.F.C., Rhead, M.M. and Brown, L. (1982), *Estuar. Coast. Shelf Sci.*, **14**, 369–389.

Ridway, I.M. and Price, N.B. (1987), *Mar. Chem.*, **21**, 229–248.

Riley, and Chester, R. (1969), *Introduction to Marine Chemistry*, Academic Press, New York, 1976.

Rohatgi, N.K. and Chen, K.Y. (1976), *J. Environ. Eng. Div. ACSE*, **102**, 675–685.

Rosenberg, R. (1977), *Mar. Pollut. Bull.*, **8**, 102–104.

Salomans, W. (1985), *Environ. Technol.*, Letters **6**, 315–326.

Salomans, W. and Forstner, U. (1984), *Metals in the Hydrocycle*, Springer-Verlag, Berlin.

Schmidt, U. and Huber, F. (1976), *Nature*, **259**, 157–158.

Sholkovitz, E.R. (1976), *Geochim. Cosmochim. Acta*, **40**, 831–845.

Sills, G.C. and Been, K. (1984), *In: Transfer Processes in Cohesive Sediment Systems. (Eds. W.R. Parker and J.J. Kinsman)*, Plenum Press, pp. 109–125.

Streeter, M.W. and Phelps, E.B. (1925), *U.S. Public Health Service Bulletin*, **146**.

Stumm, W. and Morgan, J.J. (1981), *Aquatic Chemistry*, Second Edition, Wiley Interscience, New York, p. 780.

Sulfita, J.M., Robinson, J.A. and Tiedje, J.M. (1983), *Appl. Environ. Microbiol.*, **45**, 1466–1473.

Sunda, W.G. and Gillespie, P.A. (1979), *J. Mar. Res.*, **37**, 761–777.

Sundby, B., Anderson, L.G., Hall, P.O., Iverfeldt, A., Rutgens, van der Loff, M. and Westerlund, S.F.G. (1986), *Geochim. Cosmochim. Acta*, **50**, 1281–1288.

Sundby, B. and Silverberg, N. (1985), *Limnol. Oceanogr.*, **30**, 372–381.

Sundby, B., Silverberg, N. and Chesselet, R. (1981), *Geochim. Cosmochim. Acta*, **45**, 293–307.

Tipping, E. (1981), *Geochim. Cosmochim. Acta*, **45**, 191–199.

Tipping, E. and Heaton, M.J. (1983), *Geochim. Cosmochim. Acta*, **47**, 1393–1397.

Turekian, K.K. and Wedenpohl, K.H. (1961), *Bull. Geol. Soc. Am.*, **72**, 175–192.

Vanderborght, J.P., Wollast, R. and Billen, G. (1977), *Limnol. Oceanogr.*, **22**, 787–793; 794–803.

Van der Weijden, C.H., Arnoldus, M.J.H.L. and Meurs, C.T. (1977), *Neth. J. Sea Res.*, **11**, 130–145.

Wedenpohl, K.H. (1960), *Geochim. Cosmochim. Acta*, **18**, 200–231.
Westerlund, S.F.G., Anderson, L.G., Hall, P.O., Iverfeldt, A., Rutgens, van der Loff, M. and Sundby, B. (1986), *Geochim. Cosmochim. Acta*, **50**, 1289–1296.
Williams, R.J.P. (1981), *Proc. Roy. Soc.*, (B), *213*, 361–397.
Windom, H.L., Tenore, K.T. and Rice, D.L. (1982), *Can. J. Fish. Aquatic Sci.*, **39**, 191–196.
Wong, P.T.S., Chau, Y.K. and Luxon, P.L. (1975), *Nature*, **253**, 263–264.
Wood, J.M. (1972), *Environment*, **14**, 33–39.
Xie, T.-M., Abrahamsson, K., Fogelquvist, E. and Josefsson, B. (1986), *Environ. Sci. Technol.*, **20**, 457–463.
Yamada, M. and Tonomura (1972), *J. Ferment. Technol. Osaka*, **50**, 159–166.
Zoro, J.A., Hunter, J.M., Eglington, G. and Ware, G.C. (1974), *Nature*, **247**, 235–237.

COHESIVE SEDIMENTS IN COASTAL ENGINEERING APPLICATIONS 41

A.J. Mehta

One normally associates silts with estuaries but there are many cases where they create problems in the coastal environment. Here there is the additional factor of wave action which complicates the understanding and quantification of settlement and re-erosion. It is not often realized that the problem of dealing with, say, siltation in a coastal harbour from a logistical point of view can be quite different to a similar problem in an estuary. Even programming maintenance dredging in estuarial situations is often seasonal. For example, rainfall, or lack of it, controls the discharge of rivers which in turn can affect salinity which in turn can change siltation in certain areas at certain times of the year. Programmes of maintenance can be planned to a certain extent in advance but even these can be altered by annual changes in the environmental pattern of weather. What is so different in the coastal situation, for example in a coastal harbour, is that one might have extended periods of little or no wave action when little or no siltation will take place because re-entrainment of silts does not occur. Then out of the blue comes a storm which can re-entrain large quantities of silts and sands with the consequence that the amount of siltation in a few days can equal the average annual maintenance over the years. In general, at many locations waves have their seasons too, but it is the unpredictability of storms, perhaps out of season, that can create the necessity for panic dredging to, say, re-open an access channel to an important coastal harbour or marina. Ready access to dredgers is one of the consequences of this situation.

What has been said so far is in illustration of the importance of the many factors which can affect erosion and siltation in the complex coastal environment which is the main thrust of this chapter.

CONTENTS

41.1 INTRODUCTION

Cohesive sediment transport studies in coastal engineering applications relate to port development, dredging of approach channels, coastal turbidity generation, erosion or accretion of mudflats and muddy beaches and contaminant transport associated with fine sediment movement. They require a knowledge of the basic transport principles and methodologies for measurement and analysis within a prescribed technical framework as noted in previous chapters.

41.1.1 AN APPLICATION PERSPECTIVE

As engineers and scientists we are usually called upon to investigate a macroscale problem in relation to the microscale processes which ultimately lead to such a problem. In Figure 41.1a the linkage between macroscale coastal topographic changes and the corresponding microscale near-bed processes is shown descriptively. The typical causative factor for problem occurrence is anthropogenic, e.g. dredging or shoreline modification due to port construction, which influences the topography directly and also affects hydrodynamic forcing. Both, in turn, determine the water column dynamics which are linked to muddy bed dynamics via flow–bottom interaction. Understanding flow–bottom interaction and the manner

Coastal, Estuarial and Harbour Engineer's Reference Book. Edited by M.B. Abbott and W.A. Price. Published in 1993 by Chapman & Hall, 2–6 Boundary Row, London SE1 8HN. ISBN 0 419 15430 2.

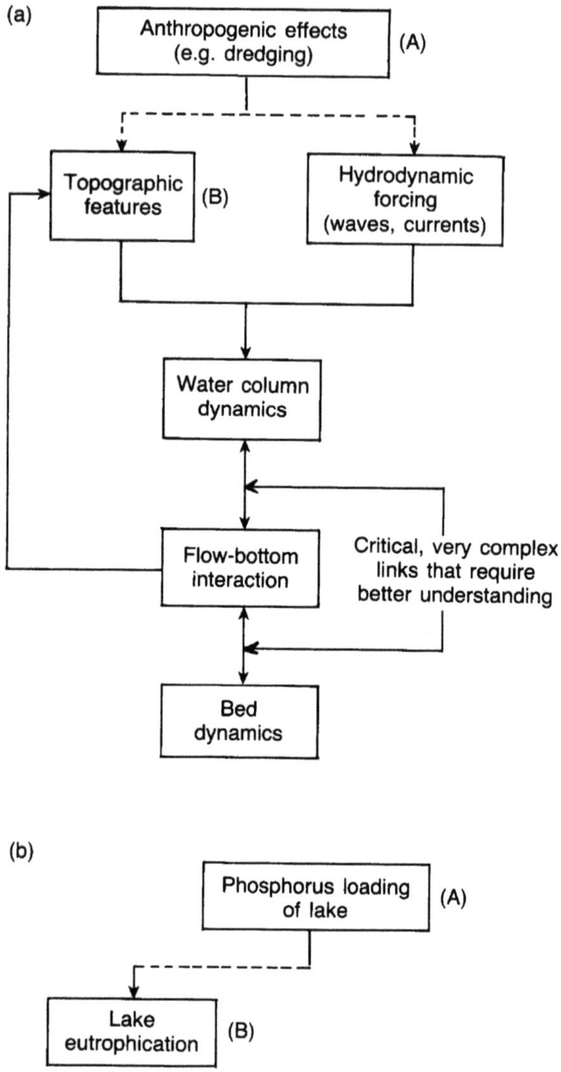

Figure 41.1 (a) Linkage between macroscale changes in coastal topography and microscale near-bed processes; (b), Replacing boxes (A) and (B) in (a) with these boxes defines the problem of lake eutrophication.

in which it links bed response to water column dynamics constitutes the core problem for cohesive sediment transport investigations. A common illustrative problem is the prediction of the rate of accumulation of 'fluff' at a channel bottom in order to estimate the frequency of maintenance dredging (Figure 41.2). The skewed nature of the concentration (C) profile, which is heavily bottom weighted, coupled with the turbulent flow velocity (u) profile, which may itself be influenced by the concentration field, can lead to a residual sediment flux largely confined to the near-bed zone. Important points to be made are that:

1. The vertical structures of the interactive concentration and velocity fields govern the horizontal transport rates.
2. Suitable measuring instrumentation must be deployed to account for the motion of the near-bed sediment slurry which may constitute the overwhelming component of the total rate of mass transport.
3. Since the rate of residual transport is important in determining the rate of channel infilling, accuracy in calculating the transport rate in a given direction must be high, and, therefore, the constitutive properties of the slurry, which dictate transport in conjunction with hydrodynamic forcing, must be carefully identified.
4. Given that the bottom response will be predominantly tide controlled in macrotidal areas but largely episodically determined in microtidal and wave dominated environments, the time scales of interest can vary from days to months or years.

A somewhat different perspective results if boxes A and B in Figure 41.1a are replaced, for example, by corresponding ones in Figure 41.1b. In the event the problem becomes one of water quality, in particular that of eutrophication. Increased trophic levels manifesting themselves as algal blooms due to external loading by phosphorus is a common problem in lakes. Although the water environment is clearly different from that of the coast, the bottom sediment re-suspension and deposition mechanisms responsible for the release of soluble reactive phosphorus due to wave action in the lake are basically the same as those that occur due to coastal waves.

The relationship between soluble reactive phosphorus and the resulting biomass is schematized in Figure 41.3 (Lijklema *et al.*, 1983). The matter of interest here is that related to sediment re-suspension, which consequently results is the desorption of suspended sediment-bound phosphorus. Re-suspension in this context is essentially inclusive of sediment settling (as distinct from biomass settling) and deposition. Response time scales of typical interest range from a few hours to a few days. These time scales are therefore typically shorter than those associated with channel infilling for example. It must be recognized that convective transport dominates over advection, and that the constitutive properties of the bottom mud play a critical role in governing suspended sediment transport.

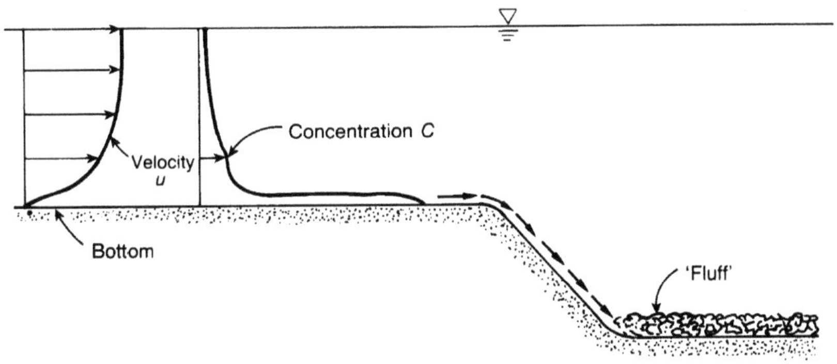

Figure 41.2 Accumulation of 'fluff' at channel bottom due to residual near-bed mud transport. Bottom response is tide controlled in macrotidal areas episodic (largely) in microtidal environment.

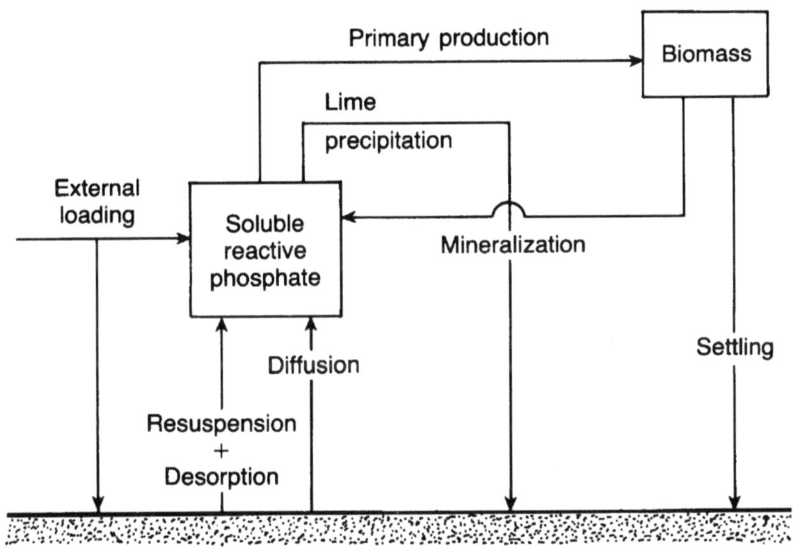

Figure 41.3 Schematic description of phosphorus loading in lakes (adapted from Lijklema *et al.*, 1983). Bottom response is entirely episodic.

Mud rheology

Fishermen as well as coastal engineers are aware of the important role played by mud flats in dissipating the energy of propagating waves as they approach the beach. Understanding the energy absorbing role of soft coastal muds requires a knowledge of the constitutive behaviour of mud as obtained through rheometric measurements. Models have been developed to calculate the degree of wave damping assuming non-Newtonian constitutive properties, e.g. pseudoplastic or viscoplastic. Similar models have also been used to calculate the horizontal mass transport of mud resulting from the non-linear ineteraction between waves and soft mud beds (Sakakiyama and Bijker, 1989).

Recent work on mud rheology points to deficiencies in the traditional approaches to rheometry for obtaining answers usually sought in mud related investigations (Williams and Williams, 1989a,b). These works clearly lay out the case for using a controlled stress (as opposed to controlled rate of strain) rheo-

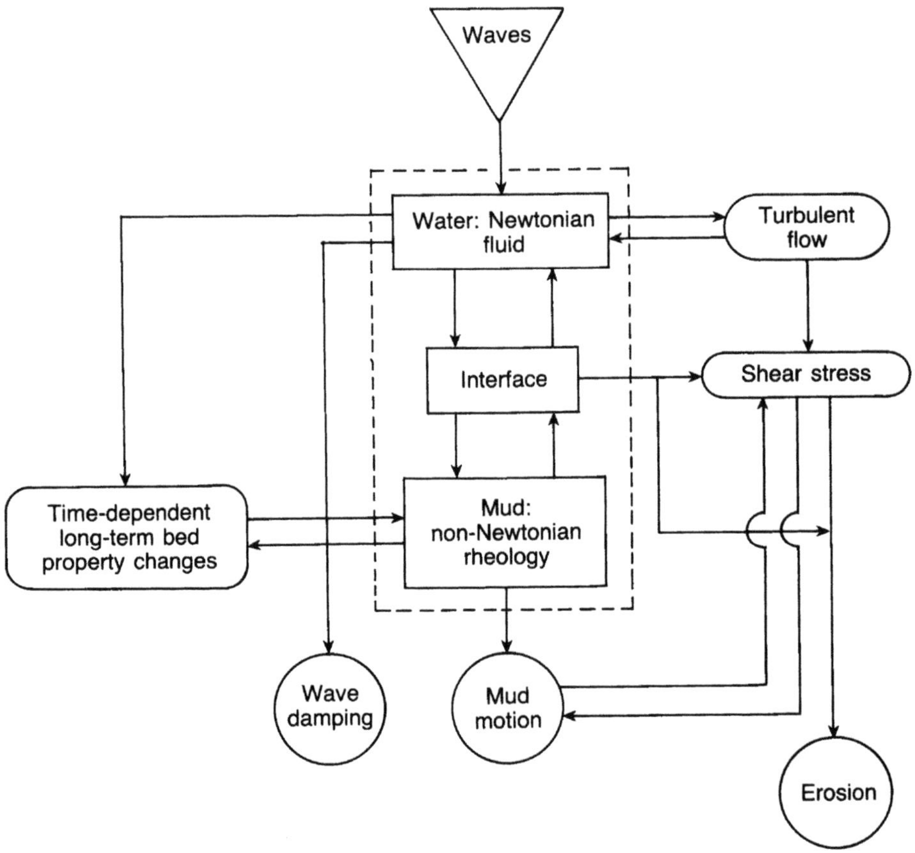

Figure 41.4 Schematic depiction of response of a mud bed to waves. Box with dashed line (---) represents two-layered system, ▽ = forcing function, □ = components of two-layered system, ○ = transfer functions and ○ = response (after Maa and Mehta, 1989).

meter, not only for determining the viscosity and the modulus of elasticity, but also for identifying a yield strength that truly defines the stress beyond which the bed structure loses its integrity. In general, the interaction between waves and muddy substrates is quite complex, particularly because wave action causes time-dependent changes in the bottom mud properties leading to its degradation over time. Erosion may occur if and when the bed becomes very weak. This is illustrated schematically in Figure 41.4. (Maa and Mehta, 1989).

41.2 PROBLEM SOLUTION STRATEGY

Simulation of fine sediment transport for validation as well as predictive purposes can range from empirical to fully three-dimensional numerical modelling. It is self-evident that while the development of solution options must rely on the outcome of the result of simulation, the latter, in turn, must be firmly grounded in physical evidence from the field and associated laboratory experiments. Among the available problem solving approaches, the so-called hybrid approach is technically meritorious in that it carefully combines the strengths of physical and mathematical modelling. A comprehensive methodology of this type has been developed over a number of years at the US Army Engineer Waterways Experiment Station, Vicksburg, Mississippi (McAnally *et al.*, 1984a). Some examples of application methodologies are very briefly mentioned below, beginning with the problem of predicting the future infilling of a coastal bay.

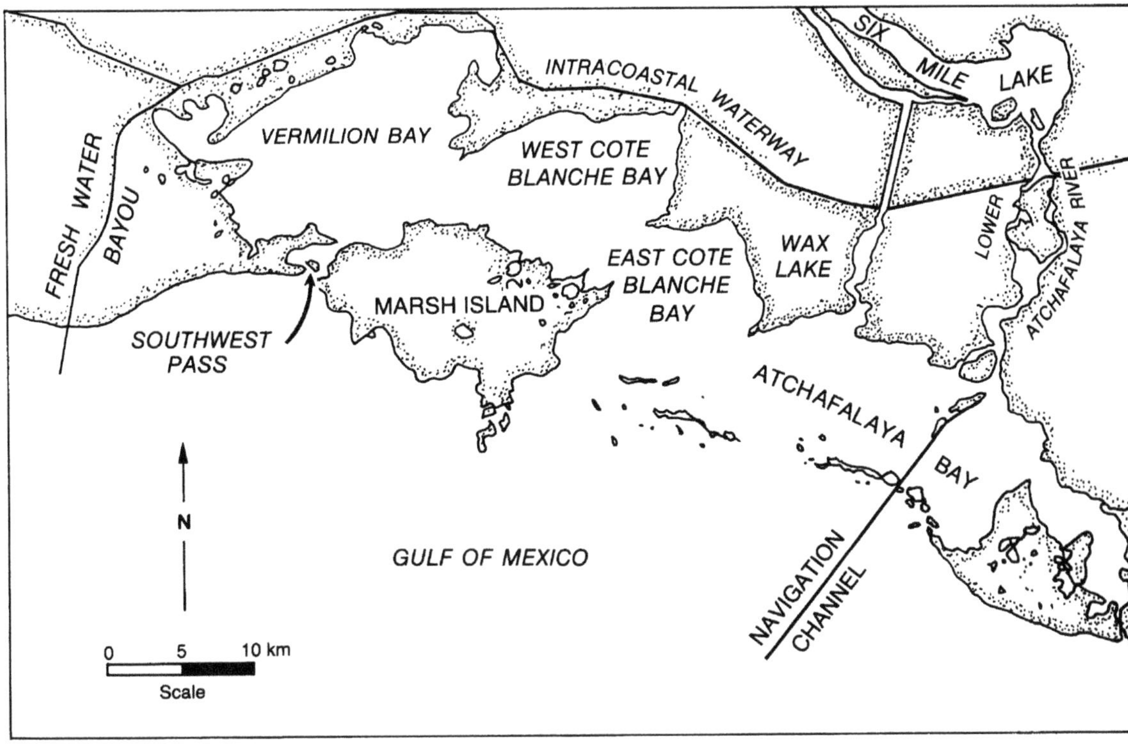

Figure 41.5 Location map of Atchafalaya Bay area, Louisiana (after McAnally *et al.*, 1984b).

41.2.1 BAY INFILLING BY FINE SEDIMENT

Prediction of sedimentation in bays can be illustrated by the case of Atchafalaya Bay in Louisiana (Figure 41.5). The Atchafalaya River, a distributary of the Mississippi River, discharges into this bay. In recent years, the delta at the mouth of the river has grown dramatically. An investigation of the bay and adjacent waters was carried out to predict the rate at which the delta will evolve in the short-term (10 years) and the long-term (50 years), and the manner by which that evolution will affect flood stages, navigation channel shoaling and the environmental resources of the area (McAnally *et al.*, 1984b; Mehta *et al.*, 1989).

A number of factors combined to make the study unusually complex. They included the long period over which predictions had to be made; the migration of the region of delta growth from lacustrine to estuarial and then to marine environments; a hydrodynamic regime that is variously dominated by river flows, wind-induced currents, tides, waves and storm surges; and the combined deposition of sediments from the sand, silt and clay classes. In order to adequately address this problem, the investigation included several predictive techniques, including:

1. Extrapolation of observed bathymetric changes into the future;
2. A generic analysis to predict future delta growth by constructing an analogy between the Atchafalaya delta and other deltas in similar environments;
3. Quasi-two-dimensional numerical modeling of hydrodynamics and sedimentation;
4. Use of extensive field and laboratory experiments.

The results showed a wide range of possible future land growth rates for 50 years in Atchafalaya Bay, and highlighted the sensitivity of delta growth to bottom subsidence rate. High subsidence rates are caused in part by the compaction of thick layers of fine sediments that have been deposited by the Mississippi River and its distributaries over millennia.

Using the extrapolation approach, Figure 41.6a, delta extent (land above 3 ft) in the year 2030 was generated by establishing a relationship between the past delta growth and the forcing phenomena of river flow and sediment supply, then using that relationship in combination with historically recorded flows to pro-

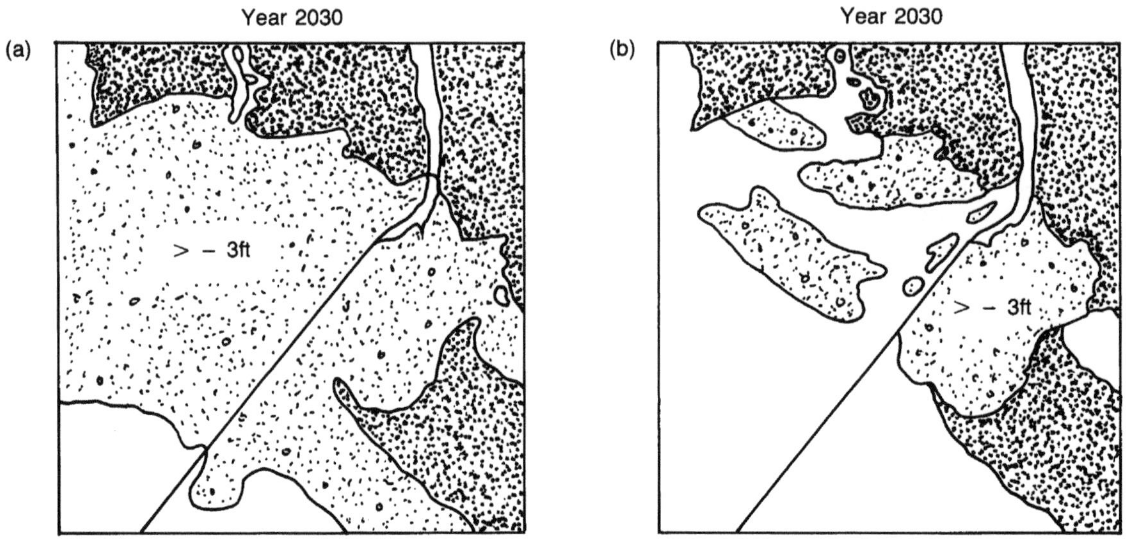

Figure 41.6 Predicted delta extent in Atchafalaya Bay in year 2030: (a) by extrapolation method; (b) by quasi-two-dimensional modelling (after McAnally *et al.*, 1984b).

ject future delta growth. It did not explicitly include subsidence effects. The quasi-two-dimensional approach (Figure 41.6b) employed a one-dimensional model with a subsidence rate of one centimetre per year. Even without any provision in the model to account for redistribution of bottom sediment by resuspension due to currents and waves, delta growth is predicted to be considerably lower than in Figure 41.6a. Recent observations in the bay suggest that the rate of emergence of the bottom is measurably lower than that suggested by the extrapolation technique (McAnally, personal communication). This outcome points to the critical importance of seemingly second-order effects, e.g. subsidence, in influencing the evolution of sediment-related macroscale phenomena.

41.2.2 SEDIMENTATION IN CLOSED-END CANALS

Closed-end channels such as pier slips, tidal docks, elongated marinas and residential canals are well known sites for fine-grained sediment deposition. In an investigation of pier slips at Mare Island naval shipyard in the San Francisco Bay area, Jenkins *et al.*, (1980) found the rate of sedimentation in the slips to be as much as 2.5 times higher than that in the main channel to which the slips were connected. Similar instances are also found at a large number of ports world-wide.

In general, advective transport due to tidal prism, wind-driven circulation and density-induced currents are processes by which sediment enters the closed-end channel. A turbidity current is driven by the difference in density between the sediment-laden outside waters and the relatively quiescent and sediment-free waters in the channel. The contribution to the total rate of sedimentation from this mechanism evidently varies with the physical conditions. In areas where tides are weak or when the suspended sediment concentration is high, turbidity currents become an important source of sediment in the channel. In many residential canals of Florida, sedimentation largely occurs during storms when the suspended sediment concentration in the waterways increases by one to three orders of magnitude over that during fair weather (Maa and Mehta, 1985). Sediment influx under these episodic conditions can be thought of as occurring in bursts due to the opening of a hypothetical gate (Figure 41.7) These bursts are separated in time by calms with relatively low-level ambient concentrations (5–10 $\mathrm{mg\,l^{-1}}$). The Orange Waterway is a residential canal off Marco River, a microtidal coastal waterbody near the southwestern coast of Florida. The rate (cm/yr) and pattern (longitudinal variation) of deposition observed in Figure 41.8, (Wanless, 1975), seem to be largely consistent with laboratory observations on turbidity transport in closed-end channels (Lin and Mehta, 1989).

The decay of mean concentration below the interface with distance which accompanies the ingress of the turbid current and the influence of an important governing parameter, namely the dimensionless

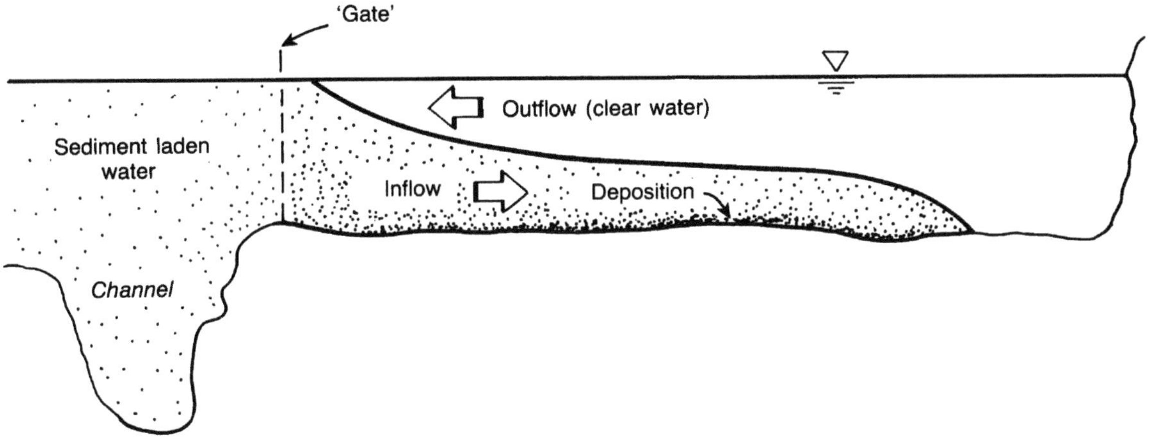

Figure 41.7 Sediment influx in a closed-end canal by turbidity current.

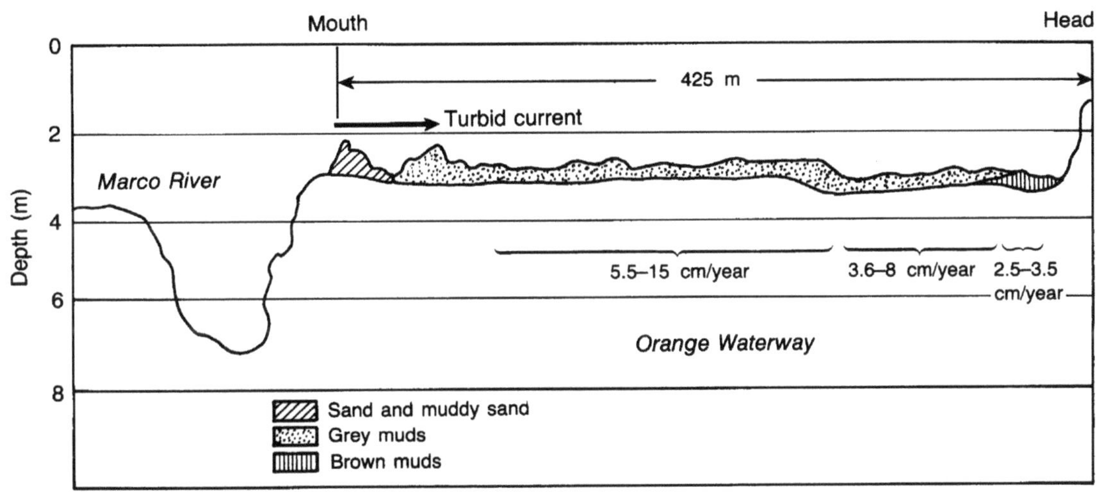

Figure 41.8 Longitudinal deposition rates and variation in Orange Waterway (adapted from Wanless, 1975).

settling velocity, w_{s1}/u_Δ, on the longitudinal concentration distribution, are apparent in Figure 41.9, in which the concentration ratio, C_b/C_{b1}, at steady state (when the rates of in-flow and out-flow become equal) is plotted against dimensionless distance, x/H. It should be noted that C_{b1} is the value of the mean concentration in the lower layer, C_b, at the mouth and H is the depth of water. The settling velocity at the mouth, w_{s1}, has been normalized by u_Δ, the interfacial celerity. In non-stratified turbulent flows, u_Δ is replaced by the friction velocity. The data were collected in a laboratory flume (Lin and Mehta, 1989). Results from seven tests using kaolinite have been included. Three lines represented by different values of w_{s1}/u_Δ are shown. These lines are based on an analytic solution for the decay of concentration with distance. The line slope increases with increasing w_{s1}/u_Δ, since, for a given u_Δ, increasing w_{s1} implies an increasing rate of deposition.

The rate of sediment influx through the mouth, S, can be practically calculated by dividing the total deposited sediment mass in the channel by the influx duration and the area of the lower half of the flow cross-section at the entrance. The quantity, S, is plotted against entrance concentration C_1 (which is practically equal to C_{b1}) in Figure 41.10. A 3/2 power

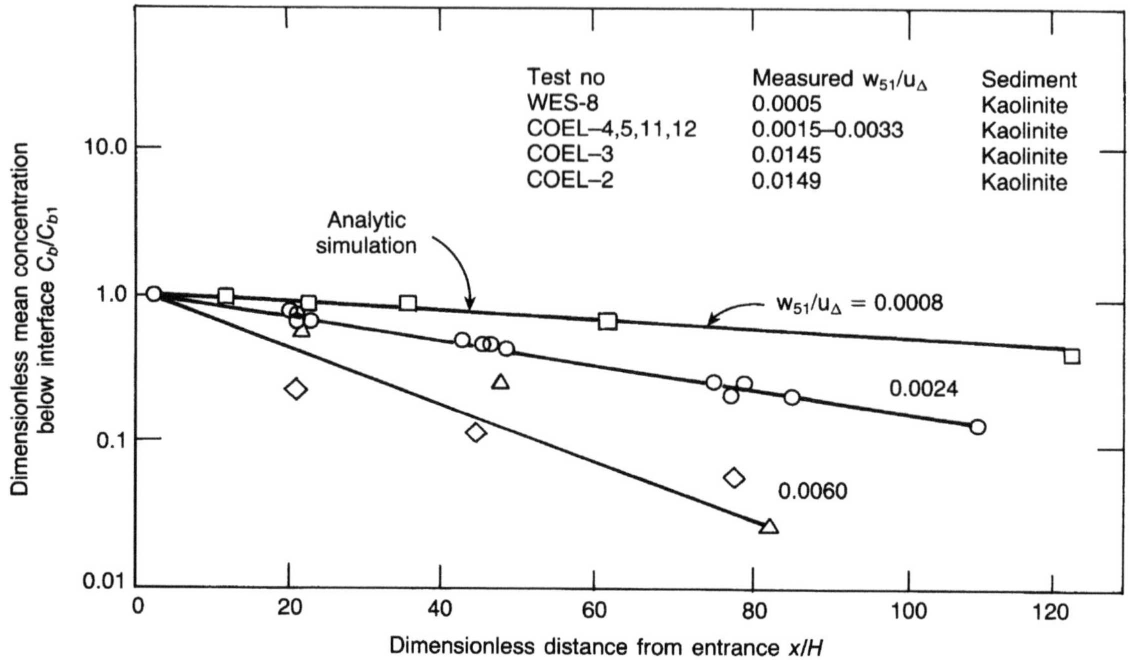

Figure 41.9 Dimensionless mean concentration below interface as a function of dimensionless distance along closed-end canal (after Lin and Mehta, 1989).

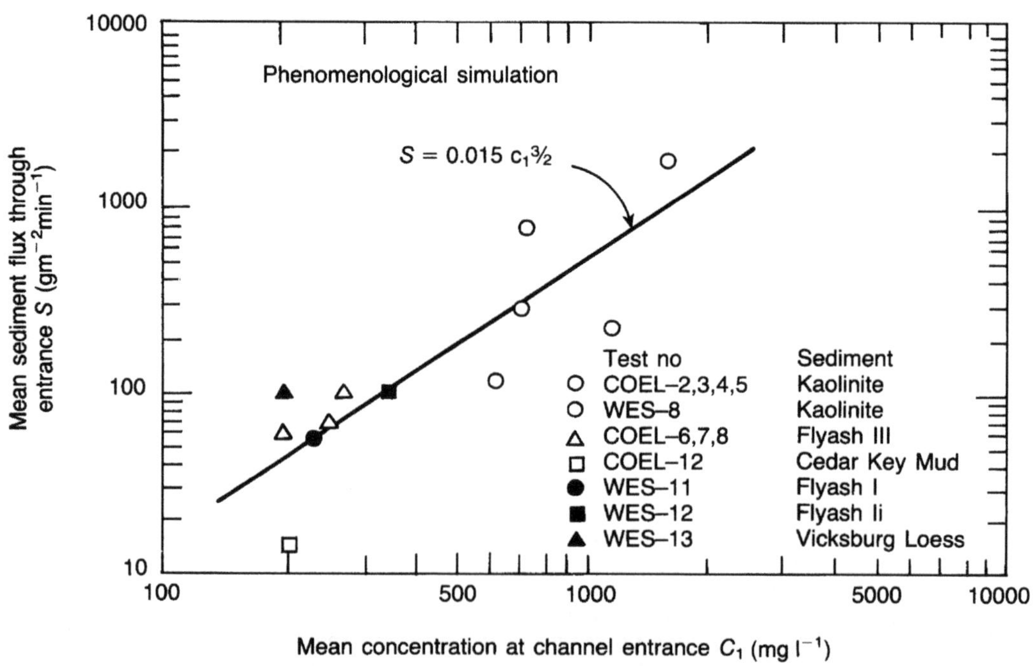

Figure 41.10 Mean sediment flux as a function of depth-mean concentration at entrance (after Lin and Mehta, 1989).

dependence of S on C_1 is evident. This relationship has a rather simple phenomenological basis (Lin and Mehta, 1989).

41.2.3 SEDIMENTATION IN SMALL MARINA BASINS

Prediction of the pattern and rate of sedimentation in marinas is required at the design stage for the estimation of the capital and maintenance costs of dredging, as well as for the implementation of measures for minimizing sediment intrusion into the basin. In general, given basis geometry and sediment load in waters outside the basin, the nature of flow circulation principally determines the degree of sedimentation within the basin.

The number of entrances and the planform shape of basins vary widely. In areas where suspended sediment loads are relatively high, such as in the San Francisco Bay system, it is desirable to construct water-tight basin walls and a single entrance in order to minimize sediment intrusion. An examination of the planform geometry of single-entrance marina basins in Florida revealed that, on average, the basins were rectangular with 100 m sides with a 55 m long entrance channel connecting the basin to the main water body (Srivastava, 1983). The corresponding mean water depth in the basin was found to be 2.2 m. In basins of this type, or even in those with more than one entrance, given the planform shape and depth, the principal factors which influence the degree of sedimentation are:

1. Range and period of tide;
2. Wind;
3. Waves;
4. Salinity gradients;
5. Concentration in outside waters,

provided that the flow environment in the basin is predominantly depositional, i.e. a basin in which there is no significant re-suspension of the deposit. If erosion does take place (for instance when the basin has two entrances of such a configuration as to cause appropriately strong currents during at least a portion of the tidal cycle), the erosional properties of the deposit must, in addition, be known.

Reference must be made here to the nature of suspended sediment concentration outside the basin. Observations have shown that in estuaries with large tidal ranges, there is an acceptable degree of correlation between the depth- and tide-averaged suspension concentration, C_0, the range of tide, and even the temperature, T. For example, measurements in the Mersey Estuary, England, yielded the following linear relationship (Halliwell and O'Dell, 1969):

$$C_0 = a + bH + dT \qquad (41.1)$$

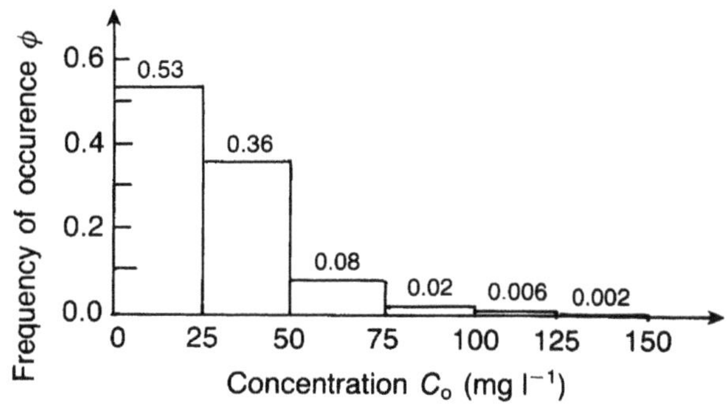

Figure 41.11 Histogram showing frequency of occurrence of concentration outside a basin (after Maa and Mehta, 1985).

where H is the high water elevation (not the tidal range) above Liverpool Bay datum, T is the temperature and a, b and d are empirical coefficients. Once evaluated for a given site, relationships such as Equation (41.1) provide a convenient means to estimate the temporal variation of C_0 at that site, given the corresponding variations of H, and T (which in many cases can be assumed to be constant, at least on a seasonal basis).

In episodically controlled environments tide-determined relationships such as Equation (41.1) are of limited use and, in general, it becomes essential to obtain extensive measurements of suspension concentration at the site. Figure 41.11 shows a concentration histogram (frequency of occurrence, of the suspension concentration, C_0) derived from measurements near Camachee Cove Yacht Harbor basin on Tolomato River near the Atlantic Coast of Florida (Srivastava, 1983).

Prediction of the pattern and rate of sedimentation in the basin involves numerical solutions of the advection–dispersion equation for sediment mass transport with appropriate descriptions of deposition and erosion, given the flow field. Results obtained in this way usually show a good degree of agreement with measurements as far as the basin-average rate of sedimentation is concerned, provided that the flow field is correctly represented. On the other hand, the pattern of deposition is complicated by the presence of docks and boat traffic, as well as by redistribution of the sediment during episodic events. Since these factors are not easily accounted for in numerical modelling, detailed patterns are often not simulated adequately; however, areas of critical shoaling, if prominent, are often identified.

For many small marina-type design problems, it is sufficient to estimate the basin-average shoaling rate without considering the depositional pattern. Simple computational procedures, even though far less ac-

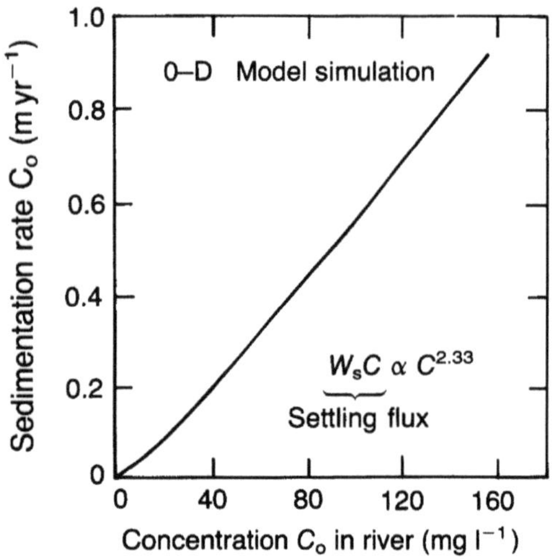

Figure 41.12 Variation of shoaling or sedimentation rate, *SR*, with outside concentration C_0 (after Maa and Mehta, 1985).

curate than sophisicated numerical approaches, can be quite useful in such cases. In Figure 41.12, the computed shoaling or sedimentation rate, *SR*, on the annual basis is plotted against the outside concentration, C_0, for Camachee Cove basin using a zero-dimensional approach i.e. an approach in which sedimentation rate is dependent on time but not on co-ordinates (Maa and Mehta, 1985). Using the observed variation of C_0 according to Figure 41.11, the actual total rate of shoaling per year may be obtained by utilizing the results of Figure 41.12. For each C_0 corresponding to a frequency of occurrence from Figure 41.11, the *SR* value is obtained from Figure 41.12. This *SR* is then weighted by multiplying it by the corresponding value of frequency. The total rate of shoaling is the sum of *SR*, which in the present case can be shown to be equal to 0.13 m year^{-1}. Bathymetric measurements in the basin over a 2.5-year period yielded 0.15 m year^{-1}. Note that in the illustration the settling velocity, w_s, was taken to be proportional to 4/3 power of concentration *C* according to the relationship:

$$W_s = K_s C^{n1} \tag{41.2}$$

thus incorporating the effect of floc aggregation on deposition. As a consequence, the settling flux, $w_s C$, becomes proportional to $C^{2.3}$. This consideration leads a higher rate of shoaling than would be the case if w_s were assumed to be constant, and thus yields an answer that is on the conservative side as far as the need for dredging is concerned.

An example of predicted sedimentation pattern (contours of deposit thickness per year) in the Camachee Cove basin is shown in Figure 41.13. A depth-averaged suspended sediment transport model was used for this purpose (Hayter and Mehta, 1986). The flow field necessary for driving the sediment model was generated through a compatible hydrodynamic model. The predicted basin-average deposit thickness is 0.146 m yr.$^{-1}$

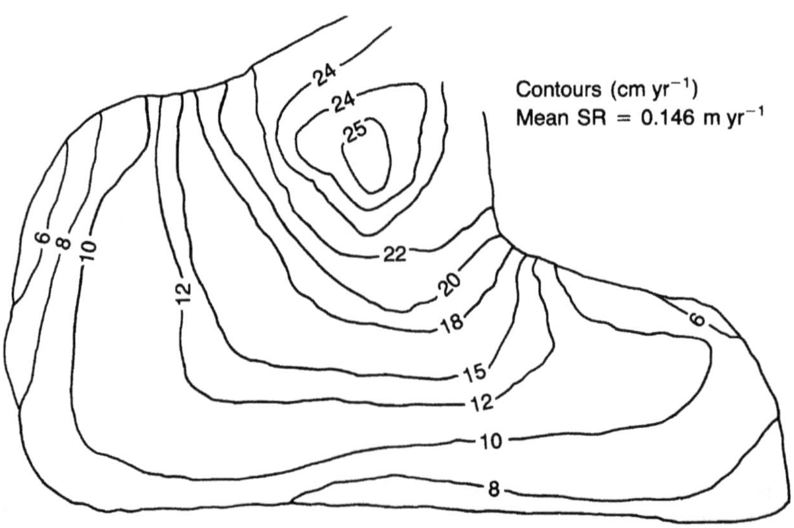

Figure 41.13 Two-dimensional model simulation of predicted sedimentation contours in Camachee Cove Yacht Harbor basin (after Hayter and Mehta, 1986).

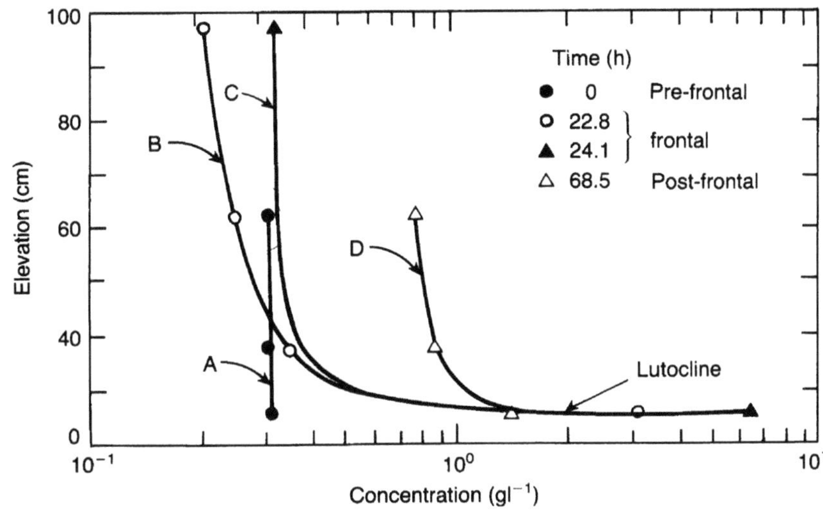

Figure 41.14 Vertical suspended sediment concentration profiles obtained before, during and after passage of a winter cold front at a coastal site in Louisiana. Times are relative to time of measurement of pre-frontal profile 1 (adapted from Kemp and Wells, 1987).

41.2.4 ERODIBLE DEPTH DUE TO WAVE ACTION

A key feature of fine-grained suspended sediment concentration profiles is the occurrence of steep vertical gradients, with concentrations which can be orders higher near the bottom than near the water surface. A representative illustration of suspension concentration profile evolution by wave action over coastal mudflats is provided by the data of Kemp and Wells (1987) in Figure 41.14. Of the four instantaneous vertical concentration profiles for suspended sediment, one represents a pre-frontal (fairweather) condition, two represent conditions during the passage of a winter cold front and one represents post-frontal conditions. The data were obtained over a three-day period at a site on the eastern margin of the Louisiana chernier plain where the tidal range is less than 0.5 m. The wave height during front passage was of the order of 13 cm and the period was 7 s. Of particular interest is the development of a near-bed, high concentration suspension layer by the wind-generated waves (profiles B and C), which was previously absent (profile A). The post-frontal profile D further suggests that this layer may have persisted following the front, conceivably due to the typically low rate at which such a layer dewaters. The suspension concentration in the upper water column was higher following the front than that during the front, possibly due to sediment advection from a neighbouring area of higher turbidity.

Concentration profiles qualitatively similar to those shown in Figure 41.14 have been reproduced in laboratory flume tests involving wave action over soft muddy deposits (Ross and Mehta, 1989). A significant lutocline develops near the bottom due to the entrainment of the mud–water interface resulting from the effects of upward, shear-induced diffusion strongly stabilized by the negative bouyancy of the high concentration suspension combined with hindered gravitational settling. The height at which the lutocline stabilizes is largely determined by a balance between the rate of turbulent kinetic energy input and the bouyancy flux determined by the sediment settling rate. In the water column above the lutocline diffusion due to the wave field is characteristically slow, so that the concentration there increases to only modest levels.

The formation of a high concentration fluidized layer of sediment at the bottom is characteristic of all wave-influenced environments having mud beds, including lakes (Wolanski *et al.*, 1989). In lakes such layers are episodically generated but due to the relatively low rates of dewatering of organic-rich sediment, they are believed to be more common and persistent than previously thought. As noted in Figure 41.1(a,b) there is therefore an obvious linkage between re-suspension and trophic level in lakes (Otsubo and Muraoka, 1987).

In Figure 41.15 simulated concentration profiles generated using a one-dimensional vertical model similar to that used by Costa (1989) are shown (Hwang, 1989). The main differences between the

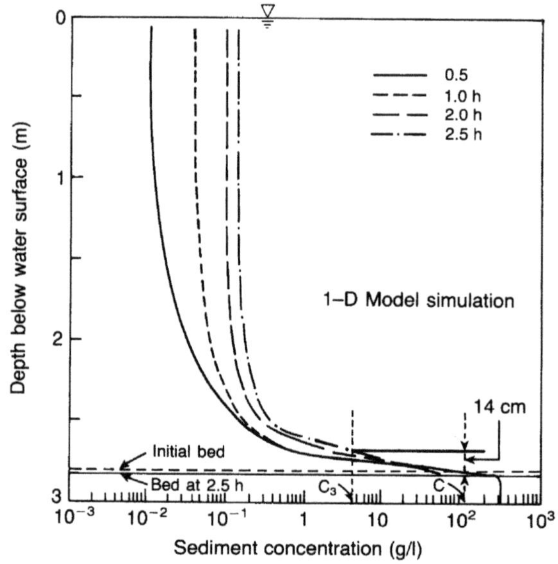

Figure 41.15 Simulated lutocline evolution under wave action (after Hwang, 1989).

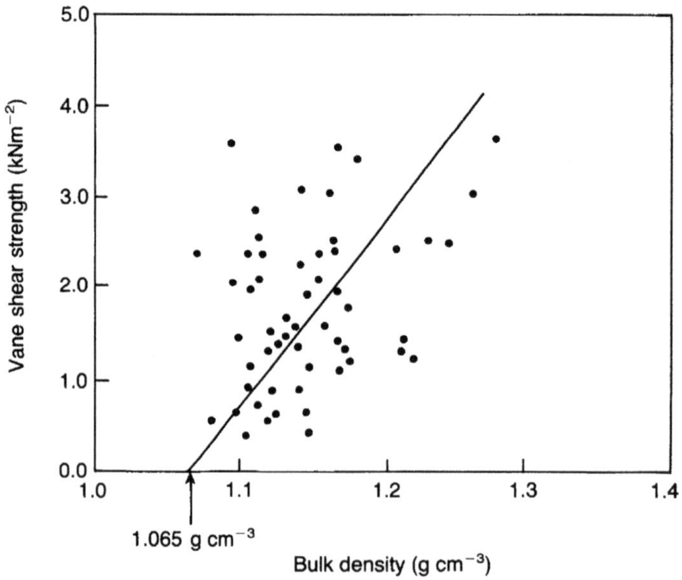

Figure 41.16 Mud vane shear strength variation with density (after Hwang, 1989).

two models is in the expression for the vertical mass diffusivity. Costa used the bouyancy stabilized form of the well known Prandtl–von Kármán expression appropriate for open channel flows, while for the simulation shown in Figure 41.15, Hwang used the corresponding expression applicable to waves. The necessary sediment-related input was obtained from laboratory tests using bottom material from Lake Okeechobee. The bed density profile required for simulation was derived though vibracoring. The water depth used for this simulation was 2.8 m, the wave height 0.8 m and a period of 3.5 s.

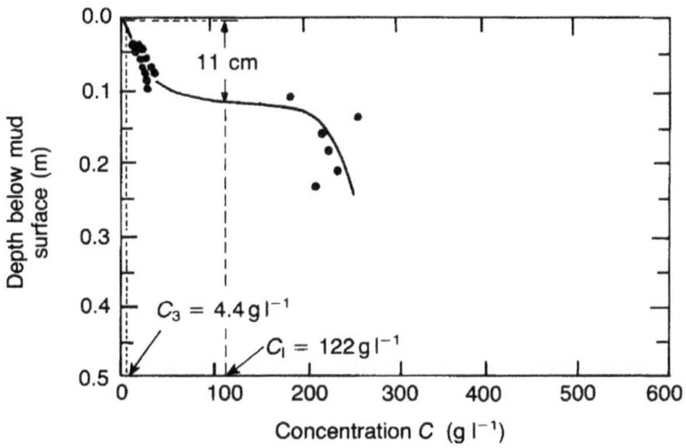

Figure 41.17 Bottom concentration profile showing 11 cm thick fluid mud (after Hwang, 1989).

The simulated evolution of the concentration profile in Figure 41.15, starting from a clear water column representing pre-storm calm, shows a strong qualitative resemblance to prototype data (e.g. Figure 41.14). The concentration $C_3 = 4.4\,\mathrm{g\,l^{-1}}$ corresponds to that value above which settling of this particular sediment was found to be hindered. The value $C_l = 122\,\mathrm{g\,l^{-1}}$ (corresponding to a bulk density of 1.065 $\mathrm{g\,cm^{-3}}$) is that above which the material possessed measurable vane shear strength (Figure 41.16), thus effectively characterizing it as a bed. Thus the range 4.4–122 $\mathrm{g\,l^{-1}}$ identifies for this sediment the lower and upper concentrations and corresponding elevations, respectively, of the fluid mud layer. At the end of 2.5 h when the surface concentration reached about $0.1\,\mathrm{g\,l^{-1}}$, a typical value during storm wave action in the lake, a 14 cm thick fluid mud layer was formed. At the same time the bed eroded 1.7 cm.

As shown in Figure 41.17, a bottom concentration (density) profile relative to the mud surface in the lake confirmed the prescence of a persistent fluid mud layer of the same order (11 cm) as that suggested by the simulation in Figure 41.15.

41.3 CONCLUSIONS

Applications of fine sediment transport principles have become quite extensive in coastal engineering and environmental work. For a brief review of the available methodologies, reference may be made to Mehta *et al.* (1989). Suffice it to note that while considerable additional work remains to be done in developing an understanding of the processes of higher than present order of accuracy, the technology has come a long way in recent decades, and continues to advance with the availability of better instrumentation for field and laboratory measurement.

A great many number of challenging engineering problems involving fine sediments actually occur in developing countries where, for instance, port development continues to remain a high priority issue owing to obvious commercial imperatives. These countries are important users and recipients of fine sediment transport-related technology, which is being pursued through research and development in a handful of industrially advanced nations. In these latter countries, research and development in fine sediment transport in the coming years will be guided by technical policy issues arising from three main concerns, the foremost being water quality and contaminant transport problems in coastal waters. This priority should also lead to new insights into the nature and the dynamics of the mud–water interface and exchange processes at this interface. Secondly, older and more conventional issues associated with sedimentation in ports and navigation channels will continue to demand attention due to the large and rising dredging costs and the need to minimize the number of shallow water dump sites for dredged material disposal. Finally, the need for considerable improvement in the prediction of the response of muddy coast shorelines and wet-lands to global climatic changes, in particular sea-level rise, will demand new thinking related to fine sediment process principles associated with small transport rates over very long time scales. There will be no shortage of challenging applications.

REFERENCES

Costa, R.C.F. (1989) Flow-fine sediment hysteresis in sediment-stratifield coastal waters. M.S. Thesis, University of Florida, Gainesville.

Halliwell, A.R. and O'Dell, M. (1969) Differences in silt patterns across an estuary. *The Dock and Harbor Authority* L(585), 125.

Hayter, E.J. and Mehta, A.J. (1986) Modelling cohesive sediment transport in estuarial waters. *Applied Mathematical Modelling* **10**, 294–303.

Hwang, K.-N. (1989) Erodibility of fine sediment in wave-dominated environments. M.S. Thesis, University of Floria, Gainesville.

Jenkins, S.A., Inman, D.L. and Bailard, J.A. (1980) Opening and maintaining tidal lagoons and estuaries. Proceedings of the Seventeenth Coastal Engineering Conference, ASCE, New York, pp. 1528–47.

Kemp, G.P. and Wells, J.T. (1987) Observations of shallow water waves over a fluid mud bottom: implications to sediment transport. Proceedings of Coastal Sediments '87, ASCE, New York, pp. 367–77.

Lijklema, L., Gelencser, P. and Szilagi, F. (1983) Sediments and sediment-water interaction, in, *Eutrophication of Shallow Lakes: Modeling and Management, The Lake Balaton Case Study,* (eds L. Somlyody, S. Herodek and J. Fischer), International Institute for Applied Systems Analysis, Laxenburg, Austria, pp. 81–100.

Lin, C.-P. and Mehta, A.J. (1989) Turbidity-induced sedimentation in closed-end channels. *Journal of Coastal Research* **5** (3), 391–401.

Maa, P.-Y. and Mehta, A.J. (1985) Fine sedimentation in small harbors, in, *Flocculation, Sedimentation and Consolidation,* (eds B.M. Moudgil and P. Somasundaran), Engineering Foundation, New York, pp. 405–14.

Maa, P.-Y. and Mehta, A.J. (1989) Considerations on soft mud response under waves, in, *Estuarine Circulation*, (eds B.J. Nielsen, A. Kuo and J. Brubaker), Humana Press, Clifton, New Jersey, pp. 309–36.

McAnally, W.H., Letter, J.V., Stewart, J.P. *et al.* (1984a) Columbia River hybrid modeling system. *Journal of Hydraulic Engineering* **110** (3), 300–11.

McAnally, W.H., Thomas, W.A., Letter, J.V. *et al.* (1984b) The Atchafalaya River delta, report 6, interim summary report of growth predictions. U.S. Army Engineer Waterways Experiment Station, Vicksburg, Mississippi. *Tech. Rep. HL-82-15.*

Mehta, A.J., McAnally, W.H., Hayter, E.J. *et al.* (1989) Cohesive sediment transport II: application. *Journal of Hydraulic Engineering* **115** (8), 1094–112.

Otsubo, K. and Muraoka, K. (1987) Field studies on physical properties of sediment resuspension in Lake Kasumigaura. *The Japanese Journal of Limnology* **48** (special issue), pp. 131–8.

Ross, M.A. and Mehta, A.J. (1989) On the mechanics of lutoclines and fluid mud. *Journal of Coastal Research* (special issue No. 5) pp. 51–62.

Sakakiyama, T and Bijker, E.C. (1989) Mass transport velocity in mud layer due to progressive waves. *Journal of Waterway, Port, Coastal and Ocean Engineering* **115** (5), 614–33.

Srivastava, M. (1983) Sediment deposition in a coastal marina. Coastal and Oceanographic Engineering Dept., University of Florida, Gainesville. *Rep. UFL/COEL/MP-83/1.*

Wanless, H.R. (1975) Sedimentation in canals. Report, Division of Marine Geology and Geophysics, Rosenstiel School of Marine and Atmospheric Science, University of Miami, Florida.

Williams, P.R. and Williams, D.A.J. (1989a) Rheometry for concentrated cohesive suspensions. *Journal of Coastal Research* (special issue No.5) pp. 151–64.

Williams, D.A.J. and Williams, P.R. (1989b) Rheology of concentrated cohesive sediments. *Journal of Coastal Research* (special issue No. 5) pp. 165–273.

Wolanski, E., Asaeda, T. and Imberger, J. (1989) Mixing across a lutocline. *Limnology and Oceanography* **34** (5), 931–8.

DETERMINING DEPTH AND NAVIGABILITY IN FINE SEDIMENT AREAS

42

W.R. Parker

It used to be the case that depths shown on hydrographic charts were those established by echo sounder or lead line. This is still the case where the sea-bed is 'hard' or consists of granular materials like sands and gravels. However, during the last 10 to 15 years it was realized that where the sea-bed consisted of 'soft' materials it should be possible to navigate through such materials and hence accept a greater depth than that established by normal survey methods. The basis of the nautical depth concept relies on the identification of the material properties of the sea-bed which influences navigation and the choice of a viable survey methodology. Dr Parker, who has been one of the leaders in this field, describes the techniques involved in establishing a safe nautical depth and illustrates the methods with practical examples.

42.1 DEPTH DETERMINATION AND NAVIGABILITY IN FINE SEDIMENT AREAS

Where only small mobile populations of fine sediment occur and deposition rates are low, the bed is usually normally consolidated with an abrupt surface density gradient. This gradient is readily and unambiguously identified by most survey techniques.

Where large populations of mobile fine sediment lead to the development of fluid muds, ambiguity may arise between the depth determined by different survey techniques. This may be further compounded if the depths so determined are referenced to the requirements of navigation or the establishment/calibration/verification of numerical models (Parker, 1989).

Depth determination is important for many reasons (Figure 42.1). It is possible to identify at least three criteria which determine the approach taken to deciding what is meant by 'depth'. These may be designated.

1. **Hydrodynamic depth:** the elevation of the interface between flowing, possibly muddy, water and stationary, possibly watery, sediment.
2. **Methodological depth:** the depth determined by some particular technique or method e.g. target strength (acoustic surveys) or shear strength (lead-line surveys).
3. **Operational depth:** e.g. nautical depth – the depth of a particular material property that affects vessel safety or handling.

The natural determinant of the level of sediment is the local hydrodynamics. The depth determinant for methodological criteria depends upon the state of evolution of the bed and the nature of the measurement technique deployed. The operational definition depends upon criteria related to safe navigation or vessel operation. An extended discussion of this concept has been given by Parker (1989). The important point to consider is that there is no specific relationship between the depths determined by various survey methods and that determined by hydrodynamics, except as a result of temporal coincidence (Parker, 1989). This has lead to a great deal of confusion and argument in relation to the navigation of vessels in muddy areas. It has been rationalized by the adoption of the 'nautical depth' concept (Kirby *et al.*, 1980; Nederlof and Vanbochove, 1981; PIANC, 1983).

The nautical depth concept accepts that different definitions of depth can co-exist and seeks to define the depth for practical navigational purposes. This also has a bearing on operational survey methods and dredging. The core of the nautical depth concept lies in the identification of the material property of the bed which influences navigation and the choice of a viable survey methodology. Current practice relies on an evaluation of the rheological properties of fluid mud and appropriate field methods for determining

Coastal, Estuarial and Harbour Engineer's Reference Book. Edited by M.B. Abbott and W.A. Price. Published in 1993 by Chapman & Hall, 2–6 Boundary Row, London SE1 8HN. ISBN 0 419 15430 2.

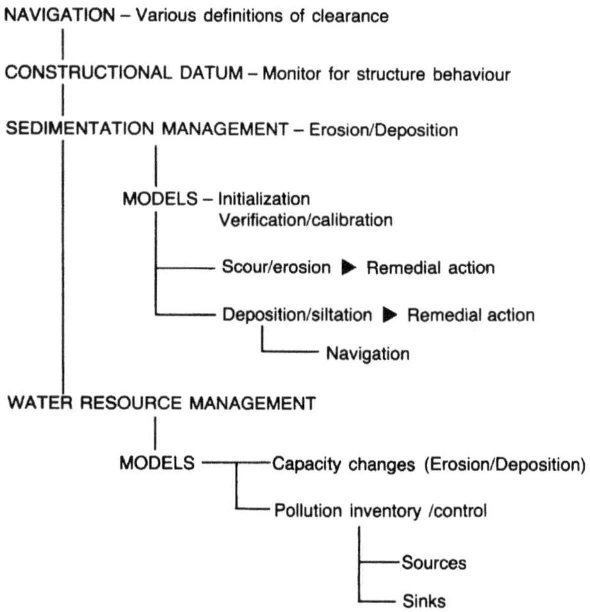

Figure 42.1 Purposes for which depths (bed levels) are determined.

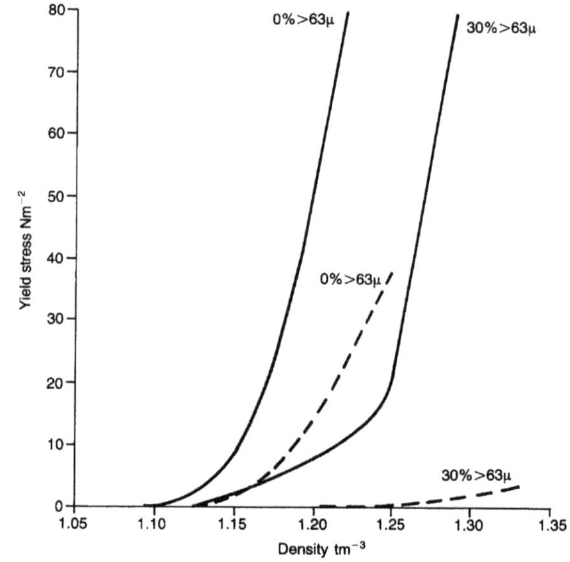

Figure 42.2 Comparison of static yield stress, as a function of density, determined by controlled stress shear rheometer (——) and imposed shear rate (---) for sediments with between 0 and 30% by weight of particles larger than 63 μm. Note great spread and under-estimate of yield arising from methodologies involved in imposed shear rate measurement.

these or a related property. The relevant rheological parameters are the static yield stress and the apparent viscosity. Extensive discussions of these parameters have been given by Barnes and Walters (1985), Cheng and Cheng (1986), James *et al.* (1987, 1988) and Williams and Williams (1989). These types of measurements also require the availability of genuinely undisturbed samples. It is apparent that *in situ* rheological probes do not satisfy the analytical criteria for rheological measurements.

An alternative has been the development of *in situ* density measurements as a survey tool to estimate navigable depth (Parker *et al.*, 1975). This methodology depends upon being able to estimate the relationship between rheological properties and solids content for a particular material. The rheological properties are generally well defined providing there is not too great a variation in the particle spectrum and mineralogy. Studies in Zeebrugge show that the rheological properties of liquid muds vary with sand and silt content such that (at a given density) the highest strength materials are those with the highest clay content (Figure 42.2). In practical terms this results in a conservative density value being defined for nautical depth. The practical advantage of density measurements lies in the availability of systems which can execute surveys at a useful rate (De Vlieger, 1988; De Vlieger and De Cloedt, 1987; Parker and De Vlieger, 1988). In The Netherlands attempts to determine nautical depth by acoustic means have taken advantage of the limited particle size spectrum of the fluid mud and consistent relationships between rheology and density. Nevertheless, nuclear measurements still form part of the survey proceedure. In the United States, attempts to develop non-nuclear nautical depth survey methods using towed plates appear to ignore the dynamic rheology of the fluff materials. Nautical depth surveys using acoustic transmissance measurements of sediment density (Chaumet–Lagrange *et al.*, 1984) require substantial programmes of particle size and other materials characterization to ensure a useable density/acoustics algorithm. A severe constraint on this type of measurement arises with the biogenic development of gas bubbles. The application of nautical depth criteria can produce considerable cost benefits by providing unambiguous, repeatable and relevant data (De Vlieger and De Cloedt, 1987; Parker and De Vlieger, 1988).

REFERENCES

Barnes, H.A. and Walters, K. (1985) *Rheological Acta*, **24**, 323–6.

Chaumet–Lagrange, M., Grandboulan, J. and Fourcassies, C. (1984) *Proceedings of 2nd International Hydrographic Technical Conference*,

Paper No 37, Plymouth, UK, pp. 1–8.
Cheng, D. and Cheng, C.-H. (1986) *Rheol Acta*, **25**, 542–54.
De Vlieger, H. (1986) *Dredging and Port Construction*, March, pp. 33–5.
De Vlieger, H. and De Cloedt (1987) *Terra Et Aqua*, **35**, 2–18.
James, A.E., Williams, D.J.A. and Williams, P.R. (1987) *Rheologica Acta*, **26**, 437–46.
James, A.E., Williams, D.J.A. and Williams, P.R. (1988) in Dronkers, J. and van Leussen, W. (eds.) *Physical Processes in Estuaries*, Springer-Verlag, 488–500.
Kirby, R., Parker, W.R. and Van Oostrum, W.H.A. (1980) in, *Hydrographic Review*, **LVIII** (1), 107–17.
Nederlof, L. and van Bochove, G. (1984) *Dock and Harbour Authority*, May, pp. 2–6.
Parker, W.R. (1989) *Journal of Coastal Research*, (Special Issue 5), pp. 175–84.
Parker, W.R. and De Vlieger, H. (1988) Proceedings of the 21st Annual Dredging Seminar, Centre for Dredging Studies. Texas A & M University, Texas, USA. *CDS Rep.*, **309**.
Parker, W.R. and Kirby, R. (1982) in, *Estuarine Comparisons*, (ed. V.S. Kennedy) Academic Press, New York, pp. 15–26.
Parker, W.R., Sills, G.C. and Paske, R.E.A. (1975) in, *First International Symposium on Dredging Technology*, (ed. S.K. Hemmings), Canterbury, U.K. BHRA, Cranfield, UK, pp. 25–41.
PIANC (1983) *Pianc Bull.*, **43**, General Secretariat of PIANC, Brussels
Williams, D.J.A. and Williams, P.R. (1989) *Journal of Coastal Research* (Special Issue 5), 165–74.
Williams, P.R. and Williams, D.J.A. (1989) *Journal of Coastal Research* (Special Issue 5), 151–64.

ESTUARIAL PROBLEMS 43

M.P. Kendrick

Estuaries have been developed for many purposes and it is important that proposals for new works should be examined both for their likely effects locally and on the hydraulic system as a whole. This chapter illustrates, often drawing experience from specific examples, how certain problems should be tackled and the studies that should be initiated to arrive at a satisfactory solution. Stated simply, this means identifying the interaction between the effects of the works and the hydraulic environment. A reader new to estuarial problems will find this chapter a useful introduction to the subject (see also Chapters 34, 38, 42 and 44).

CONTENTS

43.1 INTRODUCTION

This chapter examines civil engineering activity in estuaries, emphasizing the need to identify both the prevailing hydraulic loading on any works proposed or undertaken, and the impact of those works on existing estuary regime. Stated simply, this means recognizing the interactive effects of works and environment, and thus ensuring a sound basis for the formulation of solutions to practical problems.

Estuaries have been variously defined and classified by geomorphologists and engineers. The ideal estuary of earlier hydraulic engineers (Pilsbury, 1956) is one in which moderate tides and river flow mix in a single-channel system whose width varies exponentially with length; maximum current velocity and the near-sinusoidal tide curve shape remain virtually unchanged along the tidally-active length; bed and banks are composed of movable sediments and so the dynamically stable system readjusts fairly readily to a non-cyclic change, whether natural in origin as with a storm-surge, or man-made as with a land reclamation project. Such ideal conditions are difficult, if not impossible, to find in practice, although the Delaware River of the USA provides a good approximation, with the UK's tidal Thames not far behind.

In the present context the term estuary is taken to cover a range of tidal inlets in which one or all of the following may differ widely: climate, meteorological conditions, tidal characteristics, river flow, sediment type, marine influence. For example, Figure 43.1 shows some of the differences in tidal type and tidal range experienced round the world: diurnal tides of less than 1 m range in the Gulf of Mexico; semi-diurnal tides of 12 m range in the UK's Severn estuary; mixed, largely diurnal tides in the South China Sea; mixed, largely semi-diurnal tides in the Gulf of Aden. The influence of differences in sediment type on tidal progress along two English estuaries is illustrated by the tide curves in Figures 43.2 and 43.3. The Thames and the Mersey are characterized by semi-diurnal tides having a mean range at the entrance exceeding 4 m and similar seasonal variations in river flow, but large differences in the composition of bed and banks in the two estuaries give rise to wide dissimilarities in the frictional resistance of the sediments to flow, and hence to tidal levels up-river.

The following sections of this chapter describe some of the more common situations that confront civil engineers involved in the management and development of estuaries for a variety of activities. The topics covered include dredging, reclamation, water abstraction, marine structures associated with shipping, trade, industry and leisure, and tide-control structures such as barriers and barrages. Methods are

Coastal, Estuarial and Harbour Engineer's Reference Book. Edited by M.B. Abbott and W.A. Price. Published in 1993 by Chapman & Hall, 2–6 Boundary Row, London SE1 8HN. ISBN 0 419 15430 2.

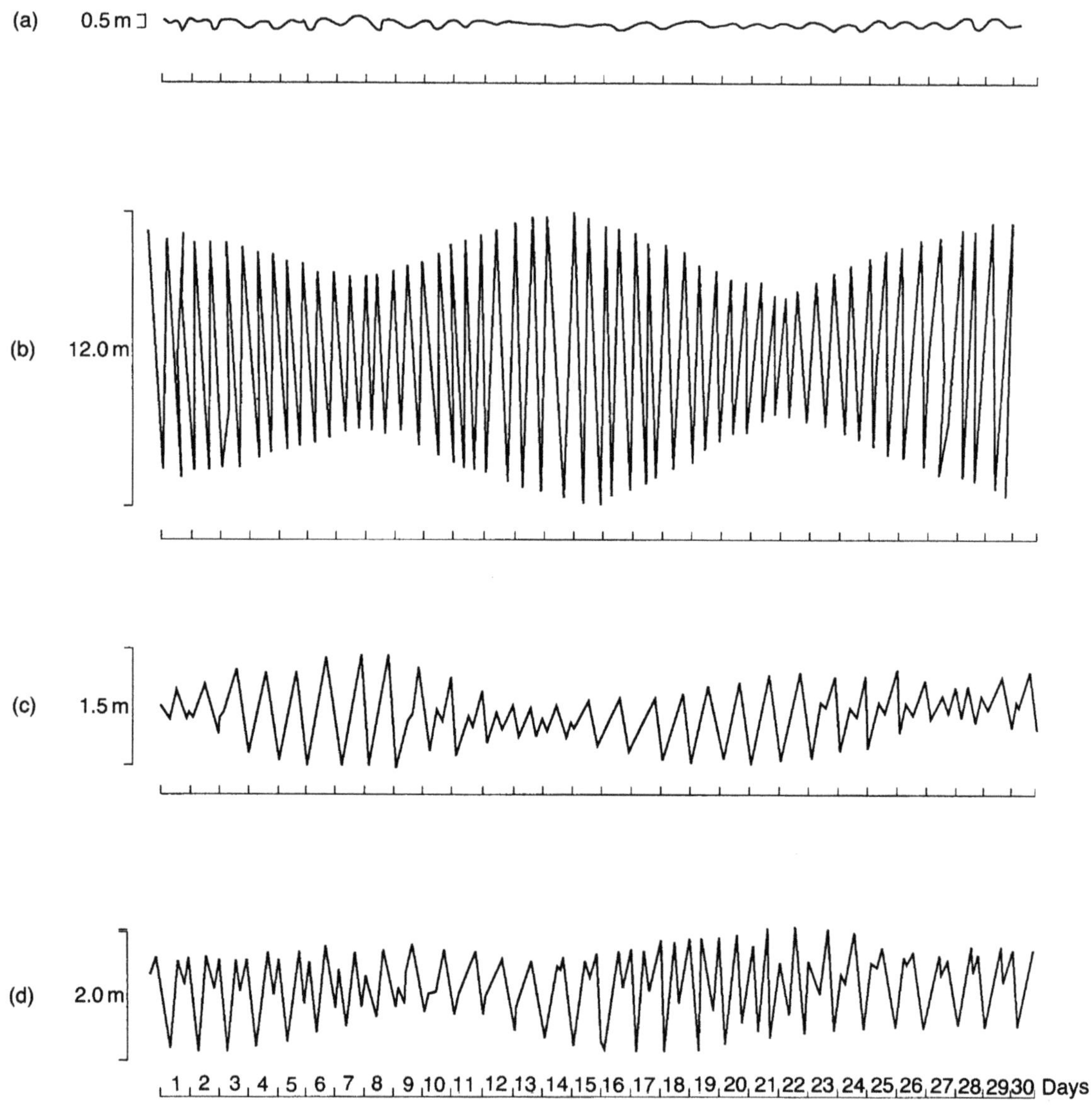

Figure 43.1 Differences in tidal characteristics (a) Galveston; (b) Avonmouth; (c) Kuala Baram; (d) Aden.

advanced for investigating problems associated with proposed or completed civil engineering projects, including an account of some of the solutions devised for rectifying past mistakes.

43.2 DREDGING

Almost all estuary works entail some dredging. The activity may be confined to a once-for-all capital dredging operation, it may involve a continuing maintenance dredging commitment, or it may take the form of commercial dredging.

43.2.1 CAPITAL DREDGING

Once it has been decided to embark on a major project involving capital dredging, such as the deepening of an estuary navigation channel by, say, 2 or more metres, there may be little scope for discriminating between alternative methods of implementation. This

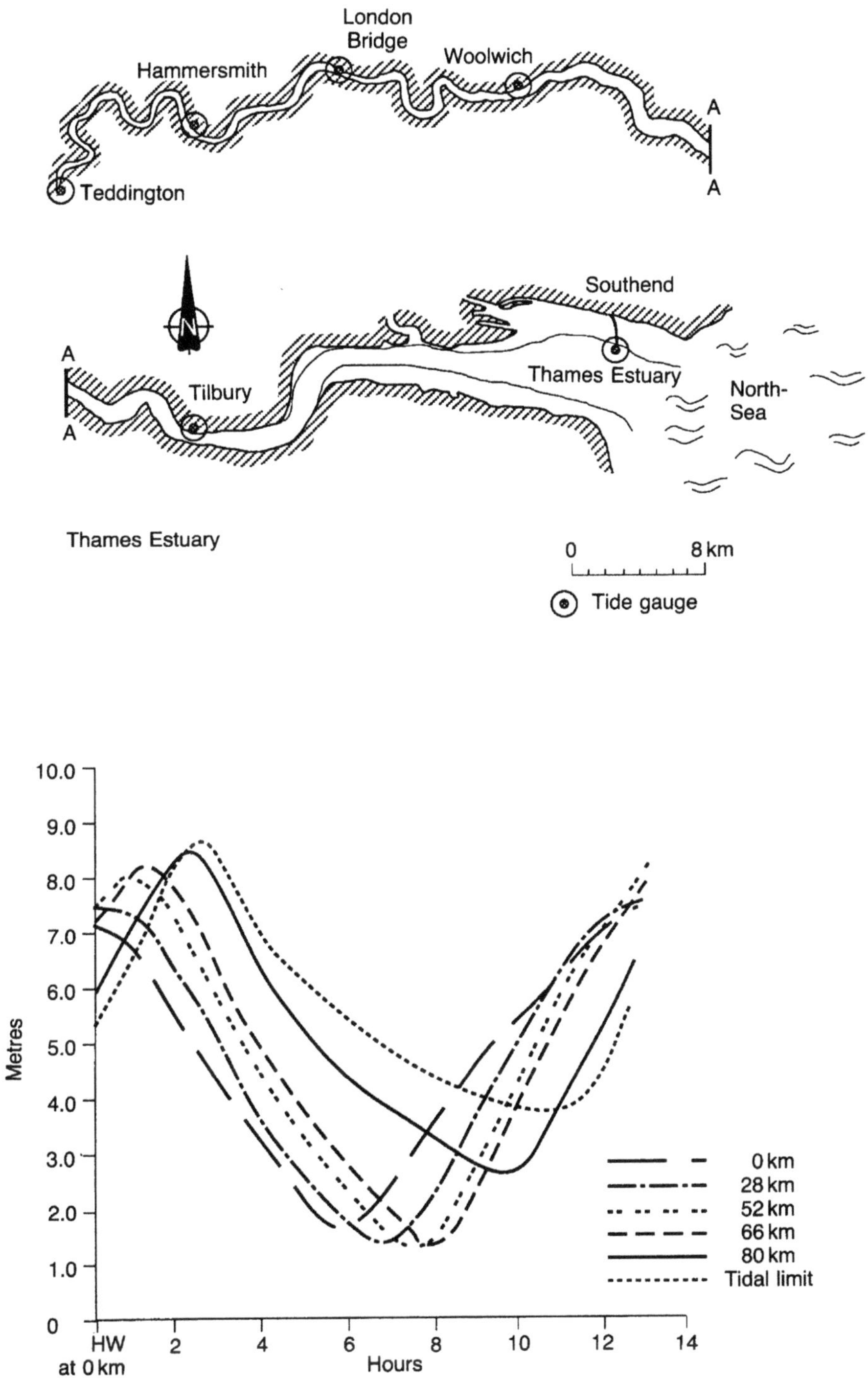

Figure 43.2 Thames tidal levels.

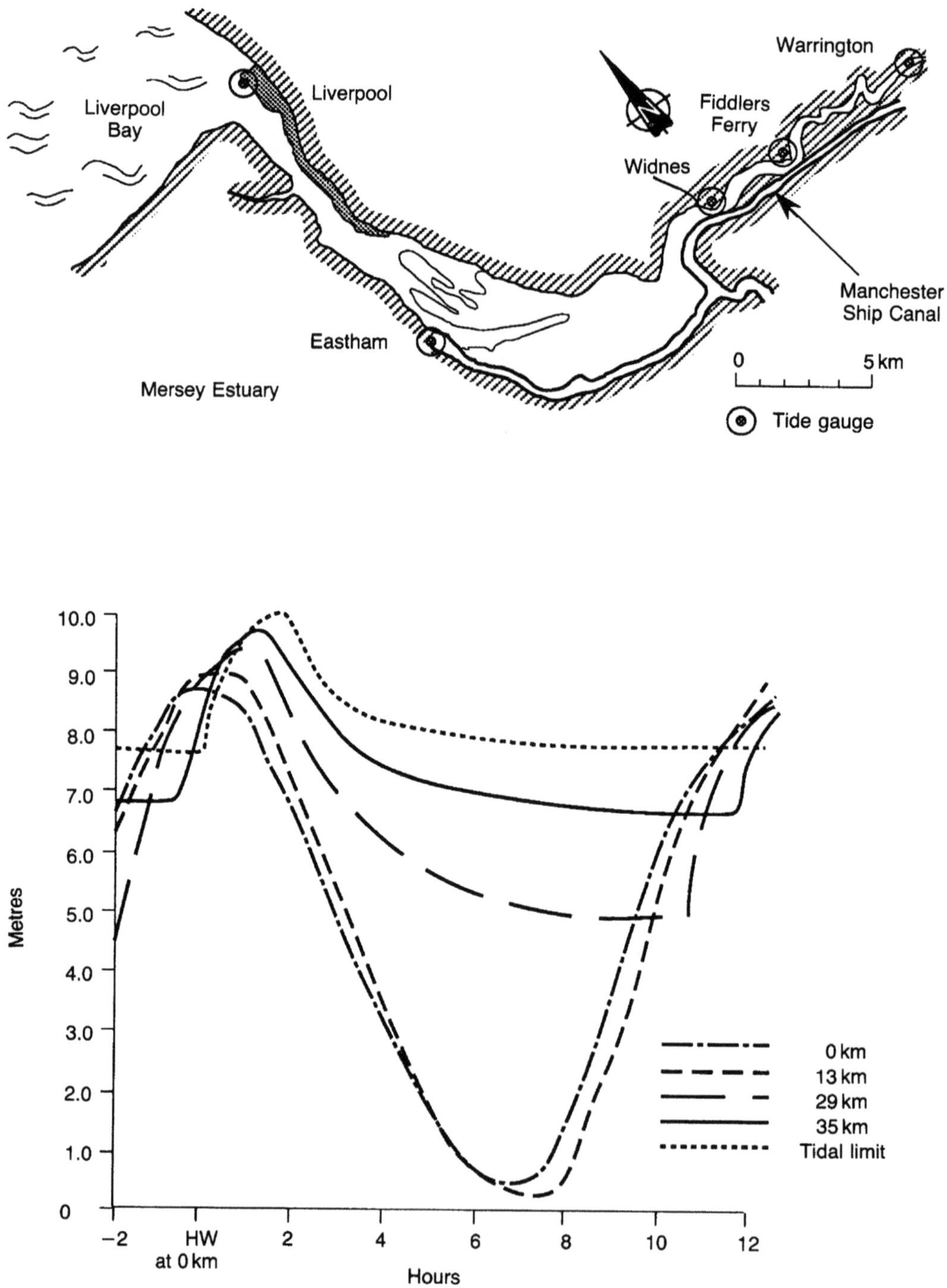

Figure 43.3 Mersey tidal levels.

does not preclude the need to ensure that the implications of the work are fully appreciated. For example, estuary channel deepening can have the following consequences:

1. Increased tidal penetration resulting in higher upriver high water levels and greater flood risk;
2. Increased tidal penetration resulting in lower upriver low water levels and depth improvement by bed erosion;
3. Increased flood and ebb current velocities giving rise to higher concentrations of sediment suspended in the flow;
4. Increased penetration of saline water leading to relocation of existing principal deposition zones;
5. Creation of a new and/or larger maintenance dredging requirement.

The extent to which any of the above consequences might apply in a given situation depends mainly on two factors: the existing state of equilibrium of the estuary and the availability of transportable sediments.

An estuarine system attempts to reach a state of dynamic stability for a specific set of hydraulic conditions. This does not mean that sediment movement ceases when that state is reached, but that over a long period (decades rather than years), fluctuations in bed-level normally occur only between certain limits and there is an average depth to which a particular part of the system tends. If the hydraulic conditions are changed significantly, the system will try to attain the new state of dynamic stability which is dictated by the new conditions, its rate of response depending on the availability of sediment from fluvial or marine sources.

It can often be established by a historical analysis whether an estuary has reached long-term stability or is in the process of readjusting to an event, such as a change in relative land/sea levels or a major civil engineering work.

Where an estuary port has been in existence for a number of years, regular survey data and dredging records may well be available from which to calculate estuary volume changes, as in Figure 43.4, which indicates that during the first half of this century the Mersey estuary was not in a state of long-term stability, the volume below high water of spring tides declining almost continuously, culminating in the accumulation of some 70 million cubic metres of sediment. Navigation channel deepening undertaken during such a period could be expected to incur heavy deposition, as indeed it did (Price and Kendrick, 1963). Even where such comprehensive bathymetric data are not available, it may still be possible to supplement what survey charts do exist with information from less obvious sources such as local historical records, geological or archaeological studies and so obtain some indication of the relative long-term stability of the area.

There is convincing evidence that a secular adjustment in land/sea levels in south-east England has effectively resulted in a rise in sea level over the last 2000 years, the present rate being about one-third of a metre a century (Dunham, 1972). The Thames estuary nevertheless virtually succeeded in maintaining some measure of long-term stability by redistributing along its course the easily moved sediments which cover large areas both in the estuary and the shallow part of the southern North Sea outside. Channel deepening by natural processes was therefore extremely slow. The system did, however, experience a relatively short-term disturbance in the form of capital dredging carried out between 1909 and 1928, when channel depths were increased by about 2 m over a distance of 16 km to provide for deeper-draughted vessels. The effect in central London was to increase the flow on a mean tide by about 4% and the tidal range by about 0.7 m (Inglis and Allen, 1957). Although only a minor factor contributing to London's complex and ever-growing flood problem (countered in the 1970s with the construction of the Thames Barrier (Gilbert and Horner, 1984)), the easier tidal access provided by the dredging certainly made the situation worse in January 1928 when a tide, higher than any that had ever occurred in the Thames before, resulted in flooding in London in which 14 people were drowned.

Probably the best-known and best-documented example of another possible effect of estuary channel deepening – an increase in saline intrusion resulting in the upriver translation of the major deposition zone – is provided by Savannah Harbour, on the Atlantic seaboard of the USA. Here, the lower 37 km of the Savannah River were dredged on four separate occasions between 1923 and 1953 to provide an overall depth increase of 3 m.

From considerations of salinity, the Savannah estuary can be classified as partly-mixed (Simmons, 1965), i.e. the vertical mixing of salt and fresh water is not complete over the saline intrusion length. This creates a transition zone where surface salinities are significantly lower than near-bed salinities. Upstream of this transition zone the net flow of water is seawards at all depths; downstream of the zone it is landwards. Transported sediments therefore tend to accumulate near the upstream limit of the zone since the net flow of water near the bed is to this region from both directions. Figure 43.5 shows the consequences of the channel deepening.

In 1923, the area of major deposition was located in the stretch of estuary labelled Zone A which needed to be dredged at a rate of about 1.5 million cubic metres a year to maintain navigation in the harbour at the depth then required. By 1953, after successive

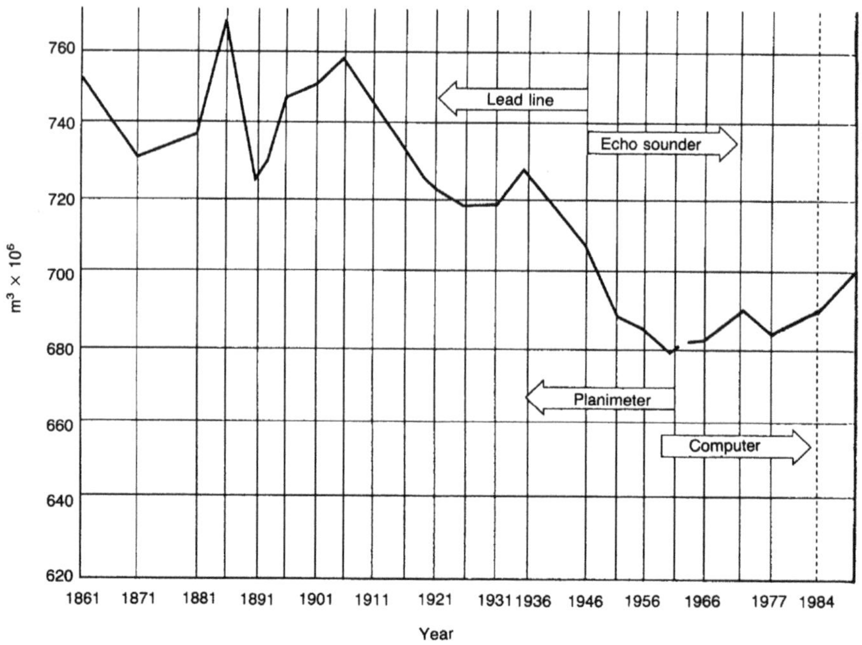

Figure 43.4 Volume of Mersey estuary below HWOST.

deepening, the deposition area had gradually been displaced upriver by some 20 km to Zone B, a reach in which average annual maintenance dredging had risen from virtually zero in 1923 to over 3.5 million cubic metres in 1953. By this time, there was almost no need for any dredging in Zone A.

Fortunately, intensive capital dredging sometimes has the desired effect of creating a permanent improvement in navigation channel depth without incurring any undesirable consequences. This was the situation in the Clyde estuary in Scotland where the only source of fine sediment was a small fluvial contribution, and once the dredged material had been deposited at sea, there was no mechanism for its recirculation in the estuary. Channel deepening in the Weser in Germany was also successful, because the removal of a large amount of material had considerably reduced low tidal levels near the head of the estuary whilst not affecting high tidal levels. The increased tidal volume flowing in and out of the system assisted the process of estuary deepening by scouring the channel.

43.2.2 MAINTENANCE DREDGING

In the earlier examples used in section 43.2.1, neither the nature nor the scale of the consequences could have been envisaged at the start of the capital dredging programmes. This is partly because at the time, the historical analyses which would have provided the background against which the changes were being effected had not been carried out, partly because the use of hydraulic models for prediction had not been widely adopted, and partly because most of the work on the properties of sediments (discussed in Chapter 38) was still to be done. This was particularly true of research on the erosion, transport and deposition of the fine, cohesive sediments which feature in a large proportion of estuarine siltation problems. With the advances of recent years, particularly in the fields of computational modelling and cohesive sediment research, the major effects of a large-scale capital dredging programme are more easily predicted. However, in respect of maintenance dredging rates, even in the 1970s a good knowledge of daily, bi-monthly, seasonal and secular changes in the distribution of flow and sediment in an estuary could not guarantee a realistic prediction: a point illustrated by events during one of the construction stages of the Thames tidal barrier.

The hydraulic investigation for the Thames barrier project was in many respects exemplary. It included the most comprehensive programme of synchronous field measurements ever before undertaken in a UK

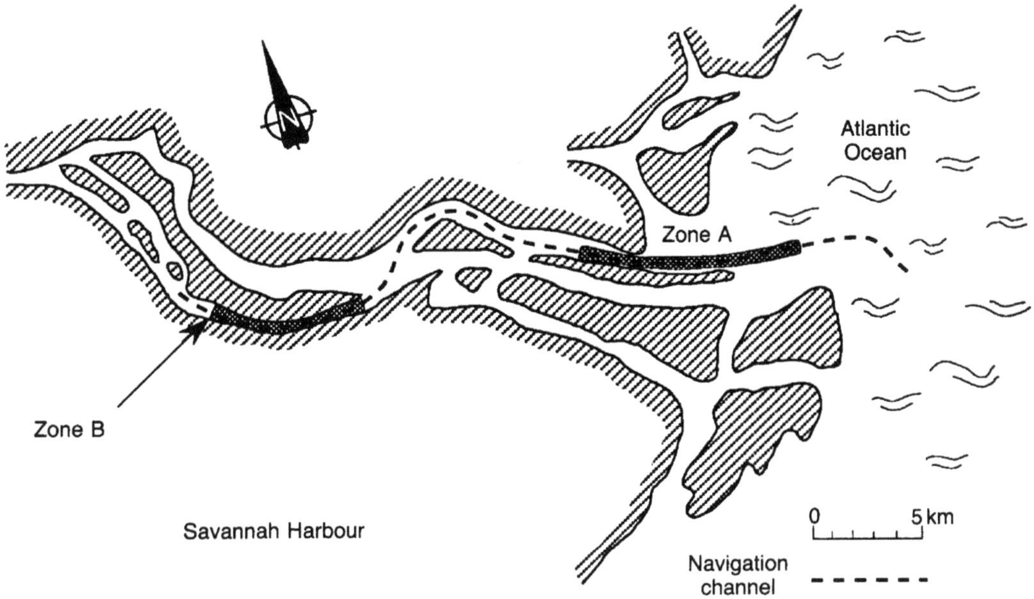

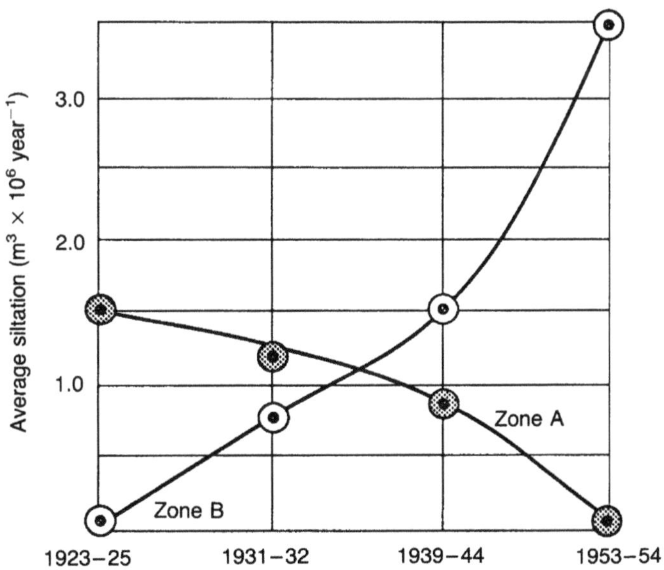

Figure 43.5 Channel deepening, Savannah Harbour.

estuary, the continuous monitoring of suspended silt concentrations from 1970 throughout a period exceeding twenty years: laboratory experiments to establish the properties of Thames mud, and studies on five physical and two mathematical models to investigate the effects of barrier site, design, construction and operation on the behaviour of flow and sediment in the estuary (Kendrick, 1972). In spite of these and other supporting studies, the maintenance dredging that was predicted following capital dredging of the northernmost span to provide deep-water navigation during one particular stage of barrier construction proved to be many times too high – an unexpected bonus for the dredging companies with fixed contracts based on the prediction.

The main factors responsible for the large discrepancy were firstly, lack of time, and secondly, other unforeseen civil engineering activities. Given the urgency to provide flood protection, time restricted the range of combined flow conditions that could be model-tested: those on which the predictions were based had therefore been selected as being seasonally typical. In the event, high fresh water flow down the River Thames was sustained during the winter of 1974–75 for an exceptionally long period. As a result, silt stored on bed and banks in the normally saline conditions became deflocculated in the consistently fresh environment and so was easily eroded and transported downriver for re-flocculation and deposition in more saline reaches, depleting upriver sediment supplies.

Dredging for a completely unrelated project was concurrently being carried out further down-estuary. As a consequence, the fine material which would normally have been carried up-estuary during the following dry season (again untypical in that it lasted for 18 months and included the drought of 1976) was limited in supply, deposited silt remaining in the new over-deep areas to seaward.

Subsequent analysis of the continuous silt monitor records confirmed estuary silt movements during the period to have been as described.

The capacity and speed of modern computers increases the range of possibilities that can be model-tested in a given time and so, by extending the information on which decisions are based, fortunately reduces the likelihood of this type of error.

So long as new ports and harbours are developed and existing facilities deepened, there will be a need for maintenance dredging. That situation is not likely to change: nor is the continuing search to reduce the cost of port and channel dredging likely to lose momentum (PIANC, 1989). What does change is the prevailing attitude to dredging and spoil disposal. In the past, the main problems associated with depth maintenance were how to extract the maximum volume with the minimum effort and to dispose of it at a safe distance, a safe distance sometimes being interpreted as somewhere far enough away to enable the dredger to regain port before the dumped sediment returned. The results of research now provide the port engineer with various options:

1. The acceptance of a bed of fluid mud having a density not exceeding $1.2\,\mathrm{g\,m^{-3}}$ as navigable;
2. The possibility of re-cycling the unwanted sediment within the system, either through side-casting or re-suspension by agitation;
3. The levelling, or bulldozing, of crests into troughs without the extensive dispersion of sediment through the water column that is involved in 2;
4. The direction of the spoil to productive uses such as beach replenishment, land fill, top soil or the making of bricks (Bray, 1989).

Clearly each of these options is dependent on local regulations, suitability of sediment and degree of contamination, as in recent years the operations of dredging and dumping have been greatly modified by environmental considerations. In some cases dredging has become closely associated with the disturbance and redistribution of pollutants, particularly where estuaries are situated at the mouths of rivers draining highly industrialized and heavily fertilized catchments. The problem then is how to avoid dredging at all, or at least how to minimize it and, if unavoidable, where to direct the dredged sediments so as not to create new problems elsewhere.

43.2.3 COMMERCIAL DREDGING

In some countries it is now no longer possible to exploit the sedimentary deposits of estuaries for commercial purposes without having regard to the consequences both for the estuary itself and for other estuary users. In such cases aggregate winning is often carried out under licence, the dredging location and rate of extraction being prescribed by the authority responsible. Even where such regulations exist, the need for awareness on the part of those charged with estuary management is underlined by an event which occurred in the UK, described below.

Sand dredging had been authorized from an estuary shoal separating two channels, one a navigable approach channel to an active port. It was stipulated that the dredging head should work along the shoal within a well defined area, excavating the material above a specified bed-level. The contractor was charged with monitoring the exercise by undertaking regular bathymetric surveys. In the event, the dredging plant remained at one station and was carving out a cut which threatened to divert the main tidal flow across the shoal, leaving the port approach channel as a silting backwater. Fortunately the offence was detected in time for the operation to be halted and remedial action taken.

43.3 RECLAMATION

As in the case of dredging, so with reclamation, an improved general understanding of natural estuarine

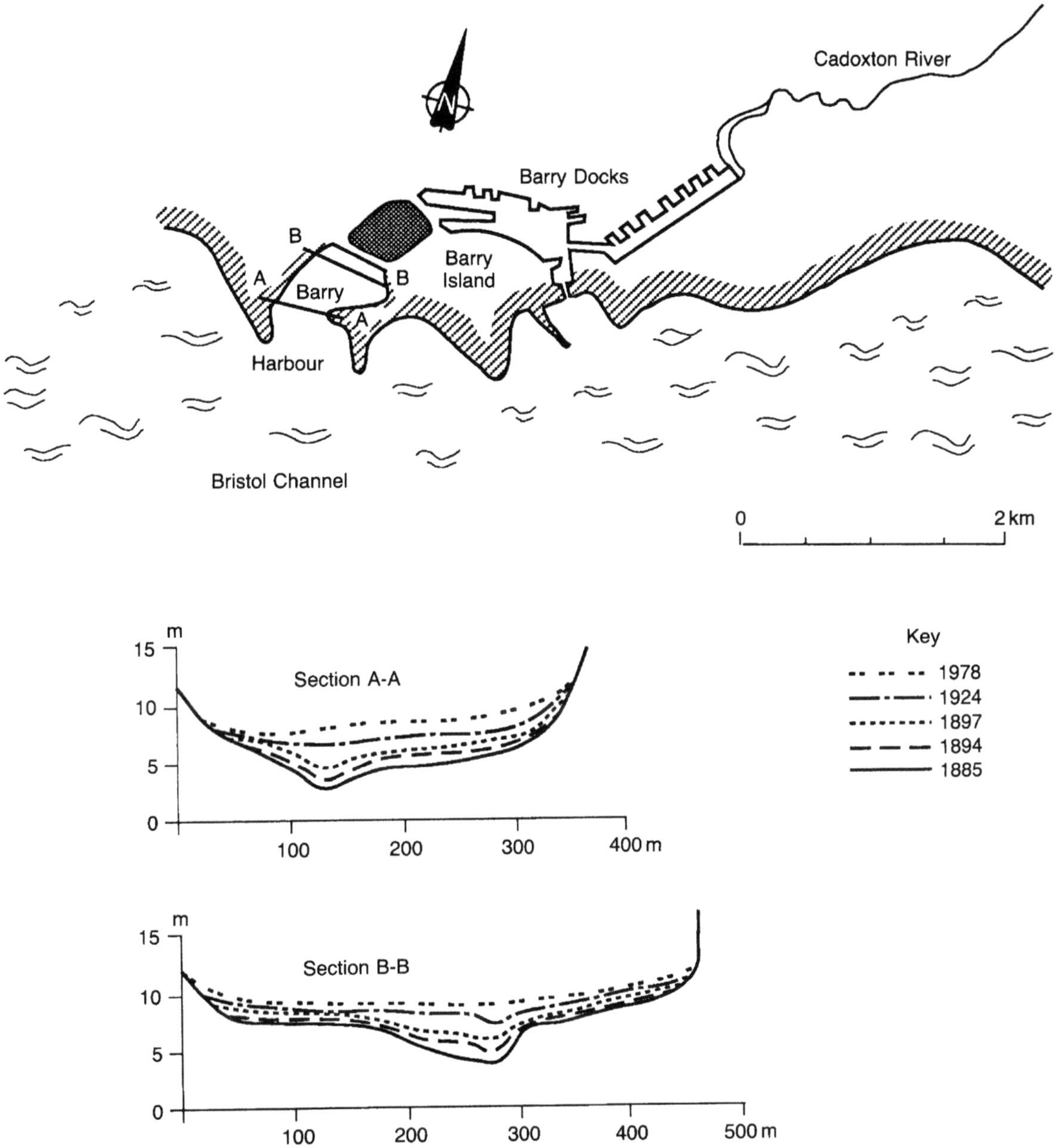

Figure 43.6 Reclamation, Barry Harbour.

processes inevitably increases the range of problems that have to be addressed when planning a project. In the past, quite ambitious schemes were undertaken without devoting too much thought to the consequences for the estuary. This was especially so during the period of rapid industrialization in Europe and the United States in the second half of the nineteenth century, when dock and harbour facilities could not be provided fast enough to meet the needs of world trade expansion. An example, small in scale but large in impact, of the type of development that occurred in some estuaries in this period can be seen in South Wales, on the Bristol Channel (Figure 43.6).

To relieve the pressure on Cardiff docks 12 km away, reclamation of the eastern part of the estuary of the Cadoxton River and dock construction were carried out in 1884. The process was repeated in 1895, the successive reclamation, its associated infrastructure and the diversion of the course of the river to the east, instead of the west, of Barry Island leaving what is now Barry Harbour as an abandoned tidal inlet at the mouth of the old estuary (Davies *et al.*, 1978). Deprived of the current velocities formerly generated through the entrance as the 8 m tide filled and emptied the tidal storage area which used to exist up-estuary, Barry Harbour shoaled by up to 5 m, as depicted by the two harbour cross-sections forming the lower half of Figure 43.6.

The example of Barry Harbour represents that end of the spectrum in which reclamation changes the estuarine system beyond all recognition. At the other end are those comparatively minor reclamations that are carried out almost unnoticed along the banks of a busy estuary whose port areas are probably already flanked with wharves, dock walls and jetties. This type of development often involves no more than the infilling of a bankside area between two existing installations, either as an extension to one of them or as a completely new enterprise. Individually such works usually have very little impact on the estuary, and can, in fact, lead to an improvement in local flow conditions provided care is taken to ensure that any upstream or downstream transitions are smooth. Cumulatively, however, minor reclamations may start to influence the behaviour of the estuary, particularly if undertaken in upper tidal reaches, where their impact on tidal discharge is proportionately greater than in reaches nearer the sea.

If it is not possible for the reduction in effective flow area following reclamation to be compensated by increased bed scour (e.g. because the bed is inerodible or the estuary not in regime), then the tidal discharge will fall, and the associated reduction in current velocities will increase the tendency for deposition to occur both above and below the works. Another possible effect of successive riparian reclamations is one which may apply only when high river flow happens to coincide with high spring tides or storm surges. If the excess water is unable to spread laterally by inundation of the flood plain, or

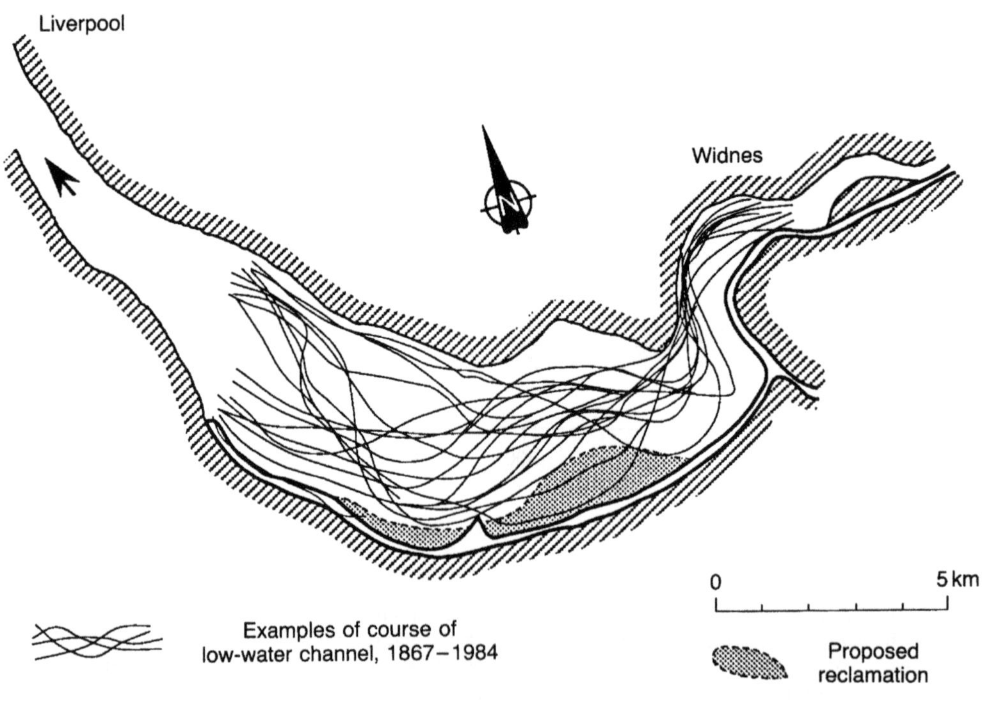

Figure 43.7 Reclamation, Mersey estuary.

by groundwater flow, maximum estuary levels could be increased and vulnerable properties flooded.

When weighed in the balance with the benefits to be derived from a given reclamation proposal, effects such as these may not be considered sufficiently significant to militate against the project. They must nevertheless be taken into account by any authority responsible for safeguarding the wider public interest.

Lying somewhere between the two reclamation extremes of complete flow blockage and riparian development, is the single, fairly ambitious proposal which might involve the creation of a completely new artificial island in the middle of an estuary or the reclamation of a relatively large area of inter-tidal bank. Nowadays the public could reasonably expect that for such a project, fairly comprehensive feasibility studies would be carried out, as was the case with the scheme proposed in the 1980s for the Ince and Stanlow Banks of the Mersey estuary in the north-west of England (Figure 43.7).

This proposal was to reclaim some 450 hectares of long-established sand and mud banks lying along one shore of the inner estuary. The banks are covered only during the highest spring tides but nevertheless still form part of the general estuary system. The remit for the hydraulic investigation therefore included the need to establish what size of reclamation, if any, would be acceptable should the scheme as originally proposed fail to satisfy regime requirements. In this respect, certain criteria had to be met. The first was that existing depths in the navigation channels to the estuary ports should not be threatened; the second, that net siltation in the inner estuary should not be increased; the third, that surge-tide levels upstream should not be increased; and the fourth, that the freedom of the low-water channel to migrate across the inner estuary should not be inhibited. The last criterion is associated with the first two in that one consequence of restraining the movement of the low-water channel would be an increase in inner estuary accretion. In an estuary like the Mersey, the meandering process of channel migration, involving the continuous erosion and deposition of sediments, keeps a check on sediment consolidation and slows down the transformation of high, inter-tidal banks into permanent features which, once established, reduce the tidal storage volume. Freedom of the channel to move can thus be an important consideration where the maintenance of adequate navigation depths further seaward is dependent on tidal discharge (Hydraulics Research, 1985).

It is important with large new projects like the Ince and Stanlow Banks reclamation, to plan studies so that the answers to problems emerge at the right stage. Time and money can be wasted in pursuing design detail if potential project-stoppers have not already been identified and satisfactory countermeasures devised. The three main impediments feared in this case were:

1. The size of acceptable reclamation would be too small for the financial viability of the project;
2. The risk of the low-water channel encroaching on the reclamation area would be too high;
3. The optimum reclamation as determined on hydraulic grounds would be incompatible with that as determined on ecological grounds.

Given a satisfactory resolution to points 1–3, the other problems to be addressed were:

4. Whether a completely different low-water channel position from that existing at the time of the study would produce any unsatisfactory changes in flow behaviour near the reclamation at particular stages in the tide;
5. How, in the (unlikely) event of the course of the low-water channel encountering the reclamation, it could be discouraged from running permanently along the reclamation boundary;
6. Which of the alternative local sources of suitable reclamation sand could be dredged without significantly affecting the estuary regime;
7. Which of possible reclamation construction methods would cause the least disturbance to existing estuary activity;
8. What the implications for sediment distribution would be following the unavoidable changes in local flow which would occur, particularly while reclamation was in progress.

Following a review of existing bathymetric, hydraulic and sedimentological data, the problems that had been identified were tackled in the following way.

Firstly, the likelihood of the low-water channel encroaching on the proposed reclamation was established by an examination of the frequency of its encroachment on the site between 1867 and 1984 (Figure 43.7). A specially developed computer programme produced a schematized outline of the inner estuary and reclamation area. On this the location of each intersection of the channel with the standard estuary survey sections was plotted, using information extracted from the chart produced each month to show the latest position of the buoyed channel.

Secondly, a 1/600 (horizontal) 1/80 (vertical) scale hydraulic model was used, first to determine the optimum size and location of reclamation from the standpoint of minimizing the effects on tidal levels and flow both generally and locally; secondly to examine flow behaviour with a completely different low-water channel course and to investigate the effectiveness of measures designed to ensure its non-stabilization along the reclamation boundary; thirdly to establish the effect on flow of sand-winning for the reclamation from nearby areas.

Finally, the results of an examination of the nature of sediments sampled from the inner estuary were used in conjunction with the model flow results to assess siltation and scour during and after reclamation (Hydraulics Research, 1986).

The investigation of the physical aspects of the Ince and Stanlow Banks reclamation project lasted about three years, during which a continuing exchange of information was maintained between parallel hydraulic and ecological studies. The problems posed by the three potential project-stoppers were satisfactorily resolved before the end of the first year.

43.4 WATER ABSTRACTION

The problems that arise in connection with water abstraction either stem directly from the abstraction itself or are associated with the subsequent return of the water to the estuary. Where water is abstracted in significant quantities over prolonged periods from the upper tidal reaches or just above, for whatever purpose, tidal penetration, saline intrusion, turbidity, and hence siltation and scour in the estuary, can all be affected.

To assess the possible implications of a specific abstraction proposal may require little more than a desk study, particularly where the estuary is already used extensively and good field records exist on bed levels and on the distribution and properties of tidal flow, river flow and sediments. Where an initial appraisal identifies the need for more detailed predictive studies, as for instance when the Thames Water Authority in the UK sought relaxation of existing legal constraints on abstractions for reservoir storage at time when river flow fell below a certain statutory minimum (Hydraulics Research, 1984), then recourse is best had to mathematical modelling. In this particular example, one of the studies involved the use of a laterally-averaged model designed to simulate the vertical structure of the flow, salinity and mud distributions in the estuary. The model was used to predict the effect on turbidity and siltation of sudden and prolonged reductions in river flow at the tidal limit (Hydraulics Research, 1985).

In the case of highly developed estuaries, a variety of river interests can be affected to a greater or lesser extent by water abstraction, *viz.* navigation, industrial processing, power station efficiency, dock and jetty operations, fishing and leisure activities. In predominantly agricultural countries which experience long periods of drought, the increase in salt water penetration following excessive use of water from upper reaches for irrigation can be more devastating in its impact, impoverishing surrounding land in the middle and lower reaches. Often in the past this type of problem has necessitated curative rather than preventative treatment. Nowadays, given adequate funding, the effects of different methods of tide and river discharge control, such as the salt-excluding barrage proposed for the Gambia River in West Africa, can be examined on mathematical models before, rather than after, the worst has happened (Hydraulics Research, 1977).

The withdrawal and return to the estuary of power station cooling water can cause changes of a very different kind, which may or may not create problems. For example, where it is necessary for a large power station situated in a highly turbid reach to clear the intake water in silt settling lagoons before passing it through the plant and back to the estuary, operation over many years can reduce local turbidity levels. The net effect will probably be of benefit to other estuary users. Some environmental advantage might also be effected by exploiting the fact that silt lagoons can make very good nature reserves if well managed.

On the other side of the balance sheet could be the combined effect created by a number of power stations discharging heated cooling water to an estuary. Mathematical modelling studies showed that the 14 power stations along the tidal Thames had produced a general rise in river temperature of about 3°C, local increases being still higher (Wood, 1982). Increased temperatures not only make a river unsuitable for sensitive organisms such as migrating fish, but also reduce the concentration of dissolved oxygen that can be maintained (Gameson and Robertson, 1955).

43.5 MARINE STRUCTURES

The main structures associated with the exploitation of estuaries for shipping, trade, industry and leisure include training works, port and harbour facilities such as jetties and terminals, dock basins and yacht marinas.

43.5.1 TRAINING WORKS

The scale of shipping activity which takes place in many estuaries has only been made possible by the construction of training works. The works are generally located in the lower estuary and have usually been designed either to stabilize the position of the entrance channel or to concentrate the available flow over a bar or shoal for the maintenance of better depths. Because the main objective of training, like that of channel deepening by dredging, has almost invariably been to further navigation, many of the schemes carried out in the past with little or no appreciation of other possible consequences would either not be undertaken at all today, or only in modified form. This is not a judgement; merely an opinion based on the increase that has occurred in the general understanding of the behaviour of estuarine systems and the imposition of constraints that has accompanied greater environmental awareness.

The UK is not unique in having several examples of training schemes which would be considered ill-conceived by today's standards. In the Dee below Chester, the main river channel was diverted from the north side of the estuary to a specially-dredged and trained channel on the south side. In a few years, sediment deposition over the remainder of the estuary had reduced the total high water area by 26% and 22 years later (1938), deterioration due to loss of tidal storage had been so severe that efforts to maintain navigation in the upper estuary were abandoned.

In promulgating a training scheme, especially one located in seaward reaches, it is necessary to be aware not only of any possible consequences for tidal storage (discussed in section 43.3 in relation to low-water channel stabilization), but also of the interaction of the scheme with the existing wave climate, entrance flow conditions and littoral current regime. The survey of Kuala Baram in East Malaysia (Figure 43.8) provides a perfect example of an offshore bar located where all of these factors had to be taken into account in assessing the feasibility of a training scheme required to maintain a 5 m navigation channel into the Baram River (Hydraulics Research, 1970, 1973). Both the local and more far-reaching effects of such a scheme vary, depending on the height, length, alignment and permeability of the structures, be they breakwater units, tipped rock, armoured bunds, steel sheet piling or bamboo screens. Furthermore, because the costs of construction vary widely, depending on the final combination of design features selected, the economies effected by early investment in hydraulic assessment can be considerable.

The procedure for identifying initial problems is basically the same as for any other major estuary work, *viz*. to obtain information both on the characteristics of flow and sediment in the area of interest and on the changes in the distribution of flow and sediment that have occurred in the long- and short-terms.

In practice there may be several study stages. The first stage will be the collection and assessment of any existing field data and relevant historical records.

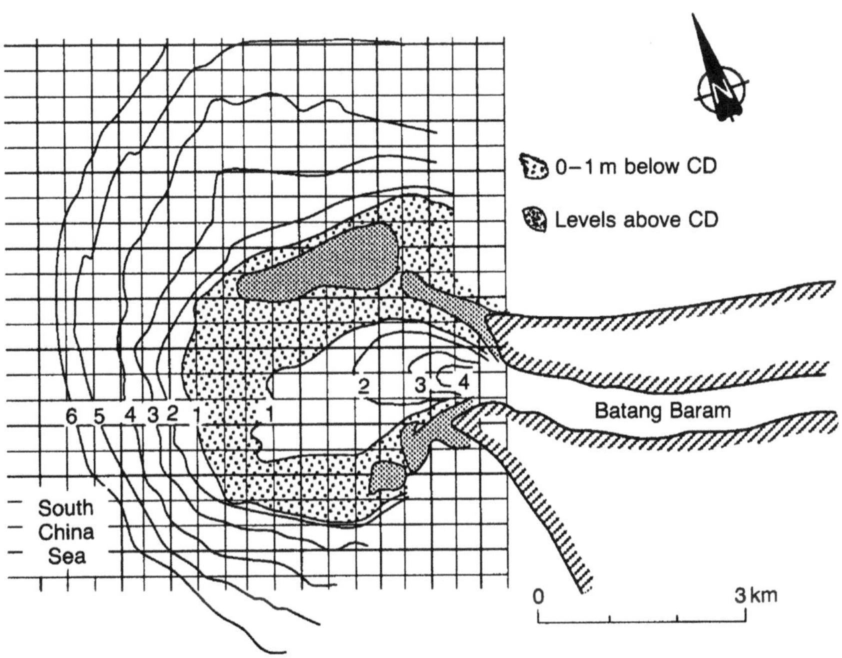

Figure 43.8 Approaches, Kuala Baram.

This may have to be supplemented in a second stage by further site observations, e.g. float tracking in the vicinity of the proposed training works; measurements of wave height, period and direction; sediment flux measurements carried out during a combination of tide and wave activity. The analysis of the new and existing field data in the third stage may provide sufficient information on which to base design decisions. If it does not, then a final stage involving physical and/or mathematical modelling will be necessary. In cases where the two modelling methods would be equally satisfactory in providing answers, then the choice may well depend on whether suitable models of either type already exist. If so, then the cost of the studies will be considerably less than if new models have to be specially designed, constructed and verified. Where no models are already available, the choice is usually dictated by the respective costs and time associated with the alternatives.

In some cases, one type of model may clearly be superior. For example, a mathematical model can be used to predict the erosion, transport and deposition of estuary muds which are impossible with a physical model. On the other hand, a physical model will reproduce detailed flow behaviour near the structures more realistically than a mathematical model. To provide a comprehensive picture of the interaction of a major training scheme and its hydraulic and sedimentary environment, it is quite usual to employ a combination of physical and mathematical modelling in the final study stage.

43.5.2 JETTIES AND TERMINALS

Much of the information of practical value to engineers designing jetties and terminals in estuaries has been acquired as a result of problems that have arisen with existing installations. Sometimes the problems occurred because structures, built to a design and located in an area quite suitable for the shipping activities of the day, became less satisfactory when modified to meet changing demands. Sometimes they occurred because lack of attention to variations in tidal flow behaviour at the site resulted in a jetty that was badly aligned in relation to the predominant flow direction, or in a pile arrangement that produced too great a head loss through the structure, or (less frequently) in the creation of pile and jetty vibration at certain flow velocities. The author has encountered several examples of such structures on the Thames alone, not surprising in view of the estuary's long history of jetty construction (Kendrick, 1984).

Irrespective of the form of structure adopted, solid or open-pile, defective jetty alignment can initiate flow separation and turbulence at certain states of the tide which make berthing hazardous and, in a turbid estuary, render both the jetty itself and adjacent areas more prone to siltation. The photographs in Figure 43.9a and b were taken during model tests on a jetty misaligned by about 10° to the main flow direction. They depict the situation at 2 h after high water when current velocities reach maximum values. The float tracks demonstrate that the disturbed conditions

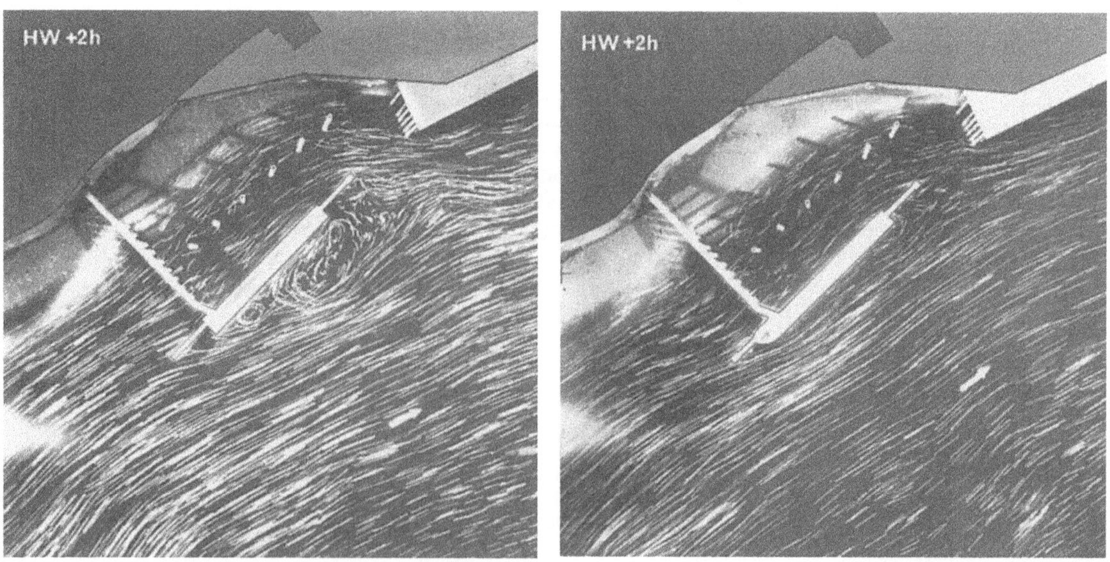

(a) (b)

Figure 43.9 (a) Jetty misalignment; (b) flow rectification.

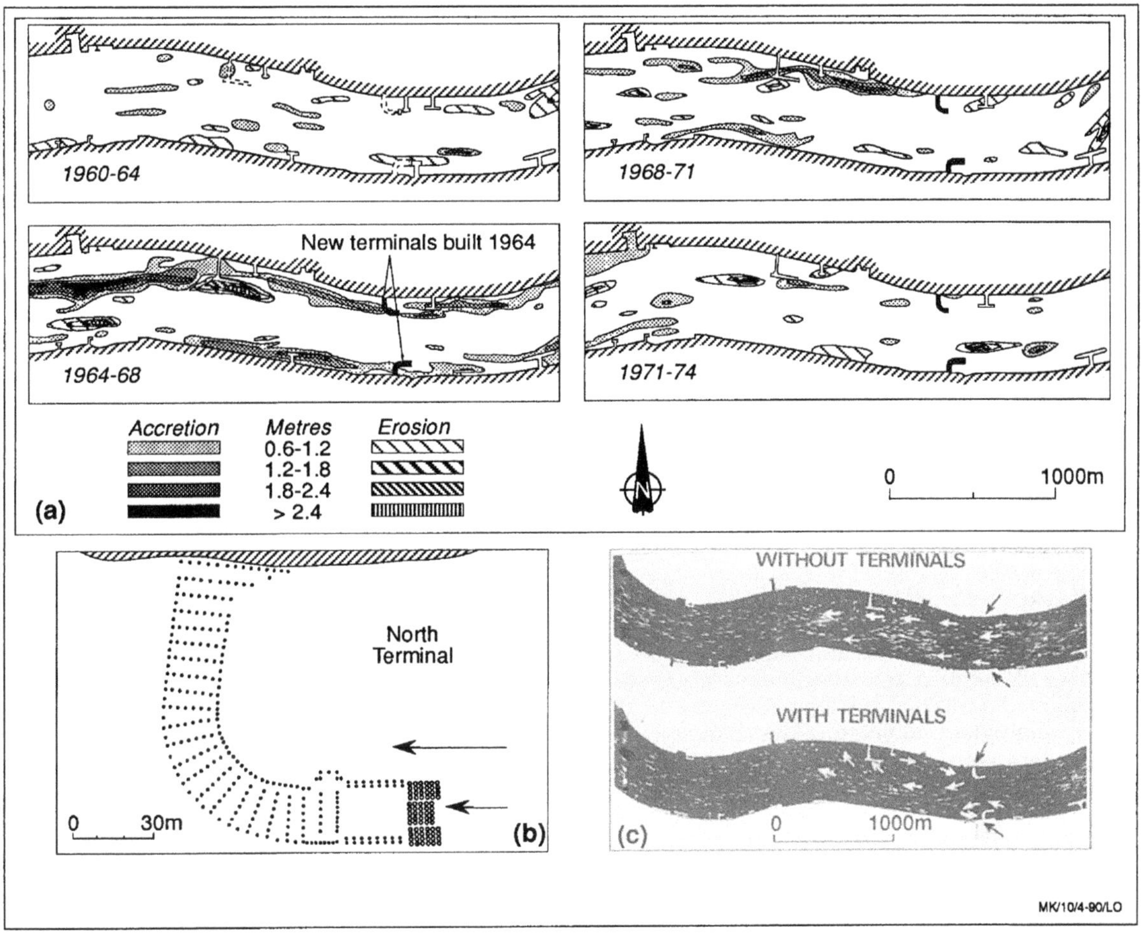

Figure 43.10 Consequences of inappropriate pile distribution.

shown on (a) at the face of the jetty could be significantly reduced by the addition of an asymmetric cutwater to the upriver end of the structure. The flow improvements on (b) make for easier berthing and reduced siltation.

With open-pile design, insufficient attention to the interaction of piles and flow can result in significant head losses through the structure. This may not only reduce current velocities on the down-flow side thereby increasing the likelihood of sediment deposition, it may also produce unwanted changes in flow direction at certain states of the tide. An example of inappropriate pile distribution on two ferry terminals, with its consequences, is given in Figure 43.10a–c.

Figure 43.10a presents accretion/erosion charts of a reach in the Thames estuary representing four periods between 1960 and 1974. From 1960 to 1964, changes in bed level rarely exceeded 1 m, areas of accretion and erosion being fairly equally distributed, indicating this part of the estuary to be in a dynamically stable state. Between 1964 and 1968 the transformation was remarkable. What occurred could almost be described as a siltation explosion. Apart from one well-defined dredging zone on each side of the channel, both banks suffered considerable loss in depth. This amounted to over 3 m near the north bank and almost 2.5 m near the south. During the next period (1968–1971), although there was still plenty of evidence of siltation on the north side, changes elsewhere had slowed down. Finally, between 1971 and 1974, the magnitude of the changes became much the same as in the first period, rarely exceeding 1 m except where some local dredging had recently taken place. This indicates that the estuary bed had more or less adjusted to a modified hydraulic environment in which natural (or regime) bed levels on both sides of the estuary were now higher than before the terminals were constructed. As a result, jetty owners were having to do more dredging to maintain the same depths that they had enjoyed in the past.

It is apparent from the layout of the north terminal piling shown on Figure 43.10b that the design made no allowance for the increase in flow obstruction that the piles would create as the structure curved through 90° on extending into the channel. This potentially undesirable situation was made worse by the fact that the silt that had deposited during construction (due to the flow blockage created when both terminals were in cofferdam for twelve months) became too well-consolidated for former depths to be even partially restored by the flow once the works were completed and all temporary obstructions removed.

Flow behaviour before and after terminal construction was demonstrated retrospectively on a model. The situation at mid-flood is shown in Figure 43.10c, the lower photograph illustrating the change in current circulation which particularly affected the north side of the channel.

In a turbid estuary, failure to assess the interaction of flow and structure can be costly in terms of remedial works, protracted litigation or, as in the last example, both. If the up-front funding for a project cannot meet the expense of model studies in addition to the cost of collecting the essential site data, then head loss calculations for alternative piling designs can be carried out for about a quarter of the modelling costs by prudent use of experimental results available in the literature on pile interference and hydrodynamic drag (Delaney and Sorensen, 1953; Hoerner, 1958; Mair and Maull, 1971). Although this will not provide the information on flow behaviour that would be obtained by modelling, it will at least indicate how to minimize potential siltation problems by minimizing the drag of the structure.

43.5.3 DOCK BASINS AND YACHTING MARINAS

Several factors influence the nature and extent of the problems associated with the construction and operation of docks and marinas on estuaries:

1. Whether the basin is fully tidal, partially tidal, or completely non-tidal;
2. Whether it is located in an area or side-arm off the main waterway, or actually sited on the estuary foreshore;
3. Whether it has permeable or impermeable boundaries;
4. Whether it is situated in a part of the estuary which experiences much salinity variation throughout the tide;
5. Whether the local sediments are predominantly sandy or muddy;
6. How the strength and direction of the main currents vary throughout the tide near the entrance to the basin.

This list could be extended, but it suffices to demonstrate the need to have adequate site information for checking present diagnoses and predicting future problems.

Where an otherwise enclosed dock or marina on a turbid estuary has its entrance open to flow at all states of the tide, siltation will be inevitable. The amount of sediment available for deposition could be a function of different combinations of river flow, tidal flow and wave activity. The actual siltation mechanism may consist of the deposition at slack high water and subsequent bed consolidation of sediment fully suspended throughout the water column; the deposition throughout a tide of mixed sediments due to the creation of eddies by strong flow past the entrance; the settlement of silt from a highly concentrated near-bed turbid layer; or the intrusion of a sediment-laden salinity/density current arising from salinity differences between the water inside and that outside the basin. The process may occur on a regular daily, monthly or seasonal basis, or be a purely random occurrence (Kendrick, 1989).

Two possible ways in which a tidal basin can become silted are shown diagramatically on Figure 43.11a and b. In (a), the deposition in a basin on the side of an estuary is due to the continuous sediment feed which occurs as a consequence of the primary and secondary eddies created by the estuary/basin water exchange during the main run of the tide past the entrance. In (b), siltation occurs as a result of the conditions prevailing in estuary and basin around the time of high tide. Although the tidal level in both is depicted as being at a maximum, half an hour remains before the main current in the estuary is seen to reverse direction from flood to ebb. The higher salinity outside the basin during the late flood can increase the entrance velocity of flow within 1 m of the bed by about $0.3\,m\,s^{-1}$, due to the water density difference inside and outside the basin. To compensate, lighter, less saline water flows out at the surface. Although the effect is reversed on the ebb, velocities inside the basin are too low to re-entrain sediment deposited on the flood. During a complete tidal cycle,

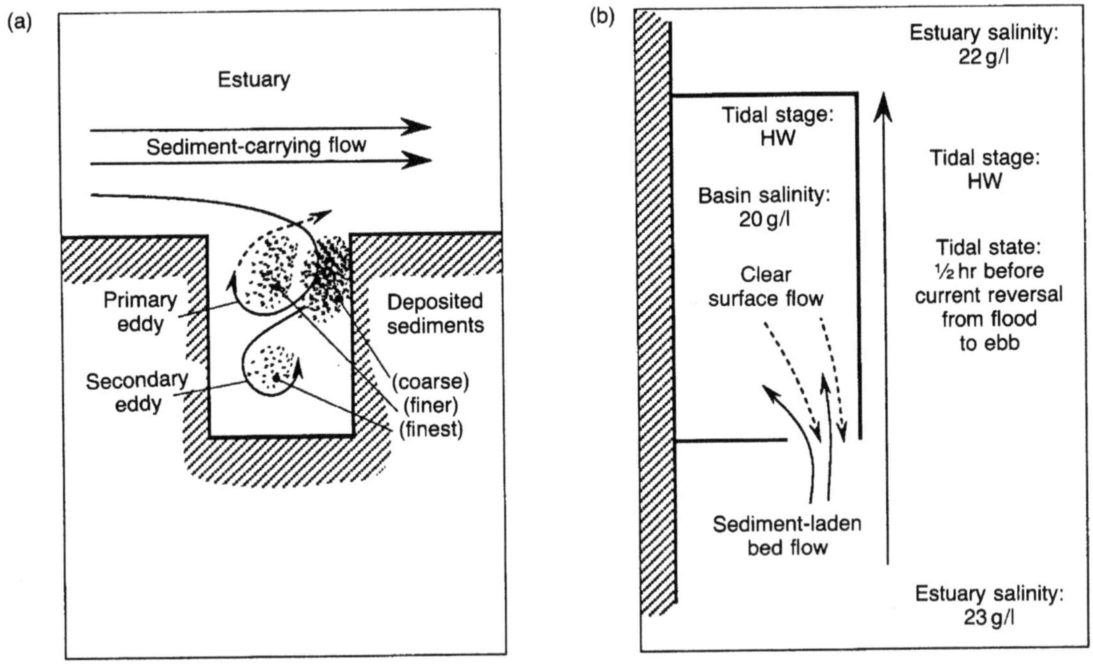

Figure 43.11 Possible causes of tidal basin siltation.

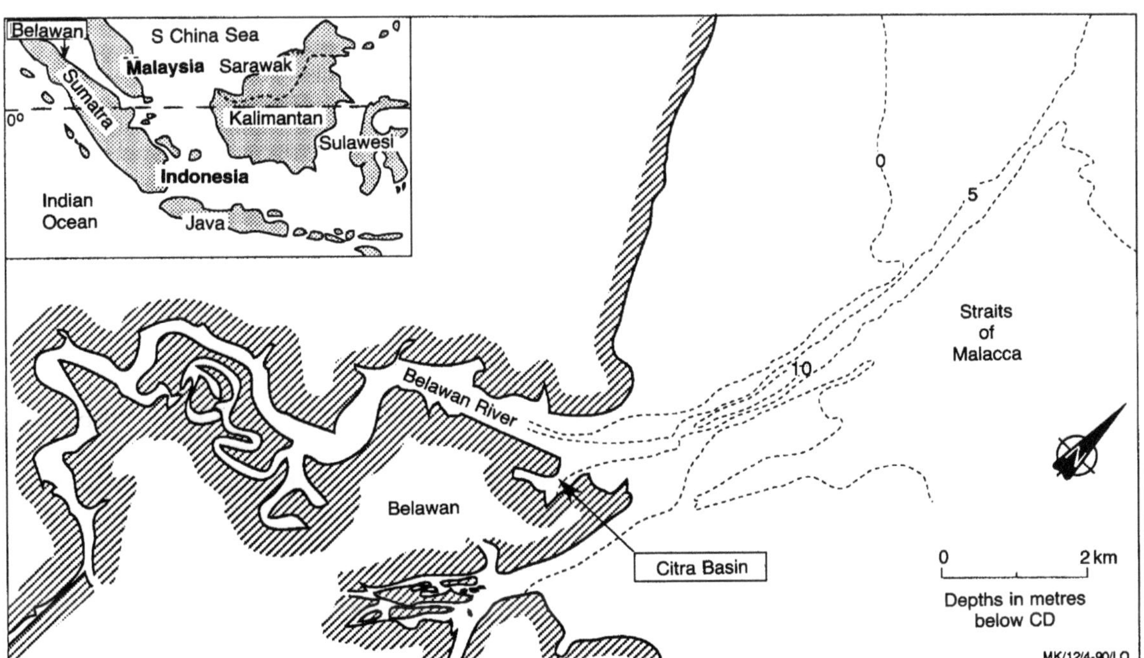

Figure 43.12 Citra Basin, Port of Belawan.

sediment transport is therefore largely one way – into the basin.

Given a basin siltation problem, it is obviously desirable to know which particular mechanism applies before attempting to design remedial measures. The construction of a sill or step to prevent the incursion of a near-bed layer of fluid mud is of little use if the main cause of the trouble is the deposition of sediments carried into the basin fully suspended throughout the flow depth. Similarly, a device to interrupt general flow circulation and eddy formation will not eliminate near-bed sediment transport due to basin/estuary salinity differences.

It is usually possible to establish the prevailing siltation mechanism by an assessment of site measurements of current velocity and suspended sediment concentration, made at intervals of water depth and time throughout a tide, at stations inside the basin, outside the basin and in the entrance. Sometimes the problem is less tractable, and prolonged field data collection may be necessary involving highly specialized instrumentation. Such was the case with the Citra Basin at the port of Belawan in Indonesia (Figure 43.12) in which siltation was thought to be due either to bankslip following dredging, or to deposition of sediment suspended throughout the flow depth. In the event, calculations based on the available field data demonstrated that neither one alone, nor the two combined, could possibly account for the annual siltation experienced.

More than twelve months of data collection in basin, estuary and port approach channel, supplemented by laboratory experiments on the properties and behaviour of local mud, were required before enough information was obtained to suggest the complicated siltation mechanism in operation at Belawan. The indications were that in this low-energy environment, a near-bed turbid layer a few centimetres thick could be formed by low but regular daily wave activity over shallow mud banks flanking the dredged port approach channel. The layer could be moved to and fro by modest currents without significant mixing throughout the flow depth and, once in the navigation channel, come under the influence of near-bed currents. Since both residual drift and the location of the major deposition zone (or turbidity maximum) in an estuary are a consequence of the interaction between fresh water flow and tides, the distribution of sediment throughout the whole tidal system of sea channel, estuary and basin was thus found to be primarily, though very indirectly, dependent on the seasonal variations in flow down the Belawan River (Kendrick *et al.*, 1985).

Where there are approach-depth restrictions and the enclosed basin has entrance gates which are kept shut for most of the tidal period except for a few hours around high water, the situation is somewhat different from that of the permanently open entrance considered above.

If the estuary is turbid, the inevitable maintenance dredging problem inside the basin can be minimized by careful control of gate design and operation (e.g. retractable high and low level sections) so that selective impounding of clearer surface water and near-bed flushing at low tidal levels can be exercised. Where estuary approach depths are adequate and 24-hour use of the basin is provided through locks, then care should be taken to ensure that the actual locking operations are conducted in such a way as to optimize sediment exchange between estuary, lock and basin. It is sometimes possible by these means to encourage deposition to occur in those areas where any necessary maintenance dredging is least likely to disrupt port or marina activities.

The problems discussed so far in this section can apply in some degree to both dock basins and yachting marinas. Chapter 31 considers marinas in general, but it is appropriate to mention here one design aspect – boundary permeability – of marinas dredged in the muddy foreshores of estuaries. This can be a problem to the conscientious designer concerned not only with the minimization of siltation in the marina itself, but also with the minimization of any impact of the works on adjacent frontages.

It is difficult, if not impossible, to satisfy both these criteria at once and the final design is usually a compromise. This is illustrated by a study in which measurements of suspended sediment concentration and current velocity throughout a tide were used in calculations to indicate how different boundary restrictions to flood and ebb flow through a dredged marina would affect siltation (Hydraulics Research, 1987).

Of the conditions examined both then and later (Kendrick, 1989), the preferred boundary options for minimizing marina siltation were, in descending order:

1. Fully impermeable;
2. Floating breakwater;
3. Open-pile;
4. Wave-screens leaving a 1-metre gap above the bed.

From the standpoint of siltation on the adjacent frontages the order of preference was:

1. Floating breakwater;
2. Open-pile;
3. Wave-screens leaving 1 m gap above the bed;
4. Fully impermeable.

If hydraulic considerations were of paramount importance, then in this case the preferred option would be the floating breakwater. In practice, other considerations frequently carry more weight.

43.6 TIDE-CONTROL STRUCTURES

Man has exercised some degree of control over the tide since early times. The extent of control may quite simply be limited to a bunded pound with tidal flaps for trapping fish on the flood tide, or it may involve the construction of machinery for harnessing tidal energy to grind cereals or saw wood. Long before the use of electricity, hydraulic power from water-wheels installed in the spans of old London Bridge provided the public water supply to the city. Sometimes saline water is excluded from upper reaches by the construction of a weir which imposes an artificial tidal limit to the estuary. Where it is necessary to navigate upriver and depths are inadequate at some low tidal levels, movable gates, to be operated in conjunction with a half-tide lock, may be installed across the channel to impound water upstream during the lower half of the tide. During these periods, navigation may be maintained through the lock.

Prominent among the larger and more complex tide-control works are the barriers of the Thames (UK) and the Oosterscheldt (The Netherlands), and the barrages of La Rance (France) and Annapolis (Canada). Chapter 44 discusses the design and construction of barriers and barrages and mentions some of the techniques employed in assessing the impact of such works on the environment.

The long-term effects of a barrier or barrage on an estuary must obviously be of primary concern. In this respect, the scale of the impact is not necessarily a function of the scale of the civil engineering works, for a flood barrier with movable gates closed only in

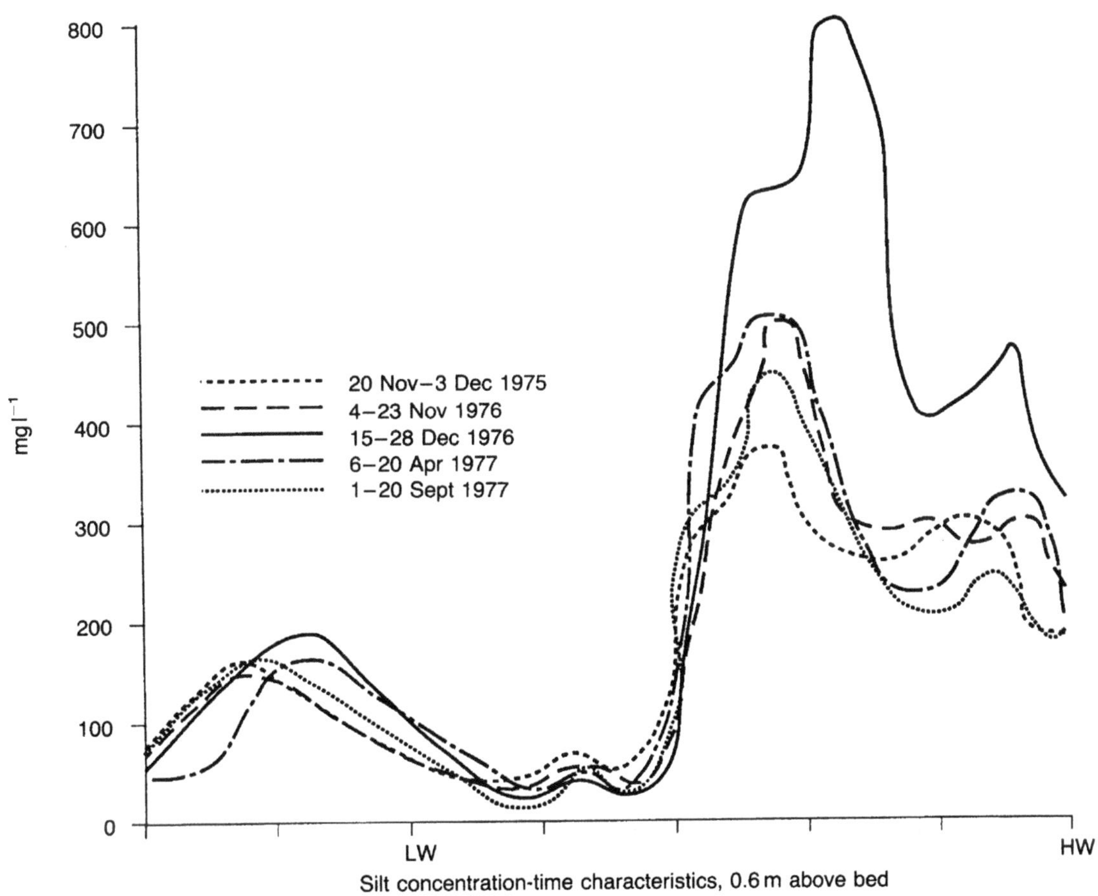

Figure 43.13 Monitoring civil engineering works.

the event of a storm surge need have no lasting effect on estuary regime. By contrast, a power-generating barrage will, by halving the tidal range above the structure, create a completely new regime which has far-reaching implications for surrounding drainage, pollutant dispersion, flora and fauna.

The short-term effects can also be important both in the case of barriers and barrages, especially where a particular construction stage produces a large change in the balance of sediment erosion, transport and deposition. Even where pre-construction flow conditions can be re-established following completion of the works, costly remedial measures may be required to restore former depths (see 43.5.2).

Solution of the practical problems which arise in the planning, implementation and subsequent operation of tide-control structures of the scale and prestige of most barrier and barrage schemes requires the application of the best analytical and predictive techniques currently available. In section 43.2.2 above, mention was made of the comprehensive field, laboratory, physical and mathematical modelling studies that were carried out in the Thames Barrier investigation. Even so, because the barrier was sited across a highly developed stretch of the estuary where adjacent banks already accommodated enclosed docks, riverside quays, jetties and various types of industrial property, environmental assessments did not have to address the detailed ecological problems that would have been posed by a barrage. Where existing activities may have to be modified considerably, or even cease altogether if the new conditions are unsuitable, then decisions must be based on sound information and relevant parameters monitored regularly.

Monitoring assumes special importance where there are claims that the presence of a barrier or barrage has caused harmful effects and evidence is required to confirm or refute the claims. One of the most frequent complaints is of siltation.

The analysis of bed levels, dredging records and the continuous measurement of suspended silt concentrations recorded at stations along the estuary before, during and after Thames Barrier construction demonstrated that it is generally possible to distinguish between those changes in silt transport and deposition that are attributable to variations in tidal range and river flow and those that arise from other causes (Hydraulics Research, 1974, 1978, 1980, 1986). Silt concentration-time characteristics during the flood of 6.1 m tides are shown in Figure 43.13 for a selection of 2/3 week recording periods between 1975 and 1977. The measurements were made at a height above the bed of 0.6 m at a station on the south side of the Thames estuary. The consistency of the position of the peaks of concentration with respect to the half-tide period is a typical feature of silt monitoring records from all stations on the Thames, the peak concentration amplitudes depending on the availability of silt in the pick-up areas. In this example, the centroid of the silt source, as determined from tidal current-speed, lay about 2 km to seaward. The marked increase in peak concentrations was a direct consequence of the disturbance of bed sediments due to new dredging operations begun in the source area on the south side of the estuary in mid-November 1976. The high values for December 1976 are particularly prominent.

The investigation of numerous allegations, sometimes quite unfounded, that increases in upriver deposition during this period were due to the construction of the Thames Barrier was assisted by this technique, employing continuous records obtained at monitoring stations distributed along the estuary.

REFERENCES

Bray, R.N. (1989) Maintenance dredging: where do we go from here?. *The Dock and Harbour Authority*, July/Aug.

Davies, C.M., Brooks, M. and Shaw, J. (1978) A geological and sedimentological investigation of Barry Harbour. Report to Vale of Glamorgan Borough Council.

Delaney, N.K. and Sorensen, N.E. (1953) Low speed drag of cylinders of various shapes. *NACA Tech Note 3038*.

Dunham, K.C. (1972) *The evidence of subsidence of south-east England, Philosophical Transactions of the Royal Society*, London, **272**.

Gameson, A.L.H. and Robertson, K.G. (1955) *Journal of Applied Chemistry*, **5**.

Gilbert, S.K. and Horner, R.W. (1984) *The Thames Barrier*, Thomas Telford Ltd, London.

Hoerner, S.F. (1958) *Fluid dynamic drag. Practical information on aerodynamic drag and hydrodynamic resistance*. Midland Park, NJ. 2nd edition, 1959.

Hydraulics Research (1970) Kuala Baram, Sarawak, Studies of bar deepening. Hydraulics Research, Wallingford, *Rep. EX 496*.

Hydraulics Research (1973) Kuala Baram, Sarawak, Effects of training walls on adjacent beaches. Hydraulics Research, Wallingford, *Rep. EX 634*.

Hydraulics Research (1974) Thames flood prevention investigation, Continuous silt monitoring 1970–1974. Hydraulics Research, Wallingford, *Rep. EX 671*.

Hydraulics Research (1977) Gambia Barrage study, Effect of barrage on tidal regime downstream. Hydraulics Research, Wallingford, *Rep. EX 795*.

Hydraulics Research (1978) Suspended solids monitoring in Long Reach, Thames estuary, September 1975–January 1978. Hydraulics Research, Wallingford, *Rep. EX 830*.

Hydraulics Research (1980) Thames flood prevention investigation, analysis of suspended solids regime prior to barrier construction. Hydraulics Research, Wallingford, *Rep. EX 934*.

Hydraulics Research (1984) Thames water abstraction, Appraisal of possible implications. Hydraulics Research, Wallingford, *Rep EX 1218*.

Hydraulics Research (1985a) Changes in the tidal regime of the River Mersey, Report of studies made in 1984–85. Hydraulics Research, Wallingford, *Rep. SR 43*.

Hydraulics Research (1985b) Teddington flow proposal, The effect on turbidity and siltation in the tidal Thames. Hydraulics Research, Wallingford, *Rep. EX 1350*.

Hydraulics Research (1986a) Thames flood prevention investigation,

continuous silt monitoring during Thames Barrier construction. Hydraulics Research, Wallingford, *Rep. EX 1279*.
Hydraulics Research (1986b) Ince and Stanlow Banks reclamation scheme, summary report on hydraulic studies. Hydraulics Research, Wallingford, *Rep. EX 1485*.
Hydraulics Research (1987) Greenhithe development, study of the hydraulic implications of a proposed marina, Hydraulics Research, Wallingford, *Rep. EX 1589*.
Inglis, C.C. and Allen, F.H. (1957) The regimen of the Thames estuary as affected by currents, salinities and river flow, *Proceedings of the Institution of Civil Engineers*, **7**.
Kendrick, M.P. (1972) Siltation problems in relation to the Thames Barrier, *Philosophical Transactions of the Royal Society, London*, **272**.
Kendrick, M.P. (1984) Impact of engineering structures on tidal flow and sediment distribution in the Thames. *Quarterly Journal of Engineering Geology*, **17**.
Kendrick, M.P., Derbyshire, B.V. and Stevenson, T.A. (1985) Harbour siltation. Physical model and computational studies to establish present cause and predict post-development siltation rates. International Conference on numerical and hydraulic modelling of Ports and Harbours, BHRA, Bedford, UK.
Kendrick, M.P. (1989) Designing marinas for efficient maintenance, *Marina Management International*, **2** (1).
Mair, W.A. and Maull, D.J. (1971) Aerodynamic behaviour of bodies in the wakes of other bodies, *Proceedings of the Royal Society*, London, **A269**.
Permanent International Association of Navigation Congresses (1989) Economic methods of channel maintenance, *Report of Working Group 14 of Permanent Tech. Ctee 11*, PIANC, Brussels.
Pilsbury, G.B. (1956) Tidal hydraulics, *Corps of Engineers*, US Army.
Price, W.A. and Kendrick, M.P. (1963) Field and model investigation into the reasons for siltation in the Mersey estuary. *Proceedings of the Institution of Civil Engineers*, **24**.
Simmons, H.B. (1965) *Committee on tidal hydraulics*, Corps of Engrs., US Army, Tech. Bull. No 8.
Wood, L.B. (1982) *The restoration of the tidal Thames*, Adam Hilger Ltd, Bristol.

BARRIERS AND BARRAGES 44

N.P. Reilly

This chapter follows naturally from Chapter 43 which dealt with many of the problems that are associated with the development of an estuary. Barriers and barrages are built for flood control, to improve amenity, to control saline intrusion and more recently to harness tidal energy. Of the many works built in rivers and estuaries they are likely to have the most fundamental effects on the hydraulic regime. Consequently before they are built they require very careful study (see also Chapters 38, 42 and 43).

CONTENTS

44.1 INTRODUCTION AND DEFINITIONS

A barrage may be defined as a low dam designed to control the flow and level of a river or estuary. Flood discharge is normally catered for by incorporating gated passages in the structure and many barrages consist substantially of gates. When the gates are fully open for the passage of a large flood, the effect of the barrage on upstream water levels may be minimal. Among the earliest barrages were the large structures constructed in the late nineteenth century on the River Nile in Egypt (Stephens, 1903; Vaughan Lee, 1941).

In modern times numerous barrages have been constructed, and many more planned, in estuarine or other marine environments for a variety of purposes. In areas having a large tidal range there is the potential to harness this to generate electricity and tidal energy barrages have been constructed in France, Russia, Canada and China. Many more are under investigation.

In some places the demands of industry and agriculture for water have so depleted the flow of rivers that saline penetration has advanced up the estuary to affect water intakes. A solution to this problem is to construct a barrage at a suitable location to prevent saline intrusion.

A barrage may be constructed purely for amenity purposes, where a tidal river flows through an urban area, to prevent exposure of unsightly mud banks at low tide or to facilitate leisure activities. In this case a fully gated structure may be adopted or, alternatively, a simple half-tide weir may suffice.

A barrier is a relatively modern concept and describes a gated structure which is normally open so that flow is unobstructed. The sole purpose of a barrier is to protect an area from being flooded by high downstream levels. Such a need for protection can arise in an estuary when high astronomical tides combine with surges caused by exceptional meteorological conditions. It can also arise when a main river floods and causes a significant backwater effect, and therefore flooding, in a tributary. A barrier across the tributary is thus an alternative to a programme of bank raising and is resorted to on grounds of either cost or amenity. Closure of a barrier is an exceptional event and, in an estuarial location, is carried out on the basis of a weather and tide forecast. Closure would generally be over a single tide and during the period of closure the fluvial flow is impounded. If there is insufficient storage volume upstream to absorb the fluvial flow during the period of closure, pumps may have to be incorporated to evacuate flow across the closed barrier.

Coastal, Estuarial and Harbour Engineer's Reference Book. Edited by M.B. Abbott and W.A. Price. Published in 1993 by Chapman & Hall, 2–6 Boundary Row, London SE1 8HN. ISBN 0 419 15430 2.

44.2 FLOOD PROTECTION AND WATER LEVELS

The design of any structure must take account of extreme water levels including the effects of waves. Generally for a structure close to the sea it will be sea-level which causes the highest water levels, but with fluvial floods often playing a significant role. As far as sea-levels are concerned, high astronomical tides in combination with surges can produce a catastrophic event. Surges arise from atmospheric conditions and are caused by two factors. A reduction in barometric pressure results in a direct elevation of the sea-level at the rate of 10 mm for 1 mbar reduction in pressure. A much greater effect can arise as a result of wind stress on the water surface and the elevation or set-up of sea-level can be several metres. (Suthons, 1963; Ackers and Ruxton, 1974).

Superimposed on these periodic and occasional variations there is evidence of a continuous secular rise in sea-level (Rossiter, 1972) of the order of 1.5 mm/year over the period 1890–1960. More recently it has come to be suspected that the increase in carbon dioxide and other gases in the atmosphere is giving rise to global warming (the so-called greenhouse effect) and a consequential gradual rise in sea level due to thermal expansion of water and melting of ice caps.

The Intergovernmental Panel on Climate Change (IPCC, 1990) has predicted a sea-level rise of 80 to 290 mm by 2030 and 210 to 710 mm by 2070 if no changes are made to the increase of emission of greenhouse gases. These are secular, or world-wide, changes of sea-level. Independently of these, there are local changes associated with tectonic movements. For example there is long-term geological evidence of a general tilting of England and Scotland about a line between Tees and Pembrokeshire with Scotland rising and SE England sinking (Dunham, 1972).

Associated with the greenhouse effect there may be an increased frequency of violent storms leading to an increase in surge levels at any given return period. For the same reason wave heights could increase.

A barrage or barrier will have a design life of not less than 50 years (and an actual life probably in excess of 100 years) and over that period these changes could be highly significant. It is clearly necessary to carry out a detailed statistical analysis of all the latest data and predictions in order to arrive at appropriate design criteria.

As already indicated, fluvial floods must also be taken into account in the design of structures and their interaction with the seaward influences is a dynamic problem requiring appropriate analysis. The nature of the interaction depends on the form and volume of the estuary, the nature of the barrier or barrage structure, and its proposed operating rules. This will normally be analysed by means of one- or two-dimensional mathematical models. The river flood magnitudes for specified return periods will normally be computed by statistical analysis of historical records using extreme value theory (Ven-te-chow, 1964).

For rivers in the UK where there are no records (i.e. smaller rivers) the Flood Studies Report (NERC, 1975) provides a method of analysis.

In general, except perhaps for small catchments, it may be assumed that the fluvial and marine events are statistically unconnected for the purposes of assessing the probability of combined events. Typically, flood defence structures in the UK are designed for a 100-year return period event where the consequences of flooding are not a threat to life. A usual standard for the design of barrier and barrage structures would be to consider a 100-year fluvial event with an average annual surge and vice versa. A check would be made of the consequences of combining the two 100-year events. However, each case would have to be considered on its merits taking into account the consequences of events of greater magnitude. In any case allowance should be made for future changes taking account of the best available predictions of long term changes of climate and sea level.

44.3 TIDAL ENERGY

Serious interest in large-scale extraction of energy from the tides dates from the start of the present century when proposals were put forward for schemes in the Bay of Fundy (Canada), the Gulf of St Malo in France and the Severn Estuary in the UK.

Figure 44.1 illustrates the method of operation of the simplest means of extracting energy from the tides, the principle of ebb generation. A barrage constructed across an estuary or inlet contains a series of turbine generators and gated sluices. During the period of the incoming tide the sluices are open and allow the upstream basin to fill. Shortly after high water the sluices are closed and as the tide ebbs a water-level difference or head develops across the barrage. When this has reached a pre-determined level the turbines are started and the basin is drawn down while electrical energy is generated. Soon after low water the head reduces and the turbines are closed down. When the levels equalize the sluices are opened and the cycle is repeated. It will be realized that a serious disadvantage of this operating cycle, in comparison with other forms of generation, is that the power supply is highly discontinuous and periods of generation do not even occur at the same time each day. Early studies of the favoured schemes

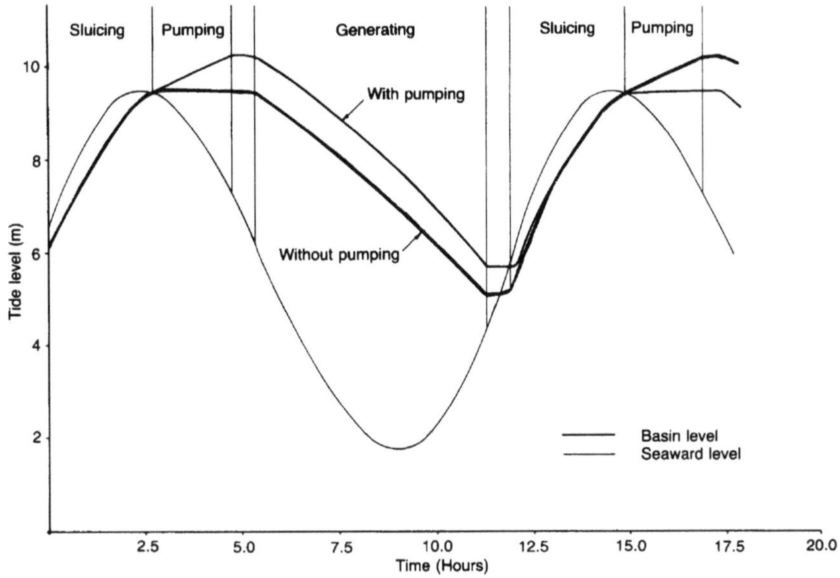

Figure 44.1 Principle of ebb-generation. (——), basin level; (——), seaward level.

devoted a considerable amount of effort to devising ingenious multi-basin arrangements incorporating facilities for pumping to provide a near uniform supply of power throughout the day. Recent studies have tended to favour the simple ebb generation principle but with the addition of a pumping capability to increase energy yield without giving any re-timing of output. This is achieved by reversing the turbine generators to act as motor driven pumps. Pumping is carried out just after high water against a low head and the increased energy capture arises because the pumped water is effectively used to generate at a significantly higher head.

The potential installed capacity and annual energy output of a tidal energy project are clearly a function of basin area and tidal range i.e. tidal volume. Baker (1986) has carried out regression analyses on these basic parameters abstracted from studies on 24 schemes world-wide and the resulting relationships form a useful guide for preliminary appraisal of a site.

44.4 DESIGN OF BARRAGES AND BARRIERS

The design and construction of barrages and barriers are closely inter-linked and it is not possible to contemplate the design without at the same time considering the method of construction. The design is also highly site and function specific and this, combined with the increasing development of new construction techniques, causes the optimum design to change with time. As a result it is almost the case that no two barrages or barriers are alike. Certainly there is a wide variety of designs which are successfully operating around the world. The following sections describe various approaches to the problems by reference to actual examples that have been constructed.

Water control gates dominate the design of most barriers and barrages and the civil works are first and foremost a support for the gates and their operating gear. It is appropriate therefore to consider first the range of gate options which currently exist.

44.5 GATES

Gates are (almost invariably) steel structures designed to obstruct (and permit) flow in an opening. When closed to obstruct the flow, large hydrostatic forces are developed on the gate which is designed to transmit these safely to the structure and hence to the support foundation. In order to control flow, the gate has to be movable and this is achieved by means of gate hoists which are designed to apply adequate forces to move the gate in the various design conditions.

Over the years a range of different gate types has been developed, each with its own characteristics and advantages in particular situations. Large gates are always designed specifically for the application and requirements of the site and so vary considerably from project to project even for the same type of gate. The main characteristics of the gate types applicable to barrages and barriers are described in the following sections. Generally the same gate types apply to both types of structure but differences in duty and detailing arise from the fact that in a barrage the gate is normally in the water and loaded while, in barriers, gates are generally out of the water.

Vertical lift gates

The oldest and simplest type of hydraulic gate is the vertical lift gate. Generally such gates span horizontally between end frames which engage in a vertical slot or gate guide in the supporting structure. In a barrage application the gate will be required to open or close in unbalanced conditions, i.e. with a hydraulic load on it, and therefore it is necessary to support it on rollers or wheels to minimize frictional forces.

A vertical lift gate normally requires an overhead gantry for the hoist gear and in the case of a barrier having to accommodate navigation, the gantry may have to be very high and may be visually unacceptable. The height of the gantry and the wind loading on the gate and structure can be reduced by adopting a turnover gate as adopted for the Hull Barrier (Figure 44.2) (Fleming *et al.*, 1980).

When used in a barrage application the vertical lift gate is normally an undershot gate, i.e. water flows under the gate. Careful design of the shape of the underside of the gate is necessary to avoid an excessive downward pull and vibration. An undershot gate is not generally suitable for passing floating trash and a variant which caters for this requirement is the double leaf gate.

(a)

Figure 44.2 River Hull tidal surge barrier (a) open; (b) closed (courtesy of Kvaerner Boring Ltd).

(b)

Figure 44.2 *Continued*

Modern gates are usually plated structures but special problems demand special solutions and the innovative design developed for the Oosterscheldt barrier in the Netherlands is a case in point (see Figure 44.8). The very long span, 39.5 m clear opening, and the nature of the duty dictated that a tubular trussed structure was the best solution. A novel form of corrugated skinplate was also developed to give maximum stiffness with minimum mass. No wheels or rollers were provided for these gates.

Radial gates

The radial gate (or Tainter gate after its American inventor) is perhaps the most widely used type of gate in hydraulic control. As illustrated in Figure 44.3, the skin plate forms part of a cylinder having a horizontal axis. Radial arms carry the hydraulic load to a pivot or trunnion located at or very close to the centre of curvature of the skin plate. As a result hydraulic forces pass directly through the pivot and (except for seal friction) they do not affect the hoisting forces. It is normally an undershot gate and even quite large gates can be used for fine flow control. A superstructure is not required and this type of gate has a considerable visual advantage over the vertical lift gate.

Flap gates

Flap gates (also called tilting gates) are steel gates hinged at either top or bottom. The bottom hinged flap gate is a useful device as a crest gate for accurate flow control, and being an overflow gate, has the capability to pass floating trash. As with any overflow gate it is necessary to ensure that the underside of the nappe is adequately ventilated otherwise unstable operation results. It is often advantageous to operate even quite long gates from one end only. In this case the gate must incorporate a torque tube. The characteristic shape of this type gives it its name of fish belly gate (See Figure 44.4). A fish belly

Figure 44.3 Radial gates at Victoria Dam, Sri Lanka (courtesy of Kvaerner Boring Ltd).

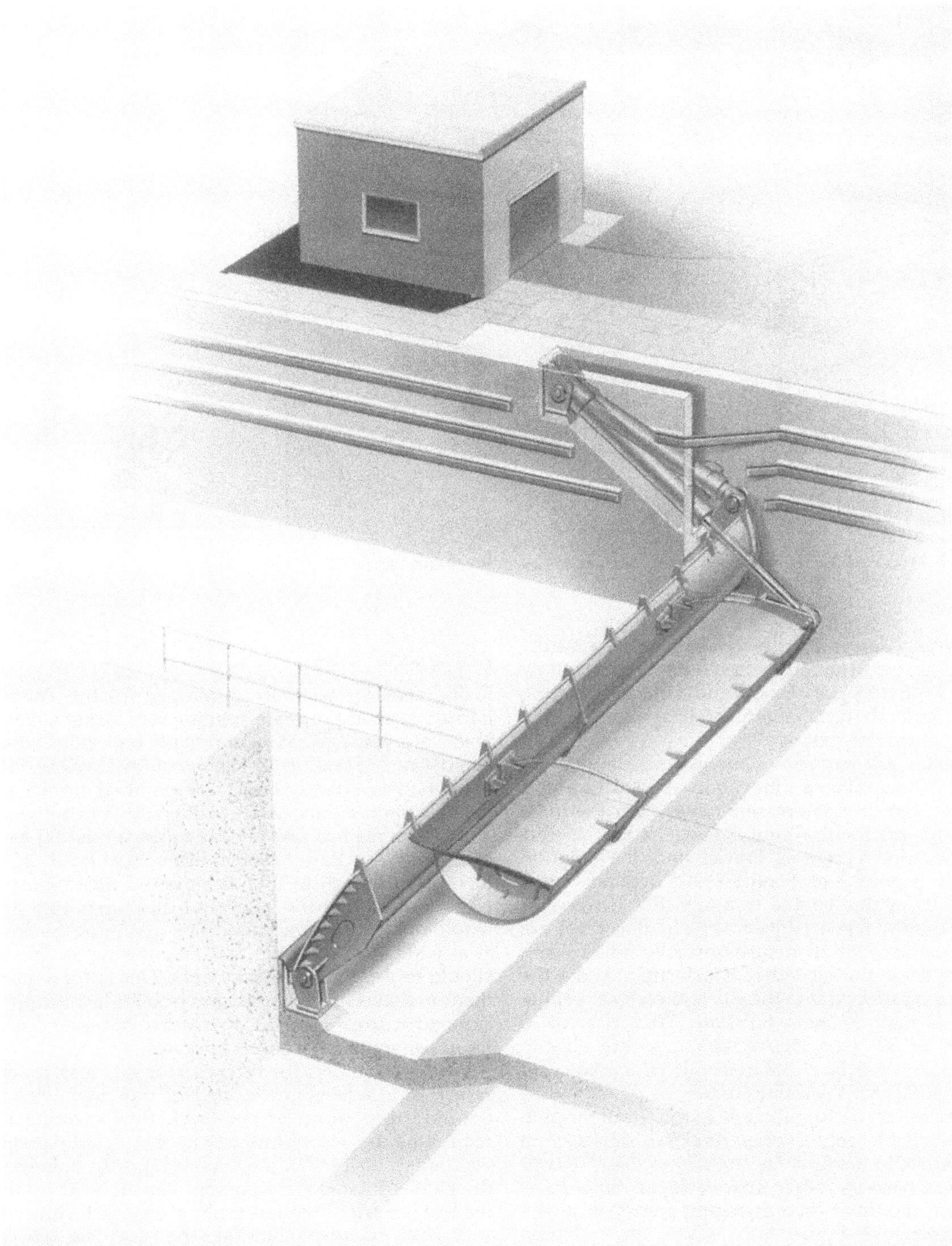

Figure 44.4 Fish belly flap gate (courtesy of Kvaerner Boring Ltd).

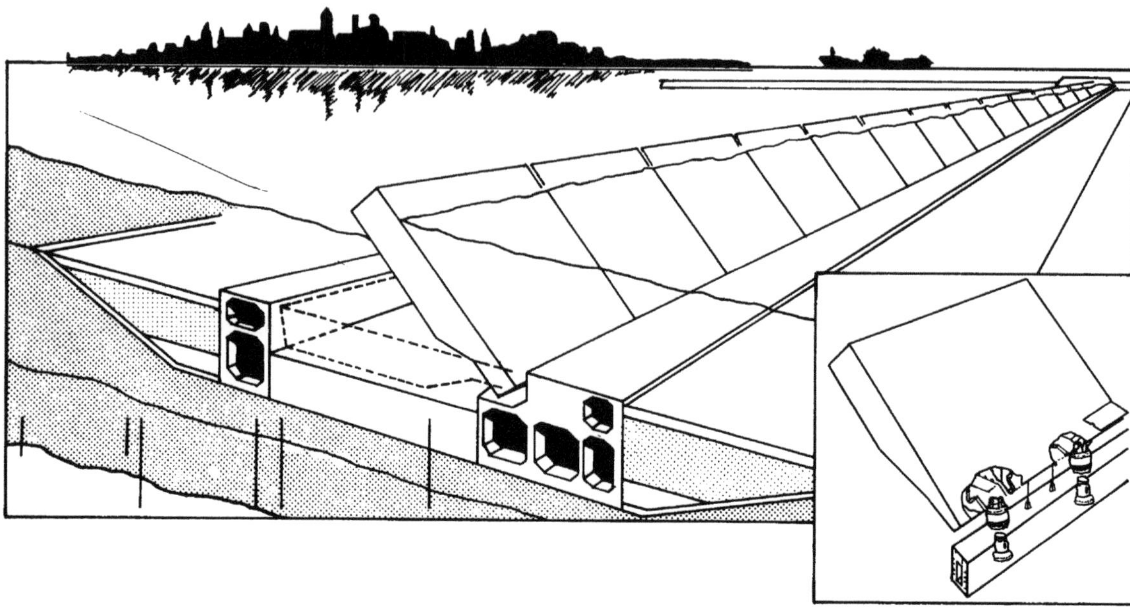

Figure 44.5 Proposed flood gates for Venice (courtesy of Lewin Fryer and Partners).

gate operated hydraulically requires no superstructure and this feature, together with accurate flow control, ability to pass trash, and fail-safe opening makes it particularly suitable as a barrage crest gate in visually sensitive locations.

Flap gates are usually of modest size but as an example of imaginative innovation applied to a particular problem it is interesting to consider the huge gates proposed for the Venice flood barrier. In this case navigation access is crucial and the solution proposed (Lewin and Scotti, 1990) depends on a series of flap gates hinged to a line of concrete sills sunk across the lagoon entrances. The flap gates are compartmented box structures normally full of water and lying flat in the sill. When it is desired to raise the barrier air is pumped into the gates, thereby expelling water and making them buoyant. They rise up to an angle of 50° (see Figure 44.5) and are able to withstand a water level difference of 1.2 m. No other hoisting equipment is required.

Flap gates can be top hinged. In this configuration they are suitable only as open or closed devices and are not suitable for flow control. Consequently they find use in barriers rather than barrages. Such gates have been mounted on a swinging gantry to give a totally unrestricted waterway when open. (Figure 44.6).

Rising sector gates

Rising sector gates were developed for the Thames Barrier as an appropriate solution for closing a navigable waterway against exceptionally high tides which could cause flooding in the heart of London. The gates had to span openings determined from navigation requirements as 60 m wide with a mean water depth of some 9 m and a mean springs tidal range of over 6 m. They have to retain differential heads of up to 9.9 m during surge tide conditions. Furthermore, it was required that the gates and their operating gear would not obstruct the passage of ocean-going vessels at any stage of the tide, and that the whole barrier should be aesthetically acceptable. This latter requirement militates against any form of high-level support superstructure, such as would be necessary with more conventional vertical lift gates.

When not in use, the rising sector gate leaf lies in a recess in a sill below the bed of the river with its flat side uppermost at bed-level, thus meeting the requirements for a completely unobstructed passageway for water flow and for navigation. In operation, the gate is rotated through approximately 90° so that the leaf is upright with its trailing edge still within the sill recess, effectively blocking the flow. The gate can be rotated further so that the gate leaf is in an inverted position above water level for maintenance.

The gate is capable of carrying differential loads in

Figure 44.6 Gantry mounted flap gates (courtesy of Peter Fraenkel BMT Ltd).

either the upriver or the downriver direction. It can also be positioned to allow some spill to pass over its upper edge or under its lower edge. The design is capable of development to cater for very large spans and differential heads.

Sealing of these gates presents problems: the Thames Barrier gates are not sealed because the leakage is not significant in the context of excluding a high surge tide from a large river basin.

44.6 EXAMPLES OF EXISTING BARRIERS AND BARRAGES

44.6.1 FLOOD PROTECTION BARRIERS

Increasing development in low lying coastal areas, coupled in some cases with settlement due to groundwater abstraction, has resulted in a need in many places for flood protection against surge events.

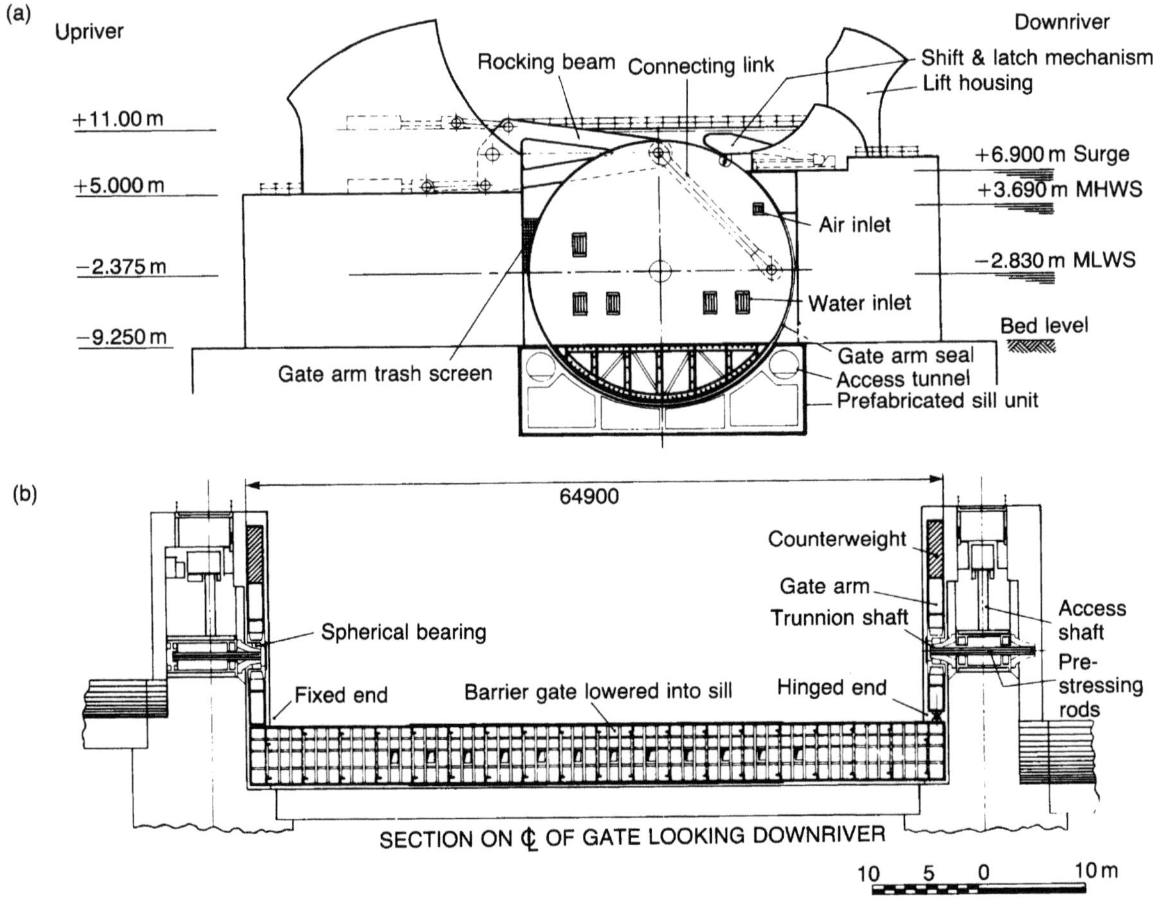

Figure 44.7 Thames Barrier (a) section through 61.0 m rising sector gate; (b) typical arrangement of rising sector gates (courtesy of Rendel Palmer and Tritton).

The Thames Barrier

Just such a need for the protection of low lying areas of London and the land bordering the Thames estuary began to be perceived in the 1960s and numerous studies were undertaken. After much study it was concluded that bank raising for the downstream estuary would be satisfactory but that this solution would not be practicable or acceptable in London itself. It was decided therefore that a movable barrier in the vicinity of Woolwich was the best option and various alternative forms of structure were investigated. At that time various docks of the Port of London upstream of the proposed site were still operational and a gated structure that would enable shipping to pass was essential. The normal solution in such circumstances would be some form of vertical lifting gates but the large span and headroom required rendered this solution unattractive. To solve this problem the designers developed a new type of gate, the rising sector gate, specially for this project as described in section 44.5.

Figure 44.7 illustrates the main features of the barrier which is 520 m long and contains 10 gates. Four of these are radial gates (at the sides) and the remainder are rising sector gates. The latter form the navigation openings which are completely unobstructed with unlimited headroom.

River Hull storm surge barrier UK

The Hull Barrier (Fleming *et al.*, 1980) is located at the confluence of the rivers Hull and Humber and protects the City of Kingston upon Hull. The navigation

Figure 44.8 Oosterscheldt Barrier **1**, pier; **2**, quarry stone dam for land abutment construction; **3**, beam supporting operating equipment; **4**, hydraulic cylinders; **5**, capping unit; **6**, upper beam; **7**, gate; **8**, sill beam; **9**, road; **10**, road box **8**, sill beam; **9**, road; **10**, road box girder and machinery for gate operation; **11**, power supply duct; **12**, sand filling of sill beam; **13**, top layer of sill; **14**, core of sill; **15**, sand filling of pier base slab; **16**, sill beam stops/bearings; **17**, upper mattress; **18**, grout filling; **19**, block mattress; **20**, bottom mattress; **21**, compacted sand under the bed of the Eastern Scheldt; **22**, gravel bag (courtesy of Dosbouw).

opening is 30 m wide and this is closed by a 10.6 m high vertical drop gate suspended from towers. A full height navigation clearance was required and in order to limit the overall height of the structure and to minimize wind loadings, a turnover gate was adopted.

Eider storm surge barrier, Germany

The Eider Barrier in Germany illustrates another solution to the problem. Here the structure is highly exposed to the North Sea and radial gates were adopted. The barrier has five openings of 40 m clear span and the gates are supported from tubular prestressed concrete girders which also carry the 9 m wide roadway inside the tube. The need for a highly secure protection was such that two sets of gates are provided, each set hinged on the underside of the tubular girder. Navigation is provided for by a lock in an adjacent structure.

Oosterscheldt Storm Surge Barrier

After the storm surge of 1953 which flooded large areas of The Netherlands claiming 1835 lives it was decided to accelerate the implementation of plans which were already in existence.

The project involved the closure of the main tidal estuaries and inlets in the southwestern part of The Netherlands with the exception of those giving access to the ports of Rotterdam and Antwerp. Not only would this shorten the country's coastline by hundreds of kilometres but, by forcing the saltwater further back towards the sea it would also provide significant improvements to freshwater management in the country.

The various parts of the project were undertaken, one after the other, without delay and in the following order:

1958 Storm surge barrier in the Hollandse Ijssel;
1960 Zandkreek Dam;
1961 Veerse Dam;
1965 Grevelingen Dam;
1970 Volkerak Dam with lock complex;
1971 Haringvliet Dam with discharge sluices;
1972 Brouwers Dam;
1985 Eastern Scheldt Barrier.

The final part of the project was the closure of the Eastern Scheldt by a storm surge barrier. This was a formidable task as this short description will illustrate.

The storm surge barrier, in all 3 km long, was built in three major channels. It has 65 prefabricated concrete piers between which slide 62 steel gates.

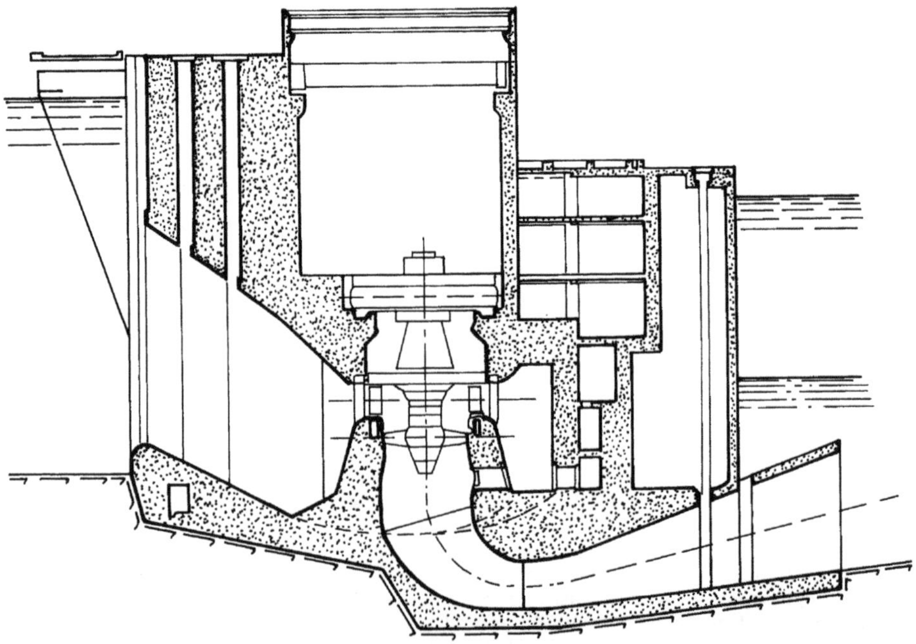

(a)

Figure 44.9 Low-head hydro-electric plant; (a) conventional; (b) incorporating bulb turbine generator; (c) incorporating 'Straflo' rim generator turbine.

(b)

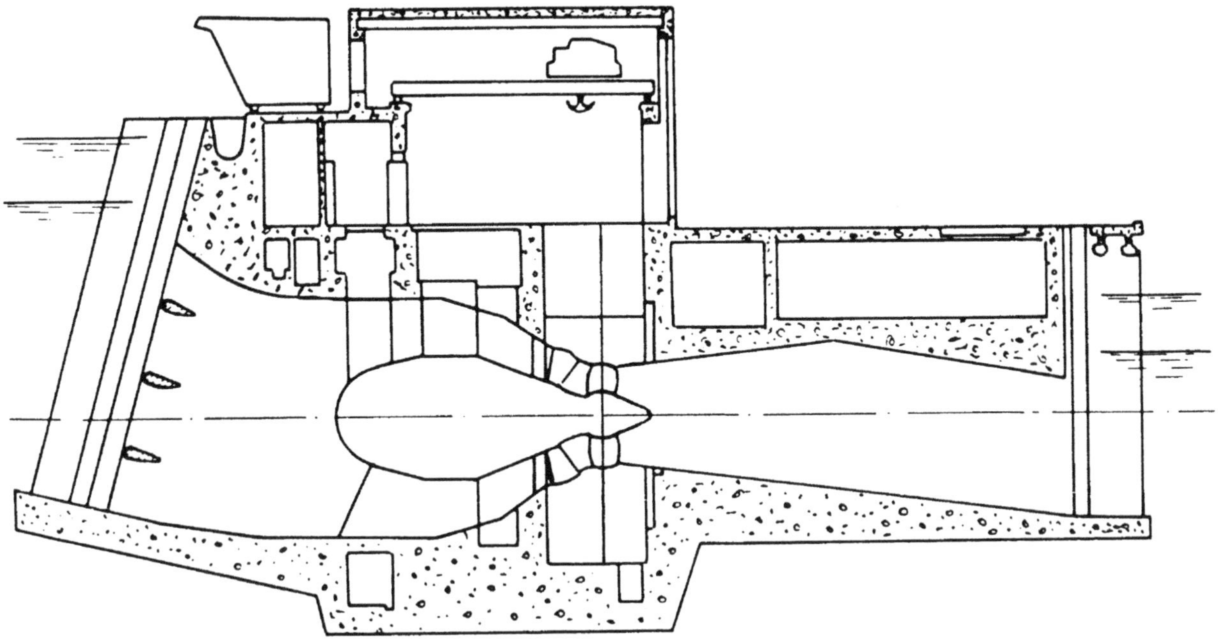

(c)

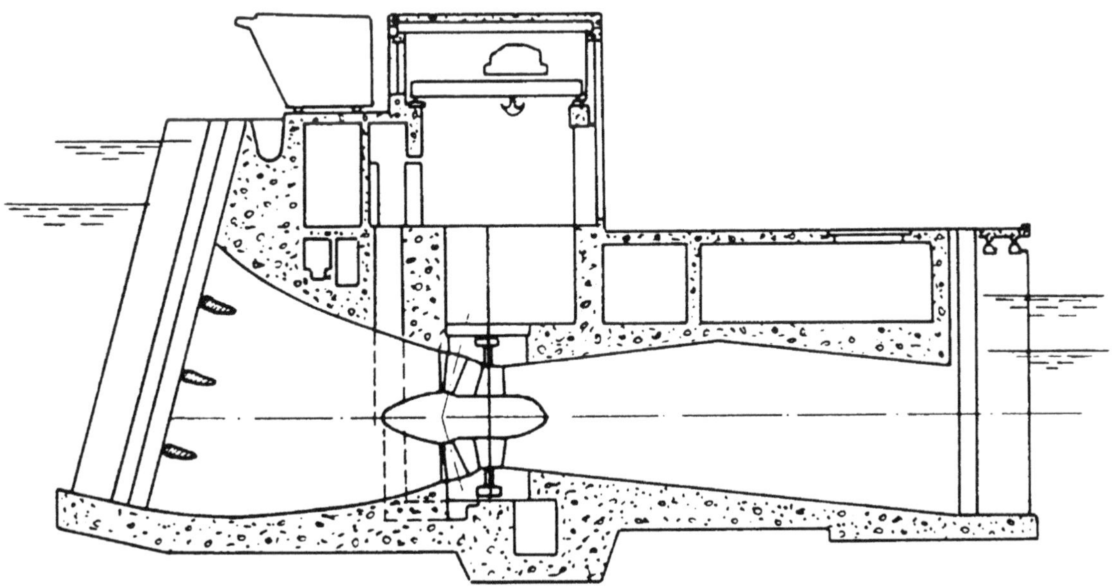

Figure 44.9 *Continued*

Each pier is a massive concrete structure, nearly 40 m high. The construction period for each pier was almost one and a half years. To transport these concrete giants weighing 18 000 t and to position them accurately to within a few centimetres in depths of 35 m and at a centre-to-centre distance of 45 m was a huge engineering task which involved the construction of two special vessels capable of transporting them out of the construction dock and placing them accurately on site.

One of the main reasons for including this short description of this major project is to illustrate that such projects generate new techniques of study, design and construction. (Further reading on the Delta project will be found in Rijkswaterstaat, 1987, 1989, 1992; Huis in't Veld *et al.*, 1987; Hydraulics Aspects of Coastal Structures, 1980; Foundation Aspects of Coastal Structures, 1978.)

44.6.2 TIDAL ENERGY BARRAGES

As stated above, man has been aware of the energy in tides for many centuries and numerous tide driven mills existed in medieval times and before. The exploitation of tidal energy on a larger scale has lagged behind that of conventional hydro-electric development for economic and technical reasons. The former relates to the hitherto cheaper alternative of energy from fossil fuel sources while the latter relates primarily to the design of water turbines. In a conventional large-scale hydro-electric plant the water flow enters the turbine horizontally (See Figure 44.9a), and turns twice through 90° before emerging horizontally from the draft tube into the tailrace downstream. While this is quite acceptable for a plant developing heads measured in tens of metres, it is extremely cumbersome, and therefore expensive, where a head of only a few metres is available.

It was the development of a water turbine with a horizontal water passage combined with prospects of higher primary energy prices which stimulated modern interest in tidal energy. There are numerous variants of turbine-generators with horizontal water passages but the two main contenders for the larger-scale developments are the bulb machine and the rim generator machine (See Figure 44.9b and c).

The following examples of existing and planned tidal developments will serve to illustrate some of the problems and possible solutions to tidal power development.

La Rance

La Rance, near St Malo in France, was the first tidal power development of significant size to be constructed. Construction commenced in 1961 and the first machines were commissioned in 1966. It has an installed capacity of 240 MW in 24 identical bulb turbine generators. It was designed to generate and pump on both ebb and flood but its operation now is generally limited to ebb generation with flood pumping.

Construction of the barrage was carried out in the dry between two lines of cellular cofferdams which almost totally blocked the estuary. This method of construction was expensive and the ecological impact on the estuary of being cut off from the sea for several years was considerable, taking some ten years to reverse. As will be seen later, the development of offshore technology and the ability to construct, float into position and place large structures has opened up new possibilities for construction which are cheaper and less environmentally damaging.

Kislogubsk

The Kislogubsk tidal power station on the Arctic coast of Russia (Bernstein, 1974a) was commissioned in 1968 with an installed capacity of 0.8 MW in two bulb turbine generators. In contrast to the La Rance project, the power station of the Kislogubsk plant was prefabricated offsite as a caisson structure, then floated and towed into place before being sunk into position. The construction site was a temporary 'dry dock' nearby and the single caisson unit containing both turbines and the sluice unit was fully equipped with its turbine generators and gates prior to float out.

Annapolis

The Annapolis tidal power station is a single unit 19 MW plant commissioned in Nova Scotia, Canada in 1984. The machine is a 'Straflo' (rim generator) type and it was built into an existing flood protection barrage as a pilot project to prove the capabilities of the machine. The project operates in an ebb-generation mode and in its early years has operated successfully with very high availability (Rice and Baker, 1987).

44.6.3 SALINITY EXCLUSION BARRAGES

The penetration of salinity up an estuary under tidal influence increases greatly during low fluvial flow conditions. If there is significant abstraction of fresh water from the river for industrial, agricultural or domestic purposes then the impact on drought flows is likely to be dramatic and salinity may penetrate far beyond its natural limit, affecting intakes or endangering groundwater supplies. This can be overcome by construction of a barrage which physically

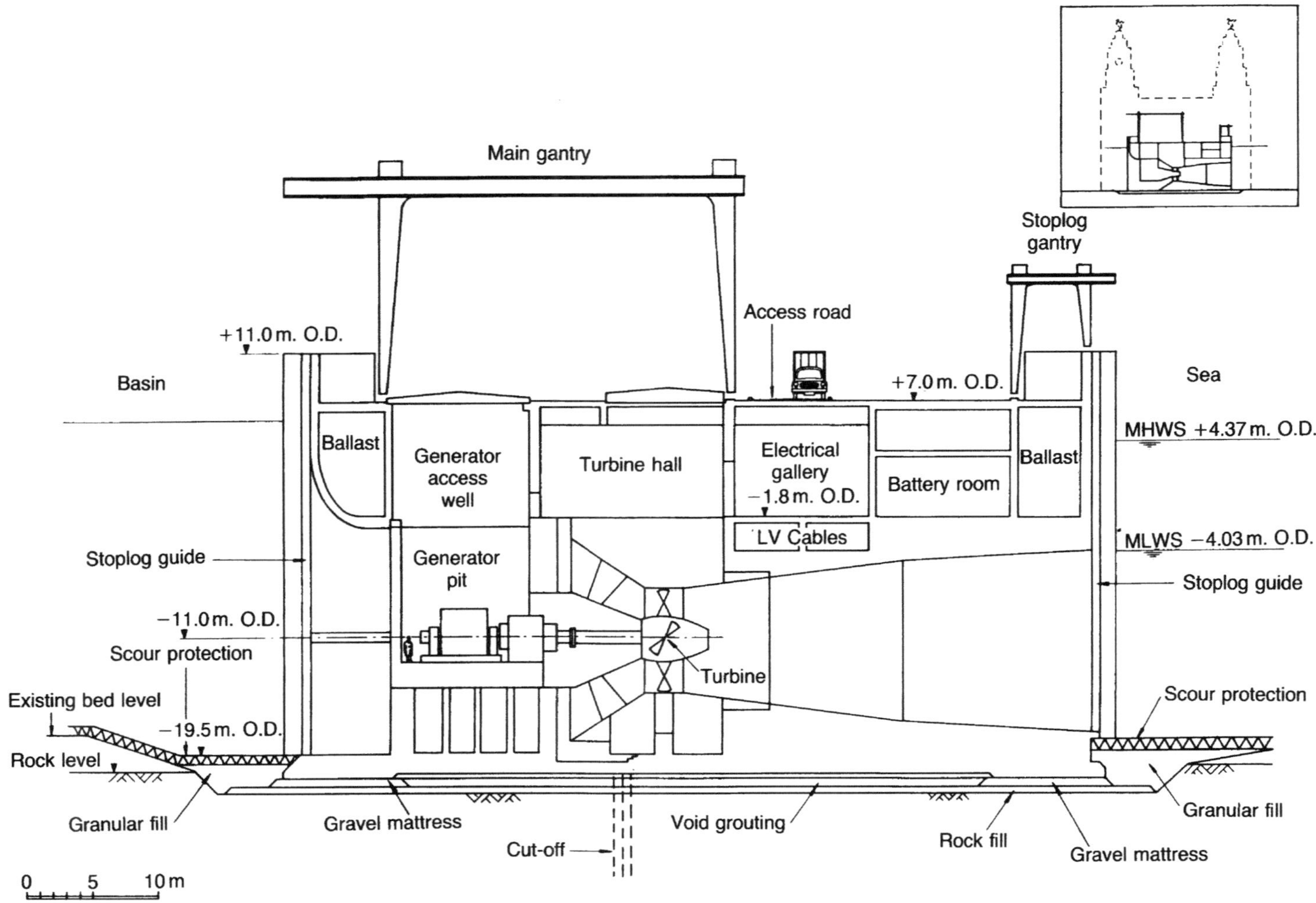

Figure 44.10 Turbine generator caisson for proposed Mersey tidal energy barrage (courtesy Rendel Parkman Ltd).

prevents the tide from penetrating above it. The upstream water level may be raised slightly above its pre-barrage maximum tidal level or, more commonly a lower level will be set in order to minimize the effect on upstream drainage. In the latter case the barrage must accept a reversal of head difference and flow control gates would be operated on every tide.

A substantial salinity exclusion barrage was constructed on the Nakdong Estuary in Korea. A smaller example is the Barmby Tidal Barrage on the River Derwent in Yorkshire in the north of England (Chadwick, 1975).

44.7 CONSTRUCTION

Traditionally barrages and barriers are constructed in the dry behind cofferdams which are often constructed of braced steel sheet piling or of cellular construction using straight webbed sheet piles backfilled with granular material. This type of construction is well established and is amply covered in the construction literature.

The present interest in very large tidal energy barrages (e.g. Bay of Fundy, Severn, Mersey etc.) has focused attention on more efficient methods of construction and it is clear that the use of conventional *in situ* methods in a hostile marine environment will tend to be expensive. Experience at La Rance, referred to earlier, showed that the environmental impact of a major blockage of an estuary is probably unacceptable. These factors, combined with the rapid development by the offshore oil industry of techniques for transporting large gravity structures, has led to a general acceptance that energy barrages will be prefabricated and floated into position. Extensive studies for the three projects mentioned above have shown that this is practicable and caissons fully equipped with turbine generators or sluice gates will be floated into position and sunk onto a prepared bed. Not only does this transfer, and possibly disperse, much construction work but it also allows completed sections of the barrage to be made permeable at an earlier stage, thereby easing the problems of final closure.

Figure 44.10 shows a section of a turbine generator caisson and a sluice caisson as developed for the proposed Mersey Barrage project. Caissons may be of concrete, steel or composite construction. The only tidal energy barrage constructed by this method to date is the small Kislogubsk scheme in Russia (Bernstein, 1974a,b) which was constructed in concrete. Steel caissons have been used for two low head hydro-electric stations in the United States, namely the 70 MW W.T Love station (Ferren–Laloc, 1987) and the 192 MW Sydney A Murry plant (Gertler and O'Connor, 1987).

Caissons are essentially self buoyant structures and are highly suitable for the power station and sluice units of tidal energy scheme. However, it is not necessary for the elements to be totally self-buoyant and the construction of the Oosterscheldt flood barrier, already described, is an example of precast construction using components which were substantially carried to the site.

Prefabrication of units offsite and transporting them to be sunk in position is clearly a feasible method for a range of structure types. Considerable development of this technique for barriers and barrages may be expected.

44.8 GEOLOGY AND FOUNDATIONS

As a part of the feasibility study for a barrage or barrier project it is necessary to have a thorough understanding of the regional and site geology. Knowledge of the regional geology and geomorphology will not only assist in understanding the site geology but will be necessary to assess the environmental impacts of the project. For example it is shown in the next section that influences on groundwater of a barrage scheme may be important. Likewise in certain areas which are prone to flooding, the abstraction of groundwater for water supply not only exacerbates the problem by causing significant settlements but may also gradually render ineffective any measures designed to give protection. Venice and Bangkok are particular examples of this problem. Rivers in alluvial beds are frequently unstable and an inadequate understanding of the river morphology may lead to a control structure being by-passed during a major flood event.

No part of the world can be regarded as totally earthquake free and all important structures should take this into account. Depending on the risk and importance of the structure either a detailed earthquake risk appraisal should be carried out or the structure should be designed to a locally specified norm.

Considering the site itself, there are three main factors which have to be taken into account in the design of a barrage or barrier structure: firstly, the ability of the foundation to support the loads; secondly, the amount of long-term settlement and differential settlements between adjacent components; and finally, the resistance to transport of material by seepage forces.

By their nature estuaries are frequently located in zones of deep fine-grained alluvial materials which can cause problems on all three fronts.

Seepage is only likely to be of concern in completed structures in relation to erosion of fine-grained granular foundation materials. The head differences developed in typical estuary barrage schemes are not great compared with those in dams and it is not difficult to control the seepage gradient to a safe level. Typical means of controlling the seepage gradients are by means of upstream blankets or cut-off walls of sheet piling or diaphragm walling either of which can be supplemented by drainage. (Casagrande, 1961). Grouting may also be considered but is more effective at reducing leakage flows than seepage gradients unless combined with drainage.

Where barrages are constructed *in situ* within cofferdams, seepage during construction may be a much greater problem because heads and gradients are much greater. In this case grouting of the foundation may well be considered to reduce inflows.

44.9 ENVIRONMENTAL ISSUES

44.9.1 INTRODUCTION

Tidal estuaries are highly complex and productive ecological systems and the consequences of interfering with an estuary may be far reaching. Before major works are undertaken it is therefore necessary to carry out detailed studies to understand the existing system and predict the consequences of the proposed works. The nature of the changes arising from a development depend on the sensitivity of the system and the type of the development. A flood barrier, for example, is open most of the time and the direct effect on the environment is minimal although indirect effects may be substantial. Energy or salinity exclusion barrages on the other hand have dramatic, although different, impacts on the environment. A fuller discussion on the nature of the environmental studies necessary to evaluate the impact of a project is given in Chapter 43 but some of the major issues which relate specifically to barrages are identified below.

44.9.2 SILTATION

One of the most important environmental (and engineering) issues in many barrage schemes is undoubtedly the effect on the sediment transport, deposition and erosion regime.

In many estuaries the tidal ebb and flow carries a considerable amount of sediment either in suspension or as bed-load. While the original source of this material may be the river flowing into the estuary, the immediate source in typical UK rivers may well be the estuary itself and the near sea-bed zone. This is because the rivers are relatively small with low silt content while tides and tidal currents are strong. (The opposite would apply in the case of large tropical rivers flowing into low tidal, oceanic regimes).

Clearly any alteration to the flow regime such as would arise with a barrage has the potential to affect the sedimentation processes. O'Brien (1931) has shown that in sandy bed estuaries there is a relation between the cross-sectional area of the entrance and the volumetric ebb and flow of the tide (cubature). A reduction in cubature, such as arises with an energy barrage, could therefore be expected to result in a gradual accumulation of sediment provided sufficient mobile material is available. Much is known about the physical processes of sediment transport and, provided adequate field data is available, mathematical models exist which can predict the morphological effects of changes to the flow regime (see also to Chapters 19–21).

44.9.3 WATER QUALITY

The water quality in an estuary may also be greatly affected by a barrage in a number of ways, some adverse and some beneficial. Considering an energy barrage, the reduced cubature will increase the average residence time of pollutants discharged upstream but may also increase the oxygen content. Reduced current velocities upstream will, in general, reduce the turbidity and thereby allow increased sunlight penetration. This, in turn, will increase the rate of mortality of faecal bacteria and increase the rate of algal growth. Algae forms the start of the food chain through invertebrates to birds and an increase in algae productivity may benefit bird life. However, it also increases the probability of excessive growth leading to 'blooms' or 'red tides' which on decay can cause a collapse of oxygen levels or release toxins. Reduced currents could also result in the formation of visible slicks of sewage in the vicinity of outfalls.

Potentially the worst water quality effects arise in the case of tidal penetrating amenity barrages (e.g. half-tide barrages). In conditions of low fluvial flow, the upstream pool can become highly stratified with a stagnant layer of saline water remaining for long periods. Such conditions arise particularly in the summer when temperatures are high and the saline bottom layer becomes anoxic, thereby generating objectionable odours of hydrogen sulphide due to decaying matter. Such situations are best avoided if possible but the problems can be overcome by measures such as low-level outlets combined, in extreme cases, with artificial mixing and an injection of oxygen.

Water quality is a complex question of interacting processes. As with sedimentation, the empirical relationships governing most of the processes are known and it is possible, given sufficient data, to model the processes and predict an outcome in given circumstances (see also Chapters 16–18).

44.9.4 GROUNDWATER AND DRAINAGE

Any scheme which affects the tidal range at a point in an estuary will alter the mean water level and consequently affect groundwater levels close to the shoreline. Generally the groundwater level will be raised and saline penetration may also increase. Foundations of structures in the area may be affected and basements may become more damp or flooded. Likewise drainage outfalls which were previously tidelocked will experience a greater degree of tide-locking and measures may have to be taken to overcome this.

44.9.5 ECOLOGY

Many aspects of the estuarine ecology will be affected by a barrage scheme. Estuaries are frequently rich in bird life and birds may find their feeding areas diminished or the supply of food affected. Fish will experience a change in water quality and may have to cross the barrage via a fish pass. In an energy barrage there may be the question of fish damage caused by transit through turbines to assess and in an anti-salinity barrage fish will experience the sudden change from fully saline to freshwater in a very short time. Salt marshes which previously experienced regular inundation during equinoctial tides may lose this and over time the flora will adjust accordingly. An environmental impact study will address all these aspects and assess the consequences of each particular change for a given development.

44.9.6 LEGAL ASPECTS

There are numerous legal obstacles to be overcome before a barrage or barrier scheme can be implemented. In the UK an Act of Parliament will be required before any project which interferes with a navigable waterway can be constructed. European Community legislation is also relevant here and EC Directive 85/337/EEC lays down the requirements for environmental statements which must be prepared for specified types of project.

Other countries have equivalent legislation and international lending agencies also lay down requirements relating to environmental impact assessments which must be carried out prior to approval of a project. The sanctioning legislation or loan agreement may also lay down requirements for post construction monitoring to compare the outcome with predictions.

REFERENCES

Ackers, P. and Ruxton, T.D. (1974) Extreme levels arising from meteorological surges. Proceedings of the 14th Coastal Engineering Conference, Copenhagen Vol 1 ASCE pp. 69–86.

Baker, A.C. (1986) The development of functions relating cost and performance of tidal power schemes and their application to small scale sites. Symposium on Tidal Power, Institution of Civil Engineers, London. October 1986.

Baker, A.C. (1987) Tidal power. *IEE Proceedings Part A* **134** (5).

Bernstein, L.B. (1974a) Kislogubsk: a small station generating great expectations. *Water Power and Dam Construction*, May.

Bernstein, L.B. (1974b) Russian tidal power station is precast offsite, floated into place. *Civil Engineering. ASCE* April.

Casagrande, A. (1961) Control of seepage through foundations and abutments of dams. 1st Rankine lecture, London, 1961. *Geotechnique* **X1** September.

Chadwick, N.R. (1975) The Barmby tidal barrage. *Journal of the Instute of Water Engineers and Scientists*, October.

Delft Hydraulics – Rijkswaterstaat (1989) Storm Surge Barrier Eastern Scheldt. Evaluation of water movement studies for design and construction of the barrier, July.

Dunham, K.C. (1972) The regional setting, in 'Discussion on problems associated with the subsidence of South Eastern England'. *Phil. Trans. R. Soc. London A*, **272**, 81–6.

European Community (1985) Council Directive of 27 June 1985 on the assessment of the effects of certain public and private projects on the environment (85/337/EEC). Official Journal of the European Communities. No L175/40 5/7/85.

Ferrer-Laloe, F.R. (1987) Prefabricated powerplant reduces civil works. *Water Power and Dam Construction Handbook.*

Fleming, J.H., McMillan, P.H. and Williams, B.P. (1980) *The River Hull Tidal Surge Barrier Proc. ICE* August.

Gertler, L. and O'Connor, E. (1987) Sydney A Murray Jr. Hydroelectric Project. *Water Power*, 1987.

Huis in't Veld, J.C. *et al.* (Eds) (1987) *The Closure of Tidal Basins, Closing of Estuaries, Tidal Inlets and Dike Breaches*, Delft University Press, The Netherlands.

Intergovernmental Panel on Climate Change (1990) *Climate change: The IPCC Assessment*, Cambridge University Press (for IPCC).

Lewin, J. and Scotti, A. (1990) The flood prevention scheme of Venice: experimental module. *Journal IWEM*. February.

NERC (1975) *Flood studies report*. Natural Environment Research Council, UK.

O'Brien, M. (1931) Estuary tidal prisms related to entrance area. *Civil Engineering* 1 8 May 1931.

Rice, R.G. and Baker, G.C. (1987) The Annapolis experience. Proceedings of Oceans '87 Conference. The Ocean, an International Workplace. Halifax, Nova Scotia. September 1987.

Rijkswaterstaat (1987) The Storm Surge Barrier in the Eastern Scheldt. For safety and environment. Rijkswaterstaat, The Netherlands.

Rijkswaterstaat (1992) Design Report of the Eastern Scheldt Storm Surge Barrier, A.A. Balkemia, Rotterdam.

Rossiter, J.R. (1972) Sea level observations and their secular variation, in Discussion on problems associated with the subsidence of South Eastern England. *Proceedings of the Royal Society*, London, 1972.

Stephens, G.H. (1903) The Barrage across the Nile at Assiut. *Proceedings of the Institution of Civil Engineers*, **158**.

Suthons, C.T. (1963) Frequency of occurrence of abnormally high sea levels on the east and south coasts of England. *Proceedings of the Institution of Civil Engineers*, **25**, 433–50.

Vaughan-Lee, A.G. (1941) The Mohammed Aly Barrages. Egypt. *Proceedings of the Institution of Civil Engineers*, **15**, 237–79.

Ven-te-chow (1964) *Handbook of Applied Hydrology. A compendium of Water Resources Technology* (Ed. Ven-te-chow), McGraw Hill.

Hydraulic aspects of coastal structures (1980) *Developments of hydraulic engineering related to the design of the Oosterschelde storm surge barrier in The Netherlands*. Delft University Press, The Netherlands.

Foundation aspects of coastal structures. (1978) International Symposium on Soil Mechanics Research and Foundation Design for the Oosterschelde Storm Surge Barrier. Delft, The Netherlands, 9–12 October 1978.

DREDGING AND DREDGERS 45

A.D. Bates

Dredging is a highly specialized subject. The author discusses the various types of dredgers to be used for specific purposes and describes their operation in different materials and situations. This theme is continued in a section on capital dredging where attention is drawn to the problems of working in difficult ground. Maintenance dredging poses its own problems, especially difficulties in measurement. Various methods of survey are discussed. Disposal of dredged material at sea is subject to increasing environmental pressures especially for materials containing pollutants and alternative methods of disposal and legislation constraints are described. The economic evaluation of a dredging, dumping or land reclamation project is obviously complex and in the final section of the chapter the author points to the many factors that have to be considered (see also Chapters 42, 43, 47, 48, 49 and 50).

CONTENTS

45.1 TYPES OF DREDGER AND APPLICATIONS

45.1.1 INTRODUCTION

There is a great variation in the ability of different types of dredger to operate in different materials and situations. In this section, a brief description is provided of each of the main types of dredger which are available (BSI, 1991) and of the types of material and operation to which each is best suited.

45.1.2 TRAILING SUCTION HOPPER DREDGER

The trailing suction hopper dredger (trailer) is a ship (see Figure 45.1) consisting of a large hopper, a propulsion unit and a suction dredging installation. This dredger is sea-going under its own power and is the most commonly used type.

Dredging takes place with the dredger moving slowly ahead under its own power at a speed over the ground of between 1 and 5 knots (0.5 to 2.5 m s^{-1}). Confined working is rarely practicable.

The dredging action is achieved by pumping with one, or two large centrifugal pumps. Upon approaching the dredging area, one or two articulated suction pipes are swung outboard on davits and lowered to the sea-bed, or to dredging level. At the suction intake a steel fabricated structure, termed a 'draghead', serves to control water flow over the sea-bed. Soft or loose bed material is forced into suspension by a combination of differential pressure and erosion. Stronger materials may be loosened by teeth, water jets, or a combination of both. A few have a mechanical cutter.

The maximum depth to which dredging may take place is limited by the vacuum of the centrifugal dredge pump. Generally, if the dredge pump is mounted within the hull, the maximum dredging depth of most medium-sized trailer dredgers is about 35 m, but exceptionally in special circumstances greater depth is possible. By mounting the dredge pump in the suction pipe, greater depth or improved production is possible. A few large trailing suction hopper dredgers are capable, with modification of dredging to a depth of 80 m below water.

Dredgers which rely on suction methods may have

Coastal, Estuarial and Harbour Engineer's Reference Book. Edited by M.B. Abbott and W.A. Price. Published in 1993 by Chapman & Hall, 2–6 Boundary Row, London SE1 8HN. ISBN 0 419 15430 2.

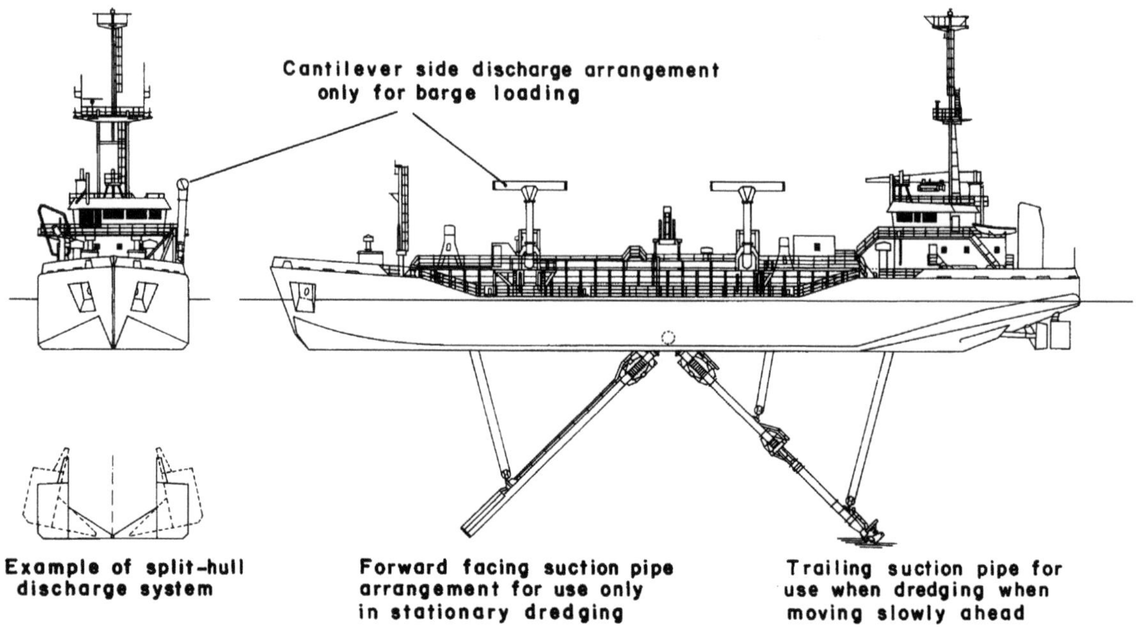

Figure 45.1 Trailing suction hopper dredger.

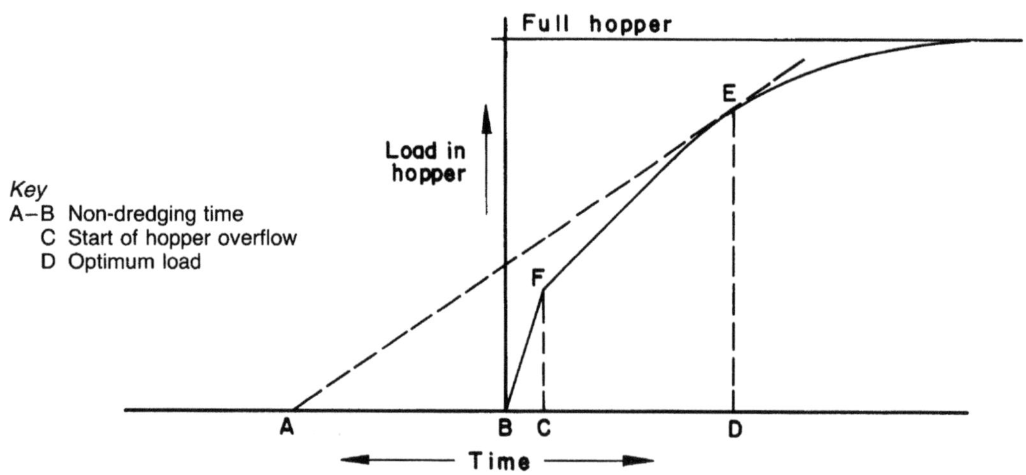

Figure 45.2 Hopper loading curve.

serious difficulty in areas heavily polluted by foreign matter. Pumping continues until an economic hopper load is achieved. In fine grained materials (less than 0.1 mm), the maximum economic load normally coincides with the commencement of hopper overflow. In coarse grained, or mixed materials, pumping may continue for some time after hopper overflow with continued improvement in hopper load (see Figure 45.2). The ship may be designed for the dredging of material of a particular density. Some dredgers are unable to carry a full load of high density materials, such as sand and gravel.

When the hopper contains an economic load, the dredger may sail to a disposal area and discharge the load via bottom doors, or valves, or by splitting the hull under hydraulic power. Alternatively, some

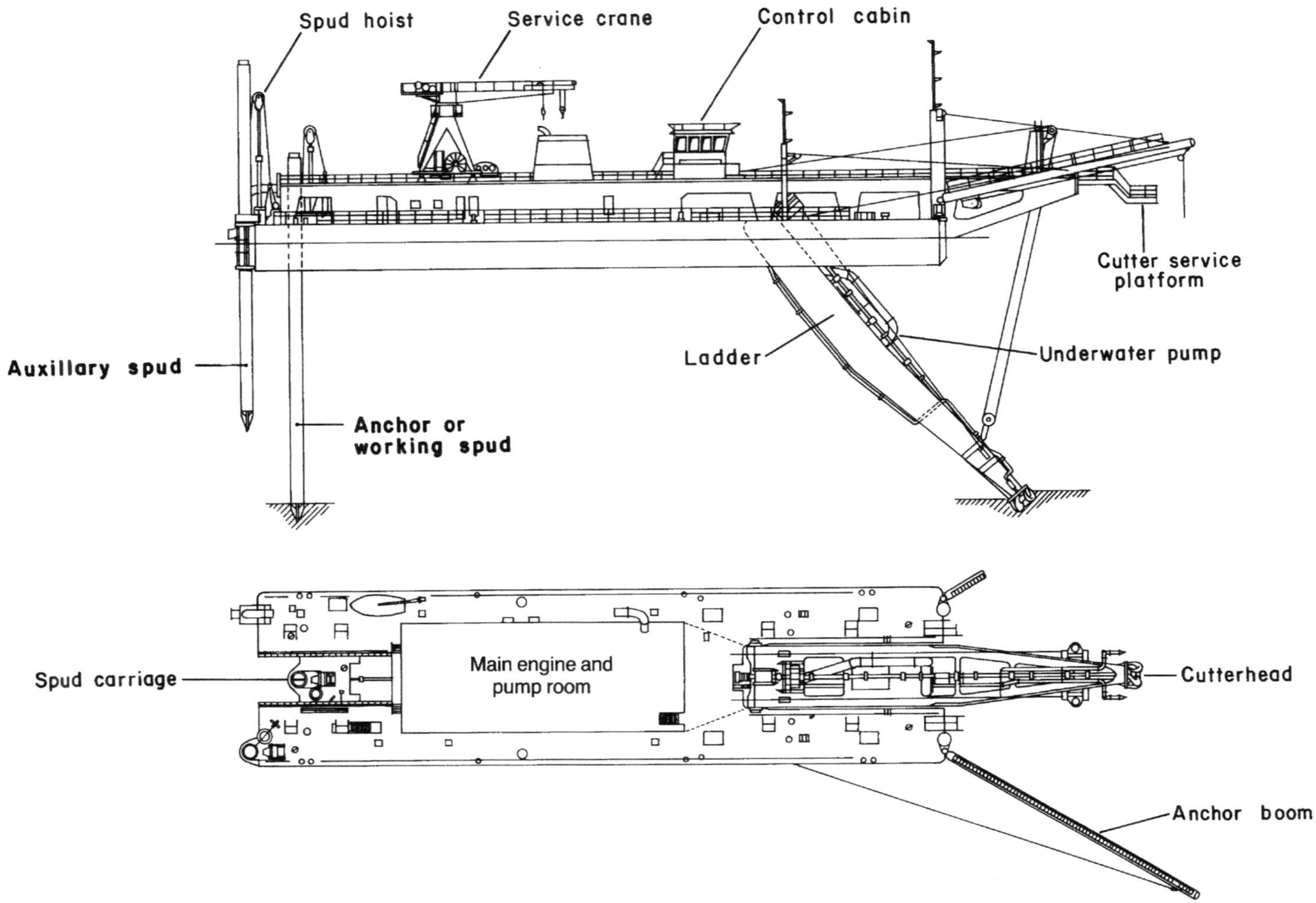

Figure 45.3 Cutter suction dredger.

dredgers are able to connect to a discharge pipeline and pump the hopper contents into a prepared containment area, or onto a beach, etc. Dredgers designed specifically for mining applications, such as marine aggregate recovery, may dry-discharge by conveyor or grab.

The main advantages of this type of dredger are its independent operation, relative immunity to rough seas and its high production. The dredger is best suited to the dredging of loose or soft materials, including, silts, sands, gravel and soft clays. Occasionally large particles can be pumped. The maximum size which the pump can pass is usually limited to about 70% of the suction intake diameter, but draghead grids are usually used to provide a much smaller limit. Suction pipe diameters may range from 250 mm to in excess of 1000 mm.

Some suction hopper dredgers are designed to work whilst at anchor. These types are suited only to the dredging of free flowing granular materials in marine mining applications. To improve production rates, fluidization of the bed material may be achieved with high pressure water jets.

45.1.3 CUTTER SUCTION DREDGER

The cutter suction dredger also relies on large centrifugal pumps for the movement of material, but has a powerful cutter surrounding the suction intake. The cutter breaks up obstinate bed materials to a size suitable for pumping.

Dredgers which rely on suction methods may have serious difficulty in areas heavily polluted by foreign matter.

The dredger is most commonly in the form of a rectangular pontoon (see Figure 45.3), but large types may have a hull in ship form. To provide a reaction to the cutting forces and to maintain accurate position, the dredger is usually secured by a spud anchor. The cutterhead is traversed across the face by on-board winches attached either to sea anchors, or to a shore fixture. Most commonly the whole dredger rotates about the working spud, but on some specialized types, the hull is anchored and only the ladder moves.

The working method is to make a series of arc shaped cuts at increasing depth until all material is removed to the required level. The dredger is then advanced into the face of excavation, either by means of a hydraulically controlled spud carriage, or if fixed spuds are fitted, by swinging the dredger on the walking spud in order to advance the working face (see Figure 45.4). Occasionally a tilting spud method may be used. The cycle is then repeated. The step distance is normally approximately equal to the front to back height of the cutterhead.

The dredger may load independent hopper barges, but most commonly discharges continuously through a combination of floating and shore pilelines.

The mechanical cutting action, which may be very powerful, allows this type of dredger to dredge a wide range of materials. Most types of material can be dredged, but if the bed material is moderately strong, as in the case of some rocks, pre-treatment of the material may be necessary. In the case of rock, this is most commonly achieved by drilling and blasting. The ability of the dredger to dredge rock directly is proportional to the power driving the cutterhead and the weight of the supporting ladder. Driving power to the cutter head may range from 35 kw to 3750 kw.

The maximum pipeline diameter and length which may be employed without the use of booster pumps, is dependent on the size of the dredge pumps and the applied power, but the characteristics of the dredged material are also very important. Materials which break down into fine particles can be pumped much further than coarse grained materials. Coarse grained materials must be transported at higher velocities and hence at higher mixture flow rates. The hydraulic head necessary to maintain flow increases with the square of velocity. The power required is proportional to the flow rate and mixture density.

The hydraulic head necessary to overcome friction when transporting solid material in a slurry mixture is determined by:

Pipe diameter;
Mixture velocity;
Roughness of the pipe wall;
Pipeline length;
Type of soil to be transported;
Specific gravity of the mixture.

The required head is not easily calculated, particularly for non-homogeneous mixtures, which will include, coarse sands, gravels, rocks, some clays and mixtures containing these materials. Formulae for determining friction head have been put forward by various researchers including, Condolios, Durand, Führböter, Gilbert and others, but the results obtained differ significantly. For nearly homogeneous mixtures containing soft clays, silts and fine sands, these formulae can be used with reasonable confidence.

Some non-propelled suction dredgers are constructed without a mechanical cutter, relying instead on high pressure water jets to fluidize or break up the bed material. These are most commonly used in marine mining applications in free flowing granular materials. The dredger remains stationary whilst dredging and hence creates a deep conical depression in the sea-bed. Such types may be adapted for the exploitation of deep, or even buried deposits.

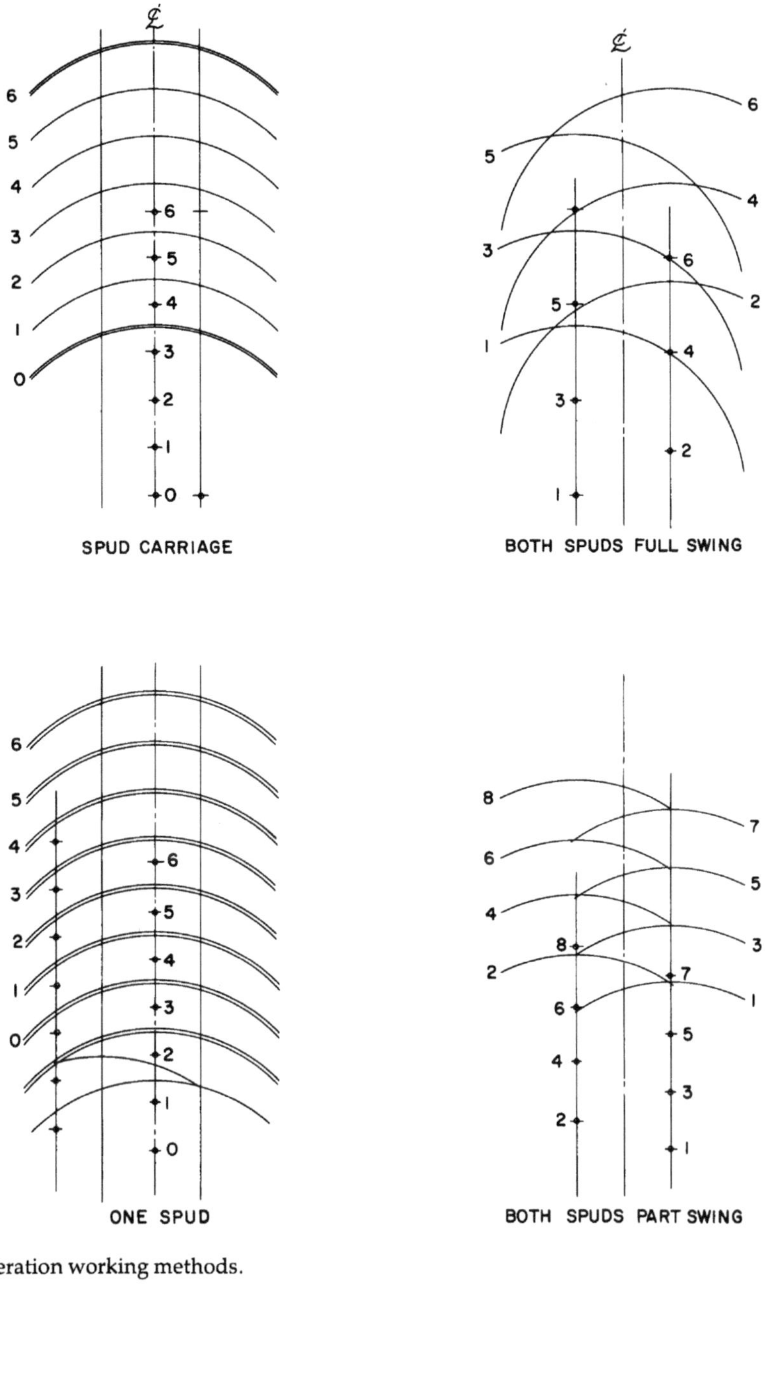

Figure 45.4 Spud operation working methods.

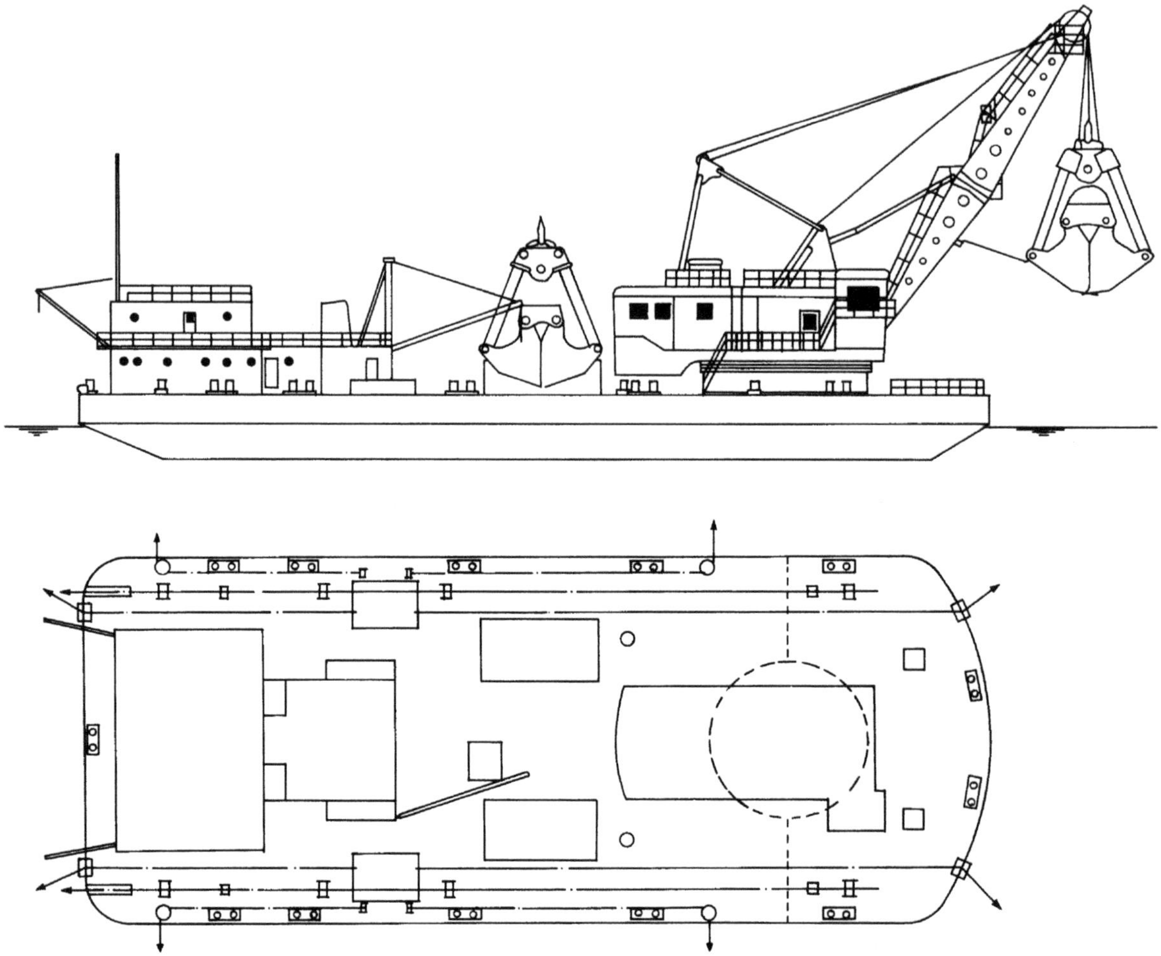

Figure 45.5 Pontoon grab dredger.

45.1.4 GRAB DREDGERS

Grab dredging cranes may be mounted on a pontoon (see Figure 45.5), or on a ship-like self-propelled hopper (see Figure 45.6). Pontoon mounted cranes are most common and the configuration usually permits larger cranes to be used. Grab hopper dredgers may have up to four independent cranes. Most are operated by ropes.

Hopper types of grab dredger are usually moored on a pattern of anchors and winches. These may interfere with navigation. Pontoon types may be similarly anchored, but those of modern construction are usually spud rigged.

Hopper types load their own hopper and sail off site to discharge via bottom doors. Pontoon types load into independent self-propelled, or towed bottom discharge hoppers.

Because the digging operation of a grab dredger is intermittent, production is relatively low and the accuracy and continuity of bottom finish is inferior. In Europe, grab dredgers are most commonly used for small works, for work in confined areas, or for trenching. In the USA and Far East, grab dredgers are favoured for a wider variety of work.

Soft or loose materials are most easily dredged, but medium to stiff clays, medium dense sands and gravels and very weak rocks, or pre-treated rocks, can all be dredged with reasonable success, depending on the size and weight of grab used. Different designs of grab bucket are used according to the type of material to be dredged.

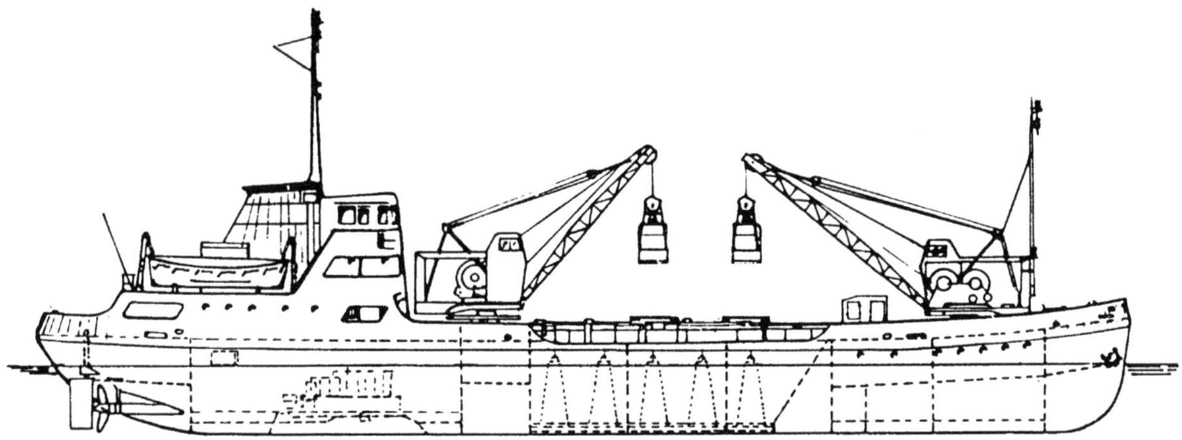

Figure 45.6 Grab hopper dredger.

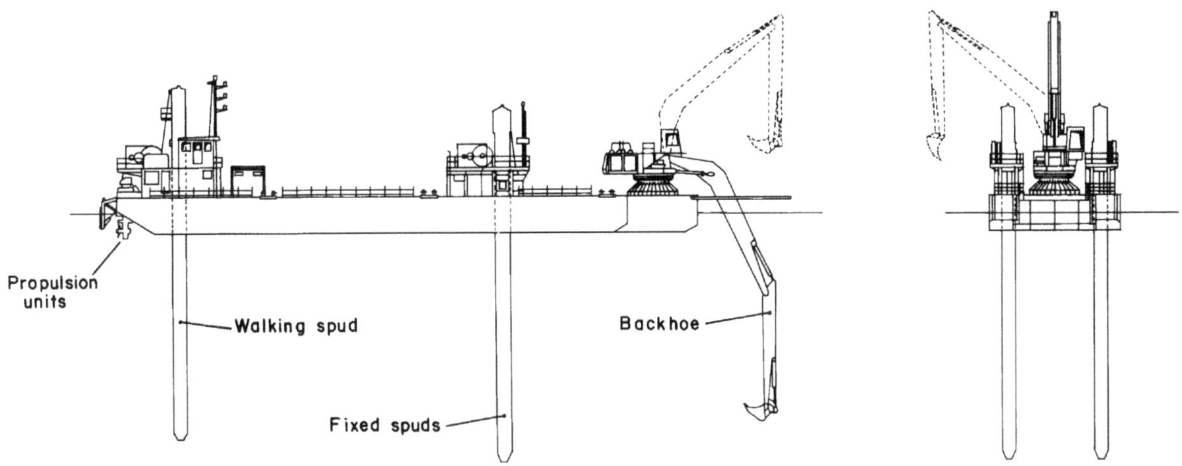

Figure 45.7 Backhoe dredger.

Grab dredgers may be particularly well suited to the dredging of areas polluted by debris, or boulders. Dredgers which rely on suction methods may have serious difficulty in such polluted areas.

45.1.5 BACKHOE DREDGERS

Backhoe dredgers are based upon conventional hydraulic backhoe excavators. Usually the excavator is mounted, without an undercarriage, onto a purpose-built pedestal on a spud rigged pontoon (see Figure 45.7). The dipper arm and monoblock are usually significantly longer than those of the equivalant land-based machine, in order to provide the greater depth of excavation which is common to dredging.

The backhoe dredger usually loads into self-propelled, or towed, bottom discharge hoppers. Very occasionally, however, smaller types may be mounted on a self-propelled hopper.

The virtues of the backhoe dredger are its powerful digging action, the ability to handle large pieces and its generally better controlled and hence more accurate digging action compared with a grab dredger. Production may also be superior due to better penetration and lower cycle times.

The backhoe dredger may be used in any type of

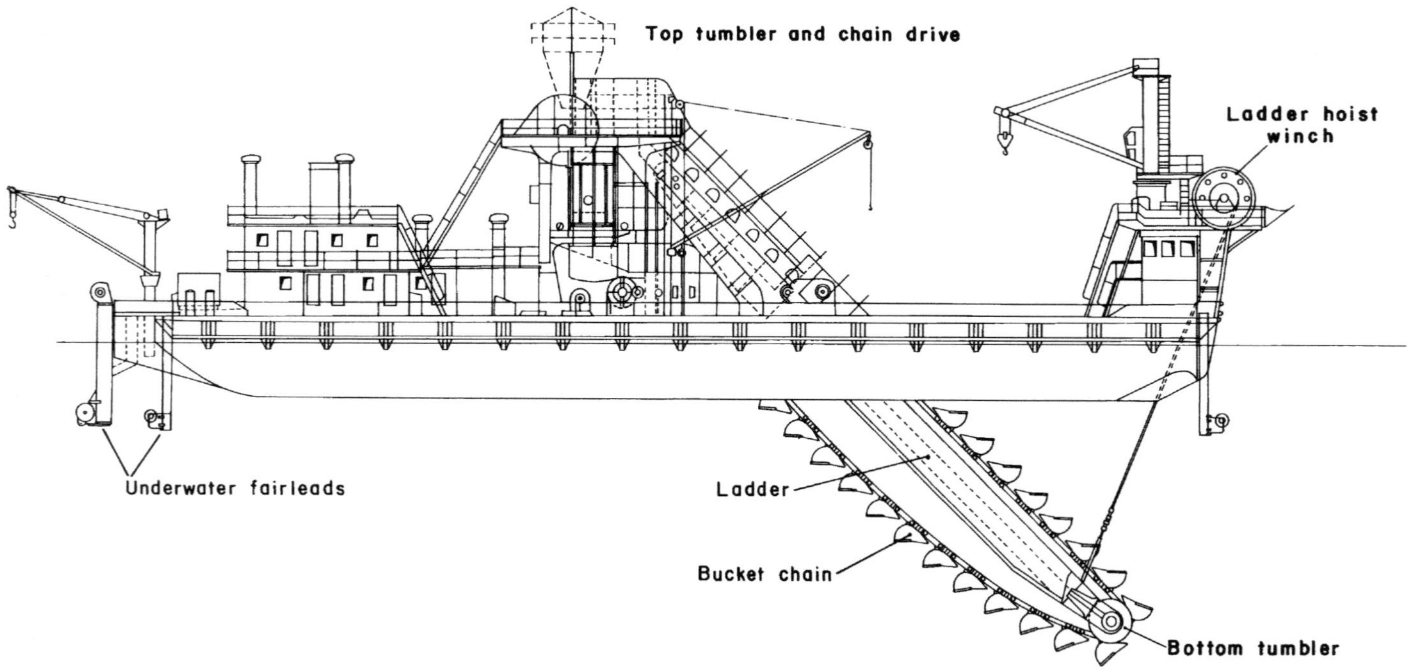

Figure 45.8 Bucket dredger.

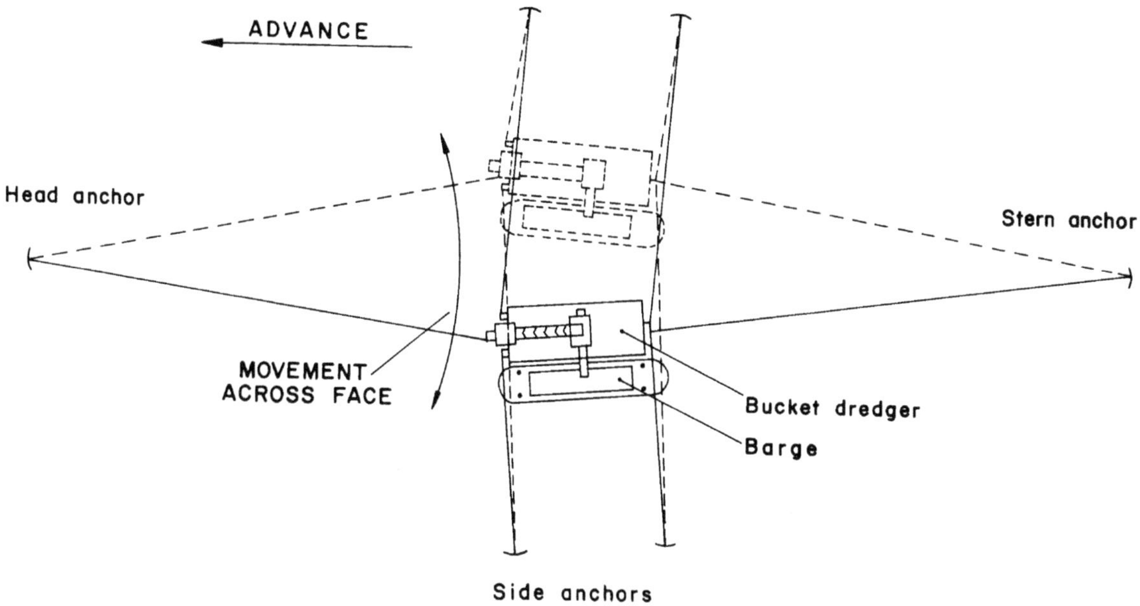

Figure 45.9 Bucket dredger working method.

material, but the maximum rock strength which can be dredged without pre-treatment is proportional to the size and power of the excavator and to the dredging depth. Moderately strong and stronger rocks will normally need pre-treatment, unless fractures are open and very closely spaced. The dredger is particularly well suited to the dredging of highly variable materials, such as boulder clays, morain and heavily fractured, or highly variable rocks.

45.1.6 BUCKET CHAIN DREDGER

The bucket dredger consists of a continuous chain of buckets supported by a heavy ladder, all mounted on a narrow rectangular pontoon (see Figure 45.8). The dredger loads hopper barges via over-side discharge chutes. Buckets discharge to the chute by inversion when passing over the top tumbler, which drives the bucket chain.

The dredger is positioned and the dredging process is controlled by a pattern of six winches connected to sea anchors, or occasionally to shore fixtures. The digging reaction is provided by a heavy head wire. The dredger swings on the head wire under the control of four side winches (see Figure 45.9). A stern winch provides stability and resistance to currents, or wind load.

Mooring wires may obstruct navigation. Noise levels may cause problems in residential areas, particularly during night work. Bucket dredgers are not capable of dredging in shallow water.

The bucket dredger is capable of dredging a wide range of materials, including some weak to moderately strong rocks, if fractures are open and very closely spaced. Strong rocks require pre-treatment. The method of working allows an accurate, level-bottom finish to be achieved. Production, which in soft materials may be high, is continuous, providing that an adequate number of attendant hopper barges is available.

45.1.7 LESS COMMON TYPES OF DREDGER

Additional suction-type dredgers include the so-called 'dustpan dredger' and the 'horizontal auger' dredger. Both are of pontoon-type construction and both work by repeatedly moving forwards and backwards under winch control. Both discharge via a floating pipeline. The dustpan has a hull width suction head with water jet assistance. The auger has a hull width twin horizontally opposed mechanical auger to dislodge and feed loose bottom sediments.

Additional bucket type dredgers include the 'dipper' dredger which may be a spud pontoon mounted version of the conventional land based rope operated dipper, or face shovel, or a hydraulic face shovel. These dredgers generally dredge only to a relatively shallow depth, but are capable of dredging difficult materials, including boulder clays, boulders, dense gravels and weak to moderately strong rocks, if fractures are open and closely spaced.

Some small suction- and bucket-type dredgers are designed for amphibious operation, mainly in the maintenance of canals and similar work.

45.1.8 BED LEVELLERS, OR PLOUGHS

The bed-leveller consists of a wide bottomless bucket, or blade, towed at a controlled height behind a tug, or workboat. Its primary use is to level the sea-bed after dredging. This may be particularly important where trailing suction hopper dredgers, or grab dredgers have been used.

It may also be used to drag material from areas inaccessible to trailer dredgers for subsequent dredging, or independently, to move soft, or loose deposits from shallow areas into deeper areas.

45.2 CAPITAL DREDGING

Capital dredging will usually require the removal of virgin undisturbed materials. Some of these materials may be strong, and in consequence the dredging plant for capital works may need to be stronger and more powerful than that which works only on maintenance.

The performance of all types of dredging plant is dependent on the nature of the ground to be dredged. A thorough and proper ground investigation in advance of dredging is, therefore, most important. Some types of ground may present special difficulty. These include:

Clays with high plasticity;
Cobbles and boulders;
Flints;
Peat;
Rock;
Sands which are highly consolidated.

These materials may each present special problems. Some clays form balls when pumped and hence require higher transport velocities. They may also be difficult to handle if pumped into contained areas. Cobbles and boulders are difficult, or impossible to handle by suction methods. Flints are highly abrasive. Peat may swell rapidly upon the removal of overburden and will not load into hopper wells by pumping. The difficulty of dredging rock increases with its strength and integrity. Pre-treatment of the rock, usually by drilling and blasting, may be necessary to permit dredging. Certain sands, usually fine and well graded, may be very dense *in situ* and hence difficult to penetrate. Such sands may reconsolidate quickly in hoppers and be difficult to discharge.

Other things which may disrupt dredging operations include abandoned munitions, debris and heavy vegetation. If possible munitions should be cleared, or rendered harmless prior to dredging. Debris may severely disrupt suction dredging. If it is known to be present, it should be cleared, if possible, by raking, or non-suction dredging methods.

45.3 MAINTENANCE DREDGING

Maintenance dredging involves the removal of recent deposits. These will usually be weak and poorly consolidated. The dredging plant employed need not be particularly strong or powerful.

The bulk of maintenance dredging is done by suction methods, primarily by trailing suction hopper dredgers, sometimes assisted by a bed-leveller. Other types are used, subject to the particular site conditions.

To maximize dredger productivity, it may be desirable to maximize the interval between dredging campaigns. This will allow greater consolidation of sea-bed deposits and, if acceptable, an increased thickness of deposit to be dredged.

Variations in the density of sea-bed materials can cause important difficulties in measurement. Materials with low density will be penetrated by low frequency echo sounding, but may reflect a high frequency signal. Hence echo sounders of different frequency will record different bed levels and the use of a sounding line, or pole, may produce further variation. In some ports where this is recognized to be a problem the vertical variation in sea-bed material density is measured. For navigational purposes, the sea-bed level is taken to be at the elevation of a specified density, on the assumption that bed materials of less density will present no hazard to navigation. Density is measured either by gamma radiation, or accoustical methods. This approach can save a great deal on maintenance.

45.4 DISPOSAL OF DREDGED MATERIAL

A range of beneficial engineering uses exist for the disposal of clean granular materials (PIANC, 19●●). These include offshore bunds, beach recharge, land reclamation and land improvement. However, problems arise when pollutants have entered coastal and estuarial zones through industrial discharges and remain trapped within the near surface sediments. This is particularly the case where the sea-bed sediments are of clay and silt. Mainly for this reason traditional methods of disposal, such as dumping at sea, are subject to rapidly increasing pressure from all concerned with the protection of the environment. For a full description of the contaminants, alternative methods of disposal and legislative constraints which apply to the disposal of such materials, the reader is referred to Chapters 40 and 47.

45.5 LEGISLATIVE CONSTRAINTS

45.5.1 EXTRACTION FROM THE SEA-BED

In the UK, formal consent is necessary before material can be removed by dredging from the sea-bed (BSI, 1991). Usually the mineral rights in respect of the sea-bed within the continental shelf limits are controlled by the Crown Estates Commissioners (CEC) whose consent is strictly required for any dredging work.

A CEC production licence may not be required when material is to be dredged from within the boundaries of a port or harbour authority for the purposes of improving navigation. If, however, the dredged material is to be used for some beneficial purpose royalties may be payable.

The Crown Estates Commissioners operate a two-stage licencing procedure. Before removing sea-bed material application must be made for a prospecting licence. Such a licence will be issued only if it is considered that the issue of a licence will not be detrimental to the interests of the Crown and other interested parties.

Unless the area in which prospecting is to take place is particularly sensitive, the issue of a prospecting licence is reasonably quick and uncomplicated. There is an established procedure of initial consultation with fishery interests (MAFF, 1981).

Samples arising from prospecting are normally subject to some laboratory analysis to determine the physical properties of the deposits. All of the information gained from any prospecting work must be supplied to the Crown Estates Commissioners.

Providing that the prospecting exercise has demonstrated that deposits of adequate quantity and quality exist, a production licence may be applied for as a second stage.

If the licence application is for a 'one-off' exercise such as beach recharge or land reclamation then the production licence may be issued specifically for the year of proposed construction.

Before the licence is issued a consultation procedure takes place to safeguard the interests of third parties. The period of consultation necessary prior to the issue of a production licence may be prolonged. If there are serious objections a licence may not be issued. There is no appeal procedure in the event of a licence application being refused.

45.5.2 PERMISSION TO DREDGE

In addition to the requirement described earlier, other permissions may be required. Dredging work which is proposed within navigable waters will also require the permission of the relevant port or harbour authority and also, in the UK, of the Department of Transport, Marine Directorate, who administer Section 34 of the Coast Protection Act 1949.

Under Section 18 of the Coast Protection Act 1949 it is unlawful to excavate or remove materials on the sea-shore or under the sea-bed. Some coastal local authorities have obtained enabling orders under the Act under which their consent for dredging within 3 miles of the coast is required.

Approval from the responsible authority may be required, under the Control of Pollution Act 1974, when dredging which may cause disturbing noise is proposed in close proximity to areas of habitation, or is to be carried out in conditions such that the suspension of significant amounts of fine materials, perhaps with toxic content, may result.

45.5.3 PERMISSION TO RECLAIM LAND

Different localities and situations may require different consents. A thorough investigation of the necessary procedures is advised before any such work commences.

In the UK, and similar rules are likely to apply elsewhere, land reclamation will require planning permission from the local authority within whose area the work is proposed and who are charged with the administration of the Town and Country Planning Act 1971, and also a licence from the Ministry of Agriculture, Fisheries and Food under the Food and Environment Protection Act (FEPA) 1985 Part II.

45.6 ECONOMICS

45.6.1 INTRODUCTION

The economics of dredging is complicated by the differences which exist between direct labour dredging and contract dredging.

45.6.2 CAPITAL CHARGES AND FIXED COSTS

Dredging plant is expensive to build. A small trailing suction dredger may cost in excess of £1 000 000 (at 1989 values). A large one may cost in excess of £10 000 000. The building costs of a few very special dredgers have been as high as £40 000 000. If a realistic commercial attitude is taken, these costs should be written off over the useful working life of the plant. This may range from ten years for a small cutter suction dredger to 30 years for a trailing suction hopper dredger. Provision should also be made for the replacement of the dredger at the end of its working life. Interest charges on capital employed should also be taken into account. These charges are independent of working hours.

The contribution of capital charges to unit costs is directly related to annual production volume, which is, in turn, directly related to working hours. It is for this reason that the majority of dredging contractors usually man their dredging plant 168 hours a week. This is less common for the port-owned dredger. The port may, however, have the advantage of substantial capital grant aid which reduces internal capital charges, but to obtain a true comparison, it can be argued that any grant should be ignored, because the money has been spent. Regardless of this, it will be clear that working hours should be maximized where possible.

Fixed charges also include insurance, permanent staff and labour, overheads, supervision, workshop expenses and, in the case of sea-going vessels, periodic survey, classification and dry-docking. To minimize unit costs, these charges should also be spread over the maximum working hours.

45.6.3 RUNNING COSTS

Running costs include fuel, oils and greases, water, various consumables, spares, repairs and overhauling, painting, catering and casual labour. Labour may be considered as a running cost rather than a fixed cost depending on policy. For a continuously operated trailing suction hopper dredger, running costs exclusive of labour, may be approximately 40% of total costs, rising to 60% if labour is included.

45.6.4 MOBILIZATION COSTS

Mobilization costs will not arise for the port which uses only its own dredging plant, but for other ports, the cost of bringing plant to the site and setting up, may be important. When quantities are small and the area of work remote, mobilization charges may exceed the cost of executing the work.

When the sailing distance for mobilization is small, charges will usually be low, or may be nil for a trailer dredger which is passing the site. For capital works which require a variety of large plant at a remote location, mobilization costs may be large.

Mobilization costs may be reduced if a flexible start date is permitted, or through co-operation with other users of contract dredging to co-ordinate works and share mobilization charges. For those works which require only the infrequent dredging of substantial quantities by trailing suction hopper dredger, mobilization costs will usually be insignificant.

45.6.5 UNIT COSTS

Unit costs can be reduced by observing 10 simple rules.

1. Understand the task.
2. Choose carefully between contract and direct labour.
3. Choose carefully the timing of the work.
4. Choose carefully the plant to be employed.
5. Minimize the haul distance for disposal.
6. Minimize the handling of dredged material.
7. Minimize the water content of the dredged material.
8. Minimize the area to be dredged.
9. Minimize the quantity to be dredged.
10. Maximize the working time.

45.7 LAND RECLAMATION

45.7.1 INTRODUCTION

Land reclamation can be divided into two quite different categories (BSI, 1991): capital works, where the primary object is to create land for development, and that which arises as a consequence of the disposal of materials from maintenance dredging. In either case, reclamation may be the result of the enclosing and filling of an area of sea-bed, or of raising the level of existing land, which is low lying in relation to the level of tides, floods, or the water table. The latter case is land improvement.

45.7.2 CAPITAL LAND CREATION

Land creation will most commonly involve only the use of free draining granular materials. The material may arise as a result of dredging of the sea-bed elsewhere on the site, or may be dredged offsite from a borrow area.

At protected sites, it may be feasible to pump coarse grained materials directly into reclamation areas without containment, or protection, but usually, prior to the commencement of filling of the area to be reclaimed, it will be necessary to construct an enclosing structure, which will serve both to retain the fill and to protect it against erosion by the sea. For beach recharge the material is not permanently contained.

45.8 CONTAINMENT

45.8.1 CAPITAL WORKS

Containment is most commonly by the prior construction of an embankment. The level of construction should provide a reasonable freeboard above the highest expected tide level plus wave set up. The embankment profile should be carefully considered so that the final foundation loading is not excessive and the seaward slope is compatible with the prevailing wave climate. The embankment may be constructed using imported fill, or by using dredged fill. In the latter case, it may be possible to pump dredged material directly to form an embankment.

The seaward slope will usually require protection against erosion by wave action. Such protection may take the form of rock rip rap, concrete module rip rap, concrete block revetment, etc. Occasionally, in protected areas, if the fill material is coarse grained, the fill may be allowed to form a natural beach profile. The beach may need occasional replenishment, but may be more acceptable for certain types of development.

Reclamation for port development may require a vertical seaward face to permit the berthing of shipping. This is most commonly achieved using steel sheet piling, but concrete may also be used in various ways, including block retaining walls or caission construction.

The enclosing structure may be permeable, or impermeable, subject to local conditions, material availability, and the intended end use of the area reclaimed. Normally some drainage is preferable.

45.8.2 CONTAINMENT OF DISPOSAL AREAS

Areas which are to contain fine grained materials which most commonly arise from maintenance dredging require a quite different approach.

Fine grained materials drain slowly and in some cases will retain high levels of moisture indefinitely. The depth of deposition should be restricted to obtain the maximum benefits from evaporation. A system of interconnected areas may be necessary to permit rotational filling. Dredged material will usually be placed by pumping. The solid material may not readily settle from suspension. Enclosed areas may therefore need to be large in relation to the volume to be contained.

The outfall which discharges drainage water from the area being filled should be as far as possible from the incoming discharge of the pumped mixture. The drainage system should be designed to remove supernatent water with the minimum risk of entraining solid particles. Under-drainage may be necessary to accelerate drainage of the fill and to improve consolidation.

The enclosing bunds should be sufficiently impervious to prevent excess seepage and the development of flow paths. Limited permeability may be advantageous for long-term drainage and consolidation. If necessary, the level of permeability, or impermeability, can be engineered by the use of geotextile lining materials, or membranes.

The enclosing bund height should be sufficient to provide adequate freeboard and to retain an adequate depth of water to facilitate deposition, even during the final stages of filling. Clay and silt-size materials will increase bulk greatly during the dredging/transport process. Bulking should be allowed when determining the required containment area capacity. Bulking can be estimated by the laboratory testing of typical samples of the material to be dredged.

45.9 CONSOLIDATION AND COMPACTION

Well graded granular materials which are placed hydraulically above the retained water-level will normally consolidate rapidly and well. For many applications, the consolidation which is achieved during well controlled hydraulic placement will be adequate.

Further compaction of near surface deposits may be achieved by vibrating roller or plate compactors. The depth of influence is dependent on the weight and frequency of the compactor, but is generally less than 2 m. Thus the fill may need to be raised progressively with intermediate compaction. Alternatively, for deeper compaction, vibro-flotation, or dynamic compaction methods should be used.

The consolidation of cohesive sediments will normally be a prolonged process directly related to the rate of release of moisture. The deposit may not achieve a strength sufficient for industrial use for many years. Consolidation can be accelerated by the more rapid release of moisture. This may be achieved by agricultural methods which employ special low ground bearing pressure plant. Close centre ditching, with a progressive increase in ditch depth, followed by reworking of the soil, is the traditional method of improvement.

Consolidation may also be accelerated by enhanced drainage. This may include, under-drainage, vertical sand drains, vacuum pumping, reduced fill thickness, thin hydraulically placed sand layers between lifts, selective planting, etc.

45.10 ENVIRONMENTAL

45.10.1 INTRODUCTION

Environmental considerations have become of major significance to dredging and associated works. An environmental impact study will be required during the planning stages of most capital dredging works and may be required when a major change of method is proposed for maintenance dredging. There are many ways in which dredging work may influence the environment. The effects may not be detrimental, but will often appear to be objectionable to the lay public.

45.10.2 POTENTIALLY ADVERSE EFFECTS

The main areas of influence are as listed below.

Release of contaminants;
Physical disturbance;
Depleted oxygen supplies;
Spillage;
Increased water turbidity;
Erosion;
Noise;
Blanket cover to sea-bed.

45.10.3 CONTAMINATED MATERIALS

When dredging polluted materials, contaminants may be released, or moved. Contamination is not a consequence of dredging, but usually of some past industrial activity. Dredging may, therefore, uncover a dormant problem. Strict controls on dredging may not be necessary, but controls on the method and place of disposal may be. Statutory controls now exist to regulate the dredging and disposal of contaminated materials.

Various measures can be adopted to eliminate, or minimize the spread of contaminated material. Where practicable, a totally confined method may be used. For example, a cutter suction dredger may be used to dredge and transport the material by pipeline to a sealed containment area. Subject to appropriate operating procedures, the release of contaminants during dredging should be negligible.

If a trailing suction dredger is used, hopper overflow should be restricted, or prohibited. Some release of material may be caused by erosion and turbulance around the draghead. This may be minimized by operating constraints, including, minimum trailing speed, use of de-gassing equipment, non-use of water jets, careful draghead selection, minimum intake flow velocity, minimum change of direction and speed, optimum draghead burial, maximum underkeel clearance, avoidance of dredging in high waves or swell, partial filling of hopper, etc. The hopper contents may be pumped into a sealed containment area. If disposal is on the sea-bed the material should be contained until release just above bed-level and on completion the area should be capped.

45.10.4 TURBIDITY

Increased turbidity may result from disturbance and turbulence in the area of the draghead, cutterhead, bucket, or blade. Subject to appropriate constraints, these are generally not a serious problem. Turbidity may also be caused by water draining from hoppers or containment areas. This may be more serious, but is easily controlled. Propeller wash may also cause the suspension of bottom sediments.

The effects of turbidity include reduced light penetration, which in very fine materials and still waters may be prolonged. Oxygen supply may be reduced.

45.10.5 PHYSICAL DISTURBANCE

Disturbance to the sea-bed will be caused by the action of dredging, the placement and movement of anchors and by the passage of vessels when underkeel clearance is small. That disturbance caused by the dredging action cannot be avoided.

If fisheries which are bottom resident and relatively immobile exist in the area to be dredged, they may be destroyed, but may re-establish later.

Anchor damage can be avoided by avoiding the use of anchors, or restricting their position, or by using only spud located plant, or plant which uses fewer anchors. A cutter suction dredger normally needs only two, a trailer suction does not need any when working.

45.10.6 EROSION

Erosion may be caused by propeller wash, already discussed, by water draining from containment/reclamation areas, or as a result of changes in nearshore sea-bed levels caused by dredging works. The latter is a design problem, not a dredging problem. For borrow areas in UK coastal waters, there are restrictions on the removal of material from nearshore. No licence to extract will be issued if coastal instability is likely to result. Drainage of water from containment areas, or from leaking pipeline joints, can be a problem, particularly in areas of saltmarsh, which may be easily eroded. Leaking pipe joints may be avoided through regular inspection and maintenance, good management and supervision. Draining water from impounded areas, or fill, can be contained or re-routed.

45.10.7 NOISE

Noise is dependent on the type of plant employed. The intrusion of noise is dependent on working hours and the closeness of habitation. Bucket dredgers are noisy, trailing suction hopper dredgers are not. Other plant falls between the two. All plant will be more intrusive at night. Noise levels of all types of dredging plant can be reduced by design or modification. Where the pre-treatment of rock by the use of explosives is necessary, detonation will cause noise, but providing that the water cover is adequate, which is desirable for maximum destructive effect, the level of noise during day time will not usually be a significant problem. The noise of an onshore pump, or generator, is as likely to upset residents as is dredging plant.

45.10.8 SPILLAGE

Most floating mechanical plants carry fuel and oils. These may be spilt, as they may be from any other craft. Good maintenance and good management will minimize the risk, which is not great. The most likely unexpected failure and source of spillage is in high pressure hydraulic systems, but the oils used may be vegetable based and non-toxic.

45.10.9 BLANKET COVER

Blanket coverage of the sea-bed may occur deliberately as a result of the concentrated discharge of dredged material, or accidently as a result of the settlement of suspended sediments which have been mobilized by the dredging process and transported by tidal currents.

If the thickness of sea-bed cover is small, damage may be minimal and recovery rapid. In some instances, the resident marine population density, or variety, may actually improve due to an improved food supply, or changed sea-bed characteristics. Some immobile shellfish species may, however, respond badly. Spawning grounds may also be affected.

There is a growing belief among marine biologists that, subject to the particular marine population, the widespread dispersion of dredged material may be preferable to the traditional concept of concentrated dumping.

45.11 CONTRACTS

45.11.1 INTRODUCTION

The guiding objectives in the preparation of a contract for any dredging work should be, to describe accurately the work to be done and the conditions under which it is to be done, and to provide a fair and equitable method of payment for that work, when satisfactorily completed.

45.11.2 INFORMATION TO BE SUPPLIED

In order to tender, a contractor should have adequate information concerning the site. This information should include the following:

Soils
Bathymetry
Currents
Services
Navigation restrictions
Debris
Wind and wave data
Visibility
Shipping

Of the matters listed, soil conditions and wave climate are the most important, as these will most affect the working of dredging plant. Bathymetry is also important (ICE, 1984), but generally this information is more readily available.

It is most important to recognize that good ground information is absolutely vital, particularly in the case of capital works (BSI, 1991; PIANC, 1984). Ground investigation over water is expensive and in consequence it is too easy to shrink from the task of carrying out a proper investigation. Failure to do so, however, may expose the employer to a high risk of dispute and over-expenditure. When a proper investigation is done, it is important to pass all relevant information gained, with explanatory notes, to all tenderers.

It is important to understand that the contractor must often prepare his tender, for work on a site of which he may be quite ignorant, in only a few weeks. In contrast, the employer, or his consultants, will usually have years of accumulated knowledge of the site and months of preparation of the particular design. The responsible employer will use this time to compile and present an adequate picture of all relevant site conditions.

45.11.3 CONDITIONS OF CONTRACT

In the UK, it is common to use the ICE Conditions of Contract for Works of Civil Engineering Construction (5th Edition January 1979) as the general conditions of contract. Internationally, the similar FIDEC conditions are more common.

45.11.4 INSURANCE

The insurance clauses in the ICE 5th and FIDEC conditions are drafted with land works in mind. Care should be taken to see that the special risks associated with marine works and craft are adequately covered. The advice of a specialist broker may be helpful.

45.11.5 MAINTENANCE PERIOD

Pure dredging work does not involve construction and there will be no structure to maintain, or show defects, beyond the period of work. Any defects in work which exist should be revealed by the post-dredge survey. For maintenance dredging works in particular, the imposition of a contract maintenance period on a site where siltation is continuous, that is most sites to some degree, will in effect be an infinite extension of the contract. This is unreasonable.

45.11.6 WEATHER DOWN TIME

Time lost due to inclement weather is common on dredging work. Given adequate records of past seasonal conditions, tenderers can estimate and make provision for lost time in a typical period, but if the conditions encountered are not typical, it may be unreasonable to expect that the contractor assumes all of the risk. This is especially true for work of short duration where, if the weather is bad, lost time may easily equal, or exceed the working time. For the tenderer, the most valuable information will be a month-by-month analysis of wind/wave conditions on a percentage exceedance basis.

45.11.7 SPECIFICATION

The specification should clearly define the works, and most important, also define any operational constraints to be placed upon the contractor.

45.11.8 MEASUREMENT

The method of measurement which will generally prove most satisfactory is that of pre-dredge and post-dredge surveys of the work. If the work is to reduce sea-bed levels, then the reduction in volume should be measured within the specified profiles. If the work is to create land, then the increase in volume should be measured. Other methods of measurement, although occasionally unavoidable, are generally less satisfactory.

45.11.9 TOLERANCES

It is only essential to specify a tolerance where structures, or some other feature, require protection. All dredging plant will require some working tolerance in order to achieve the specified dimensions, but the actual tolerance which is necessary will be affected by the type of plant used, the skill of the operators and the quality of site control. A paid over-depth tolerance is not essential. Its value is dependent on the site conditions and the objectives of the work.

REFERENCES

BS 6349 Part 5 (1991) Recommendations for dredging and for Land Reclamation. British Standards Institution, London.

ICE (1979) *Conditions of Contract for use in works of civil engineering construction*, (5th edition) June 1973 (Revised January 1979). The Institution of Civil Engineers, London.

ICE (1984) *Guidelines for the preparation of Hydrographic Surveys for Dredging*, Thomas Telford Ltd, London.

MAFF (1981) Code of practice for the extraction of marine aggregates, December 1981. Ministry of Agriculture, Fisheries and Food, London.

PIANC (1984) Classification of Soils and Rocks to be Dredged. Report of a Working Group of the Permanent Technical Committee II, Permanent International Association of Navigation Congresses, Brussels.

DREDGING IN COHESIVE SEDIMENT AREAS 46

H. de Vlieger

Perhaps an old definition of dredging would be the act of deepening an area with the aid of a dredger. However, there are a number of complementary activities which could reduce or even eliminate the need to dredge. Among others there are the ability to improve natural depths by some form of training wall or breakwater, fixed dredging installations such as silt pumping stations or by-pass plants, bed-levellers, water-injection techniques etc. Many things have to be considered in coming to an economic and environmentally acceptable solution – in silts is it possible to invoke the idea of navigable as opposed to 'natural' depths, and is there seasonality to siltation, for example, which would control the times to dredge? In other words the best plan involves considering many factors.

CONTENTS

46.1 INTRODUCTION

This chapter discusses methods for maintaining rivers, channels, harbours and maritime fairways at the prescribed depth for safe navigation.

Depending on specific local requirements dredging can be achieved by:

1. Enhancing natural conditions by, for instance bandalls or breakwaters;
2. Fixed dredging installations such as silt pumping stations or by-pass plants;
3. Mobile dredging installations such as bed-levellers, water injection plants and traditional hydraulic and mechanical dredging techniques.

Optimization of dredging operations is also site-specific but the problem can be approached by:

1. Reducing the amount of mud to be removed by introducing the navigable-depth concept;
2. Establishing the optimal maintenance dredging procedure;
3. Optimization of the maintenance dredging system on board;
4. Optimization of the disposal areas.

46.2 MAINTENANCE DREDGING

The Permanent International Association of Navigation Congresses (PIANC, 1989) defines maintenance dredging as:

> 'The removal, transport and temporary or definitive disposal of all bottom material sedimenting in harbours, access channels or other navigable or non-navigable waters, above a minimum level necessary for safe navigation in, or sufficient flow capacity through the said waters.'

Thus, maintenance dredging covers all activities involved to keep up design requirements for navigation, flood control or coastal protection. Among such activities are permanent hydrographic surveying, supply or removal, transport and temporary or definitive disposal of bottom material.

The approach of a maintenance dredging project is quite site-specific, since the necessity of maintenance dredging mainly depends on the local meteorological, hydrodynamic and sedimentological conditions. A thorough comprehension of these phenomena is thus a pre-eminent requirement for the optimum choice of available techniques.

Although sedimentation is quite site-specific, there are nevertheless some standard techniques used world-wide to solve the problem:

Coastal, Estuarial and Harbour Engineer's Reference Book. Edited by M.B. Abbott and W.A. Price. Published in 1993 by Chapman & Hall, 2–6 Boundary Row, London SE1 8HN. ISBN 0 419 15430 2.

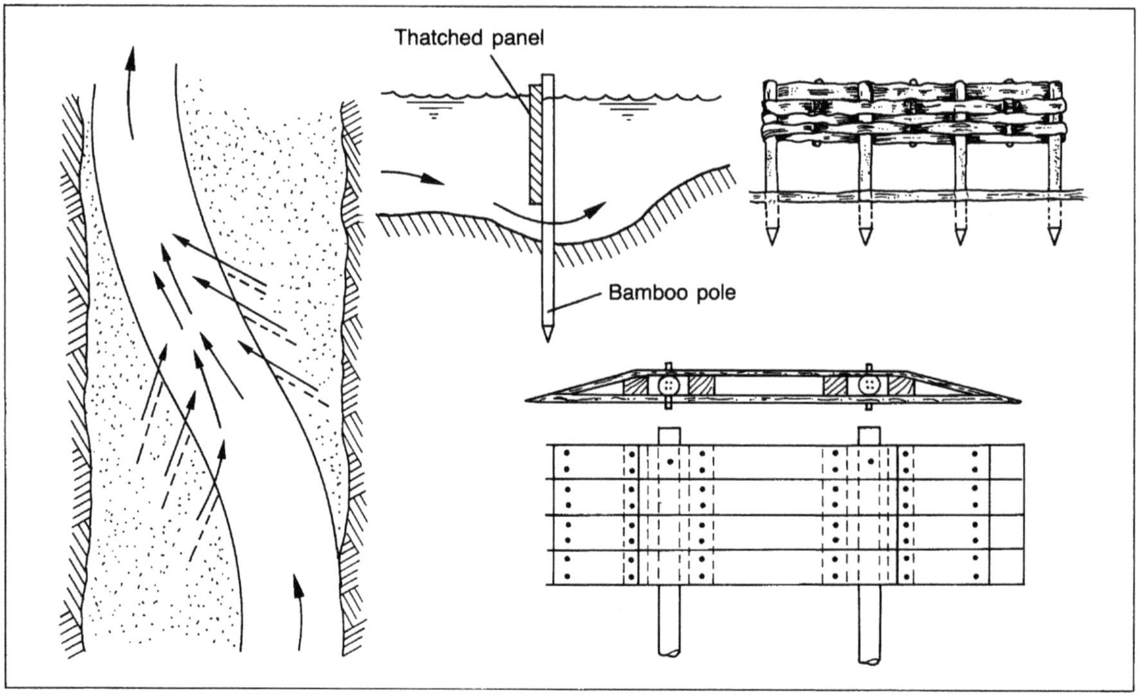

Figure 46.1 Principle and construction of bandalls (UNESCAP, 1989) (courtesy of Applied Dredging Consultancy, The Hague).

1. Enhancing natural conditions, working with nature;
2. Using stationary installations;
3. Using mobile plants.

46.2.1 WORKING WITH NATURE

The prescribed channel-profile required for safe navigation or sufficient flow can be obtained by fixed constructions, which can favour: erosion in the channel; and sedimentation out of the channel or in overdepths in the waterway. Erosion-inducing hydraulic measures such as bandalls and Potapov panels have been described by Van Ooostrum and Mackay in UNESCAP (1989). These structures (Figure 46.1) divert the water current, which results in increased turbulence and consequently in increased erosion of the bottom in the fairway.

Induced sedimentation has been discussed by PIANC (1989) and De Meyer and Malherbe (1987). The technique consists basically of designing hydraulic measures to divert sediment transport out of the zones of interest or to fix the transported sediment by structures such as tombolos and breakwaters.

Silt traps, as reported by PIANC (1988, 1989) are created by locally dredging overdepths, thus causing a reduction of the current velocity and consequently the settling of the sedimentation in specific areas. Silt traps are used to optimize the efficiency of the dredging campaigns, to optimize the maintenance dredging operations as a function of the available dredger-capacity and to restrict the dredging operations to specific areas.

46.2.2 STATIONARY INSTALLATIONS

Stationary systems for maintenance dredging have been developed on the basis of several principles: firstly, through stationary silt pumping installations fluid mud is pumped-up and reclaimed in a special mud-deposit or carried away by dumping barges; and secondly, settled mud can be re-fluidized by compressed air- or water-jets, which cause the re-suspension and consequently the removal of the sediment by existing currents.

Stationary silt pumping installations have been described by Berlamont *et al.* (1985, 1986) and reported by PIANC (1988, 1989). The method could be used for maintenance dredging of mud in certain areas, but the efficient functioning of the method is limited

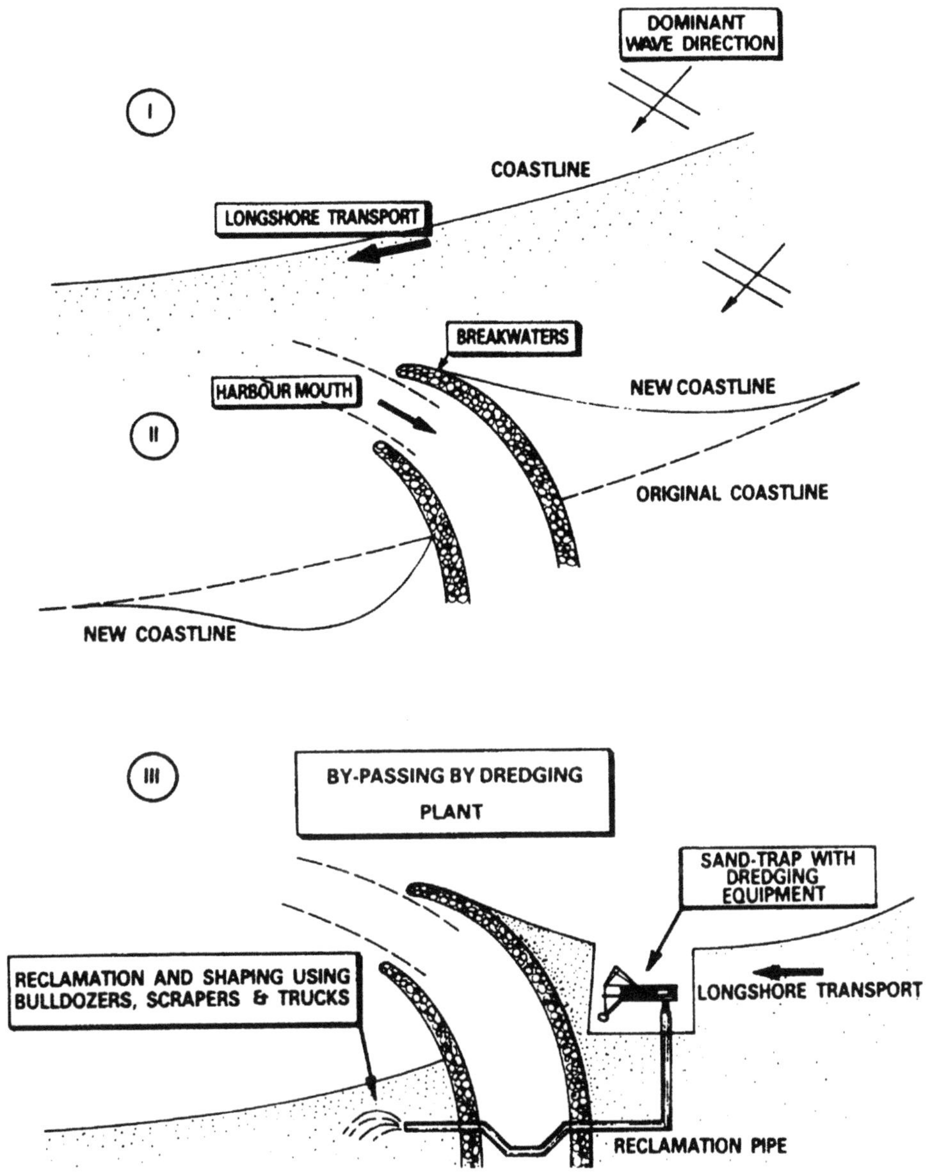

Figure 46.2 Maintenance dredging and shore protection using by-passing dredging plant (De Meyer and Malherbe, 1987) (courtesy of HAECOW N.V. Ghent).

because the radius of influence of the pump depends on the gravitational flux of the mud which is directly related to its structure and flow behaviour.

A fixed installation for re-fluidization consists of a scour jet array which is mounted along a quay wall or on the bottom of a sheltered dock. During the ebb tide, the sediment is scoured and brought into suspension by the jets and consequently removed by the existing currents. The dimensions of a scour jet array depend on the lay-out of the harbour, the intensity of the currents, the sedimentation and the characteristics of the sediments (PIANC, 1989).

By-passing plants (Figure 46.2) are primarily used in combination with sand traps to prevent the inter-

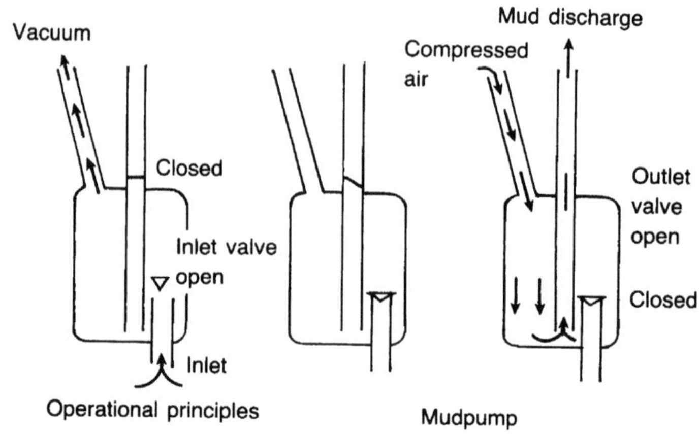

Figure 46.3 Principle of pneuma pump (courtesy of International Marine and Dredging Consultants, Antwerp).

ception of the longshore transported sand by dredging in the adjacent channel, combined with beach replenishment. This method has been applied, for example, at Vishaknapatna in India, Nerang in Australia (Wakefield, 1986), and Delaware, USA (Clausner, 1990).

46.2.3 MOBILE DREDGING PLANTS

Mobile maintenance dredging techniques have been described and discussed extensively by Turner (1984), Huston (1986), Van Oostrum in UNESCAP (1985) and Brosskamp (in PIANC (1989)), in BS 6349 and Bray (1979). Traditional dredging equipment includes mechanical, hydraulic and pneumatic dredgers, while more recently specially developed (low cost) techniques include bottom-levelling and water-injection dredging.

Mechanical dredgers excavate and lift the bottom material out of the water by means of containers (e.g. buckets) which can differ in size and form. Examples are bucket dredgers, dipper dredgers, backhoe dredgers, front shovel dredgers, dragline dredgers and grab dredgers.

Hydraulic dredgers use centrifugal pumps to create a positive suction head, which induces a waterflow in the suction pipe. The material to be dredged is scoured by the water entering the suction pipe (erosion principle) or by auxiliary devices such as cutter-head, bucketwheel, draghead, dustpan or water-jet.

Stationary suction dredgers, cutter suction dredgers, hopper suction dredgers and sidecase dredgers are hydraulic dredgers. Pneumatic dredgers are equipped with a pneuma pump the operating principle of which is the difference between the air pressure in the chambers and the head of the water over the pump (Figure 46.3). Bed-levelling techniques such as ploughs or sweepbeams have been discussed by Van Oostrum (in PIANC, 1988, 1989), van de Graaf (1987), PIANC (1989) and Claessens and Marain (1988). A bed-leveller (Figure 46.4) is usually a tug or work-boat equipped with an underwater plough or shovel which is dragged over the bed. The technique is used to level beds that cannot be reached by traditional dredging equipment and to move the soil to areas where the currents or other equipment can take it away.

The mobile water-injection dredger or Jetsed has been developed by Meublok and Van Weezenbeek (1986) and discussed by Estourgie (1988). The working principle of a mobile injection plant (Figure 46.5) is the same as for the fixed fluidization plants: water is pumped into the bed at high pressure which causes the re-suspension of the bed material, which can be removed by naturally occurring currents.

The application of one or another maintenance dredging method is highly dependent on local circumstances. Jackson (1987), Brouwer (1985) and PIANC (1989) discuss the selection of dredging equipment as a function of hydraulic conditions, navigational requirements, sedimentation characteristics and economical considerations.

46.3 OPTIMIZATION OF MAINTENANCE DREDGING

The optimization of maintenance dredging operations aims at: a reduction of the amount of material to be dredged; and an increase in the efficiency of the maintenance dredging plants by optimizing the maintenance dredging procedure.

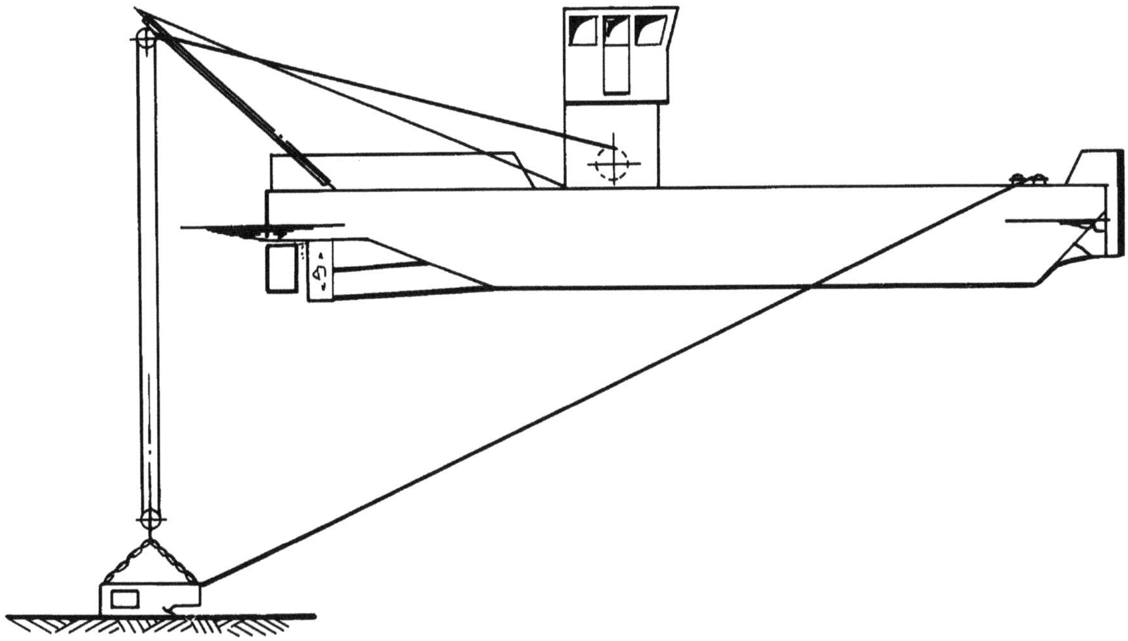

Figure 46.4 Bed-leveller (courtesy of Dredging International, Antwerp).

46.3.1 REDUCTION OF THE QUANTITY TO BE DREDGED

The quantity of sediments to be dredged can be reduced by:

1. Helping nature by applying the techniques described in section 46.1.1: induced erosion, induced sedimentation, silt traps;
2. Installing silt-screens to avoid siltation of certain areas;
3. Introducing the concept of navigable depth for muddy areas.

PIANC (1988) describes three different methods to prevent silt entering harbours: water-jet screens, air-bubble screens and impermeable flexible screens.

The concept of navigable depth (Figure 46.6) was conceived by Parker *et al.* (1975, 1979), Parker and Kirby (1977) and Kirby *et al.* (1980) and was formally introduced by PIANC in 1983 (PIANC, 1983). In most ports of the world silt covering the bottom of maritime fairways and navigation channels causes serious problems for hydrographic surveying, for maintenance dredging and for navigating deep draughted vessels:

1. The signal of a high frequency echosounder (e.g. 210 kHz) is reflected by the top of the fluid mud suspension, which has a density of about 1025–1.08 $t\,m^{-3}$ while the signal of a lower frequency echosounder (e.g. 33 kHz) passes through the fluid mud layer and is reflected by the hard bottom consisting of sand or stiff, consolidated mud.
2. In muddy areas traditional survey techniques do not measure properties related to navigability or have limitations when applied to dredging.

Several studies have indicated that the top of the mud, recorded by a high frequency echosounder, is too conservative to define the navigable depth and the required vertical clearance above the bed. PIANC (1983) established two criteria to be met while defining the navigable depth in muddy areas:

1. The ship's hull must suffer no damage even if its draught were to reach the full navigable depth.
2. The navigational response of the vessel must not be adversely affected.

Taking into account these criteria, the port of Rotterdam has established that the depth of the 1.2 $t\,m^{-3}$ density level is a safe value for navigability.

Model tests and full-scale trials with the supertanker *Lepton* leading to this conclusion have been reported by Sellmeyer and van Oortmeerssen (1983) and by van Bockhove and Nederlof (1981). In France,

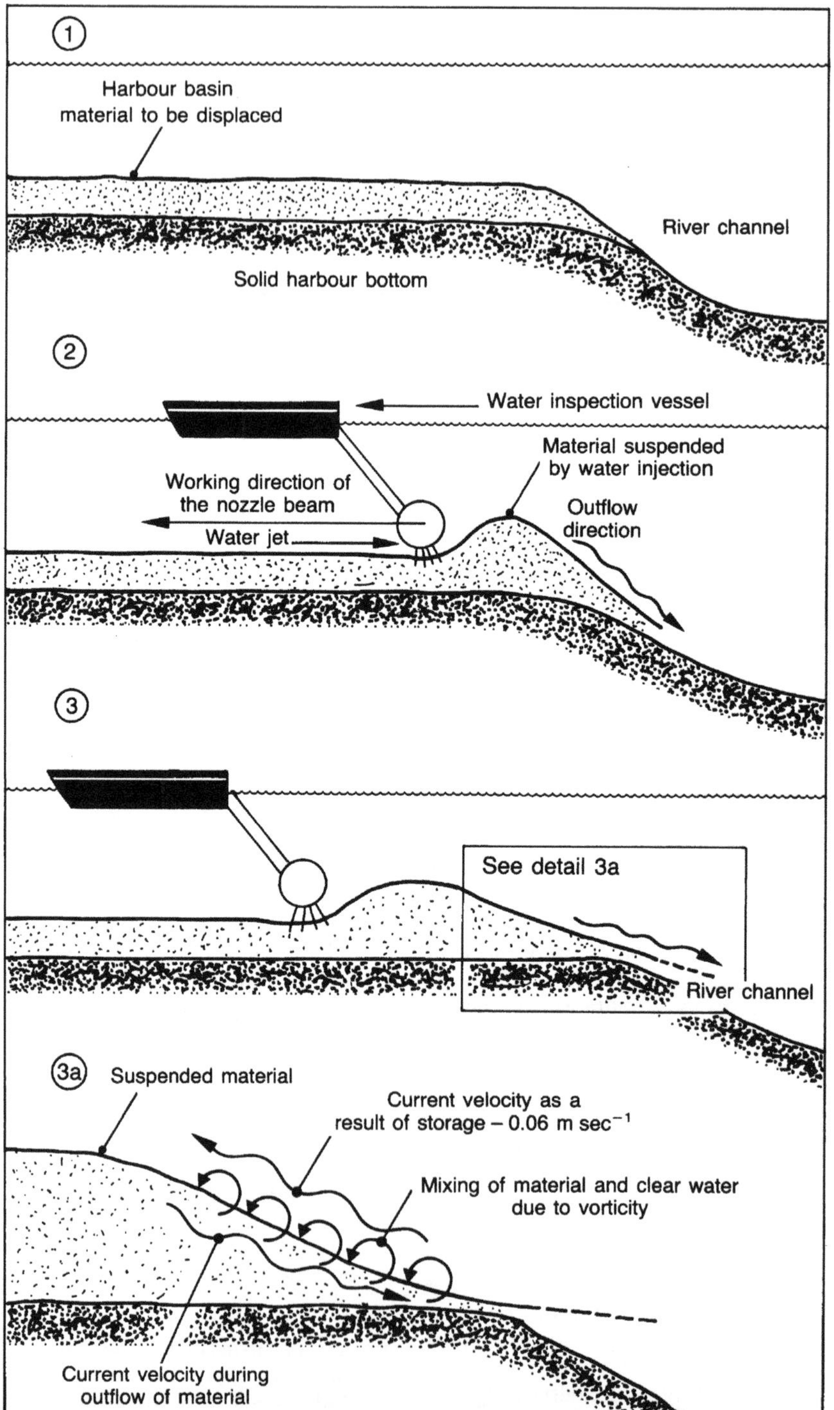

Figure 46.5 Working principle of mobile water-injection dredger Jetsed (courtesy of Volker Stevin Dredging, Rotterdam).

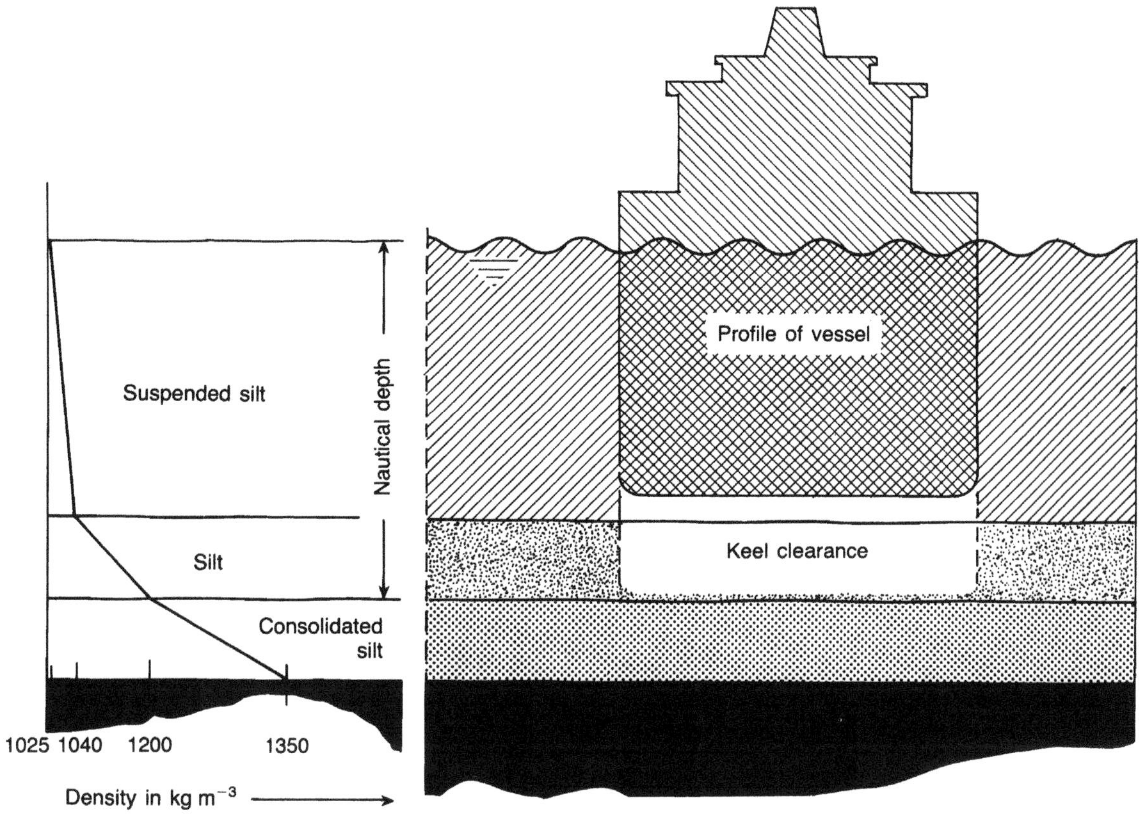

Figure 46.6 Principle of navigable depth (courtesy of PIANC, Brussels).

in the ports of Bordeaux and Nantes, the $1.2\,t\,m^{-3}$ level is also accepted as a safe level for navigability (Granboulan *et al.*, 1984). In Zeebrugge (Belgium), the $1.15\,t\,m^{-3}$ density level has been established as the safest value, after thorough research studying the various aspects related to the determination of navigable depth. Physico-chemical characteristics and structural behaviour has been evaluated by Malherbe and De Wolf (1986), De Meyer and Malherbe (1987), James *et al.* (1986, 1987), De Vlieger and De Cloedt (1987) Kerkcaert and De Vlieger (1989) and van Craenenbroeck *et al.* (1989) and PIANC (1989). The full-scale trials have been used for calibration of a model which was developed and tested by Vantorre and Coen (1988).

The establishment of the $1.15\,t\,m^{-3}$ density in Zeebrugge and the $1.2\,t\,m^{-3}$ density in Rotterdam and France as a safe physical parameter for the determination of navigable depth illustrates that neither the former or latter is a universal value to be applied world-wide, but that the appropriate density criterion depends on local circumstances. Consequently, the reduction of the amount of sediment to be dredged also depends on local circumstances.

In situ density measurements are performed by stationary or continuous measurement systems, based on the backscatter or the transmission of gamma radiation (Parker *et al.*, 1975) or the attenuation of sound (Granboulan *et al.*, 1985) in muddy sediments. The different methods have been evaluated by De Vlieger and De Cloedt (1987).

After evaluating the economical benefits derived from the introduction of the navigable depth concept, De Vlieger and De Cloedt (1987) estimated a net profit margin for the Maintenance Dredging Authority of 11% on the total yearly dredging budget.

46.3.2 OPTIMIZATION OF THE MAINTENANCE DREDGING PROCEDURE

The maintenance dredging procedure can become more effective by improving:

1. Process management;

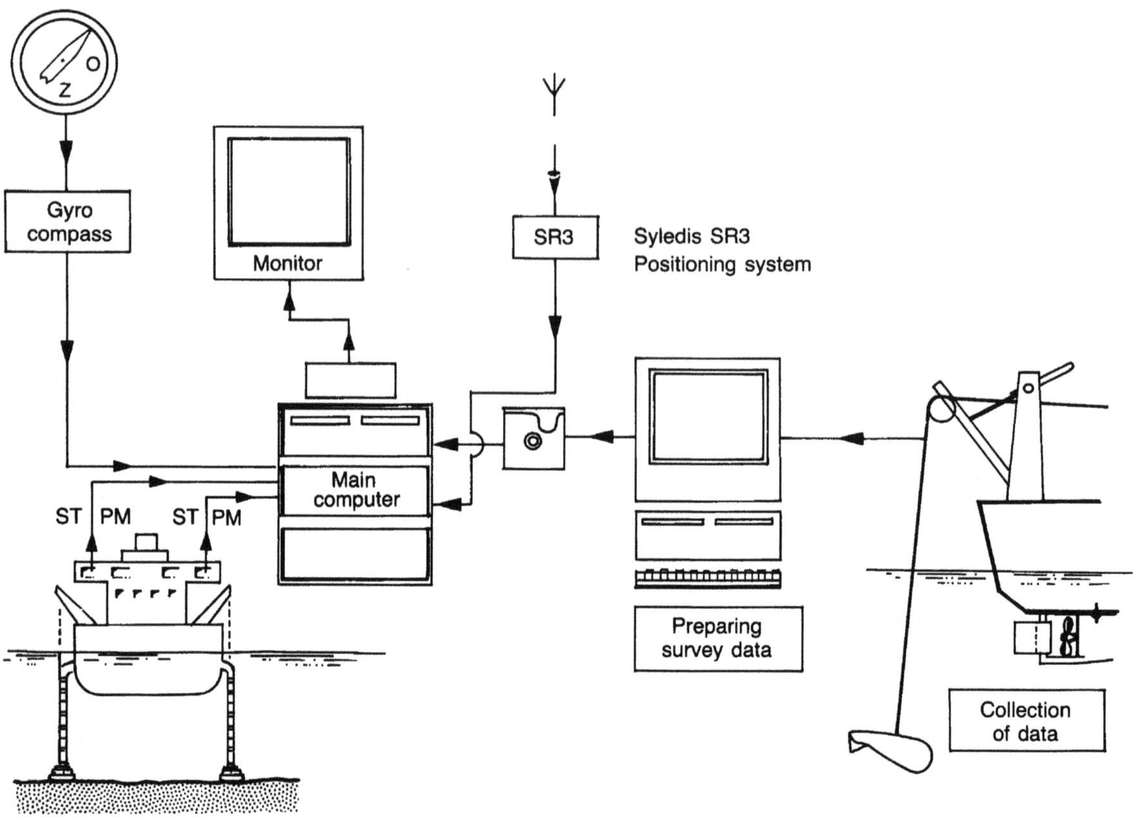

Figure 46.7 Lay-out of dredge-monitor Videoplotter (courtesy of Overseas Decloedt, Brussels).

2. Techniques used on board the dredgers;
3. Use of the disposal site.

Process management

According to Kerckaert and De Vlieger (1989) effective process management is achieved in three steps: process preparation, process control and process evaluation.

The process preparation consists of the collection and analysis of dredging-related data (hydrographic, hydrodynamic, sedimentologic and meteorologic data) which results in the prediction and calculation of maintenance dredging quantities and thus in an optimum dredging plan and organization. Of major importance here are data-collection techniques (rapid echo-sounding, densimetry, rheometry, sampling techniques.) and interpretation, as discussed by Granboulan *et al.* (in PIANC (1989)).

The implementation of the established dredging plan is controlled and evaluated by the supervisors, who are supplied with detailed information to enable the execution of the dredging program established in the process preparation phase: topographic data, maximum depths to be dredged, dredging and disposal procedures, quantities and locations where dredging has to be performed, depth charts and tidal data.

This monitoring of the dredging performances can be obtained by so-called dredge-monitors (Figure 46.7), described by Granboulan (1987), Granboulan *et al.* (1989), De Vlieger and van Craenenbroeck (1987) and PIANC (1989). These dredge-monitors visualize the required dredging-plan data and record simultaneously data related to the dredging output. (See also the next section.)

During the process evaluation, all available data are analysed, on a short-term as well as on a long-term basis. These include:

1. Dredger output and differential charts, produced after regular surveys;
2. Dredger output and pre-process production estimates;
3. Dredged volume and estimated volumes;
4. Statistical analysis of the dredger-output;
5. Sedimentation rates;
6. Natural phenomena.

Improvement of output techniques on board dredgers

Because of their economical and practical applicability for maintenance dredging in coastal and estuarine areas, the efforts to improve the dredging output on board have been pinpointed to the optimization of the dredging system on board trailing suction hopper dredgers. De Vlieger (in PIANC (1989)), van Drimmelen and van Zutphen (1987), Pary (1987) and Jansen (1989) describe the following devices:

1. For bucket dredgers: dredge monitors and the automatic bucket dredger controller for an optimum use of the bucket chain drive and the loads of the head winch.
2. For cutter suction dredgers and bucket wheel dredgers: the dredge monitor or videoplotter and the automatic cutter controller, which has been designed to control swing winches, vacuum relief valve, ladder winch and spud carriage.
3. For grab dredgers: dredge monitor, which provides the dredge master continuously with a top view of the dredger with a target plan and the actual bites taken.
4. For trailing suction hopper dredgers.
   (a) Dredge-monitors (Figure 46.7) enable the visualization of the dredger and its suction tube in the dredging zone, the registration and automatic production of the dredged tracks, the execution of the dredging performance with higher accuracy and the production of dredging intensity charts to establish the relative efficiency of the dredger (PIANC, 1989; van Craenenbroek *et al.*, 1989; De Vlieger and van Craenbroek, 1987; van Drimmelen and van Zutphen, 1987).
   (b) Suction tube position monitors indicate at any moment the exact position of the dragheads, which constantly varies due to turning manoeuvres, tidal differences, currents, waves (PIANC, 1989; van Craenenbroek *et al.*, 1989; De Vlieger and van Craenenbroek, 1987; van Drimmelen and van Zutphen, 1987).
   (c) A draught and load monitor supplies accurate data on the load of the dredger, which enables efficient and safe loading of the ship under steadily varying circumstances (PIANC, 1989; De Vlieger and van Craenenbroek, 1987; van Drimmelen and van Zutphen, 1987).
   (d) The application of degassing systems (Figure 46.8) to reduce the gas content in the dredged material results in higher mixture velocities, increased mixture density, reduced suction times (PIANC, 1989; van Drimmelen and van Zutphen, 1987).
   (e) Deep-loading systems aimed at a reduced required suction head, thus increasing mixture concentration and reducing fuel consumption (PIANC, 1989).
   (f) Automatic light mixture overboard installations increase the efficiency of the hopper loading system (PIANC, 1987; van Drimmelen and van Zutphen, 1987).
   (g) Constant flow devices maintain an appropriate velocity of the sediment in the suction tubes as a function of the mixture density (PIANC, 1989).
   (h) Different types of dragheads are deployed depending on the sediment characteristics (Kerchaert and De Vlieger, 1989; van Craenenbroek *et al.*, 1989; van Drimmelen and van Zutphen, 1987).
   (i) Adjustable overflow systems (Figure 46.9) permit the reduction of the overflow losses and are environmentally more acceptable than traditional overflow techniques (PIANC, 1989; van Drimmelen and van Zutphen, 1987).
   (j) Submerged dredging pumps result in greater dredging depths, higher mixture concentrations and more regular suction processes.
   (k) Optimization of the cycle-output of trailing suction hopper dredgers is obtained by cycle tests to establish the maximum amount of sediment dredged and removed per minute of the total dredging cycle. These tests aim at the optimum dredging time as a function of the area (and sediment) to be dredged (PIANC, 1989, Kerchaert and De Vlieger, 1989; van Craenenbroek *et al.*, 1989).

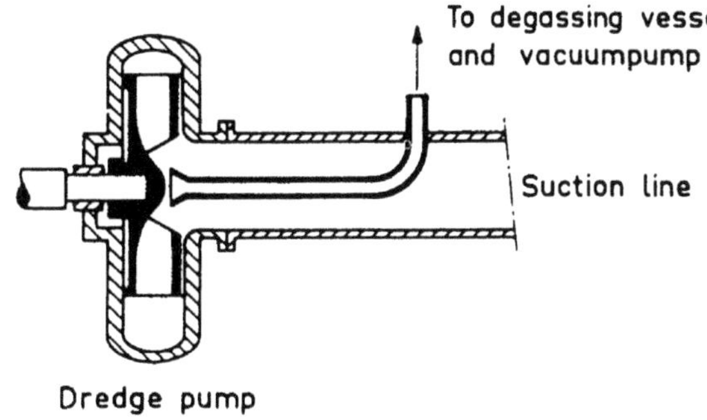

Figure 46.8 Principle of degassing systems (courtesy of IHC-Holland, Sliedrecht).

Optimization of the maintenance dredging process in relation to the natural environment

The optimization of the maintenance dredging process in the natural environment aims at minimizing disposal and overflow losses. Here too, most research has been done for maintenance dredging with trailer suction hopper dredgers.

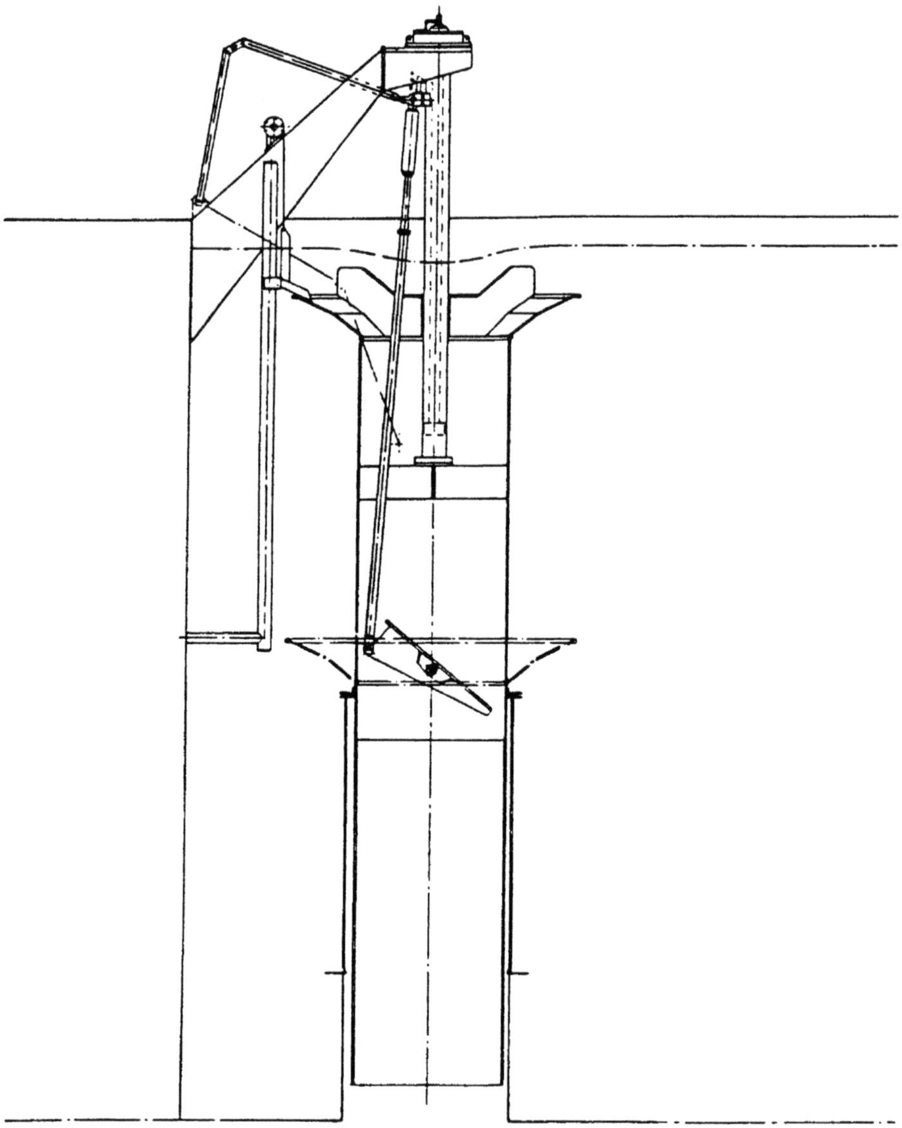

Figure 46.9 Anti-turbidity overflow (courtesy of IHC-Holland, Sliedrecht).

In the case of open water disposal, PIANC (1989) emphasises the analysis of overflow agitation efficiency and dump efficiency. An investigation which has to be conducted on several levels includes:

1. Establishment of a model depicting the overall sediment budget and balances at the site;
2. Analysis of differential charts and bottom composition in the vicinity of the dredge and dump zones;
3. Research into the dynamic behaviour of the overflow and of the disposed sediment by means of radioactive tracers.

Thorough investigations in this field have been described by Malherbe and De Vlieger (1989).

To increase the open-water disposal efficiency, several procedures have been proposed (PIANC, 1989):

1. Disposal of the sediment in deep pits;
2. Disposal of the load through the suction tube to prevent spreading of the spoil by currents;

3. Reclaiming the soil behind retaining walls;
4. Submerged diffuser (van Hoof *et al.*, (1989)).

Other disposal techniques have been described by Malherbe and De Vlieger (1989), and Barlow (in PIANC (1989)).

PIANC (1989), Gallene (1987), Vellinga (1989) and Laboyrie (1989) discuss the disposal of dredged material on land. Silty dredged material can be used for habitat development (Gallene, 1987) while contaminated dredged material needs special care. Large-scale examples of the upland disposal of contaminated sediments are the Slufter and the Papegaaiebek, or 'Parrot's beak', disposal site (Vellinga, 1989; Laboyrie, 1989).

On-board techniques to reduce the adverse environmental impact of maintenance dredging aim essentially at the reduction of turbidity generated during dredging.

For grab dredgers, watertight grabs have been developed. Van Drimmelen and Loevendie (1988) discuss the use of special dragheads and degassing anti-turbidity overflow systems in use on board trailing suction hopper dredgers (Figure 46.9), while van Doorn (1988) introduces the recirculation-of-overflow technique to increase hopperload and reduce turbidity. Pennekamp and van Raalte (1989) and van Raalte and Zwartbol (1986) report on the use of the disc bottom cutter for small cutter suction dredgers for the removal of contaminated soils during maintenance dredging operations.

REFERENCES

Berlamont, J. *et al.* (1985) A permanent mud pumping installation as an alternative for local maintenance dredging. Proceedings of the 21st IAHR-Congress Melbourne, Australia, **4**, 100–5.

Berlamont, J. *et al.* (1986) Local maintenance dredging by means of a fixed mud pumping plant. *Proceedings of the XIth World Dredging Congress.* (ed. CEDA, Delft) pp. 155–62.

Bray, R.N. (1979) *Dredging, a Handbook for Engineers*, Eduard Arnold, London.

BS 6349 (1991) *British Standard Code of Practice for Maritime Structures. Part 5. Recommendations for the Dredging of Waterways and for Land Reclamation* Part 5 to BS 6349. British Standards Institution, London.

Brouwer, K. (1985) The selection of dredging equipment. *Proceedings of the Marintec China 85 Conference*, Shanghai, 4–10 December 1985. Lloyds of London Press (Far East) Ltd., London.

Claessens, J. and Marain, J.M. (1988) Access channel to the Kallo-lock: research of alternative dredging methods. Proceedings of the 9th International Harbour Congress pp. 4.189–200. K.VIV, Antwerp.

Clausner, J.E. (1990) Jet pump sand bypassing plant Indian River Inlet, Delaware. *International Dredging Review* February 1990, pp. 10–1.

De Meyer, C.P. and Malherbe, B. (1987) Optimization of maintenance dredging in maritime and estuarine areas. *Terra et Aqua* **35**, 25–39.

De Vlieger, H. and De Cloedt, J. (1987) Navitracker: a giant step forward in tactics and economics of maintenance dredging, *Terra et Aqua* **35**, December, 2–18.

De Vlieger, H. and van Craenenbroeck, K. (1987) Computers and electronics: indispensible tools for monitoring dredging at Zeebrugge, *World Dredging and Marine Construction*, May, 8–16.

Estourgie, A.L.P. (1988) Theory and practice of water in injection dredging. *Terra and Aqua* **38**, 21–8.

Gallene, B. (1987) Case study of dumping in estuarine areas *Proceedings of the International Seminar on Environmental Aspects of Dredging Action* 27 November–December 1989, Nantes, France.

Granboulan, J. *et al.* (1984) Measuring the navigable depth in a muddy channel, Proceedings of the 2nd International Hydrographic Technical Conference September 1984, Paper 37, pp. 1–8.

Granboulan, J. (1987) Amelioration de la Productivite des Dragues *Revue Navires, Ports et Chantiers* No 441, March.

Granboulan, J. *et al.* (1985) A new instrument for optimizing the maintenance of estuary channels: the ultrasonic probe measures vertical density gradients in muddy channels. Proceedings of Portex 1985, Hamburg.

Granboulan, J. *et al.* (1989) Recent developments in data collection and processing applied to channel maintenance. The case of the Gironde Estuary, *Dredging, Technology, Environmental.* Proceedings of the XII World Dredging Congress, 1989, Orlando, Florida, 2–5 May 1989. WEDA, Fairfax, Virginia, pp. 371–96.

Huston, J.P.E. (1986) *Hydraulic Dredging Principles – Equipment – Procedures – Methods*, John Huston Inc., Corpus Christi, Texas.

Jackson, P.J. (1987) The selection of plant for maintenance dredging. *Proceedings of the Conference on Maintenance Dredging* 20–21 May 1987, Bristol UK. Thomas Telford, London.

James, A.E. *et al.* (1986) A small strain, low shear rate rheometry of cohesive sediments, in *Proceedings of the International Symposium on Physical Processes in Estuaries* 9–12 September 1986, eds van Leussen and Dronkers), Springer Verlag, Berlin.

James, A.E. *et al.* (1987) Direct measurement of static yield properties of cohesive suspensions, *Rheologica Acta*, **26**, 437–46.

Jansen, J.W.M. (1989) A Monitoring System for a Grabdredger. *Dredging, Technology, Environmental, Mining*. Proceedings of the XII World Dredging Congress, 1989, Orlando, Florida, 2–5 May 1989. WEDA, Fairfax, Virginia. pp. 185–94.

Kerckaert, P. *et al.* (1988) Maintenance dredging at the port of Zeebrugge: procedures to achieve an operational determination of the nautical bottom. Proceedings of the 9th International Harbour Congress Vol. 2, pp. 4.13–32, Royal Flemish Society of Engineers, K.VIV, Antwerp.

Kerckaert, P. and De Vlieger, J. (1989) Practical application of maintenance dredging optimization techniques in zeebrugge *Dredging, Technology, Environmental, Mining*. Proceedings XII World Dredging Congress, 1989, Orlando, Florida, 2–5 May 1989. WEDA, Fairfax, Virginia. pp. 611–32.

Kirby, R. *et al.* (1980) Definition of the seabed in navigation routes through mud areas *International Hydrographic Review* **LVII** (1) 107–17.

Laboyrie, H.P. (1989) Disposal of dredged material on land. Short Course on Dredging and the Environment: section D6-2. Stichting Postacademish Onderwijs – Civiele Techniek en Bouw-techniek, Delft.

Malherbe, B. and De Wolf, P. (1986) Nautical bottom research and survey for optimization of maintenance dredging in mud areas. Proceedings of Oceanology International 1986 Brighton, UK.

Malherbe, B. and De Vlieger, H. (1989) Case study dumping in open areas. Proceedings of the Conference on Environmental Aspects of Dredging Activities, 27 November–1 December 1989, Nantes, France.

Meulblok, W. and Van Wezenbeek, R.N. (1986) Silt dredging, water injection technique shows energy consumption advantage, *Dredging and Port Construction.* **XII** (12), 38–42.

Parry, A.J. (1987) The effective deployment of bucket type dredgers, *Maintenance Dredging*, 20–23 May 1987, Bristol. Thomas Telford London.

Parker, W.R. *et al.* (1975) In-situ nuclear density measurements in dredging practice and control. *Proceedings of the First International Symposium on Dredging Technology*. BHRA, Cranfield, UK. pp. 25–42.

Parker, W.R. and Kirby, R. (1977) Fine sediment studies relevant to dredging practice and control. *Proceedings of the Second International Symposium on Dredging Technology*, Texas A&M University. BHRA, Cranfield, UK. pp. 15–26.

Parker, W.R. *et al.* (1979) Observation of fine sediment behaviour. Proceedings of the First International Conference on Cohesive Sediment Cambridge, UK. pp. 1–10.

Pennekamp, J.G.S. and van Raalte, G.V. (1989) Technical aspects of the dredging activity in relation to the environment. Short Course on Dredging and the Environment: section D6-2. Stichting Post-academisch Onderwijs – Civiele Techniek en Bouwtechniek, Delft.

PIANC (1983) Navigation in muddy areas. Report of Working Group 3a, in *Pianc Bulletin 43*, General Secretariat of PIANC, Brussels.

PIANC (1988) Minimizing the cost of maintenance dredging. Report of the Permanent Technical Committee II – *Bull. 63* pp. 51–94, General Secretariat of PIANC, Brussels.

PIANC (1989) *Economic Methods of Channel Maintenance*. Report of Working Group 14 of the Permanent Technical Committee II – Supplement to Bulletin No 67, General Secretariat of PIANC, Brussels.

Sellmeijerb, R. and van Ootmeerssen, G. (1983) The effect of mud on tanker manoeuvres. Proceedings of the Joint Meeting of the Royal Institution of Naval Architects and the Nautical Institute 21 April 1983, pp. 105–24.

Turner, T.M. (1984) *Fundamentals of Hydraulic Dredging*, Cornwell Maritime Press, Centreville, Maryland.

Van, Bockhove, G. and Nederlof, L. (1981) Manoeuvring behaviour of ships in muddy canals and harbours. *The Dock and Harbour Authority* May, pp. 2–6.

UNESCAP (1985) *Agitation Dredging, A Low Cost System*, United Nations, New York.

UNESCAP (1989) *Low Cost Dredging Technology* United Nations, New York.

van Craenenbroeck, K. *et al.* (1989) Maintenance dredging optimization in Belgium: a review of field studies, technique development and practical application. *Proceedings of the Regional Seminar on Optimization of Maintenance Dredging and Low Cost Dredging Techniques* 12–17 February 1990, Panaji, Goa. UNESCAP, Bangkok.

van Doorn, T. (1988) Dredging polluted soil with a trailing suction hopper dredger. Environmentally Acceptable Methods of Dredging and Handling Harbour and Channel Sediments. Proceedings of the Ceda Dredging Day, 28 September 1988, Section 82.

van Drimmelen, N.J. and Loevendie, R.H. (1988) Modern dredging methods and their environmental aspects. Environmentally Acceptable Methods of Dredging and Handling Harbour and Channel Sediments. Proceedings of the Ceda Dredging Day, 29 Sept 1988. Section 84.

van Drimmelen, N.J. and van Zutphen, A.C. (1987) Current and future developments in dredging plant for maintenance dredging. *Maintenance Dredging* 20–23 May 1987, Bristol. Thomas Telford, London.

van de Graaf, C.J. (1987) The use of ploughs or bed-levellers in maintenance dredging. *Proceedings of the Congress on Maintenance Dredging*, 20–21 May 1987, Bristol, UK. Thomas Telford, London.

Van Hoof, *et al.* (1989) Underwater disposal of dredged material. *Proceedings of The Harbour, an ecological challenge* 11–15 December 1989, Public Umweltbehorde Hamburg, pp. 87–91.

van Raalte, G.H. and Zwartbol, A. (1986) The disc-bottom cutter-head, report on laboratory and field tests. Proceedings of the XIth World Dredging Congress, Brighton 1986, pp. 206–25.

Vantorre, M. and Coen, I. (1988) On sinkage and trim of vessels navigating above a mud layer. Proceedings of the 9th International Harbour Congress Vol. 2. pp. 4.149–62, Royal Flemish Society of Engineers. K.VIV, Antwerp.

Vellinga, T. (1989) Environmental effects of dredging and disposal operations. *Dredging, Technology, Environmental Mining*. Proceedings of the XII World Dredging Congress, 1989, Orlando, Florida, 2–5 May 1989. WEDA, Fairfax, Virginia. pp. 236–52.

Wakefield, A.W. (1986) The handling of sand deposits and littoral drift by submerged jet pump, with reference to the Nerang River project. Proceedings of the 10th International Conference on the Hydraulic Transport of Solids in Pipes – Hydrotransport 10, October 1986, Innsbruck, Austria. pp. 29–31.

DISPOSAL OF DREDGED MATERIAL AT SEA 47

J.H. Sargent

Far more attention is paid nowadays than used to be the case to the disposal of dredged material. Capital dredging frequently produces granular materials which are useful for land reclamation and even for beach nourishment inasmuch as deepening often takes place in virgin ground. The problem materials are the fine grained silts and clays which settle in dredged cuts and can constitute a maintenance problem. They are often polluted and many countries have passed legislation on how such materials should be dealt with. The scientific, engineering and legislative aspects of this complex subject are increasing rapidly so that apart from the chapter being a review of the fundamentals it also indicates the directions in which the topic is heading (see also Chapters 34, 38, 45 and 48).

CONTENTS

47.1 BACKGROUND

Dredging works carried out at coastal and estuarial sites will include ports, harbours, marinas, access waterways and canal entrances. Dredging is usually differentiated into capital works for new or extended construction and maintenance work required to keep a port or waterway operational.

Capital dredging frequently produces dredged material which is suitable for reclamation fill, since it often involves the dredging of new channels and berths or the deepening of existing facilities in virgin ground. Maintenance dredging, by definition, usually takes place to remove fine grained material which has collected in existing channels and berths from sedimentation processes.

In engineering terms, material which is unsuitable for use in reclamation will be very fine grained sediment including soft clays and silts. Such materials have been used as a hydraulic fill, but give rise to problems related to the long period required for effective consolidation, due to their high water content and low strength. Even after the consolidation process is effectively complete and the area has been improved by blanketing with better quality material its use will be limited to light loading unless special ground treatment is carried out.

Until relatively recently few restrictions were placed on poor quality material which required disposal. Such disposal took place at 'dump sites' at sea, was discharged into natural current streams, or placed on land in disused pits, or into specially constructed lagoons. It has been estimated that between 80% and 90% of all 'wastes' deposited at sea has arisen from dredging work (GESAMP, 1990). Volumetric estimates are difficult to determine although it seems likely that well in excess of 1 billion cubic metres is dredged globally each year. In a 1981 survey by the International Association of Ports and Harbours (IAPH) it appeared that nearly 60% was deposited in the nearshore or at sea and a repeat survey in 1986 substantiated this division. From the same surveys it appears likely that capital works account for perhaps one quarter of the volume whilst the remainder is from maintenance work.

Authorities are becoming increasingly aware of the potential value of dredged material for beneficial uses, especially in relation to the repair, protection or enhancement of coastal zones, and there appears to be an increase in the use of dredged material for beach nourishment and other coastal features (PIANC, 1992).

Coastal, Estuarial and Harbour Engineer's Reference Book. Edited by M.B. Abbott and W.A. Price. Published in 1993 by Chapman & Hall, 2–6 Boundary Row, London SE1 8HN. ISBN 0 419 15430 2.

Modern public attitudes supported in many countries by legislation now require that the treatment of dredged material is dictated by its level of cleanliness, although it is accepted that a very high percentage of dredged material is 'clean' in nature, in which case few problems should arise concerning its disposal.

However, it is essential, in present day planning, that not only a determination of the physical nature of the material to be dredged is made but also that an examination to check its chemical content is carried out so that its suitability for various uses or for its ultimate disposal can be assessed.

Whenever dredged material is of suitable quality, then every effort should be undertaken to utilize it for engineering purposes, even if a suitable construction project is not immediately available. Such material should be viewed as a valuable resource and consideration given to its retention by stock-piling on land or underwater for future use.

The comparative merits of disposal on land compared with that at sea is the subject of much debate at present. In this context the assimulative capacity of the ocean has been questioned, but the infrastructure developments which aid the economic growth of a country are themselves essential and material of all types (including dredged material) will arise and must be handled. Land deserves special environmental protection and pressures on land use are increasingly apparent from the public who will not tolerate undue environmental disturbance.

These attitudes should motivate engineers to solve disposal problems through new solutions and in this respect disposal at sea in the future may increasingly require new techniques, including the construction of artificial islands using dredging and construction techniques which have developed in the last two decades.

47.2 ENVIRONMENTAL CONCERNS ABOUT DISPOSAL

Public awareness concerning the environment was highlighted in the United Nations Conference on the Environment in 1970. This conference triggered the development of legislation concerning the control of 'wastes'.

Such legislation included the 1972 London Dumping Convention (LDC) which is global in effect and is administered by the International Maritime Organization (IMO) in London. This convention has now been re-named the 'London Convention of 1972'. Article 1 states 'Contracting parties shall individually and collectively promote the effective control of all sources of pollution of the marine environment and pledge themselves especially to take all practicable steps to prevent all pollution of the sea by the dumping of waste and other matter that is liable to create hazards to human health, to harm living resources and marine life, to damage amenities or to interfere with other legitimate uses of the sea' (IMO, 1972).

The Convention regulates the deliberate disposal of any material into the sea and has now been ratified by 67 countries (IMO, 1991).

In view of the predominantly clean nature of dredged material the term 'dumping' is unfortunate and the Permanent International Association of Navigation Congresses (PIANC) in particular has been active in promulgating an improved terminology, including the term 'disposal' in the place of 'dumping' and 'dredged material' in the place of 'spoil'. These are not just semantic niceties, but are important in separating dredged material from other 'wastes', such as industrial wastes, many of which contain toxic and harmful elements.

In 1986, after lengthy debate, the LC72 approved the publication of *Guidelines for disposal of dredged materials* (Sargent, 1987). This document effectively brought together all the requirements of the Convention relevant to the disposal of dredged material in one place and has recognized that such material should be viewed differently from other wastes.

Other conventions have developed on a regional basis with the same aims. Of particular relevance to the United Kingdom is the Oslo Convention (1974). Developments on the subject are rapid, and a recent article (Hummer *et al.*, 1993) provides the latest overview.

Dredging in coastal areas, especially maintenance dredging in ports and estuarial waterways, may encounter material which is contaminated due to industrial wastes (at the port location or upstream in the rivers and estuaries) and contaminants may find their way into dredged channels or berths as part of the siltation process.

Problems exist in some ports with contaminated sediment due to accumulation over many years from industrial misuse. In such cases special measures will have to be taken and disposal at sea is unlikely to be available as an option. Material dredged from inland waterways in industrial countries is particularly susceptible to contamination and even in rural areas sediment may be effected in waterways by the run-off from fertilizers and chemicals used in farming (PIANC, 1990).

Recent publications by IAPH (1989) and World Bank (1990) provide useful advice concerning environmental precautions to be taken by ports and harbours.

47.3 COMPARATIVE ASSESSMENTS (LAND *VERSUS* SEA)

Current practice demands that all alternatives for disposal should be considered at the planning stage to ensure minimal environmental impact. Various approaches are adopted by authorities in different countries often dictated by political attitudes. These may vary from the relatively complex procedures used in the USA where legislation is onerous and high demands are placed on the authority disposing of the waste to much more lenient measures in the developing countries or regions with less industrialization where practice is dictated principally (or solely) by consideration of costs.

A vast output of information concerning the handling of dredged material (including contaminated dredged material) and comparative assessments is available from the United States Army Engineer Waterways Experiment Station (WES) Vicksburg, Mississippi.

When making an assessment it will be necessary to consider the following factors:

1. Legal requirements;
2. Technical aspects;
3. Economic considerations;
4. Environmental aspects.

Included in the last factor will be the need to take into account the acceptability of the solution to the public, (such as the social interference that may be caused with existing amenities) and any likely impact on public health.

A suggested strategy developed in the USA for the selection of disposal options is included in a PIANC report (PIANC, 1990) and a similar but alternative strategy which is perhaps more suited to European practice was given in a paper to the 1990 PIANC Congress (Kreps-Heyndrikx *et al.*, 1990).

Once the decision has been made that disposal at sea is the most practicable solution then other factors will arise concerning the need for a licence, the best technical method of transport and disposal and the consequent need for monitoring of the disposal site.

47.4 LICENSES FOR DISPOSAL AT SEA

Increasingly (and especially in the developed countries) studies are required prior to permission for disposal at sea. Requirements pass under different names from limited testing to the all-embracing 'environmental impact assessment'. In the European Community a directive (the Environmental Assessment Directive) is in force and requires that consent for 'public and private projects which are likely to have significant effects on the environment' should only be granted after a formal environmental assessment has been carried out.

The effect of complying with legislation means that a considerable amount of information about the proposed disposal is now required.

Prior to the issue of a licence for sea disposal several criteria usually need to be satisfied:

1. Type and nature of the sediment, (engineering description) and degree, if any, of chemical contamination;
2. Assessment of material quantity;
3. Characteristics of disposal site;
4. Examination of potential impact including any interference with fishing and other users of the sea;
5. Need for monitoring, both at the time of disposal and afterwards.

Where the new dredging work consists of an extension of maintenance dredging, historical data is normally available and will be useful in reducing or eliminating the need to carry out testing.

In most cases (except in those countries where disposal at sea of any waste is not permitted), it is possible to obtain a licence providing the various steps outlined above have been followed. However, sampling, testing and analysis take time and are expensive and the cost of proving that the dredged material will not have any significant environmental impact is in some cases becoming onerous. The cost of possible monitoring programmes must also be taken into account.

Specifically in the UK it will be necessary to consider the following legislation; Environmental Protection Act (1990), London Convention of 1972, the Oslo Convention (1974), EC Environmental Assessment Directive (85/337) and other relevant regulations.

The licensing authorities are: the Ministry of Agriculture and Fisheries (MAFF), in England and Wales; the Department of Agriculture and Fisheries (DAFS), in Scotland; and the Department of the Environment for Northern Ireland (DOE NI). A useful summary of the licensing procedures for disposal in UK waters has been given by Parker (1987).

In view of the fact that problems arising in disposal of dredged material are largely, if not wholly, the result of contamination from sources such as sewage, industrial wastes, spills and other discharges (either directly into the water or as run-off from land) it is vital that legislation is increasingly enforced on the 'polluter' in order that sources of contamination are reduced or stopped.

Considerable efforts are being made in this respect by some countries, and the 'North Sea Conferences', the last being held in 1990, provide an excellent example of common actions which are in hand.

47.5 REVIEW OF DREDGING TECHNIQUES

Various publications give a good overview about plant and equipment used for the dredging and disposal of dredged material (Bray, 1979 – new edition to be published).

With particular regard to disposal at sea, the following plant is commonly employed:

Trailer dredgers

Trailer dredgers (or more correctly trailing suction hopper dredgers) are used widely for both capital and maintenance work, especially in access waterways. The technique employs a centrifugal pump either mounted within the vessel or on the suction arm, the material being pumped into a hopper which forms the body of the vessel. The hull is fitted with discharge facilities comprising either bottom doors or valves. Additionally, some trailers are equipped with 'pump out' facilities, but these will normally only be employed to discharge material onto a reclamation site. The trailer is an efficient vessel able to work and sail under difficult weather and sea conditions.

Split hull vessels

The development of the split hull vessel in which two halves of the hull open, originally took place to allow material with high adhesion to be discharged more efficiently. This design has been extended to trailer dredgers and hopper barges and has become increasingly popular.

Hopper barges

Self-propelled hopper barges or dumb barges are frequently employed for loading by bucket dredger, the hoppers then sailing to the disposal area. Large self-propelled hoppers can sail in difficult sea conditions, but tug-assisted dumb barges will be very limited in operational use at sea, and their application should be confined to work in sheltered waters.

Figure 47.1 Trailing suction hopper dredger after loading, sailing to disposal site at sea.

Figure 47.2 Split-hull trailer dredger unloading at sea.

Hydraulic dredgers (other than trailer dredgers)

Hydraulic dredgers which include a very wide range of cutter suction and suction dredging plant using floating pipeline systems are less commonly employed for discharge at sea. An exception can occur when such a dredger pumps through the pipeline into a current or tidal stream, allowing dispersion by natural processes.

Special techniques

In order to mitigate environmental impact various techniques have been developed (CEDA, 1988). For example, to reduce turbidity effects (i.e. the temporary mixing of fine grained material with water in the dredging process which produces a silt-coloured plume or trail) special overflow systems and submerged diffusers have been developed.

Dredged material quality may also be improved by other techniques including centrifuge and dewatering systems, although clearly these additional processes will add considerably to the cost of the work.

47.6 MANAGEMENT OF DISPOSAL

In order to consider the management of disposal it is necessary to review the options available which will include:

1. Ocean disposal or nearshore disposal sites;
2. Island containment nearshore or in an estuary (i.e. either an artificial island or the use of a natural island);

Figure 47.3 Bucket dredger loading hopper barges.

3. Disposal within pits in nearshore or estuarial sites;
4. Disposal into natural tidal or current streams.

Open disposal into the sea or nearshore is the most frequently used method and is usually acceptable for 'clean' material. Dispersion is unconfined and will often result in a slow accretion at the site.

Where dispersion is to be restricted or contained, other methods involving underwater dams on the sea-bed, disposal in pits or on islands will all provide environmental advantages compared with open disposal but at higher cost.

Methods involving 'capping' have taken place at various locations in which material considered to be contaminated is covered (or 'capped') with a layer of better quality material in order to both prevent dispersion and isolate contaminants from movement into the surrounding water.

Disposal directly into tidal or current streams will disperse material, but in this case, field studies, often including tests with radioactive tracers, will be required to determine the destination of the dumped material. Physical and mathematical modelling are other tools that can be invoked in this process. The method cannot be viewed favourably where there is a likelihood of a return of the material or where turbidity is markedly increased. A full treatment of the management of disposal has been given in a PIANC report (PIANC, 1986).

ACKNOWLEDGMENTS

Illustrations are reproduced by kind permission of IHC Holland, P.O. Box 208, 3350 AE Papendrecht, Holland.

REFERENCES

Bray, R.N. (1979, with new edition under publication) *Dredging*, Edward Arnold, London.

CEDA (1988) Environmentally acceptable methods of dredging and handling harbour and channel sediments. Proceedings of the Dredging Day, Hamburg, 1988.

GESAMP (1990) The state of the marine environment, *Rep. 39*, 1990 (GESAMP-United Nations Environment Programme).

Hummer, C.W. *et al.*, (1993) WODA's role at the London Dumping Convention. *Terra et Aqua*, January 1993.

IAPH (1981) Guidelines for Environmental Planning and Management in Ports and Coastal Area Developments, IAPH, Tokyo.

IMO (1972) Convention on the Prevention of Marine Pollution (London Convention of 1972), IMO, London.

IMO (1991) The London Dumping Convention – the first decade and beyond. IMO, London.

Kreps–Heyndrikx, Y. and others. Paper to Section II Subject 2 of the Proceedings of the 27th International Navigation Congress, Osaka 1990 (PIANC).

Oslo Convention (1974) Convention on the Prevention of Marine Pollution. Oslo Commission, London.

Parker, M.M. Paper. *Proceedings of the ICE/CEDA Conference on Maintenance Dredging*, Bristol 1987. Thomas Telford, London.

PIANC (1986) Disposal of dredged material at sea. Report. *Supplement to Bulletin 52*.

PIANC (1990) Management of dredged materials from inland waterways. Report. *Supplement to Bulletin*.

PIANC (1992) Beneficial uses of dredged material. Report. *Supplement to Bulletin 76*.

Sargent, J.H. (1987) PIANC and dredging – the environmental issues. *Bulletin 57*, PIANC, Brussels.

World Bank (1990) Technical Paper No. 126. Environmental Considerations for Port and Harbour Developments. World Bank, Washington D.C.

COASTAL POLLUTION, WATER QUALITY AND OIL SPILLS 48

P. Neville-Jones and G.E. Thompson

Chapter 47 discusses the disposal of dredged material at sea, and the special problems associated with the disposal of heavily polluted fine grained sediments arising from maintenance dredging are described. This chapter deals with other contaminants that can enter the marine environment from natural, domestic, industrial and agriculture sources. Many countries now have water quality standards. The UK for example, allocates sets of standards to particular areas depending on their use – fisheries, bathing and marine recreation are some of them. To meet such standards requires a strategy for waste water disposal and the design of one such scheme is described. The chapter concludes with a review of the methods available for the prediction of the movement of oil spills (see also Chapter 40).

CONTENTS

48.1 INTRODUCTION

This chapter commences with a description of the main contaminants entering the marine environment, and explains the strategy for the design of coastal waste-water disposal schemes to meet modern environmental criteria. A case history is then presented describing an environmental study for the planning of a major scheme. The chapter concludes with a summary of methods for the prediction of the processes occurring during an oil spill.

48.2 CONTAMINANT SOURCES

Pollution of the marine environment is caused by the adverse impact of contaminants released into the sea.

The main types of contaminants are;

Natural
Domestic
Industrial
Agricultural

Natural contaminants are those produced by nature without the interference of man, and include sediments, dissolved salts, chemical and organic material. Man's activities may artificially increase the supply of these contaminants, such as when loads in rivers are increased by deforestation and subsequent erosion in the catchment. Domestic contaminants are those produced in homes, and consist mainly of sewage and minor amounts of chemicals, such as detergents. Industrial contaminants cover the whole spectrum from food processing wastes to discharges from chemical plants and refineries. The heat in warm cooling water discharged from power stations is also a contaminant in this category. Agricultural contaminants are mainly nutrients released due to the run-off of fertilizers and chemicals into drainage ditches and rivers. Agriculture is also a source of pesticides which may sometimes be very persistent contaminants.

These contaminants arrive in the sea by a number of means:

River discharge;
Atmospheric transport;

Coastal, Estuarial and Harbour Engineer's Reference Book. Edited by M.B. Abbott and W.A. Price. Published in 1993 by Chapman & Hall, 2–6 Boundary Row, London SE1 8HN. ISBN 0 419 15430 2.

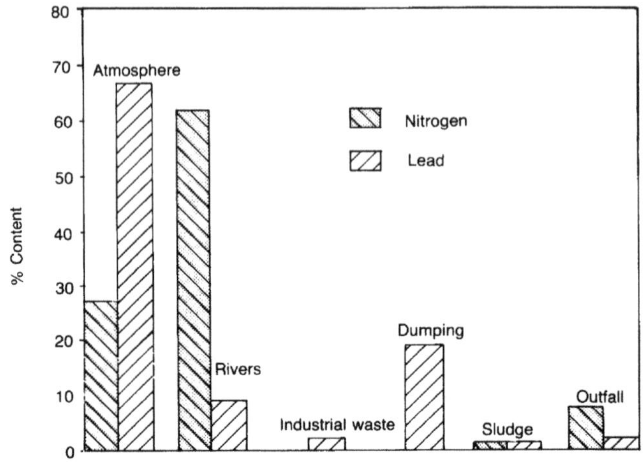

Figure 48.1 Lead and nitrogen inputs to the North Sea.

Waste-water discharges;
Dumping of dredged materials;
Industrial waste disposal;
Sewage sludge dumping.

The dumping of sludge and industrial wastes are being increasingly controlled world-wide and these practices may well be prohibited in the more environmentally concerned nations.

The relative inputs of lead and nitrogen to the North Sea may be seen in Figure 48.1. Figure 48.1 suggests that the majority of the 10 000 tonnes of lead entering the sea each year enters via the atmosphere (UK DoE, 1987a,b) which reinforces the environmental concern over lead emissions from motor vehicles. Figure 48.1 shows that the atmosphere still provides a significant contribution of nitrogen, but this is outweighed by that discharged from rivers.

The impact of contaminants is very variable and depends both on their quantity and concentration and the waters into which they are discharged. Table 48.1 gives an approximate idea of the persistence of contaminants in the environment.

The main long-term persistent contaminants are the metals and some organic chemicals. The contaminants are therefore of particular concern to the general preservation of the environment. The long-term consequences of damage to the environment are difficult to predict and may be virtually irreversible, by the time the impact becomes apparent. The avoidance of adverse environmental impact from any of the contaminants listed in Table 48.1 is obviously desirable, whether the impact is reversible or not.

Table 48.1 Contaminant persistence in the environment (illustrative)

Contaminant	*Persistence*
Metals	Indefinite
Organic chemicals	
Non-biodegradable	Years
Biodegradable	Hours to months
Nutrients	Days to months
Ammonia	Hours to days
Biochemical oxygen demand	Days
Suspended solids	Hours to days
Viruses	Hours to months (years)
Enteric bacteria	Hours to days

48.3 CONTROL OF CONTAMINANTS

The entry and dissipation of contaminants in the environment may be controlled in a number of ways, the most significant being:

1. Control contaminant production;
2. Treat contaminants at source;

3. Treat the combined contaminants by waste-water treatment methods;
4. Contain the contaminants in storage;
5. Encourage natural decay of contaminant;
6. Disperse contaminants to reduce the concentration.

Generally coastal pollution is controlled by a combination of these methods. Once the contaminants have been mixed in with a large volume of water the treatment of containment of contaminants is very difficult. Where the production of the contaminant is unavoidable, the most effective approach is treatment at source.

48.4 PROTECTION OF THE MARINE ENVIRONMENT

The marine environment in Europe is generally protected by setting emission standards and water quality standards. Emission standards are set to limit the quantity of particular contaminants which may be discharged at any one point. Water quality standards limit the maximum concentrations of contaminants released into the environment.

In Europe the general consensus has developed for a combination of emission and water quality standards as may be seen from the environmental legislation instigated by the European Commission (EC 1976a,b, 1979).

In the UK the water quality standards are set by dividing the receiving waters up into a number of use areas which are:

Fisheries
Shellfisheries
Ecosystem
Abstraction
Bathing
Marine recreation
Mixing zone

Each of these use areas has its own set of water quality standards, called environmental quality standards (EQS). The standards are taken from the UK and

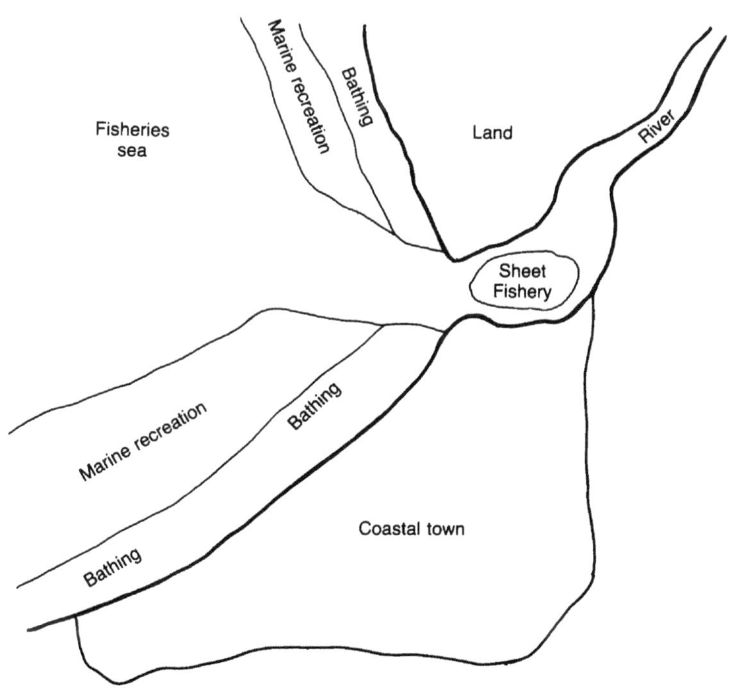

Figure 48.2 Use areas for coastal town.

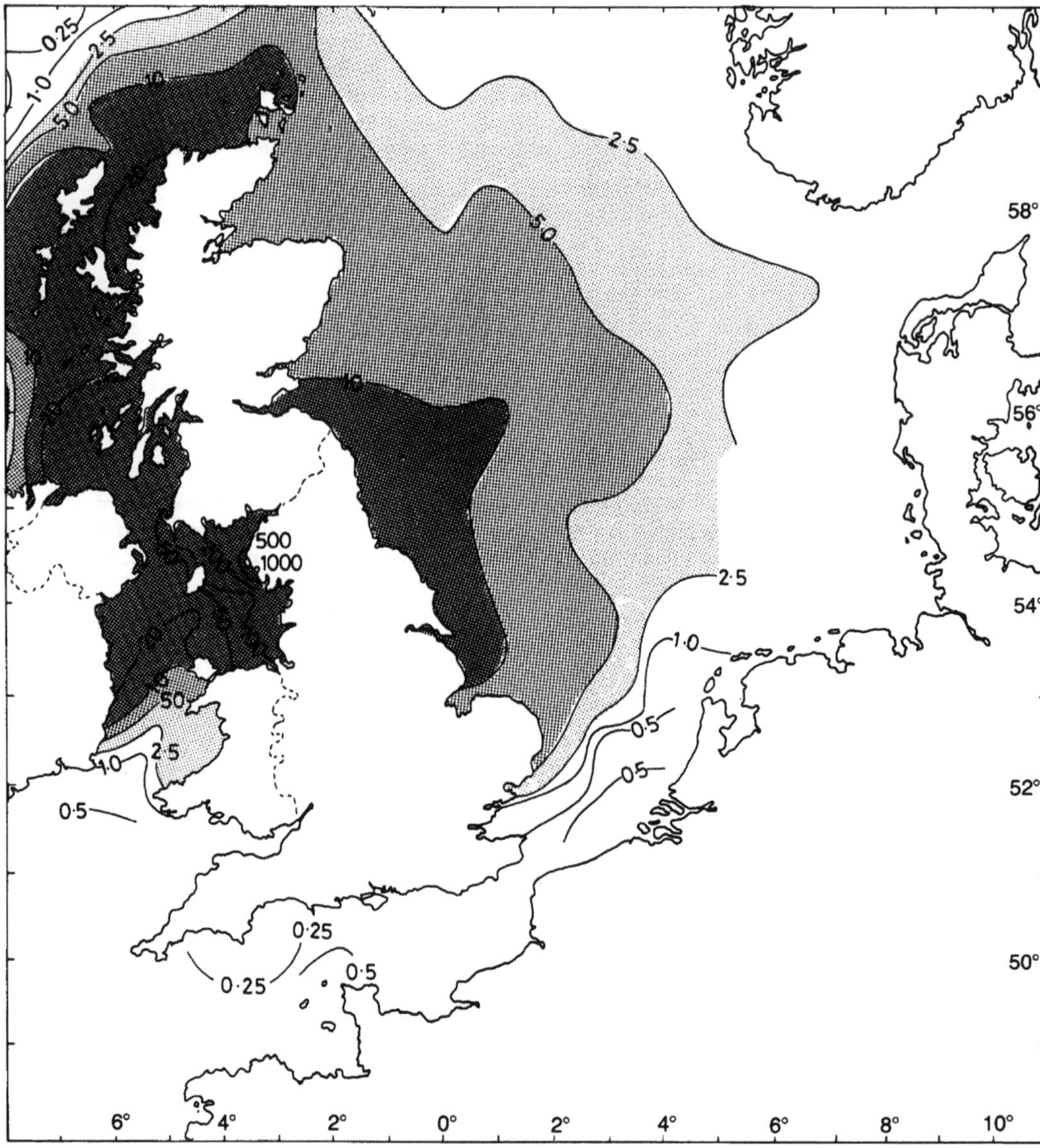

Figure 48.3 Distribution of calcium 137.

European legislation which applies to the particular use (EC, 1976a,b, 1979, UK DoE, 1988).

The use areas off a typical coastal town are shown in Figure 48.3. Bacterial standards are set to protect the inshore bathing waters and shellfisheries. The more general uses of fisheries and ecosystem are protected by standards for dangerous substances such as organic chemicals and metals mainly derived from the EC Dangerous Substances Directive (EC, 1976a).

The WRC Engineering report entitled 'The design of long sea outfalls for environmental protection' gives a more detailed account of UK environmental design practice (Neville–Jones and Dorling, 1986).

48.5 TRANSPORT OF CONTAMINANTS IN THE SEA

Contaminants in the sea may be transported as:

Dissolved substances;
Suspended solids;
Sediments;
Organic matter.

The movement and dispersion of dissolved substances is reasonably well understood and can be modelled physically or mathematically. Radioactive metals provide a particularly useful tracer for dispersion of contaminants as can be seen in Figure 48.3 which shows the concentrations of caesium emanating from the Windscale site in 1973 (MAFF, 1980). Radioactive elements may be identified when present in very small quantities. The distribution of lead in Figure 48.5 shows the emanation of lead and its dissipation in the North Sea from the Humber. Such data is extremely useful when trying to understand the long-term movement of other materials in the sea.

Much effort has been put into the development of mathematical techniques to predict contaminant dispersion, and a typical application is described later in this paper, in relation to the environmental protection of the coastal waters around Hong Kong.

The movements of suspended solids are less easy to predict, although some information may be gained by following methods similar to those for the dissolved contaminants. The real problems in prediction start when the suspended solids settle out and sediment movement has to be considered. Most settlable solids entering rivers become trapped in the estuary due to density circulation effects assisted by the accelerated flocculation of particles in sea water (McDowell and O'Connor, 1977). This explains the high contaminant loads within spoil dredged from major estuaries, as is evident in Figure 48.1 from the high concentration of lead in the North Sea from dumping.

The short-term movement of sediment can be predicted to a certain extent but the long-term fate can only be inferred from studies of existing sediment movement patterns.

48.6 CONCENTRATION, DILUTION AND DECAY OF MARINE POLLUTANTS

The atmospheric contaminants mix at low concentrations with very large volumes of water. Any environmental hazard created by atmospheric pollution can only be reduced by the natural process of mixing and decay. Consequently control at source is the only available option. Concentrations of water-borne contaminants are highest near their source. The usual sources occur downstream of urban settlements and industrial developments. For urban sewage the contaminant source is either through a direct discharge from a long sea outfall or through an indirect discharge from a sewage treatment works via a river, as shown in Figures 48.5 and 48.6, respectively. The concentration typically reduces in the direction of dispersion, except where ecological factors intervene, such as the accumulation of heavy metals and bacteria by shellfish.

Contaminants dumped from ships may either accumulate at the site of dumping or disperse, or may do both, depending on the site conditions.

The modelling of dispersion decay in the marine environment is illustrated by the case study of Hong Kong in section 48.7.

48.7 CASE STUDY: MODELLING THE COASTAL WATERS OF HONG KONG

48.7.1 INTRODUCTION

In recent years computer simulation models have been increasingly used to assist in making decisions on the appropriate strategy to be followed to attain water quality objectives in coastal waters. A specific example of such an application is the development of the WAHMO suite of models for the territorial waters of Hong Kong.

It is well understood by the Government of Hong Kong that in order to arrest the gradual deterioration of its marine waters, proper planning for the treatment and disposal of sewage is essential. The HK Government's 1988 Review says:

> By the mid-1980s it had become clear that because of the speed of growth and development in Hong Kong, and because of the changing aspirations of the community, a major review of the overall strategy for sewage disposal needed to be carried out. A new Sewage Disposal Strategy Study has therefore been undertaken. Its principal aim is to put forward a number of possible territory-wide schemes for the treatment and disposal of sewage, each of which would result in water quality objectives being met throughout the territory. The study involves the selection of a preferred option on the grounds of cost-effectiveness and, for this option, the provision of general guidelines on design criteria for the necessary infrastructure.
>
> To examine the effects of the different options and development scenarios on water quality a multilevel suite of mathematical Water Quality and Hydraulic Models was developed.
>
> With the availability of the WAHMO mathematical models, the environmental consequences of all the major developments in water quality and

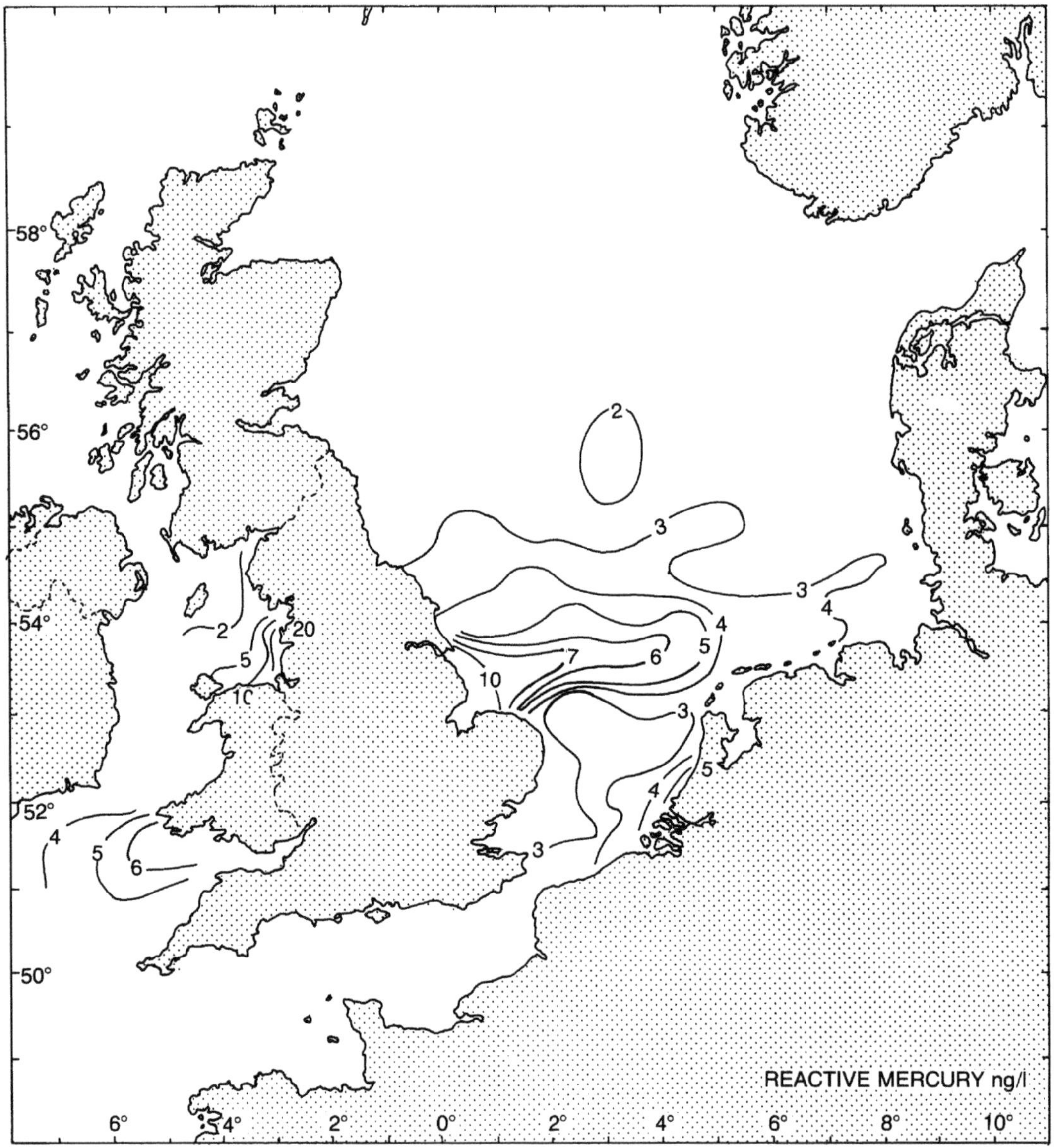

Figure 48.4 Distribution of mercury.

hydrology can be properly assessed. As a result, serious environmental decline should be avoided and any remedial actions can be planned promptly.

The development of the suite of models and the acquisition of the data base necessary for the reliable use of the models make the territorial waters of Hong Kong one of the most intensively monitored and modelled water bodies in the world.

48.7.2 OVERALL CONCEPT OF THE MODEL SUITE

The WAHMO suite of models was developed to study the effect of major reclamations on both water quality and port activities.

In developing the models there were three main areas of interest with regard to water quality which were addressed:

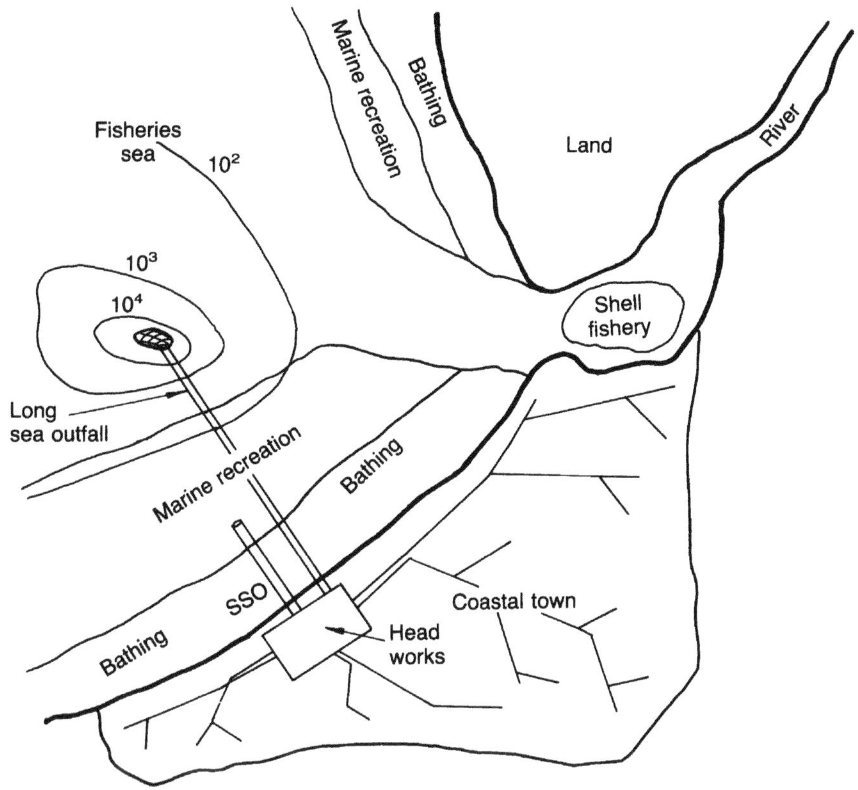

Figure 48.5 Coastal sewerage scheme with long sea outfall. 10^2 is 10^2 Ecoli/100 ml; SSO denotes storm sewage overflow.

1. The biochemical oxygen demand in the water, and the resulting dissolved oxygen levels which are indicators of the general level of organic pollution and have a major effect on marine life.
2. Bacterial levels, because of the potential health risk to humans who bathe in the water or fall in from wind-surfers or water-skis, and the health risk from eating bacterial contaminated shellfish.
3. Concentrations of algae, as measured by chlorophyll, and concentrations of inorganic nitrogen which cause algae to grow, because of the problem of unsightly and possibly toxic algal blooms and the consequent instability with regard to dissolved oxygen levels.

The water quality parameters modelled include:

Dissolved oxygen
Suspended solids
Oxidized nitrogen
Organic nitrogen
Ammoniacal nitrogen
BOD
Chlorophyll
Temperature
Salinity
Bacteria

A set of eleven complementary mathematical models were developed by Binnie and Partners, Hydraulics Research Ltd and the Water Research Centre to form the WAHMO suite. The models can be categorized into four discrete topic areas:

Water movement models;
Water quality models;
Sediment models;
Wave models.

The models range in coverage and hence detail from those which cover the whole territorial waters and the Pearl estuary as shown in Figure 48.7 to those which cover individual small coastal embankments and typhoon shelters or an individual outfall.

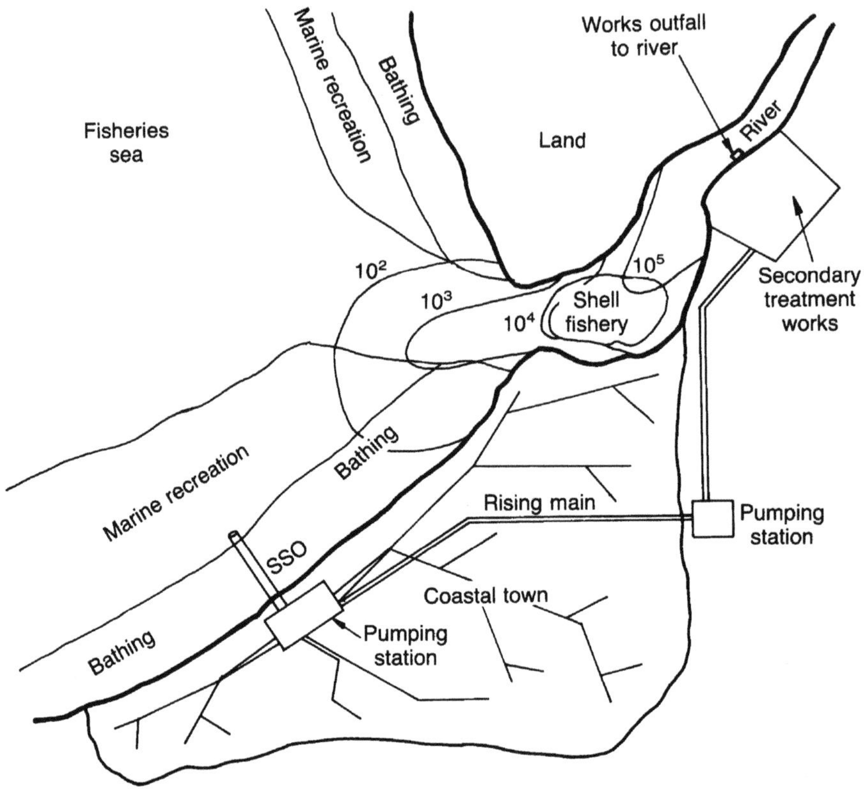

Figure 48.6 Coastal sewerage scheme with secondary treatment works. 10^2 is 10^2 Ecoli/100 ml; SSO denotes storm sewage overflow.

The models are required to provide data at two disparate time scales. They are required to simulate the minute-by-minute movement of water, pollutants and bacteria during one or more tidal cycles in order to ensure that instantaneous water quality indicators are within specified limits on a local scale. These models are referred to as intertidal models.

The second time scale is that over which nutrients and other pollutants build up and potentially create critical conditions at particular times of year. In these models the intertidal oscillation of parcels of water is of less interest than the net residual drifts and gravitational flows which vary seasonally. These models are referred to as seasonal models and are based on tidal residual flows.

48.7.3 INTERTIDAL MODELS: GENERAL

The intertidal models calculate water surface levels, water velocities, salinity, suspended sediment concentration and the concentration of a range of water quality parameters during the tidal cycle.

These models will simulate existing observed conditions which can then be used to predict the impact of large-scale reclamations or modified effluent disposal strategies on water movements, siltation and water quality in Hong Kong coastal waters. The main models on this time scale are listed below:

Intertidal flow model – global depth-averaged

This model covers the Pearl Estuary and most of the Hong Kong waters. It simulates tidal flows in the dry season only when the flows in the Pearl Estuary are low and the salinity in Hong Kong coastal water is uniform over the depth. The results from this model are used to provide input to the two-layer model of tidal flows and salt movement.

Intertidal flow model – two-layer

The two-layer model of Hong Kong coastal waters simulates tidal flows and salt movement in the wet and dry seasons where the water column has been

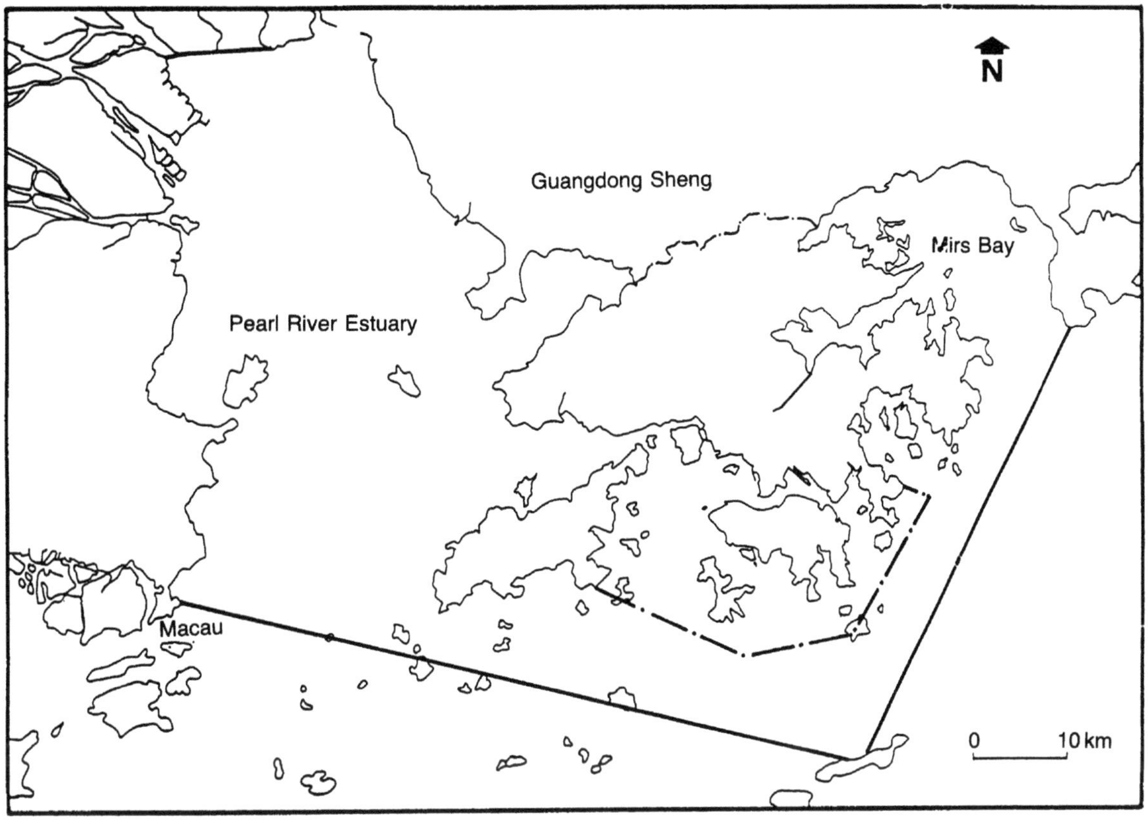

Figure 48.7 Model boundaries of Hong Kong Harbour. (——), grid 1 boundary; (–·–), grid 2 boundary.

divided into two interacting layers. This model can resolve details of the velocity variation over the water depth and can simulate the occasionally highly stratified conditions in the wet season. The results from this model provide input to the water quality and sediment transport models.

Intertidal water quality models

The intertidal water quality models, depth-averaged and two-layer, calculate the concentrations of dissolved oxygen (DO) and other substances including those that affect DO. Apart from flow and water level information, the models use estimates of loads discharged, rates of chemical reactions and physical exchanges, and concentrations around the seaward boundaries of the models.

Sediment transport model

This model of sediment transport uses the tidal flow results to simulate the movement of sediment in suspension, deposition and erosion. In the wet season, it simulates the natural influx of sediment from the Pearl Estuary. In the wet and dry seasons, it can simulate the discharge, transport and deposition of particulate effluent from the many sea outfalls.

Detailed models

The resolution afforded by any model is a compromise between the detail required and the computer resources available. The basic WAHMO tidal models have a resolution of 250 m which enables the modelling of most of the Hong Kong waters with sufficient resolution to examine the large-scale effects of reclamation schemes or effluent disposal options. It does not allow detailed simulation of conditions in typhoon shelters or narrow channels.

In order to examine these comparatively small areas, all the WAHMO tidal models can be used with appropriate model data files to simulate detailed conditions over smaller areas using finer grids. The software required to set up these detailed models forms part of the WAHMO system.

The detailed water quality model is virtually identical to the intertidal one because of the general way in which the computer code has been written. It allows analysis of small areas where the main grid is considered to be too coarse.

Outfall models

The outfall models predict bacterial concentrations near outfalls from estimates of loading, diffusion, and decay based on monitoring and experiments made in Hong Kong. They are based on the intertidal flow models. A special feature is the inclusion of dispersion resulting from a wind-induced shear flow superimposed on the depth-averaged flows, so the effects of onshore winds can be studied.

48.7.4 SEASONAL MODELS

The seasonal models are designed to simulate the long-term variation in water quality over periods of up to a year and they consist of a three-dimensional flow model and a three-dimensional water quality model. The flow model covers all of the Hong Kong coastal waters and most of the Pearl Estuary and simulates residual water movements where the main driving force is the fresh water flow from the Pearl Estuary. The water quality model uses the results from the flow model to simulate the long period seasonal changes in water quality in the Hong Kong coastal waters only. These models can be used to predict the longer term changes in water quality following large reclamation schemes or changes in effluent disposal strategies.

48.7.5 USER-FRIENDLY SOFTWARE

In order to control the way in which the models are run and in order to facilitate data modifications, an umbrella programme has been developed to:

1. Keep track of input and output files with appropriate run descriptions;
2. Ensure that runs are carried out with compatible data;
3. Facilitate data modification and output presentation;
4. Provide a menu driven system for running the models.

48.7.6 COMPUTER HARDWARE

A dedicated computer system is used for running the software. The system comprises networked workstations plus a full set of peripherals including two printers, plotters and colour hardcopy devices for obtaining colour copies of the screen graphics.

48.7.7 DATA COLLECTION

A comprehensive programme of data collection was developed to provide the data essential for calibrating and validating the mathematical models. The main part of the data collection comprised an intensive survey carried out during spring and neap tides in both summer and winter. Flow and water quality data were collected. Surveys of bacterial mortality, the dispersion of pollutants from outfalls and flows and water quality in storm and foul sewers discharging into the harbour were also carried out. Data was also collected from other studies and from Government data banks. The extensive bi-monthly monitoring data on water quality in the coastal waters, which had been collected by the government over several years, was of major importance in understanding the water chemistry and its seasonal variations.

48.7.8 VALIDITY OF THE MODELS

The successful use of the models depends upon a clear understanding of the strengths and weaknesses of the models within the suite. In particular an understanding is necessary of what the models can do, how accurate they are and what they should not be used for. This understanding has been provided by extensive training of the Hong Kong Government's officers. Not only have they been formally trained in the use of the models but they were also integrated into the team that developed the models. This on-the-job training has proved vital in establishing the knowledge base necessary not only to implement the models for increasingly more complex scenarios, but also to build the confidence for individual officers to 'champion' the use of the models.

48.7.9 CONCLUSIONS

The models described here were originally developed to allow the effect of major reclamation upon water quality to be assessed. They have subsequently been used in both the Sewage Strategy study and the Port and Airport Development Strategy study, which has involved evaluating the effect of major changes in the flow regime.

The models are currently being used to evaluate local developments and outfalls. Furthermore the models and their database are being extended to cover a larger area of the South China Sea so that the regional effects of sewage disposal policies can be evaluated.

The development of the suite of computer models

for Hong Kong has required a large commitment of resources. To date the expenditure on development, training and short-term data collection has exceeded £3 million, at 1988 prices. This excludes the cost of the long-term monitoring and the Government's internal costs in terms of monitoring the project and applying the suite of models. This sum is, however, a small percentage of the cost and indeed benefit of the schemes being considered. It is salutary to note that this scale of resources is necessary if a reasonable degree of assurance is to be obtained on a problem of this scope and coverage.

For the successful use of such a tool it has to be championed, both at the highest level of Government and at the management and operator level. The top level support can only be achieved by successful internal marketing of the model suite, its capabilities and its application, within the government organization. This requires the enthusiasm of those directly in charge of the project and attention to the detail of the presentation format of the results. The use of graphics and animation techniques is essential in such situations to achieve acceptance by both senior officials and the public.

The successful implementation and acceptance of this type of model is not an accident of fate. It requires vision on the part of the initiators, careful planning and the allocation of sufficient resources both for the initial development and the updating of the system as technology advances and more data become available.

48.8 THE OIL SPILL AGE

It is apparent that in the 1980s the popularity of 'green' politics has meant that governments everywhere have focused on limiting the production and discharge of pollutants. A major form of pollution is that of oceanic oil slicks which contaminate large areas of coastline destroying many species of animals in its wake.

This section outlines the main options available in studying oil spills whether by field investigation of real oil spills, small-scale laboratory studies or numerical prediction models.

An extensive study by the United Nations GESAMP (1977) was carried out to establish the long- and short-term effects on marine life due to major oil spills. Kuipers *et al.* (1984) outline the effect on the food chain if micro-organisms are unable to reproduce. Seip (1984) studied the physical and biological effect of the Amoco Cadiz spill in 1978. It was found that high levels of oil were still present in the fauna and flora after five years, especially in sheltered embayments.

The major transporter of oil slicks is the wind. Many oil spill models use wind transport. The oil slick speed is generally taken to be around 3% of the wind speed measured 10 m above sea-level, although Hoult (1972), Lissauer and Murphy (1979), Johansen (1983) and Kollmeyer and Thompson (1977) have used differing values within the range of 2.5% to 3.5%. The effect of the Coriolis force on the wind direction must also be considered for large-scale models. Application of wind field data needs careful consideration. Random walk analogy, Markov chain rule or a numerical meteorological model may be used. These techniques have been described by Huang (1983). Once the wind direction and speed has been established the most common method used is to apply this to an oil slick by vector addition, or a float tracking technique. Kuipers (1981), Audunson *et al.* (1980) and Cornillon *et al.* (1979) have all used this method. The modelling is simple and effective if sea conditions are calm and the oil is homogeneous.

Berridge *et al.* (1968) attempted to evaluate the evaporation and emulsification of persistent oils. Depending on the type and number of fractions within the oil evaporation can reduce the oil volume by up to 40%. Also the emulsification of water in oil can increase the spill volume by 400%. Therefore these two processes must be modelled since they are important in quantifying oil slicks as time elapses. Kuipers (1981), Aravamudan *et al.* (1981), Rasmussen (1985) and Mackay *et al.* (1985) have modelled evaporation and reasonably predicted the final oil volumes. Most evaporation formulae require the wind speed to be known. Huang (1983) reviewed many models and concluded that Mackay *et al.* (1980) produced the most appropriate evaporation rate technique. Emulsification was predicted with some success by Drapeau *et al.* (1974), Payne *et al.* (1983) and Venkatesh *et al.* (1979). Vertical dispersion, induced predominantly by breaking waves, should also be considered since many field tests have two-dimensional aerial charts with the oil slick thickness assumed to be uniform. Shonting *et al.* (1979), Audunson *et al.* (1980), Naess (1981) and Johansen (1983) have attempted to correlate dispersion with wave height and its associated mixing length.

Sedimentation of oil particles occurs over a large time span and this in turn leads to longer-term clean-up problems. Salinity, oil inertia, tidal cycle and micro-organisms all influence the rate and extent of sedimentation. McClaren (1985), Seip (1984) and Kuipers *et al.* (1984) have all investigated this phenomenon. Other processes which act on the oil chemically and biologically are dissolution, oxidation and biodegradation. These have small effects on an oil slick and are little understood. However, it is worth referring to Kolpack (1977), Berridge *et al.* (1968) and Payne *et al.* (1983) to study their methods of modelling these processes.

Coupled with advection by winds and tides the fate of an oil slick depends greatly on its ability to spread on the surface of the sea. Since the early 1960s two methods have been used to model the spread of oil: firstly, Fay's equations; and secondly, Fickian advection diffusion. Both methods are equally popular. An extensive review by Joynes (1989) examines the advantages of both methods and concludes that the Fick equation is more scientifically based and more appropriate for two-dimensional flow fields. The Fay equations concerning the balance of gravity, inertia, viscosity and surface tension simplify the oil slick and require too many empirical equations. Lehr *et al.* (1984a,b). Hoult (1972), Wang and Hwang (1974) and Buckmaster (1973) have all considered Fay's three-phrase spreading. Oil spill models using Fick's equation, which expresses the oil concentration as diffusing proportionally to the mean concentration gradient, have been used by Borthwick and Joynes (1989), Kollmeyer and Thompson (1977), Zeidler (1976) and Kolpack (1977). The use of a source or sink term, tested by Sahota and Venkatesh (1978), is useful for a time-dependent spill like an oil well blowout, Murray (1972). Solution of Fick's equation produces many numerical problems since it contains hyperbolic and parabolic components. However, Holly and Ussegliss–Polatera (1984) and Borthwick and Joynes (1989) have demonstrated techniques that accurately simulate the advected shape and diffusion of an oil slick in the open sea and coastal waters.

Another method to predict the position of oil slicks is by random walk where the oil slick is made up of many particles. These particles move a distance proportional to a random number and a diffusion coefficient. Hunter (1980) and Sahota and Venkatesh (1978) have used this technique.

In conclusion, many state-of-the-art reviews have been carried out that encompass the processes that occur during major and minor oil spillages. GESAMP (1977), Stolzenbach *et al.* (1977), Huang (1983) and Joynes (1989) are notable examples of such reviews.

REFERENCES

Aravamudan, K.S., Raj, P.K. and Marsh, G. (1981) Simplified model to predict the break-up of oil on rough seas. Proceedings of the 1981 Oil Spill Conference, A.P.I., Washington DC, pp. 153–9.

Audunson, T., Dalen, V., Mathison, J.P. *et al.* (1980) SLIKFORCAST – a simulation program for oil spill emergency tracking and long term contingency planning. Proceedings of Petromar '80, Monaco, pp. 512–41.

Berridge, S.A., Dean, Jr. A., Fallows, R.G. *et al.* (1968) The properties of persistent oils at sea. *Journal of the Institute of Petroleum*, **54** (539), 300–9.

Borthwick, A.G.L. and Joynes, S.A. (1989) Horizontal dispersion of oil pollutant in coastal waters, in *Hydraulic and Environmental Modelling of Coastal, Estuarine and River Waters* (Eds R.A. Falconer, P. Goodwin and R.G.S. Matthew) Gower Press, Aldeshot, UK, pp. 322–31.

Buckmaster, J. (1973) Viscous-gravity spreading of an oil slick. *Journal of Fluid Mechanics*, **59** (3), 481–91.

Cornillon, P.C., Spaulding, M.L. and Hansen, K. (1979) Oil spill treatment strategy modelling for Geroges Bank. Proceedings of the 1979 Oil Spill Conference. A.P.I., Washington DC, pp. 685–92.

Cooper, V.A. and Lack, T.J. (1986) Environmental effects of discharges. I.P.H.E. Symposium: Construction and maintenance of long sea outfalls, Portsmouth, UK, September 1986.

Council of European Communities (1976a) Directive 76/464/EEC on the discharge of dangerous substances. *Official Journal of the European Communities*, L54.

Council of European Communities (1976b) Directive 76/160/EEC concerning the quality of bathing water. *Official Journal of the European Communites*, L54.

Council of European Communities (1979) Directive 76/923/EEC on the quality required of shellfish waters. *Official Journal of the European Communities*, L281/4.

Drapeau, G., Harrison, W., Bien, W. *et al.* (1974) Oil slick fate in a region of strong tidal currents. Proceedings of the 14th International Conference on Coastal Engineering, ASCE, Hamburg, pp. 2245–59.

Environmental Protection Department, Environment of Hong Kong (1989) A review of 1988, Government Printer, Hong Kong.

GESAMP (1977) Impact of oil on the marine environment. Food and Agriculture Organisation of the United Nations, *Report Study GESAMP*, 6.

Holly, F.M. and Usseglio-Polatera, J. (1984) Dispersion simulation in two-dimensional tidal flow. *Journal of Hydraulics Division, ASCE*, **110** (7), 905–25.

Hoult, D. (1972) Oil spreading on the sea. *Annual Review of Fluid Mechanics*, **4**, 341–68.

Huang, J.C. (1983) A review of the state-of-the-art of oil behaviour. Proceedings of the 1983 Oil Spill Conference. A.P.I., Washington DC, pp. 313–22.

Hunter, J.R. (1980) An inter-active computer model of oil slick motion. Proceedings of Oceanology International '80, Brighton, UK, pp. 42–50.

Johansen, O. (1983) Dispersion of oil from drifing oil slicks, A.M.O.P. Technical Seminar, Edmonton, Canada, 14–16 June.

Joynes, S.A. (1989) Horizontal diffusion of a buoyant pollutant in coastal waters. PhD thesis, Department of Civil Engineering, University of Salford, UK.

Kollmeyer, R.C. and Thompson, M.E. (1977) New York Harbour oil prediction model. Proceedings of the 1977 Oil Spill Conference. A.P.I., Washington DC, pp. 441–5.

Kolpack, R.L. (1977) Fate of oil in a water environment – phase II, A dynamic model of mass balance for released oil, A.P.I., Publication 4313.

Kuipers, H.D. (1981) A simulation model for oil slicks at sea. Department of Civil Engineering, North Sea Directorate, Ministry of Transport and Public Works, The Netherlands.

Kuiper, J., Wilde, P. and Wolfe, W. (1984) Effects of an oil spill in outdoor model tidal flat exosystems. *Marine Pollution Bulletin*, **15** (3), 102–6.

Lehr, W.J., Cekirge, H.M., Fraga, R.J. *et al.* (1984a) A new technique to estimate initial spill size using a modified Fay-type spreading formula. *Marine Pollution Bulletin*, **15** (9), 326–9.

Lehr, W.J., Cekirge, H.M., Fraga, R.J. *et al.* (1984b) Empirical studies of the spreading of oil spills. *Oil and Petrol Pollution*, **2** (1), 7–11.

Lissauer, I.M. and Murphy, D.L. (1979) Oil spill forecasting – where is it going. Proceedings of the 1979 Oil Spill Conference. A.P.I., Washington DC.

Mackay, D., Paterson, S. and Trudel, K. (1980) A mathematical model of oil spill behaviour. Environmental Protection Service, Fisheries and Environment, Canada.

Mackay, D., Chau, A., Clark, B. *et al.* (1985) The dispersion and submergence of oil spills. 9th A.M.O.P. Technical Seminar, Edmonton, Canada, 10–12 June, pp. 101–11.

McClaren, P. (1985) Behaviour of diesel fuels on a high energy beach. *Marine Pollution Bulletin*, **16** (51), pp. 191–6.

McDowell, D.M. and O'Connor, B.A. (1977) *Hydraulic Behaviour of Estuaries*, McMillan, London.

Murray, A.J., and Portmann, J.E. (1986) Fisheries implications of sea outfalls – an environmental assessment and the monitoring requirements for England and Wales. Construction and maintenance of long sea outfalls symposium. I.P.H.E., Portsmouth, UK.

Murray, S.P. (1972) Turbulent diffusion of oil in the ocean. *Journal of Limnology, and Oceanography*, **17** (5), 651–9.

Naess, A. (1981) Natural dispersion of oil spills by breaking waves. Norwegian Hydrodynamics Laboratories, Trondheim, *Rep. NHL 181057*.

Neville-Jones, P. and Dorling, C. (1987) The design of long sea outfalls for environmental protection. *W.R.C. Engineering Rep. Er 2097E*, Swindon, UK.

Payne, J.R., Kirstein, B.E., McNabb, D. *et al.* (1983) Multivariate analysis of petroleum hydrocarbon weathering in the sub-arctic marine environment. Proceedings of the 1983 Oil Spill Conference. A.P.I., Washington DC, pp. 423–34.

Rasmussen, D. (1985) Oil spill modelling – a tool for clean on operations. Proceedings of the 1985 Oil Spill Conference. A.P.I., Washington DC, pp. 243–9.

Sahota, H.S. and Venkatesh, S. (1978) Prediction of the motion of oil spills in Northern Canadian waters. Atmospheric Dynamic Corporation, Elmira, Ontaria, Canada.

Seip, K.L. (1984) The Amoco Cadiz oil spill – at a glance. *Marine Pollution Bulletin*, **15** (6), 218–20.

Shonting, D.H., Temple, P.R. and Roklan, J. (1979) The windwave and turbulence observation program (WAVTOP), U.S.C.G., Washington DC, *USCG-D-68-79*.

UK Department of the Environment (1987a) Quality status of the North Sea summary. Second international conference on the protection of the North Sea, Scientific and Technical Working Group, London.

UK Department of the Environment (1987b) Inputs of dangerous substances to water proposals for a unified system of control. London.

Venkatesh, S., Sahota, H.S. and Rizkalla, A.S. (1979) Prediction of the motion of oil spills in Canadian arctic waters. Proceedings of the 1979 Oil Spill Conference. A.P.I., Washington DC, pp. 677–83.

Wang, S. and Hwang, L.S. (1974) Numerical simulation of oil slick transport in bays. Proceedings of the 14th International Coastal Engineering Conference. ASCE, pp. 2227–43.

Wood, P.C. and Abel, R. (1986) Inputs into the North Sea and their effects. Water pollution Control, *Journal of the I.W.P.C.*

Zeidler, R.B. (1976) Coastal dispersion of pollutants. *Journal of the Waterways, Harbours and Coastal Engineering Division, ASCE*, **102** (WW2), 235–54.

PART NINE

ECONOMICS AND MANAGEMENT OF COASTAL ENGINEERING WORKS

ECONOMIC APPRAISAL OF COASTAL ENGINEERING WORKS 49

Dr P.C. Barber

Considerable advances have been made in recent years in our understanding of coastal and other associated hydraulic engineering processes to the point where it is now possible to quantify, with a reasonable degree of accuracy, the effect of various schemes proposed to solve a particular problem. The result of this advance in knowledge is that alternative solutions can now be appraised on a benefit/cost basis. Taking his examples from coastal engineering, although the same techniques can be applied elsewhere, the author describes the calculations that should be made in carrying out a benefit/cost exercise (see also Chapters 22, 23 and 33).

CONTENTS

49.1 INTRODUCTION

As the demand upon available resources has increased in recent years more attention has been given to establishing the value-for-money from various investments. Scheme justification has become an important part of coastal engineering and this chapter examines the philosophy of approach to such work and sets down the basic principles for economic evaluation.

It is assumed in what follows that the coastal engineering problem has been defined and that a number of technical solution options have been identified. It remains to establish the economic worth for the particular project and then which of the solution approaches is the most cost-effective. There are a number of methods used to quantify and compare economic values. The most commonly used method in the UK is benefit cost analysis and it is this method which is given more detailed consideration below. In such an analysis the costs of the proposed scheme need to be compared against the economic value of the losses that would occur – 'benefits' – should the works not be constructed. From this the rate of return to be gained (if any) can be established for the investment. This information can be used to convince investors to finance the proposed coastal engineering works.

49.2 TYPES OF SCHEME

The most common types of scheme in coastal engineering are those associated with the alleviation of flooding or the prevention of land erosion. There are other types to do with commercial undertakings such as ports, power stations and mining but these schemes are too diverse to cover here although the same basic principles apply.

The bench-mark condition to be defined for scheme justification is the 'do-nothing' situation. The definition is different for flooding and erosion schemes and requires input from the technical investigations. Table 49.1 identifies the principal parameters that need to be defined for flooding and erosion scheme justification.

The selection of a preferred scheme involves an iteration between solution options, present day values (or similar) and timetable of benefit gain. Equivalent technical schemes by present day valuation may be different with regard to timetable for benefit gain.

The 'do-nothing' situation may not be realistic for erosion problems where localized works can ameliorate the recession. For flooding problems there is

Coastal, Estuarial and Harbour Engineer's Reference Book. Edited by M.B. Abbott and W.A. Price. Published in 1993 by Chapman & Hall, 2–6 Boundary Row, London SE1 8HN. ISBN 0 419 15430 2.

Table 49.1 Technical parameters

Flooding (A)	*Erosion (B)*
(i) Return periods of extreme water levels.	(i) Mean rate of recession (usually annual)
(ii) Wave conditions associated with (i).	(ii) Extreme event recession
(iii) Topography of flood water. Routing of flood water.	(iii) As for (i) and (ii) in 'A' Topography of immediate hinterland
(iv) Extent of flooding/ water level duration data.	(iv) Forecast rates of recession with time
(v) Design of existing works (if any).	(v) Design of existing works (if any)
(vi) Breach characteristics (length; depth; time of development etc.)	(vi) Failure characteristics (as for (iii) above)

seldom any significant gain from localized works since once a breach occurs the flood plain is inundated. For the 'do-nothing' erosion situation it may be helpful to consider the following qualifications for the responsible authority/landowner;

1. To invest the equivalent of financing charges in the protection of strategic locations across a frontage and in carrying out the work in (2).
2. To clear away unsightly debris and dangerous ground on an 'as-required' basis as conditions deteriorate.

These qualifications provide some constraints on the erosion of a frontage by compartmentalizing it between strategic locations. This prediction of recession within the embayment becomes more reliable as a consequence. The maximum depth of recession within an embayment should not exceed that derived from equilibrium beach profile dimensions which can usually be derived from selective observations of existing beach forms for a specific location.

Once the technical definition as summarized in Table 49.1 is available the evaluation of benefits can proceed.

49.3 BENEFITS

The benefits to be derived from a coastal engineering scheme are defined as the monetary value of the different behaviour of the coastal zone with or without the proposed works. Benefits are classified as tangible and intangible.

Tangible benefits are those which can be easily valued in monetary terms and therefore usually comprise physical items (property, infrastructure etc.). Generally tangible benefits will be valued in terms of market values or replacement costs.

Intangible benefits are those which cannot be easily valued in monetary terms. Such benefits include conservation, tourism, social distress and worry from a flood or erosion risk. (They would also include the effect of securing a coastline upon investment levels in the hinterland).

Tangible benefits can be grouped under three principal headings:

1. Property (private and public);
2. Infrastructure (sewerage; services – gas, electric etc. lighting; public land);
3. Transportation (diversion routes).

Property would include public and private buildings and these would be valued in different ways for flooding and erosion schemes. For a flooding situation it is necessary to establish the frequency of flooding at certain levels to be expected and then to evaluate the damage sustained by the property (structurally) on each occasion. For an erosion situation it is necessary to value the property in total and to make provision in the valuation for any depression in market value due to the erosion risk. For property valuations in both flooding and erosion schemes it is also necessary to determine whether the residual life of the property is less than or more than the proposed service life of the scheme.

Infrastructure will normally comprise sewerage and statutory undertaker services together with street lighting and furniture. For a flooding situation it will be necessary to evaluate the degree of damage for the forecast frequencies and scales of flooding events. For an erosion situation the infrastructure can usually be valued as replacement cost unless the service becomes redundant due to the erosion when it can only be valued against its residual life at the time of loss.

Transportation requires a separate consideration from infrastructure since it can be considered in a unique way. The valuation of transportation disruption is derived from the costs of diversion in the first instance with the replacement or substitution cost only becoming relevant if it is shown to be cheaper. For instances where road diversion is necessary as a result of flooding and/or erosion there are standard Department of Transport techniques in the UK for establishing the increased costs in terms of extra mileage required for vehicles to use an alternative route. For long diversions in heavily populated urban areas, these costs can be considerable when taken over the design life of the scheme. Transportation includes road, rail and canal generally, and can include airports in certain circumstances.

Intangible benefits are generally related to the

'quality of life' and can therefore vary markedly between different sites. However, these benefits can be grouped under three principal headings:

1. Community concern;
2. Conservation value;
3. Commercial value.

There are a number of different methods of defining a value against intangible benefits that can then be imported into the overall economic evaluation. The method and approach used will vary according to the 'intangible' that is being quantified. The methods of particular use in evaluating intangibles in respect of coastal engineering works are: firstly, the travel/cost method which involves determination by survey of the cost in terms of peoples' time and actual costs incurred (rail fare, petrol costs, etc.) in visiting a site. This is established by site survey. Secondly the contingent valuation method involves determination of the 'willingness-to-pay' of visitors to preserve current conditions in their present form. This can be carried out by site survey in two forms:

1. How much extra in taxes would people pay to preserve the coastline.
2. What alternative would give visitors equivalent enjoyment – location, entrance price, etc.

Community concern represents the social distress felt by people who live with the ongoing threat of flooding or erosion affecting their homes and (maybe) livelihood. There is general acceptance of community concern resulting in a cost from social behaviour and illness as a direct result of such circumstances. However, as yet there is no accepted means of quantifying the concern monetarily, and community concern is usually represented at the political level at present.

Conservation value represents the value of interesting habitats to the community, which can be valued by using public questionnaire survey techniques to establish both the nearest similar alternative location and an alternative pastime if the area were no longer available. Analysis of survey returns can provide a monetary value against these data for use in the justification of schemes.

Commercial value represents the consequential effects of 'with-scheme' or 'without-scheme' on commercial interests such as tourism. This intangible benefit can also be valued using public questionnaire survey techniques. For tourism the survey would be directed at establishing the point of coastline change when the tourist would not return. The travel/cost or contingent evaluation method can then be applied.

The evaluation of intangible benefits is a developing subject and it is likely that significant progress will be made here in the future. At present there are techniques available, as referred to above, but there are no accepted standards for data interpretation. This can lead to subjectivity in assessments. This situation will need to change for intangible benefits to gain their rightful place alongside tangible benefits in the assessment of value-for-money from scheme proposals.

There are two distinct approaches to establishing the value of benefits for coastal engineering schemes, dependent on their function. In general, losses associated with flood alleviation can be considered to be temporary in nature, in that only under extreme conditions will flooding of land and/or property result in permanent inundation of the particular elements that are affected. For erosion prevention schemes, losses will be permanent, in that land and associated infrastructure, once lost, cannot be replaced. Considerable research work has been carried out in the UK, particularly in respect of data for evaluating benefits with respect to flooding. This work has centred on Middlesex Polytechnic where, in the early 1970s, a Flood Hazard Research Centre was established. Work here has been sponsored largely by the UK Ministry of Agriculture, Fisheries and Food and by industry and two benchmark reference works for flood benefit derivation have been published.

The Benefits of Flood Alleviation produced in 1977, commonly known as the *Blue Manual* provides a volume of assessment techniques and depth damage data for the estimation of direct losses due to flood events. In 1987 a further volume was produced, *Urban Flood Protection Benefits – A Project Appraisal Guide*, which updated a considerable amount of the data contained in the *Blue Manual* and provided guidelines in respect of the evaluation of indirect losses that occur as a result of flood events, e.g. traffic disruption, loss of industrial production, emergency services costs etc. This manual has become known as the *Red Manual*. In addition the Flood Hazard Research Centre has developed a series of computer software models and programmes to cover the data available and procedures adopted.

These manuals tend to deal almost exclusively with urban benefits, damage to property, infrastructure, etc. However, it is important to recognize that, particularly in flood alleviation protection of open space (agricultural, amenity or environmentally important) may be necessary.

For the derivation of benefits to be gained from protecting agricultural land, and Hess and Morris (1985), Morris and Hess (1987) at Silsoe College in the UK have used gross margin data for different land uses – varying crops, grazing, etc. – to derive the losses to land that is subject to flood water inundation. As with the Middlesex Polytechnic work, computer models are available to cover this aspect although the basic data required can be obtained from a number of sources.

In such cases it is usual for a land use survey to be carried out for the farms affected, to establish what use different fields are put to, whether uses are rotated and what property or infrastructure would be affected.

For the valuation of losses for erosion scenarios there are no manuals of direct data available and data have to be obtained on a site specific basis. However, the degrees of data are somewhat less than for flooding in that the value of elements lost can generally be defined by a one-off lump sum or unit rate and are not dependent on differing depths of water for example. Consequently research in this respect has been carried out to a lesser degree and the engineer is required to investigate suitable data sources on a scheme-by-scheme basis.

The different types of benefit and their evaluation have been described above and it is readily evident that timescale plays a key role in the derivation of losses of benefit for a 'without-scheme' condition.

49.4 ECONOMIC ANALYSIS (UK PRACTICE)

During the planning of most coastal engineering projects it will be found that there are a number of alternative ways of achieving the same goal. The aim of economic analysis is to provide the engineer with a means of judging the relative economics of the alternatives and of ensuring that available resources are used to maximum efficiency in achieving the goal. It is important to recognize that there are a number of ways of providing an economic analysis, namely:

1. Cost-effective analysis: the least expensive method of achieving a stated goal.
2. Economically important analysis: detailed accounting of the economic benefits and costs and desegregation of them across affected parties.
3. Benefit cost analysis: organization and comparison of the value of losses – 'benefits' and costs to an equivalent datum.

The latter has become the accepted form of economic analysis in the UK for coastal engineering projects as these are largely publicly funded investments and this method of analysis is devised to determine the net impact of investment on society (in this case the nation) as a whole. The true term for this analysis is economic efficiency benefit cost analysis in that it does not take into account who contributes and who receives but analyses purely an overall net return.

The system has had its critics and is open to misuse, particularly in respect of discounting and the selected omission and inclusion of various costs and/or benefits. However, these are considered to be 'practice abuses' rather than inherent flaws in the system.

In order to economically evaluate the proposed investment in works for coastal engineering it is necessary to compare benefits and costs that will be accrued at different times over time spans (equating to design lives of schemes) of generally between 40 and 60 years.

To do this the values used for both costs and benefits are real present values which ignore inflation but are then discounted back to the present day from the time when they are incurred to provide a common base for both sets of values. The sum total of discounted benefits or costs is known as the net present value. Generally, schemes will be considered to be economically viable should the benefit–cost ratio exceed unity.

This technique is known as discounting and provides the basis not only for benefit and cost analysis for a particular scheme but also, by having a set discount rate applicable to all schemes, the means for scheme comparison. This is of use particularly when funds are limited and the rate of return of different schemes competing for such funds can be compared. In the UK coastal works have largely been funded from taxation and the discount rate is set by the Treasury. The use of this technique also provides a further additional stage in appraising schemes, particularly those whose function is erosion prevention. This involves the origin for commencement time for benefits and costs.

Should a scheme be relying on benefits that will be accrued at a time somewhat in the future it may be preferable for scheme implementation to be delayed in order that the present day value of the benefits increases. Conversely scheme costs can also rise with time, and erosion could accelerate, thereby threatening additional hinterland interests, for example, and requiring a larger scheme. Clearly there will be an optimum time for works to be carried out but this is dependent on both technical and economic considerations.

The technique of discounting is carried out, using the following formulae.

The net present value for a benefit or cost for each year in which it is accrued is calculated using the formula:

$$NPV_n = AV_n \times D_n \tag{49.1}$$

where NPV_n is the net present value of costs or benefits for year n; AV_n is the total real value of costs or benefits accrued in year n; D_n is the discount factor $= 1/(1 + r)^n$; r is the discount rate; and n is the year number.

The total net present value of the costs or benefits is then calculated by summing the values for each of the years.

$$\text{Total } NPV = \Sigma NPV_n \quad \text{for } n = 1, \ldots, n = s \qquad (49.2)$$

where s is the proposed design life of the scheme in years.

This standard method of discounting is used for all costs, and for benefits which can be valued straightforwardly by a replacement cost or known value that is not influenced by its risk from flooding or erosion, e.g. the replacement cost of a sewer or highway will be the same whether it is life-expired or is a new one just about to fall into the sea.

However, for other benefits that can be considered to be market dependent, e.g. property, Middlesex Polytechnic Flood Hazard Research Centre have recommended a method that takes into account the increased life the property realises as a result of its being protected.

The benefit is calculated using the following formula:

$$B = MV(C_{e+s} - C_e) \times r \qquad (49.3)$$

where: B is the benefit value; MV is the market value; C_n is the capitalized value factor of present value of a stream of constant annual £1 payments over a period of n years; e is the number of years until erosion; s is the design life in years of the proposed scheme, as above; r is the discount rate, as above.

This relies on the market value for the property being a typical value of a similar property in the same area that is not threatened by erosion/flooding risk. However, should the value obtained be that for the particular properties that are at risk then the reduction obtained by using Equation (49.3) is inherent in the market value used, and straightforward discounting of the values (Equation (49.1)) should be used. The total value of property benefits is summed in the same way as previously (Equation (49.2)).

49.4.1 SCHEME COSTS

Scheme costs are discounted in exactly the same manner as benefits using Equations (49.1) and (49.2) and clearly phasing of stages of construction where applicable can have a significant impact on the total net present value of costs for the scheme. For the scheme to have a cost benefit ratio that exceeds unity, clearly:

$$\Sigma NPV_{n=1,\ldots,s} \text{ (benefits)} > \Sigma NPV_{n=1,\ldots,s} \text{ (costs)} \qquad (49.4)$$

49.4.2 BENEFIT DATA SOURCES

A number of sources are available for valuing benefits. For property in the UK, local estate agents and/or the District Valuer should be approached. For infrastructure values local or water authority technical service officers will generally be able to provide replacement costs for highways/sewers etc.

For flooding damage Middlesex Polytechnic (1987), MAFF (1985), Penning Rowsell and Chatterton (1977) and Parker *et al.* (1987) provide up-to-date depth/damage data for properties and possessions whilst the annual *Farm Management Handbook* and *Farm Management Pocket Book* will provide the necessary base data for agricultural benefits.

A typical cost benefit spreadsheet is provided by way of example in table 49.1.

49.5 USEFUL ECONOMIC ANALYSIS FORMULAE

The following symbols and definitions will be used: P is the principal, a present sum of money which is lent or borrowed; r is the rate of interest per period (rate expressed as a decimal); n is the number of periods over which interest is paid; S is the sum of money at the end of n periods at interest r which is made up of the principal plus the interest; and R is the uniform series end-of-period payment which occurs for n periods.

Compound interest is used to compute sums to be paid by a borrower to a lender for the use of capital. The compounding process is described by the following formulae.

$$S = P(1 + r) \qquad (49.5)$$

The expression $(1 + r)$ is called the compound amount factor.

Rearranging Equation (49.5) to express P in terms of S, r and n gives:

$$P = S\left[\frac{1}{(1 + r)^n}\right] \qquad (49.6)$$

This equation gives the principal or initial capital, P, necessary to generate the sum S if the interest rate is r and the number of periods is n. P is the present value or present worth of S in n periods of time for the given rate of interest.

The process of calculating a present value in this way is known as discounting. The expression:

$$\left[\frac{1}{(1 + r)^n}\right]$$

is called the present value factor or present worth factor.

Any series of equal payments which are made at the ends of equal periods of time is known as an annuity.

If a payment, R, is made at the end of each of n periods and the interest rate is r, then from Equation (49.6)

$$\text{Compound amount of 1st period's payment} = R(1 + r)^{n-1}$$

Table 49.2 Example of benefit/cost spreadsheet

Discount rate (decimal) = 0.06
Scheme design life (years) = 60

Anywhere by the sea coastal works
Benefit cost appraisal

Year	*1*	*2*	*3*	*4*	*5*	*6–10*	*11–15*	*16–20*	*21–25*	*26–30*	*31–40*	*41–50*	*51–60*	*Totals*
Loss profiles														
Flooding														
Land usage	0	0	0	0	0	0	0	0	0	0	0	0	0	0
Property and infrastructure	0	0	0	0	0	0	0	0	0	0	0	0	0	0
Erosion														
Infrastructure	0	0	0	0	0	0	0	0	0	0	0	0	0	0
Traffic diversion	0	0	0	0	0	0	0	0	0	0	0	0	0	0
Property	0	0	0	0	0	0	0	0	0	0	0	0	0	0
Benefit evaluation														
Flooding benefit														
Land usage	0.000	0.000	0.000	0.000	0.000	0.000	0.000	0.000	0.000	0.000	0.000	0.000	0.000	0.000
Property and infrastructure	0.000	0.000	0.000	0.000	0.000	0.000	0.000	0.000	0.000	0.000	0.000	0.000	0.000	0.000
Yearly cost totals	0.000	0.000	0.000	0.000	0.000	0.000	0.000	0.000	0.000	0.000	0.000	0.000	0.000	0.000
(1) Discounted present value totals	0.000	0.000	0.000	0.000	0.000	0.000	0.000	0.000	0.000	0.000	0.000	0.000	0.000	0.000
Erosion benefit														
Infrastructure	0.000	0.000	0.000	0.000	0.000	0.000	0.000	0.000	0.000	0.000	0.000	0.000	0.000	0.000
Traffic diversion	0.000	0.000	0.000	0.000	0.000	0.000	0.000	0.000	0.000	0.000	0.000	0.000	0.000	0.000
Yearly cost totals	0.000	0.000	0.000	0.000	0.000	0.000	0.000	0.000	0.000	0.000	0.000	0.000	0.000	0.000
(2) Discounted present value totals	0.000	0.000	0.000	0.000	0.000	0.000	0.000	0.000	0.000	0.000	0.000	0.000	0.000	0.000
Property														
Market values	0.000	0.000	0.000	0.000	0.000	0.000	0.000	0.000	0.000	0.000	0.000	0.000	0.000	0.000
(3) Discounted present value	0.000	0.000	0.000	0.000	0.000	0.000	0.000	0.000	0.000	0.000	0.000	0.000	0.000	0.000
(1+2+3) Present value yearly totals	0.000	0.000	0.000	0.000	0.000	0.000	0.000	0.000	0.000	0.000	0.000	0.000	0.000	0.000
(4) Total N.P.V. scheme benefit	0.000													
Scheme expenditure														
Capital costs	0.000	0.000	0.000	0.000	0.000	0.000	0.000	0.000	0.000	0.000	0.000	0.000	0.000	0.000
Estimated maintenance costs	0.000	0.000	0.000	0.000	0.000	0.000	0.000	0.000	0.000	0.000	0.000	0.000	0.000	0.000
Yearly cost totals	0.000	0.000	0.000	0.000	0.000	0.000	0.000	0.000	0.000	0.000	0.000	0.000	0.000	0.000
Discounted present value totals	0.000	0.000	0.000	0.000	0.000	0.000	0.000	0.000	0.000	0.000	0.000	0.000	0.000	0.000
(5) Total net present values scheme cost	0.000													
(4/5) Scheme benefit to cost ratio	⟨DIV 0⟩													

Compound amount of 2nd period's payment $= R(1 + r)^{n-2}$

Compound amount of nth period's payment $= R(1 + r)^{n-n} = R$

$$S = R[(1 + r)^n - 1 + (1 + r)^n - 2 + \ldots + 1]$$

Multiplying both sides of the equation by $(1 + r)$,

$$S(1 + r) = R[(1 + r)^n + (1 + r)^{n-1} + \ldots + (1 + r)]$$

Subtracting the first of these equations from the second gives:

$$Sr = R[(1 + r)^n - 1]$$

or

$$S = R\left[\frac{(1 + r)^n - 1}{r}\right] \tag{49.7}$$

The expression

$$\left[\frac{(1 + r)^n - 1}{r}\right]$$

is called the uniform series compound amount factor. Equation (49.7) can be rearranged to give:

$$R = S\left[\frac{r}{(1 + r)^n - 1}\right] \tag{49.8}$$

This equation gives the uniform amount to be invested at the end of each period in order to produce a fixed amount at the end of n periods if the interest rate is r. A fund into which such payments are made is called a sinking fund. The expression:

$$\left[\frac{r}{(1 + r)^n - 1}\right]$$

is called the sinking fund deposit factor. Combining Equations (49.5) and (49.8) gives:

$$R = P(1 + r)^n\left[\frac{r}{(1 + r)^n - 1}\right]$$

or

$$R = P\left[\frac{r(1 + r)^n}{(1 + r)^n - 1}\right] \tag{49.9}$$

The expression

$$\left[\frac{r(1 + r)^n}{(1 + r)^n - 1}\right]$$

is called the uniform series capital recovery factor. By rearranging Equation (49.9) we obtain:

$$P = R\left[\frac{(1 + r)^n - 1}{r(1 + r)^n}\right]$$

The expression

$$\left[\frac{(1 + r)^n - 1}{r(1 + r)^n}\right]$$

is called the uniform series present worth factor.

The following relationships should be noted:

1. (Compound amount factor) $= \dfrac{1}{\text{(present value factor)}}$
2. (Uniform series compound amount factor) $= \dfrac{1}{\text{(sinking fund deposit factor)}}$
3. (Uniform series capital recovery factor) $= \dfrac{1}{\text{(uniform series present worth factor)}}$

If an interest rate is quoted without a time period qualification then it is taken to be the annual interest. Thus, an interest rate of 8% means an interest rate of 8% per annum. Interest can be calculated after periods of other than one year. An interest rate of 8% compounded quarterly means that there are four interest periods per year of 2%. Thus, whilst the nominal interest rate is 8% per annum, the effective interest rate is greater than 8% per annum. In general terms, if interest is compounded t times per year and the nominal interest rate per annum is r, then the effective interest rate per annum

$$= \left(\frac{1 + r}{t}\right)t - 1$$

Whilst costs (or benefits) which arise at different times may be unequal, if the time value of money is taken into account, they may be found to be equivalent. This concept is known as equivalence. Thus, it is clear that choice between alternatives should not be made by comparing absolute values of costs (or benefits).

In practice, equivalence comparisons conform to one of three generally accepted methods:

1. Annual cost method: the cash flow of each alternative is converted to an equivalent series of uniform end-of-year payments (or receipts) using a stipulated minimum acceptable rate of return as the interest rate.
2. Present worth method: the cash flow of each alternative is converted to an equivalent present worth using a stipulated minimum acceptable rate of return as the interest rate.
3. Rate of return method: the prospective rate of return on the differences between the cash flows of a pair of alternatives is compared with a stipulated minimum acceptable rate of return.

The methods are equivalent but present the results of analyses in different ways.

The annual cost method is particularly suited to situations in which annual costs and receipts are regular. In contrast, the present worth method is easier to use in cases where the cash flow is very irregular. One of the difficulties experienced with the annual cost method and the present worth method is in selecting the interest rate. Using different rates in making comparisons between the same alternatives may result in different conclusions. Thus, it may be necessary, when using these two methods, to carry out the analyses for a range of interest rates in order to determine what the effects will be of changes in the minimum acceptable rate of return. If the rate of return method is used, this procedure is unnecessary. One of the disadvantages of the rate of return method is that it is not easy to list projects in order of best return on investment when multiple alternatives exist. For this purpose, the annual cost method and the present worth method are better.

Whichever method of comparison is used, it is necessary at some stage to decide upon a minimum acceptable rate of return. This rate of return should be at least equal to the rate of return required by investors on their money. In addition, if an investment is to be acceptable, it must provide a rate of return at least as great as that of any other investment of equal risk which it displaces as a result of finance being limited.

The effects of inflation may also be considered when comparing alternatives. The real rate of return is approximately equal to the difference between the money rate of return and the rate of inflation. More accurately, if m is the money rate of return and i is the rate of inflation, both compounded at the end of each year, then the real rate of return is $((1 + m)/(1 + i)) - 1$. In carrying out comparisons between alternatives, the minimum acceptable rate of return and the cash flows may be expressed either in actual money terms or in real terms. Whichever choice is made, the same conclusions should be reached.

Methods which involve the discounting of estimated cash flows, or the use of compound interest formulae in ways that are mathematically equivalent, are often referred to as discounted cash flow (DCF) methods.

49.6 CONCLUSIONS

The economic appraisal of coastal works provides an essential part of scheme development. As available resources have reduced, the attention paid to 'value-for-money' has increased. However, the indications are that the more rigorous the appraisal the better the 'value-for-money' available from coastal engineering works. It is important that technical studies provide the right information for use in the economic analysis and both aspects of scheme development should ideally be progressed in tandem. Sound economic appraisal can often result in review of design approach especially in respect of timing of works since this can be critical to their present-day value and whether a scheme meets the necessary rate of return criteria.

REFERENCES

S.F. Edwards (1987) *An Introduction to Coastal Zone Economics; Concepts, Methods and Case Studies*, Taylor and Francis, London.

Farm Management Handbook Department of Agricultural Economics, University of Exeter, UK. (Updated annually).

Farm Management Pocketbook (Ed J. Nix) Scheme of Rural Economics and Related Studies, Wye College, Ashford, UK. (Updated annually).

Hess, T.M. and Morris, J. (1985) A computer model for agricultural land drainage scheme appraisal. MAFF Conference of River Engineers, Cranfield.

MAFF (1985) Investment appraisal of arterial flood protection and sea defence schemes – guidance for drainage authorities, UK Ministry of Agriculture Fisheries and Food, River and Coastal Engineering Group.

Middlesex Polytechnic Flood Hazard Research Centre (1987) Interim Guidelines for the Economic Evaluation of Coast Protection and Sea Defence Schemes. Middlesex Polytechnic, Enfield, UK.

Morris, J., Black, D.E. and Hess, T.M. (1987) Farmer Uptake of Drainage Benefits: A Regional Study. Report to the Ministry of Agriculture, Fisheries and Food. Silsoe College, Bedford.

Parker, D.J., Penning Rowsell, E.C. and Chatterton, J.B. (1987) *Urban Flood Protection Benefits: A Project Appraisal Guide*, Gower Technical Press, Aldershot, UK.

Penning Rowsell, E.C. and Chatterton, J.B. (1977) *The Benefits of Flood Alleviation – A Manual of Assessment Techniques*, Gower Technical Press. Aldershot, UK.

TENDERING, CONTRACT MANAGEMENT, SITE SUPERVISION AND CONTROL INCLUDING INSURANCES 50

A.G. Brinson

This chapter describes the essential elements of tendering, contract management and site supervision. Normally the first requirement on major projects is a feasibility study which more often than not includes an Environmental Impact Assessment. In fact, in the UK and certainly under EEC regulations this is a compulsory requirement. This will involve the study of existing site data and the specification of additional information required for detailed design, and the preparation of tender documents. The realistic specification of Conditions of Contract is particularly important and although these vary from country to country they form a framework within which the client can insert and amend clauses to meet his own particular requirements. Control of the works on site requires tight supervision and the chapter discusses how this can be achieved and ends with a section on insurance (see also Chapter 51).

CONTENTS

50.1 TENDERING

50.1.1 INTRODUCTION

Tendering can be by a contractor or by a consultant and the procedures for tendering are not very different in principle. If the contractor or the consultant has a quality manual then there will be quality procedures to follow.

50.1.2 TENDERING BY A CONSULTANT

A consultant may be invited by a private company operating a small port or marina to submit a bid for a study or a design in competition with other consultants. On a larger scale a consultant may be trying to win a major contract from a funding agency such as the World Bank. In this case the first requirement will be to pre-qualify. When pre-qualified then the invitation will be issued to bid. This may be of the two envelope system; one envelope being the technical proposal and the second envelope being the financial proposal. The normal procedures of the funding agency will be to judge the technical submissions in order of merit and then open the financial envelope of the company that has submitted the best technical proposal and negotiate a contract.

If a contract cannot be finalized then the financial envelope of the company that has submitted the second best technical proposal will be opened and a contract negotiated and so on.

When the consultant receives an invitation to tender the first decision which has to be made is whether to tender or not. If it is a large job then the senior partner or managing director normally would decide after being told by the department head concerned of all the points for and against.

If the decision is to proceed and prepare a tender a budget is calculated as to the expense which will be allowed for preparing the tender. This could include for instance, the cost of sending someone to visit the

Coastal, Estuarial and Harbour Engineer's Reference Book. Edited by M.B. Abbott and W.A. Price. Published in 1993 by Chapman & Hall, 2–6 Boundary Row, London SE1 8HN. ISBN 0 419 15430 2.

site, which may be overseas and therefore cost an air fare, hotel bill, etc.

The technical proposal must comply with the Scope of Work in the Terms of Reference. It should include methodology, programme, the input from personnel, the firm's experience of the type of work required and *curriculum vitae*. The financial proposal should include the costs of fees for personnel plus expenses. If it is a UK project then value-added tax (VAT) must be considered. If it is overseas then care must be taken as regards local taxes etc.

50.1.3 TENDERING BY A CONTRACTOR

As stated earlier the principles of tendering whether by a contractor or by a consultant are similar. However, a contractor must take into account, for marine civil works, additional factors such as tides and storms. Tides may be very critical in that some work may only be carried out at low tides. As regards storms the Contractor should take into account the periods when he will not be able to work due to wave action and of course possible damage to his plant/equipment, his temporary works and the permanent works. Weather records are available for any part of the World. The first decision is of course whether or not to tender. For overseas projects this may well depend on the Conditions of Contract which may be very onerous. For instance when the writer was involved in a project in the Middle East he asked a well-known British contractors if they were interested in tendering. The first response was to enquire the content of the Conditions of Contract. In this particular case the British contractors decided not to tender and the reasons given were:

1. No default of employer clause i.e. the contractor has to continue working even if he has not been paid.
2. No satisfactory war clause and there are always wars at some time or another.
3. Performance bond of 10% of the contract price. This is normal in the Middle East. Contractors take out insurance against the client calling in the bond. However, for this particular country it was not possible to obtain insurance for the total amount.

The contractor will therefore carefully study the tender documents from all aspects, legal, contractual, and, of course, technical.

The tender will be built up in a similar way in principle to that of the consultant, namely man-time, or with a contractor it will be labour, but in addition he has to add materials, equipment and plant. In order to be able to do this the contractor will normally work from first principles especially on a large contract overseas. This means he has to decide how to construct the project and the plant and labour required to do this and how long it will take to complete. He may need to construct and run a complete camp for his labour force. When the costing is carried out he must assess the cash flow situation and try to arrange payments so that he is never in a negative cash flow situation. It is said that contractors will study the tender documents to see if there are any loopholes for claims, so that they can submit a low tender price, win the job, then make a profit from claims. The last addition to the price for labour, materials and plant is overheads. This is normally decided by the managing director who will take into account all that he knows about the risks involved, the need to win the project, the actual cost of overheads which is required to fund the head office commitments and last, but not least, the profit.

50.2 CONTRACT MANAGEMENT

50.2.1 INTRODUCTION

Contract management is a very extensive subject. This is therefore a brief outline.

50.2.2 CONTRACT MANAGEMENT ALTERNATIVE APPROACHES

It is the client who decides on the type of contract management that he wishes to have for the project. For instance he, the client, will decide whether to employ a consultant or not. He may carry out the concept himself and then invite contractors to tender for detail design/construct. Again he may select a consultant whom he knows and negotiate a contract, or he may pre-qualify say six consultants and then invite them to bid. The same approach can be made with contractors.

The first requirement of any project is a feasibility report. Also, now with more awareness of the environment, there may well be an environmental impact assessment. The feasibility report may include a financial assessment as to the viability of the project. It is at this stage that the concept is developed and as stated earlier this could be by the client himself or by a consultant, but not normally by a construction contractor.

The consultant may require assistance from other companies or outside consultants for say environmental studies, or financial assessments. It is at this stage that the finance for the project should be determined. This may have been established earlier but if not then it should be part of the study to establish either public or private finance.

On completion of the study and report and after establishing finance the next step is to carry out the detail design followed by construction. The client may decide to employ a consultant to carry out the detail design and prepare the tender documents or he may invite contractors to tender for a detail design/construct package. In the latter case the client must be able to prepare the necessary documentation or he can invite the contractor to do this, in which case the final contract will almost inevitably be biased towards the contractor.

If the traditional way of employing a consultant is used, it may be decided that the consultant will carry out full working drawings or may only carry out general arrangement drawings sufficient for tendering purposes. The working drawings may then be carried out during construction whether by the consultant or by the contractor.

With the consultant carrying out the detail design and Tender documents, the documents will consist primarily of Conditions of Contract, Specification, Bills of Quantities and Drawings. At the time of tender there will also be an Invitation to Tender and a Tender Form to complete. The Tender documents will then be issued to selected pre-qualified contractors to submit tenders. During the tendering period it may be that the client decides to invite all the contractors to visit the site of the project so that they may have a better appreciation of the work involved and the general environment. One very important feature of the Tender documents is the Conditions of Contract. In the United Kingdom the most used version for heavy civil engineering works is that produced by the Institution of Civil Engineers (6th edition, February 1991). Another version is the new Engineering Contract; a consultative version of which was also issued in February 1991. This is a very new approach and, as always with something new, there is a certain amount of controversy between its supporters and critics. As regards overseas contracts, Federation Internationale des Ingenieurs Conseils (FIDIC) is the most widely known and the most recent version is FIDIC 4. However, in most countries the client will require his particular conditions which will amend, revise or omit some of the conditions in FIDIC. As stated earlier in section 50.1 this may well have a major effect on contractors invited to tender.

When the tenders are received they must be evaluated technically and commercially. Contractors may have submitted qualifications to the tender which must also be evaluated or withdrawn. In addition contractors may have submitted alternative designs, usually with the object of submitting a lower priced tender and thereby winning the contract. It is for the client to decide whether the alternative design is worth pursuing. Sometimes the alternative design has sufficient merit to pursue but normally the contractor has submitted it because it suits him i.e. he has the necessary plant or the necessary skilled labour which will be available. When the contractor is chosen the contract documents need to be prepared and the contract signed.

A letter of intent may have been issued previously or even a letter of acceptance which can be a binding contract between the client and the contractor.

Following the signing of the contract, the order to commence work is given. Other matters at this time are the request for the contractor's programme and method of construction. The contractor must be given possession of the site. The consultant should satisfy himself that the contractor's programme, method of construction, including labour, plant and materials and insurance, are all satisfactory.

The initial weeks or even months on a large project overseas will consist of mobilizing labour, plant and materials. On large projects this may consist of recruiting local labour and establishing a camp on the site consisting of accommodation, canteens, recreational facilities etc. Plant may have to be purchased and brought to site. Their equipment may need to be fabricated and then brought to the site. Most of the materials will have to be procured locally.

Also during this period it may be necessary to establish a quarry for rock fill or aggregate for concrete. Temporary access roads may be required, as well as a temporary jetty. Offices will be necessary as well as workshops. Before permanent construction starts it may well be a requirement to carry out a joint survey. The consultant should ensure that the contractor is keeping to his programme. Also he should pay close attention to the financial arrangements, in particular, he should carry out regular financial reviews, so that the client always knows the anticipated final contract price for the project. This means that the first financial review can be carried out on completion of the joint surveys during mobilization. As at that point, the final quantities can be computed for dredging, reclamation, breakwaters, etc. where the finished top levels are known, but the original ground or sea-bed levels can only be finally established when the surveys are completed.

Financial reviews at regular intervals are essential to keep the client informed of the anticipated final contract price since if the price has gone up he may have to omit something, or if the final price will be less than expected then the client may decide to make additions to the project.

The alternative to management by the client and consultant on the contractor is to employ a company to perform project management. This may be from many different types of companies such as a bank who has such expertise. The project management company will employ consultants or architects to carry out design and a contractor or contractors to carry out construction. The supervision of construction can be carried out by a consultant or directly by the project management company.

Another option is to employ a contractor to carry out the construction as a management contractor. As a management contractor he will arrange for the actual construction to be carried out by sub-contractors who have tendered and been awarded parts of the works. He will directly manage the sub-contractors and co-ordinate their activities.

The essential factor, whatever type of contract management approach is used, is for each of the individual companies concerned, client, contractor, and consultant, to know precisely their obligations and responsibilities. These must be defined from the beginning of the project.

50.3 SITE SUPERVISION AND CONTROL

Site supervision is the supervision of construction of the works and can be carried out by various concerns which will be discussed here.

50.3.1 SITE SUPERVISION BY THE CLIENT

The client may have his own engineers, inspectors and clerk of works to carry out supervision himself.

50.3.2 SUPERVISION BY THE CONSULTANT

The consulting engineer who has carried out the design and prepared the tender documents may also be employed by the client to carry out site supervision. He should be in an advantageous position compared to another consultant being employed, since he will know the background and the reasons for the designs. However, with some funding agencies, it is quite normal to have different consulting engineers for the study and report stage, the design and tender documents stage and finally the supervision stage.

Site supervision, which is the people on the ground, will always require back-up from their head office when any difficult technical or contractual matters arise.

50.3.3 SUPERVISION BY THE CONTRACTOR

With quality management becoming more and more common in the civil engineering construction market, construction by the contractor should be performed in accordance with an acredited quality manual with quality procedures. This means that the contractor employs personnel within his own organization to carry out inspection and testing. For instance, it will be the contractor who carries out concrete cube testing. The consulting engineer or client will then only check that such inspection or testing has been performed.

50.3.4 SUPERVISION BY PROJECT MANAGEMENT COMPANY

On projects in which the contract is being managed by a project management company the site supervision will be carried out by a separate company. The project management will only check and audit that this supervision is being performed satisfactorily.

50.3.5 SUPERVISION BY MANAGEMENT CONTRACTOR

This is similar to section 50.3.4.

50.3.6 SITE SUPERVISION AND CONTROL

Site supervision and control, as the words imply, mean supervision on site and control of the construction so that the project is constructed satisfactorily in accordance with the contract. This means that the personnel employed to carry out this supervision should be competent, knowledgeable and qualified in the type of work. The specification cannot always be entirely explicit as to the requirements and can be interpreted differently which can lead to arguments and claims. It is important therefore that the specification is as clear as possible to avoid such arguments and claims. It is then for the supervision staff to ensure it is met. During construction the supervision staff should know whether work is being carried out correctly at the time, rather than having to wait for the results of tests. For instance with concrete, the supervisor should check that the constituents of the mix are correct, that the placing and compaction are carried out properly, and finally, that the curing is suitable. If these are followed then the concrete quality will be satisfactory. Cubes will be taken and possibly cores but if these show that the concrete is unsatisfactory it is a major problem, possibly involving demolition and reconstruction.

Site supervision and control not only applies to all the technical matters it also includes the contractual and financial matters. Therefore programmes and method statements should be presented and updated as necessary. Progress should be checked against the programme. Any possible technical problems which can be foreseen should be overcome as early as possible. For instance, there may be an area where the piling is expected to be difficult and a test pile or piles should be attempted as early as possible so that alternatives can be arranged, if necessary, as soon as possible to avoid delays to the project.

Regarding financial matters, as discussed earlier,

there should be regular financial reviews indicating the anticipated final contract price. Civil engineering projects, especially the large ones, have a very bad record for not finishing on time and costing considerably more than the original contract price. During construction there may well be claims, for various reasons, to which the construction contractor considers he is entitled. It is important that these are dealt with as quickly as possible during the construction period and not, as on so many projects, still argued about years after completion. The contractor may be entitled to interest charges if his claim is successful, but what is more important is that relationships on site will suffer if claims are not dealt with expeditiously. Generally the projects which are carried out successfully technically, to time and to cost, are those which have been well designed and engineered and where good relationships are maintained between the client and the other parties on site, whether consultant and contractor or management contractor or project management. These relationships need to recognize the duties and responsibilities of the other party and the aim of all concerned to carry out the project to a high standard of workmanship, to time and to cost.

50.4 INSURANCES

50.4.1 INTRODUCTION

Insurances will be necessary for all those concerned with construction of a project. The client, consultants and contractor all need to take out insurance for various reasons.

50.4.2 INSURANCE BY THE CONSULTANT

The consultant will need to have professional indemnity insurance against claims for defective work, whether it be caused by design or workmanship.

50.4.3 INSURANCE BY THE CONTRACTOR

The contractor will be normally required by the client under the Conditions of Contract to take out insurances. These will be for third party, his plant and equipment, the temporary works and the permanent works.

It is the temporary works and permanent works which are most difficult for the insurance company to evaluate. The insurance company will normally employ professional engineers to assess the risks of damage to temporary works and permanent works and therefore the probabilities of claims.

An example is that of breakwater construction. The engineer has probably included in the specification certain requirements for construction. One normal requirement is for the armouring to be a limited distance behind the forward end of the breakwater at any one time. This is so that in a storm only that part unprotected by armouring will be damaged. The insurance company may require this distance to be reduced further. Also the insurance company may require the contractor to pay a large excess on any one claim. In coastal, estuarial and harbour engineering there is, by the nature of the work, more risk of damage and loss than in some other forms of civil engineering.

Some large civil engineering companies are known not to insure their plant and equipment, since they can absorb such losses when they occur and consider it is less costly in the end than paying out large premiums to insurance companies.

The contractor may well also insure against other risks such as the client cashing in the performance bond and advance payment bond.

50.4.4 INSURANCE BY THE CLIENT

The conditions of contract normally require that the client will have to insure the permanent works as soon as he takes possession or when the completion certificate is issued. This is because at the time it is no longer the contractor's responsibility to insure the works.

REFERENCES

Suggested references are papers by PIANC which cover various types of marine civil engineering works. In addition the publication by the Department of the US Army Corps of Engineers called *Shore Protection Manual Volumes 1 and 2.*

ARBITRATION 51

D. Miles

When things go wrong the parties to a dispute sometimes go to arbitration rather than to court. There are many advantages to this procedure. It can be quicker and less expensive than court, mainly because there is not so great a need for lawyer participation. Although this subject could be considered on the fringes of engineering it was felt that engineers likely to be involved in the settlement of disputes would welcome a review of the subject not only as it is applied in the UK but also when litigation is conducted under the rules of the International Chamber of Commerce.

CONTENTS

51.1 THE ARBITRATION AGREEMENT*

Arbitration occurs where the parties to a contract have agreed that their disputes will be resolved by means of arbitration. Arbitration is an alternative method of resolving disputes to that provided by the Courts. There are a number of broad similiarities between the two systems, but arbitration offers more scope for flexibility, informality and originality of approach.

*Unless otherwise stated, references in this chapter relate to the procedure under English law.

51.2 THE ADVANTAGES AND DISADVANTAGES OF ARBITRATION

51.2.1 ADVANTAGES

Arbitration is private. Members of the public are not allowed to attend the hearing (unless by agreement). This means that any adverse publicity can be avoided.

Most arbitrations in the construction industry presuppose that the dispute (if it arises) will be of a technical nature. The Arbitrator is normally selected for his technical expertise in the field of the particular dispute. If the parties are unable to agree upon an Arbitrator, the arbitration clause normally provides for an appointment by one of the professional bodies. It is unusual for construction arbitrations to be concerned principally with matters of legal interpretation.

Arbitration should be quicker than formal Court proceedings as the pace of the action should be governed by the availability of the parties and the Arbitrator, rather than dictated by the ability of the Courts to provide a Judge to hear the dispute. It should also be less expensive because there is not so great a need for lawyer participation.

In international cases where the parties to the contract may be of different nationality, arbitration avoids a party having to submit to the jurisdiction of a foreign Court with which it is not familiar.

51.2.2 DISADVANTAGES

Increasingly arbitrations are run very much upon the lines of formal hearings in the Courts, indeed it has been called 'litigation without wigs'. The lawyers

Coastal, Estuarial and Harbour Engineer's Reference Book. Edited by M.B. Abbott and W.A. Price. Published in 1993 by Chapman & Hall, 2–6 Boundary Row, London SE1 8HN. ISBN 0 419 15430 2.

who conduct the case have to be educated in the technical aspects of the dispute. Experts are normally used by both sides, even when the Arbitrator has sufficient technical knowledge.

In instances where the interpretation of the law is concerned, it is usually only a preliminary to give effect to the implications of the technical issues where matters of technical expertise are then applied.

It is only possible to enforce arbitration between parties who have signed an arbitration agreement. Some, but by no means all, arbitration agreements provide for arbitration between all parties involved in, say, an engineering project by reference to a common Arbitrator where the disputes between, for instance, the main contractor and the sub-contractor touch and concern those issues also in dispute between the main contractor and the employer. However, if one of the contracts does not contain a binding arbitration clause then, in the example cited, a main contractor may find itself in the position of having to arbitrate against its sub-contractor, but litigate against its employer. Not only is this costly and time consuming, but there is also a danger that a main contractor could obtain two different results (losing in both instances). In High Court proceedings, there is an established procedure for bringing in third parties to an action. Further, Judges are more used than Arbitrators to deciding issues between a number of defendants and third parties.

Despite the general intention that arbitration should be quicker and cheaper, it often works out as expensive as litigation and sometimes more so. In litigation the parties do not have to pay either for the hire of the room or the Judge's time. In arbitration the parties are required to arrange and pay for the rooms in which the arbitration is heard and for the Arbitrator's time. It is not uncommon for an Arbitrator to charge a cancellation fee if the matter is settled at or shortly before the hearing, which can sometimes be a disincentive to the parties to settle.

The ability of the parties to appeal against the Arbitrator's award is now strictly limited.

There is more scope for a reluctant party to delay proceedings.

51.3 THE ARBITRATION AGREEMENT

Section 32 of The Arbitration Act 1950 states:

> 'Arbitration Agreement' means a written agreement to submit present or future differences to arbitration, whether an Arbitrator is named therein or not.

Provided it is in writing, the basis of the law and practice of arbitration in England and Wales are now the Arbitration Acts 1950–1979. (There is currently underway a private initiative to introduce a new Arbitration Act sometime in 1993/1994). There can be an oral agreement to submit the differences to arbitration, but in such circumstances the Acts will not apply. It is not necessary that there should be a formal agreement or that the terms be contained in one document; the agreement may be incorporated by reference.

51.4 THE UNCITRAL MODEL LAW

Previously there were two major instruments within the United Nations system relating to international commercial arbitration. These were the 1958 United Nations Convention of the Recognition and Enforcement of Foreign Arbitral Awards (the New York Convention) and the UNCITRAL Arbitration Rules. In the late 1970s the Commission requested the preparation of the draft of a model law on arbitral procedure in consultation with interested international organizations. The result was the UNCITRAL Model Law on International Commercial Arbitration, which was adopted by the United Nations Commission on international trade law on 21st June 1985. Subsequently, the model law has been adopted in Australia, Bulgaria, Canada, Cyprus, Hong Kong, New Zealand, Nigeria and Scotland, but not in England, France, The Netherlands or Switzerland.

51.5 THE RULES APPLYING TO ARBITRATION

The individual arbitration agreements may also incorporate arbitration rules and procedures produced by the various countries, institutions and organizations which administer arbitrations. For example, rules are produced by UNCITRAL, the International Chamber of Commerce (ICC), the London Court of International Arbitration and the Chartered Institute of Arbitrators.

Particular professional bodies produce their own rules, such as the Institution of Civil Engineers and the Joint Contracts Tribunal. These rules govern the particular way in which the arbitration is to be run. They can deal with such matters as the appointment of the Arbitrator, the powers of the Arbitrator, the procedural rules before, during and after the hearing and any special proceedings, e.g. short procedures, documents only arbitrations and interim arbitrations.

51.6 TIME LIMITS AND LIMITATION PERIODS

Generally an arbitration must be begun (i.e. notice of arbitration served) within six years of the date when the claimant's cause of action arises. If the agreement is under seal, the period is twelve years. This is the general rule, but care should be taken to refer to the particular law of the contract and rules to which the arbitration is subject, for this period can sometimes be shortened dramatically. The Institution of Civil Engineers' procedure provides for a time limit of three calendar months from the date of any decision of the engineer or the failure of the engineer to give such a decision. The time limit in the FIDIC international form of contract (fourth edition 1987) is 70 days.

A further time limit is also relevant, running from the date of the award. An action to enforce the award must be brought within six years of the date of the award.

If the arbitration clause is worded so as to provide for disputes or differences to be referred to arbitration, care should be taken in the notice of arbitration to identify the particular dispute or difference which it is wished to refer to arbitration. Failure to do so may render a notice of arbitration invalid.

51.7 NOTICE TO CONCUR

Either at the same time or following notice of arbitration, it is normal for the party proposing arbitration to invite the other party to concur in the appointment of an Arbitrator (in cases where the arbitration provides for the dispute to be heard by a single Arbitrator). It is difficult to lay down specific guidelines relating to the appointment of Arbitrators. A party may propose a single Arbitrator, or offer the choice of three. Sometimes, upon receiving three proposals, the other party may counter with three of its preferred nominees – as a matter of principle. If the arbitration is being conducted under the ICC rules, these provide that each party nominate their own Arbitrator and then an attempt is made to agree the chairman of the panel.

In any event, if the parties fail to agree an Arbitrator, then most arbitration rules provide for reference to be made to a particular professional body for an appointment by the President or Vice President. In the absence of an appointing body, the Court has power to appoint an Arbitrator, who could be a Judge.

51.8 APPOINTMENT OF THE ARBITRATOR

It is unusual (and not advisable) for an Arbitrator to be named in an arbitration agreement upon the signing of the contract. This is because until the nature of the dispute has been identified, the parties will not know the particular expertise that will be needed from the Arbitrator. Where possible, it is senible for the parties to try and reach agreement upon the Arbitrator. Merely to object to the other side's nominees as a matter of principle can be counterproductive.

If failure to agree an Arbitrator results in a reerence to the President, then both parties run the risk of having an Arbitrator imposed upon them with whom neither agrees and, sometimes, he may not necessarily be sufficiently qualified or experienced to hear the dispute.

51.9 THE JURISDICTION OF THE ARBITRATOR

The jurisdiction and authority of the Arbitrator comes from the parties themselves. They give him the authrity to act by setting up in the arbitration agreement what disputes he has jurisdiction or authority to decide. They may also decide what award he can make, for example, giving him the authority to decide liability only and not the amount.

This is subject to two exceptions: firstly, the parties cannot give the Arbitrator powers which are contrary to public policy – for example, to decide an issue without regard to the substantive law; and secondly they cannot empower him to proceed against a third party without the consent of that third party.

In the case of *Northern Regional Health Authority-v-Derek Crouch Construction Company Limited* the arbiration clause gave the Arbitrator power to:

> open up, review and revise any certificate, opinion, decision, requirement or notice and to determine all matters in dispute which shall be submitted to him in the same manner as if no certificate, opinion, decision, requirement or notice had been given.

The Court of Appeal decided that the effect of such a clause was to invest these powers in the Arbitrator alone and that a Court, where such a clause existed, would not have the same powers. The effect was to require the matter to be determined by arbitration as the Courts did not have the relevant powers.

It is important at the beginning of the action that the question of the Arbitrator's jurisdiction is considered very carefully. The implications of starting an action, say in the High Court, and discovering at the hearing (many years later) that the Court has no jurisdiction to hear the case, could be very serious. Not only would there be a great deal of lost time and cost, but also, by that stage, the Limitation Acts may well apply, thereby barring the aggrieved party from commencing arbitration proceedings in lieu.

51.10 REMOVAL OF THE ARBITRATOR

The ability of the Court to remove an Arbitrator is limited. The Arbitrator must be guilty of serious and irreparable misconduct, bias or be fundamentally deficient in capability and performance.

51.11 PROCEDURE

In some cases, despite there being an arbitration clause in existence, one party may decide to proceed in the Courts. Section 4 of the Arbitration Act 1950 gives the Court power to order a stay of the proeedings until after the hearing of the arbitration. It is important that the party wishing to apply for a stay takes no step in the Court proceedings, otherwise it may be argued that the right to arbitration has been lost. Even where a case proceeds in the Courts, there may still be a fundamental problem relating to jurisdiction because of the effect of the Derek Crouch case (see section 51.9).

The manner in which the arbitration is conducted will be subject to the relevant rules to which the arbiration clause is subject. In the absence of such rules, the arbitration is conducted under the Arbitrator's directions. Generally, arbitrations follow the same pattern as litigation. However, arbitration is more informal and has greater scope for flexibility with a view to shortening the whole procedure. The arbiraion can be restricted to documents only. The tribunal can take part in the preparation for trial. There is a greater likelihood that documents will be read prior to the hearing. The hearing can be adversarial (the traditional English and Common Law procedure) or inquisitorial (as used in civil jurisdictions).

Following the hearing, the Arbitrator will make his award. Whilst the giving of an award without reasons is valid, 'the making of an award is, or should be, a rational process, formulating and recording the reason tends to accentuate its rationality'. (See the Comercial Court report on arbitration 1978) (CMND7284).

In addition, the award should contain reasons:

1. Where the arbitration agreement expressly provides that reasons are to be given;
2. Where the institutional rules require the giving of reasons; or
3. Where the parties have requested a reasoned award.

In the absence of reasons, the High Court will not be able to consider any questions of law arising out of it and accordingly determine whether there are any grounds for appeal.

51.12 APPEALS

It is the policy of the British Parliament and the Courts to ensure the finality of Arbitrators' awards. One of the main objectives of the Arbitration Act 1979 was to limit drastically the ability of a party to appeal to the High Court on what was known as the 'case stated procedure', enabling the High Court to overturn an award on the ground of errors of fact or law appearing on the face of the award.

The House of Lords in the case of *The Nema* held that judicial interference with an Arbitrator's award was only justified if it could be shown that the Arbitrator had misdirected himself in law or had reached a decision that no reasonable Arbitrator could have reached. An Arbitrator must go badly wrong for such an application to succeed.

In other countries the Courts' ability to enquire, interfere and indeed override the decisions of the Arbitrator could be radically different. Careful enquiry as to the general policy will need to be made in each particular jurisdiction.

51.13 COSTS

The rules relating to costs in the Courts apply to arbitrations (unless a contrary agreement exists). Costs, therefore, normally follow the event. This may not necessarily be so in all jurisdictions. The particular rules adopted may give the Arbitrator a jurisdiction to hear applications for security for costs (i.e., pay an amount into Court which is a pre-estimate of the costs a successful Respondent could incur), otherwise an application can be made to the Commercial Court. Further, following an award of costs, the Arbitrator has power to determine the costs of the arbitration. Alternatively, the parties may apply to the Taxing Masters to the High Court, who are well experienced in such matters. It is likely that a party will fare worse in front of a Taxing Master than in front of an Arbitrator.

It is normal for the Arbitrator only to release his award upon the payment of his own costs. Therefore, it is sometimes necessary for the parties to reach agreement as to how the costs should be paid (norally conditional upon the result) before the award is collected from the Arbitrator. If the parties wish to dispute the Arbitrator's costs, then they can require taxation in the High Court.

51.14 ALTERNATIVE DISPUTE RESOLUTION

Arbitrations, like Court actions, can prove to be both costly and time consuming. In addition, at the end, inevitably, there is a 'winner' and a 'loser'. Over the past fifteen or so years procedures have evolved whereby the confrontational aspects of arbitration are reduced by using Alternative Dispute Resolution procedures (ADR). Such techniques involve proedures such as mini-trials, conciliation and mediation.

For these techniques to work, it is vital that both parties are willing to adopt the procedures. If successful, they can result in an amicable settlement which preserves the business relationship between the parties and which has the advantage of avoiding the considerable costs and delays. Another advantage is that the parties play a more active role in resolving the dispute than otherwise would have been the case in arbitration.

In the construction field ADR techniques and proesses have been adopted world-wide. Clause 67.2 of FIDIC provides that arbitration of a dispute shall not be commenced unless an attempt has first been made by the parties to settle the dispute amicably. The South African Institution of Civil Engineers, in the early 1980s, introduced a dispute settling procedure which they call mediation into their standard civil engineering contract documentation. In Hong Kong, in consultation with the Hong Kong Contractors Asociation, the Hong Kong Government introduced a set of mediation rules for use in Government contruction contracts. In the US, at the end of 1991, the standard dispute clause in federal procurement contracts allowed the contractors and the Government, by mutual consent, to agree to use ADR.

Throughout the world there are now organizations promoting and encouraging the use of ADR. The first organization to be established was the Centre for Public Resources in New York, followed by the Australian Commercial Disputes Centre in Sydney, the Hong Kong International Arbitration Centre, the Centre for Dispute Resolution in London and the Alternative Dispute Resolution Association of South Africa based in Johannesburg. It is anticipated that the 1990s will see such organizations increasing and expanding.

REFERENCES

FIDIC (FEDERATION INTERNATIONAL DES INGENIEURS CONSEILS)

Guide to the use of FIDIC, 4th edn, by Federation International Des Ingenieurs Conseils.

Corbett, E.G. (1991) *FIDIC 4th – A practical legal guide*, Sweet and Maxwell, London.

ARBITRATION

Holzman, and Newhaus, (1989) *A guide to the UNICRAL Model Law on International Commercial Arbitration*, Kluwer.

Mustill and Boyd (1989) *Commercial Arbitration* 2nd edn, Butterworth, London.

Redfern and Hunter (1991) *Law and Practice of International Commercial Arbitration* 2nd edn, Sweet & Maxwell, London.

ADR

Donovan Leisure Newton & Irvine (1990) *ADR Practice Book*, John Wiley, Chichester.

Mackie, K. (1991) *A Handbook of Dispute Resolution – ADR in Action*, Routledge, London.

INDEX

For Product Safety Concerns and Information please contact our EU
representative GPSR@taylorandfrancis.com
Taylor & Francis Verlag GmbH, Kaufingerstraße 24, 80331 München, Germany

www.ingramcontent.com/pod-product-compliance
Ingram Content Group UK Ltd.
Pitfield, Milton Keynes, MK11 3LW, UK
UKHW050055200926
475702UK00019B/2858
* 9 7 8 0 3 6 7 8 6 4 6 4 4 *